The Environmental Dictionary

The Environmental Dictionary

and Regulatory Cross-Reference

Third Edition

Compiled by
James J. King

A Wiley-Interscience Publication
John Wiley & Sons, Inc.
New York • Chichester • Brisbane • Toronto • Singapore

This text is printed on acid-free paper.

This publication is designed to provide accurate and authoritative information in regard to the subject matter covered. It is sold with the understanding that the publisher is not engaged in rendering legal, accounting, or other professional services. If legal advice or other expert assistance is required, the services of a competent professional person should be sought.

Library of Congress Cataloging in Publication Data:

King, James J.

The environmental dictionary : and regulatory cross-reference / compiled by James J. King — 3rd ed.

p. cm.

ISBN 0-471-11995-4

1. Environmental law—United States—Dictionaries.
2. Environmental protection—United States—Dictionaries.
3. Pollution—United States—Dictionaries. I. Title.

KF3775.A68K57 1994
344.73'046'03—dc20
[347.30445603] 94-40758

Printed in the United States of America

10 9 8 7 6 5 4 3 2 1

CONTENTS

PREFACE TO THE THIRD EDITION

During the compilation of the third edition of *The Environmental Dictionary*, over the past two years nearly 1500 new and modified definitions to 40 CFR were made. This work has expanded to well over 1000 pages. And, as with the second edition the acceptance of this effort by the regulated community has been very rewarding. Thank you.

I wish the user of the 3rd edition of *The Environmental Dictionary* success in your endeavors working through 40 CFR. We are entering a very interesting legislative period. Reauthorization of environmental legislation will take on a new issues and heighten the need to keep on top of developing environmental regulatory trends. Strategic environmental management decisions will be more critical than ever. Good luck, and again, thank you.

Jim King
Knoxville, Tennessee
August, 1994

GUIDE TO USAGE

DEFINITION SOURCE:

The definitions in this publication are taken from two sources:

1. The *Code of Federal Regulations, Title 40, Protection of Environment* (40 CFR), Revised as of July 1, 1987, and

2. The *Federal Register* (updates pertaining to 40 CFR), for the period covering July 1, 1987 through June 30, 1994.

Note: Definitions in *The Environmental Dictionary* that are taken from the *Federal Register* include only those pertaining to sections of 40 CFR designated for "Definitions" or "Special definitions." Definitions in *The Environmental Dictionary* that are taken from sections of 40 CFR (7/1/87) not designated for "Definitions" or "Special Definitions" (terms defined in other sections), are not updated through June 30, 1994 by *Federal Register* entries, but are effective as of July 1, 1987. (There are relatively few definitions taken from other sections, however.) Also note that, as a rule, definitions from the appendices of 40 CFR are not included in this dictionary.

DEFINITION ENTRY FORMAT:

- Definitions taken from 40 CFR:

  Defined term, phrase, or acronym (Section number of 40 CFR where term is defined)[1] definition taken from 40 CFR.

  In most cases, where a definition is the same for two or more sections of 40 CFR, these section numbers are listed (in ascending numerical

[1]In most cases, only the section number is provided as a reference. The paragraph designation and/or other subdivisions are not included; this information is usually unnecessary since definitions are easily found within the section. Where this information is pertinent, however, it is included in the reference along with the section number.

sequence) after the defined term, phrase, or acronym, and the definition appears only once.

Example:

Defined term, phrase, or acronym (Sections xx; xxx; xxxx) definition.

However, in certain cases, definitions that are the same for two or more sections are repeated for each section number. This method is used (at the editor's discretion) where listing multiple section numbers with one definition might make it difficult for the user to locate a particular definition.

Definitions taken from the *Federal Register*:

A definition that is taken from the *Federal Register* will be distinguished by a bracketed date [indicating the publication date of the *Federal Register*] following the section number.

Example:

Defined term, phrase, or acronym (Section number [*Federal Register* date]) definition.

This format is used to designate a new definition (i.e., a definition not originally defined in 40 CFR (7/1/87)).

In the case where a definition from the *Federal Register* replaces a definition published in 40 CFR (7/1/87), the section number and bracketed date will be followed by an asterisk.

Example:

Defined term, phrase, or acronym (Section number [*Federal Register* date])* definition.

In the case where a definition from the *Federal Register* replaces a previous *Federal Register* definition (and the term, phrase, or acronym being defined is <u>not</u> in 40 CFR (7/1/87)), both *Federal Register* dates will follow the section number in brackets. The definition given is from the second *Federal Register* publication.

Example:

Defined term, phrase, or acronym (Section number [original *Federal Register* date, second *Federal Register* date]) definition of the second date.

SEQUENCE OF DEFINITIONS:

- First level:

  The definitions are arranged in alphabetical order. [Note: All defined terms, phrases or acronyms that begin with numerical values are alphabetized according to the word form of the number.]

- Second level:

  If the same term, phrase, or acronym is defined differently for different sections, the multiple definitions for that same term, phrase, or acronym are arranged by section number, in ascending numerical order. A grouping of section numbers with one definition appears last in the numerical sequence for a particular term, phrase, or acronym.

  Example:

  Article (Section 372.3) unique definition.

  Article (Section 704.3) unique definition.

  Article (Section 710.2) unique definition.

  Article (Section 725.2) unique definition.

  Article (Sections 720.3; 723.50; 747.115; 747.195; 747.200) same definition for multiple sections.

  If there are two groupings of section numbers for a defined term, phrase, or acronym, they are arranged in ascending numerical sequence according to the first section number listed in the grouping:

Flow-through (Section 797.1970) unique definition.

Flow-through (Sections 797.1300; 797.1330) same definition for multiple sections.

Flow-through (Sections 797.1400; 797.1930; 797.1950) same definition for multiple sections.

Note in the example above that the groupings of section numbers with the same definition follow the last definition of the term for an individual section.

CROSS-REFERENCING FORMAT:

Where a term, phrase, or acronym is cross-referenced, the entry for the cross-reference is as follows:

Term, phrase, or acronym being cross-referenced to defined term, phrase, or acronym (Section number). See **Defined term, phrase, or acronym (Section number)**.

Examples:

1) **Cross-reference entry:**

A/B (Section 60.471). See **Afterburner (Section 60.471)**.

Defined term/definition entry:

Afterburner (A/B) (Section 60.471) means an exhaust gas incinerator used to control emissions of particulate matter.

2) **Cross-reference entry:**

Average of daily values for thirty consecutive days (Section 407.81). See **Maximum for any one day (Section 407.81)**.

Defined term/definition entry:

Maximum for any one day and Average of daily values for thirty consecutive days (Section 407.81) shall be based on the daily average mass of final product produced during the peak thirty consecutive day production period.

Note: In a cross-reference entry, the cross-referenced term, phrase, or acronym is not repeated in the "See **defined term**" statement. (Referring to the cross-reference entry in Example No. 1 above: The acronym **A/B** is not included in the "See **Afterburner (Section 60.471)**" statement. Referring to the cross-reference entry in Example No. 2 above: The phrase **Average of daily values for thirty consecutive days** is not included in the "See **Maximum for any one day (Section 407.81)**" statement.)

This method of cross-referencing eliminates repetitive wording in the cross-reference statement. However, the user must be aware of this format when seeking a definition, since the defined term, phrase or acronym along with the cross-referenced term, phrase, or acronym are listed together as the complete entry where defined, and alphabetized accordingly.

Also note that, when searching for the defined term referred to in a cross-reference entry, it is important to look for the defined term with the correct section number (the one referred to in the cross-reference entry).

FINDING AIDS:

1) If a definition cannot be found for a particular section, the user should refer to that section in 40 CFR. There may be a reference to another part/subpart/section or an act, for an applicable definition.

2) Refer to the leading paragraph of a definition's 40 CFR section number for important general information or specific details relating to the definitions in that section.

3) In order to properly identify an "Act" being referred to in a definition, the user must read the definition of "Act" from the same section as that of the defined term, phrase, or acronym.

4) There are many inconsistencies within the CFR definitions relating to word spelling, word hyphenation, and word formation/separation. If a word cannot be located in this dictionary by using its most common form, the user should check for possible variations in spelling, hyphenation, or word formation/separation. (He/she should also look for the word in both its singular and plural form.)

Examples of inconsistencies:

byproduct	by-product
groundwater	ground water
nonprocess	non-process
onsite	on-site
wastewater	waste water

(For instance, a word that is hyphenated will fall at a different location in the alphabetical sequence than the same word when it appears without hyphenation; be sure to check for both forms of the word at the appropriate locations in the alphabetical sequence.)

5) The section number (following the defined term, phrase, or acronym) that provides the location of the definition in 40 CFR, does not include paragraph and other subdivision designations. The user must refer to 40 CFR for this information if it is needed to clarify a reference made within a definition, to another term.

6) All defined terms, phrases or acronyms that begin with numerical values are alphabetized according to the word form of the number; in some cases, the numerical value follows the word form, in parentheses (e.g., Twenty-four-hour (24-hour); Six-minute (6-minute)).

APPENDIX:

The Appendix contains a table of chapters, subchapters, parts, subparts, and sections for 40 CFR, as well as a guide for using the appendix and a brief explanation of the *Code of Federal Regulations.*

INTRODUCTION

The Environmental Dictionary is designed to be a supplement for researching environmental regulations in the *Code of Federal Regulations, Title 40, Protection of Environment* (40 CFR). 40 CFR is a maze of administrative and technical requirements, constructed with language that has very specific meanings. Locating correct terminology in the regulations can become complicated: it is not always obvious where to start looking for a term; and, the same term may be defined differently for the various regulatory areas. This book will hopefully be useful not only as a dictionary, but also as a locator for the regulation(s) that applies to a term being researched.

The Environmental Dictionary is a compilation of definitions from two sources:

1. The *Code of Federal Regulations, Title 40, Protection of Environment*, Revised as of July 1, 1987, and

2. The *Federal Register* (updates pertaining to 40 CFR), for the period covering July 1, 1987 through June 30, 1994.

Note: Definitions in *The Environmental Dictionary* that are taken from the *Federal Register* include only those pertaining to sections of 40 CFR designated for "Definitions" or "Special definitions." Definitions in *The Environmental Dictionary* that are taken from sections of 40 CFR (7/1/87) not designated for "Definitions" or "Special Definitions" (terms defined in other sections), are not updated through June 30, 1994 by *Federal Register* entries, but are effective as of July 1, 1987. (There are relatively few definitions taken from other sections, however.) Also note that, as a rule, definitions from the appendices of 40 CFR are not included in this dictionary.

The Environmental Dictionary does not guarantee the correctness (regarding content, grammar, punctuation, spelling, etc.) or validity of the definitions contained in this dictionary; they are simply reproductions of those contained in the source publications. (Refer to 40 CFR and the *Federal Register* for further information on a subject if a definition seems questionable.)

All definitions in this dictionary have been reproduced as accurately as possible. However, definitions from the dictionary should not be quoted or cited; refer to 40 CFR and the *Federal Register* for these purposes.

The user of this dictionary should consult 40 CFR and the *Federal Register* for bibliographical details on material in the definitions that is reproduced from other publications. This information is not included in the bibliography of *The Environmental Dictionary*.

It is highly recommended that the user of this dictionary take a few moments to review the information in the "Guide to Usage." It provides details on entry format, sequence of definitions, cross-referencing format, and other finding aids. Without this information, the user will be unable to utilize this dictionary to its potential.

The appendix of *The Environmental Dictionary* contains a table of chapters, subchapters, parts, subparts, and sections. It is included as a handy reference for identifying the categories under which the definitions can be found in 40 CFR. (For information on how to use the appendix, refer to the "Guide to Using the Appendix."

The final section of the book deals with a list of acronyms issued by the Environmental Protection Agency. This list is constantly expanding and changing as new regulations are promulgated or old regulations revised.

The Environmental Dictionary

A

A (Section 433.11), as in Cyanide A, shall mean amenable to alkaline chlorination.

A/B (Section 60.471). See **Afterburner (Section 60.471)**.

A/E services (Sections 33.005; 35.6015 [1/27/89]). See **Architectural or engineering services (Sections 33.005; 35.6015)**.

A_λ (Section 796.3700). See **Absorbance (Section 796.3700)**.

AAP (Section 205.151). See **Acoustical Assurance Period (Section 205.151)**.

Abandoned area (Section 61.20) means a deserted mine area in which work has ceased and in which further work is not intended. Areas which function as escapeways, and areas formerly used as lunchrooms, shops, and transformer or pumping stations are not considered abandoned areas. Except for designated ventilation passageways designed to minimize the distance to vents, worked-out mine areas are considered abandoned areas for the purpose of this subpart.

Abandoned mine (Section 434.11) means a mine where mining operations have occurred in the past and (1) The applicable reclamation bond or financial assurance has been released or forfeited or (2) If no reclamation bond or other financial assurance has been posted, no mining operations have occurred for five years or more.

Abandoned well (Section 146.3) means a well whose use has been permanently discontinued or which is in a state of disrepair such that it cannot be used for its intended purpose or for observation purposes.

Abnormally treated vehicle (Section 86.085-2 [12/16/87]) means any diesel light-duty vehicle or diesel light-duty truck that is operated for less than five miles in a 30 day period immediately prior to conducting a particulate emissions test.

Aboveground release (Section 280.12 [9/23/88]) means any release to the surface of the land or to surface water. This includes, but is not limited to, releases from the aboveground portion of an UST system and aboveground releases associated with overfills and transfer operations as the regulated substance moves to or from an UST system.

Aboveground storage facility (Section 113.3) means a tank or other container, the bottom of which is on a plane not more than 6 inches below the surrounding surface.

Aboveground tank (Section 260.10) means a device meeting the definition of tank in Section 260.10, and that is situated in such a way that the entire surface area of the tank is completely above the plane of the adjacent surrounding surface and the entire surface area of the tank (including the tank bottom) is able to be visually inspected.

Absorbance (A_λ) (Section 796.3700) is defined as the logarithm to the base 10 of the ratio of the initial intensity (I_o) of a beam of radiant energy to the intensity (I) of the same beam after passage through a sample at a fixed wavelength. Thus, $A_\lambda = \log_{10}(I_o/I)$.

ac (Section 401.11 [8/25/93])* means acre(s).

ACBM (Section 763.83 [10/30/87]). See **Asbestos-containing building material (Section 763.83)**.

Acceleration test procedure (Section 205.151) means the measurement methodologies specified in Appendix I.

Accelerator Pump (Plunger or Diaphragm) (Section 85.2122(a)(3)(ii)) means a device used to provide a supplemental supply of fuel during increasing throttle opening as required.

Acceptable of a batch (Section 204.51) means that the number of non-complying compressors in the batch sample is less than or equal to the acceptance number as determined by the appropriate sampling plan.

Acceptable Quality Level (AQL) (Sections 86.602; 205.51) means the maximum percentage of failing vehicles that, for purposes of sampling inspection, can be considered satisfactory as a process average.

Acceptable Quality Level (AQL) (Section 205.151) means the maximum allowable average percentage of vehicles or exhaust systems that can fail sampling inspection under a Selective Enforcement Audit.

Acceptable Quality Level (AQL) (Section 86.1002-84) means the maximum percentage of failing engines or vehicles, that for purposes of sampling inspection, can be considered satisfactory as a process average.

Acceptable Quality Level (AQL) (Section 204.51) means the maximum percentage of failing compressors that, for purposes of sampling inspection can be considered satisfactory as a process average.

Acceptance of a batch (Section 205.51) means that the number of noncomplying vehicles in the batch sample is less than or equal to the acceptance number as determined by the appropriate sampling plan.

Acceptance of a batch sequence (Section 205.51) means that the number of rejected batches in the sequence is less than or equal to the acceptance number as determined by the appropriate sampling plan.

Acceptance of a batch sequence (Section 204.51) means that the number of rejected batches in the sequence is less than or equal to the sequence acceptable number as determined by the appropriate sampling plan.

Acceptance of a compressor (Section 204.51) means that the measured noise emissions of the compressor, when measured in accordance with the applicable procedure, conforms to the applicable standard.

Acceptance of a vehicle (Section 205.51) means that the measured emissions of the vehicle when measured in accordance with the applicable procedure, conforms to the applicable standard.

Accessible (Section 763.83 [10/30/87]) when referring to ACM means that the material is subject to disturbance by school building occupants or custodial or maintenance personnel in the course of their normal activities.

Accessible environment (Section 191.12 [12/20/93])* means: (1) The atmosphere; (2) land surfaces; (3) surface waters; (4) oceans; and (5) all of the lithosphere that is beyond the controlled area.

Accident (Section 162.3) means an unexpected, undesirable event that adversely affects man or the environment, and that is caused by the use or presence of a pesticide.

Accident (Section 171.2) means an unexpected, undesirable event, caused by the use or presence of a pesticide, that adversely affects man or the environment.

Accidental occurrence (Section 265.141), as specified in Section 265.141(g), means an accident, including continuous or repeated exposure to conditions, which results in bodily injury or property damage neither expected nor intended from the standpoint of the insured.

Accidental occurrence (Section 264.141), as specified in Section 264.141(g), means an accident, including continuous or repeated exposure to conditions, which results in bodily injury or property damage neither expected nor intended from the standpoint of the insured.

Accidental release (Section 280.92 [10/26/88]) means any sudden or nonsudden release of petroleum from an underground storage tank that results in a need for corrective action and/or compensation for bodily injury or property damage neither expected nor intended by the tank owner or operator.

Acclimation (Section 797.1400) means the physiological compensation by test organisms to new environmental conditions (e.g., temperature, hardness, pH).

Acclimation (Section 797.1600) means the physiological or behavioral adaptation of organisms to one or more environmental conditions associated with the test method (e.g., temperature, hardness, pH).

Acclimation (Section 797.1830) is the physiological compensation by test organisms to new environmental conditions (e.g., temperature, salinity, pH).

Acclimation (Sections 797.2050; 797.2175) is the physiological or behavioral adaptation of test animals to environmental conditions and basal diet associated with the test procedure.

Acclimation (Sections 797.2130; 797.2150) means the physiological and behavioral adaptation to environmental conditions (e.g., housing and diet) associated with the test procedure.

Accreditation (Section 763.83 [10/30/87]). See **Accredited (Section 763.83)**.

Accredited or accreditation (Section 763.83 [10/30/87]) when referring to a person or laboratory means that such person or laboratory is accredited in accordance with section 206 of Title II of the Act.

Accrual date (Section 14.2) means the date of the incident causing the loss or damage or the date on which the loss or damage should have been discovered by the employee through the exercise of reasonable care.

Accumulated speculatively (Section 261.1). For purposes of Sections 261.2 and 261.6, a material is accumulated speculatively if it is accumulated before being recycled. A material is not accumulated speculatively, however, if the person accumulating it can show that the material is potentially recyclable and has a feasible means of being recycled; and that - during the calendar year (commencing on January 1) - the amount of material that is recycled, or transferred to a different site for recycling, equals at least 75 percent by weight or volume of the amount of that material accumulated at the beginning of the period. In calculating the percentage of turnover, the 75 percent requirement is to be applied to each material of the same type (e.g., slags from a single smelting process) that is recycled in the same way (i.e., from which the same material is recovered or that is used in the same way). Materials accumulating in units that would be exempt from regulation under Section 261.4(c) are not to be included in making the calculation. (Materials that are already defined as solid wastes also are not to be included in making the calculation.) Materials are no longer in this category once they are removed from accumulation for recycling, however.

Accuracy (Section 86.082-2) means the difference between a measurement and true value.

Acetylene cylinder filler (Section 763.163 [7-12-89]) means an asbestos-containing product which is intended for use as a filler for acetylene cylinders.

Acid gas (Section 60.641) means a gas stream of hydrogen sulfide (H_2S) and carbon dioxide (CO_2) that has been separated from sour natural gas by a sweetening unit.

Acid mist (Section 60.81) means sulfuric acid mist, as measured by Method 8 of Appendix A to this part or an equivalent or alternative method.

Acid or ferruginous mine drainage (Section 434.11) means mine drainage which, before any treatment, either has a pH of less than 6.0 or a total iron concentration equal to or greater than 10 mg/l.

Acid recovery (Section 420.91) means those sulfuric acid pickling operations that include processes for recovering the unreacted acid from spent pickling acid solutions.

Acid regeneration (Section 420.91) means those hydrochloric acid pickling operations that include processes for regenerating acid from spent pickling acid solutions.

ACM (Section 763.83 [10/30/87]). See **Asbestos-containing material (Section 763.83)**.

Acoustic descriptor (Section 211.102) means the numeric, symbolic, or narrative information describing a product's acoustic properties as they are determined according to the test methodology that the Agency prescribes.

Acoustical Assurance Period (AAP) (Section 205.151) means a specified period of time or miles driven after sale to the ultimate purchaser during which a newly manufactured vehicle or exhaust system, properly used and maintained, must continue in compliance with the Federal standard.

Acrylic fiber (Section 60.601) means a manufactured synthetic fiber in which the fiber-forming substance is any long-chain synthetic polymer composed of at least 85 percent by weight of acrylonitrile units.

Act (Section 104.2) means the Federal Water Pollution Control Act, as amended, 33 U.S.C. 1251 et seq., Pub. L. 92-500, 86 Stat. 816.

Act (Section 108.2) means the Federal Water Pollution Control Act, as amended.

Act (Section 110.1) means the Federal Water Pollution Control Act, as amended, 33 U.S.C. 1251, et seq., also known as the Clean Water Act.

Act (Section 113.3) means the Federal Water Pollution Control Act, as amended, 33 U.S.C. 1151, et seq.

Act (Section 121.1) means the Federal Water Pollution Control Act, 33 U.S.C. 1151, et seq.

Act (Section 123.64) is defined in Section 22.03.

Act (Section 129.2) means the Federal Water Pollution Control Act, as amended (Pub. L. 92-500, 86 Stat. 816 et seq., 33 U.S.C. 1251 et seq.). Specific references to sections within the Act will be according to Pub. L. 92-500 notation.

Act (Section 130.2) means the Clean Water Act, as amended, 33 U.S.C. 1251 et seq.

Act (Section 131.3) means the Clean Water Act (Pub. L. 92-500, as amended (33 U.S.C. 1251 et seq.).

Act (Section 133.101) means the Clean Water Act (33 U.S.C. 1251 et seq., as amended).

Act (Section 136.2) means the Clean Water Act of 1977, Pub. L. 95-217, 91 Stat. 1566, et seq. (33 U.S.C. 1251 et seq.) (The Federal Water Pollution Control Act Amendments of 1972 as amended by the Clean Water Act of 1977).

Act (Section 141.2) means the Public Health Service Act, as amended by the Safe Drinking Water Act, Pub. L. 93-523.

Act (Section 142.2) means the Public Health Service Act.

Act (Section 143.2) means the Safe Drinking Water Act as amended (42 U.S.C. 300f et seq.)

Act (Section 149.101) means the Public Health Service Act, as amended by the Safe Drinking Water Act, Pub. L. 93-523.

Act (Section 162.3) means the Federal Insecticide, Fungicide, and Rodenticide Act, as amended by the Federal Environmental Pesticide Control Act of 1972 (Pub. L. 92-516, 86 Stat. 973), and other legislation supplementary thereto and amendatory thereof.

Act (Section 163.2) means the Federal Food, Drug, and Cosmetic Act (21 U.S.C. 301 et seq.), as amended by Pub. L. 518, 83d Congress, 2d Session, "An Act to amend the Federal Food, Drug, and Cosmetic Act with respect to residues of pesticide chemicals in or on raw agricultural commodities" (68 Stat. 511).

Act (Section 164.2) means the Federal Insecticide, Fungicide, and Rodenticide Act, as amended (86 Stat. 973) and other legislation supplementary thereto and amendatory thereof.

Act (Section 165.1) means the Federal Insecticide, Fungicide, and Rodenticide

Act as amended by the Federal Environmental Pesticide Control Act of 1972 (Pub. L. 92-516, 86 Stat. 973).

Act (Section 166.3) means the Federal Insecticide, Fungicide, and Rodenticide Act, as amended, 7 U.S.C. 136 et seq.

Act (Section 167.1) as used in this part, means the Federal Insecticide, Fungicide, and Rodenticide Act, as amended (86 Stat. 973-999).

Act (Section 17.2) means section 504 of Title 5, United States Code, as amended by section 203(a)(1) of the Equal Access to Justice Act, Pub. L. No. 96-481.

Act (Section 171.2) means the Federal Insecticide, Fungicide, and Rodenticide Act, as amended (86 Stat. 973), and other legislation supplementary thereto and amendatory thereof.

Act (Section 172.1) means the Federal Insecticide, Fungicide, and Rodenticide Act, as amended (86 Stat. 973), and other legislation supplementary thereto and amendatory thereof.

Act (Section 192.00), for purposes of subparts A, B, and C of this part, means the Uranium Mill Tailings Radiation Control Act of 1978.

Act (Section 2.301) means the Clean Air Act, as amended, 42 U.S.C. 7401 et seq.

Act (Section 2.302) means the Clean Water Act, as amended, 33 U.S.C. 1251 et seq.

Act (Section 2.303) means the Noise Control Act of 1972, 42 U.S.C. 4901 et seq.

Act (Section 2.304) means the Safe Drinking Water Act, 42 U.S.C. 300f et seq.

Act (Section 2.305) means the Solid Waste Disposal Act, as amended, including amendments made by the Resource Conservation and Recovery Act of 1976, as amended, 42 U.S.C. 6901 et seq.

Act (Section 2.306) means the Toxic Substances Control Act, 15 U.S.C. 2601 et seq.

Act (Section 2.307) means the Federal Insecticide, Fungicide and Rodenticide Act, as amended, 7 U.S.C. 136 et seq., and its predecessor, 7 U.S.C. 135 et seq.

Act (Section 2.308) means the Federal Food, Drug and Cosmetic Act, as amended, 21 U.S.C. 301 et seq.

Act (Section 2.309) means the Marine Protection, Research and Sanctuaries Act of 1972, 33 U.S.C. 1401 et seq.

Act (Section 2.310) means the Comprehensive Environmental Response, Compensation, and Liability Act of 1980, 42 U.S.C. 9601 et seq.

Act (Section 2.311) means the Motor Vehicle Information and Cost Savings Act, as amended, 15 U.S.C. 1901 et seq.

Act (Section 20.2) means, when used in connection with water pollution control facilities, the Federal Water Pollution Control Act, as amended (33 U.S.C. 1251 et seq.) or, when used in connection with air pollution control facilities, the Clean Air Act, as amended (42 U.S.C. 1857 et seq.).

Act (Section 201.1) means the Noise Control Act of 1972 (Pub. L. 92-574, 86 Stat. 1234).

Act (Section 202.10) means the Noise Control Act of 1972 (Pub. L. 92-574, 86 Stat. 1234).

Act (Section 203.1) means the Noise Control Act of 1972 (Pub. L. 92-574).

Act (Section 204.2) means the Noise Control Act of 1972 (Pub. L. 92-574, 86 Stat. 1234).

Act (Section 205.2) means the Noise Control Act of 1972 (Pub. L. 92-574, 86 Stat. 1234).

Act (Section 209.3) means the Noise Control Act of 1972 (42 U.S.C. 4901, et seq.).

Act (Section 21.2) means the Federal Water Pollution Control Act, 33 U.S.C. 1151, et seq.

Act (Section 211.102) means the Noise Control Act of 1972 (Pub. L. 92-574, 86 Stat. 1234).

Act (Section 22.03) means the particular statute authorizing the institution of the proceeding at issue.

Act (Section 220.2) means the Marine Protection, Research, and Sanctuaries Act of 1972, as amended (33 U.S.C. 1401).

Act (Section 230.3) means the Clean Water Act (also known as the Federal Water Pollution Control Act or FWPCA) Pub. L. 92-500, as amended by Pub. L. 95-217, 33 U.S.C. 1251, et seq.

Act (Section 232.2 [2/11/93])* means the Clean Water Act (33 U.S.C. 1251 et seq.).

Act (Section 233.2 [6/6/88]) means the Clean Water Act (33 U.S.C. 1251 et seq.).

Act (Section 256.06) means the Solid Waste Disposal Act, as amended by the Resource Conservation and Recovery Act of 1976 (42 U.S.C. 6901 et seq.).

Act (Section 35.2005) means The Clean Water Act (33 U.S.C. 1251 et seq., as amended).

Act (Section 35.3105 [3-19-90]) The Federal Water Pollution Control Act, more commonly known as the Clean

Water Act (Pub. L. 92-500), as amended by the Water Quality Act of 1987 (Pub. L. 100-4). 33 U.S.C. 1251 et seq.

Act (Section 35.905) means the Clean Water Act (33 U.S.C. 1251 et seq., as amended.).

Act (Section 355.20) means the Superfund Amendments and Reauthorization Act of 1986.

Act (Section 39.105) means The Federal Water Pollution Control Act Amendments of 1972, as amended (Pub. L. 92-500, 33 U.S.C. 1281 et seq.).

Act (Section 401.11) means the Federal Water Pollution Control Act, as amended, 33 U.S.C. 1251 et seq., 86 Stat. 816, Pub. L. 92-500.

Act (Section 403.3) means Federal Water Pollution Control Act, also known as the Clean Water Act, as amended, 33 U.S.C. 1251, et seq.

Act (Section 50.1) means the Clean Air Act, as amended (42 U.S.C. 1857-18571, as amended by Pub. L. 91-604).

Act (Section 51.100) means the Clean Air Act (42 U.S.C. 7401 et seq., as amended by Pub. L. 91-604, 84 Stat. 1676 Pub. L. 95-95, 91 Stat., 685 and Pub. L. 95-190, 91 Stat., 1399).

Act (Section 53.1) means the Clean Air Act (42 U.S.C. 1857-18571), as amended.

Act (Section 56.1) means the Clean Air Act as amended (42 U.S.C. 7401 et seq.).

Act (Section 57.103) means the Clean Air Act, as amended.

Act (Section 58.1) means the Clean Air Act as amended (42 U.S.C. 7401 et seq.).

Act (Section 60.2 [7/21/92])* means the Clean Air Act (42 U.S.C. 7401 et seq.)

Act (Section 600.002-85) means Part I of Title V of the Motor Vehicle Information and Cost Savings Act (15 U.S.C. 1901 et seq.).

Act (Section 61.02) means the Clean Air Act (42 U.S.C. 7401 et seq.).

Act (Section 65.01) shall mean the Clean Air Act, as amended, 42 U.S.C. 7401 et seq.

Act (Section 66.3) means the Clean Air Act, 42 U.S.C. 7401 et seq. as amended on August 7, 1977, except where the context specifically indicates otherwise.

Act (Section 710.2) means the Toxic Substances Control Act, 15 U.S.C. 2601 et seq.

Act (Section 720.3) means the Toxic Substances Control Act, 15 U.S.C.

2601 et seq.

Act (Section 723.175) means the Toxic Substances Control Act, 15 U.S.C. 2601 et seq.

Act (Section 723.250) means the Toxic Substances Control Act, 15 U.S.C. 2601 et seq.

Act (Section 723.50) means the Toxic Substances Control Act, 15 U.S.C. 2601 et seq.

Act (Section 747.115) means the Toxic Substances Control Act, 15 U.S.C. 2601 et seq.

Act (Section 747.195) means the Toxic Substances Control Act, 15 U.S.C. 2601 et seq.

Act (Section 747.200) means the Toxic Substances Control Act, 15 U.S.C. 2601 et seq.

Act (Section 763.103) means the Toxic Substances Control Act (TSCA), 15 U.S.C. 2601, et seq.

Act (Section 763.83 [10/30/87]) means the Toxic Substances Control Act (TSCA), 15 U.S.C. 2601, et seq.

Act (Section 79.2) means the Clean Air Act (42 U.S.C. 1857 et seq., as amended by Pub. L. 91-604).

Act (Section 790.3) means the Toxic Substances Control Act, 15 U.S.C. 2601 et seq.

Act (Section 791.3) means the Toxic Substances Control Act (TSCA), 15 U.S.C. 2601, et seq.

Act (Section 80.2) means the Clean Air Act, as amended (42 U.S.C. 1857 et seq.).

Act (Section 85.1502 [9/25/87]) means the Clean Air Act, as amended (42 U.S.C. 7401 et seq.).

Act (Section 85.1801) shall mean the Clean Air Act, 42 U.S.C. 1857, as amended.

Act (Section 85.1902) shall mean the Clean Air Act, 42 U.S.C. 1857, as amended.

Act (Section 85.2102) means Part A of Title II of the Clean Air Act, 42 U.S.C. 7421 et seq. (formerly 42 U.S.C. 1857 et seq.), as amended.

Act (Section 85.2113) means Part A of Title II of the Clean Air Act, 42 U.S.C. 7421 et seq. (formerly 42 U.S.C. 1857 et seq.) as amended.

Act (Section 86.082-2) means Part A of Title II of the Clean Air Act, 42 U.S.C. as amended, 7521, et seq.

Act (Section 86.402-78) means Part A of Title II of the Clean Air Act, 42 U.S.C. 1857 f-1 through f-7, as amended by Pub. L. 91-604.

Act (Section 87.1) means the Clean Air Act, as amended (42 U.S.C. 7401 et seq.).

Act or CERCLA (Section 305.12) means the Comprehensive Environmental Response, Compensation, and Liability Act of 1980 (42 U.S.C. 9601 et seq.).

Act or CERCLA (Section 306.12) means the Comprehensive Environmental Response, Compensation, and Liability Act of 1980 (42 U.S.C. 9601 et seq.).

Act, or CERCLA, or Superfund (Section 302.3) means the Comprehensive Environmental Response, Compensation, and Liability Act of 1980 (Pub. L. 96-510).

Act or FIFRA (Section 152.3 [5/4/88]) means the Federal Insecticide, Fungicide, and Rodenticide Act, as amended (7 U.S.C. 136-136y).

Act or FIFRA (Section 154.3) means the Federal Insecticide, Fungicide, and Rodenticide Act, as amended, 7 U.S.C. 136 et seq.

Act or RCRA (Section 248.4 [2/17/89]) means the Solid Waste Disposal Act, as amended by the Resource Conservation and Recovery Act, as amended, 42 U.S.C. 6901 et seq.

Act or RCRA (Section 249.04) means the Solid Waste Disposal Act, as amended by the Resource Conservation and Recovery Act of 1976, as amended, 42 U.S.C. 6901 et seq.

Act or RCRA (Section 250.4 [10/6/87, 6/22/88]) means the Solid Waste Disposal Act, as amended by the Resource Conservation and Recovery Act, as amended, 42 U.S.C. 6901 et seq.

Act or RCRA (Section 252.4 [6/30/88]) means the Solid Waste Disposal Act, as amended by the Resource Conservation and Recovery Act, as amended, 42 U.S.C. 6901 et seq.

Act or RCRA (Section 253.4 [11/17/88]) means the Solid Waste Disposal Act, as amended by the Resource Conservation and Recovery Act, as amended, 42 U.S.C. 6901 et seq.

Act or RCRA (Section 260.10) means the Solid Waste Disposal Act, as amended by the Resource Conservation and Recovery Act of 1976, as amended, 42 U.S.C. section 6901 et seq.

Action level (Section 141.2 [7/17/92])* is the concentration of lead or copper in water specified in Sec. 141.80(c) which determines, in some cases, the treatment requirements contained in subpart I of this part that a water system is required to complete.

Action level (Section 763.121) means an airborne concentration of asbestos of 0.1 fiber per cubic centimeter (f/cc) of air calculated as an 8-hour time-weighted average.

Activation (Section 300.5 [3-8-90])* means notification by telephone or other expeditious manner or, when required, the assembly of some or all appropriate members of the RRT or NRT.

Active ingredient (Section 152.3 [5/4/88]) means any substance (or group of structurally similar substances if specified by the Agency) that will prevent, destroy, repel or mitigate any pest, or that functions as a plant regulator, desiccant, or defoliant within the meaning of FIFRA section 2(a).

Active ingredient (Section 158.153 [5/4/88]) means any substance (or group of structurally similar substances, if specified by the Agency) that will prevent, destroy, repel or mitigate any pest, or that functions as a plant regulator, desiccant, or defoliant within the meaning of FIFRA sec. 2(a).

Active ingredient (Section 162.3) means (1) In the case of a pesticide other than a plant regulator, defoliant, or desiccant, an ingredient which will prevent, destroy, repel, or mitigate any pest; (2) In the case of a plant regulator, an ingredient which, through physiological, biochemical action, will accelerate or retard the rate of growth or rate of maturation or otherwise alter the behavior of ornamental or crop plants or the product thereof; (3) In the case of a defoliant, an ingredient which will cause the leaves or foliage to drop from a plant; and (4) In the case of a desiccant, an ingredient which will artificially accelerate the drying of plant tissue.

Active ingredient (Section 455.10) means an ingredient of a pesticide which is intended to prevent, destroy, repel, or mitigate any pest.

Active institutional control (Section 191.12 [12/20/93])* means: (1) Controlling access to a disposal site by any means other than passive institutional controls; (2) performing maintenance operations or remedial actions at a site, (3) controlling or cleaning up releases from a site, or (4) monitoring parameters related to disposal system performance.

Active life (Section 258.2 [10-9-91]) means the period of operation beginning with the initial receipt of solid waste and ending at completion of closure activities in accordance with Sec. 258.60 of this part.

Active life of a facility (Section 260.10) means the period from the initial receipt of hazardous waste at the facility until the Regional Administrator receives certification of final closure.

Active mine (Section 61.21 [12-15-89])* means an underground uranium mine which is being ventilated to allow workers to enter the mine for any purpose.

Active mining area (Section 434.11) means the area, on and beneath land, used or disturbed in activity related to the extraction, removal, or recovery of coal from its natural deposits. This term excludes coal preparation plants, coal preparation plant associated areas and post-mining areas.

Active mining area (Section 440.132) is a place where work or other activity related to the extraction, removal, or recovery of metal ore is being conducted, except, with respect to surface mines, any area of land on or in which grading has been completed to return the earth to desired contour and reclamation work has begun.

Active portion (Section 258.2 [10-9-91]) means that part of a facility or unit that has received or is receiving wastes and that has not been closed in accordance with Sec. 258.60 of this part.

Active portion (Section 260.10) means that portion of a facility where treatment, storage, or disposal operations are being or have been conducted after the effective date of Part 261 of this chapter and which is not a closed portion. (See also closed portion and inactive portion.)

Active service (Section 60.691 [11/23/88]) means that a drain is receiving refinery wastewater from a process unit that will continuously maintain a water seal.

Active use (Section 57.103) refers to an SO_2 constant control system installed at a smelter before August 7, 1977 and not totally removed from regular service by that date.

Active waste disposal site (Section 61.141) means any disposal site other than an inactive site.

Activity (Section 35.6015 [6-5-90])* A set of CERCLA-funded tasks that makes up a segment of the sequence of events undertaken in determining, planning, and conducting a response to a release or potential release of a hazardous substance. These include Core Program, pre-remedial (i.e. preliminary assessments and site inspections), support agency, remedial investigation/feasibility studies, remedial design, remedial action, removal, and enforcement activities.

Activity (Section 61 [12-15-89]) The amount of a radioactive material. It is a measure of the transformation rate of radioactive nuclei at a given time. The customary unit of activity, the curie, is 3.7×10^{10} nuclear transformations per second.

Activity subject to regulation (Section 124.41). See **Regulated activity (Section 124.41)**.

Acts (Section 372.3 [2/16/88]) means Title III.

Actual credits (Section 86.090-2 [7-26-90]) refer to emission credits based on actual U.S. production volumes as contained in the end-of-year reports submitted to EPA. Some or all of these credits may be revoked if EPA review of the end of year reports or any subsequent audit actions uncover problems or errors.

Actual emissions (Section 51.165) means (A) the actual rate of emissions of a pollutant from an emissions unit as determined in accordance with paragraphs (a)(1)(xii) (B) through (D) of this section. (B) In general, actual emissions as of a particular date shall equal the average rate, in tons per year, at which the unit actually emitted the pollutant during a two-year period which precedes the particular date and which is representative of normal source operation. The reviewing authority shall allow the use of a different time period upon a determination that it is more representative of normal source operation. Actual emissions shall be calculated using the unit's actual operating hours, production rates, and types of materials processed, stored, or combusted during the selected time period. (C) The reviewing authority may presume that the source-specific allowable emissions for the unit are equivalent to the actual emissions of the unit. (D) For any emissions unit which has not begun normal operations on the particular date, actual emissions shall equal the potential to emit of the unit on that date.

Actual emissions (Section 51.166) means (i) the actual rate of emissions of a pollutant from an emissions unit, as determined in accordance with paragraphs (b)(21) (ii) through (iv) of this section. (ii) In general, actual emissions as of a particular date shall equal the average rate, in tons per year, at which the unit actually emitted the pollutant during a two-year period which precedes the particular date and which is representative of normal source operation. The reviewing authority may allow the use of a different time period upon a determination that it is more representative of normal source operation. Actual emissions shall be calculated using the unit's actual operating hours, production rates, and types of materials processed, stored, or combusted during the selected time period. (iii) The reviewing authority may presume that source-specific allowable emissions for the unit are equivalent to the actual emissions of the unit. (iv) For any emissions unit which has not begun normal operations on the particular date, actual emissions shall equal the potential to emit of the unit on that date.

Actual emissions (Section 52.24) means (i) the actual rate of emissions of a pollutant from an emissions unit, as determined in accordance with paragraphs (f) (ii) through (iv) of this section. (ii) In general, actual emissions as of a particular date shall equal the average rate, in tons per year, at which the unit actually emitted the pollutant during a two-year period which precedes the particular date and which is representative of normal source operation. The Administrator shall allow the use of a different time period upon a determination that it is more representative of normal source operation. Actual emissions shall be calculated using the unit's actual operating hours, production rates, and types of materials processed, stored, or combusted during the selected time period. (iii) The Administrator may presume that source-specific allowable emissions for the unit are equivalent to the actual emissions of the unit. (iv) For any emissions unit which has not begun normal operations on the particular date, actual emissions shall equal the potential to emit of the unit on that date.

Acute dermal LD_{50} (Section 152.3 [5/4/88]) means a statistically derived estimate of the single dermal dose of a substance that would cause 50 percent mortality to the test population under specified conditions.

Acute dermal LD_{50} (Section 162.3) means a single dermal dose of a substance, expressed as milligrams per kilogram of body weight, that is lethal to 50% of the test population of animals under test conditions as specified in the Registration Guidelines.

Acute inhalation LC_{50} (Section 152.3 [5/4/88]) means a statistically derived estimate of the concentration of a substance that would cause 50 percent mortality to the test population under specified conditions.

Acute inhalation toxicity (Section 798.1150) is the adverse effects caused by a substance following a single uninterrupted exposure by inhalation over a short period of time (24 hours or less) to a substance capable of being inhaled.

Acute LC_{50} (Section 162.3) means a concentration of a substance, expressed as parts per million parts of medium, that is lethal to 50% of the test population of animals under test conditions as specified in the Registration Guidelines.

Acute lethal toxicity (Section 797.1350) is the lethal effect produced on an organism within a short period of time of exposure to a chemical.

Acute oral LD_{50} (Section 152.3 [5/4/88]) means a statistically derived estimate of the single oral dose of a substance that would cause 50 percent mortality to the test population under specified conditions.

Acute oral LD_{50} (Section 162.3) means a single orally administered dose of a substance, expressed as milligrams per kilogram of body weight, that is lethal to 50 percent of the test population of animals under test conditions as specified in the Registration Guidelines.

Acute oral toxicity (Section 798.1175) is the adverse effects occurring within a short time of oral administration of a single dose of a substance or multiple doses given within 24 hours.

Acute toxicity (Section 162.3) means the property of a substance or mixture of substances to cause adverse effects in an organism through a single short-term exposure.

Acute toxicity (Section 797.1800) is the discernible adverse effects induced in an organism within a short period of time (days) of exposure to a chemical. For aquatic animals this usually refers to continuous exposure to the chemical in water for a period of up to four days. The effects (lethal or sublethal) occurring may usually be observed within the period of exposure with aquatic organisms. In this test guideline, shell deposition is used as the measure of toxicity.

Acute toxicity test (Section 797.1400) means a method used to determine the concentration of a substance that produces a toxic effect on a specified percentage of test organisms in a short period of time (e.g., 96 hours). In this guideline, death is used as the measure of toxicity.

Acutely toxic effects (Section 721.3 [7-27-89]) A chemical substance produces acutely toxic effects if it kills within a short time period (usually 14 days): (1) At least 50 percent of the exposed mammalian test animals following oral administration of a single dose of the test substance at 25 milligrams or less per kilogram of body weight (LD_{50}). (2) At least 50 percent of the exposed mammalian test animals following dermal administration of a single dose of the test substance at 50 milligrams or less per kilogram of body weight (LD_{50}). (3) At least 50 percent of the exposed mammalian test animals following administration of the test substance for 8 hours or less by continuous inhalation at a steady concentration in air at 0.5 milligrams or less per liter of air (LC_{50}).

Ad valorem tax (Sections 35.905; 35.2005) means a tax based upon the value of real property.

Adaptation (Section 796.3100) is the process by which a substance induces the synthesis of any degradative enzymes necessary to catalyze the transformation of that substance.

Added ingredients (Section 407.81) shall mean the prepared sauces (prepared from items such as dairy products, starches, sugar, tomato sauce and concentrate, spices, and other related preprocessed ingredients) which are added during the canning and freezing of fruits and vegetables.

Additional advance auction (Section 73.3 [12-17-91]) means the auction of advance allowances that were offered the previous year for sale in an advance sale.

Additions and alterations (Section 21.2) means the act of undertaking construction of any facility.

Additive (Section 79.2 [6/27/94])* means any substance that is intentionally added to a fuel named in the designation (including any added to a motor vehicle's fuel system) and that is not intentionally removed prior to sale or use.

Additive (Section 790.3) means a chemical substance that is intentionally added to another chemical substance to improve its stability or impart some other desirable quality.

Additive manufacturer (Section 79.2 [6/27/94])* means any person who produces, manufactures, or imports an additive for use as an additive and/or sells or imports for sale such additive under the person's own name.

Adequate evidence (Section 32.102) means more than mere accusation but less than substantial evidence. Consideration must be given to the amount of credible information available, reasonableness in view of surrounding circumstances, corroboration, and other inferences which may be drawn from the existence or absence of affirmative facts.

Adequate SO_2 emission limitation (Section 57.103) means a SIP emission limitation which was approved or promulgated by EPA as adequate to attain and maintain the NAAQS in the areas affected by the stack emissions without the use of any unauthorized dispersion technique.

Adequate storage (Section 165.1) means placing of pesticides in proper containers and in safe areas as per Section 165.10 as to minimize the possibility of escape which could result in unreasonable adverse effects on the environment.

Adequately wet (Section 61.141 [11-20-90]) means sufficiently mix or penetrate with liquid to prevent the release of particulates. If visible emissions are observed coming from asbestos-containing material, then that material has not been adequately wetted. However, the absence of visible emissions is not sufficient evidence of being adequately wet.

Adequately wetted (Section 61.141) means sufficiently mixed or coated with water or an aqueous solution to prevent dust emissions.

Adjacent (Section 230.3) means bordering, contiguous, or neighboring. Wetlands separated from other waters of the United States by man-made dikes or barriers, natural river berms, beach dunes, and the like are adjacent wetlands.

Adjusted configuration (Section 610.11) means the test configuration after adjustment of engine calibrations to the retrofit specifications, but excluding retrofit hardware installation.

Adjusted Loaded Vehicle Weight (Section 86.094-2 [6-5-91]) means the numerical average of vehicle curb weight and GVWR.

Administering agency (Section 8.2) means any department, agency, and establishment in the Executive Branch of the Government, including any wholly owned Government corporation, which administers a program involving federally assisted construction contracts.

Administrative action (Section 6.601) for the sake of this subpart means the issuance by EPA of an NPDES permit to discharge as a new source, pursuant to 40 CFR 124.15.

Administrative Law Judge (Section 209.3) means an administrative law judge appointed under 5 U.S.C. 3105 (see also 5 CFR Part 930, as amended by 37 FR 16787). Administrative law judge is synonymous with hearing examiner as used in Title 5 of the United States Code.

Administrative Law Judge (Section 57.103) means an administrative law judge appointed under 5 U.S.C. 3105 (see also 5 CFR Part 930, as amended by 37 FR 16787), and is synonymous with the term Hearing Examiner as formerly used in Title 5 of the U.S. Code.

Administrative Law Judge (Section 164.2) means an Administrative Law Judge appointed pursuant to 5 U.S.C. 3105 (see also 5 CFR Part 930, as amended), and such term is synonymous with the term Hearing Examiner as used in the Act or in the United States Code.

Administrative Law Judge (Sections 22.03; 123.64) means an Administrative Law Judge appointed under 5 U.S.C. 3105 (see also Pub. L. 95-251, 92 Stat. 183).

Administrative offset (Section 13.2 [9/23/88]) means the withholding of money payable by the United States to, or held by the United States for, a person to satisfy a debt the person owes the Government.

Administrator (Regional Administrator) (Section 149.101) means the Administrator (Regional Administrator) of the United States Environmental Protection Agency.

Administrator (Section 104.2) means the Administrator of the Environmental Protection Agency, or any employee of the Agency to whom the Administrator may by order delegate his authority to carry out his functions under section 307(a) of the Act, or any person who shall by operation of law be authorized to carry out such functions.

Administrator (Section 110.1) means the Administrator of the Environmental Protection Agency (EPA).

Administrator (Section 117.1) means the Administrator of the Environmental Protection Agency (EPA).

Administrator (Section 121.1) means the Administrator, Environmental Protection Agency.

Administrator (Section 122.2) means the Administrator of the United States Environmental Protection Agency, or an authorized representative.

Administrator (Section 124.2) means the Administrator of the U.S. Environmental Protection Agency, or an authorized representative.

Administrator (Section 124.41) means the Administrator of the U.S. Environmental Protection Agency, or an authorized representative.

Administrator (Section 125.58) means the EPA Administrator or a person designated by the EPA Administrator.

Administrator (Section 129.2) means the Administrator of the Environmental Protection Agency or any employee of the Agency to whom the Administrator may by order delegate the authority to carry out his functions under section 307(a) of the Act, or any person who shall by operation of law be authorized to carry out such functions.

Administrator (Section 13.2 [9/23/88]) means the Administrator of EPA or an EPA employee or official designated to act on the Administrator's behalf.

Administrator (Section 136.2) means the Administrator of the U.S. Environmental Protection Agency.

Administrator (Section 142.2) means the Administrator of the United States Environmental Protection Agency or his authorized representative.

Administrator (Section 144.3) means the Administrator of the United States Environmental Protection Agency, or an authorized representative.

Administrator (Section 146.3) means the Administrator of the United States Environmental Protection Agency, or an authorized representative.

Administrator (Section 147.2902) means the Administrator of the United States Environmental Protection Agency, or an authorized representative.

Administrator (Section 15.4) means the Administrator of the United States Environmental Protection Agency or his or her designee.

Administrator (Section 152.3 [5/4/88]) means the Administrator of the United States Environmental Protection Agency or his delegate.

Administrator (Section 154.3) means the Administrator of the Environmental Protection Agency or any officer or employee thereof to whom authority has been delegated to act for the Administrator.

Administrator (Section 162.3)

Administrator (Section 162.3) means the Administrator of the United States Environmental Protection Agency or any officer or employee of the Agency to whom authority has heretofore been delegated or to whom authority may hereafter be delegated to act in his stead.

Administrator (Section 164.2 [7/10/92])* means the Administrator of the United States Environmental Protection Agency.

Administrator (Section 165.1) means the Administrator of the Agency, or any officer or employee thereof to whom authority has been heretofore delegated or to whom authority may hereafter be delegated, to act in his stead.

Administrator (Section 17.2) means the Administrator of the Environmental Protection Agency.

Administrator (Section 171.2) means the Administrator of the Environmental Protection Agency, or any officer or employee of the Agency to whom authority has heretofore been delegated, or to whom authority may hereafter be delegated, to act in his stead.

Administrator (Section 177.3 [12-5-90]) means the Administrator of the Agency, or an officer or employee of the Agency to whom the Administrator has delegated the authority to perform functions under this part.

Administrator (Section 178.3 [12-5-90]) means the Administrator of the Agency, or any officer or employee of the Agency to whom the Administrator delegates the authority to perform functions under this part.

Administrator (Section 179.3 [12-5-90]) means the Administrator of the Agency, or any officer or employee of the Agency to whom the Administrator has delegated the authority to perform functions under this part.

Administrator (Section 191.02) means the Administrator of the Environmental Protection Agency.

Administrator (Section 2.201) means the EPA officer or employee occupying the position so titled.

Administrator (Section 20.2) means the Administrator, Environmental Protection Agency.

Administrator (Section 203.1) means the Administrator of the Environmental Protection Agency.

Administrator (Section 204.2) means the Administrator of the Environmental Protection Agency or his authorized representative.

Administrator (Section 205.2) means the Administrator of the Environmental Protection Agency or his authorized representative.

Administrator (Section 209.3) means

the Administrator of the Environmental Protection Agency or his or her delegate.

Administrator (Section 211.102) means the Administrator of the Environmental Protection Agency or his authorized representative.

Administrator (Section 22.03) means the Administrator of the United States Environmental Protection Agency or his delegate.

Administrator (Section 23.1) means the Administrator or any official exercising authority delegated by the Administrator.

Administrator (Section 232.2 [2/11/93])* means the Administrator of the Environmental Protection Agency or an authorized representative.

Administrator (Section 233.3) means the Administrator of the United States Environmental Protection Agency, or an authorized representative.

Administrator (Section 260.10) means the Administrator of the Environmental Protection Agency, or his designee.

Administrator (Section 270.2) means the Administrator of the United States Environmental Protection Agency, or an authorized representative.

Administrator (Section 302.3) means the Administrator of the United States Environmental Protection Agency (EPA).

Administrator (Section 304.12 [5/30/89]) means the EPA Administrator or his designee.

Administrator (Section 401.11) means the Administrator of the United States Environmental Protection Agency.

Administrator (Section 50.1) means the Administrator of the Environmental Protection Agency.

Administrator (Section 501.2 [5/2/89]) means the Administrator of the United States Environmental Protection Agency, or an authorized representative.

Administrator (Section 51.100) means the Administrator of the Environmental Protection Agency (EPA) or an authorized representative.

Administrator (Section 53.1) means the Administrator of the Environmental Protection Agency or his authorized representative.

Administrator (Section 57.103) means the Administrator of the U.S. Environmental Protection Agency, or the Administrator's authorized representative.

Administrator (Section 58.1) means the Administrator of the Environmental Protection Agency (EPA) or his or her authorized representative.

Administrator (Section 60.2) means the Administrator of the Environmental Protection Agency or his authorized representative.

Administrator (Section 600.002-85) means the Administrator of the Environmental Protection Agency or his authorized representative.

Administrator (Section 61.02) means the Administrator of the Environmental Protection Agency or his authorized representative.

Administrator (Section 65.01) shall mean the Administrator of the U.S. Environmental Protection Agency, or his authorized delegatee.

Administrator (Section 7.25) means the Administrator of EPA. It includes any other agency official authorized to act on his or her behalf, unless explicitly stated otherwise.

Administrator (Section 710.2) means the Administrator of the U.S. Environmental Protection Agency, any employee or authorized representative of the Agency to whom the Administrator may either herein or by order delegate his authority to carry out his functions, or any other person who shall by operation of law be authorized to carry out such functions.

Administrator (Section 723.250) has the same meaning as in section 3 of the Act (15 U.S.C. 2602).

Administrator (Section 73.3 [12-17-91]) means the Administrator of the United States Environmental Protection Agency or the Administrator's duly authorized representative.

Administrator (Section 761.3) means the Administrator of the Environmental Protection Agency, or any employee of the Agency to whom the Administrator may either herein or by order delegate his authority to carry out his functions, or any person who shall by operation of law be authorized to carry out such functions.

Administrator (Section 762.3) has the same meaning as in 15 U.S.C. 2602.

Administrator (Section 763.121) means the Administrator, U.S. Environmental Protection Agency, or designee.

Administrator (Section 79.264) means the Administrator of the Environmental Protection Agency.

Administrator (Section 8.2) means the Administrator of the Environmental Protection Agency.

Administrator (Section 8.33) means the Administrator of the Environmental Protection Agency.

Administrator (Section 80.2) means the Administrator of the Environmental Protection Agency.

Administrator (Section 82.3 [8/12/88]) means the Administrator of the

Environmental Protection Agency or his authorized representative.

Administrator (Section 85.1502 [9/25/87]) means the Administrator of the Environmental Protection Agency.

Administrator (Section 85.2102) means the Administrator of the Environmental Protection Agency or an authorized representative of the Administrator.

Administrator (Section 86.082-2) means the Administrator of the Environmental Protection Agency or his authorized representative.

Administrator (Section 86.402-78) means the Administrator of the Environmental Protection Agency or his authorized representative.

Administrator (Section 87.1) means the Administrator of the Environmental Protection Agency and any other officer or employee of the Environmental Protection Agency to whom authority involved may be delegated.

Administrator and General Counsel (Section 350.1 [7/29/88]) mean the EPA officers or employees occupying the positions so titled.

Administrator or Regional Administrator (Section 108.2) means the Administrator or a Regional Administrator of the Environmental Protection Agency.

Adsorption ratio, K_d (Section 796.2750) is the amount of test chemical adsorbed by a sediment or soil (i.e., the solid phase) divided by the amount of test chemical in the solution phase, which is in equilibrium with the solid phase, at a fixed solid/solution ratio.

Advance allowance (Section 73.3 [12-17-91]) means an allowance that may be used for purposes of compliance with a unit's sulfur dioxide emissions limitation requirements beginning no earlier than seven years following the year in which the allowance is first offered for sale.

Advance Auction (Section 73.3 [12-17-91]) means an auction of an advance allowance.

Advance Sale (Section 73.3 [12-17-91]) means a sale of an advance allowance.

Advanced air emission control devices (Section 426.11) shall mean air pollution control equipment, such as electrostatic precipitators and high energy scrubbers, that are used to treat an air discharge which has been treated initially by equipment including knockout chambers and low energy scrubbers.

Adversary adjudication (Section 17.2) means an adjudication required by statute to be held pursuant to 5 U.S.C. 554 in which the position of the United States is represented by counsel or otherwise, but excludes an adjudication for the purpose of granting or renewing a license.

Adverse impact on visibility (Section 52.21) means visibility impairment which interferes with the management, protection, preservation or enjoyment of the visitor's visual experience of the Federal Class I area. This determination must be made on a case-by-case basis taking into account the geographic extent, intensity, duration, frequency and time of visibility impairment, and how these factors correlate with (1) times of visitor use of the Federal Class I area, and (2) the frequency and timing of natural conditions that reduce visibility.

Adverse impact on visibility (Section 51.301) means, for purposes of section 307, visibility impairment which interferes with the management, protection, preservation, or enjoyment of the visitor's visual experience of the Federal Class I area. This determination must be made on a case-by-case basis taking into account the geographic extent, intensity, duration, frequency and time of visibility impairments, and how these factors correlate with (1) times of visitor use of the Federal Class I area, and (2) the frequency and timing of natural conditions that reduce visibility. This term does not include effects on integral vistas.

Adverse weather (Section 112.2 [8/25/93])* means the weather conditions that make it difficult for response equipment and personnel to cleanup or remove spilled oil, and that will be considered when identifying response systems and equipment in a response plan for the applicable operating environment. Factors to consider include significant wave height as specified in Appendix E to this part, as appropriate, ice conditions, temperatures, weather-related visibility, and currents within the area in which the systems or equipment are intended to function.

Advertised Engine Displacement (Section 205.151) means the rounded off volumetric engine capacity used for marketing purposes by the motorcycle manufacturer.

AECD (Section 600.002-85). See **Auxiliary Emission Control Device (Section 600.002-85)**.

AECD (Section 86.082-2). See **Auxiliary Emission Control Device (Section 86.082-2)**.

Aerodynamic diameter (Section 798.4350) applies to the behavioral size of particles of aerosols. It is the diameter of a sphere of unit density which behaves aerodynamically like the particles of the test substance. It is used to compare particles of different sizes, shapes, and densities and to predict where in the respiratory tract such particles may be deposited. This term is used in contrast to optical, measured or geometric diameters which are representation of actual diameters which in themselves cannot be related to deposition within the respiratory tract.

Aerodynamic diameter (Sections 798.1150; 798.2450) applies to the size of particles of aerosols. It is the diameter of a sphere of unit density which behaves aerodynamically as the particle of the test substance. It is used to compare particles of different size and densities and to predict where in the respiratory tract such particles may be deposited. This term is used in contrast to measured or geometric diameter which is representative of actual diameters which in themselves cannot be related to deposition within the respiratory tract.

Aerosol propellant (Section 762.3) means a liquefied or compressed gas in a container where the purpose of the liquefied or compressed gas is to expel from the container liquid or solid material different from the aerosol propellant.

Affected (Section 35.4010 [10/1/92])* means subject to an actual or potential health, economic or environmental threat arising from a release or a threatened release at a facility listed on the National Priorities List (NPL) or proposed for listing under the National Oil and Hazardous Substances Pollution Contingency Plan (NCP) where a response action under CERCLA has begun. Examples of affected parties include individuals who live in areas adjacent to NPL facilities whose health is or may be endangered by releases of hazardous substances at the facility, or whose economic interests are directly threatened or harmed.

Affected business (Section 2.201) means, with reference to an item of business information, a business which has asserted (and not waived or withdrawn) a business confidentiality claim covering the information, or a business which could be expected to make such a claim if it were aware that disclosure of the information to the public was proposed.

Affected facility (Section 60.2) means, with reference to a stationary source, any apparatus to which a standard is applicable.

Affecting (Section 1508.3) means will or may have an effect on.

Affiliate (Section 32.102) means any person whose governing instruments require it to be bound by the decision of another person or whose governing board includes enough voting representatives of the other person to cause or prevent action, whether or not the power is exercised. It may also include persons doing business under a variety of names, or where there is a parent/subsidiary relationship between persons.

Affiliate (Section 73.3 [12-17-91]) is defined as in section 2(a)(11) of the

Public Utility Holding Company Act of 1935, 15 U.S.C. 79b(a)(11).

Affiliated entity (Section 66.3) means a person who directly, or indirectly through one or more intermediaries, controls, is controlled by, or is under common control with the owner or operator of a source.

Afterburner (A/B) (Section 60.471) means an exhaust gas incinerator used to control emissions of particulate matter.

Aftermarket part (Section 763.163 [7-12-89]) means any part offered for sale for installation in or on a motor vehicle after such vehicle has left the manufacturer's production line.

Aftermarket Part (Section 85.2113) means any part offered for sale for installation in or on a motor vehicle after such vehicle has left the vehicle manufacturer's production line.

Aftermarket Part Manufacturer (Section 85.2113) means: (1) A manufacturer of an aftermarket part or, (2) A party that markets aftermarket parts under its own brand name, or, (3) A rebuilder of original equipment or aftermarket parts, or (4) A party that licenses others to sell its parts.

Aged Catalytic Converter (Section 85.2122(a)(15)(ii)(F)) means a converter that has been installed on a vehicle or engine stand and operated through a cycle specifically designed to chemically age, including exposure to representative lead concentrations, and mechanically stress the catalytic converter in a manner representative of in-use vehicle or engine conditions.

Agency (Section 104.2) means the Environmental Protection Agency.

Agency (Section 12.103 [8/14/87]) means the Environmental Protection Agency.

Agency (Section 13.2 [9/23/88]) means the United States Environmental Protection Agency.

Agency (Section 142.2) means the United States Environmental Protection Agency.

Agency (Section 15.4) means any department, agency, establishment, or instrumentality in the Executive Branch of the Federal Government, including corporations wholly owned by the Federal Government which award contracts, grants, or loans.

Agency (Section 152.3 [5/4/88]) means the United States Environmental Protection Agency (EPA), unless otherwise specified.

Agency (Section 162.3) means the United States Environmental Protection Agency (EPA), unless otherwise specified.

Agency (Section 163.2) means the Environmental Protection Agency.

Agency (Section 164.2) unless otherwise specified, means the United States Environmental Protection Agency.

Agency (Section 165.1) means the United States Environmental Protection Agency.

Agency (Section 166.3) means the United States Environmental Protection Agency.

Agency (Section 171.2) unless otherwise specified, means the United States Environmental Protection Agency.

Agency (Section 177.3 [12-5-90]) means the United States Environmental Protection Agency.

Agency (Section 178.3 [12-5-90]) means the United States Environmental Protection Agency.

Agency (Section 179.3 [12-5-90]) means the United States Environmental Protection Agency.

Agency (Section 191.02) means the Environmental Protection Agency.

Agency (Section 204.2) means the United States Environmental Protection Agency.

Agency (Section 205.2) means the United States Environmental Protection Agency.

Agency (Section 209.3) means the United States Environmental Protection Agency.

Agency (Section 211.102) means the United States Environmental Protection Agency.

Agency (Section 22.03) means the United States Environmental Protection Agency.

Agency (Section 34.105 [2-26-90]) as defined in 5 U.S.C. 552(f), includes Federal executive departments and agencies as well as independent regulatory commissions and Government corporations, as defined in 31 U.S.C. 9101(1).

Agency (Section 4.2) means the Federal agency, State or State agency which acquires the real property or displaces a person (see Section 4.2(f)).

Agency (Section 50.1) means the Environmental Protection Agency.

Agency (Section 51.301) means the United States Environmental Protection Agency.

Agency (Section 53.1) means the Environmental Protection Agency.

Agency (Section 761.3) means the United States Environmental Protection Agency.

Agency (Section 791.3) means the Environmental Protection Agency.

Agency (Section 8.2) means the Environmental Protection Agency.

Agency (Section 8.33) means the Environmental Protection Agency.

Agency (Section 85.2113) means the Environmental Protection Agency.

Agency trial staff (Section 124.78) means those agency employees, whether temporary or permanent, who have been designated by the Agency under Section 124.77 or Section 124.116 as available to investigate, litigate, and present the evidence, arguments, and position of the Agency in the evidentiary hearing or nonadversary panel hearing. Any EPA employee, consultant, or contractor who is called as a witness by EPA trial staff, or who assisted in the formulation of the draft permit which is the subject of the hearing, shall be designated as a member of the Agency trial staff.

Agency trial staff (Section 57.809) means those Agency employees, whether temporary or permanent, who have been designated by the Agency as available to investigate, litigate, and present the evidence arguments and position of the Agency in the evidentiary hearing or non-adversary panel hearing. Appearance as a witness does not necessarily require a person to be designated as a member of the Agency trial staff.

Aggregate costs (Section 35.9010 [10-3-89]) The total cost of all research, surveys, studies, modeling, and other technical work completed by a Management Conference during a fiscal year to develop a Comprehensive Conservation and Management Plan for the estuary.

Aggregate facility (Section 60.691 [11/23/88]) means an individual drain system together with ancillary downstream sewer lines and oil-water separators, down to and including the secondary oil-water separator, as applicable.

Agreement State (Section 191.02) means any State with which the Commission or the Atomic Energy Commission has entered into an effective agreement under subsection 274b of the Atomic Energy Act of 1954, as amended (68 Stat. 919).

Agreement State (Section 61.101 [12-15-89])* means a State with which the Atomic Energy Commission or Nuclear Regulatory Commission has entered into an effective agreement under subsection 274(b) of the Atomic Energy Act of 1954.

Agricultural commodity (Section 171.2) means any plant, or part thereof, or animal, or animal product, produced by a person (including farmers, ranchers, vineyardists, plant propagators, Christmas tree growers, aquaculturists, floriculturists, orchardists, foresters, or other comparable persons) primarily for sale, consumption, propagation, or other use by man or animals.

Agricultural employer (Section 170.3 [8/21/92])* means any person who hires or contracts for the services of workers, for any type of compensation, to perform activities related to the production of agricultural plants, or any person who is an owner of or is responsible for the management or condition of an agricultural establishment that uses such workers.

Agricultural establishment (Section 170.3 [8/21/92])* means any farm, forest, nursery, or greenhouse.

Agricultural plant (Section 170.3 [8/21/92])* means any plant grown or maintained for commercial or research purposes and includes, but is not limited to, food, feed, and fiber plants; trees; turfgrass; flowers, shrubs; ornamentals; and seedlings.

Agricultural solid waste (Sections 243.101; 246.101) means the solid waste that is generated by the rearing of animals, and the producing and harvesting of crops or trees.

Air Cleaner Filter Element (Section 85.2122(a)(16)(ii)(A)) means a device to remove particulates from the primary air that enters the air induction system of the engine.

Air emissions (Section 129.2) means the release or discharge of a toxic pollutant by an owner or operator into the ambient air either (1) by means of a stack or (2) as a fugitive dust, mist or vapor as a result inherent to the manufacturing or formulating process.

Air erosion (Section 763.83 [10/30/87]) means the passage of air over friable ACBM which may result in the release of asbestos fibers.

Air Oxidation Reactor (Section 60.611 [6-29-90; 9-7-90]) means any device or process vessel in which one or more organic reactants are combined with air, or a combination of air and oxygen, to produce one or more organic compounds. Ammoxidation and oxychlorination reactions are included in this definition.

Air Pollution Control Agency (Section 15.4) means any agency which is defined in section 302(b) or section 302(c) of the CAA.

Air quality restricted operation of a spray tower (Section 417.151) shall mean an operation utilizing formulations (e.g., those with high non-ionic content) which require a very high rate of wet scrubbing to maintain desirable quality of stack gases, and thus generate much greater quantities of waste water than can be recycled to process.

Air stripping operation (Section 264.1031 [6-21-90]) is a desorption operation employed to transfer one or more volatile components from a liquid mixture into a gas (air) either with or without the application of heat to the liquid. Packed towers, spray towers, and bubble-cap, sieve, or valve-type plate towers are among the process configurations used for contacting the air and a liquid.

Aircraft (Section 87.1) means any airplane for which a U.S. standard airworthiness certificate or equivalent foreign airworthiness certificate is issued.

Aircraft engine (Section 87.1) means a propulsion engine which is installed in or which is manufactured for installation in an aircraft.

Aircraft gas turbine engine (Section 87.1) means a turboprop, turbofan, or turbojet aircraft engine.

Airport (Section 257.3-8) means public-use airport open to the public without prior permission and without restrictions within the physical capacities of available facilities.

A_k (Section 60.741 [9-11-89]) means the area of each natural draft opening (k) in a total enclosure, in square meters.

Alaskan North Slope (Sections 60.591; 60.631) means the approximately 69,000 square mile area extending from the Brooks Range to the Arctic Ocean.

Alcohol abuse (Section 7.25) means any misuse of alcohol which demonstrably interferes with a person's health, interpersonal relations or working ability.

Aldrin/Dieldrin (Section 129.4) Aldrin means the compound aldrin as identified by the chemical name, 1,2,3,4,10,10-hexachloro-1,4,4a,5,8,8a-hexahydro-1,4-endo-5,8-exo-dimethanonaphthalene; Dieldrin means the compound dieldrin as identified by the chemical name 1,2,3,4,10,10-hexachloro-6,7-epoxy-1,4,4a,5,6,7,8,8a-octahydro-1,4-endo-5,8-exo-dimethanonaphthalene.

Aldrin/Dieldrin formulator (Section 129.100) means a person who produces, prepares or processes a formulated product comprising a mixture of either aldrin or dieldrin and inert materials or other diluents, into a product intended for application in any use registered under the Federal Insecticide, Fungicide and Rodenticide Act, as amended (7 U.S.C. 135, et seq.).

Aldrin/Dieldrin manufacturer (Section 129.100) means a manufacturer, excluding any source which is exclusively an aldrin/dieldrin formulator, who produces, prepares or processes technical aldrin or dieldrin or who uses aldrin or dieldrin as a material in the production, preparation or processing of another synthetic organic substance.

Algicidal (Section 797.1050) means having the property of killing algae.

Algistatic (Section 797.1050) means having the property of inhibiting algal growth.

Alkaline cleaning (Section 471.02) uses a solution (bath), usually detergent, to remove lard, oil, and other such compounds from a metal surface. Alkaline cleaning is usually followed by a water rinse. The rinse may consist of single or multiple stage rinsing. For the purposes of this part, an alkaline cleaning operation is defined as a bath followed by a rinse, regardless of the number of rinse stages. Each alkaline cleaning bath and rinse combination is entitled to a discharge allowance.

Alkaline cleaning bath (Section 468.02) shall mean a bath consisting of an alkaline cleaning solution through which a workpiece is processed.

Alkaline cleaning rinse (Section 468.02) shall mean a rinse following an alkaline cleaning bath through which a workpiece is processed. A rinse consisting of a series of rinse tanks is considered as a single rinse.

Alkaline cleaning rinse for forged parts (Section 468.02) shall mean a rinse following an alkaline cleaning bath through which a forged part is processed. A rinse consisting of a series of rinse tanks is considered as a single rinse.

Alkaline, mine drainage (Section 434.11) means mine drainage which, before any treatment, has a pH equal to or greater than 6.0 and total iron concentration of less than 10 mg/l.

All-electric melter (Section 60.291) means a glass melting furnace in which all the heat required for melting is provided by electric current from electrodes submerged in the molten glass, although some fossil fuel may be charged to the furnace as raw material only.

Allegation (Section 717.3) means a statement, made without formal proof or regard for evidence, that a chemical substance or mixture has caused a significant adverse reaction to health or the environment.

Allergic contact dermatitis (Section 798.4100). See **Skin sensitization (Section 798.4100)**.

Alley collection (Section 243.101) means the collection of solid waste from containers placed adjacent to or in an alley.

Allotment (Section 35.105) means an amount representing a State's share of funds requested in the President's budget or appropriated by Congress for an environmental program, as EPA determines after considering any factors indicated by this regulation. The allotment is not an entitlement but rather the objective basis for determining the range for a State's planning target.

Allowable costs (Section 30.200) means those project costs that are: eligible, reasonable, necessary, and allocable to the project; permitted by the appropriate Federal cost principles, and approved by EPA in the assistance agreement.

Allowable costs (Section 35.6015 [1/27/89]) means those project costs that are: eligible, reasonable, necessary, and allocable to the project; permitted by the appropriate Federal cost principles; and approved by EPA in the cooperative agreement and/or Superfund State Contract.

Allowable emissions (Section 51.165) means the emissions rate of a stationary source calculated using the maximum rated capacity of the source (unless the source is subject to federally enforceable limits which restrict the operating rate, or hours of operation, or both) and the most stringent of the following: (A) The applicable standards set forth in 40 CFR Part 60 or 61; (B) Any applicable State Implementation Plan emissions limitation including those with a future compliance date; or (C) The emissions rate specified as a federally enforceable permit condition, including those with a future compliance date.

Allowable emissions (Section 51.166) means the emissions rate of a stationary source calculated using the maximum rated capacity of the source (unless the source is subject to federally enforceable limits which restrict the operating rate, or hours of operation, or both) and the most stringent of the following: (i) The applicable standards as set forth in 40 CFR Parts 60 and 61; (ii) The applicable State Implementation Plan emissions limitation, including those with a future compliance date; or (iii) The emissions rate specified as a federally enforceable permit condition.

Allowable emissions (Section 52.21) means the emissions rate of a stationary source calculated using the maximum rated capacity of the source (unless the source is subject to federally enforceable limits which restrict the operating rate, or hours of operation, or both) and the most stringent of the following: (i) The applicable standards as set forth in 40 CFR Parts 60 and 61; (ii) The applicable State Implementation Plan emissions limitation, including those with a future compliance date; or (iii) The emissions rate specified as a federally enforceable permit condition, including those with a future compliance date.

Allowable emissions (Section 52.24) means the emissions rate of a stationary source calculated using the maximum rated capacity of the source (unless the source is subject to federally enforceable limits which restrict the operating rate, or hours of operation, or both) and the most stringent of the following: (i) The applicable standards set forth in 40 CFR Parts 60 and 61; (ii) Any applicable State Implementation Plan emissions limitation, including those with a future compliance date; or (iii) The emissions rate specified as a federally enforceable permit condition, including those with a future compliance date.

Allowance (Section 35.2005) means an amount based on a percentage of the project's allowable building cost, computed in accordance with Appendix B.

Allowance (Section 73.3 [12-17-91]) means an authorization, allocated by the Administrator under the Acid Rain program, to emit up to one ton of sulfur dioxide during or after a specified calendar year.

Allowance Tracking System Account (Section 73.3 [12-17-91]) means an account in the Allowance Tracking System established by the Administrator for purposes of allocating, holding, transferring, and using allowances.

Allowance Tracking System (Section 73.3 [12-17-91]) means the system by which the Administrator allocates, records, and tracks allowances.

Allowance transfer deadline (Section 73.3 [12-17-91]) means midnight of January 30 or, if January 30 is not a business day, midnight of the first business day thereafter, and is the last day on which allowances may be submitted for recordation in an affected unit's compliance subaccount for the purposes of meeting sulfur dioxide emissions limitation requirements for the previous calendar year.

Altered discharge (Section 125.58) means any discharge other than a current discharge or improved discharge, as defined in this regulation.

Alternative effluent limitations (Section 125.71) means all effluent limitations or standards of performance for the control of the thermal component of any discharge which are established under section 316(a) and this subpart.

Alternative method (Section 60.2) means any method of sampling and analyzing for an air pollutant which is not a reference or equivalent method but which has been demonstrated to the Administrator's satisfaction to, in specific cases, produce results adequate for his determination of compliance.

Alternative method (Section 61.02) means any method of sampling and analyzing for an air pollutant which is not a reference method but which has been demonstrated to the Administrator's satisfaction to produce results adequate for the Administrator's determination of compliance.

Alternative technology (Section 35.2005) means proven wastewater treatment processes and techniques which provide for the reclaiming and reuse of water, productively recycle wastewater constituents or otherwise eliminate the discharge of pollutants, or recover energy. Specifically, alternative technology includes land application of effluent and sludge; aquifer recharge; aquaculture; direct reuse (non-potable); horticulture; revegetation of disturbed land; containment ponds; sludge composting and drying prior to land application; self-sustaining incineration; and methane recovery.

Alternative to conventional treatment works for a small community (Section 35.2005), for purposes of Sections 35.2020 and 35.2032, alternative technology used by treatment works in small communities include alternative technologies defined in paragraph (b)(4), as well as, individual and onsite systems; small diameter gravity, pressure or vacuum sewers conveying treated or partially treated wastewater. These systems can also include small diameter gravity sewers carrying raw wastewater to cluster systems.

Alternative water supplies (Section 300.5 [3-8-90]) as defined by section 101(34) of CERCLA, includes, but is not limited to, drinking water and household water supplies.

Altitude performance adjustments (Section 86.1602) are adjustments or modifications made to vehicle, engine, or emission control functions in order to improve emission control performance at altitudes other than those for which the vehicles were designed.

Aluminum basis material (Section 465.02) means aluminum, aluminum alloys and aluminum coated steels which are processed in coil coating.

Aluminum Casting (Section 464.02). The remelting of aluminum or an aluminum alloy to form a cast intermediate or final product by pouring or forcing the molten metal into a mold, except for ingots, pigs, or other cast shapes related to nonferrous (primary and secondary) metals manufacturing (40 CFR Part 421) and aluminum forming (40 CFR Part 467). Processing operations following the cooling of castings not covered under aluminum forming, except for grinding scrubber operations which are covered here, are covered under the electroplating and metal finishing point source categories (40 CFR Parts 413 and 433).

Aluminum equivalent (Section 60.191) means an amount of aluminum which can be produced from a Mg of anodes produced by an anode bake plant as determined by Section 60.195(g).

Aluminum forming (Section 467.02) is a set of manufacturing operations in which aluminum and aluminum alloys are made into semifinished products by hot or cold working.

Ambient air (Section 50.1) means that

portion of the atmosphere, external to buildings, to which the general public has access.

Ambient air (Section 57.103) shall have the meaning given by 40 CFR 50.1(e), as that definition appears upon promulgation of this subpart, or as hereafter amended.

Ambient air quality (Section 57.103) refers only to concentrations of sulfur dioxide in the ambient air, unless otherwise specified.

Ambient water criterion (Section 129.2) means that concentration of a toxic pollutant in a navigable water that, based upon available data, will not result in adverse impact on important aquatic life, or on consumers of such aquatic life, after exposure of that aquatic life for periods of time exceeding 96 hours and continuing at least through one reproductive cycle; and will not result in a significant risk of adverse health effects in a large human population based on available information such as mammalian laboratory toxicity data, epidemiological studies of human occupational exposures, or human exposure data, or any other relevant data.

Ambient water criterion for toxaphene in navigable waters (Section 129.103) is 0.005 μg/l.

Ambient water criterion for aldrin/dieldrin in navigable waters (Section 129.100) is 0.003 μg/l.

Ambient water criterion for benzidine in navigable waters (Section 129.104) is 0.1 μg/l.

Ambient water criterion for DDT in navigable waters (Section 129.101) is 0.001 μg/l.

Ambient water criterion for endrin in navigable waters (Section 129.102) is 0.004 μg/l.

Ambient water criterion for PCBs in navigable waters (Section 129.105) is 0.001 μg/l.

Amendment review (Section 152.403 [5/26/88]) means review of any application requiring Agency approval to amend the registration of a currently registered product, or for which an application is pending Agency decision, not entailing a major change to the use pattern of an active ingredient.

Ammonia-N (or ammonia-nitrogen) (Section 420.02) means the value obtained by manual distillation (at pH 9.5) followed by the Nesslerization method specified in 40 CFR 136.3.

Ammonia-nitrogen (Section 420.02). See **Ammonia-N (Section 420.02)**.

Amount of pesticidal product (Section 167.3 [9/8/88]) means quantity, expressed in weight or volume of the product, and is to be reported in pounds for solid or semi-solid pesticides and active ingredients or gallons for liquid pesticides and active ingredients, or number of individual retail units for devices.

Amount of pesticide (Section 167.1) means quantity, expressed in weight or volume of the product, and is to be reported in pounds for solid or semi-solid products and gallons for liquid products.

Amount of pesticide or active ingredient (Section 169.1) means the weight or volume of the pesticide or active ingredient used in producing a pesticide expressed as weight for solid or semi-solid products and as weight or volume of liquid products.

Amphibian and reptile poisons and repellents (Section 162.3) includes all substances or mixtures of substances intended for preventing, destroying, repelling, or mitigating amphibians and reptiles declared to be pests under Section 162.14. Amphibian and reptile poisons and repellents include, but are not limited to: (i) Substances or mixtures of substances intended for use in baits or sprays for killing or repelling snakes, frogs, or lizards; and (ii) Reproductive inhibitors intended to reduce or otherwise alter the reproductive capacity or potential of amphibian or reptile pests.

Analyzer (Section 53.1). See **Automated method (Section 53.1)**.

Ancillary equipment (Section 260.10) means any device including, but not limited to, such devices as piping, fittings, flanges, valves, and pumps, that is used to distribute, meter, or control the flow of hazardous waste from its point of generation to a storage or treatment tank(s), between hazardous waste storage and treatment tanks to a point of disposal onsite, or to a point of shipment for disposal off-site.

Ancillary equipment (Section 280.12 [9/23/88]) means any devices including, but not limited to, such devices as piping, fittings, flanges, valves, and pumps used to distribute, meter, or control the flow of regulated substances to and from an UST.

Ancillary operation (Section 467.02) is a manufacturing operation that has a large flow, discharges significant amounts of pollutants, and may not be present at every plant in a subcategory, but when present is an integral part of the aluminum forming process.

Ancillary operation (Section 467.11) shall mean any operation not previously included in the core, performed on-site, following or preceding the rolling operation. The ancillary operations shall include continuous rod casting, continuous sheet casting, solution heat treatment, cleaning or etching.

Ancillary operation (Section 467.21) shall mean any operation not previously included in the core, performed on-site, following or preceding the rolling operation. The ancillary operations shall include direct chill casting, solution heat treatment, cleaning or etching, and degassing.

Ancillary operation (Section 467.31) shall mean any operation not previously included in the core, performed on-site, following or preceding the extrusion operation. The ancillary operations shall include direct chill casting, press or solution heat treatment, cleaning or etching, degassing, and extrusion press hydraulic fluid leakage.

Ancillary operation (Section 467.41) shall mean any operation not previously included in the core, performed on-site, following or preceding the forging operation. The ancillary operations shall include forging air pollution scrubbers, solution heat treatment, and cleaning or etching.

Ancillary operation (Section 468.02) shall mean any operation associated with a primary forming operation. These ancillary operations include surface and heat treatment, hydrotesting, sawing, and surface coating.

Ancillary operation (Sections 467.51; 467.61) shall mean any operation not previously included in the core, performed on-site, following or preceding the drawing operation. The ancillary operations shall include continuous rod casting, solution heat treatment, and cleaning or etching.

Ancillary operations (Section 461.2) means all of the operations specific to battery manufacturing and not included specifically within anode or cathode manufacture (ancillary operations are primarily associated with battery assembly and chemical production of anode or cathode active materials).

Anhydrous product (Sections 417.11; 417.21; 417.31; 417.41; 417.151; 417.161) shall mean the theoretical product that would result if all water were removed from the actual product.

Animal feed (Section 257.3-5) means any crop grown for consumption by animals, such as pasture crops, forage, and grain.

Animal feeding operation (Section 122.23) means a lot or facility (other than an aquatic animal production facility) where the following conditions are met: (i) Animals (other than aquatic animals) have been, are, or will be stabled or confined and fed or maintained for a total of 45 days or more in any 12-month period, and (ii) Crops, vegetation forage growth, or post-harvest residues are not sustained in the normal growing season over any portion of the lot or facility.

Animals (Section 116.3) means appropriately sensitive animals which carry out respiration by means of a lung structure permitting gaseous exchange between air and the circulatory system.

Annealing with oil (Section 468.02) shall mean the use of oil to quench a workpiece as it passes from an annealing furnace.

Annealing with water (Section 468.02) shall mean the use of a water spray or bath, of which water is the major constituent, to quench a workpiece as it passes from an annealing furnace.

Annual (Section 704.3 [12/22/88])* means the corporate fiscal year.

Annual average (Section 407.81) shall mean the maximum allowable discharge of BOD5 or TSS, as calculated by multiplying the total mass (kkg or 1000 lb) of each final product produced for the entire processing season or calendar year by the applicable annual average limitation.

Annual average (Sections 407.61; 407.71) shall mean the maximum allowable discharge of BOD5 or TSS as calculated by multiplying the total mass (kkg or 1000 lb) of each raw commodity processed for the entire processing season or calendar year by the applicable annual average limitation.

Annual capacity factor (Section 60.41b [12/16/87])* means the ratio between the actual heat input to a steam generating unit from the fuels listed in Section 60.42b(a), Section 60.43b(a), or Section 60.44b(a), as applicable, during a calendar year and the potential heat input to the steam generating unit had it been operated for 8,760 hours during a calendar year at the maximum steady state design heat input capacity. In the case of steam generating units that are rented or leased, the actual heat input shall be determined based on the combined heat input from all operations of the affected facility in a calendar year.

Annual capacity factor (Section 60.41c [9-12-90]) means the ratio between the actual heat input to a steam generating unit from an individual fuel or combination of fuels during a period of 12 consecutive calendar months and the potential heat input to the steam generating unit from all fuels had the steam generating unit been operated for 8,760 hours during that 12-month period at the maximum design heat input capacity. In the case of steam generating units that are rented or leased, the actual heat input shall be determined based on the combined heat input from all operations of the affected facility during a period of 12 consecutive calendar months.

Annual coke production (Section 61.131 [9-14-89]) means the coke produced in the batteries connected to the coke by-product recovery plant over a 12-month period. The first 12-month period concludes on the first

December 31 that comes at least 12 months after the effective date or after the date of initial startup if initial startup is after the effective date.

Annual committed effective dose (Section 191.12 [12/20/93])* means the committed effective dose resulting from one-year intake of radionuclides released plus the annual effective dose caused by direct radiation from facilities or activities subject to subparts B and C of this part.

Annual document log (Section 761.3 [12-21-89]) means the detailed information maintained at the facility on the PCB waste handling at the facility.

Annual precipitation and annual evaporation (Section 440.132) are the mean annual precipitation and mean annual lake evaporation, respectively, as established by the U.S. Department of Commerce, Environmental Science Services Administration, Environmental Data Services, or equivalent regional rainfall and evaporation data.

Annual report (Section 761.3 [12-21-89]) means the written document submitted each year by each disposer and commercial storer of PCB waste to the appropriate EPA Regional Administrator. The annual report is a brief summary of the information included in the annual document log.

Annual research period (Section 85.402) means the time period from August 1 of a previous calendar year to July 31 of the given calendar year, e.g., the 1981 annual research period would be the time period from August 1, 1980 to July 31, 1981.

Annual Work Plan (Section 35.9010 [10-3-89]) The plan, developed by the Management Conference each year, which documents projects to be undertaken during the upcoming year. The Annual Work Plan is developed within budgetary targets provided by EPA.

Annualized Cost (Section 61 [12-15-89]) A stream of annual payments for a determined time period, equal in value to a one-time payment based on a selected rate of interest.

Anode bake plant (Section 60.191) means a facility which produces carbon anodes for use in a primary aluminum reduction plant.

ANSI S3.19-1974 (Section 211.203) means a revision of the ANSI Z24.22-1957 measurement procedure using one-third octave band stimuli presented under diffuse (reverberant) acoustic field conditions.

ANSI Z24.22-1957 (Section 211.203) means a measurement procedure published by the American National Standards Institute (ANSI) for obtaining hearing protector attenuation values at nine of the one-third octave band center frequencies by using pure tone stimuli presented to ten different test subjects under anechoic conditions.

Anthracite (Section 60.41a) means coal that is classified an anthracite according to the American Society of Testing and Materials (ASTM) Standard Specification for Classification of Coals by Rank D388-77 (incorporated by reference - see Section 60.17).

Antimicrobial agents (Section 162.3) includes all substances or mixtures of substances, except those defined as fungicides in paragraph (ff)(8) of this section, and slimicides in paragraph (ff)(16) of this section, intended for inhibiting the growth of, or destroying any bacteria, fungi pathogenic to man and other animals, or viruses declared to be pests under Section 162.14 and existing in any environment except those excluded in paragraph (ff)(2)(ii) of this section. (i) Antimicrobial agents include, but are not limited to: (A) Disinfectants intended to destroy or irreversibly inactivate infectious or other undesirable bacteria, pathogenic fungi, or viruses on surfaces or inanimate objects; (B) Sanitizers intended to reduce the number of living bacteria or viable virus particles on inanimate surfaces, in water, or in air; (C) Bacteriostats intended to inhibit the growth of bacteria in the presence of moisture; (D) Sterilizers intended to destroy viruses and all living bacteria, fungi and their spores, on inanimate surfaces; (E) Fungicides and fungistats intended to inhibit the growth of, or destroy fungi (including yeasts), pathogenic to man or other animals on inanimate surfaces; and (F) Commodity preservatives and protectants intended to inhibit the growth of, or destroy bacteria in or on raw materials (such as adhesives and plastics) used in manufacturing, or manufactured products (such as fuel, textiles, lubricants, and paints). (ii) Antimicrobial agents do not include those antimicrobial substances or mixtures of substances subject to the provisions of the Federal Food, Drug and Cosmetic Act, as amended (21 U.S.C. 301 et seq.), such as: (A) Substances or mixtures of substances intended to inhibit the growth of, inactivate or destroy fungi, bacteria, or viruses in or on living man or other animals; and (B) Substances or mixtures of substances intended to inhibit the growth of, inactivate or destroy fungi, bacteria, or viruses in or on processed food, beverages, or pharmaceuticals including cosmetics.

Antimony (Section 415.661) shall mean the total antimony present in the process wastewater stream exiting the wastewater treatment system.

AOD vessel (Section 60.271a). See **Argon-oxygen decarburization vessel (Section 60.271a)**.

Apparent plateau (Section 797.1560). See **Steady-state (Section 797.1560)**.

Applicable legal requirements (Section 66.3) means any of the following: (1) In the case of any major source, any emission limitation, emission standard, or compliance schedule under any EPA-approved State implementation plan (regardless of whether the source is subject to a Federal or State consent decree); (2) In the case of any source, an emission limitation, emission standard, standard of performance, or other requirement (including, but not limited to, work practice standards) established under section 111 or 112 of the Act; (3) In the case of a source that is subject to a federal or federally approved state judicial consent decree or EPA approved extension, order, or suspension, any interim emission control requirement or schedule of compliance under that consent decree, extension, order or suspension; (4) In the case of a nonferrous smelter which has received a primary non-ferrous smelter order issued or approved by EPA under Section 119 of the Act, any interim emission control requirement (including a requirement relating to the use of supplemental or intermittent controls) or schedule of compliance under that order.

Applicable plan (Section 60.21) means the plan, or most recent revision thereof, which has been approved under Section 60.27(b) or promulgated under Section 60.27(d).

Applicable requirements (Section 300.5 [3-8-90])* means those cleanup standards, standards of control, and other substantive requirements, criteria, or limitations promulgated under federal environmental or state environmental or facility siting laws that specifically address a hazardous substance, pollutant, contaminant, remedial action, location, or other circumstances found at a CERCLA site. Only those state standards that are identified by a state in a timely manner and that are more stringent than federal requirements may be applicable.

Applicable standard (Section 130.10 [6/2/89]), for the purposes of listing waters under Section 130.10(d)(2), means a numeric criterion for a priority pollutant promulgated as part of a state water quality standard. Where a state numeric criterion for a priority pollutant is not promulgated as part of a state water quality standard, for the purposes of listing waters applicable standard means the state narrative water quality criterion to control a priority pollutant (e.g., no toxics in toxic amounts) interpreted on a chemical-by-chemical basis by applying a proposed state criterion, an explicit state policy or regulation, or an EPA national water quality criterion, supplemented with other relevant information.

Applicable standard (Section 21.2) means any requirement, not subject to an exception under Section 21.6, relating to the quality of water containing or potentially containing pollutants, if such requirement is imposed by: (1) The Act; (2) EPA regulations promulgated thereunder or permits issued by EPA or a State thereunder; (3) Regulations by any other Federal Agency promulgated thereunder; (4) Any State standard or requirement as applicable under section 510 of the Act; (5) Any requirements necessary to comply with an areawide management plan approved pursuant to section 208(b) of the Act; (6) Any requirements necessary to comply with a facilities plan developed under section 201 of the Act (see 35 CFR Subpart E); (7) Any State regulations or laws controlling the disposal of aqueous pollutants that may affect groundwater.

Applicable standards and limitations (Section 122.2 [5/2/89])* means all State, interstate, and federal standards and limitations to which a discharge, a sewage sludge use or disposal practice, or a related activity is subject under the CWA, including effluent limitations, water quality standards, standards of performance, toxic effluent standards or prohibitions, best management practices, pretreatment standards, and standards for sewage sludge use or disposal under sections 301, 302, 303, 304, 306, 307, 308, 403 and 405 of CWA.

Applicable standards and limitations (Section 124.2 [5/2/89])* means all State, interstate, and federal standards and limitations to which a discharge, a sludge use or disposal practice or a related activity is subject under the CWA, including standards for sewage sludge use or disposal, effluent limitations, water quality standards, standards of performance, toxic effluent standards or prohibitions, best management practices, and pretreatment standards under sections 301, 302, 303, 304, 306, 307, 308, 403, and 405 of CWA.

Applicable water quality standards (Section 110.1) means State water quality standards adopted by the State pursuant to section 303 of the Act or promulgated by EPA pursuant to that section.

Applicant (Section 125.58) means an applicant for a section 301(h) modified permit. Large applicants have populations contributing to their POTWs equal to or more than 50,000 people or average dry weather flows of 5.0 million gallons per day (mgd) or more; small applicants have contributing populations of less than 50,000 people and average dry weather flows of less than 5.0 mgd. For the purposes of this definition the contributing population and flows shall be based on projections for the end of the five year permit term. Average dry weather flows shall be the average daily total discharge flows for the maximum month of the dry weather season.

Applicant (Section 15.4) means any person who has applied for but has not yet received a contract, grant, or loan and includes a bidder or proposer for a contract which is not yet awarded.

Applicant (Section 152.3 [5/4/88]) means a person who applies for a registration, amended registration, or reregistration, under FIFRA section 3.

Applicant (Section 162.3) means a person who applies for a registration pursuant to section 3 of the Act.

Applicant (Section 164.2) means any person who has made application to have a pesticide registered or classified pursuant to the provisions of the Act.

Applicant (Section 172.1) means any person who applies for an experimental use permit, pursuant to section 5 of the Act.

Applicant (Section 2.307) means any person who has submitted to EPA (or to a predecessor agency with responsibility for administering the Act) a registration statement or application for registration under the Act of a pesticide or of an establishment.

Applicant (Section 20.2) means any person who files an application with the Administrator for certification that a facility is in compliance with the applicable regulations of Federal agencies and in furtherance of the general policies of the United States for cooperation with the States in the prevention and abatement of water or air pollution under the Act.

Applicant (Section 30.200) means any entity that files an application or unsolicited proposal for EPA financial assistance under this subchapter.

Applicant (Section 35.4010 [10/1/92])* means any group of individuals that files an application for a TAG.

Applicant (Section 53.1) means a person who submits an application for a reference or equivalent method determination under Section 53.4, or a person who assumes the rights and obligations of an applicant under Section 53.7.

Applicant (Section 6.501) means any individual, agency, or entity which has filed an application for grant assistance under 40 CFR Part 35, Subpart E or I.

Applicant (Section 6.601) for the sake of this subpart means any person who applies to EPA for the issuance of an NPDES permit to discharge as a new source.

Applicant (Section 7.25) means any entity that files an application or unsolicited proposal or otherwise requests EPA assistance (see definition for EPA assistance).

Applicant (Section 8.2) means an applicant for Federal assistance from the Agency involving a construction contract, or other participant in a program involving a construction contract as determined by the regulations of the Agency. The term also includes such persons after they become recipients of such Federal assistance.

Application (Section 124.2) means the EPA standard national forms for applying for a permit, including any additions, revisions or modifications to the forms; or forms approved by EPA for use in approved States, including any approved modifications or revisions. For RCRA, application also includes the information required by the Director under Sections 270.14 through 270.29 [contents of Part B of the RCRA application].

Application (Section 124.41) means an application for a PSD permit.

Application (Section 125.58) means a final application previously submitted in accordance with the June 15, 1979, section 301(h) regulations (44 FR 34784) or an application submitted between December 29, 1981 and December 29, 1982. It does not include a preliminary application submitted in accordance with the June 15, 1979, section 301(h) regulations.

Application (Section 146.3) means the EPA standard national forms for applying for a permit, including any additions, revisions or modifications to the forms; or forms approved by EPA for use in approved States, including any approved modifications or revisions. For RCRA, application also includes the information required by the Director under Section 122.25 (contents of Part B of the RCRA application).

Application (Section 2.309) means an application for a permit.

Application (Section 232.2 [2/11/93])* means a form for applying for a permit to discharge dredged or fill material into waters of the United States.

Application (Section 233.3) means the forms approved by EPA for use in approved States, including any approved modifications or revisions.

Application (Section 270.2) means the EPA standard national forms for applying for a permit, including any additions, revisions or modifications to the forms; or forms approved by EPA for use in approved States, including any approved modifications or revisions. Application also includes the information required by the Director under Sections 270.14 through 270.29 (contents of Part B of the RCRA application).

Application (Section 35.4010 [10/1/92])* means a completed formal written request for a TAG that is submitted to a State or the EPA on EPA form SF-424, Application for Federal Assistance (Non-construction Programs).

Application (Sections 122.2; 144.3) means the EPA standard national forms for applying for a permit, including any additions, revisions or modifications to the forms; or forms approved by EPA for use in approved States, including any approved modifications or revisions.

Application for research or marketing permit (Section 160.3 [8-17-89]) includes: (1) An application for registration, amended registration, or reregistration of a pesticide product under FIFRA sections 3, 4 or 24(c). (2) An application for an experimental use permit under FIFRA section 5. (3) An application for an exemption under FIFRA section 18. (4) A petition or other request for establishment or modification of a tolerance, for an exemption for the need for a tolerance, or for other clearance under FFDCA section 408. (5) A petition or other request for establishment or modification of a food additive regulation or other clearance by EPA under FFDCA section 409. (6) A submission of data in response to a notice issued by EPA under FIFRA section 3(c)(2)(B). (7) Any other application, petition, or submission sent to EPA intended to persuade EPA to grant, modify, or leave unmodified a registration or other approval required as a condition of sale or distribution of a pesticide.

Application of a pesticide (Section 162.3) means the placement for effect of a pesticide at or on the site where the pest control or other response is desired.

Application questionnaire (Section 125.58) means EPA's "Applicant Questionnaire for Modification of Secondary Treatment Requirements." Individual questionnaires for small applicants and for large applicants are published as Appendix A and Appendix B to this subpart, respectively.

Applied coating solids (Section 60.391) means the volume of dried or cured coating solids which is deposited and remains on the surface of the automobile or light-duty truck body.

Applied coating solids (Section 60.451) means the coating solids that adhere to the surface of the large appliance part being coated.

Appraisal (Section 4.103) is a written statement independently and impartially prepared by a qualified appraiser setting forth an opinion of defined value of an adequately described property as of a specific date, supported by the presentation and analysis of relevant market information.

Approach angle (Section 86.084-2) means the smallest angle in a plan side view of an automobile, formed by the level surface on which the automobile is standing and a line tangent to the front tire static loaded radius arc and touching the underside of the automobile forward of the front tire. This definition applies beginning with the 1984 model year.

Appropriate (Section 157.21) when

used with respect to child-resistant packaging, means that the packaging is chemically compatible with the pesticide contained therein.

Appropriate Act and regulations (Section 144.3) means the Solid Waste Disposal Act, as amended by the Resource Conservation and Recovery Act (RCRA); or Safe Drinking Water Act (SDWA), whichever is applicable; and applicable regulations promulgated under those statutes.

Appropriate Act and regulations (Section 124.2) means the Clean Water Act (CWA); the Solid Waste Disposal Act, as amended by the Resource Conservation Recovery Act (RCRA); or Safe Drinking Water Act (SDWA), whichever is applicable; and applicable regulations promulgated under those statutes. In the case of an approved State program appropriate Act and regulations includes program requirements.

Appropriate Act and regulations (Section 124.41) means the Clean Air Act and applicable regulations promulgated under it.

Appropriate program official (Section 6.701) means the official at each decision level within ORD to whom the Assistant Administrator has delegated responsibility for carrying out the environmental review process.

Appropriate sensitive benthic marine organisms (Section 227.27) means at least one species each representing filter-feeding, deposit-feeding, and burrowing species chosen from among the most sensitive species accepted by EPA as being reliable test organisms to determine the anticipated impact on the site; provided, however, that until sufficient species are adequately tested and documented, interim guidance on appropriate organisms available for use will be provided by the Administrator, Regional Administrator, or the District Engineer, as the case may be.

Appropriate sensitive marine organisms (Section 227.27) means at least one species each representative of phytoplankton or zooplankton, crustacean or mollusk, and fish species chosen from among the most sensitive species documented in the scientific literature or accepted by EPA as being reliable test organisms to determine the anticipated impact of the wastes on the ecosystem at the disposal site. Bioassays, except on phytoplankton or zooplankton, shall be run for a minimum of 96 hours under temperature, salinity, and dissolved oxygen conditions representing the extremes of environmental stress at the disposal site. Bioassays on phytoplankton or zooplankton may be run for shorter periods of time as appropriate for the organisms tested at the discretion of EPA, or EPA and the Corps of Engineers, as the case may be.

Appropriate treatment of the recycle water (Section 440.132) in Subpart J,

Section 440.104 includes, but is not limited to pH adjustment, settling and pH adjustment, settling, and mixed media filtration.

Approval Authority (Section 403.3) means the Director in an NPDES State with an approved State pretreatment program and the appropriate Regional Administrator in a non-NPDES State or NPDES State without an approved State pretreatment program.

Approval of the facilities plan (Section 6.501) means approval of the facilities plan for a proposed wastewater treatment works pursuant to 40 CFR Part 35, Subpart E or I.

Approved measure (Section 57.103) refers to one contained in an NSO which is in effect.

Approved permit program (Section 61.02 [3/16/94])* means a State permit program approved by the Administrator as meeting the requirements of part 70 of this chapter or a Federal permit program established in this chapter pursuant to title V of the Act (42 U.S.C. 7661).

Approved permit program (Section 60.2 [7/21/92])* means a State permit program approved by the Administrator as meeting the requirements of part 70 of this chapter or a Federal permit program established in this chapter pursuant to title V of the Act (42 U.S.C. 7661).

Approved POTW Pretreatment Program or Program or POTW Pretreatment Program (Section 403.3) means a program administered by a POTW that meets the criteria established in this regulation (Sections 403.8 and 403.9) and which has been approved by a Regional Administrator or State Director in accordance with Section 403.11 of this regulation.

Approved program (Section 124.41) means a State implementation plan providing for issuance of PSD permits which has been approved by EPA under the Clean Air Act and 40 CFR Part 51. An approved State is one administering an approved program. State Director as used in Section 124.4 means the person(s) responsible for issuing PSD permits under an approved program, or that person's delegated representative.

Approved program (Section 232.2 [2/11/93])* means a State program which has been approved by the Regional Administrator under Part 233 of this chapter or which is deemed approved under section 404(h)(3), 33 U.S.C. 1344(h)(3).

Approved program or approved State (Section 233.3) means a State or interstate program which has been approved or authorized by EPA under Subpart C.

Approved program or approved State (Section 270.2) means a State which has been approved or authorized by EPA under Part 271.

Approved program or approved State (Section 122.2) means a State or interstate program which has been approved or authorized by EPA under Part 123.

Approved Section 120 program (Section 66.3) means a State program to assess and collect Section 120 penalties that has been approved by the Administrator.

Approved State (Section 122.2). See **Approved program (Section 122.2)**.

Approved State (Section 233.3). See **Approved program (Section 233.3)**.

Approved State (Section 270.2). See **Approved program (Section 270.2)**.

Approved State primacy program (Section 142.2 [12-20-89]) consists of those program elements listed in Sec. 142.11(a) that were submitted with the initial State application for primary enforcement authority and approved by the EPA Administrator and all State program revisions thereafter that were approved by the EPA Administrator.

Approved State Program (Section 501.2 [5/2/89]) means a State program which has received EPA approval under this Part.

Approved State Program (Section 144.3 [9/26/88])* means a UIC program administered by the State or Indian Tribe that has been approved by EPA according to SDWA sections 1422 and/or 1425.

Apricots (Section 407.61) shall include the processing of apricots into the following product styles: Canned and frozen, pitted and unpitted, peeled and unpeeled, whole, halves, slices, nectar, and concentrate.

AQL (Section 204.51). See **Acceptable Quality Level (Section 204.51)**.

AQL (Section 205.151). See **Acceptable Quality Level (Section 205.151)**.

AQL (Section 86.1002-84). See **Acceptable Quality Level (Section 86.1002-84)**.

AQL (Sections 86.602; 205.51). See **Acceptable Quality Level (Sections 86.602; 205.51)**.

Aquatic animals (Section 116.3) means appropriately sensitive wholly aquatic animals which carry out respiration by means of a gill structure permitting gaseous exchange between the water and the circulatory system.

Aquatic environment and aquatic ecosystem (Section 230.3) means waters of the United States, including wetlands, that serve as habitat for interrelated and interacting communities and populations of plants and animals.

Aquatic flora (Section 116.3) means plant life associated with the aquatic eco-system including, but not limited to, algae and higher plants.

Aquifer (Section 149.101) means the Edwards Underground Reservoir.

Aquifer (Section 149.2 [2/14/89])* means a geological formation, group of formations, or part of a formation that is capable of yielding a significant amount of water to a well or spring.

Aquifer (Section 191.12 [12/20/93])* means an underground geological formation, group of formations, or part of a formation that is capable of yielding a significant amount of water to a well or spring.

Aquifer (Section 257.3-4) means a geologic formation, group of formations, or portion of a formation capable of yielding usable quantities of ground water to wells or springs.

Aquifer (Section 258.2 [10-9-91]) means a geological formation, group of formations, or porton of a formation capable of yielding significant quantities of ground water to wells or springs.

Aquifer (Section 260.10) means a geologic formation, group of formations, or part of a formation capable of yielding a significant amount of ground water to wells or springs.

Aquifer (Sections 144.3; 146.3; 147.2902; 270.2) means a geological formation, group of formations, or part of a formation that is capable of yielding a significant amount of water to a well or spring.

Aquifer Service Area (Section 149.2) means an area above the aquifer and including the area where the entire population served by the aquifer lives.

Arbitrator (Section 304.12 [5/30/89]) means the person appointed in accordance with Section 304.22 of this part and governed by the provisions of this part.

Arc chute (Section 763.163 [7-12-89]) means an asbestos-containing product that acts as a chute or guidance device and is intended to guide electric arcs in applications such as motor starter units in electric generating plants.

Architectural or engineering services (Section 35.2005) means consultation, investigations, reports, or services for design-type projects within the scope of the practice of architecture or professional engineering as defined by the laws of the State or territory in which the grantee is located.

Architectural or engineering (A/E) services (Sections 33.005; 35.6015 [1/27/89]) means consultation, investigations; reports, or services for design-type projects within the scope of the practice of architecture or professional engineering as defined by the laws of the State or territory in which the recipient is located.

Area (Section 61.151 [12-15-89]) means the vertical projection of the pile upon the earth's surface.

Area (Section 61.20) means a man-made underground void from which ore or waste has been removed.

Area (Section 61.251) means the area covered by the vertical projection of the pile upon the earth's surface.

Area coated (Section 466.02) means the area of basis material covered by each coating of enamel.

Area of review (Sections 144.3; 146.3) means the area surrounding an injection well described according to the criteria set forth in Section 146.06 or in the case of an area permit, the project area plus a circumscribing area the width of which is either 1/4 of a mile or a number calculated according to the criteria set forth in Section 146.06.

Area processed (Section 465.02) means the area actually exposed to process solutions. Usually this includes both sides of the metal strip.

Area processed (Section 466.02) means the total basis material area exposed to processing solutions.

Area source (Section 51.100) means any small residential, governmental, institutional, commercial, or industrial fuel combustion operations; onsite solid waste disposal facility; motor vehicles, aircraft vessels, or other transportation facilities or other miscellaneous sources identified through inventory techniques similar to those described in the "AEROS Manual series, Vol. II AEROS User's Manual," EPA-450/2-76-029 December 1976.

Areawide agency (Section 130.2). An agency designated under section 208 of the Act, which has responsibilities for WQM planning within a specified area of a State.

Areawide agency (Section 21.2) means an areawide management agency designated under section 208(c)(1) of the Act.

Argon-oxygen decarburization vessel (AOD vessel) (Section 60.271a) means any closed-bottom, refractory-lined converter vessel with submerged tuyeres through which gaseous mixtures containing argon and oxygen or nitrogen may be blown into molten steel for further refining.

Aromatic content (Section 80.2 [2/16/94])* is the aromatic hydrocarbon content in volume percent as determined by ASTM standard test method D 1319-88, entitled "Standard Test Method for Hydrocarbon Types in Liquid Petroleum Products by Fluorescent Indicator Adsorption". ASTM test method D 1319-88 is incorporated by reference. This incorporation by reference was approved by the Director of the Federal Register in accordance with 5 U.S.C. 552(a) and 1 CFR part 51. A copy may be obtained from the American Society for Testing and Materials, 1916 Race Street, Philadelphia, PA 19103. A copy may be inspected at the Air Docket Section (A-130), room M-1500, U.S. Environmental Protection Agency, Docket No. A-86-03, 401 M Street SW., Washington, DC 20460 or at the Office of the Federal Register, 800 North Capitol Street, NW., suite 700, Washington, DC 20408.

Arrest (Section 303.11 [5/5/88, 6/21/89]) means restraint of an arrestee's liberty or the equivalent through the service of judicial process compelling such a person to respond to a criminal accusation.

Arsenic kitchen (Section 61.181) means a baffled brick chamber where inorganic arsenic vapors are cooled, condensed, and removed in a solid form.

Arsenic-containing glass type (Section 61.161) means any glass that is distinguished from other glass solely by the weight percent of arsenic added as a raw material and by the weight percent of arsenic in the glass produced. Any two or more glasses that have the same weight percent of arsenic in the raw materials as well as in the glass produced shall be considered to belong to one arsenic-containing glass type, without regard to the recipe used or any other characteristics of the glass or the method of production.

Article (Section 372.3 [2/16/88]) means a manufactured item: (1) Which is formed to a specific shape or design during manufacture; (2) which has end use functions dependent in whole or in part upon its shape or design during end use; and (3) which does not release a toxic chemical under normal conditions of processing or use of that item at the facility or establishments.

Article (Section 704.3 [12/22/88])* means a manufactured item (1) which is formed to a specific shape or design during manufacture, (2) which has end use function(s) dependent in whole or in part upon its shape or design during end use, and (3) which has either no change of chemical composition during its end use or only those changes of composition which have no commercial purpose separate from that of the article, and that result from a chemical reaction that occurs upon end use of other chemical substances, mixtures, or articles; except that fluids and particles are not considered articles regardless of shape or design.

Article (Section 710.2) is a manufactured item: (1) Which is formed to a specific shape or design during manufacture, (2) which has end use function(s) dependent in whole or in part upon its shape or design during end use, and (3) which has either no change of chemical composition during its end use or only those changes of composition which have no commercial purpose separate from that of the article and that may occur as described in Section 710.4(d)(5); except that fluids and particles are not considered articles regardless of shape or design.

Article (Section 723.175) is a manufactured item (i) which is formed to a specific shape or design during manufacture, (ii) which has end use function(s) dependent in whole or in part upon its shape or design during end use, and (iii) which has either no change of chemical composition during its end use or only those changes of composition which have no commercial purpose separate from that of the article and that may occur as described in Section 710.2 of this chapter except that fluids and particles are not considered articles regardless of shape or design.

Article (Sections 720.3; 723.50; 747.115; 747.195; 747.200) means a manufactured item (1) which is formed to a specific shape or design during manufacture, (2) which has end use function(s) dependent in whole or in part upon its shape or design during end use, and (3) which has either no change of chemical composition during its end use or only those changes of composition which have no commercial purpose separate from that of the article and that may occur as described in Section 720.36(g)(5), except that fluids and particles are not considered articles regardless of shape or design.

As expeditiously as practicable considering technological feasibility (Section 192.31 [11/5/93])* means as quickly as possible considering: the physical characteristics of the tailings and the site; the limits of available technology; the need for consistency with mandatory requirements of other regulatory programs; and factors beyond the control of the licensee. The phrase permits consideration of the cost of compliance only to the extent specifically provided for by use of the term "available technology."

Asbestos (Section 61.141) means the asbestiform varieties of serpentinite (chrysotile), riebeckite (crocidolite), cummingtonitegrunerite, anthophyllite, and actinolite-tremolite.

Asbestos (Section 763.163 [7-12-89]) means the asbestiform varieties of: chrysotile (serpentine); crocidolite (riebeckite); amosite (cummingtonite-grunerite); tremolite; anthophyllite; and actinolite.

Asbestos (Sections 763.63; 763.83 [10/30/87]; 763.103; 763.121) means the asbestiform varieties of: chrysotile (serpentine); crocidolite (riebeckite); amosite (cummingtonite-grunerite); anthophyllite; tremolite; and actinolite.

Asbestos abatement project (Section 763.121) means any activity involving the removal, enclosure, or encapsulation of friable asbestos material.

Asbestos-cement (A/C) pipe (Section 763.163 [7-12-89]) means an asbestos-containing product made of cement and intended for use as pipe or fittings for joining pipe. Major applications of this product include: pipe used for transmitting water or sewage; conduit pipe for protection of utility or telephone cable; and pipes used for air ducts.

Asbestos clothing (Section 763.163 [7-12-89]) means an asbestos-containing product designed to be worn by persons.

Asbestos-containing waste materials (Section 61.141 [1-16-91])* means mill tailings or any waste that contains commercial asbestos and is generated by a source subject to the provisions of this subpart. This term includes filters from control devices, friable asbestos waste material, and bags or other similar packaging contaminated with commercial asbestos. As applied to demolition and renovation operations, this term also includes regulated asbestos-containing material waste and materials contaminated with asbestos including disposable equipment and clothing.

Asbestos debris (Section 763.83 [10/30/87]) means pieces of ACBM that can be identified by color, texture, or composition, or means dust, if the dust is determined by an accredited inspector to be ACM.

Asbestos diaphragm (Section 763.163 [7-12-89]) means an asbestos-containing product that is made of paper and intended for use as a filter in the production of chlorine and other chemicals, and which acts as a mechanical barrier between the cathodic and anodic chambers of an electrolytic cell.

Asbestos material (Section 61.141) means asbestos or any material containing asbestos.

Asbestos mill (Section 61.141) means any facility engaged in converting, or in any intermediate step in converting, asbestos ore into commercial asbestos. Outside storage of asbestos material is not considered a part of the asbestos mill.

Asbestos mixture (Section 763.63) means a mixture which contains bulk asbestos or another asbestos mixture as an intentional component. An asbestos mixture may be either amorphous or a sheet, cloth fabric, or other structure. This term does not include mixtures which contain asbestos as a contaminant or impurity.

Asbestos tailings (Section 61.141) means any solid waste that contains asbestos and is a product of asbestos mining or milling operations.

Asbestos waste from control devices (Section 61.141) means any waste material that contains asbestos and is collected by a pollution control device.

Asbestos-cement (A/C) corrugated sheet (Section 763.163 [7-12-89]) means an asbestos-containing product made of cement and in the form of a corrugated sheet used as a non-flat-surfaced reinforcing or insulating material. Major applications of this product include: building siding or roofing; linings for waterways; and components in cooling towers.

Asbestos-cement (A/C) flat sheet (Section 763.163 [7-12-89]) means an asbestos-containing product made of cement and in the form of a flat sheet used primarily as a flat-surfaced reinforcing or insulating material. Major applications of this product include: wall linings; partitions; soffit material; electrical barrier boards; bus bar run separators; reactance coil partitions; laboratory work surfaces; and components of vaults, ovens, safes, and broilers.

Asbestos-cement (A/C) shingle (Section 763.163 [7-12-89]) means an asbestos-containing product made of cement and intended for use as a siding, roofing, or construction shingle serving the purpose of covering and insulating the surface of building walls and roofs.

Asbestos-containing building material or ACBM (Section 763.83 [10/30/87]) means surfacing ACM, thermal system insulation ACM, or miscellaneous ACM that is found in or on interior structural members or other parts of a school building.

Asbestos-containing material (Section 763.103) means any material which contains more than 1 percent asbestos by weight.

Asbestos-containing material or ACM (Section 763.83 [10/30/87]) when referring to school buildings means any material or product which contains more than 1 percent asbestos.

Asbestos-containing product (Section 763.163 [7-12-89]) means any product to which asbestos is deliberately added in any concentration or which contains more than 1.0 percent asbestos by weight or area.

Asbestos-containing waste materials (Section 61.141) means any waste that contains commercial asbestos and is generated by a source subject to the provisions of this subpart. This term includes asbestos mill tailings, asbestos waste from control devices, friable asbestos waste material, and bags or containers that previously contained commercial asbestos. However, as applied to demolition and renovation operations, this term includes only friable asbestos waste and asbestos waste from control devices.

ASME (Section 60.51a [2-11-91]) means the American Society of Mechanical Engineers.

Asphalt processing (Section 60.471) means the storage and blowing of asphalt.

Asphalt processing plant (Section 60.471) means a plant which blows asphalt for use in the manufacture of asphalt products.

Asphalt roofing plant (Section 60.471) means a plant which produces asphalt roofing products (shingles, roll roofing, siding, or saturated felt).

Asphalt storage tank (Section 60.471) means any tank used to store asphalt at asphalt roofing plants, petroleum refineries, and asphalt processing plants. Storage tanks containing cutback asphalts (asphalts diluted with solvents to reduce viscosity for low temperature applications) and emulsified asphalts (asphalts dispersed in water with an emulsifying agent) are not subject to this regulation.

Assets (Section 144.61) means all existing and all probable future economic benefits obtained or controlled by a particular entity.

Assets (Section 264.141), as specified in Section 264.141(f), means all existing and all probable future economic benefits obtained or controlled by a particular entity.

Assets (Section 265.141), as specified in Section 265.141(f), means all existing and all probable future economic benefits obtained or controlled by a particular entity.

Assistance (Section 32.102) means EPA's contribution to a project under a grant, cooperative agreement, technical assistance award, inter-governmental agreement, or award under the Intergovernmental Personnel Act.

Assistance agreement (Section 30.200) means the legal instrument EPA uses to transfer money, property, services, or anything of value to a recipient to accomplish a public purpose. It is either a grant or a cooperative agreement and will specify: budget and project periods; the Federal share of eligible project costs; a description of the work to be accomplished; and any special conditions.

Assistant Administrator (Section 15.4) means the Assistant Administrator for Enforcement and Compliance Monitoring, United States Environmental Protection Agency, or his or her designee.

Assistant Administrator (Section 178.3 [6-24-92])* means the Agency's Assistant Administrator for Prevention, Pesticides and Toxic Substances, or any officer or employee of the Agency's Office for Prevention, Pesticides and Toxic Substances to whom the Assistant Administrator delegates the authority to perform functions under this part.

Assistant Administrator (Section 179.3 [12-5-90])* means the Agency's Assistant Administrator for Pesticides and Toxic Substances, or any officer or employee of OPTS to whom the Assistant Administrator has delegated the authority to perform functions under this part.

Assistant Administrator (Section 723.50) means the EPA Assistant Administrator for Pesticides and Toxic Substances or any employee designated by the Assistant Administrator to carry out the Assistant Administrator's functions under this section.

Assistant Administrator for Air and Radiation (Section 57.103) means the Assistant Administrator for Air and Radiation of the U.S. Environmental Protection Agency.

Assistant Attorney General (Section 12.103 [8/14/87]) means the Assistant Attorney General, Civil Rights Division, United States Department of Justice.

Assistant Attorney General (Section 7.25) is the head of the Civil Rights Division, U.S. Department of Justice.

Association (Section 304.12 [5/30/89]) means the organization offering arbitration services selected by EPA to conduct arbitrations pursuant to this part.

At retail (Section 60.531 [2/26/88]) means the sale by a commercial owner of a wood heater to the ultimate purchaser.

At-the-source (Section 421.31) means at or before the commingling of delacquering scrubber liquor blowdown with other process or nonprocess wastewaters.

Atomization (Section 471.02) is the process in which a stream of water or gas impinges upon a molten metal stream, breaking it into droplets which solidify as powder particles.

Attractants (Section 162.3) includes all substances or mixtures of substances which, through their property of attracting certain animals, are intended to mitigate a population of, or destroy any vertebrate or invertebrate animals declared to be pests under Section 162.14. (i) Attractants include, but are not limited to: (A) Sensory stimulants (such as pheromones, synthetic attractants, and certain extracts from naturally-occurring organic materials) when used alone, or when in combination with toxicants that can kill certain vertebrate or invertebrate animals, that are intended to draw certain animals into traps or away from crops or sites; these sensory stimulants are considered to be active ingredients in pesticide products; and (B) Naturally-occurring foods and certain extracts from such foods, when in combination with toxicants that can kill certain vertebrate or invertebrate animals, that are intended to draw certain animals into traps or away from crops or sites; these foods and extracts are considered to be inert ingredients in pesticide products. (ii) Attractants do not include: (A) Substances or mixtures of substances intended to attract vertebrate or invertebrate animals for survey or detection purposes only; and (B) Naturally-occurring foods, when used alone or separately and not marketed in mixtures with toxicants, for the purpose of attracting vertebrate or invertebrate animals.

Auction Subaccount (Section 73.3 [12-17-91]) means an account in the Special Allowance Reserve, as specified in section 416(b) of the Clean Air Act. The Auction Subaccount shall contain allowances to be sold at auction in the amount of 150,000 per year from 1995 through 1999, inclusive, and 200,000 per year for each year beginning in the calendar year 2000, subject to modifications noted in these regulations.

Authorized account representative (Section 73.3 [12-17-91]) means a natural person who may transfer and otherwise dispose of allowances held in an account in the Allowance Tracking System, including, in the case of a unit account, the designated representative of the owners and operators of an affected unit.

Authorized person (Section 763.121) means any person authorized by the employer and required by work duties to be present in regulated areas.

Authorized representative (Section 260.10) means the person responsible for the overall operation of a facility or an operational unit (i.e., part of a facility), e.g., the plant manager, superintendent or person of equivalent responsibility.

Automated method (or analyzer) (Section 53.1) means a method for measuring concentrations of an ambient air pollutant in which sample collection, analysis, and measurement are performed automatically.

Automated transmission component (Section 763.163 [7-12-89]) means an asbestos-containing product used as a friction material in vehicular automatic transmissions.

Automatic temperature compensator (Section 60.431) means a device that continuously senses the temperature of fluid flowing through a metering device and automatically adjusts the registration of the measured volume to the corrected equivalent volume at a base temperature.

Automobile (Section 60.391) means a motor vehicle capable of carrying no more than 12 passengers.

Automobile (Section 600.002-85) means: (i) Any four-wheel vehicle propelled by a combustion engine using onboard fuel or by an electric motor drawing current from rechargeable storage batteries or other portable energy storage devices (rechargeable using energy from a source off the vehicle such as residential electric service), (ii) Which is manufactured primarily for use on public streets, roads, or highways (except any vehicle operated on a rail or rails), (iii) Which is rated at not more than 8,500 pounds gross vehicle weight, which has a curb weight of not more than 6,000 pounds, and which has a basic vehicle frontal area of not more than 45 square feet, or (iv) Is a type of vehicle which the Secretary determines is substantially used for the same purposes.

Automobile (Section 610.11) means any four-wheeled vehicle propelled by fuel which is manufactured primarily for use on public streets, roads, and highways (except any vehicle operated exclusively on a rail or rails), and which is rated at 6,000 lbs. gross vehicle weight or less.

Automobile and light-duty truck body (Section 60.391) means the exterior surface of an automobile or light-duty truck including hoods, fenders, cargo boxes, doors, and grill opening panels.

Auxiliary aids (Section 12.103 [8/14/87]) means services or devices that enable persons with impaired sensory, manual, or speaking skills to have an equal opportunity to participate in, and enjoy the benefits of, programs or activities conducted by the agency. For example, auxiliary aids useful for persons with impaired vision include readers, Brailled materials, audio recordings, and other similar services and devices. Auxiliary aids useful for persons with impaired hearing include telephone handset amplifiers, telephones compatible with hearing aids, telecommunication devices for deaf persons (TDDs), interpreters, notetakers, written materials, and other similar services and devices.

Auxiliary Emission Control Device (AECD) (Section 600.002-85) means an element of design as defined in Part 86.

Auxiliary Emission Control Device (AECD) (Section 86.082-2) means any element of design which senses temperature, vehicle speed, engine RPM, transmission gear, manifold vacuum, or any other parameter for the purpose of activating, modulating, delaying, or deactivating the operation of any part of the emission control system.

Available purchase power (Section 60.41a) means the lesser of the following: (a) The sum of available system capacity in all neighboring companies. (b) The sum of the rated capacities of the power interconnection devices between the principal company and all neighboring companies, minus the sum of the electric power load on these interconnections. (c) The rated capacity of the power transmission lines between the power interconnection devices and the electric generating units (the unit in the principal company that has the malfunctioning flue gas desulfurization system and the unit(s) in the neighboring company supplying replacement electrical power) less the electric power load on these transmission lines.

Available system capacity (Section 60.41a) means the capacity determined by subtracting the system load and the system emergency reserves from the net system capacity.

Available technology (Section 192.31 [11/5/93])* means technologies and methods for emplacing a permanent radon barrier on uranium mill tailings piles or impoundments. This term shall not be construed to include extraordinary measures or techniques that would impose costs that are grossly excessive as measured by practice within the industry or one that is reasonably analogous, (such as, by way of illustration only, unreasonable overtime, staffing or transportation requirements, etc., considering normal practice in the industry; laser fusion, of soils, etc.), provided there is reasonable progress toward emplacement of a permanent radon barrier. To determine grossly excessive costs, the relevant baseline against which cost increases shall be compared is the cost estimate for tailings impoundment closure contained in the licensee's tailings closure plan, but costs beyond such estimates shall not automatically be considered grossly excessive.

Average concentration (Section 423.11) as it relates to chlorine discharge means the average of analyses made over a single period of chlorine release which does not exceed two hours.

Average fuel economy (Section 2.311) has the meaning given it in section 501(4) of the Act, 15 U.S.C. 2001(4).

Average fuel Economy (Section 600.002-85) means the unique fuel economy value as computed under Section 600.510 for a specific class of automobiles produced by a manufacturer that is subject to average fuel economy standards.

Average monthly discharge limitation (Section 122.2) means the highest allowable average of daily discharges over a calendar month, calculated as the sum of all daily discharges measured during a calendar month divided by the number of daily discharges measured during that month.

Average of daily values for 30 consecutive days (Section 435.11 [3/4/93])* shall be the average of the daily values obtained during any 30 consecutive day period.

Average of daily values for thirty consecutive days (Section 407.61). See **Maximum for any one day (Section 407.61)**.

Average of daily values for thirty consecutive days (Section 407.81).

See **Maximum for any one day (Section 407.81)**.

Average of daily values for thirty consecutive days (Section 407.71). See **Maximum for any one day (Section 407.71)**.

Average process water usage flow rate for a plant with more than one plastics molding and forming process that uses contact cooling and heating water (Section 463.11) is the sum of the average process water usage flow rates for the contact cooling and heating processes.

Average process water usage flow rate for a plant with more than one plastics molding and forming process that uses cleaning water (Section 463.21) is the sum of the average process water usage flow rates for the cleaning processes.

Average process water usage flow rate of a cleaning water process in liters per day (Section 463.21) is equal to the volume of process water (liters) used per year by a process divided by the number of days per year the process operates.

Average process water usage flow rate for a plant with more than one plastics molding and forming process that uses finishing water (Section 463.31) is the sum of the average process water usage flow rates for the finishing processes.

Average process water usage flow rate of a finishing water process in liters per day (Section 463.31) is equal to the volume of process water (liters) used per year by a process divided by the number of days per year the process operates.

Average process water usage flow rate of a contact cooling and heating water process in liters per day (Section 463.11) is equal to the volume of process water (liters) used per year by a process divided by the number of days per year the process operates.

Average weekly discharge limitation (Section 122.2) means the highest allowable average of daily discharges over a calendar week, calculated as the sum of all daily discharges measured during a calendar week divided by the number of daily discharges measured during that week.

Averaging for heavy-duty engines (Section 86.090-2 [7-26-90]) means the exchange of NOX and particulate emission credits among engine families within a given manufacturer's product line.

Averaging set (Section 86.090-2 [7-26-90]) means a subcategory of heavy-duty engines within which engine families can average and trade emission credits with one other.

Award (Section 35.4010 [10/1/92])* means the TAG agreement signed by both EPA and the recipient.

Award official (Section 30.200) means the EPA official with the authority to execute assistance agreements and to take other actions authorized by this subchapter and by EPA Orders.

Award official (Section 35.6015 [6-5-90])* The EPA official with the authority to execute Cooperative Agreements and Superfund State Contracts (SSCs) and to take other actions authorized by EPA Orders.

Award official (Section 7.25) means the EPA official with the authority to approve and execute assistance agreements and to take other assistance related actions authorized by this part and by other EPA regulations or delegation of authority.

Award Official (Section 35.4010 [10/1/92])* means the EPA official delegated the authority to sign grant agreements.

Axenic (Section 797.1160) means a culture of Lemna fronds free from other organisms.

Axle clearance (Section 86.084-2) means the vertical distance from the level surface on which an automobile is standing to the lowest point on the axle differential of the automobile. This definition applies beginning with the 1984 model year.

Axle Ratio (Section 600.002-85) means the number of times the input shaft to the differential (or equivalent) turns for each turn of the drive wheels.

Axle Ratio (Section 86.602) means all ratios within plus or minus 3% of the axle ratio specified in the configuration in the test order.

B

Baby foods (Section 407.81) shall mean the processing of canned fresh fruits and vegetables, meats, eggs, fruit juices, cereal, formulated entrees, desserts and snacks using fresh, preprocessed, or any combination of these and other food ingredients necessary for the production of infant foods.

Background soil pH (Section 257.3-5) means the pH of the soil prior to the addition of substances that alter the hydrogen ion concentration.

Bagging operation (Section 60.671) means the mechanical process by which bags are filled with nonmetallic minerals.

Bake oven (Sections 60.311; 60.391) means a device which uses heat to dry or cure coatings.

Balanced, indigenous community (Section 125.71) is synonymous with the term balanced, indigenous population in the Act and means a biotic community typically characterized by diversity, the capacity to sustain itself through cyclic seasonal changes, presence of necessary food chain species and by a lack of domination by pollution tolerant species. Such a community may include historically non-native species introduced in connection with a program of wildlife management and species whose presence or abundance results from substantial, irreversible environmental modifications. Normally, however, such a community will not include species whose presence or abundance is attributable to the introduction of pollutants that will be eliminated by compliance by all sources with section 301(b)(2) of the Act; and may not include species whose presence or abundance is attributable to alternative effluent limitations imposed pursuant to section 316(a).

Balanced, indigenous population (Section 125.58) means an ecological community which: (1) Exhibits characteristics similar to those of nearby, healthy communities existing under comparable but unpolluted environmental conditions; or (2) May reasonably be expected to become re-established in the polluted water body segment from adjacent waters if sources of pollution were removed.

Baler (Section 246.101) means a machine used to compress solid wastes, primary materials, or recoverable materials, with or without binding, to a density or from which will support handling and transportation as a material unit rather than requiring a disposable or reuseable container. This specifically excludes briquetters and stationary compaction equipment which is used to compact materials into disposable or reuseable containers.

Ballast (Section 419.11) shall mean the flow of waters, from a ship, that is treated along with refinery wastewaters in the main treatment system.

Bank (Section 39.105) means The Federal Financing Bank established pursuant to the Federal Financing Bank Act of 1973 (12 U.S.C. 2281 et seq.).

Banking (Section 86.090-2 [7-26-90]) means the retention of heavy-duty engine NOX and particulate emission credits, by the manufacturer generating the emission credits, for use in future model year certification programs as permitted by regulation.

Bar, billet and bloom (Section 420.91) means those acid pickling operations that pickle bar, billet or bloom products.

Barometric condensing operations (Section 409.11) shall mean those operations or processes directly associated with or related to the concentration and crystallization of sugar solutions.

Barrel (Section 113.3) means 42 United States gallons at 60 degrees Fahrenheit.

Barrel finishing (Section 471.02). See **Tumbling (Section 471.02)**.

Barrier (Section 191.12 [12/20/93])* means any material or structure that prevents or substantially delays movement of water or radionuclides toward the accessible environment. For example, a barrier may be a geologic structure, a canister, a waste form with physical and chemical characteristics that significantly decrease the mobility of radionuclides, or a material placed over and around waste, provided that the material or structure substantially delays movement of water or radionuclides.

BART (Section 51.301). See **Best Available Retrofit Technology (Section 51.310)**.

Basal diet (Section 797.2050) means the food or diet as it is prepared or received from the supplier, without the addition of any carrier, diluent, or test substance.

Basal diet (Sections 797.2130; 797.2150) means the untreated form of the diet, such as the diet obtained from a commercial source.

Base film (Section 60.711 [10/3/88]) means the substrate that is coated to produce magnetic tape.

Base Flood (Part 6, Appendix A, Section 4) means that flood which has a one percent chance of occurrence in any given year (also known as a 100-year flood). This term is used in the National Flood Insurance Program (NFIP) to indicate the minimum level of flooding to be used by a community in its floodplain management regulations.

Base Floodplain (Part 6, Appendix A, Section 4) means the land area covered by a 100-year flood (one percent chance floodplain). Also see definition of floodplain.

Base Level (Section 600.002-85) means a unique combination of basic engine inertia weight class and transmission class.

Base load (Section 60.331) means the load level at which a gas turbine is normally operated.

Base temperature (Section 60.431) means an arbitrary reference temperature for determining liquid densities or adjusting the measured volume of a liquid quantity.

Base Vehicle (Section 600.002-85) means the lowest priced version of each body style that makes up a car line.

Based flood (Section 257.3-1) means a flood that has a 1 percent or greater chance of recurring in any year or a flood of a magnitude equalled or exceeded once in 100 years on the average over a significantly long period.

Baseline area (Section 51.166) means (i) any intrastate area (and every part thereof) designated as attainment or unclassifiable under section 107(d)(1) (D) or (E) of the Act in which the major source or major modification establishing the baseline date would construct or would have an air quality impact equal to or greater than 1 $\mu g/m^3$ (annual average) of the pollutant for which the baseline date is established. (ii) Area redesignations under section 107(d)(1) (D) or (E) of the Act cannot intersect or be smaller than the area of impact of any major stationary source or major modification which: (a) Establishes a baseline date; or (b) Is subject to 40 CFR 52.21 or under regulations approved pursuant to 40 CFR 51.166, and would be constructed in the same state as the state proposing the redesignation.

Baseline area (Section 52.21) means (i) any intrastate area (and every part thereof) designated as attainment or unclassifiable under section 107(d)(1) (D) or (E) of the Act in which the major source or major modification establishing the baseline date would construct or would have an air quality impact equal to or greater than 1 $\mu g/m^3$ (annual average) of the pollutant for which the baseline date is established. (ii) Area redesignations under section 107(d)(1) (D) or (E) of the Act cannot intersect or be smaller than the area of impact of any major stationary source or major modification which: (a) Establishes a baseline date; or (b) Is subject to 40 CFR 52.21 and would be constructed in the same state as the state proposing the redesignation.

Baseline concentration (Sections 51.166; 52.21) means (i) that ambient concentration level which exists in the baseline area at the time of the applicable baseline date. A baseline concentration is determined for each pollutant for which a baseline date is established and shall include: (a) The actual emissions representative of sources in existence on the applicable baseline date, except as provided in paragraph (b)(13)(ii) of this section; (b) The allowable emissions of major stationary sources which commenced construction before January 6, 1975, but were not in operation by the applicable baseline date. (ii) The following will not be included in the baseline concentration and will affect the applicable maximum allowable increase(s): (a) Actual emissions from any major stationary source on which construction commenced after January 6, 1975; and (b) Actual emissions increases and decreases at any stationary source occurring after the baseline date.

Baseline configuration (Section 610.11) means the unretrofitted test configuration, tuned in accordance with the automobile manufacturer's specifications.

Baseline consumption allowances (Section 82.3 [12/10/93])* means the consumption allowances apportioned under Sec. 82.6.

Baseline date (Section 51.166) means (i) the earliest date after August 7, 1977, that: (a) A major stationary source or major modification subject to 40 CFR 52.21 submits a complete application under that section; or (b) A major stationary source or major modification subject to regulations approved pursuant to 40 CFR 51.166 submits a complete application under such regulations. (ii) The baseline date is established for each pollutant for which increments or other equivalent measures have been established if: (a) The area in which the proposed source or modification would construct is designated as attainment or unclassifiable under section 107(d)(i)(D) or (E) of the Act for the pollutant on the date of its complete application under 40 CFR 52.21 or under regulations approved pursuant to 40 CFR 51.166; and (b) In the case of a major stationary source, the pollutant would be emitted in significant amounts, or, in the case of a major modification, there would be a significant net emissions increase of the pollutant.

Baseline date (Section 52.21) means (i) the earliest date after August 7, 1977, on which the first complete application under 40 CFR 52.21 is submitted by a major stationary source or major modification subject to the requirements of 40 CFR 52.21. (ii) The baseline date is established for each pollutant for which increments or other equivalent measures have been established if: (a) The area in which the proposed source or modification would construct is designated as attainment or unclassifiable under section 107(d)(i) (D) or (E) of the Act for the pollutant on the date of its complete application under 40 CFR 52.21; and (b) In the case of a major stationary source, the pollutant would be emitted in significant amounts, or, in the case of a major modification, there would be a significant net emissions increase of the pollutant.

Baseline or trend assessment survey (Section 228.2) means the planned sampling or measurement of parameters at set stations or in set areas in and near disposal sites for a period of time sufficient to provide synoptic data for determining water quality, benthic, or biological conditions as a result of ocean disposal operations. The minimum requirements for such surveys are given in Section 228.13.

Baseline production allowances (Section 82.3 [12/10/93])* means the production allowances apportioned under Sec. 82.5.

Basic engine (Section 600.002-85) means a unique combination of manufacturer, engine displacement, number of cylinders, fuel system (as distinguished by number of carburetor barrels or use of fuel injection), catalyst usage, and other engine and emission control system characteristics specified by the Administrator. For electric vehicles, basic engine means a unique combination of manufacturer and electric traction motor, motor controller, battery configuration, electrical charging system, energy storage device, and other components as specified by the Administrator.

Basic engine (Section 86.082-2) means a unique combination of manufacturer, engine displacement, number of cylinders, fuel system (as distinguished by number of carburetor barrels or use of fuel injection), catalyst usage, and other engine and emission control system characteristics specified by the Administrator.

Basic oxygen furnace steelmaking (Section 420.41) means the production of steel from molten iron, steel scrap, fluxes, and various combinations thereof, in refractory lined furnaces by adding oxygen.

Basic oxygen process furnace (BOPF) (Sections 60.141; 60.141a) means any furnace with a refractory lining in which molten steel is produced by charging scrap metal, molten iron, and flux materials or alloy additions into a vessel and introducing a high volume of oxygen-rich gas. Open hearth, blast, and reverberatory furnaces are not included in this definition.

Basic vehicle frontal area (Section 86.082-2) means the area enclosed by the geometric projection of the basic vehicle along the longitudinal axis, which includes tires but excludes mirrors and air deflectors, onto a plane perpendicular to the longitudinal axis of the vehicle.

Basis material (Section 465.02) means the coiled strip which is processed.

Basis material (Section 466.02) means the metal part or base onto which porcelain enamel is applied.

BAT (Section 141.2 [7/8/87]). See **Best available technology (Section 141.2)**.

BAT (Section 467.02) means the best available technology economically achievable under section 304(b)(2)(B) of the Act.

Batch (Section 160.3 [8-17-89])* means a specific quantity or lot of a test, control, or reference substance that has been characterized according to Sec. 160.105(a).

Batch (Section 169.1) means a quantity of a pesticide product or active ingredient used in producing a pesticide made in one operation or lot or if made in a continuous or semi-continuous process or cycle, the quantity produced during an interval of time to be specified by the producer.

Batch (Section 204.51) means the collection of compressors of the same category or configuration, as designated by the Administrator in a test request, from which a batch sample is to be randomly drawn and inspected to determine conformance with the acceptability criteria.

Batch (Section 205.51) means the collection of vehicles of the same category, configuration or subgroup thereof as designated by the Administrator in a test request, from which a batch sample is to be drawn, and inspected to determine conformance with the acceptability criteria.

Batch (Section 420.111) means those alkaline cleaning operations which process steel products such as coiled wire, rods, and tubes in discrete batches or bundles.

Batch (Section 420.81) means those descaling operations in which the products are processed in discrete batches.

Batch (Section 420.91) means those pickling operations which process steel products such as coiled wire, rods, and tubes in discrete batches or bundles.

Batch (Section 792.3 [8-17-89])* means a specific quantity or lot of a test, control, or reference substance that has been characterized according to Sec. 792.105(a).

Batch distillation operation (Section 60.661 [6-29-90]) means a noncontinuous distillation operation in which a discrete quantity or batch of liquid feed is charged into a distillation unit and distilled at one time. After the initial charging of the liquid feed, no additional liquid is added during the distillation operation.

Batch MWC (Section 60.51a [2-11-91]) means an MWC unit designed such that it cannot combust MSW continuously 24 hours per day because the design does not allow waste to be fed to the unit or ash to be removed while combustion is occurring.

Batch of reformulated gasoline (Section 80.2 [2/16/94])* means a quantity of reformulated gasoline which is homogeneous with regard to those properties which are specified for reformulated gasoline certification.

Batch, pipe and tube (Section 420.81) means those descaling operations that remove surface scale from pipe and tube products in batch processes.

Batch, rod and wire (Section 420.81) means those descaling operations that remove surface scale from rod and wire products in batch processes.

Batch sample (Section 204.51) means the collection of compressors that are drawn from a batch.

Batch sample (Section 205.51) means the collection of vehicles of the same category, configuration or subgroup thereof which are drawn from a batch and from which test samples are drawn.

Batch sample size (Section 204.51) means the number of compressors of the same category or configuration which is randomly drawn from the batch sample and which will receive emissions tests.

Batch sample size (Section 205.51) means the number of vehicles of the same category or configuration in a batch sample.

Batch, sheet and plate (Section 420.81) means those descaling operations that remove surface scale from sheet and plate products in batch processes.

Batch size (Section 204.51) means the number, as designated by the Administrator in the test request, of compressors of the same category or configuration in a batch.

Batch size (Section 205.51) means the number as designated by the Administrator in the test request of vehicles of the same category or configuration in a batch.

Batt insulation (Section 248.4 [2/17/89]). See **Blanket insulation (Section 248.4)**.

Battery (Section 461.2) means a modular electric power source where part or all of the fuel is contained within the unit and electric power is generated directly from a chemical reaction rather than indirectly through a heat cycle engine. In this regulation there is no differentiation between a single cell and a battery.

Battery Configuration (Section 600.002-85) means the electrochemical type, voltage, capacity (in Watt-hours at the c/3 rate), and physical characteristics of the battery used as the tractive energy storage device.

Battery manufacturing operations (Section 461.2) means all of the specific processes used to produce a battery including the manufacture of anodes and cathodes and associated ancillary operations. These manufacturing operations are excluded from regulation under any other point source category.

Battery separator (Section 763.163 [7-12-89]) means an asbestos-containing product used as an insulator or separator between the negative and positive terminals in batteries and fuel cells.

Bauxite (Section 421.11) shall mean ore containing alumina monohydrate or alumina trihydrate which serves as the principal raw material for the production of alumina by the Bayer process or by the combination process.

BCF (Section 797.1520). See **Bioconcentration factor (Section 797.1520)**.

BCF (Section 797.1560). See **Bioconcentration factor (Section 797.1560)**.

BCF (Section 797.1830). See **Bioconcentration factor (Section 797.1830)**.

BCT (Section 467.02) means the best conventional pollutant control technology, under section 304(b)(4) of the Act.

Bead (Section 60.541 [9/15/87]) means rubber-covered strands of wire, wound into a circular form, which ensure a seal between a tire and the rim of the wheel onto which the tire is mounted.

Bead cementing operation (Section 60.541 [9/15/87]) means the system that is used to apply cement to the bead rubber before or after it is wound into its final circular form. A bead cementing operation consists of a cement application station, such as a dip tank, spray booth and nozzles, cement trough and roller or swab applicator, and all other equipment necessary to apply cement to wound beads or bead rubber and to allow evaporation of solvent from cemented beads.

Beater-add gasket (Section 763.163 [7-12-89] means an asbestos-containing product that is made of paper intended for use as a gasket, and designed to prevent leakage of liquids, solids, or gases and to seal the space between two sections of a component in circumstances not involving rotary, reciprocating, and helical motions. Major applications of beater-add gaskets include: gaskets for internal combustion engines; carburetors; exhaust manifolds; compressors; reactors; distillation columns; and other apparatus.

Beehive cokemaking (Section 420.11) means those operations in which coal is heated with the admission of air in controlled amounts for the purpose of producing coke. There are no by-product recovery operations associated with beehive cokemaking operations.

Beer-Lambert law (Section 796.3700) states that the absorbance of a solution of a given chemical species, at a fixed wavelength, is proportional to the thickness of the solution (l), or the light pathlength, and the concentration of the absorbing species (C).

Beets (Section 407.71) shall include the processing of beets into the following product styles: Canned and peeled, whole, sliced, diced, French style, sections, irregular, and other cuts but not dehydrated beets.

Begin actual construction (Section 51.165) means in general, initiation of physical on-site construction activities on an emissions unit which are of a permanent nature. Such activities include, but are not limited to, installation of building supports and foundations, laying of underground pipework, and construction of permanent storage structures. With respect to a change in method of operating this term refers to those on-site activities other than preparatory activities which mark the initiation of the change.

Begin actual construction (Sections 51.166; 52.21; 52.24) means in general, initiation of physical on-site construction activities on an emissions unit which are of a permanent nature. Such activities include, but are not limited to, installation of building supports and foundations, laying of underground pipework, and construction of permanent storage structures. With respect to a change in method of operation this term refers to those on-site activities other than preparatory activities which mark the initiation of the change.

Belowground release (Section 280.12 [9/23/88]) means any release to the subsurface of the land and to ground water. This includes, but is not limited to, releases from the belowground portions of an underground storage tank system and belowground releases associated with overfills and transfer operations as the regulated substance moves to or from an underground storage tank.

Belowground storage facility (Section 113.3) means a tank or other container located other than as defined as Aboveground.

Belt conveyor (Section 60.671) means a conveying device that transports material from one location to another by means of an endless belt that is carried on a series of idlers and routed around a pulley at each end.

Beneath the surface of the ground (Section 280.12 [9/23/88]) means beneath the ground surface or otherwise covered with earthen materials.

Beneficial organism (Section 166.3) means any pollinating insect, or any pest predator, parasite, pathogen or other biological control agent which functions naturally or as part of an integrated pest management program to control another pest.

Beneficiation (Section 60.401) means the process of washing the rock to remove impurities or to separate size fractions.

Beneficiation area (Section 440.141 [5/24/88]) means the area of land used to stockpile ore immediately before the beneficiation process, the area of land used for the beneficiation process, the area of land used to stockpile the tailings immediately after the beneficiation process, and the area of land from the stockpiled tailings to the treatment system (e.g., holding pond or settling pond, and the area of the treatment system).

Beneficiation process (Section 440.141 [5/24/88]) means the dressing or processing of gold bearing ores for the purpose of: (i) Regulating the size of, or recovering, the ore or product, (ii) Removing unwanted constituents from the ore, and (iii) Improving the quality, purity, or assay grade of a desired product.

Benzene (or priority pollutant No. 4) (Section 420.02) means the value obtained by the standard method Number 602 specified in 44 FR 69464, 69570 (December 3, 1979).

Benzene concentration (Section 61.341 [3-7-90]) means the fraction by weight of benzene in a waste as determined in accordance with the procedures specified in Sec. 61.355 of this subpart.

Benzene storage tank (Section 61.131 [9-14-89]) means any tank, reservoir, or container used to collect or store refined benzene.

Benzidine (Section 129.4) means the compound benzidine and its salts as identified by the chemical name 4,4'-diaminobiphenyl.

Benzidine Manufacturer (Section 129.104) means a manufacturer who produces benzidine or who produces benzidine as an intermediate product in the manufacture of dyes commonly used for textile, leather and paper dyeing.

Benzidine-Based Dye Applicator (Section 129.104) means an owner or operator who uses benzidine-based dyes in the dyeing of textiles, leather or paper.

Benzo(a)pyrene (or priority pollutant No. 73) (Section 420.02) means the value obtained by the standard method Number 610 specified in 44 FR 69464, 69570 (December 3, 1979).

Beryllium (Section 61.31) means the element beryllium. Where weights or concentrations are specified, such weights or concentrations apply to beryllium only, excluding the weight or concentration of any associated elements.

Beryllium alloy (Section 61.31) means any metal to which beryllium has been added in order to increase its beryllium content and which contains more than 0.1 percent beryllium by weight.

Beryllium copper alloy (Section 468.02) shall mean any copper alloy that is alloyed to contain 0.10 percent or greater beryllium.

Beryllium ore (Section 61.31) means any naturally occurring material mined or gathered for its beryllium content.

Beryllium propellant (Section 61.41) means any propellant incorporating beryllium.

Beryllium-containing waste (Section 61.31) means material contaminated with beryllium and/or beryllium compounds used or generated during any process or operation performed by a source subject to this subpart.

Best available control technology (Section 51.166) means an emissions limitation (including a visible emissions standard) based on the maximum degree of reduction for each pollutant subject to regulation under the Act which would be emitted from any proposed major stationary source or major modification which the reviewing authority, on a case-by-case basis, taking into account energy, environmental, and economic impacts and other costs, determines is achievable for such source or modification through application of production processes or available methods, systems, and techniques, including fuel cleaning or treatment or innovative fuel combination techniques for control of such pollutant. In no event shall application of best available control technology result in emissions of any pollutant which would exceed the emissions allowed by any applicable standard under 40 CFR Parts 60 and 61. If the reviewing authority determines that technological or economic limitations on the application of measurement methodology to a particular emissions unit would make the imposition of an emissions standard infeasible, a design, equipment, work practice, operational standard or combination thereof, may be prescribed instead to satisfy the requirement for the application of best available control technology. Such standard shall, to the degree possible, set forth the emissions reduction achievable by implementation of such design, equipment, work practice or operation, and shall provide for compliance by means which achieve equivalent results.

Best available control technology (Section 52.21) means an emissions limitation (including a visible emission standard) based on the maximum degree of reduction for each pollutant subject to regulation under Act which would be emitted from any proposed major stationary source or major modification which the Administrator, on a case-by-case basis, taking into account energy, environmental, and economic impacts and other costs, determines is achievable for such source or modification through application of production processes or available methods, systems, and techniques, including fuel cleaning or treatment or innovative fuel combustion techniques for control of such pollutant. In no event shall application of best available control technology result in emissions of any pollutant which would exceed the emissions allowed by any applicable standard under 40 CFR Parts 60 and 61. If the Administrator determines that technological or economic limitations on the application of measurement methodology to a particular emissions unit would make the imposition of an emissions standard infeasible, a design, equipment, work practice, operational standard, or combination thereof, may be prescribed instead to satisfy the requirement for the application of best available control technology. Such standard shall, to the degree possible, set forth the emissions reduction achievable by implementation of such design, equipment, work practice or operation, and shall provide for compliance by means which achieve equivalent results.

Best available technology (BAT) (Section 141.2 [7/8/87]) means the best technology, treatment techniques, or other means which the Administrator finds, after examination for efficacy under field conditions and not solely under laboratory conditions, are available (taking cost into consideration). For the purposes of setting MCLs for synthetic organic chemicals, any BAT must be at least as effective as granular activated carbon.

Best Available Retrofit Technology (BART) (Section 51.301) means an emission limitation based on the degree of reduction achievable through the application of the best system of continuous emission reduction for each pollutant which is emitted by an existing stationary facility. The emission limitation must be established, on a case-by-case basis, taking into consideration the technology available, the costs of compliance, the energy and nonair quality environmental impacts of compliance, any pollution control equipment in use or in existence at the source, the remaining useful life of the source, and the degree of improvement in visibility which may reasonably be anticipated to result from the use of such technology.

Best management practices (BMPs) (Section 233.3) means schedules of activities, prohibitions of practices, maintenance procedures, and other management practices to prevent or reduce the pollution of waters of the

United States, including methods, measures, practices, or design and performance standards, which facilitate compliance with section 404(b)(1) environmental guidelines (40 CFR Part 230), effluent limitations or prohibitions under section 307(a), and applicable water quality standards.

Best management practices (BMPs) (Section 122.2) means schedules of activities, prohibitions of practices, maintenance procedures, and other management practices to prevent or reduce the pollution of waters of the United States. BMPs also include treatment requirements, operating procedures, and practices to control plant site runoff, spillage or leaks, sludge or waste disposal, or drainage from raw material storage.

Best management practices (BMPs) (Section 232.2 [2/11/93])* means schedules of activities, prohibitions of practices, maintenance procedures, and other management practices to prevent or reduce the pollution of waters of the United States from discharges of dredged or fill material. BMPs include methods, measures, practices, or design and performance standards which facilitate compliance with the section 404(b)(1) Guidelines (40 CFR Part 230), effluent limitations or prohibitions under section 307(a), and applicable water quality standards.

Best Management Practice (BMP) (Section 130.2). Methods, measures or practices selected by an agency to meet its nonpoint source control needs. BMPs include but are not limited to structural and nonstructural controls and operation and maintenance procedures. BMPs can be applied before, during and after pollution-producing activities to reduce or eliminate the introduction of pollutants into receiving waters.

Best Practicable Waste Treatment Technology (BPWTT) (Section 35.2005) means the cost-effective technology that can treat wastewater, combined sewer overflows and nonexcessive infiltration and inflow in publicly owned or individual wastewater treatment works, to meet the applicable provisions of: (i) 40 CFR Part 133 - secondary treatment of wastewater; (ii) 40 CFR Part 125, Subpart G - marine discharge waivers; (iii) 40 CFR 122.44(d) - more stringent water quality standards and State standards; or (iv) 41 FR 6190 (February 11, 1976) - Alternative Waste Management Techniques for Best Practicable Waste Treatment (treatment and discharge, land application techniques and utilization practices, and reuse).

Beverage (Section 244.101) means carbonated natural or mineral waters; soda water and similar carbonated soft drinks; and beer or other carbonated malt drinks in liquid form and intended for human consumption.

Beverage can (Section 60.491) means any two-piece steel or aluminum container in which soft drinks or beer, including malt liquor, are packaged. The definition does not include containers in which fruit or vegetable juices are packaged.

Beverage container (Section 244.101) means an airtight container containing a beverage under pressure of carbonation. Cups and other open receptacles are specifically excluded from this definition.

BIA (Section 147.2902) means the Bureau of Indian Affairs, United States Department of Interior.

Binding Commitment (Section 35.3105 [3-19-90]) A legal obligation by the State to a local recipient that defines the terms for assistance under the SRF.

Bioavailability (Section 795.232 [1-8-90])* refers to the relative amount of administered test substance which reaches the systemic circulation and the rate at which this process occurs.

Bioconcentration (Section 797.1520) is the net accumulation of a substance directly from water into and onto aquatic organisms.

Bioconcentration (Section 797.1560) is the increase in concentration of test material in or on test organisms (or specified tissues thereof) relative to the concentration of test material in the ambient water.

Bioconcentration (Section 797.1830) is the net accumulation of a chemical directly from water into and onto aquatic organisms.

Bioconcentration factor (BCF) (Section 797.1520) is the quotient of the concentration of a test substance in aquatic organisms at or over a discrete time period of exposure divided by the concentration in the test water at or during the same time period.

Bioconcentration factor (BCF) (Section 797.1830) is the quotient of the concentration of a test chemical in tissues of aquatic organisms at or over a discrete time period of exposure divided by the concentration of test chemical in the test water at or during the same time period.

Bioconcentration factor (BCF) (Section 797.1560) is the ratio of the test substance concentration in the test fish (C_f) to the concentration in the test water (C_w) at steady-state.

Biological additives (Section 300.5 [3-8-90])* means microbiological cultures, enzymes, or nutrient additives that are deliberately introduced into an oil discharge for the specific purpose of encouraging biodegredation to mitigate the effects of the discharge.

Biological additives (Section 300.82), for the purposes of this subpart, are microbiological cultures, enzymes, or nutrient additives that are deliberately introduced into an oil discharge for the specific purpose of encouraging biodegradation to mitigate the effects of the discharge.

Biological control agent (Section 152.3 [5/4/88]) means any living organism applied to or introduced into the environment that is intended to function as a pesticide against another organism declared to be a pest by the Administrator.

Biologicals (Section 259.10 [3/24/89]) means preparations made from living organisms and their products, including vaccines, cultures, etc., intended for use in diagnosing, immunizing or treating humans or animals or in research pertaining thereto.

Biopolymer (Section 723.250) means a polymer directly produced by living or once-living cells or cellular components.

Bird hazard (Section 257.3-8) means an increase in the likelihood of bird/aircraft collisions that may cause damage to the aircraft or injury to its occupants.

Bird poisons and repellents (Section 162.3) includes all substances or mixtures of substances intended for preventing, destroying, repelling, or mitigating birds declared to be pests under Section 162.14. Bird poisons and repellents include, but are not limited to: (i) Toxicants intended to kill or destroy certain birds; (ii) Toxicants intended to cause, by pharmacological action, repelling of birds away from certain sites; (iii) Sensory agents utilizing taste, sight, touch, or other means, intended to repel certain bird species or populations from certain sites, to reduce their predation of certain seed and crops, or to protect other organisms or objects from injury, soiling, or harassment; and (iv) Reproductive inhibitors intended to reduce or otherwise alter the reproductive capacity or potential of certain birds.

Bitterns (Section 415.161) shall mean the saturated brine solution remaining after precipitation of sodium chloride in the solar evaporation process.

Bituminous coal (Section 60.251) means solid fossil fuel classified as bituminous coal by ASTM Designation D388-77 (incorporated by reference - see Section 60.17).

Black liquor oxidation system (Section 60.281) means the vessels used to oxidize, with air or oxygen, the black liquor, and associated storage tank(s).

Black liquor solids (Section 60.281) means the dry weight of the solids which enter the recovery furnace in the black liquor.

Blanket insulation (Section 248.4 [2/17/89]) means relatively flat and flexible insulation in coherent sheet form, furnished in units of substantial area. Batt insulation is included in this term.

Blast furnace (Section 60.131) means any furnace used to recover metal from slag.

Blast furnace (Section 60.181) means any reduction furnace to which sinter is charged and which forms separate layers of molten slag and lead bullion.

Bleached papers (Section 250.4 [10/6/87, 6/22/88]) means paper made of pulp that has been treated with bleaching agents.

Blend fertilizer (Section 418.71) shall mean a mixture of dry, straight and mixed fertilizer materials.

Blood products (Section 259.10 [3/24/89]) means any product derived from human blood, including but not limited to blood plasma, platelets, red or white blood corpuscles, and other derived licensed products, such as interferon, etc.

Blowdown (Sections 401.11; 423.11) means the minimum discharge of recirculating water for the purpose of discharging materials contained in the water, the further buildup of which would cause concentration in amounts exceeding limits established by best engineering practice.

Blowing (Section 61.171) means the injection of air or oxygen-enriched air into a molten converter bath.

Blowing still (Section 60.471) means the equipment in which air is blown through asphalt flux to change the softening point and penetration rate.

Blowing tap (Section 60.261) means any tap in which an evolution of gas forces or projects jets of flame or metal sparks beyond the ladle, runner, or collection hood.

BMP (Section 130.2). See **Best Management Practice (Section 130.2)**.

BMPs (Section 122.2). See **Best management practices (Section 122.2)**.

BMPs (Section 232.2 [6/6/88]). See **Best management practices (Section 232.2)**.

BMPs (Section 233.3). See **Best management practices (Section 233.3)**.

Board (Sections 305.12; 306.12). See **Board of Arbitrators (Sections 305.12; 306.12)**.

Board insulation (Section 248.4 [2/17/89]) means semi-rigid insulation preformed into rectangular units having a degree of suppleness, particularly related to their geometrical dimensions.

Board of Arbitrators or Board (Sections 305.12; 306.12) means a panel of one or more persons selected in accordance with section

112(b)(4)(A) of CERCLA and governed by the provisions in 40 CFR Part 305.

BOD (Section 133.101). The five day measure of the pollutant parameter biochemical oxygen demand (BOD).

BOD5 (Section 401.11) means five-day biochemical oxygen demand.

BOD5 input (Section 405.11) shall mean the biochemical oxygen demand of the materials entered into process. It can be calculated by multiplying the fats, proteins and carbohydrates by factors of 0.890, 1.031 and 0.691 respectively. Organic acids (e.g., lactic acids) should be included as carbohydrates. Composition of input materials may be based on either direct analyses or generally accepted published values.

BOD7 (Sections 417.151; 417.161) shall mean the biochemical oxygen demand as determined by incubation at 20 degrees C for a period of 7 days using an acclimated seed. Agitation employing a magnetic stirrer set at 200 to 500 rpm may be used.

Bodily injury (Section 280.92 [10/26/88]) shall have the meaning given to this term by applicable state law; however, this term shall not include those liabilities which, consistent with standard insurance industry practices, are excluded from coverage in liability insurance policies for bodily injury.

Bodily injury and property damage (Sections 264.141; 265.141). In the liability insurance requirements, the terms bodily injury and property damage shall have the meanings given these terms by applicable State law. However, these terms do not include those liabilities which, consistent with standard industry practices, are excluded from coverage in liability policies for bodily injury and property damage.

Body fluids (Section 259.10 [7-2-90])* means liquid emanating or derived from humans and limited to blood; dialysate; amniotic, cerebrospinal, synovial, pleural, peritoneal and pericardial fluids; and semen and vaginal secretions.

Body style (Section 600.002-85) means a level of commonality in vehicle construction as defined by number of doors and roof treatment (e.g., sedan, convertible, fastback, hatchback) and number of seats (i.e., front, second, or third seat) requiring seat belts pursuant to National Highway Traffic Safety Administration safety regulations. Station wagons and light trucks are identified as car lines.

Body style (Section 86.082-2) means a level of commonality in vehicle construction as defined by number of doors and roof treatment (e.g., sedan, convertible, fastback, hatchback).

Body type (Section 86.082-2) means a name denoting a group of vehicles that are either in the same car line or in different car lines provided the only reason the vehicles qualify to be considered in different car lines is that they are produced by a separate division of a single manufacturer.

Boiler (Section 260.10) means an enclosed device using controlled flame combustion and having the following characteristics: (1)(i) The unit must have physical provisions for recovering and exporting thermal energy in the form of steam, heated fluids, or heated gases; and (ii) The unit's combustion chamber and primary energy recovery sections(s) must be of integral design. To be of integral design, the combustion chamber and the primary energy recovery section(s) (such as waterwalls and superheaters) must be physically formed into one manufactured or assembled unit. A unit in which the combustion chamber and the primary energy recovery section(s) are joined only by ducts or connections carrying flue gas is not integrally designed; however, secondary energy recovery equipment (such as economizers or air preheaters) need not be physically formed into the same unit as the combustion chamber and the primary energy recovery section. The following units are not precluded from being boilers solely because they are not of integral design: process heaters (units that transfer energy directly to a process stream), and fluidized bed combustion units; and (iii) While in operation, the unit must maintain a thermal energy recovery efficiency of at least 60 percent, calculated in terms of the recovered energy compared with the thermal value of the fuel; and (iv) The unit must export and utilize at least 75 percent of the recovered energy, calculated on an annual basis. In this calculation, no credit shall be given for recovered heat used internally in the same unit. (Examples of internal use are the preheating of fuel or combustion air, and the driving of induced or forced draft fans or feed-water pumps); or (2) The unit is one which the Regional Administrator has determined, on a case-by-case basis, to be a boiler, after considering the standards in Section 260.32.

Boiler (Section 60.531 [2/26/88]) means a solid fuel burning appliance used primarily for heating spaces, other than the space where the appliance is located, by the distribution through pipes of a gas or fluid heated in the appliance. The appliance must be tested and listed as a boiler under accepted American or Canadian safety testing codes. A manufacturer may request an exemption in writing from the Administrator by stating why the testing and listing requirement is not practicable and by demonstrating that his appliance is otherwise a boiler.

Boiler (Section 60.661 [6-29-90]) means any enclosed combustion device that extracts useful energy in the form of steam.

Boiler operating day (Section 60.41a)

means a 24-hour period during which fossil fuel is combusted in a steam generating unit for the entire 24 hours.

Bond paper (Section 250.4 [10/6/87, 6/22/88]) means a generic category of paper used in a variety of end use applications such as forms (see form bond), offset printing, copy paper, stationery, etc. In the paper industry, the term was originally very specific but is now very general.

Bond release (Section 434.11) means the time at which the appropriate regulatory authority returns a reclamation or performance bond based upon its determination that reclamation work (including, in the case of underground mines, mine sealing and abandonment procedures) has been satisfactorily completed.

Book paper (Section 250.4 [10/6/87, 6/22/88]) means a generic category of papers produced in a variety of forms, weights, and finishes for use in books and other graphic arts applications, and related grades such as tablet, envelope, and converting papers.

BOPF (Sections 60.141; 60.141a). See **Basic oxygen process furnace (Sections 60.141; 60.141a)**.

Borosilicate recipe (Section 60.291) means glass product composition of the following approximate ranges of weight proportions: 60 to 80 percent silicon dioxide, 4 to 10 percent total R_2O (e.g., Na_2 and K_2O), 5 to 35 percent boric oxides, and 0 to 13 percent other oxides.

Borrower (Section 15.4) means any recipient of a loan as defined in this section under Loan.

Bottom ash (Section 240.101) means the solid material that remains on a hearth or falls off the grate after thermal processing is complete.

Bottom ash (Section 423.11) means the ash that drops out of the furnace gas stream in the furnace and in the economizer sections. Economizer ash is included when it is collected with bottom ash.

Bottom-blown furnace (Section 60.141a) means any BOPF in which oxygen and other combustion gases are introduced to the bath of molten iron through tuyeres in the bottom of the vessel or through tuyeres in the bottom and sides of the vessel.

Bottoms receiver (Section 264.1031 [6-21-90]) means a container or tank used to receive and collect the heavier bottoms fractions of the distillation feed stream that remain in the liquid phase.

Bounce (Section 85.2122(a)(5)(ii)(B)) means unscheduled point contact opening(s) after initial closure and before scheduled reopening.

BPT (Section 467.02) means the best practicable control technology currently available under section 304(b)(1) of the Act.

BPWTT (Section 35.2005). See **Best Practicable Waste Treatment Technology (Section 35.2005)**.

Brass or bronze (Section 60.131) means any metal alloy containing copper as its predominant constituent, and lesser amounts of zinc, tin, lead, or other metals.

Break block (Section 763.163 [7-12-89]) means an asbestos-containing product intended for use as a friction material in drum brake systems for vehicles rated at 26,001 pounds gross vehicle weight rating (GVWR) or more.

Breakdown Voltage (Section 85.2122(a)(6)(ii)(C)) means the voltage level at which the capacitor fails.

Breaker Point (Section 85.2122(a)(5)(ii)(A)) means a mechanical switch operated by the distributor cam to establish and interrupt the primary ignition coil current.

Breakover angle (Section 86.084-2) means the supplement of the largest angle, in the plan side view of an automobile, that can be formed by two lines tangent to the front and rear static loaded radii arcs and intersecting at a point on the underside of the automobile. This definition applies beginning with the 1984 model year.

Broccoli (Section 407.71) shall include the processing of broccoli into the following product styles: Frozen, chopped, spears, and miscellaneous cuts.

Brood stock (Sections 797.1300; 797.1330) means the animals which are cultured to produce test organisms through reproduction.

Brown papers (Section 250.4 [10/6/87, 6/22/88]) means papers usually made from unbleached kraft pulp and used for bags, sacks, wrapping paper, and so forth.

Brown stock washer system (Section 60.281) means brown stock washers and associated knotters, vacuum pumps, and filtrate tanks used to wash the pulp following the digestion system. Diffusion washers are excluded from this definition.

BTX storage tank (Section 61.131 [9-14-89]) means any tank, reservoir, or container used to collect or store benzene-toluene-xylene or other light-oil fractions.

Bubbling fluidized bed combustor (Section 60.51a [2-11-91]) means a fluidized bed combustor in which the majority of the bed material remains in a fluidized state in the primary combustion zone.

Bucket elevator (Section 60.381) means a conveying device for metallic minerals consisting of a head and foot assembly that supports and drives an endless single or double strand chain or belt to which buckets are attached.

Bucket elevator (Section 60.671) means a conveying device of nonmetallic minerals consisting of a head and foot assembly which supports and drives an endless single or double strand chain or belt to which buckets are attached.

Budget (Section 35.4010 [10/1/92])* means the financial plan for the spending of all Federal and matching funds (including in-kind contributions) for a TAG project as proposed by the applicant, and negotiated with and approved by the Award Official.

Budget period (Section 30.200) means the length of time EPA specifies in an assistance agreement during which the recipient may expend or obligate Federal Funds.

Budget period (Section 35.4010 [10/1/92])* means the length of time specified in a grant agreement during which the recipient may spend or obligate Federal funds. The budget period may not exceed three (3) years. A TAG project period may be comprised of several budget periods.

Budget period (Section 35.6015 [6-5-90])* The length of time EPA specifies in a Cooperative Agreement during which the recipient may expend or obligate Federal funds.

Building (Section 35.2005) means the erection, acquisition, alteration, remodeling, improvement or extension of treatment works.

Building (Section 60.671) means any frame structure with a roof.

Building completion (Section 35.2005) means the date when all but minor components of a project have been built, all equipment is operational and the project is capable of functioning as designed.

Building insulation (Section 248.4 [2/17/89]) means a material, primarily designed to resist heat flow, which is installed between the conditioned volume of a building and adjacent unconditioned volumes or the outside. This term includes but is not limited to insulation products such as blanket, board, spray-in-place, and loose-fill that are used as ceiling, floor, foundation, and wall insulation.

Building, structure, facility or installation (Section 52.24) means all of the pollutant-emitting activities which belong to the same industrial grouping, are located on one or more contiguous or adjacent properties, and are under the control of the same person (or persons under common control) except the activities of any vessel. Pollutant-emitting activities shall be considered as part of the same industrial grouping if they belong to the same Major Group (i.e., which have the same two-digit code) as described in the following document, *Standard Industrial Classification Manual*, 1972, as amended by the 1977 Supplement (U. S. Government Printing Office stock numbers 4101-0066 and 003-005-00176-0, respectively).

Building, structure, facility, or installation (Section 51.166) means all of the pollutant-emitting activities which belong to the same industrial grouping, are located on one or more contiguous or adjacent properties, and are under the control of the same person (or persons under common control) except the activities of any vessel. Pollutant-emitting activities shall be considered as part of the same industrial grouping if they belong to the same Major Group (i.e., which have the same two-digit code) as described in the *Standard Industrial Classification Manual*, 1972, as amended by the 1977 Supplement (U.S. Government Printing Office stock numbers 4101-0066 and 003-005-00176-0, respectively).

Building, structure, facility, or installation (Section 51.165) means all of the pollutant-emitting activities which belong to the same industrial grouping, are located on one or more contiguous or adjacent properties, and are under the control of the same person (or persons under common control) except the activities of any vessel. Pollutant-emitting activities shall be considered as part of the same industrial grouping if they belong to the same Major Group (i.e., which have the same two-digit code) as described in the *Standard Industrial Classification Manual*, 1972, as amended by the 1977 Supplement (U.S. Government Printing Office stock numbers 4101-0065 and 003-005-00176-0, respectively).

Building, structure, facility or installation (Section 52.21) means all of the pollutant-emitting activities which belong to the same industrial grouping, are located on one or more contiguous or adjacent properties, and are under the control of the same person (or persons under common control) except the activities of any vessel. Pollutant-emitting activities shall be considered as part of the same industrial grouping if they belong to the same Major Group (i.e., which have the same first two digit code) as described in the *Standard Industrial Classification Manual*, 1972, as amended by the 1977 Supplement (U. S. Government Printing Office stock numbers 4101-0066 and 003-005-00176-0, respectively).

Building, structure or facility (Section 51.301) means all of the pollutant-emitting activities which belong to the same industrial grouping, are located on one or more contiguous or adjacent properties, and are under the control of the same person (or persons under common control). Pollutant-emitting activities must be considered as part of the same industrial grouping if they belong to the same Major Group (i.e., which have the same two-digit code) as described in the *Standard Industrial Classification Manual,* 1972 as amended by the 1977 Supplement (U.S. Government Printing Office stock numbers 4101-0066 and 003-005-00176-0 respectively).

Bulk asbestos (Section 763.63) means any quantity of asbestos fiber of any type or grade, or combination of types or grades, that is mined or milled with the purpose of obtaining asbestos. This term does not include asbestos that is produced or processed as a contaminant or an impurity.

Bulk container (Section 246.101) means a large container that can either be pulled or lifted mechanically onto a service vehicle or emptied mechanically into a service vehicle.

Bulk gasoline plant (Section 60.111b) means any gasoline distribution facility that has a gasoline throughput less than or equal to 75,700 liters per day. Gasoline throughput shall be the maximum calculated design throughput as may be limited by compliance with an enforceable condition under Federal requirement or Federal, State or local law, and discoverable by the Administrator and any other person.

Bulk gasoline terminal (Section 60.501) means any gasoline facility which receives gasoline by pipeline, ship or barge, and has a gasoline throughput greater than 75,700 liters per day. Gasoline throughput shall be the maximum calculated design throughput as may be limited by compliance with an enforceable condition under Federal, State or local law and discoverable by the Administrator and any other person.

Bulk resin (Section 61.61) means a resin which is produced by a polymerization process in which no water is used.

Bulk terminal (Section 61.301 [3-7-90]) means any facility which receives liquid product containing benzene by pipelines, marine vessels, tank trucks, or railcars, and loads the product for further distribution into tank trucks, railcars, or marine vessels.

Bulkhead (Section 61.20) means an air-restraining barrier constructed for long-term control of radon-222 and radon-222 decay product levels in mine air.

Bulky waste (Section 243.101) means large items of solid waste such as household appliances, furniture, large auto parts, trees, branches, stumps, and other oversize wastes whose large size precludes or complicates their handling by normal solid wastes collection, processing, or disposal methods.

Burial operation (Section 257.3-6). See **Trenching (Section 257.3-6)**.

Burning agents (Section 300.5 [3-8-90])* means those additives that, through physical or chemical means, improve the combustibility of the materials to which they are applied.

Burning agents (Section 300.82), for the purposes of this subpart, are those additives that, through physical or chemical means, improve the combustibility of the materials to which they are applied.

Burnishing (Section 468.02). See **Tumbling (Section 468.02)**.

Burnishing (Section 471.02) is a surface finishing process in which minute surface irregularities are displaced rather than removed.

Business (Section 2.201) means any person engaged in a business, trade, employment, calling or profession, whether or not all or any part of the net earnings derived from such engagement by such person inure (or may lawfully inure) to the benefit of any private shareholder or individual.

Business (Section 4.2) means any lawful activity, except a farm operation, that is conducted: (1) Primarily for the purchase, sale, lease, and/or rental of personal and/or real property, and/or for the manufacture, processing, and/or marketing of products, commodities, and/or any other personal property; or (2) Primarily for the sale of services to the public; or (3) Solely for the purpose of Section 4.303 of these regulations, conducted primarily for outdoor advertising display purposes, when the display must be moved as a result of the project; or (4) By a nonprofit organization that has established its nonprofit status under applicable Federal or State law.

Business confidentiality claim or claim (Section 2.201) means a claim or allegation that business information is entitled to confidential treatment for reasons of business confidentiality, or a request for a determination that such information is entitled to such treatment.

Business confidentiality or confidential business information (Section 350.1 [7/29/88]) includes the concept of trade secrecy and other related legal concepts which give (or may give) a business the right to preserve the confidentiality of business information and to limit its use or disclosure by others in order that the business may obtain or retain business advantages it derives from its right in the information. The definition is meant to encompass any concept which authorizes a Federal agency to withhold business information under 5 U.S.C. 552(b)(4), as well as any concept which requires EPA to withhold information from the public for the benefit of a business under 18 U.S.C. 1905.

Business information or information (Section 2.201) means any information which pertains to the interests of any business, which was developed or acquired by that business, and (except where the context otherwise requires) which is possessed by EPA in recorded form.

Business machine (Section 60.721 [1/29/88]) means a device that uses electronic or mechanical methods to process information, perform calculations, print or copy information, or convert sound into electrical impulses for transmission, such as: (1) Products classified as typewriters under SIC Code 3572; (2) Products classified as electronic computing devices under SIC Code 3573; (3) Products classified as calculating and accounting machines under SIC Code 3574; (4) Products classified as telephone and telegraph equipment under SIC Code 3661; (5) Products classified as office machines, not elsewhere classified, under SIC Code 3579; and (6) Photocopy machines, a subcategory of products classified as photographic equipment under SIC Code 3861.

By compound (Section 60.661 [6-29-90]) means by individual stream components, not carbon equivalents.

By-pass the control device (Section 61.161) means to operate the glass melting furnace without operating the control device to which that furnace's emissions are directed routinely.

By-product (Section 261.1), for purposes of Sections 261.2 and 261.6, is a material that is not one of the primary products of a production process and is not solely or separately produced by the production process. Examples are process residues such as slags or distillation column bottoms. The term does not include a co-product that is produced for the general public's use and is ordinarily used in the form it is produced by the process.

By-product cokemaking (Section 420.11) means those cokemaking operations in which coal is heated in the absence of air to produce coke. In this process, by-products may be recovered from the gases and liquids driven from the coal during cokemaking.

By-product material (Section 61 [12-15-89]) Any radioactive material (except source material and special nuclear material) yielded in or made radioactive by exposure to the radiation incident to the process of producing or utilizing special nuclear material and wastes from the processing of ores primarily to recover their source material content.

Bypass (Section 122.41) means the intentional diversion of waste streams from any portion of a treatment facility.

Bypass (Section 60.61 [12/14/88]) means any system that prevents all or a portion of the kiln or clinker cooler exhaust gases from entering the main control device and ducts the gases through a separate control device. This does not include emergency systems designed to duct exhaust gases directly to the atmosphere in the event of a malfunction of any control device controlling kiln or clinker cooler emissions.

Bypass stack (Section 60.61 [12/14/88]) means the stack that vents exhaust gases to the atmosphere from the bypass control device.

Byproduct (Section 704.3 [12/22/88])* means a chemical substance produced without a separate commercial intent during the manufacture, processing, use, or disposal of another chemical substance(s) or mixture(s).

Byproduct (Section 712.3) means any chemical substance or mixture produced without a separate commercial intent during the manufacture, processing, use, or disposal of another chemical substance or mixture.

Byproduct (Section 716.3) means a chemical substance produced without a separate commercial intent during the manufacture, processing, use, or disposal of another chemical substance(s) or mixture(s).

Byproduct (Sections 710.2; 723.175; 761.3) means a chemical substance produced without separate commercial intent during the manufacture or processing of another chemical substance(s) or mixture(s).

Byproduct (Sections 720.3; 723.50; 747.200; 791.3) means a chemical substance produced without a separate commercial intent during the manufacture, processing, use, or disposal of another chemical substance or mixture.

Byproduct material (Sections 710.2; 720.3) shall have the meaning contained in the Atomic Energy Act of 1954, 42 U.S.C. 2014 et seq., and the regulations issued thereunder.

Byproduct/waste (Section 60.41b [12/16/87])* means any liquid or gaseous substance produced at chemical manufacturing plants or petroleum refineries (except natural gas, distillate oil, or residual oil) and combusted in a steam generating unit for heat recovery or for disposal. Gaseous substances with carbon dioxide levels greater than 50 percent or carbon monoxide levels greater than 10 percent are not byproduct/waste for the purposes of this subpart.

C

C in CT calculations (Section 141.2 [6/29/89]). See **Residual disinfectant concentration (Section 141.2)**.

Cab over axle or cab over engine (Section 205.51) means the cab which contains the operator/passenger compartment is directly above the engine and front axle and the entire cab can be tilted forward to permit access to the engine compartment.

Cab over engine (Section 205.51). See **Cab over axle (Section 205.51)**.

C_{aj} (Section 60.741 [9-11-89]) means the concentration of VOC in each gas stream (j) exiting the emission control device, in parts per million by volume.

Calcine (Section 60.161) means the solid materials produced by a roaster.

Calciner (Section 60.401) means a unit in which the moisture and organic matter of phosphate rock is reduced within a combustion chamber.

Calciner or Nodulizing Kiln (Section 61.121 [12-15-89])* means a unit in which phosphate rock is heated to high temperatures to remove organic material and/or to convert it to a nodular form. For the purpose of this subpart, calciners and nodulizing kilns are considered to be similar units.

Calcium carbide (Section 60.261) means material containing 70 to 85 percent calcium carbide by weight.

Calcium silicon (Section 60.261) means that alloy as defined by ASTM Designation A495-76 (incorporated by reference - see Section 60.17).

Calcium sulfate storage pile runoff (Section 418.11) shall mean the calcium sulfate transport water runoff from or through the calcium sulfate pile, and the precipitation which falls directly on the storage pile and which may be collected in a seepage ditch at the base of the outer slopes of the storage pile, provided such seepage ditch is protected from the incursion of surface runoff from areas outside of the outer perimeter of the seepage ditch.

Calculated level (Section 82.3 [12/10/93])* means the weighted amount of a controlled substance determined by multiplying the amount (in kilograms) of the controlled substance by that substance's ozone depletion weight listed in appendix A or appendix B of this subpart.

Calibrating gas (Section 86.082-2) means a gas of known concentration which is used to establish the response curve of an analyzer.

Calibration (Sections 86.082-2; 600.002-85) means the set of specifications, including tolerances, unique to a particular design, version, or application of a component or components assembly capable of functionally describing its operation over its working range.

Calibration of equipment (Section 171.2) means measurement of dispersal or output of application equipment and adjustment of such equipment to control the rate of dispersal, and droplet or particle size of a pesticide dispersed by the equipment.

Can (Section 465.02) means a container formed from sheet metal and consisting of a body and two ends or a body and a top.

Candidate method (Section 53.1) means a method of sampling and analyzing the ambient air for an air pollutant for which an application for a reference method determination or an equivalent method determination is submitted in accordance with Section 53.4, or a method tested at the initiative of the Administrator in accordance with Section 53.7.

Caneberries (Section 407.61) shall include the processing of the following berries: Canned and frozen blackberries, blueberries, boysenberries, currants, gooseberries, loganberries, ollalieberries, raspberries, and any other similar cane or bushberry but not strawberries or cranberries.

Canmaking (Section 465.02) means the manufacturing process or processes used to manufacture a can from a basic metal.

Canned meat processor (Section 432.91) shall mean an operation which prepares and cans meats (such as stew, sandwich spreads, or similar products) alone or in combination with other finished products at rates greater than 2730 kg (6000 lb) per day.

Canned onions (Section 407.71) shall mean the processing of onions into the following product styles: Canned, frozen, and fried (canned), peeled, whole, sliced, and any other piece size but not including frozen, battered onion rings or dehydrated onions.

Capable of Transportation of Property on a street or highway (Section 205.51) means that the vehicle: (i) Is self propelled and is capable of transporting any material or fixed apparatus, or is capable of drawing a trailer or semi-trailer; (ii) Is capable of maintaining a cruising speed of at least 25 mph over level, paved surface; (iii) Is equipped or can readily be equipped with features customarily associated with practical street or highway use, such features including but not being limited to: A reverse gear and a differential, fifth wheel, cargo platform or cargo enclosure; and (iv) Does not exhibit features which render its use on a street or highway impractical, or highly unlikely, such features including, but not being limited to, tracked road means, an inordinate size or features ordinarily associated with combat or tactical vehicles.

Capacitance (Section 85.2122(a)(6)(ii)(A)) means the property of a device which permits storage of electrically-separated charges when differences in electrical potential exist between the conductors and measured as the ratio of stored charge to the difference in electrical potential between conductors.

Capacitor (Section 761.3) means a device for accumulating and holding a charge of electricity and consisting of conducting surfaces separated by a dielectric.

Capacitor/Condenser (Section 85.2122(a)(6)(ii)(D)) means a device for the storage of electrical energy consisting of two oppositely charged conducting plates separated by a dielectric and which resists the flow of direct current.

Capacity (Section 60.671) means the cumulative rated capacity of all initial crushers that are part of the plant.

Capacity factor (Section 51.100) means the ratio of the average load on a machine or equipment for the period of time considered to the capacity rating of the machine or equipment.

Capital expenditure (Section 60.2) means an expenditure for a physical or operational change to an existing facility which exceeds the product of the applicable annual asset guideline repair allowance percentage specified in the latest edition of Internal Revenue Service (IRS) Publication 534 and the existing facility's basis, as defined by section 1012 of the Internal Revenue Code. However, the total expenditure for a physical or operational change to an existing facility must not be reduced by any excluded additions as defined in IRS Publication 534, as would be done for tax purposes.

Capital expenditure (Section 60.481) means, in addition to the definition in 40 CFR 60.2, an expenditure for a physical or operational change to an existing facility that: (a) Exceeds P, the product of the facility's replacement cost, R, and an adjusted annual asset guideline repair allowance, A, as reflected by the following equation: P = R x A, where (1) The adjusted annual asset guideline repair allowance, A, is the product of the percent of the replacement cost, Y, and the applicable basic annual asset guideline repair allowance, B, as reflected by the following equation: A = Y x (B /100); (2) The percent Y is determined from the following equation : Y = 1.0 - 0.575 log X, where X is 1982 minus the year of construction; and (3) The applicable basic annual asset guideline repair allowance, B, is selected from the table found in Section 60.481 under this definition, consistent with the applicable subpart.

Capital expenditure (Section 61.02)

means an expenditure for a physical or operational change to a stationary source which exceeds the product of the applicable annual asset guideline repair allowance percentage specified in the latest edition of Internal Revenue Service (IRS) Publication 534 and the stationary source's basis, as defined by section 1012 of the Internal Revenue Code. However, the total expenditure for a physical or operational change to a stationary source must not be reduced by any excluded additions as defined for stationary sources constructed after December 31, 1981, in IRS Publication 534, as would be done for tax purposes. In addition, annual asset guideline repair allowance may be used even though it is excluded for tax purposes in IRS Publication 534.

Capitalization Grant (Section 35.3105 [3-19-90]) The assistance agreement by which the EPA obligates and awards funds allotted to a State for purposes of capitalizing that State's revolving fund.

Capture system (Section 60.261) means the equipment (including hoods, ducts, fans, dampers, etc.) used to capture or transport particulate matter generated by an affected electric submerged arc furnace to the control device.

Capture system (Section 60.271) means the equipment (including ducts, hoods, fans, dampers, etc.) used to capture or transport particulate matter generated by an EAF to the air pollution control device.

Capture system (Section 60.271a) means the equipment (including ducts, hoods, fans, dampers, etc.) used to capture or transport particulate matter generated by an electric arc furnace or AOD vessel to the air pollution control device.

Capture system (Section 60.301) means the equipment such as sheds, hoods, ducts, fans, dampers, etc. used to collect particulate matter generated by an affected facility at a grain elevator.

Capture system (Section 60.381) means the equipment used to capture and transport particulate matter generated by one or more affected facilities to a control device.

Capture system (Section 60.671) means the equipment (including enclosures, hoods, ducts, fans, dampers, etc.) used to capture and transport particulate matter generated by one or more process operations to a control device.

Capture system (Section 60.711 [10/3/88]) means any device or combination of devices that contains or collects an airborne pollutant and directs it into a duct.

Car Coupling Sound (Section 201.1) means a sound which is heard and identified by the observer as that of car coupling impact, and that causes a sound level meter indicator (FAST) to register an increase of at least ten decibels above the level observed immediately before hearing the sound.

Car Line (Sections 86.082-2; 600.002-85) means a name denoting a group of vehicles within a make or car division which has a degree of commonality in construction (e.g., body, chassis). Car line does not consider any level of decor or opulence and is not generally distinguished by characteristics as roof line, number of doors, seats, or windows, except for station wagons or light-duty trucks. Station wagons and light-duty trucks are considered to be different car lines than passenger cars.

Car-seal (Section 61.341 [1/7/93])* means a seal that is placed on a device that is used to change the position of a valve (e.g., from opened to closed) in such a way that the position of the valve cannot be changed without breaking the seal.

Car-sealed (Section 61.301 [3-7-90]) means having a seal that is placed on the device used to change the position of a valve (e.g., from open to closed) such that the position of the valve cannot be changed without breaking the seal and requiring the replacement of the old seal, once broken, with a new seal.

Carbon (Section 420.71). See **Carbon hot forming operation (Section 420.71)**.

Carbon hot forming operation (or carbon) (Section 420.71) means those hot forming operations which produce a majority, on a tonnage basis, of carbon steel products.

Carbon regeneration unit (Section 260.10 [2-21-91]) means any enclosed thermal treatment device used to regenerate spent activated carbon.

Carbon steel (Section 420.71) means those steel products other than specialty steel products.

Carrier (Section 160.3 [8-17-89]) means any material, including but not limited to feed, water, soil, nutrient media, with which the test substance is combined for administration to a test system.

Carrier (Section 201.1) means a common carrier by railroad, or partly by railroad and partly by water, within the continental United States, subject to the Interstate Commerce Act, as amended, excluding street, suburban, and inter-urban electric railways unless operated as a part of a general railroad system of transportation.

Carrier (Section 792.226) means any material (e.g., feed, water, soil, nutrient media) with which the test substance is combined for administration to test organisms.

Carrier (Section 792.3 [8-17-89]) means any material, including but not limited to, feed, water, soil, and nutrient media, with which the test substance is combined for administration to a test system.

Carrier (Section 797.1400) means a solvent used to dissolve a test substance prior to delivery to the test chamber.

Carrier (Section 797.1520) is a solvent used to dissolve a test substance prior to delivery of the test substance to the test chamber.

Carrier (Section 797.1600) is a solvent or other agent used to dissolve or improve the solubility of the test substance in dilution water.

Carrier (Section 80.2 [2/16/94])* means any distributor who transports or stores or causes the transportation or storage of gasoline or diesel fuel without taking title to or otherwise having any ownership of the gasoline or diesel fuel, and without altering either the quality or quantity of the gasoline or diesel fuel.

Carrier of contaminant (Section 230.3) means dredged or fill material that contains contaminants.

Carrots (Section 407.71) shall include the processing of carrots into the following product styles: Canned and frozen, peeled, whole, sliced, diced, nuggets, crinkle cut, julienne, shoestrings, chunks, chips and other irregular cuts, and juices but not dehydrated carrots.

Carrying Case (Section 211.203) means the container used to store reusable hearing protectors.

Carryout collection (Section 243.101) means collection of solid waste from a storage area proximate to the dwelling unit(s) or establishment.

Cartridge filter (Section 60.621) means a discrete filter unit containing both filter paper and activated carbon that traps and removes contaminants from petroleum solvent, together with the piping and ductwork used in the installation of this device.

CAS Number (Section 704.3 [12/22/88])* means Chemical Abstracts Service Registry Number.

CAS Number (Section 721.3 [7/27/88])* means Chemical Abstracts Service Registry Number assigned to a chemical substance on the Inventory.

Case Examiner (Section 15.4) means an EPA official familiar with pollution control issues who is designated by the Assistant Administrator to conduct a listing or removal proceeding and to determine whether a facility will be placed on the List of Violating Facilities or removed from such list. The Case Examiner may not be: (1) The Listing Official; (2) the Recommending Person or anyone subordinate to the Recommending Person; or (3) closely involved in the underlying enforcement action.

Cash draw (Section 35.3105 [3-19-90]) The transfer of cash under a letter of credit (LOC) from the Federal Treasury into the State's SRF.

Casing (Section 146.3) means a pipe or tubing of appropriate material, of varying diameter and weight, lowered into a borehole during or after drilling in order to support the sides of the hole and thus prevent the walls from caving, to prevent loss of drilling mud into porous ground, or to prevent water, gas, or other fluid from entering or leaving the hole.

Casing (Section 147.2902) means a pipe or tubing of varying diameter and weight, lowered into a borehole during or after drilling in order to support the sides of the hole and, thus, prevent the walls from caving, to prevent loss of drilling mud into porous ground, or to prevent water, gas, or other fluid from entering the hole.

Casting (Section 471.02) is pouring molten metal into a mold to produce an object of desired shape.

Catalyst (Section 60.471) means a substance which, when added to asphalt flux in a blowing still, alters the penetrating-softening point relationship or increases the rate of oxidation of the flux.

Catalytic Converter (Section 85.2122(a)(15)(ii)(A)) means a device installed in the exhaust system of an internal combustion engine that utilizes catalytic action to oxidize hydrocarbon (HC) and carbon monoxide (CO) emissions to carbon dioxide (CO_2) and water (H_2O).

Catastrophic collapse (Section 146.3) means the sudden and utter failure of overlying strata caused by removal of underlying materials.

Catch basin (Section 60.691 [11/23/88]) means an open basin which serves as a single collection point for stormwater runoff received directly from refinery surfaces and for refinery wastewater from process drains.

Categorical exclusion (Section 1508.4) means a category of actions which do not individually or cumulatively have a significant effect on the human environment and which have been found to have no such effect in procedures adopted by a Federal agency in implementation of these regulations (Section 1507.3) and for which, therefore, neither an environmental assessment nor an environmental impact statement is required. An agency may decide in its procedures or otherwise, to prepare environmental assessments for the reasons stated in Section 1508.9 even though it is not required to do so. Any procedures under this section shall provide for extraordinary circumstances in which a normally excluded action may have a significant environmental effect.

Category (Section 204.51) means a

group of compressor configurations which are identical in all aspects with respect to the parameters listed in paragraph (c)(1)(i) of Section 204.55-2.

Category (Section 205.151) means a group of vehicle configurations which are identical in all material aspects with respect to the parameters listed in Section 205.157-2 of this subpart.

Category (Section 205.165) means a group of exhaust systems which are identical in all material aspects with respect to the parameters listed in Section 205.168 of this subpart.

Category (Section 205.51) means a group of vehicle configurations which are identical in all material aspects with respect to the parameters listed in Section 205.55-2.

Category (Section 211.203) means a group of hearing protectors which are identical in all aspects to the parameters listed in Section 211.210-2(c).

Category I nonfriable asbestos-containing material (ACM) (Section 61.141 [11-20 90]) means asbestos-containing packings, gaskets, resilient floor covering, and asphalt roofing products containing more than 1 percent asbestos as determined using the method specified in appendix A, subpart F, 40 CFR part 763, section 1, Polarized Light Microscopy.

Category II nonfriable ACM (Section 61.141 [11-20-90]) means any material, excluding Category I nonfriable ACM, containing more than 1 percent asbestos as determined using the methods specified in appendix A, subpart F, 40 CFR part 763, section 1, Polarized Light Microscopy that, when dry, cannot be crumbled, pulverized, or reduced to powder by hand pressure.

Category of chemical substances (Section 723.50) has the same meaning as in section 26(c)(2) of the Act (15 U.S.C. 2625(c)(2)).

Category of chemical substances (Sections 723.175; 723.250) has the same meaning as in section 26(c)(2) of the Act (15 U.S.C. 2625).

Cathode ray tubes (Section 469.31) means electronic devices in which electrons focus through a vacuum to generate a controlled image on a luminescent surface. This definition does not include receiving and transmitting tubes.

Cathodic protection (Section 280.12 [9/23/88]) is a technique to prevent corrosion of a metal surface by making that surface the cathode of an electrochemical cell. For example, a tank system can be cathodically protected through the application of either galvanic anodes or impressed current.

Cathodic protection tester (Section 280.12 [9/23/88]) means a person who can demonstrate an understanding of the principles and measurements of all common types of cathodic protection systems as applied to buried or submerged metal piping and tank systems. At a minimum, such persons must have education and experience in soil resistivity, stray current, structure-to-soil potential, and component electrical isolation measurements of buried metal piping and tank systems.

Cation exchange capacity (CEC) (Section 796.2750) is the sum total of exchangeable cations that a sediment or soil can adsorb. The CEC is expressed in milliequivalents of negative charge per 100 grams (meq/100g) or milliequivalents of negative charge per gram (meq/g) of soil or sediment.

Cation exchange capacity (CEC) (Section 796.2700) is the sum total of exchangeable cations that a soil can adsorb. The CEC is expressed in milliequivalents of negative charge per 100 grams (meq/100 g) or milliequivalents of negative charge per gram (meq/g) of soil.

Cation exchange capacity (Section 257.3-5) means the sum of exchangeable cations a soil can absorb expressed in milli-equivalents per 100 grams of soil as determined by sampling the soil to the depth of cultivation or solid waste placement, whichever is greater, and analyzing by the summation method for distinctly acid soils or the sodium acetate method for neutral, calcareous or saline soils ("Methods of Soil Analysis, Agronomy Monograph No. 9." C. A. Black , ed., American Society of Agronomy, Madison, Wisconsin. pp 891-901, 1965).

Cationic polymer (Section 723.250) means a polymer that contains one or more covalently linked subunits that bear a net positive charge.

C_{bi} (Section 60.741 [9-11-89]) means the concentration of VOC in each gas stream (i) entering the emission control device, in parts per million by volume.

$CBOD_5$ (Section 133.101). The five day measure of the pollutant parameter carbonaceous biochemical oxygen demand ($CBOD_5$).

C_{di} (Section 60.741 [9-11-89]) means the concentration of VOC in each gas stream (i) entering the emission control device from the affected coating operation, in parts per million by volume.

CEC (Section 796.2700). See **Cation exchange capacity (Section 796.2700).**

CEC (Section 796.2750). See **Cation exchange capacity (Section 796.2750).**

Ceiling Insulation (Section 248.4 [2/17/89]) means a material, primarily designed to resist heat flow, which is installed between the conditioned area of a building and an unconditioned attic as well as common ceiling floor assemblies between separately conditioned units in multi-unit structures. Where the conditioned area of a building extends to the roof, ceiling insulation includes such a material used between the underside and upperside of the roof.

Cell (Section 241.101) means compacted solid wastes that are enclosed by natural soil or cover material in a land disposal site.

Cell room (Section 61.51) means a structure(s) housing one or more mercury electrolytic chlor-alkali cells.

Cellular polyisocyanurate insulation (Section 248.4 [2/17/89]) means insulation produced principally by the polymerization of polymeric polyisocyanates, usually in the presence of polyhydroxl compounds with the addition of catalysts, cell stabilizers, and blowing agents.

Cellular polystyrene insulation (Section 248.4 [2/17/89]) means an organic foam composed principally of polymerized styrene resin processed to form a homogenous rigid mass of cells.

Cellular polyurethane insulation (Section 248.4 [2/17/89]) means insulation composed principally of the catalyzed reaction product of polyisocyanurates and polyhydroxl compounds, processed usually with a blowing agent to form a rigid foam having a predominantly closed cell structure.

Cellulose (Section 248.4 [2/17/89]) means vegetable fiber such as paper, wood, and cane.

Cellulose fiber fiberboard (Section 248.4 [2/17/89]) means insulation composed principally of cellulose fibers usually derived from paper, paperboard stock, cane, or wood, with or without binders.

Cellulose fiber loose-fill (Section 248.4 [2/17/89]) means a basic material of recycled wood-based cellulosic fiber made from selected paper, paperboard stock, or ground wood stock, excluding contaminated materials which may reasonably be expected to be retained in the finished product, with suitable chemicals introduced to provide properties such as flame resistance, processing and handling characteristics. The basic cellulosic material may be processed into a form suitable for installation by pneumatic or pouring methods.

Cementing (Sections 146.3; 147.2902) means the operation whereby a cement slurry is pumped into a drilled hole and/or forced behind the casing.

Central collection point (Section 259.10 [3/24/89]) means a location where a generator consolidates regulated medical waste brought together from original generation points prior to its transport off-site or its treatment on-site (e.g., incineration).

CEQ Regulations (Section 6.101)

means the regulations issued by the Council on Environmental Quality on November 29, 1978 (see 43 FR 55978), which implement Executive Order 11991. The CEQ Regulations will often be referred to throughout this regulation by reference to 40 CFR Part 1500 et al.

Ceramic plant (Section 61.31) means a manufacturing plant producing ceramic items.

CERCLA (Section 280.12 [9/23/88]) means the Comprehensive Environmental Response, Compensation, and Liability Act of 1980, as amended.

CERCLA (Section 300.5 [3-8-90]) is the Comprehensive Environmental Response, Compensation, and Liability Act of 1980, as amended by the Superfund Amendments and Reauthorization Act of 1986.

CERCLA (Section 304.12 [5/30/89]) means the Comprehensive Environmental Response, Compensation, and Liability Act of 1980, 42 U.S.C. 9601, et seq., as amended by the Superfund Amendments and Reauthorization Act of 1986, Pub. L. 99-499, 100 Stat. 1613 (1986).

CERCLA (Section 305.12) means the Comprehensive Environmental Response, Compensation, and Liability Act of 1980 (42 U.S.C. 9601 et seq.).

CERCLA (Section 306.12) means the Comprehensive Environmental Response, Compensation, and Liability Act of 1980 (42 U.S.C. 9601 et seq.).

CERCLA (Section 35.6015 [6-5-90])* The Comprehensive Environmental Response, Compensation, and Liability Act of 1980 (42 U.S.C. 9601-9657, Pub. L. 96-510, Dec. 11, 1980), as amended by the Superfund Amendments and Reauthorization Act of 1986 (Pub. L. 99-499, Oct. 17, 1986; 100 Stat. 1613).

CERCLA (Section 355.20) means the Comprehensive Environmental Response, Compensation, and Liability Act of 1980, as amended.

CERCLA Hazardous Substance (Section 355.20) means a substance on the list defined in Section 101(14) of CERCLA. Note: Listed CERCLA hazardous substances appear in Table 302.4 of 40 CFR Part 302.

CERCLA or Superfund (Section 300.6) is the Comprehensive Environmental Response, Compensation, and Liability Act of 1980.

CERCLA or Superfund (Section 302.3) means the Comprehensive Environmental Response, Compensation, and Liability Act of 1980 (Pub. L. 96-510).

CERCLIS (Section 300.5 [3-8-90]) is the abbreviation of the CERCLA

Information system, EPA's comprehensive data base and management system that inventories and tracks releases addressed or needing to be addressed by the Superfund program. CERCLIS contains the official inventory of CERCLA sites and supports EPA's site planning and tracking functions. Sites that EPA decides do not warrant moving further in the site evaluation process are given a "No Further Response Action Planned" (NFRAP) designation in CERCLIS. This means that no federal steps under CERCLA will be taken at the site unless future information so warrants. Sites are not removed from the data base after completion of evaluations in order to document that these evaluations too place and to preclude the possibility that they be needlessly repeated. Inclusion of a specific site or area in the CERCLIS data base does not represent a determination of any party's liability, nor does it represent a finding that any response action is necessary. Sites that are deleted from the NLP are not designated NFRAP sites. Deleted sites are listed in a separate category in the CERCLIS data base.

Cereal (Section 406.81) shall mean breakfast cereal.

Certificate holder (Section 85.1502 [9/25/87]) is the entity in whose name the certificate of conformity for a class of motor vehicles or motor vehicle engines has been issued.

Certificate of conformity (Section 85.1502 [9/25/87]) is the document issued by the Administrator under section 206(a) of the Act.

Certification (Section 163.2) means a certification by the Director that a pesticide chemical is useful for the purpose for which a tolerance or exemption is sought under the act.

Certification (Section 171.2) means the recognition by a certifying agency that a person is competent and thus authorized to use or supervise the use of restricted use pesticides.

Certification (Section 26.102 [6-18-91]) means the official notification by the institution to the supporting department or agency, in accordance with the requirements of this policy, that a research project or activity involving human subjects has been reviewed and approved by an IRB in accordance with an approved assurance.

Certification (Section 260.10) means a statement of professional opinion based upon knowledge and belief.

Certification (Section 761.3 [12-21-89]) means a written statement regarding a specific fact or representation that contains the following language: Under civil and criminal penalties of law for the making or submission of false or fraudulent statements or representations (18 U.S.C. 1001 and 15 U.S.C. 2615),

I certify that the information contained in or accompanying this document is true, accurate, and complete. As to the identified section(s) of this document for which I cannot personally verify truth and accuracy, I certify as the company official having supervisory responsibility for the persons who, acting under my direct instructions, made the verification that this information is true, accurate, and complete.

Certification Short Test (Section 86.096-2 [11/1/93])* means the test, for gasoline-fueled Otto-cycle light-duty vehicles and light-duty trucks, performed in accordance with the procedures contained in 40 CFR part 86 subpart O.

Certification Vehicle (Section 600.002-85) means a vehicle which is selected under Section 86.084-24(b)(1) and used to determine compliance under Section 86.084-30 for issuance of an original certificate of conformity.

Certified Aftermarket Part (Section 85.2113) means any aftermarket part which has been certified pursuant to this subpart.

Certified applicator (Section 171.2) means any individual who is certified to use or supervise the use of any restricted use pesticides covered by his certification.

Certified Part (Section 85.2102) means a part certified in accordance with the after-market part certification regulations contained in this subpart.

Certifying agency (Section 121.1) means the person or agency designated by the Governor of a State, by statute, or by other governmental act, to certify compliance with applicable water quality standards. If an interstate agency has sole authority to so certify for the area within its jurisdiction, such interstate agency shall be the certifying agency. Where a State agency and an interstate agency have concurrent authority to certify, the State agency shall be the certifying agency. Where water quality standards have been promulgated by the Administrator pursuant to section 10(c)(2) of the Act, or where no State or interstate agency has authority to certify, the Administrator shall be the certifying agency.

Certifying official (Section 73.3 [12-17-91]) for purposes of part 73, means: (1) for a corporation, a president, secretary, treasurer, or vice-president of the corporation in charge of a principal business function, or any other person who performs similar policy- or decision-making functions for the corporation; (2) for a partnership or sole proprietorship, a general partner or the proprietor, respectively; and (3) for a local government entity or State, Federal or other public agency, either a principal executive officer or ranking elected official.

Cetane index (Section 80.2 [2/16/94])*

or "Calculated cetane index" is a number representing the ignition properties of diesel fuel oils from API gravity and mid-boiling point as determined by ASTM standard method D 976-80, entitled "Standard Methods for Calculated Cetane Index of Distillate Fuels". ASTM test method D 976-80 is incorporated by reference. This incorporation by reference was approved by the Director of the Federal Register in accordance with 5 U.S.C. 552(a) and 1 CFR part 51. A copy may be obtained from the American Society for Testing and Materials, 1916 Race Street, Philadelphia, PA 19103. A copy may be inspected at the Air Docket Section (A-130), Room M-1500, U.S. Environmental Protection Agency, Docket No. A-86-03, 401 M Street SW., Washington, DC 20460 or at the Office of the Federal Register, 800 North Capitol Street, NW., suite 700, Washington, DC 20408.

C_{fk} (Section 60.741 [9-11-89]) means the concentration of VOC in each uncontrolled gas stream (k) emitted directly to the atmosphere from the affected coating operation, in parts per million by volume.

C_{gv} (Section 60.741 [9-11-89]) means the concentration of VOC in the gas stream entering each individual carbon adsorber vessel (v), in parts per million by volume. For purposes of calculating the efficiency of the individual adsorber vessel, C_{gv} may be measured in the carbon adsorption system's common inlet duct prior to the branching of individual inlet ducts.

Challenge exposure (Section 798.4100) is an experimental exposure of a previously treated subject to a test substance following an induction period, to determine whether the subject will react in a hypersensitive manner.

Change order (Section 35.6015 [1/27/89]) means a written order issued by a recipient, or its designated agent, to its contractor authorizing an addition to, deletion from, or revision of, a contract, usually initiated at the contractor's request.

Change room (Section 763.121). See **Equipment room (Section 763.121)**.

Changed use pattern (Section 162.3) means a significant change from a use pattern approved in connection with the registration of a pesticide product. Examples of significant changes include, but are not limited to, changes from nonfood to food use, outdoor to indoor use, ground to aerial application, terrestrial to aquatic use, and nondomestic to domestic use.

Charge (Section 60.271) means the addition of iron and steel scrap or other materials into the top of an electric arc furnace.

Charge (Section 60.271a) means the addition of iron and steel scrap or other materials into the top of an electric arc furnace or the addition of molten steel or other materials into the top of an AOD vessel.

Charge chrome (Section 60.261) means that alloy containing 52 to 70 percent by weight chromium, 5 to 8 percent by weight carbon, and 3 to 6 percent by weight silicon.

Charging (Section 61.171) means the addition of a molten or solid material to a copper converter.

Charging period (Section 60.271) means the time period commencing at the moment an EAF starts to open and ending either three minutes after the EAF roof is returned to its closed position or six minutes after commencement of opening of the roof, whichever is longer.

Chemical (Section 790.3) means a chemical substance or mixture.

Chemical Abstracts Service Registry Number (Section 721.3 [7/27/88])*. See **CAS Number (Section 721.3)**.

Chemical agents (Section 300.5 [3-8-90])* means those elements, compounds, or mixtures that coagulate, disperse, dissolve, emulsify, foam, neutralize, precipitate, reduce, solubilize, oxidize, concentrate, congeal, entrap, fix, make the pollutant mass more rigid or viscous, or otherwise facilitate the mitigation of deleterious effects or the removal of the pollutant from the water.

Chemical agents (Section 300.82), for the purposes of this subpart, in general, are those elements, compounds, or mixtures that coagulate, disperse, dissolve, emulsify, foam, neutralize, precipitate, reduce, solubilize, oxidize, concentrate, congeal, entrap, fix, make the pollutant mass more rigid or viscous, or otherwise facilitate the mitigation of deleterious effects or removal of the pollutant from the water.

Chemical composition (Section 79.2) means the name and percentage by weight of each compound in an additive and the name and percentage by weight of each element in an additive.

Chemical fate studies (Section 792.226) means studies performed for the characterization of physical, chemical, and persistence factors that may be used to evaluate transport and transformation processes.

Chemical manufacturing plant (Section 61.341 [3-7-90]) means any facility engaged in the production of chemicals by chemical, thermal, physical, or biological processes for use as a product, co-product, by-product, or intermediate including but not limited to industrial organic chemicals, organic pesticide products, pharmaceutical preparations, paint and allied products, fertilizers, and agricultural chemicals. Examples of chemical manufacturing plants include facilities at which process units are operated to produce one or more of the following chemicals: benzenesulfonic acid, benzene, chlorobenzene, cumene, cyclohexane, ethylene, ethylbenzene, hydroquinone, linear alklylbenzene,nitrobenzene, resorcinol, sulfolane, or styrene.

Chemical manufacturing plants (Section 60.41b [12/16/87])* means industrial plants which are classified by the Department of Commerce under Standard Industrial Classification (SIC) Code 28.

Chemical metal cleaning waste (Section 423.11) means any wastewater resulting from the cleaning of any metal process equipment with chemical compounds, including, but not limited to, boiler tube cleaning.

Chemical name (Section 721.3 [7-27-89]) means the scientific designation of a chemical substance in accordance with the nomenclature system developed by the International Union of Pure and Applied Chemistry or the Chemical Abstracts Service's rules of nomenclature, or a name which will clearly identify a chemical substance for the purpose of conducting a hazard evaluation.

Chemical protective clothing (Section 721.3 [7-27-89]) means items of clothing that provide a protective barrier to prevent dermal contact with chemical substances of concern. Examples can include, but are not limited to: full body protective clothing, boots, coveralls, gloves, jackets, and pants.

Chemical structure (Section 79.2) means the molecular structure of a compound in an additive.

Chemical substance (Section 723.175) has the same meaning as in section 3 of the Act (15 U.S.C. 2602).

Chemical substance (Section 2.306) has the meaning given it in section 3(2) of the Act, 15 U.S.C. 2602(2).

Chemical substance (Section 710.2) means any organic or inorganic substance of a particular molecular identity, including any combination of such substances occurring in whole or in part as a result of a chemical reaction or occurring in nature, and any chemical element or uncombined radical; except that chemical substance does not include: (1) Any mixture, (2) Any pesticide when manufactured, processed, or distributed in commerce for use as a pesticide, (3) Tobacco or any tobacco product, but not including any derivative products, (4) Any source material, special nuclear material, or byproduct material, (5) Any pistol, firearm, revolver, shells, and cartridges, and (6) Any food, food additive, drug, cosmetic, or device, when manufactured, processed, or distributed in commerce for use as a food, food additive, drug, cosmetic, or device.

Chemical substance (Section 761.3) (1) except as provided in paragraph (2) of this definition, means any organic or inorganic substance of a particular molecular identity, including: Any combination of such substances occurring in whole or part as a result of a chemical reaction or occurring in nature, and any element or uncombined radical. (2) Such term does not include: Any mixture; any pesticide (as defined in the Federal Insecticide, Fungicide, and Rodenticide Act) when manufactured, processed, or distributed in commerce for use as a pesticide; tobacco or any tobacco product; any source material, special nuclear material, or byproduct material (as such terms are defined in the Atomic Energy Act of 1954 and regulations issued under such Act); any article the sale of which is subject to the tax imposed by section 4181 of the Internal Revenue Code of 1954 (determined without regard to any exemptions from such tax provided by section 4182 or section 4221 or any provisions of such Code); and any food, food additive, drug, cosmetic, or device (as such terms are defined in section 201 of the Federal Food, Drug, and Cosmetic Act) when manufactured, processed, or distributed in commerce for use as a food, food additive, drug, cosmetic, or device.

Chemical substance (Section 762.3) has the same meaning as in 15 U.S.C. 2602.

Chemical substance (Section 763.163 [7-12-89]) has the same meaning as in section 3 of the Toxic Substances Control Act.

Chemical substance (Sections 720.3; 723.250; 747.115; 747.195; 747.200) means any organic or inorganic substance of a particular molecular identity, including any combination of such substances occurring in whole or in part as a result of a chemical reaction or occurring in nature, and any chemical element or uncombined radical, except that chemical substance does not include: (1) Any mixture. (2) Any pesticide when manufactured, processed, or distributed in commerce for use as a pesticide. (3) Tobacco or any tobacco product. (4) Any source material, special nuclear material, or byproduct material. (5) Any pistol, firearm, revolver, shells, or cartridges. (6) Any food, food additive, drug, cosmetic, or device, when manufactured, processed, or distributed in commerce for use as a food, food additive, drug, cosmetic, or device.

Chemical waste landfill (Section 761.3) means a landfill at which protection against risk of injury to health or the environment from migration of PCBs to land, water, or the atmosphere is provided from PCBs and PCB Items deposited therein by locating, engineering, and operating the landfill as specified in Section 761.75.

Chemigation (Section 170.3 [8/21/92])* means the application of pesticides through irrigation systems.

Cherries, brined (Section 407.61)

shall include the processing of all varieties of cherries into the following brined product styles: Canned, bottled and bulk, sweet and sour, pitted and unpitted, bleached, sweetened, colored and flavored, whole, halved and chopped.

Cherries, sour (Section 407.61) shall include the processing of all sour varieties of cherries into the following products styles: Frozen and canned, pitted and unpitted, whole, halves, juice and concentrate.

Cherries, sweet (Section 407.61) shall include the processing of all sweet varieties of cherries into the following products styles: Frozen and canned, pitted and unpitted, whole, halves, juice and concentrate.

Chief Executive Officer of the tribe (Section 372.3 [7-26-90]) means the person who is recognized by the Bureau of Indian Affairs as the chief elected administrative officer of the tribe.

Chief Executive Officer of the tribe (Section 370.2 [7-26-90]) means the person who is recognized by the Bureau of Indian Affairs as the chief elected administrative officer of the tribe.

Chief Executive Officer of the tribe (Section 355.20 [7-26-90]) means the person who is recognized by the Bureau of Indian Affairs as the chief elected administrative officer of the tribe.

Chief facility operator (Section 60.51a [2-11-91]) means the person in direct charge and control of the operation of an MWC and who is responsible for daily on-site supervision, technical direction, management, and overall performance of the facility. Circulating fluidized bed combustor means a fluidized bed combustor in which the majority of the fluidized bed material is carried out of the primary combustion zone and is transported back to the primary zone through a recirculation loop.

Child-resistant packaging (Section 157.21) means packaging that is designed and constructed to be significantly difficult for children under 5 years of age to open or obtain a toxic or harmful amount of the substance contained therein within a reasonable time, and that is not difficult for normal adults to use properly.

Chilled water loop (Section 749.68 [1-3-90]) means any closed cooling water system that transfers heat from air handling units or refrigeration equipment to a refrigeration machine, or chiller.

Chips, corn (Section 407.81) shall mean the processing of fried corn, made by soaking, rinsing, milling and extruding into a fryer without toasting. In terms of finished corn chips, 1 kg (lb) of finished product is equivalent to

0.9 kg (lb) of raw material.

Chips, potato (Section 407.81) shall mean the processing of fried chips, made from fresh or stored white potatoes, all varieties. In terms of finished potato chips, 1 kg (lb) of finished product is equivalent to 4 kg (lb) of raw material.

Chips, tortilla (Section 407.81) shall mean the processing of fried corn, made by soaking, rinsing, milling, rolling into sheets, toasting and frying. In terms of finished tortilla chips, 1 kg (lb) of finished product is equivalent to 0.9 kg (lb) of raw material.

Chlorinated terphenyl (Section 704.85) means a chemical substance, CAS No. 61788-33-6, comprised of chlorinated ortho-, meta-, and paraterphenyl.

Chlorine (Section 415.661) shall mean the total residual chlorine present in the process wastewater stream exiting the wastewater treatment system.

Chlorofluorocarbon (Section 82.62 [12/30/93])* means any substance listed as Class I group I or Class I group III in 40 CFR part 82, appendix A to subpart A.

Choke (Section 85.2122(a)(2)(iii)(A)) means a device to restrict air flow into a carburetor in order to enrich the air/fuel mixture delivered to the engine by the carburetor during cold-engine start and cold-engine operation.

Choke Pull-off (Section 85.2122(a)(1)(ii)(F)). See **Vacuum Break (Section 85.2122(a)(1)(ii)(F))**.

Chrome tan (Section 425.02) means the process of converting hide into leather using a form of chromium.

Chromium (Section 415.661) shall mean the total chromium present in the process wastewater stream exiting the wastewater treatment system.

Chromium (Section 420.02) means total chromium and is determined by the method specified in 40 CFR 136.3.

Chromium (Section 428.101) shall mean total chromium.

Chromium VI (Section 420.02). See **Hexavalent chromium (Section 420.02)**.

Chronic toxicity (Section 162.3) means the property of a substance or mixture of substances to cause adverse effects in an organism upon repeated or continuous exposure over a period of at least 1/2 the lifetime of that organism.

Chronic toxicity test (Section 797.1950) means a method used to determine the concentration of a substance that produces an adverse effect from prolonged exposure of an organism to that substance. In this test, mortality, number of young per female and growth are used as measures of chronic toxicity.

Chronic toxicity test (Section 797.1330) means a method used to determine the concentration of a substance in water that produces an adverse effect on a test organism over an extended period of time. In this test guideline, mortality and reproduction (and optionally, growth) are the criteria of toxicity.

C_{hv} (Section 60.741 [9-11-89]) means the concentration of VOC in the gas stream exiting each individual carbon adsorber vessel (v), in parts per million by volume.

Ci (Section 190.02). See **Curie (Section 190.02)**.

Ci (Section 192.01). See **Curie (Section 192.01)**.

City Fuel Economy (Section 600.002-85) means the fuel economy determined by operating a vehicle (or vehicles) over the driving schedule in the Federal emission test procedure.

City Fuel Economy Test (Section 610.11). See **Federal Test Procedure (Section 610.11)**.

Cladding or metal cladding (Section 471.02) is the art of producing a composite metal containing two or more layers that have been metallurgically bonded together by roll bonding (co-rolling), solder application (or brazing), or explosion bonding.

Claim (Section 14.2) means a demand for payment by an employee or his/her representative for the value or the repair cost of an item of personal property damaged, lost or destroyed as an incident to government service.

Claim (Section 2.201). See **Business confidentiality claim (Section 2.201)**.

Claim (Section 211.203) means an assertion made by a manufacturer regarding the effectiveness of his product.

Claim (Section 300.5 [3-8-90])* as defined by section 104(4) of CERCLA, means a demand in writing for a sum certain.

Claim (Section 304.12 [5/30/89]) means the amount sought by EPA as recovery of response costs incurred and to be incurred by the United States at a facility, which does not exceed $500,000, excluding interest.

Claim (Section 35.6015 [1/27/89]) means a demand or written assertion by a contractor seeking, as a matter of right, changes in contract duration, costs, or other provisions, which originally have been rejected by the recipient.

Claim (Sections 305.12; 306.12) means a demand in writing for a sum certain.

Claimant (Section 305.12) means an individual, firm, corporation, association, partnership, consortium, joint venture, commercial entity,

United States Government, State, municipality, commission, political subdivision of a State, or any interstate body who presents a claim for compensation under section 112 of CERCLA.

Claimant (Section 306.12) means any person who presents a claim for compensation under section 112 of CERCLA.

Claimant (Section 350.1 [7/29/88]) means a person submitting a claim of trade secrecy to EPA in connection with a chemical otherwise required to be disclosed in a report or other filing made under Title III.

Class (Section 205.151) means a group of vehicles which are identical in all material aspects with respect to the parameters listed in Section 205.155 of this subpart.

Class (Section 86.402-78). See Section 86.419.

Class I (Section 82.3 [12/10/93])* refers to the controlled substances listed in appendix A of this subpart.

Class I sludge management facility (Section 122.2 [5/2/89]) means any POTW identified under 40 CFR 403.8(a) as being required to have an approved pretreatment program (including such POTWs located in a State that has elected to assume local program responsibilities pursuant to 40 CFR 403.10(e)) and any other treatment works treating domestic sewage classified as a Class I sludge management facility by the Regional Administrator, or, in the case of approved State programs, the Regional Administrator in conjunction with the State Director, because of the potential for its sludge use or disposal practices to adversely affect public health and the environment.

Class I sludge management facility (Section 501.2 [5/2/89]) means any POTW identified under 40 CFR 403.8(a) as being required to have an approved pretreatment program (including such POTWs located in a State that has elected to assume local program responsibilities pursuant to 40 CFR 403.10(e)) and any other treatment works treating domestic sewage classified as a Class I sludge management facility by the Regional Administrator in conjunction with the State Program Director because of the potential for its sludge use or disposal practices to adversely affect public health or the environment.

Class II (Section 82.3 [12/10/93])* refers to the controlled substances listed in appendix B of this subpart.

Class II Substance (Section 82.62 [12/30/93])* means any substance designated as class II in 40 CFR part 82, appendix B to subpart A.

Class II Wells (Section 147.2902) means wells which inject fluids: (a) Which are brought to the surface in connection with conventional oil or natural gas production and may be commingled with waste waters from gas plants which are an integral part of production operations, unless those waters would be classified as a hazardous waste at the time of injection; (b) For enhanced recovery of oil or natural gas; and (c) For storage of hydrocarbons which are liquid at standard temperature and pressure.

Class T3 (Section 87.1) means all aircraft gas turbine engines of the JT3D model family.

Class T8 (Section 87.1) means all aircraft gas turbine engines of the JT8D model family.

Class TF (Section 87.1) means all turbofan or turbojet aircraft engines except engines of Class T3, T8, and TSS.

Class TP (Section 87.1) means all aircraft turboprop engines.

Class TSS (Section 87.1) means all aircraft gas turbine engines employed for propulsion of aircraft designed to operate at supersonic flight speeds.

Classification of Railroads (Section 201.1) means the division of railroad industry operating companies by the Interstate Commerce Commission into three categories. As of 1978, Class I railroads must have annual revenues of $50 million or greater, Class II railroads must have annual revenues of between $10 and $50 million, and Class III railroads must have less than $10 million in annual revenues.

Classified information (Section 11.4) means official information which has been assigned a security classification category in the interest of the national defense or foreign relations of the United States.

Classified material (Section 11.4) means any document, apparatus, model, film, recording, or any other physical object from which classified information can be derived by study, analysis, observation, or use of the material involved.

Classified Waste (Section 246.101) means waste material that has been given security classification in accordance with 50 U.S.C. 401 and Executive Order 11652.

Claus sulfur recovery plant (Section 60.101) means a process unit which recovers sulfur from hydrogen sulfide by a vapor-phase catalytic reaction of sulfur dioxide and hydrogen sulfide.

Clay mineral analysis (Section 796.2750) is the estimation or determination of the kinds of clay-size minerals and the amount present in a sediment or soil.

Clean air standards (Section 15.4) means any enforceable rules, regulations, guidelines, standards, limitations, orders, controls, prohibitions, or other requirements which are contained in, issued under, or otherwise adopted pursuant to the CAA or Executive Order 17738, an applicable implementation plan as described in section 110(d) of the CAA, an approved implementation procedure or plan under section 111(c) or section 111(d), respectively, of the CAA or an approved implementation procedure under section 112(d) of the CAA.

Clean room (Section 763.121) means an uncontaminated room having facilities for the storage of employees' street clothing and uncontaminated materials and equipment.

Clean water standards (Section 15.4) means any enforceable limitation, control, condition, prohibition, standard, or other requirement which is established pursuant to the CWA or contained in a permit issued to a discharger by the United States Environmental Protection Agency, or by a State under an approved program, as authorized by section 402 of the CWA, or by a local government to ensure compliance with pretreatment regulations as required by section 307 of the Clean Water Act.

Cleaning or etching (Section 467.02) is a chemical solution bath and a rinse or series of rinses designed to produce a desired surface finish on the workpiece. This term includes air pollution control scrubbers which are sometimes used to control fumes from chemical solution baths. Conversion coating and anodizing when performed as an integral part of the aluminum forming operations are considered cleaning or etching operations. When conversion coating or anodizing are covered here they are not subject to regulation under the provisions of 40 CFR Part 433, Metal Finishing.

Cleaning water (Section 463.2) is process water used to clean the surface of an intermediate or final plastic product or to clean the surfaces of equipment used in plastics molding and forming that contact an intermediate or final plastic product. It includes water used in both the detergent wash and rinse cycles of a cleaning process.

Clearance (Section 797.1560). See **Depuration (Section 797.1560)**.

Closed cooling water system (Section 749.68 [1-3-90]) means any configuration of equipment in which heat is transferred by circulating water that is contained within the equipment and not discharged to the air; chilled water loops are included.

Closed course competition event (Section 205.151) means any organized competition event covering an enclosed, repeated or confined route intended for easy viewing of the entire route by all spectators. Such events include short track, dirt track, drag race, speedway, hillclimb, ice race, and the Bonneville Speed Trials.

Closed portion (Section 260.10) means that portion of a facility which an owner or operator has closed in accordance with the approved facility closure plan and all applicable closure requirements. (See also active portion and inactive portion.)

Closed-vent system (Section 264.1031 [6-21-90]) means a system that is not open to the atmosphere and that is composed of piping, connections, and, if necessary, flow-inducing devices that transport gas or vapor from a piece or pieces of equipment to a control device.

Closed vent system (Section 60.481) means a system that is not open to the atmosphere and that is composed of piping, connections, and, if necessary, flow inducing devices that transport gas or vapor from a piece or pieces of equipment to a control device.

Closed vent system (Section 60.691 [11/23/88]) means a system that is not open to the atmosphere and is composed of piping, connections, and, if necessary, flow inducing devices that transport gas or vapor from an emission source to a control device.

Closed-vent system (Section 61.341 [3-7-90]) means a system that is not open to the atmosphere and is composed of piping, ductwork, connections, and, if necessary, flow inducing devices that transport gas or vapor from an emission source to a control device.

Closed-vent system (Section 61.241) means a system that is not open to atmosphere and that is composed of piping, connections, and, if necessary, flow-inducing devices that transport gas or vapor from a piece or pieces of equipment to a control device.

Closeout (Section 35.6015 [1/27/89]) means the final EPA or recipient actions taken to assure satisfactory completion of project work and to fulfill administrative requirements, including financial settlement, submission of acceptable required final reports, and resolution of any outstanding issues under the cooperative agreement and/or Superfund State Contract.

Closing rpm (Section 205.151) means the engine speed in Figure 2 of Appendix I.

Closurc (Section 270.2) means the act of securing a Hazardous Waste Management facility pursuant to the requirements of 40 CFR Part 264.

Closure period (Section 192.31) means the period of time beginning with the cessation, with respect to a waste impoundment, of uranium ore processing operations and ending with completion of requirements specified under a closure plan.

Closure plan (Section 192.31) means the plan required under Section 264.112 of this chapter.

Closure plan (Section 264.141) means the plan for closure prepared in accordance with the requirements of Section 264.112.

Closure plan (Section 265.141) means the plan for closure prepared in accordance with the requirements of Section 265.112.

Clutch facing (Section 763.163 [7-12-89]) means an asbestos-containing product intended for use as a friction material or lining in the clutch mechanisms or manual transmission vehicles.

CN,A (Section 413.02) shall mean cyanide amendable to chlorination as defined by 40 CFR 136.

CN,T (Section 413.02) shall mean cyanide, total.

CO (Section 58.1) means carbon monoxide.

Co-product (Section 716.3) means a chemical substance produced for a commercial purpose during the manufacture, processing, use, or disposal of another chemical substance(s) or mixture(s).

Coagulation (Section 141.2 [6/29/89]) means a process using coagulant chemicals and mixing by which colloidal and suspended materials are destabilized and agglomerated into flocs.

Coal (Section 60.251) means all solid fossil fuels classified as anthracite, bituminous, subbituminous, or lignite by ASTM Designation D388-77 (incorporated by reference - see Section 60.17).

Coal (Section 60.41) means all solid fuels classified as anthracite, bituminous, subbituminous, or lignite by the American Society of Testing and Materials, Designation D388-77 (incorporated by reference - see Section 60.17).

Coal (Section 60.41b [12/16/87])* means all solid fuels classified as anthracite, bituminous, subbituminous, or lignite by the American Society of Testing and Materials in ASTM D388-77, Standard Specification for Classification of Coals by Rank (IBR - see Section 60.17), coal refuse, and petroleum coke. Coal-derived synthetic fuels, including but not limited to solvent refined coal, gasified coal, coal-oil mixtures, and coal-water mixtures, are also included in this definition for the purposes of this subpart.

Coal (Section 60.41c [9-12-90]) means all solid fuels classified as anthracite, bituminous, subbituminous, or lignite by the American Society for Testing and Materials in ASTM D388-77, "Standard Specification for Classification of Coals by Rank" (incorporated by reference--see Sec. 60.17); coal refuse; and petroleum coke. Synthetic fuels derived from coal for the purpose of creating useful heat, including but not limited to solvent-refined coal, gasified coal, coal-oil mixtures, and coal-water mixtures, are included in this definition for the purposes of this subpart.

Coal pile runoff (Section 423.11) means the rainfall runoff from or through any coal storage pile.

Coal preparation plant (Section 60.251) means any facility (excluding underground mining operations) which prepares coal by one or more of the following processes: breaking, crushing, screening, wet or dry cleaning, and thermal drying.

Coal preparation plant (Section 434.11) means a facility where coal is subjected to cleaning, concentrating, or other processing or preparation in order to separate coal from its impurities and then is loaded for transit to a consuming facility.

Coal preparation plant associated areas (Section 434.11) means the coal preparation plant yards, immediate access roads, coal refuse piles and coal storage piles and facilities.

Coal preparation plant water circuit (Section 434.11) means all pipes, channels, basins, tanks, and all other structures and equipment that convey, contain, treat, or process any water that is used in coal preparation processes within a coal preparation plant.

Coal processing and conveying equipment (Section 60.251) means any machinery used to reduce the size of coal or to separate coal from refuse, and the equipment used to convey coal to or remove coal and refuse from the machinery. This includes, but is not limited to, breakers, crushers, screens, and conveyor belts.

Coal/RDF mixed fuel fired combustor (Section 60.51a [2-11-91]) means a combustor that fires coal and RDF simultaneously.

Coal refuse (Section 60.41) means waste-products of coal mining, cleaning, and coal preparation operations (e.g. culm, gob, etc.) containing coal, matrix material, clay, and other organic and inorganic material.

Coal refuse (Section 60.41a) means waste products of coal mining, physical coal cleaning, and coal preparation operations (e.g. culm, gob, etc.) containing coal, matrix material, clay, and other organic and inorganic material.

Coal refuse (Section 60.41b [12/16/87]) means any byproduct of coal mining or coal cleaning operations with an ash content greater than 50 percent, by weight, and a heating value less than 13,900 kJ/kg (6,000 Btu/lb) on a dry basis.

Coal refuse (Section 60.41c [9-12-90]) means any by-product of coal mining or coal cleaning operations with an ash content greater than 50 percent (by weight) and a heating value less than 13,900 kilojoules per kilogram (kJ/kg) (6,000 Btu per pound (Btu/ lb) on a dry basis.

Coal refuse disposal pile (Section 434.11) means any coal refuse deposited on the earth and intended as permanent disposal or long-term storage (greater than 180 days) of such material, but does not include coal refuse deposited within the active mining area or coal refuse never removed from the active mining area.

Coal storage system (Section 60.251) means any facility used to store coal except for open storage piles.

Coal-only heater (Section 60.531 [2/26/88]) means an enclosed, coal-burning appliance capable of space heating, or domestic water heating, which has all of the following characteristics: (a) An opening for emptying ash that is located near the bottom or the side of the appliance, (b) A system that admits air primarily up and through the fuel bed, (c) A grate or other similar device for shaking or disturbing the fuel bed or power-driven mechanical stoker, (d) Installation instructions that state that the use of wood in the stove, except for coal ignition purposes, is prohibited by law, and (e) The model is listed by a nationally recognized safety-testing laboratory for use of coal only, except for coal ignition purposes.

Coarse papers (Section 250.4 [10/6/87, 6/22/88]) means papers used for industrial purposes, as distinguished from those used for cultural or sanitary purposes.

Coastal (Section 435.41) shall mean: (1) any body of water landward of the territorial seas as defined in 40 CFR 125.1(gg), or (2) any wetlands adjacent to such waters.

Coastal waters (Section 300.5 [3-8-90])* for the purposes of classifying the size of discharges, means the waters of the coastal zone except for the Great Lakes and specified ports and harbors on inland rivers.

Coastal zone (Section 300.5 [3-8-90])* defined for the purpose of the NCP, means all United States waters subject to the tide, United States waters of the Great Lakes, specified ports and harbors on inland rivers, waters of the contiguous zone, other waters of the high seas subject to the NCP, and the land surface or land substrata, ground waters, and ambient air proximal to those waters. The term coastal zone delineates an area of federal responsibility for response action. Precise boundaries are determined by EPA/USCG agreements and identified in federal regional contingency plans.

Coating (Section 60.461) means any organic material that is applied to the surface of metal coil.

Coating application station (Section

60.461) means that portion of the metal coil surface coating operation where the coating is applied to the surface of the metal coil. Included as part of the coating application station is the flashoff area between the coating application station and the curing oven.

Coating application station (Section 60.451) means that portion of the large appliance surface coating operation where a prime coat or a top coat is applied to large appliance parts or products (e.g., dip tank, spray booth, or flow coating unit).

Coating applicator (Section 60.441) means an apparatus used to apply a surface coating to a continuous web.

Coating applicator (Section 60.711 [10/3/88]) means any apparatus used to apply a coating to a continuous base film.

Coating applicator (Section 60.741 [9-11-89]) means any apparatus used to apply a coating to a continuous substrate.

Coating blow (Section 60.471) means the process in which air is blown through hot asphalt flux to produce coating asphalt. The coating blow starts when the air is turned on and stops when the air is turned off.

Coating line (Section 60.441) means any number or combination of adhesive, release, or precoat coating applicators, flashoff areas, and ovens which coat a continuous web, located between a web unwind station and a web rewind station, to produce pressure sensitive tape and label materials.

Coating mix preparation equipment (Section 60.711 [10/3/88]) means all mills, mixers, holding tanks, polishing tanks, and other equipment used in the preparation of the magnetic coating formulation but does not include those mills that do not emit VOC because they are closed, sealed, and operated under pressure.

Coating mix preparation equipment (Section 60.741 [9-11-89]) means all mixing vessels in which solvent and other materials are blended to prepare polymeric coatings.

Coating operation (Section 60.711 [10/3/88]) means any coating applicator, flashoff area, and drying oven located between a base film unwind station and a base film rewind station that coat a continuous base film to produce magnetic tape.

Coating operation (Section 60.721 [1/29/88]) means the use of a spray booth for the application of a single type of coating (e.g., prime coat); the use of the same spray booth for the application of another type of coating (e.g., texture coat) constitutes a separate coating operation for which compliance determinations are performed separately.

Coating operation (Section 60.741 [9-11-89]) means any coating applicator(s), flashoff area(s), and drying oven(s) located between a substrate unwind station and a rewind station that coats a continuous web to produce a substrate with a polymeric coating. Should the coating process not employ a rewind station, the end of the coating operation is after the last drying oven in the process.

Coating operations (Section 466.02) means all of the operations associated with preparation and application of the vitreous coating. Usually this includes ballmilling, slip transport, application of slip to the workpieces, cleaning and recovery of faulty parts, and firing (fusing) of the enamel coat.

Coating solids applied (Section 60.721 [1/29/88]) means the coating solids that adhere to the surface of the plastic business machine part being coated.

Coating solids applied (Section 60.441) means the solids content of the coated adhesive, release, or precoat as measured by Reference Method 24.

Cobalt (Section 415.651) shall mean the total cobalt present in the process wastewater stream exiting the wastewater treatment system.

COD (Section 401.11) means chemical oxygen demand.

COD (Section 427.71) shall mean COD added to the process waste water.

Cofired combustor (Section 60.51a [2-11-91]) means a unit combusting MSW or RDF with a non-MSW fuel and subject to a Federally enforceable permit limiting the unit to combusting a fuel feed stream, 30 percent or less of the weight of which is comprised, in aggregate, of MSW or RDF as measured on a 24-hour daily basis. A unit combusting a fuel feed stream, more than 30 percent of the weight of which is comprised, in aggregate, of MSW or RDF shall be considered an MWC unit and not a cofired combustor. Cofired combustors which fire less than 30 percent segregated medical waste and no other municipal solid waste are not covered by this subpart.

Cogeneration steam generating unit (Section 60.41c [9-12-90]) means a steam generating unit that simultaneously produces both electrical (or mechanical) and thermal energy from the same primary energy source.

Cogeneration system (Section 60.41b) means a power system which simultaneously produces both electrical (or mechanical) and thermal energy from the same energy source.

Coil (Section 465.02) means a strip of basis material rolled into a roll for handling.

Coil (Section 85.2122(a)(9)(ii)(A)) means a device used to provide high voltage in an inductive ignition system.

Coil coating (Section 465.02) means the process of converting basis material strip into coated stock. Usually cleaning, conversion coating, and painting are performed on the basis material. This regulation covers processes which perform any two or more of the three operations.

Coke burn-off (Section 60.101) means the coke removed from the surface of the fluid catalytic cracking unit catalyst by combustion in the catalyst regenerator. The rate of coke burn-off is calculated by the formula specified in Section 60.106.

Coke by-product recovery plant (Section 61.131 [9-14-89]) means any plant designed and operated for the separation and recovery of coal tar derivatives (by-products) evolved from coal during the coking process of a coke oven battery.

Coke by-product recovery plant (Section 61.341 [3-7-90]) means any facility designed and operated for the separation and recovery of coal tar derivatives (by-products) evolved from coal during the coking process of a coke oven battery.

Cold rolling (Section 468.02) shall mean the process of rolling a workpiece below the recrystallization temperature of the copper or copper alloy.

Cold worked pipe and tube (Section 420.101) means those cold forming operations that process unheated pipe and tube products using either water or oil solutions for cooling and lubrication.

Collection (Sections 243.101; 246.101) means the act of removing solid waste (or materials which have been separated for the purpose of recycling) from a central storage point.

Collection frequency (Section 243.101) means the number of times collection is provided in a given period of time.

Collector sewer (Section 35.2005) means the common lateral sewers, within a publicly owned treatment system, which are primarily installed to receive wastewaters directly from facilities which convey wastewater from individual systems, or from private property, and which include service "Y" connections designed for connection with those facilities including: (i) Crossover sewers connecting more than one property on one side of a major street, road, or highway to a lateral sewer on the other side when more cost effective than parallel sewers; and (ii) Except as provided in paragraph (b)(10)(iii) of this section, pumping units and pressurized lines serving individual structures or groups of structures when such units are cost effective and are owned and maintained by the grantee. (iii) This definition excludes other facilities which convey wastewater from individual structures, from private property to the public lateral sewer, or its equivalent and also excludes facilities associated with alternatives to conventional treatment works in small communities.

Colloidal dispersion (Section 796.1840) is a mixture resembling a true solution but containing one or more substances that are finely divided but large enough to prevent passage through a semipermeable membrane. It consists of particles which are larger than molecules, which settle out very slowly with time, which scatter a beam of light, and which are too small for resolution with an ordinary light microscope.

Colony (Section 797.1160) means an aggregate of mother and daughter fronds attached to each other.

Color coat (Section 60.721 [1/29/88, 6/15/89]) means the coat applied to a part that affects the color and gloss of the part, not including the prime coat or texture coat. This definition includes fog coating, but does not include conductive sensitizers or electromagnetic interference/radio frequency interference shielding coatings.

Column dryer (Section 60.301) means any equipment used to reduce the moisture content of grain in which the grain flows from the top to the bottom in one or more continuous packed columns between two perforated metal sheets.

Combination (Section 420.101) means those cold rolling operations which include recirculation of rolling solutions at one or more mill stands, and once-through use of rolling solutions at the remaining stand or stands.

Combination acid pickling (Section 420.91) means those operations in which steel products are immersed in solutions of more than one acid to chemically remove scale and oxides, and those rinsing steps associated with such immersions.

Combined cycle gas turbine (Section 60.41a) means a stationary turbine combustion system where heat from the turbine exhaust gases is recovered by a steam generating unit.

Combined cycle gas turbine (Section 60.331) means any stationary gas turbine which recovers heat from the gas turbine exhaust gases to heat water or generate steam.

Combined cycle system (Section 60.41b [12/16/87])* means a system in which a separate source, such as a gas turbine, internal combustion engine, kiln, etc., provides exhaust gas to a heat recovery steam generating unit.

Combined cycle system (Section 60.41c [9-12-90]) means a system in which a separate source (such as a stationary gas turbine, internal combustion engine, or kiln) provides exhaust gas to a steam generating unit.

Combined Fuel Economy (Section 600.002-85) means (i) the fuel economy value determined for a vehicle (or vehicles) by harmonically averaging the city and highway fuel economy values, weighted 0.55 and 0.45 respectively, for gasoline-fueled and diesel vehicles. (ii) For electric vehicles, the term means the equivalent petroleum-based fuel economy value as determined by the calculation procedure promulgated by the Secretary of Energy.

Combined sewer (Section 35.2005) means a sewer that is designed as a sanitary sewer and a storm sewer.

Combined sewer (Section 35.905) means a sewer intended to serve as a sanitary sewer and a storm sewer, or as an industrial sewer and a storm sewer.

Combustibles (Section 240.101) means materials that can be ignited at a specific temperature in the presence of air to release heat energy.

Comfort cooling towers (Section 749.68 [1-3-90]) means cooling towers that are dedicated exclusively to and are an integral part of heating, ventilation, and air conditioning or refrigeration systems.

Commence (Section 51.165) as applied to construction of a major stationary source or major modification, means that the owner or operator has all necessary preconstruction approvals or permits and either has: (A) Begun, or caused to begin, a continuous program of actual on-site construction of the source, to be completed within a reasonable time; or (B) Entered into binding agreements or contractual obligations, which cannot be canceled or modified without substantial loss to the owner or operator, to undertake a program of actual construction of the source to be completed within a reasonable time.

Commence (Sections 51.166; 52.21; 52.24) as applied to construction of a major stationary source or major modification means that the owner or operator has all necessary preconstruction approvals or permits and either has: (i) Begun, or caused to begin, a continuous program of actual on-site construction of the source, to be completed within a reasonable time; or (ii) Entered into binding agreements or contractual obligations, which cannot be cancelled or modified without substantial loss to the owner or operator, to undertake a program of actual construction of the source to be completed within a reasonable time.

Commenced (Section 52.01) means that an owner or operator has undertaken a continuous program of construction or modification.

Commenced (Sections 60.2; 61.02) means, with respect to the definition of new source in section 111(a)(2) of the Act, that an owner or operator has undertaken a continuous program of construction or modification or that an owner or operator has entered into a contractual obligation to undertake and complete, within a reasonable time, a continuous program of construction or modification.

Commenced commercial operation (Section 73.3 [12-17-91]) means to have begun to generate electricity for sale, including test generation.

Commerce (Section 205.2) means trade, traffic, commerce, or transportation: (i) Between a place in a State and any place outside thereof, or (ii) Which affects trade, traffic, commerce, or transportation described in paragraph (a)(17)(i) of this section.

Commerce (Section 710.2) means trade, traffic, transportation, or other commerce: (1) Between a place in a State and any place outside of such State, or (2) which affects trade, traffic, transportation, or commerce described in paragraph (i)(1) of this section.

Commerce (Section 761.3) means trade, traffic, transportation, or other commerce: (1) Between a place in a State and any place outside of such State, or (2) Which affects trade, traffic, transportation, or commerce described in paragraph (1) of this definition.

Commerce (Section 762.3) has the same meaning as in 15 U.S.C. 2602.

Commerce (Section 763.163 [7-12-89]) has the same meaning as in section 3 of the Toxic Substances Control Act.

Commerce (Sections 720.3; 747.115; 747.195; 747.200) means trade, traffic, transportation, or other commerce (1) between a place in a State and any place outside of such State, or (2) which affects trade, traffic, transportation, or commerce between a place in a State and any place outside of such State.

Commercial (Section 82.62 [12/30/93])* when used to describe the purchaser of a product, means a person that uses the product in the purchaser's business or sells it to another person and has one of the following identification numbers: (1) A federal employer identification number; (2) A state sales tax exemption number; (3) A local business license number; or (4) A government contract number.

Commercial aircraft engine (Section 87.1) means any aircraft engine used or intended for use by an air carrier (including those engaged in intrastate air transportation) or a commercial operator (including those engaged in intrastate air transportation) as these terms are defined in the Federal Aviation Act and the Federal Aviation Regulations.

Commercial aircraft gas turbine

engine (Section 87.1) means a turboprop, turbofan, or turbojet commercial aircraft engine.

Commercial and industrial friction product (Section 763.163 [7-12-89]) means an asbestos-containing product, which is either molded or woven, intended for use as a friction material in braking and gear changing components in industrial and commercial machinery and consumer appliances. Major applications of this product include: hand brakes; segments; blocks; and other components used as brake linings, rings and clutches in industrial and commercial machinery and consumer appliances.

Commercial applicator (Section 171.2) means a certified applicator (whether or not he is a private applicator with respect to some uses) who uses or supervises the use of any pesticide which is classified for restricted use for any purpose or on any property other than as provided by the definition of private applicator.

Commercial arsenic (Section 61.161) means any form of arsenic that is produced by extraction from any arsenic-containing substance and is intended for sale or for intentional use in a manufacturing process. Arsenic that is a naturally occurring trace constituent of another substance is not considered commercial arsenic.

Commercial asbestos (Section 61.141) means any asbestos that is extracted from asbestos ore.

Commercial asbestos (Section 61.141 [11-20-90])* means any material containing asbestos that is extracted from ore and has value because of its asbestos content.

Commercial establishment (Section 246.101) means stores, offices, restaurants, warehouses and other non-manufacturing activities.

Commercial hexane (Section 799.2155 [2/5/88]) for purposes of this section, is a product obtained from crude oil, natural gas liquids, or petroleum refinery processing in accordance with the American Society for Testing and Materials Designation D1836-83 (ASTM D 1836), consists primarily of six-carbon alkanes or cycloalkanes, and contains at least 40 liquid volume percent n-hexane (CAS No. 110-54-3) and at least 5 liquid volume percent methylcyclopentane (MCP; CAS No. 96-37-7).

Commercial hexane test substance (Section 799.2155 [2/5/88]) for purposes of this section, is a product which conforms to the specifications of ASTM D 1836 and contains no more than 40 liquid volume percent n-hexane and no less than 10 liquid volume percent MCP.

Commercial Item Descriptions (Section 248.4 [2/17/89]) are a series of simplified item descriptions under the Federal specifications-and-standards program used in the acquisition of commercial off-the-shelf and commercial type products.

Commercial owner (Section 60.531 [2/26/88]) means any person who owns or controls a wood heater in the course of the manufacture, importation, distribution, or sale of the wood heater.

Commercial paper (Section 763.163 [7-12-89]) means an asbestos-containing product which is made of paper intended for use as general insulation paper or muffler paper. Major applications of commercial papers are insulation against fire, heat transfer, and corrosion in circumstances that require a thin, but durable, barrier.

Commercial pesticide handling establishment (Section 170.3 [8/21/92])* means any establishment, other than an agricultural establishment, that: (1) Employs any person, including a self-employed person, to apply on an agricultural establishment, pesticides used in the production of agricultural plants. (2) Employs any person, including a self-employed person, to perform on an agricultural establishment, tasks as a crop advisor.

Commercial Property (Section 201.1) means any property that is normally accessible to the public and that is used for any of the purposes described in the following standard land use codes (reference Standard Land Use Coding Manual, U.S. DOT/FHWA, reprinted March 1977): 53-59, Retail Trade; 61-64, Finance, Insurance, Real Estate, Personal, Business and Repair Services; 652-659, Legal and other professional services; 671, 672, and 673 Governmental Services: 692 and 699, Welfare, Charitable and Other Miscellaneous Services; 712 and 719, Nature exhibitions and other Cultural Activities; 721, 723, and 729, Entertainment, Public and other Public Assembly; and 74-79, Recreational, Resort, Park and other Cultural Activities.

Commercial solid waste (Section 243.101) means all types of solid wastes generated by stores, offices, restaurants, warehouses, and other non-manufacturing activities, excluding residential and industrial wastes.

Commercial solid waste (Section 246.101) means all types of solid wastes generated by stores, offices, restaurants, warehouses and other non-manufacturing activities, and non-processing wastes such as office and packing wastes generated at industrial facilities.

Commercial solid waste (Section 258.2 [10-9-91]) means all types of solid waste generated by stores, offices, restaurants, warehouses, and other nonmanufacturing activities, excluding residential and industrial wastes.

Commercial solid waste (Section 245.101) means all types of solid waste generated by stores, offices, restaurants, warehouses, and other such non-manufacturing activities, and non-processing waste generated at industrial facilities such as office and packing wastes.

Commercial storer of PCB waste (Section 761.3 [6/8/93])* means the owner or operator of each facility which is subject to the PCB storage facility standards of Sec. 761.65, and who engages in storage activities involving PCB waste generated by others, or PCB waste that was removed while servicing the equipment owned by others and brokered for disposal. The receipt of a fee or any form of compensation for storage services is not necessary to qualify as a commercial storer of PCB waste. It is sufficient under this definition that the facility stores PCB waste generated by others or the facility removed the PCB waste while servicing equipment owned by others. A generator who stores only the generator's own waste is subject to the storage requirements of Sec. 761.65, but is not required to seek approval as a commercial storer. If a facility's storage of PCB waste at no time exceeds 500 liquid gallons of PCBs, the owner or operator is not required to seek approval as a commercial storer of PCB waste.

Commercial use (Section 721.3 [7-27-89]) means the use of a chemical substance or any mixture containing the chemical substance in a commercial enterprise providing saleable goods or a service to consumers (e.g., a commercial dry cleaning establishment or painting contractor).

Commercial use request (Section 2.100 [1/5/88]) refers to a request from or on behalf of one who seeks information for a use or purpose that furthers the commercial, trade or profit interests of the requestor or the person on whose behalf the request is made. In determining whether a requestor properly belongs in this category, EPA must determine the use to which a requestor will put the documents requested. Moreover, where EPA has reasonable cause to doubt the use to which a requestor will put the records sought, or where that use is not clear from the request itself, EPA may seek additional clarification before assigning the request to a specific category.

Commission (Section 191.02) means the Nuclear Regulatory Commission.

Commission (Section 355.20 [7-26-90])* means the emergency response commission for the State in which the facility is located except where the facility is located in Indian Country, in which case, commission means the emergency response commission for the tribe under whose jurisdiction the facility is located. In absence of an emergency response commission, the Governor and the chief executive officer, respectively, shall be the commission. Where there is a cooperative agreement between a State and a Tribe, the commission shall be the entity identified in the agreement.

Commission (Section 370.2 [7-26-90])* means the emergency response commission for the State in which the facility is located except where the facility is located in Indian Country, in which case, commission means the emergency response commission for the Tribe under whose jurisdiction the facility is located. In absence of an emergency response commission, the Governor and the chief executive officer, respectively, shall be the commission. Where there is a cooperative agreement between a State and a Tribe, the commission shall be the entity identified in the agreement.

Commission (Section 61.251) means the Nuclear Regulatory Commission or its Agreement States (where applicable).

Commission finishing (Section 410.01) shall mean the finishing of textile materials, 50 percent or more of which are owned by others, in mills that are 51 percent or more independent (i.e., only a minority ownership by company(ies) with greige or integrated operations); the mills must process 20 percent or more of their commissioned production through batch, noncontinuous processing operations with 50 percent or more of their commissioned orders processed in 5000 yard or smaller lots.

Commission scouring (Section 410.11) shall mean the scouring of wool, 50 percent or more of which is owned by others, in mills that are 51 percent or more independent (i.e., only a minority ownership by company(ies) with greige or integrated operations); the mills must process 20 percent or more of their commissioned production through batch, noncontinuous processing operations.

Commitment of Federal financial assistance (Section 149.101) means a written agreement entered into by a department, agency, or instrumentality of the Federal Government to provide financial assistance as defined in paragraph (g) of this section. Renewal of a commitment which the issuing agency determines has lapsed shall not constitute a new commitment unless the Regional Administrator determines that the project's impact on the aquifer has not been previously reviewed under section 1424(e). The determination of a Federal agency that a certain written agreement constitutes a commitment shall be conclusive with respect to the existence of such a commitment.

Committee (Section 164.2) means a group of qualified scientists designated by the National Academy of Sciences according to agreement under the Act to submit an independent report to the Administrative Law Judge on questions of scientific fact referred from a hearing under Subpart B of this part.

Committee (Section 370.2 [10/15/87])

means the local emergency planning committee for the emergency planning district in which the facility is located.

Committee or local emergency planning committee (Section 370.2 [7-26-90]) means the local emergency planning committee appointed by the emergency response commission.

Committee or Local emergency planning committee (Section 355.20 [7-26-90]) means the local emergency planning committee appointed by the emergency response commission. Environment includes water, air, and land and the interrelationship which exists among and between water, air, and land and all living things.

Common carrier by motor vehicle (Section 202.10) means any person who holds himself out to the general public to engage in the transportation by motor vehicle in interstate or foreign commerce of passengers or property or any class or classes thereof for compensation, whether over regular or irregular routes.

Common emission control device (Section 60.711 [10/3/88]) means a control device controlling emissions from the coating operation as well as from another emission source within the plant.

Common emission control devise (Section 60.741 [9-11-89]) means a device controlling emissions from an affected coating operation as well as from any other emission source.

Common exposure route (Section 171.2) means a likely way (oral, dermal, respiratory) by which a pesticide may reach and/or enter an organism.

Common name (Section 721.3 [7-27-89]) means any designation or identification such as code name, code number, trade name, brand name, or generic chemical name used to identify a chemical substance other than by its chemical name.

Community relations (Section 300.5 [3-8-90]) means EPA's program to inform and encourage public participation in the Superfund process and to respond to community concerns. The term "public" includes citizens directly affected by the site, other interested citizens or parties, organized groups, elected officials, and potentially responsible parties.

Community relations coordinator (Section 300.5 [3-8-90]) means lead agency staff who work with OSC/RPM to involve and inform the public about the Superfund process and response actions in accordance with the interactive community relations requirements set forth in the NCP.

Community Relations Plan (CRP) (Section 35.6015 [1/27/89]) means a management and planning tool outlining the specific community relations activities to be undertaken during the course of a response. It is designed to provide for two-way communication between the affected community and the agencies responsible for conducting a response action, and to assure public input into the decision-making process related to the affected communities.

Community water system (Section 191.12) means a system for the provision to the public of piped water for human consumption, if such system has at least 15 service connections used by year-round residents or regularly serves at least 25 year-round residents.

Community water system (Section 141.2) means a public water system which serves at least 15 service connections used by year-round residents or regularly serves at least 25 year-round residents.

Compactor collection vehicle (Section 243.101) means a vehicle with an enclosed body containing mechanical devices that convey solid waste into the main compartment of the body and compress it into a smaller volume of greater density.

Company or firm (Section 717.3) means any person that is subject to this part, as defined in Section 717.5.

Comparable replacement dwelling (Section 4.2) means a dwelling which is: (1) Decent, safe, and sanitary as described in Section 4.2(e). (2) Functionally similar to the displacement dwelling with particular attention to the number of rooms and living space. (See Appendix A.) (3) In an area that is not subject to unreasonable adverse environmental conditions, is not generally less desirable than the location of the displaced person's dwelling with respect to public utilities and commercial and public facilities, and is reasonably accessible to the person's place of employment. (4) On a site that is typical in size for residential development with normal site improvements, including customary landscaping. The site need not include special improvements such as outbuildings, swimming pools, or greenhouses. (See also Section 4.403(a)(2).) (5) Currently available to the displaced person on the private market. However, a comparable replacement dwelling for a person receiving government housing assistance before displacement may reflect similar government housing assistance. (See Appendix A.) (6) Within the financial means of the displaced person. (i) A replacement dwelling purchased by a homeowner in occupancy for at least 180 days prior to initiation of negotiations (180-day homeowner) is considered to be within the homeowner's financial means if the homeowner is paid the full price differential as described at Section 4.401(c), all increased mortgage interest costs as described at Section 4.401(d) (for last resort housing see Appendix A, Section 4.602) and all incidental expenses as described at Section 4.401(e). (ii) A replacement dwelling rented by a displaced person is considered to be within his or her financial means if the monthly rent at the replacement dwelling does not exceed the monthly rent at the displacement dwelling, after taking into account any rental assistance which the person receives under these regulations. If the cost of any utility service is included in either rent, an appropriate adjustment must be made if necessary to ensure that like circumstances are compared. For a person who paid little or no rent before displacement, the market rent of the displacement dwelling may be used when computing costs (see Appendix A, Section 4.402(b)(1)). (iii) Whenever a $15,000 replacement housing payment under Section 4.401 or a $4,000 replacement housing payment under Section 4.402 would be insufficient to ensure that a comparable replacement dwelling is available on a timely basis to a person, the Agency shall provide additional or alternative assistance under the last resort housing provisions at Subpart G, which may include increasing the replacement housing payment so that a replacement dwelling is within the displaced person's financial means.

Compartmentalized vehicle (Section 246.101) means a collection vehicle which has two or more compartments for placement of solid wastes or recyclable materials. The compartments may be within the main truck body or on the outside of that body as in the form of metal racks.

Compatibility (Section 171.2) means that property of a pesticide which permits its use with other chemicals without undesirable results being caused by the combination.

Compatible (Section 280.12 [9/23/88]) means the ability of two or more substances to maintain their respective physical and chemical properties upon contact with one another for the design life of the tank system under conditions likely to be encountered in the UST.

Competent (Section 171.2) means properly qualified to perform functions associated with pesticide application, the degree of capability required being directly related to the nature of the activity and the associated responsibility.

Competent person (Section 763.121) means one who is capable of identifying existing asbestos hazards in the workplace and who has the authority to take prompt corrective measures to eliminate them. The duties of the competent person include at least the following: Establishing the negative-pressure enclosure, ensuring its integrity, and controlling entry to and exit from the enclosure; supervising any employee exposure monitoring required by this subpart, ensuring that all employees working within such an enclosure wear the appropriate personal protective equipment, are trained in the use of appropriate methods of exposure control, and use the hygiene facilities and decontamination procedures specified in this subpart; and ensuring that engineering controls in use are in proper operating condition and are functioning properly.

Competition motorcycle (Section 205.151) means any motorcycle designed and marketed solely for use in closed course competition events.

Complainant (Section 209.3) means the Agency acting through any person authorized by the Administrator to issue a complaint to alleged violators of the Act. The complainant shall not be the judicial officer or the Administrator.

Complainant (Section 22.03 [2-13-92])* means any person authorized to issue a complaint on behalf of the Agency to persons alleged to be in violation of the Act. The complainant shall not be a member of the Environmental Appeals Board, the Regional Judicial Officer, or any other person who will participate or advise in the decision.

Complaint (Section 22.03) means a written communication, alleging one or more violations of specific provisions of the Act, or regulations or a permit promulgated thereunder, issued by the complainant to a person under Sections 22.13 and 22.14.

Complete (Section 51.166) means, in reference to an application for a permit, that the application contains all the information necessary for processing the application. Designating an application complete for purposes of permit processing does not preclude the reviewing authority from requesting or accepting any additional information.

Complete (Section 52.21) means, in reference to an application for a permit, that the application contains all of the information necessary for processing the application.

Complete complaint (Section 12.103 [8/14/87]) means a written statement that contains the complainant's name and address and describes the agency's alleged discriminatory action in sufficient detail to inform the agency of the nature and date of the alleged violation of section 504. It shall be signed by the complainant or by someone authorized to do so on his or her behalf. Complaints filed on behalf of classes or third parties shall describe or identify (by name, if possible) the alleged victims of discrimination.

Complete destruction of pesticides (Section 165.1) means alteration by physical or chemical processes to inorganic forms.

Complete waste treatment system (Section 35.2005) consists of all the treatment works necessary to meet the requirements of title III of the Act, involving: (i) The transport of wastewater from individual homes or buildings to a plant or facility where treatment of the wastewater is accomplished; (ii) the treatment of the wastewater to remove pollutants; and (iii) the ultimate disposal, including recycling or reuse, of the treated wastewater and residues which result from the treatment process.

Complete waste treatment system (Section 35.905) consists of all the treatment works necessary to meet the requirements of title III of the Act, involved in: (a) The transport of waste waters from individual homes or buildings to a plant or facility where treatment of the wastewater is accomplished; (b) the treatment of the wastewaters to remove pollutants; and (c) the ultimate disposal, including recycling or reuse, of the treated wastewaters and residues which result from the treatment process. One complete waste treatment system would, normally, include one treatment plant or facility, but also includes two or more connected or integrated treatment plants or facilities.

Completely closed drain system (Section 60.691 [11/23/88]) means an individual drain system that is not open to the atmosphere and is equipped and operated with a closed vent system and control device complying with the requirements of Section 60.692-5.

Completely destroy (Section 82.3 [12/10/93])* means to cause the expiration of a controlled substance at a destruction efficiency of 98 percent or greater, using one of the destruction technologies approved by the Parties.

Complex (Section 112.2 [8/25/93])* means a facility possessing a combination of transportation-related and non-transportation-related components that is subject to the jurisdiction of more than one Federal agency under section 311(j) of the Clean Water Act.

Complex manufacturing operation (Section 410.61) shall mean simple unit processes (fiber preparation, dyeing and carpet backing) plus any additional manufacturing operations such as printing or dyeing and printing.

Complex manufacturing operation (Sections 410.41; 410.51) shall mean simple unit processes (desizing, fiber preparation and dyeing) plus any additional manufacturing operations such as printing, water proofing, or applying stain resistance or other functional fabric finishes.

Complex slaughterhouse (Section 432.21) shall mean a slaughterhouse that accomplishes extensive by-product processing, usually at least three of such operations as rendering, paunch and viscera handling, blood processing, hide processing, or hair processing.

Compliance (Section 15.4) means compliance with clean air standards or clean water standards. For the purpose of these regulations, compliance also shall mean compliance with a schedule or plan ordered or approved by a court of competent jurisdiction, the United States Environmental Protection Agency, or an air or water pollution control agency, in accordance with the requirements of the CAA or the CWA and regulations issued pursuant thereto.

Compliance Agency (Section 8.2) means the agency designated by the Director on a geographical, industry, or other basis to conduct compliance reviews and to undertake such other responsibilities in connection with the administration of the order as the Director may determine to be appropriate. In the absence of such a designation the Compliance Agency will be determined as follows: (1) In the case of a prime contractor not involved in construction work, the Compliance Agency will be the agency whose contracts with the prime contractor have the largest aggregate dollar value; (2) In the case of a subcontractor not involved in construction work, the Compliance Agency will be the Compliance Agency of the prime contractor with which the subcontractor has the largest aggregate value of subcontracts or purchase orders for the performance of work under contracts; (3) In the case of a prime contractor or subcontractor involved in construction work, the Compliance Agency for each construction project will be the agency providing the largest dollar value for the construction projects; and (4) In the case of a contractor who is both a prime contractor and subcontractor, the Compliance Agency will be determined as if such contractor is a prime contractor only.

Compliance cycle (Section 141.2 [7/17/92])* means the nine-year calendar year cycle during which public water systems must monitor. Each compliance cycle consists of three three-year compliance periods. The first calendar year cycle begins January 1, 1993 and ends December 31, 2001; the second begins January 1, 2002 and ends December 31, 2010; the third begins January 1, 2011 and ends December 31, 2019.

Compliance cycle (Section 141.2 [1-30-91]) means the nine-year calendar year cycle during which public water systems must monitor. Each compliance cycle consists of three three-year compliance periods. The first calendar year cycle begins January 1, 1993 and ends December 31, 2001; the second begins January 1, 2002 and ends December 31, 2010; the third begins January 1, 2011 and ends December 31, 2019.

Compliance level (Section 86.1002-84) means an emission level determined during a Production Compliance Audit pursuant to Subpart L of this part.

Compliance level (Section 86.1102-87 [11-5-90])* means the deteriorated pollutant emissions level at the 60th percentile point for a population of heavy-duty engines or heavy-duty vehicles subject to Production Compliance Audit testing pursuant to the requirements of this subpart. A compliance level for a population can only be determined for a pollutant for which an upper limit has been established in this subpart.

Compliance period (Section 141.2 [7/17/92])* means a three-year calendar year period within a compliance cycle. Each compliance cycle has three three-year compliance periods. Within the first compliance cycle, the first compliance period runs from January 1, 1993 to December 31, 1995; the second from January 1, 1996 to December 31, 1998; the third from January 1, 1999 to December 31, 2001.

Compliance schedule (Section 51.100) means the date or dates by which a source or category of sources is required to comply with specific emission limitations contained in an implementation plan and with any increments of progress toward such compliance.

Compliance schedule (Section 60.21) means a legally enforceable schedule specifying a date or dates by which a source or category of sources must comply with specific emission standards contained in a plan or with any increments of progress to achieve such compliance.

Compliance schedule (Section 61.02) means the date or dates by which a source or category of sources is required to comply with the standards of this part and with any steps toward such compliance which are set forth in a waiver of compliance under Section 61.11.

Compliance use date (Section 73.3 [12-17-91]) means the first calendar year for which an allowance may be used for purposes of meeting a unit's sulfur dioxide emissions limitation requirements.

Complying with the Protocol (Section 82.3 [12/10/93])* when referring to a foreign state not Party to the 1987 Montreal Protocol, the London Amendments, or the Copenhagen Amendments, means that the non-Party has been determined as complying with the Protocol, as indicated in appendix C of this subpart, by a meeting of the Parties as noted in the records of the directorate of the United Nations Secretariat.

Component (Section 260.10) means either the tank or ancillary equipment of a tank system.

Component (Section 270.2 [9/28/88]) means any constituent part of a unit or any group of constituent parts of a unit which are assembled to perform a specific function (e.g., a pump seal, pump, kiln liner, kiln thermocouple).

Component (Section 35.936.13) means any article, material, or supply directly incorporated in construction material.

Component (Section 60.541 [9/15/87]) means a piece of tread, combined tread/sidewall, or separate sidewall rubber, or other rubber strip that is combined into the sidewall of a finished tire.

Composite particulate standard (Section 86.090-2 [4/11/89]) for a manufacturer which elects to average light-duty vehicles and light-duty trucks together in either the petroleum-fueled or methanol-fueled light-duty particulate averaging program, means that standard calculated using the equation set forth in Section 86.090-2 under this definition. This definition applies beginning with the 1990 model year.

Composite particulate standard (Section 86.087-2 [10/31/88]) for a manufacturer which elects to average diesel light-duty vehicles and diesel light-duty trucks with a loaded vehicle weight equal to or less than 3,750 lbs (LDDT1s) together in the particulate averaging program, means that standard calculated according to the equation set forth in Section 86.087-2 under this definition.

Composite particulate standard (Section 86.090 2 [7-26-90]) for a manufacturer which elects to average light-duty vehicles and light-duty trucks together in either the petroleum-fueled or methanol-fueled light-duty particulate averaging program
< nearest one-hundredth (0.01) of a gram per mile:

(STDLDV) + (PRODLDT)
(PRODLDV) -------- - ---------
(STDLDT) (PRODLDV) +
(PRODLDT)
= Manufacturer composite particulate standard

Where: PRODLDV represents the manufacturer's total petroleum-fueled diesel or methanol-fueled diesel light-duty vehicle production for those engine families being included in the appropriate average for a given model year. STDLDV represents the light-duty vehicle particulate standard. PRODLDT represents the manufacturer's total petroleum-fueled diesel or methanol-fueled diesel light-duty truck production for those engine families being included in the appropriate average for a given model year. STDLDT represents the light-duty truck particulate standard.

Composite particulate standard (Section 86.085-2), for a manufacturer which elects to average diesel light-duty vehicles and diesel light-duty trucks together in the particulate averaging program, means that standard calculated according to the equation set forth in Section 86.085-2 under this definition and rounded to the nearest hundredth gram-per-mile. This definition applies beginning with the 1985 model year.

Composite sample (Section 471.02) is a sample composed of no less than eight grab samples taken over the compositing period.

Compressor (Section 204.51). See **Portable air compressor (Section 204.51)**.

Compressor configuration (Section 204.51) means the basic classification unit of a manufacturer's product line and is comprised of compressor lines, models or series which are identical in all material respects with regard to the parameters listed in Section 204.55-3.

Computer paper (Section 250.4 [10/6/87, 6/22/88]) means a type of paper used in manifold business forms produced in rolls and/or fan folded. It is used with computers and word processors to print out data, information, letters, advertising, etc. It is commonly called computer printout.

Computer printout (Section 250.4 [10/6/87, 6/22/88]). See **Computer paper (Section 250.4)**.

Computer program (Section 66.3) means the computer program used to calculate noncompliance penalties under section 120 of the Clean Air Act. This computer program appears as Appendix C to these regulations.

Concentrated animal feeding operation (Section 122.23) means an animal feeding operation which meets the criteria in Appendix B of this part, or which the Director designates under paragraph (c) of this section.

Concentration (Section 798.4350) refers to an exposure level. Exposure is expressed as weight or volume of test substance per volume of air (mg/l), or as parts per million (ppm).

Concentration of a solution (Section 796.1840) is the amount of solute in a given amount of solvent and can be expressed as a weight/weight or weight/volume relationship. The conversion from a weight relationship to one of volume incorporates density as a factor. For dilute aqueous solutions, the density of the solvent is approximately equal to the density of the solutions; thus, concentrations in mg/dm^3 are approximately equal to $10^{-3}g/10^3g$ or parts per million (ppm); ones in $\mu g/dm^3$ are approximately equal to $10^{-6}g/10^3g$ or parts per billion (ppb). In addition, concentration can be expressed in terms of molarity, normality, molality, and mole fraction. For example, to convert from weight/volume to molarity one incorporates molecular mass as a factor.

Concentration of a solution (Section 796.1860) is the amount of solute in a given amount of solvent or solution and can be expressed as a weight/weight or weight/volume relationship. The conversion from a weight relationship to one of volume incorporates density as a factor. For dilute aqueous solutions, the density of the solvent is approximately equal to the density of the solution; thus, concentrations in mg/L are approximately equal to $10^{-3}g/10^{3}g$ or parts per million (ppm); ones in μg/L are approximately equal to $10^{-6}g/10^{3}g$ or parts per billion (ppb). In addition, concentration can be expressed in terms of molarity, normality, molality, and mole fraction. For example, to convert from weight/volume to molarity one incorporates molecular mass as a factor.

Concentration vs. time study (Section 796.1840) results in a graph which plots the measured concentration of a given compound in a solution as a function of elapsed time. Usually, it provides a more reliable determination of equilibrium water solubility of hydrophobic compounds than can be obtained by single measurements of separate samples.

Concurrent (Section 60.711 [10/3/88]) means construction of a control device is commenced or completed within the period beginning 6 months prior to the date construction of affected coating mix preparation equipment commences and ending 2 years after the date construction of affected coating mix preparation equipment is completed.

Concurrent (Section 60.741 [9-11-89]) means the period of time in which construction of an emission control device serving an affected facility is commenced or completed, beginning 6 months prior to the date that construction of the affected facility commences and ending 2 years after the date that construction of the affected facility is completed.

Condensate (Section 60.111) means hydrocarbon liquid separated from natural gas which condenses due to changes in the temperature and/or pressure and remains liquid at standard conditions.

Condensate (Section 60.111a) means hydrocarbon liquid separated from natural gas which condenses due to changes in the temperature or pressure, or both, and remains liquid at standard conditions.

Condensate (Section 60.111b) means hydrocarbon liquid separated from natural gas that condenses due to changes in the temperature or pressure, or both, and remains liquid at standard conditions.

Condensate stripper system (Section 60.281) means a column, and associated condensers, used to strip, with air or steam, TRS compounds from condensate streams from various processes within a kraft pulp mill.

Condenser (Section 264.1031 [6-21-90]) means a heat-transfer device that reduces a thermodynamic fluid from its vapor phase to its liquid phase.

Condenser stack gases (Section 61.51) mean the gaseous effluent evolved from the stack of processes utilizing heat to extract mercury metal from mercury ore.

Conditioned (Section 248.4 [2/17/89]) means heated and/or mechanically cooled.

Conditioning (Section 797.1400) means the exposure of construction materials, test chambers, and testing apparatus to dilution water or to test solutions prior to the start of a test in order to minimize the sorption of the test substance onto the test facilities or the leaching of substances from the test facilities into the dilution water or test solution.

Conditioning (Section 797.1600) is the exposure of construction materials, test chambers, and testing apparatus to dilution water or to the test solution prior to the start of the test in order to minimize the sorption of test substance onto the test facilities or the leaching of substances from test facilities into the dilution water or the test solution.

Conductive sensitizer (Section 60.721 [1/29/88]) means a coating applied to a plastic substrate to render it conductive for purposes of electrostatic application of subsequent prime, color, texture, or touch-up coats.

Cone of influence (Section 146.61 [7/26/88]) means that area around the well within which increased injection zone pressures caused by injection into the hazardous waste injection well would be sufficient to drive fluids into an underground source of drinking water (USDW).

Confidence limits (Section 797.1350) are the limits within which, at some specified level of probability, the true value of a result lies.

Confidential (Section 11.4) refers to that national security information or material which requires protection. The test for assigning Confidential classification shall be whether its unauthorized disclosure could reasonably be expected to cause damage to the national security.

Confidential business information (Section 350.1 [7/29/88]). See **Business confidentiality (Section 350.1)**.

Confidential business information (Section 154.3) means trade secrets or confidential commercial or financial information under FIFRA section 10(b) or 5 U.S.C. 552(b) (3) or (4).

Configuration (Section 205.151) means the basic classification unit of a manufacturer's product line and is comprised of all vehicle designs, models or series which are identical in all material aspects with respect to the parameters listed in Section 205.157-3 of this subpart.

Configuration (Section 205.51) means the basic classification unit of a manufacturer's product line and is comprised of all vehicle designs, models or series which are identical in material aspects with respect to the parameters listed in Section 205.55-3.

Configuration (Section 610.11) means the mechanical arrangement, calibration and condition of a test automobile, with particular respect to carburetion, ignition timing, and emission control systems.

Configuration (Section 86.082-2) means a subclassification of an engine-system combination on the basis of engine code, inertia weight class, transmission type and gear ratios, final drive ratio, and other parameters which may be designated by the Administrator.

Configuration (Section 86.1002-84) means a subclassification, if any, of a heavy-duty engine family for which a separate projected sales figure is listed in the manufacturer's Application for Certification and which can be described on the basis of emission control system, governed speed, injector size, engine calibration, and other parameters which may be designated by the Administrator, or a subclassification of a light-duty truck engine family/emission control system combination on the basis of engine code, inertia weight class, transmission type and gear ratios, axle ratio, and other parameters which may be designated by the Administrator.

Configuration (Section 86.1102-87) means a subdivision, if any, of a heavy-duty engine family for which a separate projected sales figure is listed in the manufacturer's Application for Certification and which can be described on the basis of emission control system, governed speed, injector size, engine calibration, or other parameters which may be designated by the Administrator, or a subclassification of light-duty truck engine family emission control system combination on the basis of engine code, inertia weight class, transmission type and gear ratios, rear axle ratio, or other parameters which may be designated by the Administrator.

Configuration (Section 86.602) means a subclassification of an engine-system combination on the basis of engine code, inertia weight class, transmission type and gear ratios, axle ratio, and other parameters which may be designated by the Administrator.

Confined aquifer (Section 260.10) means an aquifer bounded above and below by impermeable beds or by beds of distinctly lower permeability than that of the aquifer itself; an aquifer containing confined ground water.

Confining bed (Sections 146.3; 147.2902) means a body of impermeable or distinctly less permeable material stratigraphically adjacent to one or more aquifers.

Confining zone (Section 146.3) means a geological formation, group of formations, or part of a formation that is capable of limiting fluid movement above an injection zone.

Confining zone (Section 147.2902)

means a geologic formation, group of formations, or part of a formation that is capable of limiting fluid movement above an injection zone.

Confluent growth (Section 141.2 [6/29/89]) means a continuous bacterial growth covering the entire filtration area of a membrane filter, or a portion thereof, in which bacterial colonies are not discrete.

Congener (Section 766.3) means any one particular member of a class of chemical substances. A specific congener is denoted by unique chemical structure, for example 2,3,7,8-tetrachlorodibenzofuran.

Connected piping (Section 280.12 [9/23/88]) means all underground piping including valves, elbows, joints, flanges, and flexible connectors attached to a tank system through which regulated substances flow. For the purpose of determining how much piping is connected to any individual UST system, the piping that joins two UST systems should be allocated equally between them.

Connector (Section 264.1031 [6-21-90]) means flanged, screwed, welded, or other joined fittings used to connect two pipelines or a pipeline and a piece of equipment. For the purposes of reporting and recordkeeping, connector means flanged fittings that are not covered by insulation or other materials that prevent location of the fittings.

Connector (Section 60.481) means flanged, screwed, welded, or other joined fittings used to connect two pipe lines or a pipe line and a piece of process equipment.

Connector (Section 61.241) means flanged, screwed, welded, or other joined fittings used to connect two pipe lines or a pipe line and a piece of equipment. For the purpose of reporting and recordkeeping, connector means flanged fittings that are not covered by insulation or other materials that prevent location of the fittings.

Consent Agreement (Section 22.03) means any written document, signed by the parties, containing stipulations or conclusions of fact or law and a proposed penalty or proposed revocation or suspension acceptable to both complainant and respondent.

Consignee (Section 262.51 [7/19/88])* means the ultimate treatment, storage or disposal facility in a receiving country to which the hazardous waste will be sent.

Consolidate (Section 29.12) means that a State may meet statutory and regulatory requirements by combining two or more plans into one document and that the State can select the format, submission date, and planning period for the consolidated plan.

Consolidated assistance (Section 30.200) means an assistance agreement awarded under more than one EPA program authority or funded together with one or more other Federal agencies. Applicants for consolidated assistance submit only one application.

Consolidated PMN (Section 700.43 [8/17/88]). See **Consolidated premanufacture notice (Section 700.43)**.

Consolidated premanufacture notice or Consolidated PMN (Section 700.43 [8/17/88]) means any PMN submitted to EPA that covers more than one chemical substance (each being assigned a separate PMN number by EPA) as a result of a prenotice agreement with EPA (See 48 FR 21734).

Consortium (Section 790.3) means an association of manufacturers and/or processors who have made an agreement to jointly sponsor testing.

Constant controls, control technology and continuous emission reduction technology (Section 57.103) mean systems which limit the quantity, rate, or concentration, excluding the use of dilution, and emissions of air pollutants on a continuous basis.

Construction (Section 124.41) has the meaning given in 40 CFR 52.21.

Construction (Section 125.91), for purposes of this subpart, includes any one of the following: Preliminary planning to determine the feasibility of treatment works; engineering, architectural, legal, fiscal, or economic investigations or studies, surveys, designs, plans, working drawings, specifications, procedures, or other necessary actions, erection, building, acquisition, alteration, remodeling, improvement, or extension of treatment works, or the inspection or supervision of any of the foregoing items, Provided: That, completion of the facility and attainment of operational level by no later than July 1, 1983, is a reasonable expectation.

Construction (Section 129.2) means any placement, assembly, or installation of facilities or equipment (including contractual obligations to purchase such facilities or equipment) at the premises where such equipment will be used, including preparation work at such premises.

Construction (Section 21.2) means the erection, building, acquisition, alteration, remodeling, modification, improvement, or extension of any facility; Provided, that it does not mean preparation or undertaking of: Plans to determine feasibility; engineering, architectural, legal, fiscal, or economic investigations or studies; surveys, designs, plans, writings, drawings, specifications or procedures.

Construction (Section 249.04) means the erection or building of new structures, or the replacement, expansion, remodeling, alteration, modernization, or extension of existing structures. It includes the engineering and architectural surveys, designs, plans, working drawings, specifications, and other actions necessary to complete the project.

Construction (Section 33.005) means erection, building, alteration, remodeling, improvement, or extension of buildings, structures or other property. Construction also includes remedial actions in response to a release, or a threat of a release, of a hazardous substance into the environment as determined by the Comprehensive Environmental Response, Compensation, and Liability Act of 1980.

Construction (Section 35.2005) means any one or more of the following: Preliminary planning to determine the feasibility of treatment works, engineering, architectural, legal, fiscal, or economic investigations or studies, surveys, designs, plans, working drawings, specifications, procedures, field testing of innovative or alternative wastewater treatment processes and techniques (excluding operation and maintenance) meeting guidelines promulgated under section 304(d)(3) of the Act, or other necessary actions, erection, building, acquisition, alteration, remodeling, improvement, or extension of treatment works, or the inspection or supervision of any of the foregoing items.

Construction (Section 35.6015 [1/27/89]) means erection, building, alteration, repair, remodeling, improvement, or extension of buildings, structures or other property.

Construction (Section 35.905) means any one or more of the following: Preliminary planning to determine the feasibility of treatment works, engineering, architectural, legal, fiscal, or economic investigations or studies, surveys, designs, plans, working drawings, specifications, procedures, or other necessary actions, erection, building, acquisition, alteration, remodeling, improvement, or extension of treatment works, or the inspection or supervision of any of the foregoing items. The phrase initiation of construction, as used in this subpart means with reference to a project for: (a) Step 1: The approval of a plan of study (see Sections 35.920-3(a)(1) and 35.925-18(a)); (b) Step 2: The award of a step 2 grant; (c) Step 3: Issuance of a notice to proceed under a construction contract for any segment of step 3 project work or, if notice to proceed is not required, execution of the construction contract.

Construction (Section 52.01) means fabrication, erection, or installation.

Construction (Sections 51.165; 51.166; 52.21; 52.24) means any physical change or change in the method of operation (including fabrication, erection, installation, demolition, or modification of an emissions unit) which would result in a change in actual emissions.

Construction (Sections 60.2; 61.02) means fabrication, erection, or installation of an affected facility.

Construction and demolition waste (Sections 243.101; 246.101) means the waste building materials, packaging, and rubble resulting from construction, remodeling, repair, and demolition operations on pavements, houses, commercial buildings, and other structures.

Construction material (Section 35.936.13) means any article, material, or supply brought to the construction site for incorporation in the building or work.

Construction work (Section 8.2) means the construction, rehabilitation, alteration, conversion, extension, demolition or repair of buildings, highways, or other changes or improvements to real property, including facilities providing utility services. The term also includes the supervision, inspection, and other on-site functions incidental to the actual construction.

Consultation with the Regional Administrator (Section 124.62(a)(2)) (Section 124.2) means review by the Regional Administrator following evaluation by a panel of the technical merits of all 301(k) applications approved by the Director. The panel (to be appointed by the Director of the Office of Water Enforcement and Permits) will consist of Headquarters, Regional, and State personnel familiar with the industrial category in question.

Consumer (Section 244.101) means any person who purchases a beverage in a beverage container for final use or consumption.

Consumer (Section 721.3 [7-27-89]) means a private individual who uses a chemical substance or any product containing the chemical substance in or around a permanent or temporary household or residence, during recreation, or for any personal use or enjoyment.

Consumer Price Index (CPI) (Section 73.3 [12-17-91]) means the United States government's primary indicator of the monetary inflation rate as published monthly by the U.S. Department of Labor, Bureau of Labor Statistics, Consumer Price Indices Branch, in the CPI Detailed Report and in the Monthly Labor Review. For purposes of part 73, the Administrator will use the "Consumer Price Index for all urban consumers for the US City Average, for all Items on the Official Reference Base" (CPI-U), or if such index is no longer published, such other index as the Administrator in his discretion determines meets the requirements of the Clean Air Act Amendments of 1990. (1) CPI (1990) means the most recently adjusted CPI for all urban consumers as of August 31, 1989. The CPI for 1990 is 124.6 (with 1982-1984 = 100). (2) CPI (year) means the most recently adjusted CPI for all urban consumers as of August 31 of the previous year.

Consumer product (Section 302.3) shall have the meaning stated in 15 U.S.C. 2052.

Consumer product (Section 721.3 [7-27-89]) means a chemical substance that is directly, or is part of a mixture, sold or made available to consumers for their use in or around a permanent or temporary household or residence, in or around a school, or in recreation.

Consumption (Section 82.3 [12/10/93])* means the production plus imports minus exports of a controlled substance (other than transhipments, or recycled or used controlled substances).

Consumption allowances (Section 82.3 [12/10/93])* means the privileges granted by this subpart to produce and import class I controlled substances; however, consumption allowances may be used to produce class I controlled substances only in conjunction with production allowances. A person's consumption allowances are the total of the allowances he obtains under Secs. 82.7, 82.6 and 82.10, as may be modified under Sec. 82.12 (transfer of allowances).

Consumptive use (Section 280.12 [9/23/88]) with respect to heating oil means consumed on the premises.

Contact cooling and heating water (Section 463.2) is process water that contacts the raw materials or plastic product for the purpose of heat transfer during the plastics molding and forming process.

Contact cooling water (Section 471.02) is any wastewater which contacts the metal workpiece or the raw materials used in forming metals for the purpose of removing heat from the metal.

Contact cooling water (Section 467.02) is any wastewater which contacts the aluminum workpiece or the raw materials used in forming aluminum.

Contact material (Section 60.101 [8-17-89]) means any substance formulated to remove metals, sulfur, nitrogen, or any other contaminant from petroleum derivatives.

Contact Resistance (Section 85.2122(a)(5)(ii)(D)) means the opposition to the flow of current between the mounting bracket and the insulated terminal.

Container (Section 165.1) means any package, can, bottle, bag, barrel, drum, tank, or other containing-device (excluding spray applicator tanks) used to enclose a pesticide or pesticide-related waste.

Container (Section 259.10 [7-2-90]) means any portable device in which a regulated medical waste is stored, transported, disposed or otherwise handled. The term container as used in this part does not include items in the Table or Regulated Medical Waste at Sec. 259.30(a) of this part.

Container (Section 260.10) means any portable device in which a material is stored, transported, treated, disposed of, or otherwise handled.

Container (Section 61.341 [3-7-90]) means any portable waste management unit in which a material is stored, transported, treated, or otherwise handled. Examples of containers are drums, barrels, tank trucks, barges, dumpsters, tank cars, dumptrucks, and ships.

Container (Section 749.68 [1-3-90]) means any bag, barrel, bottle, box, can, cylinder, drum, or the like that holds hexavalent chromium-based water treatment chemicals for use in cooling systems.

Container glass (Section 60.291) means glass made of soda-lime recipe, clear or colored, which is pressed and/or blown into bottles, jars, ampoules, and other products listed in Standard Industrial Classification 3221 (SIC 3221).

Contaminant (Section 230.3) means a chemical or biological substance in a form that can be incorporated into, onto or be ingested by and that harms aquatic organisms, consumers of aquatic organisms, or users of the aquatic environment, and includes but is not limited to the substances on the 307(a)(1) list of toxic pollutants promulgated on January 31, 1978 (43 FR 4109).

Contaminant (Section 300.6). See **Pollutant (Section 300.6)**.

Contaminant (Section 310.11 [10/21/87]). See **Pollutant (Section 310.11)**.

Contaminant (Sections 2.304; 141.2; 142.2; 143.2; 144.3; 146.3; 147.2902; 149.101) means any physical, chemical, biological, or radiological substance or matter in water.

Contaminate (Section 257.3-4) means introduce a substance that would cause: (i) The concentration of that substance in the ground water to exceed the maximum contaminant level specified in Appendix I, or (ii) An increase in the concentration of that substance in the ground water where the existing concentration of that substance exceeds the maximum contaminant level specified in Appendix I.

Contaminated nonprocess wastewater (Sections 415.91; 415.241; 415.381; 415.431; 415.441; 415.551; 415.601) shall mean any water which, during manufacturing or processing, comes into incidental contact with any raw material, intermediate product, finished product, by-product or waste product by means of (1) rainfall runoff; (2) accidental spills; (3) accidental leaks caused by the failure of process equipment, which are repaired within the shortest reasonable time not to exceed 24 hours after discovery; and (4) discharges from safety showers and related personal safety equipment: Provided, that all reasonable measures have been taken (i) to prevent, reduce and control such contact to the maximum extent feasible; and (ii) to mitigate the effects of such contact once it has occurred.

Contaminated nonprocess wastewater (Sections 418.11; 422.41; 422.51) shall mean any water including precipitation runoff which, during manufacturing or processing, comes into incidental contact with any raw material, intermediate product, finished product, by-product or waste product by means of: (1) Precipitation runoff; (2) accidental spills; (3) accidental leaks caused by the failure of process equipment and which are repaired or the discharge of pollutants therefrom contained or terminated within the shortest reasonable time which shall not exceed 24 hours after discovery or when discovery should reasonably have been made, whichever is earliest; and (4) discharges from safety showers and related personal safety equipment, and from equipment washings for the purpose of safe entry, inspection and maintenance; provided that all reasonable measures have been taken to prevent, reduce, eliminate and control to the maximum extent feasible such contact and provided further that all reasonable measures have been taken that will mitigate the effects of such contact once it has occurred.

Contaminated runoff (Section 419.11) shall mean runoff which comes into contact with any raw material, intermediate product, finished product, by-product or waste product located on petroleum refinery property.

Contiguous zone (Section 300.5 [3-8-90]) the zone of the high seas, established by the United States under Article 24 of the Convention on the Territorial sea and Contiguous Zone, which is contiguous to the territorial sea and which extends nine miles seaward from the outer limit of the territorial sea.

Contiguous zone (Sections 110.1; 116.3) means the entire zone established or to be established by the United States under article 24 of the Convention on the Territorial Sea and the Contiguous Zone.

Contiguous zone (Sections 117.1; 122.2) means the entire zone established by the United States under Article 24 of the Convention on the Territorial Sea and the Contiguous Zone.

Contingency plan (Section 260.10) means a document setting out an organized, planned, and coordinated course of action to be followed in case of a fire, explosion, or release of hazardous waste or hazardous waste constituents which could threaten human health or the environment.

Continuation award (Section 30.200) means an assistance agreement after the initial award, for a project which has more than one budget period in its approved project period, or annual awards, after the first award, to State, Interstate, or local agencies for continuing environmental programs (see Section 30.306).

Continuation award (Section 35.105) means any assistance award after the first award to a State, interstate, or local agency for a continuing environmental program.

Continuing environmental programs (Section 35.105) are those pollution control programs which will not be completed within a definable time period.

Continuous (Section 420.111) means those alkaline cleaning operations which process steel products other than in discrete batches or bundles.

Continuous (Section 420.81) means those descaling operations that remove surface scale from the sheet or wire products in continuous processes.

Continuous (Section 420.91) means those pickling operations which process steel products other than in discrete batches or bundles.

Continuous casting (Section 467.02) is the production of sheet, rod, or other long shapes by solidifying the metal while it is being poured through an open-ended mold using little or no contact cooling water. Continuous casting of rod and sheet generates spent lubricants and rod casting also generates contact cooling water.

Continuous casting (Section 471.02) is the production of sheet, rod, or other long shapes by solidifying the metal while it is being poured through an open-ended mold.

Continuous discharge (Section 122.2) means a discharge which occurs without interruption throughout the operating hours of the facility, except for infrequent shutdowns for maintenance, process changes, or other similar activities.

Continuous disposal (Section 61.251 [12-15-89])* means a method of tailings management and disposal in which tailings are dewatered by mechanical methods immediately after generation. The dried tailings are then placed in trenches or other disposal areas and immediately covered to limit emissions consistent with applicable Federal standards.

Continuous emission monitoring system or CEMS (Section 60.51a [2-11-91]) means a monitoring system for continuously measuring the emissions of a pollutant from an affected facility.

Continuous emission reduction technology (Section 57.103). See **Constant controls (Section 57.103)**.

Continuous monitoring system (Section 60.2) means the total equipment, required under the emission monitoring sections in applicable subparts, used to sample and condition (if applicable), to analyze, and to provide a permanent record of emissions or process parameters.

Continuous operations (Section 471.02) means that the industrial user introduces regulated wastewaters to the POTW throughout the operating hours of the facility, except for infrequent shutdowns for maintenance, process changes, or other similar activities.

Continuous process (Section 60.561 [3-5-91]) means a polymerization process in which reactants are introduced in a continuous manner and products are removed either continuously or intermittently at regular intervals so that the process can be operated and polymers produced essentially continuously.

Continuous recorder (Section 264.1031 [6-21-90]) means a data-recording device recording an instantaneous data value at least once every 15 minutes.

Continuous recorder (Section 60.661 [6-29-90]) means a data recording device recording an instantaneous data value at least once every 15 minutes.

Continuous vapor processing system (Section 60.501) means a vapor processing system that treats total organic compounds vapors collected from gasoline tank trucks on a demand basis without intermediate accumulation in a vapor holder.

Contract (Section 15.4) means any contract or other agreement made with an Executive Branch agency for the procurement of goods, materials, or services (including construction), and includes any subcontract made thereunder.

Contract (Section 35.4010 [10/1/92])* means a written agreement between the recipient and another party (other than a public agency) for services or supplies necessary to complete the TAG project. Contracts include contracts and subcontracts for personal and professional services or supplies necessary to complete the TAG project, and agreements with consultants, and purchase orders.

Contract (Section 35.6015 [1/27/89]) means a written agreement between an EPA recipient and another party (other than another public agency) or between the recipient's contractor and the contractor's first tier subcontractor.

Contract (Section 8.2) means any Government contract or any federally assisted construction contract.

Contract carrier by motor vehicle (Section 202.10) means any person who engages in transportation by motor vehicle of passengers or property in interstate or foreign commerce for compensation (other than transportation referred to in paragraph (b) of this section) under continuing contracts with one person or a limited number of persons either (1) for the furnishing of transportation services through the assignment of motor vehicles for a continuing period of time to the exclusive use of each person served or (2) for the furnishing of transportation services designed to meet the distinct need of each individual customer.

Contract or other approved (Section 112.2 [8/25/93])* means: (1) A written contractual agreement with an oil spill removal organization(s) that identifies and ensures the availability of the necessary personnel and equipment within appropriate response times; and/or (2) A written certification by the owner or operator that the necessary personnel and equipment resources, owned or operated by the facility owner or operator, are available to respond to a discharge within appropriate response times; and/or (3) Active membership in a local or regional oil spill removal organization(s) that has identified and ensures adequate access through such membership to necessary personnel and equipment to respond to a discharge within appropriate response times in the specified geographic areas; and/or (4) Other specific arrangements approved by the Regional Administrator upon request of the owner or operator.

Contract specifications (Section 249.04) means the set of specifications prepared for an individual construction project, which contains design, performance, and material requirements for that project.

Contractor (Section 15.4) means any person with whom an Executive Branch agency has entered into, extended, or renewed a contract as defined above, and includes subcontractors or any person holding a subcontract.

Contractor (Section 35.4010 [10/1/92])* means any party (e.g., Technical Advisor) to whom a recipient awards a contract.

Contractor (Section 35.6015 [1/27/89]) means any party to whom a recipient awards a contract.

Contractor (Section 35.936-1) means a party to whom a subagreement is awarded.

Contractor (Section 8.2) means, unless otherwise indicated, a prime contractor or subcontractor.

Contractor (Sections 30.200; 33.005) means any party to whom a recipient awards a subagreement.

Contribute materially (Section 4.2) means that during the 2 taxable years prior to the taxable year in which displacement occurs, or during such other period as the Agency determines to be more equitable, a business or farm operation: (1) Had average annual gross receipts of at least $5,000; or (2) Had average annual net earnings of at least $1,000; or (3) Contributed at least 33-1/3 percent of the owner's or operator's average annual gross income from all sources. (4) If the application of the above criteria creates an inequity or hardship in any given case, the Agency may approve the use of other criteria as determined appropriate.

Control (including the terms controlling, controlled by and under common control with) (Section 66.3) means the power to direct or cause the direction of the management and policies of a person or organization, whether by the ownership of stock, voting rights, by contract, or otherwise.

Control (Section 192.01) means any remedial action intended to stabilize, inhibit future misuse of, or reduce emissions or effluents from residual radioactive materials.

Control (Section 192.31) means any action to stabilize, inhibit future misuse of, or reduce emissions or effluents from uranium byproduct materials.

Control (Section 797.1600) is an exposure of test organisms to dilution water only or dilution water containing the test solvent or carrier (no toxic agent is intentionally or inadvertently added).

Control area (Section 80.2 [2/16/94])* means a geographic area in which only oxygenated gasoline under the oxygenated gasoline program may be sold or dispensed, with boundaries determined by Section 211(m) of the Act.

Control authority (Section 403.12) refers to: (1) The POTW if the POTW's Submission for its pretreatment program (Section 403.3(t)(1)) has been approved in accordance with the requirements of Section 403.11; or (2) the Approval Authority if the Submission has not been approved.

Control authority (Sections 413.02; 466.02; 471.02) is defined as the POTW if it has an approved pretreatment program; in the absence of such a program, the NPDES State if it has an approved pretreatment program or EPA if the State does not have an approved program.

Control device (Section 264.1031 [6-21-90]) means an enclosed combustion device, vapor recovery system, or flare. Any device the primary function of which is the recovery or capture of solvents or other organics for use, reuse, or sale (e.g., a primary condenser on a solvent recovery unit) is not a control device.

Control device (Section 60.261) means the air pollution control equipment used to remove particulate matter generated by an electric submerged arc furnace from an effluent gas stream.

Control device (Section 60.271) means the air pollution control equipment used to remove particulate matter generated by an EAF(s) from the effluent gas stream.

Control device (Section 60.271a) means the air pollution control equipment used to remove particulate matter from the effluent gas stream generated by an electric arc furnace or AOD vessel.

Control device (Section 60.381) means the air pollution control equipment used to reduce particulate matter emissions released to the atmosphere from one or more affected facilities at a metallic mineral processing plant.

Control device (Section 60.671) means the air pollution control equipment used to reduce particulate matter emissions released to the atmosphere from one or more process operations at a nonmetallic mineral processing plant.

Control device (Section 60.691 [11/23/88]) means an enclosed combustion device, vapor recovery system or flare.

Control device (Section 60.711 [10/3/88]) means any apparatus that reduces the quantity of a pollutant emitted to the air.

Control device (Section 60.741 [9-11-89]) means any apparatus that reduces the quantity of a pollutant emitted to the air.

Control device (Section 61.301 [3-7-90]) means all equipment used for recovering or oxidizing benzene vapors displaced from the affected facility.

Control device (Section 61.341 [3-7-90]) means an enclosed combustion device, vapor recovery system, or flare.

Control device (Sections 60.481; 61.241) means an enclosed combustion device, vapor recovery system, or flare.

Control device (Sections 61.171; 61.181) means the air pollution control equipment used to collect particulate matter emissions.

Control device shutdown (Section 264.1031 [6-21-90]) means the cessation of operation of a control device for any purpose.

Control period (Section 80.2 [2/16/94])* means the period during which oxygenated gasoline must be sold or dispensed in any control area, pursuant to Section 211(m)(2) of the Act.

Control period (Section 82.3 [12/10/93])* means the period from January 1, 1992 through December 31, 1992, and each twelve-month period from January 1 through December 31, thereafter.

Control strategy (Section 51.100) means a combination of measures designated to achieve the aggregate reduction of emissions necessary for attainment and maintenance of national standards including, but not limited to, measures such as: (1) Emission limitations. (2) Federal or State emission charges or taxes or other economic incentives or disincentives. (3) Closing or relocation of residential, commercial, or industrial facilities. (4) Changes in schedules or methods of operation of commercial or industrial facilities or transportation systems, including, but not limited to, short-term changes made in accordance with standby plans. (5) Periodic inspection and testing of motor vehicle emission control systems, at such time as the Administrator determines that such programs are feasible and practicable. (6) Emission control measures applicable to in-use motor vehicles, including, but not limited to, measures such as mandatory maintenance, installation of emission control devices, and conversion to gaseous fuels. (7) Any transportation control measure including those transportation measures listed in section 108(f) of the Clean Air Act as amended. (8) Any variation of, or alternative to any measure delineated herein. (9) Control or prohibition of a fuel or fuel additive used in motor vehicles, if such control or prohibition is necessary to achieve a national primary or secondary air quality standard and is approved by the Administrator under section 211(c)(4)(C) of the Act.

Control substance (Section 160.3 [8-17-89]) means any chemical substance or mixture, or any other material other than a test substance, feed, or water, that is administered to the test system in the course of a study for the purpose of establishing a basis for comparison with the test substance for known chemical or biological measurements.

Control substance (Section 792.3 [8-17-89]) means any chemical substance or mixture, or any other material other than a test substance, feed, or water, that is administered to the test system in the course of a study for the purpose of establishing a basis for comparison with the test substance for chemical or biological measurements.

Control substance (Sections 160.3; 792.3) means any chemical substance or mixture or any other material other than a test substance that is administered to the test system in the course of a study for the purpose of establishing a basis for comparison with the test substance.

Control technology (Section 57.103). See **Constant controls (Section 57.103)**.

Controlled area (Section 191.12 [12/20/93])* means: (1) A surface location, to be identified by passive institutional controls, that encompasses no more than 100 square kilometers and extends horizontally no more than five kilometers in any direction from the outer boundary of the original location of the radioactive wastes in a disposal system; and (2) the subsurface underlying such a surface location.

Controlled by (Section 66.3). See **Control (Section 66.3)**.

Controlled product (Section 82.3 [12/10/93])* means a product that contains a controlled substance listed as a Class I, Group I or II substance in appendix A of this subpart, and that belongs to one or more of the following six categories of products: (i) Automobile and truck air conditioning units (whether incorporated in vehicles or not); (ii) Domestic and commercial refrigeration and air conditioning/heat pump equipment (whether containing controlled substances as a refrigerant and/or in insulating material of the product), e.g. Refrigerators, Freezers, Dehumidifiers, Water coolers, Ice machines, Air conditioning and heat pump units; (iii) Aerosol products, except medical aerosols; (iv) Portable fire extinguishers; (v) Insulation boards, panels and pipe covers; and (vi) Pre-polymers. Controlled products include, but are not limited to, those products listed in appendix D of this subpart.

Controlled substance (Section 32.605 [5-25-90]) means a controlled substance in schedules I through V of the Controlled Substances Act (21 U.S.C. 812), and as further defined by regulation at 21 CFR 1308.11 through 1308.15.

Controlled substance (Section 82.3 [12/10/93])* means any substance listed in appendix A or appendix B of this subpart, whether existing alone or in a mixture, but excluding any such substance or mixture that is in a manufactured product other than a container used for the transportation or storage of the substance or mixture. Thus, any amount of a listed substance in appendix A or appendix B of this subpart which is not part of a use system containing the substance is a controlled substance. If a listed substance or mixture must first be transferred from a bulk container to another container, vessel, or piece of equipment in order to realize its intended use, the listed substance or mixture is a "controlled substance". The inadvertent or coincidental creation of insignificant quantities of a listed substance in appendix A or appendix B of this subpart: (1) During a chemical manufacturing process, (2) resulting from unreacted feedstock, or (3) from the listed substance's use as a process agent present as a trace quantity in the chemical substance being manufactured, is not deemed a controlled substance. Controlled substances are divided into two classes, Class I in appendix A of this subpart, and Class II listed in appendix B of this subpart. Class I substances are further divided into seven groups, Group I, Group II, Group III, Group IV, Group V, Group VI, and Group VII as set forth in appendix A of this subpart.

Controlled surface mine drainage (Section 434.11) means any surface mine drainage that is pumped or siphoned from the active mining area.

Controlling (Section 66.3). See **Control (Section 66.3)**.

Controlling interest (Section 280.92 [10/26/88]) means direct ownership of at least 50 percent of the voting stock of another entity.

Conveniently available service facility and spare parts for small-volume manufacturers (Section 86.092-2 [2-28-90]) (Note: the definitions listed in this section on this date of publication apply beginning with the 1992 model year.) means that the vehicle manufacturer has a qualified service facility at or near the authorized point of sale or delivery of its vehicles and maintains an inventory of all emission-related spare parts or has made arrangements for the part manufacturers to supply the parts by expedited shipment (e.g., utilizing overnight express or delivery services, UPS, etc.).

Conventional filtration treatment (Section 141.2 [6/29/89]) means a series of processes including coagulation, flocculation, sedimentation, and filtration resulting in substantial particulate removal.

Conventional gasoline (Section 80.2 [2/16/94])* means any gasoline which has not been certified under Sec. 80.40.

Conventional mine (Section 146.3) means an open pit or underground excavation for the production of minerals.

Conventional technology (Section 60.41c [9-12-90]) means wet flue gas desulfurization technology, dry flue gas desulfurization technology, atmospheric fluidized bed combustion technology, and oil hydrodesulfurization technology.

Conventional technology (Section 35.2005) means wastewater treatment processes and techniques involving the treatment of wastewater at a centralized treatment plant by means of biological or physical/chemical unit processes followed by direct point source discharge to surface waters.

Conventional technology (Section 60.41b [12/16/87]) means wet flue gas desulfurization (FGD) technology, dry FGD technology, atmospheric fluidized bed combustion technology, and oil hydrodesulfurization technology.

Conversion Efficiency (Section 85.2122(a)(15)(ii)(B)) means the measure of the catalytic converter's ability to oxidize HC/CO to CO_2/H_2O under fully warmed-up conditions stated as a percentage calculated by the formula set forth in Section 85.2122(a)(15)(ii)(B).

Converter (Section 60.181) means any vessel to which lead concentrate or bullion is charged and refined.

Converter arsenic charging rate (Section 61.171) means the hourly rate at which arsenic is charged to the copper converters in the copper converter department based on the arsenic content of the copper matte and of any lead matte that is charged to the copper converters.

Conveying system (Section 60.671) means a device for transporting materials from one piece of equipment or location to another location within a plant. Conveying systems include but are not limited to the following: Feeders, belt conveyors, bucket elevators and pneumatic systems.

Conveyor belt transfer point (Section 60.381) means a point in the conveying operation where the metallic mineral or metallic mineral concentrate is transferred to or from a conveyor belt except where the metallic mineral is being transferred to a stockpile.

Conviction (Section 303.11 [5/5/88, 6/21/89]) means a judgment of guilt entered in U.S. District Court, upon a verdict rendered by the court or petit jury or by a plea of guilty, including a plea of nolo contendere.

Cookstove (Section 60.531 [2/26/88]) means a wood-fired appliance that is designed primarily for cooking food and that has the following characteristics: (a) An oven, with a volume of 0.028 cubic meters (1 cubic foot) or greater, and an oven rack, (b) A device for measuring oven temperatures, (c) A flame path that is routed around the oven, (d) A shaker grate, (e) An ash pan, (f) An ash clean-out door below the oven, and (g) The absence of a fan or heat channels to dissipate heat from the appliance.

Cooling system (Section 749.68 [1-3-90]) means any cooling tower or closed cooling water system.

Cooling tower (Section 749.68 [1-3-90]) means an open water recirculating device that uses fans or natural draft to draw or force ambient air through the device to cool warm water by direct contact.

Cooperating agency (Section 1508.5) means any Federal agency other than a lead agency which has jurisdiction by law or special expertise with respect to any environmental impact involved in a proposal (or a reasonable alternative) for legislation or other major Federal action significantly affecting the quality of the human environment. The selection and responsibilities of a cooperating agency are described in Section 1501.6. A State or local agency of similar qualifications or, when the effects are on a reservation, an Indian Tribe, may by agreement with the lead agency become a cooperating agency.

Cooperative agreement (Section 35.6015 [1/27/89]) means a legal instrument EPA uses to transfer money, property, services, or anything of value to a recipient to accomplish a public purpose in which substantial EPA involvement is anticipated during the performance of the project.

Cooperative agreement (Section 30.200) means an assistance agreement in which substantial EPA involvement is anticipated during the performance of the project (does not include fellowships).

Cooperative agreement (Section 300.5 [3-8-90]) is a legal instrument EPA uses to transfer money, property, services, or anything of value to a recipient to accomplish a public purpose in which substantial EPA involvement is anticipated during the performance of the project.

Cooperator (Section 172.1) means any person who grants permission to a permittee or a permittee's designated participant for the use of an experimental use pesticide at an application site owned or controlled by the cooperator.

Copenhagen Amendments (Section 82.3 [12/10/93])* means the Montreal Protocol on Substances That Deplete the Ozone Layer, as amended at the Fourth Meeting of the Parties to the Montreal Protocol in Copenhagen in 1992.

Copper (Section 420.02) means total copper and is determined by the method specified in 40 CFR 136.3.

Copper (Sections 415.361; 415.471; 415.651) shall mean the total copper present in the process wastewater stream exiting the wastewater treatment system.

Copper Casting (Section 464.02). The remelting of copper or a copper alloy to form a cast intermediate or final product by pouring or forcing the molten metal into a mold, except for ingots, pigs, or other cast shapes related to nonferrous (primary and secondary) metals manufacturing (40 CFR Part 421). Also excluded are casting of beryllium alloys in which beryllium is present at 0.1 or greater percent by weight and precious metals alloys in which the precious metal is present at 30 or greater percent by weight. Except for grinding scrubber operations which are covered here, processing operations following the cooling of castings are covered under the electroplating and metal finishing point source categories (40 CFR Parts 413 and 433).

Copper converter (Sections 60.161; 61.171) means any vessel to which copper matte is charged and oxidized to copper.

Copper converter department (Section 61.171) means all copper converters at a primary copper smelter.

Copper matte (Section 61.171) means any molten solution of copper and iron sulfides produced by smelting copper sulfide ore concentrates or calcines.

Coproduct (Section 704.3 [12/22/88]) means a chemical substance produced for a commercial purpose during the manufacture, processing, use, or disposal of another chemical substance or mixture.

Copy of study (Section 716.3) means the written presentation of the purpose and methodology of a study and its results.

Copy paper (Section 250.4 [10/6/87, 6/22/88]). See **Xerographic paper (Section 250.4)**.

Core of the drawing with emulsions or soaps subcategory (Section 467.61) shall include drawing using emulsions or soaps, stationary casting, artificial aging, annealing, degreasing, sawing, and swaging.

Core of the drawing with neat oils subcategory (Section 467.51) shall include drawing using neat oils, stationary casting, artificial aging, annealing, degreasing, sawing, and swaging.

Core of the extrusion subcategory (Section 467.31) shall include extrusion die cleaning, dummy block cooling, stationary casting, artificial aging, annealing, degreasing, and sawing.

Core of the forging subcategory (Section 467.41) shall include forging, artificial aging, annealing, degreasing, and sawing.

Core of the rolling with emulsions subcategory (Section 467.21) shall include rolling using emulsions, roll grinding, stationary casting, homogenizing, artificial aging, annealing, and sawing.

Core of the rolling with neat oils subcategory (Section 467.11) shall include rolling using neat oils, roll grinding, sawing, annealing, stationary casting, homogenizing, artificial aging, degreasing, and stamping.

Core Program Cooperative Agreement (Section 35.6015 [6-5-90])* A Cooperative Agreement that provides funds to a State or Indian Tribe to conduct CERCLA implementation activities that are not assignable to specific sites, but are intended to support a State's ability to participate in the CERCLA response program.

Corn (Section 406.11) shall mean the shelled corn delivered to a plant before processing.

Corn, canned (Section 407.71) shall mean the processing of corn into the following product styles: Canned, yellow and white, whole kernel, cream style, and on-the-cob.

Corn, frozen (Section 407.71) shall mean the processing of corn into the following product styles: Frozen, yellow and white, whole kernel and whole cob.

Corps (Section 232.2 [2/11/93])* means the U.S. Army Corps of Engineers.

Corps (Section 233.2 [6/6/88]) means

the U.S. Army Corps of Engineers.

Corrective action management unit or CAMU (Section 260.10 [8/18/92])* means an area within a facility that is designated by the Regional Administrator under part 264 subpart S, for the purpose of implementing corrective action requirements under Sec. 264.101 and RCRA section 3008(h). A CAMU shall only be used for the management of remediation wastes pursuant to implementing such corrective action requirements at the facility.

Corrective Action Management Unit or CAMU (Section 270.2 [2/16/93])* means an area within a facility that is designated by the Regional Administrator under part 264 subpart S, for the purpose of implementing corrective action requirements under Sec. 264.101 and RCRA section 3008(h). A CAMU shall only be used for the management of remediation wastes pursuant to implementing such corrective action requirements at the facility.

Corrosion expert (Section 260.10) means a person who, by reason of his knowledge of the physical sciences and the principles of engineering and mathematics, acquired by a professional education and related practical experience, is qualified to engage in the practice of corrosion control on buried or submerged metal piping systems and metal tanks. Such a person must be certified as being qualified by the National Association of Corrosion Engineers (NACE) or be a registered professional engineer who has certification or licensing that includes education and experience in corrosion control on buried or submerged metal piping systems and metal tanks.

Corrosion expert (Section 280.12 [9/23/88]) means a person who, by reason of thorough knowledge of the physical sciences and the principles of engineering and mathematics acquired by a professional education and related practical experience, is qualified to engage in the practice of corrosion control on buried or submerged metal piping systems and metal tanks. Such a person must be accredited or certified as being qualified by the National Association of Corrosion Engineers or be a registered professional engineer who has certification or licensing that includes education and experience in corrosion control of buried or submerged metal piping systems and metal tanks.

Corrosion inhibitor (Section 141.2 [7/17/92])* means a substance capable of reducing the corrosivity of water toward metal plumbing materials, especially lead and copper, by forming a protective film on the interior surface of those materials.

Corrugated box (Section 246.101) means a container for goods which is composed of an inner fluting of material (corrugating medium) and one or two outer liners of material (linerboard).

Corrugated boxes (Section 250.4 [10/6/87, 6/22/88]) means boxes made of corrugated paperboard, which, in turn, is made from a fluted corrugating medium pasted to two flat sheets of paperboard (linerboard); multiple layers may be used.

Corrugated container waste (Section 246.101) means discarded corrugated boxes.

Corrugated paper (Section 763.163 [7-12-89]) means an asbestos-containing product made of corrugated paper, which is often cemented to a flat backing, may be laminated with foils or other materials, and has a corrugated surface. Major applications of asbestos corrugated paper include: thermal insulation for pipe coverings; block insulation; panel insulation in elevators; insulation in appliances; and insulation in low-pressure steam, hot water, and process lines.

Corrugating medium furnish subdivision mills (Section 430.51) are mills where only recycled corrugating medium is used in the production of paperboard.

Cosmetic (Sections 710.2; 720.3) shall have the meaning contained in the Federal Food, Drug, and Cosmetic Act, 21 U.S.C. 321 et seq., and the regulations issued under such Act.

Cost analysis (Section 33.005) means the review and evaluation of each element of subagreement cost to determine reasonableness, allocability and allowability.

Cost analysis (Section 35.6015 [1/27/89]) means the review and evaluation of each element of contract cost to determine reasonableness, allocability, and allowability.

Cost of production of a car line (Section 600.502-81) (applies beginning with the 1979 model year) shall mean the aggregate of the products of: (i) The average U.S. dealer wholesale price for such car line as computed from each official dealer price list effective during the course of a model year, and (ii) The number of automobiles within the car line produced during the part of the model year that the price list was in effect.

Cost share (Section 35.6015 [6-5-90])* The portion of allowable project costs that a recipient contributes toward completing its project (i.e., non-Federal share, matching share).

Cost sharing (Section 30.200) means the portion of allowable project costs that a recipient contributes toward completing its project (i.e., non-Federal share, matching share).

Cotton fiber content papers (Section 250.4 [6/22/88]) means paper that contains a minimum of 25 percent and up to 100 percent cellulose fibers derived from lint cotton, cotton linters, and cotton or linen cloth cuttings. It is also known as rag content paper or rag paper. It is used for stationery, currency, ledgers, wedding invitations, maps, and other specialty papers.

Cotton fiber furnish subdivision mills (Section 430.181) are those mills where significant quantities of cotton fibers (equal to or greater than 4 percent of the total product) are used in the production of fine papers.

Council (Section 1508.6) means the Council on Environmental Quality established by Title II of the Act.

Council (Section 1517.2) shall mean the Council on Environmental Quality established under Title II of the National Environmental Policy Act of 1969 (42 U.S.C. 4321-4347).

Cover (Section 60.711 [10/3/88]) means, with respect to coating mix preparation equipment, a device that lies over the equipment opening to prevent VOC from escaping and that meets the requirements found in Section 60.712(c)(1)-(5).

Cover (Section 60.741 [9-11-89]) with respect to coating mix preparation equipment, a device that fits over the equipment opening to prevent emissions of volatile organic compounds (VOC) from escaping.

Cover (Section 61.341 [3-7-90]) means a device or system which is placed on or over a waste placed in a waste management unit so that the entire waste surface area is enclosed and sealed to minimize air emissions. A cover may have openings necessary for operation, inspection, and maintenance of the waste management unit such as access hatches, sampling ports, and gauge wells provided that each opening is closed and sealed when not in use. Example of covers include a fixed roof installed on a tank, a lid installed on a container, and an air-supported enclosure installed over a waste management unit.

Cover material (Section 241.101) means soil or other suitable material that is used to cover compacted solid wastes in a land disposal site.

Cover paper (Section 250.4 [10/6/87, 6/22/88]). See **Cover stock (Section 250.4)**.

Cover stock or Cover paper (Section 250.4 [10/6/87, 6/22/88]) means a heavyweight paper commonly used for covers, books, brochures, pamphlets, and the like.

Coverage period (Section 704.203 [12/22/88]) means a time-span which is 1 day less than 2 years, as identified in Subpart D, and is the time-span which a person uses to determine his/her reporting year. Subject manufacturing or processing activities may or may not have occurred during the coverage period.

Covered (Section 61.251) means to cover with earth sufficient to meet Federal standards for the management of uranium byproduct materials pursuant to 40 CFR 192.32.

Covered area (Section 80.2 [2/16/94])* means each of the geographic areas specified in Sec. 80.70 in which only reformulated gasoline may be sold or dispensed to ultimate consumers.

Covered Federal Action (Section 34.105 [2-26-90]) means any of the following Federal actions: (1)The awarding of any Federal contract; (2) The making of any Federal Grant; (3) The making of any Federal loan; (4) The entering into any cooperative agreement; and, (5) The extension, continuation, renewal, amendment, or modification of any Federal contract, grant, loan, or cooperative agreement. Covered Federal action does not include receiving from any agency commitment providing for the United States to insure or guarantee a loan. Loan guarantees and loan insurance are addressed independently within this part.

Covered States (Section 259.10 [8-24-89])* means those States that are participating in the demonstration medical waste tracking program and includes: Connecticut, New Jersey, New York, Rhode Island, and Puerto Rico. Any other State is a Non-Covered State.

Cr(+6) (Section 415.171) shall mean hexavalent chromium.

Cr(T) (Section 415.171) shall mean total chromium.

Cranberries (Section 407.61) shall mean the processing of cranberries into the following product styles: Canned, bottled, and frozen, whole, sauce, jelly, juice and concentrate.

Crankcase emissions (Sections 86.082-2; 86.402-78) means airborne substances emitted to the atmosphere from any portion of the engine crankcase ventilation or lubrication systems.

Creditor agency (Section 13.2 [9/23/88]) means the Federal agency to which the debt is owed.

Criteria (Section 131.3) are elements of State water quality standards, expressed as constituent concentrations, levels, or narrative statements, representing a quality of water that supports a particular use. When criteria are met, water quality will generally protect the designated use.

Criteria (Section 220.2) means the criteria set forth in Part 227 of this Subchapter H.

Criteria (Section 256.06) means the "Criteria for Classification of Solid Waste Disposal Facilities," 40 CFR Part 257, promulgated under section 4004(a) of the Act.

Critical organ (Sections 61.91; 61.101; 191.02) means the most exposed human organ or tissue exclusive of the integumentary system (skin) and the cornea.

Critical pollutant (Part 58, Appendix G) means the pollutant or pollutant combination (TSP x SO_2) with the highest subindex during the reporting period.

Crop advisor (Section 170.3 [8/21/92])* means any person who is assessing pest numbers or damage, pesticide distribution, or the status or requirements of agricultural plants. The term does not include any person who is performing hand labor tasks.

Crops for direct human consumption (Section 257.3-6) means crops that are consumed by humans without processing to minimize pathogens prior to distribution to the consumer.

Cross recovery furnace (Section 60.281) means a furnace used to recover chemicals consisting primarily of sodium and sulfur compounds by burning black liquor which on a quarterly basis contains more than 7 weight percent of the total pulp solids from the neutral sulfite semichemical process and has a green liquor sulfidity of more than 28 percent.

CRP (Section 35.6015 [1/27/89]). See **Community Relations Plan (Section 35.6015)**.

Crude intermediate plastic material (Section 463.2) is plastic material formulated in an on-site polymerization process.

Crusher (Section 60.381) means a machine used to crush any metallic mineral and includes feeders or conveyors located immediately below the crushing surfaces. Crushers include, but are not limited to, the following types: jaw, gyratory, cone, and hammermill.

Crusher (Section 60.671) means a machine used to crush any nonmetallic minerals, and includes, but is not limited to, the following types: jaw, gyratory, cone, roll, rod mill, hammermill, and impactor.

Cr,VI (Section 413.02) shall mean hexavalent chromium.

CT or CTcalc (Section 141.2 [6/29/89]) is the product of residual disinfectant concentration (C) in mg/l determined before or at the first customer, and the corresponding disinfectant contact time (T) in minutes, i.e., C x T. If a public water system applies disinfectants at more than one point prior to the first customer, it must determine the CT of each disinfectant sequence before or at the first customer to determine the total percent inactivation or total inactivation ratio. In determining the total inactivation ratio, the public water system must determine the residual disinfectant concentration of each disinfection sequence and corresponding contact time before any subsequent disinfection application point(s). (For calculation details, refer to this definition in 54 FR 27526.)

CTcalc (Section 141.2 [6/29/89]). See **CT (Section 141.2)**.

Cubic feet or cubic meters of production (Section 429.11) in Subpart A, means the cubic feet or cubic meters of logs from which bark is removed.

Cullet (Section 426.21) shall mean any broken glass generated in the manufacturing process.

Cullet (Section 61.161) means waste glass recycled to a glass melting furnace.

Cullet water (Section 426.11) shall mean that water which is exclusively and directly applied to molten glass in order to solidify the glass.

Cumulative impact (Section 1508.7) is the impact on the environment which results from the incremental impact of the action when added to other past, present, and reasonably foreseeable future actions regardless of what agency (Federal or non-Federal) or person undertakes such other actions. Cumulative impacts can result from individually minor but collectively significant actions taking place over a period of time.

Cumulative toxicity (Section 798.2250) is the adverse effects of repeated doses occurring as a result of prolonged action on, or increased concentration of the administered test substance or its metabolites in susceptible tissues.

Cumulative toxicity (Sections 798.2450; 798.2650) is the adverse effects of repeated doses occurring as a result of prolonged action on, or increased concentration of the administered substance or its metabolites in susceptible tissues.

Curb collection (Section 243.101) means collection of solid waste placed adjacent to a street.

Curb mass (Section 86.402-78) means the actual or manufacturer's estimated mass of the vehicle with fluids at nominal capacity and with all equipment specified by the Administrator.

Curb-idle (Section 86.082-2), for manual transmission code heavy-duty engines, means the manufacturer's recommended engine speed with the transmission in neutral or with the clutch disengaged. For automatic transmission code heavy-duty engines, curb-idle means the manufacturer's recommended engine speed with the automatic transmission in gear and the output shaft stalled.

Curb-idle (Section 86.084-2) (applies beginning with the 1984 model year) means: (1) For manual transmission code light-duty trucks, the engine speed with the transmission in neutral or with the clutch disengaged and with the air conditioning system, if present, turned off. For automatic transmission code light-duty trucks, curb-idle means the engine speed with the automatic transmission in the Park position (or Neutral position if there is no Park position), and with the air conditioning system, if present, turned off. (2) For manual transmission code heavy-duty engines, the manufacturer's recommended engine speed with the clutch disengaged. For automatic transmission code heavy-duty engines, curb idle means the manufacturer's recommended engine speed with the automatic transmission in gear and the output shaft stalled. (Measured idle speed may be used in lieu of curb-idle speed for the emission tests when the difference between measured idle speed and curb idle speed is sufficient to cause a void test under either Section 86.1341 or Section 86.884-7 but not sufficient to permit adjustment in accordance with Section 86.085-25.)

Curie (Ci) (Section 190.02) means that quantity of radioactive material producing 37 billion nuclear transformations per second. (One millicurie (mCi)=0.001 Ci.)

Curie (Ci) (Section 192.01) means the amount of radioactive material that produces 37 billion nuclear transformations per second. One picocurie (pCi)=10^{-12}Ci.

Curie (Section 61.121) is a unit of radioactivity equal to 37 billion nuclear transformations (decays) per second.

Curing oven (Section 60.451) means a device that uses heat to dry or cure the coating(s) applied to large appliance parts or products.

Curing oven (Section 60.461) means the device that uses heat or radiation to dry or cure the coating applied to the metal coil.

Current assets (Section 144.61) means cash or other assets or resources commonly identified as those which are reasonably expected to be realized in cash or sold or consumed during the normal operating cycle of the business.

Current assets (Section 264.141), as specified in Section 264.141(f), means cash or other assets or resources commonly identified as those which are reasonably expected to be realized in cash or sold or consumed during the normal operating cycle of the business.

Current assets (Section 265.141), as specified in Section 265.141(f), means cash or other assets or resources commonly identified as those which are reasonably expected to be realized in cash or sold or consumed during the normal operating cycle of the business.

Current closure cost estimate (Section 265.141) means the most recent of the estimates prepared in accordance with Section 265.142(a),

(b) and (c).

Current closure cost estimate (Section 264.141) means the most recent of the estimates prepared in accordance with Section 264.142 (a), (b) and (c).

Current discharge (Section 125.58) means the volume, composition, and location of an applicant's discharge as of anytime between December 27, 1977, and December 29, 1982, as designated by the applicant.

Current liabilities (Section 144.61) means obligations whose liquidation is reasonably expected to require the use of existing resources properly classifiable as current assets or the creation of other current liabilities.

Current liabilities (Section 264.141), as specified in Section 264.141(f), means obligations whose liquidation is reasonably expected to require the use of existing resources properly classifiable as current assets or the creation of other current liabilities.

Current liabilities (Section 265.141), as specified in Section 265.141(f), means obligations whose liquidation is reasonably expected to require the use of existing resources properly classifiable as current assets or the creation of other current liabilities.

Current plugging and abandonment cost estimate (Section 265.141), as specified in Section 265.141(f), means the most recent of the estimates prepared in accordance with Section 144.62(a), (b) and (c) of this title.

Current plugging and abandonment cost estimate (Section 264.141), as specified in Section 264.141(f), means the most recent of the estimates prepared in accordance with Section 144.62(a), (b) and (c) of this title.

Current plugging cost estimate (Section 144.61) means the most recent of the estimates prepared in accordance with Section 144.62(a), (b) and (c).

Current post-closure cost estimate (Section 264.141) means the most recent of the estimates prepared in accordance with Section 264.144 (a), (b), and (c).

Current post-closure cost estimate (Section 265.141) means the most recent of the estimates prepared in accordance with Section 265.144(a), (b) and (c).

Current production (sales or distribution) (Section 167.3 [9/8/88]) means amount of planned production in the calendar year in which the pesticides report is submitted, including new pesticidal products not previously sold or distributed.

Current production (Section 167.1) means amount of planned production in the calendar year in which the pesticides report is submitted, including new products not previously sold or distributed.

Curtail (Section 61.181) means to cease operations to the extent technically feasible to reduce emissions.

Custody transfer (Section 60.111b) means the transfer of produced petroleum and/or condensate, after processing and/or treatment in the producing operations, from storage vessels or automatic transfer facilities to pipelines or any other forms of transportation.

Custody transfer (Sections 60.111; 60.111a) means the transfer of produced petroleum and/or condensate, after processing and/or treating in the producing operations, from storage tanks or automatic transfer facilities to pipelines or any other forms of transportation.

Custom blender (Section 167.3 [9/8/88]) means any establishment which provides the service of mixing pesticides to a customer's specifications, usually a pesticide(s)-fertilizer(s), pesticide-pesticide, or a pesticide-animal feed mixture, when: (1) The blend is prepared to the order of the customer and is not held in inventory by the blender; (2) the blend is to be used on the customer's property (including leased or rented property); (3) the pesticide(s) used in the blend bears end-use labeling directions which do not prohibit use of the product in such a blend; (4) the blend is prepared from registered pesticides; (5) the blend is delivered to the end-user along with a copy of the end-use labeling of each pesticide used in the blend and a statement specifying the composition of mixture; and (6) no other pesticide production activity is performed at the establishment.

Custom-molded device (Section 211.203) means a hearing protective device that is made to conform to a specific ear canal. This is usually accomplished by using a moldable compound to obtain an impression of the ear and ear canal. The compound is subsequently permanently hardened to retain this shape.

Customer (Section 704.3 [12/22/88]) means any person to whom a manufacturer, importer, or processor directly distributes any quantity of a chemical substance, mixture, mixture containing the substance or mixture, or article containing the substance or mixture, whether or not a sale is involved.

Customer (Section 721.3 [7/27/88])* means any person to whom a manufacturer, importer, or processor distributes any quantity of a chemical substance, or of a mixture containing the chemical substance, whether or not a sale is involved.

Customs territory of the United States (Section 763.163 [7-12-89]) means the 50 States, Puerto Rico, and the District of Columbia.

Customs territory of the United States (Sections 372.3 [2/16/88]; 720.3) means the 50 States, the District of Columbia, and Puerto Rico.

Cutout or by-pass or similar devices (Section 202.10) means devices which vary the exhaust system gas flow so as to discharge the exhaust gas and acoustic energy to the atmosphere without passing through the entire length of the exhaust system, including all exhaust system sound attenuation components.

Cutting (Section 61.141 [11-20-90]) means to penetrate with a sharp-edged instrument and includes sawing, but does not include shearing, slicing, or punching.

CWA (Section 122.2) means the Clean Water Act (formerly referred to as the Federal Water Pollution Control Act or Federal Water Pollution Control Act Amendments of 1972) Pub. L. 92-500, as amended by Pub. L. 95-217, Pub. L. 95-576, Pub. L. 96-483 and Pub. L. 97-117, 33 U.S.C. 1251 et seq.

CWA (Section 501.2 [5/2/89]) means the Clean Water Act (formerly referred to as the Federal Water Pollution Control Act or Federal Water Pollution Control Act Amendments of 1972), Pub. L. 92-500, as amended by Pub. L. 95-217, Pub. L. 95-576, Pub. L. 96-483, Pub. L. 97-117, and Pub. L. 100-4, 33 U.S.C. 1251 et seq.

CWA (Sections 124.2; 233.3; 270.2) means the Clean Water Act (formerly referred to as the Federal Water Pollution Control Act or Federal Water Pollution Control Act Amendments of 1972) Pub. L. 92-500, as amended by Pub. L. 95-217 and Pub. L. 95-576, 33 U.S.C. 1251 et seq.

CWA and regulations (Section 122.2) means the Clean Water Act (CWA) and applicable regulations promulgated thereunder. In the case of an approved State program, it includes State program requirements.

Cyanide (Section 420.02) means total cyanide and is determined by the method specified in 40 CFR 136.3.

Cyanide A (Sections 415.91; 415.421) means those cyanides amenable to chlorination and is determined by the methods specified in 40 CFR 136.3.

Cyanide destruction unit (Section 439.1) shall mean a treatment system designed specifically to remove cyanide.

Cyclonic flow (Section 60.251) means a spiraling movement of exhaust gases within a duct or stack.

D

Daily cover (Section 241.101) means cover material that is spread and compacted on the top and side slopes of compacted solid waste at least at the end of each operating day in order to control vectors, fire, moisture, and erosion and to assure an aesthetic appearance.

Daily discharge (Section 122.2) means the discharge of a pollutant measured during a calendar day or any 24-hour period that reasonably represents the calendar day for purposes of sampling. For pollutants with limitations expressed in units of mass, the daily discharge is calculated as the total mass of the pollutant discharged over the day. For pollutants with limitations expressed in other units of measurement, the daily discharge is calculated as the average measurement of the pollutant over the day.

Daily maximum limitation (Section 429.11) is a value that should not be exceeded by any one effluent measurement.

Daily values as applied to produced water effluent limitations and NSPS (Section 435.11 [3/4/93])* shall refer to the daily measurements used to assess compliance with the maximum for any one day.

Damage assessment claim (Section 306.12) means a claim for assessment costs described in section 111(c)(1) of CERCLA.

Damage assessment claim (Section 305.12) means a claim for assessment costs submitted to the Fund as described in section 111(c)(2) of CERCLA.

Damaged friable miscellaneous ACM (Section 763.83 [10/30/87]) means friable miscellaneous ACM which has deteriorated or sustained physical injury such that the internal structure (cohesion) of the material is inadequate or, if applicable, which has delaminated such that its bond to the substrate (adhesion) is inadequate or which for any other reason lacks fiber cohesion or adhesion qualities. Such damage or deterioration may be illustrated by the separation of ACM into layers; separation of ACM from the substrate; flaking, blistering, or crumbling of the ACM surface; water damage; significant or repeated water stains, scrapes, gouges, mars or other signs of physical injury on the ACM. Asbestos debris originating from the ACBM in question may also indicate damage.

Damaged friable surfacing ACM (Section 763.83 [10/30/87]) means friable surfacing ACM which has deteriorated or sustained physical injury such that the internal structure (cohesion) of the material is inadequate or which has delaminated such that its bond to the substrate (adhesion) is inadequate, or which, for any other reason, lacks fiber cohesion or adhesion qualities. Such damage or deterioration may be illustrated by the separation of ACM into layers; separation of ACM from the substrate; flaking, blistering, or crumbling of the ACM surface; water damage; significant or repeated water stains, scrapes, gouges, mars or other signs of physical injury on the ACM. Asbestos debris originating from the ACBM in question may also indicate damage.

Damaged or significantly damaged thermal system insulation ACM (Section 763.83 [10/30/87]) means thermal system insulation ACM on pipes, boilers, tanks, ducts, and other thermal system insulation equipment where the insulation has lost its structural integrity, or its covering, in whole or in part, is crushed, water-stained, gouged, punctured, missing, or not intact such that it is not able to contain fibers. Damage may be further illustrated by occasional punctures, gouges or other signs of physical injury to ACM; occasional water damage on the protective coverings/jackets; or exposed ACM ends or joints. Asbestos debris originating from the ACBM in question may also indicate damage.

Data Fleet (Section 610.11) means a fleet of automobiles tested at zero device-miles in baseline configuration, the retrofitted configuration and in some cases the adjusted configuration, in order to determine the changes in fuel economy and exhaust emissions due to the retrofitted configuration, and where applicable the changes due to the adjusted configuration, as compared to the fuel economy and exhaust emissions of the baseline configuration.

Data gap (Section 152.83) means the absence of any valid study or studies in the Agency's files which would satisfy a specific data requirement for a particular pesticide product.

Data Submitters List (Section 152.83) means the current Agency list, entitled "Pesticide Data Submitters by Chemical," of persons who have submitted data to the Agency.

Date of completion (Section 310.11 [1/15/93])* means the date when all field work has been completed and all deliverables (e.g., lab results, technical expert reports) have been received by the local government.

Date of discovery (Section 306.12) means the date on which the trustee became aware of the injury to the natural resource: (1) For an injury that can be visually observed, this is the date on which the trustee has available, or reasonably should have available, a document or memorandum prepared for the trustee verifying the observed injury to the natural resource, the types of injury, and which suggests that the injury may be related to the release of a hazardous substance; or (2) for an injury that cannot be visually observed, this is the date on which the trustee has available, or reasonably should have available, a document or memorandum prepared for the trustee, including such sampling and laboratory analysis as is necessary, which identifies the injured natural resource, the types of injury, and which suggests that the injury may be related to the release of a hazardous substance.

Day (Section 60.51) means 24 hours.

Day-night Sound Level (Section 201.1) means the 24-hour time of day weighted equivalent sound level, in decibels, for any continuous 24-hour period, obtained after addition of ten decibels to sound levels produced in the hours from 10 p.m. to 7 a.m. (2200-0700). It is abbreviated as L_{dn}.

Days (Section 85.1801) shall mean calendar days.

dB(A) (Section 201.1) is an abbreviation meaning A-weighted sound level in decibels, reference: 20 micropascals.

dB(A) or dBA (Sections 202.10; 204.2; 205.2) means the standard abbreviation for A-weighted sound level in decibels.

DDT (Section 129.4) means the compounds DDT, DDD, and DDE as identified by the chemical names: (DDT)-1,1,1-trichloro-2,2-bis(p-chlorophenyl) ethane and some o,p'-isomers; (DDD) or (TDE)-1,1-dichloro-2,2-bis(p-chlorophenyl) ethane and some o,p'-isomers; (DDE)-1,1-dichloro-2,2-bis(p-chlorophenyl) ethylene.

DDT formulator (Section 129.101) means a person who produces, prepares or processes a formulated product comprising a mixture of DDT and inert materials or other diluents into a product intended for application in any use registered under the Federal Insecticide, Fungicide and Rodenticide Act, as amended (7 U.S.C. 135, et seq.).

DDT Manufacturer (Section 129.101) means a manufacturer, excluding any source which is exclusively a DDT formulator, who produces, prepares or processes technical DDT, or who uses DDT as a material in the production, preparation or processing of another synthetic organic substance.

Dealer (Section 244.101) means any person who engages in the sale of beverages in beverage containers to a consumer.

Dealer (Section 600.002-85) means a person who resides or is located in the United States, any territory of the United States, or the District of Columbia and who is engaged in the sale or distribution of new automobiles to the ultimate purchaser.

Dealership (Section 171.2) means any site owned or operated by a restricted use pesticide retail dealer where any restricted use pesticide is made available for use, or where the dealer offers to make available for use any such pesticide.

Death (Section 797.1400) means the lack of opercular movement by a test fish.

Death (Sections 797.1930; 797.1950; 797.1970) means the lack of reaction of a test organism to gentle prodding.

Debarment (Section 32.102) means an action taken by the Director under Section 32.206 to deny a person the opportunity to participate in EPA assistance or subagreements.

Debris (Section 429.11) means woody material such as bark, twigs, branches, heartwood or sapwood that will not pass through a 2.54 cm (1.0 in) diameter round opening and is present in the discharge from a wet storage facility.

Debt (Section 13.2 [9/23/88]) means an amount owed to the United States from sources which include loans insured or guaranteed by the United States and all other amounts due the United States from fees, grants, contracts, leases, rents, royalties, services, sales of real or personal property, overpayments, fines, penalties, damages, interest, forfeitures (except those arising under the Uniform Code of Military Justice), and all other similar sources. As used in this regulation, the terms debt and claim are synonymous.

Debtor (Section 13.2 [9/23/88]) means an individual, organization, association, corporation, or a State or local government indebted to the United States or a person or entity with legal responsibility for assuming the debtor's obligation.

DEC system (Section 60.271a). See **Direct-shell evacuation control system (Section 60.271a)**.

Decent, safe, and sanitary dwelling (Section 4.2) means a dwelling which meets applicable housing and occupancy codes. However, any of the following standards which are not met by an applicable code shall apply, unless waived for good cause by the Federal agency funding the project. The dwelling shall: (1) Be structurally sound, weathertight, and in good repair. (2) Contain a safe electrical wiring system adequate for lighting and other electrical devices. (3) Contain a heating system capable of sustaining a healthful temperature (of approximately 70 degrees) for a displaced person, except in those areas where local climatic conditions do not require such a system. (4) Be adequate in size with respect to the number of rooms and area of living space needed to accommodate the displaced person. There shall be a separate, well-lighted and ventilated bathroom that provides privacy to the user and contains a sink, bathtub or shower stall, and a toilet, all in good working order and properly connected to appropriate sources of water and to a sewage drainage system. In the case of a housekeeping dwelling, there shall be a kitchen area that contains a fully usable sink, properly connected to potable hot and cold water and to a sewage drainage system, and adequate space and utility service connections for a stove and refrigerator. (5) Contains unobstructed egress to safe, open space at ground level. If the replacement dwelling unit is on the second story or above, with access directly from or through a common corridor, the common corridor must have at least two means of egress. (6) For a displaced person who is handicapped, be free of any barriers which would preclude reasonable ingress, egress, or use of the dwelling by such displaced person.

Decibel (Section 201.1) means the unit measure of sound level, abbreviated as dB.

Decisional body (Section 124.78) means any Agency employee who is or may reasonably be expected to be involved in the decisional process of the proceeding including the Administrator, Judicial Officer, Presiding Officer, the Regional Administrator (if he or she does not designate himself or herself as a member of the Agency trial staff), and any of their staff participating in the decisional process. In the case of a nonadversary panel hearing, the decisional body shall also include the panel members, whether or not permanently employed by the Agency.

Decisional body (Section 57.809) means any Agency employee who is or may reasonably be expected to be involved in the decisional process of the proceeding including the Administrator, Judicial Officer, Presiding Officer, the Regional Administrator (if he does not designate himself as a member of the Agency trial staff), and any of their staff participating in the decisional process. In the case of a non-adversary panel hearing, the decisional body shall also include the panel members whether or not permanently employed by the Agency.

Deck drainage (Section 435.11 [3/4/93])* shall refer to any waste resulting from deck washings, spillage, rainwater, and runoff from gutters and drains including drip pans and work areas within facilities subject to this subpart.

Declared value of imported components (Section 600.502-81) shall be the value at which components are declared by the importer to the U.S. Customs Service at the date of entry into the customs territory of the United States, or with respect to imports into Canada, the declared value of such components as if they were declared as imports into the United States at the date of entry into Canada. This definition applies beginning with the 1979 model year.

Decontamination (Section 259.10 [3/24/89]) means the process of reducing or eliminating the presence of harmful substances, such as infectious agents, so as to reduce the likelihood of disease transmission from those substances.

Decontamination area (Section 763.121) means an enclosed area adjacent and connected to the regulated area and consisting of an equipment room, shower area, and clean room, which is used for the decontamination of workers, materials, and equipment contaminated with asbestos.

Decontamination/detoxification (Section 165.1) means processes which will convert pesticides into nontoxic compounds.

Deepwater port (Section 110.1) means an offshore facility as defined in section (3)(10) of the Deepwater Port Act of 1974 (33 U.S.C. 1502(10)).

Defeat device (Section 86.094-2 [3/24/93])* means an auxilary emission control device (AECD) that reduces the effectiveness of the emission control system under conditions which may reasonably be expected to be encountered in normal vehicle operation and use, unless: (1) Such conditions are substantially included in the Federal emission test procedure; (2) The need for the AECD is justified in terms of protecting the vehicle against damage or accident; or (3) The AECD does not go beyond the requirements of engine starting.

Defeat Device (Section 86.082-2) means an AECD that reduces the effectiveness of the emission control system under conditions which may reasonably be expected to be encountered in normal urban vehicle operation and use, unless (1) such conditions are substantially included in the Federal emission test procedure, (2) the need for the AECD is justified in terms of protecting the vehicle against damage or accident, or (3) the AECD does not go beyond the requirements of engine starting.

Defoliants (Section 162.3) includes all substances or mixtures of substances intended for causing leaves or foliage to drop from plants. Defoliants include, but are not limited to, harvest-aid agents intended for defoliating plants (such as cotton) to facilitate harvesting.

Degassing (Section 467.02) is the removal of dissolved hydrogen from the molten aluminum prior to casting. Chemicals are added and gases are bubbled through the molten aluminum. Sometimes a wet scrubber is used to remove excess chlorine gas.

Degradation product (Section 162.3) means a substance resulting from the transformation of a pesticide by physiochemical, or biochemical means.

Degradation products (Section 165.1) means those chemicals resulting from partial decomposition or chemical breakdown of pesticides.

Degreasing (Section 471.02) is the removal of oils and greases from the surface of the metal workpiece. This process can be accomplished with detergents as in alkaline cleaning or by the use of solvents.

Dehydrated onions and garlic (Section 407.71) shall mean the processing of dehydrated onions and garlic into the following product styles: Air, vacuum, and freeze dried, all varieties, diced, strips, and other piece sizes ranging from large sliced to powder but not including green onions, chives, or leeks.

Dehydrated vegetables (Section 407.71) shall mean the processing of dehydrated vegetables in the following product styles: Air, vacuum and freeze dried, blanched and unblanched, peeled and unpeeled, beets, bell peppers, cabbage, carrots, celery, chili pepper, horseradish, turnips, parsnips, parsley, asparagus, tomatoes, green beans, corn, spinach, green onion tops, chives, leeks, whole, diced, and any other piece size ranging from sliced to powder.

Delayed compliance order (Section 65.01) shall mean an order issued by a State or by the Administrator to a stationary source which postpones the date by which the source is required to comply with any requirement contained in the applicable State implementation plan.

Delinquent debt (Section 13.2 [9/23/88]) means any debt which has not been paid by the date specified by the Government for payment or which has not been satisfied in accordance with a repayment agreement.

Demolition (Section 61.141 [11-20-90])* means the wrecking or taking out of any load-supporting structural member of a facility together with any related handling operations or the intentional burning of any facility.

Demolition (Section 763.121) means the wrecking or taking out of any load-supporting structural member and any related razing, removing, or stripping of asbestos products.

Density (Section 60.431) means the mass of a unit volume of liquid, expressed as grams per cubic centimeter, kilograms per liter, or pounds per gallon, at a specified temperature.

Density (Sections 796.1840; 796.1860) is the mass of a unit volume of a material. It is a function of temperature, hence the temperature at which it is measured should be specified. For a solid, it is the density of the impermeable portion rather than the bulk density. For solids and liquids, suitable units of measurement are g/cm^3. The density of a solution is the mass of a unit volume of the solution and suitable units of measurement are g/cm³.

Denuder (Section 61.51) means a horizontal or vertical container which is part of a mercury chlor-alkali cell and in which water and alkali metal amalgam are converted to alkali metal hydroxide, mercury, and hydrogen gas in a short-circuited, electrolytic reaction.

Deny or restrict the use of any defined area for specification (Section 231.2) is to deny or restrict the use of any area for the present or future discharge of any dredged or fill material.

Department (Section 191.02) means the Department of Energy.

Department or agency head (Section 26.102 [6-18-91]) means the head of any federal department or agency and any other officer or employee of any department or agency to whom authority has been delegated.

Departure angle (Section 86.084-2) means the smallest angle, in a plan side view of an automobile, formed by the level surface on which the automobile is standing and a line tangent to the rear tire static loaded radius arc and touching the underside of the automobile rearward of the rear tire. This definition applies beginning with the 1984 model year.

Deposit (Section 244.101) means the sum paid to the dealer by the consumer when beverages are purchased in returnable beverage containers, and which is refunded when the beverage container is returned.

Depository site (Section 192.01) means a disposal site (other than a processing site) selected under section 104(b) or 105(b) of the Act.

Depreciation (Section 14.2) is the reduction in value of an item caused by the elapse of time between the date of acquisition and the date of loss or damage.

Depuration (Section 797.1520) is the elimination of a test substance from a test organism.

Depuration (Section 797.1830) is the elimination of a test chemical from a test organism.

Depuration or clearance or elimination (Section 797.1560) is the process of losing test material from the test organisms.

Depuration phase (Section 797.1520) is the portion of a bioconcentration test after the uptake phase during which the organisms are in flowing water to which no test substance is added.

Depuration phase (Section 797.1830) is the portion of a bioconcentration test after the uptake phase during which the organisms are in flowing water to which no test chemical is added.

Depuration rate constant (K_2) (Section 797.1560) is the mathematically determined value that is used to define the depuration of test material from previously exposed test animals when placed in untreated dilution water, usually reported in units per hour.

Deputy Assistant Administrator (Section 85.2113) means the Deputy Assistant Administrator for Mobile Source, Noise and Radiation Enforcement of the Agency or his or her delegate.

Dermal corrosion (Section 798.4470) is the production of irreversible tissue damage in the skin following the application of the test substance.

Dermal irritation (Section 798.4470) is the production of reversible inflammatory changes in the skin following the application of a test substance.

Desiccants (Section 162.3) includes all substances or mixtures of substances intended for artificially accelerating the drying of plant tissue. Desiccants include, but are not limited to, harvest-aid agents whose use is intended to cause sufficient foliage injury so as to result in accelerated drying and death (maturation) of certain crop plants, such as cotton and soybeans.

Design capacity (Section 240.101) means the weight of solid waste of a specified gross calorific value that a thermal processing facility is designed to process in 24 hours of continuous operation; usually expressed in tons per day.

Design life (Section 35.2005) is the period during which a treatment works is planned and designed to be operated.

Designated facility (Section 260.10 [1-23-90])* means a hazardous waste treatment, storage, or disposal facility which (1) has received a permit (or interim status) in accordance with the requirements of parts 270 and 124 of this chapter, (2) has received a permit (or interim status) from a State authorized in accordance with part 271 of this chapter, or (3) is regulated under Sec. 261.6(c)(2) or subpart F of part 266 of this chapter, and (4) that has been designated on the manifest by the generator pursuant to Sec. 260.20. If a waste is destined to a facility in an authorized State which has not yet obtained authorization to regulate that particular waste as hazardous, then the designated facility must be a facility allowed by the receiving State to accept such waste.

Designated facility (Section 60.21) means any existing facility (see Section 60.2(aa)) which emits a designated pollutant and which would be subject to a standard of performance for that pollutant if the existing facility were an affected facility (see Section 60.2(e)).

Designated facility (Section 761.3 [12-21-89]) means the off-site disposer or commercial storer of PCB waste designated on the manifest as the facility that will receive a manifested shipment of PCB waste.

Designated management agency (DMA) (Section 130.2). An agency identified by a WQM plan and designated by the Governor to implement specific control recommendations.

Designated pollutant (Section 60.21) means any air pollutant, emissions of which are subject to a standard of performance for new stationary sources but for which air quality criteria have not been issued, and which is not included on a list published under section 108(a) or section 112(b)(1)(A) of the Act.

Designated State Agency (Section 172.21) means the State agency designated by State law or other authority to be responsible for registering pesticides to meet special local needs.

Designated uses (Section 131.3) are those uses specified in water quality standards for each water body or segment whether or not they are being attained.

Designated Volatility Attainment Area (Section 80.2 [2/16/94])* means an area not designated as being in nonattainment with the National Ambient Air Quality Standard for ozone pursuant to rulemaking under section 107(d)(4)(A)(ii) of the Clean Air Act.

Designated Volatility Nonattainment Area (Section 80.2 [2/16/94])* means any area designated as being in nonattainment with the National Ambient Air Quality Standard for ozone pursuant to rulemaking under section 107(d)(4)(A)(ii) of the Clean Air Act.

Desizing facilities (Section 410.41), for NSPS (Section 410.45), shall mean those facilities that desize more than 50 percent of their total production. These facilities may also perform other processing such as fiber preparation, scouring, mercerizing, functional finishing, bleaching, dyeing and printing.

Destination facility (Section 259.10 [3/24/89]) means the disposal facility, the incineration facility, or the facility that both treats and destroys regulated medical waste, to which a consignment of such is intended to be shipped, specified in Box 8 of the Medical Waste Tracking Form.

Destroyed regulated medical waste (Section 259.10 [7-2-90])* means regulated medical waste that is no longer generally recognizable as medical waste because the waste has been ruined, torn apart, or mutilated (it does not mean compaction) through: (1) Processes such as thermal treatment or melting, during which treatment and destruction could occur; or (2)

Processes such as shredding, grinding, tearing, or breaking, during which only destruction would take place.

Destruction (Section 82.3 [12/10/93])* means the expiration of a controlled substance to the destruction efficiency actually achieved, unless considered completely destroyed as defined in this section. Such destruction does not result in a commercially useful end product and uses one of the following controlled processes approved by the Parties to the Protocol: (1) Liquid injection incineration; (2) Reactor cracking; (3) Gaseous/fume oxidation; (4) Rotary kiln incineration; or (5) Cement kiln.

Destruction facility (Section 259.10 [3/24/89]) means a facility that destroys regulated medical waste by ruining or mutilating it, or tearing it apart.

Destruction or adverse modification (Section 257.3-2) means a direct or indirect alteration of critical habitat which appreciably diminishes the likelihood of the survival and recovery of threatened or endangered species using that habitat.

Desulfurization (Section 72.2 [7/30/93])* refers to various procedures whereby sulfur is removed from petroleum during or apart from the refining process. "Desulfurization" does not include such processes as dilution or blending of low sulfur content diesel fuel with high sulfur content diesel fuel from a diesel refinery not eligible under 40 CFR part 73, subpart G.

Detection limit (Section 136.2) means the minimum concentration of an analyte (substance) that can be measured and reported with a 99% confidence that the analyte concentration is greater than zero as determined by the procedure set forth at Appendix B of this part.

Development facility (Section 435.11 [3/4/93])* shall mean any fixed or mobile structure subject to this subpart that is engaged in the drilling of productive wells.

Developmental toxicity (Section 798.4350) is the property of a chemical that causes in utero death, structural or functional abnormalities or growth retardation during the period of development.

Device (Section 169.1) means any device or class of devicc as defined by the Act and determined by the Administrator to be subject to the provisions of the Act.

Device (Section 610.11). See **Retrofit device (Section 610.11)**.

Device (Sections 167.1; 167.3 [9/8/88]) means any device or class of devices as defined by the Act and determined by the Administrator pursuant to section 25(c) to be subject to the provisions of section 7 of the Act.

Device (Sections 710.2; 720.3) shall have the meaning contained in the Federal Food, Drug, and Cosmetic Act, 21 U.S.C. 321 et seq., and the regulations issued under such Act.

Device integrity (Section 610.11) means the durability of a device and effect of its malfunction on vehicle safety or other parts of the vehicle system.

Dewatered (Section 61.251 [12-15-89])* means to remove the water from recently produced tailings by mechanical or evaporative methods such that the water content of the tailings does not exceed 30 percent by weight.

Diaphragm (Section 85.2122(a)(3)(ii)). See **Accelerator Pump (Section 85.2122(a)(3)(ii))**.

Diaphragm Displacement (Section 85.2122(a)(1)(ii)(A)) means the distance through which the center of the diaphragm moves when activated. In the case of a non-modulated stem, diaphragm displacement corresponds to stem displacement.

Diatomaceous earth filtration (Section 141.2 [6/29/89]) means a process resulting in substantial particulate removal in which (1) a precoat cake of diatomaceous earth filter media is deposited on a support membrance (septum), and (2) while the water is filtered by passing through the cake on the septum, additional filter media known as body feed is continuously added to the feed water to maintain the permeability of the filter cake.

Dibenzo-p-dioxin or dioxin (Section 766.3) means any of a family of compounds which has as a nucleus a triple-ring structure consisting of two benzene rings connected through a pair of oxygen atoms.

Dibenzofuran (Section 766.3) means any of a family of compounds which has as a nucleus a triple-ring structure consisting of two benzene rings connected through a pair of bridges between the benzene rings. The bridges are a carbon-carbon bridge and a carbon-oxygen-carbon bridge at both substitution positions.

Dielectric material (Section 280.12 [9/23/88]) means a material that does not conduct direct electrical current. Dielectric coatings are used to electrically isolate UST systems from the surrounding soils. Dielectric bushings are used to electrically isolate portions of the UST system (e.g., tank from piping).

Dielectric Strength (Section 85.2122(a)(9)(ii)(C)) means the ability of the material of the coil to resist electrical breakdown.

Dielectric Strength (Section 85.2122(a)(7)(ii)(B)) means the ability of the material of the cap and/or rotor to resist the flow of electric current.

Dielectric Strength (Section 85.2122(a)(8)(ii)(E)) means the ability of the spark plug's ceramic insulator material to resist electrical breakdown.

Diesel (Section 86.090-2 [4/11/89]) means type of engine with operating characteristics significantly similar to the theoretical Diesel combustion cycle. The non-use of a throttle during normal operation is indicative of a diesel engine. This definition applies beginning with the 1990 model year.

Diesel fuel (Section 80.2 [2/16/94])* means any fuel sold in any State and suitable for use in diesel motor vehicles and diesel motor vehicle engines, and which is commonly or commercially known or sold as diesel fuel.

Diesel oil (Section 435.11 [3/4/93])* shall refer to the grade of distillate fuel oil, as specified in the American Society for Testing and Materials Standard Specification D975-81, that is typically used as the continuous phase in conventional oil-based drilling fluids. This incorporation by reference was approved by the Director of the Federal Register in accordance with 5 U.S.C. 552(a) and 1 CFR part 51. Copies may be obtained from the American Society for Testing and Materials, 1916 Race Street, Philadelphia, PA 19103. Copies may be inspected at the Office of the Federal Register, 800 North Capitol Street, NW., suite 700, Washington, DC.

Dietary LC_{50} (Section 152.161 [5/4/88]) means a statistically derived estimate of the concentration of a test substance in the diet that would cause 50 percent mortality to the test population under specified conditions.

Digester system (Section 60.281) means each continuous digester or each batch digester used for the cooking of wood in white liquor, and associated flash tank(s), below tank(s), chip steamer(s), and condenser(s).

Dike (Section 260.10) means an embankment or ridge of either natural or man-made materials used to prevent the movement of liquids, sludges, solids, or other materials.

Diluent (Section 165.1) means the material added to a pesticide by the user or manufacturer to reduce the concentration of active ingredient in the mixture.

Dilution water (Section 797.1520) is the water to which the test substance is added and in which the organisms undergo exposure.

Dilution water (Section 797.1600) is the water used to produce the flow-through conditions of the test to which the test substance is added and to which the test species is exposed.

Dioxin (Section 766.3). See **Dibenzo-p-dioxin (Section 766.3)**.

Dioxin/furan (Section 60.51a [2-11-

91]) means total tetra- through octachlorinated dibenzo-p-dioxins and dibenzofurans.

Dip coating (Section 60.311) means a method of applying coatings in which the part is submerged in a tank filled with the coatings.

Direct application (Section 420.101) means those cold rolling operations which include once-through use of rolling solutions at all mill stands.

Direct chill casting (Section 467.02) is the pouring of molten aluminum into a water-cooled mold. Contact cooling water is sprayed onto the aluminum as it is dropped into the mold, and the aluminum ingot falls into a water bath at the end of the casting process.

Direct chill casting (Section 471.02) is the pouring of molten nonferrous metal into a water-cooled mold. Contact cooling water is sprayed onto the metal as it is dropped into the mold, and the metal ingot falls into a water bath at the end of the casting process.

Direct discharge (Section 122.2) means the discharge of a pollutant.

Direct filtration (Section 141.2 [6/29/89]) means a series of processes including coagulation and filtration but excluding sedimentation resulting in substantial particulate removal.

Direct photolysis (Section 796.3700) is defined as the direct absorption of light by a chemical followed by a reaction which transforms the parent chemical into one or more products.

Direct public utility ownership (Section 72.2 [7/30/93])* means direct ownership of equipment and facilities by one or more corporations, the principal business of which is sale of electricity to the public at retail. Percentage ownership of such equipment and facilities shall be measured on the basis of book value.

Direct Sale Subaccount (Section 73.3 [12-17-91]) means an account in the Special Allowance Reserve, as defined in section 416(b) of the Clean Air Act. The Direct Sale Subaccount will contain Phase II allowances to be sold in the amount of 25,000 per year, beginnning in calendar year 1993 and of 50,000 per year beginning in the calendar year 2000.

Direct shell evacuation system (Section 60.271) means any system that maintains a negative pressure within the EAF above the slag or metal and ducts these emissions to the control device.

Direct Training (Section 5.2) means all technical and managerial training conducted directly by EPA for personnel of State and local governmental agencies, other Federal agencies, private industries, universities, and other non-EPA agencies and organizations.

Direct-shell evacuation control system (DEC system) (Section 60.271a) means a system that maintains a negative pressure within the electric arc furnace above the slag or metal and ducts emissions to the control device.

Director (Section 122.2) means the Regional Administrator or the State Director, as the context requires, or an authorized representative. When there is no approved State program, and there is an EPA administered program, Director means the Regional Administrator. When there is an approved State program, Director normally means the State Director. In some circumstances, however, EPA retains the authority to take certain actions even when there is an approved State program. (For example, when EPA has issued an NPDES permit prior to the approval of a State program, EPA may retain jurisdiction over that permit after program approval; see Section 123.1.) In such cases, the term Director means the Regional Administrator and not the State Director.

Director (Section 124.2 [9/26/88])* means the Regional Administrator, the State director or the Tribal director as the context requires, or an authorized representative. When there is no approved State or Tribal program, and there is an EPA administered program, Director means the Regional Administrator. When there is an approved State or Tribal program, Director normally means the State or Tribal director. In some circumstances, however, EPA retains the authority to take certain actions even when there is an approved State or Tribal program. (For example, when EPA has issued an NPDES permit prior to the approval of a State program, EPA may retain jurisdiction over that permit after program approval; see Section 123.1.) In such cases, the term Director means the Regional Administrator and not the State or Tribal director.

Director (Section 124.41) means the Regional Administrator.

Director (Section 136.2) means the Director of the State Agency authorized to carry out an approved National Pollutant Discharge Elimination System Program under section 402 of the Act.

Director (Section 144.3 [9/26/88])* means the Regional Administrator, the State director or the Tribal director as the context requires, or an authorized representative. When there is no approved State or Tribal program, and there is an EPA administered program, Director means the Regional Administrator. When there is an approved State or Tribal program, Director normally means the State or Tribal director. In some circumstances, however, EPA retains the authority to take certain actions even when there is an approved State or Tribal program. In such cases, the term Director means the Regional Administrator and not the State or Tribal director.

Director (Section 146.3 [9/26/88])* means the Regional Administrator, the State director or the Tribal director as the context requires, or an authorized representative. When there is no approved State or Tribal program, and there is an EPA administered program, Director means the Regional Administrator. When there is an approved State or Tribal program, Director normally means the State or Tribal director. In some circumstances, however, EPA retains the authority to take certain actions even when there is an approved State or Tribal program. (For example, when EPA has issued an NPDES permit prior to the approval of a State program, EPA may retain jurisdiction over that permit after program approval; see Section 123.69.) In such cases, the term Director means the Regional Administrator and not the State or Tribal director.

Director (Section 1517.2) means the Chairman of the Council on Environmental Quality acting as the head of the Office of Environmental Quality pursuant to the Environmental Quality Improvement Act of 1970, Pub. L. 91-224, 42 U.S.C. 4371-4374.

Director (Section 163.2) means the Director of the Pesticides Regulation Division, Environmental Protection Agency, Washington, D.C.

Director (Section 233.2 [6/6/88]). See **State Director (Section 233.2)**.

Director (Section 233.3) means the chief administrative officer of any state or interstate agency operating an approved program, or the delegated representative of the State Director. If responsibility is divided among two or more State or interstate agencies, State Director means the chief administrative officer of the State or interstate agency authorized to perform the particular procedure or function to which reference is made.

Director (Section 270.2) means the Regional Administrator or the State Director, as the context requires, or an authorized representative. When there is no approved State program, and there is an EPA administered program, Director means the Regional Administrator. When there is an approved State program, Director normally means the State Director. In some circumstances, however, EPA retains the authority to take certain actions even when there is an approved State program. In such cases, the term Director means the Regional Administrator and not the State Director.

Director (Section 32.102) means the Director, Grants Administration Division.

Director (Section 403.3) means the chief administrative officer of a State or Interstate water pollution control agency with an NPDES permit program approved pursuant to section 402(b) of the Act and an approved State pretreatment program.

Director (Section 501.2 [5/2/89]). See **State Program Director (Section 501.2)**.

Director (Section 720.3 [6-24-92])* means the Director of the EPA Office of Pollution Prevention and Toxics.

Director (Section 720.3) means the Director of the EPA Office of Toxic Substances.

Director (Section 723.250) means the Director of the EPA Office of Toxic Substances.

Director (Section 8.2) means the Director, Office of Federal Contract Compliance, U.S. Department of Labor, or any person to whom he delegates authority under the regulations of the Secretary of Labor.

Director (Section 8.33) means the Director of the Office of Civil Rights and Urban Affairs.

Director (Section 85.2113) means the Director of the Manufacturer's Operations Division of the Office of Enforcement of the Agency or his or her delegate.

Director of an approved State (Section 258.2 [10-9-91]) means the chief administrative officer of a State agency responsible for implementing the State municipal solid waste permit program or other system of prior approval that is deemed to be adequate by EPA under regulations published pursuant to sections 2002 and 4005 of RCRA.

Director of the Implementing Agency (Section 280.92 [10/26/88]) means the EPA Regional Administrator, or, in the case of a state with a program approved under section 9004, the Director of the designated state or local agency responsible for carrying out an approved UST program.

Director of the Office of Pollution Prevention and Toxics (Section 721.3 [6-24-92]) means the Director of the EPA Office of Pollution Prevention and Toxics or any EPA employee delegated by the Office Director to carry out the Office Director's functions under this part.

Director of the Office of Toxic Substances (Section 723.175) means the Director of the EPA Office of Toxic Substances or any EPA employee designated by the Office Director to carry out the Office Director's functions under this section.

Director of the Office of Toxic Substances (Section 721.3 [7-27-89]) means the Director of the EPA Office of Pollution Prevention and Toxics or any EPA employee delegated by the Office Director to carry out the Office Director's functions under this part.

Director of the Office of Toxic Substances (Section 723.50) means the Director of the EPA Office of Toxic Substances or any EPA employee designated by the Director to carry out the Director's functions under this section.

Disbursement (Section 35.3105 [3-19-90]) The transfer of cash from an SRF to an assistance recipient.

Disc break pad for light-and medium-weight vehicles (Section 763.163 [7-12-89]) means an asbestos-containing product intended for use as a friction material in disc brake systems for vehicles rated at less than 26,001 pounds gross vehicle weight rating (GVWR).

Disc break pads for heavy-weight vehicles (Section 763.163 [7-12-89]) means an asbestos-containing product intended for use as a friction material in disc brake systems for vehicles rated at 26,001 pounds gross vehicle weight rating (GVWR) or more.

Discarded material (Section 261.2) is any material which is: (i) Abandoned, as explained in paragraph (b) of this section; or (ii) Recycled, as explained in paragraph (c) of this section; or (iii) Considered inherently waste-like, as explained in paragraph (d) of this section.

Discharge (Section 110.1) when used in relation to section 311 of the Act, includes, but is not limited to, any spilling, leaking, pumping, pouring, emitting, emptying, or dumping, but excludes (A) discharges in compliance with a permit under section 402 of the Act, (B) discharges resulting from circumstances identified and reviewed and made a part of the public record with respect to a permit issued or modified under section 402 of the Act, and subject to a condition in such permit, and (C) continuous or anticipated intermittent discharges from a point source, identified in a permit or permit application under section 402 of the Act, that are caused by events occurring within the scope of relevant operating or treatment systems.

Discharge (Section 112.2 [8/25/93])* includes but is not limited to, any spilling, leaking, pumping, pouring, emitting, emptying or dumping. For purposes of this part, the term "discharge" shall not include any discharge of oil which is authorized by a permit issued pursuant to section 13 of the River and Harbor Act of 1899 (30 Stat. 1121, 33 U.S.C. 407), or sections 402 or 405 of the FWPCA Amendments of 1972 (86 Stat. 816 et seq., 33 U.S.C. 1251 et seq.).

Discharge (Section 116.3) includes, but is not limited to, any spilling, leaking, pumping, pouring, emitting, emptying or dumping, but excludes (A) discharges in compliance with a permit under section 402 of this Act, (B) discharges resulting from circumstances identified and reviewed and made a part of the public record with respect to a permit issued or modified under section 402 of this Act, and subject to a condition in such permit, and (C) continuous or anticipated intermittent discharges from a point source, identified in a permit or permit application under section 402 of this Act, which are caused by events occurring within the scope of relevant operating or treatment systems.

Discharge (Section 122.2) when used without qualification means the discharge of a pollutant.

Discharge (Section 240.101) means water-borne pollutants released to a receiving stream directly or indirectly or to a sewerage system.

Discharge (Section 300.5 [3-8-90])* as defined by section 311(a)(2) of the CWA, includes, but is not limited to, any spilling, leaking, pumping, pouring, emitting, emptying, or dumping of oil, but excludes discharges in compliance with a permit under section 402 of the CWA, discharges resulting from circumstances identified and reviewed and made a part of the public record with respect to a permit issued or modified under section 402 of the CWA, and subject to a condition in such permit, or continuous or anticipated intermittent discharges from a point source, identified in a permit or permit application under section 402 of the CWA, that are caused by events occurring within the scope of relevant operating or treatment systems. For purposes of the NCP, discharge also means threat of a discharge.

Discharge (Section 403.3). See **Indirect Discharge (Section 403.3)**.

Discharge (Sections 109.2; 113.3) includes, but is not limited to, any spilling, leaking, pumping, pouring, emitting, emptying, or dumping.

Discharge allowance (Section 461.2) means the amount of pollutant (mg per kg of production unit) that a plant will be permitted to discharge. For this category the allowances are specific to battery manufacturing operations.

Discharge in connection with activities under the Outer Continental Shelf Lands Act or the Deepwater Port Act of 1974, or which may affect natural resources belonging to, appertaining to, or under the exclusive management authority of the United States (including resources under the Fishery Conservation and Management Act of 1976) (Section 116.3) means: (1) A discharge into any waters beyond the contiguous zone from any vessel or onshore or offshore facility, which vessel or facility is subject to or is engaged in activities under the Outer Continental Shelf Lands Act or the Deepwater Port Act of 1974, and (2) any discharge into any waters beyond the contiguous zone which contain, cover, or support any natural resource belonging to, appertaining to, or under the exclusive management authority of the United States (including resources under the Fishery Conservation and Management Act of 1976).

Discharge Monitoring Report (DMR) (Section 122.2) means the EPA uniform national form, including any subsequent additions, revisions, or modifications for the reporting of self-monitoring results by permittees. DMRs must be used by approved States as well as by EPA. EPA will supply DMRs to any approved State upon request. The EPA national forms may be modified to substitute the State Agency name, address, logo, and other similar information, as appropriate, in place of EPA's.

Discharge of a pollutant (Section 122.2) means: (a) Any addition of any pollutant or combination of pollutants to waters of the United States from any point source, or (b) Any addition of any pollutant or combination of pollutants to the waters of the contiguous zone or the ocean from any point source other than a vessel or other floating craft which is being used as a means of transportation. This definition includes additions of pollutants into waters of the United States from: surface runoff which is collected or channelled by man; discharges through pipes, sewers, or other conveyances owned by a State, municipality, or other person which do not lead to a treatment works; and discharges through pipes, sewers, or other conveyances, leading into privately owned treatment works. This term does not include an addition of pollutants by any indirect discharger.

Discharge of dredged material (Section 233.3) means any addition from any point source of dredged material into waters of the United States. The term includes the addition of dredged material into waters of the United States and the runoff or overflow from a contained land or water dredge material disposal area. Discharges of pollutants into waters of the United States resulting from the subsequent onshore processing of dredged material are not included within this term and are subject to the NPDES program even though the extraction and deposit of such material may also require a permit from the Corps of Engineers or the State section 404 program.

Discharge of dredged material (Section 257.3-3) is defined in the Clean Water Act, as amended, 33 U.S.C. 1251 et seq., and implementing regulations, specifically 33 CFR Part 323 (42 FR 37122, July 19, 1977).

Discharge of dredged material (Section 232.2 [2/11/93])* (1) Except as provided below in paragraph (2), the term discharge of dredged material means any addition of dredged material into, including any redeposit of dredged material within, the waters of the United States. The term includes, but is not limited to, the following: (i) The addition of dredged material to a specified discharge site located in waters of the Untied States; (ii) The runoff or overflow, associated with a dredging operation, from a contained land or water disposal area; and (iii) Any addition, including any redeposit, of dredged material, including excavated material, into waters of the United States which is incidental to any activity, including mechanized landclearing, ditching, channelization, or other excavation. (2) The term discharge of dredged material does not include the following: (i) Discharges of pollutants into waters of the United States resulting from the onshore subsequent processing of dredged material that is extracted for any commercial use (other than fill). These discharges are subject to section 402 of the Clean Water Act even though the extraction and deposit of such material may require a permit from the Corps or applicable state. (ii) Activities that involve only the cutting or removing of vegetation above the ground (e.g., mowing, rotary cutting, and chainsawing) where the activity neither substantially disturbs the root system nor involves mechanized pushing, dragging, or other similar activities that redeposit excavated soil material. (3) Section 404 authorization is not required for the following: (i) Any incidental addition, including redeposit, of dredged material associated with any activity that does not have or would not have the effect of destroying or degrading an area of waters of the U.S. as defined in paragraphs (4) and (5) of this definition; however, this exception does not apply to any person preparing to undertake mechanized landclearing, ditching, channelization and other excavation activity in a water of the United States, which would result in a redeposit of dredged material, unless the person demonstrates to the satisfaction of the Corps, or EPA as appropriate, prior to commencing the activity involving the discharge, that the activity would not have the effect of destroying or degrading any area of waters of the United States, as defined in paragraphs (4) and (5) of this definition. The person proposing to undertake mechanized landclearing, ditching, channelization or other excavation activity bears the burden of demonstrating that such activity would not destroy or degrade any area of waters of the United States. (ii) Incidental movement of dredged material occurring during normal dredging operations, defined as dredging for navigation in navigable waters of the United States, as that term is defined in 33 CFR part 329, with proper authorization from the Congress or the Corps pursuant to 33 CFR part 322; however, this exception is not applicable to dredging activities in wetlands, as that term is defined at Sec. 232.2(r) of this Chapter. (iii) Those discharges of dredged material associated with ditching, channelization or other excavation activities in waters of the United States, including wetlands, for which Section 404 authorization was not previously required, as determined by the Corps district in which the activity occurs or would occur, provided that prior to August 25, 1993, the excavation activity commenced or was under contract to commence work and that the activity will be completed no later that August 25, 1994. This provision does not apply to discharges associated with mechanized landclearing. For those excavation activities that occur on an ongoing basis (either continuously or periodically), e.g., mining operations, the Corps retains the authority to grant, on a case-by-case basis, an extension of this 12-month grandfather provision provided that the discharger has submitted to the Corps within the 12-month period an individual permit application seeking Section 404 authorization for such excavation activity. In no event can the grandfather period under this paragraph extend beyond August 25, 1996. (iv) Certain discharges, such as those associated with normal farming, silviculture, and ranching activities, are not prohibited by or otherwise subject to regulation under Section 404. See 40 CFR 232.3 for discharges that do not require permits. (4) For purposes of this section, an activity associated with a discharge of dredged material destroys an area of waters of the United States if it alters the area in such a way that it would no longer be a water of the United States. Note: Unauthorized discharges into waters of the United States do not eliminate Clean Water Act jurisdiction, even where such unauthorized discharges have the effect of destroying waters of the United States. (5) For purposes of this section, an activity associated with a discharge of dredged material degrades an area of waters of the United States if it has more than a de minimis (i.e., inconsequential) effect on the area by causing an identifiable individual or cumulative adverse effect on any aquatic function.

Discharge of fill material (Section 233.3) means the addition from any point source of fill material into waters of the United States. The term includes the following activities in waters of the United States: Placement of fill that is necessary for the construction of any structure; the building of any structure or impoundment requiring rock, sand, dirt, or other materials for its construction; site-development fill for recreational, industrial, commercial, residential, and other uses; causeways or road fills; dams and dikes; artificial islands; property protection and/or reclamation devices such as riprap, groins, seawalls, breakwaters, and revetments; beach nourishment; levees; fill for structures such as sewage treatment facilities, intake and outfall pipes associated with power plants and subaqueous utility lines; and artificial reefs.

Discharge of fill material (Section 232.2 [2/11/93])* (1) The term discharge of fill material means the addition of fill material into waters of the United States. The term generally includes, without limitation, the following activities: Placement of fill that is necessary for the construction of any structure in a water of the United States; the building of any structure or impoundment requiring rock, sand, dirt, or other material for its construction; site-development fills for recreational, industrial, commercial, residential, and other uses; causeways or road fills; dams and dikes; artificial islands; property protection and/or reclamation devices such as riprap, groins, seawalls, breakwaters, and revetments; beach nourishment; levees; fill for structures such as sewage treatment facilities, intake and outfall pipes associated with power plants and subaqueous utility lines; and artificial reefs. (2) In addition, placement of pilings in waters of the United States constitutes a discharge of fill material and requires a Section 404 permit when such placement has or would have the effect of a discharge of fill material. Examples of such activities that have the effect of a discharge of fill material include, but are not limited to, the following: Projects where the pilings are so closely spaced that sedimentation rates would be increased; projects in which the pilings themselves effectively would replace the bottom of a waterbody; projects involving the placement of pilings that would reduce the reach or impair the flow or circulation of waters of the United States; and projects involving the placement of pilings which would result in the adverse alteration or elimination of aquatic functions. (i) Placement of pilings in waters of the United States that does not have or would not have the effect of a discharge of fill material shall not require a Section 404 permit. Placement of pilings for linear projects, such as bridges, elevated walkways, and powerline structures, generally does not have the effect of a discharge of fill material. Furthermore, placement of pilings in waters of the United States for piers, wharves, and an individual house on stilts generally does not have the effect of a discharge of fill material. All pilings, however, placed in the navigable waters of the United States, as that term is defined in 33 CFR part 329, require authorization under section 10 of the Rivers and Harbors Act of 1899 (see 33 CFR part 322). (ii) [Reserved]

Discharge of pollutant(s) (Section 401.11) means: (1) The addition of any pollutant to navigable waters from any point source and (2) any addition of any pollutant to the waters of the contiguous zone or the ocean from any point source, other than from a vessel or other floating craft. The term discharge includes either the discharge of a single pollutant or the discharge of multiple pollutants.

Discharge or hazardous waste discharge (Section 260.10) means the accidental or intentional spilling, leaking, pumping, pouring, emitting, emptying, or dumping of hazardous waste into or on any land or water.

Discharge point (Section 230.3) means the point within the disposal site at which the dredged or fill material is released.

Discharges (Section 140.1) includes, but is not limited to, any spilling, leaking, pumping, pouring, emitting, emptying, or dumping.

Disease vector (Section 257.3-6) means rodents, flies, and mosquitoes capable of transmitting disease to humans.

Disinfectant (Section 141.2) means any oxidant, including but not limited to chlorine, chlorine dioxide, chloramines, and ozone added to water in any part of the treatment or distribution process, that is intended to kill or inactivate pathogenic microorganisms.

Disinfectant contact time (T in CT calculations) (Section 141.2 [6/29/89]) means the time in minutes that it takes for water to move from the point of disinfectant application or the previous point of disinfectant residual measurement to a point before or at the point where residual disinfectant concentration (C) is measured. Where only one C is measured, T is the time in minutes that it takes for water to move from the point of disinfectant application to a point before or at where residual disinfectant concentration (C) is measured. Where more than one C is measured, T is (a) for the first measurement of C, the time in minutes that it takes for water to move from the first or only point of disinfectant application to a point before or at the point where the first C is measured and (b) for subsequent measurements of C, the time in minutes that it takes for water to move from the previous C measurement point to the C measurement point for which the particular T is being calculated. Disinfectant contact time in pipelines must be calculated based on plug flow by dividing the internal volume of the pipe by the maximum hourly flow rate through that pipe. Disinfectant contact time within mixing basins and storage reservoirs must be determined by tracer studies or an equivalent demonstration.

Disinfection (Section 141.2 [6/29/89]) means a process which inactivates pathogenic organisms in water by chemical oxidants or equivalent agents.

Dispensed fuel temperature (Section 86.098-2 [4/6/94])* means the temperature (deg.F or deg.C may be used) of the fuel being dispensed into the tank of the test vehicle during a refueling test.

Dispenser (Section 211.203) means the permanent (intended to be refilled) or disposable (discarded when empty) container designed to hold more than one complete set of hearing protector(s) for the express purpose of display to promote sale or display to promote use or both.

Dispersants (Section 300.5 [3-8-90]) means those chemical agents that emulsify, disperse, or solubilize oil into the water column or promote the surface spreading of oil slicks to facilitate dispersal of the oil into the water column.

Dispersants (Section 300.82), for the purposes of this subpart, are those chemical agents that emulsify, disperse, or solubilize oil into the water column or promote the surface spreading of oil slicks to facilitate dispersal of the oil into the water column.

Dispersion resin (Section 61.61) means a resin manufactured in such a way as to form fluid dispersions when dispersed in a plasticizer or plasticizer/diluent mixtures.

Dispersion technique (Section 51.100) (1) means any technique which attempts to affect the concentration of a pollutant in the ambient air by: (i) Using that portion of a stack which exceeds good engineering practice stack height; (ii) Varying the rate of emission of a pollutant according to atmospheric conditions or ambient concentrations of that pollutant; or (iii) Increasing final exhaust gas plume rise by manipulating source process parameters, exhaust gas parameters, stack parameters, or combining exhaust gases from several existing stacks into one stack; or other selective handling of exhaust gas streams so as to increase the exhaust gas plume rise. (2) The preceding sentence does not include: (i) The reheating of a gas stream, following use of a pollution control system, for the purpose of returning the gas to the temperature at which it was originally discharged from the facility generating the gas stream; (ii) The merging of exhaust gas streams where: (A) The source owner or operator demonstrates that the facility was originally designed and constructed with such merged gas streams; (B) After July 8, 1985, such merging is part of a change in operation at the facility that includes the installation of pollution controls and is accompanied by a net reduction in the allowable emissions of a pollutant. This exclusion from the definition of dispersion techniques shall apply only to the emission limitation for the pollutant affected by such change in operation; or (C) Before July 8, 1985, such merging was part of a change in operation at the facility that included the installation of emissions control equipment or was carried out for sound economic or engineering reasons. Where there was an increase in the emission limitation or, in the event that no emission limitation was in existence prior to the merging, an increase in the quantity of pollutants actually emitted prior to the merging, the reviewing agency shall presume that merging was significantly motivated by an intent to gain emissions credit for greater dispersion. Absent a demonstration by the source owner or operator that merging was not significantly motivated by such intent, the reviewing agency shall deny credit for the effects of such merging in calculating the allowable emissions for the source; (iii) Smoke management in agricultural or silvicultural prescribed burning programs; (iv) Episodic restrictions on residential woodburning and open burning; or (v) Techniques under Section 51.100(hh)(1)(iii) which increase final exhaust gas plume rise where the resulting allowable emissions of sulfur dioxide from the facility do not exceed 5,000 tons per year.

Displaced person (Section 4.2) means any person (defined at Section 4.2(m)) who moves from the real property or moves his or her personal property from the real property: (i) As a direct result of the Agency's acquisition of such real property in whole or in part for a project. This includes any person who moved from the real property as a result of the initiation of negotiations as described at Section 4.2(k); or (ii) As a result of a written order from the acquiring Agency to vacate such real property for the project; or (iii) As a result of the Agency's acquisition of, or written order to vacate, for a project, other real property on which the person conducts a business or farm operation. Eligibility as a displaced person under this subparagraph applies only for purposes of obtaining relocation assistance advisory services under Section 4.205 and moving expenses under Section 4.301, Section 4.302, or Section 4.303.

Displacement (Section 264.18), as used in paragraph (a)(1) of this section, means the relative movement of any two sides of a fault measured in any direction.

Displacement and Displacement Class (Section 86.402-78). See Section 86.419.

Disposable Device (Section 211.203) means a hearing protective device that is intended to be discarded after one period of use.

Disposable pay (Section 13.2 [9/23/88]) means that part of current basic pay, special pay, incentive pay, retired pay, retainer pay, or in the case of an employee not entitled to basic pay, other authorized pay remaining after the deduction of any amount described in 5 CFR 581.105(b) through (f). These deductions include, but are not limited to: Social security withholdings; Federal, State and local tax withholdings; health insurance premiums; retirement contributions; and life insurance premiums.

Disposal (Section 191.02) means permanent isolation of spent nuclear fuel or radioactive waste from the accessible environment with no intent of recovery, whether or not such isolation permits the recovery of such fuel or waste. For example, disposal of waste in a mined geologic repository occurs when all of the shafts to the repository are back-filled and sealed.

Disposal (Section 245.101) means the collection, storage, treatment, utilization, processing, or final disposal of solid waste.

Disposal (Section 270.2) means the discharge, deposit, injection, dumping, spilling, leaking, or placing of any hazardous waste into or on any land or water so that such hazardous waste or any constituent thereof may enter the environment or be emitted into the air or discharged into any waters, including ground water.

Disposal (Section 373.4 [4-16-90]) means the discharge, deposit, injection, dumping, spilling, leaking or placing of any hazardous substance into or on any land or water so that such hazardous substance or any constituent thereof may enter the environment or be emitted into the air or discharged into any waters, including groundwater.

Disposal (Section 761.3) means intentionally or accidentally to discard, throw away, or otherwise complete or terminate the useful life of PCBs and PCB Items. Disposal includes spills, leaks, and other uncontrolled discharges of PCBs as well as actions related to containing, transporting, destroying, degrading, decontaminating, or confining PCBs and PCB Items.

Disposal (Sections 257.2; 260.10) means the discharge, deposit, injection, dumping, spilling, leaking, or placing of any solid waste or hazardous waste into or on any land or water so that such solid waste or hazardous waste or any constituent thereof may enter the environment or be emitted into the air or discharged into any waters, including ground waters.

Disposal area (Section 192.31) means the region within the perimeter of an impoundment or pile containing uranium byproduct materials to which the post-closure requirements of Section 192.32(b)(1) of this subpart apply.

Disposal facility (Section 260.10 [8/18/92])* means a facility or part of a facility at which hazardous waste is intentionally placed into or on any land or water, and at which waste will remain after closure. The term disposal facility does not include a corrective action management unit into which remediation wastes are placed.

Disposal facility (Section 270.2 [2/16/93])* means a facility or part of a facility at which hazardous waste is intentionally placed into or on the land or water, and at which hazardous waste will remain after closure. The term disposal facility does not include a corrective action management unit into which remediation wastes are placed.

Disposal site (Section 192.01) means the region within the smallest perimeter of residual radioactive material (excluding cover materials) following completion of control activities.

Disposal site (Section 192.31) means a site selected pursuant to section 83 of the Act.

Disposal site (Section 228.2) means an interim or finally approved and precise geographical area within which ocean dumping of wastes is permitted under conditions specified in permits issued under sections 102 and 103 of the Act. Such sites are identified by boundaries established by (1) coordinates of latitude and longitude for each corner, or by (2) coordinates of latitude and longitude for the center point and a radius in nautical miles from that point. Boundary coordinates shall be identified as precisely as is warranted by the accuracy with which the site can be located with existing navigational aids or by the implantation of transponders, buoys or other means of marking the site.

Disposal site (Section 230.3) means that portion of the waters of the United States where specific disposal activities are permitted and consist of a bottom surface area and any overlying volume of water. In the case of wetlands on which surface water is not present, the disposal site consists of the wetland surface area.

Disposal site (Section 233.3) means that portion of the waters of the United States enclosed within fixed boundaries consisting of a bottom surface area and any overlaying volume of water. In the case of wetlands on which water is not present, the disposal site consists of the wetland surface area. Fixed boundaries may consist of fixed geographic point(s) and associated dimensions, or of a discharge point and specific associated dimensions.

Disposal site designation study (Section 228.2) means the collection, analysis and interpretation of all available pertinent data and information on a proposed disposal site prior to use, including but not limited to, that from baseline surveys, special purpose surveys of other Federal agencies, public data archives, and social and economic studies and records of areas which would be affected by use of the proposed site.

Disposal site evaluation study (Section 228.2) means the collection, analysis, and interpretation of all pertinent information available concerning an existing disposal site, including but not limited to, data and information from trend assessment surveys, monitoring surveys, special purpose surveys of other Federal agencies, public data archives, and social and economic studies and records of affected areas.

Disposal system (Section 191.12 [12/20/93])* means any combination of engineered and natural barriers that isolate spent nuclear fuel or radioactive waste after disposal.

Disposal well (Sections 146.3; 147.2902) means a well used for the disposal of waste into a subsurface stratum.

Disposer of PCB waste (Section 761.3 [12-21-89]) means the off-site disposer or commercial storer of PCB waste designated on the manifest as the facility that will receive a manifested shipment of PCB waste.

Dispute (Section 791.3) refers to a present controversy between parties subject to a test rule over the amount or method of reimbursement for the cost of developing health and environmental data on the test chemical.

Distance piece (Section 60.481) means an open or enclosed casing through which the piston rod travels, separating the compressor cylinder from the crankcase.

Distillate oil (Section 60.41b [12/16/87])* means fuel oils that contain 0.05 weight percent nitrogen or less and comply with the specifications for fuel oil numbers 1 and 2, as defined by the American Society of Testing and Materials in ASTM D396-78, Standard Specifications for Fuel Oils (incorporated by reference - see Section 60.17).

Distillate oil (Section 60.41c [9-12-90]) means fuel oil that complies with the specifications for fuel oil numbers 1 or 2, as defined by the American Society for Testing and Materials in ASTM D396-78, "Standard Specification for Fuel Oils" (incorporated by reference--see Sec. 60.17).

Distillate receiver (Section 264.1031 [6-21-90]) means a container or tank used to receive and collect liquid material (condensed) from the overhead condenser of a distillation unit and from which the condensed liquid is pumped to larger storage tanks or other process units.

Distillation operation (Section 264.1031 [6-21-90]) means an operation, either batch or continuous, separating one or more feed stream(s) into two or more exit streams, each exit stream having component concentrations different from those in the feed stream(s). The separation is achieved by the redistribution of the components between the liquid and vapor phase as they approach equilibrium within the distillation unit.

Distillation operation (Section 60.661 [6-29-90]) means an operation separating one or more feed stream(s) into two or more exit stream(s), each exit stream having component concentrations different from those in the feed stream(s). The separation is achieved by the redistribution of the components between the liquid and vapor-phase as they approach equilibrium within the distillation unit.

Distillation unit (Section 60.661 [6-29-90]) means a device or vessel in which distillation operations occur, including all associated internals (such as trays or packing) and accessories (such as reboiler, condenser, vacuum pump, steam jet, etc.), plus any associated recovery system.

Distribute in commerce (Section 762.3) has the same meaning as in 15 U.S.C. 2602.

Distribute in commerce (Section 720.3) means to sell in commerce, to introduce or deliver for introduction into commerce, or to hold after introduction into commerce.

Distribute in commerce (Section 763.163 [7-12-89]) has the same meaning as in section 3 of the Act, but the term does not include actions taken with respect to an asbestos-containing product (to sell, resale, deliver, or hold) in connection with the end use of the product by persons who are users (persons who use the product for its intended purpose after it is manufactured or processed). The term also does not include distribution by manufacturers, importers, and processors, and other persons solely for purposes of disposal of an asbestos-containing product.

Distribute in commerce (Section 205.2) means sell in, offer for sale in, or introduce or deliver for introduction into, commerce.

Distribute in commerce and distribution in commerce (Sections 710.2; 761.3) when used to describe an action taken with respect to a chemical substance or mixture or article containing a substance or mixture, mean to sell or the sale of, the substance, mixture, or article in commerce; to introduce or deliver for introduction into commerce, or the introduction or delivery for introduction into commerce of, the substance, mixture, or article; or to hold, or the holding of, the substance, mixture, or article after its introduction into commerce.

Distribute or sell, distributed or sold, or distribution or sale (Section 152.3 [5/4/88]) means the acts of distributing, selling, offering for sale, holding for sale, shipping, holding for shipment, delivering for shipment, or receiving and (having so received) delivering or offering to deliver, or releasing for shipment to any person in any State.

Distributed or sold (Section 152.3 [5/4/88]). See **Distribute or sell (Section 152.3)**.

Distribution or sale (Section 152.3 [5/4/88]). See **Distribute or sell (Section 152.3)**.

Distributor (Section 244.101) means any person who engages in the sale of beverages, in beverage containers, to a dealer, including any manufacturer who engages in such sale.

Distributor (Section 749.68 [1-3-90]) means any person who distributes in commerce water treatment chemicals for use in cooling systems.

Distributor (Section 80.2 [2/16/94])* means any person who transports or stores or causes the transportation or storage of gasoline or diesel fuel at any point between any gasoline or diesel fuel refinery or importer's facility and any retail outlet or wholesale purchaser-consumer's facility.

Distributor (Section 82.62 [12/30/93])* when used to describe a person taking action with regard to a product means: (1) The seller of a product to a consumer or another distributor; or (2) A person who sells or distributes that product in interstate commerce for export from the United States.

Distributor (Section 85.2122 (a) (11) (ii) (A)) means a device for directing the secondary current from the induction coil to the spark plugs at the proper intervals and in the proper firing order.

Distributor Firing Angle (Section 85.2122(a)(11)(ii)(B)) means the angular relationship of breaker point opening from one opening to the next in the firing sequence.

Diurnal breathing losses (Section 86.082-2) means evaporative emissions as a result of the daily range in temperature.

Diurnal breathing losses (Section 86.096-2 [11/1/93])* means diurnal emissions.

Diurnal emissions (Section 86.096-2 [11/1/93])* means evaporative emissions resulting from the daily cycling of ambient temperatures.

DMA (Section 130.2). See **Designated management agency (Section 130.2)**.

DMR (Section 122.2) means Discharge Monitoring Report.

DOC (Section 300.5) is the Department of Commerce.

DOD (Section 300.5) is the Department of Defense.

DOE (Section 300.5) is the Department of Energy.

DOI (Section 300.5) is the Department of the Interior.

Doilies (Section 250.4 [10/6/87, 6/22/88]) means paper place mats used on food service trays in hospitals and other institutions.

DOJ (Section 300.5) is the Department of Justice.

DOL (Section 300.5) is the Department of Labor.

Domestic (Section 704.3 [12/22/88])* means within the geographical boundaries of the 50 United States, including the District of Columbia, the Commonwealth of Puerto Rico, the Virgin Islands, Guam, American

Samoa, the Northern Mariana Islands, and any other territory or possession of the United States.

Domestic application (Section 162.3) means application of a pesticide directly to humans or pets, or application of a pesticide in, on or around all structures, vehicles or areas associated with the household or home life, patient care areas of health related institutions, or areas where children spend time including but not limited to: (1) Gardens, non-commercial greenhouses, yards, patios, houses, pleasure marine craft, mobile homes, campers and recreational vehicles, non-commercial campsites, home swimming pools and kennels: (2) Articles, objects, devices or surfaces handled or contacted by humans or pets in all structures, vehicles or areas listed above; (3) Patient care areas of nursing homes, mental institutions, hospitals, and convalescent homes; (4) Educational, lounging and recreational areas of preschools, nurseries and day camps.

Domestic construction material (Section 35.936.13) means an unmanufactured construction material which has been mined or produced in the United States, or a manufactured construction material which has been manufactured in the United States if the cost of its components which are mined, produced, or manufactured in the United States exceeds 50 percent of the cost of all its components.

Domestic or other non-distribution system plumbing problem (Section 141.2 [6/29/89]) means a coliform contamination problem in a public water system with more than one service connection that is limited to the specific service connection from which the coliform-positive sample was taken.

Domestic septage (Section 257.2 [3/19/93])* is either liquid or solid material removed from a septic tank, cesspool, portable toilet, Type III marine sanitation device, or similar treatment works that receives only domestic sewage. Domestic septage does not include liquid or solid material removed from a septic tank, cesspool, or similar treatment works that receives either commercial wastewater or industrial wastewater and does not include grease removed from a grease trap at a restaurant.

Domestic waste (Section 435.11 [3/4/93])* shall refer to materials discharged from sinks, showers, laundries, safety showers, eye-wash stations, hand-wash stations, fish cleaning stations, and galleys located within facilities subject to this subpart.

DOS (Section 300.5) is the Department of State.

Dose (Section 798.1175) is the amount of test substance administered. Dose is expressed as weight of test substance (g, mg) per unit weight of test animal (e.g., mg/kg).

Dose (Section 798.2250) in a dermal test, is the amount of test substance applied to the skin (applied daily in subchronic tests). Dose is expressed as weight of the substance (g, mg) per unit weight of test animal (e.g., mg/kg).

Dose (Section 798.2450) refers to an exposure level. Exposure is expressed as weight or volume of test substance per volume of air (mg/l), or as parts per million (ppm).

Dose (Section 798.2650) is the amount of test substance administered. Dose is expressed as weight of test substance (g, mg) per unit weight of test animal (e.g., mg/kg), or as weight of test substance per unit weight of food or drinking water.

Dose equivalent (Section 141.2) means the product of the absorbed dose from ionizing radiation and such factors as account for differences in biological effectiveness due to the type of radiation and its distribution in the body as specified by the International Commission on Radiological Units and Measurements (ICRU).

Dose equivalent (Section 190.02) means the product of absorbed dose and appropriate factors to account for differences in biological effectiveness due to the quality of radiation and its spatial distribution in the body. The unit of dose equivalent is the rem. (One millirem (mrem)=0.001 rem.)

Dose equivalent (Section 191.12 [12/20/93])* means the product of absorbed dose and appropriate factors to account for differences in biological effectiveness due to the quality of radiation and its spatial distribution in the body; the unit of dose equivalent is the "rem" ("sievert" in SI units).

Dose equivalent (Sections 61.91; 61.101) means the product of absorbed dose and appropriate factors to account for differences in biological effectiveness due to the quality of radiation and its distribution in the body. The unit of dose equivalent is the rem.

Dose response (Section 798.1150) is the relationship between the dose (or concentration) and the proportion of a population sample showing a defined effect.

Dose Standard (Section 61 [12-15-89]) A regulatory standard that requires a regulated facility to limit its emissions to the level necessary to ensure that no individual receives an effective dose equivalent greater than the specified level.

Dose-response (Section 798.1175) is the relationship between the dose and the proportion of a population sample showing a defined effect.

DOT (Section 300.5) is the Department of Transportation.

Double block and bleed system

(Section 264.1031 [6-21-90]) means two block valves connected in series with a bleed valve or line that can vent the line between the two block valves.

Double block and bleed system (Sections 60.481; 61.241) means two block valves connected in series with a bleed valve or line that can vent the line between the two block valves.

Double wash/rinse (Section 761.123) means a minimum requirement to cleanse solid surfaces (both impervious and nonimpervious) two times with an appropriate solvent or other material in which PCBs are at least 5 percent soluble (by weight). A volume of PCB-free fluid sufficient to cover the contaminated surface completely must be used in each wash/rinse. The wash/rinse requirement does not mean the mere spreading of solvent or other fluid over the surface, nor does the requirement mean a once-over wipe with a soaked cloth. Precautions must be taken to contain any runoff resulting from the cleansing and to dispose properly of wastes generated during the cleansing.

Draft permit (Section 144.3) means a document prepared under Section 124.6 indicating the Director's tentative decision to issue or deny, modify, revoke and reissue, terminate, or reissue a permit. A notice of intent to terminate a permit, and a notice of intent to deny a permit, as discussed in Section 124.5 are types of draft permits. A denial of a request for modification, revocation and reissuance, or termination, as discussed in Section 124.5 is not a draft permit.

Draft permit (Sections 122.2; 124.2; 124.41; 233.3; 270.2) means a document prepared under Section 124.6 indicating the Director's tentative decision to issue or deny, modify, revoke and reissue, terminate, or reissue a permit. A notice of intent to terminate a permit, and a notice of intent to deny a permit, as discussed in Section 124.5, are types of draft permits. A denial of a request for modification, revocation and reissuance, or termination, as discussed in Section 124.5, is not a draft permit. A proposed permit is not a draft permit.

Drainage water (Section 440.141 [5/24/88]) means incidental surface waters from diverse sources such as rainfall, snow melt or permafrost melt.

Drawing (Section 467.02) is the process of pulling metal through a die or succession of dies to reduce the metal's diameter or alter its shape. There are two aluminum forming subcategories based on the drawing process. In the drawing with neat oils subcategory, the drawing process uses a pure or neat oil as a lubricant. In the drawing with emulsions or soaps subcategory, the drawing process uses an emulsion or soap solution as a lubricant.

Drawing (Section 468.02) shall mean

pulling the workpiece through a die or succession of dies to reduce the diameter or alter its shape.

Drawing (Section 471.02) is the process of pulling a metal through a die or succession of dies to reduce the metal's diameter or alter its cross-sectional shape.

Dredge (Section 440.141 [5/24/88]) means a self-contained combination of an elevating excavator (e.g., bucket line dredge), the beneficiation or gold-concentrating plant, and a tailings disposal plant, all mounted on a floating barge.

Dredged material (Section 232.2 [2/11/93])* means material that is excavated or dredged from waters of the United States.

Dredged Material Permit (Section 220.2) means a permit issued by the Corps of Engineers under section 103 of the Act (see 33 CFR 209.120) and any Federal projects reviewed under section 103(e) of the Act (see 33 CFR 209.145).

Dried fruit (Section 407.61) shall mean the processing of various fruits into the following products styles: Air, vacuum, and freeze dried, pitted and unpitted, blanched and unblanched, whole, halves, slices and other similar styles of apples, apricots, figs, peaches, pears, prunes, canned extracted prune juice and pulp from rehydrated and cooked dehydrated prunes; but not including dates or raisins.

Drift (Section 162.3) means movement of a pesticide during or immediately after application or use through air to a site other than the intended site of application or use.

Drill cuttings (Section 435.11 [3/4/93])* shall refer to the particles generated by drilling into subsurface geologic formations and carried to the surface with the drilling fluid.

Drilling and production facility (Section 60.111) means all drilling and servicing equipment, wells, flow lines, separators, equipment, gathering lines, and auxiliary nontransportation-related equipment used in the production of petroleum but does not include natural gasoline plants.

Drilling fluid (Section 435.11 [3/4/93])* shall refer to the circulating fluid (mud) used in the rotary drilling of wells to clean and condition the hole and to counterbalance formation pressure. A water-based drilling fluid is the conventional drilling mud in which water is the continuous phase and the suspending medium for solids, whether or not oil is present. An oil-based drilling fluid has diesel oil, mineral oil, or some other oil as its continuous phase with water as the dispersed phase.

Drilling mud (Section 144.3) means a heavy suspension used in drilling an injection well, introduced down the drill pipe and through the drill bit.

Drinking water supply (Section 300.5 [3-8-90])* as defined by section 101(7) of CERCLA, means any raw or finished water source that is or may be used by a public water system (as defined in the Safe Drinking Water Act) or as drinking water by one or more individuals.

Drip pad (Section 260.10 [12-6-90]) is an engineered structure consisting of a curbed, free-draining base, constructed of non-earthen materials and designed to convey preservative kick-back or drippage from treated wood, precipitation, and surface water run-on to an associated collection system at wood preserving plants.

Drive System (Section 600.002-85) is determined by the number and location of drive axles (e.g., front wheel drive, rear wheel drive, four wheel drive) and any other feature of the drive system if the Administrator determines that such other features may result in a fuel economy difference.

Drive train configuration (Section 86.082-2) means a unique combination of engine code, transmission configuration, and axle ratio.

Dross reverberatory furnace (Section 60.181) means any furnace used for the removal or refining of impurities from lead bullion.

Drug (Sections 710.2; 720.3) shall have the meaning contained in the Federal Food, Drug, and Cosmetic Act, 21 U.S.C. 321 et seq., and the regulations issued under such Act.

Drug abuse (Section 7.25) means: (a) The use of any drug or substance listed by the Department of Justice in 21 CFR 1308.11, under authority of the Controlled Substances Act, 21 U.S.C. 801, as a controlled substance unavailable for prescription because: (1) The drug or substance has a high potential for abuse, (2) The drug or other substance has no currently accepted medical use in treatment in the United States, or (3) There is a lack of accepted safety for use of the drug or other substance under medical supervision. (b) The misuse of any drug or substance listed by the Department of Justice in 21 CFR 1308.12-1308.15 under authority of the Controlled Substances Act as a controlled substance available for prescription.

Drum break lining (Section 763.163 [7-12-89]) means any asbestos-containing product intended for use as a friction material in drum brake systems for vehicles rated at less than 26,001 pounds gross vehicle weight rating (GVWR).

Dry beans (Section 407.71) shall mean the production of canned pinto, kidney, navy, great northern, red, pink or related type, with and without formulated sauces, meats and gravies.

Dry flue gas desulfurization technology (Section 60.41b [12/16/87]) means a sulfur dioxide control system that is located downstream of the steam generating unit and removes sulfur oxides from the combustion gases of the steam generating unit by contacting the combustion gases with an alkaline slurry or solution and forming a dry powder material. This definition includes devices where the dry powder material is subsequently converted to another form. Alkaline slurries or solutions used in dry flue gas desulfurization technology include but are not limited to lime and sodium.

Dry flue gas desulfurization technology (Section 60.41c [9-12-90]) means a sulfur dioxide (SO2) control system that is located between the steam generating unit and the exhaust vent or stack, and that removes sulfur oxides from the combustion gases of the steam generating unit by contacting the combustion gases with an alkaline slurry or solution and forming a dry powder material. This definition includes devices where the dry powder material is subsequently converted to another form. Alkaline reagents used in dry flue gas desulfurization systems include, but are not limited to, lime and sodium compounds.

Dry lot (Section 412.21) shall mean a confinement facility for growing ducks in confinement with a dry litter floor cover and no access to swimming areas.

Dry solid form (Section 721.3 [7/27/88])*. See **Powder (Section 721.3)**.

Dryer (Section 60.161) means any facility in which a copper sulfide ore concentrate charge is heated in the presence of air to eliminate a portion of the moisture from the charge, provided less than 5 percent of the sulfur contained in the charge is eliminated in the facility.

Dryer (Section 60.401) means a unit in which the moisture content of phosphate rock is reduced by contact with a heated gas stream.

Dryer (Section 60.621) means a machine used to remove petroleum solvent from articles of clothing or other textile or leather goods, after washing and removing of excess petroleum solvent, together with the piping and ductwork used in the installation of this device.

Drying area (Section 60.541 [9/15/87]) means the area where VOC from applied cement or green tire sprays is allowed to evaporate.

Drying oven (Section 60.711 [10/3/88]) means a chamber in which heat is used to bake, cure, polymerize, or dry a surface coating.

Drying oven (Section 60.741 [9-11-89]) means a chamber within which heat is used to dry a surface coating; drying may be the only process or one of multiple processes performed in the chamber.

Duct burner (Section 60.41b [12/16/87])* means a device that combusts fuel and that is placed in the exhaust duct from another source, such as a stationary gas turbine, internal combustion engine, kiln, etc., to allow the firing of additional fuel to heat the exhaust gases before the exhaust gases enter a heat recovery steam generating unit.

Duct burner (Section 60.41c [9-12-90]) means a device that combusts fuel and that is placed in the exhaust duct from another source (such as a stationary gas turbine, internal combustion engine, kiln, etc.) to allow the firing of additional fuel to heat the exhaust gases before the exhaust gases enter a steam generating unit.

Dumping (Section 220.2) means a disposition of material: Provided, That it does not mean a disposition of any effluent from any outfall structure to the extent that such disposition is regulated under the provisions of the FWPCA, under the provisions of section 13 of the River and Harbor Act of 1899, as amended (33 U.S.C. 407), or under the provisions of the Atomic Energy Act of 1954, as amended (42 U.S.C. 2011), nor does it mean a routine discharge of effluent incidental to the propulsion of, or operation of motor-driven equipment on, vessels: Provided further, That it does not mean the construction of any fixed structure or artificial island nor the intentional placement of any device in ocean waters or on or in the submerged land beneath such waters, for a purpose other than disposal, when such construction or such placement is otherwise regulated by Federal or State law or occurs pursuant to an authorized Federal or State program; And provided further, That it does not include the deposit of oyster shells, or other materials when such deposit is made for the purpose of developing, maintaining, or harvesting fisheries resources and is otherwise regulated by Federal or State law or occurs pursuant to an authorized Federal or State program.

Duplication (Section 2.100 [1/5/88]) refers to the process of making a copy of a document necessary to respond to an FOIA request. Such copies can take the form of paper copy, microform, audio-visual materials, or machine readable documentation (e.g., magnetic tape or disk), among others. The copy provided must be in a form that is reasonably usable by requesters.

Duplicator paper (Section 250.4 [10/6/87, 6/22/88]) means writing papers used for masters or copy sheets in the aniline ink or hectograph process of reproduction (commonly called spirit machines).

Durability fleet (Section 610.11) means a fleet of automobiles operated for mileage accumulation used to assess deterioration effects associated with the retrofit device.

Durability useful life (Section 86.094-2 [3/24/93])* means the highest useful life mileage out of the set of all useful life mileages that apply to a given vehicle. The durability useful life determines the duration of service accumulation on a durability data vehicle. The determination of durability useful life shall reflect any alternative useful life mileages approved by the Administrator under Sec. 86.094-21(f). The determination of durability useful life shall exclude any standard and related useful life mileage for which the manufacturer has obtained a waiver of emission data submission requirements under Sec. 86.094-23(c)

Dust-handling equipment (Section 60.271) means any equipment used to handle particulate matter collected by the control device and located at or near the control device for an EAF subject to this subpart.

Dust-handling equipment (Section 60.261) means any equipment used to handle particulate matter collected by the air pollution control device (and located at or near such device) serving any electric submerged arc furnace subject to this subpart.

Dust-handling system (Section 60.271a) means equipment used to handle particulate matter collected by the control device for an electric arc furnace or AOD vessel subject to this subpart. For the purposes of this subpart, the dust-handling system shall consist of the control device dust hoppers, the dust-conveying equipment, any central dust storage equipment, the dust-treating equipment (e.g., pug mill, pelletizer), dust transfer equipment (from storage to truck), and any secondary control devices used with the dust transfer equipment.

Dwell Angle (Section 85.2122(a)(11)(ii)(C)) means the number of degrees of distributordtr mechanical rotation during which the breaker points are capable of conducting current.

Dwell Angle (Section 85.2122(a)(5)(ii)(C)) means the number of degrees of distributor mechanical rotation during which the breaker points are conducting current.

Dwelling (Section 4.2) means the place of permanent or customary and usual residence of a person, according to local custom or law, including a single family house; a single family unit in a two-family, multi-family, or multi-purpose property; a unit of a condominium or cooperative housing project; a non-housekeeping unit; a mobile home; or any other residential unit.

Dye penetrant testing (Section 471.02) is a nondestructive method for finding discontinuities that are open to the surface of the metal. A dye is applied to the surface of metal and the excess is rinsed off. Dye that penetrates surface discontinuities will not be rinsed away thus marking these discontinuities.

Dynamometer-idle (Section 86.082-2) for automatic transmission code heavy-duty engines means the manufacturer's recommended engine speed without a transmission that simulates the recommended engine speed with a transmission and with the transmission in neutral.

E

E (Section 60.641) is the sulfur emission rate expressed as elemental sulfur, kilograms per hour (kg/hr) rounded to one decimal place.

E (Section 60.741 [9-11-89]) means the control device efficiency achieved for the duration of the emission test (expressed as a fraction).

EAF (Section 60.271). See **Electric arc furnace (Section 60.271)**.

EAF (Section 60.271a). See **Electric arc furnace (Section 60.271a)**.

Ear Insert Device (Section 211.203) means a hearing protective device that is designed to be inserted into the ear canal, and to be held in place principally by virtue of its fit inside the ear canal.

Ear Muff Device (Section 211.203) means a hearing protective device that consists of two acoustic enclosures which fit over the ears and which are held in place by a spring-like headband to which the enclosures are attached.

Early entry (Section 170.3 [8/21/92])* means entry by a worker into a treated area on the agricultural establishment after a pesticide application is complete, but before any restricted-entry interval for the pesticide has expired.

Early life stage toxicity test (Section 797.1600) is a test to determine the minimum concentration of a substance which produces a statistically significant observable effect on hatching, survival, development and/or growth of a fish species continuously exposed during the period of their early development.

EC-X, ECX, or ECx (Sections 797.1050; 797.1060; 797.1160; 797.2750; 797.2800; 797.2850) means the experimentally derived chemical concentration that is calculated to effect X percent of the test criterion.

EC_{50} (Section 797.1800) is that experimentally derived concentration of a chemical in water that is calculated to induce shell deposition 50 percent less than that of the controls in a test batch of organisms during continuous exposure within a particular exposure period which should be stated.

EC_{50} (Section 797.1830) is that experimentally derived concentration of a chemical in water that is calculated to induce shell deposition 50 percent less than that of the controls in a test batch of organisms during continuous exposure within a particular period of exposure (which should be stated).

EC_{50} (Sections 797.1300; 797.1330) means that experimentally derived concentration of test substance in dilution water that is calculated to affect 50 percent of a test population during continuous exposure over a specified period of time. In this guideline, the effect measured is immobilization.

Ecological effects studies (Section 792.226) means studies performed for the development of information on non-human toxicity and potential ecological impact of test substances and their degradation and activation products.

Economic poison (Section 163.2) shall have the same meaning as it has under the Federal Insecticide, Fungicide, and Rodenticide Act (7 U.S.C. 135-135k) and the regulations issued thereunder (Part 162 of this chapter).

EDP (Section 60.311). See **Electrodeposition (Section 60.311)**.

EDP (Section 60.391). See **Electrodeposition (Section 60.391)**.

EDP (Section 60.451). See **Electrodeposition (Section 60.451)**.

Educational institution (Section 2.100 [1/5/88]) refers to a preschool, a public or private elementary or secondary school, an institution of graduate higher education, an institution of undergraduate higher education, an institution or professional education, and an institution of vocational education, which operates a program or programs of scholarly research.

Effective corrosion inhibitor residual (Section 141.2 [7/17/92])* for the purpose of subpart I of this part only, means a concentration sufficient to form a passivating film on the interior walls of a pipe.

Effective date (Section 61.02) is the date of promulgation in the *Federal Register* of an applicable standard or other regulation under this part.

Effective date of a UIC program (Section 146.3) means the date that a State UIC program is approved or established by the Administrator.

Effective date of an NSO (Section 57.103) means the effective date listed in the *Federal Register* publication of EPA's issuance or approval of an NSO.

Effective dose (Section 191.12 [12/20/93])* means the sum over specified tissues of the products of the dose equivalent received following an exposure of, or an intake of radionuclides into, specified tissues of the body, multiplied by appropriate weighting factors. This allows the various tissue-specific health risks to be summed into an overall health risk. The method used to calculate effective dose is described in Appendix B of this part.

Effective dose equivalent (Section 61.91, 61.101 [12-15-89])* means the sum of the products of absorbed dose and appropriate factors to account for differences in biological effectiveness due to the quality of radiation and its distribution in the body of reference man. The unit of the effective dose equivalent is the rem. For purposes of this subpart, doses caused by radon-222 and its respective decay products formed after the radon is released from the facility are not included. The method for calculating effective dose equivalent and the definition of reference man are outlined in the International Commission on Radiological Protection's Publication No. 26.

Effective dose equivalent (Section 61.21 [12-15-89]) means the sum of the products of absorbed dose and appropriate factors to account for differences in biological effectiveness due to the quality of radiation and its distribution in the body of reference man. The unit of the effective dose equivalent is the rem. The method for calculating effective dose equivalent and the definition of reference man are outlined in the International Commission on Radiological Protection's Publication No. 26.

Effects (Section 1508.8) include: (a) Direct effects, which are caused by the action and occur at the same time and place. (b) Indirect effects, which are caused by the action and are later in time or farther removed in distance, but are still reasonably foreseeable. Indirect effects may include growth inducing effects and other effects related to induced changes in the pattern of land use, population density or growth rate, and related effects on air and water and other natural systems, including ecosystems. Effects and impacts as used in these regulations are synonymous. Effects includes ecological (such as the effects on natural resources and on the components, structures, and functioning of affected ecosystems), aesthetic, historic, cultural, economic, social, or health, whether direct, indirect, or cumulative. Effects may also include those resulting from actions which may have both beneficial and detrimental effects, even if on balance the agency believes that the effect will be beneficial.

Efficacy (Section 162.3) means the capacity of a pesticide product when used according to label directions to control, kill, or induce the desired action in the target pest.

Efficiency (Section 85.2122(a)(16)(ii)(C)) means the ability of the air cleaner or the unit under test to remove contaminant.

Efficiency (Section 60.331) means the gas turbine manufacturer's rated heat rate at peak load in terms of heat input per unit of power output based on the lower heating value of the fuel.

Effluent (Section 232.2 [2/11/93])* means dredged material or fill material, including return flow from confined sites.

Effluent concentrations consistently achievable through proper operation and maintenance (Section 133.101). (1) For a given pollutant parameter, the 95th percentile value for the 30-day average effluent quality achieved by a treatment works in a period of at least two years, excluding values attributable to upsets, bypasses, operational errors, or other unusual conditions, and (2) a 7-day average value equal to 1.5 times the value derived under paragraph (f)(1) of this section.

Effluent data (Section 2.302) means (i) with reference to any source of discharge of any pollutant (as that term is defined in section 502(6) of the Act, 33 U.S.C. 1362(6)) (A) Information necessary to determine the identity, amount, frequency, concentration, temperature, or other characteristics (to the extent related to water quality) of any pollutant which has been discharged by the source (or of any pollutant resulting from any discharge from the source), or any combination of the foregoing; (B) Information necessary to determine the identity, amount, frequency, concentration, temperature, or other characteristics (to the extent related to water quality) of the pollutants which, under an applicable standard or limitation, the source was authorized to discharge (including, to the extent necessary for such purpose, a description of the manner or rate of operation of the source); and (C) A general description of the location and/or nature of the source to the extent necessary to identify the source and to distinguish it from other sources (including, to the extent necessary for such purposes, a description of the device, installation, or operation constituting the source). (ii) Notwithstanding paragraph (a)(2)(i) of this section, the following information shall be considered to be effluent data only to the extent necessary to allow EPA to disclose publicly that a source is (or is not) in compliance with an applicable standard or limitation, or to allow EPA to demonstrate the feasibility, practicability, or attainability (or lack thereof) of an existing or proposed standard or limitation: (A) Information concerning research, or the results of research, on any product, method, device, or installation (or any component thereof) which was produced, developed, installed, and used only for research purposes; and (B) Information concerning any product, method, device, or installation (or any component thereof) designed and intended to be marketed or used commercially but not yet so marketed or used.

Effluent limitation (Section 108.2) means any effluent limitation which is established as a condition of a permit issued or proposed to be issued by a State or by the Environmental Protection Agency pursuant to section 402 of the Act; any toxic or pretreatment effluent standard established under section 307 of the Act; any standard of performance established under section 306 of the

Act; and any effluent limitation established under section 302, section 316, or section 318 of the Act.

Effluent limitation (Section 122.2) means any restriction imposed by the Director on quantities, discharge rates, and concentrations of pollutants which are discharged from point sources into waters of the United States, the waters of the contiguous zone, or the ocean.

Effluent limitation (Section 401.11) means any restriction established by the Administrator on quantities, rates, and concentrations of chemical, physical, biological and other constituents which are discharged from point sources, other than new sources, into navigable waters, the waters of the contiguous zone or the ocean.

Effluent limitations guidelines (Section 122.2) means a regulation published by the Administrator under section 304(b) of CWA to adopt or revise effluent limitations.

Effluent limitations guidelines (Section 401.11) means any effluent limitations guidelines issued by the Administrator pursuant to section 304(b) of the Act.

Effluent standard (Section 104.2) means any effluent standard or limitation, which may include a prohibition of any discharge, established or proposed to be established for any toxic pollutant under section 307(a) of the Act.

Effluent standard (Section 129.2) means, for purposes of section 307, the equivalent of effluent limitation as that term is defined in section 502(11) of the Act with the exception that it does not include a schedule of compliance.

Effluents (Section 233.3) means dredged material or fill material, including return flow from confined sites.

Eggs cracked (Sections 797.2130; 797.2150) are eggs determined to have cracked shells when inspected with a candling lamp. Fine cracks cannot be detected without using a candling lamp and if undetected will bias data by adversely affecting embryo development. Values are expressed as a percentage of eggs laid by all hens during the test.

Eggs laid (Sections 797.2130; 797.2150) refers to the total egg production during the test, which normally includes 10 weeks of laying. Values are expressed as numbers of eggs per pen per season (or test).

Eggs set (Sections 797.2130; 797.2150) are all eggs placed under incubation, i.e., total eggs minus cracked eggs and those selected for analysis of eggshell thickness. The number of eggs set, itself, is an artificial number, but it is essential for the statistical analysis of other development parameters.

Eggshell thickness (Sections 797.2130;

797.2150) is the thickness of the shell and the membrane of the egg at several points around the girth after the egg has been opened, washed out, and the shell and membrane dried for at least 48 hours at room temperature. Values are expressed as the average thickness of the several measured points in millimeters.

Eight-hour (8-hour) time weighted average (Section 704.142) means the cumulative exposure for an 8-hour work shift computed as set forth in paragraph (a)(5) of this section.

Electric arc furnace (EAF) (Section 60.271a) means a furnace that produces molten steel and heats the charge materials with electric arcs from carbon electrodes. For the purposes of this subpart, an EAF shall consist of the furnace shell and roof and the transformer. Furnaces that continuously feed direct-reduced iron ore pellets as the primary source of iron are not affected facilities within the scope of this definition.

Electric arc furnace (EAF) (Section 60.271) means a furnace that produces molten steel and heats the charge materials with electric arcs from carbon electrodes. Furnaces that continuously feed direct-reduced iron ore pellets as the primary source of iron are not affected facilities within the scope of this definition.

Electric arc furnace steelmaking (Section 420.41) means the production of steel principally from steel scrap and fluxes in refractory lined furnaces by passing an electric current through the scrap or steel bath.

Electric furnace (Section 60.131) means any furnace which uses electricity to produce over 50 percent of the heat required in the production of refined brass or bronze.

Electric smelting furnace (Section 60.181) means any furnace in which the heat necessary for smelting of the lead sulfide ore concentrate charge is generated by passing an electric current through a portion of the molten mass in the furnace.

Electric submerged arc furnace (Section 60.261) means any furnace wherein electrical energy is converted to heat energy by transmission of current between electrodes partially submerged in the furnace charge.

Electric Traction Motor (Section 600.002-85) means an electrically powered motor which provides tractive energy to the wheels of a vehicle.

Electric utility combined cycle gas turbine (Section 60.41a) means any combined cycle gas turbine used for electric generation that is constructed for the purpose of supplying more than one-third of its potential electric output capacity and more than 25 MW electrical output to any utility power distribution system for sale. Any steam distribution system that is constructed for the purpose of providing steam to a steam electric generator that would produce electrical power for sale is also considered in determining the electrical energy output capacity of the affected facility.

Electric utility company (Section 60.41a) means the largest interconnected organization, business, or governmental entity that generates electric power for sale (e.g., a holding company with operating subsidiary companies).

Electric utility stationary gas turbine (Section 60.331) means any stationary gas turbine constructed for the purpose of supplying more than one-third of its potential electric output capacity to any utility power distribution system for sale.

Electric utility steam generating unit (Section 60.41a) means any steam electric generating unit that is constructed for the purpose of supplying more than one-third of its potential electric output capacity and more than 25 MW electrical output to any utility power distribution system for sale. Any steam supplied to a steam distribution system for the purpose of providing steam to a steam-electric generator that would produce electrical energy for sale is also considered in determining the electrical energy output capacity of the affected facility.

Electrical capacitor manufacturer (Section 129.105) means a manufacturer who produces or assembles electrical capacitors in which PCB or PCB-containing compounds are part of the dielectric.

Electrical Charging System (Section 600.002-85) means a device to convert 60Hz alternating electric current, as commonly available in residential electric service in the United States, to a proper form for recharging the energy storage device.

Electrical equipment (Section 280.12 [9/23/88]) means underground equipment that contains dielectric fluid that is necessary for the operation of equipment such as transformers and buried electrical cable.

Electrical transformer manufacturer (Section 129.105) means a manufacturer who produces or assembles electrical transformers in which PCB or PCB-containing compounds are part of the dielectric.

Electrically-heated Choke (Section 85.2122(a)(2)(iii)(C)) means a device which contains a means for applying heat to the thermostatic coil by electrical current.

Electrocoating (Section 471.02) is the electrodeposition of a metallic or nonmetallic coating onto the surface of a workpiece.

Electrodeposition (EDP) (Section 60.451) means a method of coating application in which the large appliance part or product is submerged in a tank filled with coating material suspended in water and an electrical potential is used to enhance deposition of the material on the part or product.

Electrodeposition (EDP) (Section 60.391) means a method of applying a prime coat by which the automobile or light-duty truck body is submerged in a tank filled with coating material and an electrical field is used to effect the deposition of the coating material on the body.

Electrodeposition (EDP) (Section 60.311) means a method of applying coatings in which the part is submerged in a tank filled with the coatings and in which an electrical potential is used to enhance deposition of the coatings on the part.

Electroless plating (Section 413.71) shall mean the deposition of conductive material from an autocatalytic plating solution without application of electrical current.

Electromagnetic interference/radio frequency interference (EMI/RFI) shielding coating (Section 60.721 [6/15/89]) means a conductive coating that is applied to a plastic substrate to attenuate EMI/RFI signals.

Electronic crystals (Section 469.22) means crystals or crystalline material which because of their unique structural and electronic properties are used in electronic devices. Examples of these crystals are crystals comprised of quartz, ceramic, silicon, gallium arsenide, and idium arsenide.

Electroplating process wastewater (Section 413.02) shall mean process wastewater generated in operations which are subject to regulation under any of Subparts A through H of this Part.

Electrostatic precipitator (ESP) (Section 60.471) means an air pollution control device in which solid or liquid particulates in a gas stream are charged as they pass through an electric field and precipitated on a collection surface.

Electrostatic spray application (Section 60.391) means a spray application method that uses an electrical potential to increase the transfer efficiency of the coating solids. Electrostatic spray application can be used for prime coat, guide coat, or topcoat operations.

Electrostatic spray application (Section 60.311) means a spray application method that uses an electrical potential to increase the transfer efficiency of the coatings.

Element of design (Section 86.094-2 [3/24/93])* means any control system (i.e., computer software, electronic control system, emission control system, computer logic), and/or control system calibrations, and/or the results of systems interaction, and/or hardware items on a motor vehicle or motor vehicle engine.

Elemental phosphorus plant or plant (Section 61.121 [12-15-89])* means any facility that processes phosphate rock to produce elemental phosphorus. A plant includes all buildings, structures, operations, calciners and nodulizing kilns on one contiguous site.

Elementary neutralization unit (Section 270.2) means a device which: (a) Is used for neutralizing wastes which are hazardous wastes only because they exhibit the corrosivity characteristic defined in Section 261.22 of this chapter, or are listed in Subpart D of Part 261 of this chapter only for this reason; and (b) Meets the definition of tank, container, transport vehicle, or vessel in Section 260.10 of this chapter.

Elementary neutralization unit (Section 260.10 [9/2/88])* means a device which: (1) Is used for neutralizing wastes that are hazardous only because they exhibit the corrosivity characteristic defined in Section 261.22 of this chapter, or they are listed in Subpart D of Part 261 of the chapter only for this reason; and (2) Meets the definition of tank, tank system, container, transport vehicle, or vessel in Section 260.10 of this chapter.

Eleven-AA (11-AA) (Section 704.25) means the chemical substance 11-aminoundecanoic acid, CAS Number 2432-99-7.

Eligible Indian Tribe (Section 35.105 [3/23/94])* means for purposes of the Clean Water Act, any federally recognized Indian Tribe that meets the requirements set forth at 40 CFR 130.6(d).

Elimination (Section 797.1560). See **Depuration (Section 797.1560)**.

ELWK (equivalent live weight killed) (Sections 432.11; 432.21; 432.31; 432.41) shall mean the total weight of the total number of animals slaughtered at locations other than the slaughterhouse or packinghouse, which animals provide hides, blood, viscera or renderable materials for processing at that slaughterhouse, in addition to those derived from animals slaughtered on site.

Embryo (Section 797.2750) means the young sporophytic plant before the start of germination.

Embryo cup (Section 797.1600) is a small glass jar or similar container with a screened bottom in which the embryos of some species (i.e., minnow) are placed during the incubation period and which is normally oscillated to ensure a flow of water through the cup.

Emergency condition (Section 60.41a) means that period of time when: (a)

The electric generation output of an affected facility with a malfunctioning flue gas desulfurization system cannot be reduced or electrical output must be increased because: (1) All available system capacity in the principal company interconnected with the affected facility is being operated, and (2) All available purchase power interconnected with the affected facility is being obtained, or (b) The electric generation demand is being shifted as quickly as possible from an affected facility with a malfunctioning flue gas desulfurization system to one or more electrical generating units held in reserve by the principal company or by a neighboring company, or (c) An affected facility with a malfunctioning flue gas desulfurization system becomes the only available unit to maintain a part or all of the principal company's system emergency reserves and the unit is operated in spinning reserve at the lowest practical electric generation load consistent with not causing significant physical damage to the unit. If the unit is operated at a higher load to meet load demand, an emergency condition would not exist unless the conditions under (a) of this definition apply.

Emergency condition (Section 166.3) means an urgent, non-routine situation that requires the use of a pesticide(s) and shall be deemed to exist when: (1) No effective pesticides are available under the Act that have labeled uses registered for control of the pest under the conditions of the emergency; and (2) No economically or environmentally feasible alternative practices which provide adequate control are available; and (3) The situation: (i) Involves the introduction or dissemination of a pest new to or not theretofore known to be widely prevalent or distributed within or throughout the United States and its territories; or (ii) Will present significant risks to human health; or (iii) Will present significant risks to threatened or endangered species, beneficial organisms, or the environment; or (iv) Will cause significant economic loss due to: (A) An outbreak or an expected outbreak of a pest; or (B) A change in plant growth or development caused by unusual environmental conditions where such change can be rectified by the use of a pesticide(s).

Emergency fuel (Section 60.331) is a fuel fired by a gas turbine only during circumstances, such as natural gas supply curtailment or breakdown of delivery system, that make it impossible to fire natural gas in the gas turbine.

Emergency gas turbine (Section 60.331) means any stationary gas turbine which operates as a mechanical or electrical power source only when the primary power source for a facility has been rendered inoperable by an emergency situation.

Emergency permit (Section 144.3) means a UIC permit issued in accordance with Section 144.34.

Emergency permit (Section 233.3) means a State 404 permit issued in accordance with Section 233.38.

Emergency permit (Section 270.2) means a RCRA permit issued in accordance with Section 270.61.

Emergency Planning and Community Right-To-Know Act of 1986 (Section 310.11 [1/15/93])* means Title III--Emergency Planning and Community Right-To-Know Act of the Superfund Amendments and Reauthorization Act of 1986 (EPCRA) (Pub. L. 99-499, 42 U.S.C. 960).

Emergency project (Section 763.121) means a project involving the removal, enclosure, or encapsulation of friable asbestos-containing material that was not planned but results from a sudden unexpected event.

Emergency renovation operation (Section 61.141 [11-20-90]) means a renovation operation that was not planned but results from a sudden, unexpected event that, if not immediately attended to, presents a safety or public health hazard, is necessary to protect equipment from damage, or is necessary to avoid imposing an unreasonable financial burden. This term includes operations necessitated by nonroutine failures of equipment.

Emergency Situation (Section 761.3 [7/19/88]) for continuing use of a PCB Transformer exists when: (1) Neither a non-PCB Transformer nor a PCB-Contaminated transformer is currently in storage for reuse or readily available (i.e., available within 24 hours) for installation. (2) Immediate replacement is necessary to continue service to power users.

Emergency vent stream (Section 60.561 [3-22-91]) means, for the purposes of these standards, an intermittent emission that results from a decomposition, attempts to prevent decompositions, power failure, equipment failure, or other unexpected cause that requires immediate venting of gases from process equipment in order to avoid safety hazards or equipment damage. This includes intermittent vents that occur from process equipment where normal operating parameters (e.g., pressure or temperature) are exceeded such that the process equipment can not be returned to normal operating conditions using the design features of the system and venting must occur to avoid equipment failure or adverse safety personnel consequences and to minimize adverse effects of the runaway reaction. This does not include intermittent vents that are designed into the process to maintain normal operating conditions of process vessels including those vents that regulate normal process vessel pressure.

Emerging technology (Section 60.41b [12/16/87]) means any sulfur dioxide control system that is not defined as a conventional technology under this section, and for which the owner or operator of the facility has applied to the Administrator and received approval to operate as an emerging technology under Section 60.49b(a)(4).

Emerging technology (Section 60.41c [9-12-90]) means any SO2 control system that is not defined as a conventional technology under this section, and for which the owner or operator of the affected facility has received approval from the Administrator to operate as an emerging technology under Sec. 60.48c(a)(4).

EMI/RFI shielding coating (Section 60.721 [6/15/89]). See **Electromagnetic interference/radio frequency interference shielding coating (Section 60.721)**.

Emission (Section 240.101) means gas-borne pollutants released to the atmosphere.

Emission control device (Section 60.581) means any solvent recovery or solvent destruction device used to control volatile organic compounds (VOC) emissions from flexible vinyl and urethane rotogravure printing lines.

Emission control system (Section 60.581) means the combination of an emission control device and a vapor capture system for the purpose of reducing VOC emissions from flexible vinyl and urethane rotogravure printing lines.

Emission credits (Section 86.090-2 [7-26-90]) mean the amount of emission reductions or exceedances, by a heavy-duty engine family, below or above the emission standard, respectively. Emission credits below the standard are considered as "positive credits," while emission credits above the standard are considered as "negative credits." In addition, "projected credits" refer to emission credits based on the projected U.S. production volume of the engine family.

Emission data (Section 2.301) means (i) with reference to any source of emission of any substance into the air (A) Information necessary to determine the identity, amount, frequency, concentration, or other characteristics (to the extent related to air quality) of any emission which has been emitted by the source (or of any pollutant resulting from any emission by the source), or any combination of the foregoing; (B) Information necessary to determine the identity, amount, frequency, concentration, or other characteristics (to the extent related to air quality) of the emissions which, under an applicable standard or limitation, the source was authorized to emit (including, to the extent necessary for such purposes, a description of the manner or rate of operation of the source); and (C) A general description of the location and/or nature of the source to the extent necessary to identify the source and to distinguish it from other sources (including, to the extent necessary for such purposes, a description of the device, installation, or operation constituting the source). (ii) Notwithstanding paragraph (a)(2)(i) of this section, the following information shall be considered to be emission data only to the extent necessary to allow EPA to disclose publicly that a source is (or is not) in compliance with an applicable standard or limitation, or to allow EPA to demonstrate the feasibility, practicability, or attainability (or lack thereof) of an existing or proposed standard or limitation: (A) Information concerning research, or the results of research, on any project, method, device or installation (or any component thereof) which was produced, developed, installed, and used only for research purposes; and (B) Information concerning any product, method, device, or installation (or any component thereof) designed and intended to be marketed or used commercially but not yet so marketed or used.

Emission guideline (Section 60.21) means a guideline set forth in Subpart C of this part, or in a final guideline document published under Section 60.22(a), which reflects the degree of emission reduction achievable through the application of the best system of emission reduction which (taking into account the cost of such reduction) the Administrator has determined has been adequately demonstrated for designated facilities.

Emission limitation and emission standard (Section 51.100) mean a requirement established by a State, local government, or the Administrator which limits the quantity, rate, or concentration of emissions of air pollutants on a continuous basis, including any requirements which limit the level of opacity, prescribe equipment, set fuel specifications, or prescribe operation or maintenance procedures for a source to assure continuous emission reduction.

Emission measurement system (Section 87.1) means all of the equipment necessary to transport and measure the level of emissions. This includes the sample system and the instrumentation system.

Emission Performance Warranty (Section 85.2102) means that warranty given pursuant to this subpart and section 207(b) of the Act.

Emission Short Test (Section 85.2102). See **EPA-Approved Emission Test (Section 85.2102)**.

Emission standard (Section 51.100). See **Emission limitation (Section 51.100)**.

Emission standard (Section 60.21) means a legally enforceable regulation setting forth an allowable rate of emissions into the atmosphere, or prescribing equipment specifications for control of air pollution emissions.

Emission Warranty (Section 85.2113) means those warranties given by vehicle manufacturers pursuant to section 207 of the Act.

Emission-Critical Parameters (Section 85.2113). See **Emission-Related Standards (Section 85.2113)**.

Emission-related defect (Section 85.1902) shall mean a defect in design, materials, or workmanship in a device, system, or assembly described in the approved Application for Certification (required by 40 CFR 86.077-22 and like provisions of Part 85 and Part 86 of Title 40 of the Code of Federal Regulations) which affects any parameter or specification enumerated in Appendix VIII.

Emission-related maintenance (Section 86.084-2) means that maintenance which does substantially affect emissions or which is likely to affect the deterioration of the vehicle or engine with respect to emissions, even if the maintenance is performed at some time other than that which is recommended. This definition applies beginning with the 1984 model year.

Emission-Related Standards or Emission-Critical Parameters (Section 85.2113) means those critical parameters and tolerances which, if equivalent from one part to another, will not cause the vehicle to exceed applicable emission standards with such parts installed.

Emissions unit (Sections 51.165; 51.166; 52.21; 52.24) means any part of a stationary source which emits or would have the potential to emit any pollutant subject to regulation under the Act.

Employee (Section 13.2 [9/23/88]) means a current employee of the Federal Government including a current member of the Armed Forces.

Employee (Section 14.2) means a person appointed to a position with EPA.

Employee (Section 3.102) means any officer or employee of the Environmental Protection Agency, Public Health Service commissioned officers assigned to EPA, employees detailed to EPA from other federal agencies and employees detailed or assigned to EPA under the Intergovernmental Personnel Act. The term does not include special Government employees.

Employee (Section 311.2 [6/23/89]) in Section 311.1 is defined as a compensated or non-compensated worker who is controlled directly by a State or local government, as contrasted to an independent contractor.

Employee (Section 32.605 [5-25-90]) means the employee of a grantee directly engaged in the performance of work under the grant, including: (i) All "direct charge" employees; (ii) All "indirect charge" employees, unless their impact or involvement is insignificant to the performance of the grant; and, (iii) Temporary personnel and consultants who are directly engaged in the performance of work under the grant and who are on the grantee's payroll. This definition does not include workers not on the payroll of the grantee (e.g., volunteers, even if used to meet a matching requirement; consultants or independent contractors not on the payroll; or employees of subrecipients or subcontractors in covered workplaces).

Employee exposure (Section 763.121) means that exposure to airborne asbestos would occur if the employee were not using respiratory protective equipment.

Employee salary offset (Section 13.2 [9/23/88]) means the administrative collection of a debt by deductions at one or more officially established pay intervals from the current pay account of an employee without the employee's consent.

Employer (Section 721.3 [7-27-89]) means any manufacturer, importer, processor, or user of chemical substances or mixtures.

Employer (Section 763.121) means the public department, agency, or entity which hires an employee. The term includes, but is not limited to, any State, County, City, or other local governmental entity which operates or administers schools, a department of health or human services, a library, a police department, a fire department, or similar public service agencies or offices.

Emulsions (Section 467.02) are stable dispersions of two immiscible liquids. In the aluminum forming category this is usually an oil and water mixture.

Emulsions (Section 471.02) are stable dispersions of two immiscible liquids. In the Nonferrous Metals Forming and Metal Powders Point Source category, this is usually an oil and water mixture.

Encapsulate (Section 165.1) means to seal a pesticide, and its container if appropriate, in an impervious container made of plastic, glass, or other suitable material which will not be chemically degraded by the contents. This container then should be sealed within a durable container made from steel, plastic, concrete, or other suitable material of sufficient thickness and strength to resist physical damage during and subsequent to burial or storage.

Encapsulation (Section 763.83 [10/30/87]) means the treatment of ACBM with a material that surrounds or embeds asbestos fibers in an adhesive matrix to prevent the release of fibers, as the encapsulant creates a membrane over the surface (bridging encapsulant) or penetrates the material and binds its components together (penetrating encapsulant).

Enclosed process (Section 704.3 [12/22/88]) means a manufacturing or processing operation that is designed and operated so that there is no intentional release into the environment of any substance present in the operation. An operation with fugitive, inadvertent, or emergency pressure relief releases remains an enclosed process so long as measures are taken to prevent worker exposure to and environmental contamination from the releases.

Enclosed process (Section 721.347 [10/27/87]) means a process that is designed and operated so that there is no intentional release of any substance present in the process. A process with fugitive, inadvertent, or emergency relief releases remains an enclosed process so long as measures are taken to prevent worker exposure to and environmental contamination from the releases.

Enclosed process (Sections 704.25; 704.104 [10/27/87]) means a process that is designed and operated so that there is no intentional release of any substance present in the process. A process with fugitive, inadvertent, or emergency pressure relief releases remains an enclosed process so long as measures are taken to prevent worker exposure to and environmental contamination from the releases.

Enclosed storage area (Section 60.381) means any area covered by a roof under which metallic minerals are stored prior to further processing or loading.

Enclosed truck or railcar loading station (Section 60.671) means that portion of a nonmetallic mineral processing plant where nonmetallic minerals are loaded by an enclosed conveying system into enclosed trucks or railcars.

Enclosure (Section 60.441). See **Hood (Section 60.441)**.

Enclosure (Section 60.541 [9/15/87]) means a structure that surrounds a VOC (cement, solvent, or spray) application area and drying area, and that captures and contains evaporated VOC and vents it to a control device. Enclosures may have permanent and temporary openings.

Enclosure (Section 763.83 [10/30/87]) means an airtight, impermeable, permanent barrier around ACBM to prevent the release of asbestos fibers into the air.

End box (Section 61.51) means a container(s) located on one or both ends of a mercury chlor-alkali electrolyzer which serves as a connection between the electrolyzer and denuder for rich and stripped amalgam.

End box ventilation system (Section 61.51) means a ventilation system which collects mercury emissions from the end-boxes, the mercury pump sumps, and their water collection systems.

End use product (Section 152.3 [5/4/88]) means a pesticide product whose labeling (1) Includes directions for use of the product (as distributed or sold, or after combination by the user with other substances) for controlling pests or defoliating, desiccating, or regulating the growth of plants, and (2) Does not state that the product may be used to manufacture or formulate other pesticide products.

End use product (Section 158.153 [5/4/88]) means a pesticide product whose labeling (1) Includes directions for use of the product (as distributed or sold, or after combination by the user with other substances) for controlling pests or defoliating, desiccating or regulating growth of plants, and (2) Does not state that the product may be used to manufacture or formulate other pesticide products.

Endangered or threatened species (Section 257.3-2) means any species listed as such pursuant to section 4 of the Endangered Species Act.

Endrin (Section 129.4) means the compound endrin as identified by the chemical name 1,2,3,4,10,10-hexachloro-6,7-epoxy-1,4,4a,5,6,7,8,8a-octahydro-1,4-endo-5,8-endodimethanonaphthalene.

Endrin (Section 704.142) means the pesticide 2,7:3,6-Dimethanonaphth[2,3-b]oxirene,3,4,5,6,9,9-hexachloro-1a,2,2a,3,6,6a,7,7a-octahydro-, (1aalpha, 2beta, 2abeta, 3alpha, 6alpha, 6abeta, 7beta, 7aalpha)-, CAS No. 72-20-8.

Endrin Formulator (Section 129.102) means a person who produces, prepares or processes a formulated product comprising a mixture of endrin and inert materials or other diluents into a product intended for application in any use registered under the Federal Insecticide, Fungicide and Rodenticide Act, as amended (7 U.S.C. 135, et seq.).

Endrin Manufacturer (Section 129.102) means a manufacturer, excluding any source which is exclusively an endrin formulator, who produces, prepares or processes technical endrin or who uses endrin as a material in the production, preparation or processing of another synthetic organic substance.

Energy Average Level (Section 201.1) means a quantity calculated by taking ten times the common logarithm of the arithmetic average of the antilogs of one-tenth of each of the levels being averaged. The levels may be of any consistent type, e.g., maximum sound levels, sound exposure levels, and day-night sound levels.

Energy Storage Device (Section 600.002-85) means a rechargeable means of storing tractive energy on board a vehicle such as storage batteries or a flywheel.

Energy Summation of Levels (Section 201.1) means a quantity calculated by taking ten times the common logarithm of the sum of the antilogs of one-tenth of each of the levels being summed. The levels may be of any consistent type, e.g., day-night sound level or equivalent sound level.

Enforceable requirements of the Act (Section 35.905) are those conditions or limitations of section 402 or 404 permits which, if violated, could result in the issuance of a compliance order or initiation of a civil or criminal action under section 309 of the Act. If a permit has not been issued, the term shall include any requirement which, in the Regional Administrator's judgment, would be included in the permit when issued. Where no permit applies, the term shall include any requirement which the Regional Administrator determines is necessary to meet applicable criteria for best practicable waste treatment technology (BPWTT).

Enforceable requirements of the Act (Section 35.2005) means those conditions or limitations of section 402 or 404 permits which, if violated, could result in the issuance of a compliance order or initiation of a civil or criminal action under section 309 of the Act or applicable State laws. If a permit has not been issued, the term shall include any requirement which, in the Regional Administrator's judgment, would be included in the permit when issued. Where no permit applies, the term shall include any requirement which the Regional Administrator determines is necessary for the best practicable waste treatment technology to meet applicable criteria.

Engine code (Section 600.002-85) means, for gasoline-fueled and diesel vehicles, a unique combination, within an engine-system combination (as defined in Part 86 of this chapter), of displacement, carburetor (or fuel injection) calibration, distributor calibration, choke calibration, auxiliary emission control devices, and other engine and emission control system components specified by the Administrator. For electric vehicles, engine code means a unique combination of manufacturer, electric traction motor, motor configuration, motor controller, and energy storage device.

Engine code (Section 86.082-2) means a unique combination, within an engine-system combination, of displacement, carburetor (or fuel injection) calibration, choke calibration, distributor calibration, auxiliary emission control devices, and other engine and emission control system components specified by the Administrator.

Engine Configuration (Section

85.2113). See **Vehicle Configuration (Section 85.2113)**.

Engine displacement (Section 205.151) means volumetric engine capacity as defined in Section 205.153.

Engine family (Section 85.2113) means the basic classification unit of a vehicle's product line for a single model year used for the purpose of emission-data vehicle or engine selection and as determined in accordance with 40 CFR 86.078-24.

Engine family (Section 86.082-2) means the basic classification unit of a manufacturer's product line used for the purpose of test fleet selection and determined in accordance with Section 86.082-24.

Engine family (Section 86.402-78) means the basic classification unit of a manufacturer's product line used for the purpose of test fleet selection and determined in accordance with Section 86.420.

Engine family group (Section 86.082-2) means a combination of engine families for the purpose of determining a minimum deterioration factor under the Alternative Durability program.

Engine lubricating oils (Section 252.4 [6/30/88]) means petroleum-based oils used for reducing friction in engine parts.

Engine Model (Section 87.1) means all commercial aircraft turbine engines which are of the same general series, displacement, and design characteristics and are usually approved under the same type certificate.

Engine warm-up cycle (Section 86.094-2 [3/24/93])* means sufficient vehicle operation such that the coolant temperature has risen by at least 40 deg.F from engine starting and reaches a minimum temperature of 160 deg.F.

Engine-displacement-system combination (Section 86.402-78) means an engine family-displacement-emission control system combination.

Engine-system combination (Section 86.082-2) means an engine family-exhaust emission control system combination.

Envelopes (Section 250.4 [10/6/87, 6/22/88]) means brown, manila, padded, or other mailing envelopes not included with stationery.

Environment (Section 171.2) means water, air, land, and all plants and man and other animals living therein, and the interrelationships which exist among them.

Environment (Section 300.5 [3-8-90])* as defined by section 101(8) of CERCLA, means the navigable waters, the waters of the contiguous zone, and the ocean waters of which the natural resources are under the exclusive management authority of the United

States under the Magnuson Fishery Conservation and Management Act; and any other surface water, ground water, drinking water supply, land surface or subsurface strata, or ambient air within the United States or under the jurisdiction of the United States.

Environment (Section 302.3) means (1) the navigable waters, the waters of the contiguous zone, and the ocean waters of which the natural rcsources are under the exclusive management authority of the United States under the Fishery Conservation and Management Act of 1976, and (2) any other surface water, ground water, drinking water supply, land surface or subsurface strata, or ambient air within the United States or under the jurisdiction of the United States.

Environment (Section 355.20) includes water, air, and land and the interrelationship which exists among and between water, air, and land and all living things.

Environment (Section 370.2 [10/15/87]) includes water, air, and land and the interrelationship that exists among and between water, air, and land and all living things.

Environment (Section 6.1003) as used in this subpart, means the natural and physical environment and excludes social, economic and other environments. An action significantly affects the environment if it does significant harm to the environment even though on balance the action may be beneficial to the environment. To the extent applicable, the responsible official shall address the considerations set forth in the CEQ Regulations under 40 CFR 1508.27 in determining significant effect.

Environment (Section 723.50) has the same meaning as in section 3 of the Act (15 U.S.C. 2602).

Environmental Appeals Board (Section 209.3 [2-13-92]) means the Board within the Agency described in Sec. 1.25 of this title. The Administrator delegates authority to the Environmental Appeals Board to issue final decisions in appeals filed under this part. An appeal directed to the Administrator, rather than to the Environmental Appeals Board, will not be considered. This delegation of authority to the Environmental Appeals Board does not preclude the Environmental Appeals Board from referring an appeal or a motion filed under this part to the Administrator for decision when the Environmental Appeals Board, in its discretion, deems it appropriate to do so. When an appeal or motion is referred to the Administrator, all parties shall be so notified and the rules in this part referring to the Environmental Appeals Board shall be interpreted as referring to the Administrator.

Environmental Appeals Board (Section 22.03 [2-13-92]) means the Board within the Agency described in

Sec. 1.25 of this title, located at U.S. Environmental Protection Agency, A-110, 401 M St. SW., Washington, DC 20460.

Environmental Appeals Board (Section 66.3 [2-13-92]) shall mean the Board within the Agency described in Sec. 1.25 of this title. The Administrator delegates authority to the Environmental Appeals Board to issue final decisions in appeals filed under this part. Appeals directed to the Administrator, rather than to the Environmental Appeals Board, will not be considered. This delegation of authority to the Environmental Appeals Board does not preclude the Environmental Appeals Board from referring an appeal or a motion filed under this part to the Administrator for decision when the Environmental Appeals Board, in its discretion, deems it appropriate to do so. When an appeal or motion is referred to the Administrator, all parties shall be so notified and the rules in this part referring to the Environmental Appeals Board shall be interpreted as referring to the Administrator.

Environmental Appeals Board (Section 27.2 [2-13-92]) means the Board within the Agency described in Sec. 1.25 of this title.

Environmental Appeals Board (Section 124.72 [2-13-92]) shall mean the Board within the Agency described in Sec. 1.25 of this title. The Administrator delegates authority to the Environmental Appeals Board to issue final decisions in NPDES appeals filed under this subpart. An appeal directed to the Administrator, rather than to the Environmental Appeals Board, will not be considered. This delegation does not preclude the Environmental Appeals Board from referring an appeal or a motion to the Administrator when the Environmental Appeals Board, in its discretion, deems it appropriate to do so. When an appeal or motion is referred to the Administrator by the Environmental Appeals Board, all parties shall be so notified and the rules in this subpart referring to the Environmental Appeals Board shall be interpreted as referring to the Administrator.

Environmental Appeals Board (Section 124.2 [12/22/93])* shall mean the Board within the Agency described in Sec. 1.25(e) of this title. The Administrator delegates authority to the Environmental Appeals Board to issue final decisions in RCRA, PSD, UIC, or NPDES permit appeals filed under this subpart, including informal appeals of denials of requests for modification, revocation and reissuance,or termination of permits under Section 124.5(b). An appeal directed to the Administrator, rather than to the Environmental Appeals Board, will not be considered. This delegation does not preclude the EnvironmentalAppeals Board from referring an appeal or a motion under this subpart to the Administrator when the Environmental Appeals Board, in its discretion, deems it appropriate to do so. When anappeal or motion is referred to the Administrator by the Environmental Appeals Board, all parties shall be so notified and the rules in this subpart referring to the Environmental Appeals Board shall be interpreted as referring to the Administrator.

Environmental Appeals Board (Section 164.2 [7/10/92])* shall mean the Board within the Agency described in Sec. 1.25 of this title. The Administrator delegates authority to the Environmental Appeals Board to issue final decisions in appeals filed under subparts B and C of this part. An appeal directed to the Administrator, rather than to the Environmental Appeals Board, will not be considered. This delegation does not preclude the Environmental Appeals Board from referring an appeal or a motion under subparts B and C to the Administrator when the Environmental Appeals Board, in its discretion, deems it appropriate to do so. When an appeal or motion is referred to the Administrator, all of the parties shall be so notified and the rules in subparts B and C referring to the Environmental Appeals Board shall be interpreted as referring to the Administrator.

Environmental assessment (Section 1508.9): (a) Means a concise public document for which a Federal agency is responsible that serves to: (1) Briefly provide sufficient evidence and analysis for determining whether to prepare an environmental impact statement or a finding of no significant impact. (2) Aid an agency's compliance with the Act when no environmental impact statement is necessary. (3) Facilitate preparation of a statement when one is necessary. (b) Shall include brief discussions of the need for the proposal, of alternatives as required by section 102(2)(E), of the environmental impacts of the proposed action and alternatives, and a listing of agencies and persons consulted.

Environmental document (Section 1508.10) includes the documents specified in Section 1508.9 (environmental assessment), Section 1508.11 (environmental impact statement), Section 1508.13 (finding of no significant impact), and Section 1508.22 (notice of intent).

Environmental education and environmental education and training (Section 47.105 [3-9-92]) mean educational activities and training activities involving elementary, secondary, and postsecondary students, as such terms are defined in the State in which they reside, and environmental education personnel, but does not include technical training activities directed toward environmental management professionals or activities primarily directed toward the support of noneducational research and development.

Environmental impact statement (Section 1508.11) means a detailed written statement as required by section 102(2)(C) of the Act.

Environmental information document (Section 6.101) means any written analysis prepared by an applicant, grantee or contractor describing the environmental impacts of a proposed action. This document will be of sufficient scope to enable the responsible official to prepare an environmental assessment as described in the remaining subparts of this regulation.

Environmental noise (Section 205.2) means the intensity, duration, and the character of sounds from all sources.

Environmental Protection Agency (EPA) (Sections 122.2; 144.3; 146.3; 233.3; 270.2; 401.11) means the United States Environmental Protection Agency.

Environmental review (Section 6.101) means the process whereby an evaluation is undertaken by EPA to determine whether a proposed Agency action may have a significant impact on the environment and therefore require the preparation of the EIS.

Environmental studies (Section 792.226) refers to either ecological effects studies or chemical fate studies or both.

Environmental transformation product (Section 723.50) means any chemical substance resulting from the action of environmental processes on a parent compound that changes the molecular identity of the parent compound.

Environmentally related measurements (Section 30.200) means any data collection activity or investigation involving the assessment of chemical, physical, or biological factors in the environment which affect human health or the quality of life. The following are examples of environmentally related measurements: (a) A determination of pollutant concentrations from sources or in the ambient environment, including studies of pollutant transport and fate; (b) a determination of the effects of pollutants on human health and on the environment; (c) a determination of the risk/benefit of pollutants in the environment; (d) a determination of the quality of environmental data used in economic studies; and (c) a determination of the environmental impact of cultural and natural processes.

Environmentally transformed (Section 721.3 [7-27-89]) A chemical substance is "environmentally transformed" when its chemical structure changes as a result of the action of environmental processes on it.

EPA (Section 124.41) shall have the meaning set forth in Section 124.2, except when EPA has delegated authority to administer those regulations to another agency under the applicable subsection of 40 CFR 52.21, the term EPA shall mean the delegate agency.

EPA (Section 35.4010 [10/1/92])* means the Environmental Protection Agency. Where a State administers the TAG Program, the term "EPA" may mean a State agency.

EPA (Sections 2.100; 7.25; 16.2; 17.2; 21.2; 122.2; 124.2; 144.3; 146.3; 147.2902; 160.3; 233.3; 270.2; 300.5; 372.3 [2/16/88]; 704.3 [12/22/88]*; 707.63; 710.2; 712.3; 716.3; 720.3; 723.50; 723.250; 763.63; 790.3; 792.3) means the United States Environmental Protection Agency.

EPA Acknowledgement of Consent (Section 262.51 [7/19/88])* means the cable sent to EPA from the U.S. Embassy in a receiving country that acknowledges the written consent of the receiving country to accept the hazardous waste and describes the terms and conditions of the receiving country's consent to the shipment.

EPA and the Agency (Section 57.103) means the Administrator of the U.S. Environmental Protection Agency, or the Administrator's authorized representative.

EPA assistance (Section 7.25) means any grant or cooperative agreement, loan, contract (other than a procurement contract or a contract of insurance or guaranty), or any other arrangement by which EPA provides or otherwise makes available assistance in the form of: (1) Funds; (2) Services of personnel; or (3) Real or personal property or any interest in or use of such property, including: (i) Transfers or leases of such property for less than fair market value or for reduced consideration; and (ii) Proceeds from a subsequent transfer or lease of such property if EPA's share of its fair market value is not returned to EPA.

EPA Claims Officer (Section 14.2) is the Agency official delegated the responsibility by the Administrator to carry out the provisions of the Act.

EPA Enforcement Officer (Section 86.402-78) means any officer or employee of the Environmental Protection Agency so designated in writing by the Administrator (or by his designee).

EPA Enforcement Officer (Section 86.082-2) means any officer or employee of the Environmental Protection Agency so designated in writing by the Administrator (or by his designee).

EPA hazardous waste number (Section 260.10) means the number assigned by EPA to each hazardous waste listed in Part 261, Subpart D, of this chapter and to each characteristic identified in Part 261, Subpart C, of this chapter.

EPA identification number (Section

260.10) means the number assigned by EPA to each generator, transporter, and treatment, storage, or disposal facility.

EPA identification number (Section 761.3 [12-21-89]) means the 12-digit number assigned to a facility by EPA upon notification of PCB waste activity under Sec. 761.205.

EPA legal office (Section 2.201) means the EPA General Counsel and any EPA office over which the General Counsel exercises supervisory authority, including the various Offices of Regional Counsel. (See paragraph (m) of this section.)

EPA office (Section 2.201) means any organizational element of EPA, at any level or location. (The terms EPA office and EPA legal office are used in this subpart for the sake of brevity and ease of reference. When this subpart requires that an action be taken by an EPA office or by an EPA legal office, it is the responsibility of the officer or employee in charge of that office to take the action or ensure that it is taken.)

EPA record or record (Section 2.100 [1/5/88])* means any document, writing, photograph, sound or magnetic recording, drawing, or other similar thing by which information has been preserved, from which the information can be retrieved and copied, and over which EPA has possession or control. It may include copies of the records of other Federal agencies (see Section 2.111(d)). The term includes informal writings (such as drafts and the like), and also includes information preserved in a form which must be translated or deciphered by machine in order to be intelligible to humans. The term includes documents and the like which were created or acquired by EPA, its predecessors, its officers, and its employees by use of Government funds or in the course of transacting official business. However, the term does not include materials which are the personal records of an EPA officer or employee. Nor does the term include materials published by non-Federal organizations which are readily available to the public, such as books, journals, and periodicals available through reference libraries, even if such materials are in EPA's possession.

EPA region (Section 260.10) means the states and territories found in any one of the following ten regions: Region I - Maine, Vermont, New Hampshire, Massachusetts, Connecticut, and Rhode Island; Region II - New York, New Jersey, Commonwealth of Puerto Rico, and the U. S. Virgin Islands; Region III - Pennsylvania, Delaware, Maryland, West Virginia, Virginia, and the District of Columbia; Region IV - Kentucky, Tennessee, North Carolina, Mississippi, Alabama, Georgia, South Carolina, and Florida; Region V - Minnesota, Wisconsin, Illinois, Michigan, Indiana and Ohio; Region VI - New Mexico, Oklahoma, Arkansas, Louisiana, and Texas;

Region VII - Nebraska, Kansas, Missouri, and Iowa; Region VIII - Montana, Wyoming, North Dakota, South Dakota, Utah, and Colorado; Region IX - California, Nevada, Arizona, Hawaii, Guam, American Samoa, Commonwealth of the Northern Mariana Islands; Region X - Washington, Oregon, Idaho, and Alaska.

EPA-Approved Emission Test or Emission Short Test (Section 85.2102) means any test prescribed under 40 CFR 85.2201 et seq., and meeting all of the requirements thereunder.

Ephippium (Sections 797.1300; 797.1330) means a resting egg which develops under the carapace in response to stress conditions in daphnids.

Equal opportunity clause (Section 8.2) means the contract provisions set forth in section 4(a) or (b), as appropriate.

Equipment (Section 264.1031 [6-21-90]) means each valve, pump, compressor, pressure relief device, sampling connection system, open-ended valve or line, or flange, and any control devices or systems required by this subpart.

Equipment (Section 35.6015 [1/27/89]) means tangible, nonexpendable, personal property having a useful life of more than one year and an acquisition cost of $5,000 or more per unit.

Equipment (Section 60.481) means each pump, compressor, pressure relief device, sampling connection system, open-ended valve or line, valve, and flange or other connector in VOC service and any devices or systems required by this subpart.

Equipment (Section 60.591) means each valve, pump, pressure relief device, sampling connection system, open-ended valve or line, and flange or other connector in VOC service. For the purposes of recordkeeping and reporting only, compressors are considered equipment.

Equipment (Section 60.631) means each pump, pressure relief device, open-ended valve or line, valve, compressor, and flange or other connector that is in VOC service or in wet gas service, and any device or system required by this subpart.

Equipment (Section 61.131 [9-14-89]) means each pump, valve, exhauster, pressure relief device, sampling connection system, open-ended valve or line, and flange or other connector in benzene service.

Equipment (Section 61.241) means each pump, compressor, pressure relief device, sampling connection system, open-ended valve or line, valve, flange or other connector, product accumulator vessel in VHAP service, and any control devices or systems required by this subpart.

Equipment room (change room) (Section 763.121) means a contaminated room located within the decontamination area that is supplied with impermeable bags or containers for the disposal of contaminated protective clothing and equipment.

Equivalency projects (Section 35.3105 [3-19-90]) Those section 212 wastewater treatment projects constructed in whole or in part before October 1, 1994, with funds "directly made available by" the capitalization grant. These projects must comply with the requirements of section 602(b)(6) of the Act.

Equivalent (Section 790.3) means that a chemical substance or mixture is able to represent or substitute for another in a test or series of tests, and that the data from one substance can be used to make scientific and regulatory decisions concerning the other substance.

Equivalent diameter (Section 60.711 [10/3/88]) means four times the area of an opening divided by its perimeter.

Equivalent diameter (Section 60.741 [9-11-89]) means four times the area of an opening divided by its perimeter.

Equivalent live weight killed (Sections 432.11; 432.21; 432.31; 432.41). See **ELWK (Sections 432.11; 432.21; 432.31; 432.41)**.

Equivalent method (Section 260.10) means any testing or analytical method approved by the Administrator under Sections 260.20 and 260.21.

Equivalent method (Section 50.1) means a method of sampling and analyzing the ambient air for an air pollutant that has been designated as an equivalent method in accordance with Part 53 of this chapter; it does not include a method for which an equivalent method designation has been cancelled in accordance with Section 53.11 or Section 53.16 of this chapter.

Equivalent method (Section 53.1) means a method of sampling and analyzing the ambient air for an air pollutant that has been designated as an equivalent method in accordance with this part; it does not include a method for which an equivalent method designation has been cancelled in accordance with Section 53.11 or Section 53.16.

Equivalent method (Section 60.2) means any method of sampling and analyzing for an air pollutant which has been demonstrated to the Administrator's satisfaction to have a consistent and quantitatively known relationship to the reference method, under specified conditions.

Equivalent P_2O_5 feed (Sections 60.211; 60.221; 60.231) means the quantity of phosphorus, expressed as phosphorous pentoxide, fed to the process.

Equivalent P_2O_5 stored (Section 60.241) means the quantity of phosphorus, expressed as phosphorus pentoxide, being cured or stored in the affected facility.

Equivalent petroleum-based fuel economy value (Section 600.502-81) means a number which represents the average number of miles traveled by an electric vehicle per gallon of gasoline. This definition applies beginning with the 1979 model year.

Equivalent Sound Level (Section 201.1) means the level, in decibels, of the mean-square A-weighted sound pressure during a stated time period, with reference to the square of the standard reference sound pressure of 20 micropascals. It is the level of the sound exposure divided by the time period and is abbreviated as L_{eq}.

Equivalent test weight (Section 86.094-2 [6-5-91]) means the weight, within an inertia weight class, which is used in the dynamometer testing of a vehicle and which is based on its loaded vehicle weight or adjusted loaded vehicle weight in accordance with the provisions of subparts A and B of this part.

ERT (Section 300.5) is the Environmental Response Team.

ESP (Section 60.471). See **Electrostatic precipitator (Section 60.471)**.

Establishment (Section 167.1) for purposes of this part, means each site where a pesticide, as defined by this Act, or a device is produced, regardless of whether such site is independently owned or operated and regardless of whether such site is domestic and producing any pesticide or device for export only or whether the site is foreign and producing any pesticide or device for import into the United States.

Establishment (Section 167.3 [9/8/88]) means any site where a pesticidal product, active ingredient, or device is produced, regardless of whether such site is independently owned or operated, and regardless of whether such site is domestic and producing a pesticidal product for export only, or whether the site is foreign and producing any pesticidal product for import into the United States.

Establishment (Section 372.3 [2/16/88]) means an economic unit, generally at a single physical location, where business is conducted or where services or industrial operations are performed.

Etching (Section 467.02). See **Cleaning (Section 467.02)**.

Ethanol blender (Section 80.2 [3/22/89]) means any person who owns, leases, operates, controls, or supervises an ethanol blending plant.

Ethanol blending plant (Section 80.2 [3/22/89]) means any refinery at which gasoline is produced solely through the addition of ethanol to gasoline, and at which the quality or quantity of gasoline is not altered in any other manner.

Ethnic foods (Section 407.81) shall mean the production of canned and frozen Chinese and Mexican specialties utilizing fresh and pre-processed bean sprouts, bamboo shoots, water chestnuts, celery, cactus, tomatoes, and other similar vegetables necessary for the production of the various characteristic product styles.

Ethylene dichloride plant (Section 61.61) includes any plant which produces ethylene dichloride by reaction of oxygen and hydrogen chloride with ethylene.

Ethylene dichloride purification (Section 61.61 [2-10-90])* includes any part of the process of ethylene dichloride purification following ethylene dichloride formation, but excludes crude, intermediate, and final ethylene dichloride storage tanks.

Evaluation program or program (Section 610.11) means the sequence of analyses and tests prescribed by the Administrator as described in Section 610.13 in order to evaluate the performance of a retrofit device.

Evaporative emission code (Section 86.082-2) means a unique combination, in an evaporative emission family-evaporative emission control system combination, of purge system calibrations, fuel tank and carburetor bowl vent calibrations and other fuel system and evaporative emission control system components and calibrations specified by the Administrator.

Evaporative emissions (Section 86.082-2) means hydrocarbons emitted into the atmosphere from a motor vehicle, other than exhaust and crankcase emissions.

Evaporative/refueling emission control system (Section 86.098-2 [4/6/94])* means a unique combination within an evaporative/refueling family of canister adsorptive material, purge system configuration, purge strategy, and other parameters determined by the Administrator to affect evaporative and refueling emission control system durability or deterioration factors.

Evaporative/refueling emission family (Section 86.098-2 [4/6/94])* means the basic classification unit of a manufacturers' product line used for the purpose of evaporative and refueling emissions test fleet selection and determined in accordance with Sec. 86.098-24.

Evaporative vehicle configuration (Section 86.082-2) means a unique combination of basic engine, engine code, body type, and evaporative emission code.

Ex parte communication (Section 305.12) means any communication, written or oral, relating to the merits of the proceeding between the Arbitrator and any party, or other interested person which was not originally filed or stated in the administrative record or in the hearing.

Ex parte communication (Section 304.12 [5/30/89]) means any communication, written or oral, relating to the merits of the arbitral proceeding, between the Arbitrator and any interested person, which was not originally filed or stated in the administrative record of the proceeding. Such communication is not ex parte communication if all parties to the proceeding have received prior written notice of the proposed communication and have been given the opportunity to be present and to participate therein.

Ex parte communication (Sections 57.809; 124.78) means any communication, written or oral, relating to the merits of the proceeding between the decisional body and an interested person outside the Agency or the Agency trial staff which was not originally filed or stated in the administrative record or in the hearing. Ex parte communications do not include: (i) Communications between Agency employees other than between the Agency trial staff and the members of the decisional body; (ii) Discussions between the decisional body and either: (A) Interested persons outside the Agency, or (B) The Agency trial staff, if all parties have received prior written notice of the proposed communications and have been given the opportunity to be present and participate therein.

Excavation zone (Section 280.12 [9/23/88]) means the volume containing the tank system and backfill material bounded by the ground surface, walls, and floor of the pit and trenches into which the UST system is placed at the time of installation.

Excess ammonia-liquor storage tank (Section 61.131 [9-14-89]) means any tank, reservoir, or container used to collect or store a flushing liquor solution prior to ammonia or phenol recovery.

Excess emissions (Section 51.100) means emissions of an air pollutant in excess of an emission standard.

Excess Emissions and Monitoring Systems Performance Report (Section 60.2 [12-13 90]) is a report that must be submitted periodically by a source in order to provide data on its compliance with stated emission limits and operating parameters, and on the performance of its monitoring systems.

Excess pesticides (Section 165.1) means all pesticides which cannot be legally sold pursuant to the Act or which are to be discarded.

Excess property (Section 35.6015 [1/27/89]) means any property under the control of a Federal agency that is not required for immediate or foreseeable needs and thus is a candidate for disposal.

Excessive concentration (Section 51.100) is defined for the purpose of determining good engineering practice stack height under Section 51.100(ii)(3) and means: (1) For sources seeking credit for stack height exceeding that established under Section 51.100(ii)(2) a maximum ground-level concentration due to emissions from a stack due in whole or part to downwash, wakes, and eddy effects produced by nearby structures or nearby terrain features which individually is at least 40 percent in excess of the maximum concentration experienced in the absence of such downwash, wakes, or eddy effects and which contributes to a total concentration due to emissions from all sources that is greater than an ambient air quality standard. For sources subject to the prevention of significant deterioration program (40 CFR 51.166 and 52.21), an excessive concentration alternatively means a maximum ground-level concentration due to emissions from a stack due in whole or part to downwash, wakes, or eddy effects produced by nearby structures or nearby terrain features which individually is at least 40 percent in excess of the maximum concentration experienced in the absence of such downwash, wakes, or eddy effects and greater than a prevention of significant deterioration increment. The allowable emission rate to be used in making demonstrations under this part shall be prescribed by the new source performance standard that is applicable to the source category unless the owner or operator demonstrates that this emission rate is infeasible. Where such demonstrations are approved by the authority administering the State implementation plan, an alternative emission rate shall be established in consultation with the source owner or operator. (2) For sources seeking credit after October 11, 1983, for increases in existing stack heights up to the heights established under Section 51.100(ii)(2), either (i) a maximum ground-level concentration due in whole or part to downwash, wakes or eddy effects as provided in paragraph (kk)(1) of this section, except that the emission rate specified by any applicable State implementation plan (or, in the absence of such a limit, the actual emission rate) shall be used, or (ii) the actual presence of a local nuisance caused by the existing stack, as determined by the authority administering the State implementation plan; and (3) For sources seeking credit after January 12, 1979, for a stack height determined under Section 51.100(ii)(2) where the authority administering the State implementation plan requires the use of a field study or fluid model to verify GEP stack height, for sources seeking stack height credit after November 9, 1984, based on the aerodynamic influence of cooling towers, and for sources seeking stack height credit after December 31, 1970, based on the aerodynamic influence of structures not adequately represented by the equations in Section 51.100(ii)(2), a maximum ground-level concentration due in whole or part to downwash, wakes or eddy effects that is at least 40 percent in excess of the maximum concentration experienced in the absence of such downwash, wakes, or eddy effects.

Excessive infiltration/inflow (Section 35.905) means the quantities of infiltration/inflow which can be economically eliminated from a sewerage system by rehabilitation, as determined in a cost-effectiveness analysis that compares the costs for correcting the infiltration/inflow conditions to the total costs for transportation and treatment of the infiltration/inflow, subject to the provisions in Section 35.927.

Excessive infiltration/inflow (Section 35.2005) means the quantities of infiltration/inflow which can be economically eliminated from a sewer system as determined in a cost-effectiveness analysis that compares the costs for correcting the infiltration/inflow conditions to the total costs for transportation and treatment of the infiltration/inflow. (See Sections 35.2005(b) (28) and (29) and 35.2120.)

Excluded manufacturing process (Section 761.3) means a manufacturing process in which quantities of PCBs, as determined in accordance with the definition of inadvertently generated PCBs, calculated as defined, and from which releases to products, air, and water meet the requirements of paragraphs (1) through (5) of this definition, or the importation of products containing PCBs as unintentional impurities, which products meet the requirements of paragraph (1) and (2) of this definition. (1) The concentration of inadvertently generated PCBs in products leaving any manufacturing site or imported into the United States must have an annual average of less than 25 ppm, with a 50 ppm maximum. (2) The concentration of inadvertently generated PCBs in the components of detergent bars leaving the manufacturing site or imported into the United States must be less than 5 ppm. (3) The release of inadvertently generated PCBs at the point at which emissions are vented to ambient air must be less than 10 ppm. (4) The amount of inadvertently generated PCBs added to water discharged from a manufacturing site must be less than 100 micrograms per resolvable gas chromatographic peak per liter of water discharged. (5) Disposal of any other process wastes above concentrations of 50 ppm PCB must be in accordance with Subpart D of this part.

Excluded PCB products (Section 761.3 [6/27/88]) means PCB materials which appear at concentrations less than 50 ppm, including but not limited to: (1) Non-Aroclor inadvertently generated PCBs as a byproduct or impurity resulting from a chemical manufacturing process. (2) Products contaminated with Aroclor or other PCB materials from historic PCB uses (investment casting waxes are one example). (3) Recycled fluids and/or equipment contaminated during use involving the products described in paragraphs (1) and (2) of this definition (heat transfer and hydraulic fluids and equipment and other electrical equipment components and fluids are examples). (4) Used oils, provided that in the cases of paragraphs (1) through (4) of this definition: (i) The products or source of the products containing < 50 ppm concentration PCBs were legally manufactured, processed, distributed in commerce, or used before October 1, 1984. (ii) The products or source of the products containing < 50 ppm concentrations PCBs were legally manufactured, processed, distributed in commerce, or used, i.e., pursuant to authority granted by EPA regulation, by exemption petition, by settlement agreement, or pursuant to other Agency-approved programs; (iii) The resulting PCB concentration (i.e., below 50 ppm) is not a result of dilution, or leaks and spills of PCBs in concentrations over 50 ppm.

Exclusive use study (Section 152.83) means a study that meets each of the following requirements: (1) The study pertains to a new active ingredient (new chemical) or new combination of active ingredients (new combination) first registered after September 30, 1978; (2) The study was submitted in support of, or as a condition of approval of, the application resulting in the first registration of a product containing such new chemical or new combination (first registration), or an application to amend such registration to add a new use; and (3) The study was not submitted to satisfy a data requirement imposed under FIFRA section 3(c)(2)(B); Provided that, a study is an exclusive use study only during the 10-year period following the date of the first registration.

Executive Order (Section 8.33) means Executive Order 11246, 30 FR 12319, as amended.

Exempted aquifer (Section 144.3) means an aquifer or its portion that meets the criteria in the definition of underground source of drinking water but which has been exempted according to the procedures in Section 144.7.

Exempted aquifer (Section 146.3) means an aquifer or its portion that meets the criteria in the definition of underground source of drinking water but which has been exempted according to the procedures of Section 144.8(b).

Exempted on-highway diesel fuel (Section 80.2 [8-21-90]) means any diesel fuel which is produced by a small refinery under the provisions of sections 80.29(a)(2) and 80.29(c).

Exemption (Section 790.3) means an exemption from a testing requirement of a test rule promulgated under section 4 of the Act and Part 799 of this chapter.

Exemption application (Section 700.43 [8/17/88]) means any application submitted to EPA under section 5(h)(2) of the Act.

Exemption category (Section 723.175) means a category of chemical substances for which a person(s) has applied for or been granted an exemption under section 5(h)(4) of the Act (15 U.S.C. 2604).

Exemption holder (Section 791.3) refers to a manufacturer or processor, subject to a test rule, that has received an exemption under sections 4(c)(1) or 4(c)(2) of TSCA from the requirement to conduct a test and submit data.

Exemption notice (Section 700.43 [8/17/88]) means any notice submitted to EPA under Section 723.175 of this chapter.

Exhaust emissions (Section 87.1) means substances emitted to the atmosphere from the exhaust discharge nozzle of an aircraft or aircraft engine.

Exhaust emissions (Sections 86.082-2; 86.402-78) means substances emitted to the atmosphere from any opening downstream from the exhaust port of a motor vehicle engine.

Exhaust gas (Section 61.61 [2-10-90])* means any offgas (the constituents of which may consist of any fluids, either as a liquid and/or gas) discharged directly or ultimately to the atmosphere that was initially contained in or was in direct contact with the equipment for which gas limits are prescribed in Secs. 61.62(a) and (b); 61.63(a); 61.64 (a)(1), (b), (c), and (d); 61.65 (b)(1)(ii), (b)(2), (b)(3), (b)(5), (b)(6)(ii), (b)(7), and (b)(9)(ii); and 61.65(d). A leak as defined in paragraph (w) of this section is not an exhaust gas. Equipment which contains exhaust gas is subject to Sec. 61.65(b)(8), whether or not that equipment contains 10 percent by volume vinyl chloride.

Exhaust header pipe (Section 205.165) means any tube of constant diameter which conducts exhaust gas from an engine exhaust port to other exhaust system components which provide noise attenuation. Tubes with cross connections or internal baffling are not considered to be exhaust header pipes.

Exhaust system (Section 202.10) means the system comprised of a combination of components which provides for enclosed flow of exhaust gas from engine parts to the atmosphere.

Exhaust system (Section 205.151) means the combination of components which provides for the enclosed flow of exhaust gas from the engine exhaust port to the atmosphere. Exhaust system further means any constituent components of the combination which conduct exhaust gases and which are sold as separate products. Exhaust system does not mean any of the constituent components of the combination, alone, which do not conduct exhaust gases, such as brackets and other mounting hardware.

Exhaust system (Section 205.51) means the system comprised of a combination of components which provides for enclosed flow of exhaust gas from engine exhaust port to the atmosphere.

Exhauster (Section 61.131 [9-14-89]) means a fan located between the inlet gas flange and outlet gas flange of the coke oven gas line that provides motive power for coke oven gases.

Existing Class II Wells (Section 147.2902) means wells that were authorized by BIA and constructed and completed before the effective date of this program.

Existing component (Section 260.10). See **Existing tank system (Section 260.10)**.

Existing facility (Section 260.10). See **Existing hazardous waste management (HWM) facility (Section 260.10)**.

Existing facility (Section 270.2). See **Existing hazardous waste management (HWM) facility (Section 270.2)**.

Existing facility (Section 60.2) means, with reference to a stationary source, any apparatus of the type for which a standard is promulgated in this part, and the construction or modification of which was commenced before the date of proposal of that standard; or any apparatus which could be altered in such a way as to be of that type.

Existing hazardous waste management (HWM) facility or existing facility (Section 260.10) means a facility which was in operation or for which construction commenced on or before November 19, 1980. A facility has commenced construction if: (1) The owner or operator has obtained the Federal, State and local approvals or permits necessary to begin physical construction; and either (2)(i) A continuous on-site, physical construction program has begun; or (ii) The owner or operator has entered into contractual obligations - which cannot be cancelled or modified without substantial loss - for physical construction of the facility to be completed within a reasonable time.

Existing hazardous waste management (HWM) facility or existing facility (Section 270.2) means a facility which was in operation or for which construction commenced on or before November 19, 1980. A facility has commenced construction if: (a) The owner or operator has obtained the Federal, State and local approvals or permits necessary to begin physical construction; and either (b)(1) A continuous on-site, physical construction program has begun; or (2) The owner or operator has entered into contractual obligations which cannot be cancelled or modified without substantial loss - for physical construction of the facility to be completed within a reasonable time.

Existing HWM facility (Section 270.2). See **Existing hazardous waste management (HWM) facility (Section 270.2)**.

Existing HWM facility (Section 260.10). See **Existing hazardous waste management (HWM) facility (Section 260.10)**.

Existing impoundment (Section 61.251 [12-15-89]) means any uranium mill tailings impoundment which is licensed to accept additional tailings and is in existence as of December 15, 1989.

Existing indirect dischargers (Section 420.31) means only those two iron blast furnace operations with discharges to publicly owned treatment works prior to May 27, 1982.

Existing injection well (Sections 144.3; 146.3) means an injection well other than a new injection well.

Existing portion (Section 192.31 [11/5/93])* means that land surface area of an existing surface impoundment on which significant quantities of uranium byproduct materials have been placed prior to promulgation of this standard.

Existing portion (Section 260.10) means that land surface area of an existing waste management unit, included in the original Part A permit application, on which wastes have been placed prior to the issuance of a permit.

Existing source (Section 122.29) means any source which is not a new source or a new discharger.

Existing source (Section 129.2) means any source which is not a new source as defined in 40 CFR Section 129.2.

Existing source (Section 61.02) means any stationary source which is not a new source.

Existing Stationary Facility (Section 51.301) means any of the following stationary sources of air pollutants, including any reconstructed source, which was not in operation prior to August 7, 1962, and was in existence on August 7, 1977, and has the potential to emit 250 tons per year or more of any air pollutant. In determining potential to emit, fugitive emissions, to the extent quantifiable, must be counted. (1) Fossil-fuel fired steam electric plants of more than 250 million British thermal units per hour heat input, (2) Coal cleaning plants (thermal dryers), (3) Kraft pulp mills, (4) Portland cement plants, (5) Primary zinc smelters, (6) Iron and steel mill plants, (7) Primary aluminum ore reduction plants, (8) Primary copper smelters, (9) Municipal incinerators capable of charging more than 250 tons of refuse per day, (10) Hydrofluoric, sulfuric, and nitric acid plants, (11) Petroleum refineries, (12) Lime plants, (13) Phosphate rock processing plants, (14) Coke oven batteries, (15) Sulfur recovery plants, (16) Carbon black plants (furnace process), (17) Primary lead smelters, (18) Fuel conversion plants, (19) Sintering plants, (20) Secondary metal production facilities, (21) Chemical process plants, (22) Fossil-fuel boilers of more than 250 million British thermal units per hour heat input, (23) Petroleum storage and transfer facilities with a capacity exceeding 300,000 barrels, (24) Taconite ore processing facilities, (25) Glass fiber processing plants, and (26) Charcoal production facilities.

Existing tailings pile (Section 61.251) means a tailings pile that is in operation on the effective date of this rule.

Existing tank system (Section 280.12 [9/23/88]) means a tank system used to contain an accumulation of regulated substances or for which installation has commenced on or before December 22, 1988. Installation is considered to have commenced if: (a) The owner or operator has obtained all federal, state, and local approvals or permits necessary to begin physical construction of the site or installation of the tank system; and if, (b)(1) Either a continuous on-site physical construction or installation program has begun; or, (2) The owner or operator has entered into contractual obligations - which cannot be cancelled or modified without substantial loss - for physical construction at the site or installation of the tank system to be completed within a reasonable time.

Existing tank system or existing component (Section 260.10) means a tank system or component that is used for the storage or treatment of hazardous waste and that is in operation, or for which installation has commenced on or prior to July 14, 1986. Installation will be considered to have commenced if the owner or operator has obtained all Federal, State, and local approvals or permits necessary to begin physical construction of the site or installation of the tank system and if either (1) a continuous on-site physical construction or installation program has begun, or (2) the owner or operator has entered into contractual obligations - which cannot be canceled or modified without substantial loss - for physical construction of the site or installation of the tank system to be completed within a reasonable time.

Existing unit (Section 72.2 [7/30/93])* means a unit (including a unit subject to section 111 of the Act) that commenced commercial operation before November 15, 1990 and that on or after November 15, 1990 served a generator with nameplate capacity of greater than 25 MWe. "Existing unit" does not include simple combustion turbines or any unit that on or after November 15, 1990 served only generators with a nameplate capacity of 25 MWe or less. Any "existing unit" that is modified, reconstructed, or repowered after November 15, 1990 shall continue to be an "existing unit."

Existing uses (Section 131.3) are those uses actually attained in the water body on or after November 28, 1975, whether or not they are included in the water quality standards.

Existing vapor processing system (Section 60.501) means a vapor processing system (capable of achieving emissions to the atmosphere no greater than 80 milligrams of total organic compounds per liter of gasoline loaded), the construction or refurbishment of which was commenced before December 17, 1980, and which was not constructed or refurbished after that date.

Existing vessel (Section 140.1) refers to any vessel on which construction was initiated before January 30, 1975.

Existing well (Section 146.61 [7/26/88]) means a Class I well which was authorized prior to August 25, 1988 by an approved State program, or an EPA-administered program or a well which has become a Class I well as a result of a change in the definition of the injected waste which would render the waste hazardous under Section 261.3 of this Part.

Expedited Hearing (Section 164.2 [7/10/92])* means a hearing commenced as the result of the issuance of a notice of intention to suspend or the suspension of a registration of a pesticide by an emergency order, and is limited to a consideration as to whether a pesticide presents an imminent hazard which justifies such suspension.

Expendable personal property (Section 30.200) means all tangible personal property other than nonexpendable personal property.

Experimental animals (Section 172.1) means individual animals or groups of animals, regardless of species, intended for use and used solely for research purposes and does not include animals intended to be used for any food purposes.

Experimental furnace (Section 60.291) means a glass melting furnace with the sole purpose of operating to evaluate glass melting processes, technologies, or glass products. An experimental furnace does not produce glass that is sold (except for further research and development purposes) or that is used as a raw material for nonexperimental furnaces.

Experimental start date (Section 792.3 [8-17-89]) means the first date the test substance is applied to the test system.

Experimental start date (Section 160.3 [8-17-89]) means the first date the test substance is applied to the test system.

Experimental technology (Section 146.3) means a technology which has not been proven feasible under the conditions in which it is being tested.

Experimental termination date (Section 160.3 [8-17-89]) means the last date on which data are collected directly from the study.

Experimental termination date (Section 792.3 [8-17-89] means the last date on which data are collected directly from the study.

Experimental use permit review (Section 152.403 [5/26/88]) means review of an application for a permit pursuant to section 5 of FIFRA to apply a limited quantity of a pesticide in order to accumulate information necessary to register the pesticide. The application may be for a new chemical or for a new use of an old chemical. The fee applies to such experimental uses of a single unregistered active ingredient (no limit on the number of other active ingredients, in a tank mix, already registered for the crops involved) and no more than three crops. This fee does not apply to experimental use permits required for small-scale field testing of microbial pest control agents (40 CFR 172.3).

Exploratory facility (Section 435.11 [3/4/93])* shall mean any fixed or mobile structure subject to this subpart that is engaged in the drilling of wells to determine the nature of potential hydrocarbon reservoirs.

Explosive gas (Section 257.3-8) means methane (CH_4).

Export (Section 82.3 [12/10/93])* means the transport of virgin, used, or recycled controlled substances from inside the United States or its territories to persons outside the United States or its territories, excluding United States military bases and ships for on-board use.

Export (Section 82.3 [6-15-90])* means the transport of virgin, used or recycled controlled substances from inside the United States or its territories to persons outside the United States or its territories, excluding United States military bases and ships for on-board use.

Export exemption (Section 211.102) means an exemption from the prohibitions of section 10(a)(3) and (4) of the Act; this type of exemption is granted by statute under section 10(b)(2) of the Act for the purpose of exporting regulated products.

Export exemption (Section 85.1702) means an exemption granted by statute under section 203(b)(3) of the Act for the purpose of exporting new motor vehicles or new motor vehicle engines.

Export exemption (Sections 204.2; 205.2) means an exemption from the prohibitions of section 10(a) (1), (2), (3), and (4) of the Act, granted by statute under section 10(b)(2) of the Act for the purpose of exporting regulated products.

Exporter (Section 707.63) means the person who, as the principal party in interest in the export transaction, has the power and responsibility for determining and controlling the sending of the chemical substance or mixture to a destination out of the customs territory of the United States.

Exporter (Section 82.3 [12/10/93])* means the person who contracts to sell controlled substances for export or transfers controlled substances to his affiliate in another country.

Exporter (Section 82.3 [8/12/88]) means the person who contracts to sell controlled substances for export, or transfers controlled substances to his affiliate in another country.

Exposure period (Section 797.2050) is the 5-day period during which test birds are offered a diet containing the test substance.

Extent of chlorination (Section 704.83) means the percent by weight of chlorine.

Extent of chlorination (Section 704.85) means the percent by weight of chlorine for each isomer (ortho, meta, and para).

Exterior base coating operation (Section 60.491) means the system on each beverage can surface coating line used to apply a coating to the exterior of a two-piece beverage can body. The exterior base coat provides corrosion resistance and a background for lithography or printing operations. The exterior base coat operation consists of the coating application station, flashoff area, and curing oven. The exterior base coat may be pigmented or clear (unpigmented).

External floating roof (Section 61.341 [3-7-90]) means a pontoon-type or double-deck type cover with certain rim sealing mechanisms that rests on the liquid surface in a waste management unit with no fixed roof.

Extraction plant (Section 61.31) means a facility chemically processing beryllium ore to beryllium metal, alloy, or oxide, or performing any of the intermediate steps in these processes.

Extraction site (Section 230.3) means the place from which the dredged or fill material proposed for discharge is to be removed.

Extractor column (Section 796.1860) is used to extract the solute from the saturated solutions produced by the generator column. After extraction onto a chromatographic support, the solute is eluted with a solvent/water mixture and subsequently analyzed by high pressure liquid chromatography (HPLC). A detailed description of the preparation of the extractor column is given in paragraph (b)(1)(i) of this section.

Extremely hazardous substance (Section 355.20 [7-26-90])* means a substance listed in Appendices A and B of this part.

Extremely hazardous substance (Section 370.2 [10/15/87]) means a substance listed in the Appendices to 40 CFR Part 355, Emergency Planning and Notification.

Extrusion (Section 467.02) is the application of pressure to a billet of aluminum, forcing the aluminum to flow through a die orifice. The extrusion subcategory is based on the extrusion process.

Extrusion (Section 468.02) shall mean the application of pressure to a copper workpiece, forcing the copper to flow through a die orifice.

Extrusion (Section 471.02) is the application of pressure to a billet of metal, forcing the metal to flow through a die orifice.

Extrusion die cleaning (Section 467.31) shall mean the process by which the steel dies used in extrusion of aluminum are cleaned. The term includes a dip into a concentrated caustic bath to dissolve the aluminum followed by a water rinse. It also includes the use of a wet scrubber with the die cleaning operation.

Extrusion heat treatment (Section 468.02) shall mean the spray application of water to a workpiece immediately following extrusions for the purpose of heat treatment.

F

F (Section 60.741 [9-11-89]) means the VOC emission capture efficiency of the vapor capture system achieved for the duration of the emission test (expressed as a fraction).

Fabricating (Section 61.141 [11-20-90])* means any processing (e.g., cutting, sawing, drilling) of a manufactured product that contains commercial asbestos, with the exception of processing at temporary sites (field fabricating) for the construction or restoration of facilities. In the case of friction products, fabricating includes bonding, debonding, grinding, sawing, drilling, or other similar operations performed as part of fabricating.

Facial tissue (Section 250.4 [10/6/87, 6/22/88]) means a class of soft absorbent papers in the sanitary tissue group.

Facilities (Section 8.2) includes, but is not limited to, waiting rooms, work areas, restaurants and other eating areas, time clocks, restrooms, washrooms, locker rooms and other storage or dressing areas, parking lots, drinking fountains, recreation or entertainment areas, transportation, and housing facilities provided for employees.

Facilities eligible for treatment equivalent to secondary treatment (Section 133.101). Treatment works shall be eligible for consideration for effluent limitations described for treatment equivalent to secondary treatment (Section 133.105), if: (1) The BOD_5 and SS effluent concentrations consistently achievable through proper operation and maintenance (Section 133.101(f)) of the treatment works exceed the minimum level of the effluent quality set forth in Sections 133.102(a) and 133.102(b); (2) A trickling filter or waste stabilization pond is used as the principal process; and (3) The treatment works provide significant biological treatment of municipal wastewater.

Facilities or equipment (Section 122.29) means buildings, structures, process or production equipment or machinery which form a permanent part of the new source and which will be used in its operation, if these facilities or equipment are of such value as to represent a substantial commitment to construct. It excludes facilities or equipment used in connection with feasibility, engineering, and design studies regarding the source or water pollution treatment for the source.

Facility (Section 12.103 [8/14/87]) means all or any portion of buildings, structures, equipment, roads, walks, parking lots, rolling stock or other conveyances, or other real or personal property.

Facility (Section 15.4) means any building, plant, installation, structure, mine, vessel or other floating craft, location or site of operations owned, leased, or supervised by an applicant, contractor, grantee, or borrower to be used in the performance of a contract, grant, or loan. Where a location or site of operations contains or includes more than one building, plant, installation, or structure, the entire location or site shall be deemed to be a facility, except where the Assistant Administrator determines that independent facilities are located in one geographic area.

Facility (Section 2.310) has the meaning given it in section 101(9) of the Act, 42 U.S.C. 9601(9).

Facility (Section 20.2) means property comprising any new identifiable treatment facility which removes, alters, disposes of, stores, or prevents the creation of pollutants, contaminants, wastes, or heat.

Facility (Section 21.2) means any building, structure, installation or vessel, or portion thereof.

Facility (Section 240.101) means all thermal processing equipment, buildings, and grounds at a specific site.

Facility (Section 245.101) means any building, installation, structure, or public work owned by or leased to the Federal Government. Ships at sea, aircraft in the air, land forces on maneuvers, other mobile facilities, and U.S. Government installations located on foreign soil are not considered Federal facilities for the purpose of these guidelines.

Facility (Section 256.06) refers to any resource recovery system or component thereof, any system, program or facility for resource conservation, and any facility for collection, source separation, storage, transportation, transfer, processing, treatment or disposal of solid waste, including hazardous waste, whether such facility is associated with facilities generating such wastes or not.

Facility (Section 257.2 [10-9-91])* means all contiguous land and structures, other appurtenances, and improvements on the land used for the disposal of solid waste.

Facility (Section 257.2) means any land and appurtenances thereto used for the disposal of solid wastes.

Facility (Section 258.2 [10-9-91]) means all contiguous land and structures, other appurtenances, and improvements on the land used for the disposal of solid waste.

Facility (Section 259.10 [3/24/89]) means all contiguous land and structures, other appurtenances, and improvements on the land, used for treating, destroying, storing, or disposing of regulated medical waste. A facility may consist of several treatment, destruction, storage, or disposal operational units.

Facility (Section 260.10) means all contiguous land, and structures, other appurtenances, and improvements on the land, used for treating, storing, or disposing of hazardous waste. A facility may consist of several treatment, storage, or disposal operational units (e.g., one or more landfills, surface impoundments, or combinations of them).

Facility (Section 300.5 [3-8-90])* as defined by section 101(9) of CERCLA, means any building, structure, installation, equipment, pipe or pipeline (including any pipe into a sewer or publically owned treatment works), well, pit, pond, lagoon, impoundment, ditch, landfill, storage container, motor vehicle, rolling stock, or aircraft, or any site or area, where a hazardous substance has been deposited, stored, dispose of, or placed, or otherwise come to be located; but does not include any consumer use or any vessel.

Facility (Section 302.3) means (1) any building, structure, installation, equipment, pipe or pipeline (including any pipe into a sewer or publicly owned treatment works), well, pit, pond, lagoon, impoundment, ditch, landfill, storage container, motor vehicle, rolling stock, or aircraft, or (2) any site or area where a hazardous substance has been deposited, stored, disposed of, or placed, or otherwise come to be located; but does not include any consumer product in consumer use or any vessel.

Facility (Section 355.20 [7-26-90])* means all buildings, equipment, structure, and other stationary items that are located on a single site or on contiguous or adjacent sites and which are owned or operated by the same person (or by any person which controls, is controlled by, or under common control with, such person). Facility shall include manmade structures in which chemicals are purposefully placed or removed through human means such that it functions as a containment structure for human use. For purposes of emergency release notification, the term includes motor vehicles, rolling stock, and aircraft.

Facility (Section 370.2 [10/15/87]) means all buildings, equipment, structures, and other stationary items that are located on a single site or on contiguous or adjacent sites and that are owned or operated by the same person (or by any person which controls, is controlled by, or under common control with, such person). For purposes of emergency release notification, the term includes motor vehicles, rolling stock, and aircraft.

Facility (Section 372.3 [2/16/88])

means all buildings, equipment, structures, and other stationary items which are located on a single site or on contiguous or adjacent sites and which are owned or operated by the same person (or by any person which controls, is controlled by, or under common control with such person). A facility may contain more than one establishment.

Facility (Section 51.165). See **Building, structure, facility, or installation (Section 51.165)**.

Facility (Section 51.166). See **Building, structure, facility, or installation (Section 51.166)**.

Facility (Section 51.301). See **Building, structure or facility (Section 51.301)**.

Facility (Section 52.21). See **Building, structure, facility, or installation (Section 52.21)**.

Facility (Section 52.24). See **Building, structure, facility, or installation (Section 52.24)**.

Facility (Section 61.101 [12-15-89]) means all buildings, structures and operations on one contiguous site.

Facility (Section 61.141 [11-20-90])* means any institutional, commercial, public, industrial, or residential structure, installation, or building (including any structure, installation, or building containing condominiums or individual dwelling units operated as a residential cooperative, but excluding residential buildings having four or fewer dwelling units); any ship; and any active or inactive waste disposal site. For purposes of this definition, any building, structure, or installation that contains a loft used as a dwelling is not considered a residential structure, installation, or building. Any structure, installation or building that was previously subject to this subpart is not excluded, regardless of its current use or function.

Facility (Section 61.341 [3-7-90]) means all process units and product tanks that generate waste within a stationary source, and all waste management units that are used for waste treatment, storage, or disposal within a stationary source.

Facility (Section 61.91 [12-15-89]) means all buildings, structures and operations on one contiguous site.

Facility (Section 7.25) means all, or any part of, or any interests in structures, equipment, roads, walks, parking lots, or other real or personal property.

Facility (Section 82.3 [8/12/88]) means any process equipment (e.g., reactor, distillation column) to convert raw materials or feedstock chemicals into controlled substances.

Facility component (Section 61.141 [11-20-90])* means any part of a facility including equipment.

Facility mailing list (Section 270.2 [9/28/88]) means the mailing list for a facility maintained by EPA in accordance with 40 CFR 124.10(c)(viii).

Facility or activity (Section 122.2) means any NPDES point source or any other facility or activity (including land or appurtenances thereto) that is subject to regulation under the NPDES program.

Facility or activity (Section 124.2 [5/2/89])* means any HWM facility, UIC injection well, NPDES point source or treatment works treating domestic sewage or State 404 dredge or fill activity, or any other facility or activity (including land or appurtenances thereto) that is subject to regulation under the RCRA, UIC, NPDES, or 404 programs.

Facility or activity (Section 124.41) means a major PSD stationary source or major PSD modification.

Facility or activity (Section 144.3) means any UIC injection well, or another facility or activity that is subject to regulation under the UIC program.

Facility or activity (Section 146.3) means any HWM facility, UIC injection well, NPDES point source, or State 404 dredge and fill activity, or any other facility or activity (including land or appurtenances thereto) that is subject to regulation under the RCRA, UIC, NPDES, or 404 programs.

Facility or activity (Section 233.3) means any State 404 dredge or fill activity, or any other facility or activity (including land or appurtenances thereto) that is subject to regulation under the 404 program.

Facility or activity (Section 270.2) means any HWM facility or any other facility or activity (including land or appurtenances thereto) that is subject to regulation under the RCRA program.

Facility personnel (Section 260.10). See **Personnel (Section 260.10)**.

Facility structures (Section 257.3-8) means any building and sheds or utility or drainage lines on the facility.

Failing compressor (Section 204.51) means that the measured noise emissions of the compressor, when measured in accordance with the applicable procedure, exceeds the applicable standard.

Failing exhaust system (Section 205.165) means that, when installed on any Federally regulated motorcycle for which it is designed and marketed, that motorcycle and exhaust system exceed the applicable standards.

Failing vehicle (Section 205.151) means a vehicle whose noise level is in excess of the applicable standard.

Failing vehicle (Section 205.51) means that the measured emissions of the vehicle, when measured in accordance with the applicable procedure, exceeds the applicable standard.

Fair market value (Section 35.6015 [1/27/89]) means the amount at which property would change hands between a willing buyer and a willing seller, neither being under any compulsion to buy or sell and both having reasonable knowledge of the relevant facts. Fair market value is the price in cash, or its equivalent, for which the property would have been sold on the open market.

Family emission limit (FEL) (Section 86.090-2 [7-26-90]) means an emission level declared by the manufacturer which serves in lieu of an emission standard for certification purposes in any of the averaging, trading, or banking programs. FELs must be expressed to the same number of decimal places as the applicable emission standard. The FEL for an engine family using NOX or particulate NCPs must equal the value of the current NOX or particulate emission standard.

Family particulate emission limit (Section 86.085-2) means the diesel particulate emission level to which an engine family is certified in the particulate averaging program, expressed to an accuracy of one hundredth gram-per-mile. This definition applies beginning with the 1985 model year.

Farm operation (Section 4.2) means any activity conducted solely or primarily for the production of one or more agricultural products or commodities, including timber, for sale or home use, and customarily producing such products or commodities in sufficient quantity to be capable of contributing materially to the operator's support.

Farm tank (Section 280.12 [9/23/88]) is a tank located on a tract of land devoted to the production of crops or raising animals, including fish, and associated residences and improvements. A farm tank must be located on the farm property. Farm includes fish hatcheries, rangeland and nurseries with growing operations.

Fast meter response (Section 201.1) means that the fast response of the sound level meter shall be used. The fast dynamic response shall comply with the meter dynamic characteristics in paragraph 5.3 of the American National Standard Specification for Sound Level Meters, ANSI S1.4-1971. This publication is available from the American National Standards Institute, Inc., 1430 Broadway, New York, New York 10018.

Fast meter response (Sections 202.10; 205.2) means that the fast dynamic response of the sound level meter shall be used. The fast dynamic response shall comply with the meter dynamic characteristics in paragraph 5.3 of the American National Standard Specification for Sound Level Meters, ANSI S1.4-1971. This publication is available from the American National Standards Institute, Inc., 1430 Broadway, New York, New York 10018.

Fast turnaround operation of a spray drying tower (Section 417.151) shall mean operation involving more than 6 changes of formulation in a 30 consecutive day period that are of such degree and type (e.g., high phosphate to no phosphate) as to require cleaning of the tower to maintain minimal product quality.

Fast turnaround operation of automated fill lines (Section 417.161) shall mean an operation involving more than 8 changes of formulation in a 30 consecutive day period that are of such degree and type as to require thorough purging and washing of the fill line to maintain minimal product quality.

Fault (Section 264.18), as used in paragraph (a)(1) of this section, means a fracture along which rocks on one side have been displaced with respect to those on the other side.

Fault (Sections 146.3; 147.2902) means a surface or zone of rock fracture along which there has been displacement.

FCO (Section 300.5) is the Federal Coordinating Officer.

FDA (Section 160.3 [8-17-89])* means the U.S. Food and Drug Administration.

FDA (Section 792.3 [8-17-89])* means the U.S. Food and Drug Administration.

FDA (Sections 160.3; 792.3) means the U.S. Food and Drug Administration.

Feasibility study (FS) (Section 300.5 [3-8-90])* means a study undertaken by the lead agency to develop and evaluate options for remedial action. The FS emphasizes data analysis and is generally performed concurrently and in an interactive fashion with the remedial investigation (RI), using data gathered during the RI. The RI data are used to define the objectives of the response action, to develop remedial action alternatives, and to undertake an initial screening and detailed analysis of the alternatives. The term also refers to a report that describes the results of the study.

Fecal coliform bacteria (Section 140.1) are those organisms associated with the intestines of warm-blooded animals that are commonly used to indicate the presence of fecal material and the potential presence of organisms capable of causing human disease.

Federal Act (Section 109.2) means the Federal Water Pollution Control Act, as amended, 33 U.S.C. 1151, et seq.

Federal agency (Section 142.2) means any department, agency, or instrumentality of the United States.

Federal agency (Section 1508.12) means all agencies of the Federal Government. It does not mean the Congress, the Judiciary, or the President, including the performance of staff functions for the President in his Executive Office. It also includes for purposes of these regulations States and units of general local government and Indian tribes assuming NEPA responsibilities under section 104(h) of the Housing and Community Development Act of 1974.

Federal agency (Section 205.2) means an executive agency (as defined in section 105 of Title 5, United States Code) and includes the United States Postal Service.

Federal agency (Section 244.101) means any department, agency, establishment, or instrumentality of the executive branch of the United States Government.

Federal agency (Section 248.4 [2/17/89]) means any department, agency, or other instrumentality of the Federal government; any independent agency or establishment of the Federal government including any government corporation; and the Government Printing Office.

Federal agency (Section 249.04) means any department, agency, or other instrumentality of the Federal Government, any independent agency or establishment of the Federal Government including any Government corporation, and the Government Printing Office (Pub. L. 94-580, 90 Stat. 2799, 42 U.S.C. 6903).

Federal agency (Section 250.4 [10/6/87, 6/22/88]) means any department, agency, or other instrumentality of the Federal Government, any independent agency or establishment of the Federal Government including a government corporation, and the Government Printing Office.

Federal agency (Section 252.4 [6/30/88]) means any department, agency, or other instrumentality of the Federal Government; and independent agency or establishment of the Federal Government including any Government corporation; and the Government Printing Office.

Federal agency (Section 253.4 [11/17/88]) means any department, agency or other instrumentality of the Federal Government, any independent agency or establishment of the Federal Government including any Government corporation, and the Government Printing Office.

Federal agency (Section 260.10) means any department, agency, or other instrumentality of the Federal Government, any independent agency or establishment of the Federal

Government including any Government corporation, and the Government Printing Office.

Federal agency (Section 4.2) means any department, agency, or instrumentality in the Executive Branch of the Government, any wholly owned Government corporation, and the Architect of the Capitol, the Federal Reserve Banks and branches thereof.

Federal agency or agency of the United States (Section 47.105 [3-9-92]) means any department, agency or other instrumentality of the Federal Government, any independent agency or establishment of the Federal Government including any Government corporation

Federal Class I area (Section 51.301) means any Federal land that is classified or reclassified Class I.

Federal contract (Section 34.105 [2-26-90]) means an acquisition contract awarded by an agency, including those subject to the Federal Acquisition Regulation (FAR), and any other acquisition contract for real or personal property or services not subject to the FAR.

Federal cooperative agreement (Section 34.105 [2-26-90]) means a cooperative agreement entered into by an agency.

Federal delayed compliance order (Section 65.01) shall mean a delayed compliance order issued by the Administrator under section 113(d) (1), (3), (4) or (5) of the Act.

Federal Emission Test Procedure (Section 600.002-85) refers to the dynamometer driving schedule, dynamometer procedure, and sampling and analytical procedures described in Part 86 for the respective model year, which are used to derive city fuel economy data for gasoline-fueled or diesel vehicles.

Federal facility (Section 244.101) means any building, installation, structure, land, or public work owned by or leased to the Federal Government. Ships at sea, aircraft in the air, land forces on maneuvers, and other mobile facilities; and United States Government installations located on foreign soil or on land outside the jurisdiction of the United States Government are not considered Federal facilities for the purpose of these guidelines.

Federal facility (Section 61.101 [12-15-89]) means any facility owned or operated by any department, commission, agency, office, bureau or other unit of the government of the United States of America except for facilities owned or operated by the Department of Energy.

Federal facility (Sections 243.101; 246.101) means any building, installation, structure, land, or public work owned by or leased to the

Federal Government. Ships at sea, aircraft in the air, land forces on maneuvers, and other mobile facilities are not considered Federal facilities for the purpose of these guidelines. United States Government installations located on foreign soil or on land outside the jurisdiction of the United States Government are not considered Federal facilities for the purpose of these guidelines.

Federal financial assistance (Section 149.101) means any financial benefits provided directly as aid to a project by a department, agency, or instrumentality of the Federal government in any form including contracts, grants, and loan guarantees. Actions or programs carried out by the Federal government itself such as dredging performed by the Army Corps of Engineers do not involve Federal financial assistance. Actions performed for the Federal government by contractors, such as construction of roads on Federal lands by a contractor under the supervision of the Bureau of Land Management, should be distinguished from contracts entered into specifically for the purpose of providing financial assistance, and will not be considered programs or actions receiving Federal financial assistance. Federal financial assistance is limited to benefits earmarked for a specific program or action and directly awarded to the program or action. Indirect assistance, e.g., in the form of a loan to a developer by a lending institution which in turn receives Federal assistance not specifically related to the project in question is not Federal financial assistance under section 1424(e).

Federal financial assistance (Section 4.2) means any Federal grant, loan, or contribution, except a Federal guarantee or insurance.

Federal Government (Section 203.1) includes the legislative, executive, and judicial branches of the Government of the United States, and the government of the District of Columbia.

Federal grant (Section 34.105 [2-26-90]) means an award of financial assistance in the form of money, or property in lieu of money, by the Federal Government or a direct appropriation made by law to any person. The term does not include technical assistance which provides services instead of money, or other assistance in the form of revenue sharing, loans, loan guarantees, loan insurance, interest subsidies, insurance, or direct United States cash assistance to an individual.

Federal Highway Fuel Economy Test Procedure (Section 600.002-85) refers to the dynamometer driving schedule, dynamometer procedure, and sampling and analytical procedures described in Subpart B of this part and which are used to derive highway fuel economy data for gasoline-fueled or diesel vehicles.

Federal Indian reservation (Section 35.105 [4/11/89]) means for purposes of Clean Water Act, all land within the limits of any Indian reservation under the jurisdiction of the United States Government, notwithstanding the issuance of any patent, and including rights-of-way running through the reservation.

Federal Indian Reservation, Indian Reservation, or Reservation (Section 131.3 [12 12-91]) means all land within the limits of any Indian reservation under the jurisdiction of the United States Government, notwithstanding the issuance of any patent, and including rights-of-way running through the reservation.

Federal Land Manager (Section 124.41) has the meaning given in 40 CFR 52.21.

Federal Land Manager (Section 51.301) means the Secretary of the department with authority over the Federal Class I area or, with respect to Roosevelt-Campobello International Park, the Chairman of the Roosevelt-Campobello International Park Commission.

Federal Land Manager (Sections 51.166; 52.21) means, with respect to any lands in the United States, the Secretary of the department with authority over such lands.

Federal loan (Section 34.105 [2-26-90]) means a loan made by an agency. The term does not include loan guarantee or loan insurance.

Federal Register **document** (Section 23.1) means a document intended for publication in the *Federal Register* and bearing in its heading an identification code including the letters "FRL."

Federal standards (Section 205.165) means, for the purpose of this subpart, the standards specified in Section 205.152(a)(1), (2) and (3).

Federal, State and local approvals or permits necessary to begin physical construction (Sections 260.10; 270.2) means permits and approvals required under Federal, State or local hazardous waste control statutes, regulations or ordinances.

Federal Test Procedure (Section 85.1502 [9/25/87]). See **FTP (Section 85.1502)**.

Federal Test Procedure or City Fuel Economy Test (Section 610.11) means the test procedures specified in 40 CFR Part 86, except as those procedures are modified in these protocols.

Federally assisted construction contract (Section 8.2) means any agreement or modification thereof between any applicant and any person for construction work which is paid for in whole or in part with funds obtained from the Agency or borrowed on the credit of the Agency pursuant to any Federal program involving a grant, contract, loan, insurance, or guarantee, or undertaken pursuant to any Federal program involving such grant, contract, loan, insurance, or guarantee, or any application or modification thereof approved by the Agency for a grant, contract, loan, insurance, or guarantee under which the applicant itself participates in the construction work.

Federally enforceable (Part 51, Appendix S, paragraph II.A.12 [6/28/89]) means all limitations and conditions which are enforceable by the Administrator, including those requirements developed pursuant to 40 CFR Parts 60 and 61, requirements within any applicable State implementation plan, any permit requirements established pursuant to 40 CFR 52.21 or under regulations approved pursuant to 40 CFR Part 51, Subpart I, including operating permits issued under an EPA-approved program that is incorporated into the State implementation plan and expressly requires adherence to any permit issued under such program.

Federally enforceable (Section 51.165 [6/28/89])* means all limitations and conditions which are enforceable by the Administrator, including those requirements developed pursuant to 40 CFR Parts 60 and 61, requirements within any applicable State implementation plan, any permit requirements established pursuant to 40 CFR 52.21 or under regulations approved pursuant to 40 CFR Part 51, Subpart I, including operating permits issued under an EPA-approved program that is incorporated into the State implementation plan and expressly requires adherence to any permit issued under such program.

Federally enforceable (Section 51.166 [6/28/89])* means all limitations and conditions which are enforceable by the Administrator, including those requirements developed pursuant to 40 CFR Parts 60 and 61, requirements within any applicable State implementation plan, any permit requirements established pursuant to 40 CFR 52.21 or under regulations approved pursuant to 40 CFR Part 51, Subpart I, including operating permits issued under an EPA-approved program that is incorporated into the State implementation plan and expressly requires adherence to any permit issued under such program.

Federally enforceable (Section 51.301) means all limitations and conditions which are enforceable by the Administrator under the Clean Air Act including those requirements developed pursuant to Parts 60 and 61 of this title, requirements within any applicable State Implementation Plan, and any permit requirements established pursuant to Section 52.21 of this chapter or under regulations approved pursuant to Part 51, 52, or 60 of this title.

Federally enforceable (Section 52.21 [6/28/89])* means all limitations and conditions which are enforceable by the Administrator, including those requirements developed pursuant to 40 CFR Parts 60 and 61, requirements within any applicable State implementation plan, any permit requirements established pursuant to 40 CFR 52.21 or under regulations approved pursuant to 40 CFR Part 51, Subpart I, including operating permits issued under an EPA-approved program that is incorporated into the State implementation plan and expressly requires adherence to any permit issued under such program.

Federally enforceable (Section 52.24) means all limitations and conditions which are enforceable by the Administrator, including those requirements developed pursuant to 40 CFR Parts 60 and 61, requirements within any applicable State Implementation Plan, and any permit requirements established pursuant to 40 CFR 52.21 or under regulations approved pursuant to 40 CFR Subpart I and 51.166.

Federally enforceable (Section 60.41b [12/16/87])* means all limitations and conditions that are enforceable by the Administrator, including the requirements of 40 CFR Parts 60 and 61, requirements within any applicable State Implementation Plan, and any permit requirements established under 40 CFR 52.21 or under 40 CFR 51.18 and 40 CFR 51.24.

Federally enforceable (Section 60.41c [9-12-90]) means all limitations and conditions that are enforceable by the Administrator, including the requirements of 40 CFR Parts 60 and 61, requirements within any applicable State implementation plan, and any permit requirements established under 40 CFR 52.21 or under 40 CFR 51.18 and 40 CFR 51.24.

Federally-enforceable (Section 60.51a [2-11-91]) means all limitations and conditions that are enforceable by the Administrator including the requirements of 40 CFR parts 60 and 61, requirements within any applicable State implementation plan, and any permit requirements established under 40 CFR 52.21 or under 40 CFR 51.18 and 40 CFR 51.24.

Federally permitted release (Section 300.6) as defined by section 101(10) of CERCLA, means discharges in compliance with a permit under section 402 of the Federal Water Pollution Control Act; discharges resulting from circumstances identified and reviewed and made part of the public record with respect to a permit issued or modified under section 402 of the Federal Water Pollution Control Act and subject to a condition of such permit; continuous or anticipated intermittent discharges from a point source, identified in a permit or permit application under section 402 of the Federal Water Pollution Control Act, which are caused by events occurring within the scope of relevant operating or treatment systems; discharges in compliance with a legally enforceable permit under section 404 of the Federal Water Pollution Control Act; releases in compliance with a legally enforceable final permit issued pursuant to section 3005 (a) through (d) of the Solid Waste Disposal Act from a hazardous waste treatment, storage, or disposal facility when such permit specifically identifies the hazardous substances and makes such substances subject to a standard of practice, control procedure or bioassay limitation or condition, or other control on the hazardous substances in such releases; any release in compliance with a legally enforceable permit issued under section 102 or 103 of the Marine Protection, Research, and Sanctuaries Act of 1972; any injection of fluids authorized under Federal underground injection control programs or State programs submitted for Federal approval (and not disapproved by the Administrator of EPA) pursuant to Part C of the Safe Drinking Water Act; any emission into the air subject to a permit or control regulation under section 111, section 112, Title 1 Part C, Title 1 Part D, or State implementation plans submitted in accordance with section 110 of the Clean Air Act (and not disapproved by the Administrator of EPA), including any schedule or waiver granted, promulgated, or approved under these sections; any injection of fluids or other materials authorized under applicable State law for the purpose of stimulating or treating wells for the production of crude oil, natural gas, or water, for the purpose of secondary, tertiary, or other enhanced recovery of crude oil or natural gas, or which are brought to the surface in conjunction with the production of crude oil or natural gas and which are reinjected; the introduction of any pollutant into a publicly owned treatment works when such pollutant is specified in and in compliance with applicable pretreatment standards of section 307(b) or (c) of the CWA and enforceable requirements in a pretreatment program submitted by a State or municipality for Federal approval under section 402 of such Act; and any release of source, special nuclear, or by-product material, as those terms are defined in the Atomic Energy Act of 1954, in compliance with a legally enforceable license, permit, regulation, or order issued pursuant to the Atomic Energy Act of 1954.

Federally registered (Section 162.152) means currently registered under sec. 3 of the Act, after having been initially registered under the Federal Insecticide, Fungicide, and Rodenticide Act of 1947 (Pub. L. 86-139; 73 Stat. 286; June 25, 1947) by the Secretary of Agriculture or under FIFRA by the Administrator.

Federally regulated motorcycle (Section 205.165) means, for the purpose of this subpart, any motorcycle subject to the noise standards of Subpart D of this part.

Feed Gas (Section

85.2122(a)(15)(ii)(E)) means the chemical composition of the exhaust gas measured at the converter inlet.

Feedlot (Sections 412.11; 412.21) shall mean a concentrated, confined animal or poultry growing operation for meat, milk or egg production, or stabling, in pens or houses wherein the animals or poultry are fed at the place of confinement and crop or forage growth or production is not sustained in the area of confinement.

Feedstock (Section 419.11) shall mean the crude oil and natural gas liquids fed to the topping units.

FEMA (Section 300.5) is the Federal Emergency Management Agency.

Ferrochrome silicon (Section 60.261) means that alloy as defined by ASTM Designation A482-76 (incorporated by reference - see Section 60.17).

Ferromanganese blast furnace (Section 420.31) means those blast furnaces which produce molten iron containing more than fifty percent manganese.

Ferromanganese silicon (Section 60.261) means that alloy containing 63 to 66 percent by weight manganese, 28 to 32 percent by weight silicon, and a maximum of 0.08 percent by weight carbon.

Ferrosilicon (Section 60.261) means that alloy as defined by ASTM Designation A100-69 (Reapproved 1974) (incorporated by reference - see Section 60.17) grades A, B, C, D, and E, which contains 50 or more percent by weight silicon.

Ferrous Casting (Section 464.02). The remelting of ferrous metals to form a cast intermediate or finished product by pouring the molten metal into a mold. Except for grinding scrubber operations which are covered here, processing operations following the cooling of castings are covered under the electroplating and metal finishing point source categories (40 CFR Parts 413 and 433).

FFDCA (Section 160.3 [8-17-89])* means the Federal Food, Drug and Cosmetic Act, as amended (21 U.S.C. 321 et seq).

FFDCA (Section 177.3 [12-5-90]) means the Federal Food, Drug, and Cosmetic Act, as amended, 21 U.S.C. 301-392.

FFDCA (Section 178.3 [12-5-90]) means the Federal Food, Drug, and Cosmetic Act, as amended, 21 U.S.C. 301-392.

FFDCA (Section 179.3 [12-5-90]) means the Federal Food, Drug, and Cosmetic Act, as amended, 21 U.S.C. 301-392.

Fiber (Section 410.21) shall mean the dry wool and other fibers as received at the wool finishing mill for processing into wool and blended products.

Fiber (Section 763.121) means a particulate form of asbestos, 5 micrometers or longer, with a length-to-diameter ratio of at least 3 to 1.

Fiber or fiberboard boxes (Section 250.4 [10/6/87, 6/22/88]) means boxes made from containerboard, either solid fiber or corrugated paperboard (general term); or boxes made from solid paperboard of the same material throughout (specific term).

Fiber release episode (Section 763.83 [10/30/87]) means any uncontrolled or unintentional disturbance of ACBM resulting in visible emission.

Fiberboard boxes (Section 250.4 [10/6/87, 6/22/88]). See **Fiber boxes (Section 250.4)**.

Fiberglass insulation (Section 248.4 [2/17/89]) means insulation which is composed principally of glass fibers, with or without binders.

Field gas (Section 60.631) means feedstock gas entering the natural gas processing plant.

Field testing (Section 35.2005) means practical and generally small-scale testing of innovative or alternative technologies directed to verifying performance and/or refining design parameters not sufficiently tested to resolve technical uncertainties which prevent the funding of a promising improvement in innovative or alternative treatment technology.

FIFRA (Section 160.3 [8-17-89])* means the Federal Insecticide, Fungicide and Rodenticide Act as amended (7 U.S.C. 136 et seq).

FIFRA (Section 177.3 [12-5-90]) means the Federal Insecticide, Fungicide, and Rodenticide Act, 7 U.S.C. 136-136y.

FIFRA (Section 179.3 [12-5-90]) means the Federal Insecticide, Fungicide, and Rodenticide Act, 7 U.S.C. 136-136y.

FIFRA or Act (Section 152.3 [5/4/88]) means the Federal Insecticide, Fungicide, and Rodenticide Act, as amended (7 U.S.C. 136-136y).

FIFRA or Act (Section 154.3) means the Federal Insecticide, Fungicide, and Rodenticide Act, as amended, 7 U.S.C. 136 et seq.

Fill (Section 60.111b) means the introduction of VOL into a storage vessel but not necessarily to complete capacity.

Fill material (Sections 232.2 [6/6/88]; 233.3) means any pollutant which replaces portions of the waters of the United States with dry land or which changes the bottom elevation of a water body for any purpose.

Filtration (Section 141.2 [6/29/89]) means a process for removing particulate matter from water by passage through porous media.

Final authorization (Section 270.2) means approval by EPA of a State program which has met the requirements of section 3006(b) of RCRA and the applicable requirements of Part 271, Subpart A.

Final closure (Section 260.10) means the closure of all hazardous waste management units at the facility in accordance with all applicable closure requirements so that hazardous waste management activities under Parts 264 and 265 of this chapter are no longer conducted at the facility unless subject to the provisions in Section 262.34.

Final cover (Section 241.101) means cover material that serves the same functions as daily cover but, in addition, may be permanently exposed on the surface.

Final Order (Section 22.03) means (a) an order issued by the Administrator after an appeal of an initial decision, accelerated decision, decision to dismiss, or default order, disposing of a matter in controversy between the parties, or (b) an initial decision which becomes a final order under Section 22.27(c).

Final printed labeling (Section 152.3 [5/4/88]) means the label or labeling of the product when distributed or sold. Final printed labeling does not include the package of the product, unless the labeling is an integral part of the package.

Final printed labeling (Section 162.3) means the printed label and the labeling which will appear on or will accompany the pesticide product.

Final product (Section 700.43 [8/17/88]) means a new chemical substance (as new chemical substance is defined in Section 720.3 of this chapter) that is manufactured by a person for distribution in commerce, or for use by the person other than as an intermediate.

Financial reporting year (Section 280.92 [10/26/88]) means the latest consecutive twelve-month period for which any of the following reports used to support a financial test is prepared: (1) a 10-K report submitted to the SEC; (2) an annual report of tangible net worth submitted to Dun and Bradstreet; or (3) annual reports submitted to the Energy Information Administration or the Rural Electrification Administration. Financial reporting year may thus comprise a fiscal or a calendar year period.

Finding of no significant impact (Section 1508.13) means a document by a Federal agency briefly presenting the reasons why an action, not otherwise excluded (Section 1508.4), will not have a significant effect on the human environment and for which an environmental impact statement therefore will not be prepared. It shall include the environmental assessment or a summary of it and shall note any other environmental documents related to it (Section 1501.7(a)(5)). If the assessment is included, the finding need not repeat any of the discussion in the assessment but may incorporate it by reference.

Finish coat operation (Section 60.461) means the coating application station, curing oven, and quench station used to apply and dry or cure the final coating(s) on the surface of the metal coil. Where only a single coating is applied to the metal coil, that coating is considered a finish coat.

Finished product (Section 432.51) shall mean the final manufactured product as fresh meat cuts, hams, bacon or other smoked meats, sausage, luncheon meats, stew, canned meats or related products.

Finished product (Section 432.61) shall mean the final manufactured product as fresh meat cuts including, but not limited to, steaks, roasts, chops, or boneless meats.

Finished product (Section 432.71) shall mean the final manufactured product as fresh meat cuts including steaks, roasts, chops or boneless meat, bacon or other smoked meats (except hams) such as sausage, bologna or other luncheon meats, or related products (except canned meats).

Finished products (Section 432.81) shall mean the final manufactured product as fresh meat cuts including steaks, roasts, chops or boneless meat, smoked or cured hams, bacon or other smoked meats, sausage, bologna or other luncheon meats (except canned meats).

Finished products (Section 432.91) shall mean the final manufactured product as fresh meat cuts including steaks, roasts, chops or boneless meat, hams, bacon or other smoked meats, sausage, bologna or other luncheon meats, stews, sandwich spreads or other canned meats.

Finishing water (Section 463.2) is processed water used to remove waste plastic material generated during a finishing process or to lubricate a plastic product during a finishing process. It includes water used to machine or to assemble intermediate or final plastic products.

Fire-fighting turbine (Section 60.331) means any stationary gas turbine that is used solely to pump water for extinguishing fires.

Firm or company (Section 717.3) means any person that is subject to this part, as defined in Section 717.5.

First attempt at repair (Section 264.1031 [6-21-90]) means to take rapid action for the purpose of stopping or reducing leakage of organic material to the atmosphere using best practices.

First attempt at repair (Sections 60.481; 61.241) means to take rapid action for the purpose of stopping or reducing leakage of organic material to atmosphere using best practices.

First draw sample (Section 141.2 [6-7-91]) means a one-liter sample of tap water, collected in accordance with Sec. 141.86(b)(2), that has been standing in plumbing pipes at least 6 hours and is collected without flushing the tap.

First federal official (Section 300.5 [3-8-90])* means the first federal representative of a participating agency of the National Response Team to arrive at the scene of a discharge or a release. This official coordinates activities under the NCP and may initiate, in consultation with the OSC, any necessary actions until the arrival of the predesignated OSC. A state with primary jurisdiction over a site covered by a cooperative agreement will act in the stead of the first federal official for any incident at the site.

First food use (Section 166.3) refers to the use of a pesticide on a food or in a manner which otherwise would be expected to result in residues in a food, if no permanent tolerance, exemption from the requirement of a tolerance, or food additive regulation for residues of the pesticide on any food has been established for the pesticide under section 408 (d) or (e) or 409 of the Federal Food, Drug, and Cosmetic Act.

First-order reaction (Section 796.3700) is defined as a reaction in which the rate of disappearance of a chemical is directly proportional to the concentration of the chemical and is not a function of the concentration of any other chemical present in the reaction mixture.

First-tier subcontractor (Section 8.2) refers to a subcontractor holding a subcontract with a prime contractor.

Fish poisons and repellents (Section 162.3) includes all substances or mixtures of substances intended for destroying, repelling, or mitigating fish declared to be pests under Section 162.14. Fish poisons and repellents include, but are not limited to: (i) Toxicants intended to kill fish in lakes, ponds, or streams; (ii) Repellents intended to repel species dangerous to man or injurious to aquatic organisms which man wishes to protect; and (iii) Sex influence agents intended to control sexual development of fish, such as to cause young to develop into all-female populations.

Five Year State/EPA Conference Agreement. (Section 35.9010 [10-3-89]) Agreement negotiated among the States represented in a Management Conference and the EPA shortly after the Management Conference is convened. The agreement identifies milestones to be achieved during the term of the Management Conference.

Five-year (5-year), six-hour (6-hour) precipitation event (Section 440.141 [5/24/88]) means the maximum 6-hour precipitation event with a probable recurrence interval of once in 5 years as established by the U.S. Department of Commerce, National Oceanic and Atmospheric Administration, National Weather Service, or equivalent regional or rainfall probability information.

Fixed capital cost (Section 51.301) means the capital needed to provide all of the depreciable components.

Fixed plant (Section 60.671) means any nonmetallic mineral processing plant at which the processing equipment specified in Section 60.670(a) is attached by a cable, chain, turnbuckle, bolt or other means (except electrical connections) to any anchor, slab, or structure including bedrock.

Fixed roof (Section 60.691 [11/23/88]) means a cover that is mounted to a tank or chamber in a stationary manner and which does not move with fluctuations in wastewater levels.

Fixed roof (Section 61.341 [3-7-90]) means a cover that is mounted on a waste management unit in a stationary manner and that does not move with fluctuations in liquid level.

Fixed source (Section 247.101) means, for the purpose of these guidelines, a stationary facility that converts fossil fuel into energy, such as steam, hot water, electricity, etc.

Flame zone (Section 264.1031 [6-21-90]) means the portion of the combustion chamber in a boiler occupied by the flame envelope.

Flame zone (Section 60.661 [6-29-90]) means the portion of the combustion chamber in a boiler occupied by the flame envelope.

Flash-off area (Section 60.311) means the portion of a surface coating operation between the coating application area and bake oven.

Flash-off area (Section 60.391) means the structure on automobile and light-duty truck assembly lines between the coating application system (dip tank or spray booth) and the bake oven.

Flashoff area (Section 60.441) means the portion of a coating line after the coating applicator and usually before the oven entrance.

Flashoff area (Section 60.451) means the portion of a surface coating line between the coating application station and the curing oven.

Flashoff area (Section 60.711 [10/3/88]) means the portion of a coating operation between the coating applicator and the drying oven where solvent begins to evaporate from the coated base film.

Flashoff area (Section 60.741 [9-11-89]) means the portion of a coating operation between the coating applicator and the drying oven where VOC begins to evaporate from the coated substrate.

Flashover (Section 85.2122(a)(7)(ii)(A)) means the discharge of ignition voltage across the surface of the distributor cap and/or rotor rather than at the spark plug gap.

Flashover (Section 85.2122(a)(8)(ii)(F)) means the discharge of ignition voltage at any point other than at the spark plug gap.

Flashover (Section 85.2122(a)(9)(ii)(B)) means the discharge of ignition voltage across the coil.

Flat glass (Section 60.291) means glass made of soda-lime recipe and produced into continuous flat sheets and other products listed in SIC 3211.

Flat mill (Section 420.71) means those steel hot forming operations that reduce heated slabs to plates, strip and sheet, or skelp.

Flexible fuel vehicle (or engine) (Section 86.090-2 [7-26-90])* means any motor vehicle (or motor vehicle engine) engineered and designed to be operated on a petroleum fuel, a methanol fuel, or any mixture of the two.

Flexible vinyl and urethane products (Section 60.581) means those products, except for resilient floor coverings (1977 Standard Industry Code 3996) and flexible packaging, that are more than 50 micrometers (0.002 inches) thick, and that consist of or contain a vinyl or urethane sheet or a vinyl or urethane coated web.

Floating roof (Section 60.111) means a storage vessel cover consisting of a double deck, pontoon single deck, internal floating cover or covered floating roof, which rests upon and is supported by the petroleum liquid being contained, and is equipped with a closure seal or seals to close the space between the roof edge and tank wall.

Floating roof (Section 60.691 [11/23/88]) means a pontoon-type or double-deck type cover that rests on the liquid surface.

Floating roof (Section 61.341 [3-7-90]) means a cover with certain rim sealing mechanisms consisting of a double deck, pontoon single deck, internal floating cover or covered floating roof, which rests upon and is supported by the liquid being contained, and is equipped with a closure seal or seals to close the space between the roof edge and unit wall.

Flocculation (Section 141.2 [6/29/89]) means a process to enhance agglomeration or collection of smaller floc particles into larger, more easily settleable particles through gentle stirring by hydraulic or mechanical means.

Flood or Flooding (Part 6, Appendix A, Section 4) means a general and temporary condition of partial or complete inundation of normally dry land areas from the overflow of inland and/or tidal waters, and/or the unusual and rapid accumulation or runoff of surface waters from any source, or flooding from any other source.

Floodplain (Part 6, Appendix A, Section 4) means the lowland and relatively flat areas adjoining inland and coastal waters and other floodprone areas such as offshore islands, including at a minimum, that area subject to a one percent or greater chance of flooding in any given year. The base floodplain shall be used to designate the 100-year floodplain (one percent chance floodplain). The critical action floodplain is defined as the 500-year floodplain (0.2 percent chance floodplain).

Floodplain (Section 257.3-1) means the lowland and relatively flat areas adjoining inland and coastal waters, including flood-prone areas of offshore islands, which are inundated by the base flood.

Floodproofing (Part 6, Appendix A, Section 4) means modification of individual structures and facilities, their sites, and their contents to protect against structural failure, to keep water out or to reduce effects of water entry.

Floor insulation (Section 248.4 [2/17/89]) means a material, primarily designed to resist heat flow, which is installed between the first level conditioned area of a building and an unconditioned basement, a crawl space, or the outside beneath it. Where the first level conditioned area of a building is on a ground level concrete slab, floor insulation includes such a material installed around the perimeter of or on the slab. In the case of mobile homes, floor insulation also means skirting to enclose the space between the building and the ground.

Flooring felt (Section 763.163 [7-12-89]) means an asbestos-containing product which is made of paper felt intended for use as an underlayer for floor coverings, or to be bonded to the underside of vinyl sheet flooring.

Flow channels (Section 60.291) means appendages used for conditioning and distributing molten glass to forming apparatuses and are a permanently separate source of emissions such that no mixing of emissions occurs with emissions from the melter cooling system prior to their being vented to the atmosphere.

Flow coating (Section 60.311) means a method of applying coatings in which the part is carried through a chamber containing numerous nozzles which direct unatomized streams of coatings from many different angles onto the surface of the part.

Flow indicator (Section 264.1031 [6-21-90]) means a device that indicates whether gas flow is present in a vent stream.

Flow indicator (Section 60.661 [6-29-90]) means a device which indicates whether gas flow is present in a vent stream.

Flow proportional composite sample (Section 471.02) is composed of grab samples collected continuously or discretely in proportion to the total flow at time of collection or to the total flow since collection of the previous grab sample. The grab volume or frequency of grab collection may be varied in proportion to flow.

Flow rate (Section 146.3) means the volume per time unit given to the flow of gases or other fluid substance which emerges from an orifice, pump, turbine or passes along a conduit or channel.

Flow through (Section 797.1600) refers to the continuous or very frequent passage of fresh test solution through a test chamber with no recycling.

Flow-through (Section 797.1970) means a continuous passage of test solution or dilution water through a test chamber, holding or acclimation tank with no recycling.

Flow-through (Sections 797.1300; 797.1330) means a continuous or an intermittent passage of test solution or dilution water through a test chamber or culture tank with no recycling.

Flow-through (Sections 797.1400; 797.1930; 797.1950) means a continuous or an intermittent passage of test solution or dilution water through a test chamber, or a holding or acclimation tank with no recycling.

Flow-through process tank (Section 280.12 [9/23/88]) is a tank that forms an integral part of a production process through which there is a steady, variable, recurring, or intermittent flow of materials during the operation of the process. Flow-through process tanks do not include tanks used for the storage of materials prior to their introduction into the production process or for the storage of finished products or by-products from the production process.

Flow-through test (Section 797.1350) is a toxicity test in which water is renewed continuously in the test chambers, the test chemical being transported with the water used to renew the test medium.

Fluid (Sections 144.3; 146.3; 147.2902) means any material or substance which flows or moves whether in a semisolid, liquid, sludge, gas, or any other form or state.

Fluid catalytic cracking unit (Section 60.101 [8-17-89]) means a refinery process unit in which petroleum derivatives are continuously charged;

hydrocarbon molecules in the presence of a catalyst suspended in a fluidized bed are fractured into smaller molecules, or react with a contact material suspended in a fluidized bed to improve feedstock quality for additional processing; and the catalyst or contact material is continuously regenerated by burning off coke and other deposits. The unit includes the riser, reactor, regenerator, air blowers, spent catalyst or contact material stripper, catalyst or contact material recovery equipment, and regenerator equipment for controlling air pollutant emissions and for heat recovery.

Fluid catalytic unit catalyst regenerator (Section 60.101 [8-17-89]) means one or more regenerators (multiple regenerators) which comprise that portion of the fluid catalytic cracking unit in which coke burn-off and catalyst or contact material regeneration occurs, and includes the regenerator combustion air blower(s).

Fluidized bed combustion steam generating unit (Section 60.41b) means a device wherein fuel and solid sorbent are distributed onto or into a bed, or series of beds, of aggregate for combustion and these materials together with solid products of combustion are forced upward in the device by the flow of combustion air and the gaseous products of combustion.

Fluidized bed combustion technology (Section 60.41c [9-12-90]) means a device wherein fuel is distributed onto a bed (or series of beds) of limestone aggregate (or other sorbent materials) for combustion; and these materials are forced upward in the device by the flow of combustion air and the gaseous products of combustion. Fluidized bed combustion technology includes, but is not limited to, bubbling bed units and circulating bed units.

Fluidized bed combustion technology (Section 60.41b [12/16/87]) means combustion of fuel in a bed or series of beds (including but not limited to bubbling bed units and circulating bed units) of limestone aggregate (or other sorbent materials) in which these materials are forced upward by the flow of combustion air and the gaseous products of combustion.

Fluorescent light ballast (Section 761.3) means a device that electrically controls fluorescent light fixtures and that includes a capacitor containing 0.1 kg or less of dielectric.

Flushing-liquor circulation tank (Section 61.131 [9-14-89]) means any vessel that functions to store or contain flushing liquor that is separated from the tar in the tar decanter and is recirculated as the cooled liquor to the gas collection system.

Flux standard (Section 61 [12-15-89]) A regulatory standard that limits the amount of radon that can emanate per square meter of regulated material per second, averaged over a single source.

Fly ash (Section 240.101) means suspended particles, charred paper, dust, soot, and other partially oxidized matter carried in the products of combustion.

Fly ash (Section 249.04) means the component of coal which results from the combustion of coal, and is the finely divided mineral residue which is typically collected from boiler stack gases by electrostatic precipitator or mechanical collection devices.

Fly ash (Section 423.11) means the ash that is carried out of the furnace by the gas stream and collected by mechanical precipitators, electrostatic precipitators, and/or fabric filters. Economizer ash is included when it is collected with fly ash.

Foam-in-place insulation (Section 248.4 [2/17/89]) is rigid cellular foam produced by catalyzed chemical reactions that hardens at the site of the work. The term includes spray-applied and injected applications such as spray-in-place foam and pour-in-place.

Fog coating (also known as mist coating and uniforming) (Section 60.721 [1/29/88]) means a thin coating applied to plastic parts that have molded-in color or texture or both to improve color uniformity.

Folding boxboard (Section 250.4 [10/6/87, 6/22/88]) means a paperboard suitable for the manufacture of folding cartons.

Food (Section 166.3) means any article used for food or drink for man or animals.

Food (Sections 710.2; 720.3) shall have the meaning contained in the Federal Food, Drug, and Cosmetic Act, 21 U.S.C. 321 et seq., and the regulations issued under such Act. In addition, the term food includes poultry and poultry products, as defined in the Poultry Products Inspection Act, 21 U.S.C. 453 et seq.; meats and meat food products, as defined in the Federal Meat Inspection Act, 21 U.S.C. 60 et seq.; and eggs and egg products, as defined in the Egg Products Inspection Act, 21 U.S.C. 1033 et seq.

Food additive (Section 177.3 [12-5-90]) means any substance the intended use of which results or may reasonably be expected to result, directly or indirectly, in its becoming a component of or otherwise affecting the characteristics of any food (including any such substance intended for use in producing, manufacturing, packing, processing, preparing, treating, packaging, transporting, or holding food), except that such term does not include: (1) A pesticide chemical in or on a raw agricultural commodity. (2) A pesticide chemical to the extent that it is intended for use or is used in the production, storage, or transportation of any raw agricultural commodity. (3) A color additive. (4) Any substance used in accordance with a sanction or approval granted prior to September 6, 1958, pursuant to the FFDCA, the

Poultry Products Inspection Act, or the Federal Meat Inspection Act. (5) A new animal drug. (6) A substance that is generally recognized, among experts qualified by scientific training and experience to evaluate its safety, as having been adequately shown through scientific procedures (or, in the case of a substance used in food prior to January 1, 1958, through either scientific procedures or experience based on common use in food) to be safe under the conditions of its intended use.

Food additive (Sections 710.2; 720.3) shall have the meaning contained in the Federal Food, Drug, and Cosmetic Act, 21 U.S.C. 321 et seq., and the regulations issued under such Act.

Food additive regulation (Section 177.3 [12-5-90]) means a regulation issued pursuant to FFDCA section 409 that states the conditions under which a food additive may be safely used. A food additive regulation under this part ordinarily establishes a tolerance for pesticide residues in or on a particular processed food or a group of such foods. It may also specify: (1) The particular food or classes of food in or on which a food additive may be used. (2) The maximum quantity of the food additive which may be used in or on such food. (3 The manner in which the food additive may be added to or used in or on such food. (4) Directions or other labeling or packaging requirements for the food additive.

Food waste (Sections 243.101; 246.101) means the organic residues generated by the handling, storage, sale, preparation, cooking, and serving of foods, commonly called garbage.

Food-chain crops (Section 257.3-5) means tobacco, crops grown for human consumption, and animal feed for animals whose products are consumed by humans.

Food-chain crops (Section 260.10) means tobacco, crops grown for human consumption, and crops grown for feed for animals whose products are consumed by humans.

Force account work (Section 30.200) means the use of the recipient's own employees or equipment for construction, construction-related activities (including A and E services), or for repair or improvement to a facility.

Foreign awards (Section 30.200) means an EPA award of assistance when all or part of the project is performed in a foreign country by (a) a U.S. recipient, (b) a foreign recipient, or (c) an international organization.

Forest (Section 171.2) means a concentration of trees and related vegetation in non-urban areas sparsely inhabited by and infrequently used by humans; characterized by natural terrain and drainage patterns.

Forging (Section 467.02) is the

exertion of pressure on dies or rolls surrounding heated aluminum stock, forcing the stock to change shape and in the case where dies are used to take the shape of the die. The forging subcategory is based on the forging process.

Forging (Section 471.02) is deforming metal, usually hot, with compressive force into desired shapes, with or without dies. Where dies are used, the metal is forced to take the shape of the die.

Form bond (Section 250.4 [10/6/87, 6/22/88]) means a lightweight commodity paper designed primarily for business forms including computer printout and carbonless paper forms. (See manifold business forms.)

Formal amendment (Section 30.200) means a written modification of an assistance agreement signed by both the authorized representative of the recipient and the award official.

Formal hearing (Section 124.2) means any evidentiary hearing under Subpart E or any panel hearing under Subpart F but does not mean a public hearing conducted under Section 124.12.

Formation (Section 144.3) means a body of consolidated or unconsolidated rock characterized by a degree of lithologic homogeneity which is prevailingly, but not necessarily, tabular and is mappable on the earth's surface or traceable in the subsurface.

Formation (Sections 146.3; 147.2902) means a body of rock characterized by a degree of lithologic homogeneity which is prevailingly, but not necessarily, tabular and is mappable on the earth's surface or traceable in the subsurface.

Formation fluid (Sections 144.3; 146.3) means fluid present in a formation under natural conditions as opposed to introduced fluids, such as drilling mud.

Former employee (Section 3.102) means a former Environmental Protection Agency employee, or a former special Government employee.

Forming (Section 471.02) is a set of manufacturing operations in which metals and alloys are made into semifinished products by hot or cold working.

Formulation (Section 158.153 [5/4/88]) means (1) The process of mixing, blending, or dilution of one or more active ingredients with one or more other active or inert ingredients, without an intended chemical reaction, to obtain a manufacturing use product or an end use product, or (2) The repackaging of any registered product.

Fossil fuel (Sections 60.41; 60.41a; 60.161) means natural gas, petroleum, coal, and any form of solid, liquid, or gaseous fuel derived from such materials for the purpose of creating useful heat.

Fossil fuel and wood residue-fired steam generating unit (Section 60.41) means a furnace or boiler used in the process of burning fossil fuel and wood residue for the purpose of producing steam by heat transfer.

Fossil fuel-fired steam generator (Section 51.100) means a furnace or boiler used in the process of burning fossil fuel for the primary purpose of producing steam by heat transfer.

Fossil-fuel fired steam generating unit (Section 60.41) means a furnace or boiler used in the process of burning fossil fuel for the purpose of producing steam by heat transfer.

Foundation insulation (Section 248.4 [2/17/89]) means a material, primarily designed to resist heat flow, which is installed in foundation walls between conditioned volumes and unconditioned volumes and the outside or surrounding earth, at the perimeters of concrete slab-on-grade foundations, and at common foundation wall assemblies between conditioned basement volumes.

Foundry (Section 61.31) means a facility engaged in the melting or casting of beryllium metal or alloy.

Foundry coke (Section 61.131 [9-14-89]) means coke that is produced from raw materials with less than 26 percent volatile material by weight and that is subject to a coking period of 24 hours or more. Percent volatile material of the raw materials (by weight) is the weighted average percent volatile material of all raw materials (by weight) charged to the coke oven per coking cycle.

Foundry coke by-product recovery plant (Section 61.131 [9-14-89]) means a coke by-product recovery plant that is not a foundry coke by-product recovery plant.

Four-hour block average or 4-hour block average (Section 60.51a [2-11-91]) means the average of all hourly emission rates when the affected facility is operating and combusting MSW measured over 4-hour periods of time from 12 midnight to 4 a.m., 4 a.m. to 8 a.m., 8 a.m. to 12 noon, 12 noon to 4 p.m., 4 p.m. to 8 p.m., and 8 p.m. to 12 midnight.

Four-Wheel-Drive General Utility Vehicle (Section 600.002-85) means a four-wheel-drive, general purpose automobile capable of off-highway operation that has a wheelbase not more than 110 inches and that has a body shape similar to a 1977 Jeep CJ-5 or CJ-7, or the 1977 Toyota Land Cruiser, as defined by the Secretary of Transportation at 49 CFR 553.4.

Fourteen-day (14-day) old survivors (Sections 797.2130; 797.2150) are birds that survive for 2 weeks following hatch. Values are expressed both as a percentage of hatched eggs and as the number per pen per season (test).

Fractionation operation (Section 264.1031 [6-21-90]) means a distillation operation or method used to separate a mixture of several volatile components of different boiling points in successive stages, each stage removing from the mixture some proportion of one of the components.

Free available chlorine (Section 423.11) shall mean the value obtained using the amperometric titration method for free available chlorine described in "Standard Methods for the Examination of Water and Wastewater," page 112 (13th edition).

Free liquids (Section 260.10) means liquids which readily separate from the solid portion of a waste under ambient temperature and pressure.

Free moisture (Sections 240.101; 241.101) means liquid that will drain freely by gravity from solid materials.

Free product (Section 280.12 [9/23/88]) refers to a regulated substance that is present as a non-aqueous phase liquid (e.g., liquid not dissolved in water).

Free stall barn (Section 412.11) shall mean specialized facilities wherein producing cows are permitted free movement between resting and feeding areas.

Freeboard (Section 260.10) means the vertical distance between the top of a tank or surface impoundment dike, and the surface of the waste contained therein.

Fresh feed (Section 60.101 [8-17-89]) means any petroleum derivative feedstock stream charged directly into the riser or reactor of a fluid catalytic cracking unit except for petroleum derivatives recycled within the fluid catalytic cracking unit, fractionator, or gas recovery unit.

Fresh granular triple superphosphate (Section 60.241) means granular triple superphosphate produced no more than 10 days prior to the date of the performance test.

Freshwater (Section 147.2902) means underground source of drinking water.

Friable (Section 763.83 [10/30/87]) when referring to material in a school building means that the material, when dry, may be crumbled, pulverized, or reduced to powder by hand pressure, and includes previously nonfriable material after such previously nonfriable material becomes damaged to the extent that when dry it may be crumbled, pulverized, or reduced to powder by hand pressure.

Friable asbestos material (Section 61.141 [11-20-90])* means any material containing more than 1 percent asbestos as determined using the method specified in appendix A, subpart F, 40 CFR part 763 section 1, Polarized Light Microscopy, that, when dry, can be crumbled, pulverized, or reduced to powder by hand pressure. If the asbestos content is less than 10 percent as determined by a method other than point counting by polarized light microscopy (PLM), verify the asbestos content by point counting using PLM.

Friable asbestos material (Section 763.121) means any material containing more than 1 percent asbestos by weight which, when dry, may be crumbled, pulverized, or reduced to powder by hand pressure.

Friable material (Section 763.103) means any material applied onto ceilings, walls, structural members, piping, ductwork, or any other part of the building structure which, when dry, may be crumbled, pulverized, or reduced to powder by hand pressure.

Frond (Section 797.1160) means a single Lemna leaf-like structure.

Frond mortality (Section 797.1160) means dead fronds which may be identified by a total discoloration (yellow, white, black or clear) of the entire frond.

Front panel (Section 162.3) means that portion of the label of a pesticide product that is ordinarily visible to the purchaser under the usual conditions of display for sale.

FTP (Section 85.1502 [9/25/87]) is the Federal Test Procedure at Part 86.

Fuel (Section 600.002-85) means (i) gasoline and diesel fuel for gasoline- or diesel-powered automobiles or (ii) electrical energy for electrically powered automobiles.

Fuel (Section 79.2) means any material which is capable of releasing energy or power by combustion or other chemical or physical reaction.

Fuel economy (Section 2.311) has the meaning given it in section 501(6) of the Act, 15 U.S.C. 2001(6).

Fuel economy (Section 600.002-85) means (i) the average number of miles traveled by an automobile or group of automobiles per gallon of gasoline or diesel fuel consumed as computed in Section 600.113 or Section 600.207 or (ii) the equivalent petroleum-based fuel economy for an electrically powered automobile as determined by the Secretary of Energy.

Fuel economy (Section 610.11) means the average number of miles traveled by an automobile per gallon of gasoline (or equivalent amount of other fuel) consumed, as determined by the Administrator in accordance with procedures established under Subparts D or F.

Fuel economy data (Section 2.311) means any measurement or calculation of fuel economy for any model type and average fuel economy of a manufacturer under section 503(d) of the Act, 15 U.S.C. 2003(d).

Fuel Economy Data Vehicle (Section 600.002-85) means a vehicle used for the purpose of determining fuel economy which is not a certification vehicle.

Fuel evaporative emissions (Section 86.082-2) means vaporized fuel emitted into the atmosphere from the fuel system of a motor vehicle.

Fuel gas (Section 60.101) means any gas which is generated at a petroleum refinery and which is combusted. Fuel gas also includes natural gas when the natural gas is combined and combusted in any proportion with a gas generated at a refinery. Fuel gas does not include gases generated by catalytic cracking unit catalyst regenerators and fluid coking burners.

Fuel gas combustion device (Section 60.101) means any equipment, such as process heaters, boilers and flares used to combust fuel gas, except facilities in which gases are combusted to produce sulfur or sulfuric acid.

Fuel manufacturer (Section 79.2) means any person who, for sale or introduction into commerce, produces or manufactures a fuel or causes or directs the alteration of the chemical composition of, or the mixture of chemical compounds in, a bulk fuel by adding to it an additive.

Fuel pretreatment (Section 60.41b [12/16/87]) means a process that removes a portion of the sulfur in a fuel before combustion of the fuel in a steam generating unit.

Fuel pretreatment (Section 60.41c [9-12-90]) means a process that removes a portion of the sulfur in a fuel before combustion of the fuel in a steam generating unit.

Fuel supply agreement (Section 73.3 [12-17-91]) means a legally binding document between a firm associated with a new independent power production facility (IPPF) or a new IPPF and a fuel supplier that establishes the terms and conditions under which the fuel supplier commits to provide fuel to be delivered to a specific new IPPE.

Fuel system (Section 86.082-2) means the combination of fuel tank(s), fuel pump, fuel lines, and carburetor or fuel injection components, and includes all fuel system vents and fuel evaporative emission control system components.

Fuel system (Section 86.402-78) means the combination of fuel tank, fuel pump, fuel lines, oil injection metering system, and carburetor or fuel injection components, and includes all fuel system vents.

Fuel venting emissions (Section 87.1) means raw fuel, exclusive of hydrocarbons in the exhaust emissions, discharged from aircraft gas turbine engines during all normal ground and flight operations.

Fugitive dust, mist or vapor (Section 129.2) means dust, mist or vapor containing a toxic pollutant regulated under this part which is emitted from any source other than through a stack.

Fugitive emission (Section 60.301) means the particulate matter which is not collected by a capture system and is released directly into the atmosphere from an affected facility at a grain elevator.

Fugitive emission (Section 60.671) means particulate matter that is not collected by a capture system and is released to the atmosphere at the point of generation.

Fugitive emissions (Section 57.103) means any air pollutants emitted to the atmosphere other than from a stack.

Fugitive emissions (Sections 51.165; 51.166; 51.301; 52.21; 52.24) means those emissions which could not reasonably pass through a stack, chimney, vent or other functionally equivalent opening.

Fugitive source (Section 61.141 [11-20-90]) means any source of emissions not controlled by an air pollution control device.

Fugitive volatile organic compounds (Section 60.441) means any volatile organic compounds which are emitted from the coating applicator and flashoff areas and are not emitted in the oven.

Full capacity (Section 60.41b [12/16/87])* means operation of the steam generating unit at 90 percent or more of the maximum steady-state design heat input capacity.

Full-time employee (Section 372.3 [2/16/88]) means 2,000 hours per year of full-time equivalent employment. A facility would calculate the number of full-time employees by totaling the hours worked during the calendar year by all employees, including contract employees, and dividing that total by 2,000 hours.

Full-time fellow (Section 46.120) means an individual enrolled in an academic educational program directly related to pollution abatement and control, and taking a minimum of 30 credit hours or an academic workload otherwise defined by the institution as a full-time curriculum for a school year. The fellow need not be pursuing a degree.

Fume scrubber (Section 420.121) means wet air pollution control devices used to remove and clean fumes originating from hot coating operations.

Fume scrubber (Section 420.91) means those pollution control devices used to remove and clean fumes originating in pickling operations.

Fume suppression system (Section 60.141a) means the equipment comprising any system used to inhibit the generation of emissions from steelmaking facilities with an inert gas, flame, or steam blanket applied to the surface of molten iron or steel.

Functional space (Section 763.83 [10/30/87]) means a room, group of rooms, or homogeneous area (including crawl spaces or the space between a dropped ceiling and the floor or roof deck above), such as classroom(s), a cafeteria, gymnasium, hallway(s), designated by a person accredited to prepare management plans, design abatement projects, or conduct response actions.

Functionally equivalent component (Section 270.2 [9/28/88]) means a component which performs the same function or measurement and which meets or exceeds the performance specifications of another component.

Fund (Sections 305.12; 306.12) means the Hazardous Substance Response Trust Fund established under section 221 of CERCLA.

Fund or Trust Fund (Section 300.5 [3-8-90])* means the Hazardous Substance Superfund established by section 9507 of the Internal Revenue Code of 1986.

Funds "directly made available by" capitalization grants (Section 35.3105 [3-19-90]) Funds equaling the amount of the grant.

Fungicides (Section 162.3) includes all substances or mixtures of substances intended for preventing or inhibiting the growth of, or destroying any fungi declared to be pests under Section 162.14, except those substances defined as slimicides in paragraph (ff)(16) of this section and those fungicides and fungistats defined as antimicrobial agents in paragraph (ff)(2)(i)(E) of this section and those antimicrobial substances or mixtures of substances subject to the provisions of the Federal Food, Drug, and Cosmetic Act, as amended (21 U.S.C. 301 et seq.), as delineated in paragraph (ff)(2)(ii) of this section. Fungicides include, but are not limited to: (i) Seed, plant, and soil treatment materials intended to prevent, mitigate, or cure fungal, bacterial, or viral diseases of plants; (ii) Substances intended for use in inhibiting the growth of fungi on inanimate surfaces, in water or in air, including those intended for control of mold and mildew on surfaces and inanimate objects; (iii) Commodity preservatives and protectants intended for use in inhibiting the growth of, or destroying fungi (including yeasts) in or on raw materials (such as adhesives and plastics) used in manufacturing, in or on manufactured products (such as fuels, textiles, lubricants, and paints), or in or on containers and equipment (such as for storage and transportation of commodities); (iv) Wood preservatives intended to prevent or inhibit growth of, or destroying organisms which cause staining, decay, or rotting of wood; and (v) Fumigants and certain other fungicidal agents intended to destroy fungi in the air of enclosed spaces and/or in or on objects within such spaces.

Furnace (Section 240.101) means the chambers of the combustion train where drying, ignition, and combustion of waste material and evolved gases occur.

Furnace (Section 60.531 [2/26/88]) means a solid fuel burning appliance that is designed to be located outside of ordinary living areas and that warms spaces other than the space where the appliance is located, by the distribution of air heated in the appliance through ducts. The appliance must be tested and listed as a furnace under accepted American or Canadian safety testing codes unless exempted from this provision by the Administrator. A manufacturer may request an exemption in writing from the Administrator by stating why the testing and listing requirement is not practicable and by demonstrating that his appliance is otherwise a furnace.

Furnace charge (Section 60.261) means any material introduced into the electric submerged arc furnace, and may consist of, but is not limited to, ores, slag, carbonaceous material, and limestone.

Furnace coke (Section 61.131 [9-14-89]) means coke produced in by-product ovens that is not foundry coke.

Furnace coke by-product recovery plant (Section 61.131 [9-14-89]) means a coke by-product recovery plant that is not a foundry coke by-product recovery plant.

Furnace cycle (Section 60.261) means the time period from completion of a furnace product tap to the completion of the next consecutive product tap.

Furnace power input (Section 60.261) means the resistive electrical power consumption of an electric submerged arc furnace as measured in kilowatts.

Furnace pull (Sections 426.81; 426.121) shall mean that amount of glass drawn from the glass furnace or furnaces.

FV (Section 60.741 [9-11-89]) means the average inward face velocity across all natural draft openings in a total enclosure, in meters per hour.

FWPCA (Section 116.3) means the Federal Water Pollution Control Act, as amended by the Federal Water Pollution Control Act Amendments of 1972 (Pub. L. 92-500), and as further amended by the Clean Water Act of 1977 (Pub. L. 95-217), 33 U.S.C. 1251 et seq.; and as further amended by the Clean Water Act Amendments of 1978 (Pub. L. 95-676).

FWPCA (Section 220.2) means the Federal Water Pollution Control Act, as amended (33 U.S.C. 1251).

FWS (Section 233.2 [6/6/88]) means

FWS

the U.S. Fish and Wildlife Service.

G

G1 (Generation 1) (Section 797.1950) means those mysids which are used to begin the test, also referred to as adults.

G2 (Generation 2) (Section 797.1950) are the young produced by G1.

Galvanized basis material (Section 465.02) means zinc coated steel, galvalum, brass and other copper base strip which is processed in coil coating.

Galvanizing (Section 420.121) means coating steel products with zinc by the hot dip process including the immersion of the steel product in a molten bath of zinc metal, and the related operations preceding and subsequent to the immersion phase.

Gap Location (Section 85.2122(a)(8)(ii)(D)) means the position of the electrode gap in the combustion chamber.

Gap Spacing (Section 85.2122(a)(8)(ii)(C)) means the distance between the center electrode and the ground electrode where the high voltage ignition arc is discharged.

Garrison facility (Section 60.331) means any permanent military installation.

Gas phase process (Section 60.561 [3-5-91]) means a polymerization process in which the polymerization rection is carried out in the gas phase; i.e., the monomer(s) are gases in a fluidized bed of catalyst particles and granular polymer.

Gas plant (Section 60.631). See **Natural gas processing plant (Section 60.631)**.

Gas turbine model (Section 60.331) means a group of gas turbines having the same nominal air flow, combuster inlet pressure, combuster inlet temperature, firing temperature, turbine inlet temperature and turbine inlet pressure.

Gas well (Section 435.61) shall mean any well which produces natural gas in a ratio to the petroleum liquids produced greater than 15,000 cubic feet of gas per 1 barrel (42 gallons) of petroleum liquids.

Gas-tight (Section 60.691 [11/23/88]) means operated with no detectable emissions.

Gasoline (Section 60.501) means any petroleum distillate or petroleum distillate/alcohol blend having a Reid vapor pressure of 27.6 kilopascals or greater which is used as a fuel for internal combustion engines.

Gasoline (Section 80.2) means any fuel sold in any State for use in motor vehicles and motor vehicle engines, and commonly or commercially known or sold as gasoline.

Gasoline blending stock or component (Section 80.2) means any liquid compound which is blended with other liquid compounds or with lead additives to produce gasoline.

Gasoline service station (Section 60.111b) means any site where gasoline is dispensed to motor vehicle fuel tanks from stationary storage tanks.

Gasoline tank truck (Section 60.501) means a delivery tank truck used at bulk gasoline terminals which is loading gasoline or which has loaded gasoline on the immediately previous load.

Gathering lines (Section 280.12 [9/23/88]) means any pipeline, equipment, facility, or building used in the transportation of oil or gas during oil or gas production or gathering operations.

GCWR (Sections 202.10; 205.51). See **Gross Combination Weight Rating (Sections 202.10; 205.51)**.

Gear oils (Section 252.4 [6/30/88]) means petroleum-based oils used for lubricating machinery gears.

General Counsel (Section 15.4) means the General Counsel of the U.S. Environmental Protection Agency, or his or her designee.

General Counsel (Section 23.1 [8/3/88]) means the General Counsel of EPA or any official exercising authority delegated by the General Counsel.

General Counsel and Administrator (Section 350.1 [7/29/88]) mean the EPA officers or employees occupying the positions so titled.

General environment (Section 190.02) means the total terrestrial, atmospheric and aquatic environments outside sites upon which any operation which is part of a nuclear fuel cycle is conducted.

General environment (Section 191.02) means the total terrestrial, atmospheric, and aquatic environments outside sites within which any activity, operation, or process associated with the management and storage of spent nuclear fuel or radioactive waste is conducted.

General permit (NPDES and 404) (Section 124.2 [5/2/89])* means an NPDES or 404 permit authorizing a category of discharges or activities under the CWA within a geographical area. For NPDES, a general permit means a permit issued under Section 122.28. For 404, a general permit means a permit issued under Section 233.37.

General permit (Section 122.2) means an NPDES permit issued under Section

122.28 authorizing a category of discharges under the CWA within a geographical area.

General permit (Section 232.2 [6/6/88]) means a permit authorizing a category of discharges of dredged or fill material under the Act. General permits are permits for categories of discharge which are similar in nature, will cause only minimal adverse environmental effects when performed separately, and will have only minimal cumulative adverse effect on the environment.

General permit (Section 233.3) means 404 permit issued under Section 233.37 authorizing a category of discharges under the CWA within a geographical area.

General processing (Section 410.31) shall mean the internal subdivision of the low water use processing subcategory for facilities described in Section 410.30 that do not qualify under the water jet weaving subdivision.

General purpose unit of local government (Section 310.11 [10/21/87]) means the governing body of a county, parish, municipality, city, town, township, Federally-recognized Indian tribe or similar governing body.

Generation (Sections 243.101; 246.101) means the act or process of producing solid waste.

Generation 1 (G1) (Section 797.1950) means those mysids which are used to begin the test, also referred to as adults.

Generation 2 (G2) (Section 797.1950) are the young produced by G1.

Generator (Section 259.10 [3/24/89]) means any person, by site, whose act or process produces regulated medical waste as defined in Subpart D of this part, or whose act first causes a regulated medical waste to become subject to regulation. In the case where more than one person (e.g., doctors with separate medical practices) are located in the same building, each individual business entity is a separate generator for the purposes of this part.

Generator (Section 260.10) means any person, by site, whose act or process produces hazardous waste identified or listed in Part 261 of this chapter or whose act first causes a hazardous waste to become subject to regulation.

Generator (Sections 144.3; 146.3; 270.2) means any person, by site location, whose act or process produces hazardous waste identified or listed in 40 CFR Part 261.

Generator column (Section 796.1860) is used to produce or generate saturated solutions of a solute in a solvent. The column (see Figure 1 under paragraph (b)(1)(i)(A) of this section) is packed with a solid support coated with the solute, i.e., the organic compound whose solubility is to be determined. When water (the solvent) is pumped through the column, saturated solutions of the solute are generated. Preparation of the generator column is described under paragraph (b)(1)(i) of this section.

Generator of PCB waste (Section 761.3 [12-21-89]) means any person whose act or process produces PCBs that are regulated for disposal under subpart D of this part, or whose act first causes PCBs or PCB Items to become subject to the disposal requirements of subpart D of this part, or who has physical control over the PCBs when a decision is made that the use of the PCBs has been terminated and therefore is subject to the disposal requirements of subpart D of this part. Unless another provision of this part specifically requires a site-specific meaning, "generator of PCB waste" includes all of the sites of PCB waste generation owned or operated by the person who generates PCB waste.

Geometric mean diameter or median diameter (Sections 798.1150; 798.4350) is the calculated aerodynamic diameter which divides the particles of an aerosol in half based on the weight of the particles. Fifty percent of the particles by weight will be larger than the median diameter and 50 percent of the particles will be smaller than the median diameter. The median diameter and its geometeric standard deviation are used to statistically describe the particle size distribution of any aerosol based on the weight and size of the particles.

Geometric mean diameter or the median diameter (Section 798.2450) is the calculated aerodynamic diameter which divides the particles of an aerosol in half based on the weight of the particles. Fifty percent of the particles by weight will be larger than the median diameter and 50 percent of the particles will be smaller than the median diameter. The median diameter describes the particle size distribution of any aerosol based on the weight and size of the particles.

GEP stack height (Section 51.100). See **Good engineering practice stack height (Section 51.100)**.

Germination (Section 797.2750) means the resumption of active growth by an embryo. The primary root should attain a length of 5 mm for the seed to be counted as having germinated.

Germination (Section 797.2800) means the resumption of active growth by an embryo.

Gigawatt-year (Section 190.02) refers to the quantity of electrical energy produced at the busbar of a generating station. A gigawatt is equal to one billion watts. A gigawatt-year is equivalent to the amount of energy output represented by an average electric power level of one gigawatt sustained for one year.

Glass fiber reinforced polyisocyanurate/polyurethane foam (Section 248.4 [2/17/89]) means cellular polyisocyanurate or cellular polyurethane insulation made with glass fibers within the foam core.

Glass melting furnace (Section 60.291) means a unit comprising a refractory vessel in which raw materials are charged, melted at high temperature, refined, and conditioned to produce molten glass. The unit includes foundations, superstructure and retaining walls, raw material charger systems, heat exchangers, melter cooling system, exhaust system, refractory brick work, fuel supply and electrical boosting equipment, integral control systems and instrumentation, and appendages for conditioning and distributing molten glass to forming apparatuses. The forming apparatuses, including the float bath used in flat glass manufacturing and flow channels in wool fiberglass and textile fiberglass manufacturing, are not considered part of the glass melting furnace.

Glass melting furnace (Section 61.161) means a unit comprising a refractory vessel in which raw materials are charged, melted at high temperature, refined, and conditioned to produce molten glass. The unit includes foundations, superstructure and retaining walls, raw material charger systems, heat exchangers, melter cooling system, exhaust system, refractory brick work, fuel supply and electrical boosting equipment, integral control systems and instrumentation, and appendages for conditioning and distributing molten glass to forming apparatuses. The forming apparatuses, including the float bath used in flat glass manufacturing, are not considered part of the glass melting furnace.

Glass produced (Section 60.291) means the weight of the glass pulled from the glass melting furnace.

Glass produced (Section 61.161) means the glass pulled from the glass melting furnace.

Glass pull rate (Section 60.681) means the mass of molten glass utilized in the manufacture of wool fiberglass insulation at a single manufacturing line in a specified time period.

Global commons (Section 6.1003) is that area (land, air, water) outside the jurisdiction of any nation.

Glove bag (Section 61.141 [11-20-90]) means a sealed compartment with attached inner gloves used for the handling of asbestos-containing materials. properly installed and used, glove bags provide a small work area enclosure typically used for small-scale asbestos stripping operations. Information on glove-bag installation, equipment and supplies, and work practices is contained in the Occupational Safety and Health

Administration's (OSHA's) final rule on occupational exposure to asbestos (appendix G to 29 CFR 1926.58).

Good engineering practice (GEP) stack height (Section 51.100) means the greater of: (1) 65 meters, measured from the ground-level elevation at the base of the stack; (2)(i) For stacks in existence on January 12, 1979, and for which the owner or operator had obtained all applicable permits or approvals required under 40 CFR Parts 51 and 52. $H_g = 2.5H$, provided the owner or operator produces evidence that this equation was actually relied on in establishing an emission limitation; (ii) For all other stacks, $H_g = H + 1.5L$ where H_g = good engineering practice stack height, measured from the ground-level elevation at the base of the stack, H = height of nearby structure(s) measured from the ground-level elevation at the base of the stack, and L = lesser dimension, height or projected width, of nearby structure(s), provided that the EPA, State or local control agency may require the use of a field study or fluid model to verify GEP stack height for the source; or (3) The height demonstrated by a fluid model or a field study approved by the EPA State or local control agency, which ensures that the emissions from a stack do not result in excessive concentrations of any air pollutant as a result of atmospheric downwash, wakes, or eddy effects created by the source itself, nearby structures or nearby terrain features.

Governing instruments (Section 32.102) means those legal documents which establish the existence of an organization and define its powers and parameters of operation. They include such documents as the Articles of Incorporation or Association, Constitution, Charter and By-Laws.

Government (Section 8.2) means the Government of the United States of America.

Government contract (Section 8.2) means any agreement or modification thereof between any contracting agency and any person for the furnishing of supplies or services or for the use of real or personal property, including lease arrangements. The term services, as used in this definition includes, but is not limited to, the following services: Utility, construction, transportation, research, insurance, and fund depository. The term government contract does not include (1) agreements in which the parties stand in the relationship of employer and employee, and (2) federally assisted construction contracts.

Governor (Section 15.4) means the governor or principal executive officer of a state.

Grab sample (Section 471.02) is a single sample which is collected at a time and place most representative of total discharge.

Grade of resin (Section 61.61) means

the subdivision of resin classification which describes it as a unique resin, i.e., the most exact description of a resin with no further subdivision.

Grain (Section 60.301) means corn, wheat, sorghum, rice, rye, oats, barley, and soybeans.

Grain elevator (Section 60.301) means any plant or installation at which grain is unloaded, handled, cleaned, dried, stored, or loaded.

Grain handling operations (Section 60.301) include bucket elevators or legs (excluding legs used to unload barges or ships), scale hoppers and surge bins (garners), turn heads, scalpers, cleaners, trippers, and the headhouse and other such structures.

Grain loading station (Section 60.301) means that portion of a grain elevator where the grain is transferred from the elevator to a truck, railcar, barge, or ship.

Grain storage elevator (Section 60.301) means any grain elevator located at any wheat flour mill, wet corn mill, dry corn mill (human consumption), rice mill, or soybean oil extraction plant which has a permanent grain storage capacity of 35,200 m^3 (ca. 1 million bushels).

Grain terminal elevator (Section 60.301) means any grain elevator which has a permanent storage capacity of more than 88.100 m^3 (ca. 2.5 million U.S. bushels), except those located at animal food manufacturers, pet food manufacturers, cereal manufacturers, breweries, and livestock feedlots.

Grain unloading station (Section 60.301) means that portion of a grain elevator where the grain is transferred from a truck, railcar, barge, or ship to a receiving hopper.

Grant (Section 15.4) means any grant or cooperative agreement awarded by an Executive Branch agency including all subagreements awarded thereunder. This includes grants-in-aid, except where such assistance is solely in the form of general revenue sharing funds, distributed under the State and Local Fiscal Assistance Act of 1972, 31 U.S.C. 1221 et seq.

Grant (Section 32.605 [5-25-90]) means an award of financial assistance, including a cooperative agreement, in the form of money, or property in lieu of money, by a Federal agency directly to a grantee. The term grant includes block grant and entitlement grant programs, whether or not exempted from coverage under the grants management government-wide common rule on uniform administrative requirements for grants and cooperative agreements. The term does not include technical assistance that provides services instead of money, or other assistance in the form of loans, loan guarantees, interest subsidies, insurance, or direct appropriations; or any veterans' benefits to individuals, i.e., any benefit to veterans, their families, or survivors by virtue of the service of a veteran in the Armed Forces of the United States.

Grant (Section 6.101) as used in this part, means an award of funds or other assistance by a written grant agreement or cooperative agreement under 40 CFR Chapter I, Subpart B.

Grant agreement (Section 30.200) means an assistance agreement that does not substantially involve EPA in the project and where the recipient has the authority and capability to complete all elements of the program (does not include fellowships).

Grant agreement (Section 35.936-1) means the written agreement and amendments thereto between EPA and a grantee in which the terms and conditions governing the grant are stated and agreed to by both parties under Section 30.345 of this subchapter.

Grantee (Section 15.4) means any person with whom an Executive Branch agency has entered into, extended, or renewed a grant, subgrant, or other assistance agreement defined under grant in this section.

Grantee (Section 32.605 [5-25-90]) means a person who applies for or receives a grant directly from a Federal agency (except another Federal agency).

Grantee (Section 35.936-1) means any municipality which has been awarded a grant for construction of a treatment works under this subpart. In addition, where appropriate in Sections 35.936 through 35.939, grantee may also refer to an applicant for a grant.

Grantee (Section 6.501) means any individual, agency, or entity which has been awarded wastewater treatment construction grant assistance under 40 CFR Part 35, Subpart E or I.

Granular diammonium phosphate plant (Section 60.221) means any plant manufacturing granular diammonium phosphate by reacting phosphoric acid with ammonia.

Granular triple superphosphate storage facility (Section 60.241) means any facility curing or storing granular triple superphosphate.

Grape juice canning (Section 407.61) shall mean the processing of grape juice into the following products and product styles: Canned and frozen, fresh and stored, natural grape juice for the manufacture of juices, drinks, concentrates, jams, jellies, and other related finished products but not wine or other spirits. In terms of raw material processed 1000 kg (1000 lb) of grapes are equivalent to 834 liters (100 gallons) of grape juice.

Grape pressing (Section 407.61) shall mean the washing and subsequent handling including pressing, heating, and filtration of natural juice from all varieties of grapes for the purpose of manufacturing juice, drink, concentrate, and jelly but not wine or other spirits. In terms of raw material processed 1000 kg (1000 lb) of grapes are equivalent to 834 liters (100 gallons) of grape juice.

Grate siftings (Section 240.101) means the materials that fall from the solid waste fuel bed through the grate openings.

Gravity separation methods (Section 440.141 [5/24/88]) means the treatment of mineral particles which exploits differences between their specific gravities. The separation is usually performed by means of sluices, jigs, classifiers, spirals, hydrocyclones, or shaking tables.

Gravure cylinder (Section 60.431) means a printing cylinder with an intaglio image consisting of minute cells or indentations specially engraved or etched into the cylinder's surface to hold ink when continuously revolved through a fountain of ink.

Gravure cylinder (Section 60.581) means a plated cylinder with a printing image consisting of minute cells or indentations, specifically engraved or etched into the cylinder's surface to hold ink when continuously revolved through a fountain of ink.

Green liquor sulfidity (Section 60.281) means the sulfidity of the liquor which leaves the smelt dissolving tank.

Green tire (Section 60.541 [9/15/87]) means an assembled, uncured tire.

Green tire spraying operation (Section 60.541 [9/15/87]) means the system used to apply a mold release agent and lubricant to the inside and/or outside of green tires to facilitate the curing process and to prevent rubber from sticking to the curing press. A green tire spraying operation consists of a booth where spraying is performed, the spray application station, and related equipment, such as the lubricant supply system.

Grid casting facility (Section 60.371) means the facility which includes all lead melting pots and machines used for casting the grid used in battery manufacturing.

Grinder (Section 60.401) means a unit which is used to pulverize dry phosphate rock to the final product size used in the manufacture of phosphate fertilizer and does not include crushing devices used in mining.

Grinding (Section 471.02) is the process of removing stock from a workpiece by the use of a tool consisting of abrasive grains held by a rigid or semi-rigid grinder. Grinding includes surface finishing, sanding, and slicing.

Grinding (Section 61.141 [11-20-90])

means to reduce to powder or small fragments and includes mechanical chipping or drilling.

Grinding mill (Section 60.671) means a machine used for the wet or dry fine crushing of any nonmetallic mineral. Grinding mills include, but are not limited to, the following types: hammer, roller, rod, pebble and ball, and fluid energy. The grinding mill includes the air conveying system, air separator, or air classifier, where such systems are used.

Gross alpha particle activity (Section 141.2) means the total radioactivity due to alpha particle emission as inferred from measurements on a dry sample.

Gross boat particle activity (Section 141.2) means the total radioactivity due to beta particle emission as inferred from measurements on a dry sample.

Gross calorific value (Section 240.101) means heat liberated when waste is burned completely and the products of combustion are cooled to the initial temperature of the waste. Usually expressed in British thermal units per pound.

Gross cane (Sections 409.61; 409.81) shall mean that amount of crop material as harvested, including field trash and other extraneous material.

Gross Combination Weight Rating (GCWR) (Sections 202.10; 205.51) means the value specified by the manufacturer as the loaded weight of a combination vehicle.

Gross production of fiberboard products (Section 429.11) means the air dry weight of hardboard or insulation board following formation of the mat and prior to trimming and finishing operations.

Gross vehicle weight (Section 86.082-2) means the manufacturer's gross weight rating for the individual vehicle.

Gross vehicle weight rating (GVWR) (Section 86.082-2) means the value specified by the manufacturer as the maximum design loaded weight of a single vehicle.

Gross vehicle weight rating (GVWR) (Sections 202.10; 205.51) means the value specified by the manufacturer as the loaded weight of a single vehicle.

Gross vehicle weight rating (GVWR) (Section 763.163 [7-12-89]) means the value specified by the manufacturer as the maximum design loaded weight of a single vehicle.

Gross vehicle weight rating (Section 600.002-85) means the manufacturer's gross weight rating for the individual vehicle.

Grotthus-Draper law (Section 796.3700), the first law of photochemistry, states that only light which is absorbed can be effective in producing a chemical transformation.

Ground phosphate rock handling and storage system (Section 60.401) means a system which is used for the conveyance and storage of ground phosphate rock from grinders at phosphate rock plants.

Ground water (Section 258.2 [10-9-91]) means water below the land surface in a zone of saturation.

Ground water (Section 300.5 [3-8-90])* as defined by section 101(12) of CERCLA, means water in a saturated zone or stratum beneath the surface of land or water.

Ground water (Sections 144.3; 146.3; 147.2902; 191.12; 257.3-4; 260.10; 270.2) means water below the land surface in a zone of saturation.

Ground water under the direct influence of surface water (Section 141.2 [6/29/89]) means any water beneath the surface of the ground with (1) significant occurrence of insects or other macroorganisms, algae, or large-diameter pathogens such as Giardia lamblia, or (2) significant and relatively rapid shifts in water characteristics such as turbidity, temperature, conductivity, or pH which closely correlate to climatological or surface water conditions. Direct influence must be determined for individual sources in accordance with criteria established by the State. The State determination of direct influence may be based on site-specific measurements of water quality and/or documentation of well construction characteristics and geology with field evaluation.

Groundwater (Section 241.101) means water present in the saturated zone of an aquifer.

Groundwater infiltration (Section 440.132) in Section 440.131 means that water which enters the treatment facility as a result of the interception of natural springs, aquifers, or run-off which percolates into the ground and seeps into the treatment facility's tailings pond or wastewater holding facility and that cannot be diverted by ditching or grouting the tailings pond or wastewater holding facility.

Group I storm water discharge (Section 122.26) means any storm water point source which is: (i) Subject to effluent limitations guidelines, new source performance standards, or toxic pollutant effluent standards; (ii) Designated under paragraph (c) of this section; or (iii) Located at an industrial plant or in plant associated areas. Plant associated areas means industrial plant yards, immediate access roads, drainage ponds, refuse piles, storage piles or areas and material or products loading and unloading areas. The term excludes areas located on plant lands separate from the plant's industrial activities, such as office buildings and accompanying parking lots.

Group II storm water discharge

(Section 122.26) means any storm water point source not included in paragraph (b)(2) of this section. (See Section 122.21(g)(10) for exemption from certain application requirements.)

Growth (Section 797.1050) means a relative measure of the viability of an algal population based on the number and/or weight of algal cells per volume of nutrient medium or test solution in a specified period of time.

Growth rate (Section 797.1060) means an increase in biomass or cell numbers of algae per unit time.

Guaranteed Loan Program (Section 39.105) means the program established pursuant to Pub. L. 94-558 which amended the Act by adding section 213.

Guide coat operation (Section 60.391) means the guide coat spray booth, flash-off area and bake oven(s) which are used to apply and dry or cure a surface coating between the prime coat and topcoat operation on the components of automobile and light-duty truck bodies.

Guide specification (Section 249.04) means a general specification - often referred to as a design standard or design guideline - which is a model standard and is suggested or required for use in the design of all of the construction projects of an agency.

Guidelines (Section 766.3) means the Midwest Research Institute (MRI) publication *Guidelines for the Determination of Polyhalogenated Dioxins and Dibenzofurans in Commercial Products*, EPA contract No. 68- 02-3938; MRI Project No. 8201-A(41), 1985.

GVWR (Section 86.082-2). See **Gross vehicle weight rating (Section 86.082-2)**.

GVWR (Sections 202.10; 205.51). See **Gross vehicle weight rating (Sections 202.10; 205.51)**.

H

Hair pulp (Section 425.02) means the removal of hair by chemical dissolution.

Hair save (Section 425.02) means the physical or mechanical removal of hair which has not been chemically dissolved, and either selling the hair as a by-product or disposing of it as a solid waste.

Half-life ($t_{1/2}$) of a chemical (Section 796.3700) is defined as the time required for the concentration of the chemical being tested to be reduced to one-half its initial value.

Half-Life (Section 61 [12-15-89]) The time which half the atoms of a particular radioactive substance transform, or decay, to another nuclear form.

Halogen (Section 141.2) means one of the chemical elements chlorine, bromine or iodine.

Halogenated organic compounds or HOCs (Section 268.2 [6-1-90])* means those compounds having a carbon-halogen bond which are listed under appendix III to this part.

Halogenated vent stream (Section 60.661 [6-29-90]) means any vent stream determined to have a total concentration (by volume) of compounds containing halogens of 20 ppmv (by compound) or greater.

Ham processor (Section 432.81) shall mean an operation which manufactures hams alone or in combination with other finished products at rates greater than 2730 kg (6000 lb) per day.

Hand glass melting furnace (Section 60.291) means a glass melting furnace where the molten glass is removed from the furnace by a glass worker using a blowpipe or a pontil.

Handicapped person (Section 7.25) (a) means any person who (1) has a physical or mental impairment which substantially limits one or more major life activities, (2) has a record of such an impairment, or (3) is regarded as having such an impairment. For purposes of employment, the term handicapped person does not include any person who is an alcoholic or drug abuser whose current use of alcohol or drugs prevents such individual from performing the duties of the job in question or whose employment, by reason of such current drug or alcohol abuse, would constitute a direct threat to property or the safety of others. (b) As used in this paragraph, the phrase (1) Physical or mental impairment means (i) any physiological disorder or condition, cosmetic disfigurement, or anatomical loss affecting one or more of the following body systems: Neurological; musculoskeletal; special sense organs; respiratory, including speech organs; cardiovascular; reproductive; digestive; genito-urinary; hemic and lymphatic; skin; and endocrine; and (ii) any mental or psychological disorder, such as mental retardation, organic brain syndrome, emotional or mental illness, and specific learning disabilities. (2) Major life activities means functions such as caring for one's self, performing manual tasks, walking, seeing, hearing, speaking, breathing, learning, and working. (3) Has a record of such an impairment means has a history of, or has been misclassified as having, a mental or physical impairment that substantially limits one or more major life activities. (4) Is regarded as having an impairment means: (i) Has a physical or mental impairment that does not substantially limit major life activities but that is treated by a recipient as constituting such a limitation; (ii) Has a physical or mental impairment that substantially limits major life activities only as a result of the attitudes of others toward such impairment; or (iii) Has none of the impairments defined above but is treated by a recipient as having such an impairment.

Hang-up (Section 86.082-2) refers to the process of hydrocarbon molecules being adsorbed, condensed, or by any other method removed from the sample flow prior to reaching the instrument detector. It also refers to any subsequent desorption of the molecules into the sample flow when they are assumed to be absent.

Hardboard (Section 429.11) means a panel manufactured from interfelted ligno-cellulosic fibers consolidated under heat and pressure to a density of 0.5 g/cu cm (31 lb/cu ft) or greater.

Hardness (Section 797.1600) is the total concentration of the calcium and magnesium ions in water expressed as calcium carbonate (mg $CaCO_3$/liter).

Hatch (Section 797.2050) means eggs or young birds that are the same age and that are derived from the same adult breeding population, where the adults are of the same strain and stock.

Hatch (Section 797.2175). Eggs or birds that are the same age and that are derived from the same adult breeding population, where the adults are of the same strain and stock.

Hatchability (Section 797.2130) means embryos that mature, pip the shell, and liberate themselves from the eggs on day 23 or 24 of incubation. Values are expressed as percentage of viable embryos (fertile eggs).

Hatchability (Section 797.2150) means embryos that mature, pip the shell, and liberate themselves from their eggs on day 25, 26, or 27 of incubation. Values are expressed as a percentage of viable embryos (fertile eggs).

Hatchback (Section 600.002-85) means a passenger automobile where the conventional luggage compartment, i.e., trunk, is replaced by a cargo area which is open to the passenger compartment and accessed vertically by a rear door which encompasses the rear window.

Hazard (Section 162.3) means the likelihood that use of a pesticide would result in an adverse effect on man or the environment in a given situation.

Hazard (Section 171.2) means a probability that a given pesticide will have an adverse effect on man or the environment in a given situation, the relative likelihood of danger or ill effect being dependent on a number of interrelated factors present at any given time.

Hazard Category (Section 370.2 [10/15/87]) means any of the following: (1) Immediate (acute) health hazard, including highly toxic, toxic, irritant, sensitizer, corrosive (as defined under Section 1910.1200 of Title 29 of the Code of Federal Regulations) and other hazardous chemicals that cause an adverse effect to a target organ and which effect usually occurs rapidly as a result of short term exposure and is of short duration; (2) Delayed (chronic) health hazard, including carcinogens (as defined under Section 1910.1200 of Title 29 of the Code of Federal Regulations) and other hazardous chemicals that cause an adverse effect to a target organ and which effect generally occurs as a result of long term exposure and is of long duration; (3) Fire hazard, including flammable, combustible liquid, pyrophoric, and oxidizer (as defined under Section 1910.1200 of Title 29 of the Code of Federal Regulations); (4) Sudden release of pressure, including explosive and compressed gas (as defined under Section 1910.1200 of Title 29 of the Code of Federal Regulations); and (5) Reactive, including unstable reactive, organic peroxide, and water reactive (as defined under Section 1910.1200 of Title 29 of the Code of Federal Regulations).

Hazardous chemical (Section 355.20 [7-26-90])* means any hazardous chemical as defined under Sec. 1910.1200(c) of Title 29 of the Code of Federal Regulations, except that such term does not include the following substances: (1) Any food, food additive, color additive, drug, or cosmetic regulated by the Food and Drug Administration. (2) Any substance present as a solid in any manufactured item to the extent exposure to the substance does not occur under normal conditions of use. (3) Any substance to the extent it is used for personal, family, or household purposes, or is present in the same form and concentration as a product packaged for distribution and use by the general public. (4) Any substance to the extent it is used in a research laboratory or a hospital or other medical facility under the direct supervision of a technically qualified individual. (5) Any substancc to the extent it is used in routine agricultural operations or is a fertilizer held for sale by a retailer to the ultimate customer.

Hazardous Chemical (Section 370.2 [10/15/87]) means any hazardous chemical as defined under Section 1910.1200(c) of Title 29 of the Code of Federal Regulations, except that such term does not include the following substances: (1) Any food, food additive, color additive, drug, or cosmetic regulated by the Food and Drug Administration. (2) Any substance present as a solid in any manufactured item to the extent exposure to the substance does not occur under normal conditions of use. (3) Any substance to the extent it is used for personal, family, or household purposes, or is present in the same form and concentration as a product packaged for distribution and use by the general public. (4) Any substance to the extent it is used in a research laboratory or a hospital or other medical facility under the direct supervision of a technically qualified individual. (5) Any substance to the extent it is used in routine agricultural operations or is a fertilizer held for sale by a retailer to the ultimate customer.

Hazardous constituent or constituents (Section 268.2 [6-1-90])* means those constituents listed in appendix VIII to part 261 of this chapter.

Hazardous Ranking System (HRS) (Section 300.5 [3-8-90]) means the method used by EPA to evaluate the relative potential of hazardous substance releases to cause health or safety problems, or ecological or environmental damage.

Hazardous substance (Section 122.2) means any substance designated under 40 CFR Part 116 pursuant to section 311 of CWA.

Hazardous substance (Section 2.310) has the meaning given it in section 101(14) of the Act, 42 U.S.C. 9601(14).

Hazardous substance (Section 300.5 [3-8-90])* as defined by section 101(14) of CERCLA, means: Any substance designated pursuant to section 311(b)(2)(A) of the CWA; any element, compound, mixture, solution, or substance designated pursuant to section 102 of CERCLA; any hazardous waste having the characteristics identified under or listed pursuant to section 3001 of the Solid Waste Disposal Act (but not including any waste the regulation of which under the Solid Waste Disposal Act has been suspended by Act of Congress); any toxic pollutant listed under section 307(a) of the CWA; any hazardous air pollutant listed under section 112 of the Clean Air Act; and any imminently hazardous chemical substance or mixture with respect to which the EPA Administrator has taken action pursuant to section 7 of the

Toxic Substances Control Act. The term does not include petroleum, including crude oil or any fraction thereof which is not otherwise specifically listed or designated as a hazardous substance in the first sentence of this paragraph, and the term does not include natural gas, natural gas liquids, liquified natural gas, or synthetic gas usable for fuel (or mixtures of natural gas and such synthetic gas).

Hazardous substance (Section 302.3) means any substance designated pursuant to 40 CFR Part 302.

Hazardous substance (Section 310.11 [10/21/87]) as defined by section 101(14) of CERCLA, means: (1) Any substance designated pursuant to section 311(b)(2)(A) of the Federal Water Pollution Control Act, (2) Any element, compound, mixture, solution, or substance designated pursuant to section 102 of CERCLA, (3) Any hazardous waste having the characteristics identified under or listed pursuant to section 3001 of the Solid Waste Disposal Act (but not including any waste the regulation of which under the Solid Waste Disposal Act has been suspended by Act of Congress), (4) Any toxic pollutant listed under section 307(a) of the Federal Water Pollution Control Act, (5) Any hazardous air pollutant listed under section 112 of the Clean Air Act, and (6) Any imminently hazardous chemical substance or mixture with respect to which the Administrator has taken action pursuant to section 7 of the Toxic Substances Control Act. The term does not include petroleum, including crude oil or any fraction thereof that is not otherwise specifically listed or designated as a hazardous substance under paragraphs (d)(1) through (d)(6) of this paragraph, and the term does not include natural gas, natural gas liquids, liquefied natural gas, or synthetic gas usable for fuel (or mixtures or natural gas and such synthetic gas).

Hazardous substance (Sections 305.12; 306.12) means (1) any substance designated pursuant to section 311(b)(2)(A) of the Federal Water Pollution Control Act, (2) any element, compound, mixture, solution, or substance designated pursuant to section 102 of this Act, (3) any hazardous waste having the characteristics identified under or listed pursuant to section 3001 of the Solid Waste Disposal Act (but not including any waste the regulation of which under the Solid Waste Disposal Act has been suspended by Act of Congress), (4) any toxic pollutant listed under section 307(a) of the Federal Water Pollution Control Act, (5) any hazardous air pollutant listed under section 112 of the Clean Air Act, and (6) any imminently hazardous chemical substance or mixture with respect to which the Administrator has taken action pursuant to section 7 of the Toxic Substances Control Act. The term does not include petroleum, including crude oil or any fraction thereof which is not otherwise specifically listed or designated as a hazardous substance under paragraphs (h)(1) through (6) of this section, and the term does not include natural gas, natural gas liquids, liquefied natural gas, or synthetic gas usable for fuel (or mixtures of natural gas and such synthetic gas).

Hazardous substance UST system (Section 280.12 [9/23/88]) means an underground storage tank system that contains a hazardous substance defined in section 101(14) of the Comprehensive Environmental Response, Compensation and Liability Act of 1980 (but not including any substance regulated as a hazardous waste under subtitle C) or any mixture of such substances and petroleum, and which is not a petroleum UST system.

Hazardous substances (Section 373.4 [4-16-90]) means that group of substances defined as hazardous under CERCLA 101(14), and that appear at 40 CFR 302.4.

Hazardous waste (Section 2.305) has the meaning given it in section 1004(5) of the Act, 42 U.S.C. 6903(5).

Hazardous waste (Section 240.101) means any waste or combination of wastes which pose a substantial present or potential hazard to human health or living organisms because such wastes are non-degradable or persistent in nature or because they can be biologically magnified, or because they can be lethal, or because they may otherwise cause or tend to cause detrimental cumulative effects.

Hazardous waste (Section 243.101) means a waste or combination of wastes of a solid, liquid, contained gaseous, or semisolid form which may cause, or contribute to, an increase in mortality or an increase in serious irreversible, or incapacitating reversible illness, taking into account the toxicity of such waste, its persistence and degradability in nature, its potential for accumulation or concentration in tissue, and other factors that may otherwise cause or contribute to adverse acute or chronic effects on the health of persons or other organisms.

Hazardous waste (Section 260.10) means a hazardous waste as defined in Section 261.3 of this chapter.

Hazardous waste (Section 261.3). A solid waste, as defined in Section 261.2, is a hazardous waste if: (1) It is not excluded from regulation as a hazardous waste under Section 261.4(b); and (2) It meets the criteria set forth in Section 261.3(a)(2) (i) through (iv).

Hazardous waste (Section 302.3) shall have the meaning provided in 40 CFR 261.3.

Hazardous waste (Sections 144.3; 146.3; 270.2) means a hazardous waste as defined in 40 CFR 261.3.

Hazardous waste constituent (Section 260.10) means a constituent that caused the Administrator to list the hazardous waste in Part 261, Subpart D, of this chapter, or a constituent listed in Table 1 of Section 261.24 of this chapter.

Hazardous waste discharge (Section 260.10). See **Discharge (Section 260.10)**.

Hazardous waste management (Section 260.10). See **Management (Section 260.10)**.

Hazardous waste management facility (HWM facility) (Sections 144.3; 146.3; 270.2) means all contiguous land, and structures, other appurtenances, and improvements on the land used for treating, storing, or disposing of hazardous waste. A facility may consist of several treatment, storage, or disposal operational units (for example, one or more landfills, surface impoundments, or combination of them).

Hazardous waste management unit (Section 260.10) is a contiguous area of land on or in which hazardous waste is placed, or the largest area in which there is significant likelihood of mixing hazardous waste constituents in the same area. Examples of hazardous waste management units include a surface impoundment, a waste pile, a land treatment area, a landfill cell, an incinerator, a tank and its associated piping and underlying containment system and a container storage area. A container alone does not constitute a unit; the unit includes containers and the land or pad upon which they are placed.

Hazardous waste management unit shutdown (Section 264.1031 [6-21-90]) means a work practice or operational procedure that stops operation of a hazardous waste management unit or part of a hazardous waste management unit. An unscheduled work practice or operational procedure that stops operation of a hazardous waste management unit or part of a hazardous waste management unit for less than 24 hours is not a hazardous waste management unit shutdown. The use of spare equipment and technically feasible bypassing of equipment without stopping operation are not hazardous waste management unit shutdowns.

Hazardous wastes (Section 241.101) means any waste or combination of wastes which pose a substantial present or potential hazard to human health or living organisms because such wastes are nondegradable or persistent in nature or because they can be biologically magnified, or because they can be lethal, or because they may otherwise cause or tend to cause detrimental cumulative effects.

HDD or 2,3,7,8-HDD (Section 766.3) means any of the dibenzo-p-dioxins totally chlorinated or totally brominated at the following positions on the molecular structure: 2,3,7,8; 1,2,3,7,8; 1,2,3,4,7,8; 1,2,3,6,7,8; 1,2,3,7,8,9; and 1,2,3,4,7,8,9.

HDF or 2,3,7,8-HDF (Section 766.3) means any of the dibenzofurans totally chlorinated or totally brominated at the following positions on the molecular structure: 2,3,7,8; 1,2,3,7,8; 2,3,4,7,8; 1,2,3,4,7,8; 1,2,3,6,7,8; 1,2,3,7,8,9; 2,3,4,6,7,8; 1,2,3,4,6,7,8; and 1,2,3,4,7,8,9.

Headband (Section 211.203) means the component of hearing protective device which applies force to, and holds in place on the head, the component which is intended to acoustically seal the ear canal.

Health and safety data (Section 2.306) means (i) the information described in paragraphs (a)(3)(i) (A), (B), and (C) of this section with respect to any chemical substance or mixture offered for commercial distribution (including for test marketing purposes and for use in research and development), any chemical substance included on the inventory of chemical substances under section 8 of the Act (15 U.S.C. 2607), or any chemical substance or mixture for which testing is required under section 4 of the Act (15 U.S.C. 2603) or for which notification is required under section 5 of the Act (15 U.S.C. 2604). (A) Any study of any effect of a chemical substance or mixture on health, on the environment, or on both, including underlying data and epidemiological studies; studies of occupational exposure to a chemical substance or mixture; and toxicological, clinical, and ecological studies of a chemical substance or mixture; (B) Any test performed under the Act; and (C) Any data reported to, or otherwise obtained by, EPA from a study described in paragraph (a)(3)(i)(A) of this section or a test described in paragraph (a)(3)(i)(B) of this section. (ii) Notwithstanding paragraph (a)(3)(i) of this section, no information shall be considered to be health and safety data if disclosure of the information would (A) In the case of a chemical substance or mixture, disclose processes used in the manufacturing or processing the chemical substance or mixture or, (B) In the case of a mixture, disclose the portion of the mixture comprised by any of the chemical substances in the mixture.

Health and safety plan (Section 35.6015 [6-5-90])* A plan that specifies the procedures that are sufficient to protect on-site personnel and surrounding communities from the physical, chemical, and/or biological hazards of the site. The health and safety plan outlines: (i) Site hazards; (ii) Work areas and site control procedures; (iii) Air surveillance procedures; (iv) Levels of protection; (v) Decontamination and site emergency plans; (vi) Arrangements for weather-related problems; and (vii) Responsibilities for implementing the health and safety plan.

Health and safety study or study (Section 716.3) means any study of any effect of a chemical substance or mixture on health or the environment or on both, including underlying data and epidemiological studies, studies of occupational exposure to a chemical substance or mixture, toxicological, clinical, and ecological or other studies of a chemical substance or mixture, and any test performed under TSCA. (1) It is intended that the term health and safety study be interpreted broadly. Not only is information which arises as a result of a formal, disciplined study included, but other information relating to the effects of a chemical substance or mixture on health or the environment is also included. Any data that bear on the effects of a chemical substance on health or the environment would be included. Chemical identity is part of, or underlying data to, a health and safety study. (2) Examples are: (i) Long- and short-term tests of mutagenicity, carcinogenicity, or teratogenicity; data on behavioral disorders; dermatoxicity; pharmacological effects; mammalian absorption, distribution, metabolism, and excretion; cumulative, additive, and synergistic effects; and acute, subchronic, and chronic effects. (ii) Tests for ecological or other environmental effects on invertebrates, fish, or other animals, and plants, including: Acute toxicity tests, chronic toxicity tests, critical life-stage tests, behavioral tests, algal growth tests, seed germination tests, plant growth or damage tests, microbial function tests, bioconcentration or bioaccumulation tests, and model ecosystem (microcosm) studies. (iii) Assessments of human and environmental exposure, including workplace exposure, and impacts of a particular chemical substance or mixture on the environment, including surveys, tests, and studies of: Biological, photochemical, and chemical degradation; structure/activity relationships; air, water, and soil transport; biomagnification and bioconcentration; and chemical and physical properties, e.g., boiling point, vapor pressure, evaporation rates from soil and water, octanol/water partition coefficient, and water solubiiity. (iv) Monitoring data, when they have been aggregated and analyzed to measure the exposure of humans or the environment to a chemical substance or mixture.

Health and safety study or study (Sections 720.3; 723.50) means any study of any effect of a chemical substance or mixture on health or the environment or on both, including underlying data and epidemiological studies, studies of occupational exposure to a chemical substance or mixture, toxicological, clinical, and ecological, or other studies of a chemical substance or mixture, and any test performed under the Act. Chemical identity is always part of a health and safety study. (1) Not only is information which arises as a result of a formal, disciplined study included, but other information relating to the effects of a chemical substance or mixture on health or the environment is also included. Any data that bear on the effects of a chemical substance on health or the environment would be included. (2) Examples include: (i) Long- and short-term tests of mutagenicity, carcinogenicity, or teratogenicity; data on behavioral disorders; dermatoxicity; pharmacological effects; mammalian absorption, distribution, metabolism, and excretion; cumulative, additive, and synergistic effects; acute, subchronic, and chronic effects; and structure/activity analyses. (ii) Tests for ecological or other environmental effects on invertebrates, fish, or other animals, and plants, including: Acute toxicity tests, chronic toxicity tests, critical life stage tests, behavioral tests, algal growth tests, seed germination tests, plant growth or damage tests, microbial function tests, bioconcentration or bioaccumulation tests, and model ecosystem (microcosm) studies. (iii) Assessments of human and environmental exposure, including workplace exposure, and impacts of a particular chemical substance or mixture on the environment, including surveys, tests, and studies of: Biological, photochemical, and chemical degradation; air, water, and soil transport; biomagnification and bioconcentration; and chemical and physical properties, e.g., boiling point, vapor pressure, evaporation rates from soil and water, octanol/water partition coefficient, and water solubility. (iv) Monitoring data, when they have been aggregated and analyzed to measure the exposure of humans or the environment to a chemical substance or mixture. (v) Any assessments of risk to health and the environment resulting from the manufacture, processing, distribution in commerce, use, or disposal of the chemical substance.

Hearing (Section 164.2 [2-13-92])* means a public hearing which is conducted pursuant to the provisions of Chapter 5, Subchapter II of Title 5 of the United States Code and the regulations of this part.

Hearing (Section 22.03) means a hearing on the record open to the public and conducted under these rules of practice.

Hearing (Section 8.33) means a hearing conducted as specified in this subpart to enable the Agency to decide whether to impose sanctions on a respondent for violations of the Executive Order and rules, regulations, and orders thereunder.

Hearing Clerk (Section 164.2 [2-13-92])* means the Hearing Clerk, Environmental Protection Agency, Washington, DC 20460.

Hearing Clerk (Sections 104.2; 124.72) means The Hearing Clerk, U.S. Environmental Protection Agency, 401 M Street, SW, Washington DC 20460.

Hearing Clerk (Sections 22.03; 123.64) means the Hearing Clerk, A-110, United States Environmental Protection Agency, 401 M St. SW, Washington DC 20460.

Hearing Clerk (Sections 85.1807; 86.614; 86.1014-84; 86.1115-87; 209.3) shall mean the Hearing Clerk of the Environmental Protection Agency.

Hearing examiner (Section 8.33) means a hearing examiner appointed by the Assistant Administrator for Enforcement and General Counsel.

Hearing officer (Section 8.2) means the individual or board of individuals designated to conduct hearings.

Hearing Officer (Section 142.202 [1-30-91]) means an Environmental Protection Agency employee who has been delegated by the Administrator the authority to preside over a public hearing held pursuant to section 1414(g)(2) of the Safe Drinking Water Act, 42 U.S.C. 300g-3(g)(2).

Hearing Protective Device (Section 211.203) means any device or material, capable of being worn on the head or in the ear canal, that is sold wholly or in part on the basis of its ability to reduce the level of sound entering the ear. This includes devices of which hearing protection may not be the primary function, but which are nonetheless sold partially as providing hearing protection to the user. This term is used interchangeably with the terms, hearing protector and device.

Heat cycle (Section 60.271a) means the period beginning when scrap is charged to an empty EAF and ending when the EAF tap is completed or beginning when molten steel is charged to an empty AOD vessel and ending when the AOD vessel tap is completed.

Heat input (Section 52.01) means the total gross calorific value (where gross calorific value is measured by ASTM Method D2015-66, D240-64, or D1826-64) of all fuels burned.

Heat input (Section 60.41b [12/16/87])* means heat derived from combustion of fuel in a steam generating unit and does not include the heat input from preheated combustion air, recirculated flue gases, or exhaust gases from other sources, such as gas turbines, internal combustion engines, kilns, etc.

Heat input (Section 60.41c [9-12-90]) means heat derived from combustion of fuel in a steam generating unit and does not include the heat derived from preheated combustion air, recirculated flue gases, or exhaust gases from other sources (such as stationary gas turbines, internal combustion engines, and kilns).

Heat Rating (Section 85.2122(a)(8)(ii)(B)) means that measurement of engine indicated mean effective pressure (IMEP) value obtained on the engine at a point when the supercharge pressure is 25.4 mm (one inch) Hg below the preignition point of the spark plug, as rated according to SAE J549A Recommended Practice.

Heat release rate (Section 60.41b [12/16/87])* means the steam generating unit design heat input capacity (in MW or Btu/hour) divided by the furnace volume (in cubic meters or cubic feet); the furnace volume is that volume bounded by the front furnace wall where the burner is located, the furnace side waterwall, and extending to the level just below or in front of the first row of convection pass tubes.

Heat time (Section 60.271) means the period commencing when scrap is charged to an empty EAF and terminating when the EAF tap is completed.

Heat transfer medium (Section 60.41b [12/16/87])* means any material that is used to transfer heat from one point to another point.

Heat transfer medium (Section 60.41c [9-12-90]) means any material that is used to transfer heat from one point to another point.

Heat treatment (Section 467.02) is the application of heat of specified temperature and duration to change the physical properties of the metal.

Heat treatment (Section 468.02) shall mean the application or removal of heat to a workpiece to change the physical properties of the metal.

Heat treatment (Section 471.02) is the application of heat of specified temperature and duration to change the physical properties of the metal.

Heating Oil (Section 280.12 [9/23/88]) means petroleum that is No. 1, No. 2, No. 4 - light, No. 4 - heavy, No. 5 - light, No. 5 - heavy, and No. 6 technical grades of fuel oil; other residual fuel oils (including Navy Special Fuel Oil and Bunker C); and other fuels when used as substitutes for one of these fuel oils. Heating oil is typically used in the operation of heating equipment, boilers, or furnaces.

Heavy light-duty truck (Section 86.094-2 [6-5-91]) means any light-duty truck rated greater than 6000 lbs GVWR.

Heavy metal (Section 191.12) means all uranium, plutonium, or thorium placed into a nuclear reactor.

Heavy metals (Section 165.1) means metallic elements of higher atomic weights, including but not limited to arsenic, cadmium, copper, lead, mercury, manganese, zinc, chromium, tin, thallium, and selenium.

Heavy-duty engine (Section 86.082-2) means any engine which the engine manufacturer could reasonably expect to be used for motive power in a heavy-duty vehicle.

Heavy-duty vehicle (Section 86.082-2) means any motor vehicle rated at more than 8,500 pounds GVWR or that has a vehicle curb weight of more than 6,000 pounds or that has a basic vehicle frontal area in excess of 45 square feet.

Heavy-passenger cars (Section 86.084-2) means, for the 1984 model year only, a passenger car or passenger car derivative capable of seating 12 passengers or less, rated at 6,000 pounds GVW or more and having an equivalent test weight of 5,000 pounds or more.

HEPA (Section 763.83 [10/30/87]). See **High-efficiency particulate air (Section 763.83)**.

HEPA filter (Section 763.121). See **High-efficiency particulate air filter (Section 763.121)**.

Herbicides (Section 162.3) includes all substances or mixtures of substances, except defoliants as defined in paragraph (ff)(5) of this section, desiccants as defined in paragraph (ff)(6) of this section, plant regulators as defined in paragraph (ff)(14) of this section, and slimicides as defined in paragraph (ff)(16) of this section, intended for use in preventing or inhibiting the growth of, or killing or destroying plants and plant parts which are declared to be pests under Section 162.14. Herbicides include, but are not limited to: (i) Direct contact herbicides intended to kill or destroy weeds, unwanted brush and trees, or unwanted plant parts, or to mitigate their adverse effects on desirable plants; (ii) Soil treatment herbicides intended to kill or destroy weeds, unwanted brush and trees, or unwanted plant parts, or to prevent the establishment of any or all plants; (iii) Pre-emergence herbicides intended to prevent or inhibit the germination or growth of weed seeds or seedlings; (iv) Root control herbicides intended to prevent the growth of, or kill roots in certain sites such as sewer lines and drainage tiles; (v) Aquatic herbicides intended to prevent, inhibit, or control the growth of, or kill aquatic weeds; (vi) Algaecides, except slimicides as defined in paragraph (ff)(16) of this section, intended to prevent or inhibit the multiplication of, or destroy algae in ponds, swimming pools, aquaria or similar confined sites; (vii) Debarking agents intended to kill trees by treatment of bark on trunks; and (viii) Biological weed-control agents such as specific pathogenic organisms or entities prepared and utilized by man.

HEX-BCH (Section 704.142) means the chemical substance 1,2,3,4,7,7-hexachloronorbornadiene, CAS No. 3389-71-7.

Hexavalent chromium (or chromium VI) (Section 420.02) means the value obtained by the method specified in 40 CFR 136.3.

Hexavalent chromium chemicals (Section 749.68 [1-3-90]) means any combination of chemical substances containing hexavalent chromium and includes hexavalent chromium-based water treatment chemicals.

Hexavalent chromium-based water treatment chemicals (Section 749.68 [1-3-89]) means any hexavalent chromium, alone or in combination with other water treatment chemicals, used to treat water.

HFPO (Section 704.104 [10/27/87]) means the chemical substance hexafluoropropylene oxide, CAS Number 428-59-1. (Listed in TSCA Inventory as oxirane, trifluoro(trifluoromethyl)-.)

HHS (Section 300.5) is the Department of Health and Human Services.

Hide (Section 425.02) means any animal pelt or skin as received by a tannery as raw material to be processed.

High altitude (Section 86.082-2) means any elevation over 1,219 meters (4,000 feet).

High dose (Section 795.232 [1-8-90]) shall not exceed the lower explosive limit (LEL) and ideally should induce minimal toxicity.

High heat release rate (Section 60.41b [12/16/87])* means a heat release rate greater than 730,000 J/sec-m^3 (70,000 Btu/hour-ft^3).

High level of volatile impurities (Section 60.161) means a total smelter charge containing more than 0.2 weight percent arsenic, 0.1 weight percent antimony, 4.5 weight percent lead or 5.5 weight percent zinc, on a dry basis.

High terrain (Sections 51.166; 52.21) means any area having an elevation 900 feet or more above the base of the stack of a source.

High velocity air filter (HVAF) (Section 60.471) means an air pollution control filtration device for the removal of sticky, oily, or liquid aerosol particulate matter from exhaust gas streams.

High-altitude conditions (Section 86.082-2) means a test altitude of 1,620 meters (5,315 feet), plus or minus 100 meters (328 feet), or equivalent observed barometric test conditions of 83.3 plus or minus 1 kilopascals.

High-altitude reference point (Section 86.082-2) means an elevation of 1,620 meters (5,315 feet) plus or minus 100 meters (328 feet), or equivalent observed barometric test conditions of 83.3 kPa (24.2 inches Hg), plus or minus 1 kPa (0.30 Hg).

High-carbon ferrochrome (Section 60.261) means that alloy as defined by ASTM Designation A101-73 (incorporated by reference - see Section 60.17) grades HC1 through HC6.

High-concentration PCBs (Section 761.123) means PCBs that contain 500 ppm or greater PCBs, or those materials which EPA requires to be assumed to contain 500 ppm or greater PCBs in the absence of testing.

High-contact industrial surface (Section 761.123) means a surface in an industrial setting which is repeatedly touched, often for relatively long periods of time. Manned machinery and control panels are examples of high-contact industrial surfaces. High-contact industrial surfaces are generally of impervious solid material. Examples of low-contact industrial surfaces include ceilings, walls, floors, roofs, roadways and sidewalks in the industrial area, utility poles, unmanned machinery, concrete pads beneath electrical equipment, curbing, exterior structural building components, indoor vaults, and pipes.

High-contact residential/commercial surface (Section 761.123) means a surface in a residential/commercial area which is repeatedly touched, often for relatively long periods of time. Doors, wall areas below 6 feet in height, uncovered flooring, windowsills, fencing, bannisters, stairs, automobiles, and children's play areas such as outdoor patios and sidewalks are examples of high-contact residential/commercial surfaces. Examples of low-contact residential/commercial surfaces include interior ceilings, interior wall areas above 6 feet in height, roofs, asphalt roadways, concrete roadways, wooden utility poles, unmanned machinery, concrete pads beneath electrical equipment, curbing, exterior structural building components (e.g., aluminum/vinyl siding, cinder block, asphalt tiles), and pipes.

High-efficiency particulate air (HEPA) filter (Section 763.121) means a filter capable of trapping and retaining at least 99.97 percent of all monodispersed particles of 0.3 micrometer in diameter or larger.

High-efficiency particulate air (HEPA) (Section 763.83 [10/30/87]) refers to a filtering system capable of trapping and retaining at least 99.97 percent of all monodispersed particles 0.3 μm in diameter or larger.

High-grade electrical paper (Section 763.163 [7-12-89]) means an asbestos-containing product that is made of paper and consisting of asbestos fibers and high-temperature resistant organic binders and used in or with electrical devices for purposes of insulation or protection. Major applications of this product include insulation for high-temperature, low voltage applications such as in motors, generators, transformers, switch gears, and other heavy electrical apparatus.

High-grade paper (Section 246.101) means letterhead, dry copy papers, miscellaneous business forms, stationery, typing paper, tablet sheets, and computer printout paper and cards, commonly sold as white ledger, computer printout and tab card grade by the wastepaper industry.

High-level radioactive waste (Section 191.02) as used in this part, means high-level radioactive waste as defined in the Nuclear Waste Policy Act of 1982 (Pub. L. 97-425).

High-processing packinghouse (Section 432.41) shall mean a packinghouse which processes both animals slaughtered at the site and additional carcasses from outside sources.

Highway (Sections 202.10; 205.2) means the streets, roads, and public ways in any State.

Highway Fuel Economy (Section 600.002-85) means the fuel economy determined by operating a vehicle (or vehicles) over the driving schedule in the Federal highway fuel economy test procedure.

Highway Fuel Economy Test (Section 610.11) means the test procedure described in Section 600.111(b).

HOCs (Section 268.2 [7/8/87]). See **Halogenated organic compounds (Section 268.2)**.

Holding of a copper converter (Section 61.171) means suspending blowing operations while maintaining in a heated state the molten bath in the copper converter.

Holocene (Section 264.18), as used in paragraph (a)(1) of this section, means the most recent epoch of the Quarternary period, extending from the end of the Pleistocene to the present.

Homogeneous area (Section 763.83 [10/30/87]) means an area of surfacing material, thermal system insulation material, or miscellaneous material that is uniform in color and texture.

Homolog (Section 766.3) means a group of isomers that have the same degree of halogenation. For example, the homologous class of tetrachlorodibenzo-p-dioxins consists of all dibenzo-p-dioxins containing four chlorine atoms. When the homologous classes discussed in this Part are referred to, the following abbreviations for the prefix denoting the number of halogens are used: tetra-, T (4 atoms); penta-, Pe (5 atoms); hexa-, Hx (6 atoms); hepta-, Hp (7 atoms).

Hood or enclosure (Section 60.441) means any device used to capture fugitive volatile organic compounds.

Hosiery products (Section 410.51), for NSPS (Section 410.55), shall mean the internal subdivision of the knit fabric finishing subcategory for facilities that are engaged primarily in dyeing or finishing hosiery of any type.

Host (Section 171.2) means any plant or animal on or in which another lives for nourishment, development, or protection.

Hot forming (Section 420.71) means those steel operations in which solidified, heated steel is shaped by rolls.

Hot metal transfer station (Section 60.141a) means the facility where molten iron is emptied from the railroad torpedo car or hot metal car to the shop ladle. This includes the transfer of molten iron from the torpedo car or hot metal car to a mixer (or other intermediate vessel) and from a mixer (or other intermediate vessel) to the ladle. This facility is also known as the reladling station or ladle transfer station.

Hot mix asphalt facility (Section 60.91) means any facility, as described in Section 60.90, used to manufacture hot mix asphalt by heating and drying aggregate and mixing with asphalt cements.

Hot pressing (Section 471.02) is forming a powder metallurgy compact at a temperature high enough to effect concurrent sintering.

Hot strip and sheet mill (Section 420.71) means those steel hot forming operations that produce flat hot-rolled products other than plates.

Hot water seal (Section 467.02 [12/27/88]) is a heated water bath (heated to approximately 180 degrees F) used to seal the surface coating on formed aluminum which has been anodized and coated. In establishing an effluent allowance for this operation, the hot water seal shall be classified as a cleaning or etching rinse.

Hot well (Section 264.1031 [6-21-90]) means a container for collecting condensate as in a steam condenser serving a vacuum-jet or steam-jet ejector.

Hot-soak losses (Section 86.082-2) means evaporative emissions after termination of engine operation.

Housed lot (Section 412.11) shall mean totally roofed buildings which may be open or completely enclosed on the sides wherein animals or poultry are housed over solid concrete or dirt floors, slotted (partially open) floors over pits or manure collection areas in pens, stalls or cages, with or without bedding materials and mechanical ventilation. For the purposes hereof, the term housed lot is synonymous with the terms slotted floor buildings (swine, beef), barn (dairy cattle) or stable (horses), houses (turkeys, chickens), which are terms widely used in the industry.

Household waste (Section 258.2 [10-9-91]) means any solid waste (including garbage, trash, and sanitary waste in septic tanks) derived from households (including single and multiple residences, hotels and motels, bunkhouses, ranger stations, crew quarters, campgrounds, picnic grounds, and day-use recreation areas).

HRGC (Section 766.3) means high resolution gas chromatography.

HRMS (Section 766.3) means high resolution mass spectrometry.

H_{sys} (Section 60.741 [9-11-89]) means the carbon adsorption system efficiency calculated when each adsorber vessel has an individual exhaust stack.

Human environment (Section 1508.14) shall be interpreted comprehensively to include the natural and physical environment and the relationship of people with that environment. (See the definition of effects (Section 1508.8).) This means that economic or social effects are not intended by themselves to require preparation of an environmental impact statement. When an environmental impact statement is prepared and economic or social and natural or physical environmental effects are interrelated, then the environmental impact statement will discuss all of these effects on the human environment.

Human subject (Section 26.102 [6-18-91]) means a living individual about whom an investigator (whether professional or student) conducting research obtains (1) data through intervention or interaction with the individual, or (2) identifiable private information. Intervention includes both physical procedures by which data are gathered (for example, venipuncture) and manipulations of the subject or the subject's environment that are performed for research purposes. Interaction includes communication or interpersonal contact between investigator and subject. "Private information" includes information about behavior that occurs in a context in which an individual can reasonably expect that no observation or recording is taking place, and information which has been provided for specific purposes by an individual and which the individual can reasonably expect will not be made public (for example, a medical record). Private information must be individually identifiable (i.e., the identity of the subject is or may readily be ascertained by the investigator or associated with the information) in order for obtaining the information to constitute research involving human subjects.

H_v (Section 60.741 [9-11-89]) means the individual carbon adsorber vessel (v) efficiency achieved for the duration of the emission test (expressed as a fraction).

HVAF (Section 60.471). See **High velocity air filter (Section 60.471)**.

HWM facility (Sections 144.3; 146.3; 270.2). See **Hazardous waste**

management facility (Sections 144.3; 146.3; 270.2).

Hydraulic barking (Section 429.11) means a wood processing operation that removes bark from wood by the use of water under a pressure of 6.8 atm (100 psia) or greater.

Hydraulic fluids (Section 252.4 [6/30/88]) means petroleum-based hydraulic fluids.

Hydraulic lift tank (Section 280.12 [9/23/88]) means a tank holding hydraulic fluid for a closed-loop mechanical system that uses compressed air or hydraulic fluid to operate lifts, elevators, and other similar devices.

Hydrocarbon (Section 60.111) means any organic compound consisting predominantly of carbon and hydrogen.

Hydrochloric acid pickling (Section 420.91) means those operations in which steel products are immersed in hydrochloric acid solutions to chemically remove oxides and scale, and those rinsing operations associated with such immersions.

Hydrogen gas stream (Section 61.51) means a hydrogen stream formed in the chlor-alkali cell denuder.

Hydrotesting (Section 471.02) is the testing of piping or tubing by filling with water and pressurizing to test for integrity.

Hypocotyl (Section 797.2750) means that portion of the axis of an embryo or seedling situated between the cotyledons (seed leaves) and the radicle.

I

Ice fog (Section 60.331) means an atmospheric suspension of highly reflective ice crystals.

ICS (Section 51.100). See **Intermittent control system (Section 51.100)**.

Identity (Section 721.3 [7-27-89]) means any chemical or common name used to identify a chemical substance or a mixture containing that substance.

Idle (Section 201.1) means that condition where all engines capable of providing motive power to the locomotive are set at the lowest operating throttle position; and where all auxiliary non-motive power engines are not operating.

Immediate container (Section 162.3) means that container which is directly in contact with the pesticide or device.

Immediate container (Section 167.1) means the individual innermost package holding the pesticide.

Immediate family (Section 170.3 [8/21/92])* includes only spouse, children, stepchildren, foster children, parents, stepparents, foster parents, brothers, and sisters.

Immediate use (Section 721.3 [7-27-89]) A chemical substance is for the "immediate use" of a person if it is under the control of, and used only by, the person who transferred it from a labeled container and will only be used by that person within the work shift in which it is transferred from the labeled container.

Imminent hazard (Section 165.1) means a situation which exists when the continued use of a pesticide during the time required for cancellation proceedings would be likely to result in unreasonable adverse effects on the environment or will involve unreasonable hazard to the survival of a species declared endangered by the Secretary of the Interior under Pub. L. 91-135.

Immobilization (Section 797.1300) means the lack of movement by the test organisms except for minor activity of the appendages.

Immobilization (Section 797.1330) means the lack of movement by daphnids except for minor activity of the appendages.

Impervious (Section 721.3 [7-27-89]) Chemical protective clothing is "impervious" to a chemical substance if the substance causes no chemical or mechanical degradation, permeation, or penetration of the chemical protective clothing under the conditions of, and the duration of, exposure.

Impervious solid surfaces (Section 761.123) means solid surfaces which are nonporous and thus unlikely to absorb spilled PCBs within the short period of time required for cleanup of spills under this policy. Impervious solid surfaces include, but are not limited to, metals, glass, aluminum siding, and enameled or laminated surfaces.

Implementation (Section 249.04) means putting a plan into practice by carrying out planned activities, or ensuring that these activities are carried out.

Implementation (Section 256.06) means putting the plan into practice by carrying out planned activities, including compliance and enforcement activities, or ensuring such activities are carried out.

Implementing agency (Section 191.12 [12/20/93])* means: (1) The Commission for facilities licensed by the Commission; (2) The Agency for those implementation responsibilities for the Waste Isolation Pilot Plant, under this part, given to the Agency by the Waste Isolation Pilot Plant Land Withdrawal Act (Pub. L. 102-579, 106 Stat. 4777) which, for the purposes of this part, are: (i) Determinations by the Agency that the Waste Isolation Pilot Plant is in compliance with subpart A of this part; (ii) Issuance of criteria for the certifications of compliance with subparts B and C of this part of the Waste Isolation Pilot Plant's compliance with subparts B and C of this part; (iii) Certifications of compliance with subparts B and C of this part of the Waste Isolation Pilot Plant's compliance with subparts B and C of this part; (iv) If the initial certification is made, periodic recertification of the Waste Isolation Pilot Plant's continued compliance with subparts B and C of this part; (v) Review and comment on performance assessment reports of the Waste Isolation Pilot Plant; and (vi) Concurrence by the Agency with the Department's determination under Sec. 191.02(i) that certain wastes do not need the degree of isolation required by subparts B and C of this part; and (3) The Department of Energy for any other disposal facility and all other implementation responsibilities for the Waste Isolation Pilot Plant, under this part, not given to the Agency.

Implementing agency (Section 280.12 [9/23/88]) means EPA, or, in the case of a state with a program approved under section 9004 (or pursuant to a memorandum of agreement with EPA), the designated state or local agency responsible for carrying out an approved UST program.

Import (Section 372.3 [2/16/88]) means to cause a chemical to be imported into the customs territory of the United States. For purposes of this definition, to cause means to intend that the chemical be imported and to control the identity of the imported chemical and the amount to be imported.

Import (Section 704.3 [12/22/88]) means to import for commercial purposes.

Import (Section 704.83) means to import in bulk form or as part of a mixture.

Import (Section 716.3) means to import for commercial purposes.

Import (Section 763.163 [7-12-89]) means to bring into the customs territory of the United States, except for: (1) Shipment through the customs territory of the United States for export without any use, processing, or disposal within the customs territory of the United States; or (2) entering the customs territory of the United States as a component of a product during normal personal or business activities involving use of the product.

Import (Section 82.3 [12/10/93])* means to land on, bring into, or introduce into, or attempt to land on, bring into, or introduce into any place subject to the jurisdiction of the United States whether or not such landing, bringing, or introduction constitutes an importation within the meaning of the customs laws of the United States, with the following exemptions: (1) Off-loading used or excess controlled substances or controlled products from a ship during servicing; (2) Bringing controlled substances into the U.S. from Mexico where the controlled substance had been admitted into Mexico in bond and was of U.S.origin; and (3) Bringing a controlled product into the U.S. when transported in a consignment of personal or household effects or in a similar non-commercial situation normally exempted from U.S. Customs attention.

Import for commercial purposes (Section 716.3) means to import with the purpose of obtaining an immediate or eventual commercial advantage for the importer, and includes the importation of any amount of a chemical substance or mixture. If a chemical substance or mixture containing impurities is imported for commercial purposes, then those impurities are also imported for commercial purposes.

Import for commercial purposes (Section 704.3 [12/22/88]) means to import with the purpose of obtaining an immediate or eventual commercial advantage for the importer, and includes the importation of any amount of a chemical substance or mixture. If a chemical substance or mixture containing impurities is imported for commercial purposes, then those impurities also are imported for commercial purposes.

Import in bulk form (Section 704.3 [12/22/88])* means to import a chemical substance (other than as part of a mixture or article) in any quantity, in cans, bottles, drums, barrels, packages, tanks, bags, or other containers, if the chemical substance is intended to be removed from the container and the substance has an end use or commercial purpose separate from the container.

Import in bulk form (Section 712.3) means to import a chemical substance (other than as part of a mixture or article) in any quantity, in cans, bottles, drums, barrels, packages, tanks, bags, or other containers used for purposes of transportation or containment, if the chemical substance has an end use or commercial purpose separate from the container.

Importer (Section 704.3 [12/22/88])* means (1) any person who imports any chemical substance or any chemical substance as part of a mixture or article into the customs territory of the United States, and includes: (i) The person primarily liable for the payment of any duties on the merchandise, or (ii) An authorized agent acting on his behalf (as defined in 19 CFR 1.11). (2) Importer also includes, as appropriate: (i) The consignee. (ii) The importer of record. (iii) The actual owner if an actual owner's declaration and superseding bond have been filed in accordance with 19 CFR 141.20. (iv) The transferee, if the right to draw merchandise in a bonded warehouse has been transferred in accordance with Subpart C of 19 CFR Part 144. (3) For the purposes of this definition, the customs territory of the United States consists of the 50 States, Puerto Rico, and the District of Columbia.

Importer (Section 710.2) means any person who imports any chemical substance or any chemical substance as part of a mixture or article into the customs territory of the U.S. and includes: (1) The person primarily liable for the payment of any duties on the merchandise, or (2) an authorized agent acting on his behalf (as defined in 19 CFR 1.11).

Importer (Section 712.3) means anyone who imports a chemical substance, including a chemical substance as part of a mixture or article, into the customs territory of the U.S. and includes the person liable for the payment of any duties on the merchandise, or an authorized agent on his behalf. Importer also includes, as appropriate: (1) The consignee. (2) The importer of record. (3) The actual owner if an actual owner's declaration and superseding bond has been filed in accordance with 19 CFR 141.20. (4) The transferee, if the right to withdraw merchandise in a bonded warehouse has been transferred in accordance with Subpart C of 19 CFR Part 144. For the purposes of this definition, the customs territory of the U.S. consists of the 50 states, Puerto Rico, and the District of Columbia.

Importer (Section 716.3) means any person who imports a chemical substance, including a chemical substance as a part of a mixture or article, into the customs territory of the

United States and includes the person primarily liable for the payment of any duties on the merchandise or an authorized agent acting on his behalf (as defined in 19 CFR 1.11). Importer also includes, as appropriate: (1) The consignee. (2) The importer of record. (3) The actual owner, if an actual owner's declaration and superseding bond has been filed in accordance with 19 CFR 141.20. (4) The transferee, if the right to draw merchandise in a bonded warehouse has been transferred in accordance with Subpart C of 19 CFR Part 144. For the purpose of this definition, the customs territory of the United States consists of the 50 States, Puerto Rico, and the District of Columbia.

Importer (Section 763.163 [7-12-89]) means anyone who imports a chemical substance, including a chemical substance as part of a mixture or article, into the customs territory of the United States. "Importer" includes the person primarily liable for the payment of any duties on the merchandise or an authorized agent acting on his or her behalf. The term includes as appropriate: (1) The consignee. (2) The importer of record. (3) The actual owner if an actual owner's declaration and superseding bond has been filed in accordance with 19 CFR 141.20. (4) The transferee, if the right to withdraw merchandise in a bonded warehouse has been transferred in accordance with Subpart C of 19 CFR Part 144.

Importer (Section 763.63) means anyone who imports any chemical substance, including a chemical substance as part of a mixture or article, into the customs territory of the U.S. and includes the person liable for the payment of any duties on the merchandise, or an authorized agent on his behalf. Importer also includes, as appropriate: (1) The consignee. (2) The importer of record. (3) The actual owner if an actual owner's declaration and superseding bond has been filed in accordance with 19 CFR 141.20. (4) The transferee, if the right to draw merchandise in a bonded warehouse has been transferred in accordance with Subpart C of 19 CFR Part 144. For the purpose of this definition, the customs territory of the U.S. consists of the 50 states, Puerto Rico, and the District of Columbia.

Importer (Section 80.2 [2/16/94])* means a person who imports gasoline, gasoline blending stocks or components, or diesel fuel from a foreign country into the United States (including the Commonwealth of Puerto Rico, the Virgin Islands, Guam, American Samoa, and the Northern Mariana Islands).

Importer (Section 82.3 [12/10/93])* means any person who imports a controlled substance or a controlled product into the United States. "Importer" includes the person primarily liable for the payment of any duties on the merchandise or an authorized agent acting on his or her behalf. The term also includes, as appropriate: (1) The consignee; (2) The importer of record; (3) The actual owner; or (4) The transferee, if the right to draw merchandise in a bonded warehouse has been transferred.

Importer (Sections 720.3; 723.50; 723.250; 747.115; 747.195; 747.200) means any person who imports a chemical substance, including a chemical substance as part of a mixture or article, into the customs territory of the United States. Importer includes the person primarily liable for the payment of any duties on the merchandise or an authorized agent acting on his or her behalf. The term also includes, as appropriate: (1) The consignee. (2) The importer of record. (3) The actual owner if an actual owner's declaration and superseding bond has been filed in accordance with 19 CFR 141.20; or (4) The transferee, if the right to draw merchandise in a bonded warehouse has been transferred in accordance with Subpart C of 19 CFR Part 144. (See principal importer.)

Impoundment (Section 260.10). See **Surface impoundment (Section 260.10)**.

Impregnation (Section 471.02) is the process of filling pores of a formed powder part, usually with a liquid such as a lubricant, or mixing particles of a nonmetallic substance in a matrix of metal powder.

Improved discharge (Section 125.58) means the volume, composition, and location of an applicant's discharge following: (1) Construction of planned outfall improvements, including, without limitation, outfall relocation, outfall repair, or diffuser modification; or (2) Construction of planned treatment system improvements to treatment levels or discharge characteristics; or (3) Implementation of a planned program to improve operation and maintenance of an existing treatment system or to eliminate or control the introduction of pollutants into the applicant's treatment works.

Impulsive Noise (Section 211.203) means an acoustic event characterized by very short rise time and duration.

Impurity (Section 158.153 [5/4/88]) means any substance (or group of structurally similar substances if specified by the Agency) in a pesticide product other than an active ingredient or an inert ingredient, including unreacted starting materials, side reaction products, contaminants, and degradation products.

Impurity (Section 704.3 [12/22/88])* means a chemical substance which is unintentionally present with another chemical substance.

Impurity (Section 79.2 [6/27/94])* means any chemical element present in an additive that is not included in the chemical formula or identified in the breakdown by element in the chemical composition of such additive.

Impurity (Sections 710.2; 716.3; 720.3; 723.50; 723.250; 747.115; 747.195; 747.200; 761.3; 790.3) means a chemical substance which is unintentionally present with another chemical substance.

Impurity (Sections 712.3; 791.3) means a chemical substance unintentionally present with another chemical substance or mixture.

Impurity associated with an active ingredient (Section 158.153 [5/4/88]) means: (1) Any impurity present in the technical grade of active ingredient; and (2) Any impurity which forms in the pesticide product through reactions between the active ingredient and any other component of the product or packaging of the product.

In benzene service (Section 61.111) means that a piece of equipment either contains or contacts a fluid (liquid or gas) that is at least 10 percent benzene by weight as determined according to the provisions of Section 61.245(d). The provisions of Section 61.245(d) also specify how to determine that a piece of equipment is not in benzene service.

In benzene service (Section 61.131 [9-14-89]) means a piece of equipment, other than an exhauster, that either contains or contacts a fluid (liquid or gas) that is at least 10 percent benzene by weight or any exhauster that either contains or contacts a fluid (liquid or gas) at least 1 percent benzene by weight as determined by the provisions of Sec. 61.137(b). The provisions of Sec. 61.137(b) also specify how to determine that a piece of equipment is not in benzene service.

In existence (Section 51.301) means that the owner or operator has obtained all necessary preconstruction approvals or permits required by Federal, State, or local air pollution emissions and air quality laws or regulations and either has (1) begun, or caused to begin, a continuous program of physical on-site construction of the facility or (2) entered into binding agreements or contractual obligations, which cannot be cancelled or modified without substantial loss to the owner or operator, to undertake a program of construction of the facility to be completed in a reasonable time.

In gas/vapor service (Section 264.1031 [6-21-90]) means that the piece of equipment contains or contacts a hazardous waste stream that is in the gaseous state at operating conditions.

In gas/vapor service (Sections 60.481; 61.241) means that the piece of equipment contains process fluid that is in the gaseous state at operating conditions.

In heavy liquid service (Section 264.1031 [6-21-90]) means that the piece of equipment is not in gas/vapor service or in light liquid service.

In heavy liquid service (Section 60.481) means that the piece of equipment is not in gas/vapor service or in light liquid service.

In hydrogen service (Section 60.591) means that a compressor contains a process fluid that meets the conditions specified in Section 60.593(b).

In-kind contribution (Section 35.6015 [6-5-90]) The value of a non-cash contribution (generally from third parties) to meet a recipient's cost sharing requirements. An in-kind contribution may consist of charges for real property and equipment or the value of goods and services directly benefiting the CERCLA-funded project.

In light liquid service (Section 60.591) means that the piece of equipment contains a liquid that meets the conditions specified in Section 60.593(c).

In light liquid service (Section 60.631) means that the piece of equipment contains a liquid that meets the conditions specified in Section 60.485(e) or Section 60.633(h)(2).

In light liquid service (Section 60.481) means that the piece of equipment contains a liquid that meets the conditions specified in Section 60.485(e).

In liqht liquid service (Section 264.1031 [6-21-90]) means that the piece of equipment contains or contacts a waste stream where the vapor pressure of one or more of the components in the stream is greater than 0.3 kilopascals (kPa) at 20 deg.C, the total concentration of the pure components having a vapor pressure greater than 0.3 kPa at 20 deg.C is equal to or greater than 20 percent by weight, and the fluid is a liquid at operating conditions.

In liquid service (Section 61.241) means that a piece of equipment is not in gas/vapor service.

In operation (Section 51.301) means engaged in activity related to the primary design function of the source.

In operation (Sections 260.10; 270.2) refers to a facility which is treating, storing, or disposing of hazardous waste.

In or Near Commercial Buildings (Section 761.3) means within the interior of, on the roof of, attached to the exterior wall of, in the parking area serving, or within 30 meters of a non-industrial non-substation building. Commercial buildings are typically accessible to both members of the general public and employees, and include: (1) Public assembly properties, (2) educational properties, (3) institutional properties, (4) residential properties, (5) stores, (6) office buildings, and (7) transportation centers (e.g., airport terminal buildings, subway stations, bus stations, or train stations).

In poor condition (Section 61.141 [11-20-90]) means the binding of the material is losing its integrity as indicated by peeling, cracking, or crumbling of the material.

In situ sampling systems (Section 264.1031 [6-21-90]) means nonextractive samplers or in-line samplers.

In the Hands of the Manufacturer (Section 86.602) means that vehicles are still in the possession of the manufacturer and have not had their bills of lading transferred to another person for the purpose of transporting.

In vacuum service (Section 264.1031 [6-21-90]) means that equipment is operating at an internal pressure that is at least 5 kPa below ambient pressure.

In vacuum service (Sections 60.481; 61.241) means that equipment is operating at an internal pressure which is at least 5 kilopascals (kPa) below ambient pressure.

In VHAP service (Section 61.241) means that a piece of equipment either contains or contacts a fluid (liquid or gas) that is at least 10 percent by weight a volatile hazardous air pollutant (VHAP) as determined according to the provisions of Section 61.245(d). The provisions of Section 61.245(d) also specify how to determine that a piece of equipment is not in VHAP service.

In vinyl chloride service (Section 61.61) means that a piece of equipment either contains or contacts a liquid that is at least 10 percent vinyl chloride by weight or a gas that is at least 10 percent by volume vinyl chloride as determined according to the provisions of Section 61.67(h). The provisions of Section 61.67(h) also specify how to determine that a piece of equipment is not in vinyl chloride service. For the purposes of this subpart, this definition must be used in place of the definition of in VHAP service in Subpart V of this part.

In VOC service (Section 60.481) means that the piece of equipment contains or contacts a process fluid that is at least 10 percent VOC by weight. (The provisions of Section 60.485(d) specify how to determine that a piece of equipment is not in VOC service.)

In VOC service (Section 61.241) means, for the purposes of this subpart, that (a) the piece of equipment contains or contacts a process fluid that is at least 10 percent VOC by weight (see 40 CFR 60.2 for the definition of volatile organic compound or VOC and 40 CFR 60.485(d) to determine whether a piece of equipment is not in VOC service) and (b) the piece of equipment is not in heavy liquid service as defined in 40 CFR 60.481.

In wet gas service (Section 60.631) means that a piece of equipment contains or contacts the field gas before the extraction step in the process.

In-kind contribution (Section 30.200) means the value of a non-cash contribution to meet a recipient's cost sharing requirements. An in-kind contribution may consist of charges for real property and equipment or the value of goods and services directly benefiting the EPA funded project.

In-kind contribution (Section 35.6015 [1/27/89]) means the value of a non-cash contribution (generally from third parties) to meet a recipient's cost sharing requirements in a cooperative agreement only. An in-kind contribution may consist of charges for real property and equipment or the value of goods and services directly benefiting the CERCLA-funded project.

In-process control technology (Sections 467.02; 471.02) is the conservation of chemicals and water throughout the production operations to reduce the amount of wastewater to be discharged.

In-situ leach methods (Section 440.132) means the processes involving the purposeful introduction of suitable leaching solutions into a uranium ore body to dissolve the valuable minerals in place and the purposeful leaching of uranium ore in a static or semistatic condition either by gravity through an open pile, or by flooding a confined ore pile. It does not include the natural dissolution of uranium by ground waters, the incidental leaching of uranium by mine drainage, nor the rehabilitation of aquifers and the monitoring of these aquifers.

In-situ sampling systems (Sections 60.481; 61.241) means nonextractive samplers or in-line samplers.

In-use aircraft gas turbine engine (Section 87.1) means an aircraft gas turbine engine which is in service.

Inability (Section 169.1) means the incapacity of any person to maintain, furnish or permit access to any records under this Act and regulations, where such incapacity arises out of causes beyond the control and without the fault or negligence of such person. Such causes may include, but are not restricted to acts of God or of the public enemy, fires, floods, epidemics, quarantine restrictions, strikes, and unusually severe weather, but in every case, the failure must be beyond the control and without the fault or negligence of said person.

Inactive facility (Section 256.06) means a facility which no longer receives solid waste.

Inactive mine (Section 61.20) is a mine from which uranium ore has been previously removed but which is not an active mine as of the effective date of the standard. Inactive mines which become active mines after the effective date of the standard are considered new sources under the provisions of subparts A and B of this part.

Inactive portion (Section 260.10) means that portion of a facility which is not operated after the effective date of Part 261 of this chapter. (See also active portion and closed portion.)

Inactive waste disposal site (Section 61.141 [11-20-90])* means any disposal site or portion of it where additional asbestos-containing waste material has not been deposited within the past year.

Incidence (Section 61 12-15-89]) This term denotes the predicted number of fatal cancers in a population from exposure to a pollutant. Other health effects (non-fatal cancers, genetic, and developmental) are noted separately.

Incineration (Section 240.101) means the controlled process which combustible solid, liquid, or gaseous wastes are burned and changed into noncombustible gases.

Incinerator (Section 240.101) means a facility consisting of one or more furnaces in which wastes are burned.

Incinerator (Section 260.10 [2-21-91])* means any enclosed device that: (1) Uses controlled flame combustion and neither meets the criteria for classification as a boiler, sludge dryer, or carbon regeneration unit, nor is listed as an industrial furnace; or (2) Meets the definition of infrared incinerator or plasma arc incinerator.

Incinerator (Section 60.51) means any furnace used in the process of burning solid waste for the purpose of reducing the volume of the waste by removing combustible matter.

Incinerator (Section 60.661 [6-29-90]) means any enclosed combustion device that is used for destroying organic compounds and does not extract energy in the form of steam or process heat.

Incinerator (Section 61.301 [3-7-90]) means any enclosed combustion device that is used for destroying organic compounds and that does not extract energy in the form of steam or process heat. These devices do not rely on the heating value of the waste gas to sustain efficient combustion. Auxiliary fuel is burned in the device and the heat from the fuel flame heats the waste gas to combustion temperature. Temperature is controlled by controlling combustion air or fuel.

Incinerator (Section 61.31) means any furnace used in the process of burning waste for the primary purpose of reducing the volume of the waste by removing combustible matter.

Incinerator (Section 761.3) means an engineered device using controlled flame combustion to thermally degrade PCBs and PCB Items. Examples of devices used for incineration include rotary kilns, liquid injection incinerators, cement kilns, and high temperature boilers.

Incipient LC_{50} (Section 797.1400) means that test substance concentration, calculated from experimentally-derived mortality data, that is lethal to 50 percent of a test population when exposure to the test substance is continued until the mean increase in mortality does not exceed 10 percent in any concentration over a 24-hour period.

Incompatible waste (Section 260.10) means a hazardous waste which is unsuitable for: (1) Placement in a particular device or facility because it may cause corrosion or decay of containment materials (e.g., container inner liners or tank walls); or (2) Commingling with another waste or material under uncontrolled conditions because the commingling might produce heat or pressure, fire or explosion, violent reaction, toxic dusts, mists, fumes, or gases, or flammable fumes or gases. (See Part 265, Appendix V, of this chapter for examples.)

Incomplete gasoline-fueled heavy-duty vehicle (Section 86.085-2) means any gasoline-fueled heavy-duty vehicle which does not have the primary load-carrying device, or passenger compartment, or engine compartment or fuel system attached. This definition applies beginning with the 1985 model year.

Incomplete truck (Section 86.082-2) means any truck which does not have the primary load carrying device or container attached.

Incorporated into the soil (Sections 257.3-5; 257.3-6) means the injection of solid waste beneath the surface of the soil or the mixing of solid waste with the surface soil.

Increments of progress (Section 51.100) means steps toward compliance which will be taken by a specific source, including: (1) Date of submittal of the source's final control plan to the appropriate air pollution control agency; (2) Date by which contracts for emission control systems or process modifications will be awarded; or date by which orders will be issued for the purchase of component parts to accomplish emission control or process modification; (3) Date of initiation of on-site construction or installation of emission control equipment or process change; (4) Date by which on-site construction or installation of emission control equipment or process modification is to be completed; and (5) Date by which final compliance is to be achieved.

Increments of progress (Section 60.21) means steps to achieve compliance which must be taken by an owner or operator of a designated facility, including: (1) Submittal of a final control plan for the designated facility to the appropriate air pollution control agency; (2) Awarding of contracts for emission control systems or for process modifications, or issuance of orders for the purchase of component parts to accomplish emission control or process modification; (3) Initiation of on-site construction or installation of emission control equipment or process change; (4) Completion of on-site construction or installation of emission control equipment or process change; and (5) Final compliance.

Independent commercial importer (ICI) (Section 85.1502 [9/25/87]) is an importer who is not an original equipment manufacturer (OEM) or does not have a contractual agreement with an OEM to act as its authorized representative for the distribution of motor vehicles or motor vehicle engines in the U.S. market.

Independent laboratory (Section 610.11) means a test facility operated independently of any motor vehicle, motor vehicle engine, or retrofit device manufacturer capable of performing retrofit device evaluation tests. Additionally, the laboratory shall have no financial interests in the outcome of these tests other than a fee charged for each test performed.

Independent Power Production Facility (IPP) (Section 72.2 [7/30/93])* means a source that: (1) Is nonrecourse project financed, as defined by the Secretary of Energy at 10 CFR part 715; (2) Is used for the generation of electricity, eighty percent or more of which is sold at wholesale; and (3) Is a new unit required to hold allowances under Title IV of the Clean Air Act; but only if direct public utility ownership of the equipment comprising the facility does not exceed 50 percent.

Independent printed circuit board manufacturer (Section 433.11) shall mean a facility which manufactures printed circuit boards principally for sale to other companies.

Independently audited (Section 265.141), as specified in Section 265.141(f), refers to an audit performed by an independent certified public accountant in accordance with generally accepted auditing standards.

Independently audited (Section 264.141), as specified in Section 264.141(f), refers to an audit performed by an independent certified public accountant in accordance with generally accepted auditing standards.

Independently audited (Section 144.61) refers to an audit performed by an independent certified public accountant in accordance with generally accepted auditing standards.

Index Mark (Section 85.2122(a)(2)(iii)(H)). See **Index (Section 85.2122(a)(2)(iii)(H))**.

Index or Index Mark (Section 85.2122(a)(2)(iii)(H)) means a mark on a choke thermostat housing, located in a fixed relationship to the thermostatic coil tang position to aid in assembly and service adjustment of the choke.

Indian Country (Section 355.20 [7-26-90]) means Indian country as defined in 18 U.S.C. 1151. That section defines Indian country as: (a) All land within the limits of any Indian reservation under the jurisdiction of the United States government, notwithstanding the issuance of any patent, and including rights-of-way running through the reservation; (b) All dependent Indian communities within the borders f the United States whether within the original or subsequently acquired territory thereof, and whether within or without the limits of a State; and (c) All Indian allotments, the Indian titles to which have not been extinguished, including rights-of-way running through the same.

Indian Country (Section 372.3 [7-26-90]) means "Indian country" as defined in 18 U.S.C. 1151. That section defines Indian country as: (a) All land within the limits of any Indian reservation under the jurisdiction of the United States government, notwithstanding the issuance of any patent, and includng rights-of-way running through the reservation; (b) All dependent Indian communities within the borders of the United States whether within the original or subsequently acquired territory thereof, and whether within or without the limits of a State; and (c) All Indian allotments, the Indian titles to which have not been extinguished, including rights-of-way running through the same.

Indian Governing Body (Sections 51.166; 52.21; 58.1; 124.41) means the governing body of any tribe, band, or group of Indians subject to the jurisdiction of the United States and recognized by the United States as possessing power of self-government.

Indian lands (Section 144.3) means Indian country as defined in 18 U.S.C. 1151. That section defines Indian country as: (a) All land within the limits of any Indian reservation under the jurisdiction of the United States government, notwithstanding the issuance of any patent, and, including rights-of-way running through the reservation; (b) All dependent Indian communities within the borders of the United States whether within the original or subsequently acquired territory thereof, and whether within or without the limits of a State; and (c) All Indian allotments, the Indian titles to which have not been extinguished, including rights-of-way running through the same.

Indian lands or Indian country (Section 258.2 [10/1/93])* means: (1) All land within the limits of any Indian reservation under the jurisdiction of the United States Government, notwithstanding the issuance of any patent, and including rights-of-way running throughout the reservation; (2) All dependent Indian communities within the borders of the United States whether within the original or subsequently acquired territory thereof, and whether within or without the limits of the State; and (3) All Indian allotments, the Indian titles to which have not been extinguished, including rights of way running through the same.

Indian Reservation (Sections 51.166; 52.21; 58.1) means any federally recognized reservation established by Treaty, Agreement, Executive Order, or Act of Congress.

Indian tribe (Section 300.5 [3-8-90]) as defined by section 101(36) of CERCLA, means any Indian tribe, band, nation, or other organized group or community, including any Alaska Native village but not including any Alaska Native regional or village corporation which is recognized as eligible for the special programs and services provided by the United States to Indians because of their status as Indians.

Indian tribe (Section 355.20 [7-26-90]) means those tribes federally recognized by the Secretary of the Interior.

Indian tribe (Section 372.3 [7-26-90]) means those tribes federally recognized by the Secretary of the Interior.

Indian tribe and tribal organization (Section 34.105 [2-26-90]) have the meaning provided in section 4 of the Indian Self-Determination and Education Assistance Act (25U.S.C. 450B). Alaskan Natives are included under the definitions of Indian tribes in that act.

Indian Tribe (Section 122.1 [12/22/93])* means any Indian Tribe, band, group, or community recognized by the Secretary of the Interior and exercising governmental authority over a Federal Indian reservation.

Indian Tribe (Section 124.2 [12/22/93])* means (in the case of UIC) any Indian Tribe having a federally recognized governing body carrying out substantial governmental duties and powers over a defined area. For the NPDES program, the term "Indian Tribe" means any Indian Tribe, band, group, or community recognized by the Secretary of the Interior and exercising governmental authority over a Federal Indian reservation.

Indian Tribe (Section 130.2 [4/11/89]) means any Indian Tribe, band, group, or community recognized by the Secretary of the Interior and exercising governmental authority over a Federal Indian reservation.

Indian Tribe (Section 142.2 [9/26/88]) means any Indian Tribe having a Federally recognized governing body carrying out substantial governmental duties and powers over a defined area.

Indian Tribe (Section 144.3 [9/26/88]) means any Indian Tribe having a Federally recognized governing body carrying out substantial governmental duties and powers over a defined area.

Indian Tribe (Section 146.3 [9/26/88]) means any Indian Tribe having a Federally recognized governing body carrying out substantial governmental duties and powers over a defined area.

Indian Tribe (Section 232.2 [2/11/93])* means any Indian Tribe, band, group, or community recognized by the Secretary of the Interior and exercising governmental authority over a Federal Indian reservation.

Indian Tribe (Section 35.105 [9/26/88, 4/11/89]) means for purposes of the Clean Water Act, any Indian Tribe, band, group, or community recognized by the Secretary of the Interior and exercising governmental authority over a Federal Indian reservation.

Indian Tribe (Section 35.6015 [1/27/89]) means, as defined by section 101(36) of CERCLA, any Indian Tribe, band, nation, or other organized group or community, including any Alaska Native Village but not including any Alaska Native regional or village corporation, which is recognized as eligible for the special programs and services provided by the United States to Indians because of their status as Indians.

Indian Tribe or Tribe (Section 131.3 [12-12-91]) means any Indian Tribe, band, group, or community recognized by the Secretary of the Interior and exercising governmental authority over a Federal Indian reservation.

Indian Tribe or Tribe (Section 258.2 [10/1/93])* means any Indian tribe, band, nation, or community recognized by the Secretary of the Interior and exercising substantial governmental duties and powers on Indian lands.

Indian waters (Section 300.5 [3-8-90]) for the purposes of classifying the size of discharges, means those waters of the United States in the inland zone, waters of the Great Lakes, and specified ports and harbors on inland rivers.

Indirect ammonia recovery system (Section 420.11) means those systems which recover ammonium hydroxide as a by-product from coke oven gases and waste ammonia liquors.

Indirect Discharge or Discharge (Section 403.3) means the introduction of pollutants into a POTW from any non-domestic source regulated under section 307(b), (c) or (d) of the Act.

Indirect discharger (Section 122.2) means a nondomestic discharger introducing pollutants to a publicly owned treatment works.

Individual (Section 1516.2) means a citizen of the United States or an alien lawfully admitted for permanent residence.

Individual (Section 16.2) shall have

the meaning given to it by 5 U.S.C. 552a (a)(2).

Individual (Section 303.11 [5/5/88, 6/21/89]) means a natural person, not a corporation or other legal entity nor an association of persons.

Individual (Section 32.605 [5-25-90]) means a natural person.

Individual drain system (Section 60.691 [11/23/88]) means all process drains connected to the first common downstream junction box. The term includes all such drains and common junction box, together with their associated sewer lines and other junction boxes, down to the receiving oil-water separator.

Individual drain system (Section 61.341 [1/7/93])* means the system used to convey waste from a process unit, product storage tank, or waste management unit to a waste management unit. The term includes all process drains and common junction boxes, together with their associated sewer lines and other junction boxes, down to the receiving waste management unit.

Individual generation site (Section 260.10 [2-21-91])* means the contiguous site at or on which one or more hazardous wastes are generated. An individual generation site, such as a large manufacturing plant, may have one or more sources of hazardous waste but is considered a single or individual generation site if the site or property is contiguous.

Individual systems (Section 35.2005) means privately owned alternative wastewater treatment works (including dual waterless/gray water systems) serving one or more principal residences, or small commercial establishments. Normally these are onsite systems with localized treatment and disposal of wastewater, but may be systems utilizing small diameter gravity, pressure or vacuum sewers conveying treated or partially treated wastewater. These systems can also include small diameter gravity sewers carrying raw wastewater to cluster systems.

Individual with handicaps (Section 12.103 [8/14/87]) means any person who has a physical or mental impairment that substantially limits one or more major life activities, has a record of such an impairment, or is regarded as having such an impairment. As used in this definition, the phrase: (1) Physical or mental impairment includes (i) Any physiological disorder or condition, cosmetic disfigurement, or anatomical loss affecting one or more of the following body systems: Neurological; musculoskeletal; special sense organs; respiratory, including speech organs; cardiovascular, reproductive, digestive, genitourinary, hemic and lymphatic; skin, and endocrine; or (ii) Any mental or psychological disorder, such as mental retardation, organic brain syndrome, emotional or mental illness, and specific learning disabilities. The term physical or mental impairment includes, but is not limited to, such diseases and conditions as orthopedic, visual, speech, and hearing impairments, cerebral palsy, epilepsy, muscular dystrophy, multiple sclerosis, cancer, heart disease, diabetes, mental retardation, emotional illness, and drug addiction and alcoholism. (2) Major life activities includes functions such as caring for one's self, performing manual tasks, walking, seeing, hearing, speaking, breathing, learning, and working. (3) Has a record of such an impairment means has a history of, or has been misclassified as having, a mental or physical impairment that substantially limits one or more major life activities. (4) Is regarded as having an impairment means (i) Has a physical or mental impairment that does not substantially limit major life activities but is treated by the agency as constituting such a limitation; (ii) Has a physical or mental impairment that substantially limits major life activities only as a result of the attitudes of others toward such impairment; or (iii) Has none of the impairments defined in subparagraph (1) of this definition but is treated by the agency as having such an impairment.

Induction exposure (Section 798.4100) is an experimental exposure of a subject to a test substance with the intention of inducing a hypersensitive state.

Induction period (Section 798.4100) is a period of at least 1 week following a sensitization exposure during which a hypersensitive state is developed.

Industrial building (Section 761.3) means a building directly used in manufacturing or technically productive enterprises. Industrial buildings are not generally or typically accessible to other than workers. Industrial buildings include buildings used directly in the production of power, the manufacture of products, the mining of raw materials, and the storage of textiles, petroleum products, wood and paper products, chemicals, plastics, and metals.

Industrial cooling tower (Section 749.68 [1-3-90]) means any cooling tower used to remove heat form industrial processes, chemical reactions, or plants producing electrical power.

Industrial cost recovery (Section 35.905) means (a) The grantee's recovery from the industrial users of a treatment works of the grant amount allocable to the treatment of waste from such users under section 204(b) of the Act and this subpart. (b) The grantee's recovery from the commercial users of an individual system of the grant amount allocable to the treatment of waste from such users under section 201(h) of the Act and this subpart.

Industrial cost recovery period

(Section 35.905) means that period during which the grant amount allocable to the treatment of wastes from industrial users is recovered from the industrial users of such works.

Industrial furnace (Section 260.10 [2-21-91])* means any of the following enclosed devices that are integral components of manufacturing processes and that use thermal treatment to accomplish recovery of materials or energy: (1) Cement kilns (2) Lime kilns (3) Aggregate kilns (4) Phosphate kilns (5) Coke ovens (6) Blast furnaces (7) Smelting, melting and refining furnaces (including pyrometallurgical devices such as cupolas, reverberator furnaces, sintering machine, roasters, and foundry furnaces) (8) Titanium dioxide chloride process oxidation reactors (9) Methane reforming furnaces (10) Pulping liquor recovery furnaces (11) Combustion devices used in the recovery of sulfur values from spent sulfuric acid (12) Halogen acid furnaces (HAFs) for the production of acid from halogenated hazardous waste generated by chemical production facilities where the furnace is located on the site of a chemical production facility, the acid product has a halogen acid content of at least 3%, the acid product is used in a manufacturing process, and, except for hazardous waste burned as fuel, hazardous waste fed to the furnace has a minimum halogen content of 20% as-generated (13) Such other devices as the Administrator may, after notice and comment, add to this list on the basis of one or more of the following factors: (i) The design and use of the device primarily to accomplish recovery of material products; (ii) The use of the device to burn or reduce raw materials to make a material product; (iii) The use of the device to burn or reduce secondary materials as effective substitutes for raw materials, in processes using raw materials as principal feedstocks; (iv) The use of the device to burn or reduce secondary materials as ingredients in an industrial process to make a material product; (v) The use of the device in common industrial practice to produce a material product; and (vi) Other factors, as appropriate.

Industrial solid waste (Section 258.2 [10-9-91]) means solid waste generated by manufacturing or industrial processes that is not a hazardous waste regulated under subtitle C of RCRA. Such waste may include, but is not limited to, waste resulting from the following manufacturing processes: Electric power generation; fertilizer/agricultural chemicals; food and related products/by-products; inorganic chemicals; iron and steel manufacturing; leather and leather products; nonferrous metals manufacturing/foundries; organic chemicals; plastics and resins manufacturing; pulp and paper industry; rubber and miscellaneous plastic products; stone, glass, clay, and concrete products; textile manufacturing; transportation equipment; and water treatment. This term does not include mining waste or oil and gas waste.

Industrial solid waste (Sections 243.101; 246.101) means the solid waste generated by industrial processes and manufacturing.

Industrial source (Section 125.58) means any source of nondomestic pollutants regulated under section 307 (b) or (c) of the Clean Water Act which discharges into a POTW.

Industrial user (Section 35.2005) means any nongovernmental, nonresidential user of a publicly owned treatment works which is identified in the *Standard Industrial Classification Manual*, 1972, Office of Management and Budget, as amended and supplemented, under one of the following divisions: Division A: Agriculture, Forestry, and Fishing. Division B: Mining. Division D: Manufacturing. Division E: Transportation, Communications, Electric, Gas, and Sanitary Services. Division I: Services.

Industrial user (Section 35.905) means (a) Any nongovernmental, nonresidential user of a publicly owned treatment works which discharges more than the equivalent of 25,000 gallons per day (gpd) of sanitary wastes and which is identified in the *Standard Industrial Classification Manual*, 1972, Office of Management and Budget, as amended and supplemented under one of the following divisions: Division A: Agriculture, Forestry, and Fishing. Division B: Mining. Division D: Manufacturing. Division E: Transportation, Communications, Electric, Gas, and Sanitary Services. Division I: Services. (1) In determining the amount of a user's discharge for purposes of industrial cost recovery, the grantee may exclude domestic wastes or discharges from sanitary conveniences. (2) After applying the sanitary waste exclusion in paragraph (1) of this paragraph (b) (if the grantee chooses to do so), dischargers in the above divisions that have a volume exceeding 25,000 gpd or the weight of biochemical oxygen demand (BOD) or suspended solids (SS) equivalent to that weight found in 25,000 gpd of sanitary waste are considered industrial users. Sanitary wastes, for purposes of this calculation of equivalency, are the wastes discharged from residential users. The grantee, with the Regional Administrator's approval, shall define the strength of the residential discharges in terms of parameters including, as a minimum, BOD and SS per volume of flow. (b) Any nongovernmental user of a publicly owned treatment works which discharges wastewater to the treatment works which contains toxic pollutants or poisonous solids, liquids, or gases in sufficient quantity either singly or by interaction with other wastes, to contaminate the sludge of any municipal systems, or to injure or to interfere with any sewage treatment process, or which constitutes a hazard to humans or animals, creates a public nuisance, or creates any hazard in or has an adverse effect on the waters receiving any discharge from the treatment works. (c) All commercial users of an individual system constructed with grant assistance under section 201(h) of the Act and this subpart. (See Section 35.918(a)(3).)

Industrial User or User (Section 403.3) means a source of Indirect Discharge.

Industrial wipers (Section 250.4 [10/6/87, 6/22/88]) means paper towels especially made for industrial cleaning and wiping.

Inert ingredient (Section 152.3 [5/4/88]) means any substance (or group of structurally similar substances if designated by the Agency), other than an active ingredient, which is intentionally included in a pesticide product.

Inert ingredient (Section 158.153 [5/4/88]) means any substance (or group of structurally similar substances if designated by the Agency), other than an active ingredient, which is intentionally included in a pesticide product.

Inert ingredients (Section 162.3) means all ingredients which are not active ingredients as defined in Section 162.3(c), and includes, but is not limited to, the following types of ingredients (except when they have pesticidal efficacy of their own): Solvents such as water; baits such as sugar, starches, and meat scraps; dust carriers such as talc and clay; fillers; wetting and spreading agents; propellents in aerosol dispensers; emulsifiers.

Inertia Weight Class (Sections 86.082-2; 600.002-85) means the class, which is a group of test weights, into which a vehicle is grouped based on its loaded vehicle weight in accordance with the provisions of Part 86.

Infectious agent (Section 259.10 [3/24/89]) means any organism (such as a virus or a bacteria) that is capable of being communicated by invasion and multiplication in body tissues and capable of causing disease or adverse health impacts in humans.

Infectious waste (Sections 240.101; 241.101; 243.101) means: (1) Equipment, instruments, utensils, and fomites of a disposable nature from the rooms of patients who are suspected to have or have been diagnosed as having a communicable disease and must, therefore, be isolated as required by public health agencies; (2) laboratory wastes such as pathological specimens (e.g., all tissues, specimens of blood elements, excreta, and secretions obtained from patients or laboratory animals) and disposable fomites (any substance that may harbor or transmit pathogenic organisms) attendant thereto; (3) surgical operating room pathologic specimens and disposable fomites attendant thereto and similar disposable materials from out-patient areas and emergency rooms.

Infectious waste (Sections 245.101; 246.101) means: (1) Equipment, instruments, utensils, and fomites (any substance that may harbor or transmit pathogenic organisms) of a disposable nature from the rooms of patients who are suspected to have or have been diagnosed as having a communicable disease and must, therefore, be isolated as required by public health agencies; (2) laboratory wastes, such as pathological specimens (e.g., all tissues, specimens of blood elements, excreta, and secretions obtained from patients or laboratory animals) and disposable fomites attendant thereto; (3) surgical operating room pathologic specimens and disposable fomites attendent thereto and similar disposable materials from outpatient areas and emergency rooms.

Infiltration (Section 35.2005) means water other than wastewater that enters a sewer system (including sewer service connections and foundation drains) from the ground through such means as defective pipes, pipe joints, connections, or manholes. Infiltration does not include, and is distinguished from, inflow.

Infiltration (Section 35.905) means water other than wastewater that enters a sewerage system (including sewer service connections) from the ground through such means as defective pipes, pipe joints, connections, or manholes. Infiltration does not include, and is distinguished from, inflow.

Infiltration/inflow (Section 35.905) means the total quantity of water from both infiltration and inflow without distinguishing the source.

Infiltration water (Section 440.141 [5/24/88]) means that water which permeates through the earth into the plant site.

Inflow (Section 35.2005) means water other than wastewater that enters a sewer system (including sewer service connections) from sources such as, but not limited to, roof leaders, cellar drains, yard drains, area drains, drains from springs and swampy areas, manhole covers, cross connections between storm sewers and sanitary sewers, catch basins, cooling towers, storm waters, surface runoff, street wash waters, or drainage. Inflow does not include, and is distinguished from, infiltration.

Inflow (Section 35.905) means water other than wastewater that enters a sewerage system (including sewer service connections) from sources such as roof leaders, cellar drains, yard drains, area drains, foundation drains, drains from springs and swampy areas, manhole covers, cross connections between storm sewers and sanitary sewers, catch basins, cooling towers, storm waters, surface runoff, street wash waters, or drainage. Inflow does not include, and is distinguished from, infiltration.

Influencing or attempting to influence (Section 34.105 [2-26-90]) means making, with the intent to influence, any communication to or appearance before an officer or employee or any agency, a Member of Congress, an officer or employee of Congress, or an employee of a Member of Congress in connection with any covered Federal action.

Information (Section 2.201). See **Business information (Section 2.201)**.

Information which is available to the public (Section 2.201) is information in EPA's possession which EPA will furnish to any member of the public upon request and which EPA may make public, release or otherwise make available to any person whether or not its disclosure has been requested.

Infrared incinerator (Section 260.10 [8/18/92])* means any enclosed device that uses electric powered resistance heaters as a source of radiant heat followed by an afterburner using controlled flame combustion and which is not listed as an industrial furnace.

Inground tank (Section 260.10) means a device meeting the definition of tank in Section 260.10, whereby a portion of the tank wall is situated to any degree within the ground, thereby preventing visual inspection of that external surface area of the tank that is in the ground.

Inhalable diameter (Section 798.4350) refers to that aerodynamic diameter of a particle which is considered to be inhalable for the organism. It is used to refer to particles which are capable of being inhaled and may be deposited anywhere within the respiratory tract from the trachea to the deep lung (the alveoli). For man, the inhalable diameter is considered here as 15 micrometers or less.

Inhalable diameter (Sections 798.1150; 798.2450) refers to that aerodynamic diameter of a particle which is considered to be inhalable for the organism. It is used to refer to particles which are capable of being inhaled and may be deposited anywhere within the respiratory tract from the trachea to the alveoli. For man, the inhalable diameter is considered as 15 micrometers or less.

Inhalation LC_{50} (Section 162.3) means a concentration of a substance, expressed as milligrams per liter of air or parts per million parts of air, that is lethal to 50% of the test population of animals under test conditions as specified in the Registration Guidelines.

Inhibition (Section 797.1060) means any decrease in the growth rate of the test algae compared to the control algae.

Initial compliance period (Section

141.2 [7/17/92])* means the first full three-year compliance period which begins at least 18 months after promulgation, except for contaminants listed at Sec. 141.61(a) (19)-(21), (c)(19)-(33), and 141.62(b) (11)-(15), initial compliance period means the first full three-year compliance period after promulgation for systems with 150 or more service connections (January 1993- December 1995), and first full three-year compliance period after the effective date of the regulation (January 1996-December 1998) for systems having fewer than 150 service connections.

Initial crusher (Section 60.671) means any crusher into which nonmetallic minerals can be fed without prior crushing in the plant.

Initial Decision (Section 164.2 [7/10/92])* means the decision of the Administrative Law Judge supported by findings of fact and conclusions regarding all material issues of law, fact, or discretion, as well as reasons therefor. Such decision shall become the final decision and order of the Administrator without further proceedings unless an appeal therefrom is taken or the Administrator orders review thereof as herein provided.

Initial Decision (Section 22.03) means the decision issued by the Presiding Officer based upon the record of the proceedings out of which it arises.

Initiation of negotiations (Section 4.2) means the delivery of the initial written offer by the Agency to the owner or the owner's representative to purchase real property for a project for the amount determined to be just compensation, unless applicable. Federal program regulations specify a different action to serve this purpose. However: (1) If the Agency issues a notice of its intent to acquire the real property, and a person moves after that notice, but before delivery of the initial written purchase offer, the initiation of negotiations means the date the person moves from the property. (See also Section 4.505(c).) (2) In the case of a permanent relocation to protect the public health and welfare, under the Comprehensive Environmental Response Compensation and Liability Act of 1980 (Pub. L. 96-510, or Superfund) the initiation of negotiations means the formal announcement of such relocation or the Federal or federally-coordinated health advisory where the Federal Government later decides to conduct a permanent relocation.

Initiation of operation (Section 35.2005) means the date specified by the grantee on which use of the project begins for the purpose for which it was planned, designed, and built.

Injection interval (Section 146.61 [7/26/88]) means that part of the injection zone in which the well is screened, or in which the waste is otherwise directly emplaced.

Injection interval (Section 148.2 [7/26/88]) means that part of the injection zone in which the well is screened, or in which the waste is otherwise directly emplaced.

Injection well (Section 260.10) means a well into which fluids are injected. (See also underground injection.)

Injection well (Sections 144.3; 146.3; 147.2902; 270.2) means a well into which fluids are being injected.

Injection zone (Sections 144.3; 147.2902) means a geological formation, group of formations, or part of a formation receiving fluids through a well.

Injury (Section 112.2 [8/25/93])* means a measurable adverse change, either long- or short-term, in the chemical or physical quality or the viability of a natural resource resulting either directly or indirectly from exposure to a discharge of oil, or exposure to a product of reactions resulting from a discharge of oil.

Ink (Section 60.581) means any mixture of ink, coating solids, organic solvents including dilution solvent, and water that is applied to the web of flexible vinyl or urethane on a rotogravure printing line.

Ink solids (Section 60.581) means the solids content of an ink as determined by Reference Method 24, ink manufacturer's formulation data, or plant blending records.

Inland waters (Section 300.6) for the purposes of classifying the size of discharges, means those waters of the U.S. in the inland zone, waters of the Great Lakes, and specified ports and harbors on inland rivers.

Inland zone (Section 300.5 [3-8-90])* means the environment inland of the coastal zone excluding the Great Lakes and specified ports and harbors on inland rivers. The term inland zone delineates an area of federal responsibility for response action. Precise boundaries are determined by EPA/USCG agreements and identified in federal regional contingency plans.

Inner liner (Section 260.10) means a continuous layer of material placed inside a tank or container which protects the construction materials of the tank or container from the contained waste or reagents used to treat the waste.

Innovative control technology (Sections 51.166; 52.21) means any system of air pollution control that has not been adequately demonstrated in practice, but would have a substantial likelihood of achieving greater continuous emissions reduction than any control system in current practice or of achieving at least comparable reductions at lower cost in terms of energy, economics, or nonair quality environmental impacts.

Innovative technology (Section 125.22) means a production process, a pollution control technique, or a combination of the two which satisfies one of the criteria in Section 125.23 and which has not been commercially demonstrated in the industry of which the requesting discharger is a part.

Innovative technology (Section 35.2005) means developed wastewater treatment processes and techniques which have not been fully proven under the circumstances of their contemplated use and which represent a significant advancement over the state of the art in terms of significant reduction in life cycle cost or significant environmental benefits through the reclaiming and reuse of water, otherwise eliminating the discharge of pollutants, utilizing recycling techniques such as land treatment, more efficient use of energy and resources, improved or new methods of waste treatment management for combined municipal and industrial systems, or the confined disposal of pollutants so that they will not migrate to cause water or other environmental pollution.

Inorganic arsenic (Sections 61.161; 61.171; 61.181) means the oxides and other noncarbon compounds of the element arsenic included in particulate matter, vapors, and aerosols.

Inorganic pesticides (Section 165.1) means noncarbon-containing substances used as pesticides.

Inorganic Solid Debris (Section 268.2 [1-31-91])* means nonfriable inorganic solids contaminated with D004-D011 hazardous wastes that are incapable of passing through a 9.5 mm standard sieve; and that require cutting, or crushing and grinding in mechanical sizing equipment prior to stabilization; and, are limited to the following inorganic or metal materials: (1) Metal slags (either dross or scoria); (2) Glassified slag; (3) Glass; (4) Concrete (excluding cementitious or pozzolanic stabilized hazardous wastes); (5) Masonry and refractory bricks; (6) Metal cans, containers, drums, or tanks; (7) Metal nuts, bolts, pipes, pumps, valves, appliances, or industrial equipment; (8) Scrap metal as defined in 40 CFR 261.1(c)(6).

Inprocess wastewater (Section 61.61) means any water which, during manufacturing or processing, comes into direct contact with vinyl chloride or polyvinyl chloride or results from the production or use of any raw material, intermediate product, finished product, by-product, or waste product containing vinyl chloride or polyvinyl chloride but which has not been discharged to a wastewater treatment process or discharged untreated as wastewater. Gas-holder seal water is not inprocess wastewater until it is removed from the gasholder.

Insecticides (Section 162.3) includes all substances or mixtures of substances intended for preventing or inhibiting the establishment, reproduction, development, or growth of, destroying or repelling any member of the Class Insecta or other allied classes in the Phylum Arthropoda declared to be pests under Section 162.14. Insecticides include, but are not limited to: (i) Plant protection insecticides intended for use directly or indirectly against insects or allied organisms that attack or infest plants or plant parts, to prevent or mitigate their injury, debilitation, or destruction; (ii) Animal protection insecticides intended for use directly or indirectly against insects or allied organisms that attack or infest man, other mammals, birds, or certain other animals, to prevent or mitigate their injury, irritation, harassment, or debilitation; (iii) Premise and indoor insecticides intended for use directly or indirectly against insects or allied organisms to prevent or mitigate their decimation or contamination of man's stored food and animal feeds, injury to raw or manufactured goods, or weakening or destruction of buildings and building materials; and (iv) Biological insect control agents such as specific pathogenic organisms or entities prepared and utilized by man.

Inside spray coating operation (Section 60.491) means the system on each beverage can surface coating line used to apply a coating to the interior of a two-piece beverage can body. This coating provides a protective film between the contents of the beverage can and the metal can body. The inside spray coating operation consists of the coating application station, flashoff area, and curing oven. Multiple applications of an inside spray coating are considered to be a single coating operation.

Inspection Criteria (Sections 204.51; 205.51) means the rejection and acceptance numbers associated with a particular sampling plan.

Inspection Criteria (Sections 86.602; 86.1002-84) means the pass and fail numbers associated with a particular sampling plan.

Installation (Section 51.165). See **Building, structure, facility, or installation (Section 51.165)**.

Installation (Section 51.166). See **Building, structure, facility, or installation (Section 51.166)**.

Installation (Section 51.301) means an identifiable piece of process equipment.

Installation (Section 52.21). See **Building, structure, facility, or installation (Section 52.21)**.

Installation (Section 52.24). See **Building, structure, facility, or installation (Section 52.24)**.

Installation (Section 61.141 [11-20-90]) means any building or structure or any group of buildings or structures at a single demolition or renovation site that are under the control of the same owner or operator (or owner or operator under common control).

Installation inspector (Section 260.10) means a person who, by reason of his knowledge of the physical sciences and the principles of engineering, acquired by a professional education and related practical experience, is qualified to supervise the installation of tank systems.

Instant photographic film aritcle (Section 723.175) means a self-developing photographic film article designed so that all the chemical substances contained in the article, including the chemical substances required to process the film, remain sealed during distribution and use.

Institution (Section 26.102 [6-18-91]) means any public or private entity or agency (including federal, state, and other agencies).

Institutional solid waste (Section 245.101) means solid wastes originating from educational, health care, correctional, and other institutional facilities.

Institutional solid waste (Sections 243.101; 246.101) means solid wastes generated by educational, health care, correctional, and other institutional facilities.

Institutional use (Section 152.3 [5/4/88]) means any application of a pesticide in or around any property or facility that functions to provide a service to the general public or to public or private organizations, including but not limited to: (1) Hospitals and nursing homes. (2) Schools other than preschools and day care facilities. (3) Museums and libraries. (4) Sports facilities. (5) Office buildings.

Instructions (Section 205.2). See **Maintenance instructions (Section 205.2)**.

Insulation board (Section 429.11) means a panel manufactured from interfelted ligno-cellulosic fibers consolidated to a density of less than 0.5 g/cu cm (less than 31 lb/cu ft).

Integral vista (Section 51.301) means a view perceived from within the mandatory Class I Federal area of a specific landmark or panorama located outside the boundary of the mandatory Class I Federal area.

Integrated facility (Section 413.02) is defined as a facility that performs electroplating as only one of several operations necessary for manufacture of a product at a single physical location and has significant quantities of process wastewater from non-electroplating manufacturing operations. In addition, to qualify as an integrated facility one or more plant electroplating process wastewater lines must be combined prior to or at the point of treatment (or proposed treatment) with one or more plant sewers carrying process wastewater from non-electroplating manufacturing operations.

Integrated refueling emission control system (Section 86.098-2 [4/6/94])* means a system where vapors resulting from refueling are stored in a common vapor storage unit(s) with other evaporative emissions of the vehicle and are purged through a common purge system.

Integrated system (Section 158.153 [5/4/88]) means a process for producing a pesticide product that: (1) Contains any active ingredient derived from a source that is not an EPA-registered product; or (2) Contains any active ingredient that was produced or acquired in a manner that does not permit its inspection by the Agency under FIFRA section 9(a) prior to its use in the process.

Interceptor sewer (Section 35.2005) means a sewer which is designed for one or more of the following purposes: (i) To intercept wastewater from a final point in a collector sewer and convey such wastes directly to a treatment facility or another interceptor. (ii) To replace an existing wastewater treatment facility and transport the wastes to an adjoining collector sewer or interceptor sewer for conveyance to a treatment plant. (iii) To transport wastewater from one or more municipal collector sewers to another municipality or to a regional plant for treatment. (iv) To intercept an existing major discharge of raw or inadequately treated wastewater for transport directly to another interceptor or to a treatment plant.

Interceptor sewer (Sections 21.2; 35.905) means a sewer whose primary purpose is to transport wastewaters from collector sewers to a treatment facility.

Interconnected (Section 60.41a) means that two or more electric generating units are electrically tied together by a network of power transmission lines, and other power transmission equipment.

Interested person (Section 304.12 [5/30/89]) means the Administrator, any EPA employee, any party to the proceeding, any potentially responsible party associated with the facility concerned, any person who filed written comments in the proceeding, any participant or intervenor in the proceeding, all officers, directors, employees, consultants, and agents of any party, and any attorney of record for any of the foregoing persons.

Interested person (Section 305.12) means the Administrator, any EPA employee, any party, any potentially responsible party associated with the site, any person who filed written comments in the proceeding, any person who requested the hearing, any person who requested to participate or intervene in the hearing, any participant in the hearing, all officers, directors, employees, consultants, and agents of the claimant and the persons represented by the claimant, and any attorney of record for those persons.

Interested person outside the Agency (Section 124.78) includes the permit applicant, any person who filed written comments in the proceeding, any person who requested the hearing, any person who requested to participate or intervene in the hearing, any participant in the hearing and any other interested person not employed by the Agency at the time of the communications, and any attorney of record for those persons.

Interested person outside the Agency (Section 57.809) includes the smelter owner, any person who filed written comments in the proceeding, any person who requested the hearing, any person who requested to participate or intervene in the hearing, any participant or party in the hearing and any other interested person not employed by the Agency at the time of the communications, and the attorney of record for such persons.

Interference (Section 403.3) means a discharge which, alone or in conjunction with a discharge or discharges from other sources, both: (1) Inhibits or disrupts the POTW, its treatment processes or operations, or its sludge processes, use or disposal; and (2) Therefore is a cause of a violation of any requirement of the POTW's NPDES permit (including an increase in the magnitude or duration of a violation) or of the prevention of sewage sludge use or disposal in compliance with the following statutory provisions and regulations or permits issued thereunder (or more stringent State or local regulations): Section 405 of the Clean Water Act, the Solid Waste Disposal Act (SWDA) (including Title II, more commonly referred to as the Resource Conservation and Recovery Act (RCRA), and including State regulations contained in any State sludge management plan prepared pursuant to Subtitle D of the SWDA), the Clean Air Act, the Toxic Substances Control Act, and the Marine Protection, Research and Sanctuaries Act.

Interference (Section 425.02) means the discharge of sulfides in quantities which can result in human health hazards and/or risks to human life, and an inhibition or disruption of POTW as defined in 40 CFR 403.3(i).

Intergovernmental agreement (Sections 32.102; 33.005) means any written agreement between units of government under which one public agency performs duties for or in concert with another public agency using EPA assistance. This includes substate and interagency agreements.

Intergovernmental Agreement (Section 35.6015 [6-5-90]) Any written agreement between units of government under which one public agency performs duties for or in concert with another public agency using EPA assistance. This includes substate and interagency agreements.

Interim authorization (Section 270.2) means approval by EPA of a State hazardous waste program which has met the requirements of section 3006(c) of RCRA and applicable requirements of Part 271, Subpart B.

Intermediate (Section 704.3 [12/22/88])* means any chemical substance that is consumed, in whole or in part, in chemical reactions used for the intentional manufacture of other chemical substances or mixtures, or that is intentionally present for the purpose of altering the rates of such chemical reactions.

Intermediate (Section 710.2) means any chemical substance: (1) Which is intentionally removed from the equipment in which it is manufactured, and (2) which either is consumed in whole or in part in chemical reaction(s) used for the intentional manufacture of other chemical substance(s) or mixture(s), or is intentionally present for the purpose of altering the rate of such chemical reaction(s). Note: The equipment in which it was manufactured includes the reaction vessel in which the chemical substance was manufactured and other equipment which is strictly ancillary to the reaction vessel, and any other equipment through which the chemical substance may flow during a continuous flow process, but does not include tanks or other vessels in which the chemical substance is stored after its manufacture.

Intermediate (Section 712.3) means any chemical substance that is consumed, in whole or in part, in chemical reactions used for the intentional manufacture of other chemical substances or mixtures, or that is intentionally present for the purpose of altering the rates of such chemical reactions. (See also paragraph (j) of this section.)

Intermediate (Section 720.3) means any chemical substance that is consumed, in whole or in part, in chemical reactions used for the intentional manufacture of another chemical substance(s) or mixture(s), or that is intentionally present for the purpose of altering the rates of such chemical reactions.

Intermediate (Section 723.175) means any chemical substance which is consumed in whole or in part in a chemical reaction(s) used for the intentional manufacture of another chemical substance.

Intermediate cover (Section 241.101) means cover material that serves the same functions as daily cover, but must resist erosion for a longer period of time, because it is applied on areas where additional cells are not to be constructed for extended periods of time.

Intermediate handler (Section 259.10 [3/24/89]) is a facility that either treats regulated medical waste or destroys regulated medical waste but does not do both. The term, as used in this Part, does not include transporters.

Intermediate PMN (Section 700.43 [8/17/88]). See **Intermediate premanufacture notice (Section 700.43)**.

Intermediate premanufacture notice or intermediate PMN (Section 700.43 [8/17/88]) means any PMN submitted to EPA for a chemical substance which is an intermediate (as intermediate is defined in Section 720.3 of this chapter) in the production of a final product, provided that the PMN for the intermediate is submitted to EPA at the same time as, and together with, the PMN for the final product and that the PMN for the intermediate identifies the final product and describes the chemical reactions leading from the intermediate to the final product. If PMNs are submitted to EPA at the same time for several intermediates used in the production of a final product, each of those is an intermediate PMN if they all identify the final product and every other associated intermediate PMN and are submitted to EPA at the same time as, and together with, the PMN for the final product.

Intermediate speed (Section 86.082-2) means peak torque speed if peak torque speed occurs between 60 and 75 percent of rated speed. If the peak torque speed is less than 60 percent of rated speed, intermediate speed means 60 percent of rated speed. If the peak torque speed is greater than 75 percent of rated speed, intermediate speed means 75 percent of rated speed.

Intermediate Temperature Cold Testing (Section 86.094-2 [3/24/93])* means testing done pursuant to the driving cycle and testing conditions contained in 40 CFR part 86, subpart C, at temperatures between 25 deg.F (-4 deg.C) and 68 deg.F (20 deg.C).

Intermediate useful life (Section 86.098-2 [4/6/94])* is a period of use of 5 years or 50,000 miles, whichever occurs first. Full useful life is a period of use of 10 years or 100,000 miles, whichever occurs first, except as otherwise noted in Sec. 86.094-9. The useful life of evaporative and/or refueling emission control systems on the portion of these vehicles subject to the evaporative emission test requirements of Sec. 86.130-96, and/or the refueling emission test requirements of Sec. 86.151-98, is defined as a period of use of 10 years or 100,000 miles, whichever occurs first. For light light-duty trucks subject to the Tier 0 standards of Sec. 86.094-9(a), and for heavy light-duty truck engine families, intermediate and/or full useful life. Intermediate useful life is a period of use of 5 years or 50,000 miles, whichever occurs first.

Intermittent control system (ICS)

(Section 51.100) means a dispersion technique which varies the rate at which pollutants are emitted to the atmosphere according to meteorological conditions and/or ambient concentrations of the pollutant, in order to prevent ground-level concentrations in excess of applicable ambient air quality standards. Such a dispersion technique is an ICS whether used alone, used with other dispersion techniques, or used as a supplement to continuous emission controls (i.e., used as a supplemental control system).

Intermittent operations (Section 471.02) means the industrial user does not have a continuous operation.

Intermittent vapor processing system (Section 60.501) means a vapor processing system that employs an intermediate vapor holder to accumulate total organic compounds vapors collected from gasoline tank trucks, and treats the accumulated vapors only during automatically controlled cycles.

Internal floating roof (Section 61.341 [3-7-90]) means a cover that rests or floats on the liquid surface inside a waste management unit that has a fixed roof.

Internal subunit (Section 704.25) means a subunit that is covalently linked to at least two other subunits. Internal subunits of polymer molecules are chemically derived from monomer molecules that have formed covalent links between two or more other molecules.

Internal subunit (Section 723.250) means a subunit that is covalently linked to at least two other subunits. Internal subunits of polymer molecules are chemically derived from monomer molecules that have formed covalent links between two or more other subunits.

International shipment (Section 260.10) means the transportation of hazardous waste into or out of the jurisdiction of the United States.

International System of Units (Section 191.12 [12/20/93])* is the version of the metric system which has been established by the International Bureau of Weights and Measures and is administered in the United States by the National Institute of Standards and Technology. The abbreviation for this system is "SI."

Interstate agency (Section 122.2) means an agency of two or more States established by or under an agreement or compact approved by the Congress, or any other agency of two or more States having substantial powers or duties pertaining to the control of pollution as determined and approved by the Administrator under the CWA and regulations.

Interstate agency (Section 124.2) means an agency of two or more States established by or under an agreement or compact approved by the Congress, or any other agency of two or more States having substantial powers or duties pertaining to the control of pollution as determined and approved by the Administrator under the appropriate Act and regulations.

Interstate agency (Section 142.2 [9/26/88]) means an agency of two or more States established by or under an agreement or compact approved by the Congress, or any other agency of two or more States or Indian Tribes having substantial powers or duties pertaining to the control of pollution as determined and approved by the Administrator.

Interstate agency (Section 144.3 [9/26/88])* means an agency of two or more States established by or under an agreement or compact approved by the Congress, or any other agency of two or more States or Indian Tribes having substantial powers or duties pertaining to the control of pollution as determined and approved by the Administrator under the appropriate Act and regulations.

Interstate agency (Section 232.2 [2/11/93])* means an agency of two or more States established by or under an agreement or compact approved by the Congress, or any other agency of two or more States having substantial powers or duties pertaining to the control of pollution.

Interstate agency (Section 233.2 [6/6/88]) means an agency of two or more States established by or under an agreement or compact approved by the Congress, or any other agency of two or more States having substantial powers or duties pertaining to the control of pollution.

Interstate agency (Section 233.3) means an agency of two or more States established by or under an agreement or compact approved by the Congress, or any other agency of two or more States having substantial powers or duties pertaining to the control of pollution as determined and approved by the Administrator under the CWA.

Interstate agency (Sections 35.905; 35.2005) means an agency of two or more States established under an agreement or compact approved by the Congress, or any other agency of two or more States, having substantial powers or duties pertaining to the control of water pollution.

Interstate commerce (Section 201.1) means the commerce between any place in a State and any place in another State, or between places in the same State through another State, whether such commerce moves wholly by rail or partly by rail and partly by motor vehicle, express, or water. This definition of interstate commerce for purposes of this regulation is similar to the definition of interstate commerce in section 203(a) of the Interstate Commerce Act (49 U.S.C. 303(a)).

Interstate commerce (Section 202.10) means the commerce between any place in a State and any place in another State or between places in the same State through another State, whether such commerce moves wholly by motor vehicle or partly by motor vehicle and partly by rail, express, water or air. This definition of interstate commerce for purposes of these regulations is the same as the definition of interstate commerce in section 203(a) of the Interstate Commerce Act (49 U.S.C. 303(a)).

Intervener (Section 209.3) means a person who files a motion to be made a party under Section 209.15 or Section 209.16, and whose motion is approved.

Intervener (Section 85.1807) shall mean a person who files a petition to be made an intervener pursuant to paragraph (g) of this section and whose petition is approved.

Inventory (Sections 720.3; 723.250; 747.115; 747.195; 747.200) means the list of chemical substances manufactured or processed in the United States that EPA compiled and keeps current under section 8(b) of the Act.

Inventory form (Section 370.2 [10/15/87]) means the Tier I and Tier II emergency and hazardous chemical inventory forms set forth in Subpart D of this Part.

Inventory of open dumps (Section 256.06) means the inventory required under section 4005(b) and is defined as the list published by EPA of those disposal facilities which do not meet the criteria.

Inventory system (Section 60.581) means a method of physically accounting for the quantity of ink, solvent, and solids used at one or more affected facilities during a time period. The system is based on plant purchase or inventory records.

Invertebrate animal poisons and repellents (Section 162.3) includes all substances or mixtures of substances intended for preventing the establishment of, destroying, repelling, or mitigating invertebrate animals declared to be pests under Section 162.14, except those pesticides defined as insecticides in paragraph (ff)(10) of this section or nematicides in paragraph (ff)(13) of this section. (i) Invertebrate animal poisons and repellents include, but are not limited to: (A) Antifouling agents intended for use on boat and ship bottoms, pier and dock pilings, and similar submerged structures to prevent attachment or damage and destruction by marine invertebrates; (B) Mollusk control agents intended to repel or destroy snails or slugs; and (C) Protozoa control agents intended to destroy disease-inducing and/or parasitic protozoa in aquatic situations. (ii) Invertebrate animal poisons and repellents do not include those substances or mixtures of substances subject to the provisions of the Federal Food, Drug, and Cosmetic Act, as amended (21 U.S.C. 301 et seq.), intended for use in controlling or killing parasitic invertebrates on or in living man or other animals.

Invitation For Bids (Section 248.4 [2/17/89]) is the solicitation for prospective suppliers by a purchaser requesting their competitive price quotations.

IRB (Section 26.102 [6-18-91]) means an institutional review board established in accord with and for the purposes expressed in this policy.

IRB approval (Section 26.102 [6-18-91]) means the determination of the IRB that the research has been reviewed and may be conducted at an institution within the constraints set forth by the IRB and by other institutional and federal requirements.

Iron and steel (Section 420.11) means those by-product cokemaking operations other than merchant cokemaking operations.

Iron blast furnace (Section 420.31) means all blast furnaces except ferromanganese blast furnaces.

Irreparable harm (Section 125.121) means significant undesirable effects occurring after the date of permit issuance which will not be reversed after cessation or modification of the discharge.

ISO standard day conditions (Section 60.331) means 288 degrees Kelvin, 60 percent relative humidity and 101.3 kilopascals pressure.

Isodrin (Section 704.142) means the pesticide 1,4:5,8-Dimethanonaphthalene,1,2,3,4,10,10-hexacholoro-1,4,4a,5,8,8a-hexahydro-, (1alpha, 4alpha, 4abeta, 5beta, 8beta, 8abeta)-, CAS No. 465-73-6.

Isokinetic sampling (Section 60.2) means sampling in which the linear velocity of the gas entering the sampling nozzle is equal to that of the undisturbed gas stream at the sample point.

Isomeric ratio (Section 704.83) means the relative amounts of each isomeric chlorinated naphthalene that composes the chemical substance; and for each isomer the relative amounts of each chlorinated naphthalene designated by the position of the chlorine atom(s) on the naphthalene.

Isomeric ratio (Section 704.85) means the ratios of ortho-, meta-, and parachlorinated terphenyls.

Issuance (Section 60.2 [7/21/92])* of a part 70 permit will occur, if the State is the permitting authority, in accordance with the requirements of part 70 of this chapter and the applicable, approved State permit program. When the EPA is the permitting authority, issuance of a title V permit occurs immediately after the EPA takes final action on the final permit.

Issuance (Section 61.02 [3/16/94])* of a part 70 permit will occur, if the State is the permitting authority, in accordance with the requirements of part 70 of this chapter and the applicable, approved State permit program. When the EPA is the permitting authority, issuance of a title V permit occurs immediately after the EPA takes final action on the final permit.

Issuance of an NSO (Section 57.103) means the final transmittal of the NSO pursuant to Section 57.107(a) by an issuing agency (other than EPA) to EPA for approval, or the publication of an NSO issued by EPA in the *Federal Register.*

Issuing agency (Section 57.103), unless otherwise specifically indicated, means the state or local air pollution control agency to which a smelter's owner has applied for an NSO, or which has issued the NSO, or EPA, when the NSO application has been made to EPA. Any showings or demonstrations required to be made under this part to the issuing agency, when not EPA, are subject to independent determinations by EPA.

J

Jams and jellies (Section 407.81) shall include the production of jams, jellies and preserves defined as follows: The combination of fruit and fruit concentrate, sugar, pectin, and other additives in an acidic medium resulting in a gelatinized and thickened finished product.

Job shop (Section 433.11) shall mean a facility which owns not more than 50% (annual area basis) of the materials undergoing metal finishing.

Joint sponsor (Section 790.3) means a person who sponsors testing pursuant to section 4(b)(3)(A) of the Act.

Joint sponsorship (Section 790.3) means the sponsorship of testing by two or more persons in accordance with section 4(b)(3)(A) of the Act.

Joint submitters (Section 700.43 [8/17/88]) means two or more persons who submit a section 5 notice together.

Judicial Officer (Section 124.72) means a permanent or temporary employee of the Agency appointed as a Judicial Officer by the Administrator under these regulations and subject to the following conditions: (a) A Judicial Officer shall be a licensed attorney. A Judicial Officer shall not be employed in the Office of Enforcement or the Office of Water and Waste Management, and shall not participate in the consideration or decision of any case in which he or she performed investigative or prosecutorial functions, or which is factually related to such a case. (b) The Administrator may delegate any authority to act in an appeal of a given case under this subpart to a Judicial Officer who, in addition, may perform other duties for EPA, provided that the delegation shall not preclude a Judicial Officer from referring any motion or case to the Administrator when the Judicial Officer decides such action would be appropriate. The Administrator, in deciding a case, may consult with and assign the drafting of preliminary findings of fact and conclusions and/or a preliminary decision to any Judicial Officer.

Judicial Officer (Section 164.2 [7/10/92])* means an officer or employee of the Agency designated as a judicial officer, pursuant to these rules, who shall meet the qualifications and perform functions as herein provided.

Judicial Officer (Section 179.3 [12-5-90]) means a person who has been designated by the Administrator under Sec.179.117 to serve as a judicial officer.

Judicial Officer (Section 209.3) means an officer or employee of the Agency appointed as a judicial officer by the

Administrator under this section who shall meet the qualifications and perform functions as follows: (1) Position. There may be designated for the purposes of this section one or more judicial officers. As work requires, there may be a judicial officer designated to act for a particular case. (2) Qualifications. A judicial officer shall be a permanent or temporary employee of the Agency, and may perform other duties for the Agency. The judicial officer shall not be employed by the Office of Enforcement or have any connection with the preparation or presentation of evidence for a hearing held under this subpart in which he or she participates as judicial officer. (3) Functions. The Administrator may consult with a judicial officer or delegate all or part of his or her authority to act under these rules of practice to a judicial officer in a given case. The judicial officer may refer any motion or case to the Administrator, even after this delegation.

Judicial Officer (Section 22.03) means the person designated by the Administrator under Section 22.04(b) to serve as the Judicial Officer.

Judicial Officer (Section 57.103) means an attorney who is a permanent or temporary employee of the U.S. Environmental Protection Agency.

Judicial Officer (Section 85.1807) shall mean an officer or employee of the Agency appointed as a Judicial Officer by the Administrator pursuant to this section who shall meet the qualifications and perform functions as follows: (i) Officer. There may be designated for the purposes of this section one or more Judicial Officers. As work requires, there may be a Judicial Officer designated to act for the purposes of a particular case. (ii) Qualifications. A Judicial Officer may be a permanent or temporary employee of the Agency who performs other duties for the Agency. Such Judicial Officer shall not be employed by the Office of Enforcement and General Counsel or the Office of Mobile Source and Noise Enforcement or have any connection with the preparation or presentation of evidence for a hearing held pursuant to this subpart. (iii) Functions. The Administrator may consult with a Judicial Officer or delegate all or part of his authority to act in a given case under this section to a Judicial Officer: Provided, That this delegation shall not preclude the Judicial Officer from referring any motion or case to the Administrator when the Judicial Officer determines such referral to be appropriate.

Judicial Officer (Section 86.1014-84) means an officer or employee of the Agency appointed as a Judicial Officer by the Administrator pursuant to this section who shall meet the qualifications and perform functions as follows: (i) Officer. There may be designated for the purposes of this section one or more Judicial Officers. As work requires, there may be a

Judicial Officer designated to act for the purposes of a particular case. (ii) Qualifications. A Judicial Officer may be a permanent or temporary employee of the Agency who performs other duties for the Agency. The Judicial Officer shall not be employed by the Office of Enforcement or have any connection with the preparation or presentation of evidence for a hearing held pursuant to this subpart. The Judicial Officer shall be a graduate of an accredited law school and a member in good standing of a recognized Bar Association of any state or the District of Columbia. (iii) Functions. The Administrator may consult with Judicial Officer or delegate all or part of his authority to act in a given case under this section to a Judicial officer: Provided, That this delegation does not preclude the Judicial Officer from referring any motion or case to the Administrator when the Judicial Officer determines such referral to be appropriate.

Judicial Officer (Section 86.1115-87) shall mean an officer or employee of the Agency appointed as a Judicial Officer by the Administrator pursuant to this section who shall meet the qualifications and perform functions as follows: (i) Officer. There may be designated for purposes of this section one or more Judicial Officers. As work requires, there may be a Judicial Officer designated to act for the purposes of a particular case. (ii) Qualifications. A Judicial Officer may be a permanent or temporary employee of the Agency who performs other duties for the Agency. Such Judicial Officer shall not be employed by the Office of Air and Radiation or have any connection with the preparation or presentation of evidence for a hearing held pursuant to this subpart. (iii) Functions. The Administrator may consult with a Judicial Officer or delegate all or part of his authority to act in a given case under this section to a Judicial Officer, Provided, That this delegation shall not preclude the Judicial Officer from referring any motion or case to the Administrator when the Judicial Officer determines such referral to be appropriate.

Judicial Officer (Section 86.614) shall mean an officer or employee of the Agency appointed as a Judicial Officer by the Administrator pursuant to this section who shall meet the qualifications and perform functions as follows: (i) Officer. There may be designated for the purposes of this section one or more Judicial Officers. As work requires, there may be a Judicial Officer designated to act for the purposes of a particular case. (ii) Qualifications. A Judicial Officer may be a permanent or temporary employee of the Agency who performs other duties for the Agency. Such Judicial Officer shall not be employed by the Office of Enforcement or have any connection with the preparation or presentation of evidence for a hearing held pursuant to this subpart. (iii) Functions. The Administrator may consult with a Judicial Officer or delegate all or part of his authority to act in a given case under this section to a Judicial Officer: Provided, That this delegation shall not preclude the Judicial Officer from referring any motion or case to the Administrator when the Judicial Officer determines such referral to be appropriate.

Junction box (Section 60.691 [11/23/88]) means a manhole or access point to a wastewater sewer system line.

Jurisdiction by law (Section 1508.15) means agency authority to approve, veto, or finance all or part of the proposal.

K

K_1 (Section 797.1560). See **Uptake rate constant (Section 797.1560)**.

K_2 (Section 797.1560). See **Depuration rate constant (Section 797.1560)**.

K_d (Section 796.2750). See **Adsorption ratio (Section 796.2750)**.

Known human effects (Section 717.3) means a commonly recognized human health effect of a particular substance or mixture as described either in: (i) Scientific articles or publications abstracted in standard reference sources. (ii) The firm's product labeling or material safety data sheets (MSDS). (2) However, an effect is not a known human effect if it: (i) Was a significantly more severe toxic effect than previously described. (ii) Was a manifestation of a toxic effect after a significantly shorter exposure period or lower exposure level than described. (iii) Was a manifestation of a toxic effect by an exposure route different from that described.

Known to or reasonably ascertainable by (Section 712.3) means all information in a person's possession or control, plus all information that a reasonable person similarly situated might be expected to possess, control, or know, or could obtain without unreasonable burden.

Known to or reasonably ascertainable by (Section 723.175) means all information in a person's possession or control, plus all information that a reasonable person similarly situated might be expected to possess, control, or know, or could obtain without unreasonable burden or cost.

Known to or reasonably ascertainable by (Section 763.63) means all information in a person's possession or control, plus all information that a reasonable person might be expected to possess, control, or know, or could obtain without unreasonable burden or cost.

Known to or reasonably ascertainable by (Section 704.3 [12/22/88])* means all information in a person's possession or control, plus all information that a reasonable person similarly situated might be expected to possess, control, or know.

Known to or reasonably ascertainable by (Sections 720.3; 723.50; 723.250) means all information in a person's possession or control, plus all information that a reasonable person similarly situated might be expected to possess, control, or know.

Kraft pulp mill (Section 60.281) means any stationary source which produces pulp from wood by cooking (digesting) wood chips in a water solution of sodium hydroxide and sodium sulfide (white liquor) at high temperature and pressure. Regeneration of the cooking chemicals through a recovery process is also considered part of the kraft pulp mill.

L

L_λ (Section 796.3700). See **Solar irradiance in water (Section 796.3700)**.

LA (Section 130.2). See **Load allocation (Section 130.2)**.

Label (Section 211.203) means that item, as described in this regulation, which is inscribed on, affixed to or appended to a product, its packaging, or both for the purpose of giving noise reduction effectiveness information appropriate to the product.

Label (Section 600.002-85) means a sticker that contains fuel economy information and is affixed to new automobiles in accordance with Subpart D of this part.

Label (Section 749.68 [1-3-90]) means any written, printed, or graphic material displayed on or affixed to containers of hexavalent chromium-based water treatment chemicals that are to be used in cooling systems.

Laboratory (Section 259.10 [3/24/89]) means any research, analytical, or clinical facility that performs health care related analysis or service. This includes medical, pathological, pharmaceutical, and other research, commercial, or industrial laboratories.

Laboratory (Section 761.3 [12-21-89]) means a facility that analyzes samples for PCBs and is unaffiliated with any entity whose activities involve PCBs.

Land (Section 192.11) means any surface or subsurface land that is not part of a disposal site and is not covered by an occupiable building.

Land application unit (Section 257.2 [3/19/93])* means an area where wastes are applied onto or incorporated into the soil surface (excluding manure spreading operations) for agricultural purposes or for treatment and disposal.

Land disposal (Section 268.2 [5/24/93])* means placement in or on the land, except in a corrective action management unit, and includes, but is not limited to, placement in a landfill, surface impoundment, waste pile, injection well, land treatment facility, salt dome formation, salt bed formation, underground mine or cave, or placement in a concrete vault, or bunker intended for disposal purposes.

Land treatment facility (Section 260.10) means a facility or part of a facility at which hazardous waste is applied onto or incorporated into the soil surface; such facilities are disposal facilities if the waste will remain after closure.

Landfill (Section 257.2 [3/19/93])* means an area of land or an excavation in which wastes are placed for permanent disposal, and that is not a land application unit, surface impoundment, injection well, or waste pile.

Landfill (Section 259.10 [3/24/89]) means a disposal facility of part of a facility where regulated medical waste is placed in or on the land and which is not a land treatment facility, a surface impoundment, or an injection well.

Landfill (Section 260.10 [8/18/92])* means a disposal facility or part of a facility where hazardous waste is placed in or on land and which is not a pile, a land treatment facility, a surface impoundment, an underground injection well, a salt dome formation, a salt bed formation, an underground mine, a cave, or a corrective action management unit.

Landfill cell (Section 260.10) means a discrete volume of a hazardous waste landfill which uses a liner to provide isolation of wastes from adjacent cells or wastes. Examples of landfill cells are trenches and pits.

Large (Sections 407.61; 407.71; 407.81) shall mean a point source that processes a total annual raw material production of fruits, vegetables, specialties and other products that exceeds 9,080 kkg (10,000 tons) per year.

Large appliance part (Section 60.451) means any organic surface-coated metal lid, door, casing, panel, or other interior or exterior metal part or accessory that is assembled to form a large appliance product. Parts subject to in-use temperatures in excess of 250 degrees F are not included in this definition.

Large appliance product (Section 60.451) means any organic surface-coated metal range, oven, microwave oven, refrigerator, freezer, washer, dryer, dishwasher, water heater, or trash compactor manufactured for household, commercial, or recreational use.

Large appliance surface coating line (Section 60.451) means that portion of a large appliance assembly plant engaged in the application and curing of organic surface coatings on large appliance parts or products.

Large high voltage capacitor (Section 761.3) means a capacitor which contains 1.36 kg (3 lbs.) or more of dielectric fluid and which operates at 2,000 volts (a.c. or d.c.) or above.

Large low voltage capacitor (Section 761.3) means a capacitor which contains 1.36 kg (3 lbs.) or more of dielectric fluid and which operates below 2,000 volts (a.c. or d.c.).

Large MWC plant (Section 60.31a [2-11-91]) means an MWC plant with an MWC plant capacity greater than 225 megagrams per day (250 tons per day) but less than or equal to 1,000 megagrams per day (1,100 tons per day) of MSW.

Large MWC plant (Section 60.51a [2-11-91]) means an MWC plant with an MWC plant capacity greater than 225 megagrams per day (250 tons per day) of MSW.

Large water system (Section 141.2 [7/17/92])* for the purpose of subpart I of this part only, means a water system that serves more than 50,000 persons.

Large-sized plants (Section 428.71) shall mean plants which process more than 10,430 kg/day (23,000 lbs/day) of raw materials.

Lateral expansion (Section 258.2 [10-9-91]) means a horizontal expansion of the waste boundaries of an existing MSWLF unit.

Lateral sewer (Section 21.2) means a sewer which connects the collector sewer to the interceptor sewer.

Latex resin (Section 61.61) means a resin which is produced by a polymerization process which initiates from free radical catalyst sites and is sold undried.

LC_{50} (Section 116.3) means that concentration of material which is lethal to one-half of the test population of aquatic animals upon continuous exposure for 96 hours or less.

LC_{50} (Section 797.1350) is the median lethal concentration, i.e., that concentration of a chemical in air or water killing 50 percent of a test batch of organisms within a particular period of exposure (which shall be stated).

LC_{50} (Section 797.1400) means that test substance concentration, calculated from experimentally-derived mortality data, that is lethal to 50 percent of a test population during continuous exposure over a specified period of time.

LC_{50} (Section 797.1970) means that experimentally derived concentration of test substance that is calculated to have killed 50 percent of a test population during continuous exposure over a specified period of time.

LC_{50} (Section 797.2050) is the empirically derived concentration of the test substance in the diet that is expected to result in mortality of 50 percent of a population of birds which is exposed exclusively to the treated diet under the conditions of the test.

LC_{50} (Sections 797.1930; 797.1950) means that experimentally derived concentration of test substance that is calculated to kill 50 percent of a test population during continuous exposure over a specified period of time.

LD_{50} (Section 797.2175) is the empirically derived dose of the test substance that is expected to result in mortality of 50 percent of a population of birds which is treated with a single oral dose under the conditions of the test.

Leach (Section 162.3) means to undergo the process by which pesticides in the soil are moved into a lower layer of soil or are dissolved and carried through soil by water.

Leachate (Section 241.101) means liquid that has percolated through solid waste and has extracted dissolved or suspended materials from it.

Leachate (Section 257.2) means liquid that has passed through or emerged from solid waste and contains soluble, suspended or miscible materials removed from such wastes.

Leachate (Section 258.2 [10-9-91]) means a liquid that has passed through or emerged from solid waste and contains soluble, suspended, or miscible materials removed from such waste.

Leachate (Section 260.10) means any liquid, including any suspended components in the liquid, that has percolated through or drained from hazardous waste.

Lead (Section 415.61) shall mean total lead.

Lead (Section 420.02) means total lead and is determined by the method specified in 40 CFR 136.3.

Lead additive (Section 80.2) means any substance containing lead or lead compounds.

Lead additive manufacturer (Section 80.2) means any person who produces a lead additive or sells a lead additive under his own name.

Lead agency (Section 1508.16) means the agency or agencies preparing or having taken primary responsibility for preparing the environmental impact statement.

Lead agency (Section 300.5 [3-8-90])* means the agency that provides the OSC/RPM to plan and implement response action under the NCP, EPA, the USCG, another federal agency, or state (or political subdivision of a state) operating pursuant to a contract or cooperative agreement executed pursuant to section 104(d)(1) of CERCLA, or designated pursuant to a Superfund Memorandum of Agreement (SMOA) entered into pursuant to subpart F of the NCP or other agreements may be the lead agency for a response action. In the case of a release of a hazardous substance, pollutant, or contaminant, where the release is on, or the sole source of the release is from, any facility or vessel under the jurisdiction, custody, or control of the Department of Defense (DOD) or Department of Energy (DOE), then DOD or DOE will be the lead agency. Where the release is on, or the sole source of the release is from, any facility or vessel under the jurisdiction, custody, or control of a federal agency other than EPA, the USCG, DOD, or DOE, then that agency will be the lead agency for remedial actions and removal actions other than emergencies. The federal agency maintains its lead agency responsibilities whether the remedy is selected by the federal agency for non-NPL sites or by EPA and the federal agency or by EPA alone under CERCLA section 120. The lead agency will consult with the support agency, if one exists, throughout the response process.

Lead agency (Section 35.6015 [6-5-90])* The Federal agency, State agency, political subdivision, or Indian Tribe that has primary responsibility for planning and implementing a response action under CERCLA.

Lead matte (Section 61.171) means any molten solution of copper and other metal sulfides produced by reduction of sinter product from the oxidation of lead sulfide ore concentrates.

Lead oxide manufacturing facility (Section 60.371) means a facility that produces lead oxide from lead, including product recovery.

Lead recipe (Section 60.291) means glass product composition of the following ranges of weight proportions: 50 to 60 percent silicon dioxide, 18 to 35 percent lead oxides, 5 to 20 percent total R_2O (e.g., Na_2M and K_2O), 0 to 8 percent total R_2O_3 (e.g., Al_2O_3), 0 to 15 percent total RO (e.g., CaO, MgO), other than lead oxide, and 5 to 10 percent other oxides.

Lead reclamation facility (Section 60.371) means the facility that remelts lead scrap and casts it into lead ingots for use in the battery manufacturing process, and which is not a furnace affected under Subpart L of this part.

Lead service line (Section 141.2 [7/17/92])* means a service line made of lead which connects the water main to the building inlet and any lead pigtail, gooseneck or other fitting which is connected to such lead line.

Lead trustee (Section 306.12) means a trustee authorized to act on behalf of all affected trustees where there are multiple trustees because of co-existing or contiguous natural resources or concurrent jurisdiction.

Lead-acid battery manufacturing plant (Section 60.371) means any plant that produces a storage battery using lead and lead compounds for the plates and sulfuric acid for the electrolyte.

Leaded gasoline (Section 80.2) means gasoline which is produced with the use of any lead additive or which contains more than 0.05 gram of lead per gallon or more than 0.005 gram of phosphorus per gallon.

Leak (Section 61.301 [3-7-90]) means any instrument reading of 10,000 ppmv or greater using method 21 of 40 CFR part 60, appendix A.

Leak (Section 61.61 [2-10-90])* means any of several events that indicate interruption of confinement of vinyl chloride within process equipment. Leaks include events regulated under subpart V of this part such as: (1) An instrument reading of 10,000 ppm or greater measured according to Method 21 (see appendix A of 40 CFR part 60); (2) A sensor detection of failure of a seal system, failure of a barrier fluid system, or both; (3) Detectable emissions as indicated by an instrument reading of greater than 500 ppm above background for equipment designated for no detectable emissions measured according to Test Method 21 (see appendix A of 40 CFR part 60); and (4) In the case of pump seals regulated under Sec. 61.242-2, indications of liquid dripping constituting a leak under Sec. 61.242-2. Leaks also include events regulated under Sec. 61.65(b)(8)(i) for detection of ambient concentrations in excess of background concentrations. A relief valve discharge is not a leak.

Leak or leaking (Section 761.3) means any instance in which a PCB Article, PCB Container, or PCB Equipment has any PCBs on any portion of its external surface.

Leak-tight (Section 61.141 [11-20-90]) means that solids or liquids cannot escape or spill out. It also means dust-tight.

Leak-detection system (Section 260.10) means a system capable of detecting the failure of either the primary or secondary containment structure or the presence of a release of hazardous waste or accumulated liquid in the secondary containment structure. Such a system must employ operational controls (e.g., daily visual inspections for releases into the secondary containment system of aboveground tanks) or consist of an interstitial monitoring device designed to detect continuously and automatically the failure of the primary or secondary containment structure or the presence of a release of hazardous waste into the secondary containment structure.

Ledger paper (Section 250.4 [10/6/87, 6/22/88]) means a type of paper generally used in a broad variety of recordkeeping type applications such as in accounting machines.

Legal defense cost (Section 280.92 [10/26/88]) is any expense that an owner or operator or provider of financial assurance incurs in defending against claims or actions brought, (1) By EPA or a state to require corrective action or to recover the costs of corrective action; (2) By or on behalf of a third party for bodily injury or property damage caused by an accidental release; or (3) By any person to enforce the terms of a financial assurance mechanism.

Legal defense costs (Section 264.141), as specified in Section 264.141(g), means any expenses that an insurer incurs in defending against claims of third parties brought under the terms and conditions of an insurance policy.

Legal defense costs (Section 265.141), as specified in Section 265.141(g), means any expenses that an insurer incurs in defending against claims of third parties brought under the terms and conditions of an insurance policy.

Legally authorized representative (Section 26.102 [6-18-91]) means an individual or judicial or other body authorized under applicable law to consent on behalf of a prospective subject to the subject's participation in the procedure(s) involved in the research.

Legionella (Section 141.2 [6/29/89]) means a genus of bacteria, some species of which have caused a type of pneumonia called Legionnaires Disease.

Legislation (Section 1508.17) includes a bill or legislative proposal to Congress developed by or with the significant cooperation and support of a Federal agency, but does not include requests for appropriations. The test for significant cooperation is whether the proposal is in fact predominantly that of the agency rather than another source. Drafting does not by itself constitute significant cooperation. Proposals for legislation include requests for ratification of treaties. Only the agency which has primary responsibility for the subject matter involved will prepare a legislative environmental impact statement.

Level of Detection (Section 761.3 [6/27/88]). See **Quantifiable Level (Section 761.3)**.

Level of quantitation or LOQ (Section 766.3) means the lowest concentration at which HDDs/HDFs can be reproducibly measured in a specific chemical substance within specified confidence limits, as described in this Part.

Liabilities (Section 144.61) means probable future sacrifices of economic benefits arising from present obligations to transfer assets or provide services to other entities in the future as a result of past transactions or events.

Liabilities (Section 264.141), as specified in Section 264.141(f), means probable future sacrifices of economic benefits arising from present obligations to transfer assets or provide services to other entities in the future as a result of past transactions or events.

Liabilities (Section 265.141), as specified in Section 265.141(f), means probable future sacrifices of economic benefits arising from present obligations to transfer assets or provide services to other entities in the future as a result of past transactions or events.

License or permit (Section 121.1) means any license or permit granted by an agency of the Federal Government to conduct any activity which may result in any discharge into the navigable waters of the United States.

Licensed site (Section 192.31) means the area contained within the boundary of a location under the control of persons generating or storing uranium byproduct materials under a license issued pursuant to section 84 of the Act. For purposes of this subpart, licensed site is equivalent to regulated unit in Subpart F of Part 264 of this chapter.

Licensed site (Section 61.251) means the area contained within the boundary of a location under the control of persons generating or storing uranium byproduct materials under a license issued by the Commission. This includes such areas licensed by Agreement States, i.e., those States which have entered into an effective agreement under section 274(b) of the Atomic Energy Act of 1954, as amended.

Licensing or permitting agency (Section 121.1) means any agency of the Federal Government to which application is made for a license or permit.

Light-duty truck 1 (Section 86.094-2 [6-5-91]) means any light light-duty truck up through 3750 lbs loaded vehicle weight.

Light-duty truck 2 (Section 86.094-2 [6-5-91]) means any light light-duty truck greater than 3750 lbs loaded vehicle weight.

Light-duty truck 3 (Section 86.094-2 [6-5-91]) means any heavy light-duty truck up through 5750 lbs adjusted loaded vehicle weight.

Light-duty truck 4 (Section 86.094-2 [6-5-91]) means any heavy light-duty truck greater than 5750 lbs adjusted loaded vehicle weight.

Light light-duty truck (Section 86.094-2 [6-5-91]) means any light-duty truck rated up through 6000 lbs GVWR.

Light-duty truck (Section 60.391) means any motor vehicle rated at 3,850 kilograms gross vehicle weight or less, designed mainly to transport property.

Light-duty truck (Section 86.082-2) means any motor vehicle rated at 8,500 pounds GVWR or less which has a vehicle curb weight of 6,000 pounds or less and which has a basic vehicle frontal area of 45 square feet or less, which is: (1) Designed primarily for purposes of transportation of property or is a derivation of such a vehicle, or (2) Designed primarily for transportation of persons and has a capacity of more than 12 persons, or

(3) Available with special features enabling off-street or off-highway operation and use.

Light-duty vehicle (Section 86.082-2) means a passenger car or passenger car derivative capable of seating 12 passengers or less.

Light-off Time or LOT (Section 85.2122(a)(15)(ii)(C)) means the time required for a catalytic converter (at ambient temperature 68-86 degrees F) to warm-up sufficiently to convert 50% of the incoming HC and CO to CO_2 and H_2O.

Light-oil condenser (Section 61.131 [9-14-89]) means any unit in the light-oil recovery operation that functions to condense benzene-containing vapors.

Light-oil decanter (Section 61.131 [9-14-89]) means any vessel, tank, or other type of device in the light-oil recovery operation that functions to separate light oil from water downstream of the light-oil condenser. A light-oil decanter also may be known as a light-oil separator.

Light-oil storage tank (Section 61.131 [9-14-89]) means any tank, reservoir, or container used to collect or store crude or refined light-oil.

Light-oil sump (Section 61.131 [9-14-89]) means any tank, pit, enclosure, or slop tank in light-oil recovery operations that functions as a wastewater separation device for hydrocarbon liquids on the surface of the water.

Lignite (Section 60.41a) means coal that is classified as lignite A or B according to the American Society of Testing and Materials (ASTM) Standard Specification for Classification of Coals by Rank D388-77 (incorporated by reference - see Section 60.17).

Lignite (Section 60.41b [12/16/87])* means a type of coal classified as lignite A or lignite B by the American Society of Testing and Materials in ASTM D388-77, Standard Specification for Classification of Coals by Rank (IBR - see Section 60.17).

Lima beans (Section 180.1) means the beans and the pod.

Lima beans (Section 407.71) shall mean the processing of lima beans into the following product styles: Canned and frozen, green and white, all varieties and sizes.

Lime kiln (Section 60.281) means a unit used to calcine lime mud, which consists primarily of calcium carbonate, into quicklime, which is calcium oxide.

Lime manufacturing plant (Section 60.341) means any plant which uses a rotary lime kiln to produce lime product from limestone by calcination.

Lime product (Section 60.341) means the product of the calcination process including, but not limited to, calcitic lime, dolomitic lime, and dead-burned dolomite.

Limitation (Section 2.301). See **Standard (Section 2.301)**.

Limitation (Section 2.302). See **Standard (Section 2.302)**.

Limited water-soluble substances (Section 797.1060) means chemicals which are soluble in water at less than 1,000 mg/l.

Limiting permissible concentration (LPC) of the liquid phase of a material (Section 227.27) is: (1) That concentration of a constituent which, after allowance for initial mixing as provided in Section 227.29, does not exceed applicable marine water quality criteria; or, when there are no applicable marine water quality criteria, (2) That concentration of waste or dredged material in the receiving water which, after allowance for initial mixing, as specified in Section 227.29, will not exceed a toxicity threshold defined as 0.01 of a concentration shown to be acutely toxic to appropriate sensitive marine organisms in a bioassay carried out in accordance with approved EPA procedures. (3) When there is reasonable scientific evidence on a specific waste material to justify the use of an application factor other than 0.01 as specified in paragraph (a)(2) of this section, such alternative application factor shall be used in calculating the LPC.

Limiting permissible concentration of the suspended particulate and solid phases of a material (Section 227.27) means that concentration which will not cause unreasonable acute or chronic toxicity or other sublethal adverse effects based on bioassay results using appropriate sensitive marine organisms in the case of the suspended particulate phase, or appropriate sensitive benthic marine organisms in the case of the solid phase; and which will not cause accumulation of toxic materials in the human food chain. These bioassays are to be conducted in accordance with procedures approved by EPA, or, in the case of dredged material, approved by EPA and the Corps of Engineers. (An implementation manual is being developed jointly by EPA and the Corps of Engineers, and announcement of the availability of the manual will be published in the *Federal Register*. Until this manual is available, interim guidance on the appropriate procedures can be obtained from the Marine Protection Branch, WH-548, Environmental Protection Agency, 401 M Street SW, Washington, DC 20460, or the Corps of Engineers, as the case may be.)

Line purge (Section 60.391). See **Purge (Section 60.391)**.

Liner (Section 260.10) means a continuous layer of natural or man-made materials, beneath or on the sides of a surface impoundment, landfill, or landfill cell, which restricts the downward or lateral escape of hazardous waste, hazardous waste constituents, or leachate.

Liquid trap (Section 280.12 [9/23/88]) means sumps, well cellars, and other traps used in association with oil and gas production, gathering, and extraction operations (including gas production plants), for the purpose of collecting oil, water, and other liquids. These liquid traps may temporarily collect liquids for subsequent disposition or reinjection into a production or pipeline stream, or may collect and separate liquids from a gas stream.

Liquid-mounted seal (Section 60.111a) means a foam or liquid-filled primary seal mounted in contact with the liquid between the tank wall and the floating roof continuously around the circumference of the tank.

Liquid-mounted seal (Section 61.341 [3-7-90]) means a foam or liquid-filled primary seal mounted in contact with the liquid between the waste management unit wall and the floating roof continuously around the circumference.

Liquids dripping (Section 60.481) means any visible leakage from the seal including spraying, misting, clouding, and ice formation.

List of Violating Facilities (Section 15.4) means a list of facilities which are ineligible for any agency contract, grant or loan.

List Official (Section 15.4) means an EPA official designated by the Assistant Administrator to maintain the List of Violating Facilities.

Listed mixture (Section 716.3) means any mixture listed in Section 716.120.

Listing proceeding (Section 15.4) means an informal hearing, conducted by the Case Examiner, held to determine whether a facility should be placed on the List of Violating Facilities.

Lithology (Sections 146.3; 147.2902) means the description of rocks on the basis of their physical and chemical characteristics.

Lithosphere (Section 191.12 [12/20/93])* means the solid part of the Earth below the surface, including any ground water contained within it.

Live 18-day embryos (Section 797.2130) are embryos that are developing normally after 18 days of incubation. This is determined by candling the eggs. Values are expressed as a percentage of viable embryos (fertile eggs).

Live 21-day embryos (Section 797.2150) are embryos that are developing normally after 21 days of incubation. This is determined by candling the eggs. Values are expressed as a percentage of viable embryos (fertile eggs).

Live weight killed (Sections 432.11; 432.21; 432.31; 432.41). See **LWK (Sections 432.11; 432.21; 432.31; 432.41)**.

Load allocation (LA) (Section 130.2). The portion of a receiving water's loading capacity that is attributed either to one of its existing or future nonpoint sources of pollution or to natural background sources. Load allocations are best estimates of the loading, which may range from reasonably accurate estimates to gross allotments, depending on the availability of data and appropriate techniques for predicting the loading. Wherever possible, natural and nonpoint source loads should be distinguished.

Load Cell (Section 201.1) means a device external to the locomotive, of high electrical resistance, used in locomotive testing to simulate engine loading while the locomotive is stationary. (Electrical energy produced by the diesel generator is dissipated in the load cell resistors instead of the traction motors.)

Load or loading (Section 130.2). An amount of matter or thermal energy that is introduced into a receiving water; to introduce matter or thermal energy into a receiving water. Loading may be either man-caused (pollutant loading) or natural (natural background loading).

Loaded vehicle mass (Section 86.402-78) means curb mass plus 80 kg (176 lb.), average driver mass.

Loaded vehicle weight (Section 86.082-2) means the vehicle curb weight plus 300 pounds.

Loading (Section 61.341 [3-7-90]) means the introduction of waste into a waste management unit but not necessarily to complete capacity (also referred to as filling).

Loading (Section 797.1400) means the ratio of fish biomass (grams, wet weight) to the volume (liters) of test solution in a test chamber or passing through it in a 24-hour period.

Loading (Section 797.1520) is the ratio of fish biomass (grams, wet weight) to the volume (liters) of test solution passing through the test chamber during a 24-hour period.

Loading (Section 797.1600) is the ratio of biomass (grams of fish, wet weight) to the volume (liters) of test solution passing through the test chamber during a specific interval (normally a 24-hour period).

Loading (Section 797.1830) is the ratio of the number of oysters to the volume (liters) of test solution passing through the test chamber per hour.

Loading (Sections 797.1300; 797.1330) means the ratio of daphnid biomass (grams, wet weight) to the volume (liters) of test solution in a test chamber at a point in time, or passing through the test chamber during a specific interval.

Loading (Sections 797.1930; 797.1950; 797.1970) means the ratio of test organism biomass (grams, wet weight) to the volume (liters) of test solution in a test chamber.

Loading capacity (Section 130.2). The greatest amount of loading that a water can receive without violating water quality standards.

Loading cycle (Section 61.301 [3-7-90]) means the time period from the beginning of filling a tank truck, railcar, or marine vessel until flow to the control device ceases, as measured by the flow indicator.

Loading rack (Section 60.501) means the loading arms, pumps, meters, shutoff valves, relief valves, and other piping and valves necessary to fill delivery tank trucks.

Loading rack (Section 61.301 [3-7-90]) means the loading arms, pumps, meters, shutoff valves, relief valves, and other piping and valves necessary to fill tank trucks, railcars, or marine vessels.

Loan (Section 15.4) means an agreement or other arrangement under which any portion of a business, activity, or program is assisted under a loan issued by an agency and includes any subloan issued under a loan issued by an agency.

Loan agreement (Section 39.105) means a written agreement between the Bank and the guaranteed borrower stating the terms of the loan.

Loan guarantee agreement (Section 39.105) means a written agreement between EPA and the guaranteed borrower stating the terms of the loan guarantee.

Loan guarantee and loan insurance (Section 34.105 [2-26-90]) means an agency's guarantee or insurance of a loan made by a person.

Local agency (Section 60.21) means any local governmental agency.

Local agency (Sections 51.100; 58.1) means any local government agency other than the State agency, which is charged with responsibility for carrying out a portion of the plan.

Local comprehensive emergency response plan (Section 310.11 [10/21/87]) means the emergency plan prepared by the local emergency planning committee as required by section 303 of the Emergency Planning and Community Right-To-Know Act of 1986 (SARA Title III).

Local education agency (Section

47.105 [3-9-92]) means any education agency as defined in section 198 of the Elementary and Secondary Education Act of 1965 (20 U.S.C. 3381) and shall include any tribal education agency, as defined in Sec. 47.105(f).

Local education agency (Section 763.103) means: (1) Any local education agency as defined in section 198(a)(10) of the Elementary and Secondary Education Act of 1965 (20 U.S.C. 2854). (2) The governing authority of any nonprofit elementary or secondary school, where the term nonprofit means owned and operated by one or more nonprofit corporations or associations no part of the net earnings of which inures, or may lawfully inure, to the benefit of any private shareholder or individual.

Local education agency (Section 763.83 [10/30/87]) means: (1) Any local educational agency as defined in section 198 of the Elementary and Secondary Education Act of 1965 (20 U.S.C. 3381). (2) The owner of any nonpublic, nonprofit elementary, or secondary school building. (3) The governing authority of any school operated under the defense dependents' education system provided for under the Defense Dependents' Education Act of 1978 (20 U.S.C. 921, et seq.).

Local emergency response plan (Section 310.11 [1/15/93])* means the emergency plan prepared by the Local Emergency Planning Committee (LEPC) as required by section 303 of the Emergency Planning and Community Right-To-Know Act of 1986 (SARA Title III) (EPCRA).

Local government (Section 34.105 [2-26-90]) means a unit of government in a State and, if chartered, established, or otherwise recognized by a State for the performance of a governmental duty, including a local public authority, a special district, an intrastate district, a council of governments, a sponsor group representative organization, and any other instrumentality of a local government.

Local share (Section 39.105) means the amount of the total grant eligible and allowable project costs which a public body is obligated to pay under the grant.

Locomotive (Section 201.1) means for the purpose of this regulation, a self-propelled vehicle designed for and used on railroad tracks in the transport of rail cars, including self-propelled rail passenger vehicles.

Locomotive Load Cell Test Stand (Section 201.1) means the load cell (Section 201.1(o)) and associated structure, equipment, trackage and locomotive being tested.

Log sorting and log storage facilities (Section 122.27) means facilities whose discharges result from the holding of unprocessed wood, for example, logs or roundwood with bark or after removal of bark held in self-contained bodies of water (mill ponds or log ponds) or stored on land where water is applied intentionally on the logs (wet decking). (See 40 CFR Part 429, Subpart I, including the effluent limitations guidelines.)

London Amendments (Section 82.3 [12/10/93])* means the Montreal Protocol, as amended at the Second Meeting of the Parties to the Montreal Protocol in London in 1990.

Long term stabilization (Section 61.221 [12-15-89]) means the addition of material on a uranium mill tailings pile for purpose of ensuring compliance with the requirements of 40 CFR 192.02(a) or 192.32(b)(i). These actions shall be considered complete when the Nuclear Regulatory Commission determines that the requirements of 40 CFR 192.02(a) or 192.32(b)(i) have been met.

Loose-fill insulation (Section 248.4 [2/17/89]) means insulation in granular, nodular, fibrous, powdery, or similar form, designed to be installed by pouring, blowing or hand placement.

LOQ (Section 766.3). See **Level of quantitation (Section 766.3)**.

LOT (Section 85.2122(a)(15)(ii)(C)). See **Light-off Time (Section 85.2122(a)(15)(ii)(C))**.

Low altitude (Sections 86.082-2; 86.1602) means any elevation equal to or less than 1,219 meters (4,000 feet).

Low altitude conditions (Section 86.082-2) means a test altitude less than 549 meters (1,800 feet).

Low dose (Section 795.232 [1-8-90]) should correspond to 1/10 of the high dose.

Low heat release rate (Section 60.41b [12/16/87])* means a heat release rate of 730,000 J/sec-m^3 (70,000 Btu/hour-ft^3) or less.

Low processing packinghouse (Section 432.31) shall mean a packinghouse that processes no more than the total animals killed at that plant, normally processing less than the total kill.

Low terrain (Sections 51.166; 52.21) means any area other than high terrain.

Low volume waste sources (Section 423.11) means, taken collectively as if from one source, wastewater from all sources except those for which specific limitations are otherwise established in this part. Low volume wastes sources include, but are not limited to: wastewaters from wet scrubber air pollution control systems, ion exchange water treatment system, water treatment evaporator blowdown, laboratory and sampling streams, boiler blowdown, floor drains, cooling tower basin cleaning wastes, and recirculating house service water systems. Sanitary and air conditioning wastes are not included.

Low-concentration PCBs (Section 761.123) means PCBs that are tested and found to contain less than 500 ppm PCBs, or those PCB-containing materials which EPA requires to be assumed to be at concentrations below 500 ppm (i.e., untested mineral oil dielectric fluid).

Low-Noise-Emission Product Determination (Section 203.1) means the Administrator's determination whether or not a product, for which a properly filed application has been received, meets the low-noise-emission product criterion.

Lower explosive limit (Section 257.3-8) means the lowest percent by volume of a mixture of explosive gases which will propagate a flame in air at 25 degrees C and atmospheric pressure.

Lowest achievable emission rate (Section 51.165) means, for any source, the more stringent rate of emissions based on the following: (A) The most stringent emissions limitation which is contained in the implementation plan of any State for such class or category of stationary source, unless the owner or operator of the proposed stationary source demonstrates that such limitations are not achievable; or (B) The most stringent emissions limitation which is achieved in practice by such class or category of stationary sources. This limitation, when applied to a modification, means the lowest achievable emissions rate for the new or modified emissions units within or stationary source. In no event shall the application of the term permit a proposed new or modified stationary source to emit any pollutant in excess of the amount allowable under an applicable new source standard of performance.

LPC (Section 227.27) means limiting permissible concentration.

Luminescent materials (Section 469.41) shall mean materials that emit light upon excitation by such energy sources as photons, electrons, applied voltage, chemical reactions or mechanical energy and which are specifically used as coatings in fluorescent lamps and cathode ray tubes. Luminescent materials include, but are not limited to, calcium halophosphate, yttrium oxide, zinc sulfide, and zinc-cadmium sulfide.

LWK (live weight killed) (Sections 432.11; 432.21; 432.31; 432.41) shall mean the total weight of the total number of animals slaughtered during the time to which the effluent limitations apply; i.e., during any one day or any period of thirty consecutive days.

M

M10 (Section 435.11 [3/4/93])* shall mean those offshore facilities continuously manned by ten (10) or more persons.

M10 (Section 435.41) shall mean those coastal facilities continuously manned by ten (10) or more persons.

M9IM (Section 435.11 [3/4/93])* shall mean those offshore facilities continuously manned by nine (9) or fewer persons or only intermittently manned by any number of persons.

M9IM (Section 435.41) shall mean those coastal facilities continuously manned by nine (9) or fewer persons or intermittently manned by any number of persons.

Machine shop (Section 61.31) means a facility performing cutting, grinding, turning, honing, milling, deburring, lapping, electrochemical machining, etching, or other similar operations.

Magnetic tape (Section 60.711 [10/3/88]) means any flexible substrate that is covered on one or both sides with a coating containing magnetic particles and that is used for audio or video recording or information storage.

Maintain (Section 1516.2) means maintain, collect, use or disseminate.

Maintain (Section 16.2) shall have the meaning given to it by 5 U.S.C. 552a (a)(3).

Maintenance (Section 280.12 [9/23/88]) means the normal operational upkeep to prevent an underground storage tank system from releasing product.

Maintenance instructions (Section 204.2) means those instructions for maintenance, use, and repair, which the Administrator is authorized to require pursuant to section 6(c)(1) of the Act.

Maintenance instructions or instructions (Section 205.2) means those instructions for maintenance, use, and repair, which the Administrator is authorized to require pursuant to section 6(c)(1) of the Act.

Major disaster (Section 109.2) means any hurricane, tornado, storm, flood, high water, wind-driven water, tidal wave, earthquake, drought, fire, or other catastrophe in any part of the United States which, in the determination of the President, is or threatens to become of sufficient severity and magnitude to warrant disaster assistance by the Federal Government to supplement the efforts and available resources of States and local governments and relief organizations in alleviating the damage, loss, hardship, or suffering caused thereby.

Major discharge (Section 300.6). See **Size classes of discharges (Section 300.6)**.

Major facility (Section 122.2) means any NPDES facility or activity classified as such by the Regional Administrator, or, in the case of approved State programs, the Regional Administrator in conjunction with the State Director.

Major facility (Section 124.2) means any RCRA, UIC, NPDES, or 404 facility or activity classified as such by the Regional Administrator, or, in the case of approved State programs, the Regional Administrator in conjunction with the State Director.

Major facility (Section 144.3) means any UIC facility or activity classified as such by the Regional Administrator, or, in the case of approved State programs, the Regional Administrator in conjunction with the State Director.

Major facility (Section 233.3) means any 404 facility or activity classified as such by the Regional Administrator in conjunction with the State Director.

Major facility (Section 270.2) means any facility or activity classified as such by the Regional Administrator, or, in the case of approved State programs, the Regional Administrator in conjunction with the State Director.

Major Federal action (Section 1508.18) includes actions with effects that may be major and which are potentially subject to Federal control and responsibility. Major reinforces but does not have a meaning independent of significantly (Section 1508.27). Actions include the circumstance where the responsible officials fail to act and that failure to act is reviewable by courts or administrative tribunals under the Administrative Procedure Act or other applicable law as agency action. (a) Actions include new and continuing activities, including projects and programs entirely or partly financed, assisted, conducted, regulated, or approved by Federal agencies; new or revised agency rules, regulations, plans, policies, or procedures; and legislative proposals (Sections 1506.8, 1508.17). Actions do not include funding assistance solely in the form of general revenue sharing funds, distributed under the State and Local Fiscal Assistance Act of 1972, 31 U.S.C. 1221 et seq., with no Federal agency control over the subsequent use of such funds. Actions do not include bringing judicial or administrative civil or criminal enforcement actions. (b) Federal actions tend to fall within one of the following categories: (1) Adoption of official policy, such as rules, regulations, and interpretations adopted pursuant to the Administrative Procedure Act, 5 U.S.C. 551 et seq.; treaties and international conventions or agreements; formal documents establishing an agency's policies which will result in or substantially alter agency programs. (2) Adoption of formal plans, such as official documents prepared or approved by federal agencies which guide or prescribe alternative uses of federal resources, upon which future agency actions will be based. (3) Adoption of programs, such as a group of concerted actions to implement a specific policy or plan; systematic and connected agency decisions allocating agency resources to implement a specific statutory program or executive directive. (4) Approval of specific projects, such as construction or management activities located in a defined geographic area. Projects include actions approved by permit or other regulatory decision as well as federal and federally assisted activities.

Major modification (Section 51.165) means (A) Any physical change in or change in the method of operation of a major stationary source that would result in a significant net emissions increase of any pollutant subject to regulation under the Act. (B) Any net emissions increase that is considered significant for volatile organic compounds shall be considered significant for ozone. (C) A physical change or change in the method of operation shall not include: (1) Routine maintenance, repair and replacement; (2) Use of an alternative fuel or raw material by reason of an order under sections 2 (a) and (b) of the Energy Supply and Environmental Coordination Act of 1974 (or any superseding legislation) or by reason of a natural gas curtailment plan pursuant to the Federal Power Act; (3) Use of an alternative fuel by reason of an order or rule under section 125 of the Act; (4) Use of an alternative fuel at a steam generating unit to the extent that the fuel is generated from municipal solid waste; (5) Use of an alternative fuel or raw material by a stationary source which; (i) The source was capable of accommodating before December 21, 1976, unless such change would be prohibited under any federally enforceable permit condition which was established after December 12, 1976, pursuant to 40 CFR 52.21 or under regulations approved pursuant to 40 CFR Subpart I or Section 51.166, or (ii) The source is approved to use under any permit issued under regulations approved pursuant to this section; (6) An increase in the hours of operation or in the production rate, unless such change is prohibited under any federally enforceable permit condition which was established after December 21, 1976, pursuant to 40 CFR 52.21 or regulations approved pursuant to 40 CFR Part 51 Subpart I or 40 CFR 51.166; (7) Any change in ownership at a stationary source.

Major modification (Section 52.24) means (i) any physical change in or change in the method of operation of a major stationary source that would result in a significant net emissions increase of any pollutant subject to regulation under the Act. (ii) Any net emissions increase that is considered significant for volatile organic compounds shall be considered significant for ozone. (iii) A physical change or change in the method of operation shall not include: (a) Routine maintenance, repair, and replacement; (b) Use of an alternative fuel or raw material by reason of an order under sections 2 (a) and (b) of the Energy Supply and Environmental Coordination Act of 1974 (or any superseding legislation) or by reason of a natural gas curtailment plan pursuant to the Federal Power Act; (c) Use of an alternative fuel by reason of an order or rule under section 125 of the Act; (d) Use of an alternative fuel at a steam generating unit to the extent that the fuel is generated from municipal solid waste; (e) Use of an alternative fuel or raw material by a stationary source which: (1) The source was capable of accommodating before July 1, 1979, unless such change would be prohibited under any federally enforceable permit condition which was established after July 1, 1979, pursuant to 40 CFR 52.21 or under regulations approved pursuant to 40 CFR Subpart I or 40 CFR 51.166; or (2) The source is approved to use under any permit issued under regulations approved pursuant to 40 CFR Subpart I; (f) An increase in the hours of operation or in the production rate, unless such change is prohibited under any federally enforceable permit condition which was established after July 1, 1979, pursuant to 40 CFR 52.21 or under regulations approved pursuant to 40 CFR Subpart I or 40 CFR 51.166. (g) Any change in ownership at a stationary source.

Major modification (Sections 51.166; 52.21) means (i) any physical change in or change in the method of operation of a major stationary source that would result in a significant net emissions increase of any pollutant subject to regulation under the Act. (ii) Any net emissions increase that is significant for volatile organic compounds shall be considered significant for ozone. (iii) A physical change or change in the method of operation shall not include: (a) Routine maintenance, repair, and replacement; (b) Use of an alternative fuel or raw material by reason of any order under section 2 (a) and (b) of the Energy Supply and Environmental Coordination Act of 1974 (or any superseding legislation) or by reason of a natural gas curtailment plan pursuant to the Federal Power Act; (c) Use of an alternative fuel by reason of an order or rule under section 125 of the Act; (d) Use of an alternative fuel at a steam generating unit to the extent that the fuel is generated from municipal solid waste; (e) Use of an alternative fuel or raw material by a stationary source which: (1) The source was capable of accommodating before January 6, 1975, unless such change would be prohibited under any federally enforceable permit condition which was established after January 6, 1975, pursuant to 40 CFR 52.21 or under regulations approved pursuant to 40 CFR Subpart I or Section 51.166; or (2) The source is approved to use under any permit issued under 40 CFR 52.21 or under regulations approved pursuant to 40 CFR 51.166; (f) An increase in the hours of operation or in the production rate, unless such change would be prohibited under any federally enforceable permit condition which was established after January 6, 1975, pursuant to 40 CFR 52.21 or under regulations approved pursuant to 40 CFR Subpart I or Section 51.166; (g) Any change in ownership at a stationary source.

Major PSD modification (Section 124.41) means a major modification as defined in 40 CFR 52.21.

Major PSD stationary source (Section 124.41) means a major stationary source as defined in 40 CFR 52.21(b)(1).

Major release (Section 300.6). See **Size classes of releases (Section 300.6)**.

Major stationary source (Section 65.01) shall mean any stationary source which directly emits, or has the potential to emit, 100 tons per year or more of any air pollutant for which a national ambient air quality standard under section 109 of the Act is in effect (including any major stationary source of fugitive emissions of any such pollutant, as determined by rule by the Administrator).

Major stationary source (Section 66.3 [2-13-92])* means any stationary facility or source of air pollutants which directly emits, or has the potential to emit, one hundred tons per year or more of any air pollutant regulated by EPA under the Clean Air Act.

Major stationary source (Section 52.24) means: (i)(a) Any stationary source of air pollutants which emits, or has the potential to emit, 100 tons per year or more of any pollutant subject to regulation under the Act; or (b) Any physical change that would occur at a stationary source not qualifying under paragraph (f)(5)(i)(a) of this section, as a major stationary source, if the change would constitute a major stationary source by itself. (ii) A major stationary source that is major for volatile for organic compounds shall be considered major for ozone. (iii) The fugitive emissions of a stationary source shall not be included in determining for any of the purposes of this section whether it is a major stationary source, unless the source belongs to one of the following categories of stationary sources: (a) Coal cleaning plants (with thermal dryers); (b) Kraft pulp mills; (c) Portland cement plants; (d) Primary zinc smelters; (e) Iron and steel mills; (f) Primary aluminum ore reduction plants; (g) Primary copper smelters; (h) Municipal incinerators capable of charging more than 250 tons of refuse per day; (i) Hydrofluoric, sulfuric, or nitric acid plants; (j) Petroleum refineries; (k) Lime plants; (l) Phosphate rock processing plants; (m) Coke oven batteries; (n) Sulfur recovery plants; (o) Carbon black plants (furnace process); (p) Primary lead smelters; (q) Fuel conversion plants; (r) Sintering plants; (s) Secondary metal production plants; (t) Chemical process plants; (u) Fossil-fuel boilers (or combination thereof) totaling more than 250 million British thermal units per hour heat input; (v) Petroleum storage and transfer units with a total storage capacity exceeding 300,000 barrels; (w) Taconite ore processing plants; (x) Glass fiber processing plants; (y) Charcoal production plants; (z) Fossil fuel-fired steam electric plants of more than 250 million British thermal units per hour heat input; (z) Fossil fuel-fired steam electric plants of more than 250 million British thermal units per hour heat input; (aa) Any other stationary source category which, as of August 7, 1980, is being regulated under section 111 or 112 of the Act.

Major stationary source (Section 51.165) means: (A)(1) Any stationary source of air pollutants which emits, or has the potential to emit 100 tons per year or more of any pollutant subject to regulation under the Act, or (2) Any physical change that would occur at a stationary source not qualifying under paragraph (a)(1)(iv)(A)(1) as a major stationary source, if the change would constitute a major stationary source by itself. (B) A major stationary source that is major for volatile organic compounds shall be considered major for ozone. (C) The fugitive emissions of a stationary source shall not be included in determining for any of the purposes of this paragraph whether it is a major stationary source, unless the source belongs to one of the following categories of stationary sources: (1) Coal cleaning plants (with thermal dryers); (2) Kraft pulp mills; (3) Portland cement plants; (4) Primary zinc smelters; (5) Iron and steel mills; (6) Primary aluminum ore reduction plants; (7) Primary copper smelters; (8) Municipal incinerators capable of charging more than 250 tons of refuse per day; (9) Hydrofluoric, sulfuric, or nitric acid plants; (10) Petroleum refineries; (11) Lime plants; (12) Phosphate rock processing plants; (13) Coke oven batteries; (14) Sulfur recovery plants; (15) Carbon black plants (furnace process); (16) Primary lead smelters; (17) Fuel conversion plants; (18) Sintering plants; (19) Secondary metal production plants; (20) Chemical process plants; (21) Fossil-fuel boilers (or combination thereof) totaling more than 250 million British thermal units per hour heat input; (22) Petroleum storage and transfer units with a total storage capacity exceeding 300,000 barrels; (23) Taconite ore processing plants; (24) Glass fiber processing plants; (25) Charcoal production plants; (26) Fossil fuel-fired steam electric plants of more than 250 million British thermal units per hour heat input; and (27) Any other stationary source category which, as of August 7, 1980, is being regulated under section 111 or 112 of the Act.

Major stationary source (Sections 51.166; 52.21) means: (i)(a) Any of the following stationary sources of air pollutants which emits, or has the potential to emit, 100 tons per year or more of any pollutant subject to regulation under the Act: Fossil fuel-fired steam electric plants of more than 250 million British thermal units per hour heat input, coal cleaning plants (with thermal dryers), kraft pulp mills, portland cement plants, primary zinc smelters, iron and steel mill plants, primary aluminum ore reduction plants, primary copper smelters, municipal incinerators capable of charging more than 250 tons of refuse per day, hydrofluoric, sulfuric, and nitric acid plants, petroleum refineries, lime plants, phosphate rock processing plants, coke oven batteries, sulfur recovery plants, carbon black plants (furnace process), primary lead smelters, fuel conversion plants, sintering plants, secondary metal production plants, chemical process plants, fossil fuel boilers (or combinations thereof) totaling more than 250 million British thermal units per hour heat input, petroleum storage and transfer units with a total storage capacity exceeding 300,000 barrels, taconite ore processing plants, glass fiber processing plants, and charcoal production plants; (b) Notwithstanding the stationary source size specified in paragraph (b)(1)(i)(a) of this section, any stationary source which emits, or has the potential to emit, 250 tons per year or more of any air pollutant subject to regulation under the Act; or (c) Any physical change that would occur at a stationary source not otherwise qualifying under paragraph (b)(1) of this section, as a major stationary source if the change would constitute a major stationary source by itself. (ii) A major source that is major for volatile organic compounds shall be considered major for ozone. (iii) The fugitive emissions of a stationary source shall not be included in determining for any of the purposes of this section whether it is a major stationary source, unless the source belongs to one of the following categories of stationary sources: (a) Coal cleaning plants (with thermal dryers); (b) Kraft pulp mills; (c) Portland cement plants; (d) Primary zinc smelters; (e) Iron and steel mills; (f) Primary aluminum ore reduction plants; (g) Primary copper smelters; (h) Municipal incinerators capable of charging more than 250 tons of refuse per day; (i) Hydrofluoric, sulfuric, or nitric acid plants; (j) Petroleum refineries; (k) Lime plants; (l) Phosphate rock processing plants; (m) Coke oven batteries; (n) Sulfur recovery plants; (o) Carbon black plants (furnace process); (p) Primary lead smelters; (q) Fuel conversion plants; (r) Sintering plants; (s) Secondary metal production plants; (t) Chemical process plants; (u) Fossil-fuel boilers (or combination thereof) totaling more than 250 million British thermal units per hour heat input; (v) Petroleum storage and transfer units with a total storage capacity exceeding 300,000 barrels; (w) Taconite ore processing plants; (x) Glass fiber processing plants; (y) Charcoal production plants; (z) Fossil fuel-fired steam electric plants of more than 250 million British thermal units per hour heat input; (aa) Any other stationary source category which, as of August 7, 1980, is being regulated under section 111 or 112 of the Act.

Make available for use (Section 171.2) means to distribute, sell, ship, deliver for shipment, or receive and (having so received) deliver, to any person. However, the term excludes transactions solely between persons who are pesticide producers, registrants, wholesalers, or retail sellers, acting only in those capacities.

Makeup solvent (Section 60.601) means the solvent introduced into the affected facility that compensates for solvent lost from the affected facility during the manufacturing process.

Malfunction (Section 264.1031 [6-21-90]) means any sudden failure of a control device or a hazardous waste management unit or failure of a hazardous waste management unit to operate in a normal or usual manner, so that organic emissions are increased.

Malfunction (Section 57.103 [2-13-92])* means any unanticipated and unavoidable failure of air pollution control equipment or process equipment or of a process to operate in a normal or usual manner. Failures that are caused entirely or in part by poor design, poor maintenance, careless operation, or any other preventable upset condition or preventable equipment breakdown shall not be considered malfqnctions. A malfunction exists only for the minimum time necessary to implement corrective measures.

Malfunction (Section 60.2 [7/21/92])* means any sudden, infrequent, and not reasonably preventable failure of air pollution control equipment, process equipment, or a process to operate in a normal or usual manner. Failures that are caused in part by poor maintenance or careless operation are not malfunctions.

Malfunction (Section 61.141 [11-20-90]) means any sudden and unavoidable failure of air pollution control equipment or process equipment or of a process to operate in a normal or usual manner so that emissions of asbestos are increased. Failures of equipment shall not be considered malfunctions if they are caused in any way by poor maintenance, careless operation, or any other preventable upset conditions, equipment breakdown, or process failure.

Malfunction (Section 61.161) means any sudden failure of air pollution control equipment or process equipment or of a process to operate in a normal or usual manner so that emissions of arsenic are increased.

Malfunction (Section 86.082-2) means not operating according to specifications (e.g., those specifications listed in the application for certification).

Malfunction (Sections 61.171; 61.181) means any sudden failure of air pollution control equipment or process equipment or of a process to operate in a normal or usual manner so that emissions of inorganic arsenic are increased.

Mammal poisons and repellents (Section 162.3) includes all substances or mixtures of substances, except rodenticides, as defined in paragraph (ff)(15) of this section, intended for preventing, destroying, repelling, or mitigating mammals declared to be pests under Section 162.14. Mammal poisons and repellents include, but are not limited to: (i) Taste, odor, and irritant repellents intended to repel mammals or their adverse, undesired, or destructive activities such as attacking, foraging, chewing, gnawing, urinating, or defacating in or on specific sites or on or near specific objects, persons, plants, or animals; (ii) Predacides intended to kill certain mammals that prey upon other vertebrate animals which man deems necessary to protect; (iii) Toxicants, baits, and poisons intended to kill certain mammals causing injury or destruction to crops, stored foods, or other organisms and objects which man deems necessary to protect; and (iv) Reproductive inhibitors intended to reduce or otherwise alter the reproductive capacity or potential of certain mammals.

Man-made beta particle and photon emitters (Section 141.2) means all radionuclides emitting beta particles and/or photons listed in Maximum Permissible Body Burdens and Maximum Permissible Concentration of Radionuclides in Air or Water for Occupational Exposure, NBS Handbook 69, except the daughter products of thorium-232, uranium-235 and uranium-238.

Management (Section 191.02) means any activity, operation, or process (except for transportation) conducted to prepare spent nuclear fuel or radioactive waste for storage or disposal, or the activities associated with placing such fuel or waste in a disposal system.

Management authority (Section 228.2) means the EPA organizational entity assigned responsibility for implementing the management functions identified in Section 228.3.

Management Conference (Section 35.9010 [10-3-89]) A Management Conference convened by the Administrator under Section 320 of the CWA for an estuary in the NEP (National Estuary Program).

Management of migration (Section 300.5 [3-8-90])* means actions that are taken to minimize and mitigate the migration of hazardous substances or pollutants or contaminants and the effects of such migration. Measures may include, but are not limited to, management of a plume of contamination, restoration of a drinking water aquifer, or surface water restoration.

Management or hazardous waste management (Section 260.10) means the systematic control of the collection, source separation, storage, transportation, processing, treatment, recovery, and disposal of hazardous waste.

Mandatory Class I Federal Area (Section 51.301) means any area identified in Part 81, Subpart D of this title.

Manifest (Section 260.10) means the shipping document EPA form 8700-22 and, if necessary, EPA form 8700-22A, originated and signed by the generator in accordance with the instructions included in the Appendix to Part 262.

Manifest (Section 761.3 [12-21-89]) means the shipping document EPA form 8700-22 and any continuation sheet attached to EPA form 8700-22, originated and signed by the generator of PCB waste in accordance with the instructions included with the form and subpart K of this part.

Manifest (Sections 144.3; 270.2) means the shipping document originated and signed by the generator which contains the information required by Subpart B of 40 CFR Part 262.

Manifest document number (Section 260.10) means the U.S. EPA twelve digit identification number assigned to the generator plus a unique five digit document number assigned to the Manifest by the generator for recording and reporting purposes.

Manifold business forms (Section 250.4 [10/6/87, 6/22/88]) means a type of product manufactured by business forms manufacturers that is commonly produced as marginally punched continuous forms in small rolls or fan folded sets with or without carbon paper interleaving. It has a wide variety of uses such as invoices, purchase orders, office memoranda, shipping orders, and computer printout.

Manned Control Center (Section 761.3) means an electrical power distribution control room where the operating conditions of a PCB Transformer are continuously monitored during the normal hours of operation (of the facility), and, where the duty engineers, electricians, or other trained personnel have the capability to deenergize a PCB Transformer completely within 1 minute of the receipt of a signal indicating abnormal operating conditions such as an overtemperature condition or overpressure condition in a PCB Transformer.

Manual (Section 66.3 [2-13-92])* means the "Noncompliance Penalties

Instruction Manual" which accompanies these regulations. This Manual appears as Appendix B to these regulations.

Manual method (Section 53.1) means a method for measuring concentrations of an ambient air pollutant in which sample collection, analysis, or measurement, or some combination thereof, is performed manually.

Manufacture (Section 125.101) means to produce as an intermediate or final product, or by-product.

Manufacture (Section 372.3 [2/16/88]) means to produce, prepare, import, or compound a toxic chemical. Manufacture also applies to a toxic chemical that is produced coincidentally during the manufacture, processing, use, or disposal of another chemical or mixture of chemicals, including a toxic chemical that is separated from that other chemical or mixture of chemicals as a byproduct, and a toxic chemical that remains in that other chemical or mixture of chemicals as an impurity.

Manufacture (Section 704.3 [12/22/88])* means to manufacture for commercial purposes.

Manufacture (Section 716.3) means to manufacture for commercial purposes.

Manufacture (Section 761.3) means to produce, manufacture, or import into the customs territory of the United States.

Manufacture (Section 762.3) has the same meaning as in 15 U.S.C. 2602.

Manufacture (Section 763.163 [7-12-89]) means to produce or manufacture in the United States.

Manufacture (Sections 710.2; 720.3; 723.50; 723.250) means to produce or manufacture in the United States or import into the customs territory of the United States.

Manufacture for commercial purposes (Section 717.3) means (1) to import, produce, or manufacture with the purpose of obtaining an immediate or eventual commercial advantage for the manufacturer, and includes, among other things, such manufacture of any amount of a chemical substance or mixture: (i) For distribution in commerce, including for test marketing. (ii) For use by the manufacturer, including use for product research and development, or as an intermediate. (2) Manufacture for commercial purposes also applies to substances that are produced coincidentally during the manufacture, processing, use, or disposal of another substance or mixture, including both byproducts that are separated from that other substance or mixture and impurities that remain in that substance or mixture. Such byproducts and impurities may, or may not, in themselves have commercial value. They are nonetheless produced for the purpose of obtaining a commercial advantage since they are part of the manufacture of a chemical product for a commercial purpose.

Manufacture for commercial purposes (Section 704.3 [12/22/88])* means: (1) To import, produce, or manufacture with the purpose of obtaining an immediate or eventual commercial advantage for the manufacturer, and includes among other things, such manufacture of any amount of a chemical substance or mixture: (i) For commercial distribution, including for test marketing. (ii) For use by the manufacturer, including use for product research and development, or as an intermediate. (2) Manufacture for commercial purposes also applies to substances that are produced coincidentally during the manufacture, processing, use, or disposal of another substance or mixture, including both byproducts that are separated from that other substance or mixture and impurities that remain in that substance or mixture. Such byproducts and impurities may, or may not, in themselves have commercial value. They are nonetheless produced for the purpose of obtaining a commercial advantage since they are part of the manufacture of a chemical product for a commercial purpose.

Manufacture for commercial purposes (Section 763.63) means to import, produce, or manufacture with the purpose of obtaining an immediate or eventual commercial advantage for the manufacturer and includes, among other things, such manufacture of any amount of a chemical substance or mixture: (1) For commercial distribution, including for test marketing, and (2) For use by the manufacturer, including use for product research and development, or as an intermediate. Manufacture for commercial purposes also applies to substances that are produced coincidentally during the manufacture, processing, use, or disposal of another substance or mixture, including both byproducts and coproducts that are separated from that other substance or mixture, and impurities that remain in that substance or mixture. Byproducts and impurities may not in themselves have commercial value. They are nonetheless produced for the purpose of obtaining a commercial advantage since they are part of the manufacture of a chemical product for a commercial purpose.

Manufacture for commercial purposes (Section 712.3) means to import, produce, or manufacture with the purpose of obtaining an immediate or eventual commercial advantage for the manufacturer and includes, among other things, such manufacture of any amount of a chemical substance or mixture: (1) For commercial distribution, including for test marketing. (2) For use by the manufacturer, including use for product research and development, or as an intermediate. Manufacture for commercial purposes also applies to substances that are produced coincidentally during the manufacture, processing, use, or disposal of another substance or mixture, including byproducts and coproducts that are separated from that other substance or mixture, and impurities that remain in that substance or mixture. Byproducts and impurities may not in themselves have commercial value. They are nonetheless produced for the purpose of obtaining a commercial advantage since they are part of the manufacture of a chemical produced for a commercial purpose.

Manufacture for commercial purposes (Section 716.3) means (1) To produce, with the purpose of obtaining an immediate or eventual commercial advantage for the manufacturer, and includes among other things such manufacture of any amount of a chemical substance or mixture: (i) For commercial distribution, including for test marketing. (ii) For use by the manufacturer, including use for product research and development, or as an intermediate. (2) Manufacture for commercial purposes also applies to substances that are produced coincidentally during the manufacture, processing, use, or disposal of another substance or mixture, including byproducts and impurities. Such byproducts and impurities may, or may not, in themselves have commercial value. They are nonetheless produced for the purpose of obtaining a commercial advantage since they are part of the manufacture of a chemical product for a commercial purpose.

Manufacture of electronic crystals (Section 469.22) means the growing of crystals and/or the production of crystal wafers for use in the manufacture of electronic devices.

Manufacture of semiconductors (Section 469.12) means those processes, beginning with the use of crystal wafers, which lead to or are associated with the manufacture of semiconductor devices.

Manufacture or import for commercial purposes (Section 710.2) means to manufacture or import: (1) For distribution in commerce, including for test marketing purposes, or (2) For use by the manufacturer, including for use as an intermediate.

Manufacture or import for commercial purposes (Sections 720.3; 747.200) means (1) To import, produce, or manufacture with the purpose of obtaining an immediate or eventual commercial advantage for the manufacturer or importer, and includes, among other things, manufacture of any amount of a chemical substance or mixture: (i) For commercial distribution, including for test marketing. (ii) For use by the manufacturer, including use for product research and development or as an intermediate. (2) The term also applies to substances that are produced coincidentally during the manufacture, processing, use, or disposal of another substance or mixture, including byproducts that are separated from that other substance or mixture and impurities that remain in that substance or mixture. Byproducts and impurities without separate commercial value are nonetheless produced for the purpose of obtaining a commercial advantage, since they are part of the manufacture of a chemical substance for commercial purposes.

Manufacture or process (Section 717.3) means to manufacture or process for commercial purposes.

Manufacture solely for export (Sections 720.3; 747.200) means to manufacture or import for commercial purposes a chemical substance solely for export from the United States under the following restrictions on activities in the United States: (1) Distribution in commerce is limited to purposes of export or processing solely for export as defined in Section 721.3 of this chapter. (2) The manufacturer or importer, and any person to whom the substance is distributed for purposes of export or processing solely for export (as defined in Section 721.3 of this chapter), may not use the substance except in small quantities solely for research and development in accordance with Section 720.36.

Manufactured (Section 60.531 [2/26/88]) means completed and ready for shipment (whether or not packaged).

Manufacturer (Section 129.2) means any establishment engaged in the mechanical or chemical transformation of materials or substances into new products including but not limited to the blending of materials such as pesticidal products, resins, or liquors.

Manufacturer (Section 2.301) has the meaning given it in section 216(1) of the Act, 42 U.S.C. 7550(1).

Manufacturer (Section 2.303) has the meaning given it in 42 U.S.C. 4902(6).

Manufacturer (Section 2.311) has the meaning given it in section 501(9) of the Act, 15 U.S.C. 2001(9).

Manufacturer (Section 205.2) means any person engaged in the manufacturing or assembling of new products, or the importing of new products for resale, or who acts for and is controlled by any such person in connection with the distribution of such products.

Manufacturer (Section 211.203) as stated in the Act, means any person engaged in the manufacturing or assembling of new products, or the importing of new products for resale, or who acts for, and is controlled by, any such person in connection with the distribution of such products.

Manufacturer (Section 60.531 [2/26/88]) means any person who constructs or imports a wood heater.

Manufacturer (Section 610.11) means a person or company which is engaged in the business of producing or assembling, and which has primary control over the design specifications, of a retrofit device for which a fuel economy improvement claim is made.

Manufacturer (Section 704.3 [12/22/88])* means a person who imports, produces, or manufactures a chemical substance. A person who extracts a component chemical substance from a previously existing chemical substance or a complex combination of substances is a manufacturer of that component chemical substance.

Manufacturer (Section 716.3) means a person who produces or manufactures a chemical substance. A person who extracts a component chemical substance from a previously existing chemical substance or a complex combination of substances is a manufacturer of that component chemical substance.

Manufacturer (Section 763.163 [7-12-89]) means a person who produces or manufactures in the United States.

Manufacturer (Section 85.1807) refers to a manufacturer contesting a recall order directed at that manufacturer.

Manufacturer (Section 85.1902) shall be given the meaning ascribed to it by section 214 of the Act.

Manufacturer (Section 86.1115-87) means a manufacturer contesting a compliance level or penalty determination sent to the manufacturer.

Manufacturer (Sections 720.3; 723.250; 747.115; 747.195; 747.200) means a person who imports, produces, or manufactures a chemical substance. A person who extracts a component chemical substance from a previously existing chemical substance or a complex combination of substances is a manufacturer of that component chemical substance. A person who contracts with a manufacturer to manufacture or produce a chemical substance is also a manufacturer if (1) the manufacturer manufactures or produces the substance exclusively for that person, and (2) that person specifies the identity of the substance and controls the total amount produced and the basic technology for the plant process.

Manufacturer (Sections 86.614; 86.1014-84) means a manufacturer contesting a suspension or revocation order directed at the manufacturer.

Manufacturer parts (Section 86.1602) are parts produced or sold by the manufacturer of the motor vehicle or motor vehicle engine.

Manufacturers' rated dryer capacity (Section 60.621) means the dryer's rated capacity of articles, in pounds or kilograms of clothing articles per load, dry basis, that is typically found on each dryer on the manufacturer's name-plate or in the manufacturer's equipment specifications.

Manufacturing (Section 61.141 [11-20-90])* means the combining of commercial asbestos--or, in the case of woven friction products, the combining of textiles containing commercial asbestos--with any other material(s), including commercial asbestos, and the processing of this combination into a product. Chlorine production is considered a part of manufacturing.

Manufacturing activities (Section 704.203 [12/22/88]) means all those activities at one site which are necessary to produce a substance identified in Subpart D of this Part and make it ready for sale or use as the listed substance, including purifying or importing the substance.

Manufacturing line (Section 60.681) means the manufacturing equipment comprising the forming section, where molten glass is fiberized and a fiberglass mat is formed; the curing section, where the binder resin in the mat is thermally set; and the cooling section, where the mat is cooled.

Manufacturing process (Section 761.3) means all of a series of unit operations operating at a site, resulting in the production of a product.

Manufacturing stream (Section 721.3 [7-27-89]) means all reasonably anticipated transfer, flow, or disposal of a chemical substance, regardless of physical state or concentration, through all intended operations of manufacture, including the cleaning of equipment.

Manufacturing use product (Section 152.3 [5/4/88]) means any pesticide product that is not an end-use product.

Manufacturing use product (Section 158.153 [5/4/88]) means any pesticide product other than an end use product. A product may consist of the technical grade of active ingredient only, or may contain inert ingredients, such as stabilizers or solvents.

Manufacturing-use product (Section 162.152) means any pesticide product other than a product to be labeled with directions for end use. This term includes any product intended for use as a pesticide after re-formulation or repackaging.

Marine bays and estuaries (Section 35.2005) are semi-enclosed coastal waters which have a free connection to the territorial sea.

Marine environment (Section 125.121) means that territorial seas, the contiguous zone and the oceans.

Marine sanitation device (Section 140.1) includes any equipment for installation onboard a vessel and which is designed to receive, retain, treat, or discharge sewage and any process to treat such sewage.

Marine vessel (Section 61.301 [3-7-90]) means any tank ship or tank barge which transports liquid product such as benzene.

Mark (Section 761.3) means the descriptive name, instructions, cautions, or other information applied to PCBs and PCB Items, or other objects subject to these regulations.

Marked (Section 761.3) means the marking of PCB Items and PCB storage areas and transport vehicles by means of applying a legible mark by painting, fixation of an adhesive label, or by any other method that meets the requirements of these regulations.

Market/Marketers (Section 761.3 [6/27/88]) means the processing or distributing in commerce, or the person who processes or distributes in commerce, used oil fuels to burners or other marketers, and may include the generator of the fuel if it markets the fuel directly to the burner.

Marking (Section 11.4) means the act of physically indicating the classification assignment on classified material.

MARPOL 73/78 (Section 110.1) means the International Convention for the Prevention of Pollution from Ships, 1973, as modified by the Protocol of 1978 relating thereto, Annex I, which regulates pollution from oil and which entered into force on October 2, 1983.

Mass balance (Section 797.2850) means a quantitative accounting of the distributions of chemical in plant components, support medium, and test solutions. It also means a quantitative determination of uptake as the difference between the quantity of gas entering an exposure chamber, the quantity leaving the chamber, and the quantity adsorbed to the chamber walls.

Mass burn refractory MWC (Section 60.51a [2-11-91]) means a combustor that combusts MSW in a refractory wall furnace. This does not include rotary combustors without waterwalls.

Mass burn rotary waterwall MWC (Section 60.51a [2-11-91]) means a combustor that combusts MSW in a cylindrical rotary waterwall furnace. This does not include rotary combustors without waterwalls.

Mass burn waterwall MWC (Section 60.51a [2-11-91]) means a combustor that combusts MSW in a conventional waterwall furnace.

Mass of pollutant that can be discharged (Section 463.2) is the pollutant mass calculated by multiplying the pollutant concentration times the average process water usage flow rate.

Mass-feed stoker steam generating unit (Section 60.41b [12/16/87])* means a steam generating unit where solid fuel is introduced directly into a retort or is fed directly onto a grate where it is combusted.

Master Inventory File (Section 710.23) means EPA's comprehensive list of chemical substances which constitute the Chemical Substances Inventory compiled under section 8(b) of the Act. It includes substances reported under Subpart A of this Part and substances reported under Part 720 of this chapter for which a Notice of Commencement of Manufacture or Import has been received under Section 720.120 of this chapter.

MATC (Sections 797.1330; 797.1950). See **Maximum Acceptable Toxicant Concentration (Sections 797.1330; 797.1950)**.

Matching funds (Section 35.4010 [10/1/92])* means the portion of allowable project costs that a recipient contributes toward completing the TAG project using non-Federal funds or Federal funds if expressly authorized by statute. The match may include in-kind as well as cash contributions.

Material (Section 220.2) means matter of any kind or description, including, but not limited to, dredged material, solid waste, incinerator residue, garbage, sewage, sewage sludge, munitions, radiological, chemical, and biological warfare agents, radioactive materials, chemicals, biological and laboratory waste, wreck or discarded equipment, rock, sand, excavation debris, industrial, municipal, agricultural, and other waste, but such term does not mean sewage from vessels within the meaning of section 312 of the FWPCA. Oil within the meaning of section 311 of the FWPCA shall constitute material for purposes of this Subchapter H only to the extent that it is taken on board a vessel or aircraft for the primary purpose of dumping.

Material recovery section (Section 60.561 [3-22-91]) means the equipment that recovers unreacted or by-product materials from any process section for return to the process line, off-site purification or treatment, or sale. Equipment designed to separate unreacted or by-product material from the polymer product are to be included in this process section, provided at least some of the material is recovered for reuse in the process, off-site purification or treatment, or sale, at the time the process section becomes an affected facility. Otherwise such equipment are to be assigned to one of the other process sections, as appropriate. Equipment that treats recovered materials are to be included in this process section, but equipment that also treats raw materials are not to be included in this process section. The latter equipment are to be included in the raw materials preparation section. If equipment is used to return unreacted or by-product material directly to the same piece of process equipment from which it was emitted, then that equipment is considered part of the process section that contains the process equipment. If equipment is used to recover unreacted or by-product material from a process section and return it to another process section or a different piece of process equipment in the same process section or sends it off-site for purification, treatment, or sale, then such equipment are considered part of a material recovery section. Equipment used for the on-site recovery of ethylene glycol from poly(ethylene terephthalate) plants, however, are not included in the material recovery section, but are covered under the standards applicable to the polymerization reaction section (Sec. 60.562-1(c)(1)(ii)(A) or (2)(ii)(A)).

Material Safety Data Sheet (MSDS) (Section 370.2 [10/15/87]) means the sheet required to be developed under Section 1910.1200(g) of Title 29 of the Code of Federal Regulations.

Material specification (Sections 247.101; 249.04) means a specification that stipulates the use of certain materials to meet the necessary performance requirements.

Matter (Section 1508.19) includes for purposes of Part 1504: (a) With respect to the Environmental Protection Agency, any proposed legislation, project, action or regulation as those terms are used in section 309(a) of the Clean Air Act (42 U.S.C. 7609). (b) With respect to all other agencies, any proposed major federal action to which section 102(2)(C) of NEPA applies.

Maximum (Section 435.11 [3/4/93])* as applied to BAT effluent limitations and NSPS for drilling fluids and drill cuttings shall mean the maximum concentration allowed as measured in any single sample of the barite. (l) The term maximum for any one day as applied to BPT, BCT and BAT effluent limitations and NSPS for oil and grease in produced water shall mean the maximum concentration allowed as measured by the average of four grab samples collected over a 24-hour period that are analyzed separately. Alternatively, for BAT and NSPS the maximum concentration allowed may be determined on the basis of physical composition of the four grab samples prior to a single analysis.

Maximum 30-day average (Section 439.1) shall mean the maximum average of daily values for 30 consecutive days.

Maximum Acceptable Toxicant Concentration (MATC) (Sections 797.1330; 797.1950) means the maximum concentration at which a chemical can be present and not be toxic to the test organism.

Maximum contaminant level (Section 142.2) means the maximum permissible level of a contaminant in water which is delivered to the free flowing outlet of the ultimate user of a public water system, except in the case of turbidity where the maximum permissible level is measured at the point of entry to the distribution system. Contaminants added to the water under circumstances controlled by the user, except those resulting from corrosion of piping and plumbing caused by water quality, are excluded from this definition.

Maximum contaminant level (Section 141.2 [7/17/92])* means the maximum permissable level of a contaminant in water which is delivered to any user of a public water system.

Maximum contaminant level goal (MCLG) (Section 141.2) means the maximum level of a contaminant in drinking water at which no known or anticipated adverse effect on the health of persons would occur, and which allows an adequate margin of safety. Maximum contaminant level goals are nonenforceable health goals.

Maximum daily discharge limitation (Section 122.2) means the highest allowable daily discharge.

Maximum demonstrated MWC unit load (Section 60.51a [2-11-91]) means the maximum 4-hour block average MWC unit load achieved during the most recent dioxin/furan test demonstrating compliance with the applicable standard for MWC organics specified under Sec. 60.53a.

Maximum demonstrated particulate matter control device temperature (Section 60.51a [2-11-91]) means the maximum 4-hour block average temperature measured at the final particulate matter control device inlet during the most recent dioxin/furan test demonstrating compliance with the applicable standard for MWC organics specified under Sec. 60.53a. If more than one particulate matter control device is used in series at the affected facility, the maximum 4-hour block average temperature is measured at the final particulate matter control device.

Maximum design heat input capacity (Section 60.41c [9-12-90]) means the ability of a steam generating unit to combust a stated maximum amount of fuel (or combination of fuels) on a steady state basis as determined by the physical design and characteristics of the steam generating unit.

Maximum extent practicable (Section 112.2 [8/25/93])* means the limitations used to determine oil spill planning resources and response times for on-water recovery, shoreline protection, and cleanup for worst case discharges from onshore non-transportation-related facilities in adverse weather. It considers the planned capability to respond to a worst case discharge in adverse weather, as contained in a response plan that meets the requirements in Sec. 112.20 or in a specific plan approved by the Regional Administrator.

Maximum for any one day and Average of daily values for thirty consecutive days (Section 407.61) shall be based on the daily average mass of material processed during the peak thirty consecutive day production period.

Maximum for any one day and Average of daily values for thirty consecutive days (Section 407.81) shall be based on the daily average mass of final product produced during the peak thirty consecutive day production period.

Maximum for any one day and Average of daily values for thirty consecutive days (Section 407.71) shall be based on the daily average mass of raw material processed during the peak thirty consecutive day production period.

Maximum heat input capacity (Section 60.41b [12/16/87])* means the ability of a steam generating unit to combust a stated maximum amount of fuel on a steady state basis, as determined by the physical design and characteristics of the steam generating unit.

Maximum Individual-Risk (Section 61 [12-15-89]) The maximum additional cancer risk of a person due to exposure to an emitted pollutant for a 70-year lifetime.

Maximum organic vapor pressure (Section 61.341 [1/7/93])* means the equilibrium partial pressure exerted by the waste at the temperature equal to the highest calendar-month average of the waste storage temperature for waste stored above or below the ambient temperature or at the local maximum monthly average temperature as reported by the National Weather Service for waste stored at the ambient temperature, as determined: (1) In accordance with Sec. 60.17(c); or (2) As obtained from standard reference texts; or (3) In accordance with Sec. 60.17(a)(37); or (4) Any other method approved by the Administrator.

Maximum potential NOx emission rate (Section 72.2 [7/30/93])* means the emission rate of nitrogen oxides (in lb/mmBtu) calculated in accordance with section 3 of appendix F of part 75 of this chapter, using the maximum potential nitrogen oxides concentration and either the minimum oxygen concentration (in % O2) or the maximum carbon dioxide concentration (in % CO2) as defined in section 2 of appendix A of part 75 of this chapter.

Maximum production capacity (Section 57.103) means either the maximum demonstrated rate at which a smelter has produced its principal metallic final product under the process equipment configuration and operating procedures prevailing on or before August 7, 1977, or a rate which the smelter is able to demonstrate by calculation is attainable with process equipment existing on August 7, 1977. The rate may be expressed as a concentrate feed rate to the smelter.

Maximum rated horsepower (Section 86.082-2) means the maximum brake horsepower output of an engine as stated by the manufacturer in his sales and service literature and his application for certification under Section 86.082-21.

Maximum rated RPM (Section 205.151) means the engine speed measured in revolutions per minute (RPM) at which peak net brake power (SAE J-245) is developed for motorcycles of a given configuration.

Maximum rated torque (Section 86.082-2) means the maximum torque produced by an engine as stated by the manufacturer in his sales and service literature and his application for certification under Section 86.082-21.

Maximum Rated Capacity (Section 204.51) means that the portable air compressor, operating at the design full speed with the compressor on load, delivers its rated cfm output and pressure, as defined by the manufacturer.

Maximum Sound Level (Section 201.1) means the greatest A-weighted sound level in decibels measured during the designated time interval or during the event, with either fast meter response (Section 201.1(l)) or slow meter response (Section 201.1(ii)) as specified. It is abbreviated as L_{max}.

Maximum Total Trihalomethane Potential (MTP) (Section 141.2) means the maximum concentration of total trihalomethanes produced in a given water containing a disinfectant residual after 7 days at a temperature of 25 degrees C or above.

Maximum true vapor pressure (Section 60.111b) means the equilibrium partial pressure exerted by the stored liquid at the temperature equal to the highest calendar-month average of the liquid storage temperature for liquids stored above or below the ambient temperature or at the local maximum monthly average temperature as reported by the National Weather Service for liquids stored at the ambient temperature, as determined: (1) In accordance with methods described in American Petroleum institute Bulletin 2517, Evaporation Loss From External Floating Roof Tanks (incorporated by reference - see Section 60.17); or (2) As obtained from standard reference texts; or (3) As determined by ASTM Method D2879-83 (incorporated by reference - see Section 60.17); (4) Any other method approved by the Administrator.

Maximum true vapor pressure (Section 60.11 [8-11-89]) means the equilibrium partial pressure exerted by the stored VOL at the temperature equal to the highest calendar-month average of the VOL storage temperature for VOL's stored above or below the ambient temperature or at the local maximum monthly average temperature reported by the National Weather Service for VOL's stored at the ambient temperature, as determined.

Mayonnaise and salad dressings (Section 407.81) shall be defined as the emulsified and non-emulsified semisolid food prepared from the combining of edible vegetable oil with acidifying, and egg yolk containing ingredients, or gum and starch combinations to which certain colorings, spices, and flavorings have been added.

Mbbl (Section 419.11) means one thousand barrels (one barrel is equivalent to 42 gallons).

MBE (Section 35.6015 [1/27/89]). See **Minority Business Enterprise (Section 35.6015)**.

M_{ci} (Section 60.741 [9-11-89]) means the total mass (kg) of each coating (i) applied to the substrate at an affected coating operation during a nominal 1-month period as determined from facility records.

MCLG (Section 141.2). See **Maximum contaminant level goal (Section 141.2)**.

MCW unit capacity (Section 60.51a [2-11-91]) means the maximum design charging rate of an MWC unit expressed in megagrams per day (tons per day) of MSW combusted, calculated according to the procedures under Sec. 60.58a, paragraph (j). Municipal waste combustor unit capacity is calculated using a design heating value of 10,500 kilojoules per kilogram (4,500 British thermal units per pound) for MSW and 19,800 kilojoules per kilogram (8,500 British thermal units per pound) for medical waste. The calculational procedures under Sec. 60.58a(j) include procedures for determining MWC unit capacity for batch MWC's and cofired combustors and combustors firing mixtures of medical waste and other MSW.

Measurement Period (Section 201.1) means a continuous period of time during which noise of railroad yard operations is assessed, the beginning and finishing times of which may be selected after completion of the measurements.

Meat cutter (Section 432.61) shall mean an operation which fabricates, cuts, or otherwise produces fresh meat cuts and related finished products from livestock carcasses, at rates greater than 2730 kg (6000 lb) per day.

Mechanical and Thermal Integrity (Section 85.2122(a)(15)(ii)(G)) means the ability of a converter to continue to operate at its previously determined efficiency and light-off time and be free from exhaust leaks when subject to thermal and mechanical stresses representative of the intended application.

Mechanical Torque Rate (Section 85.2122(a)(2)(iii)(F)) means a term applied to a thermostatic coil, defined as the torque accumulation per angular degree of deflection of a thermostatic coil.

Mechanism (Section 56.1) means an administrative procedure, guideline, manual, or written statement.

Median diameter (Section 798.2450). See **Geometric mean diameter (Section 798.2450)**.

Median diameter (Sections 798.1150; 798.4350). See **Geometric mean diameter (Sections 798.1150; 798.4350)**.

Medical emergency (Section 350.40 [7/29/88]) means any unforeseen condition which a health professional would judge to require urgent and unscheduled medical attention. Such a condition is one which results in sudden and/or serious symptom(s) constituting a threat to a person's physical or psychological well-being and which requires immediate medical attention to prevent possible deterioration, disability, or death.

Medical waste (Section 259.10 [3/24/89]) means any solid waste which is generated in the diagnosis, treatment (e.g., provision of medical services), or immunization of human beings or animals, in research pertaining thereto, or in the production or testing of biologicals. The term does not include any hazardous waste identified or listed under Part 261 of this chapter or any household waste as defined in Section 261.4(b)(I) of this chapter.

Medical waste (Section 60.51a [2-11-91]) means any solid waste which is generated in the diagnosis, treatment, or immunization of human beings or animals, in research pertaining thereto, or in production or testing of biologicals. Medical waste does not include any hazardous waste identified under subtitle C of the Resource Conservation and Recovery Act or any household waste as defined in regulations under subtitle C of the Resource Conservation and Recovery Act.

Medium (Sections 407.61; 407.71; 407.81) shall mean a point source that processes a total annual raw material production of fruits, vegetables, specialties and other products that is between 1,816 kkg (2,000 tons) per year and 9,080 kkg (10,000 tons) per year.

Medium discharge (Section 300.6). See **Size classes of discharges (Section 300.6)**.

Medium release (Section 300.6). See **Size classes of releases (Section 300.6)**.

Medium-size water system (Section 141.2 [7/17/92])* for the purpose of subpart I of this part only, means a water system that serves greater than 3,300 and less than or equal to 50,000 persons.

Medium-sized plants (Section 428.61) shall mean plants which process between 3,720 kg/day (8,200 lbs/day) and 10,430 kg/day (23,000 lbs/day) of raw materials.

Meeting (Section 1517.2) means the deliberations of at least two Council members where such deliberations determine or result in the joint conduct or disposition of official collegial Council business, but does not include deliberations to take actions to open or close a meeting under Sections 1517.4 and 1517.5 or to release or withhold information under Sections 1517.4 and 1517.7. Meeting shall not be construed to prevent Council members from considering individually Council business that is circulated to them sequentially in writing.

Melt (Section 409.21) shall mean that amount of raw material (raw sugar) contained within aqueous solution at the beginning of the process for production of refined cane sugar.

Meltdown and refining (Section 60.271) means that phase of the steel production cycle when charge material is melted and undesirable elements are removed from the metal.

Meltdown and refining period (Section 60.271) means the time period commencing at the termination of the initial charging period and ending at the initiation of the tapping period, excluding any intermediate charging periods.

Melting (Section 60.271a) means that phase of steel production cycle during which the iron and steel scrap is heated to the molten state.

Member of the public (Section 191.02) means any individual except during the time when that individual is a worker engaged in any activity, operation, or process that is covered by the Atomic Energy Act of 1954, as amended.

Member of the public (Section 190.02) means any individual that can receive a radiation dose in the general environment, whether he may or may not also be exposed to radiation in an occupation associated with a nuclear fuel cycle. However, an individual is not considered a member of the public during any period in which he is engaged in carrying out any operation which is part of a nuclear fuel cycle.

Merchant (Section 420.11) means those by-product cokemaking operations which provide more than fifty percent of the coke produced to operations, industries, or processes other than iron making blast furnaces associated with steel production.

Mercury (Section 415.61) shall mean the total mercury present in the process wastewater stream exiting the mercury treatment system.

Mercury (Section 61.51) means the element mercury, excluding any associated elements, and includes mercury in particulates, vapors, aerosols, and compounds.

Mercury chlor-alkali cell (Section 61.51) means a device which is basically composed of an electrolyzer section and a denuder (decomposer) section and utilizes mercury to produce chlorine gas, hydrogen gas, and alkali metal hydroxide.

Mercury chlor-alkali electrolyzer (Section 61.51) means an electrolytic device which is part of a mercury chlor-alkali cell and utilizes a flowing mercury cathode to produce chlorine gas and alkali metal amalgam.

Mercury ore (Section 61.51) means a mineral mined specifically for its mercury content.

Mercury ore processing facility (Section 61.51) means a facility processing mercury ore to obtain mercury.

Metabolism (Section 795.232 [1-8-90]) means the sum of the enzymatic and nonenzymatic process by which a particular substance is handled in the body.

Metabolite (Section 162.3) means any substance produced in or by living organisms by biological processes and derived from a pesticide.

Metabolite (Section 723.50) means a chemical entity produced by one or more enzymatic or nonenzymatic reactions as a result of exposure of an organism to a chemical substance.

Metal cladding (Section 471.02). See **Cladding (Section 471.02)**.

Metal cleaning waste (Section 423.11) means any wastewater resulting from cleaning (with or without chemical cleaning compounds) any metal process equipment including, but not limited to, boiler tube cleaning, boiler fireside cleaning, and air preheater cleaning.

Metal coil surface coating operation (Section 60.461) means the application system used to apply an organic coating to the surface of any continuous metal strip with thickness of 0.15 millimeter (mm) (0.006 in.) or more that is packaged in a roll or coil.

Metal powder production operations (Section 471.02) are mechanical process operations which convert metal to a finely divided form.

Metal preparation (Section 466.02) means any and all of the metal processing steps preparatory to applying the enamel slip. Usually this includes cleaning, pickling and applying a nickel flash or chemical coating.

Metallic mineral concentrate (Section 60.381) means a material containing metallic compounds in concentrations higher than naturally occurring in ore but requiring additional processing if pure metal is to be isolated. A metallic mineral concentrate contains at least one of the following metals in any of its oxidation states and at a concentration that contributes to the concentrate's commercial value: Aluminum, copper, gold, iron, lead, molybdenum, silver, titanium, tungsten, uranium, zinc, and zirconium. This definition shall not be construed as requiring that material containing metallic compounds be refined to a pure metal in order for the material to be considered a metallic mineral concentrate to be covered by the standards.

Metallic mineral processing plant (Section 60.381) means any combination of equipment that produces metallic mineral concentrates from ore. Metallic mineral processing commences with the mining of ore and includes all operations either up to and including the loading of wet or dry concentrates or solutions of metallic minerals for transfer to facilities at non-adjacent locations that will subsequently process metallic concentrates into purified metals (or other products), or up to and including all material transfer and storage operations that precede the operations that produce refined metals (or other products) from metallic mineral concentrates at facilities adjacent to the metallic mineral processing plant. This definition shall not be construed as requiring that mining of ore be conducted in order for the combination of equipment to be considered a metallic mineral processing plant. (See also the definition of metallic mineral concentrate.)

Metallic shoe seal (Section 60.111a) includes but is not limited to a metal sheet held vertically against the tank wall by springs or weighted levers and is connected by braces to the floating roof. A flexible coated fabric (envelope) spans the annular space between the metal sheet and the floating roof.

Metallo-organic active ingredients (Section 455.31) means carbon containing active ingredients containing one or more metallic atoms in the structure.

Metallo-organic pesticides (Section 165.1) means a class of organic pesticides containing one or more metal or metalloid atoms in the structure.

Metalworking fluid (Section 721.3 [7-27-89])* means a liquid of any viscosity or color containing intentionally added water and used in metal machining operations for the purpose of cooling, lubricating, or rust inhibition.

Metalworking fluid (Section 747.200) means a liquid of any viscosity or color containing intentionally added water used in metal machining operations for the purpose of cooling or lubricating.

Metalworking fluid (Sections 747.115; 747.195) means a liquid of any viscosity or color containing intentionally added water used in metal machining operations for the purpose of cooling, lubricating, or rust inhibition.

Meteorological measurements (Section 58.1 [2/12/93])* means measurements of wind speed, wind direction, barometric pressure, temperature, relative humidity, and solar radiation.

Methanol-fueled (Section 86.090-2 [4/11/89]) means any motor vehicle or motor vehicle engine that is engineered and designed to be operated using methanol fuel (i.e., a fuel that contains at least 50 percent methanol (CH_3OH) by volume) as fuel. Flexible fuel vehicles are methanol-fueled vehicles. This definition applies beginning with the 1990 model year.

Method 101 (Part 61, Appendix B). Reference method for determination of particulate and gaseous mercury emissions from chlor-alkali plants - air streams.

Method 101A (Part 61, Appendix B). Determination of particulate and gaseous mercury emissions from sewage sludge incinerators.

Method 102 (Part 61, Appendix B). Determination of particulate and gaseous mercury emissions from chlor-alkali plants - hydrogen streams.

Method 103 (Part 61, Appendix B). Beryllium screening method.

Method 104 (Part 61, Appendix B). Reference method for determination of beryllium emissions from stationary sources.

Method 105 (Part 61, Appendix B). Method for determination of mercury in wastewater treatment plant sewage sludges.

Method 106 (Part 61, Appendix B). Determination of vinyl chloride from stationary sources.

Method 107 (Part 61, Appendix B). Determination of vinyl chloride content of inprocess wastewater samples, and vinyl chloride content of polyvinyl chloride resin, slurry, wet cake, and latex samples.

Method 107A (Part 61, Appendix B). Determination of vinyl chloride content of solvents, resin-solvent solution, polyvinyl chloride resin, resin slurry, wet resin, and latex samples.

Method 111 (Part 61, Appendix B). Determination of Polonium-210 emissions from stationary sources.

Methods of operation (Section 21.2) means the installation, emplacement, or introduction of materials, including those involved in construction, to achieve a process or procedure to control: Surface water pollution from non-point sources - that is, agricultural, forest practices, mining, construction; ground or surface water pollution from well, subsurface, or surface disposal operations; activities resulting in salt water intrusion; or changes in the movement, flow, or circulation of navigable or ground waters.

Metropolitan Statistical Area or MSA (Section 60.331) as defined by the Department of Commerce.

mg/l (Section 133.101). Milligrams per liter.

Mgal (Section 419.11) means one thousand gallons.

Michelin-A operation (Section 60.541 [9/15/87]) means the operation identified as Michelin-A in the Emission Standards and Engineering Division confidential file as referenced in Docket A-80-9, Entry II-B-12.

Michelin-B operation (Section 60.541 [9/15/87]) means the operation identified as Michelin-B in the Emission Standards and Engineering Division confidential file as referenced in Docket A-80-9, Entry II-B-12.

Michelin-C-automatic operation (Section 60.541 [9/15/87]) means the operation identified as Michelin-C-automatic in the Emission Standards and Engineering Division confidential file as referenced in Docket A-80-9, Entry II-B-12.

Milestone (Section 192.31 [11/5/93])* means an enforceable date by which action, or the occurrence of an event, is required for purposes of achieving compliance with the 20 pCi/m2-s flux standard.

Military engine (Section 86.082-2) means any engine manufactured solely for the Department of Defense to meet military specifications.

Milking center (Section 412.11) shall mean a separate milking area with storage and cooling facilities adjacent to a free stall barn or cowyard dairy operation.

Milkroom (Section 412.11) shall mean milk storage and cooling rooms normally used for stall barn dairies.

Mill (Section 440.132) is a preparation facility within which the metal ore is cleaned, concentrated, or otherwise processed before it is shipped to the customer, refiner, smelter, or manufacturer. A mill includes all ancillary operations and structures necessary to clean, concentrate, or otherwise process metal ore, such as ore and gangue storage areas and loading facilities.

Mill broke (Section 250.4 [10/6/87, 6/22/88]) means any paper waste generated in a paper mill prior to completion of the papermaking process. It is usually returned directly to the pulping process. Mill broke is excluded from the definition of recovered materials.

Millboard (Section 763.163 [7-12-89]) means an asbestos-containing product made of paper and similar in consistency to cardboard produced in sections rather than as a continuous sheet. Major applications of this product include: thermal protection for large circuit breakers; barriers from flame or heat; linings in floors, partitions, and fire doors; linings for stoves and heaters; gaskets; table pads; trough liners; covers for operations involving molten metal; and stove mats.

Milling (Section 471.02) is the mechanical treatment of a nonferrous metal to produce powder, or to coat one component of a powder mixture with another.

Mimeo paper (Section 250.4 [10/6/87, 6/22/88]) means a grade of writing paper used for making copies on stencil duplicating machines.

Mine (Section 436.181) shall mean an area of land, surface or underground, actively used for or resulting from the extraction of a mineral from natural deposits.

Mine (Section 436.21) shall mean an area of land, surface or underground, actively mined for the production of crushed and broken stone from natural deposits.

Mine (Section 440.132) is an active mining area, including all land and property placed under, or above the surface of such land, used in or resulting from the work of extracting metal ore or minerals from their natural deposits by any means or method, including secondary recovery of metal ore from refuse or other storage piles, wastes, or rock dumps and mill tailings derived from the mining, cleaning, or concentration of metal ores.

Mine (Section 440.141 [5/24/88]) means a place where work or other activity related to the extraction or recovery of ore is performed.

Mine (Sections 436.31; 436.41) shall mean an area of land, surface or underground, actively mined for the production of sand and gravel from natural deposits.

Mine area (Section 440.141 [5/24/88]) means the land area from which overburden is stripped and ore is removed prior to moving the ore to the beneficiation area.

Mine dewatering (Section 436.181) shall mean any water that is impounded or that collects in the mine and is pumped, drained or otherwise removed from the mine through the efforts of the mine operator.

Mine dewatering (Section 436.21) shall mean any water that is impounded or that collects in the mine and is pumped, drained or otherwise removed from the mine through the efforts of the mine operator. However, if a mine is also used for treatment of process generated waste water, discharges of commingled water from the facilities shall be deemed discharges of process generated waste water.

Mine dewatering (Sections 436.31; 436.41) shall mean any water that is impounded or that collects in the mine and is pumped, drained or otherwise removed from the mine through the efforts of the mine operator. This term shall also include wet pit overflows caused solely by direct rainfall and ground water seepage. However, if a mine is also used for treatment of process generated waste water, discharges of commingled water from the mine shall be deemed discharges of process generated waste water.

Mine drainage (Section 434.11) means any drainage, and any water pumped or siphoned, from an active mining area or a post-mining area.

Mine drainage (Section 436.381) shall mean any water drained, pumped or siphoned from a mine.

Mine drainage (Sections 440.132; 440.141 [5/24/88]) means any water drained, pumped or siphoned from a mine.

Miner of asbestos (Section 763.63) is a person who produces asbestos by mining or extracting asbestos-containing ore so that it may be further milled to produce bulk asbestos for distribution in commerce, and includes persons who conduct milling operations to produce bulk asbestos by processing asbestos-containing ore. Milling involves the separation of the fibers from the ore, grading and sorting the fibers, or fiberizing crude asbestos ore. To mine or mill is to manufacture for commercial purposes under TSCA.

Mineral fiber insulation (Section 248.4 [2/17/89]) means insulation (rock wool or fiberglass) which is composed principally of fibers manufactured from rock, slag or glass, with or without binders.

Mineral handling and storage facility (Section 60.471) means the areas in asphalt roofing plants in which minerals are unloaded from a carrier, the conveyor transfer points between the carrier and the storage silos, and the storage silos.

Mineral Oil PCB Transformer (Section 761.3 [7/19/88]) means any transformer originally designed to contain mineral oil as the dielectric fluid and which has been tested and found to contain 500 ppm or greater PCBs.

Minimal risk (Section 26.102 [6-18-91]) means that the probability and magnitude of harm or discomfort anticipated in the research are not greater in and of themselves than those ordinarily encountered in daily life or during the performance of routine physical or psychological examinations or tests.

Minimize (Part 6, Appendix A, Section 4) means to reduce to the smallest possible amount or degree.

Minimum (Section 435.11 [3/4/93])* as applied to BAT effluent limitations and NSPS for drilling fluids and drill cuttings shall mean the minimum 96-hour LC50 value allowed as measured in any single sample of the discharged waste stream. The term minimum as applied to BPT and BCT effluent limitations and NSPS for sanitary wastes shall mean the minimum concentration value allowed as measured in any single sample of the discharged waste stream.

Mining overburden returned to the mine site (Section 260.10) means any material overlying an economic mineral deposit which is removed to gain access to that deposit and is then used for reclamation of a surface mine.

Mining wastes (Sections 243.101; 246.101) means residues which result from the extraction of raw materials from the earth.

Minor discharge (Section 300.6). See **Size classes of discharges (Section 300.6)**.

Minor release (Section 300.6). See **Size classes of releases (Section 300.6)**.

Minority business enterprise (Section 33.005) is a business which is: (1) Certified as socially and economically disadvantaged by the Small Business Administration, (2) certified as a minority business enterprise by a State or Federal agency, or (3) an independent business concern which is at least 51 percent owned and controlled by minority group member(s). A minority group member is an individual who is a citizen of the United States and one of the following: (i) Black American; (ii) Hispanic American (with origins from Puerto Rico, Mexico, Cuba, South or Central America); (iii) Native American (American Indian, Eskimo, Aleut, native Hawaiian), or (iv) Asian-Pacific American (with origins from Japan, China, the Philippines, Vietnam, Korea, Samoa, Guam, the U.S. Trust Territories of the Pacific, Northern Marianas, Laos, Cambodia, Taiwan or the Indian subcontinent).

Minority Business Enterprise (MBE) (Section 35.6015 [6-5-90])* A business which is: (i) Certified as socially and economically disadvantaged by the Small Business Administration; (ii) Certified as a minority business enterprise by a State or Federal agency; or (iii) An independent business concern which is at least 51 percent owned and controlled by minority group member(s). A minority group member is an individual who is a citizen of the United States and one of the following: (A) Black American; (B) Hispanic American (with origins from Puerto Rico, Mexico, Cuba, South or Central America); (C) Native American (American Indian, Eskimo,

Aleut, native Hawaiian); or (D) Asian-Pacific American (with origins from Japan, China, the Philippines, Vietnam, Korea, Samoa, Guam, the U.S. Trust Territories of the Pacific, Northern Marianas, Laos, Cambodia, Taiwan or the Indian subcontinent).

Minority group (Section 8.2) as used herein shall include, where appropriate, female employees and prospective female employees.

Miscellaneous ACM (Section 763.83 [10/30/87]) means miscellaneous material that is ACM in a school building.

Miscellaneous material (Section 763.83 [10/30/87]) means interior building material on structural components, structural members or fixtures, such as floor and ceiling tiles, and does not include surfacing material or thermal system insulation.

Miscellaneous oil spill control agent (Section 300.5 [3-8-90]) is any product, other than a dispersant, sinking agent, surface collecting agent, biological additive, or burning agent, that can be used to enhance oil spill cleanup, removal, treatment, or mitigation.

Miscellaneous unit (Section 260.10 [8/18/92])* means a hazardous waste management unit where hazardous waste is treated, stored, or disposed of and that is not a container, tank, surface impoundment, pile, land treatment unit, landfill, incinerator, boiler, industrial furnace, underground injection well with appropriate technical standards under 40 CFR part 146, containment building, corrective action management unit, or unit eligible for research, development, and demonstration permit under Sec. 270.65.

Miscellaneous waste stream (Section 468.02) shall mean the following additional waste streams related to forming copper: hydrotesting, sawing, surface milling, and maintenance.

Miscellaneous wastewater streams (Section 461.2) shall mean the combined wastewater streams from the process operations listed below for each subcategory. If a plant has one of these streams then the plant receives the entire miscellaneous waste stream allowance. (1) Cadmium subcategory. Cell wash, electrolyte preparation, floor and equipment wash, and employee wash. (2) Lead subcategory. Floor wash, wet air pollution control, battery repair, laboratory, hand wash, and respirator wash. (3) Lithium subcategory. Floor and equipment wash, cell testing, and lithium scrap disposal. (4) Zinc subcategory. Cell wash, electrolyte preparation, employee wash, reject cell handling, floor and equipment wash.

Mischmetal (Section 421.271) refers to a rare earth metal alloy comprised of the natural mixture of rare earths to about 94-99 percent. The balance of the alloy includes traces of other elements and one to two percent iron.

Missile liner (Section 763.163 [7-12-89]) means an asbestos-containing product used as a liner for coating the interior surfaces of rocket motors.

Mist coating (Section 60.721 [1/29/88]). See **Fog coating (Section 60.721)**.

Mitigation (Section 1508.20) includes: (a) Avoiding the impact altogether by not taking a certain action or parts of an action. (b) Minimizing impacts by limiting the degree or magnitude of the action and its implementation. (c) Rectifying the impact by repairing, rehabilitating, or restoring the affected environment. (d) Reducing or eliminating the impact over time by preservation and maintenance operations during the life of the action. (e) Compensating for the impact by replacing or providing substitute resources or environments.

Mixed fertilizer (Section 418.71) shall mean a mixture of wet and/or dry straight fertilizer materials, mixed fertilizer materials, fillers and additives prepared through chemical reaction to a given formulation.

Mixing zone (Section 125.121) means the zone extending from the sea's surface to seabed and extending laterally to a distance of 100 meters in all directions from the discharge point(s) or to the boundary of the zone of initial dilution as calculated by a plume model approved by the director, whichever is greater, unless the director determines that the more restrictive mixing zone or another definition of the mixing zone is more appropriate for a specific discharge.

Mixing zone (Section 230.3) means a limited volume of water serving as a zone of initial dilution in the immediate vicinity of a discharge point where receiving water quality may not meet quality standards or other requirements otherwise applicable to the receiving water. The mixing zone should be considered as a place where wastes and water mix and not as a place where effluents are treated.

Mixture (Section 116.3) means any combination of two or more elements and/or compounds in solid, liquid, or gaseous form except where such substances have undergone a chemical reaction so as to become inseparable by physical means.

Mixture (Section 2.306) has the meaning given it in section 3(8) of the Act, 15 U.S.C. 2602(8).

Mixture (Section 355.20) means a heterogeneous association of substances where the various individual substances retain their identities and can usually be separated by mechanical means. Includes solutions or compounds but does not include alloys or amalgams.

Mixture (Section 372.3 [2/16/88]) means any combination of two or more chemicals, if the combination is not, in whole or in part, the result of a chemical reaction. However, if the combination was produced by a chemical reaction but could have been produced without a chemical reaction, it is also treated as a mixture. A mixture also includes any combination which consists of a chemical and associated impurities.

Mixture (Section 710.2) means any combination of two or more chemical substances if the combination does not occur in nature and is not, in whole or in part, the result of a chemical reaction; except that mixture does include: (1) Any combination which occurs, in whole or in part, as a result of a chemical reaction if the combination could have been manufactured for commercial purposes without a chemical reaction at the time the chemical substances comprising the combination were combined and if, after the effective date or premanufacture notification requirements, none of the chemical substances comprising the combination is a new chemical substance, and (2) hydrates of a chemical substance or hydrated ions formed by association of a chemical substance with water.

Mixture (Section 712.3) means any combination of two or more chemical substances if the combination does not occur in nature and is not, in whole or in part, the result of a chemical reaction; except that mixture does include (1) any combination which occurs, in whole or in part, as a result of a chemical reaction if the combination could have been manufactured for commercial purposes without a chemical reaction at the time the chemical substances comprising the combination were combined, and if all of the chemical substances comprising the combination are included in the EPA, TSCA Chemical Substance Inventory after the effective date of the premanufacture notification requirement under 40 CFR Part 720, and (2) hydrates of a chemical substance or hydrated ions formed by association of a chemical substance with water. The term mixture includes alloys, inorganic glasses, ceramics, frits, and cements, including Portland cement.

Mixture (Section 761.3) means any combination of two or more chemical substances if the combination does not occur in nature and is not, in whole or in part, the result of a chemical reaction; except that such term does include any combination which occurs, in whole or in part, as a result of a chemical reaction if none of the chemical substances comprising the combination is a new chemical substance and if the combination could have been manufactured for commercial purposes without a chemical reaction at the time the chemical substances comprising the combination were combined.

Mixture (Sections 720.3; 723.250) means any combination of two or more chemical substances if the combination does not occur in nature and is not, in whole or in part, the result of a chemical reaction; except mixture does include (1) any combination which occurs, in whole or in part, as a result of a chemical reaction if the combination could have been manufactured for commercial purposes without a chemical reaction at the time the chemical substances comprising the combination were combined, and if all of the chemical substances comprising the combination are not new chemical substances, and (2) hydrates of a chemical substance or hydrated ions formed by association of a chemical substance with water, so long as the nonhydrated form is itself not a new chemical substance.

ml/l (Section 434.11) means milliliters per liter.

Mobile source (Section 117.1) means any vehicle, rolling stock, or other means of transportation which contains or carries a reportable quantity of a hazardous substance.

Model (Section 86.082-2) means a specific combination of car line, body style, and drive-train configuration.

Model line (Section 60.531 [2/26/88]) means all wood heaters offered for sale by a single manufacturer that are similar in all material respects.

Model specific code (Section 205.151) means the designation used for labeling purposes in Sections 205.158 and 205.169 for identifying the motorcycle manufacturer, class, and advertised engine displacement, respectively.

Model type (Section 2.311) has the meaning given it in section 501(11) of the Act, 15 U.S.C. 2001(11).

Model type (Sections 86.082-2; 600.002-85) means a unique combination of car line, basic engine, and transmission class.

Model year (Section 205.151) means the manufacturer's annual production period, which includes January 1 of any calendar year, or if the manufacturer has no annual production period, the term model year shall mean the calendar year.

Model year (Section 600.002-85) means the manufacturer's annual production period (as determined by the Administrator) which includes January 1 of such calendar year. If a manufacturer has no annual production period, the term model year means the calendar year.

Model year (Section 763.163 [7-12-89]) means the manufacturer's annual production period which includes January 1 of such calendar year, provided that if the manufacturer has no annual production period, the term "model year" shall mean the calendar year.

Model year (Section 85.1502

[9/25/87]) means the manufacturer's annual production period (as determined by the Administrator) which includes January 1 of such calendar year; Provided, That if the manufacturer has no annual production period, the term model year shall mean the calendar year in which a vehicle is modified. A certificate holder shall be deemed to have produced a vehicle or engine when the certificate holder has modified the nonconforming vehicle or engine.

Model year (Section 86.082-2) means the manufacturer's annual production period (as determined by the Administrator) which includes January 1 of such calendar year: Provided, That if the manufacturer has no annual production period, the term model year shall mean the calendar year.

Model year (Section 86.402-78) means the manufacturer's annual production period (as determined by the Administrator) which includes January first of such calendar year. If the manufacturer has no annual production period, the term model year shall mean the calendar year.

Model year (Sections 204.51; 205.51) means the manufacturer's annual production period which includes January 1 of such calendar year; Provided, that if the manufacturer has no annual production period, the term model year shall mean the calendar year.

Model Year (Section 85.2102 [6-25-90])* means the manufacturer's annual production period (as determined by the Office Director) which includes January 1 of such calendar year; however, if the manufacturer has no annual production period, the term "model year" shall mean the calendar year.

Modification (Section 60.2) means any physical change in, or change in the method of operation of, an existing facility which increases the amount of any air pollutant (to which a standard applies) emitted into the atmosphere by that facility or which results in the emission of any air pollutant (to which a standard applies) into the atmosphere not previously emitted.

Modification (Section 61.20) as applied to an active underground uranium mine means any major change in the method of operation or mining procedure which will result in an increase in the amount of radon-222 emitted to air. The normal development or operation of an active mine, even though it results in an increase in emissions, is not considered a modification for the purposes of this subpart.

Modification (Section 8.2) means any alteration in the terms and conditions of a contract, including supplemental agreements, amendments and extensions.

Modification or modified source

(Section 52.01) mean any physical change in, or change in the method of operation of, a stationary source which increases the emission rate of any pollutant for which a national standard has been promulgated under Part 50 of this chapter or which results in the emission of any such pollutant not previously emitted, except that: (1) Routine maintenance, repair, and replacement shall not be considered a physical change, and (2) The following shall not be considered a change in the method of operation: (i) An increase in the production rate, if such increase does not exceed the operating design capacity of the source; (ii) An increase in the hours of operation; (iii) Use of an alternative fuel or raw material, if prior to the effective date of a paragraph in this part which imposes conditions on or limits modifications, the source is designed to accommodate such alternative use.

Modified discharge (Section 125.58) means the volume, composition, and location of the discharge proposed by the applicant for which a modification under section 301(h) of the Act is requested. A modified discharge may be a current discharge, improved discharge, or altered discharge.

Modified source (Section 52.01). See **Modification (Section 52.01)**.

Modular excess air MWC (Section 60.51a [2-11-91]) means a combustor that combusts MSW and that is not field-erected and has multiple combustion chambers, all of which are designed to operate at conditions with combustion air amounts in excess of theoretical air requirements.

Modular starved air MWC (Section 60.51a [2-11-91]) means a combustor that combusts MSW and that is not field-erected and has multiple combustion chambers in which the primary combustion chamber is designed to operate at substoichiometric conditions.

Modulated Stem (Section 85.2122(a)(1)(ii)(G)) means a stem attached to the vacuum break diaphragm in such a manner as to allow stem displacement independent of diaphragm displacement.

Modulated Stem Displacement (Section 85.2122(a)(1)(ii)(C)) means the distance through which the modulated stem may move when actuated independent of diaphragm displacement.

Modulated Stem Displacement Force (Section 85.2122(a)(1)(ii)(D)) means the amount of force required at start and finish of a modulated stem displacement.

Molar absorptivity (Section 796.3700) is defined as the proportionality constant in the Beer-Lambert law when the concentration is given in terms of moles per liter (i.e., molar concentration). (Refer to Section 796.3700(2)(iv) for mathematical details.)

Monitoring device (Section 60.2) means the total equipment, required under the monitoring of operations sections in applicable subparts, used to measure and record (if applicable) process parameters.

Monitoring system (Section 61.02) means any system, required under the monitoring sections in applicable subparts, used to sample and condition (if applicable), to analyze, and to provide a record of emissions or process parameters.

Monomer (Sections 704.25; 723.250) means a chemical substance that has the capacity to form links between two or more other molecules.

Monovent (Section 60.61 [12/14/88]) means an exhaust configuration of a building or emission control device (e.g., positive-pressure fabric filter) that extends the length of the structure and has a width very small in relation to its length (i.e., length to width ratio is typically greater than 5:1). The exhaust may be an open vent with or without a roof, louvered vents, or a combination of such features.

Month (Section 60.541 [9/15/87]) means a calendar month or a prespecified period of 28 days or 35 days (utilizing a 4-4-5-week recordkeeping and reporting schedule).

Monthly average (Section 425.02) means the arithmetic average of eight (8) individual data points from effluent sampling and analysis during any calendar month.

Monthly average regulatory values (Section 461.3) shall be the basis for the monthly average discharge in direct discharge permits and for pretreatment standards. Compliance with the monthly discharge limit is required regardless of the number of samples analyzed and averaged.

Montreal Protocol (Section 82.3 [12/10/93])* means the Montreal Protocol on Substances that Deplete the Ozone Layer, a protocol to the Vienna Convention for the Protection of the Ozone Layer, including adjustments adopted by the Parties thereto and amendments that have entered into force.

Motor carrier (Section 202.10) means a common carrier by motor vehicle, a contract carrier by motor vehicle, or a private carrier of property by motor vehicle as those terms are defined by paragraphs (14), (15), and (17) of section 203(a) of the Interstate Commerce Act (49 U.S.C. 303(a)).

Motor Controller (Section 600.002-85) means an electronic or electro-mechanical device to convert energy stored in an energy storage device into a form suitable to power the traction motor.

Motor fuel (Section 280.12 [9/23/88])

means petroleum or a petroleum-based substance that is motor gasoline, aviation gasoline, No. 1 or No. 2 diesel fuel, or any grade of gasohol, and is typically used in the operation of a motor engine.

Motor vehicle (Section 202.10) means any vehicle, machine, tractor, trailer, or semitrailer propelled or drawn by mechanical power and used upon the highways in the transportation of passengers or property, or any combination thereof, but does not include any vehicle, locomotive, or car operated exclusively on a rail or rails.

Motorcycle (Section 205.151) means any motor vehicle, other than a tractor, that: (i) Has two or three wheels; (ii) Has a curb mass less than or equal to 680 kg (1499 lb); and (iii) Is capable, with an 80 kg (176 lb) driver, of achieving a maximum speed of at least 24 km/h (15 mph) over a level paved surface.

Motorcycle (Section 86.402-78) means any motor vehicle with a headlight, taillight, and stoplight and having: two wheels, or three wheels and a curb mass less than or equal to 680 kilograms (1499 pounds).

Motorcycle noise level (Section 205.151) means the A-weighted noise level of a motorcycle as measured by the acceleration test procedure.

Move laterally (in soils) (Section 162.3) means to undergo transfer through soil generally in a horizontal plane from the original site of application or use by physical, chemical, or biological means.

Movement (Section 260.10) means that hazardous waste transported to a facility in an individual vehicle.

mpc (Section 427.71) shall mean 1000 pieces of floor tile.

M_r (Section 60.741 [9-11-89]) means the total mass (kg) of VOC recovered for a nominal 1-month period.

MSA (Section 60.331). See **Metropolitan Statistical Area (Section 60.331)**.

MSBu (Sections 406.11; 406.41) shall mean 1000 standard bushels.

MSDS (Section 370.2 [10/15/87]). See **Material Safety Data Sheet (Section 370.2)**.

MSDS (Section 721.3 [7-27-89]) means material safety data sheet, the written listing of data for the chemical substance as required under Sec. 721.72(c).

MTP (Section 141.2). See **Maximum Total Trihalomethane Potential (Section 141.2)**.

Muffler (Section 202.10) means a device for abating the sound of escaping gases of an internal combustion engine.

Multiple stands (Section 420.101) means those recirculation or direct application cold rolling mills which include more than one stand of work rolls.

Multiple-effect evaporator system (Section 60.281) means the multiple-effect evaporators and associated condenser(s) and hotwell(s) used to concentrate the spent cooking liquid that is separated from the pulp (black liquor).

Municipal solid waste landfill (MSWLF) unit (Section 257.2 [3/19/93])* means a discrete area of land or an excavation that receives household waste, and that is not a land application unit, surface impoundment, injection well, or waste pile, as those terms are defined in this section. A MSWLF unit also may receive other types of RCRA Subtitle D wastes, such as commercial solid waste, nonhazardous sludge, and industrial solid waste. Such a landfill may be publicly or privately owned. An MSWLF unit may be a new MSWLF unit, an existing MSWLF unit or a lateral expansion.

Municipal solid waste landfill unit (Section 258.2 [10/1/93])* means a discrete area of land or an excavation that receives household waste, and that is not a land application unit, surface impoundment, injection well, or waste pile, as those terms are defined under Sec. 257.2. A MSWLF unit also may receive other types of RCRA subtitle D wastes, such as commercial solid waste, nonhazardous sludge, conditionally exempt small quantity generator waste and industrial solid waste. Such a landfill may be publicly or privately owned. A MSWLF unit may be a new MSWLF unit, an existing MSWLF unit or a lateral expansion.

Municipal solid wastes (Section 761.3) means garbage, refuse, sludges, wastes, and other discarded materials resulting from residential and non-industrial operations and activities, such as household activities, office functions, and commercial housekeeping wastes.

Municipal solid wastes (Sections 240.101; 241.101) means normally, residential and commercial solid wastes generated within a community.

Municipal-type solid waste or MSW (Section 60.51a [2-11-91]) means household, commercial/retail, and/or institutional waste. Household waste includes material discarded by single and multiple residential dwellings, hotels, motels, and other similar permanent or temporary housing establishments or facilities. Commercial/retail waste includes material discarded by stores, offices, restaurants, warehouses, nonmanufacturing activities at industrial facilities, and other similar establishments or facilities. Institutional waste includes material discarded by schools, hospitals, nonmanufacturing activities at prisons and government facilities and other similar establishments or facilities. Household, commercial/retail, and institutional waste do not include sewage, wood pallets, construction and demolition wastes, industrial process or manufacturing wastes, or motor vehicles (including motor vehicle parts or vehicle fluff). Municipal-type solid waste does include motor vehicle maintenance materials, limited to vehicle batteries, used motor oil, and tires. Municipal type solid waste does not include wastes that are solely segregated medical wastes. However, any mixture of segregated medical wastes and other wastes which contains more than 30 percent waste medical waste discards, is considered to be municipal-type solid waste.

Municipal waste combustor or MWC or MWC unit (Section 60.51a [2-11-91]) means any device that combusts, solid, liquid, or gasified MSW including, but not limited to, field-erected incinerators (with or without heat recovery), modular incinerators (starved air or excess air), boilers (i.e., steam generating units), furnaces (whether suspension-fired, grate-fired, mass-fired, or fluidized bed-fired) and gasification/combustion units. This does not include combustion units, engines, or other devices that combust landfill gases collected by landfill gas collection systems.

Municipal-type solid waste (Section 60.41b [12/16/87])* means refuse, more than 50 percent of which is waste consisting of a mixture of paper, wood, yard wastes, food wastes, plastics, leather, rubber, and other combustible materials, and noncombustible materials such as glass and rock.

Municipality (Section 122.2) means a city, town, borough, county, parish, district, association, or other public body created by or under State law and having jurisdiction over disposal of sewage, industrial wastes, or other wastes, or an Indian tribe or an authorized Indian tribal organization, or a designated and approved management agency under section 208 of CWA.

Municipality (Section 142.2 [9/26/88])* means a city, town, or other public body created by or pursuant to State law, or an Indian Tribe which does not meet the requirements of Subpart H of this part.

Municipality (Section 35.2005) is a city, town, borough, county, parish, district, association, or other public body (including an intermunicipal agency of two or more of the foregoing entities) created under State law, or an Indian tribe or an authorized Indian tribal organization, having jurisdiction over disposal of sewage, industrial wastes, or other waste, or a designated and approved management agency under section 208 of the Act. (i) This definition includes a special district created under State law such as a water district, sewer district, sanitary district, utility district, drainage district or similar entity or an integrated waste management facility, as defined in section 201(e) of the Act, which has as one of its principal responsibilities the treatment, transport, or disposal of domestic wastewater in a particular geographic area. (ii) This definition excludes the following: (A) Any revenue producing entity which has as its principal responsibility an activity other than providing wastewater treatment services to the general public, such as an airport, turnpike, port facility or other municipal utility. (B) Any special district (such as school district or a park district) which has the responsibility to provide wastewater treatment services in support of its principal activity at specific facilities, unless the special district has the responsibility under State law to provide wastewater treatment services to the community surrounding the special district's facility and no other municipality, with concurrent jurisdiction to serve the community, serves or intends to serve the special district's facility or the surrounding community.

Municipality (Section 35.905) means a city, town, borough, county, parish, district, association, or other public body (including an intermunicipal agency of two or more of the foregoing entities) created under State law, or an Indian tribe or an authorized Indian tribal organization, having jurisdiction over disposal of sewage, industrial wastes, or other waste, or a designated and approved management agency under section 208 of the Act. (a) This definition includes a special district created under State law such as a water district, sewer district, sanitary district, utility district, drainage district, or similar entity or an integrated waste management facility, as defined in section 201(e) of the Act, which has as one of its principal responsibilities the treatment, transport, or disposal of liquid wastes of the general public in a particular geographic area. (b) This definition excludes the following: (1) Any revenue producing entity which has as its principal responsibility an activity other than providing wastewater treatment services to the general public, such as an airport, turnpike, port facility, or other municipal utility. (2) Any special district (such as school district or a park district) which has the responsibility to provide wastewater treatment services in support of its principal activity at specific facilities, unless the special district has the responsibility under State law to provide waste water treatment services to the community surrounding the special district's facility and no other municipality, with concurrent jurisdiction to serve the community, serves or intends to serve the special district's facility or the surrounding community.

Municipality (Section 501.2 [5/2/89]) means a city, town, borough, county, parish, district, association, or other public body (including an intermunicipal agency of two or more of the foregoing entities) created under State law (or an Indian tribe or an authorized Indian tribal organization), or a designated and approved management agency under section 208 of the Clean Water Act. This definition includes a special district created under State law such as a water district, sewer district, sanitary district, utility district, drainage district, or similar entity, or an integrated waste management facility as defined in section 201(e) of the CWA, as amended, that has as one of its principal responsibilities the treatment, transport, or disposal of sewage sludge.

Mushrooms (Section 407.71) shall mean the processing of mushrooms into the following product styles: Canned, frozen, dehydrated, all varieties, shapes and sizes.

Mutagenic (Section 162.3) means the property of a substance or mixture of substances to induce changes in the genetic complement of either somatic or germinal tissue in subsequent generations.

MWC acid gases (Section 60.51a [2-11-91]) means all acid gases emitted in the exhaust gases from MWC units including, but not limited to, sulfur dioxide and hydrogen chloride gases.

MWC metals (Section 60.51a [2-11-91]) means metals and metal compounds emitted in the exhaust gases from MWC units.

MWC organics (Section 60.51a [2-11-91]) means organic compounds emitted in the exhaust gases from MWC units and includes total tetra- through octa-chlorinated dibenzo-p-dioxins and dibenzofurans.

MWC plant (Section 60.31a [2-11-91]) means one or more MWC units at the same location for which construction, modification, or reconstruction is commenced on or before December 20, 1989.

MWC plant (Section 60.51a [2-11-91]) means one or more MWC units at the same location for which construction, modification, or reconstruction is commenced after December 20, 1989.

MWC plant capacity (Section 60.31a [2-11-91]) means the aggregate MWC unit capacity of all MWC units at an MWC plant for which construction, modification, or reconstruction is commenced on or before December 20, 1989.

MWC plant capacity (Section 60.51a [2-11-91]) means the aggregate MWC unit capacity of all MWC units at an MWC plant for which construction, modification, or reconstruction commenced after December 20, 1989. Any MWC units for which construction, modification, or reconstruction is commenced on or before December 20, 1989, are not included for determining applicability under this subpart.

Mwh (Section 424.11) shall mean megawatt hour(s) of electrical energy consumed in the smelting process (furnace power consumption).

N

NAAQS (Section 57.103). See **National Ambient Air Quality Standards (Section 57.103).**

NAAQS and National Ambient Air Quality Standards (Section 57.103 [2-13-92])* unless otherwise specified, refer only to the National Primary and Secondary Ambient Air Quality Standards for sulfur dioxide.

NAMS (Section 58.1) means National Air Monitoring Station(s). Collectively the NAMS are a subset of the SLAMS ambient air quality monitoring network.

Naphthalene (or priority pollutant No. 55) (Section 420.02) means the value obtained by the standard method Number 610 specified in 44 FR 69464, 69571 (December 3, 1979).

Naphthalene processing (Section 61.131 [9-14-89]) means any operations required to recover naphthalene including the separation, refining, and drying of crude or refined naphthalene.

National Air Monitoring Station(s) (Section 58.1). See **NAMS (Section 58.1).**

National Ambient Air Quality Standards (NAAQS) (Section 57.103), unless otherwise specified, refers only to the National Primary and Secondary Ambient Air Quality Standards for sulfur dioxide.

National Contingency Plan (Section 310.11 [10/21/87]) means the National Oil and Hazardous Substances Pollution Contingency Plan (40 CFR Part 300).

National Contingency Plan or NCP (Section 306.12) means the National Oil and Hazardous Substances Pollution Contingency Plan developed under section 311(c) of the Clean Water Act and revised pursuant to section 105 of CERCLA (40 CFR Part 300).

National Contingency Plan or NCP (Section 305.12) means the National Oil and Hazardous Substances Pollution Contingency Plan, developed under section 331(c) of the Clean Water Act and revised pursuant to section 105 of CERCLA (40 CFR Part 300).

National Contingency Plan or NCP (Section 304.12 [5/30/89]) means the National Oil and Hazardous Substances Pollution Contingency Plan, developed under section 311(c)(2) of the Federal Water Pollution Control Act, 33 U.S.C. 1251, et seq., as amended, revised periodically pursuant to section 105 of CERCLA, 42 U.S.C. 9605, and published at 40 CFR Part 300.

National Panel of Environmental Arbitrators or Panel (Section 304.12 [5/30/89]) means a panel of environmental arbitrators selected and maintained by the Association to arbitrate cost recovery claims under this part.

National Pollutant Discharge Elimination System (NPDES) (Section 136.2) means the national system for the issuance of permits under section 402 of the Act and includes any State or interstate program which has been approved by the Administrator, in whole or in part, pursuant to section 402 of the Act.

National Pollutant Discharge Elimination System (NPDES) (Sections 122.2; 270.2) means the national program for issuing, modifying, revoking and reissuing, terminating, monitoring and enforcing permits, and imposing and enforcing pretreatment requirements, under sections 307, 402, 318, and 405 of CWA. The term includes an approved program.

National Pretreatment Standard or Pretreatment Standard (Section 117.1) means any regulation containing pollutant discharge limits promulgated by the EPA in accordance with section 307(b) and (c) of the Act, which applies to industrial users of a publicly owned treatment works. It further means any State or local pretreatment requirement applicable to a discharge and which is incorporated into a permit issued to a publicly owned treatment works under section 402 of the Act.

National Pretreatment Standard or Pretreatment Standard or Standard (Section 403.3) means any regulation containing pollutant discharge limits promulgated by the EPA in accordance with section 307 (b) and (c) of the Act, which applies to Industrial Users. This term includes prohibitive discharge limits established pursuant to Section 403.5.

National primary drinking water regulation (Section 142.2) means any primary drinking water regulation contained in Part 141 of this chapter.

National Priorities List (NPL) (Section 300.5 [3-8-90]) means the list, complied by EPA pursuant to CERCLA section 105, of uncontrolled hazardous substance releases in the United States that arc priorities for long-term remedial evaluation and response.

National Priorities List (NPL) (Section 35.6015 [6-5-90]) EPA's list of the most serious uncontrolled or abandoned hazardous waste sites identified for possible long-term remedial action under Superfund. A site must be on the NPL to receive money from the Trust Fund for remedial action. The list is based primarily on the score a site receives from the Hazard Ranking System.

National Program Assistance

Agreements (Section 35.9010 [10-3-89]) Assistance Agreements approved by the EPA Assistant Administrator for Water for work undertaken to accomplish broad NEP goals and objectives.

National Response Center (Section 310.11 [1/15/93])* means the national communications center located in Washington, DC, that receives and relays notice of oil discharge or releases of hazardous substances to appropriate Federal officials.

National security exemption (Sections 204.2: 205.2) means an exemption from the prohibitions of section 10(a) (1), (2), (3), and (5) of the Act, which may be granted under section 10(b)(1) of the Act for the purpose of national security.

National security exemption (Section 85.1702) means an exemption which may be granted under section 203(b)(1) of the Act for the purpose of national security.

National security exemption (Section 211.102) means an exemption from the prohibitions of section 10(a)(3) and (5) of the Act, which may be granted under section 10(b)(1) of the Act in cases involving national security.

National security information (Section 11.4) as used in this order, is synonymous with classified information. It is any information which must be protected against unauthorized disclosure in the interest of the national defense or foreign relations of the United States.

National standard (Section 51.100) means either a primary or secondary standard.

Nations complying with, but not joining, the Protocol (Section 82.3 [12/10/93])* means any nation listed in appendix C , Annex 2, of this subpart.

Natural barrier (Section 61.141 [11-20-90]) means a natural object that effectively precludes or deters access. Natural barriers include physical obstacles such as cliffs, lakes or other large bodies of water, deep and wide ravines, and mountains. Remoteness by itself is not a natural barrier.

Natural Conditions (Section 51.301) includes naturally occurring phenomena that reduce visibility as measured in terms of visual range, contrast, or coloration.

Natural draft opening (Section 60.741 [9-11-89]) means any opening in a room, building, or total enclosure that remains open during operation of the facility and that is not connected to a duct in which a fan is installed. The rate and direction of the natural draft across such an opening is a consequence of the difference in pressures on either side of the wall or barrier containing the opening.

Natural draft opening (Section 60.711 [10/3/88]) means any opening in a room, building, or total enclosure that remains open during operation of the facility and that is not connected to a duct in which a fan is installed. The rate and direction of the natural draft across such an opening is a consequence of the difference in pressures on either side of the wall containing the opening.

Natural gas (Section 60.41b [12/16/87])* means (1) a naturally occurring mixture of hydrocarbon and nonhydrocarbon gases found in geologic formations beneath the earth's surface, of which the principal constituent is methane; or (2) liquid petroleum gas, as defined by the American Society for Testing and Materials in ASTM D1835-82, "Standard Specification for Liquid Petroleum Gases" (IBR - see Section 60.17).

Natural gas (Section 60.41c [9-12-90]) means (1) a naturally occurring mixture of hydrocarbon and nonhydrocarbon gases found in geologic formations beneath the earth's surface, of which the principal constituent is methane, or (2) liquefied petroleum (LP) gas, as defined by the American Society for Testing and Materials in ASTM D1835-86, "Standard Specification for Liquefied Petroleum Gases" (incorporated by reference--see Sec. 60.17).

Natural gas (Section 60.641) means a naturally occurring mixture of hydrocarbon and nonhydrocarbon gases found in geologic formations beneath the earth's surface. The principal hydrocarbon constituent is methane.

Natural gas liquids (Section 60.631) means the hydrocarbons, such as ethane, propane, butane, and pentane, that are extracted from field gas.

Natural gas processing plant (gas plant) (Section 60.631) means any processing site engaged in the extraction of natural gas liquids from field gas, fractionation of mixed natural gas liquids to natural gas products, or both.

Natural resources (Section 300.5 [3-8-90])* means land, fish, wildlife, biota, air, water, ground water, drinking water supplies, and other such resources belonging to, managed by, held in trust by appertaining to, or otherwise controlled by the United States (including the resources of the exclusive economic zone defined by the Magnuson Fishery Conservation and Management Act of 1976), any state or local government, any foreign government, any Indian tribe, or, if such resources are subject to a trust restriction on alienation, any member of an Indian tribe.

Natural resources (Section 305.12) means land, fish, wildlife, biota, air, water, ground water, drinking water supplies, and other such resources belonging to, managed by, held in trust by, appertaining to, or otherwise controlled by the United States (including the resources of the fishery conservation zone established by the Magnuson Fishery Conservation and Management Act of 1976), any State or local government, or any foreign government.

Natural resources (Section 306.12) means land, fish, wildlife, biota, air, water, ground water, drinking water supplies, and other such resources belonging to, managed by, held in trust by, appertaining to, or otherwise controlled by the United States (including the resources of the fishery conservation zone established by the Magnuson Fishery Conservation and Management Act), any State or local government, or any foreign government.

Navigable water (Section 300.82), for the purposes of this subpart, means the water of the United States, including the territorial seas.

Navigable waters (Section 110.1) means the waters of the United States, including the territorial seas. The term includes: (a) All waters that are currently used, were used in the past, or may be susceptible to use in interstate or foreign commerce, including all waters that are subject to the ebb and flow of the tide; (b) Interstate waters, including interstate wetlands; (c) All other waters such as intrastate lakes, rivers, streams (including intermittent streams), mudflats, sandflats, and wetlands, the use, degradation, or destruction of which would affect or could affect interstate or foreign commerce including any such waters: (1) That are or could be used by interstate or foreign travelers for recreational or other purposes; (2) From which fish or shellfish are or could be taken and sold in interstate or foreign commerce; (3) That are used or could be used for industrial purposes by industries in interstate commerce; (d) All impoundments of waters otherwise defined as navigable waters under this section; (e) Tributaries of waters identified in paragraphs (a) through (d) of this section, including adjacent wetlands; and (f) Wetlands adjacent to waters identified in paragraphs (a) through (e) of this section: Provided, That waste treatment systems (other than cooling ponds meeting the criteria of this paragraph) are not waters of the United States.

Navigable waters (Section 112.2 [8/25/93])* of the United States means "navigable waters" as defined in section 502(7) of the FWPCA, and includes: (1) All navigable waters of the United States, as defined in judicial decisions prior to passage of the 1972 Amendments to the FWPCA (Pub. L. 92-500), and tributaries of such waters; (2) Interstate waters; (3) Intrastate lakes, rivers, and streams which are utilized by interstate travelers for recreational or other purposes; and (4) Intrastate lakes, rivers, and streams from which fish or shellfish are taken and sold in interstate commerce. Navigable waters do not include prior converted cropland. Notwithstanding the determination of an area's status as prior converted cropland by any other federal agency, for the purposes of the Clean Water Act, the final authority regarding Clean Water Act jurisdiction remains with EPA.

Navigable waters (Section 116.3 [8/25/93])* do not include prior converted cropland. Notwithstanding the determination of an area's status as prior converted cropland by any other federal agency, for the purposes of the Clean Water Act, the final authority regarding Clean Water Act jurisdiction remains with EPA.

Navigable waters (Section 117.1) means waters of the United States, including the territorial seas. This term includes: (1) All waters which are currently used, were used in the past, or may be susceptible to use in interstate or foreign commerce, including all waters which are subject to the ebb and flow of the tide; (2) Interstate waters, including interstate wetlands; (3) All other waters such as intrastate lakes, rivers, streams (including intermittent streams), mudflats, sandflats, and wetlands, the use, degradation or destruction of which would affect or could affect interstate or foreign commerce including any such waters: (i) which are or could be used by interstate or foreign travelers for recreational or other purposes; (ii) From which fish or shellfish are or could be taken and sold in interstate or foreign commerce; (iii) Which are used or could be used for industrial purposes by industries in interstate commerce; (4) All impoundments of waters otherwise defined as navigable waters under this paragraph; (5) Tributaries of waters identified in paragraphs (i)(1) through (4) of this section, including adjacent wetlands; and (6) Wetlands adjacent to waters identified in paragraphs (i)(1) through (5) of this section. (Wetlands means those areas that are inundated or saturated by surface or ground water at a frequency and duration sufficient to support, and that under normal circumstances do support, a prevalence of vegetation typically adapted for life in saturated soil conditions. Wetlands generally included playa lakes, swamps, marshes, bogs, and similar areas such as sloughs, prairie potholes, wet meadows, prairie river overflows, mudflats, and natural ponds.): Provided, That waste treatment systems (other than cooling ponds meeting the criteria of this paragraph) are not waters of the United States.

Navigable waters (Section 300.5 [3-8-90]) as defined by 40 CFR 110.1, means the waters of the United States, including the territorial seas. The term includes: (a) All waters that are currently used, were used in the past, or may be susceptible to use in interstate or foreign commerce, including all waters that are subject to the ebb and flow of the tide; (b) interstate waters, including interstate wetlands; (c) All other waters such as intrastate lakes, rivers, streams (including intermittent streams), mudflats, sandflats, and wetlands, the use, degradation, or destruction of which would affect or could affect interstate or foreign commerce including any such wastes: (1) That are or could be used by interstate or foreign travelers for recreational or other purposes; (2) From which fish or shellfish are or could be taken and sold in interstate or foreign commerce; (3) That are used or could be used for industrial purposes by industries in interstate commerce; (d) All impoundments of waters otherwise defined as navigable waters under this section; (e) Tributaries of waters identified in paragraphs (a) through (d) of this definition, including adjacent wetlands; and (f) Wetlands adjacent to waters identified in paragraphs (a) through (e) of this definition: Provided, that waste treatment systems (other than cooling ponds meeting the criteria of this paragraph) are not waters of the United States.

Navigable waters (Section 401.11 [8/25/93])* includes: All navigable waters of the United States; tributaries of navigable waters of the United States; interstate waters; intrastate lakes, rivers, and streams which are utilized by interstate travelers for recreational or other purposes; intrastate lakes, rivers, and streams from which fish or shellfish are taken and sold in interstate commerce; and intrastate lakes, rivers, and streams which are utilized for industrial purposes by industries in interstate commerce. Navigable waters do not include prior converted cropland. Notwithstanding the determination of an area's status as prior converted cropland by any other federal agency, for the purposes of the Clean Water Act, the final authority regarding Clean Water Act jurisdiction remains with EPA.

Navigable waters or navigable waters of the United States (Section 302.3) means waters of the United States, including the territorial seas.

NCP (Section 304.12 [5/30/89]). See **National Contingency Plan (Section 304.12)**.

NCP (Section 305.12). See **National Contingency Plan (Section 305.12)**.

NCP (Section 306.12). See **National Contingency Plan (Section 306.12)**.

NCP (Section 86.1102-87) means a nonconformance penalty as described in section 206(g) of the Clean Air Act and in this subpart.

Near the first service connection (Section 141.2 [6/29/89]) means at one of the 20 percent of all service connections in the entire system that are nearest the water supply treatment facility, as measured by water transport time within the distribution system.

Nearby (Section 51.100) as used in Section 51.100(ii) of this part, is defined for a specific structure or terrain feature and (1) For purposes of applying the formulae provided in Section 51.100(ii)(2) means that distance up to five times the lesser of the height or the width dimension of a structure, but not greater than 0.8 km (1/2 mile), and (2) For conducting demonstrations under Section 51.100(ii)(3) means not greater than 0.8 km (1/2 mile), except that the portion of a terrain feature may be considered to be nearby which falls within a distance of up to 10 times the maximum height (H_t) of the feature, not to exceed 2 miles if such feature achieves a height (H_t) 0.8 km from the stack that is at least 40 percent of the GEP stack height determined by the formulae provided in Section 51.100(ii)(2)(ii) of this part or 26 meters, whichever is greater, as measured from the ground-level elevation at the base of the stack. The height of the structure or terrain feature is measured from the ground-level elevation at the base of the stack.

Neat oil (Section 467.02) is a pure oil with no or few impurities added. In aluminum forming its use is mostly as a lubricant.

Neat oil (Section 471.02) is a pure oil with no or few impurities added. In nonferrous metals forming, its use is mostly as a lubricant.

Neat soap (Sections 417.11; 417.31) shall mean the solution of completely saponified and purified soap containing about 20-30 percent water which is ready for final formulation into a finished product.

Necessary and adequate (Section 21.2), for purposes of paragraph 7(g)(2) of the Small Business Act, refers to additions, alterations, or methods of operation in the absence of which a small business concern could not comply with one or more applicable standards. This can be determined with reference to design specifications provided by manufacturers, suppliers, or consulting engineers; including, without limitations, additions, alterations, or methods of operation the design specifications of which will provide a measure of treatment or abatement of pollution in excess of that required by the applicable standard.

Necessary preconstruction approvals or permits (Section 51.165) means those Federal air quality control laws and regulations and those air quality control laws and regulations which are part of the applicable State Implementation Plan.

Necessary preconstruction approvals or permits (Sections 51.166; 52.21; 52.24) means those permits or approvals required under federal air quality control laws and regulations and those air quality control laws and regulations which are part of the applicable State Implementation Plan.

Negative-pressure fabric filter (Section 60.271a) means a fabric filter with the fans on the downstream side of the filter bags.

Negligible residue (Section 180.1) means any amount of a pesticide chemical remaining in or on a raw agricultural commodity or group of raw agricultural commodities that would result in a daily intake regarded as toxicologically insignificant on the basis of scientific judgment of adequate safety data. Ordinarily this will add to the diet an amount which will be less than 1/2,000th of the amount that has been demonstrated to have no effect from feeding studies on the most sensitive animal species tested. Such toxicity studies shall usually include at least 90-day feeding studies in two species of mammals.

Neighboring company (Section 60.41a) means any one of those electric utility companies with one or more electric power interconnections to the principal company and which have geographically adjoining service areas.

Nematicides (Section 162.3) includes all substances or mixtures of substances intended for preventing or inhibiting the multiplication or establishment of, preventing or mitigating the adverse effects of, repelling, or destroying any members of the Class Nematoda of the Phylum Nemathelminthes declared to be pests under Section 162.14. (i) Nematicides include but are not limited to, plant parasitic nematode control agents intended for use in or on plants, plant parts, soil, or certain infested agricultural commodities or articles. (ii) Nematicides do not include those substances or mixtures of substances subject to the provisions of the Federal Food, Drug and Cosmetic Act, as amended (21 U.S.C. 301 et seq.), such as substances or mixtures of substances intended for use in preventing reproduction of, inactivating, or destroying nematodes in living man or other animals.

NEPA process (Section 1508.21) means all measures necessary for compliance with the requirements of section 2 and Title I of NEPA.

Net (Section 409.21) shall mean the addition of pollutants.

Net cane (Section 409.61) shall mean that amount of gross cane less the weight of extraneous material.

Net emissions increase (Section 52.24) means (i) the amount by which the sum of the following exceeds zero: (a) Any increase in actual emissions from a particular physical change or change in the method of operation at a stationary source; and (b) Any other increases and decreases in actual emissions at the source that are contemporaneous with the particular change and are otherwise creditable. (ii) An increase or decrease in actual emissions is contemporaneous with the increase from the particular change only if it occurs between: (a) The date five years before construction on the particular change commences and (b) The date that the increase from the particular change occurs. (iii) An increase or decrease in actual emissions is creditable only if the Administrator has not relied on it in issuing a permit for the source under regulations approved pursuant to 40 CFR Subpart I which permit is in effect when the increase in actual emissions from the particular change occurs. (iv) An increase in actual emissions is creditable only to the extent that the new level of actual emissions exceeds the old level. (v) A decrease in actual emissions is creditable only to the extent that: (a) The old level of actual emissions or the old level of allowable emissions, whichever is lower, exceeds the new level of actual emissions; (b) It is federally enforceable at and after the time that construction on the particular change begins; and (c) The Administrator or reviewing authority has not relied on it in issuing any permit under regulations approved pursuant to 40 CFR Subpart I or the State has not relied on it in demonstrating attainment or reasonable further progress. (d) It has approximately the same qualitative significance for public health and welfare as that attributed to the increase from the particular change. (vi) An increase that results from a physical change at a source occurs when the emissions unit on which construction occurred becomes operational and begins to emit a particular pollutant. Any replacement unit that requires shakedown becomes operational only after a reasonable shakedown period, not to exceed 180 days.

Net emissions increase (Section 51.165) means (A) The amount by which the sum of the following exceeds zero: (1) Any increase in actual emissions from a particular physical change or change in the method of operation at a stationary source; and (2) Any other increases and decreases in actual emissions at the source that are contemporaneous with the particular change and are otherwise creditable. (B) An increase or decrease in actual emissions is contemporaneous with the increase from the particular change only if it occurs before the date that the increase from the particular change occurs; (C) An increase or decrease in actual emissions is creditable only if: (1) It occurs within a reasonable period to be specified by the reviewing authority; and (2) The reviewing authority has not relied on it in issuing a permit for the source under regulations approved pursuant to this section which permit is in effect when the increase in actual emissions from the particular change occurs; (D) An increase in actual emissions is creditable only to the extent that the new level of actual emissions exceeds the old level; (E) A decrease in actual emissions is creditable only to the extent that: (1) The old level of actual emission or the old level of allowable emissions whichever is lower, exceeds the new level of actual emissions; (2) It is federally enforceable at and after the time that actual construction on the particular change begins; and (3) The reviewing authority has not relied on it in issuing any permit under regulations approved pursuant to 40 CFR Part 51 Subpart I or the state has not relied on it in demonstrating attainment or reasonable further progress; (4) It has approximately the same qualitative significance for public health and welfare as that attributed to the increase from the particular change; (F) An increase that results from a physical change at a source occurs when the emissions unit on which construction occurred becomes operational and begins to emit a particular pollutant. Any replacement unit that requires shakedown becomes operational only after a reasonable shakedown period, not to exceed 180 days.

Net emissions increase (Section 52.21) means (i) the amount by which the sum of the following exceeds zero: (a) Any increase in actual emissions from a particular physical change or change in method of operation at a stationary source; and (b) Any other increases and decreases in actual emissions at the source that are contemporaneous with the particular change and are otherwise creditable. (ii) An increase or decrease in actual emissions is contemporaneous with the increase from the particular change only if it occurs between: (a) The date five years before construction on the particular change commences; and (b) The date that the increase from the particular change occurs. (iii) An increase or decrease in actual emissions is creditable only if the Administrator has not relied on it in issuing a permit for the source under this section, which permit is in effect when the increase in actual emissions from the particular change occurs. (iv) An increase or decrease in actual emissions of sulfur dioxide or particulate matter which occurs before the applicable baseline date is creditable only if it is required to be considered in calculating the amount of maximum allowable increases remaining available. (v) An increase in actual emissions is creditable only to the extent that the new level of actual emissions exceeds the old level. (vi) A decrease in actual emissions is creditable only to the extent that: (a) The old level of actual emissions or the old level of allowable emissions, whichever is lower, exceeds the new level of actual emissions; (b) It is federally enforceable at and after the time that actual construction on the particular change begins; and (c) It has approximately the same qualitative significance for public health and welfare as that attributed to the increase from the particular change. (viii) An increase that results from a physical change at a source occurs when the emissions unit on which construction occurred becomes operational and begins to emit a particular pollutant. Any replacement unit that requires shakedown becomes operational only after a reasonable shakedown period, not to exceed 180 days.

Net emissions increase (Section 51.166) means (i) the amount by which the sum of the following exceeds zero: (a) Any increase in actual emissions from a particular physical change or change in the method of operation at a stationary source; and (b) Any other increases and decreases in actual emissions at the source that are contemporaneous with the particular change and are otherwise creditable. (ii) An increase or decrease in actual emissions is contemporaneous with the increase from the particular change only if it occurs within a reasonable period (to be specified by the state) before the date that the increase from the particular change occurs. (iii) An increase or decrease in actual emissions is creditable only if the reviewing authority has not relied on it in issuing a permit for the source under regulations approved pursuant to this section, which permit is in effect when the increase in actual emissions from the particular change occurs. (iv) An increase or decrease in actual emissions of sulfur dioxide or particulate matter which occurs before the applicable baseline date is creditable only if it is required to be considered in calculating the amount of maximum allowable increases remaining available. (v) An increase in actual emissions is creditable only to the extent that the new level of actual emissions exceeds the old level. (vi) A decrease in actual emissions is creditable only to the extent that: (a) The old level of actual emissions or the old level of allowable emissions, whichever is lower, exceeds the new level of actual emissions; (b) It is federally enforceable at and after the time that actual construction on the particular change begins; and (c) It has approximately the same qualitative significance for public health and welfare as that attributed to the increase from the particular change. (vii) An increase that results from a physical change at a source occurs when the emissions unit on which construction occurred becomes operational and begins to emit a particular pollutant. Any replacement unit that requires shakedown becomes operational only after a reasonable shakedown period, not to exceed 180 days.

Net system capacity (Section 60.41a) means the sum of the net electric generating capability (not necessarily equal to rated capacity) of all electric generating equipment owned by an electric utility company (including steam generating units, internal combustion engines, gas turbines, nuclear units, hydroelectric units, and all other electric generating equipment) plus firm contractual purchases that are interconnected to the affected facility that has the malfunctioning flue gas desulfurization system. The electric generating capability of equipment under multiple ownership is prorated based on ownership unless the proportional entitlement to electric output is otherwise established by contractual arrangement.

Net working capital (Section 264.141), as specified in Section 264.141(f), means current assets minus current liabilities.

Net working capital (Section 265.141), as specified in Section 265.141(f), means current assets minus current liabilities.

Net working capital (Section 144.61) means current assets minus current liabilities.

Net worth (Section 144.61) means total assets minus total liabilities and is equivalent to owner's equity.

Net worth (Section 264.141), as specified in Section 264.141(f), means total assets minus total liabilities and is equivalent to owner's equity.

Net worth (Section 265.141), as specified in Section 265.141(f), means total assets minus total liabilities and is equivalent to owner's equity.

Neutral sulfite semichemical pulping operation (Section 60.281) means any operation in which pulp is produced from wood by cooking (digesting) wood chips in a solution of sodium sulfite and sodium bicarbonate, followed by mechanical defibrating (grinding).

Neutralization (Section 420.91) means those acid pickling operations that do not include acid recovery or acid regeneration processes.

New aircraft turbine engine (Section 87.1) means an aircraft gas turbine engine which has never been in service.

New biochemical and microbial registration review (Section 152.403 [5/26/88]) means review of an application for registration of a biochemical or microbial pesticide product containing a biochemical or microbial active ingredient not contained in any other pesticide product that is registered under FIFRA at the time the application is made. For purposes of this Subpart, the definitions of biochemical and microbial pesticides contained in Section 158.65 (a) and (b) of this chapter shall apply.

New chemical (Section 166.3) means an active ingredient not contained in any currently registered pesticide.

New chemical registration review (Section 152.403 [5/26/88]) means review of an application for registration of a pesticide product containing a chemical active ingredient which is not contained as an active ingredient in any other pesticide product that is registered under FIFRA at the time the application is made.

New chemical substance (Section

710.2) means any chemical substance which is not included in the inventory compiled and published under subsection 8(b) of the Act.

New chemical substance (Sections 720.3; 723.50; 723.250; 747.200) means any chemical substance which is not included on the Inventory.

New Class II Wells (Section 147.2902) means wells constructed or converted after the effective date of this program, or which are under construction on the effective date of this program.

New discharger (Sections 122.2; 122.29) means any building, structure, facility, or installation: (a) From which there is or may be a discharge of pollutants; (b) That did not commence the discharge of pollutants at a particular site prior to August 13, 1979; (c) Which is not a new source; and (d) Which has never received a finally effective NPDES permit for discharges at that site. This definition includes an indirect discharger which commences discharging into waters of the United States after August 13, 1979. It also includes any existing mobile point source (other than an offshore or coastal oil and gas exploratory drilling rig or a coastal oil and gas developmental drilling rig) such as a seafood processing rig, seafood processing vessel, or aggregate plant, that begins discharging at a site for which it does not have a permit; and any offshore or coastal mobile oil and gas exploratory drilling rig or coastal mobile oil and gas developmental drilling rig that commences the discharge of pollutants after August 13, 1979, at a site under EPA's permitting jurisdiction for which it is not covered by an individual or general permit and which is located in an area determined by the Regional Administrator in the issuance of a final permit to be an area of biological concern. In determining whether an area is an area of biological concern, the Regional Administrator shall consider the factors specified in 40 CFR 125.122(a)(1) through (10). An offshore or coastal mobile exploratory drilling rig or coastal mobile developmental drilling rig will be considered a new discharger only for the duration of its discharge in an area of biological concern.

New facility (Section 260.10). See **New hazardous waste management facility (Section 260.10)**.

New hazardous waste management facility or new facility (Section 260.10) means a facility which began operation, or for which construction commenced after October 21, 1976. (See also Existing hazardous waste management facility.)

New HWM facility (Section 270.2) means a Hazardous Waste Management facility which began operation or for which construction commenced after November 19, 1980.

New independent power production facility (Section 73.3 [12-17-91]) means, for purposes of this part, a unit(s) that: (1) Commences commercial operation on or after November 15, 1990; (2) Is nonrecourse project-financed, as defined in 10 CFR part 715; (3) Sells 80% of electricity generated at wholesale; and (4) Does not sell electricity to any affiliate or, if it does, demonstrates it cannot obtain the required allowances from such an affiliate.

New injection wells (Section 144.3) means an injection well which began injection after a UIC program for the State applicable to the well is approved or prescribed.

New MSWLF unit (Section 258.2 [10/1/93])* means any municipal solid waste landfill unit that has not received waste prior to October 9, 1993, or prior to October 9, 1995 if the MSWLF unit meets the conditions of Sec. 258.1(f)(1).

New product (Section 162.152) means a pesticide product which is not a federally registered product.

New product (Section 205.2) means (i) a product the equitable or legal title of which has never been transferred to an ultimate purchaser, or (ii) a product which is imported or offered for importation into the United States and which is manufactured after the effective date of a regulation under section 6 or 8 which would have been applicable to such product had it been manufactured in the United States.

New source (Section 129.2) means any source discharging a toxic pollutant, the construction of which is commenced after proposal of an effluent standard or prohibition applicable to such source if such effluent standard or prohibition is thereafter promulgated in accordance with section 307.

New source (Section 401.11) means any building, structure, facility or installation from which there is or may be the discharge of pollutants, the construction of which is commenced after the publication of proposed regulations prescribing a standard of performance under Section 306 of the Act which will be applicable to such source if such standard is thereafter promulgated in accordance with Section 306 of the Act.

New source (Section 403.3 [10/17/88])* means any building, structure, facility or installation from which there is or may be a discharge of pollutants, the construction of which commenced after the publication of proposed Pretreatment Standards under section 307(c) of the Act which will be applicable to such source if such Standards are thereafter promulgated in accordance with that section, provided that: (i) The building, structure, facility or installation is constructed at a site at which no other source is located; or (ii) The building, structure, facility or installation totally replaces the process or production equipment that causes the discharge of pollutants at an existing source; or (iii) The production or wastewater generating processes of the building, structure, facility or installation are substantially independent of an existing source at the same site. In determining whether these are substantially independent, factors such as the extent to which the new facility is integrated with the existing plant, and the extent to which the new facility is engaged in the same general type of activity as the existing source should be considered.

New source (Section 435.11 [3/4/93]*; 435.41) means any facility or activity of this subcategory that meets the definition of "new source" under 40 CFR 122.2 and meets the criteria for determination of new sources under 40 CFR 122.29(b) applied consistently with all of the following definitions.

New source (Section 61.02) means any stationary source, the construction or modification of which is commenced after the publication in the *Federal Register* of proposed national emission standards for hazardous air pollutants which will be applicable to such source.

New source (Sections 122.2; 122.29) means any building, structure, facility, or installation from which there is or may be a discharge of pollutants, the construction of which commenced: (a) After promulgation of standards of performance under section 306 of CWA which are applicable to such source, or (b) After proposal of standards of performance in accordance with section 306 of CWA which are applicable to such source, but only if the standards are promulgated in accordance with section 306 within 120 days of their proposal.

New source coal mine (Section 434.11) (1) Notwithstanding any other provision of this Chapter, subject to paragraph (j)(2) of this section means a coal mine (excluding coal preparation plants and coal preparation plant associated areas) including an abandoned mine which is being re-mined: (i) The construction of which is commenced after May 4, 1984; or (ii) Which is determined by the EPA Regional Administrator to constitute a major alteration. In making this determination, the Regional Administrator shall take into account whether one or more of the following events resulting in a new, altered or increased discharge of pollutants has occurred after May 4, 1984 in connection with the mine for which the NPDES permit is being considered: (A) Extraction of a coal seam not previously extracted by that mine; (B) Discharge into a drainage area not previously affected by wastewater discharge from the mine; (C) Extensive new surface disruption at the mining operation; (D) A construction of a new shaft, slope, or drift; and (E) Such other factors as the Regional Administrator deems relevant. (2) No provision in this part shall be deemed to affect the classification as a new source of a facility which was classified as a new source coal mine under previous EPA regulations, but would not be classified as a new source under this section, as modified. Nor shall any provision in this part be deemed to affect the standards applicable to such facilities, except as provided in Section 434.65 of this chapter.

New tailings (Section 61.251) means uranium tailings produced after the effective date of this rule.

New tailings impoundment (Section 61.251) means any location or structure at which uranium mill tailings are temporarily or permanently stored and which is placed in operation after the promulgation of this rule.

New tank component (Section 260.10). See **New tank system (Section 260.10)**.

New tank system (Section 280.12 [9/23/88]) means a tank system that will be used to contain an accumulation of regulated substances and for which installation has commenced after December 22, 1988. (See also Existing Tank System.)

New tank system or new tank component (Section 260.10) means a tank system or component that will be used for the storage or treatment of hazardous waste and for which installation has commenced after July 14, 1986; except, however, for purposes of Section 264.193(g)(2) and Section 265.193(g)(2), a new tank system is one for which construction commences after July 14, 1986. (See also existing tank system.)

New use (Section 152.3 [5/4/88]) when used with respect to a product containing a particular active ingredient, means: (1) Any proposed use pattern that would require the establishment of, the increase in, or the exemption from the requirement of, a tolerance or food additive regulation under section 408 or 409 of the Federal Food, Drug and Cosmetic Act; (2) Any aquatic, terrestrial, outdoor, or forestry use pattern, if no product containing the active ingredient is currently registered for that use pattern; or (3) Any additional use pattern that would result in a significant increase in the level of exposure, or a change in the route of exposure, to the active ingredient of man or other organisms.

New use pattern registration review (Section 152.403 [5/26/88]) means review of an application for registration, or for amendment of a registration entailing a major change to the use pattern of an active ingredient contained in a product registered under FIFRA or pending Agency decision on a prior application at the time of application. For purposes of this paragraph, examples of major changes include but are not limited to, changes from non-food to food use, outdoor to indoor use, ground to aerial application, terrestrial to aquatic use, and non-residential to residential use.

New uses of asbestos (Section 763.163 [7-12-89]) means commercial uses of asbestos not identified in Sec. 763.165 the manufacture, importation or processing of which would be initiated for the first time after August 25, 1989. The following products are also not new uses of asbestos: acetylene cylinders, arc chutes, asbestos diaphragms, battery separators, high grade electrical paper, missile liner, packing, reinforced plastic, sealant tape, specialty industrial gaskets, and textiles.

New vessel (Section 140.1) refers to any vessel on which construction was initiated on or after January 30, 1975.

New water (Section 440.141 [5/24/88]) means water from any discrete source such as a river, creek, lake or well which is deliberately allowed or brought into the plant site.

New well (Section 146.61 [7/26/88]) means any Class I hazardous waste injection well which is not an existing well.

Newsprint (Section 250.4 [10/6/87, 6/22/88]) means paper of the type generally used in the publication of newspapers or special publications like the *Congressional Record.* It is made primarily from mechanical wood pulps combined with some chemical wood pulp.

Nickel (Section 420.02) means total nickel and is determined by the method specified in 40 CFR 136.3.

Nickel (Sections 415.361; 415.471; 415.651) shall mean the total nickel present in the process wastewater stream exiting the wastewater treatment system.

NIOSH (Section 300.5) is the National Institute for Occupational Safety and Health.

NIOSH (Section 721.3 [7-27-89]) means the National Institute for Occupational Safety and Health of the U.S. Department of Health and Human Services.

Nitric acid plant (Section 51.100) means any facility producing nitric acid 30 to 70 percent in strength by either the pressure or atmospheric pressure process.

Nitric acid production unit (Section 60.71) means any facility producing weak nitric acid by either the pressure or atmospheric pressure process.

Nitrogen oxides (Section 60.2) means all oxides of nitrogen except nitrous oxide, as measured by test methods set forth in this part.

Nitrosating agent (Section 747.200) means any substance that has the potential to transfer a nitrosyl group (-

NO) to a secondary or tertiary amine to form the corresponding nitrosamine.

Nitrosating agent (Sections 747.115; 747.195) means any substance that has the potential to transfer a nitrosyl group (-NO) to a primary, secondary, or tertiary amine to form the corresponding nitrosamine.

NMFS (Section 232.2 [2/11/93])* means the National Marine Fisheries Service.

NMFS (Section 233.2 [6/6/88]) means the National Marine Fisheries Service.

No detectable emissions (Section 61.341 [3-7-90]) means less than 500 parts per million by volume (ppmv) above background levels, as measured by a detection instrument reading in accordance with the procedures specified in Sec. 61.355(h) of this subpart.

No detectable emissions (Section 60.691 [11/23/88]) means less than 500 ppm above background levels, as measured by a detection instrument in accordance with Method 21 in Appendix A of 40 CFR Part 60.

No discernible adverse effect (Section 162.3) means no adverse effect observable within the limitations and sensitivity specified in the Registration Guidelines.

No discharge of free oil (Section 435.11 [3/4/93])* shall mean that waste streams may not be discharged when they would cause a film or sheen upon or a discoloration of the surface of the receiving water or fail the static sheen test defined in Appendix 1 to 40 CFR 435, subpart A.

No observed effect concentration (NOEC) (Section 797.1600) is the highest tested concentration in an acceptable early life stage test: (i) which did not cause the occurrence of any specified adverse effect (statistically different from the control at the 95 percent level); and (ii) below which no tested concentration caused such an occurrence.

No reasonable alternatives (Section 125.121) means: (1) No land-based disposal sites, discharge point(s) within internal waters, or approved ocean dumping sites within a reasonable distance of the site of the proposed discharge the use of which would not cause unwarranted economic impacts on the discharger, or, notwithstanding the availability of such sites, (2) On-site disposal is environmentally preferable to other alternative means of disposal after consideration of: (i) The relative environmental harm of disposal on-site, in disposal sites located on land, from discharge point(s) within internal waters, or in approved ocean dumping sites, and (ii) The risk to the environment and human safety posed by the transportation of the pollutants.

No-effect level/No-toxic-effect level/No-adverse-effect level/No-

observed-effect level (Section 798.2650) is the maximum dose used in a test which produces no observed adverse effects. A no-observed-effect level is expressed in terms of the weight of a substance given daily per unit weight of test animal (mg/kg). When administered to animals in food or drinking water the no-observed-effect level is expressed as mg/kg of food or mg/ml of water.

No-effect level/No-toxic-effect level/No-adverse-effect level/No-observed-effect level (Section 798.2450) is the maximum dose used in a test which produces no observed adverse effects. A no-observed-effect level is expressed in terms of weight or volume of test substance given daily per unit volume of air (mg/l or ppm).

No-effect level/No-toxic-effect level/No-adverse-effect level/No-observed-effect level (Section 798.2250) is the maximum dose used in a test which produces no observed adverse effects. A no-observed-effect level is expressed in terms of the weight of a test substance given daily per unit weight of test animal (mg/kg).

No-observed-effect level (Section 798.4350) is the maximum concentration in a test which produces no observed adverse effects. A no-observed-effect level is expressed in terms of weight or volume of test substance given daily per unit volume of air.

NO2 (Section 58.1 [2/12/93])* means nitrogen dioxide. NO means nitrogen oxide. NOX means oxides of nitrogen and is defined as the sum of the concentrations of NO2 and NO.

NOAA (Section 300.5) is the National Oceanic and Atmospheric Administration.

Nodulizing kiln (Section 61.121). See **Calciner (Section 61.121)**.

NOEC (Section 797.1600). See **No observed effect concentration (Section 797.1600)**.

Noise control system (Section 205.51) includes any vehicle part, component or system the primary purpose of which is to control or cause the reduction of noise emitted from a vehicle.

Noise control system (Section 205.151) means any vehicle part, component or system, the purpose of which includes control or the reduction of noise emitted from a vehicle, including all exhaust system components.

Noise emission standard (Section 205.151) means the noise levels in Section 205.152 or Section 205.166.

Noise emission test (Section 204.51) means a test conducted pursuant to the measurement methodology specified in Section 204.54.

Noise emission test (Sections 205.51; 205.151) means a test conducted pursuant to a measurement methodology specified in this subpart.

Noise Reduction Rating (NRR) (Section 211.203) means a single number noise reduction factor in decibels, determined by an empirically derived technique which takes into account performance variation of protectors in noise reducing effectiveness due to differing noise spectra, fit variability and the mean attenuation of test stimuli at the one-third octave band test frequencies.

Nominal 1-month period (Section 60.741 [9-11-89]) means a calendar month or, if established prior to the performance test in a statement submitted with notification of anticipated startup pursuant to 40 CFR 60.7(a)(2), a similar monthly time period (e.g., 30-day month or accounting month).

Nominal 1-month period (Section 60.711 [10/3/88]) means a calendar month or, if established prior to the performance test in a statement submitted with notification of anticipated startup pursuant to 40 CFR 60.7(a)(2), a similar monthly time period (e.g., 30-day month or accounting month).

Nominal 1-month period (Section 60.721 [1/29/88]) means either a calendar month, 30-day month, accounting month, or similar monthly time period that is established prior to the performance test (i.e., in a statement submitted with notification of anticipated actual startup pursuant to 40 CFR 60.7(2)).

Nominal concentration (Section 158.153 [5/4/88]) means the amount of an ingredient which is expected to be present in a typical sample of a pesticide product at the time the product is produced, expressed as a percentage by weight.

Nominal fuel tank capacity (Section 86.082-2) means the volume of the fuel tank(s), specified by the manufacturer to the nearest tenth of a U.S. gallon, which may be filled with fuel from the fuel tank filler inlet.

Non-integrated refueling emission control system (Section 86.098-2 [4/6/94])* means a system where fuel vapors from refueling are stored in a vapor storage unit assigned solely to the function of storing refueling vapors.

Non-regenerative carbon adsorber (Section 61.131 [9-19-91]) means a series, over time, of non-regenerative carbon beds applied to a single source or group of sources, where non-regenerative carbon beds are carbon beds that are either never regenerated or are moved from their location for regeneration.

Non-commercial scientific institution (Section 2.100 [1/5/88]) refers to an institution that is not operated on a commercial basis as that term is referenced in paragraph (e) of this section, and which is operated solely for the purpose of conducting scientific research the results of which are not intended to promote any particular product or industry.

Non-community water system (Section 141.2) means a public water system that is not a community water system.

Non-contact cooling water (Sections 418.21; 418.51) shall mean water which is used in a cooling system designed so as to maintain constant separation of the cooling medium from all contact with process chemicals but which may on the occasion of corrosion, cooling system leakage or similar cooling system failures contain small amounts of process chemicals: Provided, that all reasonable measures have been taken to prevent, reduce, eliminate and control to the maximum extent feasible such contamination: And provided further, that all reasonable measures have been taken that will mitigate the effects of such contamination once it has occurred.

Non-contact cooling water system (Section 60.691 [11/23/88]) means a once-through drain, collection and treatment system designed and operated for collecting cooling water which does not come into contact with hydrocarbons or oily wastewater and which is not recirculated through a cooling tower.

Non-continuous discharger (Section 431.11) is a mill which is prohibited by the NPDES authority from discharging pollutants during specific periods of time for reasons other than treatment plant upset control, such periods being at least 24 hours in duration. A mill shall not be deemed a non-continuous discharger unless its permit, in addition to setting forth the prohibition described above, requires compliance with the effluent limitations established by this subpart for non-continuous dischargers and also requires compliance with maximum day and average of 30 consecutive days effluent limitations. Such maximum day and average of 30 consecutive days effluent limitations for non-continuous dischargers shall be established by the NPDES authority in the form of concentrations which reflect wastewater treatment levels that are representative of the application of the best practicable control technology currently available, the best conventional pollutant control technology, or new source performance standards in lieu of the maximum day and average of 30 consecutive days effluent limitations for conventional pollutants set forth in this subpart.

Non-continuous discharger (Section 464.02) is a plant which does not discharge pollutants during specific periods of time for reasons other than treatment plant upset, such periods being at least 24 hours in duration. A typical example of a non-continuous discharger is a plant where wastewaters are routinely stored for periods in excess of 24 hours to be treated on a batch basis. For non-continuous discharging direct discharging plants, NPDES permit authorities shall apply the mass-based annual average effluent limitations or standards and the concentration-based maximum day and maximum for monthly average effluent limitations or standards established in the regulations. POTWs may elect to establish concentration-based standards for non-continuous discharges to POTWs. They may do so by establishing concentration-based pretreatment standards equivalent to the mass-based standards provided in Sections 464.15, 464.16, 464.25, 464.26, 464.35, 464.36, 464.45, and 464.46 of the regulations. Equivalent concentration standards may be established by following the procedures outlined in Section 464.03(b).

Non-continuous discharger (Section 430.01) is a mill which is prohibited by the NPDES authority from discharging pollutants during specific periods of time for reasons other than treatment plant upset control, such periods being at least 24 hours in duration. A mill shall not be deemed a non-continuous discharger unless its permit, in addition to setting forth the prohibition described above, requires compliance with the effluent limitations established for non-continuous dischargers and also requires compliance with maximum day and average of 30 consecutive days effluent limitations. Such maximum day and average of 30 consecutive days effluent limitations for non-continuous dischargers shall be established by the NPDES authority in the form of concentrations which reflect wastewater treatment levels that are representative of the application of the best practicable control technology currently available, the best conventional pollutant control technology, or new source performance standards in lieu of the maximum day and average of 30 consecutive days effluent limitations for conventional pollutants set forth in each subpart.

Non-emission related maintenance (Section 86.084-2) means that maintenance which does not substantially affect emissions and which does not have a lasting effect on the deterioration of the vehicle or engine with respect to emissions once the maintenance is performed at any particular date. This definition applies beginning with the 1984 model year.

Non-enclosed process (Section 721.3 [7-27-89]) means any equipment system (such as an open-top reactor, storage tank, or mixing vessel) in which a chemical substance is manufactured, processed, or otherwise used where significant direct contact of the bulk chemical substance and the workplace air may occur.

Non-industrial use (Section 721.3 [7-27-89]) means use other than at a facility where chemical substances or mixtures are manufactured, imported, or processed.

Non-isolated intermediate (Section 712.3) means any intermediate that is not intentionally removed from the equipment in which it is manufactured, including the reaction vessel in which it is manufactured, equipment which is ancillary to the reaction vessel, and any equipment through which the substance passes during a continuous flow process, but not including tanks or other vessels in which the substance is stored after its manufacture. (See also paragraph (f) of this section.)

Non-isolated intermediate (Section 704.3 [12/22/88])* means any intermediate that is not intentionally removed from the equipment in which it is manufactured, including the reaction vessel in which it is manufactured, equipment which is ancillary to the reaction vessel, and any equipment through which the substance passes during a continuous flow process, but not including tanks or other vessels in which the substance is stored after its manufacture. Mechanical or gravity transfer through a closed system is not considered to be intentional removal, but storage or transfer to shipping containers isolates the substance by removing it from process equipment in which it is manufactured.

Non-oxygenated hydrocarbon (Section 86.090-2 [4/11/89]) means organic emissions measured by a flame ionization detector excluding methanol. This definition applies beginning with the 1990 model year.

Non-PCB Transformer (Section 761.3 [7/19/88]) means any transformer that contains less than 50 ppm PCB; except that any transformer that has been converted from a PCB Transformer or a PCB-Contaminated transformer cannot be classified as a non-PCB Transformer until reclassification has occurred, in accordance with the requirements of Section 761.30(a)(2)(v).

Non-roof coating (Section 763.163 [7-12-89]) means an asbestos-containing product intended for use as a coating, cement, adhesive, or sealant and not intended for use on roofs. Major applications of this product include: liquid sealants; semi-liquid glazing, caulking and patching compounds; asphalt-based compounds; epoxy adhesives; butyl rubber sealants; vehicle undercoatings; vinyl sealants; and compounds containing asbestos fibers that are used for bonding, weather proofing, sound deadening, sealing, coating; and other such applications.

Non-target organism (Section 171.2) means a plant or animal other than the one against which the pesticide is applied.

Non-transient non-community water system (NTNCWS) (Section 141.2

[7/8/87]) means a public water system that is not a community water system and that regularly serves at least 25 of the same persons over 6 months per year.

Non-transportation-related and Transportation-related (Section 112.2) as applied to an onshore or offshore facility, are defined in the Memorandum of Understanding between the Secretary of Transportation and the Administrator of the Environmental Protection Agency, dated November 24, 1971, 36 FR 24080.

Noncommercial education broadcasting entities (Section 47.105 [3-9-92]) means any noncommercial educational broadcasting station (and/or its legal nonprofit affiliates) as defined and licensed by the Federal Communications Commission.

Noncommercial purposes (Section 280.12 [9/23/88]) with respect to motor fuel means not for resale.

Nonconformance penalty (Section 86.1102-87). See **NCP (Section 86.1102-87)**.

Nonconforming engine (Section 85.1502 [9/25/87]). See **Nonconforming vehicle (Section 85.1502)**.

Nonconforming vehicle or engine (Section 85.1502 [9/25/87]) is a motor vehicle or motor vehicle engine which is not covered by a certificate of conformity prior to final or conditional importation and which has not been finally admitted into the United States under the provisions of Section 85.1505, Section 85.1509, or the applicable provisions of Section 85.1512. Excluded from this definition are vehicles admitted under provisions of Section 85.1512 covering EPA approved manufacturer and U.S. Government Agency catalyst and O_2 sensor control programs.

Nonconsumer article (Section 762.3) means any article subject to TSCA which is not a consumer product within the meaning of the Consumer Product Safety Act (CPSA), 15 U.S.C. 2052.

Noncontact cooling water (Section 401.11) means water used for cooling which does not come into direct contact with any raw material, intermediate product, waste product or finished product.

Noncontact cooling water pollutants (Section 401.11) means pollutants present in noncontact cooling waters.

Noncontinental area (Section 60.41c [9-12-90]) means the State of Hawaii, the Virgin Islands, Guam, American Samoa, the Commonwealth of Puerto Rico, or the Northern Mariana Islands.

Noncontinental area (Sections 60.41a; 60.41b [12/16/87]) means the State of Hawaii, the Virgin Islands, Guam, American Samoa, the Commonwealth of Puerto Rico, or the Northern Mariana Islands.

Noncorrugating medium furnish subdivision mills (Section 430.51) are mills where recycled corrugating medium is not used in the production of paperboard.

Nondomestic construction material (Section 35.936.13) means a construction material other than a domestic construction material.

Nonexcessive infiltration (Section 35.2005) means the quantity of flow which is less than 120 gallons per capita per day (domestic base flow and infiltration) or the quantity of infiltration which cannot be economically and effectively eliminated from a sewer system as determined in a cost-effectiveness analysis. (See Sections 35.2005(b)(16) and 35.2120.)

Nonexcessive inflow (Section 35.2005) means the maximum total flow rate during storm events which does not result in chronic operational problems related to hydraulic overloading of the treatment works or which does not result in a total flow of more than 275 gallons per capita per day (domestic base flow plus infiltration plus inflow). Chronic operational problems may include surcharging, backups, bypasses, and overflows. (See Sections 35.2005(b)(16) and 35.2120.)

Nonexpendable personal property (Section 30.200) means personal property with a useful life of at least two years and an acquisition cost of $500 or more.

Nonferrous metal (Section 471.02) is any pure metal other than iron or any metal alloy for which a metal other than iron is its major constituent in percent by weight.

Nonfractionating plant (Section 60.631) means any gas plant that does not fractionate mixed natural gas liquids into natural gas products.

Nonfriable (Section 763.83 [10/30/87]) means material in a school building which when dry may not be crumbled, pulverized, or reduced to powder by hand pressure.

Nonfriable asbestos-containing material (Section 61.141 [11-20-90]) means any material containing more than 1 percent asbestos as determined using the method specified in appendix A, subpart F, 40 CFR part 763, section 1, Polarized Light Microscopy, that, when dry, cannot be crumbled, pulverized, or reduced to powder by hand pressure.

Nongaseous losses (Section 60.601) means the solvent that is not volatilized during fiber production, and that escapes the process and is unavailable for recovery, or is in a form or concentration unsuitable for economical recovery.

Nonimpervious solid surfaces (Section 761.123) means solid surfaces which are porous and are more likely to absorb spilled PCBs prior to completion of the cleanup requirements prescribed in this policy. Nonimpervious solid surfaces include, but are not limited to, wood, concrete, asphalt, and plasterboard.

Nonindustrial source (Section 125.58) means any source of pollutants which is not an industrial source.

Nonisolated intermediate (Section 720.3) means any intermediate that is not intentionally removed from the equipment in which it is manufactured, including the reaction vessel in which it is manufactured, equipment which is ancillary to the reaction vessel, and any equipment through which the chemical substance passes during a continuous flow process, but not including tanks or other vessels in which the substance is stored after its manufacture.

Nonisolated intermediate (Section 710.23) means any intermediate that is not intentionally removed from the equipment in which it is manufactured, including the reaction vessel in which it is manufactured, equipment which is ancillary to the reaction vessel, and any equipment through which the substance passes during a continuous flow process, but not including tanks or other vessels in which the substance is stored after its manufacture.

Nonmetallic mineral (Section 60.671) means any of the following minerals or any mixture of which the majority is any of the following minerals: (a) Crushed and broken stone, including limestone, dolomite, granite, traprock, sandstone, quartz, quartzite, marl, marble, slate, shale, oil shale, and shell (b) Sand and gravel (c) Clay including kaolin, fireclay, bentonite, fuller's earth, ball clay, and common clay (d) Rock salt (e) Gypsum (f) Sodium compounds, including sodium carbonate, sodium chloride, and sodium sulfate (g) Pumice (h) Gilsonite (i) Talc and pyrophyllite (j) Boron, including borax, kernite, and colemanite (k) Barite (l) Fluorospar (m) Feldspar (n) Diatomite (o) Perlite (p) Vermiculite (q) Mica (r) Kyanite, including andalusite, sillimanite, topaz, and dumortierite.

Nonmetallic mineral processing plant (Section 60.671) means any combination of equipment that is used to crush or grind any nonmetallic mineral wherever located, including lime plants, power plants, steel mills, asphalt concrete plants, portland cement plants, or any other facility processing nonmetallic minerals except as provided in Section 60.670(b) and (c).

Nonpassenger Automobile (Section 600.002-85) means an automobile that is not a passenger automobile, as defined by the Secretary of Transportation at 49 CFR 523.5.

Nonperishable raw agricultural

commodity (Section 180.1) means any raw agricultural commodity not subject to rapid decay or deterioration that would render it unfit for consumption. Examples are cocoa beans, coffee beans, field-dried beans, field-dried peas, grains, and nuts. Not included are eggs, milk, meat, poultry, fresh fruits, and vegetables such as onions, parsnips, potatoes, and carrots.

Nonprocess waste water (Section 428.11) shall be classified as water used only for tread cooling.

Nonrestricted access areas (Section 761.123) means any area other than restricted access, outdoor electrical substations, and other restricted access locations, as defined in this section. In addition to residential/commercial areas, these areas include unrestricted access rural areas (areas of low density development and population where access is uncontrolled by either man-made barriers or naturally occurring barriers, such as rough terrain, mountains, or cliffs).

Nonscheduled renovation operation (Section 61.141 [11-20-90]) means a renovation operation necessitated by the routine failure of equipment, which is expected to occur within a given period based on past operating experience, but for which an exact date cannot be predicted.

Nonsudden accidental occurrence (Section 264.141), as specified in Section 264.141(g), means an occurrence which takes place over time and involves continuous or repeated exposure.

Nonsudden accidental occurrence (Section 265.141), as specified in Section 265.141(g), means an occurrence which takes place over time and involves continuous or repeated exposure.

Nontarget organisms (Section 162.3) means those flora and fauna (including man) that are not intended to be controlled, injured, killed or detrimentally affected in any way by a pesticide.

Nonvapor tight (Section 61.301 [3-7-90]) means any tank truck, railcar, or marine vessel that does not pass the required vapor-tightness test.

Nonwastewaters (Section 268.2 [6-1-90]) arc wastes that do not meet the criteria for wastewaters in paragraph (g)(6) of this section.

Normal ambient value (Section 228.2) means that concentration of a chemical species reasonably anticipated to be present in the water column, sediments, or biota in the absence of disposal activities at the disposal site in question.

Normal liquid detergent operations (Section 417.161) shall mean all such operations except those defined as fast turnaround operation of automated fill lines.

Normal operation of a spray tower (Section 417.151) shall mean operation utilizing formulations that present limited air quality problems from stack gases and associated need for extensive wet scrubbing, and without more than 6 turnarounds in a 30 consecutive day period, thus permitting essentially complete recycle of waste water.

Not detectable emissions (Section 61.341 [3-7-90]) means less than 500 parts per million by volume (ppmv) above background levels, as measured by a detection instrument reading in accordance with the procedures specified in section 61.355(h) of this subpart.

Not-for-profit organization (Section 47.105 [3-9-92]) means an organization, association, or institution described in section 501(c)(3) of the Internal Revenue Code of 1986, which is exempt from taxation pursuant to the provisions of section 501(a) of such Code.

Note (Section 39.105) means an evidence of the debt, including a bond, obligation to pay, or other evidence of indebtedness where appropriate.

Notice (Section 8.33) means a notice of hearing.

Notice of claim (Section 306.12) means a written notice of intent to file a claim in accordance with Section 306.22 of this part.

Notice of intent (Section 1508.22) means a notice that an environmental impact statement will be prepared and considered. The notice shall briefly: (a) Describe the proposed action and possible alternatives. (b) Describe the agency's proposed scoping process including whether, when, and where any scoping meeting will be held. (c) State the name and address of a person within the agency who can answer questions about the proposed action and the environmental impact statement.

NPDES (Section 110.1) means National Pollutant Discharge Elimination System.

NPDES (Section 124.2) means National Pollutant Discharge Elimination System.

NPDES (Section 133.101). National Pollutant Discharge Elimination System.

NPDES (Section 136.2). See **National Pollutant Discharge Elimination System (Section 136.2)**.

NPDES (Sections 122.2; 270.2). See **National Pollutant Discharge Elimination System (Sections 122.2; 270.2)**.

NPDES Permit or Permit (Section 403.3) means a permit issued to a POTW pursuant to section 402 of the Act.

NPDES State (Section 403.3) means a State (as defined in 40 CFR 122.2) or Interstate water pollution control agency with an NPDES permit program approved pursuant to section 402(b) of the Act.

NPL (Section 306.12) means the National Priorities List established under the NCP.

NPL (Section 35.6015 [1/27/89]). See **National Priorities List (Section 35.6015)**.

NRC (Section 300.5) is the National Response Center.

NRC-licensed facility (Section 61.101 [12-14-89])* means any facility licensed by the Nuclear Regulatory Commission (NRC) or any Agreement State to receive title to, receive, possess, use, transfer, or deliver any source, by-product, or special nuclear material.

NRR (Section 211.203). See **Noise Reduction Rating (Section 211.203)**.

NRT (Section 300.5) is the National Response Team.

NSF (Section 300.5) is the National Strike Force.

NSPS (Section 467.02) means new source performance standards under section 306 of the Act.

NTNCWS (Section 141.2 [7/8/87]). See **Non-transient non-community water system (Section 141.2)**.

Nuclear fuel cycle (Section 190.02) means the operations defined to be associated with the production of electrical power for public use by any fuel cycle through utilization of nuclear energy.

Number-average molecular weight (Section 723.250) means the arithmetic average (mean) of the molecular weight of all molecules in a polymer.

Nursery (Section 170.3 [8/21/92])* means any operation engaged in the outdoor production of any agricultural plant to produce cut flowers and ferns or plants that will be used in their entirety in another location. Such plants include, but are not limited to, flowering and foliage plants or trees; tree seedlings; live Christmas trees; vegetable, fruit, and ornamental transplants; and turfgrass produced for sod.

NWPA (Section 191.02) means the Nuclear Waste Policy Act of 1982 (Pub. L. 97-425).

O

O&G (Section 420.02). See **Oil and grease (Section 420.02)**.

O&M (Section 35.6015 [1/27/89]). See **Operation and maintenance (Section 35.6015)**.

O_3 (Section 58.1) means ozone.

Observation period (Section 797.2175) is the portion of the test that begins after the test birds have been dosed and extends at least 14 days.

Observed effect concentration (OEC) (Section 797.1600) is the lowest tested concentration in an acceptable early life stage test: (i) which caused the occurrence of any specified adverse effect (statistically different from the control at the 95 percent level); and (ii) above which all tested concentrations caused such an occurrence.

Occurrence (Section 280.92 [10/26/88]) means an accident, including continuous or repeated exposure to conditions, which results in a release from an underground storage tank. Note: This definition is intended to assist in the understanding of these regulations and is not intended either to limit the meaning of occurrence in a way that conflicts with standard insurance usage or to prevent the use of other standard insurance terms in place of occurrence.

Ocean dumping (Section 165.1) means the disposal of pesticides in or on the oceans and seas, as defined in Pub. L. 92-532.

Ocean or ocean waters (Section 220.2) means those waters of the open seas lying seaward of the baseline from which the territorial sea is measured, as provided for in the Convention on the Territorial Sea and the Contiguous Zone (15 UST 1606; TIAS 5639); this definition includes the waters of the territorial sea, the contiguous zone and the oceans as defined in section 502 of the FWPCA.

Ocean waters (Section 125.58) means those coastal waters landward of the baseline of the territorial seas, the deep waters of the territorial seas, or the waters of the contiguous zone.

Ocean waters (Section 220.2). See **Ocean (Section 220.2)**.

OCR (Section 7.25). See **Office of Civil Rights (Section 7.25)**.

Octane number, (R + M) /2 method (Section 80.2) means measurement of a gasoline's antiknock characteristics which is obtained by dividing the sum of the Research Octane Number and the Motor Octane Number by two, as explained by the American Society for Testing and Materials (ASTM) in ASTM D 439-81, entitled "Standard

Specifications for Automotive Gasoline." The Research Octane Number is determined by ASTM standard test method D 2699-80 and the Motor Octane Number is determined by ASTM standard test method D 2700-81. ASTM standards D 439-81, D 2699-80 and D 2700-81 are incorporated by reference. They are available from ASTM, 1916 Race Street, Philadelphia, PA, 19103, and are also available for inspection as part of Docket EN- 81-11, located at the Central Docket Section, EPA, Gallery I, West Tower, 401 M Street, SW, Washington DC 20460. Standard D 439-81 is contained in the Annual Book of ASTM Standards, Part 23; standards D 2699-80 and D 2700-81 are contained in Part 47. This incorporation by reference was approved by the Director of the *Federal Register* on January 19, 1983. These materials are incorporated as they exist on the date of the approval and a notice of any change in these materials will be published in the *Federal Register*.

Octave Band Attenuation (Section 211.203) means the amount of sound reduction determined according to the measurement procedure of Section 211.206 for one-third octave bands of noise.

OEC (Section 797.1600). See **Observed effect concentration (Section 797.1600)**.

OEM (Section 85.1502 [9/25/87]). See **Original equipment manufacturer (Section 85.1502)**.

Off-kg (off-lb) (Section 471.02) means the mass of metal or metal alloy removed from a forming operation at the end of a process cycle for transfer to a different machine or process.

Off-kilogram (off-pound) (Section 468.02) shall mean the mass of copper or copper alloy removed from a forming or ancillary operation at the end of a process cycle for transfer to a different machine or process.

Off-kilogram (off-pound) (Section 467.02) shall mean the mass of aluminum or aluminum alloy removed from a forming or ancillary operation at the end of a process cycle for transfer to a different machine or process.

Off-lb (Section 471.02). See **Off-kg (Section 471.02)**.

Off-pound (Section 467.02). See **Off-kilogram (Section 467.02)**.

Off-pound (Section 468.02). See **Off-kilogram (Section 468.02)**.

Off-road motorcycle (Section 205.151) means any motorcycle that is not a street motorcycle or competition motorcycle.

Off-site (Section 270.2) means any site which is not on-site.

Office Director (Section 85.2102 [6-25-90]) means the Director for the Office of Mobile Sources--Office of Air and Radiation of the Environmental Protection Agency or other authorized representative of the Office Director.

Office of Civil Rights (OCR) (Section 7.25) means the Director of the Office of Civil Rights, EPA Headquarters or his/her designated representative.

Office of Civil Rights (Section 8.33) means the Office of Civil Rights and Urban Affairs in the Agency.

Office of Federal Contract Compliance (Section 8.33) means the Office of Federal Contract Compliance, U.S. Department of Labor.

Office of the Administrator (Section 179.3 [12-5-90]) means the Agency's Administrator and Deputy Administrator and their immediate staff, including the judicial officer.

Office of the Assistant Administrator for Enforcement and General Counsel (Section 8.33) means the Office of the Assistant Administrator for Enforcement and General Counsel in the Agency.

Office papers (Section 250.4 [10/6/87, 6/22/88]) means note pads, loose-leaf fillers, tablets, and other papers commonly used in offices, but not defined elsewhere.

Officer or employee of an agency (Section 34.105 [2-26-90]) includes the following individuals who are employed by an agency: (1) An individual who is appointed to a position in the Government under title 5 U.S. Code, including a position under a temporary appointment; (2) A member of the uniformed services as defined in section 101(3), title 37, U.S. Code; (3) A special Government employee a defined in section 202, title 18, U.S. Code; and, (4) An individual who is a member of a Federal advisory committee, as defined by the Federal Advisory Committee Act, title 5 U.S. Code appendix 2.

Offset printing paper (Section 250.4 [10/6/87, 6/22/88]) means an uncoated or coated paper designed for offset lithography.

Offshore facility (Section 112.2 [8/25/93])* means any facility of any kind located in, on, or under any of the navigable waters of the United States, which is not a transportation-related facility.

Offshore facility (Section 300.5 [3-8-90])* as defined by section 101(17) of CERCLA and section 311(a)(11) of the CWA, means any facility of any kind located in, on, or under any of the navigable waters of the United States and any facility of any kind which is subject to the jurisdiction of the United States and is located in, on, or under any other waters, other than a vessel or a public vessel.

Offshore facility (Sections 110.1; 116.3; 302.3) means any facility of any kind located in, on, or under any of the navigable waters of the United States, and any facility of any kind that is subject to the jurisdiction of the United States and is located in, on, or under any other waters, other than a vessel or a public vessel.

Offshore platform gas turbines (Section 60.331) means any stationary gas turbine located on a platform in an ocean.

Oil (Section 110.1) when used in relation to section 311 of the Act, means oil of any kind or in any form, including, but not limited to, petroleum, fuel oil, sludge, oil refuse, and oil mixed with wastes other than dredged spoil. Oil, when used in relation to section 18(m)(3) of the Deepwater Port Act of 1974, has the meaning provided in section 3(14) of the Deepwater Port Act of 1974.

Oil (Section 112.2 [8/25/93])* means oil of any kind or in any form, including, but not limited to petroleum, fuel oil, sludge, oil refuse and oil mixed with wastes other than dredged spoil.

Oil (Section 300.5 [3-8-90])* as defined by section 311(a)(1) of the CWA, means oil of any kind or in any form, including, but not limited to, petroleum, fuel oil, sludge, oil refuse, and oil mixed with wastes other than dredged spoil.

Oil (Section 60.41b [12/16/87])* means crude oil or petroleum or a liquid fuel derived from crude oil or petroleum, including distillate and residual oil.

Oil (Section 60.41c [9-12-90]) means crude oil or petroleum, or a liquid fuel derived from crude oil or petroleum, including distillate oil and residual oil.

Oil (Sections 109.2; 113.3) means oil of any kind or in any form, including, but not limited to, petroleum, fuel oil, sludge, oil refuse, and oil mixed with wastes other than dredged spoil.

Oil (Sections 426.81; 426.121) shall mean those components of a waste water amenable to measurement by the technique or techniques described in the most recent edition of "Standard Methods" for the analysis of grease in polluted waters, waste waters, and effluents, such as "Standard Methods," 13th Edition, 2nd Printing, page 407.

Oil and grease (O&G) (Section 420.02) means the value obtained by the method specified in 40 CFR 136.3.

Oil and grease (Section 410.11) shall mean total recoverable oil and grease as measured by the procedure listed in 40 CFR Part 136.

Oil and grease (Sections 408.11; 408.21; 408.31; 408.41; 408.51) shall mean those components of a waste water amenable to measurement by the method described in "Methods for

Chemical Analysis of Water and Wastes," 1971, Environmental Protection Agency, Analytical Quality Control Laboratory, page 217.

Oil and grease (Sections 432.11; 432.21; 432.31; 432.41) shall mean those components of process waste water amenable to measurement by the method described in "Methods for Chemical Analysis of Water and Wastes," 1971, EPA, Analytical Quality Control Laboratory, page 217.

Oil pollution fund (Section 300.5 [3-8-90])* means the fund established by section 311(k) of the CWA.

Oil Spill Removal Organization (Section 112.2 [8/25/93])* means an entity that provides oil spill response resources, and includes any for-profit or not-for-profit contractor, cooperative, or in-house response resources that have been established in a geographic area to provide required response resources.

Oil-water separator (Section 61.341 [3-7-90]) means a waste management unit, generally a tank or surface impoundment, used to separate oil from water. An oil-water separator consists of not only the separation unit but also the forebay and other separator basins, skimmers, weirs, grit chambers, sludge hoppers, and bar screens that are located directly after the individual drain system and prior to additional treatment units such as an air flotation unit, clarifier, or biological treatment unit. Examples of an oil-water separator incude an API separator, parallel-plate interceptor, and corrugated-plate interceptor with the associated ancillary equipment.

Oil-water separator (Section 60.691 [11/23/88]) means wastewater treatment equipment used to separate oil from water consisting of a separation tank, which also includes the forebay and other separator basins, skimmers, weirs, grit chambers, and sludge hoppers. Slop oil facilities, including tanks, are included in this term along with storage vessels and auxiliary equipment located between individual drain systems and the oil-water separator. This term does not include storage vessels or auxiliary equipment which do not come in contact with or store oily wastewater.

Oily wastewater (Section 60.691 [11/23/88]) means wastewater generated during the refinery process which contains oil, emulsified oil, or other hydrocarbons. Oily wastewater originates from a variety of refinery processes including cooling water, condensed stripping steam, tank draw-off, and contact process water.

Old chemical registration review (Section 152.403 [5/26/88]) means review of an application for registration of a new product containing active ingredients and uses which are substantially similar or identical to those currently registered or for which an application is pending

Agency decision.

Olives (Section 407.61) shall mean the processing of olives into the following product styles: Canned, all varieties, fresh and stored, green ripe, black ripe, spanish, sicilian, and any other styles to which spices, acids, and flavorings may have been added.

On site (Section 761.3) means within the boundaries of a contiguous property unit.

On the premises where stored (Section 280.12 [9/23/88]) with respect to heating oil means UST systems located on the same property where the stored heating oil is used.

On-Premise Sales (Section 244.101) means sales transactions in which beverages are purchased by a consumer for immediate consumption within the area under control of the dealer.

On-scene coordinator (OSC) (Section 300.5 [3-8-90])* means the federal official predesignated be EPA or the USCG to coordinate and direct federal responses under subpart D, or the official designated by the lead agency to coordinate and direct removal actions under subpart E of the NCP.

On-Scene Coordinator (Section 113.3) is the single Federal representative designated pursuant to the National Oil and Hazardous Substances Pollution Contingency Plan and identified in approved Regional Oil and Hazardous Substances Pollution Contingency Plans.

On-site (Section 260.10) means the same or geographically contiguous property which may be divided by public or private right-of-way, provided the entrance and exit between the properties is at a cross-roads intersection, and access is by crossing as opposed to going along, the right-of-way. Non-contiguous properties owned by the same person but connected by a right-of-way which he controls and to which the public does not have access, is also considered on-site property.

On-site (Section 270.2) means on the same or geographically contiguous property which may be divided by public or private right(s)-of-way, provided the entrance and exit between the properties is at a cross-roads intersection, and access is by crossing as opposed to going along, the right(s)-of-way. Non-contiguous properties owned by the same person but connected by a right-of-way which the person controls and to which the public does not have access, is also considered on-site property.

On-site (Section 300.5 [3-8-90]) means the areal extent of contamination and all suitable areas in very close proximity to the contamination necessary for implementation of the response action.

Once through cooling water (Section

423.11) means water passed through the main cooling condensers in one or two passes for the purpose of removing waste heat.

Once-through cooling water (Section 419.11) shall mean those waters discharged that are used for the purpose of heat removal and that do not come into direct contact with any raw material, intermediate, or finished product.

Oncogenic (Section 162.3) means the property of a substance or a mixture of substances to produce or induce benign or malignant tumor formations in living animals.

One hundred-year (100-year) floodplain (Section 264.18), as used in paragraph (b)(1) of this section, means any land area which is subject to a one percent or greater chance of flooding in any given year from any source.

One hundred-year (100-year) flood (Section 264.18), as used in paragraph (b)(1) of this section, means a flood that has a one percent chance of being equalled or exceeded in any given year.

One-hour (1-hour) period (Section 60.2) means any 60-minute period commencing on the hour.

One-year (1-year), Twenty-four-hour (24-hour) precipitation event (Section 434.11) means the maximum 24-hour precipitation event with a probable recurrence interval of once in one year as defined by the National Weather Service and Technical Paper No. 40, "Rainfall Frequency Atlas of the U.S.," May 1961, or equivalent regional or rainfall probability information developed therefrom.

One-year (1-year), Two-year (2-year), and Ten-year (10-year), Twenty-four-hour (24-hour) precipitation events (Section 434.11) means the maximum 24-hour precipitation event with a probable recurrence interval of once in one, two, and ten years respectively as defined by the National Weather Service and Technical Paper No. 40, "Rainfall Frequency Atlas of the U.S.," May 1961, or equivalent regional or rainfall probability information developed therefrom.

Onground tank (Section 260.10) means a device meeting the definition of tank in Section 260.10 and that is situated in such a way that the bottom of the tank is on the same level as the adjacent surrounding surface so that the external tank bottom cannot be visually inspected.

Onshore (Section 435.51) shall mean all land areas landward of the territorial seas as defined in 40 CFR 125.1(gg).

Onshore (Section 435.61) shall mean all land areas landward of the inner boundary of the territorial seas as defined in 40 CFR 125.1(gg).

Onshore (Sections 60.631; 60.641) means all facilities except those that are located in the territorial seas or on the outer continental shelf.

Onshore facility (Section 112.2 [8/25/93])* means any facility of any kind located in, on, or under any land within the United States, other than submerged lands, which is not a transportation-related facility.

Onshore facility (Section 300.5 [3-8-90])* as defined by section 101(18) of CERCLA, means any facility (including, but not limited to, motor vehicles and rolling stock) of any kind located in, on, or under any land or non-navigable waters within the United States; and, as defined by section 311(a)(10) of the CWA, means any facility including, but not limited to, motor vehicles and rolling stock of any kind located in, on, or under any land within the United States other than submerged land.

Onshore facility (Section 302.3) means any facility (including, but not limited to, motor vehicles and rolling stock) of any kind located in, on, or under, any land or non-navigable waters within the United States.

Onshore facility (Sections 110.1; 116.3) means any facility (including, but not limited to, motor vehicles and rolling stock) of any kind located in, on, or under any land within the United States, other than submerged land.

Onshore Oil Storage Facility (Section 113.3) means any facility (excluding motor vehicles and rolling stock) of any kind located in, on, or under, any land within the United States, other than submerged land.

Onsite coating mix preparation equipment (Section 60.741 [9-11-89]) means those pieces of coating mix preparation equipment located at the same plant as the coating operation they serve.

OP year (Section 85.1502 [9/25/87]). See **Original production year (Section 85.1502)**.

OP years old (Section 85.1502 [9/25/87]). See **Original production years old (Section 85.1502)**.

Opacity (Section 60.2) means the degree to which emissions reduce the transmission of light and obscure the view of an object in the background.

Opacity (Section 86.082-2) means the fraction of a beam of light, expressed in percent, which fails to penetrate a plume of smoke.

Opacity (Sections 61.171; 61.181) means the degree to which emissions reduce the transmission of light.

Open burning (Section 165.1) means the combustion of a pesticide or pesticide container in any fashion other than incineration.

Open burning (Section 258.2 [10-9-91]) means the combustion of solid waste without: (1) Control of combustion air to maintain adequate temperature for efficient combustion, (2) Containment of the combustion reaction in an enclosed device to provide sufficient residence time and mixing for complete combustion, and (3) Control of the emission of the combustion products.

Open burning (Section 260.10) means the combustion of any material without the following characteristics: (1) Control of combustion air to maintain adequate temperature for efficient combustion, (2) Containment of the combustion-reaction in an enclosed device to provide sufficient residence time and mixing for complete combustion, and (3) Control of emission of the gaseous combustion products. (See also incineration and thermal treatment.)

Open burning (Sections 240.101; 241.101) means burning of solid wastes in the open, such as in an open dump.

Open combustion (Section 420.41) means those basic oxygen furnace steelmaking wet air cleaning systems which are designed to allow excess air to enter the air pollution control system for the purpose of combusting the carbon monoxide in furnace gases.

Open cut mine (Section 440.141 [5/24/88]) means any form of recovery of ore from the earth except by a dredge.

Open dump (Section 257.2) means a facility for the disposal of solid waste which does not comply with this part.

Open dump (Sections 240.101; 241.101) means a land disposal site at which solid wastes are disposed of in a manner that does not protect the environment, are susceptible to open burning, and are exposed to the elements, vectors, and scavengers.

Open dumping (Section 165.1) means the placing of pesticides or containers in a land site in a manner which does not protect the environment and is exposed to the elements, vectors, and scavengers.

Open-ended valve or line (Section 264.1031 [6-21-90]) means any valve, except pressure relief valves, having one side of the valve seat in contact with process fluid and one side open to the atmosphere, either directly or through open piping.

Open hearth furnace steelmaking (Section 420.41) means the production of steel from molten iron, steel scrap, fluxes, and various combinations thereof, in refractory lined fuel-fired furnaces equipped with regenerative chambers to recover heat from the flue and combustion gases.

Open lot (Section 412.11) shall mean pens or similar confinement areas with dirt, or concrete (or paved or hard) surfaces wherein animals or poultry are substantially or entirely exposed to the outside environment except for possible small portions affording some protection by windbreaks, small shed-type shade areas. For the purposes hereof the term open lot is synonymous with the terms cowyard (dairy cattle), pasture lot (swine), and dirt lot (swine, sheep or turkeys), dry lot (swine, cattle, sheep, or turkeys) which are terms widely used in the industry.

Open site (Section 202.10) means an area that is essentially free of large sound-reflecting objects, such as barriers, walls, board fences, signboards, parked vehicles, bridges, or buildings.

Open-ended valve or line (Section 60.481) means any valve, except safety relief valves, having one side of the valve seat in contact with process fluid and one side open to the atmosphere, either directly or through open piping.

Open-ended valve or line (Section 61.241) means any valve, except pressure relief valves, having one side of the valve seat in contact with process fluid and one side open to atmosphere, either directly or through open piping.

Operable treatment works (Section 35.905) is a treatment works that: (a) Upon completion of construction will treat waste water, transport waste water to or from treatment, or transport and dispose of waste water in a manner which will significantly improve an objectionable water quality situation or health hazard, and (b) Is a component part of a complete waste treatment system which, upon completion of construction for the complete waste treatment system (or completion of construction of other treatment works in the system in accordance with a schedule approved by the Regional Administrator) will comply with all applicable statutory and regulatory requirements.

Operable unit (Section 300.5 [3-8-90]) means a discrete action that comprises an incremental step toward comprehensively addressing site problems. This discrete portion of a remedial response manages migration, or eliminates or mitigates a release, threat of a release, or pathway of exposure. The cleanup of a site can be divided into a number of operable units, depending on the complexity of the problems associated with the site. Operable units may address geographical portions of a site, specific site problems, or initial phases of an action, or may consist of any set of actions performed over time or any actions that are concurrent but located in different parts of a site.

Operable unit (Section 35.4010 [10/1/92])* means a discrete action that comprises an incremental step toward comprehensively addressing site problems.

Operable unit (Section 35.6015 [6-5-90]) A discrete action, as described in the Cooperative Agreement or SSC, that comprises an incremental step toward comprehensively addressing site problems. The cleanup of a site can be divided into a number of operable units, depending on the complexity of the problems associated with the site. Operable units may address geographical portions of a site, specific site problems, or initial phases of an action, or may consist of any set of actions performed over time or any actions that are concurrent but located in different parts of a site.

Operated by the same producer (Section 162.3) means (1) another registered establishment owned by the registrant of the pesticide product or (2) another registered establishment operated under contract with the registrant of the pesticide either to package the pesticide product or to use the pesticide as a constituent part of another pesticide product, provided that the final pesticide product is registered by the transferor establishment.

Operated by the same producer (Section 152.3 [5/4/88]) when used with respect to two establishments, means that each such establishment is either owned by, or leased for operation by and under the control of, the same person. The term does not include establishments owned or operated by different persons, regardless of contractural agreement between such persons.

Operating Hours (Section 86.078-7), where vehicle, component, or engine storage areas or facilities are concerned, shall mean all times during which personnel other than custodial personnel are at work in the vicinity of the area or facility and have access to it. Where facilities or areas other than those covered by paragraph (c)(7)(ii) of this section are concerned, operating hours shall mean all times during which an assembly line is in operation or all times during which testing, maintenance, mileage (or service) accumulation, production or compilation of records, or any other procedure or activity related to certification testing, to translation of designs from the test stage to the production stage, or to vehicle (or engine) manufacture or assembly is being carried out in a facility.

Operation (Section 413.11) shall mean any step in the electroplating process in which a metal is electrodeposited on a basis material and which is followed by a rinse; this includes the related operations of alkaline cleaning, acid pickle, stripping, and coloring when each operation is followed by a rinse.

Operation (Section 413.71) shall mean any step in the electroless plating process in which a metal is deposited on a basis material and which is followed by a rinse; this includes the related operations of alkaline cleaning, acid pickle, and stripping, when each operation is followed by a rinse.

Operation (Section 61.251 [12-15-89])* means that an impoundment is being used for the continued placement of new tailings or is in standby status for such placement. An impoundment is in operation from the day that tailings are first placed in the impoundment until the day that final closure begins.

Operation and maintenance (O&M) (Section 35.6015 [6-5-90])* Measures required to maintain the effectiveness of response actions.

Operation and maintenance (O&M) (Section 300.5 [3-8-90]) means measures required to maintain the effectiveness of response actions.

Operation and Maintenance (Section 35.2005) are activities required to assure the dependable and economical function of treatment works. (i) Maintenance: Preservation of functional integrity and efficiency of equipment and structures. This includes preventive maintenance, corrective maintenance and replacement of equipment (See Section 35.2005(b)(36)) as needed. (ii) Operation: Control of the unit processes and equipment which make up the treatment works. This includes financial and personnel management; records, laboratory control, process control, safety and emergency operation planning.

Operational (Section 192.31 [11/5/93])* means that a uranium mill tailings pile or impoundment is being used for the continued placement of uranium byproduct material or is in standby status for such placement. A tailings pile or impoundment is operational from the day that uranium byproduct material is first placed in the pile or impoundment until the day final closure begins.

Operational (Section 61.221 [12-15-89]) means a uranium mill tailings pile that is licensed to accept additional tailings, and those tailings can be added without violating subpart W or any other Federal, state or local rule or law. A pile cannot be considered operational if it is filled to capacity or the mill it accepts tailings from has been dismantled or otherwise decommissioned.

Operational life (Section 280.12 [9/23/88]) refers to the period beginning when installation of the tank system has commenced until the time the tank system is properly closed under Subpart G.

Operations and maintenance program (Section 763.83 [10/30/87]) means a program of work practices to maintain friable ACBM in good condition, ensure clean up of asbestos fibers previously released, and prevent further release by minimizing and controlling friable ACBM disturbance or damage.

Operator (Section 256.06) includes facility owners and operators.

Operator (Section 258.2 [10-9-91]) means the person(s) responsible for the overall operation of a facility or part of a facility.

Operator (Section 260.10) means the person responsible for the overall operation of a facility.

Operator (Section 280.1) means any person in control of, or having responsibility for, the daily operation of the underground storage tank.

Operator (Section 280.12 [9/23/88]) means any person in control of, or having responsibility for, the daily operation of the UST system.

Operator (Section 610.11) means any person who installs, services or maintains a retrofit device in an automobile or who operates an automobile with a retrofit device installed.

OPPTS (Section 179.3 [6-24-92]) means the Agency's Office of Prevention, Pesticides and Toxic Substances.

Optimal corrosion control treatment (Section 141.2 [7/17/92])* for the purpose of subpart I of this part only, means the corrosion control treatment that minimizes the lead and copper concentrations at users' taps while insuring that the treatment does not cause the water system to violate any national primary drinking water regulations.

Option (Section 86.082-2) means any available equipment or feature not standard equipment on a model.

OPTS (Section 179.3 [12-5-90]) means the Agency's Office of Pesticides and Toxic Substances.

ORD (Section 6.700) means Office of Research and Development.

Order (Section 108.2) means any order issued by the Administrator under section 309 of the Act; any order issued by a State to secure compliance with a permit, or condition thereof, issued under a program approved pursuant to section 402 of the Act; or any order issued by a court in an action brought pursuant to section 309 or section 505 of the Act.

Order (Section 8.2) means Parts II, III, and IV of Executive Order 11246, dated September 24, 1965 (30 FR 12319), and any Executive Order amending or superseding such orders.

Ore (Section 440.141 [5/24/88]) means gold placer deposit consisting of metallic gold-bearing gravels, which may be: residual, from weathering of rocks in-situ; river gravels in active streams; river gravels in abandoned and often buried channels; alluvial fans; sea-beaches; and sea-beaches now elevated and inland. Ore is the raw bank run material measured in place, before being moved by mechanical or hydraulic means to a beneficiation process.

Organ (Section 190.02) means any human organ exclusive of the dermis, the epidermis, or the cornea.

Organic active ingredients (Section 455.21) means carbon-containing active ingredients used in pesticides, excluding metalloorganic active ingredients.

Organic chlorine (Section 797.1520) is the chlorine associated with all chlorine-containing compounds that elute just before lindane to just after mirex during gas chromatographic analysis using a halogen detector.

Organic coating (Section 60.311) means any coating used in a surface coating operation, including dilution solvents, from which volatile organic compound emissions occur during the application or the curing process. For the purpose of this regulation, powder coatings are not included in this definition.

Organic coating (Section 60.451) means any coating used in a surface coating operation, including dilution solvents, from which VOC emissions occur during the application or the curing process. For the purpose of this regulation, powder coatings are not included in this definition.

Organic Material Hydrocarbon Equivalent (Section 86.090-2 [7-26-90])* means the sum of the carbon mass contributions of non-oxygenated hydrocarbons, methanol and formaldehyde as contained in a gas sample, expressed as gasoline fueled vehicle hydrocarbons. In the case of exhaust emissions, the hydrogen-to-carbon ratio of the equivalent hydrocarbon is 1.85:1. In the case of diurnal and hot soak emissions, the hydrogen-to-carbon ratios of the equivalent hydrocarbons are 2.33:1 and 2.2:1, respectively.

Organic matter (Section 796.2750) is the organic fraction of the sediment or soil; it includes plant and animal residues at various stages of decomposition, cells and tissues of soil organisms, and substances synthesized by the microbial population.

Organic pesticide chemicals (Section 455.21 [9/28/93])* means the sum of all organic active ingredients listed in Sec. 455.20(b) which are manufactured at a facility subject to this subpart.

Organic pesticides (Section 165.1) means carbon-containing substances used as pesticides, excluding metallo-organic compounds.

Organic solvent-based green tire spray (Section 60.541 [9/15/87]) means any mold release agent and lubricant applied to the inside or outside of green tires that contains more than 12 percent, by weight, of VOC as sprayed.

Organochlorine pesticides (Section 797.1520) are those pesticides which contain carbon and chlorine such as aldrin, DDD, DDE, DDT, dieldrin, endrin, and heptachlor.

Original data submitter (Section 152.83) means the person who possesses all rights to exclusive use or compensation under FIFRA section 3(c)(1)(D) in a study originally submitted in support of an application for registration, amended registration, reregistration, or experimental use permit, or to maintain an existing registration in effect. The term includes the person who originally submitted the study, any person to whom the rights under FIFRA section 3(c)(1)(D) have been transferred, or the authorized representative of a group of joint data developers.

Original equipment manufacturer (OEM) (Section 85.1502 [9/25/87]) is the entity which originally manufactured the motor vehicle or motor vehicle engine prior to conditional importation.

Original equipment market part (Section 763.163 [7-12-89]) means any part installed in or on a motor vehicle in the manufacturer's production line.

Original Equipment Part (Section 85.2102) means a part present in or on a vehicle at the time the vehicle is sold to the ultimate purchaser, except for components installed by a dealer which are not manufactured by the vehicle manufacturer or are not installed at the direction of the vehicle manufacturer.

Original production (OP) year (Section 85.1502 [9/25/87]) is the calendar year in which the motor vehicle or motor vehicle engine was originally produced by the OEM.

Original production (OP) years old (Section 85.1502 [9/25/87]) means the age of a vehicle as determined by subtracting the original production year of the vehicle from the calendar year of importation.

Ornamental (Section 171.2) means trees, shrubs, and other plantings in and around habitations generally, but not necessarily located in urban and suburban areas, including residences, parks, streets, retail outlets, industrial and institutional buildings.

OSC (Section 300.5) is the On-Scene Coordinator.

Other coatings (Section 420.121) means coating steel products with metals other than zinc or terne metal by the hot dip process including the immersion of the steel product in a molten bath of metal, and the related operations preceding and subsequent to the immersion phase.

Other lead-emitting operation (Section 60.371) means any lead-acid battery manufacturing plant operation from which lead emissions are collected and ducted to the atmosphere and which is not part of a grid casting, lead oxide manufacturing, lead reclamation, paste mixing, or three-process operation facility, or a furnace affected under Subpart L of this part.

Other restricted access (nonsubstation) locations (Section 761.123) means areas other than electrical substations that are at least 0.1 kilometer (km) from a residential/commercial area and limited by man-made barriers (e.g., fences and walls) to substantially limited by naturally occurring barriers such as mountains, cliffs, or rough terrain. These areas generally include industrial facilities and extremely remote rural locations. (Areas where access is restricted but are less than 0.1 km from a residential/commercial area are considered to be residential/commercial areas.)

Other significant evidence (Section 154.3) means factually significant information that relates to the uses of the pesticide and their adverse risk to man or to the environment but does not include evidence based only on misuse of the pesticide unless such misuse is widespread and commonly recognized practice.

Otherwise subject to the jurisdiction of the United States (Section 116.3) means subject to the jurisdiction of the United States by virtue of United States citizenship, United States vessel documentation or numbering, or as provided for by international agreement to which the United States is a party.

Otherwise use or use (Section 372.3 [2/16/88]) means any use of a toxic chemical that is not covered by the terms manufacture or process and includes use of a toxic chemical contained in a mixture or trade name product. Relabeling or redistributing a container of a toxic chemical where no repackaging of the toxic chemical occurs does not constitute use or processing of the toxic chemical.

Otto-cycle (Section 86.090-2 [4/11/89]) means type of engine with operating characteristics significantly similar to the theoretical Otto combustion cycle. The use of a throttle during normal operation is indicative of an Otto-cycle engine. This definition applies beginning with the 1990 model year.

Outdoor application (Section 162.3) means any pesticide application or use that occurs outside enclosed manmade structures or the consequences of which extend beyond enclosed manmade structures, including, but not limited to, pulp and paper mill water treatments and industrial cooling water treatments.

Outdoor electrical substations (Section 761.123) means outdoor, fenced-off, and restricted access areas used in the transmission and/or distribution of electrical power. Outdoor electrical substations restrict public access by being fenced or walled off as defined under Section 761.30(l)(1)(ii). For purposes of this TSCA policy, outdoor electrical substations are defined as being located at least 0.1 km from a residential/commercial area. Outdoor fenced-off and restricted access areas used in the transmission and/or distribution of electrical power which are located less than 0.1 km from a residential/commercial area are considered to be residential/commercial areas.

Outdoor use (Section 152.161 [5/4/88]) means any pesticide application that occurs outside enclosed manmade structures or the consequences of which extend beyond enclosed manmade structures, including, but not limited to, pulp and paper mill water treatments and industrial cooling water treatments.

Output (Section 35.105) means an activity or product which the applicant agrees to complete during the budget period.

Outside air (Section 61.141 [11-20-90])* means the air outside buildings and structures, including, but not limited to, the air under a bridge or in an open air ferry dock.

Outside employment or other outside activity (Section 3.500) is any work or service performed by an employee other than the performance of official duties. It includes such activities as writing and editing, publishing, teaching, lecturing, consulting, self-employment and other work or services. Employees must ensure that their outside activities may not reasonably be construed as implying official EPA endorsement of any statement, activity, product or service.

Oven (Section 60.441) means a chamber which uses heat or irradiation to bake, cure, polymerize, or dry a surface coating.

Over-the-Head Position (Section 211.203) means the mode of use of a device with a headband, in which the headband is worn such that it passes over the user's head. This is contrast to the behind-the-head and under-the-chin positions.

Overfill release (Section 280.12 [9/23/88]) is a release that occurs when a tank is filled beyond its capacity, resulting in a discharge of the regulated substance to the environment.

Overfire air (Section 240.101) means air, under control as to quantity and direction, introduced above or beyond a fuel bed by induced or forced draft.

Oversaturated (supersaturated) solution (Section 796.1840) is a solution that contains a greater concentration of a solute than is possible at equilibrium under fixed conditions of temperature and pressure.

Oversubscription payment deadline (Section 73.3 [12-17-91]) means 30 calendar days prior to the allowance transfer deadline.

Overvarnish coating operation (Section 60.491) means the system on each beverage can surface coating line used to apply a coating over ink which reduces friction for automated beverage can filling equipment, provides gloss, and protects the finished beverage can body from abrasion and corrosion. The overvarnish coating is applied to two-piece beverage can bodies. The overvarnish coating operation consists of the coating application station, flashoff area, and curing oven.

Own or control (Section 704.3 [12/22/88])* means ownership of 50 percent or more of a company's voting stock or other equity rights, or the power to control the management and policies of that company. A company may own or control one or more sites. A company may be owned or controlled by a foreign or domestic parent company.

Owned or controlled by the parent company (Section 712.3) means the parent owns or controls 50 percent or more of the other company's voting stock or other equity rights, or has the power to control the management and policies of the other company.

Owner (Section 170.3 [8/21/92])* means any person who has a present possessory interest (fee, leasehold, rental, or other) in an agricultural establishment covered by this part. A person who has both leased such agricultural establishment to another person and granted that same person the right and full authority to manage and govern the use of such agricultural establishment is not an owner for purposes of this part.

Owner (Section 258.2 [10-9-91]) means the person(s) who owns a facility or part of a facility.

Owner (Section 260.10) means the person who owns a facility or part of a facility.

Owner (Section 280.1) means (a) in the case of an underground storage tank in use on November 8, 1984, or brought into use after that date, any person who owns an underground storage tank used for the storage, use, or dispensing of regulated substances, and (b) in the case of any underground storage tank in use before November 8, 1984, but no longer in use on that date, any person who owned such tank immediately before discontinuation of its use.

Owner (Section 280.12 [9/23/88]) means (a) In the case of an UST system in use on November 8, 1984, or brought into use after that date, any person who owns an UST system used for storage, use, or dispensing of regulated substances; and (b) In the case of any UST system in use before November 8, 1984, but no longer in use on that date, any person who owned such UST immediately before the discontinuation of its use.

Owner (Section 61.251) means any

person who owns or operates a uranium mill or an existing tailings pile or a new impoundment.

Owner (Section 73.3 [12-17-91]) means any of the following persons: (1) Any holder of any portion of the legal or equitable title in an affected unit; or (2) Any holder of a leasehold interest in an affected unit; or (3) Any purchaser of power from an affected unit under a life-of-the-unit, firm power contractual arrangement as that term is used in section 408(i) of the Act. However, unless expressly provided for in a leasehold agreement, owner shall not include a passive lessor, or a person who has an equitable interest through such lessor, whose rental payments are not based, either directly or indirectly, upon the revenues or income from the affected unit.

Owner (Section 85.2102) means the original purchaser or any subsequent purchaser of a vehicle.

Owner of a boat or marine vessel (Section 82.62 [12/30/93])* means any person who possesses a title, registration or other documentation that indicates that the person presenting this documentation is in possession of a marine vessel as defined in 33 CFR part 177.

Owner of a noncommercial aircraft (Section 82.62 [12/30/93])* means any person who possesses a title, registration or other documentation that indicates that the person presenting this documentation is in possession of a noncommercial aircraft.

Owner of displacement dwelling (Section 4.2). A displaced person is considered to have met the requirement to own a displacement dwelling if the person holds any of the following interests in real property acquired for a project: (1) Fee title, a life estate, a 99-year lease, or a lease, including any options for extension, with at least 50 years to run from the date of acquisition; or (2) An interest in a cooperative housing project which includes the right to occupy a dwelling; or (3) A contract to purchase any of the interests or estates described in paragraphs (l) (1) or (2) of this section; or (4) Any other interest, including a partial interest, which in the judgment of the Agency warrants consideration as ownership.

Owner/operator (Section 147.2902) means the owner or operator of any facility or activity subject to regulation under the Osage UIC program.

Owner or operator (Section 112.2 [8/25/93])* means any person owning or operating an onshore facility or an offshore facility, and in the case of any abandoned offshore facility, the person who owned or operated such facility immediately prior to such abandonment.

Owner or operator (Section 122.2) means the owner or operator of any facility or activity subject to regulation under the NPDES program.

Owner or operator (Section 124.41) means the owner or operator of any facility or activity subject to regulation under 40 CFR 52.21 or by an approved State.

Owner or operator (Section 129.2) means any person who owns, leases, operates, controls, or supervises a source as defined in 40 CFR Section 129.2.

Owner or operator (Section 144.3) means the owner or operator of any facility or activity subject to regulation under the UIC program.

Owner or operator (Section 232.2 [2/11/93])* means the owner or operator of any activity subject to regulation under the 404 program.

Owner or operator (Section 233.3) means the owner or operator of any facility or activity subject to regulation under the 404 program.

Owner or operator (Section 270.2) means the owner or operator of any facility or activity subject to regulation under RCRA.

Owner or operator (Section 280.92 [10/26/88]) when the owner or operator are separate parties, refers to the party that is obtaining or has obtained financial assurances.

Owner or operator (Section 51.100) means any person who owns, leases, operates, controls, or supervises a facility, building, structure, or installation which directly or indirectly result or may result in emissions of any air pollutant for which a national standard is in effect.

Owner or operator (Section 60.2) means any person who owns, leases, operates, controls, or supervises an affected facility or a stationary source of which an affected facility is a part.

Owner or operator (Section 61.02) means any person who owns, leases, operates, controls, or supervises a stationary source.

Owner or operator (Section 66.3 [2-13-92])* means any person who owns, leases, operates or supervises a facility, building, structure or installation which emits or has the potential to emit any air pollutant regulated by EPA under the Act.

Owner or operator (Section 73.3 [12-17-91]) means any person who is an owner or who operates, controls, or supervises in any way an affected unit or affected source of which an affected unit is a part, and shall include, but not be limited to any holding company, operating company, utility system, designated representative, or plant manager of an affected unit or affected source.

Owner or operator (Sections 124.2;

146.3) means owner or operator of any facility or activity subject to regulation under the RCRA, UIC, NPDES, or 404 programs.

Owner or operator of a demolition or renovation activity (Section 61.141 [11-20 90]) means any person who owns, leases, operates, controls, or supervises the facility being demolished or renovated or any person who owns, leases, operates, controls, or supervises the demolition or renovation operation, or both.

Owner's Manual (Section 85.2102) means the instruction booklet normally provided to the purchaser of a vehicle.

Oxidation control system (Section 60.101) means an emission control system which reduces emissions from sulfur recovery plants by converting these emissions to sulfur dioxide.

Oxides of nitrogen (Sections 86.082-2; 86.402-78) means the sum of the nitric oxide and nitrogen dioxide contained in a gas sample as if the nitric oxide were in the form of nitrogen dioxide.

Oxygenate (Section 80.2 [2/16/94])* means any substance which, when added to gasoline, increases the oxygen content of that gasoline. Lawful use of any of the substances or any combination of these substances requires that they be "substantially similar" under section 211(f)(1) of the Clean Air Act, or be permitted under a waiver granted by the Administrator under the authority of section 211(f)(4) of the Clean Air Act.

Oxygenate blender (Section 80.2 [2/16/94])* means any person who owns, leases, operates, controls, or supervises an oxygenate blending facility, or who owns or controls the blendstock or gasoline used or the gasoline produced at an oxygenate blending facility.

Oxygenate blending facility (Section 80.2 [2/16/94])* means any facility (including a truck) at which oxygenate is added to gasoline or blendstock, and at which the quality or quantity of gasoline is not altered in any other manner except for the addition of deposit control additives.

Oxygenated fuels program reformulated gasoline, or OPRG (Section 80.2 [2/16/94])* means reformulated gasoline which is intended for use in an oxygenated fuels program control area, as defined at paragraph (pp) of this section, during an oxygenated fuels program control period, as defined at paragraph (qq) of this section.

Oxygenated gasoline (Section 80.2 [2/16/94])* means gasoline which contains a measurable amount of oxygenate.

P

P-TBB (Section 704.33) means the substance p-tert-butylbenzaldehyde, also identified as 4-(1,1-dimethylethyl) benzaldehyde, CAS No. 939-97-9.

P-TBBA (Section 704.33) means the substance p-tert-butylbenzoic acid , also identified as 4-(1,1-dimethylethyl)benzoic acid, CAS No. 98-73-7.

P-TBT (Section 704.33) means the substance p-tert-butyltoluene, also identified as 1-(1,1-dimethylethyl)-4-methylbenzene, CAS No. 98-51-1.

PAAT (Section 300.5) is the Public Affairs Assist Team.

Package (Section 211.203) means the container in which a hearing protective device is presented for purchase or use. The package in some cases may be the same as the carrying case.

Package (Section 259.10 [7-2-90]) means the packaging/containers and its contents.

Package or packaging (Sections 152.3 [5/4/88]; 157.21) means the immediate container or wrapping, including any attached closure(s), in which the pesticide is contained for distribution, sale, consumption, use, or storage. The term does not include any shipping or bulk container used for transporting or delivering the pesticide unless it is the only such package.

Packaging (Section 259.10 [7-2-90]) means the assembly of one or more containers and any other components necessary to assure minimum compliance with Sec. 259.41 of this part.

Packaging (Sections 152.3 [5/4/88]; 157.21). See **Package (Sections 152.3; 157.21)**.

Packer (Section 146.3) means a device lowered into a well to produce a fluid-tight seal.

Packer (Section 147.2902) means a device lowered into a well to produce a fluid-tight seal within the casing.

Packing (Section 763.163 [7-12-89]) means an asbestos-containing product intended for use as a mechanical seal in circumstances involving rotary, reciprocating, and helical motions, and which are intended to restrict fluid or gas leakage between moving and stationary surfaces. Major applications of this product include: seals in pumps; seals in valves; seals in compressors; seals in mixers; seals in swing joints; and seals in hydraulic cylinders.

Packinghouse (Sections 432.31; 432.41) shall mean a plant that both slaughters animals and subsequently processes carcasses into cured, smoked, canned or other prepared meat products.

PAMS (Section 58.1 [2/12/93])* means Photochemical Assessment Monitoring Stations.

Panel (Section 304.12 [5/30/89]). See **National Panel of Environmental Arbitrators (Section 304.12)**.

Paper (Section 250.4 [10/6/87, 6/22/88]) means one of two broad subdivisions of paper products, the other being paperboard. Paper is generally lighter in basis weight, thinner, and more flexible than paperboard. Sheets 0.012 inch or less in thickness are generally classified as paper. Its primary uses are for printing, writing, wrapping, and sanitary purposes. However, in this guideline, the term paper is also used as a generic term that includes both paper and paperboard. It includes the following types of papers: Bleached paper, bond paper, book paper, brown paper, coarse paper, computer paper, cotton fiber content paper, cover stock or cover paper, duplicator paper, form bond, ledger paper, manifold business forms, mimeo paper, newsprint, office papers, offset printing paper, printing paper, stationery, tabulating paper, unbleached papers, writing paper, and xerographic/copy paper.

Paper napkins (Section 250.4 [10/6/87, 6/22/88]) means special tissues, white or colored, plain or printed, usually folded, and made in a variety of sizes for use during meals or with beverages.

Paper product (Section 250.4 [10/6/87, 6/22/88]) means any item manufactured from paper or paperboard. The term paper product is used in this guideline to distinguish such items as boxes, doilies, and paper towels from printing and writing papers. It includes the following types of products: Corrugated boxes, doilies, envelopes, facial tissue, fiberboard boxes, folding boxboard, industrial wipers, paper napkins, paper towels, tabulating cards, and toilet tissue.

Paper towels (Section 250.4 [10/6/87, 6/22/88]) means paper toweling in folded sheets, or in raw form, for use in drying or cleaning, or where quick absorption is required.

Paperboard (Section 250.4 [10/6/87, 6/22/88]) means one of the two broad subdivisions of paper, the other being paper itself. Paperboard is usually heavier in basis weight and thicker than paper. Sheets 0.012 inch or more in thickness are generally classified as paperboard. The broad classes of paperboard are containerboard, which is used for corrugated boxes; boxboard, which is principally used to make cartons; and all other paperboard.

Parent company (Section 704.3 [12/22/88])* is a company that owns or controls another company.

Parent corporation (Section 144.61)

means a corporation which directly owns at least 50 percent of the voting stock of the corporation which is the injection well owner or operator; the latter corporation is deemed a subsidiary of the parent corporation.

Parent corporation (Sections 264.141; 265.141) means a corporation which directly owns at least 50 percent of the voting stock of the corporation which is the facility owner or operator; the latter corporation is deemed a subsidiary of the parent corporation.

Part 70 permit (Section 60.2 [7/21/92])* means any permit issued, renewed, or revised pursuant to part 70 of this chapter.

Part 70 permit (Section 61.02 [3/16/94])* means any permit issued, renewed, or revised pursuant to part 70 of this chapter.

Part-time fellow (Section 46.120) means an individual enrolled in an academic educational program directly related to pollution abatement and control and taking at least 6 credit hours but less than 30 credit hours per school year, or an academic workload otherwise defined by the institution as less than a full-time curriculum. The fellow need not be pursuing a degree.

Partial closure (Section 260.10) means the closure of a hazardous waste management unit in accordance with the applicable closure requirements of Parts 264 and 265 of this chapter at a facility that contains other active hazardous waste management units. For example, partial closure may include the closure of a tank (including its associated piping and underlying containment systems), landfill cell, surface impoundment, waste pile, or other hazardous waste management unit, while other units of the same facility continue to operate.

Participant (Section 172.1) means any person acting as a representative of the permittee and responsible for making available for use, or supervising the use or evaluation of, an experimental use pesticide to be applied at a specific application site.

Participating PRP (Section 304.12 [5/30/89]) is any potentially responsible party who has agreed, pursuant to Section 304.21 of this part, to submit one or more issues arising in an EPA claim for resolution pursuant to the procedures established by this part.

Particle size analysis (Section 796.2750) is the determination of the various amounts of the different particle sizes in a sample (i.e., sand, silt, clay), usually by sedimentation, sieving, micrometry, or combinations of these methods. The names and diameter range commonly used in the United States are set forth in paragraph (a)(2)(iv) of this section.

Particle size analysis (Section 796.2700) is the determination of the various amounts of the different particle sizes in a soil sample (i.e., sand, silt, clay) usually by sedimentation, sieving, micrometry, or combinations of these methods. The names and size limits of these particles as widely used in the United States are set forth in paragraph (a)(2)(ii) of this section.

Particulate asbestos material (Section 61.141 [11-20-90])* means finely divided particles of asbestos or material containing asbestos.

Particulate matter (Section 51.100) means any airborne finely divided solid or liquid material with an aerodynamic diameter smaller than 100 micrometers.

Particulate matter (Section 60.2) means any finely divided solid or liquid material, other than uncombined water, as measured by the reference methods specified under each applicable subpart, or an equivalent or alternative method.

Particulate matter (Section 60.51a [2-11-91]) means total particulate matter emitted from MWC units as measured by Method 5 (see Sec. 60.58a).

Particulate matter (Section 61.171) means any finely divided solid or liquid material, other than uncombined water, as measured by the specified reference method.

Particulate matter emissions (Section 51.100) means all finely divided solid or liquid material, other than uncombined water, emitted to the ambient air as measured by applicable reference methods, or an equivalent or alternative method, specified in this chapter, or by a test method specified in an approved State implementation plan.

Party (Section 104.2) means the Environmental Protection Agency as the proponent of an effluent standard or standards, and any person who files an objection pursuant to Section 104.3 hereof.

Party (Section 108.2) means an employee filing a request under Section 108.3, any employee similarly situated, the employer of any such employee, and the Regional Administrator or his designee.

Party (Section 123.64) means the petitioner, the State, the Agency, and any other person whose request to participate as a party is granted.

Party (Section 124.72) means the EPA trial staff under Section 124.78 and any person whose request for a hearing under Section 124.74 or whose request to be admitted as a party or to intervene under Section 124.79 or Section 124.117 has been granted.

Party (Section 142.202 [1-30-91]) means any "person" or "supplier of water" as defined in section 1401 of the SDWA, 42 U.S.C. 300f, alleged to have violated any regulation implementation section 1412 of the SDWA, 42 U.S.C. 300g-1, any schedule or other requirement imposed pursuant to section 1415 or section 1416 of the SDWA, 42 U.S.C. 300g-4 and 300g-5, or section 1445 of the SDWA, 42 U.S.C. 300j-4, or any regulation implementing section 1445.

Party (Section 164.2 [7/10/92])* means any person, group, organization, or Federal agency or department that participates in a hearing.

Party (Section 209.3) means the Environmental Protection Agency, the respondent(s) and any interveners.

Party (Section 22.03) means any person that participates in a hearing as complainant, respondent, or intervenor.

Party (Section 304.12 [5/30/89]) means EPA and any person who has agreed, pursuant to Section 304.21 of this part, to submit one or more issues arising in an EPA claim for resolution pursuant to the procedures established by this part, and any person who has been granted leave to intervene pursuant to Section 304.24(a) of this part.

Party (Section 305.12) means EPA or a claimant.

Party (Section 791.3) refers to a person subject to a section 4 test rule, who: (1) Seeks reimbursement from another person under these rules, or (2) From whom reimbursement is sought under these rules.

Party (Section 8.33) means a respondent; the Director; and any person or organization participating in a proceeding pursuant to section 8.

Party (Section 82.3 [12/10/93])* means any foreign state that is listed in appendix C of this subpart (pursuant to instruments of ratification, acceptance, or approval deposited with the Depositary of the United Nations Secretariat), as having ratified the specified control measure in effect under the Montreal Protocol. Thus, for purposes of the trade bans specified in Sec. 82.4(d)(2) pursuant to the London Amendments, only those foreign states that are listed in appendix C of this subpart as having ratified both the 1987 Montreal Protocol and the London Amendments shall be deemed to be Parties.

Party (Section 85.1807) shall include the Environmental Protection Agency, the manufacturer, and any interveners.

Party (Sections 86.614; 86.1014-84; 86.1115.87) shall include the Agency and the manufacturer.

Pass Through (Section 403.3) means a Discharge which exits the POTW into waters of the United States in quantities or concentrations which, alone or in conjunction with a discharge or discharges from other sources, is a cause of a violation of any requirement of the POTW's

NPDES permit (including an increase in the magnitude or duration of a violation).

Passenger Automobile (Section 600.002-85) means any automobile which the Secretary determines is manufactured primarily for use in the transportation of no more than 10 individuals.

Passive institutional control (Section 191.12 [12/20/93])* means: (1) Permanent markers placed at a disposal site, (2) public records and archives, (3) government ownership and regulations regarding land or resource use, and (4) other methods of preserving knowledge about the location, design, and contents of a disposal system.

Past year (Sections 167.1; 167.3 [9/8/88]) means the calendar year immediately prior to that in which the report is submitted.

Paste mixing facility (Section 60.371) means the facility including lead oxide storage, conveying, weighing, metering, and charging operations; paste blending, handling, and cooling operations; and plate pasting, takeoff, cooling, and drying operations.

Pasture crops (Section 257.3-5) means crops such as legumes, grasses, grain stubble and stover which are consumed by animals while grazing.

Pathway (Section 61 [12-15-89]) A way that radionuclides might contaminate the environment or reach people, e.g. air, water food.

Payment (Section 35.3105 [3-19-90]) An action by the EPA to increase the amount of capitalization grant funds available for cash draw from an LOC.

Pb (Section 58.1) means lead.

PBBs (Section 704.195). See **Polybrominated biphenyls (Section 704.195)**.

PBDD (Section 766.3). See **Polybrominated dibenzo-p-dioxin (Section 766.3)**.

PCA (Section 86.1102-87 [11-5-90])* means Production Compliance Audit as described in Sec. 86.1106-87 of this subpart.

PCB and PCBs (Section 761.3) means any chemical substance that is limited to the biphenyl molecule that has been chlorinated to varying degrees or any combination of substances which contains such substance. Refer to Section 761.1(b) for applicable concentrations of PCBs. PCB and PCBs as contained in PCB items are defined in Section 761.3. For any purposes under this part, inadvertently generated non-Aroclor PCBs are defined as the total PCBs calculated following division of the quantity of monochlorinated biphenyls by 50 and dichlorinated biphenyls by 5.

PCB Article (Section 761.3) means any manufactured article, other than a PCB Container, that contains PCBs and whose surface(s) has been in direct contact with PCBs. PCB Article includes capacitors, transformers, electric motors, pumps, pipes and any other manufactured item (1) which is formed to a specific shape or design during manufacture, (2) which has end use function(s) dependent in whole or in part upon its shape or design during end use, and (3) which has either no change of chemical composition during its end use or only those changes of composition which have no commercial purpose separate from that of the PCB Article.

PCB Article Container (Section 761.3) means any package, can, bottle, bag, barrel, drum, tank, or other device used to contain PCB Articles or PCB Equipment, and whose surface(s) has not been in direct contact with PCBs.

PCB Container (Section 761.3) means any package, can, bottle, bag, barrel, drum, tank, or other device that contains PCBs or PCB Articles and whose surface(s) has been in direct contact with PCBs.

PCB Equipment (Section 761.3) means any manufactured item, other than a PCB Container or a PCB Article Container, which contains a PCB Article or other PCB Equipment, and includes microwave ovens, electronic equipment, and fluorescent light ballasts and fixtures.

PCB Item (Section 761.3) is defined as any PCB Article, PCB Article Container, PCB Container, or PCB Equipment, that deliberately or unintentionally contains or has a part of it any PCB or PCBs.

PCB Manufacturer (Section 129.105) means a manufacturer who produces polychlorinated biphenyls.

PCB Transformer (Section 761.3) means any transformer that contains 500 ppm PCB or greater.

PCB waste(s) Section 761.3 [12-21-89]) means those PCBs and PCB Items that are subject to the disposal requirements of subpart D of this part.

PCB-Contaminated Electrical Equipment (Section 761.3) means any electrical equipment, including but not limited to transformers (including those used in railway locomotives and self-propelled cars), capacitors, circuit breakers, reclosers, voltage regulators, switches (including sectionalizers and motor starters), electromagnets, and cable, that contain 50 ppm or greater PCB, but less than 500 ppm PCB. Oil-filled electrical equipment other than circuit breakers, reclosers, and cable whose PCB concentration is unknown must be assumed to be PCB-Contaminated Electrical Equipment. (See Section 761.30(a) and (h) for provisions permitting reclassification of electrical equipment containing 500 ppm or greater PCBs to PCB-Contaminated Electrical Equipment.)

PCBs (Section 129.4). See **Polychlorinated Biphenyls (Section 129.4)**.

PCBs (Section 268.2 [7/8/87]). See **Polychlorinated biphenyls (Section 268.2)**.

PCBs (Section 761.123) means polychlorinated biphenyls as defined under Section 761.3. As specified under Section 761.1(b), no requirements may be avoided through dilution of the PCB concentration.

PCDD (Section 766.3). See **Polychlorinated dibenzo-p-dioxin (Section 766.3)**.

PCV Valve (Section 85.2122(a)(4)(ii)) means a device to control the flow of blow-by gasses and fresh air from the crankcase to the fuel induction system of the engine.

PCW (Section 246.101). See **Post-consumer waste (Section 246.101)**.

PCW (Section 247.101). See **Post consumer waste (Section 247.101).**

Peaches (Section 407.61) shall mean the processing of peaches into the following product styles: Canned or frozen, all varieties, peeled, pitted and unpitted, whole, halves, sliced, diced, and any other cuts, nectar, and concentrate but not dehydrated.

Peak Air Flow (Section 85.2122(a)(15)(ii)(D)) means the maximum engine intake mass air flow rate measure during the 195 second to 202 second time interval of the Federal Test Procedure.

Peak load (Section 60.331) means 100 percent of the manufacturer's design capacity of the gas turbine at ISO standard day conditions.

Peak torque speed (Section 86.082-2) means the speed at which an engine develops maximum torque.

Peanuts (Section 180.1) means the peanut meat after removal of the hulls.

Pears (Section 407.61) shall mean the processing of pears into the following product styles: Canned, peeled, halved, sliced, diced, and any other cuts, nectar and concentrate but not dehydrated.

Peas (Section 407.71) shall mean the processing of peas into the following product styles: Canned and frozen, all varieties and sizes, whole.

Peel-apart film article (Section 723.175) means a self-developing photographic film article consisting of a positive image receiving sheet, a light sensitive negative sheet, and a sealed reagent pod containing a developer reagent and designed so that all the chemical substances required to develop or process the film will not remain sealed within the article during and after the development of the film.

Percent load (Section 86.082-2) means the fraction of the maximum available torque at a specified engine speed.

Percent removal (Section 133.101). A percentage expression of the removal efficiency across a treatment plant for a given pollutant parameter, as determined from the 30-day average values of the raw wastewater influent pollutant concentrations to the facility and the 30-day average values of the effluent pollutant concentrations for a given time period.

Perceptible leaks (Section 60.621) means any petroleum solvent vapor or liquid leaks that are conspicuous from visual observation or that bubble after application of a soap solution, such as pools or droplets of liquid, open containers of solvent, or solvent laden waste standing open to the atmosphere.

Perfected (Section 306.12) means the point at which EPA determines that the filing requirements for a claim have been met.

Performance assessment (Section 191.12 [12/20/93])* means an analysis that: (1) Identifies the processes and events that might affect the disposal system; (2) examines the effects of these processes and events on the performance of the disposal system; and (3) estimates the cumulative releases of radionuclides, considering the associated uncertainties, caused by all significant processes and events. These estimates shall be incorporated into an overall probability distribution of cumulative release to the extent practicable.

Performance averaging period (Section 60.431) means 30 calendar days, one calendar month, or four consecutive weeks as specified in sections of this subpart.

Performance evaluation sample (Section 141.2) means a reference sample provided to a laboratory for the purpose of demonstrating that the laboratory can successfully analyze the sample within limits of performance specified by the Agency. The true value of the concentration of the reference material is unknown to the laboratory at the time of the analysis.

Performance specification (Section 247.101) means a specification that states the desired operation or function of a product but does not specify the materials from which the product must be constructed.

Periodic application of cover material (Section 257.3-8) means the application and compaction of soil or other suitable material over disposed solid waste at the end of each operating day or at such frequencies and in such a manner as to reduce the risk of fire and to impede disease vectors' access to the waste.

Periodic application of cover material (Section 257.3-6) means the application and compaction of soil or other suitable material over disposed solid waste at the end of each operating day or at such frequencies and in such a manner as to reduce the risk of fire and to impede vectors access to the waste.

Perlite composite board (Section 248.4 [2/17/89]) means insulation board composed of expanded perlite and fibers formed into rigid, flat, rectangular units with a suitable sizing material incorporated in the product. It may have on one or both surfaces a facing or coating to prevent excessive hot bitumen strike-in during roofing installation.

Permanent opening (Section 60.541 [9/15/87]) means an opening designed into an enclosure to allow tire components to pass through the enclosure by conveyor or other mechanical means, to provide access for permanent mechanical or electrical equipment, or to direct air flow into the enclosure. A permanent opening is not equipped with a door or other means of obstruction of air flow.

Permanent Radon Barrier (Section 192.31 [11/5/93])* means the final radon barrier constructed to achieve compliance with, including attainment of, the limit on releases of radon-222 in Sec. 192.32(b)(1)(ii).

Permanent storage capacity (Section 60.301) means grain storage capacity which is inside a building, bin, or silo.

Permit (Section 122.2) means an authorization, license or equivalent control document issued by EPA or an approved State to implement the requirements of this part and Parts 123 and 124. Permit includes an NPDES general permit (Section 122.28). Permit does not include any permit which has not yet been the subject of final agency action, such as a draft permit or a proposed permit.

Permit (Section 124.2) means an authorization, license, or equivalent control document issued by EPA or an approved State to implement the requirements of this part and Parts 122, 123, 144, 145, 233, 270 and 271. Permit includes RCRA permit by rule (Section 270.60), UIC area permit (Section 144.33), NPDES or 404 general permit (Sections 270.61, 144.34, and 233.38). Permit does not include RCRA interim status (Section 270.70), UIC authorization by rule (Section 144.21), or any permit which has not yet been the subject of final agency action, such as a draft permit or a proposed permit.

Permit (Section 129.2) means a permit for the discharge of pollutants into navigable waters under the National Pollutant Discharge Elimination System established by section 402 of the Act and implemented in regulations in 40 CFR Parts 124 and 125.

Permit (Section 144.3) means an authorization, license, or equivalent control document issued by EPA or an approved State to implement the requirements of this part, Parts 145, 146 and 124. Permit includes an area permit (Section 144.33) and an emergency permit (Section 144.34). Permit does not include UIC authorization by rule (Section 144.21), or any permit which has not yet been the subject of final agency action, such as a draft permit.

Permit (Section 146.3) means an authorization, license, or equivalent control document issued by EPA or an approved State to implement the requirements of this part and Parts 124, 144, and 145. Permit does not include RCRA interim status (Section 122.23), UIC authorization by rule (Sections 144.21 to 144.26 and 144.15), or any permit which has not yet been the subject of final agency action, such as a draft permit or a proposed permit.

Permit (Section 147.2902) means an authorization issued by EPA to implement UIC program requirements. Permit does not include the UIC authorization by rule or any permit which has not yet been the subject of final Agency action.

Permit (Section 2.309) means any permit applied for or granted under the Act.

Permit (Section 21.2) means any permit issued by either EPA or a State under the authority of section 402 of the Act; or by the Corps of Engineers under section 404 of the Act.

Permit (Section 22.03) means a permit issued under section 102 of the Marine Protection, Research, and Sanctuaries Act.

Permit (Section 232.2 [2/11/93])* means a written authorization issued by an approved State to implement the requirements of Part 233, or by the Corps under 33 CFR Parts 320-330. When used in these regulations, "permit" includes "general permit" as well as individual permit.

Permit (Section 233.3) means an authorization, license, or equivalent control document issued by an approved State to implement the requirements of this part and Part 124. Permit includes 404 general permit (Section 233.37), and 404 emergency permit (Section 233.38).

Permit (Section 256.06) is an entitlement to commence and continue operation of a facility as long as both procedural and performance standards are met. The term permit includes any functional equivalent such as a registration or license.

Permit (Section 270.2) means an authorization, license, or equivalent control document issued by EPA or an approved State to implement the requirements of this part and Parts 271 and 124. Permit includes permit by rule (Section 270.60), and emergency permit (Section 270.61). Permit does not include RCRA interim status (Subpart G of this part), or any permit which has not yet been the subject of final agency action, such as a draft permit or a proposed permit.

Permit (Section 403.3). See **NPDES Permit (Section 403.3)**.

Permit (Section 501.2 [5/2/89]) means an authorization, license, or equivalent control document issued by EPA or an approved State program to implement the requirements of this Part.

Permit Area (Section 440.141 [5/24/88]) means the area of land specified or referred to in an NPDES permit in which active mining and related activities may occur that result in the discharge regulated under the terms of the permit. Usually this is specifically delineated in an NPDES permit or permit application, but in other cases may be ascertainable from an Alaska Tri-agency permit application or similar document specifying the mine location, mining plan and similar data.

Permit or PSD permit (Section 124.41) means a permit issued under 40 CFR 52.21 or by an approved State.

Permit program (Section 60.2 [7/21/92])* means a comprehensive State operating permit system established pursuant to title V of the Act (42 U.S.C. 7661) and regulations codified in part 70 of this chapter and applicable State regulations, or a comprehensive Federal operating permit system established pursuant to title V of the Act and regulations codified in this chapter.

Permit program (Section 61.02 [3/16/94])* means a comprehensive State operating permit system established pursuant to title V of the Act (42 U.S.C. 7661) and regulations codified in part 70 of this chapter and applicable State regulations, or a comprehensive Federal operating permit system established pursuant to title V of the Act and regulations codified in this chapter.

Permit-by-rule (Section 270.2) means a provision of these regulations stating that a facility or activity is deemed to have a RCRA permit if it meets the requirements of the provision.

Permittee (Section 172.1) means any applicant to whom an experimental use permit has been granted.

Permitting authority (Section 230.3) means the District Engineer of the U.S. Army Corps of Engineers or such other individual as may be designated by the Secretary of the Army to issue or deny permits under section 404 of the Act; or the State Director of a permit program approved by EPA under section 404(g) and section 404(h) or his delegated representative.

Permitting authority (Section 60.2 [7/21/92])* means: (1) The State air pollution control agency, local agency, other State agency, or other agency authorized by the Administrator to carry out a permit program under part 70 of this chapter; or (2) The Administrator, in the case of EPA-implemented permit programs under title V of the Act (42 U.S.C. 7661).

Permitting authority (Section 61.02 [3/16/94])* means: (1) The State air pollution control agency, local agency, other State agency, or other agency authorized by the Administrator to carry out a permit program under part 70 of this chapter; or (2) The Administrator, in the case of EPA-implemented permit programs under title V of the Act (42 U.S.C. 7661).

Person (Section 104.2) means an individual, corporation, partnership, association, state, municipality or other political subdivision of a state, or any interstate body.

Person (Section 110.1) includes an individual, firm, corporation, association, and a partnership.

Person (Section 112.2 [8/25/93])*" includes an individual, firm, corporation, association, and a partnership.

Person (Section 122.2) means an individual, association, partnership, corporation, municipality, State or Federal agency, or an agent or employee thereof.

Person (Section 123.64) means the Agency, the State and any individual or organization having an interest in the subject matter of the proceeding.

Person (Section 124.2 [9/26/88])* means an individual, association, partnership, corporation, municipality, State, Federal, or Tribal agency, or an agency or employee thereof.

Person (Section 124.41) includes an individual, corporation, partnership, association, State, municipality, political subdivision of a State, and any agency, department, or instrumentality of the United States and any officer, agent or employee thereof.

Person (Section 13.2 [9/23/88]) means an individual, firm, partnership, corporation, association and, except for purposes of administrative offsets under Subpart C and interest, penalty and administrative costs under Subpart B of this regulation, includes State and local governments and Indian tribes and components of tribal governments.

Person (Section 141.2 [9/26/88])* means an individual; corporation; company; association; partnership; municipality; or State, Federal, or tribal agency.

Person (Section 142.2 [9/26/88])* means an individual; corporation; company; association; partnership; municipality; or State, Federal, or Tribal agency.

Person (Section 144.3 [9/26/88])*

means an individual, association, partnership, corporation, municipality, State, Federal, or Tribal agency, or an agency or employee thereof.

Person (Section 149.101) means an individual, corporation, company, association, partnership, State, or municipality.

Person (Section 15.4) means any natural person, corporation, partnership, unincorporated association, State or local government, or any agency, instrumentality, or subdivision of such a government or any interstate body.

Person (Section 154.3) means an applicant, registrant, manufacturer, pesticide user, environmental group, labor union, or other individual or group of individuals interested in pesticide regulation.

Person (Section 160.3 [8-17-89])* includes an individual, partnership, corporation, association, scientific or academic establishment, government agency, or organizational unit thereof, and any other legal entity.

Person (Section 163.2) means individuals, partnerships, corporations, and associations.

Person (Section 164.2 [7/10/92])* includes any individual, partnership, association, corporation, and any organized group of persons, whether incorporated or not.

Person (Section 2.201) means an individual, partnership, corporation, association, or other public or private organization or legal entity, including Federal, State or local governmental bodies and agencies and their employees.

Person (Section 2.305) has the meaning given it in section 1004(15) of the Act, 42 U.S.C. 6903(15).

Person (Section 2.310) has the meaning given it in section 101(21) of the Act, 42 U.S.C. 9601 (21).

Person (Section 205.2) means an individual, corporation, partnership, or association, and except as provided in sections 11(e) and 12(a) of the Act includes any officer, employee, department, agency or instrumentality of the United States, a State or any political subdivision of a State.

Person (Section 209.3) means any individual, corporation, partnership, or association, and includes any officer, employee, department, agency or instrumentality of the United States, a State, or any political subdivision of a State.

Person (Section 22.03) includes any individual, partnership, association, corporation, and any trustee, assignee, receiver or legal successor thereof; any organized group of persons whether incorporated or not; and any officer, employee, agent, department, agency or instrumentality of the Federal

Government, of any State or local unit of government, or of any foreign government.

Person (Section 231.2) means an individual, corporation, partnership, association, Federal agency, state, municipality, or commission, or political subdivision of a state, or any interstate body.

Person (Section 232.2 [2/11/93])* means an individual, association, partnership, corporation, municipality, State or Federal agency, or an agent or employee thereof.

Person (Section 232.2 [6/6/88]) means an individual, association, partnership, corporation, municipality, State or Federal agency, or an agent or employee thereof.

Person (Section 233.3) means an individual, association, partnership, corporation, municipality, State or Federal agency, or an agent or employee thereof.

Person (Section 248.4 [2/17/89]) means an individual, trust, firm, joint stock company, corporation (including a government corporation), partnership, association, Federal agency, State, municipality, commission, political subdivision of a State, or any interstate body.

Person (Section 249.04) means an individual, trust, firm, joint stock company, Federal agency, corporation (including a government corporation), partnership, association, State, municipality, commission, political subdivision of a State, or any interstate body.

Person (Section 250.4 [6/22/88]) means an individual, trust, firm, joint stock company, corporation (including a government corporation), partnership, association, State, municipality, commission, political subdivision of a State, or any interstate body.

Person (Section 252.4 [6/30/88]) means an individual, trust, firm, joint stock company, corporation (including a government corporation), partnership, association, Federal agency, State, municipality, commission, political subdivision of a State, or any interstate body.

Person (Section 253.4 [11/17/88]) means an individual, trust, firm, joint stock company, corporation (including a government corporation), partnership, association, Federal agency, State, municipality, commission, political subdivision of a State, or any interstate body.

Person (Section 259.10 [3/24/89]) means an individual, trust, firm, joint stock company, corporation (including a government corporation), partnership, association, State, municipality, commission, political subdivision of a State, any interstate body, or any department, agency or instrumentality of the United States.

Person (Section 260.10)

Person (Section 260.10) means an individual, trust, firm, joint stock company, Federal Agency, corporation (including a government corporation), partnership, association, State, municipality, commission, political subdivision of a State, or any interstate body.

Person (Section 270.2) means an individual, association, partnership, corporation, municipality, State or Federal agency, or an agent or employee thereof.

Person (Section 280.1) has the same meaning as provided in section 1004(15) of the Resource Conservation and Recovery Act, as amended, except that such term includes a consortium, a joint venture, a commercial entity, and the United States Government.

Person (Section 280.12 [9/23/88]) means an individual, trust, firm, joint stock company, Federal agency, corporation, state, municipality, commission, political subdivision of a state, or any interstate body. Person also includes a consortium, a joint venture, a commercial entity, and the United States Government.

Person (Section 300.5 [3-8-90])* as defined by section 101(21) of CERCLA, means an individual, firm, corporation, association, partnership, consortium, joint venture, commercial entity, United States government, state, municipality, commission, political subdivision of a state, or any interstate body.

Person (Section 302.3) means an individual, firm, corporation, association, partnership, consortium, joint venture, commercial entity, United States Government, State, municipality, commission, political subdivision of a State, or any interstate body.

Person (Section 32.102) means any individual, organization or unit of government that is or may become eligible to receive EPA assistance, or its employee. It also means any individual or organization that is or may become eligible to receive a subagreement or intergovernmental agreement.

Person (Section 34.105 [2-26-90]) means an individual, corporation, company, association, authority, firm, partnership, society, State, and local government, regardless of whether such entity is operated for profit or not for profit. This term excludes an Indian tribe, tribal organization, or any other Indian organization with respect to expenditures specifically permitted by other Federal law.

Person (Section 355.20) means any individual, trust, firm, joint stock company, corporation (including a government corporation), partnership, association, State, municipality, commission, political subdivision of a State, or interstate body.

Person (Section 370.2 [10/15/87]) means any individual, trust, firm, joint stock company, corporation (including a government corporation), partnership, association, State, municipality, commission, political subdivision of State, or interstate body.

Person (Section 4.2) means any individual, family, partnership, corporation, or association.

Person (Section 501.2 [5/2/89]) is an individual, association, partnership, corporation, municipality, State or Federal Agency, or an agent or employee thereof.

Person (Section 700.43 [8/17/88]) means a manufacturer, importer, or processor.

Person (Section 704.3 [12/22/88])* includes any individual, firm, company, corporation, joint venture, partnership, sole proprietorship, association, or any other business entity; any State or political subdivision thereof; and municipality; any interstate body; and any department, agency, or instrumentality of the Federal Government.

Person (Section 710.2) means any natural or juridicial person including any individual, corporation, partnership, or association, any State or political subdivision thereof, or any municipality, any interstate body and any department, agency, or instrumentality of the Federal Government.

Person (Section 712.3) means any natural person, firm, company, corporation, joint venture, partnership, sole proprietorship, association, or any other business entity, any State or political subdivision thereof, any municipality, any interstate body, and any department, agency, or instrumentality of the Federal government.

Person (Section 716.3) includes any individual, firm, company, corporation, joint-venture, partnership, sole proprietorship, association, or any other business entity, any State or political subdivision thereof, any municipality, any interstate body, and any department, agency, or instrumentality of the Federal government.

Person (Section 717.3) includes any individual, firm, company, corporation, joint venture, partnership, sole proprietorship, association, or any other business entity, any State or political subdivision thereof, and any department, agency, or instrumentality of the Federal Government.

Person (Section 720.3) means any natural person, firm, company, corporation, joint-venture, partnership, sole proprietorship, association, or any other business entity, any State or political subdivision thereof, any municipality, any interstate body, and any department, agency or instrumentality of the Federal

Government.

Person (Section 723.250) means any natural person, firm, company, corporation, joint-venture, partnership, sole proprietorship, association, or any other business entity, any State or political subdivision thereof, any municipality, any interstate body, and any department, agency or instrumentality of the Federal Government.

Person (Section 723.50) means any natural person, firm, company, corporation, joint-venture, partnership, sole proprietorship, association, or any other business entity, any State or political subdivision thereof, any municipality, any interstate body, and any department, agency or instrumentality of the Federal Government.

Person (Section 747.115) means any natural person, firm, company, corporation, joint-venture, partnership, sole proprietorship, association, or any other business entity, any State or political subdivision thereof, any municipality, any interstate body, and any department, agency or instrumentality of the Federal Government.

Person (Section 747.195) means any natural person, firm, company, corporation, joint-venture, partnership, sole proprietorship, association, or any other business entity, any State or political subdivision thereof, any municipality, any interstate body, and any department, agency or instrumentality of the Federal Government.

Person (Section 747.200) means any natural person, firm, company, corporation, joint-venture, partnership, sole proprietorship, association, or any other business entity, any State or political subdivision thereof, any municipality, any interstate body, and any department, agency or instrumentality of the Federal Government.

Person (Section 749.68 [1-3-90]) means ny natural person, firm, company, corporation, joint venture, partnership, sole proprietorship, association, or any other business entity; any State or political subdivision thereof; any municipality; any interstate body; and any department, agency, or instrumentality of the Federal Government.

Person (Section 761.3) means any natural or judicial person including any individual, corporation, partnership, or association; any State or political subdivision thereof; any interstate body; and any department, agency, or instrumentality of the Federal Government.

Person (Section 762.3) includes any natural person, corporation, firm, company, joint venture, partnership, sole proprietorship, association, or any other business entity, any State or political subdivision thereof, any municipality, any interstate body and any department, agency, or instrumentality of the Federal Government.

Person (Section 763.163 [7-12-89]) means any natural person, firm, company, corporation, joint-venture, partnership, sole proprietorship, association, or any other business entity; any State or political subdivision thereof, or any municipality; any interstate body and any department, agency, or instrumentality of the Federal Government.

Person (Section 763.63) means any natural person, firm, company, corporation, joint venture, partnership, sole proprietorship, association, or any other business entity, any State or political subdivision thereof, any municipality, any interstate body, and any department, agency, or instrumentality of the Federal Government.

Person (Section 790.3) means an individual, partnership, corporation, association, scientific or academic establishment, or organizational unit thereof, and any other legal entity.

Person (Section 792.3 [8-17-89])* includes an individual, partnership, corporation, association, scientific or academic establishment, government agency, or organizational unit thereof, and any other legal entity.

Person (Section 8.2) means any natural person, corporation, partnership, unincorporated association, State or local government, and any agency, instrumentality, or subdivision of such a government.

Person (Section 8.33) means any natural person, corporation, partnership, unincorporated association, State or local government, and any agency, instrumentality or subdivision of such a government.

Person (Section 82.3 [12/10/93])* means any individual or legal entity, including an individual, corporation, partnership, association, state, municipality, political subdivision of a state, Indian tribe; any agency, department, or instrumentality of the United States; and any officer, agent, or employee thereof.

Personal property (Section 247.101) means any property that is not real property and that is movable or not attached to the land. Personal property can be a single or multi-component or multi-material product.

Personal property (Section 30.200) means property other than real property. It may be tangible (having physical existence), such as equipment and supplies, or intangible (having no physical existence), such as patents, inventions, and copyrights.

Personal property (Section 35.6015 [6-5-90])* Property other than real property. It includes both supplies and equipment.

Personal protective equipment (Section 721.3 [7-27-89]) means any chemical protective clothing or device placed on the body to prevent contact with, and exposure to, an identified chemical substance or substances in the work area. Examples include, but are not limited to, chemical protective clothing, aprons, hoods, chemical goggles, face splash shields, or equivalent eye protection, and various types of respirators. Barrier creams are not included in this definition.

Personnel or facility personnel (Section 260.10) means all persons who work, at, or oversee the operations of, a hazardous waste facility, and whose actions or failure to act may result in noncompliance with the requirements of Part 264 or 265 of this chapter.

Persons (Section 304.12 [5/30/89]) means an individual, firm, corporation, association, partnership, consortium, joint venture, commercial entity, United States Government, State, municipality, commission, political subdivision of a State, or any interstate body.

Persons not displaced (Section 4.2). The following is a nonexclusive listing of persons who do not qualify as a displaced person under these regulations. (i) A person who moves before the initiation of negotiations (see also Section 4.403(e)); or (ii) A person who initially enters into occupancy of the property after the date of its acquisition for the project; or (iii) A person who is not required to relocate permanently as a direct result of a project. Such determination shall be made by the Agency in accordance with any guidelines established by the Federal agency funding the project (see also Appendix A); or (iv) A person whom the Agency determines is not displaced as a direct result of a partial acquisition; or (v) A person who, after receiving a notice of relocation eligibility (described at Section 4.203), is notified in writing that he or she will not be displaced for a project. Such notice shall not be issued unless the person has not moved and the Agency agrees to reimburse the person for any expenses incurred to satisfy any binding contractual relocation obligations entered into after the effective date of the notice of relocation eligibility; or (vi) An owner-occupant who voluntarily sells his or her property (as described at Section 4.101(a) in Appendix A) after being informed in writing that if a mutually satisfactory agreement of sale cannot be reached, the Agency will not acquire the property. In such cases, however, any resulting displacement of a tenant is subject to these regulations; or (vii) A person who retains the right of use and occupancy of the real property for life following its acquisition by the Agency; or (viii) A person who retains the right of use and occupancy of the real property for a fixed term after its acquisition by the Department of Interior under Pub. L. 93-477 or Pub. L. 93-303.

Pest (Section 162.3) means (1) any insect, rodent, nematode, fungus, weed, or (2) any other form of terrestrial or aquatic plant or animal life or virus, bacterial organism or other microorganism (except viruses, bacteria, or other microorganisms on or in living man or other living animals) which the Administrator declares to be a pest under section 25(c)(1) of the Act and Section 162.14 as being injurious to health or environment.

Pest (Section 455.10) means (1) Any insect, rodent, nematode, fungus, weed, or (2) any other form of terrestrial or aquatic plant or animal life or virus, bacteria, or other micro-organism (except virusus, bacteria, or other micro-organisms on or in living man or other living animals) which the Administrator declares to be a pest under section 25(c)(1) of Pub. L. 94-140, Federal Insecticide, Fungicide and Rodenticide Act.

Pest problem (Section 162.152) means (1) a pest infestation and its consequences, or (2) any condition for which the use of plant regulators, defoliants, or desiccants would be appropriate.

Pesticidal product (Section 167.3 [9/8/88]) means a pesticide, active ingredient, or device.

Pesticidal product report (Section 167.3 [9/8/88]) means information showing the types and amounts of pesticidal products which were: (1) Produced in the past calendar year; (2) produced in the current calendar year; and, (3) sold or distributed in the past calendar year. For active ingredients, the pesticidal product report must include information on the types and amounts of an active ingredient for which there is actual or constructive knowledge of its use or intended use as a pesticide. This pesticidal product report also pertains to those products produced for export only which must also be reported. A positive or a negative annual report is required in order to maintain registration for the establishment.

Pesticide (Section 152.3 [5/4/88]) means any substance or mixture of substances intended for preventing, destroying, repelling, or mitigating any pest, or intended for use as a plant regulator, defoliant, or desiccant, other than any article that: (1) Is a new animal drug under FFDCA section 201(w), or (2) Is an animal drug that has been determined by regulation of the Secretary of Health and Human Services not to be a new animal drug, or (3) Is an animal feed under FFDCA section 201(x) that bears or contains any substances described by paragraph (s)(1) or (2) of this section.

Pesticide (Section 162.3) means any substance or mixture of substances intended for preventing, destroying, repelling, or mitigating any pest, and any substance or mixture of substances intended for use as a plant regulator, defoliant, or desiccant. The term pesticide when not specifically modified or delimited by other words, shall include any one or combination of the following aspects of the term: the active ingredient (chemical or biological); the pesticide formulation; and the pesticide product. Examples of classes of pesticides are defined in Section 162.3(ff).

Pesticide (Section 165.1) means (1) any substance or mixture of substances intended for preventing, destroying, repelling, or mitigating any pest, or (2) any substance or mixture of substances intended for use as a plant regulator, defoliant, or desiccant.

Pesticide (Section 455.10) means any substance or mixture of substances intended for preventing, destroying, repelling, or mitigating any pest.

Pesticide (Sections 710.2; 720.3) shall have the meaning contained in the Federal Insecticide, Fungicide, and Rodenticide Act, 7 U.S.C. 136 et seq., and the regulations issued thereunder.

Pesticide chemical (Section 163.2) shall have the same meaning as it has in paragraph (q) of section 201 of the Act.

Pesticide chemical (Section 177.3 [12-5-90]) means any substance which alone, or in chemical combination with or in formulation with one or more other substances, is a "pesticide" within the meaning of FIFRA and which is used in the production, storage, or transportation of any raw agricultural commodity or processed food. The term includes any substance that is an active ingredient, intentionally-added inert ingredient, or impurity of such a "pesticide."

Pesticide chemical (Section 180.1) as defined in section 201(q) of the act, means any substance which, alone, in chemical combination, or in formulation with one or more other substances, is an economic poison within the meaning of the Federal Insecticide, Fungicide, and Rodenticide Act (7 U.S.C. 135-135k) and as defined in Section 362.2 of regulations for its enforcement (7 CFR 362.2), as now in force or as hereafter amended, and which is used in the production, storage, or transportation of raw agricultural commodities.

Pesticide chemicals (Section 455.10) means the sum of all active ingredients manufactured at each facility covered by this part.

Pesticide formulation (Section 162.3) means the substance or mixture of substances comprised of all active and inert (if any) ingredients of a pesticide product.

Pesticide incinerator (Section 165.1) means any installation capable of the controlled combustion of pesticides, at a temperature of 1000 degrees C (1832 degrees F) for two seconds dwell time in the combustion zone, or lower temperatures and related dwell times that will assure complete conversion of the specific pesticide to inorganic gases and solid ash residues. Such installation complies with the Agency Guidelines for the Thermal Processing of Solid Wastes as prescribed in 40 CFR Part 240.

Pesticide product (Section 152.3 [5/4/88]) means a pesticide in the particular form (including composition, packaging, and labeling) in which the pesticide is, or is intended to be, distributed or sold. The term includes any physical apparatus used to deliver or apply the pesticide if distributed or sold with the pesticide.

Pesticide product (Section 162.152). See **Product (Section 162.152)**.

Pesticide product (Section 162.3) means a pesticide offered for distribution and use, and includes any labeled container and any supplemental labeling.

Pesticide residue (Section 177.3 [12-5-90]) means a residue of a pesticide chemical or of any metabolite or degradation product of a pesticide chemical.

Pesticide use (Section 154.3) means a use of a pesticide (described in terms of the application site and other applicable identifying factors) that is included in the labeling of a pesticide product which is registered, or for which an application for registration is pending, and the terms and conditions (or proposed terms and conditions) of registration for the use.

Pesticide-related wastes (Section 165.1) means all pesticide-containing wastes or by-products which are produced in the manufacturing or processing of a pesticide and which are to be discarded, but which, pursuant to acceptable pesticide manufacturing or processing operations, are not ordinarily a part of or contained within an industrial waste stream discharged into a sewer or the waters of a state.

Pesticides (Section 125.58) means demeton, guthion, malathion, mirex, methoxychlor, and parathion.

Pesticides report (Section 167.1) means information showing the types and amounts of pesticides or devices which are being produced in the current calendar year, have been produced in the past calendar year, and which have been sold or distributed in the past calendar year.

Petition (Section 163.2) means a petition filed with the Administrator, Environmental Protection Agency pursuant to section 408(d)(1) of the Act.

Petition (Section 2.308) means a petition for the issuance of a regulation establishing a tolerance for a pesticide chemical or exempting the pesticide chemical from the necessity of a tolerance, pursuant to section 408(d) of the Act, 21 U.S.C. 346a(d).

Petitioner (Section 123.64) means any person whose petition for commencement of withdrawal proceedings has been granted by the Administrator.

Petitioner (Section 164.2 [7/10/92])* means any person adversely affected by a notice of the Administrator who requests a public hearing.

Petitioner (Section 2.308) means a person who has submitted a petition to EPA (or to a predecessor agency).

Petitioner (Section 350.1 [7/29/88]) is any person who submits a petition under this regulation requesting disclosure of a chemical identity claimed as trade secret.

Petrochemical operations (Section 419.31) shall mean the production of second-generation petrochemicals (i.e., alcohols, ketones, cumene, styrene, etc.) or first generation petrochemicals and isomerization products (i.e., BTX, olefins, cyclohexane, etc.) when 15 percent or more of refinery production is as first-generation petrochemicals and isomerization products.

Petroleum (Section 61.341 [3-7-90]) means the crude oil removed from the earth and the oils derived from tar sands, and coal.

Petroleum (Sections 60.101; 60.111; 60.111a; 60.111b; 60.591; 60.691 [11/23/88]) means the crude oil removed from the earth and the oils derived from tar sands, shale, and coal.

Petroleum dry cleaner (Section 60.621) means a dry cleaning facility that uses petroleum solvent in a combination of washers, dryers, filters, stills, and settling tanks.

Petroleum liquids (Section 60.111b) means petroleum, condensate, and any finished or intermediate products manufactured in a petroleum refinery.

Petroleum liquids (Sections 60.111; 60.111a) means petroleum, condensate, and any finished or intermediate products manufactured in a petroleum refinery but does not mean Nos. 2 through 6 fuel oils as specified in ASTM D396-78, gas turbine fuel oils Nos. 2-GT through 4-GT as specified in ASTM D2880-78, or diesel fuel oils Nos. 2-D and 4-D as specified in ASTM D975-78. (These three methods are incorporated by reference - see Section 60.17.)

Petroleum marketing facilities (Section 280.92 [10/26/88]) include all facilities at which petroleum is produced or refined and all facilities from which petroleum is sold or transferred to other petroleum marketers or to the public.

Petroleum marketing firms (Section 280.92 [10/26/88]) are all firms owning petroleum marketing facilities. Firms owning other types of facilities with USTs as well as petroleum marketing facilities are considered to be petroleum marketing firms.

Petroleum refinery (Section 60.101) means any facility engaged in producing gasoline, kerosene, distillate fuel oils, residual fuel oils, lubricants, or other products through distillation of petroleum or through redistillation, cracking or reforming of unfinished petroleum derivatives.

Petroleum refinery (Section 60.41b [12/16/87])* means industrial plants as classified by the Department of Commerce under Standard Industrial Classification (SIC) Code 29.

Petroleum refinery (Section 60.591) means any facility engaged in producing gasoline, kerosene, distillate fuel oils, residual fuel oils, lubricants, or other products through the distillation of petroleum, or through the redistillation, cracking, or reforming of unfinished petroleum derivatives.

Petroleum refinery (Section 60.691 [11/23/88]) means any facility engaged in producing gasoline, kerosene, distillate fuel oils, residual fuel oils, lubricants, or other products through the distillation of petroleum, or through the redistillation of petroleum, cracking, or reforming unfinished petroleum derivatives.

Petroleum refinery (Section 61.341 [3-7-90]) means any facility engaged in producing gasoline, kerosene, distillate fuel oils, residual fuel oils, lubricants, or other products through the distillation of petroleum, or through the redistillation, cracking, or reforming of unfinished petroleum derivatives. Petroleum means the crude oil removed from the earth and the oils derived from tar sands, shale, and coal.

Petroleum refinery (Sections 60.111; 60.111a) means each facility engaged in producing gasoline, kerosene, distillate fuel oils, residual fuel oils, lubricants, or other products through distillation of petroleum or through redistillation, cracking, extracting, or reforming of unfinished petroleum derivatives.

Petroleum UST system (Section 280.12 [9/23/88]) means an underground storage tank system that contains petroleum or a mixture of petroleum with de minimis quantities of other regulated substances. Such systems include those containing motor fuels, jet fuels, distillate fuel oils, residual fuel oils, lubricants, petroleum solvents, and used oils.

pH (Section 257.3-5) means the logarithm of the reciprocal of hydrogen ion concentration.

pH (Section 420.02) means the value obtained by the standard method specified in 40 CFR 136.3.

pH (Section 796.2750) of a sediment

or soil is the negative logarithm to the base ten of the hydrogen ion activity of the sediment or soil suspension. It is usually measured by a suitable sensing electrode coupled with a suitable reference electrode at a 1/1 solid/solution ratio by weight.

Pharmacokinetics (Section 795.232 [1-8-90]) means the study of the rates of absorption, tissue distribution, biotransformation, and excretion.

Phase I (Section 270.2) means that phase of the Federal hazardous waste management program commencing on the effective date of the last of the following to be initially promulgated: 40 CFR Parts 260, 261, 262, 263, 265, 270, and 271. Promulgation of Phase I refers to promulgation of the regulations necessary for Phase I to begin.

Phase II (Section 270.2) means that phase of Federal hazardous waste management program commencing on the effective date of the first subpart of 40 CFR Part 264, Subparts F through R to be initially promulgated. Promulgation of Phase II refers to promulgation of the regulations necessary for Phase II to begin.

Phased disposal (Section 61.251 [12-15-89])* means a method of tailings management and disposal which uses lined impoundments which are filled and then immediately dried and covered to meet all applicable Federal standards.

PHDD (Section 766.3). See **Polyhalogenated dibenzo-p-dioxin (Section 766.3)**.

PHDF (Section 766.3). See **Polyhalogenated dibenzofuran (Section 766.3)**.

Phenolic compounds (Section 420.02). See **Phenols 4AAP (Section 420.02)**.

Phenolic insulation (Section 248.4 [2/17/89]) means insulation made with phenolic plastics which are plastics based on resins made by the condensation of phenols, such as phenol or cresol, with aldehydes.

Phenols (Section 410.01) shall mean total phenols as measured by the procedure listed in 40 CFR Part 136.

Phenols 4AAP (or phenolic compounds) (Section 420.02) means the value obtained by the method specified in 40 CFR 136.3.

Phosphate rock feed (Section 60.401) means all material entering the process unit including, moisture and extraneous material as well as the following ore minerals: Fluorapatite, hydroxylapatite, chlorapatite, and carbonateapatite.

Phosphate rock plant (Section 60.401) means any plant which produces or prepares phosphate rock product by any or all of the following processes: Mining, beneficiation, crushing, screening, cleaning, drying, calcining, and grinding.

Photographic article (Section 723.175) means any article which will become a component of an instant photographic or peel-apart film article.

Photoperiod (Sections 797.2130; 797.2150) means the light and dark periods in a 24 hour day. This is usually expressed in a form such as 17 hours light/7 hours dark or 17L/7D.

Physical and Thermal Integrity (Section 85.2122(a)(7)(ii)(C)) means the ability of the material of the cap and/or rotor to resist physical and thermal breakdown.

Physical chemical treatment system (Section 420.11) means those full scale coke plant wastewater treatment systems incorporating full scale granular activated carbon adsorption units which were in operation prior to January 7, 1981, the date of proposal of this regulation.

Physical construction (Section 270.2) means excavation, movement of earth, erection of forms or structures, or similar activity to prepare an HWM facility to accept hazardous waste.

PIAT (Section 300.5) is the Public Information Assist Team.

Pickles, fresh (Section 407.61) shall mean the processing of fresh cucumbers and other vegetables, all varieties, all sizes from whole to relish, all styles, cured after packing.

Pickles, processed (Section 407.61) shall mean the processing of pickles, cucumbers and other vegetables, all varieties, sizes and types, made after fermentation and storage.

Pickles, salt stations (Section 407.61) shall mean the handling and subsequent preserving of cucumbers and other vegetables at salting stations or tankyards, by salt and other chemical additions necessary to achieve proper fermentation for the packing of processed pickle products. Limitations include allowances for the discharge of spent brine, tank wash, tank soak, and cucumber wash waters. At locations where both salt station and process pack operations (Section 407.61(o)) occur, additive allowances shall be made for both of these sources in formulation of effluent limitations. The effluent limitations are to be calculated based upon the total annual weight (1000 lb, kkg) of raw product processed at each of the salt station and process pack operations. Allowances for contaminated stormwater runoff should be considered in NPDES permit formulation on a case-by-case basis.

Pickling bath (Section 468.02) shall mean any chemical bath (other than alkaline cleaning) through which a workpiece is processed.

Pickling fume scrubber (Section 468.02) shall mean the process of using an air pollution control device to remove particulates and fumes from air above a pickling bath by entraining the pollutants in water.

Pickling rinse (Section 468.02) shall mean a rinse, other than an alkaline cleaning rinse, through which a workpiece is processed. A rinse consisting of a series of rinse tanks is considered as a single rinse.

Pickling rinse for forged parts (Section 468.02) shall mean a rinse, other than an alkaline cleaning rinse, through which forged parts are processed. A rinse consisting of a series of rinse tanks is considered as a single rinse.

Pickup Truck (Section 600.002-85) means a non-passenger automobile which has a passenger compartment and an open cargo bed.

Picocurie (pCi) (Section 141.2) means the quantity of radioactive material producing 2.22 nuclear transformations per minute.

Pieces (Section 427.71) shall mean floor tile measured in the standard size of 12" x 12" x 3/32".

Pile (Section 260.10) means any non-containerized accumulation of solid, nonflowing hazardous waste that is used for treatment or storage.

Pineapples (Section 407.61) shall mean the processing of pineapple into the following product styles: Canned, peeled, sliced, chunk, tidbit, diced, crushed, and any other related piece size, juice and concentrate. It also specifically includes the on-site production of by-products such as alcohol, sugar or animal feed.

Pipe and tube mill (Section 420.71) means those steel hot forming operations that produce butt welded or seamless tubular steel products.

Pipe or Piping (Section 280.12 [9/23/88]) means a hollow cylinder or tubular conduit that is constructed of non-earthen materials.

Pipe, tube and other (Section 420.91) means those acid pickling operations that pickle pipes, tubes or any steel product other than those included in paragraphs (k), (l) and (m) of Section 420.91.

Pipeline facilities (including gathering lines) (Section 280.12 [9/23/88]) are new and existing pipe rights-of-way and any associated equipment, facilities, or buildings.

Pipeline wrap (Section 763.163 [7-12-89]) Means an asbestos-containing product made of paper felt intended for use in wrapping or coating pipes for insulation purposes.

Piping (Section 280.12 [9/23/88]). See **Pipe (Section 280.12)**.

Plan (Section 300.6) means the

National Oil and Hazardous Substances Pollution Contingency Plan published under section 311(c) of the CWA and revised pursuant to section 105 of CERCLA.

Plan (Section 51.100) means an implementation plan approved or promulgated under section 110 or 172 of the Act.

Plan (Section 58.1) means an implementation plan, approved or promulgated pursuant to section 110 of the Clean Air Act.

Plan (Section 60.21) means a plan under section 111(d) of the Act which establishes emission standards for designated pollutants from designated facilities and provides for the implementation and enforcement of such emission standards.

Planned renovation operations (Section 61.141 [11-20-90])* means a renovation operation, or a number of such operations, in which some RACM will be removed or stripped within a given period of time and that can be predicted. Individual nonscheduled operations are included if a number of such operations can be predicted to occur during a given period of time based on operating experience.

Planning (Section 256.06) includes identifying problems, defining objectives, collecting information, analyzing alternatives and determining necessary activities and courses of action.

Planning target (Section 35.105) means the amount of Federal financial assistance which the Regional Administrator suggests that an applicant consider in developing its application and work program.

Plans (Section 240.101) means reports and drawings, including a narrative operating description, prepared to describe the facility and its proposed operation.

Plans (Section 241.101) means reports and drawings, including a narrative operating description, prepared to describe the land disposal site and its proposed operation.

Plant (Section 82.3 [12/10/93])* means one or more facilities at the same location owned by or under common control of the same person.

Plant blending records (Section 60.581) means those records which document the weight fraction of organic solvents and solids used in the formulation or preparation of inks at the vinyl or urethane printing plant where they are used.

Plant regulators (Section 162.3) includes all substances or mixtures of substances, except defoliants as defined in paragraph (ff)(5) of this section, desiccants as defined in paragraph (ff)(16) of this section, herbicides as defined in paragraph (ff)(9) of this section, and nutrients, intended to cause, through physiological and biochemical action, plant responses of benefit to man. Plant regulators include, but are not limited to, substances or mixtures of substances intended to cause fruit thinning, fruit setting, stem elongation, stimulation or retardation, abscission inhibition, branch structure modification, sucker control, flower induction or inhibition, increased flowering, altered sex expression, extended flowering periods, fruit ripening stimulation, physiological disease inhibition, rooting of cuttings, or dormancy induction or release.

Plant site (Section 440.141 [5/24/88]) means the area occupied by the mine, necessary haulage ways from the mine to the beneficiation process, the beneficiation area, the area occupied by the wastewater treatment facilities and the storage areas for waste materials and solids removed from the wastewaters during treatment.

Plasma arc incinerator (Section 260.10 [8/18/92])* means any enclosed device using a high intensity electrical discharge or arc as a source of heat followed by an afterburner using controlled flame combustion and which is not listed as an industrial furnace.

Plastic body (Section 60.391) means an automobile or light-duty truck body constructed of synthetic organic material.

Plastic body component (Section 60.391) means any component of an automobile or light-duty truck exterior surface constructed of synthetic organic material.

Plastic material (Section 463.2) is a synthetic organic polymer (i.e., a thermoset polymer, a thermoplastic polymer, or a combination of a natural polymer and a thermoset or thermoplastic polymer) that is solid in its final form and that was shaped by flow. The material can be either a homogeneous polymer or a polymer combined with fillers, plasticizers, pigments, stabilizers, or other additives.

Plastic parts (Section 60.721 [1/29/88]) means panels, housings, bases, covers, and other business machine components formed of synthetic polymers.

Plastic rigid foam (Section 248.4 [2/17/89]) means cellular polyurethane insulation, cellular polyisocyanurate insulation, glass fiber reinforced polyisocyanurate/polyurethane foam insulation, cellular polystyrene insulation, phenolic foam insulation, spray-in-place foam and foam-in-place insulation.

Plastics molding and forming (Section 463.2) is a manufacturing process in which plastic materials are blended, molded, formed, or otherwise processed into intermediate or final products.

Plate mill (Section 420.71) means

those steel hot forming operations that produce flat hot-rolled products which are (1) between 8 and 48 inches wide and over 0.23 inches thick; or (2) greater than 48 inches wide and over 0.18 inches thick.

Plate soak (Section 461.2) shall mean the process operation of soaking or reacting lead subcategory battery plates, that are more than 2.5 mm (0.100 in) thick, in sulfuric acid.

Plugging (Sections 144.3; 146.3; 147.2902) means the act or process of stopping the flow of water, oil or gas into or out of a formation through a borehole or well penetrating that formation.

Plugging and abandonment plan (Section 144.61) means the plan for plugging and abandonment prepared in accordance with the requirements of Section 144.28 and Section 144.51.

Plugging record (Section 146.3) means a systematic listing of permanent or temporary abandonment of water, oil, gas, test, exploration and waste injection wells, and may contain a well log, description of amounts and types of plugging material used, the method employed for plugging, a description of formations which are sealed and a graphic log of the well showing formation location, formation thickness, and location of plugging structures.

Plums (Section 407.61) shall mean the processing of plums into the following product styles: Canned and frozen, pitted and unpitted, peeled and unpeeled, blanched and unblanched, whole, halved, and other piece size.

Plunger (Section 85.2122(a)(3)(ii)). See **Accelerator Pump (Section 85.2122(a)(3)(ii))**.

PM_{10} (Sections 51.100; 58.1) means particulate matter with an aerodynamic diameter less than or equal to a nominal 10 micrometers as measured by a reference method based on Appendix J of Part 50 of this chapter and designated in accordance with Part 53 of this chapter or by an equivalent method designated in accordance with Part 53 of this chapter.

PM_{10} emissions (Section 51.100) means finely divided solid or liquid material, with an aerodynamic diameter less than or equal to a nominal 10 micrometers emitted to the ambient air as measured by an applicable reference method, or an equivalent or alternative method, specified in this chapter or by a test method specified in an approved State implementation plan.

PM_{10} sampler (Section 53.1) means a device, associated with a manual method for measuring PM_{10}, designed to collect PM_{10} from an ambient air sample, but lacking the ability to automatically analyze or measure the collected sample to determine the mass concentration of PM_{10} in the sampled air.

PMN (Section 700.43 [8/17/88]). See **Premanufacture notice (Section 700.43)**.

Pneumatic coal-cleaning equipment (Section 60.251) means any facility which classifies bituminous coal by size or separates bituminous coal from refuse by application of air stream(s).

Point of disinfectant application (Section 141.2 [6/29/89]) is the point where the disinfectant is applied and water downstream of that point is not subject to recontamination by surface water runoff.

Point of waste generation (Section 61.341 [1/7/93])* means the location where the waste stream exits the process unit component or storage tank prior to handling or treatment in an operation that is not an integral part of the production process, or in the case of waste management units that generate new wastes after treatment, the location where the waste stream exits the waste management unit component.

Point source (Section 122.2 [1/4/89])* means any discernible, confined, and discrete conveyance, including but not limited to, any pipe, ditch, channel, tunnel, conduit, well, discrete fissure, container, rolling stock, concentrated animal feeding operation, landfill leachate collection system, vessel or other floating craft from which pollutants are or may be discharged. This term does not include return flows from irrigated agriculture or agricultural storm water runoff. (See Section 122.3.)

Point source (Section 257.3-3) is defined in the Clean Water Act, as amended, 33 U.S.C. 1251 et seq., and implementing regulations, specifically 33 CFR Part 323 (42 FR 37122, July 19, 1977).

Point source (Section 401.11) means any discernible, confined and discrete conveyance, including but not limited to any pipe, ditch, channel, tunnel, conduit, well, discrete fissure, container, rolling stock, concentrated animal feeding operation, or vessel or other floating craft, from which pollutants are or may be discharged.

Point source (Section 51.100) means the following: (1) For particulate matter, sulfur oxides, carbon monoxide, volatile organic compounds (VOC) and nitrogen dioxide: (i) Any stationary source the actual emissions of which are in excess of 90.7 metric tons (100 tons) per year of the pollutant in a region containing an area whose 1980 urban place population, as defined by the U.S. Bureau of the Census, was equal to or greater than 1 million. (ii) Any stationary source the actual emissions of which are in excess of 22.7 metric tons (25 tons) per year of the pollutant in a region containing an area whose 1980 urban place population, as defined by the U.S. Bureau of the Census, was less than 1 million; or (2) For lead or lead compounds measured as elemental lead, any stationary source that actually emits a total of 4.5 metric tons (5 tons) per year or more.

Point source (Sections 233.3; 260.10) means any discernible, confined, and discrete conveyance, including but not limited to any pipe, ditch, channel, tunnel, conduit, well, discrete fissure, container, rolling stock, concentrated animal feeding operation, vessel, or other floating craft from which pollutants are or may be discharged. This term does not include return flows from irrigated agriculture.

Point-of-entry treatment device (Section 141.2 [7/8/87]) is a treatment device applied to the drinking water entering a house or building for the purpose of reducing contaminants in the drinking water distributed throughout the house or building.

Point-of-use treatment device (Section 141.2 [7/8/87]) is a treatment device applied to a single tap used for the purpose of reducing contaminants in drinking water at that one tap.

Political subdivision (Section 35.6015 [6-5-90])* The unit of government that the State determines to have met the State's legislative definition of a political subdivision.

Pollutant (Section 122.2) means dredged spoil, solid waste, incinerator residue, filter backwash, sewage, garbage, sewage sludge, munitions, chemical wastes, biological materials, radioactive materials (except those regulated under the Atomic Energy Act of 1954, as amended (42 U.S.C. 2011 et seq.)), heat, wrecked or discarded equipment, rock, sand, cellar dirt and industrial, municipal, and agricultural waste discharged into water. It does not mean: (a) Sewage from vessels; or (b) Water, gas, or other material which is injected into a well to facilitate production of oil or gas, or water derived in association with oil and gas production and disposed of in a well, if the well used either to facilitate production or for disposal purposes is approved by authority of the State in which the well is located, and if the State determines that the injection or disposal will not result in the degradation of ground or surface water resources. Note: Radioactive materials covered by the Atomic Energy Act are those encompassed in its definition of source, byproduct, or special nuclear materials. Examples of materials not covered include radium and accelerator-produced isotopes. See Train v. Colorado Public Interest Research Group, Inc., 426 U.S. 1 (1976).

Pollutant (Section 21.2) means dredged spoil, solid waste, incinerator residue, sewage, garbage, sewage sludge, munitions, chemical wastes, biological materials, radioactive materials, heat, wrecked or discarded equipment, rock, sand, cellar dirt and industrial, municipal, and agricultural waste discharged into water. For the purposes of this section, the term also means sewage from vessels within the meaning of section 312 of the Act.

Pollutant (Section 230.3) means dredged spoil, solid waste, incinerator residue, sewage, garbage, sewage sludge, munitions, chemical wastes, biological materials, radioactive materials not covered by the Atomic Energy Act, heat, wrecked or discarded equipment, rock, sand, cellar dirt, and industrial, municipal, and agricultural waste discharged into water. The legislative history of the Act reflects that radioactive materials as included within the definition of pollutant in section 502 of the Act means only radioactive materials which are not encompassed in the definition of source, byproduct, or special nuclear materials as defined by the Atomic Energy Act of 1954, as amended, and regulated under the Atomic Energy Act. Examples of radioactive materials not covered by the Atomic Energy Act and, therefore, included within the term pollutant, are radium and accelerator produced isotopes. See Train v. Colorado Public Interest Research Group, Inc., 426 U.S. 1 (1976).

Pollutant (Section 233.3) means dredged spoil, solid waste, incinerator residue, filter backwash, sewage, garbage, sewage sludge, munitions, chemical wastes, biological materials, radioactive material (except those regulated under the Atomic Energy Act of 1954, as amended (42 U.S.C. 2011 et seq.)), heat, wrecked or discarded equipment, rock, sand, cellar dirt and industrial, municipal, and agricultural waste discharged into water. It does not mean: (a) Sewage from vessels; or (b) Water, gas or other material which is injected into a well to facilitate production of oil or gas, or water derived in association with oil and gas production and disposed of in a well, if the well used either to facilitate production or for disposal purposes is approved by authority of the State in which the well is located, and if the State determines that injection or disposal will not result in the degradation of ground or surface water resource.

Pollutant (Section 257.3-3) is defined in the Clean Water Act, as amended, 33 U.S.C. 1251 et seq., and implementing regulations, specifically 33 CFR Part 323 (42 FR 37122, July 19, 1977).

Pollutant (Section 401.11) means dredged spoil, solid waste, incinerator residue, sewage, garbage, sewage sludge, munitions, chemical wastes, biological materials, radioactive materials, heat, wrecked or discarded equipment, rock, sand, cellar dirt and industrial, municipal and agricultural waste discharged into water. It does not mean (1) sewage from vessels or (2) water, gas or other material which is injected into a well to facilitate production of oil or gas, or water derived in association with oil or gas production and disposed of in a well, if the well, used either to facilitate production or for disposal purposes, is approved by authority of the State in which the well is located, and if such State determines that such injection or disposal will not result in degradation of ground or surface water resources.

Pollutant or contaminant (Section 300.5 [3-8-90])* as defined by section 101(33) of CERCLA, shall include, but not be limited to, any element, substance, compound, or mixture, including disease-causing agents, which after release into the environment and upon exposure, ingestion, inhalation, or assimilation into any organism, either directly from the environment or indirectly by ingestion through food chains, will or may reasonably be anticipated to cause death, disease, behavioral abnormalities, cancer, genetic mutation, physiological malfunctions (including malfunctions in reproduction) or physical deformations, in such organisms or their offspring. The term does not include petroleum, including crude oil or any fraction thereof which is not otherwise specifically listed or designated as a hazardous substance under section 101(14)(A) through (F) of CERCLA, nor does it include natural gas, liquified natural gas, or synthetic gas of pipeline quality (or mixtures of natural gas and such synthetic gas). For purposes of the NCP, the term pollutant or contaminant means any pollutant or contaminant that may present an imminent and substantial danger to public health or welfare.

Pollutant or contaminant (Section 310.11 [1/15/93])* as defined by section 104(a)(2) of CERCLA, includes, but is not limited to, any element, substance, compound, or mixture, including disease-causing agents, which after release into the environment and upon exposure, ingestion, inhalation, or assimilation into any organism, either directly from the environment or indirectly by ingestion through food chains, will or may reasonably be anticipated to cause death, disease, behavioral abnormalities, cancer, genetic mutation, physiological malfunctions (including malfunctions in reproduction) or physical deformations, in such organisms or their offspring. The term does not include petroleum, including crude oil and any fraction thereof that is not otherwise specifically listed or designated as a hazardous substance under section 101(14) (A) through (F) of CERCLA, not does it include natural gas, liquefied natural gas, or synthetic gas of pipeline quality (or mixtures of natural gas and such synthetic gas).

Pollution (Section 230.3) means the man-made or man-induced alteration of the chemical, physical, biological or radiological integrity of an aquatic ecosystem.

Pollution (Sections 130.2; 401.11). The man-made or man-induced alteration of the chemical, physical, biological, and radiological integrity of water.

Poly(ethylene terephthalate) (PET) manufacture using dimethyl terephthalate (Section 60.561 [3-22-91]) means the manufacturing of poly(ethylene terephthalate) based on the esterification of dimethyl terephthalate (DMT) with ethylene glycol to form the intermediate monomer bis-(2-hydroxyethyl)-terephthalate (BHET) that is subsequently polymerized to form PET.

Poly(ethylene terephthalate) (PET) manufacture using dimethyl terephthalic (Section 60.561 [3-5-91]) means the manufacturing of poly(ethylene terephthalate) based on the esterification of dimethyl terephthalate (DMT) with ethylene glycol to form the intermediate monomer bis-(2-hydroxyethyl)-terephthalate (BHET) that is subsequently polymerized to form PET.

Polybrominated biphenyls (PBBs) (Section 704.195) means chemical substances the compositions of which, without regard to impurities, consist of brominated biphenyl molecules having the molecular formula $C_{12}H_xBr_y$ where $x + y = 10$ and y ranges from 1 to 10.

Polybrominated dibenzo-p-dioxin or PBDD (Section 766.3) refers to any member of a class of dibenzo-p-dioxins with two to eight bromine substituents.

Polybrominated dibenzofurans (Section 766.3) refers to any member of a class of dibenzofurans with two to eight bromine substituents.

Polychlorinated biphenyl (Sections 704.83; 704.85) means any chemical substance that is limited to the biphenyl molecule and that has been chlorinated to varying degrees.

Polychlorinated biphenyls or PCBs (Section 268.2 [6-1-90])* are halogenated organic compounds defined in accordance with 40 CFR 761.3.

Polychlorinated Biphenyls (PCBs) (Section 129.4) means a mixture of compounds composed of the biphenyl molecule which has been chlorinated to varying degrees.

Polychlorinated dibenzo-p-dioxin or PCDD (Section 766.3) means any member of a class of dibenzo-p-dioxins with two to eight chlorine substituents.

Polychlorinated dibenzofuran (Section 766.3) means any member of a class of dibenzofurans with two to eight chlorine substituents.

Polyester (Section 723.250) means a chemical substance that meets the definition of polymer and whose polymer molecules contain at least two carboxylic acid ester linkages, at least one of which links internal subunits together.

Polyhalogenated dibenzo-p-dioxin or PHDD (Section 766.3) means any member of a class of dibenzo-p-dioxins containing two to eight chlorine substituents or two to eight bromine substituents.

Polyhalogenated dibenzofuran or PHDF (Section 766.3) means any member of a class of dibenzofurans containing two to eight chlorine, bromine, or a combination of chlorine and bromine substituents.

Polymer (Section 60.601) means any of the natural or synthetic compounds of usually high molecular weight that consist of many repeated links, each link being a relatively light and simple molecule.

Polymer (Sections 704.25; 723.250) means a chemical substance that consists of at least a simple weight majority of polymer molecules but consists of less than a simple weight majority of molecules with the same molecular weight. Collectively, such polymer molecules must be distributed over a range of molecular weights wherein differences in molecular weight are primarily attributable to differences in the number of internal subunits.

Polymer molecule (Sections 704.25; 723.250) means a molecule which includes at least four covalently linked subunits, at least two of which are internal subunits.

Polymeric coating of supporting substrates (Section 60.741 [9-11-89]) means a web coating process that applies elastomers, polymers, or prepolymers to a supporting web other than paper, plastic film, metallic foil, or metal coil.

Polyvinyl chloride plant (Section 61.61) includes any plant where vinyl chloride alone or in combination with other materials is polymerized.

Pond water surface area (Section 421.61) when used for the purpose of calculating the volume of wastewater which may be discharged shall mean the water surface area of the pond created by the impoundment for storage of process wastewater at normal operating level. This surface shall in no case be less than one-third of the surface area of the maximum amount of water which could be contained by the impoundment. The normal operating level shall be the average level of the pond during the preceding calendar month.

Pond water surface area (Section 421.11) for the purpose of calculating the volume of waste water shall mean the area within the impoundment for rainfall and the actual water surface area for evaporation.

Porcelain enameling (Section 466.02) means the entire process of applying a fused vitreous enamel coating to a metal basis material. Usually this includes metal preparation and coating operations.

Portable air compressor or compressor (Section 204.51) means any wheel, skid, truck, or railroad car mounted, but not self-propelled, equipment designed to activate pneumatic tools. This consists of an air compressor (air end), and a reciprocating rotary or turbine engine rigidly connected in permanent alignment and mounted on a common frame. Also included are all cooling, lubricating, regulating, starting, and fuel systems, and all equipment necessary to constitute a complete, self-contained unit with a rated capacity of 75 cfm or greater which delivers air at pressures greater than 50 psig, but does not include any pneumatic tools themselves.

Portable plant (Section 60.671) means any nonmetallic mineral processing plant that is mounted on any chassis or skids and may be moved by the application of a lifting or pulling force. In addition, there shall be no cable, chain, turnbuckle, bolt or other means (except electrical connections) by which any piece of equipment is attached or clamped to any anchor, slab, or structure, including bedrock that must be removed prior to the application of a lifting or pulling force for the purpose of transporting the unit.

Portland cement plant (Section 60.61) means any facility manufacturing portland cement by either the wet or dry process.

Posing an exposure risk to food or feed (Section 761.3) means being in any location where human food or animal feed products could be exposed to PCBs released from a PCB Item. A PCB Item poses an exposure risk to food or feed if PCBs released in any way from the PCB Item have a potential pathway to human food or animal feed. EPA considers human food or animal feed to include items regulated by the U.S. Department of Agriculture or the Food and Drug Administration as human food or animal feed; this includes direct additives. Food or feed is excluded from this definition if it is used or stored in private homes.

Positive test result (Section 766.3) means: (1) Any resolvable gas chromatographic peak for any 2,3,7,8-HDD or HDF which exceeds the LOQ listed under Section 766.27 for that congener, or (2) exceeds LOQs approved by EPA under Section 766.28.

Positive-pressure fabric filter (Sections 60.271a; 60.341) means a fabric filter with the fans on the upstream side of the filter bags.

Possession or control (Section 704.3 [12/22/88]) means in the possession or control of any person, or of any subsidiary, partnership in which the person is a general partner, parent company, or any company or partnership which the parent company owns or controls, if the subsidiary, parent company, or other company or partnership is associated with the person in the research, development, test marketing, or commercial marketing of the substance in question. Information is in the possession or control of a person if it is: (1) In the person's own files including files maintained by employees of the person in the course of their employment. (2) In commercially available data bases to which the person has purchased access. (3) Maintained in the files in the course of employment by other agents of the person who are associated with research, development, test marketing, or commercial marketing of the chemical substance in question.

Possession or control (Sections 720.3; 723.50; 723.250) means in possession or control of the submitter, or of any subsidiary, partnership in which the submitter is a general partner, parent company, or any company or partnership which the parent company owns or controls, if the subsidiary, parent company, or other company or partnership is associated with the submitter in the research, development, test marketing, or commercial marketing of the chemical substance in question. (A parent company owns or controls another company if the parent owns or controls 50 percent or more of the other company's voting stock. A parent company owns or controls any partnership in which it is a general partner.) Information is included within this definition if it is: (1) In files maintained by submitter's employees who are: (i) Associated with research, development, test marketing, or commercial marketing of the chemical substance in question. (ii) Reasonably likely to have such data. (2) Maintained in the files of other agents of the submitter who are associated with research, development, test marketing, or commercial marketing of the chemical substance in question in the course of their employment as such agents.

Post consumer waste (**PCW**) (Section 247.101) means a material or product that has served its intended use and has been discarded for disposal after passing through the hands of a final user. Post consumer waste is a part of the broader category, Recycled material.

Post-closure plan (Section 264.141) means the plan for post-closure care prepared in accordance with the requirements of Sections 264.117 through 264.120.

Post-closure plan (Section 265.141) means the plan for post-closure care prepared in accordance with the requirements of Sections 265.117 through 265.120.

Post-consumer waste (**PCW**) (Section 246.101) means a material or product that has served its intended use and has been discarded for disposal or recovery after passing through the hands of a final consumer.

Post-exposure period (Section

797.2050) is the portion of the test that begins with the test birds being returned from a treated diet to the basal diet. This period is typically 3 days in duration, but may be extended if birds continue to die or demonstrate other toxic effects.

Post-mining area (Section 434.11) means: (1) A reclamation area or (2) the underground workings of an underground coal mine after the extraction, removal, or recovery of coal from its natural deposit has ceased and prior to bond release.

Post-removal site control (Section 300.5 [3-8-90]) means those activities that are necessary to sustain the integrity of a Fund-financed removal action following its conclusion. Post-removal site control may be a removal or remedial action under CERCLA. The term includes, without being limited to, activities such as relighting gas flares, replacing filters, and collecting leachate.

Postconsumer recovered paper (Section 248.4 [2/17/89]) means: (1) Paper, paperboard and fibrous wastes from retail stores, office buildings, homes and so forth, after they have passed through their end-usage as a consumer item including: used corrugated boxes; old newspapers; old magazines; mixed waste paper; tabulating cards and used cordage; and (2) All paper, paperboard and fibrous wastes that enter and are collected from municipal solid waste.

Pot furnace (Section 61.161) means a glass melting furnace that contains one or more refractory vessels in which glass is melted by indirect heating. The openings of the vessels are in the outside wall of the furnace and are covered with refractory stoppers during melting.

Potatoes (Section 407.71) shall mean the processing of sweet potatoes into the following product styles: Canned, peeled, solid, syrup, and vacuum packed. The following white potato product styles are also included: Canned, peeled, white, all varieties, whole and sliced.

Potential combustion concentration (Section 60.41a) means the theoretical emissions (ng/J, lb/million Btu heat input) that would result from combustion of a fuel in an uncleaned state without emission control systems and: (a) For particulate matter is: (1) 3,000 ng/J (7.0 lb/million Btu) heat input for solid fuel; and (2) 75 ng/J (0.17 lb/million Btu) heat input for liquid fuels. (b) For sulfur dioxide is determined under Section 60.48a(b). (c) For nitrogen oxides is: (1) 290 ng/J (0.67 lb/million Btu) heat input for gaseous fuels; (2) 310 ng/J (0.72 lb/million Btu) heat input for liquid fuels; and (3) 990 ng/J (2.30 lb/million Btu) heat input for solid fuels.

Potential damage (Section 763.83 [10/30/87]) means circumstances in which: (1) Friable ACBM is in an area regularly used by building occupants, including maintenance personnel, in the course of their normal activities. (2) There are indications that there is a reasonable likelihood that the material or its covering will become damaged, deteriorated, or delaminated due to factors such as changes in building use, changes in operations and maintenance practices, changes in occupancy, or recurrent damage.

Potential electrical output capacity (Section 60.41a) is defined as 33 percent of the maximum design heat input capacity of the steam generating unit (e.g., a steam generating unit with a 100-MW (340 million Btu/hr) fossil-fuel heat input capacity would have a 33-MW potential electrical output capacity). For electric utility combined cycle gas turbines the potential electrical output capacity is determined on the basis of the fossil-fuel firing capacity of the steam generator exclusive of the heat input and electrical power contribution by the gas turbine.

Potential electrical output capacity (Section 72.2 [7/30/93])* means the MWe capacity rating for the units which shall be equal to 33 percent of the maximum design heat input capacity of the steam generating unit, as calculated according to appendix D of part 72.

Potential for industry-wide application (Section 125.22) means that an innovative technology can be applied in two or more facilities which are in one or more industrial categories.

Potential hydrogen chloride emission rate (Section 60.51a [2-11-91]) means the hydrogen chloride emission rate that would occur from combustion of MSW in the absence of any hydrogen chloride emissions control.

Potential production allowances (Section 82.3 [12/10/93])* means the production allowances obtained under Sec. 82.9(a).

Potential significant damage (Section 763.83 [10/30/87]) means circumstances in which: (1) Friable ACBM is in an area regularly used by building occupants, including maintenance personnel, in the course of their normal activities. (2) There are indications that there is a reasonable likelihood that the material or its covering will become significantly damaged, deteriorated, or delaminated due to factors such as changes in building use, changes in operations and maintenance practices, changes in occupancy, or recurrent damage. (3) The material is subject to major or continuing disturbance, due to factors including, but not limited to, accessibility or, under certain circumstances, vibration or air erosion.

Potential sulfur dioxide emission rate (Section 60.41b [12/16/87]) means the theoretical sulfur dioxide emissions (ng/J, lb/million Btu heat input) that would result from combusting fuel in an uncleaned state and without using emission control systems.

Potential sulfur dioxide emission rate (Section 60.41c [9-12-90]) means the theoretical SO 2 emissions (nanograms per joule [ng/J], or pounds per million Btu [lb/million Btu] heat input) that would result from combusting fuel in an uncleaned state and without using emission control systems.

Potential sulfur dioxide emission rate (Section 60.51a [2-11-91]) means the sulfur dioxide emission rate that would occur from combustion of MSW in the absence of any sulfur dioxide emissions control.

Potential to emit (Section 51.165) means the maximum capacity of a stationary source to emit a pollutant under its physical and operational design. Any physical or operational limitation on the capacity of the source to emit a pollutant, including air pollution control equipment and restrictions on hours of operation or on the type or amount of material combusted, stored, or processed, shall be treated as part of its design only if the limitation or the effect it would have on emissions is federally enforceable. Secondary emissions do not count in determining the potential to emit of a stationary source.

Potential to emit (Section 52.24) means the maximum capacity of a stationary source to emit a pollutant under its physical and operational design. Any physical or operational limitation on the capacity of the source to emit a pollutant, including air pollution control equipment and restrictions on hours of operation or on amount of material combusted, stored, or processed, shall be treated as part of its design only if the limitation or the effect it would have on emissions is federally enforceable. Secondary emissions do not count in determining the potential to emit of a stationary source.

Potential to emit (Section 66.3 [2-13-92])* means the capability at maximum design capacity to emit a pollutant after the application of air pollution control equipment. Annual potential shall be based on the larger of the maximum annual rated capacity of the stationary source assuming continuous operation, or on a projection of actual annual emissions. Enforceable permit conditions on the type of materials combusted or processed may be used in determining the annual potential. Fugitive emissions, to the extent quantifiable, will be considered in determining annual potential for those stationary sources whose fugitive emissions are regulated by the applicable state implementation plan.

Potential to emit (Sections 51.301; 51.166; 52.21) means the maximum capacity of a stationary source to emit a pollutant under its physical and operational design. Any physical or operational limitation on the capacity of the source to emit a pollutant including air pollution control equipment and restrictions on hours of operation or on the type or amount of material combusted, stored, or processed, shall be treated as part of its design if the limitation or the effect it would have on emissions is federally enforceable. Secondary emissions do not count in determining the potential to emit of a stationary source.

Potentially responsible party (Section 306.12) means either: (1) An owner, or operator of the vessel or facility from which there is a release or threatened release of a hazardous substance; or (2) any other person who may be liable under section 107 of CERCLA.

Potentially responsible party (PRP) (Section 304.12 [5/30/89]) means any person who may be liable pursuant to section 107(a) of CERCLA, 42 U.S.C. 9607(a), for response costs incurred and to be incurred by the United States not inconsistent with NCP.

Potentially Responsible Party (PRP) (Section 35.6015 [6-5-90])* Any individual(s), or company(ies) identified as potentially liable under CERCLA for cleanup or payment for costs of cleanup of Hazardous Substance sites. PRPs may include individual(s), or company(ies) identified as having owned, operated, or in some other manner contributed wastes to Hazardous Substance sites.

Potroom (Section 60.191) means a building unit which houses a group of electrolytic cells in which aluminum is produced.

Potroom group (Section 60.191) means an uncontrolled potroom, a potroom which is controlled individually, or a group of potrooms or potroom segments ducted to a common control system.

POTW (Section 117.1). See **Publicly owned treatment works (Section 117.1)**.

POTW (Section 125.58). See **Publicly owned treatment works (Section 125.58)**.

POTW (Section 260.10). See **Publicly owned treatment works (Section 260.10)**.

POTW (Section 403.3). See **Publicly owned treatment works (Section 403.3)**.

POTW (Section 464.02) shall mean publicly owned treatment works.

POTW (Section 501.2 [5/2/89]) means a publicly owned treatment works.

POTW (Sections 122.2; 270.2). See **Publicly owned treatment works (Sections 122.2; 270.2)**.

POTW Pretreatment Program (Section 403.3). See **Approved POTW Pretreatment Program**

(**Section 403.3**).

POTW Treatment Plant (Section 403.3) means that portion of the POTW which is designed to provide treatment (including recycling and reclamation) of municipal sewage and industrial waste.

Pouring (Section 61.171) means the removal of blister copper from the copper converter bath.

Powder coating (Sections 60.311; 60.451) means any surface coating which is applied as a dry powder and is fused into a continuous coating film through the use of heat.

Powder forming (Section 471.02) includes forming and compressing powder into a fully dense finished shape, and is usually done within closed dies.

Powder or dry solid form (Section 721.3 [7-27-89])* means a state where all or part of the substance would have the potential to become fine, loose, solid particles.

Power purchase commitment (Section 72.2 [7/30/93])* means a commitment or obligation of a utility to purchase electric power from a facility pursuant to: (1) A power sales agreement; (2) A state regulatory authority order requiring a utility to: (i) Enter into a power sales agreement with the facility; (ii) Purchase from the facility; or (iii) Enter into arbitration concerning the facility for the purpose of establishing terms and conditions of the utility's purchase of power; (3) A letter of intent or similar instrument committing to purchase power (actual electrical output or generator output capacity) from the source at a previously offered or lower price and a power sales agreement applicable to the source is executed within the time frame established by the terms of the letter of intent but no later than November 15, 1992 or, where the letter of intent does not specify a timeframe, a power sales agreement applicable to the source is executed on or before November 15, 1992; or (4) A utility competitive bid solicitation that has resulted in the selection of the qualifying facility or independent power production facility as the winning bidder.

Power sales agreement (Section 72.2 [7/30/93])* is a legally binding agreement between a QF, IPP, new IPP, or firm associated with such facility and a regulated electric utility that establishes the terms and conditions for the sale of power from the facility to the utility.

Power sales agreement (Section 73.3 [12-17-91]) is a legally-binding document between a firm associated with a new independent power production facility (IPPF) or a new IPPF and a regulated electric utility that establishes the terms and conditions for the sale of power from a specific new IPPF to the utility.

Power setting (Section 87.1) means the power or thrust output of an engine in terms of kilonewtons thrust for turbojet and turbofan engines and shaft power in terms of kilowatts for turboprop engines.

Practicable (Part 6, Appendix A, Section 4) means capable of being done within existing constraints. The test of what is practicable depends upon the situation and includes consideration of the pertinent factors such as environment, community welfare, cost, or technology.

Practicable (Section 157.21) when used with respect to child-resistant packaging, means that the packaging can be mass produced and can be used in assembly line production.

Practicable (Section 230.3) means available and capable of being done after taking into consideration cost, existing technology, and logistics in light of overall project purposes.

Practicable (Section 248.4 [2/17/89]) means capable of being used consistent with: performance in accordance with applicable specifications, availability at a reasonable price, availability within a reasonable period of time, and maintenance of a satisfactory level of competition.

Practicable (Section 250.4 [10/6/87, 6/22/88]) means capable of being used consistent with: Performance in accordance with applicable specifications, availability at a reasonable price, availability within a reasonable period of time, and maintenance of a satisfactory level of competition.

Practicable (Section 252.4 [6/30/88]) means capable of being used consistent with: Performance in accordance with applicable specifications, availability at a reasonable price, availability within a reasonable period of time, and maintenance of a satisfactory level of competition.

Practicable (Section 253.4 [11/17/88]) means capable of being used consistent with: performance in accordance with applicable specifications, availability at a reasonable price, availability within a reasonable period of time, and maintenance of a satisfactory level of competition.

Practical knowledge (Section 171.2) means the possession of pertinent facts and comprehension together with the ability to use them in dealing with specific problems and situations.

Practice (Section 257.2) means the act of disposal of solid waste.

Pre-certification vehicle (Section 85.1702) means an uncertified vehicle which a manufacturer employs in fleets from year to year in the ordinary course of business for product development, production method assessment, and market promotion purposes, but in a manner not involving lease or sale.

Pre-certification vehicle engine (Section 85.1702) means an uncertified heavy-duty engine owned by a manufacturer and used in a manner not involving lease or sale in a vehicle employed from year to year in the ordinary course of business for product development, production method assessment, and market promotion purposes.

Preauthorization (Sections 305.12; 306.12) means EPA's approval to submit a claim for reimbursement to the Fund.

Precious metal (Section 466.02) means gold, silver, or platinum group metals and the principal alloys of those metals.

Precious metals (Section 421.261) shall mean gold, platinum, palladium, rhodium, iridium, osmium, and ruthenium.

Precious metals (Section 468.02) shall mean gold, platinum, palladium and silver and their alloys. Any alloy containing 30 or greater percent by weight of precious metals is considered a precious metal.

Precious metals (Section 471.02) include gold, platinum, palladium, and silver and their alloys. Any alloy containing 30 or greater percent by weight of precious metals is considered a precious metal alloy.

Precipitation bath (Section 60.601) means the water, solvent, or other chemical bath into which the polymer or prepolymer (partially reacted material) solution is extruded, and that causes physical or chemical changes to occur in the extruded solution to result in a semihardened polymeric fiber.

Precision (Section 86.082-2) means the standard deviation of replicated measurements.

Precoat (Section 60.441) means a coating operation in which a coating other than an adhesive or release is applied to a surface during the production of a pressure sensitive tape or label product.

Preconditioning (Section 610.11) means the operation of an automobile through one (1) EPA Urban Dynamometer Driving Schedule, described in 40 CFR Part 86.

Precursor (Section 766.3) means a chemical substance which is not contaminated due to the process conditions under which it is manufactured, but because of its molecular structure, and under favorable process conditions, it may cause or aid the formation of HDDs/HDFs in other chemicals in which it is used as a feedstock or intermediate.

Preliminary analysis (Section 610.11) means the engineering analysis performed by EPA prior to testing prescribed by the Administrator based on data and information submitted by a manufacturer or available from other sources.

Preliminary assessment (PA) (Section 300.5 [3-8-90]) means review of existing information and an off-site reconnaissance, if appropriate, to determine if a release may require additional investigation or action. A PA may include an on-site reconnaissance, if appropriate.

Premanufacture notice or PMN (Section 700.43 [8/17/88]) means any notice submitted to EPA pursuant to section 5(a)(1)(A) of the Act in accordance with Part 720 of this chapter or Section 723.250 of this chapter.

Present in the same form and concentration as a product packaged for distribution and use by the general public (Section 370.2 [10/15/87]) means a substance packaged in a similar manner and present in the same concentration as the substance when packaged for use by the general public, whether or not it is intended for distribution to the general public or used for the same purpose as when it is packaged for use by the general public.

Presentation of credentials (Section 86.078-7) shall mean display of the document designating a person as an EPA Enforcement Officer.

Preserve (Part 6, Appendix A, Section 4) means to prevent modification to the natural floodplain environment or to maintain it as closely as possible to its natural state.

Presiding Officer (Section 104.2) means the Chief Administrative Law Judge of the Agency or a person designated by the Chief Administrative Law Judge or by the Administrator to preside at a hearing under this part, in accordance with Section 104.6 hereof.

Presiding Officer (Section 123.64) is defined in Section 22.03.

Presiding Officer (Section 124.72) for the purposes of this subpart means an Administrative Law Judge appointed under 5 U.S.C. 3105 and designated to preside at the hearing. Under Subpart F other persons may also serve as hearing officers. See Section 124.119.

Presiding Officer (Section 164.2 [7/10/92])* means any person designated by the Administrator to conduct an expedited hearing.

Presiding Officer (Section 17.2) means the official, without regard to whether he is designated as an administrative law judge or a hearing officer or examiner, who presides at the adversary adjudication.

Presiding Officer (Section 22.03) means the Administrative Law Judge designated by the Chief Administrative Law Judge to serve as Presiding

Officer, unless otherwise specified by any Supplemental Rules.

Presiding Officer (Sections 85.1807; 86.614; 86.1014-84; 86.1115-87) shall mean an Administrative Law Judge appointed pursuant to 5 U.S.C. 3105 (see also 5 CFR Part 930 as amended).

Pressed and blown glass (Section 60.291) means glass which is pressed, blown, or both, including textile fiberglass, noncontinuous flat glass, noncontainer glass, and other products listed in SIC 3229. It is separated into: (1) Glass of borosilicate recipe, (2) Glass of soda-lime and lead recipes, (3) Glass of opal, fluoride, and other recipes.

Pressure (Sections 146.3; 147.2902) means the total load or force per unit area acting on a surface.

Pressure Drop (Section 85.2122(a)(16)(ii)(B)) means a measure, in kilopascals, of the difference in static pressure measured immediately upstream and downstream of the air filter element.

Pressure release (Section 264.1031 [6-21-90]) means the emission of materials resulting from the system pressure being greater than the set pressure of the pressure relief device.

Pressure release (Sections 60.481; 61.241) means the emission of materials resulting from system pressure being greater than set pressure of the pressure relief device.

Pretreatment (Section 403.3) means the reduction of the amount of pollutants, the elimination of pollutants, or the alteration of the nature of pollutant properties in wastewater prior to or in lieu of discharging or otherwise introducing such pollutants into a POTW. The reduction or alteration may be obtained by physical, chemical or biological processes, process changes or by other means, except as prohibited by Section 403.6(d). Appropriate pretreatment technology includes control equipment, such as equalization tanks or facilities, for protection against surges or slug loadings that might interfere with or otherwise be incompatible with the POTW. However, where wastewater from a regulated process is mixed in an equalization facility with unregulated wastewater or with wastewater from another regulated process, the effluent from the equalization facility must meet an adjusted pretreatment limit calculated in accordance with Section 403.6(e).

Pretreatment requirements (Section 403.3) means any substantive or procedural requirement related to Pretreatment, other than a National Pretreatment Standard, imposed on an Industrial User.

Pretreatment Standard (Section 403.3). See **National Pretreatment Standard (Section 403.3)**.

Pretreatment Standard (Section 117.1). See **National Pretreatment Standard (Section 117.1)**.

Preventive measures (Section 763.83 [10/30/87]) means actions taken to reduce disturbance of ACBM or otherwise eliminate the reasonable likelihood of the material's becoming damaged or significantly damaged.

Price analysis (Section 33.005) means the process of evaluating a prospective price without regard to the contractor's separate cost elements and proposed profit. Price analysis determines the reasonableness of the proposed subagreement price based on adequate price competition, previous experience with similar work, established catalog or market price, law, or regulation.

Price analysis (Section 35.6015 [6-5-90])* The process of evaluating a prospective price without regard to the contractor's separate cost elements and proposed profit. Price analysis determines the reasonableness of the proposed contract price based on adequate price competition, previous experience with similar work, established catalog or market price, law, or regulation.

Primary aluminum reduction plant (Section 60.191) means any facility manufacturing aluminum by electrolytic reduction.

Primary control system (Section 60.191) means an air pollution control system designed to remove gaseous and particulate fluorides from exhaust gases which are captured at the cell.

Primary copper smelter (Section 61.171) means any installation or intermediate process engaged in the production of copper from copper-bearing materials through the use of pyrometallurgical techniques.

Primary copper smelter (Section 60.161) means any installation or any intermediate process engaged in the production of copper from copper sulfide ore concentrates through the use of pyrometallurgical techniques.

Primary emission control system (Section 61.171) means the hoods, ducts, and control devices used to capture, convey, and collect process emissions.

Primary emission control system (Section 61.181) means the hoods, enclosures, ducts, and control devices used to capture, convey, and remove particulate matter from exhaust gases which are captured directly at the source of generation.

Primary emission control system (Section 60.141a) means the combination of equipment used for the capture and collection of primary emissions (e.g., an open hood capture system used in conjunction with a particulate matter cleaning device such as an electrostatic precipitator or a closed hood capture system used in conjunction with a particulate matter cleaning device such as a scrubber).

Primary emissions (Section 60.141) means particulate matter emissions from the BOPF generated during the steel production cycle and captured by the BOPF primary control system.

Primary emissions (Section 60.141a) means particulate matter emissions from the BOPF generated during the steel production cycle which are captured by, and do not thereafter escape from, the BOPF primary control system.

Primary enforcement responsibility (Section 142.2) means the primary responsibility for administration and enforcement of primary drinking water regulations and related requirements applicable to public water systems within a State.

Primary Exporter (Section 262.51 [7/19/88])* means any person who is required to originate the manifest for a shipment of hazardous waste in accordance with 40 CFR Part 262, Subpart B, or equivalent State provision, which specifies a treatment, storage, or disposal facility in a receiving country as the facility to which the hazardous waste will be sent and any intermediary arranging for the export.

Primary industry category (Section 122.2) means any industry category listed in the NRDC settlement agreement (Natural Resources Defense Council et al. v. Train, 8 E.R.C. 2120 (D.D.C. 1976), modified 12 E.R.C. 1833 (D.D.C. 1979)); also listed in Appendix A of Part 122.

Primary intended service class (Section 86.085-2) (applies beginning with the 1985 model year) means: (a) The primary service application group for which a heavy-duty diesel engine is designed and marketed, as determined by the manufacturer. The primary intended service classes are designated as light, medium, and heavy heavy-duty diesel engines. The determination is based on factors such as vehicle GVW, vehicle usage and operating patterns, other vehicle design characteristics, engine horsepower, and other engine design and operating characteristics. (1) Light heavy-duty diesel engines usually are non-sleeved and not designed for rebuild; their rated horsepower generally ranges from 70 to 170. Vehicle body types in this group might include any heavy-duty vehicle built for a light-duty truck chassis, van trucks, multi-stop vans, recreational vehicles, and some single axle straight trucks. Typical applications would include personal transportation, light-load commercial hauling and delivery, passenger service, agriculture, and construction. The GVWR of these vehicles is normally less than 19,500 lbs. (2) Medium heavy-duty diesel engines may be sleeved or non-sleeved and may be designed for rebuild. Rated horsepower generally ranges from 170 to 250. Vehicle body types in this group would typically include school buses, tandem axle straight trucks, city tractors, and a variety of special purpose vehicles such as small dump trucks, and trash compactor trucks. Typical applications would include commercial short haul and intra-city delivery and pickup. Engines in this group are normally used in vehicles whose GVWR varies from 19,500-33,000 lbs. (3) Heavy heavy-duty diesel engines are sleeved and designed for multiple rebuilds. Their rated horsepower generally exceeds 250. Vehicles in this group are normally tractors, trucks, and buses used in inter-city, long-haul applications. These vehicles normally exceed 33,000 lbs. GVWR.

Primary intended service class (Section 86.090-2 [7-26-90])* means: (a) The primary service application group for which a heavy-duty diesel engine is designed and marketed, as determined by the manufacturer. The primary intended service classes are designated as light, medium, and heavy heavy-duty diesel engines. The determination is based on factors such as vehicle GVW, vehicle usage and operating patterns, other vehicle design characteristics, engine horsepower, and other engine design and operating characteristics. (1) Light heavy-duty diesel engines usually are non-sleeved and not designed for rebuild; their rated horsepower generally ranges from 70 to 170. Vehicle body types in this group might include any heavy-duty vehicle built for a light-duty truck chassis, van trucks, multi-stop vans, recreational vehicles, and some single axle straight trucks. Typical applications would include personal transportation, light-load commercial hauling and delivery, passenger service, agriculture, and construction. The GVWR of these vehicles is normally less than 19,500 lbs. (2) Medium heavy-duty diesel engines may be sleeved or non-sleeved and may be designed for rebuild. Rated horsepower generally ranges from 170 to 250. Vehicle body types in this group would typically include school buses, tandem axle straight trucks, city tractors, and a variety of special purpose vehicles such as small dump trucks, and trash compactor trucks. Typical applications would include commercial short haul and intra-city delivery and pickup. Engines in this group are normally used in vehicles whose GVWR varies from 19,500-33,000 lbs. (3) Heavy heavy-duty diesel engines are sleeved and designed for multiple rebuilds. Their rated horsepower generally exceeds 250. Vehicles in this group are normally tractors, trucks, and buses used in inter-city, long-haul applications. These vehicles normally exceed 33,000 lbs. GVWR.

Primary lead smelter (Section 60.181) means any installation or any intermediate process engaged in the production of lead from lead sulfide ore concentrates through the use of pyrometallurgical techniques.

Primary mill (Section 420.71) means those steel hot forming operations that reduce ingots to blooms or slabs by passing the ingots between rotating steel rolls. The first hot forming operation performed on solidified steel after it is removed from the ingot molds is carried out on a primary mill.

Primary oxygen blow (Section 60.141a) means the period in the steel production cycle of a BOPF during which a high volume of oxygen-rich gas is introduced to the bath of molten iron by means of a lance inserted from the top of the vessel. This definition does not include any additional, or secondary, oxygen blows made after the primary blow.

Primary oxygen blow (Section 60.141) means the period in the steel production cycle of a BOPF during which a high volume of oxygen-rich gas is introduced to the bath of molten iron by means of a lance inserted from the top of the vessel or through tuyeres in the bottom or through the bottom and sides of the vessel. This definition does not include any additional or secondary oxygen blows made after the primary blow or the introduction of nitrogen or other inert gas through tuyeres in the bottom or bottom and sides of the vessel.

Primary Panel (Section 211.203) means the surface that is considered to be the front surface or that surface which is intended for initial viewing at the point of ultimate sale or the point of distribution for use.

Primary processor of asbestos (Section 763.63) is a person who processes for commercial purposes bulk asbestos.

Primary Resistor (Section 85.2122(a)(10)(ii)) means a device used in the primary circuit of an inductive ignition system to limit the flow of current.

Primary standard (Section 51.100) means a national primary ambient air quality standard promulgated pursuant to section 109 of the Act.

Primary zinc smelter (Section 60.171) means any installation engaged in the production, or any intermediate process in the production, of zinc or zinc oxide from zinc sulfide ore concentrates through the use of pyrometallurgical techniques.

Prime coat (Section 60.721 [1/29/88]) means the initial coat applied to a part when more than one coating is applied, not including conductive sensitizers or electromagnetic interference/radio frequency interference shielding coatings.

Prime coat operation (Section 60.461) means the coating application station, curing oven, and quench station used to apply and dry or cure the initial coating(s) on the surface of the metal coil.

Prime coat operation (Section 60.391) means the prime coat spray booth or dip tank, flash-off area, and bake oven(s) which are used to apply and dry or cure the initial coating on components of automobile or light-duty truck bodies.

Prime contractor (Section 8.2) means any person holding a contract, and for the purposes of Subpart B (General Enforcement, Compliance Review, and Complaint Procedure) of the rules, regulations, and relevant orders of the Secretary of Labor, any person who has held a contract subject to the order.

Principal company (Section 60.41a) means the electric utility company or companies which own the affected facility.

Principal importer (Section 720.3) means the first importer who, knowing that a new chemical substance will be imported rather than manufactured domestically, specifies the identity of the chemical substance and the total amount to be imported. Only persons who are incorporated, licensed, or doing business in the United States may be principal importers.

Principal importer (Section 721.3 [7/27/88])* means the first importer who, knowing that a chemical substance will be imported for a significant new use rather than manufactured in the United States, specifies the chemical substance and the amount to be imported. Only persons who are incorporated, licensed, or doing business in the United States may be principal importers.

Principal place of business (Section 171.2) means the principal location, either residence or office, in the State in which an individual, partnership, or corporation applies pesticides.

Principal residence (Section 35.2005) means, for the purposes of Section 35.2034, the habitation of a family or household for at least 51 percent of the year. Second homes, vacation or recreation residences are not included in this definition.

Principal Source Aquifer (Section 149.2 [2/14/89])*. See **Sole Source Aquifer (Section 149.2)**.

Principal Source Aquifer (Section 146.3). See **Sole Source Aquifer (Section 146.3)**.

Principal sponsor (Section 790.3) means an individual sponsor or the joint sponsor who assumes primary responsibility for the direction of a study and for oral and written communication with EPA.

Printing paper (Section 250.4 [10/6/87, 6/22/88]) means paper designed for printing, other than newsprint, such as offset and book paper.

Priority pollutant No. 4 (Section 420.02). See **Benzene (Section**

420.02).

Priority pollutant No. 55 (Section 420.02). See **Naphthalene (Section 420.02**).

Priority pollutant No. 73 (Section 420.02). See **Benzo(a)pyrene (Section 420.02**).

Priority pollutant No. 85 (Section 420.02). See **Tetrachloroethylene (Section 420.02**).

Priority Pollutants (Section 455.10 [9/28/93])* means the toxic pollutants listed in 40 CFR part 423, appendix A.

Priority water quality areas (Section 35.2005) means, for the purposes of Section 35.2015, specific stream segments or bodies of water, as determined by the State, where municipal discharges have resulted in the impairment of a designated use or significant public health risks, and where the reduction of pollution from such discharges will substantially restore surface or groundwater uses.

Private applicator (Section 171.2) means a certified applicator who uses or supervises the use of any pesticide which is classified for restricted use for purposes of producing any agricultural commodity on property owned or rented by him or his employer or (if applied without compensation other than trading of personal services between producers of agricultural commodities) on the property of another person.

Private carrier of property by motor vehicle (Section 202.10) means any person not included in terms common carrier by motor vehicle or contract carrier by motor vehicle, who or which transports in interstate or foreign commerce by motor vehicle property of which such person is the owner, lessee, or bailee, when such transportation is for sale, lease, rent or bailment, or in furtherance of any commercial enterprise.

Privately owned treatment works (Section 122.2) means any device or system which is (a) used to treat wastes from any facility whose operator is not the operator of the treatment works and (b) not a POTW.

Proceeding (Section 17.2) means an adversary adjudication as defined in Section 17.2(b).

Proceeding (Section 2.302) means any rulemaking, adjudication, or licensing conducted by EPA under the Act or under regulations which implement the Act, except for determinations under this part.

Proceeding (Section 2.304) means any rulemaking, adjudication, or licensing process conducted by EPA under the Act or under regulations which implement the Act, except for any determination under this part.

Proceeding (Section 2.305) means any

rulemaking, adjudication, or licensing conducted by EPA under the Act or under regulations which implement the Act including the issuance of administrative orders and the approval or disapproval of plans (e.g. closure plans) submitted by persons subject to regulation under the Act, but not including determinations under this subpart.

Proceeding (Section 2.310) means any rulemaking or adjudication conducted by EPA under the Act or under regulations which implement the Act (including the issuance of administrative orders under section 106 of the Act), or any administrative determination made under section 104 of the Act, but not including determinations under this subpart.

Proceeding (Sections 2.301; 2.303; 2.306) means any rulemaking, adjudication, or licensing conducted by EPA under the Act or under regulations which implement the Act, except for determinations under this subpart.

Process (Section 372.3 [2/16/88]) means the preparation of a toxic chemical, after its manufacture, for distribution in commerce: (1) In the same form or physical state as, or in a different form or physical state from, that in which it was received by the person so preparing such substance, or (2) As part of an article containing the toxic chemical. Process also applies to the processing of a toxic chemical contained in a mixture or trade name product.

Process (Section 704.3 [12/22/88]) means to process for commercial purposes.

Process (Section 716.3) means to process for commercial purposes.

Process (Section 761.3) means the preparation of a chemical substance or mixture, after its manufacture, for distribution in commerce: (1) In the same form or physical state as, or in a different form or physical state from that in which it was received by the person so preparing such substance or mixture, or (2) As part of an article containing the chemical substance or mixture.

Process (Section 762.3) has the same meaning as in 15 U.S.C. 2602.

Process (Section 763.163 [7-12-89]) has the same meaning as in section 3 of the Act.

Process (Sections 710.2; 720.3; 723.250; 747.115; 747.195; 747.200) means the preparation of a chemical substance or mixture, after its manufacture, for distribution in commerce (1) in the same form or physical state as, or in a different form or physical state from, that in which it was received by the person so preparing such substance or mixture, or (2) as part of a mixture or article containing the chemical substance or mixture.

Process emission (Section 60.301) means the particulate matter which is collected by a capture system.

Process emissions (Section 61.171) means inorganic arsenic emissions from copper converters that are captured directly at the source of generation.

Process emissions (Section 61.181) means inorganic arsenic emissions that are captured and collected in a primary emission control system.

Process for commercial purposes (Sections 716.3; 717.3) means the preparation of a chemical substance or mixture, after its manufacture, for distribution in commerce with the purpose of obtaining an immediate or eventual commercial advantage for the processor. Processing of any amount of a chemical substance or mixture is included. If a chemical substance or mixture containing impurities is processed for commercial purposes, then those impurities are also processed for commercial purposes.

Process for commercial purposes (Sections 712.3; 763.63) means the preparation of a chemical substance or mixture, after its manufacture, for distribution in commerce with the purpose of obtaining an immediate or eventual commercial advantage for the processor. Processing of any amount of a chemical substance or mixture is included. If a chemical or mixture containing impurities is processed for commercial purposes, then those impurities are also processed for commercial purposes.

Process for commercial purposes (Section 721.3 [7/27/88])* means the preparation of a chemical substance or mixture containing the chemical substance, after manufacture of the substance, for distribution in commerce with the purpose of obtaining an immediate or eventual commercial advantage for the processor. Processing of any amount of a chemical substance or mixture containing the chemical substance is included in this definition. If a chemical substance or mixture containing impurities is processed for commercial purposes, the impurities also are processed for commercial purposes.

Process for commercial purposes (Section 710.2) means to process (1) for distribution in commerce, including for test marketing purposes, or (2) for use as an intermediate.

Process for commercial purposes (Section 704.3 [12/22/88])* means the preparation of a chemical substance or mixture after its manufacture for distribution in commerce with the purpose of obtaining an immediate or eventual commercial advantage for the processor. Processing of any amount of a chemical substance or mixture is included in this definition. If a chemical substance or mixture containing impurities is processed for commercial purposes, then the impurities also are processed for commercial purposes.

Process fugitive emissions (Section 60.381) means particulate matter emissions from an affected facility that are not collected by a capture system.

Process gas (Section 60.101) means any gas generated by a petroleum refinery process unit, except fuel gas and process upset gas as defined in this section.

Process generated waste water (Sections 436.31; 436.41) shall mean any waste water used in the slurry transport of mined material, air emissions control, or processing exclusive of mining. The term shall also include any other water which becomes commingled with such waste water in a pit, pond, lagoon, mine or other facility used for treatment of such waste water. The term does not include waste water used for the suction dredging of deposits in a body of water and returned directly to the body of water without being used for other purposes or combined with other waste water.

Process generated waste water (Sections 412.11; 412.21) shall mean water directly or indirectly used in the operation of a feedlot for any or all of the following: Spillage or overflow from animal or poultry watering systems; washing, cleaning or flushing pens, barns, manure pits or other feedlot facilities; direct contact swimming, washing or spray cooling of animals; and dust control.

Process generated waste water (Section 436.21) shall mean any waste water used in the slurry transport of mined material, air emissions control, or processing exclusive of mining. The term shall also include any other water which becomes commingled with such waste water in a pit, pond, lagoon, mine, or other facility used for treatment of such waste water.

Process generated waste water (Section 436.181) shall mean any waste water used in the slurry transport of mined material, air emissions control, or processing exclusive of mining. The term shall also include any other water which becomes commingled with such waste water in a pit, pond, lagoon, mine, or other facility used for settling or treatment of such waste water.

Process heater (Section 264.1031 [6-21-90]) means a device that transfers heat liberated by burning fuel to fluids contained in tubes, including all fluids except water that are heated to produce steam.

Process heater (Section 60.41b [12/16/87])* means a device that is primarily used to heat a material to initiate or promote a chemical reaction in which the material participates as a reactant or catalyst.

Process heater (Section 60.41c [9-12-90]) means a device that is primarily used to heat a material to initiate or promote a chemical reaction in which the material participates as a reactant or catalyst.

Process heater (Section 60.661 [6-29-90]) means a device that transfers heat liberated by burning fuel to fluids contained in tubes, including all fluids except water that is heated to produce steam.

Process heater (Section 61.301 [3-7-90]) means a device that transfers heat liberated by burning fuel to fluids contained in tubes, except water that is heated to produce steam.

Process improvement (Section 60.481) means routine changes made for safety and occupational health requirements, for energy savings, for better utility, for ease of maintenance and operation, for correction of design deficiencies, for bottleneck removal, for changing product requirements, or for environmental control.

Process or distribute in commerce solely for export (Sections 747.115; 747.195) means to process or distribute in commerce solely for export from the United States under the following restrictions on domestic activity: (i) Processing must be performed at sites under the control of the processor. (ii) Distribution in commerce is limited to purposes of export. (iii) The processor or distributor may not use the substance except in small quantities solely for research and development.

Process solely for export (Section 721.3 [7/27/88])* means to process for commercial purposes solely for export from the United States under the following restrictions on activity in the United States: Processing must be performed at sites under the control of the processor; distribution in commerce is limited to purposes of export; and the processor may not use the chemical substance except in small quantities solely for research and development.

Process stream (Section 721.3 [7-27-89]) means all reasonably anticipated transfer, flow, or disposal of a chemical substance, regardless of physical state or concentration, through all intended operations of processing, including the cleaning of equipment.

Process unit (Section 60.481) means components assembled to produce, as intermediate or final products, one or more of the chemicals listed in Section 60.489 of this part. A process unit can operate independently if supplied with sufficient feed or raw materials and sufficient storage facilities for the product.

Process unit (Section 60.591) means components assembled to produce intermediate or final products from petroleum, unfinished petroleum derivatives, or other intermediates; a process unit can operate independently if supplied with sufficient feed or raw materials and sufficient storage facilities for the product.

Process unit (Section 60.631) means equipment assembled for the extraction of natural gas liquids from field gas, the fractionation of the liquids into natural gas products, or other operations associated with the processing of natural gas products. A process unit can operate independently if supplied with sufficient feed or raw materials and sufficient storage facilities for the products.

Process unit (Section 60.661 [6-29-90]) means equipment assembled and connected by pipes or ducts to produce, as intermediates or final products, one or more of the chemicals in Sec. 60.667. A process unit can operate independently if supplied with sufficient fuel or raw materials and sufficient product storage facilities.

Process unit (Section 61.241) means equipment assembled to produce a VHAP or its derivatives as intermediates or final products, or equipment assembled to use a VHAP in the production of a product. A process unit can operate independently if supplied with sufficient feed or raw materials and sufficient product storage facilities.

Process unit (Section 61.341 [3-7-90]) means equipment assembled and connected by pipes or ducts to produce intermediate or final products. A process unit can be operated independently if supplied with sufficient fuel or raw materials and sufficient product storage facilities.

Process unit shutdown (Sections 60.481; 61.241) means a work practice or operational procedure that stops production from a process unit or part of a process unit. An unscheduled work practice or operational procedure that stops production from a process unit or part of a process unit for less than 24 hours is not a process unit shutdown. The use of spare equipment and technically feasible by-passing of equipment without stopping production are not process unit shutdowns.

Process unit turnaround (Section 61.341 [1/7/93])* means the shutting down of the operations of a process unit, the purging of the contents of the process unit, the maintenance or repair work, followed by restarting of the process.

Process unit turnaround waste (Section 61.341 [1/7/93])* means a waste that is generated as a result of a process unit turnaround.

Process upset gas (Section 60.101) means any gas generated by a petroleum refinery process unit as a result of start-up, shut-down, upset or malfunction.

Process vent (Section 264.1031 [6-21-90]) means any open-ended pipe or stack that is vented to the atmosphere either directly, through a vacuum-producing system, or through a tank (e.g., distillate receiver, condenser, bottoms receiver, surge control tank, separator tank, or hot well) associated with hazardous waste distillation, fractionation, thin-film evaporation, solvent extraction, or air or steam stripping operations.

Process vessel (Section 61.131 [9-14-89]) means each tar decanter, flushing-liquor circulation tank, light-oil condenser, light-oil decanter, wash-oil decanter, or wash-oil circulation tank.

Process waste water (Section 117.1 [8/25/93])* means any water which, during manufacturing or processing, comes into direct contact with or results from the production or use of any raw material, intermediate product, finished product, byproduct, or waste product.

Process waste water (Section 422.41) means any water which, during manufacturing or processing, comes into direct contact with or results from the production or use of any raw material, intermediate product, finished product, by-product, or waste product. The term process waste water does not include contaminated nonprocess waste water, as defined in Section 422.41.

Process waste water (Section 422.51) means any water which, during manufacturing or processing, comes into direct contact with or results from the production or use of any raw material, intermediate product, finished product, by-product, or waste product. The term process waste water does not include contaminated non-process waste water, as defined in Section 422.51.

Process waste water (Section 428.11) shall mean, in the case of tire and inner tube plants constructed before 1959, discharges from the following: Soapstone solution applications; steam cleaning operations; air pollution control equipment; unroofed process oil unloading areas; mold cleaning operations; latex applications; and air compressor receivers. Discharges from other areas of such plants shall not be classified as process waste water for the purposes of this section.

Process waste water (Sections 458.11; 458.21) shall mean waters which result from baghouse operations or thermal quench operations.

Process waste water (Sections 412.11; 412.21) shall mean any process generated waste water and any precipitation (rain or snow) which comes into contact with any manure, litter or bedding, or any other raw material or intermediate or final material or product used in or resulting from the production of animals or poultry or direct products (e.g. milk, eggs).

Process waste water (Sections 401.11)

means any water which, during manufacturing or processing, comes into direct contact with or results from the production or use of any raw material, intermediate product, finished product, byproduct, or waste product.

Process waste water pollutants (Section 401.11) means pollutants present in process waste water.

Process wastes (Section 129.2) means any designated toxic pollutant, whether in wastewater or otherwise present, which is inherent to or unavoidably resulting from any manufacturing process, including that which comes into direct contact with or results from the production or use of any raw material, intermediate product, finished product, by-product or waste product and is discharged into the navigable waters.

Process wastewater (Section 122.2) means any water which, during manufacturing or processing, comes into direct contact with or results from the production or use of any raw material, intermediate product, finished product, byproduct, or waste product.

Process wastewater (Section 415.241) means any water which, during manufacturing or processing, comes into direct contact with or results from the production or use of any raw material, intermediate product, finished product, by-product, or waste product. The term process wastewater does not include contaminated nonprocess wastewater, as defined in Section 415.241.

Process wastewater (Section 415.381) means any water which, during manufacturing or processing, comes into direct contact with or results from the production or use of any raw material, intermediate product, finished product, by-product, or waste product. The term process wastewater does not include contaminated nonprocess wastewater, as defined in Section 415.381.

Process wastewater (Section 415.431) means any water which, during manufacturing or processing, comes into direct contact with or results from the production or use of any raw material, intermediate product, finished product, by-product, or waste product. The term process wastewater does not include contaminated non-process wastewater, as defined in Section 415.431.

Process wastewater (Section 415.441) means any water which, during manufacturing or processing, comes into direct contact with or results from the production or use of any raw material, intermediate product, finished product, by-product, or waste product. The term process wastewater does not include contaminated non-process wastewater, as defined in Section 415.441.

Process wastewater (Section 415.551) means any water which, during manufacturing or processing, comes into direct contact with or results from the production or use of any raw material, intermediate product, finished product, by-product, or waste product. The term process wastewater does not include contaminated non-process wastewater, as defined in Section 415.551.

Process wastewater (Section 415.601) means any water which, during manufacturing or processing, comes into direct contact with or results from the production or use of any raw material, intermediate product, finished product, by-product, or waste product. The term process wastewater does not include contaminated nonprocess wastewater, as defined in Section 415.601.

Process wastewater (Section 415.91) means any water which, during manufacturing or processing, comes into direct contact with or results from the production or use of any raw material, intermediate product, finished product, by-product, or waste product. The term process wastewater does not include contaminated non-process wastewater, as defined in Section 415.91.

Process wastewater (Section 418.11) means any water which, during manufacturing or processing, comes into direct contact with or results from the production or use of any raw material, intermediate product, finished product, by-product, or waste product. The term process wastewater does not include contaminated non-process wastewater, as defined in Section 418.11.

Process wastewater (Section 418.21) shall mean any water which, during manufacturing or processing, comes into direct contact with or results from the production or use of any raw material, intermediate product, finished product, by-product, or waste product. The term process wastewater does not include non-contact cooling water, as defined in Section 418.21.

Process wastewater (Section 429.11) specifically excludes noncontact cooling water, material storage yard runoff (either raw material or processed wood storage), and boiler blowdown. For the dry process hardboard, veneer, finishing, particleboard, and sawmills and planing mills subcategories, fire control water is excluded from the definition.

Process wastewater (Section 440.141 [5/24/88]) means all water used in and resulting from the beneficiation process, including but not limited to the water used to move the ore to and through the beneficiation process, the water used to aid in classification, and the water used in gravity separation, mine drainage, and infiltration and drainage waters which commingle with mine drainage or waters resulting from the beneficiation process.

Process wastewater (Section 443.21)

shall mean any water which, during the manufacturing process, comes into direct contact with any raw material, intermediate product, by-product, or product used in or resulting from the production of paving asphalt concrete.

Process wastewater (Section 443.31) shall mean any water which, during the manufacturing process, comes into direct contact with any raw material, intermediate product, by-product, or product used in or resulting from the production of asphalt roofing materials.

Process wastewater (Section 61.341 [1/7/93])* means water which comes in contact with benzene during manufacturing or processing operations conducted within a process unit. Process wastewater is not organic wastes, process fluids, product tank drawdown, cooling tower blowdown, steam trap condensate, or landfill leachate.

Process wastewater flow (Section 455.21 [9/28/93])* means the sum of the average daily flows from the following wastewater streams: Process stream and product washes, equipment and floor washes, water used as solvent for raw materials, water used as reaction medium, spent acids, spent bases, contact cooling water, water of reaction, air pollution control blowdown, steam jet blowdown, vacuum pump water, pump seal water, safety equipment cleaning water, shipping container cleanout, safety shower water, contaminated storm water, and product/process laboratory quality control wastewater. Notwithstanding any other regulation, process wastewater flow for the purposes of this subpart does not include wastewaters from the production of intermediate chemicals.

Process wastewater pollutants (Sections 415.91; 415.241; 415.381; 415.431; 415.441; 415.551; 415.601; 443.21; 443.31) means pollutants present in process wastewater.

Process wastewater pollutants (Section 443.11) shall mean any pollutants present in the process wastewaters and rainwater runoff.

Process wastewater pollutants (Section 455.21 [9/28/93])* means those pollutants present in process wastewater flow.

Process wastewater stream (Section 61.341 [3-7-90]) means a waste stream that contains only process wastewater.

Process water (Section 463.2) is any raw, service, recycled, or reused water that contacts the plastic product or contacts shaping equipment surfaces such as molds and mandrels that are, or have been, in contact with the plastic product.

Processing activities (Section 704.203 [12/22/88]) means all those activities which include (1) preparation of a substance identified in Subpart D of this Part after its manufacture to make another substance for sale or use, (2) repackaging of the identified substance, or (3) purchasing and preparing the identified substance for use or distribution in commerce.

Processing site (Section 192.10) means (a) Any site, including the mill, containing residual radioactive materials at which all or substantially all of the uranium was produced for sale to any Federal agency prior to January 1, 1971, under a contract with any Federal agency, except in the case of a site at or near Slick Rock, Colorado, unless: (1) Such site was owned or controlled as of January 1, 1978, or is thereafter owned or controlled, by any Federal agency, or (2) A license (issued by the (Nuclear Regulatory) Commission or its predecessor agency under the Atomic Energy Act of 1954 or by a State as permitted under section 274 of such Act) for the production at site of any uranium or thorium product derived from ores is in effect on January 1, 1978, or is issued or renewed after such date; and (b) Any other real property or improvement thereon which: (1) Is in the vicinity of such site, and (2) Is determined by the Secretary, in consultation with the Commission, to be contaminated with residual radioactive materials derived from such site.

Processor (Section 704.3 [12/22/88]) means any person who processes a chemical substance or mixture.

Processor (Section 762.3) has the same meaning as in 15 U.S.C. 2602.

Processor (Section 763.163 [7-12-89]) has the same meaning as in section 3 of the Act.

Processor (Sections 710.2; 720.3; 747.115; 747.195; 747.200) means any person who processes a chemical substance or mixture.

Procurement item (Section 248.4 [2/17/89]) means any device, good, substance, material, product, or other item, whether real or personal property, which is the subject of any purchase, barter, or other exchange made to procure such item.

Procurement item (Section 249.04) means any device, goods, substance, material, product, or other item whether real or personal property which is the subject of any purchase, barter, or other exchange made to procure such item (Pub. L. 94-580, 90 Stat. 2800, 42 U.S.C. 6903).

Procurement item (Section 250.4 [10/6/87, 6/22/88]) means any device, good, substance, material, product, or other item, whether real or personal property, that is the subject of any purchase, barter, or other exchange made to procure such item.

Procurement item (Section 252.4 [6/30/88]) means any device, good, substance, material, product,or other item, whether real or personal property, which is the subject of any purchase, barter, or other exchange made to procure such item.

Procurement item (Section 253.4 [11/17/88]) means any device, good, substance, material, product, or other item, whether real or personal property, which is the subject of any purchase, barter, or other exchange made to procure such item.

Procuring agency (Section 248.4 [2/17/89]) means any Federal agency, or any State agency or agency of a political subdivision of a State which is using appropriated Federal funds for such procurement, or any person contracting with any such agency with respect to work performed under such contract.

Procuring agency (Section 249.04) means any Federal agency, or any State agency or agency of a political subdivision of a State which is using appropriated Federal funds for such procurement, or any person contracting with any such agency with respect to work performed under such contract (Pub. L. 94-580, 90 Stat. 2800, 42 U.S.C. 6903).

Procuring agency (Section 250.4 [10/6/87, 6/22/88]) means any Federal agency, or any State agency or agency of a political subdivision of a State that is using appropriated Federal funds for such procurement, or any person contracting with any such agency with respect to work performed under such contract.

Procuring agency (Section 252.4 [6/30/88]) means any Federal agency, or any State agency or agency of a political subdivision of a State which is using appropriated Federal funds for such procurement, or any person contracting with any such agency with respect to work performed under such contract.

Procuring agency (Section 253.4 [11/17/88]) means any Federal agency, or any State agency or agency of a political subdivision of a State which is using appropriated Federal funds for such procurement, or any person contracting with any such agency with respect to work performed under such contract.

Produce (Section 167.1) means to manufacture, prepare, propagate, compound, or process any pesticide, including any pesticide produced pursuant to Section 5, or device, or to repackage or otherwise change the container of any pesticide or device.

Produce (Section 167.3 [9/8/88]) means to manufacture, prepare, propagate, compound, or process any pesticide, including any pesticide produced pursuant to section 5 of the Act, any active ingredient or device, or to package, repackage, label, relabel, or otherwise change the container of any pesticide or device.

Produced sand (Section 435.11

[3/4/93])* shall refer to slurried particles used in hydraulic fracturing, the accumulated formation sands and scales particles generated during production. Produced sand also includes desander discharge from the produced water waste stream, and blowdown of the water phase from the produced water treating system.

Produced water (Section 435.11 [3/4/93])* shall refer to the water (brine) brought up from the hydrocarbon-bearing strata during the extraction of oil and gas, and can include formation water, injection water, and any chemicals added downhole or during the oil/water separation process.

Producer (Section 167.1) means any person, as defined by the Act, who produces any pesticide or device.

Producer (Section 167.3 [9/8/88]) means any person, as defined by the Act, who produces any pesticide, active ingredient, or device (including packaging, repackaging, labeling and relabeling).

Producer (Section 169.1) means the person, as defined by the Act, who produces or imports any pesticide or device or active ingredient used in producing a pesticide.

Product (Section 2.303) has the meaning given it in 42 U.S.C. 4902(3).

Product (Section 203.1) means any manufactured article or goods or component thereof; except that such term does not include (i) Any aircraft, aircraft engine, propeller or appliance, as such terms are defined in section 101 of the Federal Aviation Act of 1958; or (ii)(a) Any military weapons or equipment which are designed for combat use; (b) any rockets or equipment which are designed for research, experimental or developmental work to be performed by the National Aeronautics and Space Administration; or (c) to the extent provided by regulations of the Administrator, any other machinery or equipment designed for use in experimental work done by or for the Federal Government.

Product (Section 204.2) means any construction equipment for which regulations have been promulgated under this part and includes test product.

Product (Section 205.2) means any transportation equipment for which regulations have been promulgated under this part and includes test product.

Product (Section 211.102) means any noise-producing or noise-reducing product for which regulations have been promulgated under Part 211; the term includes test product.

Product (Section 408.291) shall mean the weight of the scallop meat after processing.

Product (Section 408.51) shall mean the weight of the oyster meat after shucking.

Product (Section 409.11) shall mean crystallized refined sugar.

Product (Section 410.01) except where a specialized definition is included in the subpart, shall mean the final material produced or processed at the mill.

Product (Section 410.61) shall mean the final carpet produced or processed including the primary backing but excluding the secondary backing.

Product (Section 415.161) shall mean sodium chloride.

Product (Section 415.171) shall mean sodium dichromate.

Product (Section 415.201) shall mean sodium sulfite.

Product (Section 415.221) shall mean titanium dioxide.

Product (Section 415.231) means aluminum fluoride produced by the dry process in which partially dehydrated alumina hydrate is reacted with hydrofluoric acid gas.

Product (Section 415.241) shall mean ammonium chloride.

Product (Section 415.281) shall mean boric acid.

Product (Section 415.301) shall mean calcium carbonate.

Product (Section 415.361) shall mean copper salts.

Product (Section 415.41) shall mean calcium chloride.

Product (Section 415.421) means hydrogen cyanide.

Product (Section 415.451) shall mean lithium carbonate.

Product (Section 415.471) shall mean nickel salts.

Product (Section 415.61) shall mean chlorine.

Product (Section 415.91) shall mean hydrogen peroxide as a one hundred percent hydrogen peroxide solution.

Product (Section 418.21) shall mean the anhydrous ammonia content of the compound manufactured.

Product (Section 418.31) shall mean the 100 percent urea content of the material manufactured.

Product (Section 418.51) shall mean nitric acid on the basis of 100 percent HNO3.

Product (Section 421.11) shall mean alumina.

Product (Section 421.21) shall mean

hot aluminum metal.

Product (Section 421.31) shall mean hot aluminum metal.

Product (Section 421.51) means electrolytically refined copper.

Product (Section 421.81) shall mean zinc metal.

Product (Section 439.11) shall mean pharmaceutical products derived from fermentation processes.

Product (Section 439.21) shall mean biological and natural extraction products. This subcategory shall include blood fractions, vaccines, serums, animal bile derivatives, endocrine products, and isolation of medicinal products, such as alkaloids, from botanical drugs and herbs.

Product (Section 439.31) shall mean pharmaceutical products derived from chemical synthesis processes.

Product (Section 439.41) shall mean products from plants which blend, mix, compound, and formulate pharmaceutical ingredients. Pharmaceutical preparations for human and veterinary use such as ampules, tablets, capsules, vials, ointments, medicinal powders, solutions, and suspensions are included.

Product (Section 439.51) shall mean products or services resulting from pharmaceutical research, which includes microbiological, biological, and chemical operations.

Product (Section 454.11) shall mean char and charcoal briquets.

Product (Section 454.21) shall mean gum rosin and turpentine.

Product (Section 454.31) shall mean products from wood rosin, turpentine and pine oil.

Product (Section 454.41) shall mean tall oil rosin, pitch and fatty acids.

Product (Section 454.51) shall mean essential oils.

Product (Section 454.61) shall mean rosin-based derivatives.

Product (Section 457.11) shall mean dynamite, nitroglycerin, cyclotrimethylene trinitramine (RDX), cyclotetramethylene tetranitramine (HMX), and trinitrotoluene (TNT).

Product (Section 457.31) shall mean products from plants which blend explosives and market a final product, and plants that fill shells and blasting caps. Examples of such installations would be plants manufacturing ammonium nitrate and fuel oil (ANFO), nitrocarbonitrate (NCN), slurries, water gels, and shells.

Product (Section 458.11) shall mean carbon black manufactured by the furnace process.

Product (Section 458.21) shall mean carbon black manufactured by the thermal process.

Product (Section 458.31) shall mean carbon black manufactured by the channel process.

Product (Section 458.41) shall mean carbon black manufactured by the lamp process.

Product (Section 459.11) shall mean articles developed or printed by photographic processes, such as paper prints, slides, negatives, enlargements, movie film and other sensitized materials.

Product (Section 460.11) shall mean service resulting from the hospital activity in terms of 1,000 occupied beds.

Product (Section 60.661 [6-29-90]) means any compound or chemical listed in Sec. 60.667 that is produced for sale as a final product as that chemical, or for use in the production of other chemicals or compounds. By-products, co-products, and intermediates are considered to be products.

Product accumulator vessel (Section 61.241) means any distillate receiver, bottoms receiver, surge control vessel, or product separator in VHAP service that is vented to atmosphere either directly or through a vacuum-producing system. A product accumulator vessel is in VHAP service if the liquid or the vapor in the vessel is at least 10 percent by weight VHAP.

Product change (Section 60.261) means any change in the composition of the furnace charge that would cause the electric submerged arc furnace to become subject to a different mass standard applicable under this subpart.

Product finishing section (Section 60.561 [3-5-91]) means the equipment that treats, shapes, or modifies the polymer or resin to produce the finished end product of the particular facility, including equipment that prepares the product for product finishing. For the purposes of these standards, the product finishing section begins with the equipment used to transfer the polymerized product from the polymerization reaction section and ends with the last piece of equipment that modifies the characteristics of the polymer. Product finishing equipment may accomplish product separation, extruding and pelletizing, cooling and drying, blending, additives introduction, curing, or annealing. Equipment used to separate unreacted or by-product material from the product are to be included in this process section, provided the material separated from the polymer product is not recovered at the time the process section becomes an affected facility. If the material is being recovered, then the separation equipment are to be included in the material recovery section. Product finishing does not include polymerization, the physical mixing of the pellets to obtain a homogenous mixture of the polymer (except as noted below), or the shaping (such as fiber spinning, molding, or fabricating) or modification (such as fiber stretching and crimping) of the finished end product. If physical mixing occurs in equipment located between product finishing equipment (i.e., before all the chemical and physical characteristics have been "set" by virtue of having passed through the last piece of equipment in the product finishing section), then such equipment are to be included in this process section. Equipment used to physically mix the finished product that are located after the last piece of equipment in the product finishing section are part of the product storage section.

Product frosted (Section 426.121) shall mean that portion of the furnace pull associated with the fraction of finished incandescent lamp envelopes which is frosted; this quantity shall be calculated by multiplying furnace pull by the fraction of finished incandescent lamp envelopes which is frosted.

Product or pesticide product (Section 162.152) means a pesticide offered for distribution and use, and includes any labeled container and any supplemental labeling.

Product packaging station (Section 60.381) means the equipment used to fill containers with metallic compounds or metallic mineral concentrates.

Product tank (Section 61.341 [3-7-90]) means a stationary unit that is designed to contain an accumulation of materials that are fed to or produced by a process unit, and is constructed primarily of non-earthen materials (e.g., wood, concrete, steel, plastic) which provide structural support.

Product tank drawdown (Section 61.341 [3-7-90]) means any material or mixture of materials discharged from a product tank for the purpose of removing water or other contaminants from the product tank.

Product testing (Section 471.02) includes operations such as dye penetrant testing, hydrotesting, and ultrasonic testing.

Production (Section 430.01) shall be defined as the annual off-the-machine production (including off-the-machine coating where applicable) divided by the number of operating days during that year. Paper and paperboard production shall be measured at the off-the-machine moisture content, except for Subparts A, B, D, and E where paper and paperboard production shall be measured in air-dry-tons (10% moisture content). Market pulp shall be measured in air-dry-tons (10% moisture). Production shall be determined for each mill based upon past production practices, present trends, or committed growth.

Production (Section 431.11) shall be defined as the annual off-the-machine production (including off-the-machine coating where applicable) divided by the number of operating days during that year. Production shall be measured at the off-the-machine moisture content. Production shall be determined for each mill based upon past production practices, present trends, or committed growth.

Production (Section 82.3 [12/10/93])* means the manufacture of a controlled substance from any raw material or feedstock chemical, but does not include: (1) The manufacture of a controlled substance that is subsequently transformed; (2) The reuse or recycling of a controlled substance; (3) Amounts that are destroyed by the approved technologies; or (4) Amounts that are spilled or vented unintentionally.

Production allowances (Section 82.3 [12/10/93])* means the privileges granted by this subpart to produce controlled substances; however, production allowances may be used to produce controlled substances only in conjunction with consumption allowances. A person's production allowances are the total of the allowances he obtains under Secs. 82.7, 82.5 and 82.9 as may be modified under Sec.82.12 (transfer of allowances).

Production area size (Section 443.11) shall mean that area in which the oxidation towers, loading facilities, and all buildings that house product processes are located.

Production Compliance Audit (Section 86.1102-87). See **PCA (Section 86.1102-87)**.

Production facility (Section 435.11 [3/4/93])* shall mean any fixed or mobile structure subject to this subpart that is either engaged in well completion or used for active recovery of hydrocarbons from producing formations.

Production line (Section 60.671) means all affected facilities (crushers, grinding mills, screening operations, bucket elevators, belt conveyors, bagging operations, storage bins, and enclosed truck and railcar loading stations) which are directly connected or are connected together by a conveying system.

Production normalizing mass (/kkg) for each core or ancillary operation (Section 467.02) is the mass (off-kkg or off-lb) processed through that operation.

Production volume (Section 600.002-85) means, for a domestic manufacturer, the number of vehicle units domestically produced in a particular model year but not exported, and for a foreign manufacturer, means the number of vehicle units of a particular model imported into the United States.

Production volume (Section 704.3 [12/22/88])* means the quantity of a substance which is produced by a manufacturer, as measured in kilograms or pounds.

Production weighted particulate average (Section 86.090-2 [7-26-90]) means the manufacturer's production-weighted average particulate emission level, for certification purposes, of all of its diesel engine families included in the light-duty particulate averaging program. It is calculated at the end of the model year by multiplying each family particulate emission limit by its respective production, summing those terms, and dividing the sum by the total production of the effected families. Those vehicles produced for sale in California or at high altitude shall each be averaged separately from those produced for sale in any other area.

Production-weighted average (Section 86.085-2) means the manufacturer's production-weighted average particulate emission level, for certification purposes, of all of its diesel engine families included in the particulate averaging program. It is calculated at the end of the model year by multiplying each family particulate emission limit by its respective production, summing these terms, and dividing the sum by the total production of the effected families. Those vehicles produced for sale in California or at high altitude shall each be averaged separately from those produced for sale in any other area. This definition applies beginning with the 1985 model year.

Production-weighted average (Section 86.087-2 [10/31/88]) means the manufacturer's production-weighted average particulate emission level, for certification purposes, of all of its diesel engine families included in the particulate averaging program. It is calculated at the end of the model year by multiplying each family particulate emission limit by its respective production, summing these terms, and dividing the sum by the total production of the affected families. Those vehicles produced for sale in California or at high altitude shall each be averaged separately from those produced for sale in any other area. Diesel light-duty trucks with a loaded vehicle weight equal to or greater than 3,751 lbs (LDDT2s) shall only be averaged with other diesel light-duty trucks with a loaded vehicle weight equal to or greater than 3,751 lbs produced by that manufacturer.

Production-weighted NO_x average, for heavy-duty engines (Section 86.091-2) means the average of the manufacturer's family NO_x emission limits within the subclass (gasoline-fueled; or, light, medium, or heavy diesel) being averaged, weighted to account for differences in production volume and rated BHP. It is calculated at the end of the model year for determining compliance with the standard by summing, for all engine families in the subclass being averaged, the products (per engine family) of production volume, BHP rating, and the family NO_x emission limit, and dividing by the sum, for these engine families, of the products (per engine family) of production volume and BHP rating. The calculation is set forth in Section 86.091-2 under this definition. Those engines produced for sale in California or in 49-state areas shall each be averaged separately. Gasoline-fueled engines shall be averaged separately from diesel engines. Engines for use in urban buses may be averaged with other engines of the same subclass. This definition applies beginning with the 1991 model year.

Production-weighted particulate average, for heavy-duty diesel engines (Section 86.091-2) means the average of the manufacturer's family particulate emission limits within the subclass (light, medium, or heavy) being averaged, weighted to account for differences in production volume and rated BHP. It is calculated at the end of the model year for determining compliance with the standard by summing, for all engine families in the subclass being averaged, the products (per engine family) of production volume, BHP rating, and family particulate emission limit, and dividing by the sum, for these engine families, of the products (per engine family) of production volume and BHP rating. The calculation is set forth in Section 86.091-2 under this definition. Those engines produced for sale in California or in 49-state areas shall each be averaged separately. Engines for use in urban buses shall be excluded from participation in the particulate averaging program. This definition applies beginning with the 1991 model year.

Profit (Section 35.6015 [6-5-90])* The net proceeds obtained by deducting all allowable costs (direct and indirect) from the price. (Because this definition of profit is based on applicable Federal cost principles, it may vary from many firms' definition of profit, and may correspond to those firms' definition of "fee.")

Profit (Sections 33.005 [1/27/89]) means the net proceeds obtained by deducting all allowable costs (direct and indirect) from the price. (Because this definition of profit is based on applicable Federal cost principles, it may vary from many firms' definition of profit, and may correspond to those firms' definition of fee.)

Program (Section 403.3). See **Approved POTW Pretreatment Program (Section 403.3)**.

Program (Section 610.11). See **Evaluation program (Section 610.11)**.

Program directive (Section 56.1) means any formal written statement by the Administrator, the Deputy Administrator, the Assistant Administrator, a Staff Office Director, the General Counsel, a Deputy

Assistant Administrator, an Associate General Counsel, or a division Director of an Operational Office that is intended to guide or direct Regional Offices in the implementation or enforcement of the provisions of the act.

Program element (Section 35.105) means one of the major groupings of outputs of a continuing environmental program (e.g., administration, enforcement, monitoring).

Program income (Section 30.200) means gross income the recipient earns during its project period from charges for the project. This may include income from service fees, sale of commodities, trade-in allowances, or usage or rental fees. Fees from royalties are program income only if the assistance agreement so states. Revenue generated under the governing powers of a State or local government which could have been generated without an award is not considered program income. Such revenues include fines or penalties levied under judicial or penal power and used as a means to enforce laws. (Revenue from wastewater treatment construction grant projects under Title II of the Clean Water Act, as amended, is not program income. It must be used for operation and maintenance costs of the recipient's wastewater facilities.)

Program of requirements (Section 6.901) means a comprehensive document (booklet) describing program activities to be accomplished in the new special purpose facility or improvement. It includes architectural, mechanical, structural, and space requirements.

Prohibit specification (Section 231.2) means to prevent the designation of an area as a present or future disposal site.

Prohibited (Section 129.2) means that the constituent shall be absent in any discharge subject to these standards, as determined by any analytical method.

Project (Section 149.101) means a program or action for which an application for Federal financial assistance has been made.

Project (Section 30.200) means the activities or tasks EPA identifies in the assistance agreement.

Project (Section 35.2005) means the activities or tasks the Regional Administrator identifies in the grant agreement for which the grantee may expend, obligate or commit funds.

Project (Section 35.6015 [6-5-90])* The activities or tasks EPA identifies in the Cooperative Agreement and/or Superfund State Contract.

Project (Section 35.905) means the scope of work for which a grant or grant amendment is awarded under this subpart. The scope of work is defined as step 1, step 2, or step 3 of treatment works construction or segments (see definition of treatment works segment and Section 35.930-4).

Project (Sections 144.3; 146.3) means a group of wells in a single operation.

Project costs (Section 30.200) means all costs the recipient incurs in carrying out the project. EPA considers all allowable project costs to include the Federal share.

Project manager (Section 35.6015 [6-5-90]) The recipient official designated in the Cooperative Agreement or SSC as the program contact with EPA.

Project officer (Section 30.200) means the EPA official designated in the assistance agreement as EPA's program contact with the recipient. Project officers are responsible for monitoring the project.

Project officer (Section 35.6015 [6-5-90])* The EPA official designated in the Cooperative Agreement as EPA's program contact with the recipient. Project officers are responsible for monitoring the project.

Project officer (Section 7.25) means the EPA official designated in the assistance agreement (as defined in EPA assistance) as EPA's program contact with the recipient; Project Officers are responsible for monitoring the project.

Project performance standards (Section 35.2005) means the performance and operations requirements applicable to a project including the enforceable requirements of the Act and the specifications, including the quantity of excessive infiltration and inflow proposed to be eliminated, which the project is planned and designed to meet.

Project period (Section 30.200) means the length of time EPA specifies in the assistance agreement for completion of all project work. It may be composed of more than one budget period.

Project period (Section 35.6015 [6-5-90])* The length of time EPA specifies in the Cooperative Agreement and/or Superfund State Contract for completion of all project work. It may be composed of more than one budget period.

Project schedule (Section 35.2005) means a timetable specifying the dates of key project events including public notices of proposed procurement actions, subagreement awards, issuance of notice to proceed with building, key milestones in the building schedule, completion of building, initiation of operation and certification of the project.

Proof press (Section 60.431) means any device used only to check the quality of the image formation of newly engraved or etched gravure cylinders and prints only nonsaleable items.

Propellant (Section 162.3) means a gas or volatile liquid used in a pressurized pesticide product for the purpose of expelling the contents of the container.

Propellant (Section 61.31) means a fuel and oxidizer physically or chemically combined which undergoes combustion to provide rocket propulsion.

Propellant plant (Section 61.31) means any facility engaged in the mixing, casting, or machining of propellant.

Property damage (Section 280.92 [10/26/88]) shall have the meaning given this term by applicable state law. This term shall not include those liabilities which, consistent with standard insurance industry practices, are excluded from coverage in liability insurance policies for property damage. However, such exclusions for property damage shall not include corrective action associated with releases from tanks which are covered by the policy.

Property damage (Sections 264.141; 265.141). See **Bodily injury (Sections 264.141; 265.141)**.

Proportional sampling (Section 60.2) means sampling at a rate that produces a constant ratio of sampling rate to stack gas flow rate.

Proposal (Section 1508.23) exists at that stage in the development of an action when an agency subject to the Act has a goal and is actively preparing to make a decision on one or more alternative means of accomplishing that goal and the effects can be meaningfully evaluated. Preparation of an environmental impact statement on a proposal should be timed (Section 1502.5) so that the final statement may be completed in time for the statement to be included in any recommendation or report on the proposal. A proposal may exist in fact as well as by agency declaration that one exists.

Propose to manufacture, import, or process (Section 704.3 [12/22/88])* means that a person has made a firm management decision to commit financial resources for the manufacture, import, or processing of a specified chemical substance or mixture.

Propose to manufacture, import, or process (Section 716.3) means that a person has made a management decision to commit financial resources toward the manufacture, importation, or processing of a substance or mixture.

Proposed permit (Section 122.2) means a State NPDES permit prepared after the close of the public comment period (and, when applicable, any public hearing and administrative appeals) which is sent to EPA for review before final issuance by the State. A proposed permit is not a draft permit.

Protective equipment (Section 171.2) means clothing or any other materials or devices that shield against unintended exposure to pesticides.

Protocol (Section 790.3) means the plan and procedures which are to be followed in conducting a test.

Protocol 1 gas (Section 72.2 [7/30/93])* means a calibration gas mixture prepared and analyzed according to the "Procedure for NBS-Traceable Certification of Compressed Gas Working Standards Used for Calibration and Audit of Continuous Emission Monitors ("Revised Traceability Protocol No. 1")," Quality Assurance Handbook for Air Pollution Measurement Systems, Volume III, Stationary Source Specific Methods, Section 3.04, EPA-600/4-77-027b, June 1987 (set forth in Appendix H of part 75 of this chapter) or such revised procedure as approved by the Administrator.

Proven emission control systems (Section 86.092-2 [2-28-90]) are emission control components or systems (and fuel metering systems) that have completed full durability testing evaluation over a vehicle's useful life in some other certified engine family, or have completed bench or road testing demonstrated to be equal or more severe than certification mileage accumulation requirements. Alternatively, proven components or systems are those that are determined by EPA to be of comparable functional quality and manufactured using comparable materials and production techniques as components or systems which have been durability demonstrated in some other certified engine family. In addition, the components or systems must be employed in an operating environment (e.g., temperature, exhaust flow, etc.,) similar to that experienced by the original or comparable components or systems in the original certified engine family.

Provide for (Section 256.06) in the phrase "the plan shall (should) provide for" means explain, establish or set forth steps or courses of action.

Provider of financial assurance (Section 280.92 [10/26/88]) means an entity that provides financial assurance to an owner or operator of an underground storage tank through one of the mechanisms listed in Sections 280.95-280.103, including a guarantor, insurer, risk retention group, surety, issuer of a letter of credit, issuer of a state-required mechanism, or a state.

PRP (Section 304.12 [5/30/89]). See **Potentially responsible party (Section 304.12)**.

PRP (Section 35.6015 [1/27/89]). See **Potentially responsible party (Section 35.6015)**.

PSD permit (Section 124.41). See **Permit (Section 124.41)**.

PSD station (Section 58.1) means any station operated for the purpose of establishing the effect on air quality of the emissions from a proposed source for purposes of prevention of significant deterioration as required by Section 51.24(n) of Part 51 of this chapter.

PSES (Section 467.02) means pretreatment standards for existing sources, under section 307(b) of the Act.

PSNS (Section 467.02) means pretreatment standards for new sources, under section 307(c) of the Act.

PTC Type Choke Heaters (Section 85.2122(a)(2)(iii)(I)) means a positive temperature coefficient resistant ceramic disc capable of providing heat to the thermostatic coil when electrically energized.

Public body (Section 39.105) means a State, interstate agency, a municipality, or an intermunicipal agency, defined as follows: (1) State. A State, the District of Columbia, the Commonwealth of Puerto Rico, the Virgin Islands, Guam, American Samoa, and the Trust Territory of the Pacific Islands. (2) Interstate agency. An agency of two or more States established by or pursuant to an agreement or compact approved by the Congress, or any other agency of two or more States, having substantial powers or duties pertaining to the control of water pollution. (3) Municipality. A city, town, borough, county, parish, district, association, or other public body (including an intermunicipal agency of two or more of the foregoing entities) created by or pursuant to State law, or an Indian Tribe or an authorized Indian tribal organization, having jurisdiction over disposal of sewage, industrial wastes, or other wastes, or a designated and approved management agency under section 208 of the Act. This definition excludes a special district, such as a school district, which does not have as one of its principal responsibilities the treatment, transport, or disposal of liquid wastes.

Public or Private Agricultural Research Agency or Educational Institution (Section 172.21) means any organization engaged in research pertaining to the agricultural use of pesticides, or any educational institution engaged in pesticides research. Any research agency or educational institution whose principal function is to promote, or whose principal source of income is directly derived from, the sale or distribution of pesticides (or their active ingredients) does not come within the meaning of this term.

Public record (Section 117.1) means the NPDES permit application or the NPDES permit itself and the record for final permit as defined in Section 124.122.

Public vessel (Sections 110.1; 116.3)

means a vessel owned or bareboat-chartered and operated by the United States, or a State or political subdivision thereof, or by a foreign nation, except when such vessel is engaged in commerce.

Public water supplies (Section 125.58) means water distributed from a public water system.

Public water system (Section 125.58) means a system for the provision to the public of piped water for human consumption, if such system has at least fifteen (15) service connections or regularly serves at least twenty-five (25) individuals. This term includes (1) any collection, treatment, storage and distribution facilities under the control of the operator of the system and used primarily in connection with the system, and (2) any collection or pretreatment storage facilities not under the control of the operator of the system which are used primarily in connection with the system.

Public water system (Section 142.2) means a system for the provision to the public of piped water for human consumption, if such system has at least fifteen service connections or regularly serves an average of at least twenty-five individuals daily at least 60 days out of the year. Such term includes (1) any collection, treatment, storage, and distribution facilities under control of the operator of such system and used primarily in connection with such system, and (2) any collection or pretreatment storage facilities not under such control which are used primarily in connection with such system.

Public water system (Sections 141.2; 143.2) means a system for the provision to the public of piped water for human consumption, if such a system has at least fifteen service connections or regularly serves an average of at least twenty-five individuals daily at least 60 days out of the year. Such term includes (1) any collection, treatment, storage, and distribution facilities under control of the operator of such system and used primarily in connection with such system, and (2) any collection or pretreatment storage facilities not under such control which are used primarily in connection with such system. A public water system is either a community water system or a non-community water system.

Publication rotogravure printing press (Section 60.431) means any number of rotogravure printing units capable of printing simultaneously on the same continuous web or substrate and includes any associated device for continuously cutting and folding the printed web, where saleable paper products are printed, as specified in Section 60.431(a), under the definition of publication rotogravure printing press.

Publicly owned treatment works (POTW) (Section 260.10) means any device or system used in the treatment (including recycling and reclamation) of municipal sewage or industrial wastes of liquid nature which is owned by a State or municipality (as defined by section 502(4) of the CWA). This defintion includes sewers, pipes, or other conveyances only if they convey wastewater to a POTW providing treatment.

Publicly owned treatment works (POTW) (Section 125.58) means a treatment works, as defined in section 212(2) of the Act, which is owned by a State, municipality or intermunicipal or interstate agency.

Publicly owned treatment works (POTW) (Section 403.3) means a treatment works as defined by section 212 of the Act, which is owned by a State or municipality (as defined by section 502(4) of the Act). This definition includes any devices and systems used in the storage, treatment, recycling and reclamation of municipal sewage or industrial wastes of a liquid nature. It also includes sewers, pipes and other conveyances only if they convey wastewater to a POTW Treatment Plant. The term also means the municipality as defined in section 502(4) of the Act, which has jurisdiction over the indirect discharges to and the discharges from such a treatment works.

Publicly owned treatment works (POTW) (Section 117.1) means a treatment works as defined by section 212 of the Act, which is owned by a State or municipality (as defined by section 502(4) of the Act). This definition includes any sewers that convey wastewater to such a treatment works, but does not include pipes, sewers or other conveyances not connected to a facility providing treatment. The term also means the municipality as defined in section 502(4) of the Act, which has jurisdiction over the indirect discharges to and the discharges from such a treatment works.

Publicly owned treatment works (Section 501.2 [5/2/89]) means a treatment works treating domestic sewage that is owned by a municipality or State.

Publicly owned treatment works (POTW) (Sections 122.2; 270.2) means any device or system used in the treatment (including recycling and reclamation) of municipal sewage or industrial wastes of a liquid nature which is owned by a State or municipality. This definition includes sewers, pipes, or other conveyances only if they convey wastewater to a POTW providing treatment.

Pulverized coal-fired steam generating unit (Section 60.41b [12/16/87])* means a steam generating unit in which pulverized coal is introduced into an air stream that carries the coal to the combustion chamber of the steam generating unit where it is fired in suspension. This includes both conventional pulverized coal-fired and micropulverized coal-fired steam generating units.

Purchasing (Section 248.4 [2/17/89]) means the act of and the function of responsibility for the acquisition of equipment, materials, supplies, and services, including: buying, determining the need, selecting the supplier, arriving at a fair and reasonable price and terms and conditions, preparing the contract or purchase order, and follow up.

Purchasing activities (Section 248.4 [2/17/89]) means all activities included in the purchasing function.

Purge or line purge (Section 60.391) means the coating material expelled from the spray system when clearing it.

Putrescible wastes (Section 257.3-8) means solid waste which contains organic matter capable of being decomposed by microorganisms and of such a character and proportion as to be capable of attracting or providing food for birds.

Pyrolytic gas and oil (Section 245.101) means gas or liquid products that possess useable heating value that is recovered from the heating of organic material (such as that found in solid waste), usually in an essentially oxygen-free atmosphere.

Q

QA (Section 766.3) means quality assurance.

Q_{aj} (Section 60.741 [9-11-89]) means the volumetric flow rate of each gas stream (j) exiting the emission control device, in dry standard cubic meters per hour when Method 18 or 25 is used to measure VOC concentration or in standard cubic meters per hour (wet basis) when Method 25A is used to measure VOC concentration.

Q_{bi} (Section 60.741 [9-11-89]) means the volumetric flow rate of each gas stream (i) entering the emission control device, in dry standard cubic meters per hour when Method 18 or 25 is used to measure VOC concentration or in standard cubic meters per hour (wet basis) when Method 25A is used to measure VOC concentration.

QC (Section 766.3) means quality control.

Q_{di} (Section 60.741 [9-11-89]) means the volumetric flow rate of each gas stream (i) entering the emission control device from the affected coating operation, in dry standard cubic meters per hour when Method 18 or 25 is used to measure VOC concentration or in standard cubic meters per hour (wet basis) when Method 25A is used to measure VOC concentration.

Q_{fk} (Section 60.741 [9-11-89]) means the volumetric flow rate of each uncontrolled gas stream (k) emitted directly to the atmosphere from the affected coating operation, in dry standard cubic meters per hour when Method 18 or 25 is used to measure VOC concentration or in standard cubic meters per hour (wet basis) when Method 25A is used to measure VOC concentration.

Q_{gv} (Section 60.741 [9-11-89]) means the volumetric flow rate of the gas stream entering each individual carbon adsorber vessel (v), in dry standard cubic meters per hour when Method 18 or 25 is used to measure VOC concentration or in standard cubic meters per hour (wet basis) when Method 25A is used to measure VOC concentration. For purposes of calculating the efficiency of the individual adsorber vessel, the value of Qgv can be assumed to equal the value of Qhv measured for that adsorber vessel.

Q_{hv} (Section 60.741 [9-11-89]) means the volumetric flow rate of the gas stream exiting each individual carbon adsorber vessel (v), in dry standard cubic meters per hour when Method 18 or 25 is used to measure VOC concentration or in standard cubic meters per hour (wet basis) when Method 25A is used to measure VOC concentration.

Q_{ini} (Section 60.741 [9-11-89]) means the volumetric flow rate of each gas stream (i) entering the total enclosure through a forced makeup air duct, in standard cubic meters per hour (wet basis).

Q_{outj} (Section 60.741 [9-11-89]) means the volumetric flow rate of each gas stream (j) exiting the total enclosure through an exhaust duct or hood, in standard cubic meters per hour (wet basis).

Qualified Ground-Water Scientist (Section 260.10 [12-23-91]) means a scientist or engineer who has received a baccalaureate or post-graduate degree in the natural sciences or engineering, and has sufficient training and experience in ground-water hydrology and related fields as may be demonstrated by state registration, professional certifications, or completion of accredited university courses that enable that individual to make sound professional judgements regarding ground-water monitoring and contaminant fate and transport.

Qualified handicapped person (Section 7.25) means: (a) With respect to employment: A handicapped person who, with reasonable accommodation, can perform the essential functions of the job in question. (b) With respect to services: A handicapped person who meets the essential eligibility requirements for the receipt of such services.

Qualified incinerator (Section 761.3 [6/8/93])* means one of the following: (1) An incinerator approved under the provisions of Sec. 761.70. Any level of PCB concentration can be destroyed in an incinerator approved under Sec. 761.70. (2) A high efficiency boiler which complies with the criteria of Sec. 761.60(a)(2)(iii)(A), and for which the operator has given written notice to the appropriate EPA Regional Administrator in accordance with the notification requirements for the burning of mineral oil dielectric fluid under Sec. 761.60(a)(2)(iii)(B). (3) An incinerator approved under section 3005(c) of the Resource Conservation and Recovery Act (42 U.S.C. 6925(c)) (RCRA). (4) Industrial furnaces and boilers which are identified in 40 CFR 260.10 and 40 CFR 279.61(a)(1) and (2) when operating at their normal operating temperatures (this prohibits feeding fluids, above the level of detection, during either startup or shutdown operations).

Qualified individual with handicaps (Section 12.103 [8/14/87]) means (1) With respect to any agency program or activity under which a person is required to perform services or to achieve a level of accomplishment, an individual with handicaps who meets the essential eligibility requirements and who can achieve the purpose of the program or activity, without modifications in the program or activity that the agency can demonstrate would result in a fundamental alteration in its nature; or

(2) With respect to any other program or activity an individual with handicaps who meets the essential eligibility requirements for participation in, or receipt of benefits from, that program or activity. (3) Qualified handicapped person as that term is defined for purposes of employment in 29 CFR 1613.702(f), which is made applicable to this part by Section 12.140.

Qualifying facility (QF) (Section 72.2 [7/30/93])* means a "qualifying small power production facility" within the meaning of section 3(17)(C) of the Federal Power Act or a "qualifying cogeneration facility" within the meaning of section 3(18)(B) of the Federal Power Act.

Qualifying power purchase commitment (Section 72.2 [7/30/93])* means a power purchase commitment in effect as of November 15, 1990 without regard to changes to that commitment so long as: (1) The identity of the electric output purchaser; or (2) The identity of the steam purchaser and the location of the facility, remain unchanged as of the date the facility commences commercial operation; and (3) The terms and conditions of the power purchase commitment are not changed in such a way as to allow the costs of compliance with the Acid Rain Program to be shifted to the purchaser.

Quality assurance narrative statement (Section 30.200) means a description of how precision, accuracy, representativeness, completeness, and compatibility will be assessed, and which is sufficiently detailed to allow an unambiguous determination of the quality assurance practices to be followed throughout a research project.

Quality assurance program plan (Section 30.200) means a formal document which describes an orderly assembly of management policies, objectives, principles, organizational responsibilities, and procedures by which an agency or laboratory specifies how it intends to: (a) Produce data of documented quality, and (b) Provide for the preparation of quality assurance project plans and standard operating procedures.

Quality assurance project plan (Section 30.200) means an organization's written procedures which delineate how it produces quality data for a specific project or measurement method.

Quality assurance unit (Section 160.3 [8-17-89]) means any person or organizational element, except the study director, designated by testing facility management to perform the duties relating to quality assurance of the studies.

Quality assurance unit (Section 792.3 [8-17-89]) means any person or organizational element, except the study director, designated by testing facility management to perform the duties relating to quality assurance of the studies.

Quality Assurance Project Plan (Section 35.6015 [6-5-90])* A written document, associated with remedial site sampling, which presents in specific terms the organization (where applicable), objectives, functional activities, and specific quality assurance and quality control activities and procedures designed to achieve the data quality objectives of a specific project(s) or continuing operation(s).

Quantifiable Level/Level of Detection (Section 761.3 [6/27/88]) means 2 micrograms per gram from any resolvable gas chromatographic peak, i.e. 2 ppm.

Quarter (Section 60.481) means a 3-month period; the first quarter concludes on the last day of the last full month during the 180 days following initial startup.

Quench station (Section 60.461) means that portion of the metal coil surface coating operation where the coated metal coil is cooled, usually by a water spray, after baking or curing.

R

R (Section 60.641) is the sulfur emission reduction efficiency achieved in percent, carried to one decimal place.

R (Section 60.741 [9-11-89]) means the overall VOC emission reduction achieved for the duration of the emission test (expressed as a fraction).

Racial classifications (Section 7.25) (a) American Indian or Alaskan native. A person having origins in any of the original peoples of North America, and who maintains cultural identification through tribal affiliation or community recognition. (b) Asian or Pacific Islander. A person having origins in any of the original peoples of the Far East, Southeast Asia, the Indian subcontinent, or the Pacific Islands. This area includes, for example, China, Japan, Korea, the Philippine Islands, and Samoa. (c) Black and not of Hispanic origin. A person having origins in any of the black racial groups of Africa. (d) Hispanic. A person of Mexican, Puerto Rican, Cuban, Central or South American or other Spanish culture or origin, regardless of race. (e) White, not of Hispanic origin. A person having origins in any of the original peoples of Europe, North Africa, or the Middle East.

Rack dryer (Section 60.301) means any equipment used to reduce the moisture content of grain in which the grain flows from the top to the bottom in a cascading flow around rows of baffles (racks).

RACT (Section 51.100). See **Reasonably available control technology (Section 51.100)**.

Radiant energy or radiation (Section 796.3700) is defined as the energy traveling as a wave unaccompanied by transfer of matter. Examples include x-rays, visible light, ultraviolet light, radio waves, etc.

Radiation (Section 190.02) means any or all of the following: Alpha, beta, gamma, or X-rays; neutrons; and high-energy electrons, protons, or other atomic particles; but not sound or radio waves, nor visible, infrared, or ultraviolet light.

Radiation (Section 796.3700). See **Radiant energy (Section 796.3700)**.

Radicle (Section 797.2750) means that portion of the plant embryo which develops into the primary root.

Radioactive material (Section 190.02) means any material which spontaneously emits radiation.

Radioactive material (Section 191.12 [12/20/93])* means matter composed of or containing radionuclides, with radiological half-lives greater than 20 years, subject to the Atomic Energy Act of 1954, as amended.

Radioactive waste (Section 191.02) as used in this part, means the high-level and transuranic radioactive waste covered by this part.

Radioactive waste (Sections 144.3; 146.3) means any waste which contains radioactive material in concentrations which exceed those listed in 10 CFR Part 20, Appendix B, Table II, Column 2.

Radionuclide (Section 61.101 [12-15-89])* means any type of atom which spontaneously undergoes radioactive decay.

Radionuclide (Section 61.91 [12-15-89]) means a type of atom which spontaneously undergoes radioactive decay.

Rail Car (Section 201.1) means a non-self-propelled vehicle designed for and used on railroad tracks.

Railcar (Section 60.301) means railroad hopper car or boxcar.

Railcar loading station (Section 60.381) means that portion of a metallic mineral processing plant where metallic minerals or metallic mineral concentrates are loaded by a conveying system into railcars.

Railcar unloading station (Section 60.381) means that portion of a metallic mineral processing plant where metallic ore is unloaded from a railcar into a hopper, screen, or crusher.

Railroad (Section 201.1) means all the roads in use by any common carrier operating a railroad, whether owned or operated under a contract, agreement, or lease.

Raisins (Section 407.61) shall mean the production of raisins from the following products: Dried grapes, all varieties, bleached and unbleached, which have been cleaned and washed prior to packaging.

Random Incident Field (Section 211.203) means a sound field in which the angle of arrival of sound at a given point in space is random in time.

Range of concentration (Section 79.2) means the highest concentration, the lowest concentration, and the average concentration of an additive in a fuel.

Rare earth metals (Section 421.271) refers to the elements scandium, yttrium, and lanthanum to lutetium, inclusive.

Rated output (rO) (Section 87.1) means the maximum power/thrust available for takeoff at standard day conditions as approved for the engine by the Federal Aviation Administration, including reheat contribution where applicable, but excluding any contribution due to water injection.

Rated pressure ratio (rPR) (Section 87.1) means the ratio between the combustor inlet pressure and the engine inlet pressure achieved by an engine operating at rated output.

Rated speed (Section 86.082-2) means the speed at which the manufacturer specifies the maximum rated horsepower of an engine.

Raw agricultural commodity (Section 163.2) shall have the same meaning as it has in paragraph (r) of section 201 of the Act.

Raw data (Section 160.3 [8-17-89]) means any laboratory worksheets, records, memoranda, notes, or exact copies thereof, that are the result of original observations and activities of a study and are necessary for the reconstruction and evaluation of the report of that study. In the event that exact transcripts of raw data have been prepared (e.g., tapes which have been transcribed verbatim, dated, and verified accurate by signature), the exact copy or exact transcript may be substituted for the original source as raw data. "Raw data" may include photographs, microfilm or microfiche copies, computer printouts, magnetic media, including dictated observations, and recorded data from automated instruments.

Raw data (Section 792.3 [8-17-89]) means any laboratory worksheets, records, memoranda, notes, or exact copies thereof, that are the result of original observations and activities of a study and are necessary for the reconstruction and evaluation of the report of that study. In the event that exact transcripts of raw data have been prepared (e.g., tapes which have been transcribed verbatim, dated, and verified accurate by signature), the exact copy or exact transcript may be substituted for the original source as raw data. "Raw data" may include photographs, microfilm or microfiche copies, computer printouts, magnetic media, including dictated observations, and recorded data from automated instruments.

Raw ink (Section 60.431) means all purchased ink.

Raw material (Section 425.02) means the hides received by the tannery except for facilities covered by Subpart D and Subpart I where raw material means the hide or split in the condition in which it is first placed into a wet process.

Raw material (Section 428.101) shall mean all latex solids used in the manufacture of latex-dipped, latex-extruded, and latex-molded products.

Raw material (Section 428.11) shall mean all natural and synthetic rubber, carbon black, oils, chemical compounds, fabric and wire used in the manufacture of pneumatic tires and inner tubes or components thereof.

Raw material (Section 428.111) shall mean all latex solids used in the manufacture of latex foam.

Raw material (Sections 428.61; 428.71) shall mean all natural and synthetic rubber, carbon black, oils, chemical compounds, and fabric used in the manufacture of general molded, extruded, and fabricated rubber products.

Raw material equivalent (Sections 428.61; 428.71) shall be equal to the raw material usage multiplied by the volume of air scrubbed via wet scrubbers divided by the total volume of air scrubbed.

Rayon fiber (Section 60.601) means a manufactured fiber composed of regenerated cellulose, as well as manufactured fibers composed of regenerated cellulose in which substituents have replaced not more than 15 percent of the hydrogens of the hydroxyl groups.

RCRA (Section 144.3) means the Solid Waste Disposal Act as amended by the Resource Conservation and Recovery Act of 1976 (Pub. L. 94-580, as amended by Pub. L. 95-609, Pub. L. 96-510, 42 U.S.C. 6901 et seq.).

RCRA (Section 248.4 [2/17/89]) means the Solid Waste Disposal Act, as amended by the Resource Conservation and Recovery Act, as amended, 42 U.S.C. 6901 et seq.

RCRA (Section 249.04) means the Solid Waste Disposal Act, as amended by the Resource Conservation and Recovery Act of 1976, as amended, 42 U.S.C. 6901 et seq.

RCRA (Section 250.4 [10/6/87, 6/22/88]) means the Solid Waste Disposal Act, as amended by the Resource Conservation and Recovery Act, as amended, 42 U.S.C. 6901 et seq.

RCRA (Section 252.4 [6/30/88]) means the Solid Waste Disposal Act, as amended by the Resource Conservation and Recovery Act, as amended, 42 U.S.C. 6901 et seq.

RCRA (Section 253.4 [11/17/88]) means the Solid Waste Disposal Act, as amended by the Resource Conservation and Recovery Act, as amended, 42 U.S.C. 6901 et seq.

RCRA (Section 260.10) means the Solid Waste Disposal Act, as amended by the Resource Conservation and Recovery Act of 1976, as amended, 42 U.S.C. section 6901 et seq.

RCRA (Section 270.2) means the Solid Waste Disposal Act as amended by the Resource Conservation and Recovery Act of 1976 (Pub. L. 94-580, as amended by Pub. L. 95-609 and Pub. L. 96-482, 42 U.S.C. 6901 et seq.).

RCRA (Sections 124.2; 146.3) means the Solid Waste Disposal Act as amended by the Resource Conservation and Recovery Act of 1976 (Pub. L. 94-580, as amended by Pub. L. 95-609, 42 U.S.C. 6901 et seq.).

RDF stoker (Section 60.51a [2-11-91]) means a steam generating unit that combusts RDF in a semi-suspension firing mode using air-fed distributors.

Re-refined oils (Section 252.4 [6/30/88]) means used oils from which the physical and chemical contaminants acquired through previous use have been removed through a refining process.

Reactant (Section 723.250) means a chemical substance that is used intentionally in the manufacture of a polymer to become chemically a part of the polymer composition.

Reaction quantum yield for an excited-state process (Section 796.3700) is defined as the fraction of absorbed light that results in photoreaction at a fixed wavelength. It is the ratio of the number of molecules that photoreact to the number of quanta of light absorbed or the ratio of the number of moles that photoreact to the number of einsteins of light absorbed at a fixed wavelength.

Reaction spinning process (Section 60.601) means the fiber-forming process where a prepolymer is extruded into a fluid medium and solidification takes place by chemical reaction to form the final polymeric material.

Reactive functional group (Section 723.250) means an atom or associated group of atoms in a chemical substance that is intended or can reasonably be anticipated to undergo facile chemical reaction.

Reactor (Section 61.61) includes any vessel in which vinyl chloride is partially or totally polymerized into polyvinyl chloride.

Reactor opening loss (Section 61.61) means the emissions of vinyl chloride occurring when a reactor is vented to the atmosphere for any purpose other than an emergency relief discharge as defined in Section 61.65(a).

Readily water-soluble substances (Section 797.1060) means chemicals which are soluble in water at a concentration equal to or greater than 1,000 mg/l.

Ready Biodegradability (Section 796.3100) is an expression used to describe those substances which, in certain biodegradation test procedures, produce positive results that are unequivocal and which lead to the reasonable assumption that the substance will undergo rapid and ultimate biodegradation in aerobic aquatic environments.

Real property (Section 247.101) means any property that is immovable and attached to the land.

Real property (Section 30.200) means land, including land improvements, and structures and appurtenances, excluding movable machinery and equipment.

Real property (Section 35.6015 [6-5-90])* Land, including land improvements, structures, and appurtenances thereto, excluding movable machinery and equipment.

Real-Ear Protection at Threshold (Section 211.203) is the mean value in decibels of the occluded threshold of audibility (hearing protector in place) minus the open threshold of audibility (ears open and uncovered) for all listeners on all trials under otherwise identical test conditions.

Reasonable assistance (Section 86.078-7) includes, but is not limited to, clerical, copying, interpretation and translation services, the making available on request of personnel of the facility being inspected during their working hours to inform the EPA Enforcement Officer of how the facility operates and to answer his questions, and the performance on request of emissions tests on any vehicle (or engine) which is being, has been, or will be used for certification testing. Such tests shall be nondestructive, but may require appropriate mileage (or service) accumulation. A manufacturer may be compelled to cause the personal appearance of any employee at such a facility before an EPA Enforcement Officer by written request for his appearance, signed by the Assistant Administrator for Enforcement, served on the manufacturer. Any such employee who has been instructed by the manufacturer to appear will be entitled to be accompanied, represented, and advised by counsel.

Reasonable assistance (Sections 204.2; 205.2) means providing timely and unobstructed access to test products or products and records required by this part and opportunity for copying such records or testing such test products.

Reasonable compensation (Section 34.105 [2-26-90]) means, with respect to a regularly employed officer or employee of any person, compensation that is consistent with the normal compensation for such officer or employee for work that is not furnished to, not funded by, or not furnished in cooperation with the Federal Government.

Reasonable payment (Section 34.105 [2-26-90]) means, with respect to perfessional and other technical services, a payment in an amount that is consistent with the amount normally paid for such services in the private sector.

Reasonable terms (Section 39.105) means rates determined by the Secretary of the Treasury with relationship to the current average yield on outstanding marketable obligations of municipalities of comparable maturity.

Reasonably anticipated (Section 723.250) means that a knowledgeable person would expect a given physical or chemical composition or characteristic to occur based on such factors as the nature of the precursors used to manufacture the polymer, the type of reaction, the type of manufacturing process, the products produced in polymerization, the intended uses of the substance, or associated use conditions.

Reasonably attributable (Section 51.301) means attributable by visual observation or any other technique the State deems appropriate.

Reasonably available control technology (RACT) (Section 51.100) means devices, systems process modifications, or other apparatus or techniques that are reasonably available taking into account (1) the necessity of imposing such controls in order to attain and maintain a national ambient air quality standard, (2) the social, environmental and economic impact of such controls, and (3) alternative means of providing for attainment and maintenance of such standard. (This provision defines RACT for the purposes of Sections 51.110(c)(2) and 51.341(b) only.)

Reasons of business confidentiality (Section 2.201) include the concept of trade secrecy and other related legal concepts which give (or may give) a business the right to preserve the confidentiality of business information and to limit its use or disclosure by others in order that the business may obtain or retain business advantages it derives from its rights in the information. The definition is meant to encompass any concept which authorizes a Federal agency to withhold business information under 5 U.S.C. 552(b)(4), as well as any concept which requires EPA to withhold information from the public for the benefit of a business under 18 U.S.C. 1905 or any of the various statutes cited in Section 2.301 through Section 2.309.

Rebricking (Sections 60.291; 61.161) means cold replacement of damaged or worn refractory parts of the glass melting furnace. Rebricking includes replacement of the refractories comprising the bottom, sidewalls, or roof of the melting vessel; replacement of refractory work in the heat exchanger; and replacement of refractory portions of the glass conditioning and distribution system.

Receiving country (Section 262.51 [7/19/88])* means a foreign country to which a hazardous waste is sent for the purpose of treatment, storage or disposal (except short-term storage incidental to transportation).

Receiving Property (Section 201.1) means any residential or commercial property that receives the sound from railroad facility operations, but that is not owned or operated by a railroad; except that occupied residences located on property owned or controlled by the railroad are included in the definition of receiving property. For purposes of this definition railroad crew sleeping quarters located on property owned or controlled by the railroad are not considered as residences. If, subsequent to the publication date of these regulations, the use of any property that is currently not applicable to this regulation changes, and it is newly classified as either residential or commercial, it is not receiving property until four years have elapsed from the date of the actual change in use.

Receiving Property Measurement Location (Section 201.1) means a location on receiving property that is on or beyond the railroad facility boundary and that meets the receiving property measurement location criteria of Subpart C.

Recharge (Section 149.2 [2/14/89])* means a process, natural or artificial, by which water is added to the saturated zone of an aquifer.

Recharge Area (Section 149.2 [2/14/89])* means an area in which water reaches the zone of saturation (ground water) by surface infiltration; in addition, a major recharge area is an area where a major part of the recharge to an aquifer occurs.

Recharge zone (Section 149.101) means the area through which water enters the Edwards Underground Reservoir as defined in the December 16, 1975, Notice of Determination.

Recipient (Section 30.200) means any entity which has been awarded and accepted an EPA assistance agreement.

Recipient (Section 344.105 [2-26-90]) includes all contractors, subcontractors at any tier, and subgrantees at any tier of the recipient of funds received in connection with a Federal contract, grant, loan, or cooperative agreement. The term excludes an Indian tribe, tribal organization, or any other Indian organization with respect to expenditures specifically permitted by other Federal law.

Recipient (Section 35.4010 [10/1/92])* means any group of individuals that has been awarded a TAG.

Recipient (Section 35.6015 [6-5-90])* Any State, political subdivision thereof, or Indian Tribe which has been awarded and has accepted an EPA Cooperative Agreement.

Recipient (Section 7.25) means, for the purposes of this regulation, any state or its political subdivision, any instrumentality of a state or its political subdivision, any public or private agency, institution, organization, or other entity, or any person to which Federal financial assistance is extended directly or through another recipient, including any successor, assignee, or transferee of a recipient, but excluding the ultimate beneficiary of the assistance.

Recipient (Section 721.3 [7/27/88]) means any person who purchases or otherwise obtains a chemical substance directly from a person who manufacturers, imports, or processes the substance.

Reciprocating compressor (Section 60.631) means a piece of equipment that increases the pressure of a process gas by positive displacement, employing linear movement of the driveshaft.

Recirculated cooling water (Section 423.11) means water which is passed through the main condensers for the purpose of removing waste heat, passed through a cooling device for the purpose of removing such heat from the water and then passed again, except for blowdown, through the main condenser.

Recirculation (Section 420.101) means those cold rolling operations which include recirculation of rolling solutions at all mill stands.

Reclaimed (Section 261.1). For purposes of Sections 261.2 and 261.6, a material is reclaimed if it is processed to recover a usable product, or if it is regenerated. Examples are recovery of lead values from spent batteries and regeneration of spent solvents.

Reclamation area (Section 434.11) means the surface area of a coal mine which has been returned to required contour and on which revegetation (specifically, seeding or planting) work has commenced.

Recommencing discharger (Section 122.2) means a source which recommences discharge after terminating operations.

Recommendation to list (Section 15.4) means a written request which has been signed and sent by recommending person to the Listing Official asking that EPA place a facility on the List of Violating Facilities.

Recommended Decision (Section 164.2 [7/10/92])* means the recommended findings and conclusions of the Presiding Officer in an expedited hearing.

Recommending person (Section 15.4) means a Regional Administrator, the Associate Enforcement Counsel for Air or the Associate Enforcement Counsel for Water or their successors, the Assistant Administrator for Air and Radiation or the Assistant Administrator for Water or their successors, a Governor, or a member of the public.

Reconfigured emission-data vehicle (Section 86.082-2) means an emission-data vehicle obtained by modifying a previously used emission-data vehicle to represent another emission-data vehicle.

Reconstruction (Section 51.301) will be presumed to have taken place where the fixed capital cost of the new component exceeds 50 percent of the fixed capital cost of a comparable entirely new source. Any final decision as to whether reconstruction has occurred must be made in accordance with the provisions of Section 60.15 (f) (1) through (3) of this title.

Record (Section 1516.2) means any item or collection or grouping of information about an individual that is maintained by the Council (including, but not limited to, his or her employment history, payroll information, and financial transactions), and that contains his or her name, or an identifying number, symbol, or other identifying particular assigned to the individual such as a social security number.

Record (Section 16.2) shall have the meaning given to it by 5 U.S.C. 552a (a)(4).

Record (Section 2.100 [1/5/88])*. See **EPA record (Section 2.100)**.

Recorded (Section 2.201) means written or otherwise registered in some form for preserving information, including such forms as drawings, photographs, videotape, sound recordings, punched cards, and computer tape or disk.

Recoverable resources (Sections 245.101; 246.101) means materials that still have useful physical, chemical, or biological properties after serving their original purpose and can, therefore, be reused or recycled for the same or other purposes.

Recovered material (Section 249.04) means waste material and byproducts which have been recovered or diverted from solid waste, but such term does not include those materials and byproducts generated from, and commonly reused within, an original manufacturing process (Pub. L. 94-580, 90 Stat. 2800, 42 U.S.C. 6903, as amended by Pub. L. 96-482).

Recovered materials (Section 248.4 [2/17/89]) means waste material and byproducts which have been recovered or diverted from solid waste, but such term does not include those materials and byproducts generated from, and commonly reused within, an original manufacturing process.

Recovered materials (Section 250.4 [10/6/87, 6/22/88]) means waste material and by-products that have been recovered or diverted from solid waste, but such term does not include those materials and by-products generated from, and commonly reused within, an original manufacturing process. In the case of paper and paper products, the term recovered materials includes: (1) Postconsumer materials such as: (i) Paper, paperboard, and fibrous wastes from retail stores, office buildings, homes, and so forth, after they have passed through their end usage as a consumer item, including: Used corrugated boxes, old newspapers, old magazines, mixed waste paper, tabulating cards, and used cordage, and (ii) All paper, paperboard, and fibrous wastes that enter and are collected from municipal solid waste; and (2) Manufacturing, forest residues, and other wastes such as: (i) Dry paper and paperboard waste generated after completion of the papermaking process (that is, those manufacturing operations up to and including the cutting and trimming of the paper machine reel into smaller rolls or rough sheets) including envelope cuttings, bindery trimmings, and other paper and paperboard waste, resulting from printing, cutting, forming, and other converting operations; bag, box and carton manufacturing wastes; and butt rolls, mill wrappers, and rejected unused stock; and (ii) Finished paper and paperboard from obsolete inventories of paper and paperboard manufacturers, merchants, wholesalers, dealers, printers, converters, or others; (iii) Fibrous by-products of harvesting, manufacturing, extractive, or wood-cutting processes, flax, straw, linters, bagasse, slash, and other forest residues; (iv) Wastes generated by the conversion of goods made from fibrous material (e.g., waste rope from cordage manufacture, textile mill waste, and cuttings); and (v) Fibers recovered from waste water that otherwise would enter the waste stream.

Recovered solvent (Section 60.601) means the solvent captured from liquid and gaseous process streams that is concentrated in a control device and that may be purified for reuse.

Recovery (Sections 245.101; 246.101) means the process of obtaining materials or energy resources from solid waste.

Recovery device (Section 60.661 [6-29-90]) means an individual unit of equipment, such as an absorber, carbon adsorber, or condenser, capable of and used for the purpose of recovering chemicals for use, reuse, or sale.

Recovery furnace (Section 60.281) means either a straight kraft recovery furnace or a cross recovery furnace, and includes the direct-contact evaporator for a direct-contact furnace.

Recovery system (Section 60.661 [6-29-90]) means an individual recovery device or series of such devices applied to the same vent stream.

Recruiting and training agency (Section 8.2) means any person who refers workers to any contractor or subcontractor, or who provides or supervises apprenticeship or training for employment by any contractor or subcontractor.

Recurrent expenditures (Section 35.105) are those expenses associated with the activities of a continuing environmental program. All expenditures, except those for equipment purchases with a unit acquisition cost of $5,000 or more, are considered recurrent unless justified by the applicant as unique and approved as such by the Regional Administrator in the assistance award.

Recyclable paper (Section 250.4 [10/6/87, 6/22/88]) means any paper separated at its point of discard or from the solid waste stream for utilization as a raw material in the manufacture of a new product. It is often called waste paper or paper stock. Not all paper in the waste stream is recyclable; it may be heavily contaminated or otherwise unusable.

Recycled (Section 261.1). For purposes of Sections 261.2 and 261.6, a material is recycled if it is used, reused, or reclaimed.

Recycled material (Section 247.101) means a material that can be utilized in place of a raw or virgin material in manufacturing a product and consists of materials derived from post consumer waste, industrial scrap, material derived from agricultural wastes and other items, all of which can be used in the manufacture of new products.

Recycled material (Sections 245.101; 246.101) means a material that is utilized in place of a primary raw, or virgin material in manufacturing a product.

Recycled PCBs (Section 761.3 [6/27/88])* means those PCBs which appear in the processing of paper products or asphalt roofing materials from PCB-contaminated raw materials. Processes which recycle PCBs must meet the following requirements: (1) There are no detectable concentrations of PCBs in asphalt roofing material products leaving the processing site. (2) The concentration of PCBs in paper products leaving any manufacturing site processing paper products, or in paper products imported into the United States, must have an annual average of less than 25 ppm with a 50 ppm maximum. (3) The release of PCBs at the point at which emissions are vented to ambient air must be less than 10 ppm. (4) The amount of Aroclor PCBs added to water discharged from an asphalt roofing processing site must at all times be less than 3 micrograms per liter (μg/L) for total Aroclors (roughly 3 parts per billion (3ppb)). Water discharges from the processing of paper products must at all times be less than 3 micrograms per liter (μg/l) for total Aroclors (roughly 3 ppb), or comply with the equivalent mass-based limitation. (5) Disposal of any other process wastes at concentrations of 50 ppm or greater must be in accordance with Subpart D of this part.

Recycling (Sections 244.101; 245.101;

246.101) means the process by which recovered materials are transformed into new products.

Reduced sulfur compounds (Sections 60.101; 60.641) means hydrogen sulfide (H_2S), carbonyl sulfide (COS) and carbon disulfide (CS_2).

Reduction control system (Section 60.101) means an emission control system which reduces emissions from sulfur recovery plants by converting these emissions to hydrogen sulfide.

Reentry (Section 162.3) means the action of entering an area or site at, in, or on which a pesticide has been applied.

Reference method (Section 50.1) means a method of sampling and analyzing the ambient air for an air pollutant that is specified as a reference method in an appendix to this part, or a method that has been designated as a reference method in accordance with Part 53 of this chapter; it does not include a method for which a reference method designation has been cancelled in accordance with Section 53.11 or Section 53.16 of this chapter.

Reference method (Section 53.1) means a method of sampling and analyzing the ambient air for an air pollutant that is specified as a reference method in an appendix to Part 50 of this chapter, or a method that has been designated as a reference method in accordance with this part; it does not include a method for which a reference method designation has been cancelled in accordance with Section 53.11 or Section 53.16.

Reference method (Section 60.2) means any method of sampling and analyzing for an air pollutant as described in Appendix A to this part.

Reference method (Section 61.02) means any method of sampling and analyzing for an air pollutant, as described in Appendix B to this part.

Reference substance (Section 797.1350) is a chemical used to access the constancy of response of a given species of test organisms to that chemical, usually by use of the acute LC_{50} . (It is assumed that any change in sensitivity to the reference substance will indicate the existence of some similar change in degree of sensitivity to other chemicals whose toxicity is to be determined.)

Reference substance (Section 792.226) means any chemical substance or mixture or material other than a test substance that is administered to or used in analyzing the test system in the course of a study for the purposes of establishing a basis for comparison with the test substance. For purposes of this Subpart all references to control substance in Subparts B through J also include reference substance as defined herein.

Reference substance (Section 160.3 [8-17-89]) means any chemical substance or mixture, or analytical standard, or material other than a test substance, feed, or water, that is administered to or used in analyzing the test system in the course of a study for the purposes of establishing a basis for comparison with the test substance for known chemical or biological measurements.

Reference substance (Section 792.3 [8-17-89]) means any chemical substance or mixture, or analytical standard, or material other than a test substance, feed, or water, that is administered to or used in analyzing the test system in the course of a study for the purposes of establishing a basis for comparison with the test substance for known chemical or biological measurements.

Referring agency (Section 1508.24) means the federal agency which has referred any matter to the Council after a determination that the matter is unsatisfactory from the standpoint of public health or welfare or environmental quality.

Refillable Beverage Container (Section 244.101) means a beverage container that when returned to a distributor or bottler is refilled with a beverage and reused.

Refinery (Section 80.2 [2/16/94])* means a plant at which gasoline or diesel fuel is produced.

Refinery (Section 80.2 [8-21-90])* means a plant at which gasoline or diesel fuel is produced.

Refinery process unit (Section 60.101) means any segment of the petroleum refinery in which a specific processing operation is conducted.

Refining (Section 60.271a) means that phase of the steel production cycle during which undesirable elements are removed from the molten steel and alloys are added to reach the final metal chemistry.

Reformulated gasoline (Section 80.2 [2/16/94])* means any gasoline whose formulation has been certified under Sec. 80.40, which meets each of the standards and requirements prescribed under Sec. 80.41,and which contains less than the maximum concentration of the marker specified in Sec. 80.82 that is allowed for reformulated gasoline under Sec. 80.82.

Reformulated gasoline blendstock for oxygenate blending, or RBOB (Section 80.2 [2/16/94])* means a petroleum product which, when blended with a specified type and percentage of oxygenate, meets the definition of reformulated gasoline, and to which the specified type and percentage of oxygenate is added other than by the refiner or importer of the RBOB at the refinery or import facility where the RBOB is produced or imported.

Reformulated gasoline credit (Section 80.2 [2/16/94])* means the unit of measure for the paper transfer of oxygen or benzene content resulting from reformulated gasoline which contains more than 2.1 weight percent of oxygen or less than 0.95 volume percent benzene.

Refractory metals (Section 471.02) includes the metals of columbium, tantalum, molybdenum, rhenium, tungsten and vanadium and their alloys.

Refueling emissions canister(s)(Section 86.098-2 [4/6/94])* means any vapor storage unit(s) that is exposed to the vapors generated during refueling.

Refueling emissions (Section 86.098-2 [4/6/94])* means evaporative emissions that emanate from a motor vehicle fuel tank(s) during a refueling operation.

Refund (Section 244.101) means the sum, equal to the deposit, that is given to the consumer or the dealer or both in exchange for empty returnable beverage containers.

Refurbishment (Section 60.501) means, with reference to a vapor processing system, replacement of components of, or addition of components to, the system within any 2-year period such that the fixed capital cost of the new components required for such component replacement or addition exceeds 50 percent of the cost of a comparable entirely new system.

Refuse-derived fuel or RDF (Section 60.51a [2-11-91]) means a type of MSW produced by processing MSW through shredding and size classification. This includes all classes of RDF including low density fluff RDF through densified RDF and RDF fuel pellets.

Regenerative carbon adsorber (Section 61.131 [9-19-91]) means a carbon adsorber applied to a single source or group of sources, in which the carbon beds are regenerated without being moved from their location.

Regenerative cycle gas turbine (Section 60.331(c)) means any stationary gas turbine which recovers heat from the gas turbine exhaust gases to preheat the inlet combustion air to the gas turbine.

Regenerative cycle gas turbine (Section 60.331(s)) means any stationary gas turbine that recovers thermal energy from the exhaust gases and utilizes the thermal energy to preheat air prior to entering the combustor.

Region (Section 51.100) means an area designated as an air quality control region (AQCR) under section 107(c) of the Act.

Region (Section 60.21) means an air

quality control region designated under section 107 of the Act and described in Part 81 of this chapter.

Regional Administrator (Section 20.2) means the Regional designee appointed by the Administrator to certify facilities under this part.

Regional Administrator (Section 21.2) means the Regional Administrator of EPA for the region including the State in which the facility or method of operation is located, or his designee.

Regional Administrator (Section 260.10) means the Regional Administrator for the EPA Region in which the facility is located, or his designee.

Regional Administrator (Section 165.1) means the Administrator of a Regional Office of the Agency or his delegatee.

Regional Administrator (Section 22.03) means the Administrator of any Regional Office of the Agency or any officer or employee thereof to whom his authority is duly delegated. Where the Regional Administrator has authorized the Regional Judicial Officer to act, the term Regional Administrator shall include the Regional Judicial Officer. In a case where the complainant is the Assistant Administrator for Enforcement or his delegate, the term Regional Administrator as used in these rules shall mean the Administrator.

Regional Administrator (Section 124.41) shall have the meaning set forth in Section 124.2, except when EPA has delegated authority to administer those regulations to another agency under the applicable subsection of 40 CFR 52.21, the term Regional Administrator shall mean the chief administrative officer of the delegate agency.

Regional Administrator (Section 108.2). See **Administrator (Section 108.2)**.

Regional Administrator (Section 121.1) means the Regional designee appointed by the Administrator, Environmental Protection Agency.

Regional Administrator (Section 136.2) means one of the EPA Regional Administrators.

Regional Administrator (Section 147.2902) means the Regional Administrator of Region 6 of the United States Environmental Protection Agency, or an authorized representative.

Regional Administrator (Section 65.01) shall mean the Regional Administrator for an EPA regional office whose duties include those set forth at 40 CFR 1.41, or his authorized delegatee.

Regional Administrator (Section

58.1) means the Administrator of one of the ten EPA Regional Offices or his or her authorized representative.

Regional Administrator (Section 403.3) means the appropriate EPA Regional Administrator.

Regional Administrator (Section 112.2 [8/25/93])*", means the Regional Administrator of the Environmental Protection Agency, or his designee, in and for the Region in which the facility is located.

Regional Administrator (Section 232.2 [2/11/93])* means the Regional Administrator of the appropriate Regional Office of the Environmental Protection Agency or the authorized representative of the Regional Administrator.

Regional Administrator (Sections 122.2; 144.3; 233.3; 270.2) means the Regional Administrator of the appropriate Regional Office of the Environmental Protection Agency or the authorized representative of the Regional Administrator.

Regional Hearing Clerk (Section 22.03) means an individual duly authorized by the Regional Administrator to serve as hearing clerk for a given region. Correspondence may be addressed to the Regional Hearing Clerk, United States Environmental Protection Agency (address of Regional Office - see Appendix). In a case where the complainant is the Assistant Administrator for Enforcement or his delegate, the term Regional Hearing Clerk as used in these rules shall mean the Hearing Clerk.

Regional Hearing Clerk (Section 124.72 [2-13-92])* means an employee of the Agency designated by a Regional Administrator to establish a repository for all books, records, documents, and other materials relating to hearings under this subpart.

Regional Judicial Officer (Section 22.03) means a person designated by the Regional Administrator under Section 22.04(b) to serve as a Regional Judicial Officer.

Regional Office (Section 51.100) means one of the ten (10) EPA Regional Offices.

Regional Office of AAA (Section 305.12) means the Regional Office of AAA in Washington, D.C.

Registrant (Section 164.2 [7/10/92])* means any person who has registered a pesticide pursuant to the provisions of the Act.

Registrant (Section 2.307) means any person who has obtained registration under the Act of a pesticide or of an establishment.

Registration office (Section 5.2) means any of the several offices in EPA which have been designated to receive applications for attendance at direct training courses. (See Section 5.4 for a listing of such courses.)

Regularly employed (Section 34.105 [2-26-90]) means, with respect to an officer or employee of a person requesting or receiving a Federal contract, grant, loan, or cooperative agreement or a commitment providing for the United States to insure or guarantee a loan, an officer or employee who is employed by such person for at least 130 working days within one year immediately preceding the date of the submission that initiates agency consideration of such person for receipt of such contract, grant, loan, cooperative agreement, loan insurance commitment, or loan guarantee commitment. An officer or employee who is employed by such person for less than 130 working days within one year immediately preceding the date of the submission that initiates agency consideration of such person shall be considered to be regularly employed as soon as he or she is employed by such person for 130 working days.

Regulated activity or activity subject to regulation (Section 124.41) means a major PSD stationary source or major PSD modification.

Regulated area (Section 763.121) means an area established by the employer to demarcate areas where airborne concentrations of asbestos exceed or can reasonably be expected to exceed the permissible exposure limit. The regulated area may take the form of: (1) A temporary enclosure, as required by paragraph (e)(6) of this section, or (2) an area demarcated in any manner that minimizes the number of employees exposed to asbestos.

Regulated asbestos-containing material (RACM) (Section 61.141 [11-20-90]) means (a) Friable asbestos material, (b) Category I nonfriable ACM that has become friable, (c) Category I nonfriable ACM that will be or has been subjected to sanding, grinding, cutting, or abrading, or (d) Category II nonfriable ACM that has a high probability of becoming or has become crumbled, pulverized, or reduced to powder by the forces expected to act on the material in the course of demolition or renovation operations regulated by this subpart.

Regulated chemical (Section 707.63) means any chemical substance or mixture for which export notice is required under Section 707.60.

Regulated pest (Section 171.2) means a specific organism considered by a State or Federal agency to be a pest requiring regulatory restrictions, regulations, or control procedures in order to protect the host, man and/or his environment.

Regulated substance (Section 280.1) means (a) Any substance defined in section 101(14) of the Comprehensive Environmental Response, Compensation and Liability Act of

1980 (but not including any substance regulated as a hazardous waste under Subtitle C of the Resource Conservation and Recovery Act, as amended), and (b) Petroleum, including crude oil or any fraction thereof which is liquid at standard conditions of temperature and pressure (60 degrees Fahrenheit and 14.7 pounds per square inch absolute).

Regulated substance (Section 280.12 [9/23/88]) means: (a) Any substance defined in section 101(14) of the Comprehensive Environmental Response, Compensation and Liability Act (CERCLA) of 1980 (but not including any substance regulated as a hazardous waste under subtitle C), and (b) Petroleum, including crude oil or any fraction thereof that is liquid at standard conditions of temperature and pressure (60 degrees Fahrenheit and 14.7 pounds per square inch absolute). The term regulated substance includes but is not limited to petroleum and petroleum-based substances comprised of a complex blend of hydrocarbons derived from crude oil through processes of separation, conversion, upgrading, and finishing, such as motor fuels, jet fuels, distillate fuel oils, residual fuel oils, lubricants, petroleum solvents, and used oils.

Regulations published under this part (Section 211.102) means all subparts to Part 211.

Regulatory agency (Section 190.02) means the government agency responsible for issuing regulations governing the use of sources of radiation or radioactive materials or emissions therefrom and carrying out inspection and enforcement activities to assure compliance with such regulations.

Regulatory agency (Section 192.31) means the U.S. Nuclear Regulatory Commission.

Reid vapor pressure (Section 60.111a) is the absolute vapor pressure of volatile crude oil and nonviscous petroleum liquids, except liquified petroleum gases, as determined by ASTM D323-82 (incorporated by reference - see Section 60.17).

Reid vapor pressure (Sections 60.111; 60.111b) is the absolute vapor pressure of volatile crude oil and volatile nonviscous petroleum liquids, except liquified petroleum gases, as determined by ASTM D323-82 (incorporated by reference - see Section 60.17).

Reimbursement period (Section 791.3) refers to a period that begins when the data from the last non-duplicative test to be completed under a test rule is submitted to EPA and ends after an amount of time equal to that which had been required to develop that data or after 5 years, whichever is later.

Reimbursement period (Section 766.3) means the period that begins when the data from the last test to be completed under this Part for a specific chemical substance listed in Section 766.25 is submitted to EPA, and ends after an amount of time equal to that which had been required to develop that data or 5 years, whichever is later.

Reimbursement period (Section 790.3) refers to a period that begins when the data from the last non-duplicative test to be completed under a test rule are submitted to EPA and ends after an amount of time equal to that which had been required to develop data or after five years, whichever is later.

Reinforced plastic (Section 763.163 [7-12-89]) means an asbestos-containing product made of plastic. Major applications of this product include: electro-mechanical parts in the automotive and appliance industries; components of printing plates; and as high-performance plastics in the aerospace industry.

Rejection of a batch (Section 204.51) means that the number of non-complying compressors in the batch sample is greater than or equal to the rejection number as determined by the appropriate sampling plan.

Rejection of a batch (Section 205.51) means the number of noncomplying vehicles in the batch sample is greater than or equal to the rejection number as determined by the appropriate sampling plan.

Rejection of a batch sequence (Section 205.51) means that the number of rejected batches in a sequence is equal to or greater than the rejection number as determined by the appropriate sampling plan.

Rejection of a batch sequence (Section 204.51) means that the number of rejected batches in a sequence is greater than or equal to the sequence rejection number as determined by the appropriate sampling plan.

Related coatings (Section 60.431) means all non-ink purchased liquids and liquid-solid mixtures containing VOC solvent, usually referred to as extenders or varnishes, that are used at publication rotogravure printing presses.

Release (Section 2.310) has the meaning given it in section 101(22) of the Act, 42 U.S.C. 9601(22).

Release (Section 280.1) means any spilling, leaking, emitting, discharging, escaping, leaching, or disposing from an underground storage tank into ground water, surface water, or subsurface soils.

Release (Section 280.12 [9/23/88]) means any spilling, leaking, emitting, discharging, escaping, leaching or disposing from an UST into ground water, surface water or subsurface soils.

Release (Section 300.6)

Release (Section 300.6) as defined by section 101(22) of CERCLA, means any spilling, leaking, pumping, pouring, emitting, emptying, discharging, injection, escaping, leaching, dumping, or disposing into the environment, but excludes: Any release which results in exposure to persons solely within a workplace, with respect to a claim which such persons may assert against the employer of such persons; emissions from the engine exhaust of a motor vehicle, rolling stock, aircraft, vessel, or pipeline pumping station engine; release of source, byproduct or special nuclear material from a nuclear incident, as those terms are defined in the Atomic Energy Act of 1954, if such release is subject to requirements with respect to financial protection established by the Nuclear Regulatory Commission under section 170 of such act, or, for the purpose of section 104 of CERCLA or any other response action, any release of source, byproduct, or special nuclear material from any processing site designated under section 122(a)(1) or 302(a) of the Uranium Mill Tailings Radiation Control Act of 1978; and the normal application of fertilizer. For the purpose of this Plan, release also means substantial threat of release.

Release (Section 302.3) means any spilling, leaking, pumping, pouring, emitting, emptying, discharging, injecting, escaping, leaching, dumping, or disposing into the environment, but excludes (1) any release which results in exposure to persons solely within a workplace, with respect to a claim which such persons may assert against the employer of such persons, (2) emissions from the engine exhaust of a motor vehicle, rolling stock, aircraft, vessel, or pipeline pumping station engine, (3) release of source, byproduct, or special nuclear material from a nuclear incident, as those terms are defined in the Atomic Energy Act of 1954, if such release is subject to requirements with respect to financial protection established by the Nuclear Regulatory Commission under section 170 of such Act, or for the purposes of section 104 of the Comprehensive Environmental Response, Compensation, and Liability Act or any other response action, any release of source, byproduct, or special nuclear material from any processing site designated under section 102(a)(1) or 302(a) of the Uranium Mill Tailings Radiation Control Act of 1978, and (4) the normal application of fertilizer.

Release (Section 310.11 [1/15/93])* as defined by section 101(22) of CERCLA, means any spilling, leaking, pumping, pouring, emitting, emptying, discharging, injection, escaping, leaching, dumping, or disposing into the environment, but excludes: any release that results in exposure to persons solely within a workplace, with respect to a claim that such persons may assert against the employer ofsuch persons; emissions from the engine exhaust of a motor vehicle, rolling stock, aircraft, vessel, or pipeline pumping station engine; release of source, by-product or special nuclear material from a nuclear incident, as those terms are defined in the Atomic Energy Act of 1954, if such release is subject to requirements with respect to financial protection established by the Nuclear Regulatory Commission under section 170 of such act, or, for the purpose of section 104 of CERCLA or any other response action, any release of source, by-product, or special nuclear material from any processing site designated under section 122(a)(1) or 302(a) of the Uranium Mill Tailings Radiation Control Act of 1978; and the normal application of fertilizer. For the purposes of this part, release also means threat of release.

Release (Section 355.20) means any spilling, leaking, pumping, pouring, emitting, emptying, discharging, injecting, escaping, leaching, dumping, or disposing into the environment (including the abandonment or discarding of barrels, containers, and other closed receptacles) of any hazardous chemical, extremely hazardous substance, or CERCLA hazardous substance.

Release (Section 372.3 [2/16/88]) means any spilling, leaking, pumping, pouring, emitting, emptying, discharging, injecting, escaping, leaching, dumping, or disposing into the environment (including the abandonment or discarding of barrels, containers, and other closed receptacles) of any toxic chemical.

Release (Section 373.4 [4-16-90]) is defined as specified by CERCLA 101(22).

Release detection (Section 280.12 [9/23/88]) means determining whether a release of a regulated substance has occurred from the UST system into the environment or into the interstitial space between the UST system and its secondary barrier or secondary containment around it.

Relevant and appropriate requirements (Section 300.6) are those Federal requirements that, while not applicable, are designed to apply to problems sufficiently similar to those encountered at CERCLA sites that their application is appropriate. Requirements may be relevant and appropriate if they would be applicable but for jurisdictional restrictions associated with the requirement.

Relief valve (Section 61.61) means each pressure relief device including pressure relief valves, rupture disks and other pressure relief systems used to protect process components from over-pressure conditions. Relief valve does not include polymerization shortstop systems, refrigerated water systems or control valves or other devices used to control flow to an incinerator or other air pollution control device.

Relief valve discharge (Section 61.61) means any nonleak discharge through a relief valve. Relief valve discharge does not include discharges ducted to a control system from which the concentration of vinyl chloride in the exhaust gases does not exceed 10 ppm (average for 3-hour period), or equivalent as provided in Section 61.66.

Rem (Section 141.2) means the unit of dose equivalent from ionizing radiation to the total body or any internal organ or organ system. A millirem (mrem) is 1/1000 of a rem.

Remedial action (Section 192.01) means any action performed under section 108 of the Act.

Remedial action (Section 300.6). See **Remedy (Section 300.6)**.

Remedial investigation (Section 300.6) is a process undertaken by the lead agency (or responsible party if the responsible party will be developing a cleanup proposal) which emphasizes data collection and site characterization. The remedial investigation is generally performed concurrently and in an interdependent fashion with the feasibility study. However, in certain situations, the lead agency may require potentially responsible parties to conclude initial phases of the remedial investigation prior to initiation of the feasibility study. A remedial investigation is undertaken to determine the nature and extent of the problem presented by the release. This includes sampling and monitoring, as necessary, and includes the gathering of sufficient information to determine the necessity for and proposed extent of remedial action. Part of the remedial investigation involves assessing whether the threat can be mitigated or minimized by controlling the source of the contamination at or near the area where the hazardous substances or pollutants or contaminants were originally located (source control remedial actions) or whether additional actions will be necessary because the hazardous substances or pollutants or contaminants have migrated from the area of their original location (management of migration).

Remedial Project Manager (RPM) (Section 300.6) means the Federal official designated by EPA (or the USCG for vessels) to coordinate, monitor, or direct remedial or other response activities under Subpart F of this Plan; or the Federal official DOD designates to coordinate and direct Federal remedial or other response actions resulting from releases of hazardous substances, pollutants, or contaminants from DOD facilities or vessels.

Remediation waste (Section 260.10 [8/18/92])* means all solid and hazardous wastes, and all media (including groundwater, surface water, soils, and sediments) and debris, which contain listed hazardous wastes or which themselves exhibit a hazardous waste characteristic, that are managed for the purpose of implementing corrective action requirements under

Sec. 264.101 and RCRA section 3008(h). For a given facility, remediation wastes may originate only from within the facility boundary, but may include waste managed in implementing RCRA sections 3004(v) or 3008(h) for releases beyond the facility boundary.

Remedy or remedial action (Section 300.6) as defined by section 101(24) of CERCLA, means those actions consistent with permanent remedy taken instead of, or in addition to, removal action in the event of a release or threatened release of a hazardous substance into the environment, to prevent or minimize the release of hazardous substances so that they do not migrate to cause substantial danger to present or future public health or welfare or the environment. The term includes, but is not limited to, such actions at the location of the release as storage, confinement, perimeter protection using dikes, trenches or ditches, clay cover, neutralization, cleanup of released hazardous substances or contaminated materials, recycling or reuse, diversion, destruction, segregation of reactive wastes, dredging or excavations, repair or replacement of leaking containers, collection of leachate and runoff, on-site treatment or incineration, provision of alternative water supplies, and any monitoring reasonably required to assure that such actions protect the public health and welfare and the environment. The term includes the costs of permanent relocation of residents and businesses and community facilities where the President determines that, alone or in combination with other measures, such relocation is more cost-effective than and environmentally preferable to the transportation, storage, treatment, destruction, or secured disposition off-site of such hazardous substances, or may otherwise be necessary to protect the public health or welfare. The term does not include off-site transport of hazardous substances or contaminated materials unless the President determines that such actions: are more cost-effective than other remedial actions; will create new capacity to manage in compliance with Subtitle C of the Solid Waste Disposal Act, hazardous substances in addition to those located at the affected facility; or are necessary to protect public health or welfare or the environment from a present or potential risk which may be created by further exposure to the continued presence of such substances or materials.

Removal (Section 403.7) means a reduction in the amount of a pollutant in the POTW's effluent or alteration of the nature of a pollutant during treatment at the POTW. The reduction or alteration can be obtained by physical, chemical or biological means and may be the result of specifically designed POTW capabilities or may be incidental to the operation of the treatment system. Removal as used in this subpart shall not mean dilution of a pollutant in the POTW.

Removal (Section 763.121) means the taking out or stripping of asbestos or materials containing asbestos.

Removal (Section 763.83 [10/30/87]) means the taking out or the stripping of substantially all ACBM from a damaged area, a functional space, or a homogeneous area in a school building.

Remove (Section 61.141 [11-20-90])* means to take out RACM or facility components that contain or are covered with RACM from any facility.

Remove or removal (Section 109.2) refers to the removal of the oil from the water and shorelines or the taking of such other actions as may be necessary to minimize or mitigate damage to the public health or welfare, including, but not limited to, fish, shellfish, wildlife, and public and private property, shorelines, and beaches.

Remove or removal (Section 113.3) means the removal of the oil from the water and shorelines or the taking of such other actions as the Federal On-Scene Coordinator may determine to be necessary to minimize or mitigate damage to the public health or welfare, including but not limited to, fish, shellfish, wildlife, and public and private property, shorelines, and beaches.

Remove or removal (Section 117.1) refers to removal of the oil or hazardous substances from the water and shoreline or the taking of such other actions as may be necessary to minimize or mitigate damage to the public health or welfare, including, but not limited to, fish, shellfish, wildlife, and public and private property, shorelines, and beaches.

Remove or removal (Section 300.6) as defined by section 311(a)(8) of the CWA, refers to removal of oil or hazardous substances from the water and shorelines or the taking of such other actions as may be necessary to minimize or mitigate damage to the public health, welfare, or the environment. As defined by section 101(23) of CERCLA, remove or removal means the cleanup or removal of released hazardous substances from the environment; such actions as may be necessary to monitor, assess, and evaluate the release or threat of release of hazardous substances; the disposal of removal material or the taking of such other actions as may be necessary to prevent, minimize, or mitigate damage to the public health or welfare or the environment which may otherwise result from such release or threat of release. The term includes, in addition, without being limited to, security fencing or other measures to limit access, provision of alternative water supplies, temporary evacuation and housing of threatened individuals not otherwise provided for, action taken under section 104(b) of CERCLA, and any emergency assistance which may be provided under the Disaster Relief Act of 1974.

Renderer (Section 432.101) shall mean an independent or off-site rendering operation, conducted separate from a slaughterhouse, packinghouse or poultry dressing or processing plant, which manufactures at rates greater than 75,000 pounds of raw material per day of meat meal, tankage, animal fats or oils, grease, and tallow, and may cure cattle hides, but excluding marine oils, fish meal, and fish oils.

Renewal system (Section 797.1330) means the technique in which test organisms are periodically transferred to fresh test solution of the same composition.

Renewal test (Section 797.1350) is a test without continuous flow of solution, but with occasional renewal of test solutions after prolonged periods, e.g., 24 hours.

Renovation (Section 61.141 [11-20-90])* means altering a facility or one or more facility components in any way, including the stripping or removal of RACM from a facility component. Operations in which load-supporting structural members are wrecked or taken out are demolitions.

Renovation (Section 763.121) means the modifying of any existing structure, or portion thereof, where exposure to airborne asbestos may result.

Repackager (Section 704.203 [12/22/88]) means a person who buys a substance identified in Subpart D of this Part or mixture, removes the substance or mixture from the container in which it was bought, and transfers this substance, as is, to another container for sale.

Repair (Section 280.12 [9/23/88]) means to restore a tank or UST system component that has caused a release of product from the UST system.

Repair (Section 763.121) means overhauling, rebuilding, reconstructing, or reconditioning of structures or substrates where asbestos is present.

Repair (Section 763.83 [10/30/87]) means returning damaged ACBM to an undamaged condition or to an intact state so as to prevent fiber release.

Repaired (Section 264.1031 [6-21-90]) means that equipment is adjusted, or otherwise altered, to eliminate a leak.

Repaired (Section 60.481) means that equipment is adjusted, or otherwise altered, in order to eliminate a leak as indicated by one of the following: an instrument reading of 10,000 ppm or greater, indication of liquids dripping, or indication by a sensor that a seal or barrier fluid system has failed.

Repaired (Section 61.241) means that equipment is adjusted, or otherwise altered, to eliminate a leak as indicated by one of the following: an instrument reading of 10,000 ppm or greater, indication of liquids dripping, or indication by a sensor that a seal or barrier fluid system has failed.

Repeat compliance period (Section 141.2 [7/17/92])* means any subsequent compliance period after the initial compliance period.

Replacement (Section 35.2005) means obtaining and installing equipment, accessories, or appurtenances which are necessary during the design or useful life, whichever is longer, of the treatment works to maintain the capacity and performance for which such works were designed and constructed.

Replacement (Section 35.905) means expenditures for obtaining and installing equipment, accessories, or appurtenances which are necessary during the useful life of the treatment works to maintain the capacity and performance for which such works were designed and constructed. The term operation and maintenance includes replacement.

Replacement cost (Section 60.481) means the capital needed to purchase all the depreciable components in a facility.

Replacement unit (Section 260.10 [1-29-92]) means a landfill, surface impoundment, or waste pile unit (1) from which all or substantially all of the waste is removed, and (2) that is subsequently reused to treat, store, or dispose of hazardous waste. Replacement unit does not apply to a unit from which waste is removed during closure, if the subsequent reuse solely involves the disposal of waste from that unit and other closing units or corrective action areas at the facility, in accordance with an approved closure plan or EPA or State approved corrective action.

Replicate (Section 797.1600) is two or more duplicate tests, samples, organisms, concentrations, or exposure chambers.

Reportable quantities (Section 117.1) means quantities that may be harmful as set forth in Section 117.3, the discharge of which is a violation of section 311(b)(3) and requires notice as set forth in Section 117.21.

Reportable quantity (Section 302.3) means that quantity, as set forth in this part, the release of which requires notification pursuant to this part.

Reportable quantity (Section 355.20) means, for any CERCLA hazardous substance, the reportable quantity established in Table 302.4 of 40 CFR Part 302, for such substance, for any other substance, the reportable quantity is one pound.

Reporting agency (Part 58, Appendix G) means the applicable State agency or, in metropolitan areas, a local air pollution control agency designated by the State to carry out the provisions of Section 58.40.

Reporting area (Part 58, Appendix G) means the geographical area for which the daily index is representative for the reporting period. This area(s) may be the total urban area (or subpart thereof) or each of any number of distinct geographical subregions of the urban area deemed necessary by the reporting agency for adequate presentation of local air quality conditions.

Reporting day (Part 58, Appendix G) means the calendar day during which the daily report is given.

Reporting period (Part 58, Appendix G) means the time interval for which the daily report is representative. Normally, the reporting period is the 24-hour period immediately preceding the time of the report and should coincide to the extent practicable with the reporting day. In cases where the index will be forecasted the reporting period will include portions of the reporting day for which no monitoring data are available at the time of the report.

Reporting period (Section 704.203 [12/22/88]) means the time period during which CAIR reporting forms are to be submitted to EPA.

Reporting Year (Section 704.203 [12/22/88]) means the most recent complete corporate fiscal year during which a person manufactures, imports, or processes the listed substance, and which falls within a coverage period identified with a substance in Subpart D of this Part.

Representative affected facility (Section 60.531 [2/26/88]) means an individual wood heater that is similar in all material respects to other wood heaters within the model line it represents.

Representative important species (Section 125.71) means species which are representative, in terms of their biological needs, of a balanced, indigenous community of shellfish, fish and wildlife in the body of water into which a discharge of heat is made.

Representative of the news media (Section 2.100 [1/5/88]) refers to any person actively gathering news for an entity that is organized and operated to publish or broadcast news to the public. The term news means information that is about current events or that would be of current interest to the public. Examples of news media entities include television or radio stations broadcasting to the public at large, and publishers of periodicals (but only in those instances when they can qualify as disseminators of news) who make their products available for purchase or subscription by the general public. These examples are not intended to be all-inclusive. Moreover, as traditional methods of news delivery evolve (e.g., electronic dissemination of newspapers through telecommunications services), such alternative media would be included in this category. In the case of freelance journalists, they may be regarded as working for a news organization if they can demonstrate a solid basis for expecting publication through that organization, even though not actually employed by it. A publication contract would be the clearest proof, but EPA may also look to the past publication record of a requestor in making this determination.

Representative sample (Section 260.10) means a sample of a universe or whole (e.g., waste pile, lagoon, ground water) which can be expected to exhibit the average properties of the universe or whole.

Request (Section 2.100) means a request to inspect or obtain a copy of one or more records.

Request for Proposal (Section 248.4 [2/17/89]) is a request for an offer by one party to another of terms and conditions with references to some work or undertaking; the initial overture or preliminary statement for consideration by the other party to a proposed agreement.

Requester (Section 403.13) means an Industrial User or a POTW or other interested person seeking a variance from the limits specified in a categorical Pretreatment Standard.

Requestor (Section 2.100) means any person who has submitted a request to EPA.

Requirements and standards (Section 761.123) means: (1) Requirements as used in this policy refers to both the procedural responses and numerical decontamination levels set forth in this policy as constituting adequate cleanup of PCBs. (2) Standards refers to the numerical decontamination levels set forth in this policy.

Research (Section 26.102 [6-18-91]) means a systematic investigation, including research development, testing and evaluation, designed to develop or contribute to generalizable knowledge. Activities which meet this definition constitute research for purposes of this policy, whether or not they are conducted or supported under a program which is considered research for other purposes. For example, some demonstration and service programs may include research activities. Research subject to regulation, and similar terms are intended to encompass those research activities for which a federal department or agency has specific responsibility for regulating as a research activity, (for example, Investigational New Drug requirements administered by the Food and Drug Administration). It does not include research activities which are incidentally regulated by a federal department or agency solely as part of the department's or agency's broader responsibility to regulate certain types of activities whether research or non-research in nature (for example, Wage and Hour requirements administered by the Department of

Labor).

Reseller (Section 80.2 [2/16/94])* means any person who purchases gasoline or diesel fuel identified by the corporate, trade, or brand name of a refiner from such refiner or a distributor and resells or transfers it to retailers or wholesale purchaser-consumers displaying the refiner's brand, and whose assets or facilities are not substantially owned, leased, or controlled by such refiner.

Reserve (Section 35.105) means a portion of the State's construction grant allotment which the State proposes to set aside to use for construction or permit program management or water quality management planning activities.

Reserved credits (Section 86.090-2 [7-26-90]) are emission credits generated within a model year waiting to be reported to EPA at the end of the model year.

Residence (Section 61.91 [12-15-89]) means any home, house, apartment building, or other place of dwelling which is occupied during any portion of the relevant year.

Residential/commercial areas (Section 761.123) means those areas where people live or reside, or where people work in other than manufacturing or farming industries. Residential areas include housing and the property on which housing is located, as well as playgrounds, roadways, sidewalks, parks, and other similar areas within a residential community. Commercial areas are typically accessible to both members of the general public and employees and include public assembly properties, institutional properties, stores, office buildings, and transportation centers.

Residential Property (Section 201.1) means any property that is used for any of the purposes described in the following standard land use codes (ref. Standard Land Use Coding Manual, U.S. DOT/FHWA Washington D.C., reprinted March 1977): 1, Residential; 651, Medical and other Health Services; 68, Educational Services; 691, Religious Activities; and 711, Cultural Activities.

Residential solid waste (Section 245.101) means the garbage, rubbish, trash, and other solid waste resulting from the normal activities of households.

Residential solid waste (Sections 243.101; 246.101) means the wastes generated by the normal activities of households, including, but not limited to, food wastes, rubbish, ashes, and bulky wastes.

Residential tank (Section 280.12 [9/23/88]) is a tank located on property used primarily for dwelling purposes.

Residential use (Section 152.3 [5/4/88]) means use of a pesticide directly: (1) On humans or pets, (2) In, on, or around any structure, vehicle, article, surface, or area associated with the household, including but not limited to areas such as non-agricultural outbuildings, non-commercial greenhouses, pleasure boats and recreational vehicles, or (3) In any preschool or day care facility.

Residential use (Section 157.21) means use of a pesticide or device: (1) Directly on humans or pets; (2) In, on, or around any structure, vehicle, article, surface or area associated with the household, including but not limited to areas such as non-agricultural outbuildings, non-commercial greenhouses, pleasure boats and recreational vehicles; or (3) In or around any preschool or day care facility.

Residual disinfectant concentration (C in CT calculations) (Section 141.2 [6/29/89]) means the concentration of disinfectant measured in mg/l in a representative sample of water.

Residual oil (Section 60.41b [12/16/87])* means crude oil, fuel oil numbers 1 and 2 that have a nitrogen content greater than 0.05 weight percent, and all fuel oil numbers 4, 5 and 6, as defined by the American Society of Testing and Materials in ASTM D396-78, Standard Specifications for Fuel Oils (IBR - see Section 60.17).

Residual oil (Section 60.41c [9-12-90]) means crude oil, fuel oil that does not comply with the specifications under the definition of distillate oil, and all fuel oil numbers 4, 5, and 6, as defined by the American Society for Testing and Materials in ASTM D396-78, "Standard Specification for Fuel Oils" (incorporated by reference--see Sec. 60.17).

Residue (Section 162.3) means the active ingredient(s), metabolite(s) or degradation product(s) that can be detected in the crops, soil, water, or other component of the environment, including man, following the use of the pesticide.

Residue (Sections 240.101; 241.101) means all the solids that remain after completion of thermal processing, including bottom ash, fly ash, and grate siftings.

Resilient floor covering (Section 61.141 [11-20-90]) means asbestos-containing floor tile, including asphalt and vinyl floor tile, and sheet vinyl floor covering containing more than 1 percent asbestos as determined using polarized light microscopy according to the method specified in appendix A, subpart F, 40 CFR part 763, Section 1, Polarized Light Microscopy.

Resource recovery facility (Section 245.101) means any physical plant that processes residential, commercial, or institutional solid wastes biologically, chemically, or physically, and recovers useful products, such as shredded fuel, combustible oil or gas, steam, metal, glass, etc. for recycling.

Resource recovery unit (Section 60.41a) means a facility that combusts more than 75 percent non-fossil fuel on a quarterly (calendar) heat input basis.

Respond or response (Section 300.6) as defined by section 101(25) of CERCLA, means remove, removal, remedy, or remedial action.

Respondent (Section 164.2 [7/10/92])* means the Assistant Administrator of the Office of Hazardous Materials Control of the Agency.

Respondent (Section 209.3) means any person against whom a complaint has been issued under this subpart.

Respondent (Section 22.03 [2-13-92])* means any person proceeded against in the complaint.

Respondent (Section 8.33) means a person against whom sanctions are proposed because of alleged violations of the Executive Order and rules, regulations, and orders thereunder.

Response (Section 300.6). See **Respond (Section 300.6)**.

Response action (Section 304.12 [5/30/89]) means remove, removal, remedy and remedial action, as those terms are defined by section 101 of CERCLA, 42 U.S.C. 9601, including enforcement activities related thereto.

Response action (Section 763.83 [10/30/87]) means a method, including removal, encapsulation, enclosure, repair, operations and maintenance, that protects human health and the environment from friable ACBM.

Response action (Sections 305.12; 306.12) means remove, removal, remedy, and remedial action.

Response claim (Sections 305.12; 306.12) means a preauthorized demand in writing for a sum certain for response costs referred to in section 111(a)(2) of CERCLA.

Response costs (Section 304.12 [5/30/89]) means all costs of removal or remedial action incurred and to be incurred by the United States at a facility pursuant to section 104 of CERCLA, 42 U.S.C. 9604, including, but not limited to, all costs of investigation and information gathering, planning and implementing a response action, administration, enforcement, litigation, interest and indirect costs.

Response factor (RF) (Section 796.1860) is the solute concentration required to give a one unit area chromatographic peak or one unit output from the HPLC recording integrator at a particular recorder attenuation. The factor is required to convert from units of area to units of concentration. The determination of the response factor is given in paragraph (b)(3)(i)(B)(2) of this section.

Responsible agency (Section 240.101) means the organizational element that has the legal duty to ensure that owners, operators, or users of facilities comply with these guidelines.

Responsible agency (Section 241.101) means the organizational element that has the legal duty to ensure that owners, operators or users of land disposal sites comply with these guidelines.

Responsible agency (Section 243.101) means the organizational element that has the legal duty to ensure compliance with these guidelines.

Responsible agency (Section 247.101) means a department, agency, establishment or instrumentality of the executive branch of the Federal Government or the organizational element within such responsible agency that has the primary responsibility for procurement of materials or products or the preparation of specifications for the procurement of materials or products.

Responsible official (Section 56.1) means the EPA Administrator or any EPA employee who is accountable to the Administrator for carrying out a power or duty delegated under section 301(a)(1) of the act, or is accountable in accordance with EPA's formal organization for a particular program or function as described in Part 1 of this title.

Responsible official (Section 6.1003) is either the EPA Assistant Administrator or Regional Administrator as appropriate for the particular EPA program. To the extent applicable, the responsible official shall address the considerations set forth in the CEQ Regulations under 40 CFR 1508.27 in determining significant effect.

Responsible official (Section 6.501) means a Federal or State official authorized to fulfill the requirements of this subpart. The responsible federal official is the EPA Regional Administrator and the responsible State official is as defined in a delegation agreement under 205(g) of the Clean Water Act. The responsibilities of the State official are subject to the limitations in Section 6.514 of this subpart.

Responsible party (Section 761.123) means the owner of the PCB equipment, facility, or other source of PCBs of his/her designated agent (e.g., a facility manager or foreman).

Resting losses (Section 86.096-2 [11/1/93])* means evaporative emissions that may occur continuously, that are not diurnal emissions, hot soak emissions, running losses, or spitback emissions.

Resting losses (Section 86.098-2 [4/6/94])* means evaporative emissions that may occur continuously, that are not diurnal emissions, hot soak emissions, refueling emissions, running losses, or spitback emissions.

Restoration claim (Section 305.12) means a preauthorized or emergency claim for restoring, rehabilitating, replacing or acquiring the equivalent of any natural resources injured by the release of a hazardous substance.

Restoration claim (Section 306.12) means a preauthorized demand in writing for a sum certain for the cost of restoring, rehabilitating, replacing or acquiring the equivalent of any natural resource injured, destroyed, or lost as a result of the release of a hazardous substance.

Restoration or Restore (Section 305.12) means the restoration, rehabilitation, replacement, or acquiring the equivalent of any natural resources injured, destroyed or lost as a result of a release of a hazardous substance.

Restoration or Restoring (Section 306.12) means the restoration, rehabilitation, or replacement, or acquiring the equivalent of any natural resources injured, destroyed, or lost as a result of a release of a hazardous substance.

Restore (Part 6, Appendix A, Section 4) means to re-establish a setting or environment in which the natural functions of the floodplain can again operate.

Restore (Section 305.12). See **Restoration (Section 305.12).**

Restoring (Section 306.12). See **Restoration (Section 306.12).**

Restricted-entry interval (Section 170.3 [8/21/92])* means the time after the end of a pesticide application during which entry into the treated area is restricted.

Restricted use pesticide (Section 171.2) means a pesticide that is classified for restricted use under the provisions of section 3(d)(1)(C) of the Act.

Restricted use pesticide retail dealer (Section 171.2) means any person who makes available for use any restricted use pesticide, or who offers to make available for use any such pesticide.

Restricted wastes (Section 263 [9-6-89]) are those categories of hazardous wastes that are prohibited from one or more methods of land disposal either by regulation or statute (regardless of whether subcategories of such wastes are subject to a Section 268.5 extension, Section 268.8 "no migration" exemption, or any other exemption, any of which makes them currently eligible for one or more methods of land disposal). In other words, a hazardous waste is "restricted" no later than the date of the first automatic prohibition established in, or pursuant to, RCRA section 3004(d),(e),(f), or (g).

Retail outlet (Section 80.2 [2/16/94])* means any establishment at which gasoline or diesel fuel is sold or offered for sale for use in motor vehicles.

Retailer (Section 717.3) means a person who distributes in commerce a chemical substance, mixture, or article to ultimate purchasers who are not commercial entities.

Retailer (Section 80.2) means any person who owns, leases, operates, controls, or supervises a retail outlet.

Retan-wet finish (Section 425.02) means the final processing steps performed on a tanned hide including, but not limited to, the following wet processes: retan, bleach, color, and fatliquor.

Retarder (Active) (Section 201.1) means a device or system for decelerating rolling rail cars and controlling the degree of deceleration on a car by car basis.

Retarder Sound (Section 201.1) means a sound which is heard and identified by the observer as that of a retarder, and that causes a sound level meter indicator at fast meter response Section 201.1(l) to register an increase of at least ten decibels above the level observed immediately before hearing the sound.

Retention chamber (Section 797.1930) means a structure within a flow-through test chamber which confines the test organisms, facilitating observation of test organisms and eliminating loss of organisms in outflow water.

Retention chamber (Section 797.1950) means a structure within a flow-through test chamber which confines the test organisms, facilitating observation of test organisms and eliminating washout from test chambers.

Retread tire (Section 253.4 [11/17/88]) means a worn automobile, truck, or other motor vehicle tire whose tread has been replaced.

Retrofill (Section 761.3 [7/19/88]) means to remove PCB or PCB-contaminated dielectric fluid and to replace it with either PCB, PCB-contaminated, or non-PCB dielectric fluid.

Retrofit (Section 610.11) means the addition of a new item, modification or removal of an existing item of equipment beyond that of regular maintenance, on an automobile after its initial manufacture.

Retrofit device or device (Section 610.11) means: (i) Any component, equipment, or other device (except a flow measuring instrument or other driving aid, or lubricant or lubricant additive) which is designed to be installed in or on an automobile as an addition to, as a replacement for, or through alteration or modification of, any original component, or other devices; or (ii) Any fuel additive which is to be added to the fuel supply of an automobile by means other than fuel dispenser pumps; and (iii) Which any manufacturer, dealer, or distributor of such device represents will provide higher fuel economy than would have resulted with the automobile as originally equipped, as determined under rules of the Administrator.

Retrofitted configuration (Section 610.11) means the test configuration after adjustment of engine calibrations to the retrofit specifications and after all retrofit hardware has been installed.

Returnable Beverage Container (Section 244.101) means a beverage container for which a deposit is paid upon purchase and for which a refund of equal value is payable upon return.

Reverberation Time (Section 211.203) is the time that would be required for the mean-square sound pressure level, originally in a steady state, to fall 60 dB after the source is stopped.

Reverberatory furnace (Section 60.131) includes the following types of reverberatory furnaces: Stationary, rotating, rocking, and tilting.

Reverberatory smelting furnace (Section 60.161) means any vessel in which the smelting of copper sulfide ore concentrates or calcines is performed and in which the heat necessary for smelting is provided primarily by combustion of a fossil fuel.

Review (Section 2.100 [1/5/88]) refers to the process of examining documents located in response to a request that is for a commercial use (see paragraph (e) of this section) to determine whether any portion of any document located is permitted to be withheld. It also includes processing any documents for disclosure, e.g., doing all that is necessary to excise them and otherwise prepare them for release. Review does not include time spent resolving legal or policy issues regarding the application of exemptions. (Documents must be reviewed in responding to all requests; however, review time may only be charged to Commercial Use Requesters.)

Reviewing official (Section 27.2 [2-13-92]) means the General Counsel of the Authority or his designee who is--(a) Not subject to supervision by, or required to report to, the investigating official; (b) Not employed in the organizational unit of the Authority in which the investigating official is employed; and (c) Serving in a position for which the rate of basic pay is not less than the minimum rate of basic pay for grade GS-16 under the General Schedule.

R_f (Section 796.2700) is the furthest distance traveled by a test material on a thin-layer chromatography plate divided by the distance traveled by a solvent front (arbitrarily set at 10.0 cm in soil TLC studies).

RF (Section 796.1860). See **Response factor (Section 796.1860)**.

Rise Time (Section 85.2122(a)(9)(ii)(D)) means the time required for the spark voltage to increase from 10% to 90% of its maximum value.

rO (Section 87.1). See **Rated output (Section 87.1)**.

Roadways (Section 61.141 [11-20-90])* means surfaces on which vehicles travel. This term includes public and private highways, roads, streets, parking areas, and driveways.

Roaster (Section 60.161) means any facility in which a copper sulfide ore concentrate charge is heated in the presence of air to eliminate a significant portion (5 percent or more) of the sulfur contained in the charge.

Roaster (Section 60.171) means any facility in which a zinc sulfide ore concentrate charge is heated in the presence of air to eliminate a significant portion (more than 10 percent) of the sulfur contained in the charge.

Roasting (Section 61.181) means the use of a furnace to heat arsenic plant feed material for the purpose of eliminating a significant portion of the volatile materials contained in the feed.

Rock crushing and gravel washing facilities (Section 122.27) means facilities which process crushed and broken stone, gravel, and riprap (See 40 CFR Part 436, Subpart B, including the effluent limitations guidelines).

Rock wool insulation (Section 248.4 [2/17/89]) means insulation which is composed principally from fibers manufactured from slag or natural rock, with or without binders.

Rocket motor test site (Section 61.41) means any building, structure, facility, or installation where the static test firing of a beryllium rocket motor and/or the disposal of beryllium propellant is conducted.

Rod, wire and coil (Section 420.91) means those acid pickling operations that pickle rod, wire or coiled rod and wire products.

Rodenticides (Section 162.3) includes all substances or mixtures of substances, except mammal poisons and repellents as defined in paragraph (ff)(12) of this section, intended for preventing, destroying, repelling, or mitigating animals belonging to the Order Rodentia of the Class Mammalia, and closely related species, declared to be pests under Section 162.14. Rodenticides include, but are not limited to: (i) Baits, tracking powders, and fumigants intended to kill or repel rodents; (ii) Repellents intended for use on plants, surfaces, in premises, or in or on packaging or other materials such as food containers, plastic pipe, telephone cables, and building materials, for the purpose of repelling rodents; and (iii) Reproductive inhibitors intended to reduce or otherwise alter the reproductive capacity or potential of rodents.

Roll bonding (Section 471.02) is the process by which a permanent bond is created between two metals by rolling under high pressure in a bonding mill (co-rolling).

Rollboard (Section 763.163 [7-12-89]) means an asbestos-containing product made of paper that is produced in a continuous sheet, is flexible, and is rolled to achieve a desired thickness. Asbestos rollboard consists of two sheets of asbestos paper laminated together. Major applications of this product include: office partitioning; garage paneling; linings for stoves and electric switch boxes; and fire-proofing agent for security boxes, safes, and files.

Rolling (Section 467.02) is the reduction in thickness or diameter of a workpiece by passing it between lubricated steel rollers. There are two subcategories based on the rolling process. In the rolling with neat oils subcategory, pure or neat oils are used as lubricants for the rolling process. In the rolling with emulsions subcategory, emulsions are used as lubricants for the rolling process.

Rolling (Section 468.02) shall mean the reduction in the thickness or diameter of a workpiece by passing it between rollers.

Rolling (Section 471.02) is the reduction in thickness or diameter of a workpiece by passing it between lubricated steel rollers.

Roof coating (Section 763.163 [7-12-89]) means an asbestos-containing product intended for use as a coating, cement, adhesive, or sealant on roofs. Major applications of this product include: waterproofing; weather resistance; sealing; repair; and surface rejuvenation.

Roof monitor (Section 60.191) means that portion of the roof of a potroom where gases not captured at the cell exit from the potroom.

Roofing felt (Section 763.163 [7-12-89]) means an asbestos-containing product that is made of paper felt intended for use on building roofs as a covering or underlayer for other roof coverings.

Root crops (Section 257.3-5) means plants whose edible parts are grown below the surface of the soil.

Rotary lime kiln (Section 60.341) means a unit with an inclined rotating drum that is used to produce a lime product from limestone by calcination.

Rotary spin (Section 60.681) means a process used to produce wool fiberglass insulation by forcing molten glass through numerous small orifices in the side wall of a spinner to form continuous glass fibers that are then broken into discrete lengths by high velocity air flow.

Rotogravure print station (Section 60.581) means any device designed to print or coat inks on one side of a continuous web or substrate using the intaglio printing process with a gravure cylinder.

Rotogravure printing line (Section 60.581) means any number of rotogravure print stations and associated dryers capable of printing or coating simultaneously on the same continuous vinyl or urethane web or substrate, which is fed from a continuous roll.

Rotogravure printing unit (Section 60.431) means any device designed to print one color ink on one side of a continuous web or substrate using a gravure cylinder.

Rounded (Section 600.002-85) means a number shortened to the specific number of decimal places in accordance with the "Round Off Method" specified in ASTM E 29-67.

Routine maintenance area (Section 763.83 [10/30/87]) means an area, such as a boiler room or mechanical room, that is not normally frequented by students and in which maintenance employees or contract workers regularly conduct maintenance activities.

Routine use (Section 1516.2) means with respect to the disclosure of a record, the use of such record for a purpose which is compatible with the purpose for which it was collected.

Routine use (Section 16.2) shall have the meaining given to it by 5 U.S.C. 552a (a)(7).

RPM (Section 300.6). See **Remedial Project Manager (Section 300.6)**.

RRC (Section 300.5) is the Regional Response Center.

RRT (Section 300.5) is the Regional Response Team.

RS_i (Section 60.741 [9-11-89]) means the total mass (kg) of VOC retained on the coated substrate after oven drying or contained in waste coating for a given combination of coating and substrate.

Rubbish (Section 243.101) means a general term for solid waste, excluding food wastes and ashes, taken from residences, commercial establishments, and institutions.

Rules, regulations, and relevant orders of the Secretary of Labor

(Section 8.2) used in both paragraph (4) of the equal opportunity clause and elsewhere herein means rules, regulations, and relevant orders of the Secretary of Labor or his designee issued pursuant to the Order.

Run (Section 61.61) means the net period of time during which an emission sample is collected.

Run (Sections 60.2; 61.02) means the net period of time during which an emission sample is collected. Unless otherwise specified, a run may be either intermittent or continuous within the limits of good engineering practice.

Run-off (Section 258.2 [10-9-91]) means any rainwater, leachate, or other liquid that drains over land from any part of a facility.

Run-on (Section 258.2 [10-9-91]) means any rainwater, leachate, or other liquid that drains over land onto any part of a facility.

Run-of-pile triple superphosphate (Section 60.231) means any triple superphosphate that has not been processed in a granulator and is composed of particles at least 25 percent by weight of which (when not caked) will pass through a 16 mesh screen.

Run-off (Section 260.10) means any rainwater, leachate, or other liquid that drains over land from any part of a facility.

Run-on (Section 260.10) means any rainwater, leachate, or other liquid that drains over land onto any part of a facility.

Running changes (Section 85.1502 [9/25/87]) are those changes in vehicle or engine configuration, equipment or calibration which are made by an OEM or ICI in the course of motor vehicle or motor vehicle engine production.

Running loss (Section 86.082-2) means fuel evaporative emissions resulting from an average trip in an urban area or the simulation of such a trip.

Running losses (Section 86.096-2 [11/1/93])* means evaporative emissions that occur during vehicle operation.

Runoff (Section 241.101) means the portion of precipitation that drains from an area as surface flow.

Runoff (Section 419.11) shall mean the flow of storm water resulting from precipitation coming into contact with petroleum refinery property.

Rupture of a PCB Transformer (Section 761.3) means a violent or non-violent break in the integrity of a PCB Transformer caused by an overtemperature and/or overpressure condition that results in the release of PCBs.

S

S (Section 60.641) is the sulfur production rate in kilograms per hour (kg/hr) rounded to one decimal place.

S1S (Section 429.11) See **Smooth-one-side hardboard (Section 429.11)**.

S2S (Section 429.11). See **Smooth-two-sides hardboard (Section 429.11)**.

Safe disposal (Section 165.1) means discarding pesticides or containers, in a permanent manner so as to comply with these proposed procedures and so as to avoid unreasonable adverse effects on the environment.

Salad dressings (Section 407.81). See **Mayonnaise and salad dressings (Section 407.81)**.

Sale (Section 60.531 [2/26/88]) means the transfer of ownership or control, except that transfer of control shall not constitute a sale for purposes of Section 60.530(f).

Sale for purposes other than resale (Section 761.3) means sale of PCBs for purposes of disposal and for purposes of use, except where use involves sale for distribution in commerce. PCB Equipment which is first leased for purposes of use any time before July 1, 1979, will be considered sold for purposes other than resale.

Saline estuarine waters (Section 125.58) means those semi-enclosed coastal waters which have a free connection to the territorial sea, undergo net seaward exchange with ocean waters, and have salinities comparable to those of the ocean. Generally, these waters are near the mouth of estuaries and have cross-sectional annual mean salinities greater than twenty-five (25) parts per thousand.

Salt bath descaling, oxidizing (Section 420.81) means the removal of scale from semi-finished steel products by the action of molten salt baths other than those containing sodium hydride.

Salt bath descaling, reducing (Section 420.81) means the removal of scale from semi-finished steel products by the action of molten salt baths containing sodium hydride.

Salvage value (Section 4.2) means the probable sale price of an item, if offered for sale on the condition that it will be removed from the property at the buyer's expense, allowing a reasonable period of time to find a person buying with knowledge of the uses and purposes for which it is adaptable and capable of being used, including separate use of serviceable components and scrap when there is no reasonable prospect of sale except on that basis.

Salvaging (Section 241.101) means the controlled removal of waste materials for utilization.

Same location (Section 60.51a [2-11-91]) means the same or contiguous property that is under common ownership or control, including properties that are separated only by a street, road, highway, or other public right-of-way. Common ownership or control includes properties that are owned, leased, or operated by the same entity, parent entity, subsidiary, subdivision, or any combination thereof, including any municipality or other governmental unit, or any quasigovernmental authority (e.g., a public utility district or regional waste disposal authority).

Sample loop (Section 796.1860) is a 1/16 in. O.D. (1.6 mm) stainless steel tube with an internal volume between 20 and 50 μL. The loop is attached to the sample injection valve of the HPLC and is used to inject standard solutions into the mobile phase of the HPLC when determining the response factor for the recording integrator. The exact volume of the loop must be determined as described in paragraph (b)(3)(i)(B)(1) of this section when the HPLC method is used.

Sample system (Section 87.1) means the system which provides for the transportation of the gaseous emission sample from the sample probe to the inlet of the instrumentation system.

Sampling area (Section 763.103) means any area, whether contiguous or not, within a school building which contains friable material that is homogeneous in texture and appearance.

Sanitary landfill (Section 165.1) means a disposal facility employing an engineered method of disposing of solid wastes on land in a manner which minimizes environmental hazards by spreading the solid wastes in thin layers, compacting the solid wastes to the smallest practical volume, and applying cover material at the end of each working day. Such facility complies with the Agency Guidelines for the Land Disposal of Solid Wastes as prescribed in 40 CFR Part 241.

Sanitary landfill (Section 257.2) means a facility for the disposal of solid waste which complies with this part.

Sanitary landfill (Sections 240.101; 241.101) means a land disposal site employing an engineered method of disposing of solid wastes on land in a manner that minimizes environmental hazards by spreading the solid wastes in thin layers, compacting the solid wastes to the smallest practical volume, and applying and compacting cover material at the end of each operating day.

Sanitary sewer (Section 35.2005) means a conduit intended to carry liquid and water-carried wastes from residences, commercial buildings, industrial plants and institutions together with minor quantities of ground, storm and surface waters that are not admitted intentionally.

Sanitary sewer (Section 35.905) means a sewer intended to carry only sanitary or sanitary and industrial waste waters from residences, commercial buildings, industrial plants, and institutions.

Sanitary survey (Section 141.2) means an onsite review of the water source, facilities, equipment, operation and maintenance of a public water system for the purpose of evaluating the adequacy of such source, facilities, equipment, operation and maintenance for producing and distributing safe drinking water.

Sanitary survey (Section 142.2) means an onsite review of the water source, facilities, equipment, operation and maintenance of a public water system for the purpose of evaluating the adequacy of such source, facilities, equipment, operation and maintenance for producing and distributing safe drinking water.

Sanitary waste (Section 435.11 [3/4/93])* shall refer to human body waste discharged from toilets and urinals located within facilities subject to this subpart.

Sanitized (Section 350.1 [7/29/88]) means a version of a document from which information claimed as trade secret or confidential has been omitted or withheld.

SARA (Section 280.12 [9/23/88]) means the Superfund Amendments and Reauthorization Act of 1986.

SAROAD (Section 58.1). See **Storage and Retrieval of Aerometric Data system (Section 58.1)**.

SAROAD site identification form (Section 58.1) is one of the several forms in the SAROAD system. It is the form which provides a complete description of the site (and its surroundings) of an ambient air quality monitoring station.

Satellite vehicle (Section 243.101) means a small collection vehicle that transfers its load into a larger vehicle operating in conjunction with it.

Saturated solution (Section 796.1860) is a solution in which the dissolved solute is in equilibrium with an excess of undissolved solute; or a solution in equilibrium such that at a fixed temperature and pressure, the concentration of the solute in the solution is at its maximum value and will not change even in the presence of an excess of solute.

Saturated solution (Section 796.1840) is a solution in which the dissolved solute is in equilibrium with an excess of undisolved solute; or a solution in equilibrium such that at a fixed temperature and pressure, the concentration of the solution is at its maximum value and will not change even in the presence of an excess of solute.

Saturated zone (Section 258.2 [10-9-91]) means that part of the earth's crust in which all voids are filled with water.

Saturated zone or zone of saturation (Section 260.10) means that part of the earth's crust in which all voids are filled with water.

Saturator (Section 60.471) means the equipment in which asphalt is applied to felt to make asphalt roofing products. The term saturator includes the saturator, wet looper, and coater.

Sauerkraut canning (Section 407.71) shall mean the draining and subsequent filling and canning of fermented cabbage and juice.

Sauerkraut cutting (Section 407.71) shall mean the trimming, cutting, and subsequent preparatory handling of cabbage necessary for and including brining and fermentation, and subsequent tank soaking.

Sausage and luncheon meat processor (Section 432.71) shall mean an operation which cuts fresh meats, grinds, mixes, seasons, smokes or otherwise produces finished products such as sausage, bologna and luncheon meats at rates greater than 2730 kg (6000 lb) per day.

Sawing (Section 471.02) is cutting a workpiece with a band, blade, or circular disc having teeth.

SBA (Section 21.2) means the Small Business Administration.

Scarfing (Section 420.71) means those steel surface conditioning operations in which flames generated by the combustion of oxygen and fuel are used to remove surface metal imperfections from slabs, billets, or blooms.

Scavenging (Section 241.101) means uncontrolled removal of solid waste materials.

Scavenging (Section 243.101) means the uncontrolled and unauthorized removal of materials at any point in the solid waste management system.

SCF (Section 464.02) shall mean standard cubic feet.

Schedule of compliance (Section 270.2) means a schedule of remedial measures included in a permit, including an enforceable sequence of interim requirements (for example, actions, operations, or milestone events) leading to compliance with the Act and regulations.

Schedule of compliance (Section 233.3) means a schedule of remedial measures included in a permit, including an enforceable sequence of interim requirements (for example, actions, operations, or milestone events) leading to compliance with the CWA and its regulations.

Schedule of compliance (Section 122.2) means a schedule of remedial measures included in a permit, including an enforceable sequence of interim requirements (for example, actions, operations, or milestone events) leading to compliance with the CWA and regulations.

Schedule of compliance (Sections 124.2; 144.3) means a schedule of remedial measures included in a permit, including an enforceable sequence of interim requirements (for example, actions, operations, or milestone events) leading to compliance with the appropriate Act and regulations.

Scheduled maintenance (Section 86.402-78) means any adjustment, repair, removal, disassembly, cleaning, or replacement of vehicle components or systems which is performed on a periodic basis to prevent part failure or vehicle malfunction, or anticipated as necessary to correct an overt indication of vehicle malfunction or failure for which periodic maintenance is not appropriate.

Scheduled maintenance (Section 57.103 [2-13-92])* means any periodic procedure, necessary to maintain the integrity or reliability of emissions control performance, which can be anticipated and scheduled in advance. In sulfuric acid plants, it includes among other items the screening or replacement of catalyst, the re-tubing of heat exchangers, and the routine repair and cleaning of gas handling/cleaning equipment.

Scheduled maintenance (Section 86.082-2) means any adjustment, repair, removal, disassembly, cleaning, or replacement of vehicle components or systems which is performed on a periodic basis to prevent part failure or vehicle (if the engine were installed in a vehicle) malfunction.

Scheduled maintenance (Section 86.084-2) means any adjustment, repair, removal, disassembly, cleaning, or replacement of vehicle components or systems which is performed on a periodic basis to prevent part failure or vehicle (if the engine were installed in a vehicle) malfunction, or anticipated as necessary to correct an overt indication of vehicle malfunction or failure for which periodic maintenance is not appropriate. This definition applies beginning with the 1984 model year.

School (Section 763.103) means any public or private day or residential school that provides elementary or secondary education for grade 12 or under as determined under State law, or any school of any Agency of the United States.

School (Section 763.83 [10/30/87])

means any elementary or secondary school as defined in section 198 of the Elementary and Secondary Education Act of 1965 (20 U.S.C. 2854).

School building (Section 763.83 [10/30/87]) means: (1) Any structure suitable for use as a classroom, including a school facility such as a laboratory, library, school eating facility, or facility used for the preparation of food. (2) Any gymnasium or other facility which is specially designed for athletic or recreational activities for an academic course in physical education. (3) Any other facility used for the instruction or housing of students or for the administration of educational or research programs. (4) Any maintenance, storage, or utility facility, including any hallway, essential to the operation of any facility described in this definition of school building under paragraphs (1), (2), or (3). (5) Any portico or covered exterior hallway or walkway. (6) Any exterior portion of a mechanical system used to condition interior space.

School buildings (Section 763.103) means: (1) Structures used for the instruction of school children, including classrooms, laboratories, libraries, research facilities and administrative facilities. (2) School eating facilities, and school kitchens. (3) Gymnasiums or other facilities used for athletic or recreational activities, or for courses in physical education. (4) Dormitories or other living areas of residential schools. (5) Maintenance, storage, or utility facilities essential to the operation of the facilities described in paragraphs (h)(1) through (4) of this section.

Scope (Section 1508.25) consists of the range of actions, alternatives, and impacts to be considered in an environmental impact statement. The scope of an individual statement may depend on its relationships to other statements (Sections 1502.20 and 1508.28). To determine the scope of environmental impact statements, agencies shall consider 3 types of actions, 3 types of alternatives, and 3 types of impacts. They include: (a) Actions (other than unconnected single actions) which may be: (1) Connected actions, which means that they are closely related and therefore should be discussed in the same impact statement. Actions are connected if they: (i) Automatically trigger other actions which may require environmental impact statements. (ii) Cannot or will not proceed unless other actions are taken previously or simultaneously. (iii) Are interdependent parts of a larger action and depend on the larger action for their justification. (2) Cumulative actions, which when viewed with other proposed actions have cumulatively significant impacts and should therefore be discussed in the same impact statement. (3) Similar actions, which when viewed with other reasonably foreseeable or proposed agency actions, have similarities that provide a basis for evaluating their environmental consequences together, such as common timing or geography. An agency may wish to analyze these actions in the same impact statement. It should do so when the best way to assess adequately the combined impacts of similar actions or reasonable alternatives to such actions is to treat them in a single impact statement. (b) Alternatives, which include: (1) No action alternative. (2) Other reasonable courses of actions. (3) Mitigation measures (not in the proposed action). (c) Impacts, which may be: (1) Direct; (2) Indirect; (3) Cumulative.

Scope of work (Section 6.901) means a document similar in content to the program of requirements but substantially abbreviated. It is usually prepared for small-scale projects.

Scrap metal (Section 261.1), for purposes of Sections 261.2 and 261.6, is bits and pieces of metal parts (e.g., bars, turnings, rods, sheets, wire) or metal pieces that may be combined together with bolts or soldering (e.g., radiators, scrap automobiles, railroad box cars), which when worn or superfluous can be recycled.

Screen (Section 60.381) means a device for separating material according to size by passing undersize material through one or more mesh surfaces (screens) in series and retaining oversize material on the mesh surfaces (screens).

Screening operation (Section 60.671) means a device for separating material according to size by passing undersize material through one or more mesh surfaces (screens) in series, and retaining oversize material on the mesh surfaces (screens).

Scrubbing (Section 165.1) means the washing of impurities from any process gas stream.

SCS (Section 57.103). See **Supplementary control system (Section 57.103)**.

SDWA (Section 124.2) means the Safe Drinking Water Act (Pub. L. 95-523, as amended by Pub. L. 95-1900; 42 U.S.C. 300f et seq.).

SDWA (Section 144.3) means the Safe Drinking Water Act (Pub. L. 93-523, as amended; 42 U.S.C. 300f et seq.).

SDWA (Section 146.3) means the Safe Drinking Water Act (Pub. L. 95-523, as amended by Pub. L. 95-190, 42 U.S.C. 300(f) et seq.).

SDWA (Section 270.2) means the Safe Drinking Water Act (Pub. L. 95-523, as amended by Pub. L. 95-1900; 42 U.S.C. 3001 et seq.).

Seafood (Sections 408.11; 408.21; 408.31; 408.41; 408.51) shall mean the raw material, including freshwater and saltwater fish and shellfish, to be processed, in the form in which it is received at the processing plant.

Sealant tape (Section 763.163 [7-12-89]) means an asbestos-containing product which is initially a semi-liquid mixture of butyl rubber and asbestos, but which solidifies when exposed to air, and which is intended for use as a sealing agent. Major applications of this product include: sealants for building and automotive windows, sealants for aerospace equipment components, and sealants for insulated glass.

Search (Section 2.100 [1/5/88]) includes all time spent looking for material that is responsive to a request, including page-by-page or line-by-line identification of material within documents. Searching for material must be done in the most efficient and least expensive manner so as to minimize costs for both the EPA and the requestor. For example, EPA will not engage in line-by-line search when merely duplicating an entire document would prove the less expensive and quicker method of complying with a request. Search will be distinguished, moreover, from review of material in order to determine whether the material is exempt from disclosure (see paragraph (j) of this section). Searches may be done manually or by computer using existing programming.

Secondary emission control system (Section 60.141a) means the combination of equipment used for the capture and collection of secondary emissions (e.g., (1) An open hood system for the capture and collection of primary and secondary emissions from the BOPF, with local hooding ducted to a secondary emission collection device such as a baghouse for the capture and collection of emissions from the hot metal transfer and skimming station; or (2) An open hood system for the capture and collection of primary and secondary emissions from the furnace, plus a furnace enclosure with local hooding ducted to a secondary emission collection device, such as a baghouse, for additional capture and collection of secondary emissions from the furnace, with local hooding ducted to a secondary emission collection device, such as a baghouse, for the capture and collection of emissions from hot metal transfer and skimming station; or (3) A furnace enclosure with local hooding ducted to a secondary emission collection device such as a baghouse for the capture and collection of secondary emissions from a BOPF controlled by a closed hood primary emission control system, with local hooding ducted to a secondary emission collection device, such as a baghouse, for the capture and collection of emissions from hot metal transfer and skimming stations).

Secondary emissions (Section 60.141a) means particulate matter emissions that are not captured by the BOPF primary control system, including emissions from hot metal transfer and skimming stations. This definition also includes particulate matter emissions that escape from openings in the primary emission control system, such as from lance hole openings, gaps or tears in the ductwork of the primary emission control system, or leaks in hoods.

Secondary emissions (Section 51.301) means emissions which occur as a result of the construction or operation of an existing stationary facility but do not come from the existing stationary facility. Secondary emissions may include, but are not limited to, emissions from ships or trains coming to or from the existing stationary facility.

Secondary emissions (Section 52.21) means emissions which would occur as a result of the construction or operation of a major stationary source or major modification, but do not come from the major stationary source or major modification itself. Secondary emissions include emissions from any offsite support facility which would not be constructed or increase its emissions except as a result of the construction or operation of the major stationary source or major modification. Secondary emissions do not include any emissions which come directly from a mobile source, such as emissions from the tailpipe of a motor vehicle, from a train, or from a vessel. (i) Emissions from ships or trains coming to or from the new or modified stationary source; and (ii) Emissions from any offsite support facility which would not otherwise be constructed or increase its emissions as a result of the construction or operation of the major stationary source or major modification.

Secondary emissions (Section 52.24) means emissions which would occur as a result of the construction or operation of a major stationary source or major modification, but do not come from the major stationary source or major modification itself. For the purpose of this section, secondary emissions must be specific, well defined, quantifiable, and impact the same general area as the stationary source or modification which causes the secondary emissions. Secondary emissions include emissions from any offsite support facility which would otherwise not be constructed or increase its emissions except as a result of the construction or operation of the major stationary source or major modification. Secondary emissions do not include any emissions which come directly from a mobile source, such as emissions from the tailpipe of a motor vehicle, from a train, or from a vessel.

Secondary emissions (Sections 61.171; 61.181) means inorganic arsenic emissions that escape capture by a primary emission control system.

Secondary emissions (Sections 51.165; 51.166) means emissions which would occur as a result of the construction or operation of a major stationary source or major modification, but do not come from the major stationary source or major modification itself. For the purpose of this section, secondary emissions must be specific, well defined, quantifiable, and impact the same general area as the stationary source or modification which causes the secondary emissions. Secondary emissions include emissions from any offsite support facility which would not be constructed or increase its emissions except as a result of the construction or operation of the major stationary source or major modification. Secondary emissions do not include any emissions which come directly from a mobile source such as emissions from the tailpipe of a motor vehicle, from a train, or from a vessel.

Secondary hood system (Section 61.171) means the equipment (including hoods, ducts, fans, and dampers) used to capture and transport secondary inorganic arsenic emissions.

Secondary industry category (Section 122.2) means any industry category which is not a primary industry category.

Secondary materials (Section 261.2 [8-27-91]) fed to a halogen acid furnace that exhibit a characteristic of a hazardous waste or are listed as a hazardous waste as defined in subparts C or D of this part, except for brominated material that meets the following criteria: (i) The material must contain a bromine concentration of at least 45%; and (ii) The material must contain less than a total of 1% of toxic organic compounds listed in appendix VIII; and (iii) The material is processed on-site in the halogen acid furnace via direct conveyance (hard piping).

Secondary maximum contaminant levels (Section 143.2) means SMCLs which apply to public water systems and which, in the judgment of the Administrator, are requisite to protect the public welfare. The SMCL means the maximum permissible level of a contaminant in water which is delivered to the free flowing outlet of the ultimate user of public water system. Contaminants added to the water under circumstances controlled by the user, except those resulting from corrosion of piping and plumbing caused by water quality, are excluded from this definition.

Secondary processor of asbestos (Section 763.63) is a person who processes for commercial purposes an asbestos mixture.

Secondary standard (Section 51.100) means a national secondary ambient air quality standard promulgated pursuant to section 109 of the Act.

Secondary treatment (Section 125.58) means the term as defined in 40 CFR Part 133.

Secret (Section 11.4) refers to that national security information or material which requires a substantial degree of protection. The test for assigning Secret classification shall be whether its unauthorized disclosure could reasonably be expected to cause serious damage to the national security. Examples of serious damage include disruption of foreign relations significantly affecting the national security; significant impairment of a program or policy directly related to the national security; revelation of significant military plans or intelligence operations; and compromise of scientific or technological developments relating to national security. The classification Secret shall be sparingly used.

Secretary (Section 232.2 [2/11/93])* means the Secretary of the Army acting through the Chief of Engineers.

Secretary (Section 600.002-85) means the Secretary of Transportation or his authorized representative.

Secretary (Section 87.1) means the Secretary of Transportation and any other officer or employee of the Department of Transportation to whom the authority involved may be delegated.

Secretary (Sections 122.2; 233.3) means the Secretary of the Army, acting through the Chief of Engineers.

Secretary of Energy (Section 600.002-85) means the Secretary of Energy or his authorized representative.

Section 13 (Section 7.25) refers to section 13 of the Federal Water Pollution Control Act Amendments of 1972.

Section 304(a) criteria (Section 131.3) are developed by EPA under authority of section 304(a) of the Act based on the latest scientific information on the relationship that the effect of a constituent concentration has on particular aquatic species and/or human health. This information is issued periodically to the States as guidance for use in developing criteria.

Section 404 program or State 404 program or 404 (Sections 124.2; 233.3) means an approved State program to regulate the discharge of dredged material and the discharge of fill material under section 404 of the Clean Water Act in State regulated waters.

Section 5 notice (Section 700.43 [8/17/88]) means any PMN, consolidated PMN, intermediate PMN, significant new use notice, exemption notice, or exemption application.

Section 504 (Section 12.103 [8/14/87]) means section 504 of the Rehabilitation Act of 1973 (Pub. L. 93-112, 87 Stat. 394 (29 U.S.C. 794)), as amended by the Rehabilitation Act Amendments of 1974 (Pub. L. 93-516, 88 Stat. 1617); and the Rehabilitation, Comprehensive Services, and Developmental Disabilities Amendments of 1978 (Pub. L. 95-602, 92 Stat. 2955); and the Rehabilitation Act Amendments of 1986 (Pub. L. 99-506, 100 Stat. 1810). As used in this part, section 504 applies only to programs or activities conducted by Executive agencies and not to federally assisted programs.

Section mill (Section 420.71) means those steel hot forming operations that produce a variety of finished and semi-finished steel products other than the products of those mills specified in paragraphs (d), (e), (g), and (h) of Section 420.71.

Security classification assignment (Section 11.4) means the prescription of a specific security classification for a particular area or item of information. The information involved constitutes the sole basis for determining the degree of classification assigned.

Security classification category (Section 11.4) means the specific degree of classification (Top Secret, Secret or Confidential) assigned to classified information to indicate the degree of protection required.

Sediment (Section 796.2750) is the unconsolidated inorganic and organic material that is suspended in and being transported by surface water, or has settled out and has deposited into beds.

Sedimentation (Section 141.2 [6/29/89]) means a process for removal of solids before filtration by gravity or separation.

Segregated stormwater sewer system (Section 61.341 [3-7-90]) means a drain and collection system designed and operated for the sole purpose of collecting rainfall runoff at a facility, and which is segregated from all other individual drain systems.

Selenium (Section 415.361) shall mean the total selenium present in the process wastewater stream exiting the wastewater treatment system.

Semi-wet (Section 420.41) means those steelmaking air cleaning systems that use water for the sole purpose of conditioning the temperature and humidity of furnace gases such that the gases may be cleaned in dry air pollution control systems.

Semiannual (Section 61.111) means a 6-month period; the first semiannual period concludes on the last day of the last month during the 180 days following initial startup for new sources; and the first semiannual period concludes on the last day of the last full month during the 180 days after June 6, 1984 for existing sources.

Semiannual (Section 61.131 [9-14-89]) means a 6-month period; the first semiannual period concludes on the last day of the last full month during the 180 days following initial startup for new sources; the first semiannual period concludes on the last day of the last full month during the 180 days after the effective date of the regulation for existing sources.

Semiannual (Section 61.241) means a

6-month period; the first semiannual period concludes on the last day of the last month during the 180 days following initial startup for new sources; and the first semiannual period concludes on the last day of the last full month during the 180 days after the effective date of a specific subpart that references this subpart for existing sources.

Semiconductors (Section 469.12) means solid state electrical devices which perform functions such as information processing and display, power handling, and interconversion between light energy and electrical energy.

Senior management official (Section 372.3 [2/16/88]) means an official with management responsibility for the person or persons completing the report, or the manager of environmental programs for the facility or establishments, or for the corporation owning or operating the facility or establishments responsible for certifying similar reports under other environmental regulatory requirements.

Senior management official (Section 350.1 [7/29/88]) means an official with management responsibility for the person or persons completing the report, or the manager of environmental programs for the facility or establishments, or for the corporation owning or operating the facility or establishments responsible for certifying similar reports under other environmental regulatory requirements.

Sensor (Section 264.1031 [6-21-90]) means a device that measures a physical quantity or the change in a physical quantity, such as temperature, pressure, flow rate, pH, or liquid level.

Sensor (Sections 60.481; 61.241) means a device that measures a physical quantity or the change in a physical quantity such as temperature, pressure, flow rate, pH, or liquid level.

Separate collection (Section 246.101) means collecting recyclable materials which have been separated at the point of generation and keeping those materials separate from other collected solid waste in separate compartments of a single collection vehicle or through the use of separate collection vehicles.

Separator tank (Section 264.1031 [6-21-90]) means a device used for separation of two immiscible liquids.

Septage (Section 122.2 [5/2/89]) means the liquid and solid material pumped from a septic tank, cesspool, or similar domestic sewage treatment system, or a holding tank when the system is cleaned or maintained.

Septage (Section 501.2 [5/2/89]) means the liquid and solid material pumped from a septic tank, cesspool, or similar domestic sewage treatment system, or a holding tank, when the system is cleaned or maintained.

Septic tank (Section 280.12 [9/23/88]) is a water-tight covered receptacle designed to receive or process, through liquid separation or biological digestion, the sewage discharged from a building sewer. The effluent from such receptacle is distributed for disposal through the soil and settled solids and scum from the tank are pumped out periodically and hauled to a treatment facility.

Serial number (Section 205.151) means the identification number assigned by the manufacturer to a specific production unit.

Series Resistance (Section 85.2122(a)(6)(ii)(B)) means the sum of resistances from the condenser plates to the condenser's external connections.

Serious acute effects (Section 721.3 [7-27-89]) means human injury or human disease processes that have a short latency period for development, result from short-term exposure to a chemical substance, or are a combination of these factors and which are likely to result in death or severe or prolonged incapacitation.

Serious acute effects (Section 723.50) means human disease processes or other adverse effects that have a short latency period for development, result from short-term exposure, or are a combination of these factors and that are likely to result in death, severe or prolonged incapacitation, disfigurement, or severe or prolonged loss of the ability to use a normal bodily or intellectual function with a consequent impairment of normal activities.

Serious chronic effects (Section 723.50) means human disease processes or other adverse effects that have a long latency period for development, result from long-term exposure, are long-term illnesses, or are a combination of these factors and that are likely to result in death, severe or prolonged incapacitation, disfigurement, or severe or prolonged loss of the ability to use a normal bodily or intellectual function with a consequent impairment of normal activities.

Serious chronic effects (Section 721.3 [7-27-89]) means human injury or human disease processes that have a long latency period for development, result from long-term exposure to a chemical substance, or are a combination of these factors and which are likely to result in death or severe or prolonged incapacitation.

Service line sample (Section 141.2 [7/17/92])* means a one-liter sample of water collected in accordance with Sec. 141.86(b)(3), that has been standing for at least 6 hours in a service line.

Services (Section 35.6015 [6-5-90])* A recipient's in-kind or a contractor's labor, time, or efforts which do not involve the delivery of a specific end item, other than documents (e.g., reports, design drawings, specifications). This term does not include employment agreements or collective bargaining agreements.

Services (Sections 33.005; 35.2005) means a contractor's labor, time, or efforts which do not involve the delivery of a specific end item, other than documents (e.g., reports, design drawing, specifications). This term does not include employment agreements or collective bargaining agreements.

Settle (Section 14.2) means the act of considering, ascertaining, adjusting, determining or otherwise resolving a claim.

Settleable solids (Section 434.11) is that matter measured by the volumetric method specified in Section 434.64.

Settleable solids (Section 440.141 [5/24/88]) means the particulate material (both organic or inorganic) which will settle in one hour expressed in milliliters per liter (ml/l) as determined using an Imhoff cone and the method described for Residue - Settleable in 40 CFR Part 136.

Settling tank (Section 60.621) means a container that gravimetrically separates oils, grease, and dirt from petroleum solvent, together with the piping and ductwork used in the installation of this device.

Seven-day average (7-day average) (Section 133.101). The arithmetic mean of pollutant parameter values for samples collected in a period of 7 consecutive days.

Severe property damage (Section 122.41) means substantial physical damage to property, damage to the treatment facilities which causes them to become inoperable, or substantial and permanent loss of natural resources which can reasonably be expected to occur in the absence of a bypass. Severe property damage does not mean economic loss caused by delays in production.

Sewage (Section 140.1) means human body wastes and the wastes from toilets and other receptacles intended to receive or retain body wastes.

Sewage collection system (Section 35.905) means, for the purpose of Section 35.925-13, each, and all, of the common lateral sewers, within a publicly owned treatment system, which are primarily installed to receive waste waters directly from facilities which convey waste water from individual structures or from private property, and which include service connection "Y" fittings designed for connection with those facilities. The facilities which convey waste water from individual structures, from private property to the public lateral sewer, or its equivalent, are specifically excluded from the definition, with the exception of pumping units, and pressurized lines, for individual structures or groups of structures when such units are cost effective and are owned and maintained by the grantee.

Sewage from vessels (Section 122.2) means human body wastes and the wastes from toilets and other receptacles intended to receive or retain body wastes that are discharged from vessels and regulated under section 312 of CWA, except that with respect to commercial vessels on the Great Lakes this term includes graywater. For the purposes of this definition, graywater means galley, bath, and shower water.

Sewage sludge (Section 122.2 [5/2/89])* means any solid, semi-solid, or liquid residue removed during the treatment of municipal waste water or domestic sewage. Sewage sludge includes, but is not limited to, solids removed during primary, secondary, or advanced waste water treatment, scum, septage, portable toilet pumpings, type III marine sanitation device pumpings (33 CFR Part 159), and sewage sludge products. Sewage sludge does not include grit or screenings, or ash generated during the incineration of sewage sludge.

Sewage sludge (Section 257.2 [3/19/93])* means solid, semi-solid, or liquid residue generated during the treatment of domestic sewage in a treatment works. Sewage sludge includes, but is not limited to, domestic septage; scum or solids removed in primary, secondary, or advanced wastewater treatment processes; and a material derived from sewage sludge. Sewage sludge does not include ash generated during the firing of sewage sludge in a sewage sludge incinerator or grit and screenings generated during preliminary treatment of domestic sewage in a treatment works.

Sewage sludge (Section 501.2 [5/2/89]) means any solid, semi-solid, or liquid residue removed during the treatment of municipal waste water or domestic sewage. Sewage sludge includes, but is not limited to, solids removed during primary, secondary, or advanced waste water treatment, scum, septage, portable toilet pumpings, type III marine sanitation device pumpings (33 CFR Part 159), and sewage sludge products. Sewage sludge does not include grit or screenings, or ash generated during the incineration of sewage sludge.

Sewage sludge use or disposal practice (Section 122.2 [5/2/89]) means the collection, storage, treatment, transportation, processing, monitoring, use, or disposal of sewage sludge.

Sewage Treatment Works (Section 220.2) means municipal or domestic waste treatment facilities of any type which are publicly owned or regulated to the extent that feasible compliance schedules are determined by the availability of funding provided by Federal, State, or local governments.

Sewer line (Section 60.691 [11/23/88]) means a lateral, trunk line, branch line, ditch, channel, or other conduit used to convey refinery wastewater to downstream components of a refinery wastewater treatment system. This term does not include buried, below-grade sewer lines.

Sewer line (Section 61.341 [3-7-90]) means a lateral, trunk line, branch line, or other enclosed conduit used to convey waste to a downstream waste management unit.

Shaft power (Section 87.1) means only the measured shaft power output of a turboprop engine.

Shall (Section 256.06) denotes requirements for the development and implementation of the State plan.

Sheen (Section 110.1) means an iridescent appearance on the surface of water.

Sheet gasket (Section 763.163 [7-12-89]) means either (1) an asbestos-containing product consisting of asbestos and elastomeric or other binders rolled in homogeneous sheets at some point in its manufacture and intended for use as a gasket, or (2) any asbestos-containing product made from braided or twisted rope, slit or woven tape, yarn, or other textile products intended for use as a gasket. Sheet gaskets are used to seal the space between two sections of a component and thereby prevent leakage in such applications as: exhaust, cylinder head, and intake manifolds; pipe flanges; autoclaves; vulcanizers; pressure vessels; cooling towers; turbochargers; and gear cases. This category includes flange, spiralwound, tadpole, manhole, handhole, door, and other gaskets or seals.

Shell deposition (Section 797.1800) is the measured length of shell growth that occurs between the time the shell is ground at test initiation and test termination 96 hours later.

Shellfish, fish and wildlife (Section 125.58) means any biological population or community that might be adversely affected by the applicant's modified discharge.

Shift (Sections 204.51; 205.51) means the regular production work period for one group of workers.

Shift supervisor (Section 60.51a [2-11-91]) means the person in direct charge and control of the operation of an MWC and who is responsible for on-site supervision, technical direction, management, and overall performance of the facility during an assigned shift.

Shipment (Section 749.68 [1-3-90]) means the act or process of shipping goods by any form of conveyance.

Shipped liquid ammonia (Section 418.51) shall mean liquid ammonia commercially shipped for which the Department of Transportation requires 0.2 percent minimum water content.

Shipping losses (Sections 418.21; 418.51) shall mean: Discharges resulting from loading tank cars or tank trucks; discharges resulting from cleaning tank cars or tank trucks; and discharges from air pollution control scrubbers designed to control emissions from loading or cleaning tank cars or tank trucks.

Shop (Section 60.271) means the building which houses one or more EAFs.

Shop (Section 60.271a) means the building which houses one or more EAFs or AOD vessels.

Shop opacity (Section 60.271) means the arithmetic average of 24 or more opacity observations of emissions from the shop taken in accordance with Method 9 of Appendix A of this part for the applicable time periods.

Shop opacity (Section 60.271a) means the arithmetic average of 24 observations of the opacity of emissions from the shop taken in accordance with Method 9 of Appendix A of this part.

Short term test indicative of the potential to cause a developmentally toxic effect (Section 721.3 [7-27-89]) means either any in vivo preliminary development toxicity screen conducted in a mammalian species, or any in vitro developmental toxicity screen, including any test system other than the intact pregnant mammal, that has been extensively evaluated and judged reliable for its ability to predict the potential to cause developmentally toxic effects in intact systems across a broad range of chemicals or within a class of chemicals that includes the substance of concern.

Short-term test indicative of carcinogenic potential (Section 721.3 [7-27-89]) means either any limited bioassay that measures tumor or preneoplastic induction, or any test indicative of interaction of a chemical substance with DNA (i.e., positive response in assays for gene mutation, chromosomal aberrations, DNA damage and repair, or cellular transformation).

Shot casting (Section 471.02) is the production of shot by pouring molten metal in finely divided streams to form spherical particles.

Should (Section 256.06) denotes recommendations for the development and implementation of the State plan.

Shutdown (Section 60.2) means the cessation of operation of an affected facility for any purpose.

Shutdown (Section 61.161) means the cessation of operation of an affected source for any purpose.

Shutdown (Section 61.171) means the cessation of operation of a stationary source for any reason.

Shutdown (Section 61.181) means the cessation of operation of a stationary source for any purpose.

SI unit (Section 191.12 [12/20/93])* means a unit of measure in the International System of Units (Sievert).

Sidewall cementing operation (Section 60.541 [9/15/87]) means the system used to apply cement to a continuous strip of sidewall component or any other continuous strip component (except combined tread/sidewall component) that is incorporated into the sidewall of a finished tire. A sidewall cementing operation consists of a cement application station and all other equipment, such as the cement supply system and feed and takeaway conveyors, necessary to apply cement to sidewall strips or other continuous strip component (except combined tread/sidewall component) and to allow evaporation of solvent from the cemented rubber.

Sievert (Section 191.12 [12/20/93])* is the SI unit of effective dose and is equal to 100 rem or one joule per kilogram. The abbreviation is "Sv."

Significant (Section 51.166) means (i) in reference to a net emissions increase or the potential of a source to emit any of the pollutants specified in Section 51.166 (b)(23)(i), a rate of emissions that would equal or exceed any of the specified rates. (ii) Significant means, in reference to a net emissions increase or the potential of a source to emit a pollutant subject to regulation under the Act that paragraph (b)(23)(i) of this section, does not list, any emissions rate. (iii) Notwithstanding paragraph (b)(23)(i) of this section, significant means any emissions rate or any net emissions increase associated with a major stationary source or major modification, which would construct within 10 kilometers of a Class I area, and have an impact on such area equal to or greater than 1 $\mu g/m^3$ (24-hour average).

Significant (Section 52.21) means (i) in reference to a net emissions increase or the potential of a source to emit any of the pollutants specified in Section 52.21 (b)(23)(i), a rate of emissions that would equal or exceed any of the specified rates. (ii) Significant means, in reference to a net emissions increase or the potential of a source to emit a pollutant subject to regulation under the Act that paragraph (b)(23)(i) of this section, does not list, any emissions rate. (iii) Notwithstanding paragraph (b)(23)(i) of this section, significant means any emissions rate or any net emissions increase associated with a major stationary source or major modification, which would construct within 10 kilometers of a Class I area, and have an impact on such area equal to or greater than 1 μg/m^3 (24-hour average).

Significant (Section 52.24) means, in reference to a net emissions increase or the potential of a source to emit any of the pollutants specified in Section 52.24(f)(10), a rate of emissions that would equal or exceed any of the rates specified in Section 52.24 (f)(10).

Significant adverse environmental effects (Section 721.3 [7-27-89]) means injury to the environment by a chemical substance which reduces or adversely affects the productivity, utility, value, or function of biological, commercial, or agricultural resources, or which may adversely affect a threatened or endangered species. A substance will be considered to have the potential for significant adverse environmental effects if it has one of the following: (1) An acute aquatic EC_{50} of 1 mg/L or less. (2) An acute aquatic EC_{50} of 20 mg/L or less where the ratio of aquatic vertebrate 24-hour to 48-hour EC_{50} is greater than or equal to 2.0. (3) A Maximum Acceptable Toxicant Concentration (MATC) of less than or equal to 100 parts per billion (100 ppb). (4) An acute aquatic EC_{50} of 20 mg/L or less coupled with either a measured bioconcentration factor (BCF) equal to or greater than 1,000x or in the absence of bioconcentration data a log P value equal to or greater than 4.3.

Significant adverse reactions (Section 717.3) are reactions that may indicate a substantial impairment of normal activities, or long-lasting or irreversible damage to health or the environment.

Significant biological treatment (Section 133.101). The use of an aerobic or anaerobic biological treatment process in a treatment works to consistently achieve a 30-day average of at least 65 percent removal of BOD_5.

Significant economic loss (Section 166.3) means that, under the emergency conditions: for a productive activity, the profitability would be substantially below the expected profitability for that activity; or, for other types of activities, where profits cannot be calculated, the value of public or private fixed assets would be substantially below the expected value for those assets. Only losses caused by the emergency conditions, specific to the impacted site, and specific to the geographic area affected by the emergency conditions are included. The contribution of obvious mismanagement to the loss will not be considered in determining loss. In evaluating the significance of an economic loss for productive activities, the Agency will consider whether the expected reduction in profitability exceeds what would be expected as a result of normal fluctuations over a number of years, and whether the loss would affect the long-term financial viability expected from the productive activity. In evaluating the significance of an economic loss for situations other than productive activities, the Agency will consider reasonable measures of expected loss.

Significant environmental effects (Section 723.50) means either: (i) Any irreversible damage to biological, commercial, or agricultural resources of importance to society, (ii) Any reversible damage to biological, commercial, or agricultural resources of importance to society if the damage persists beyond a single generation of the damaged resource or beyond a single year, or (iii) Any known or reasonably anticipated loss of members of an endangered or threatened species. Endangered or threatened species are those species identified as such by the Secretary of the Interior in accordance with the Endangered Species Act, as amended (16 U.S.C. 1531).

Significant hazard to public health (Section 149.101) means any level of contaminant which causes or may cause the aquifer to exceed any maximum contaminant level set forth in any promulgated National Primary Drinking Water Standard at any point where the water may be used for drinking purposes or which may otherwise adversely affect the health of persons, or which may require a public water system to install additional treatment to prevent such adverse effect.

Significant impairment (Section 51.301) means, for purposes of section 303, visibility impairment which, in the judgment of the Administrator, interferes with the management, protection, preservation, or enjoyment of the visitor's visual experience of the mandatory Class I Federal area. This determination must be made on a case-by-case basis taking into account the geographic extent, intensity, duration, frequency and time of the visibility impairment, and how these factors correlate with (1) times of visitor use of the mandatory Class I Federal area, and (2) the frequency and timing of natural conditions that reduce visibility.

Significant Industrial User (Section 403.3 [7-24-90]) (1) Except as provided in paragraph (t)(2) of this section, the term Significant Industrial User means: (i) All industrial users subject to Categorical Pretreatment Standards under 40 CFR 403.6 and 40 CFR Chapter I, Subchapter N; and (ii) Any other industrial user that: discharges an average of 25,000 gallons per day or more of process wastewater to the POTW (excluding sanitary, noncontact cooling and boiler blowdown wastewater); contributes a process wastestream which makes up 5 percent or more of the average dry weather hydraulic or organic capacity of the POTW treatment plant; or is designated as such by the Control Authority as defined in 40 CFR 403.12(a) on the basis that the industrial user has a reasonable potential for adversely affecting the POTW's operation or for violating any pretreatment standard or requirement (in accordance with 40 CFR

403.8(f)(6)). (2) Upon a finding that an industrial user meeting the criteria in paragraph (t)(1)(ii) of this section has no reasonable potential for adversely affecting the POTW's operation or for violating any pretreatment standard or requirement, the Control Authority (as defined in 40 CFR 403.12(a)) may at any time, on its own initiative or in response to a petition received from an industrial user or POTW, and in accordance with 40 CFR 403.8(f)(6), determine that such industrial user is not a significant industrial user.

Significant new use notice (Section 700.43 [8/17/88]) means any notice submitted to EPA pursuant to section 5(a)(1)(B) of the Act in accordance with Part 721 of this chapter.

Significant site preparation work (Section 435.11 [3/4/93])* as used in 40 CFR 122.29 shall mean the process of surveying, clearing or preparing an area of the ocean floor for the purpose of constructing or placing a development or production facility on or over the site. "New Source" does not include facilities covered by an existing NPDES permit immediately prior to the effective date of these guidelines pending EPA issuance of a new source NPDES permit.

Significant source of ground water (Section 191.12) as used in this part, means: (1) An aquifer that: (i) Is saturated with water having less than 10,000 milligrams per liter of total dissolved solids; (ii) is within 2,500 feet of the land surface; (iii) has a transmissivity greater than 200 gallons per day per foot, Provided, That any formation or part of a formation included within the source of ground water has a hydraulic conductivity greater than 2 gallons per day per square foot; and (iv) is capable of continuously yielding at least 10,000 gallons per day to a pumped or flowing well for a period of at least a year; or (2) an aquifer that provides the primary source of water for a community water system as of the effective date of this subpart.

Significantly (Section 1508.27) as used in NEPA requires considerations of both context and intensity: (a) Context. This means that the significance of an action must be analyzed in several contexts such as society as a whole (human, national), the affected region, the affected interests, and the locality. Significance varies with the setting of the proposed action. For instance, in the case of a site-specific action, significance would usually depend upon the effects in the locale rather than in the world as a whole. Both short- and long-term effects are relevant. (b) Intensity. This refers to the severity of impact. Responsible officials must bear in mind that more than one agency may make decisions about partial aspects of a major action. Section 1508.27(b) (1) through (10) should be considered in evaluating intensity.

Significantly damaged friable

surfacing ACM (Section 763.83 [10/30/87]) means damaged friable surfacing ACM in a functional space where the damage is extensive and severe.

Significantly damaged friable miscellaneous ACM (Section 763.83 [10/30/87]) means damaged friable miscellaneous ACM where the damage is extensive and severe.

Significantly greater effluent reduction than BAT (Section 125.22) means that the effluent reduction over BAT produced by an innovative technology is significant when compared to the effluent reduction over best practicable control technology currently available (BPT) produced by BAT.

Significantly lower cost (Section 125.22) means that an innovative technology must produce a significant cost advantage when compared to the technology used to achieve BAT limitations in terms of annual capital costs and annual operation and maintenance expenses over the useful life of the technology.

Significantly more stringent limitation (Section 133.101) means BOD_5 and SS limitations necessary to meet the percent removal requirements of at least 5 mg/l more stringent than the otherwise applicable concentration-based limitations (e.g. less than 25 mg/l in the case of the secondary treatment limits for BOD_5 and SS), or the percent removal limitations in Sections 133.102 and 133.105, if such limits would, by themselves, force significant construction or other significant capital expenditure.

Silicomanganese (Section 60.261) means that alloy as defined by ASTM Designation A483-64 (Reapproved 1974) (incorporated by reference - see Section 60.17).

Silicomanganese zirconium (Section 60.261) means that alloy containing 60 to 65 percent by weight silicon, 1.5 to 2.5 percent by weight calcium, 5 to 7 percent by weight zirconium, 0.75 to 1.25 percent by weight aluminum, 5 to 7 percent by weight manganese, and 2 to 3 percent by weight barium.

Silicon metal (Section 60.261) means any silicon alloy containing more than 96 percent silicon by weight.

Silvery iron (Section 60.261) means any ferrosilicon, as defined by ASTM Designation A100-69 (Reapproved 1974) (incorporated by reference - see Section 60.17), which contains less than 30 percent silicon.

Silvicultural point source (Section 122.27) means any discernible, confined and discrete conveyance related to rock crushing, gravel washing, log sorting, or log storage facilities which are operated in connection with silvicultural activities and from which pollutants are discharged into waters of the United

States. The term does not include non-point source silvicultural activities such as nursery operations, site preparation, reforestation and subsequent cultural treatment, thinning, prescribed burning, pest and fire control, harvesting operations, surface drainage, or road construction and maintenance from which there is natural runoff. However, some of these activities (such as stream crossing for roads) may involve point source discharges of dredged or fill material which may require a CWA section 404 permit (See 33 CFR 209.120 and Part 233).

Similar composition (Section 162.152) refers to a pesticide product which contains only the same active ingredient(s), or combination of active ingredients, and which is in the same category of toxicity, as a federally registered pesticide product.

Similar in all material respects (Section 60.531 [2/26/88]) means that the construction materials, exhaust and inlet air system, and other design features are within the allowed tolerances for components identified in Section 60.533(k).

Similar product (Section 162.152) means a pesticide product which, when compared to a federally registered product, has a similar composition and a similar use pattern.

Similar systems (Section 86.092-2 [2-28-90]) (Note: the definitions listed in this section on this date of publication apply beginning with the 1992 model year.) are engines , fuel metering and emission control system combinations which use the same fuel (e.g., gasoline, diesel, etc.), combustion cycle (i.e., two or four stroke), general type of fuel system (i.e., carburetor or fuel injection), catalyst system (e.g., none, oxidation, three-way plus oxidation, three-way only, etc.), fuel control system (i.e., feedback or non-feedback), secondary air system (i.e., equipment or not equipment), and EGR (i.e., equipped or not equipped).

Similar use pattern (Section 162.152) refers to a use of a pesticide product which, when compared to a federally registered use of a product with a similar composition, does not require a change in precautionary labeling under Section 162.10(h), and which is substantially the same as the federally registered use. Registrations involving changed use patterns are not included in this term.

Simple combustion turbine (Section 72.2 [7/30/93])* means a unit that is a rotary engine driven by a gas under pressure that is created by the combustion of any fuel. This term includes combined cycle units without auxiliary firing. This term excludes combined cycle units with auxiliary firing, unless the unit did not use the auxiliary firing from 1985 through 1987 and does not use auxiliary firing at any time after November 15, 1990.

Simple cycle gas turbine (Section

60.331) means any stationary gas turbine which does not recover heat from the gas turbine exhaust gases to preheat the inlet combustion air to the gas turbine, or which does not recover heat from the gas turbine exhaust gases to heat water or generate steam.

Simple manufacturing (Section 410.61) shall mean the following unit processes: fiber preparation and dyeing with or without carpet backing.

Simple manufacturing operation (Sections 410.41; 410.51) shall mean all the following unit processes: Desizing, fiber preparation and dyeing.

Simple slaughterhouse (Section 432.11) shall mean a slaughterhouse which accomplishes very limited by-product processing, if any, usually no more than two of such operations as rendering, paunch and viscera handling, blood processing, hide processing, or hair processing.

Simplify (Section 29.12) means that a State may develop its own format, choose its own submission date, and select the planning period for a State plan.

Single family structure (Section 141.2 [7/17/92])* for the purpose of subpart I of this part only, means a building constructed as a single-family residence that is currently used as either a residence or a place of business.

Single response (Section 310.11 [1/15/93])* means all of the concerted activities conducted in response to a single episode, incident or threat causing or contributing to a release or threatened release of hazardous substances of pollutants or contaminants.

Single stand (Section 420.101) means those recirculation or direct application cold rolling mills which include only one stand of work rolls.

Sinking agents (Section 300.82), for the purposes of this subpart, are those additives applied to oil discharges to sink floating pollutants below the water surface.

Sinter bed (Section 60.181) means the lead sulfide ore concentrate charge within a sintering machine.

Sintering machine (Section 60.171) means any furnace in which calcines are heated in the presence of air to agglomerate the calcines into a hard porous mass called sinter.

Sintering machine (Section 60.181) means any furnace in which a lead sulfide ore concentrate charge is heated in the presence of air to eliminate sulfur contained in the charge and to agglomerate the charge into a hard porous mass called sinter.

Sintering machine discharge end (Section 60.181) means any apparatus which receives sinter as it is discharged from the conveying grate of a sintering machine.

Site (Section 124.41) means the land or water area upon which a major PSD stationary source or major PSD modification is physically located or conducted, including but not limited to adjacent land used for utility systems; as repair, storage, shipping or processing areas; or otherwise in connection with the major PSD stationary source or major PSD modification.

Site (Section 190.02) means the area contained within the boundary of a location under the control of persons possessing or using radioactive material on which is conducted one or more operations covered by this part.

Site (Section 191.02) means an area contained within the boundary of a location under the effective control of persons possessing or using spent nuclear fuel or radioactive waste that are involved in any activity, operation, or process covered by this subpart.

Site (Section 704.3 [12/22/88])* means a contiguous property unit. Property divided only by a public right-of-way shall be considered one site. There may be more than one plant on a single site. The site for a person who imports a substance is the site of the operating unit within the person's organization which is directly responsible for importing the substance and which controls the import transaction and may in some cases be the organization's headquarters office in the United States.

Site (Section 710.2) means a contiguous property unit. Property divided only by a public right-of-way shall be considered one site. There may be more than one manufacturing plant on a single site. For the purposes of imported chemical substances, the site shall be the business address of the importer.

Site (Section 717.3) means a contiguous property unit. Property divided only by a public right-of-way is considered one site. There may be multiple manufacturing, processing, or distribution activities occurring within a single site.

Site (Section 721.3 [7/27/88]) means a contiguous property unit. Property divided only by a public right-of-way is one site. There may be more than one manufacturing plant on a single site.

Site (Sections 122.2; 122.29; 124.2; 144.3; 146.3; 233.3; 270.2) means the land or water area where any facility or activity is physically located or conducted, including adjacent land used in connection with the facility or activity.

Site (Sections 712.3; 723.50; 763.63) means a contiguous property unit. Property divided only by a public right-of-way shall be considered one site. There may be more than one manufacturing plant on a single site.

Site lease (Section 73.3 [12-17-91]) is a legally-binding document signed between a firm associated with a new independent power production facility (IPPF) or a new IPPF and a site owner that establishes the term and conditions under which the firm associated with the new IPPE has the binding right to utilize a specific site for the purposes of operating or constructing the new IPPF.

Site of construction (Section 8.2) means the general physical location of any building, highway, or other change or improvement to real property which is undergoing construction, rehabilitation, alteration, conversion, extension, demolition, and repair and any temporary location or facility at which a contractor, subcontractor, or other participating party meets a demand or performs a function relating to the contract or subcontract.

Site Quality Assurance and Sampling Plan (Section 300.6) is a written document, associated with site sampling activities, which presents in specific terms the organization (where applicable), objectives, functional activities, and specific quality assurance (QA) and quality control (QC) activities designed to achieve the data quality goals of a specific project(s) or continuing operation(s). The QA Project Plan is prepared for each specific project or continuing operation (or group of similar projects or continuing operations). The QA Project Plan will be prepared by the responsible program office, regional office, laboratory, contractor, recipient of an assistance agreement, or other organization.

Site-limited (Section 710.23) means a chemical substance is manufactured and processed only within a site and is not distributed for commercial purposes as a substance or as part of a mixture or article outside the site. Imported substances are never site-limited.

Site-limited intermediate (Section 721.3 [7/27/88]) means an intermediate manufactured, processed, and used only within a site and not distributed in commerce other than as an impurity or for disposal. Imported intermediates cannot be site-limited.

Six-minute (6-minute) period (Section 60.2) means any one of the 10 equal parts of a one-hour period.

Size (Section 60.671) means the rated capacity in tons per hour of a crusher, grinding mill, bucket elevator, bagging operation, or enclosed truck or railcar loading station; the total surface area of the top screen of a screening operation; the width of a conveyor belt; and the rated capacity in tons of a storage bin.

Size classes of discharges (Section 300.6) refers to the following size classes of oil discharges which are provided as guidance to the OSC and serve as the criteria for the actions delineated in Subpart E. They are not meant to imply associated degrees of hazard to public health or welfare, nor are they a measure of environmental damage. Any oil discharge that poses a substantial threat to the public health or welfare or results in critical public concern shall be classified as a major discharge regardless of the following quantitative measures: (a) Minor discharge means a discharge to the inland waters of less than 1,000 gallons of oil or a discharge to the coastal waters of less than 10,000 gallons of oil. (b) Medium discharge means a discharge of 1,000 to 10,000 gallons of oil to the inland waters or a discharge of 10,000 to 100,000 gallons of oil to the coastal waters. (c) Major discharge means a discharge of more than 10,000 gallons of oil to the inland waters or more than 100,000 gallons of oil to the coastal waters.

Size classes of releases (Section 300.6) refers to the following size classifications which are provided as guidance to the OSC for meeting pollution reporting requirements in Subpart C. The final determination of the appropriate classification of a release will be made by the OSC based on consideration of the particular release (e.g., size, location, impact, etc.): (a) Minor release means a release of a quantity of hazardous substance(s), pollutant(s), or contaminant(s) that poses minimal threat to public health or welfare or the environment. (b) Medium release means all releases not meeting the criteria for classification as a minor or major release. (c) Major release means a release of any quantity of hazardous substance(s), pollutant(s), or contaminant(s) that poses a substantial threat to public health or welfare or the environment or results in significant public concern.

Skimming (Section 61.171) means the removal of slag from the molten converter bath.

Skimming station (Section 60.141a) means the facility where slag is mechanically raked from the top of the bath of molten iron.

Skin sensitization (allergic contact dermatitis) (Section 798.4100) is an immunologically mediated cutaneous reaction to a substance. In the human, the responses may be characterized by pruritis, erythema, edema, papules, vesicles, bullae, or a combination of these. In other species the reactions may differ and only erythema and edema may be seen.

Slag (Section 60.261) means the more or less completely fused and vitrified matter separated during the reduction of a metal from its ore.

SLAMS (Section 58.1) means State or Local Air Monitoring Station(s). The SLAMS make up the ambient air quality monitoring network which is required by Section 58.20 to be provided for in the State's implementation plan. This definition places no restrictions on the use of the physical structure or facility housing the SLAMS. Any combination of SLAMS and any other monitors (Special Purpose, NAMS, PSD) may occupy the same facility or structure without affecting the respective definitions of those monitoring stations.

Slaughterhouse (Sections 432.11; 432.21) shall mean a plant that slaughters animals and has as its main product fresh meat as whole, half or quarter carcasses or smaller meat cuts.

Slimicides (Section 162.3) includes all substances or mixtures of substances, except antimicrobial agents as defined in paragraph (ff)(2) of this section, fungicides as defined in paragraph (ff)(8) of this section, and herbicides as defined in paragraph (ff)(9) of this section, intended for use in preventing or inhibiting the growth of, or destroying biological slimes composed of combinations of algae, bacteria or fungi declared to be pests under Section 162.14. Slimicides include, but are not limited to, slime control agents for use in industrial water cooling systems and in pulp and paper mill wet end systems.

Slip gauge (Section 61.61) means a gauge which has a probe that moves through the gas/liquid interface in a storage or transfer vessel and indicates the level of vinyl chloride in the vessel by the physical state of the material the gauge discharges.

Slop oil (Section 60.691 [11/23/88]) means the floating oil and solids that accumulate on the surface of an oil-water separator.

Slop oil (Section 61.341 [3-7-90]) means the floating oil and solids that accumulate on the surface of an oil-water separator.

Slow meter response (Section 201.1) means that the slow response of the sound level meter shall be used. The slow dynamic response shall comply with the meter dynamic characteristics in paragraph 5.4 of the American National Standard Specification for Sound Level Meters, ANSI S1.4-1971. This publication is available from the American National Standards Institute Inc., 1430 Broadway, New York, New York 10018.

Slow meter response (Section 204.2) means the meter ballistics of meter dynamic characteristics as specified by American National Standard S1.4-1971 or subsequent approved revisions.

Slow sand filtration (Section 141.2 [6/29/89]) means a process involving passage of raw water through a bed of sand at low velocity (generally less than 0.4 m/h) resulting in substantial particulate removal by physical and biological mechanisms.

Sludge (Section 110.1) means an

aggregate of oil or oil and other matter of any kind in any form other than dredged spoil having a combined specific gravity equivalent to or greater than water.

Sludge (Section 257.2) means any solid, semisolid, or liquid waste generated from a municipal, commercial, or industrial wastewater treatment plant, water supply treatment plant, or air pollution control facility or any other such waste having similar characteristics and effect.

Sludge (Section 258.2 [10-9-91]) means any solid, semi-solid, or liquid waste generated from a municipal, commercial, or industrial wastewater treatment plant, water supply treatment plant, or air pollution control facility exclusive of the treated effluent from a wastewater treatment plant.

Sludge (Section 260.10) means any solid, semi-solid, or liquid waste generated from a municipal, commercial, or industrial wastewater treatment plant, water supply treatment plant, or air pollution control facility exclusive of the treated effluent from a wastewater treatment plant.

Sludge (Section 261.1), for purposes of Sections 261.2 and 261.6, has the same meaning used in Section 260.10 of this chapter.

Sludge (Section 61.51) means sludge produced by a treatment plant that processes municipal or industrial waste waters.

Sludge (Sections 240.101; 241.101) means the accumulated semiliquid suspension of settled solids deposited from wastewaters or other fluids in tanks or basins. It does not include solids or dissolved material in domestic sewage or other significant pollutants in water resources, such as silt, dissolved or suspended solids in industrial wastewater effluents, dissolved materials in irrigation return flows or other common water pollutants.

Sludge (Sections 243.101; 246.101) means the accumulated semiliquid suspension of settled solids deposited from wastewaters or other fluids in tanks or basins. It does not include solids or dissolved material in domestic sewage or other significant pollutants in water resources, such as silt, dissolved materials in irrigation return flows or other common water pollutants.

Sludge dryer (Section 260.10 [2-21-91]) means any enclosed thermal treatment device that is used to dehydrate sludge and that has a maximum total thermal input, excluding the heating value of the sludge itself, of 2,500 Btu/lb of sludge treated on a wet-weight basis.

Sludge dryer (Section 61.51) means a device used to reduce the moisture content of sludge by heating to temperatures above 65 degrees C (ca. 150 degrees F) directly with combustion gases.

Sludge Requirements (Section 403.7) shall mean the following statutory provisions and regulations or permits issued thereunder (or more stringent State or local regulations): Section 405 of the Clean Water Act; the Solid Waste Disposal Act (SWDA) (including Title II more commonly referred to as the Resource Conservation Recovery Act (RCRA) and State regulations contained in any State sludge management plan prepared pursuant to Subtitle D of SWDA); the Clean Air Act; the Toxic Substances Control Act; and the Marine Protection, Research and Sanctuaries Act.

Sludge-only facility (Section 122.2 [5/2/89]) means any treatment works treating domestic sewage whose methods of sewage sludge use or disposal are subject to regulations promulgated pursuant to section 405(d) of the CWA, and is required to obtain a permit under Section 122.1(b)(3) of this Part.

Sm^3 (Section 464.02) shall mean standard cubic meters.

Small business (Section 33.005) means a business as defined in section 3 of the Small Business Act, as amended (15 U.S.C. 632).

Small business (Section 35.6015 [6-5-90])* A business as defined in section 3 of the Small Business Act, as amended (15 U.S.C. 632).

Small business (Section 704.142) means any manufacturer, importer, or processor who meets either paragraph (a)(4)(i) or (ii) of this section. (i) A business is small if its total annual sales, when combined with those of its parent (if any), are less than $40 million. However, if the annual manufacture, importation, or processing volume of a particular chemical substance at any individual site owned or controlled by the business is greater than 45,400 kilograms (100,000 pounds), the business shall not qualify as small for purposes of reporting on the manufacture, importation, or processing of that chemical substance at that site, unless the business qualifies as small under paragraph (a)(4)(ii) of this section. (ii) A business is small if its total annual sales, when combined with those of its parent company (if any), are less than $4 million, regardless of the quantity of the particular chemical substance manufactured, imported, or processed by that business. (iii) For imported and processed mixtures containing HEX-BCH, the 45,400 kilograms (100,000 pounds) standard in paragraph (a)(4)(i) of this section applies only to the amount of HEX-BCH in a mixture and not the other components of the mixture.

Small business (Section 791.3) refers to a manufacturer or importer whose annual sales, when combined with those of its parent company (if any) are less than $30 million.

Small business concern (Section 700.43 [8/17/88]) means any person whose total annual sales in the person's fiscal year preceding the date of the submission of the applicable section 5 notice, when combined with those of the parent company (if any), are less than $40 million.

Small business concern (Section 21.2) means a concern defined by section 2[3] of the Small Business Act, 15 U.S.C. 632, 13 CFR Part 121, and regulations of the Small Business Administration promulgated thereunder.

Small capacitor (Section 761.3) means a capacitor which contains less than 1.36 kg (3 lbs.) of dielectric fluid. The following assumptions may be used if the actual weight of the dielectric fluid is unknown. A capacitor whose total volume is less than 1,639 cubic centimeters (100 cubic inches) may be considered to contain less than 1.36 kgs (3 lbs.) of dielectric fluid and a capacitor whose total volume is more than 3,278 cubic centimeters (200 cubic inches) must be considered to contain more than 1.36 kg (3 lbs.) of dielectric fluid. A capacitor whose volume is between 1,639 and 3,278 cubic centimeters may be considered to contain less than 1.36 kg (3 lbs.) of dielectric fluid if the total weight of the capacitor is less than 4.08 kg (9 lbs.).

Small commercial establishments (Section 35.2005) means, for purposes of Section 35.2034, private establishments such as restaurants, hotels, stores, filling stations, or recreational facilities and private, non-profit entities such as churches, schools, hospitals, or charitable organizations with dry weather wastewater flows less than 25,000 gallons per day.

Small community (Section 35.2005) means, for purposes of Sections 35.2020(b) and 35.2032, any municipality with a population of 3,500 or less or highly dispersed sections of larger municipalities, as determined by the Regional Administrator.

Small Diesel Refinery (Section 72.2 [7/30/93])* means a domestic motor diesel fuel refinery or portion of a refinery that, as an annual average of calendar years 1988 through 1990 and as reported to the Department of Energy on Form 810, had bona fide crude oil throughput less than 18,250,000 barrels per year, and the refinery or portion of a refinery is owned or controlled by a refiner with a total combined bona fide crude oil throughput of less than 50,187,500 barrels per year.

Small manufacturer (Section 704.3) means a manufacturer (or importer) that meets either of the standards set forth below. Small manufacturers should read the introductory paragraph of Section 704.5 and paragraph (d) of Section 704.5 to obtain complete information on the TSCA section 8(a) small manufacturer exemption. (1) First standard. A manufacturer of a chemical substance is small if its total annual sales, when combined with those of its parent company (if any), are less than $40 million. However, if the annual production volume of a particular chemical substance at any individual site owned or controlled by the manufacturer is greater than 45,400 kilograms (100,000 pounds), the manufacturer shall not qualify as small for purposes of reporting on the production of that chemical substance at that site, unless the manufacturer qualifies as small under small manufacturer (2) of this section. (2) Second standard. A manufacturer of a chemical substance is small if its total annual sales, when combined with those of its parent company (if any), are less than $4 million, regardless of the quantity of chemicals produced by that manufacturer. (3) Inflation index. EPA shall make use of the Producer Price Index for Chemicals and Allied Products, as compiled by the U.S. Bureau of Labor Statistics, for purposes of determining the need to adjust the total annual sales values and for determining new sales values when adjustments are made. EPA may adjust the total annual sales values whenever the Agency deems it necessary to do so, provided that the Producer Price Index for Chemicals and Allied Products has changed more than 20 percent since either the most recent previous change in sales values or the date of promulgation of this rule, whichever is later. EPA shall provide *Federal Register* notification when changing the total annual sales values.

Small manufacturer (Section 704.83) means a manufacturer (including importers) who meets either paragraph (a)(5)(i) or (ii) of this section: (i) A manufacturer of a chemical substance is small if its total annual sales, when combined with those of its parent company (if any), are less than $40 million. However, if the annual production volume of a particular chemical substance at any individual site owned or controlled by the manufacturer is greater than 45,400 kilograms (100,000 pounds), the manufacturer shall not qualify as small for purposes of reporting on the production of that chemical substance at that site, unless the manufacturer qualifies as small under paragraph (a)(5)(ii) of this section. (ii) A manufacturer of a chemical substance is small if its total annual sales, when combined with those of its parent company (if any), are less than $4 million, regardless of the quantity of the particular chemical substance produced by that manufacturer. (iii) For imported mixtures containing a chemical substance identified in paragraph (b) of this section, the 45,400 kilograms (100,000 pounds) standard in paragraph (a)(5)(i) of this section applies only to the amount of the chemical substance in a mixture and not the other components of the mixture.

Small manufacturer (Section 704.85) means a manufacturer (importers are defined as manufacturers under TSCA) who meets either of the standards under this rule as set forth in paragraph (a)(5)(i) and (a)(5)(ii) of this section.

Small manufacturer or importer (Section 704.3 [12/22/88]) means a manufacturer or importer that meets either of the following standards: (1) First standard. A manufacturer or importer of a substance is small if its total annual sales, when combined with those of its parent company (if any), are less than $40 million. However, if the annual production or importation volume of a particular substance at any individual site owned or controlled by the manufacturer or importer is greater than 45,400 kilograms (100,000 pounds), the manufacturer or importer shall not qualify as small for purposes of reporting on the production or importation of that substance at that site, unless the manufacturer or importer qualifies as small under standard (2) of this definition. (2) Second standard. A manufacturer or importer of a substance is small if its total annual sales, when combined with those of its parent company (if any), are less than $4 million, regardless of the quantity of substances produced or imported by that manufacturer or importer. (3) Inflation index. EPA shall make use of the Producer Price Index for Chemicals and Allied Products, as compiled by the U.S. Bureau of Labor Statistics, for purposes of determining the need to adjust the total annual sales values and for determining new sales values when adjustments are made. EPA may adjust the total annual sales values whenever the Agency deems it necessary to do so, provided that the Producer Price Index for Chemicals and Allied Products has changed more than 20 percent since either the most recent previous change in sales values or the date of promulgation of this rule, whichever is later. EPA shall provide *Federal Register* notification when changing the total annual sales values.

Small manufacturer or importer (Section 710.2) means a manufacturer or importer whose total annual sales are less than $5,000,000, based upon the manufacturer's or importer's latest complete fiscal year as of January 1, 1978, except that no manufacturer or importer is a small manufacturer or importer with respect to any chemical substance which such person manufactured at one site or imported in quantities greater than 100,000 pounds during calendar year 1977. In the case of a company which is owned or controlled by another company, total annual sales shall be based on the total annual sales of the owned or controlled company, the parent company, and all companies owned or controlled by the parent company taken together. Note: The purpose of the exception to the definition is to ensure that manufacturing and importers report production volumes for all chemical substances which they manufactured at one site or imported in quantities equal to or greater than 100,000 pounds during calendar year 1977.

Small manufacturer, processor, or importer (Section 763.63) means a manufacturer or processor who employed no more than 10 full-time employees at any one time in 1981.

Small processor (Section 432.51) shall mean an operation that produces up to 2730 kg (6000 lb) per day of any type or combination of finished products.

Small processor (Section 704.104 [10/27/87]) means a processor that meets either the standard in paragraph (a)(3)(i) of this section or the standard in paragraph (a)(3)(ii) of this section.

Small processor (Section 704.203 [12/22/88]) means a processor that meets either of the following standards: (1) First Standard. A processor of a substance is small if its total annual sales, when combined with those of its parent company (if any), are less than $40 million. However, if the annual processing volume of a particular substance at any individual site owned or controlled by the processor is greater than 45,400 kilograms (100,000 pounds), the processor shall not qualify as small for purposes of reporting on that substance at that site, unless the processor qualifies as small under standard (2) of this definition. (2) Second standard. A processor of a substance is small if its total annual sales, when combined with those of its parent company (if any), are less than $4 million, regardless of the quantity of substances processed by that processor. (3) Inflation index. EPA shall make use of the Producer Price Index for Chemicals and Allied Products, as compiled by the U.S. Bureau of Labor Statistics, for purposes of determining the need to adjust the total annual sales values and for determining new sales when adjustments are made. EPA may adjust the total annual sales values whenever the Agency deems it necessary to do so, provided that the Producer Price Index for Chemicals and Allied Products has changed more than 20 percent since either the most recent previous change in sales values or the date of promulgation of this rule, whichever is later. EPA shall provide *Federal Register* notification when changing the total annual sales values.

Small processor (Section 704.25) means a processor that meets either the standard in paragraph (a)(7)(i) of this section or the standard in paragraph (a)(7)(ii) of this section: (i) First standard. A processor of a chemical substance is small if its total annual sales, when combined with those of its parent company, if any, are less than $40 million. However, if the annual processing volume of a particular chemical substance at any individual site owned or controlled by the processor is greater than 45,400 kilograms (100,000 pounds), the processor shall not qualify as small for purposes of reporting on the processing of that chemical substance at that site, unless the processor qualifies as small under paragraph (a)(7)(ii) of this section. (ii) Second standard. A processor of a chemical substance is small if its total annual sales, when combined with those of its parent company (if any), are less than $4 million, regardless of the quantity of the particular chemical substance processed by that company. (iii) Inflation index. EPA will use the Inflation Index described in the definition of small manufacturer set forth in Section 704.3, for purposes of adjusting the total annual sales values of this small processor definition. EPA will provide notice in the *Federal Register* when changing the total annual sales values of this definition.

Small processor (Section 704.33) means a processor that meets either the standard in paragraph (a)(4)(i) of this section or the standard in paragraph (a)(4)(ii) of this section. (i) First standard. A processor of a chemical substance is small if its total annual sales, when combined with those of its parent company, if any, are less than $40 million. However, if the annual processing volume of a particular chemical substance at any individual site owned or controlled by the processor is greater than 45,400 kilograms (100,000 pounds), the processor shall not qualify as small for purposes of reporting on the processing of that chemical substance at that site, unless the processor qualifies as small under paragraph (a)(1)(ii) of this section. (ii) Second standard. A processor of a chemical substance is small if its total annual sales, when combined with those of its parent company (if any), are less than $4 million, regardless of the quantity of the particular chemical substance processed by that company. (iii) Inflation index. EPA shall use the Inflation Index described in the definition of small manufacturer that is set forth in Section 704.3, for purposes of adjusting the total annual sales values of this small processor definition. EPA shall provide *Federal Register* notification when changing the total annual sales values of this definition.

Small quantities for purposes of scientific experimentation or analysis or chemical research on, or analysis of, such substance or another substance, including any such research or analysis for the development of a product (hereinafter sometimes shortened to small quantities for research and development (Section 710.2) means quantities of a chemical substance manufactured, imported, or processed or proposed to be manufactured, imported, or processed that (1) are no greater than reasonably necessary for such purposes and (2) after the publication of the revised inventory, are used by, or directly under the supervision of, a technically qualified individual(s). Note: Any chemical substances manufactured, imported or processed in quantities less than 1,000 pounds annually shall be presumed to be manufactured, imported or processed for research and development purposes. No person may report for the inventory any chemical substance in such quantities unless that person can certify, that the substance was not manufactured, imported, or processed solely in small quantities for research and development, as defined in this section.

Small quantities for research and development (Section 710.2). See **Small quantities for purposes of scientific experimentation or analysis or chemical research on, or analysis of, such substance or another substance, including any such research or analysis for the development of a product (Section 710.2)**.

Small quantities for research and development (Section 761.3) means any quantity of PCBs (1) that is originally packaged in one or more hermetically sealed containers of a volume of no more than five (5.0) milliliters, and (2) that is used only for purposes of scientific experimentation or analysis, or chemical research on, or analysis of, PCBs, but not for research or analysis for the development of a PCB product.

Small quantities solely for purposes of scientific experimentation or analysis or chemical research on, or analysis of, such substance or another substance, including such research or analysis of the development of a product (Section 747.200). See **Small quantities solely for research and development (Section 747.200)**.

Small quantities solely for purposes of scientific experimentation or analysis or chemical research on, or analysis of, such substance or another substance, including such research or analysis of the development of a product (Section 747.195). See **Small quantities solely for research and development (Section 747.195)**.

Small quantities solely for purposes of scientific experimentation or analysis or chemical research on, or analysis of, such substance or another substance, including such research or analysis of the development of a product (Section 704.3 [12/22/88]). See **Small quantities solely for research and development (Section 704.3)**.

Small quantities solely for purposes of scientific experimentation or analysis or chemical research on, or analysis of, such substance or another substance, including such research or analysis of the development of a product (Section 720.3). See **Small quantities solely for research and development (Section 720.3)**.

Small quantities solely for purposes

of scientific experimentation or analysis or chemical research on, or analysis of, such substance or another substance, including such research or analysis of the development of a product (Section 747.115). See **Small quantities solely for research and development (Section 747.115)**.

Small quantities solely for research and development (or small quantities solely for purposes of scientific experimentation or analysis or chemical research on, or analysis of, such substance or another substance, including such research or analysis for the development of a product) (Section 747.195) means quantities of a chemical substance manufactured, imported, or processed or proposed to be manufactured, imported, or processed solely for research and development that are not greater than reasonably necessary for such purposes.

Small quantities solely for research and development (or small quantities solely for purposes of scientific experimentation or analysis or chemical research on, or analysis of, such substance or another substance, including such research or analysis for the development of a product) (Section 704.3 [12/22/88]) means quantities of a chemical substance manufactured, imported, or processed or proposed to be manufactured, imported, or processed solely for research and development that are not greater than reasonably necessary for such purposes.

Small quantities solely for research and development (or small quantities solely for purposes of scientific experimentation or analysis or chemical research on, or analysis of, such substance or another substance, including such research or analysis for the development of a product) (Section 747.200) means quantities of a chemical substance manufactured, imported, or processed or proposed to be manufactured, imported, or processed solely for research and development that are not greater than reasonably necessary for such purposes.

Small quantities solely for research and development (or small quantities solely for purposes of scientific experimentation or analysis or chemical research on, or analysis of, such substance or another substance, including such research or analysis for the development of a product) (Section 720.3) means quantities of a chemical substance manufactured, imported, or processed or proposed to be manufactured, imported, or processed solely for research and development that are not greater than reasonably necessary for such purposes.

Small quantities solely for research and development (or small quantities solely for purposes of scientific experimentation or analysis or

chemical research on, or analysis of, such substance or another substance, including such research or analysis for the development of a product) (Section 747.115) means quantities of a chemical substance manufactured, imported, or processed or proposed to be manufactured, imported, or processed solely for research and development that are not greater than reasonably necessary for such purposes.

Small Quantity Generator (Section 260.10) means a generator who generates less than 1000 kg of hazardous waste in a calendar month.

Small refinery (Section 80.2 [2/16/94])* means a domestic diesel fuel refinery (1) Which has a crude oil or bonafide feedstock capacity of 50,000 barrels per day or less, and (2) Which is not owned or controlled by any refiner with a total combined crude oil or bonafide feedstock capacity greater than 137,500 barrels per day. The above capacities shall be measured in terms of the average of the actual daily utilization rates of the affected refiners or refineries during the period January 1, 1988 to December 31, 1990. These averages will be calculated as barrels per calendar day.

Small water system (Section 141.2 [7/17/92])* for the purpose of subpart I of this part only, means a water system that serves 3,300 persons or fewer.

SMCL (Section 143.2). See **Secondary maximum contaminant levels (Section 143.2)**.

Smelt dissolving tank (Section 60.281) means a vessel used for dissolving the smelt collected from the recovery furnace.

Smelter owner and operator (Section 57.103 [2-13-92])* means the owner or operator of the smelter, without distinction.

Smelting (Section 60.161) means processing techniques for the melting of a copper sulfide ore concentrate or calcine charge leading to the formation of separate layers of molten slag, molten copper, and/or copper matte.

Smelting furnace (Section 60.161) means any vessel in which the smelting of copper sulfide ore concentrates or calcines is performed and in which the heat necessary for smelting is provided by an electric current, rapid oxidation of a portion of the sulfur contained in the concentrate as it passes through an oxidizing atmosphere, or the combustion of a fossil fuel.

Smoke (Sections 86.082-2; 87.1) means the matter in the exhaust emission which obscures the transmission of light.

Smoke number (SN) (Section 87.1) means the dimensionless term quantifying smoke emissions.

Smooth-one-side (S1S) hardboard (Section 429.11) means hardboard which is produced by the wet-matting, wet-pressing process.

Smooth-two-sides (S2S) hardboard (Section 429.11) means hardboard which is produced by the wet-matting, dry-pressing process.

SN (Section 87.1). See **Smoke number (Section 87.1)**.

Snap beans (Section 407.71) shall mean the processing of snap beans into the following product styles: Canned and frozen green, Italian, wax, string, bush, and other related varieties, whole, French, fancy, Extra Standard, Standard, and other cuts.

SO_2 (Section 58.1) means sulfur dioxide.

Soda-lime recipe (Section 60.291) means glass product composition of the following ranges of weight proportions: 60 to 75 percent silicon dioxide, 10 to 17 percent total R_2O (e.g., Na_2O and K_2O), 8 to 20 percent total RO but not to include any PbO (e.g., CaO, and MgO), 0 to 8 percent total R_2O_3 (e.g., Al_2O_3), and 1 to 5 percent other oxides.

Soil (Section 192.11) means all unconsolidated materials normally found on or near the surface of the earth including, but not limited to, silts, clays, sands, gravel, and small rocks.

Soil (Section 761.123) means all vegetation, soils and other ground media, including but not limited to, sand, grass, gravel, and oyster shells. It does not include concrete and asphalt.

Soil (Sections 796.2700; 796.2750) is the unconsolidated mineral material on the immediate surface of the earth that serves as a natural medium for the growth of land plants; its formation and properties are determined by various factors such as parent material, climate, macro- and microorganisms, topography, and time.

Soil aggregate (Section 796.2700) is the combination or arrangement of soil separates (sand, silt, clay) into secondary units. These units may be arranged in the profile in a distinctive characteristic pattern that can be classified on the basis of size, shape, and degree of distinctness into classes, type, and grades.

Soil aggregate (Section 796.2750) is the combination or arrangement of soil separates (sand, silt, clay) into secondary units. These units may be arranged in the soil profile in a distinctive characteristic pattern that can be classified according to size, shape, and degree of distinctness into classes, types, and grades.

Soil classification (Section 796.2700) is the systematic arrangement of soils into groups or categories. Broad groupings are made on the basis of general characteristics, subdivisions, on the basis of more detailed differences in specific properties. The soil classification system used today in the United States is the 7th Approximation Comprehensive System. The ranking of subdivisions under the system is: order, suborder, great group, family and series.

Soil classification (Section 796.2750) is the systematic arrangement of soils into groups or categories. Broad groupings are based on general soil characteristics while subdivisions are based on more detailed differences in specific properties. The soil classification system used in this standard and the one used today in the United States is the 7th Approximation-Comprehensive System. The ranking of subdivisions under this system is: Order, Suborder, Great group, family, and series.

Soil horizon (Section 796.2700) is a layer of soil approximately parallel to the land surface. Adjacent layers differ in physical, chemical, and biological properties or characteristics such as color, structure, texture, consistency, kinds, and numbers of organisms present, and degree of acidity or alkalinity.

Soil horizon (Section 796.2750) is a layer of soil approximately parallel to the land surface. Adjacent layers differ in physical, chemical, and biological properties such as color, structure, texture, consistency, kinds and numbers of organisms present, and degree of acidity or alkalinity.

Soil injection (Section 165.1) means the emplacement of pesticides by ordinary tillage practices within the plow layer of a soil.

Soil order (Section 796.2700) is the broadest category of soil classification and is based on general similarities of physical/chemical properties. The formation by similar genetic processes causes these similarities. The soil orders found in the United States are: Alfisol, Aridisol, Entisol, Histosol, Inceptisol, Mollisol, Oxisol, Spodosol, Ultisol, and Vertisol.

Soil order (Section 796.2750) is the broadest category of soil classification and is based on the general similarities of soil physical/chemical properties. The formation of soil by similar general genetic processes causes these similarities. The Soil Orders found in the United States are: Alfisol, Aridisol, Entisol, Histosol, Inceptisol, Mollisol, Oxisol, Spodosol, Ultisol, and Vertisol.

Soil organic matter (Section 796.2700) is the organic fraction of the soil; it includes plant and animal residues at various stages of decomposition, cells and tissues of soil organisms, and substances synthesized by the microbial population.

Soil pH (Section 257.3-5) is the value obtained by sampling the soil to the depth of cultivation or solid waste placement, whichever is greater, and analyzing by the electrometric method. ("Methods of Soil Analysis, Agronomy Monograph No. 9," C. A. Black, ed., American Society of Agronomy, Madison, Wisconsin, pp. 914-926, 1965.)

Soil pH (Section 796.2700) is the negative logarithm to the base 10 of the hydrogen ion activity of a soil as determined by means of a suitable sensing electrode coupled with a suitable reference electrode at a 1:1 soil:water ratio.

Soil series (Sections 796.2700; 796.2750) is the basic unit of soil classification and is a subdivision of a family. A series consists of soils that were developed under comparable climatic and vegetational conditions. The soils comprising a series are essentially alike in all major profile characteristics except for the texture of the "A" horizon (i.e., the surface layer of soil).

Soil texture (Section 796.2700) refers to the classification of soils based on the relative proportions of the various soil separates present. The soil textural classes are: clay, sandy clay, silty clay, clay loam, silty clay loam, sandy clay loam, loam, silt loam, silt, sandy loam, loamy sand, and sand.

Soil texture (Section 796.2750) is a classification of soils that is based on the relative proportions of the various soil separates present. The soil textural classes are: clay, sandy clay, silty clay, clay loam, silty clay loam, sandy clay loam, loam, silt loam, silt, sandy loam, loamy sand, and sand.

Solar irradiance in water (L_λ) (Section 796.3700) is related to the sunlight intensity in water and is proportional to the average light flux (in the units of 10^{-3} einsteins cm^{-2} day^{-1}) that is available to cause photoreaction in a wavelength interval centered at λ over a 24-hour day at a specific latitude and season date.

Sold or distributed (Section 167.1) means the aggregate amount of a product released for shipment by the establishment in which the pesticide or device was produced.

Sold or distributed (Section 167.3 [9/8/88]) means the aggregate amount of a pesticidal product released for shipment by the establishment in which the pesticidal product was produced.

Sole or principal source aquifer (Section 146.3) means an aquifer which has been designated by the Administrator pursuant to section 1424 (a) or (e) of the SDWA.

Sole or Principal Source Aquifer (SSA) (Section 149.2 [2/14/89])* means an aquifer which is designated as an SSA under section 1424(e) of the SDWA.

Solid waste (Section 247.101) means garbage, refuse, sludges, and other discarded solid materials, including solid waste materials resulting from industrial, commercial, and agricultural operations, and from community activities, but does not include solids or dissolved materials in domestic sewage or other significant pollutants in water resources, such as silt, dissolved or suspended solids in industrial waste water effluents, dissolved materials in irrigation return flow, or other common water pollutants.

Solid waste (Section 257.2) means any garbage, refuse, sludge from a waste treatment plant, water supply treatment plant, or air pollution control facility and other discarded material, including solid, liquid, semisolid, or contained gaseous material resulting from industrial, commercial, mining, and agricultural operations, and from community activities, but does not include solid or dissolved materials in domestic sewage, or solid or dissolved material in irrigation return flows or industrial discharges which are point sources subject to permits under section 402 of the Federal Water Pollution Control Act, as amended (86 Stat. 880), or source, special nuclear, or byproduct material as defined by the Atomic Energy Act of 1954, as amended (68 Stat. 923).

Solid waste (Section 258.2 [10-9-91]) means any garbage, or refuse, sludge from a wastewater treatment plant, water supply treatment plant, or air pollution control facility and other discarded material, including solid, liquid, semi-solid, or contained gaseous material resulting from industrial, commercial, mining, and agricultural operations, and from community activities, but does not include solid or dissolved materials in domestic sewage, or solid or dissolved materials in irrigation return flows or industrial discharges that are point sources subject to permit under 33 U.S.C. 1342, or source, special nuclear, or by-product material as defined by the Atomic Energy Act of 1954, as amended (68 Stat. 923).

Solid waste (Section 259.10 [3/24/89]) means a solid waste defined in Section 1004(27) of RCRA.

Solid waste (Section 260.10) means a solid waste as defined in Section 261.2 of this chapter.

Solid waste (Section 261.2) (a)(1) is any discarded material (as defined in Section 261.2) that is not excluded by Section 261.4(a) or that is not excluded by variance granted under Sections 260.30 and 260.31.

Solid waste (Section 261.3 [6-1-92]) as defined in Sec. 261.2, is a hazardous waste if: (1) It is not excluded from regulation as a hazardous waste under Sec. 261.4(b); and (2) It meets any of the following criteria: (i) It exhibits any of the characteristics of hazardous waste identified in subpart C except that any mixture of a waste from the extraction, beneficiation, and processing of ores and minerals excluded under Sec. 261.4(b)(7) and any other solid waste exhibiting a characteristic of hazardous waste under subpart C of this part only if it exhibits a characteristic that would not have been exhibited by the excluded waste alone if such mixture had not occurred or if it continues to exhibit any of the characteristics exhibited by the non-excluded wastes prior to mixture. Further, for the purposes of applying the Toxicity Characteristic to such mixtures, the mixture is also a hazardous waste if it exceeds the maximum concentration for any contaminant listed in table I to Sec. 261.24 that would not have been exceeded by the excluded waste alone if the mixture had not occurred or if it continues to exceed the maximum concentration for any contaminant exceeded by the nonexempt waste prior to mixture.

Solid waste (Section 60.51) means refuse, more than 50 percent of which is municipal type waste consisting of a mixture of paper, wood, yard wastes, food wastes, plastics, leather, rubber, and other combustibles, and noncombustible materials such as glass and rock.

Solid waste (Sections 243.101; 246.101) means garbage, refuse, sludges, and other discarded solid materials, including solid waste materials resulting from industrial, commercial, and agricultural operations, and from community activities, but does not include solids or dissolved materials in domestic sewage or other significant pollutants in water resources, such as silt, dissolved or suspended solids in industrial wastewater effluents, dissolved materials in irrigation return flows or other common water pollutants. Unless specifically noted otherwise, the term solid waste as used in these guidelines shall not include mining, agricultural, and industrial solid wastes; hazardous wastes; sludges; construction and demolition wastes; and infectious wastes.

Solid waste boundary (Section 257.3-4) means the outermost perimeter of the solid waste (projected in the horizontal plane) as it would exist at completion of the disposal activity.

Solid waste storage container (Section 243.101) means a receptacle used for the temporary storage of solid waste while awaiting collection.

Solid Waste Incinerator (Section 72.2 [7/30/93])* means a source as defined in section 129(g)(1) of the Act.

Solid wastes (Sections 240.101; 241.101) means garbage, refuse, sludges, and other discarded solid materials resulting from industrial and commercial operations and from community activities. It does not include solids or dissolved material in domestic sewage or other significant pollutants in water resources, such as silt, dissolved or suspended solids in industrial wastewater effluents, dissolved materials in irrigation return flows or other common water pollutants.

Solid-derived fuel (Section 60.41a) means any solid, liquid, or gaseous fuel derived from solid fuel for the purpose of creating useful heat and includes, but is not limited to, solvent refined coal, liquified coal, and gasified coal.

Solid-waste-derived fuel (Section 247.101) means a fuel that is produced from solid waste that can be used as a primary or supplementary fuel in conjunction with or in place of fossil fuels. The solid-waste-derived fuel can be in the form of raw (unprocessed) solid waste, shredded (or pulped) and classified solid waste, gas or oil derived from pyrolyzed solid waste, or gas derived from the biodegradation of solid waste.

Solution (Sections 796.1840; 796.1860) is a homogeneous mixture of two or more substances constituting a single phase.

Solution heat treatment (Section 468.02) shall mean the process introducing a workpiece into a quench bath for the purpose of heat treatment following rolling, drawing or extrusion.

Solvent applied in the coating (Section 60.441) means all organic solvent contained in the adhesive, release, and precoat formulations that is metered into the coating applicator from the formulation area.

Solvent extraction operation (Section 264.1031 [6-21-90]) means an operation or method of separation in which a solid or solution is contacted with a liquid solvent (the two being mutually insoluble) to preferentially dissolve and transfer one or more components into the solvent.

Solvent feed (Section 60.601) means the solvent introduced into the spinning solution preparation system or precipitation bath. This feed stream includes the combination of recovered solvent and makeup solvent.

Solvent filter (Section 60.621) means a discrete solvent filter unit containing a porous medium that traps and removes contaminants from petroleum solvent, together with the piping and ductwork used in the installation of this device.

Solvent inventory variation (Section 60.601) means the normal changes in the total amount of solvent contained in the affected facility.

Solvent recovery dryer (Section 60.621) means a class of dry cleaning dryers that employs a condenser to condense and recover solvent vapors evaporated in a closed-loop stream of heated air, together with the piping and ductwork used in the installation of this device.

Solvent recovery system (Section 60.601) means the equipment associated with capture, transportation, collection, concentration, and purification of organic solvents. It may include enclosures, hoods, ducting, piping, scrubbers, condensers, carbon adsorbers, distillation equipment, and associated storage vessels.

Solvent recovery system (Section 60.431) means an air pollution control system by which VOC solvent vapors in air or other gases are captured and directed through a condenser(s) or a vessel(s) containing beds of activated carbon or other adsorbents. For the condensation method, the solvent is recovered directly from the condenser. For the adsorption method, the vapors are adsorbed, then desorbed by steam or other media, and finally condensed and recovered.

Solvent-borne (Section 60.391) means a coating which contains five percent or less water by weight in its volatile fraction.

Solvent-borne ink systems (Section 60.431) means ink and related coating mixtures whose volatile portion consists essentially of VOC solvent with not more than five weight percent water, as applied to the gravure cylinder.

Solvent-spun synthetic fiber (Section 60.601) means any synthetic fiber produced by a process that uses an organic solvent in the spinning solution, the precipitation bath, or processing of the sun fiber.

Solvent-spun synthetic fiber process (Section 60.601) means the total of all equipment having a common spinning solution preparation system or a common solvent recovery system, and that is used in the manufacture of solvent-spun synthetic fiber. It includes spinning solution preparation, spinning, fiber processing and solvent recovery, but does not include the polymer production equipment.

Sorbent (Section 260.10 [8/18/92])* means a material that is used to soak up free liquids by either adsorption or absorption, or both. Sorb means to either adsorb or absorb, or both.

Sound Exposure Level (Section 201.1) means the level in decibels calculated as ten times the common logarithm of time integral of squared A-weighted sound pressure over a given time period or event divided by the square of the standard reference sound pressure of 20 micropascals and a reference duration of one second.

Sound level (Section 201.1) means the level, in decibels, measured by instrumentation which satisfies the requirements of American National Standard Specification for Sound Level Meters S1.4-1971 Type 1 (or S1A) or Type 2 if adjusted as shown in Table 1. This publication is available from the American National Standards

Institute, Inc., 1430 Broadway, New York, New York 10018. For the purpose of these procedures the sound level is to be measured using the A-weighting of spectrum and either the FAST or SLOW dynamic averaging characteristics, as designated. It is abbreviated as L_A.

Sound level (Section 202.10) means the quantity in decibels measured by a sound level meter satisfying the requirements of American National Standards Specification for Sound Level Meters S1.4-1971. This publication is available from the American National Standards Institute, Inc., 1430 Broadway, New York, New York 10018. Sound level is the frequency-weighted sound pressure level obtained with the standardized dynamic characteristic fast or slow and weighting A, B, or C; unless indicated otherwise, the A-weighting is understood.

Sound level (Section 204.2) means the weighted sound pressure level measured by the use of a metering characteristic and weighting A, B, or C as specified in American National Standard Specification for Sound Level Meters S1.4-1971 or subsequent approved revision. The weighting employed must be specified, otherwise A-weighting is understood.

Sound level (Section 205.2) means 20 times the logarithm to base 10 of the ratio of pressure of a sound to the reference pressure. The reference pressure is 20 micropascals (20 micronewtons per square meter). Note: Unless otherwise explicitly stated, it is to be understood that the sound pressure is the effective (rms) sound pressure, per American National Standards Institute, Inc., 1430 Broadway, New York, New York 10018.

Sound pressure level (Section 201.1) (in stated frequency band) means the level, in decibels, calculated as 20 times the common logarithm of the ratio of a sound pressure to the reference sound pressure of 20 micropascals.

Sound pressure level (Section 204.2) means, in decibels, 20 times the logarithm to the base ten of the ratio of a sound pressure to the reference sound pressure of 20 micropascals (20 micronewtons per square meter). In the absence of any modifier, the level is understood to be that of a root-mean-square pressure.

Sound pressure level (Section 205.2) means in decibels, 20 times the logarithm to the base 10 of the ratio of a sound pressure to the reference sound pressure of 20 micropascals (20 micronewtons per square meter). In the absence of any modifier, the level is understood to be that of a root-mean-square pressure. The unit of any sound level is the decibel, having the unit symbol dB.

Soups (Section 407.81) shall mean the

combination of various fresh and pre-processed meats, fish, dairy products, eggs, flours, starches, vegetables, spices, and other similar raw ingredients into a variety of finished mixes and styles but not including dehydrated soups.

Sour water stream (Section 61.341 [1/7/93])* means a stream that: (1) Contains ammonia or sulfur compounds (usually hydrogen sulfide) at concentrations of 10 ppm by weight or more; (2) is generated from separation of water from a feed stock, intermediate, or product that contained ammonia or sulfur compounds; and (3) requires treatment to remove the ammonia or sulfur compounds.

Sour water stripper (Section 61.341 [1/7/93])* means a unit that: (1) Is designed and operated to remove ammonia or sulfur compounds (usually hydrogen sulfide) from sour water streams; (2) has the sour water streams transferred to the stripper through hard piping or other enclosed system; and (3) is operated in such a manner that the offgases are sent to a sulfur recovery unit, processing unit, incinerator, flare, or other combustion device.

Source (Section 122.29) means any building, structure, facility, or installation from which there is or may be a discharge of pollutants.

Source (Section 129.2) means any building, structure, facility, or installation from which there is or may be the discharge of toxic pollutants designated as such by the Administration under section 307(a)(1) of the Act.

Source (Section 66.3 [2-13-92])* means any source of air pollution subject to applicable legal requirements as defined in paragraph (c).

Source control remedial action (Section 300.6) means measures that are intended to contain the hazardous substances or pollutants or contaminants where they are located or eliminate potential contamination by transporting the hazardous substances or pollutants or contaminants to a new location. Source control remedial actions may be appropriate if a substantial concentration or amount of hazardous substances or pollutants or contaminants remains at or near the area where they are originally located and inadequate barriers exist to retard migration of hazardous substances or pollutants or contaminants into the environment. Source control remedial actions may not be appropriate if most hazardous substances or pollutants or contaminants have migrated from the area where originally located or if the lead agency determines that the hazardous substances or pollutants or contaminants are adequately contained.

Source material (Sections 710.2; 720.3) has the meaning contained in the Atomic Energy Act of 1954, 42 U.S.C. 2014 et seq., and the regulations issued thereunder.

Source separation (Section 246.101) means the setting aside of recyclable materials at their point of generation by the generator.

Source Term (Section 61 [12-15-89]) The amount of radioactive material emitted to the atmosphere from a source, either estimated, measured or reported, that is used in the risk assessment.

SOW (Section 35.6015 [1/27/89]). See **Statement of Work (Section 35.6015)**.

Span gas (Section 86.082-2) means a gas of known concentration which is used routinely to set the output level of an analyzer.

Span gas (Section 86.402-78) means a gas of known concentration which is used routinely to set the output level of any analyzer.

Spandex fiber (Section 60.601) means a manufactured fiber in which the fiber-forming substance is a long chain synthetic polymer comprised of at least 85 percent of a segmented polyurethane.

Spare flue gas desulfurization system module (Section 60.41a) means a separate system of sulfur dioxide emission control equipment capable of treating an amount of flue gas equal to the total amount of flue gas generated by an affected facility when operated at maximum capacity divided by the total number of nonspare flue gas desulfurization modules in the system.

Spark Plug (Section 85.2122(a)(8)(ii)(A)) means a device to suitably deliver high tension electrical ignition voltage to the spark gap in the engine combustion chamber.

Special aquatic sites (Section 230.3) means those sites identified in Subpart E. They are geographic areas, large or small, possessing special ecological characteristics of productivity, habitat, wildlife protection, or other important and easily disrupted ecological values. These areas are generally recognized as significantly influencing or positively contributing to the general overall environmental health or vitality of the entire ecosystem of a region. (See Section 230.10(a)(3).)

Special expertise (Section 1508.26) means statutory responsibility, agency mission, or related program experience.

Special features enabling off-street or off-highway operation and use (Section 86.084-2) (applies beginning with the 1984 model year) means a vehicle: (1) That has 4-wheel drive; and (2) That has at least four of the following characteristics calculated when the automobile is at curb weight, on a level surface, with the front wheels parallel to the vehicle's longitudinal centerline, and the tires inflated to the manufacturer's recommended pressure; (i) Approach angle of not less than 28 degrees. (ii) Breakover angle of not less than 14 degrees. (iii) Departure angle of not less than 20 degrees. (iv) Running clearance of not less than 8 inches. (v) Front and rear axle clearances of not less than 7 inches each.

Special fellow (Section 46.120) means an individual enrolled in an educational program relating to environmental sciences, engineering, professional schools, and allied sciences.

Special Government employee (Section 3.102) means an officer or employee of the Environmental Protection Agency who is retained, designated, appointed or employed to perform, with or without compensation, temporary duties either on a full-time or intermittent basis, for not to exceed 130 days during any period of 365 consecutive days.

Special industrial gaskets (Section 763.163 [7-12-89]) means sheet or beater-add gaskets designed for industrial uses in either (1) environments where temperatures are 750 degrees Fahrenheit or greater, or (2) corrosive environments. An industrial gasket is one designed for use in an article which is not a "consumer product" within the meaning of the Consumer Product Safety Act (CPSA), 15 U.S.C. 2052, or for use in a "motor vehicle" or "motor vehicle equipment" within the meaning of the National Traffic and Motor Vehicle Safety Act of 1966, as amended, 15 U.S.C. 1381. A corrosive environment is one in which the gasket is exposed to concentrated (pH less than 2), highly oxidizing mineral acids (e.g., sulfuric, nitric, or chromic acid) at temperatures above ambient.

Special local need (Section 162.152) means an existing or imminent pest problem within a State for which the State lead agency, based upon satisfactory supporting information, has determined that an appropriate federally registered pesticide product is not sufficiently available.

Special nuclear material (Sections 710.2; 720.3) has the meaning contained in the Atomic Energy Act of 1954, 42 U.S.C. 2014 et seq., and the regulations issued thereunder.

Special production area (Section 723.175) means a demarcated area within which all manufacturing, processing, and use of a new chemical substancc takes place, except as provided in paragraph (f) of this section, in accordance with the requirements of paragraph (e) of this section.

Special purpose facility (Section 6.901) means a building or space, including land incidental to its use, which is wholly or predominantly utilized for the special purpose of an agency and not generally suitable for other uses, as determined by the General Services Administration.

Special Purpose Equipment (Section 201.1) means maintenance-of-way equipment which may be located on or operated from rail cars including: Ballast cribbing machines, ballast regulators, conditioners and scarifiers, bolt machines, brush cutters, compactors, concrete mixers, cranes and derricks, earth boring machines, electric welding machines, grinders, grouters, pile drivers, rail heaters, rail layers, sandblasters, snow plows, spike drivers, sprayers and other types of such maintenance-of-way equipment.

Special Review (Section 166.3) refers to any interim administrative review of the risks and benefits of the use of a pesticide conducted pursuant to the provisions of EPA's Rebuttable Presumption Against Registration rules, 40 CFR 162.11(a), or any subsequent version of those rules.

Special source of ground water (Section 191.12) as used in this part, means those Class I ground waters identified in accordance with the Agency's Ground-Water Protection Strategy published in August 1984 that: (1) Are within the controlled area encompassing a disposal system or are less than five kilometers beyond the controlled area; (2) are supplying drinking water for thousands of persons as of the date that the Department chooses a location within that area for detailed characterization as a potential site for a disposal system (e.g., in accordance with section 112(b)(1)(B) of the NWPA); and (3) are irreplaceable in that no reasonable alternative source of drinking water is available to that population.

Special Track Work (Section 201.1) means track other than normal tie and ballast bolted or welded rail or containing devices such as retarders or switching mechanisms.

Special wastes (Section 240.101) means nonhazardous solid wastes requiring handling other than that normally used for municipal solid waste.

Specially designated landfill (Section 165.1) means a landfill at which complete long term protection is provided for the quality of surface and subsurface waters from pesticides, pesticide containers, and pesticide-related wastes deposited therein, and against hazard to public health and the environment. Such sites should be located and engineered to avoid direct hydraulic continuity with surface and subsurface waters, and any leachate or subsurface flow into the disposal area should be contained within the site unless treatment is provided. Monitoring wells should be established and a sampling and analysis program conducted. The location of the disposal site should be permanently recorded in the appropriate local office of legal jurisdiction. Such facility complies with the Agency Guidelines for the Land Disposal of Solid Wastes as prescribed in 40 CFR Part 241.

Specialty (Section 420.71). See **Specialty hot forming operation**

(**Section 420.71**).

Specialty hot forming operation (or specialty) (Section 420.71) applies to all hot forming operations other than carbon hot forming operations.

Specialty paper (Section 763.163 [7-12-89]) means an asbestos-containing product that is made of paper intended for use as filters for beverages or other fluids or as paper fill for cooling towers. Cooling tower fill consists of asbestos paper that is used as a cooling agent for liquids from industrial processes and air conditioning systems.

Specialty steel (Section 420.71) means those steel products containing alloying elements which are added to enhance the properties of the steel product when individual alloying elements (e.g., aluminum, chromium, cobalt, columbium, molybdenum, nickel, titanium, tungsten, vanadium, zirconium) exceed 3% or the total of all alloying elements exceed 5%.

Specific chemical identity (Section 350.1 [7/29/88]) means the chemical name, Chemical Abstracts Service (CAS) Registry Number, or any other information that reveals the precise chemical designation of the substance. Where the trade name is reported in lieu of the specific chemical identity, the trade name will be treated as the specific chemical identity for purposes of this part.

Specification (Section 246.101) means a clear and accurate description of the technical requirements for materials, products or services, identifying the minimum requirements for quality and construction of materials and equipment necessary for an acceptable product. In general, specifications are in the form of written descriptions, drawings, prints, commercial designations, industry standards, and other descriptive references.

Specification (Section 248.4 [2/17/89]) means a description of the technical requirements for a material, product, or service that includes the criteria for determining whether these requirements are met. In general, specifications are in the form of written commercial designations, industry standards, and other descriptive references.

Specification (Section 250.4 [10/6/87, 6/22/88]) means a detailed description of the technical requirements for materials, products, or services that specifies the minimum requirement for quality and construction of materials and equipment necessary for an acceptable product. Specifications are generally in the form of a written description, drawings, prints, commercial designations, industry standards, and other descriptive references.

Specification (Section 252.4 [6/30/88]) means a description of the technical requirements for a material, product, or service that includes the criteria for determining whether these requirements are met. In general, specifications are in the form of written commercial designations, industry standards, and other descriptive references.

Specification (Section 253.4 [11/17/88]) means a description of the technical requirements for a material, product, or service that includes the criteria for determining whether these requirements are met. In general, specifications are in the form of written commercial designations, industry standards, and other descriptive references.

Specification (Sections 247.101; 249.04) means a clear and accurate description of the technical requirement for materials, products or services, which specifies the minimum requirement for quality and construction of materials and equipment necessary for an acceptable product. In general, specifications are in the form of written descriptions, drawings, prints, commercial designations, industry standards, and other descriptive references.

Specified ports and harbors (Section 300.6) means those port and harbor areas on inland rivers, and land areas immediately adjacent to those waters, where the USCG acts as predesignated on-scene coordinator. Precise locations are determined by EPA/USCG regional agreements and identified in Federal regional contingency plans.

Specimen (Section 792.3 [8-17-89])* means any material derived from a test system for examination or analysis.

Specimen (Sections 160.3; 792.3) means any material derived from a test system for examination or analysis.

Specimens (Section 160.3 [8-17-89])* means any material derived from a test system for examination or analysis.

Spectral uncertainty (Section 211.203) means possible variation in exposure to the noise spectra in the workplace. (To avoid the underprotection that would result from these variations relative to the assumed Pink Noise used to determine the NRR, an extra three decibel reduction is included when computing the NRR.)

Spent acid solution (or spent pickle liquor) (Section 420.91) means those solutions of steel pickling acids which have been used in the pickling process and are discharged or removed therefrom.

Spent lubricant (Section 468.02) shall mean water or an oil-water mixture which is used in forming operations to reduce friction, heat and wear and ultimately discharged.

Spent material (Section 261.1), for purposes of Sections 261.2 and 261.6, is any material that has been used and as a result of contamination can no longer serve the purpose for which it was produced without processing.

Spent nuclear fuel (Section 191.02) means fuel that has been withdrawn from a nuclear reactor following irradiation, the constituent elements of which have not been separated by reprocessing.

Spent pickle liquor (Section 420.91). See **Spent acid solution (Section 420.91)**.

Spill (Section 761.123) means both intentional and unintentional spills, leaks, and other uncontrolled discharges where the release results in any quantity of PCBs running off or about to run off the external surface of the equipment or other PCB source, as well as the contamination resulting from those releases. This policy applies to spills of 50 ppm or greater PCBs. The concentration of PCBs spilled is determined by the PCB concentration in the material spilled as opposed to the concentration of PCBs in the material onto which the PCBs were spilled. Where a spill of untested mineral oil occurs, the oil is presumed to contain greater than 50 ppm, but less than 500 ppm PCBs and is subject to the relevant requirements of this policy.

Spill area (Section 761.123) means the area of soil on which visible traces of the spill can be observed plus a buffer zone of 1 foot beyond the visible traces. Any surface or object (e.g., concrete sidewalk or automobile) within the visible traces area or on which visible traces of the spilled material are observed is included in the spill area. This area represents the minimum area assumed to be contaminated by PCBs in the absence of precleanup sampling data and is thus the minimum area which must be cleaned.

Spill boundaries (Section 761.123) means the actual area of contamination as determined by postcleanup verification sampling or by precleanup sampling to determine actual spill boundaries. EPA can require additional cleanup when necessary to decontaminate all areas within the spill boundaries to the levels required in this policy (e.g., additional cleanup will be required if postcleanup sampling indicates that the area decontaminated by the responsible party, such as the spill area as defined in this section, did not encompass the actual boundaries of PCB contamination).

Spill event (Section 112.2 [8/25/93])* means a discharge of oil into or upon the navigable waters of the United States or adjoining shorelines in harmful quantities, as defined at 40 CFR Part 110.

Spinach (Section 407.71) shall mean the processing of spinach and leafy greens into the following product styles: Canned or frozen, whole leaf, chopped, and other related cuts.

Spinning reserve (Section 60.41a) means the sum of the unutilized net generating capability of all units of the electric utility company that are synchronized to the power distribution system and that are capable of immediately accepting additional load. The electric generating capability of equipment under multiple ownership is prorated based on ownership unless the proportional entitlement to electric output is otherwise established by contractual arrangement.

Spinning solution (Section 60.601) means the mixture of polymer, prepolymer, or copolymer and additives dissolved in solvent. The solution is prepared at a viscosity and solvent-to-polymer ratio that is suitable for extrusion into fibers.

Spinning solution preparation system (Section 60.601) means the equipment used to prepare spinning solutions; the system includes equipment for mixing, filtering, blending, and storage of the spinning solutions.

Spitback emissions (Section 86.096-2 [11/1/93])* means evaporative emissions resulting from the loss of liquid fuel that is emitted from a vehicle during a fueling operation.

Sponsor (Section 160.3 [8-17-89])* means: (1) A person who initiates and supports, by provision of financial or other resources, a study; (2) A person who submits a study to the EPA in support of an application for a research or marketing permit; or (3) A testing facility, if it both initiates and actually conducts the study.

Sponsor (Section 790.3) means the person or persons who design, direct and finance the testing of a substance or mixture.

Sponsor (Section 792.3 [8-17-89])* means: (1) A person who initiates and supports, by provision of financial or other resources, a study; (2) A person who submits a study to the EPA in response to a TSCA section 4(a) test rule and/or a person who submits a study under a TSCA section 4 testing consent agreement or a TSCA section 5 rule or order to the extent the agreement, rule or order references this part; or (3) A testing facility, if it both initiates and actually conducts the study.

Spot allowance (Section 73.3 [12-17-91]) means an allowance that may be used for purposes of compliance with a unit's sulfur dioxide emissions limitations requirements beginning in the year in which the allowance is offered for sale.

Spot Auction (Section 73.3 [12-17-91]) means an auction of a spot allowance.

Spot Sale (Section 73.3 [12-17-91]) means a sale of a spot allowance.

Spray application (Section 60.311) means a method of applying coatings by atomizing and directing the atomized spray toward the part to be coated.

Spray application (Section 60.391) means a method of applying coatings by atomizing the coating material and directing the atomized material toward the part to be coated. Spray applications can be used for prime coat, guide coat, and topcoat operations.

Spray application (Section 721.3 [7/27/88])* means any method of projecting a jet of vapor of finely divided liquid onto a surface to be coated; whether by compressed air, hydraulic pressure, electrostatic forces, or other methods of generating a spray.

Spray booth (Section 60.391) means a structure housing automatic or manual spray application equipment where prime coat, guide coat, or topcoat is applied to components of automobile or light-duty truck bodies.

Spray booth (Section 60.451) means the structure housing automatic or manual spray application equipment where a coating is applied to large appliance parts or products.

Spray booth (Section 60.721 [1/29/88]) means the structure housing automatic or manual spray application equipment where a coating is applied to plastic parts for business machines.

Spray-in-place foam (Section 248.4 [2/17/89]) is rigid cellular polyurethane or polyisocyanurate foam produced by catalyzed chemical reactions that hardens at the site of the work. The term includes spray-applied and injected applications.

Spray-in-place insulation (Section 248.4 [2/17/89]) means insulation material that is sprayed onto a surface or into cavities and includes cellulose fiber spray-on as well as plastic rigid foam products.

Spreader stoker steam generating unit (Section 60.41b [12/16/87])* means a steam generating unit in which solid fuel is introduced to the combustion zone by a mechanism that throws the fuel onto a grate from above. Combustion takes place both in suspension and on the grate.

sq m (sq ft) (Sections 413.11; 413.71) shall mean the area plated expressed in square meters (square feet).

Squash (Section 407.71) shall include the processing of pumpkin and squash into canned and frozen styles.

SRF (Section 35.3105 [3-19-90]) State water pollution control revolving fund.

SS (Section 133.101). The pollutant parameter total suspended solids.

SSA (Section 149.2 [2/14/89])*. See **Sole Source Aquifer (Section 149.2)**.

SSC (Section 300.5) is the Scientific Support Coordinator.

Stack (Section 129.2) means any chimney, flue, conduit, or duct arranged to conduct, emissions to the ambient air.

Stack (Section 51.100) means any point in a source designed to emit solids, liquids, or gases into the air, including a pipe or duct but not including flares.

Stack emission (Section 60.671) means the particulate matter that is released to the atmosphere from a capture system.

Stack emissions (Section 60.381) means the particulate matter captured and released to the atmosphere through a stack, chimney, or flue.

Stack in existence (Section 51.100) means that the owner or operator had (1) begun, or caused to begin, a continuous program of physical on-site construction of the stack or (2) entered into binding agreements or contractual obligations, which could not be cancelled or modified without substantial loss to the owner or operator, to undertake a program of construction of the stack to be completed within a reasonable time.

Stall barn (Section 412.11) shall mean specialized facilities wherein producing cows and replacement cows are milked and fed in a fixed location.

Standard (Section 171.2) means the measure of knowledge and ability which must be demonstrated as a requirement for certification.

Standard (Section 403.3). See **National Pretreatment Standard (Section 403.3)**.

Standard (Section 60.2) means a standard of performance proposed or promulgated under this part.

Standard (Section 61.02) means a national emission standard including a design, equipment, work practice or operational standard for a hazardous air pollutant proposed or promulgated under this part.

Standard bushel (Section 406.11) shall mean a bushel of shelled corn weighing 56 pounds.

Standard bushel (Section 406.41) shall mean a bushel of wheat weighing 60 pounds.

Standard conditions (Section 60.2) means a temperature of 293 K (68 degrees F) and a pressure of 101.3 kilopascals (29.92 in Hg).

Standard conditions (Section 60.51a [2-11-91]) means a temperature of 293 deg. Kelvin (68 deg. Fahrenheit) and a pressure of 101.3 kilopascals (29. 92 inches of mercury).

Standard day conditions (Section 87.1) means standard ambient conditions as described in the United States Standard Atmosphere, 1976, (i.e., Temperature = 15 degrees C, specific humidity = 0.00 kg/H_2O/kg dry air, and pressure = 101325 Pa).

Standard equipment (Section 86.082-2) means those features or equipment which are marketed on a vehicle over which the purchaser can exercise no choice.

Standard ferromanganese (Section 60.261) means that alloy as defined by ASTM Designation A99-76 (incorporated by reference - see Section 60.17).

Standard of performance (Section 401.11) means any restriction established by the Administrator pursuant to section 306 of the Act on quantities, rates, and concentrations of chemical, physical, biological, and other constituents which are or may be discharged from new sources into navigable waters, the waters of the contiguous zone or the ocean.

Standard operating procedure (Section 61.61) means a formal written procedure officially adopted by the plant owner or operator and available on a routine basis to those persons responsible for carrying out the procedure.

Standard operating procedure (Section 30.200) means a document which describes in detail an operation, analysis, or action which is commonly accepted as the preferred method for performing certain routine or repetitive tasks.

Standard or limitation (Section 2.302) means any prohibition, any effluent limitation, or any toxic, pre-treatment or new source performance standard established or publicly proposed pursuant to the Act or pursuant to regulations under the Act, including limitations or prohibitions in a permit issued or proposed by EPA or by a State under section 402 of the Act, 33 U.S.C. 1342.

Standard or limitation (Section 2.301) means any emission standard or limitation established or publicly proposed pursuant to the Act or pursuant to any regulation under the Act.

Standard pressure (Section 61.61) means a pressure of 760 mm of Hg (29.92 in. of Hg).

Standard sample (Section 141.2) means the aliquot of finished drinking water that is examined for the presence of coliform bacteria.

Standard temperature (Section 61.61) means a temperature of 20 degrees C (69 degrees F).

Standard wipe test (Section 761.123) means, for spills of high-concentration PCBs on solid surfaces, a cleanup to numerical surface standards and sampling by a standard wipe test to verify that the numerical standards have been met. This definition constitutes the minimum requirements for an appropriate wipe testing protocol. A standard-size template (10 centimeters (cm) x 10 cm) will be used to delineate the area of cleanup; the wiping medium will be a gauze pad or glass wool of known size which has been saturated with hexane. It is important that the wipe be performed very quickly after the hexane is exposed to air. EPA strongly recommends that the gauze (or glass wool) be prepared with hexane in the laboratory and that the wiping medium be stored in sealed glass vials until it is used for the wipe test. Further, EPA requires the collection and testing of field blanks and replicates.

Standards for sewage sludge use or disposal (Section 501.2 [5/2/89]) means the regulations promulgated at 40 CFR Part 503 pursuant to section 405(d) of the CWA which govern minimum requirements for sludge quality, management practices, and monitoring and reporting applicable to the generation or treatment of sewage sludge from a treatment works treating domestic sewage or use or disposal of that sewage sludge by any person.

Standards for sewage sludge use or disposal (Section 122.2 [5/2/89]) means the regulations promulgated pursuant to section 405(d) of the CWA which govern minimum requirements for sludge quality, management practices, and monitoring and reporting applicable to sewage sludge or the use or disposal of sewage sludge by any person.

Stark-Einstein law (Section 796.3700), the second law of photochemistry, states that only one molecule is activated to an excited state per photon or quantum of light absorbed.

Starting material (Section 158.153 [5/4/88]) means a substance used to synthesize or purify a technical grade of active ingredient (or the practical equivalent of the technical grade ingredient if the technical grade cannot be isolated) by chemical reaction.

Startup (Section 264.1031 [6-21-90]) means the setting in operation of a hazardous waste management unit or control device for any purpose.

Startup (Section 52.01) means the setting in operation of a source for any purpose.

Startup (Section 60.2) means the setting in operation of an affected facility for any purpose.

Startup (Section 61.02) means the setting in operation of a stationary source for any purpose.

State (Section 122.1 [12/22/93])* means any of the 50 States, the District of Columbia, Guam, the Commonwealth of Puerto Rico, the Virgin Islands, American Samoa, the Commonwealth of the Northern Mariana Islands, the Trust Territory of the Pacific Islands, or an Indian Tribe as defined in these regulations which meets the requirements of Sec. 123.31 of this chapter.

State (Section 122.2 [1/4/89])* means any of the 50 States, the District of Columbia, Guam, the Commonwealth of Puerto Rico, the Virgin Islands, American Samoa, the Commonwealth of the Northern Mariana Islands, and the Trust Territory of the Pacific Islands.

State (Section 124.2 [9/26/88])* means one of the States of the United States, the District of Columbia, the Commonwealth of Puerto Rico, the Virgin Islands, Guam, American Samoa, the Trust Territory of the Pacific Islands (except in the case of RCRA), the Commonwealth of the Northern Mariana Islands, or an Indian Tribe treated as a State (except in the case or RCRA).

State (Section 124.41) means a State, the District of Columbia, the Commonwealth of Puerto Rico, the Virgin Islands, Guam, and American Samoa and includes the Commonwealth of the Northern Mariana Islands.

State (Section 141.2 [9/26/88])* means the agency of the State or Tribal government which has jurisdiction over public water systems. During any period when a State or Tribal government does not have primary enforcement responsibility pursuant to Section 1413 of the Act, the term State means the Regional Administrator, U.S. Environmental Protection Agency.

State (Section 142.2 [9/26/88])* means one of the States of the United States, the District of Columbia, the Commonwealth of Puerto Rico, the Virgin Islands, Guam, American Samoa, the Commonwealth of the Northern Mariana Islands, the Trust Territory of the Pacific Islands, or an Indian Tribe treated as a State.

State (Section 143.2 [9/26/88])* means the agency of the State or Tribal government which has jurisdiction over public water systems. During any period when a State does not have responsibility pursuant to section 1443 of the Act, the term State means the Regional Administrator, U.S. Environmental Protection Agency.

State (Section 144.3 [9/26/88])* means any of the 50 States, the District of Columbia, Guam, the Commonwealth of Puerto Rico, the Virgin Islands, American Samoa, the Trust Territory of the Pacific Islands, the Commonwealth of the Northern Mariana Islands, or an Indian Tribe treated as a State.

State (Section 15.4) means a State, the District of Columbia, the Commonwealth of Puerto Rico, the Virgin Islands, Guam, American Samoa, the Commonwealth of the Northern Mariana Islands, or the Trust Territories of the Pacific Islands.

State (Section 171.2) means a State, the District of Columbia, the Commonwealth of Puerto Rico, the Virgin Islands, Guam, the Trust

Territory of the Pacific Islands, and American Samoa.

State (Section 20.2) means the States, the District of Columbia, the Commonwealth of Puerto Rico, the Canal Zone, Guam, American Samoa, the Virgin Islands, and the Trust Territory of the Pacific Islands.

State (Section 205.2) includes the District of Columbia, the Commonwealth of Puerto Rico, the Virgin Islands, American Samoa, Guam, and the Trust Territory of the Pacific Islands.

State (Section 21.2) means a State, the District of Columbia, the Commonwealth of Puerto Rico, the Virgin Islands, Guam, American Samoa, and the Trust Territory of the Pacific Islands.

State (Section 231.2) means any state agency administering a 404 program which has been approved under section 404(h).

State (Section 232.2 [2/11/93])* means any of the 50 States, the District of Columbia, Guam, the Commonwealth of Puerto Rico, the Virgin Islands, American Samoa, the Commonwealth of the Northern Mariana Islands, the Trust Territory of the Pacific Islands, or an Indian Tribe, as defined in this part, which meet the requirements of Sec. 233.60. For purposes of this part, the word State also includes any interstate agency requesting program approval or administering an approved program.

State (Section 233.2 [6/6/88]) means any of the 50 States, the District of Columbia, Guam, the Commonwealth of Puerto Rico, the Virgin Islands, American Samoa, the Commonwealth of the Northern Mariana Islands, and the Trust territory of the Pacific Islands. For purposes of this regulation, the word State also includes any interstate agency requesting program approval or administering an approved program.

State (Section 233.3) means any of the 50 States, the District of Columbia, Guam, the Commonwealth of Puerto Rico, the Virgin Islands, American Samoa, and the Trust Territory of the Pacific Islands.

State (Section 250.4 [10/6/87, 6/22/88]) means any of the several States, the District of Columbia, the Commonwealth of Puerto Rico, the Virgin Islands, Guam, American Samoa, and the Commonwealth of the Northern Mariana Islands.

State (Section 257.2) means any of the several States, the District of Columbia, the Commonwealth of Puerto Rico, the Virgin Islands, Guam, American Samoa, and the Commonwealth of the Northern Mariana Islands.

State (Section 258.2 [10-9-91]) means any of the several States, the District of Columbia, the Commonwealth of Puerto Rico, the Virgin Islands, Guam, American Samoa, and the Commonwealth of the Northern Mariana Islands.

State (Section 260.10) means any of the several States, the District of Columbia, the Commonwealth of Puerto Rico, the Virgin Islands, Guam, American Samoa, and the Commonwealth of the Northern Mariana Islands.

State (Section 270.2) means any of the 50 States, the District of Columbia, Guam, the Commonwealth of Puerto Rico, the Virgin Islands, American Samoa, and the Commonwealth of the Northern Mariana Islands.

State (Section 32.605 [5-25-90]) means any of the States of the United States, the District of Columbia, the Commonwealth of Puerto Rico, any territory or possession of the United States, or any agency of a State, exclusive of institutions of higher education, hospitals, and units of local government. A State instrumentality will be considered part of the State government if it has a written determination from a State government that such State considers the instrumentality to be an agency of the State government.

State (Section 34.105 [2-26-90]) means a State of the United States, the District of Columbia, the Commonwealth of Puerto Rico, a territory or possession of the United States, an agency or instrumentality of a State, and a multi-State, regional, or interstate entity having governmental duties and powers.

State (Section 35.105 [3/23/94])* means within the context of Public Water Systems Supervision and Underground Water Source Protection grants or of financial assistance programs under the Clean Water Act, one of the States of the United States, the District of Columbia, the Commonwealth of Puerto Rico, the Virgin Islands, Guam, American Samoa, the Commonwealth of the Northern Mariana Islands, the Trust Territories of the Pacific Islands or an eligible Indian Tribe.

State (Section 35.2005) means a State, the District of Columbia, the Commonwealth of Puerto Rico, the Virgin Islands, Guam, American Samoa, the Trust Territory of the Pacific Islands, and the Commonwealth of the Northern Marianas. For the purposes of applying for a grant under section 201(g)(1) of the act, a State (including its agencies) is subject to the limitations on revenue producing entities and special districts contained in Section 35.2005(b)(27)(ii).

State (Section 35.6015 [6-5-90])* The several States of the United States, the District of Columbia, the Commonwealth of Puerto Rico, Guam, American Samoa, the Virgin Islands, the Commonwealth of Northern

Marianas, and any territory or possession over which the United States has jurisdiction.

State (Section 35.905) means a State, the District of Columbia, the Commonwealth of Puerto Rico, the Virgin Islands, Guam, American Samoa, the Trust Territory of the Pacific Islands, and the Commonwealth of the Northern Marianas.

State (Section 370.2 [10/15/87]) means any State of the United States, the District of Columbia, the Commonwealth of Puerto Rico, Guam, American Samoa, the United States Virgin Islands, the Northern Mariana Islands, and any other territory or possession over which the United States has jurisdiction.

State (Section 372.3 [7-26-90]) means any State of the United States, the District of Columbia, the Commonwealth of Puerto Rico, Guam, American Samoa, the United States Virgin Islands, the Commonwealth of the Northern Mariana Islands, and any other territory or possession over which the United States has jurisdiction and Indian Country.

State (Section 4.2) means any of the several States of the United States, the District of Columbia, the Commonwealth of Puerto Rico, any territory or possession of the United States, the Trust Territories of the Pacific Islands, or a political subdivision of any of these jurisdictions.

State (Section 501.2 [5/2/89]) means a State, the District of Columbia, the Commonwealth of Puerto Rico, the Virgin Islands, Guam, American Samoa, the Trust Territory of the Pacific Islands, and the Commonwealth of the Northern Mariana Islands, and an Indian Tribe eligible for treatment as a State pursuant to regulations promulgated under the authority of section 518(e) of the CWA.

State (Section 60.2 [7/21/92])* means all non-Federal authorities, including local agencies, interstate associations, and State-wide programs, that have delegated authority to implement: (1) The provisions of this part; and/or (2) the permit program established under part 70 of this chapter. The term State shall have its conventional meaning where clear from the context.

State (Section 61.02 [3/16/94])* means all non-Federal authorities, including local agencies, interstate associations, and State-wide programs, that have delegated authority to implement: (1) The provisions of this part; and/or (2) The permit program established under part 70 of this chapter. The term State shall have its conventional meaning where clear from the context.

State (Section 710.2) means any State of the United States, the District of Columbia, the Commonwealth of Puerto Rico, the Virgin Islands, Guam, the Canal Zone, American Samoa, the

Northern Mariana Islands, or any other territory or possession of the United States.

State (Section 720.3) means any State of the United States and the District of Columbia, the Commonwealth of Puerto Rico, the Virgin Islands, Guam, the Canal Zone, American Samoa, the Northern Mariana Islands, and any other territory or possession of the United States.

State (Section 762.3) has the same meaning as in 15 U.S.C. 2602.

State (Section 763.163 [7-12-89]) has the same meaning as in section 3 of the Act.

State (Section 763.83 [10/30/87]) means a State, the District of Columbia, the Commonwealth of Puerto Rico, Guam, American Samoa, the Northern Marianas, the Trust Territory of the Pacific Islands, and the Virgin Islands.

State 404 program (Sections 124.2; 233.3). See **Section 404 program (Sections 124.2; 233.3)**.

State 404 program or State program (Section 233.2 [6/6/88]) means a State program which has been approved by EPA under Section 404 of the Act to regulate the discharge of dredged or fill material into certain waters as defined in Section 232.2(p).

State 404 program or State program (Section 232.2 [2/11/93])* means a State program which has been approved by EPA under section 404 of the Act to regulate the discharge of dredged or fill material into certain waters as defined in Sec. 232.2(p).

State agency (Section 35.2005) means the State agency designated by the Governor having responsibility for administration of the construction grants program under section 205(g) of the Act.

State agency (Section 35.905) means the State water pollution control agency designated by the Governor having responsibility for enforcing State laws relating to the abatement of pollution.

State agency (Section 4.2) means any department, agency or instrumentality of a State or of a political subdivision of a State, or two or more States, or of two or more political subdivisions of a State or States.

State agency (Sections 51.100; 58.1) means the air pollution control agency primarily responsible for development and implementation of a plan under the Act.

State certifying authority (Section 20.2) means: (1) For water pollution control facilities, the State pollution control agency as defined in section 502 of the Act; (2) For air pollution control facilities, the air pollution control agency designated pursuant to section 302(b)(1) of the Act; or (3) For both air and water pollution control facilities, any interstate agency authorized to act in place of the certifying agency of a State.

State delayed compliance order (Section 65.01) shall mean a delayed compliance order issued by a State or by a political subdivision of a State.

State Director (Director) (Section 232.2 [2/11/93])* means the chief administrative officer of any State or interstate agency operating an approved program, or the delegated representative of the Director. If responsibility is divided among two or more State or interstate agencies, Director means the chief administrative officer of the State or interstate agency authorized to perform the particular procedure or function to which reference is made.

State Director (Section 122.2) means the chief administrative officer of any State or interstate agency operating an approved program, or the delegated representative of the State Director. If responsibility is divided among two or more State or interstate agencies, State Director means the chief administrative officer of the State or interstate agency authorized to perform the particular procedure or function to which reference is made.

State Director (Section 124.2 [9/26/88])* means the chief administrative officer of any State, interstate, or Tribal agency operating an approved program, or the delegated representative of the State director. If the responsibility is divided among two or more States, interstate, or Tribal agencies, State Director means the chief administrative officer of the State, interstate, or Tribal agency authorized to perform the particular procedure or function to which reference is made.

State Director (Section 129.2) means the chief administrative officer of a State or interstate water pollution control agency operating an approved HPDES permit program. In the event responsibility for water pollution control and enforcement is divided among two or more state or interstate agencies, the term State Director means the administrative officer authorized to perform the particular procedure to which reference is made.

State Director (Section 133.101) means the chief administrative officer of any State or interstate agency operating an approved program, or the delegated representative of the State Director.

State Director (Section 144.3 [9/26/88])* means the chief administrative officer of any State, interstate, or Tribal agency operating an approved program, or the delegated representative of the State director. If the responsibility is divided among two or more States, interstate, or Tribal agencies, State Director means the chief administrative officer of the State, interstate, or Tribal agency authorized to perform the particular procedure or function to which reference is made.

State Director (Section 146.3 [9/26/88])* means the chief administrative officer of any State, interstate, or Tribal agency operating an approved program, or the delegated representative of the State Director. If the responsibility is divided among two or more State, interstate, or Tribal agencies, State Director means the chief administrative officer of the State, interstate, or Tribal agency authorized to perform the particular procedure or function to which reference is made.

State Director (Section 258.2 [10-9-91]) means the chief administrative officer of the State agency responsible for implementing the State municipal solid waste permit program or other system of prior approval.

State Director (Section 270.2) means the chief administrative officer of any State agency operating an approved program, or the delegated representative of the State Director. If responsibility is divided among two or more State agencies, State Director means the chief administrative officer of the State agency authorized to perform the particular procedure or function to which reference is made.

State Director or Director (Section 233.2 [6/6/88]) means the chief administrative officer of any State or interstate agency operating an approved program, or the delegated representative of the Director. If responsibility is divided among two or more State or interstate agencies, Director means the chief administrative officer of the State or interstate agency authorized to perform the particular procedure or function to which reference is made.

State/EPA Agreement (Section 233.3) means an agreement between the Regional Administrator and the State which coordinates EPA and State activities, responsibilities and programs including those under the CWA.

State/EPA Agreement (Sections 122.2; 144.3; 270.2) means an agreement between the Regional Administrator and the State which coordinates EPA and State activities, responsibilities and programs including those under the CWA programs.

State implementation plan (Section 65.01) shall mean the plan, including the most recent revision thereof, which has been approved or promulgated by the Administrator under section 110 of the Act, and which implements the requirements of section 110.

State lead agency (Section 162.152). See **State or State lead agency (Section 162.152)**.

State or Local Air Monitoring

Station(s) (Section 58.1). See **SLAMS (Section 58.1)**.

State or State lead agency (Section 162.152) as used in this subpart means the State agency designated by the State to be responsible for registering pesticides to meet special local needs under sec. 24(c) of the Act.

State primary drinking water regulation (Section 142.2) means a drinking water regulation of a State which is comparable to a national primary drinking water regulation.

State program (Section 233.2 [6/6/88]). See **State 404 program (Section 233.2)**.

State program revision (Section 142.2 [12-20-89]) means a change in an approved State primacy program.

State Program Director or Director (Section 501.2 [5/2/89]) means the chief executive officer of the State sewage sludge management agency.

State regulated waters (Section 233.3) means those waters of the United States in which the Corps of Engineers suspends the issuance of section 404 permits upon approval of a State's section 404 permit program by the Administrator under section 404(h). These waters shall be identified in the program description as required by Section 233.22(h)(1). The Secretary shall retain jurisdiction over the following waters (see CWA section 404(g)(1): (a) Waters which are subject to the ebb and flow of the tide; (b) Waters which are presently used, or are susceptible to use in their natural condition or by reasonable improvement as a means to transport interstate or foreign commerce shoreward to their ordinary high water mark; and (c) Wetlands adjacent to waters in paragraphs (a) and (b) of this definition.

State regulated waters (Section 232.2 [2/11/93])* means those waters of the United States in which the Corps suspends the issuance of section 404 permits upon approval of a State's section 404 permit program by the Administrator under section 404(h). The program cannot be transferred for those waters which are presently used, or are susceptible to use in their natural condition or by reasonable improvement as a means to transport interstate or foreign commerce shoreward to their ordinary high water mark, including all waters which are subject to the ebb and flow of the tide shoreward to the high tide line, including wetlands adjacent thereto. All other waters of the United States in a State with an approved program shall be under jurisdiction of the State program, and shall be identified in the program description as required by Part 233.

State sewage sludge management agency (Section 501.2 [5/2/89]) means the agency designated by the Governor as having the lead responsibility for managing or coordinating the approved State program under this Part.

Statement (Section 21.2) means a written approval by EPA, or if appropriate, a State, of the application.

Statement (Section 27.2 [2-13-92]) means any representation, certification, affirmation, document, record, or accounting or bookkeeping entry made--(a) With respect to a claim or to obtain the approval or payment of a claim (including relating to eligibility to make a claim); or (b) With respect to (including relating to eligibility for)--(1) A contract with, or a bid or proposal for a contract with; or (2) A grant, loan, or benefit from, the Authority, or any State, political subdivision of a State, or other party, if the United States Government provides any portion of the money or property under such contract or for such grant, loan, or benefit, or if the Government will reimburse such State, political subdivision, or party for any portion of the money or property under such contract or for such grant, loan, or benefit.

Statement of Work (SOW) (Section 35.6015 [6-5-90])* The portion of the Cooperative Agreement application and/or Superfund State Contract that describes the purpose and scope of activities and tasks to be carried out as a part of the proposed project.

States (Section 131.3 [12-12-91])* include: The 50 States, the District of Columbia, Guam, the Commonwealth of Puerto Rico, Virgin Islands, American Samoa, the Trust Territory of the Pacific Islands, the Commonwealth of the Northern Mariana Islands, and Indian Tribes that EPA determines qualify for treatment as States for purposes of water quality standards.

Static (Section 797.1400) means the test solution is not renewed during the period of the test.

Static loaded radius arc (Section 86.084-2) means a portion of a circle whose center is the center of a standard tire-rim combination of an automobile and whose radius is the distance from that center to the level surface on which the automobile is standing, measured with the automobile at curb weight, the wheel parallel to the vehicle's longitudinal centerline, and the tire inflated to the manufacturer's recommended pressure. This definition applies beginning with the 1984 model year.

Static sheen test (Section 435.11 [3/4/93])* shall refer to the standard test procedure that has been developed for this industrial subcategory for the purpose of demonstrating compliance with the requirement of no discharge of free oil. The methodology for performing the static sheen test is presented in Appendix 1 to 40 CFR 435, subpart A.

Static system (Section 797.1050)

means a test container in which the test solution is not renewed during the period of the test.

Static system (Section 797.1300) means a test system in which the test solution and test organisms are placed in the test chamber and kept there for the duration of the test without renewal of the test solution.

Static system (Section 797.1930) means a test chamber in which the test solution is not renewed during the period of the test.

Static test (Section 797.1350) is a toxicity test with aquatic organisms in which no flow of test solution occurs. Solutions may remain unchanged throughout the duration of the test.

Static-replacement test (Section 797.1160) means a test method in which the test solution is periodically replaced at specific intervals during the test.

Station Wagon (Section 600.002-85) means a passenger automobile with an extended roof line to increase cargo or passenger capacity, cargo compartment open to the passenger compartment, a tailgate, and one or more rear seats readily removed or folded to facilitate cargo carrying.

Stationary casting (Section 467.02) is the pouring of molten aluminum into molds and allowing the metal to air cool.

Stationary casting (Section 471.02) is the pouring of molten metal into molds and allowing the metal to cool.

Stationary compactor (Sections 243.101; 246.101) means a powered machine which is designed to compact solid waste or recyclable materials, and which remains stationary when in operation.

Stationary gas turbine (Section 60.331) means any simple cycle gas turbine, regenerative cycle gas turbine or any gas turbine portion of a combined cycle steam/electric generating system that is not self propelled. It may, however, be mounted on a vehicle for portability.

Stationary source (Section 51.301) means any building, structure, facility, or installation which emits or may emit any air pollutant.

Stationary source (Section 52.01) means any building, structure, facility, or installation which emits or may emit an air pollutant for which a national standard is in effect.

Stationary source (Section 60.2 [7/21/92])*" means any building, structure, facility, or installation which emits or may emit any air pollutant.

Stationary source (Section 61.02) means any building, structure, facility, or installation which emits or may emit any air pollutant which has been designated as hazardous by the

Administrator.

Stationary source (Section 65.01) shall mean any stationary building, facility, equipment, installation, or operation (or combination thereof) which is located on one or more contiguous or adjacent properties and which is owned or operated by the same person (or by persons under common control), and which emits an air pollutant for which a national ambient air quality standard promulgated under section 109 of the Act is in effect.

Stationary source (Sections 51.165; 51.166; 52.21; 52.24) means any building, structure, facility, or installation which emits or may emit any air pollutant subject to regulation under the Act.

Stationery (Section 250.4 [10/6/87, 6/22/88]) means writing paper suitable for pen and ink, pencil, or typing. Matching envelopes are included in this definition.

Statistical significance (Section 228.2) shall mean the statistical significance determined by using appropriate standard techniques of multivariate analysis with results interpreted at the 95 percent confidence level and based on data relating species which are present in sufficient numbers at control areas to permit a valid statistical comparison with the areas being tested.

Statistical Sound Level (Section 201.1) means the level in decibels that is exceeded in a stated percentage (x) of the duration of the measurement period. It is abbreviated as L_x.

Steady-state (Section 797.1520) is the time period during which the amounts of test substance being taken up and depurated by the test organisms are equal, i.e., equilibrium.

Steady-state (Section 797.1830) is the time period during which the amounts of test chemical being taken up and depurated by the test oysters are equal, i.e., equilibrium.

Steady-state bioconcentration factor (Section 797.1830) is the mean concentration of the test chemical in test organisms during steady-state divided by the mean concentration of the test chemical in the test solution during the same period.

Steady-state bioconcentration factor (Section 797.1520) is the mean concentration of the test substance in test organisms during steady-state divided by the mean concentration in the test solution during the same period.

Steady-state or apparent plateau (Section 797.1560) is a condition in which the amount of test material being taken up and depurated is equal at a given water concentration.

Steam generating unit (Section 60.41c [9-12-90]) means a device that combusts any fuel and produces steam or heats water or any other heat transfer medium. This term includes any duct burner that combusts fuel and is part of a combined cycle system. This term does not include process heaters as defined in this subpart.

Steam generating unit (Section 61.301 [3-7-90]) means any enclosed combustion device that uses fuel energy in the form of steam.

Steam generating unit (Section 60.41a) means any furnace, boiler, or other device used for combusting fuel for the purpose of producing steam (including fossil-fuel-fired steam generators associated with combined cycle gas turbines; nuclear steam generators are not included).

Steam generating unit (Section 60.41b [12/16/87])* means a device that combusts any fuel or byproduct/waste to produce steam or to heat water or any other heat transfer medium. This term includes any municipal-type solid waste incinerator with a heat recovery steam generating unit or any steam generating unit that combusts fuel and is part of a cogeneration system or a combined cycle system. This term does not include process heaters as they are defined in this subpart.

Steam generating unit operating day (Section 60.41b [12/16/87])* means a 24-hour period between 12:00 midnight and the following midnight during which any fuel is combusted at any time in the steam generating unit. It is not necessary for fuel to be combusted continuously for the entire 24-hour period.

Steam generating unit operating day (Section 60.41c [9-12-90]) means a 24-hour period between 12:00 midnight and the following midnight during which any fuel is combusted at any time in the steam generating unit. It is not necessary for fuel to be combusted continuously for the entire 24-hour period.

Steam sales agreement (Section 72.2 [7/30/93])* is a legally binding agreement between a QF, IPP, new IPP, or firm associated with such facility and an industrial or commercial establishment requiring steam that establishes the terms and conditions under which the facility will supply steam to the establishment. Total planned net output capacity means the planned generator output capacity, excluding that portion of the electrical power which is designed to be used at the power production facility, as specified under one or more qualifying power purchase commitments or contemporaneous documents as of November 15, 1990; "Total installed net output capacity" shall be the generator output capacity, excluding that portion of the electrical power actually used at the power production facility, as installed.

Steam sales agreement (Section 73.3

[12-17-91]) is a legally-binding document between a firm associated with a new independent power production facility (IPPF) or a new IPPF and an industrial or commercial establishment requiring steam that sets the terms and conditions under which a specific new IPPF will provide steam to the establishment.

Steam stripping operation (Section 264.1031 [6-21-90]) means a distillation operation in which vaporization of the volatile constituents of a liquid mixture takes place by the introduction of steam directly into the charge.

Steel basis material (Section 465.02) means cold rolled steel, hot rolled steel, and chrome, nickel and tin coated steel which are processed in coil coating.

Steel production cycle (Section 60.141a) means the operations conducted within the BOPF steelmaking facility that are required to produce each batch of steel, including the following operations: scrap charging, preheating (when used), hot metal charging, primary oxygen blowing, sampling (vessel turndown and turnup), additional oxygen blowing (when used), tapping, and deslagging. Hot metal transfer and skimming operations for the next steel production cycle are also included when the hot metal transfer station or skimming station is an affected facility.

Steel production cycle (Section 60.141) means the operations conducted within the BOPF steelmaking facility that are required to produce each batch of steel and includes the following operations: scrap charging, preheating (when used), hot metal charging, primary oxygen blowing, sampling (vessel turndown and turnup), additional oxygen blowing (when used), tapping, and deslagging. This definition applies to an affected facility constructed, modified, or reconstructed after January 20, 1983. For an affected facility constructed, modified, or reconstructed after June 11, 1973, but on or before January 20, 1983, steel production cycle means the operations conducted within the BOPF steelmaking facility that are required to produce each batch of steel and includes the following operations: scrap charging, preheating (when used), hot metal charging, primary oxygen blowing, sampling (vessel turndown and turnup), additional oxygen blowing (when used), and tapping.

Step 1 (Section 35.2005). Facilities planning.

Step 1 facilities planning (Section 6.501) means preparation of a plan for facilities as described in 40 CFR Part 35, Subpart E or I.

Step 2 (Section 35.2005). Preparation of design drawings and specifications.

Step 2 (Section 6.501) means a project to prepare design drawings and specifications as described in 40 CFR Part 35, Subpart E or I.

Step 2 + 3 (Section 35.2005). Design and building of a treatment works and building related services and supplies.

Step 2 + 3 (Section 6.501) means a project which combines preparation of design drawings and specifications as described in Section 6.501(b) and building as described in Section 6.501(c).

Step 3 (Section 35.2005). Building of a treatment works and related services and supplies.

Step 3 (Section 6.501) means a project to build a publicly owned treatment works as described in 40 CFR Part 35, Subpart E or I.

Step 7 (Section 35.2005 [6-29-90]) Design/building of treatment works wherein a grantee awards a single contract for designing and building certain treatment works.

Still (Section 60.621) means a device used to volatilize, separate, and recover petroleum solvent from contaminated solvent, together with the piping and ductwork used in the installation of this device.

Stipend (Section 45.115) means supplemental financial assistance, other than tuition and fees, paid directly to the trainee by the recipient organization.

Stipend (Section 46.120) means supplemental financial assistance other than tuition, fees, and book allowance, paid directly to the fellow.

Stock configuration (Section 205.165) means that no modifications have been made to the original equipment motorcycle that would affect the noise emissions of the vehicle when measured according to the acceleration test procedure.

Stock solution (Section 797.1520) is the concentrated solution of the test substance which is dissolved and introduced into the dilution water.

Stock solution (Section 797.1600) is the source of the test solution prepared by dissolving the test substance in dilution water or a carrier which is then added to dilution water at a specified, selected concentration by means of the test substance delivery system.

Stock-on-hand (Section 763.163 [7-12-89]) means the products which are in the possession, direction, or control of a person and are intended for distribution in commerce.

Stone feed (Section 60.341) means limestone feedstock and millscale or other iron oxide additives that become part of the product.

Storage (Section 191.02) means retention of spent nuclear fuel or radioactive wastes with the intent and capability to readily retrieve such fuel or waste for subsequent use, processing, or disposal.

Storage (Section 259.10 [3/24/89]) means the temporary holding of regulated medical wastes at a designated accumulation area before treatment, disposal, or transport to another location.

Storage (Section 373.4 [4-16-90]) means the holding of hazardous substances for a temporary period, at the end of which the hazardous substance is either used, neutralized, disposed of, or stored elsewhere.

Storage (Sections 243.101; 246.101) means the interim containment of solid waste after generation and prior to collection for ultimate recovery or disposal.

Storage (Sections 260.10; 270.2) means the holding of hazardous waste for a temporary period, at the end of which the hazardous waste is treated, disposed of, or stored elsewhere.

Storage and Retrieval of Aerometric Data (SAROAD) system (Section 58.1) is a computerized system which stores and reports information relating to ambient air quality.

Storage bin (Section 60.381) means a facility for storage (including surge bins and hoppers) of metallic minerals prior to further processing or loading.

Storage bin (Section 60.671) means a facility for storage (including surge bins) of nonmetallic minerals prior to further processing or loading.

Storage for disposal (Section 761.3) means temporary storage of PCBs that have been designated for disposal.

Storage vessel (Section 60.111) means any tank, reservoir, or container used for the storage of petroleum liquids, but does not include: (1) Pressure vessels which are designed to operate in excess of 15 pounds per square inch gauge without emissions to the atmosphere except under emergency conditions, (2) Subsurface caverns or porous rock reservoirs, or (3) Underground tanks if the total volume of petroleum liquids added to and taken from a tank annually does not exceed twice the volume of the tank.

Storage vessel (Section 60.111a) means each tank, reservoir, or container used for the storage of petroleum liquids, but does not include: (1) Pressure vessels which are designed to operate in excess of 204.9 kPa (15 psig) without emissions to the atmosphere except under emergency conditions. (2) Subsurface caverns or porous rock reservoirs, or (3) Underground tanks if the total volume of petroleum liquids added to and taken from a tank annually does not exceed twice the volume of the tank.

Storage vessel (Section 60.111b) means each tank, reservoir, or container used for the storage of volatile organic liquids but does not include: (1) Frames, housing, auxiliary supports, or other components that are not directly involved in the containment of liquids or vapors; or (2) Subsurface caverns or porous rock reservoirs.

Storage vessel (Section 60.691 [11/23/88]) means any tank, reservoir, or container used for the storage of petroleum liquids, including oily wastewater.

Storm sewer (Section 35.2005 [6-29-90])* A sewer designed to carry only storm waters, surface run-off, street wash waters, and drainage.

Storm sewer (Section 35.905) means a sewer intended to carry only storm waters, surface runoff, street wash waters, and drainage.

Storm water point source (Section 122.26) means a conveyance or system of conveyances (including pipes, conduits, ditches, and channels) primarily used for collecting and conveying storm water runoff and which: (i) Is located at an urbanized area as designated by the Bureau of the Census according to the criteria in 39 FR 15202 (May 1, 1974); or (ii) Discharges from lands or facilities used for industrial or commercial activities; or (iii) Is designated under paragraph (c) of this section. Conveyances that discharge storm water runoff combined with municipal sewage are point sources that must obtain NPDES permits, but are not storm water point sources.

Storm-water or wastewater collection system (Section 280.12 [9/23/88]) means piping, pumps, conduits, and any other equipment necessary to collect and transport the flow of surface water run-off resulting from precipitation, or domestic, commercial, or industrial wastewater to and from retention areas or any areas where treatment is designated to occur. The collection of storm water and wastewater does not include treatment except where incidental to conveyance.

Stormwater sewer system (Section 60.691 [11/23/88]) means a drain and collection system designed and operated for the sole purpose of collecting stormwater and which is segregated from the process wastewater collection system.

Straight kraft recovery furnace (Section 60.281) means a furnace used to recover chemicals consisting primarily of sodium and sulfur compounds by burning black liquor which on a quarterly basis contains 7 weight percent or less of the total pulp solids from the neutral sulfite semichemical process or has green liquor sulfidity of 28 percent or less.

Stratum (plural strata) (Sections 144.3; 146.3) means a single sedimentary bed or layer, regardless of thickness, that consists of generally the same kind of rock material.

Strawberries (Section 407.61) shall mean the processing of strawberries into the following product styles: Canned and frozen, whole, sliced, and pureed.

Streamflow source zone (Section 149.101) means the upstream headwaters area which drains into the recharge zone as defined in the December 16, 1975, Notice of Determination.

Street motorcycle (Section 205.151) means: (i) Any motorcycle that: (A) With an 80 kg (176 lb) driver, is capable of achieving a maximum speed of at least 40 km/h (25 mph) over a level paved surface; and (B) Is equipped with features customarily associated with practical street or highway use, such features including but not limited to any of the following: stoplight, horn, rear view mirror, turn signals; or (ii) Any motorcycle that: (A) Has an engine displacement less than 50 cubic centimeters; (B) Produces no more than two brake horse power; (C) With a 80 kg (176 lb) driver, cannot exceed 48 km/h (30 mph) over a level paved surface.

Street wastes (Section 243.101) means materials picked up by manual or mechanical sweepings of alleys, streets, and sidewalks; wastes from public waste receptacles; and material removed from catch basins.

Stressed waters (Section 125.58) means those receiving environments in which an applicant can demonstrate to the satisfaction of the Administrator, that the absence of a balanced, indigenous population is caused solely by human perturbations other than the applicant's modified discharge.

Strip (Section 61.141 [11-20-90])* means to take off RACM from any part of a facility or facility components.

Strip, sheet, and miscellaneous products (Section 420.121) means steel products other than wire products and fasteners.

Strip, sheet and plate (Section 420.91) means those acid pickling operations that pickle strip, sheet or plate products.

Stripper (Section 61.61) includes any vessel in which residual vinyl chloride is removed from polyvinyl chloride resin, except bulk resin, in the slurry form by the use of heat and/or vacuum. In the case of bulk resin, stripper includes any vessel which is used to remove residual vinyl chloride from polyvinyl chloride resin immediately following the polymerization step in the plant process flow.

Strong chelating agents (Section 413.02) is defined as all compounds which, by virtue of their chemical structure and amount present, form soluble metal complexes which are not removed by subsequent metals control techniques such as pH adjustment followed by clarification or filtration.

Structural member (Section 61.141) means any load-supporting member of a facility, such as beams and load supporting walls; or any nonload-supporting member, such as ceilings and nonload-supporting walls.

Structure (Section 51.165). See **Building, structure, facility, or installation (Section 51.165)**.

Structure (Section 51.166). See **Building, structure, facility, or installation (Section 51.166)**.

Structure (Section 51.301). See **Building, structure or facility (Section 51.301)**.

Structure (Section 52.21). See **Building, structure, facility, or installation (Section 52.21)**.

Structure (Section 52.24). See **Building, structure, facility, or installation (Section 52.24)**.

Study (Section 160.3 [8-17-89])* means any experiment at one or more test sites, in which a test substance is studied in a test system under laboratory conditions or in the environment to determine or help predict its effects, metabolism, product performance (efficacy studies only as required by 40 CFR 158.640), environmental and chemical fate, persistence and residue, or other characteristics in humans, other living organisms, or media. The term "study" does not include basic exploratory studies carried out to determine whether a test substance or a test method has any potential utility.

Study (Section 716.3). See **Health and safety study (Section 716.3)**.

Study (Section 792.3 [8-17-89])* means any experiment at one or more test sites, in which a test substance is studied in a test system under laboratory conditions or in the environment to determine or help predict its effects, metabolism, environmental and chemical fate, persistence, or other characteristics in humans, other living organisms, or media. The term "study" does not include basic exploratory studies carried out to determine whether a test substance or a test method has any potential utility.

Study (Sections 720.3; 723.50). See **Health and safety study (Sections 720.3; 723.50)**.

Study completion date (Section 160.3 [8-17-89]) means the date the final report is signed by the study director.

Study completion date (Section 792.3 [8-17-89]) means the date the final report is signed by the study director.

Study director (Section 160.3 [8-17-89])* means the individual responsible for the overall conduct of a study.

Study director (Section 792.3 [8-17-89])* means the individual responsible for the overall conduct of a study.

Study initiation date (Section 160.3 [8-17-89]) means the date the protocol is signed by the study director.

Study initiation date (Section 792.3 [8-17-89]) means the date the protocol is signed by the study director.

Subacute dietary LC_{50} (Section 162.3) means a concentration of a substance, expressed as parts per million in food that is lethal to 50 percent of the test population of animals under test conditions as specified in the Registration Guidelines.

Subacute toxicity (Section 162.3) means the property of a substance or mixture of substances to cause adverse effects in an organism upon repeated or continuous exposure within less than 1/2 the lifetime of that organism.

Subagreement (Section 32.102) means a written agreement between a recipient of EPA assistance and another party (other than a public agency) and any tier of agreement thereunder for the furnishing of supplies, services, or equipment necessary to complete the project for which the assistance was awarded. These agreements include contracts and subcontracts for personal and professional services, agreements with consultants, and purchase orders.

Subagreement (Section 35.936-1) means a written agreement between an EPA grantee and another party (other than another public agency) and any tier of agreement thereunder for the furnishing of services, supplies, or equipment necessary to complete the project for which a grant was awarded, including contracts and subcontracts for personal and professional services, agreements with consultants and purchase orders, but excluding employment agreements subject to State or local personnel systems. (See Sections 35.937-12 and 35.938-9 regarding subcontracts of any tier under prime contracts for architectural or engineering services or construction awarded by the grantee - generally applicable only to subcontracts in excess of $10,000.)

Subagreement (Sections 30.200; 33.005) means a written agreement between an EPA recipient and another party (other than another public agency) and any lower tier agreement for services, supplies, or construction necessary to complete the project. Subagreements include contracts and subcontracts for personal and professional services, agreements with consultants, and purchase orders.

Subbituminous coal (Section 60.41a) means coal that is classified as subbituminous A, B, or C according to the American Society of Testing and Materials (ASTM) Standard Specification for Classification of Coals by Rank D388-77 (incorporated by reference - see Section 60.17).

Subchronic dermal toxicity (Section 798.2250) is the adverse effects occurring as a result of the repeated daily exposure of experimental animals to a chemical by dermal application for part (approximately 10 percent) of a life span.

Subchronic inhalation toxicity (Section 798.2450) is the adverse effects occurring as a result of the repeated daily exposure of experimental animals to a chemical by inhalation for part (approximately 10 percent) of a life span.

Subchronic oral toxicity (Section 798.2650) is the adverse effects occurring as a result of the repeated daily exposure of experimental animals to a chemical by the oral route for a part (approximately 10 percent for rats) of a life span.

Subclass (Section 86.1102-87 [11-5-90])* means a classification of heavy-duty engines of heavy-duty vehicles based on such factors as gross vehicle weight rating, fuel usage (gasoline-, diesel-, and methanol-fueled), vehicle usage, engine horsepower or additional criteria that the Administrator shall apply. Subclasses include, but are not limited to: (i) Light-duty gasoline-fueled Otto cycle trucks (6,001-8,500 lb. GVW); (ii) Light-duty methanol-fueled Otto cycle trucks (6,001-8,500 lb. GVW); (iii) Light-duty petroleum-fueled diesel trucks (6,001-8,500 lb. GVW); (iv) Light-duty methanol-fueled diesel trucks (6,001-8,500 lb. GVW); (v) Light heavy-duty gasoline-fueled Otto cycle engines (for use in vehicles of 8,501-14,000 lb. GVW); (vi) Light heavy-duty methanol-fueled Otto cycle engines (for use in vehicles of 8,501-14,000 lb. GVW); (vii) Heavy heavy-duty gasoline-fueled Otto cycle engines (for use in vehicles of 14,001 lb and above GVW); (viii) Heavy heavy-duty methanol-fueled Otto cycle engines (for use in vehicles of 14,001 lb. and above GVW); (ix) Light heavy-duty petroleum-fueled diesel engines (see Sec. 86.085-2(a)(1)); Light heavy-duty methanol-fueled diesel engines (see Sec. 86.085-2(a)(1)); Medium heavy-duty petroleum-fueled diesel engines (see Sec. 86.085-2(a)(2)); (xii) Medium heavy-duty methanol-fueled dieselengines (see Sec. 86.085-2(a)(2)); (xiii) Heavy heavy-duty petroleum-fueled diesel engines (see Sec. 86.085-2(a)(3)); (xiv) Heavy heavy-duty methanol-fueled diesel engines (see Sec. 86.085-2(a)(3)); (xv) Petroleum-fueled Urban Bus engines (see Sec. 86.091-2); (xvi) Methanol-fueled Urban Bus engines (see Sec. 86.091-2). For NCP purposes, all optionally certified engines and/or vehicles (engines certified in accordance with Sec. 86.087-10(a)(3) and vehicles certified in accordance with Sec. 86.085-1(b)) shall be considered part of, and included in the FRAC calculation of, the subclass for which they are optionally certified.

Subconfiguration (Section 600.002-85) means a unique combination, within a vehicle configuration of equivalent test weight, road-load horsepower, and any other operational characteristics or parameters which the Administrator determines may significantly affect fuel economy within a vehicle configuration.

Subcontract (Section 8.2) means any agreement or arrangement between a contractor and any person (in which the parties do not stand in the relationship of any employer and an employee): (1) For the furnishing of supplies or services or for the use of real or personal property, including lease arrangements, which, in whole or in part, is necessary to the performance of any one or more contracts; or (2) Under which any portion of the contractor's obligation under any one or more contracts is performed, undertaken or assumed.

Subcontractor (Section 35.6015 [6-5-90])* Any first tier party that has a contract with the recipient's prime contractor.

Subcontractor (Section 8.2) means any person holding a subcontract and, for the purposes of Subpart B (General Enforcement; Compliance Review; and Complaint Procedure) of the rules, regulations, and relevant orders of the Secretary of Labor any person who had held a subcontract subject to the order.

Subindex (Part 58, Appendix G) means the calculated index value for a single pollutant as described in section 7.

Submission (Section 403.3 [7-24-90])* means: (1) A request by a POTW for approval of a Pretreatment Program to the EPA or a Director; (2) A request by a POTW to the EPA or a Director for authority to revise the discharge limits in categorical Pretreatment Standards to reflect POTW pollutant removals; or (3) A request to the EPA by an NPDES State for approval of its State pretreatment program.

Submitter (Section 350.1 [7/29/88]) means a person filing a required report or making a claim of trade secrecy to EPA under sections 303(d)(2) and (d)(3), 311, 312, and 313 of Title III.

Subsidence (Sections 146.3; 147.2902) means the lowering of the natural land surface in response to: Earth movements; lowering of fluid pressure; removal of underlying supporting material by mining or solution of solids, either artificially or from natural causes; compaction due to wetting (Hydrocompaction); oxidation of organic matter in soils; or added load on the land surface.

Substance (Section 704.3 [12/22/88]) means either a chemical substance or mixture unless otherwise indicated.

Substance (Section 716.3) means chemical substance as defined at section 3(2)(A) of TSCA, 15 U.S.C.

2602(2)(A).

Substance (Section 717.3) means a chemical substance or mixture unless otherwise indicated.

Substantial business relationship (Section 280.92 [10/26/88]) means the extent of a business relationship necessary under applicable state law to make a guarantee contract issued incident to that relationship valid and enforceable. A guarantee contract is issued incident to that relationship if it arises from and depends on existing economic transactions between the guarantor and the owner or operator.

Substantial business relationship (Section 264.141 [9/1/88]) means the extent of a business relationship necessary under applicable State law to make a guarantee contract issued incident to that relationship valid and enforceable. A substantial business relationship must arise from a pattern of recent or ongoing business transactions, in addition to the guarantee itself, such that a currently existing business relationship between the guarantor and the owner or operator is demonstrated to the satisfaction of the applicable EPA Regional Administrator.

Substantial evidence (Section 32.102) means such relevant evidence as a reasonable person might accept as sufficient to support a particular conclusion.

Substantiation (Section 350.1 [7/29/88]) means the written answers submitted to EPA by a submitter to the specific questions set forth in this regulation in support of a claim that chemical identity is a trade secret.

Substate (Section 256.06) refers to any public regional, local, county, municipal, or intermunicipal agency, or regional or local public (including interstate) solid or hazardous waste management authority, or other public agency below the State level.

Substitute (Section 29.12) means that a State may use a plan or other document that is has developed for its own purposes to meet Federal requirements.

Substrate (Section 60.741 [9-11-89]) means the surface to which a coating is applied.

Subunit (Sections 704.25; 723.250) means an atom or group of associated atoms chemically derived from corresponding reactants.

Sudden accidental occurrence (Section 265.141), as specified in Section 265.141(g), means an occurrence which is not continuous or repeated in nature.

Sudden accidental occurrence (Section 264.141), as specified in Section 264.141(g), means an occurrence which is not continuous or repeated in nature.

Suitable Substitute Decision (Section 203.1) means the Administrator's decision whether a product which the Administrator has determined to be a low-noise-emission product is a suitable substitute for a product or products presently being purchased by the Federal Government.

Sulfide (Section 410.01) shall mean total sulfide (dissolved and acid soluble) as measured by the procedures listed in 40 CFR Part 136.

Sulfide (Section 425.02 [3/21/88])* shall mean total sulfide as measured by the potassium ferricyanide titration method described in Appendix A or the modified Monier-Williams method described in Appendix B.

Sulfite cooking liquor (Sections 430.101; 430.211) shall be defined as bisulfite cooking liquor when the pH of the liquor is between 3.0 and 6.0 and as acid sulfite cooking liquor when the pH is less than 3.0.

Sulfur percentage (Section 80.2 [2/16/94])* is the percentage of sulfur as determined by ASTM standard test method D 2622-87, entitled "Standard Test Method for Sulfur in Petroleum Products by X-Ray Spectrometry". ASTM test method D 2622-87 is incorporated by reference. This incorporation by reference was approved by the Director of the Federal Register in accordance with 5 U.S.C. 552(a) and 1 CFR part 51. A copy may be obtained from the American Society for Testing and Materials, 1916 Race Street, Philadelphia, PA 19103. A copy may be inspected at the Air Docket Section (A-130), room M-1500, U.S. Environmental Protection Agency, Docket No. A-86-03, 401 M Street SW., Washington DC 20460 or at the Office of the Federal Register, 800 North Capitol Street, NW., suite 700, Washington, DC 20408.

Sulfur production rate (Section 60.641) means the rate of liquid sulfur accumulation from the sulfur recovery unit.

Sulfur recovery unit (Section 60.641) means a process device that recovers element sulfur from acid gas.

Sulfuric acid pickling (Section 420.91) means those operations in which steel products are immersed in sulfuric acid solutions to chemically remove oxides and scale, and those rinsing operations associated with such immersions.

Sulfuric acid plant (Section 51.100) means any facility producing sulfuric acid by the contact process by burning elemental sulfur, alkylation acid, hydrogen sulfide, or acid sludge, but does not include facilities where conversion to sulfuric acid is utilized primarily as a means of preventing emissions to the atmosphere of sulfur dioxide or other sulfur compounds.

Sulfuric acid plant (Sections 60.161; 60.171; 60.181) means any facility producing sulfuric acid by the contact process.

Sulfuric acid production unit (Section 60.81) means any facility producing sulfuric acid by the contact process by burning elemental sulfur, alkylation acid, hydrogen sulfide, organic sulfides and mercaptans, or acid sludge, but does not include facilities where conversion to sulfuric acid is utilized primarily as a means of preventing emissions to the atmosphere of sulfur dioxide or other sulfur compounds.

Sump (Section 260.10 [1-29-92])* means any pit or reservoir that meets the definition of tank and those troughs/trenches connected to it that serve to collect hazardous waste for transport to hazardous waste storage, treatment, or disposal facilities; except that as used in the landfill, surface impoundment, and waste pile rules, "sump" means any lined pit or reservoir that serves to collect liquids drained from a leachate collection and removal system or leak detection system for subsequent removal from the system.

Sunlight direct aqueous photolysis rate constant (k_{pE}) (Section 796.3700) is the first-order rate constant in the units of day^{-1} and is a measure of the rate of disappearance of a chemical dissolved in a water body in sunlight.

Superfund (Section 300.6). See **CERCLA (Section 300.6)**.

Superfund (Section 302.3). See **CERCLA (Section 302.3)**.

Superfund State Contract (SSC) (Section 35.6015 [6-5-90])* A joint, legally binding agreement between EPA and another party(s) to obtain the necessary assurances before an EPA-lead remedial action or any political subdivision-lead activities can begin at a site, and to ensure State or Indian Tribe involvement as required under CERCLA section 121(f).

Superphosphoric acid plant (Section 60.211) means any facility which concentrates wet-process phosphoric acid to 66 percent or greater P_2O_5 content by weight for eventual consumption as a fertilizer.

Supersaturated solution (Section 796.1840). See **Oversaturated solution (Section 796.1840)**.

Supplementary control system (SCS) (Section 57.103 [2-13-92])* means any technique for limiting the concentration of a pollutant in the ambient air by varying the emissions of that pollutant according to atmospheric conditions. For the purposes of this part, the term supplementary control system does not include any dispersion technique based solely on the use of a stack the height of which exceeds good engineering practice (as determined under regulations implementing section 123 of the Act).

Supplier of water (Sections 141.2;

142.2; 143.2) means any person who owns or operates a public water system.

Supplies (Section 33.005) means all property, including equipment, materials, printing, insurances, and leases of real property, but excluding land or a permanent interest in land.

Supplies (Section 35.6015 [6-5-90])* All tangible personal property other than equipment as defined in this subpart.

Support agency (Section 35.6015 [6-5-90])* The agency that furnishes necessary data to the lead agency, reviews response data and documents, and provides other assistance to the lead agency.

Support media (Section 797.2800) means the quartz sand or glass beads used to support the plant.

Support media (Section 797.2850) means the sand or glass beads used to support the plant.

Suppressed combustion (Section 420.41) means those basic oxygen furnace steelmaking wet air cleaning systems which are designed to limit or suppress the combustion of carbon monoxide in furnace gases by restricting the amount of excess air entering the air pollution control system.

Surface casing (Section 146.3) means the first string of well casing to be installed in the well.

Surface coating (Section 468.02) shall mean the process of coating a copper workpiece as well as the associated surface finishing and flattening.

Surface coating operation (Section 60.391) means any prime coat, guide coat, or topcoat operation on an automobile or light-duty truck surface coating line.

Surface coating operation (Section 60.451) means the system on a large appliance surface coating line used to apply and dry or cure an organic coating on the surface of large appliance parts or products. The surface coating operation may be a prime coat or a topcoat operation and includes the coating application station(s), flashoff area, and curing oven.

Surface coating operation (Section 60.311) means the system on a metal furniture surface coating line used to apply and dry or cure an organic coating on the surface of the metal furniture part or product. The surface coating operation may be a prime coat or a top coat operation and includes the coating application station(s), flash-off area, and curing oven.

Surface collecting agents (Section 300.82), for the purposes of this subpart, are those chemical agents that form a surface film to control the layer thickness of oil.

Surface impoundment (Section 280.12 [9/23/88]) is a natural topographic depression, man-made excavation, or diked area formed primarily of earthen materials (although it may be lined with man-made materials) that is not an injection well.

Surface impoundment (Section 61.341 [3-7-90]) means a waste management unit which is a natural topographic depression, man-made excavation, or diked area formed primarily of earthen materials (although it may be lined with man-made materials), which is designed to hold an accumulation of liquid wastes or waste containing free liquids, and which is not an injection well. Examples of surface impoundments are holding, storage, settling, and aeration pits, ponds, and lagoons.

Surface impoundment or impoundment (Section 257.2 [3/19/93])* means a facility or part of a facility that is a natural topographic depression, human-made excavation, or diked area formed primarily of earthern materials (although it may be lined with human-made materials), that is designed to hold an accumulation of liquid wastes or wastes containing free liquids and that is not an injection well. Examples of surface impoundments are holding storage, settling, and aeration pits, ponds, and lagoons.

Surface impoundment or impoundment (Section 260.10) means a facility or part of a facility which is a natural topographic depression, man-made excavation, or diked area formed primarily of earthen materials (although it may be lined with man-made materials), which is designed to hold an accumulation of liquid wastes or wastes containing free liquids, and which is not an injection well. Examples of surface impoundments are holding, storage, settling, and aeration pits, ponds, and lagoons.

Surface moisture (Section 60.381) means water that is not chemically bound to a metallic mineral or metallic mineral concentrate.

Surface treatment (Section 471.02) is a chemical or electrochemical treatment applied to the surface of a metal. Such treatments include pickling, etching, conversion coating, phosphating, and chromating. Surface treatment baths are usually followed by a water rinse. The rinse may consist of single or multiple stage rinsing. For the purposes of this part, a surface treatment operation is defined as a bath followed by a rinse, regardless of the number of stages. Each surface treatment bath, rinse combination is entitled to discharge allowance.

Surface water (Section 141.2 [6/29/89]) means all water which is open to the atmosphere and subject to surface runoff.

Surfacing ACM (Section 763.83 [10/30/87]) means surfacing material that is ACM.

Surfacing material (Section 763.83 [10/30/87]) means material in a school building that is sprayed-on, troweled-on, or otherwise applied to surfaces, such as acoustical plaster on ceilings and fireproofing materials on structural members, or other materials on surfaces for acoustical, fireproofing, or other purposes.

Surfactant (Sections 417.151; 417.161) shall mean those methylene blue active substances amenable to measurement by the method described in "Methods for Chemical Analysis of Water and Wastes," 1971, Environmental Protection Agency, Analytical Quality Control Laboratory, page 131.

Surge control tank (Section 264.1031 [6-21-90]) means a large-sized pipe or storage reservoir sufficient to contain the surging liquid discharge of the process tank to which it is connected.

Susceptibility (Section 171.2) means the degree to which an organism is affected by a pesticide at a particular level of exposure.

Suspension (Section 32.102) means an action taken by the Director under Section 32.300 to disqualify a person temporarily from receiving any EPA assistance or subagreement.

Swaging (Section 471.02) is a process in which a solid point is formed at the end of a tube, rod, or bar by the repeated blows of one or more pairs of opposing dies.

Sweet water (Section 417.41) shall mean the solution of 8-10 percent crude glycerine and 90-22 percent water that is a by-product of saponification or fat splitting.

Sweetening unit (Section 60.641) means a process device that separates the H_2S and CO_2 contents from the sour natural gas stream.

Switcher Locomotive (Section 201.1) means any locomotive designated as a switcher by the builder or reported to the ICC as a switcher by the operator-owning-railroad and including, but not limited to, all locomotives of the builder/model designations listed in Appendix A to this subpart.

Synthetic fiber (Section 60.601) means any fiber composed partially or entirely of materials made by chemical synthesis, or made partially or entirely from chemically-modified naturally-occurring materials.

Synthetic organic chemicals manufacturing industry (Section 60.481) means the industry that produces, as intermediates or final products, one or more of the chemicals listed in Section 60.489.

System (Section 86.082-2) includes

any motor vehicle engine modification which controls or causes the reduction of substances emitted from motor vehicles.

System (Section 86.402-78) includes any motor vehicle modification which controls or causes the reduction of substances emitted from motor vehicles.

System emergency reserves (Section 60.41a) means an amount of electric generating capacity equivalent to the rated capacity of the single largest electric generating unit in the electric utility company (including steam generating units, internal combustion engines, gas turbines, nuclear units, hydroelectric units, and all other electric generating equipment) which is interconnected with the affected facility that has the malfunctioning flue gas desulfurization system. The electric generating capability of equipment under multiple ownership is prorated based on ownership unless the proportional entitlement to electric output is otherwise established by contractual arrangement.

System load (Section 60.41a) means the entire electric demand of an electric utility company's service area interconnected with the affected facility that has the malfunctioning flue gas desulfurization system plus firm contractual sales to other electric utility companies. Sales to other electric utility companies (e.g., emergency power) not on a firm contractual basis may also be included in the system load when no available system capacity exists in the electric utility company to which the power is supplied for sale.

System of records (Section 1516.2) means a group of any records under the control of the Council from which information is retrieved by the name of the individual or by some identifying number, symbol, or other identifying particular assigned to the individual.

System of records (Section 16.2) shall have the meaning given to it by 5 U.S.C. 552a (a)(5).

System with a single service connection (Section 141.2 [6/29/89]) means a system which supplies drinking water to consumers via a single service line.

T

T (Section 433.11), as in Cyanide, T, shall mean total.

T in CT calculations (Section 141.2 [6/29/89]). See **Disinfectant contact time (Section 141.2)**.

Tabulating cards (Section 250.4 [10/6/87, 6/22/88]) means cards used in automatic tabulating machines.

Tabulating paper (Section 250.4 [10/6/87, 6/22/88]) means paper used in tabulating forms for use on automatic data processing equipment.

Tag (Section 211.203) means stiff paper, metal or other hard material that is tied or otherwise affixed to the packaging of a protector.

Tailings (Section 61.251). See **Uranium byproduct material (Section 61.251)**.

Tailings Closure Plan (Radon) (Section 192.31 [11/5/93])* means the Nuclear Regulatory Commission or Agreement State approved plan detailing activities to accomplish timely emplacement of a permanent radon barrier. A tailings closure plan shall include a schedule for key radon closure milestone activities such as wind blown tailings retrieval and placement on the pile, interim stabilization (including dewatering or the removal of freestanding liquids and recontouring), and emplacement of a permanent radon barrier constructed to achieve compliance with the 20 pCi/m2-s flux standard as expeditiously as practicable considering technological feasibility (including factors beyond the control of the licensee).

Taking (Section 257.3-2) means harassing, harming, pursuing, hunting, wounding, killing, trapping, capturing, or collecting or attempting to engage in such conduct.

Tampering (Section 205.151) means the removal or rendering inoperative by any person, other than for purposes of maintenance, repair, or replacement, of any device or element of design incorporated into any product in compliance with regulations under section 6, prior to its sale or delivery to the ultimate purchaser or while it is in use; or the use of a product after such device or element of design has been removed or rendered inoperative by any person.

Tampering (Sections 204.2; 204.51; 205.2; 205.51) means those acts prohibited by section 10(a)(2) of the Act.

Tangible net worth (Section 264.141), as specified in Section 264.141(f), means the tangible assets that remain after deducting liabilities; such assets would not include intangibles such as goodwill and rights to patents or royalties.

Tangible net worth (Section 265.141), as specified in Section 265.141(f), means the tangible assets that remain after deducting liabilities; such assets would not include intangibles such as goodwill and rights to patents or royalties.

Tangible net worth (Section 144.61) means the tangible assets that remain after deducting liabilities; such assets would not include intangibles such as goodwill and rights to patents or royalties.

Tangible net worth (Section 280.92 [10/26/88]) means the tangible assets that remain after deducting liabilities; such assets do not include intangibles such as goodwill and rights to patents or royalties. For purposes of this definition, assets means all existing and all probable future economic benefits obtained or controlled by a particular entity as a result of past transactions.

Tank (Section 260.10) means a stationary device, designed to contain an accumulation of hazardous waste which is constructed primarily of non-earthen materials (e.g., wood, concrete, steel, plastic) which provide structural support.

Tank (Section 280.12 [9/23/88]) is a stationary device designed to contain an accumulation of regulated substances and constructed of non-earthen materials (e.g., concrete, steel, plastic) that provide structural support.

Tank (Section 61.341 [3-7-90]) means a stationary waste management unit that is designed to contain an accumulation of waste and is constructed primarily of nonearthen materials (e.g., wood, concrete, steel, plastic) which provide structural support.

Tank fuel volume (Section 86.082-2) means the volume of fuel in the fuel tank(s), which is determined by taking the manufacturer's nominal fuel tank(s) capacity and multiplying by 0.40, the result being rounded using ASTM E 29-67 to the nearest tenth of a U.S. gallon.

Tank system (Section 260.10) means a hazardous waste storage or treatment tank and its associated ancillary equipment and containment system.

Tank system (Section 280.12 [9/23/88]). See **UST system (Section 280.12)**.

Tankage (Section 432.101) shall mean dried animal by-product residues used in feedstuffs.

Tap (Section 60.271) means the pouring of molten steel from an EAF.

Tap (Section 60.271a) means the pouring of molten steel from an EAF or AOD vessel.

Tapping (Section 60.261) means the removal of slag or product from the electric submerged arc furnace under normal operating conditions such as removal of metal under normal pressure and movement by gravity down the spout into the ladle.

Tapping period (Section 60.261) means the time duration from initiation of the process of opening the tap hole until plugging of the tap hole is complete.

Tapping period (Section 60.271) means the time period commencing at the moment an EAF begins to tilt to pour and ending either three minutes after an EAF returns to an upright position or six minutes after commencing to tilt, whichever is longer.

Tapping station (Section 60.261) means that general area where molten product or slag is removed from the electric submerged arc furnace.

Tar decanter (Section 61.131 [9-14-89]) means any vessel, tank, or container that functions to separate heavy tar and sludge from flushing liquor by means of gravity, heat, or chemical emulsion breakers. A tar decanter also may be known as a flushing-liquor decanter.

Tar storage tank (Section 61.131 [9-14-89]) means any vessel, tank, reservoir, or other type of container used to collect or store crude tar or tar-entrained naphthalene, except for tar products obtained by distillation, such as coal tar pitch, creosotes, or carbolic oil. This definition also includes any vessel, tank, reservoir, or container used to reduce the water content of the tar by means of heat, residence time, chemical emulsion breakers, or centrifugal separation. A tar storage tank also may be known as a tar-dewatering tank.

Tar-intercepting sump (Section 61.131 [9-14-89]) means any tank, pit, or enclosure that serves to receive or separate tars and aqueous condensate discharged from the primary cooler. A tar-intercepting sump also may be known as a primary-cooler decanter.

Task (Section 35.6015 [6-5-90])* An element of a Superfund response activity identified in the Statement of Work of a Superfund Cooperative Agreement or a Superfund State Contract.

Taxi/idle (in) (Section 87.1) means those aircraft operations involving taxi and idle between the time of landing roll-out and final shutdown of all propulsion engines.

Taxi/idle (out) (Section 87.1) means those aircraft operations involving taxi and idle between the time of initial starting of the propulsion engine(s) used for the taxi and turn on to duty runway.

TDS (Section 146.3). See **Total**

dissolved solids (Section 146.3).

TDS (Section 401.11) means total dissolved solids.

Technical grade of active ingredient (Section 158.153 [5/4/88]) means a material containing an active ingredient: (1) Which contains no inert ingredient, other than one used for purification of the active ingredient; and (2) Which is produced on a commercial or pilot-plant production scale (whether or not it is ever held for sale).

Technical Support Document (Section 66.3 [2-13-92])* means the "Noncompliance Penalties Technical Support Document" which accompanies these regulations. The Technical Support Document appears as Appendix A to these regulations.

Technically feasible (Section 157.21) when applied to child-resistant packaging, means that the technology exists to produce the child-resistant packaging for a particular pesticide.

Technically qualified individual (Section 720.3) means a person or persons (1) who, because of education, training, or experience, or a combination of these factors, is capable of understanding the health and environmental risks associated with the chemical substance which is used under his or her supervision, (2) who is responsible for enforcing appropriate methods of conducting scientific experimentation, analysis, or chemical research to minimize such risks, and (3) who is responsible for the safety assessments and clearances related to the procurement, storage, use, and disposal of the chemical substance as may be appropriate or required within the scope of conducting a research and development activity.

Technically qualified individual (Section 710.2) means a person: (1) Who because of his education, training, or experience, or a combination of these factors, is capable of appreciating the health and environmental risks associated with the chemical substance which is used under his supervision, (2) who is responsible for enforcing appropriated methods of conducting scientific experimentation, analysis, or chemical research in order to minimize such risks, and (3) who is responsible for the safety assessments and clearances related to the procurement, storage, use, and disposal of the chemical substance as may be appropriate or required within the scope of conducting the research and development activity. The responsibilities in paragraph (aa)(3) of this section may be delegated to another individual, or other individuals, as long as each meets the criteria in paragraph (aa)(1) of this section.

Tempering (Section 426.61) shall mean the process whereby glass is heated near the melting point and then rapidly cooled to increase its mechanical and thermal endurance.

Temporarily abandoned area (Section 61.20) means a mine area in which further work is not intended for at least six months. Areas which function as escapeways, formerly-used lunchrooms, shops, and transformer or pumping stations are not considered abandoned areas. Except for designated ventilation passageways designed to minimize the distance to vents. worked-out mine areas are considered temporarily abandoned areas for the purpose of this subpart if work is not intended in the area for at least six months.

Temporary enclosure (Section 60.711 [10/3/88]) means a total enclosure that is constructed for the sole purpose of measuring the fugitive emissions from an affected facility. A temporary enclosure must be constructed and ventilated (through stacks suitable for testing) so that it has minimal impact on the performance of the permanent capture system. A temporary enclosure will be assumed to achieve total capture of fugitive VOC emissions if it conforms to the requirements found in Section 60.713(b)(5)(i) and if all natural draft openings are at least four duct or hood equivalent diameters away from each exhaust duct or hood. Alternatively, the owner or operator may apply to the Administrator for approval of a temporary enclosure on a case-by-case basis.

Temporary enclosure (Section 60.741 [9-11-89]) means a total enclosure that is constructed for the sole purpose of measuring the fugitive VOC emissions from an affected facility.

Temporary opening (Section 60.541 [9/15/87]) means an opening into an enclosure that is equipped with a means of obstruction, such as a door, window, or port, that is normally closed.

Ten-year (10-year) Twenty-four-hour (24-hour) rainfall event (Section 423.11) means a rainfall event with a probable recurrence interval of once in ten years as defined by the National Weather Service in Technical Paper No. 40, "Rainfall Frequency Atlas of the United States," May 1961 or equivalent regional rainfall probability information developed therefrom.

Ten-year (10-year) Twenty-four-hour (24-hour) rainfall event (Section 418.11) shall mean the maximum 24-hour precipitation event with a probable recurrence interval of once in 10 years as defined by the National Weather Service in technical paper No. 40, "Rainfall Frequency Atlas of the United States," May 1961, and subsequent amendments in effect as of the effective date of this regulation.

Ten-year (10-year), Twenty-four-hour (24-hour) rainfall event and Twenty-five-year (25-year), Twenty-four-hour (24-hour) rainfall event (Sections 412.11; 412.21) shall mean a rainfall event with a probable recurrence interval of once in ten years or twenty-five years, respectively, as defined by the National Weather Service in Technical Paper Number 40, "Rainfall Frequency Atlas of the United States," May 1961, and subsequent amendments, or equivalent regional or state rainfall probability information developed therefrom.

Ten-year (10-year), Twenty-four-hour (24-hour) precipitation event (Section 440.132) is the maximum 24-hour precipitation event with a probable recurrence interval of once in 10 years as established by the U.S. Department of Commerce, National Oceanic and Atmospheric Administration, National Weather Service, or equivalent regional or rainfall probability information.

Ten-year (10-year) Twenty-four-hour (24-hour) rainfall event (Sections 422.41, 422.51) shall mean the maximum precipitation event with a probable recurrence interval of once in 10 years as defined by the National Weather Service in technical paper no. 40, "Rainfall Frequency Atlas of the United States," May 1961, and subsequent amendments or equivalent regional or State rainfall probability information developed therefrom.

Ten-year (10-year), Twenty-four-hour (24-hour) rainfall event (Section 411.31) shall mean a rainfall event with a probable recurrence interval of once in ten years as defined by the National Weather Service in Technical Paper No. 40, "Rainfall Frequency Atlas of the United States," May 1961, and subsequent amendments, or equivalent regional or state rainfall probability information developed therefrom.

Ten-year (10-year), Twenty-four-hour (24-hour) precipitation event (Sections 436.21; 436.31; 436.41; 436.181) shall mean the maximum 24-hour precipitation event with a probable reoccurrence interval of once in 10 years. This information is available in "Weather Bureau Technical Paper No. 40," May 1961 and "NOAA Atlas 2," 1973 for the 11 Western States, and may be obtained from the National Climatic Center of the Environmental Data Service, National Oceanic and Atmospheric Administration, U.S. Department of Commerce.

Ten-year twenty-four-hour rainfall event (Section 129.2) means the maximum precipitation event with a probable recurrence interval of once in 10 years as defined by the National Weather Service in Technical Paper No. 40, "Rainfall Frequency Atlas of the United States," May 1961, and subsequent amendments or equivalent regional or State rainfall probability information developed therefrom.

Tenant (Section 4.2) means a person who has the temporary use and occupancy of real property owned by another.

Teratogenic (Section 162.3) means the property of a substance or mixture of substances to produce or induce functional deviations or developmental anomalies, not heritable, in or on an animal embryo or fetus.

Termination (Section 280.92 [11-9-89]) under Sec. 280.97(b)(1) and Sec. 280.97(b)(2) means only those changes that could result in a gap in coverage as where the insured has not obtained substitute coverage or has obtained substitute coverage with a different retroactive date than the retroactive date of the original policy.

Terminology (Section 6.101). All terminology used in this part will be consistent with the terms as defined in 40 CFR Part 1508 (the CEQ Regulations). Any qualifications will be provided in the definitions set forth in each subpart of this regulation.

Terms and conditions of registration (Section 154.3) means the terms and conditions governing lawful sale, distribution, and use approved in conjunction with registration, including labeling, use classification, composition, and packaging.

Terne coating (Section 420.121) means coating steel products with terne metal by the hot dip process including the immersion of the steel product in a molten bath of lead and tin metals, and the related operations preceding and subsequent to the immersion phase.

Territorial sea (Section 230.3) means the belt of the sea measured from the baseline as determined in accordance with the Convention on the Territorial Sea and the Contiguous Zone and extending seaward a distance of three miles.

Territorial seas (Section 116.3) means the belt of the seas measured from the line of ordinary low water along that portion of the coast which is in direct contact with the open sea and the line marking the seaward limit of inland waters, and extending seaward a distance of 3 miles.

Territorial seas (Section 300.82), for the purposes of this subpart, means the belt of the seas measured from the line of ordinary low water along that portion of the coast which is in direct contact with the open sea and the line marking the seaward limit of inland waters, and extending seaward a distance of three miles.

Test analyzer (Section 53.1) means an analyzer subjected to testing as a candidate method in accordance with Subparts B, C, and/or D of this part, as applicable.

Test chamber (Section 797.1520) is the container in which the test organisms are maintained during the test period.

Test chamber (Section 797.1600) is defined as the individual containers in which test organisms are maintained during exposure to test solution.

Test Compressor (Section 204.51) means a compressor used to demonstrate compliance with the applicable noise emissions standard.

Test data (Section 610.11) means any information which is a quantitative measure of any aspect of the behavior of a retrofit device.

Test data (Section 723.175) means: (i) Data from a formal or informal study, test, experiment, recorded observation, monitoring, or measurement. (ii) Information concerning the objectives, experimental methods and materials, protocols, results, data analyses (including risk assessments), and conclusions from a study, test, experiment, recorded observation, monitoring, or measurement.

Test data (Sections 720.3; 723.50; 723.250) means data from a formal or informal test or experiment, including information concerning the objectives, experimental methods and materials, protocols, results, data analyses, recorded observations, monitoring data, measurements, and conclusions from a test or experiment.

Test Engine (Section 86.1002-84) means an engine in a test sample.

Test exhaust system (Section 205.165) means an exhaust system in Selective Enforcement Audit test sample.

Test Facility (Section 211.203) for this subpart, a laboratory that has been set up and calibrated to conduct ANSI Std S3.19-1974 tests on hearing protective devices. It must meet the applicable requirements of these regulations.

Test Hearing Protector (Section 211.203) means a hearing protector that has been selected for testing to verify the value to be put on the label, or which has been designated for testing to determine compliance of the protector with the labeled value.

Test marketing (Section 704.3 [12/22/88]) means the distribution in commerce of no more than a predetermined amount of a chemical substance, mixture, article containing that chemical substance or mixture, or a mixture containing that substance, by a manufacturer or processor, to no more than a defined number of potential customers to explore market capability in a competitive situation during a predetermined testing period prior to the broader distribution of that chemical substance, mixture, or article in commerce.

Test marketing (Section 712.3) means distributing in commerce a limited amount of a chemical substance or mixture, or article containing such substance or mixture, to a defined number of potential customers, during a predetermined testing period, to explore market capability prior to broader distribution in commerce.

Test marketing (Sections 710.2;

720.3) means the distribution in commerce of no more than a predetermined amount of a chemical substance, mixture, or article containing that chemical substance or mixture, by a manufacturer or processor, to no more than a defined number of potential customers to explore market capability in a competitive situation during a predetermined testing period prior to the broader distribution of that chemical substance, mixture or article in commerce.

Test period (Section 797.2050) is the combination of the exposure period and the post-exposure period; or, the entire duration of the test.

Test product (Section 211.102) means any product that must be tested according to regulations published under Part 211.

Test product (Sections 204.2; 205.2) means any product that is required to be tested pursuant to this part.

Test Request (Section 211.203) means a request submitted to the manufacturer by the Administrator that will specify the hearing protector category, and test sample size to be tested according to Section 211.212-1, and other information regarding the audit.

Test rule (Section 791.3) refers to a regulation ordering the development of data on health or environmental effects or chemical fate for a chemical substance or mixture pursuant to TSCA sec. 4(a).

Test sample (Section 204.51) means the collection of compressors from the same category or configuration which is randomly drawn from the batch sample and which will receive emissions tests.

Test sample (Section 205.51) means the collection of vehicles from the same category, configuration or subgroup thereof which is drawn from the batch sample and which will receive noise emissions tests.

Test sample (Section 86.1002-84) means the collection of vehicles or engines of the same configuration which have been drawn from the population of engines or vehicles of that configuration and which will receive exhaust emission testing.

Test sample (Section 86.602) means the collection of vehicles of the same configuration which have been drawn from the population of vehicles of that configuration and which will receive exhaust emission testing.

Test sample size (Section 204.51) means the number of compressors of the same configuration in a test sample.

Test sample size (Section 205.51) means the number of vehicles of the same category or configuration in a test sample.

Test Sample (Section 86.1102-87 [11-5-90])* means a group of heavy-duty engines or heavy-duty vehicles of the same configuration which have been selected for emission testing.

Test sampler (Section 53.1) means a sampler subjected to testing as part of a candidate method in accordance with Subpart C or D of this part.

Test solution (Section 797.1400) means the test substance and the dilution water in which the test substance is dissolved or suspended.

Test solution (Section 797.1520) is dilution water containing the dissolved test substance to which test organisms are exposed.

Test solution (Section 797.1600) is dilution water with a test substance dissolved or suspended in it.

Test solution (Section 797.2750) means the test chemical and the dilution water in which the test chemical is dissolved or suspended.

Test substance (Section 160.3 [8-17-89]) means a substance or mixture administered or added to a test system in a study, which substance or mixture: (1) Is the subject of an application for a research or marketing permit supported by the study, or is the contemplated subject of such an application; or (2) Is an ingredient, impurity, degradation product, metabolite, or radioactive isotope of a substance described by paragraph (1) of this definition, or some other substance related to a substance described by that paragraph, which is used in the study to assist in characterizing the toxicity, metabolism, or other characteristics of a substance described by that paragraph.

Test substance (Section 790.3) means the form of chemical substance or mixture that is specified for use in testing.

Test substance (Section 792.3 [8-17-89]) means a substance or mixture administered or added to a test system in a study, which substance or mixture is used to develop data to meet the requirements of a TSCA section 4(a) test rule and/or is developed under a TSCA section 4 testing consent agreement or section 5 rule or order to the extent the agreement, rule or order references this part.

Test substance (Section 795.232 [1-8-90]) refers to the unlabeled and both radiolabeled mixtures (^{14}C-n-hexane and ^{14}C-methylcyclopentane) of commercial hexane used in testing.

Test substance (Section 797.1600) is the specific form of a chemical substance or mixture that is used to develop data.

Test substance (Sections 797.2050; 797.2130; 797.2150; 797.2175) is the specific form of a chemical or mixture of chemicals that is used to develop the data.

Test substance or mixture (Section 792.3) means a substance or mixture administered or added to a test system in a study, which substance or mixture is used to develop data to meet the requirements of a TSCA section 4(a) test rule and/or is developed under a negotiated testing agreement or section 5 rule/order to the extent the agreement or rule/order references this part.

Test substance or mixture (Section 160.3) means a substance or mixture administered or added to a test system in a study, which substance or mixture: (1) Is the subject of an application for a research or marketing permit supported by the study, or is the contemplated subject of such an application; or (2) Is an ingredient, impurity, degradation product, metabolite, or radioactive isotope of a substance described by paragraph (o)(1) of this section, or some other substance related to a substance described by that paragraph, which is used in the study to assist in characterizing the toxicity, metabolism, or other characteristics of a substance described by that paragraph.

Test system (Section 160.3 [8-17-89])* means any animal, plant, microorganism, chemical or physical matrix, including but not limited to soil or water, or subparts thereof, to which the test, control, or reference substance is administered or added for study. "Test system" also includes appropriate groups or components of the system not treated with the test, control, or reference substance.

Test system (Section 792.226) includes, in addition to those systems listed in Section 792.3(p), chemical or physical matrices (e.g., soil or water), or subparts thereof, to which the test or control substance or mixture is administered or added for study.

Test system (Section 792.3 [8-17-89])* means any animal, plant, microorganism, chemical or physical matrix, including but not limited to, soil or water, or components thereof, to which the test, control, or reference substance is administered or added for study. "Test system" also includes appropriate groups or components of the system not treated with the test, control, or reference substance.

Test vehicle (Section 205.151) means a vehicle in a Selective Enforcement Audit test sample.

Test vehicle (Section 205.51) means a vehicle selected and used to demonstrate compliance with the applicable noise emission standards.

Test vehicle (Sections 86.602; 86.1002-84) means a vehicle in a test sample.

Test weight (Sections 86.082-2; 600.002-85) means the weight, within an inertia weight class, which is used in the dynamometer testing of a vehicle, and which is based on its loaded vehicle weight in accordance with the provisions of Part 86.

Test weight basis (Section 86.094-2 [6-5-91]) means the basis on which equivalent test weight is determined in accordance with Sec. 86.129-94 of subpart B of this part.

Testing agent (Section 610.11) means any person who develops test data on a retrofit device.

Testing exemption (Section 204.2) means an exemption from the prohibitions of section 10(a) (1), (2), (3), and (5) of the Act, which may be granted under section 10(b)(1) of the Act for the purpose of research, investigations, studies, demonstrations, or training, but not including national security where lease or sale of the exempted product is involved.

Testing exemption (Section 205.2) means an exemption from the prohibitions of section 10(a)(1), (2), (3), and (5) of the Act, which may be granted under section 10(b)(1) of the Act for the purpose of research, investigations, studies, demonstrations, or training, but not including national security.

Testing exemption (Section 211.102) means an exemption from the prohibitions of section 10(a)(1), (2), (3), and (5) of the Act, which may be granted under section 10(b)(1) of the Act for research, investigations, studies, demonstrations, or training, but not for national security.

Testing exemption (Section 85.1702) means an exemption which may be granted under section 203(b)(1) for the purpose of research investigations, studies, demonstrations or training, but not including national security.

Testing facility (Section 160.3 [8-17-89])* means a person who actually conducts a study, i.e., actually uses the test substance in a test system. "Testing facility" encompasses only those operational units that are being or have been used to conduct studies.

Testing facility (Section 792.3 [8-17-89])* means a person who actually conducts a study, i.e., actually uses the test substance in a test system. "Testing facility" encompasses only those operational units that are being or have been used to conduct studies.

Tetrachloroethylene (or priority pollutant No. 85) (Section 420.02) means the value obtained by the standard method Number 610 specified in 44 FR 69464, 69571 (December 3, 1979).

Textile fiberglass (Section 60.291) means fibrous glass in the form of continuous strands having uniform thickness.

Textiles (Section 763.163 [7-12-89]) means an asbestos-containing product such as: yarn; thread; wick; cord;

braided and twisted rope; braided and woven tubing; mat; roving; cloth; slit and woven tape; lap; felt; and other bonded or non-woven fabrics.

Texture coat (Section 60.721 [1/29/88, 6/15/89]) means the rough coat that is characterized by discrete, raised spots on the exterior surface of the part. This definition does not include conductive sensitizers or EMI/RFI shielding coatings.

Theoretical arsenic emissions factor (Section 61.161) means the amount of inorganic arsenic, expressed in grams per kilogram of glass produced, as determined based on a material balance.

Thermal Deflection Rate (Section 85.2122(a)(2)(iii)(G)) means the angular degrees of rotation per degree of temperature change of the thermostatic coil.

Thermal dryer (Section 60.251) means any facility in which the moisture content of bituminous coal is reduced by contact with a heated gas stream which is exhausted to the atmosphere.

Thermal dryer (Section 60.381) means a unit in which the surface moisture content of a metallic mineral or a metallic mineral concentrate is reduced by direct or indirect contact with a heated gas stream.

Thermal processing (Section 240.101) means processing of waste material by means of heat.

Thermal system insulation (Section 763.83 [10/30/87]) means material in a school building applied to pipes, fittings, boilers, breeching, tanks, ducts, or other interior structural components to prevent heat loss or gain, or water condensation, or for other purposes.

Thermal system insulation ACM (Section 763.83 [10/30/87]) means thermal system insulation that is ACM.

Thermal treatment (Section 260.10) means the treatment of hazardous waste in a device which uses elevated temperatures as the primary means to change the chemical, physical, or biological character or composition of the hazardous waste. Examples of thermal treatment processes are incineration, molten salt, pyrolysis, calcination, wet air oxidation, and microwave discharge. (See also incinerator and open burning.)

Thermostat (Section 85.2122(a)(2)(iii)(B)) means a temperature-actuated device.

Thermostatic Coil (Section 85.2122(a)(2)(iii)(D)) means a spiral-wound coil of thermally-sensitive material which provides rotary force (torque) and/or displacement as a function of applied temperature.

Thermostatic Switch (Section 85.2122(a)(2)(iii)(E)) means an element of thermally-sensitive material which acts to open or close an electrical circuit as a function of temperature.

Thin-film evaporation (Section 264.1031 [6-21-90]) operation means a distillation operation that employs a heating surface consisting of a large diameter tube that may be either straight or tapered, horizontal or vertical. Liquid is spread on the tube wall by a rotating assembly of blades that maintain a close clearance from the wall or actually ride on the film of liquid on the wall.

Thirty-day (30-day) average (Section 133.101). The arithmetic mean of pollutant parameter values of samples collected in a period of 30 consecutive days.

Thirty-day (30-day) limitation (Section 429.11) is a value that should not be exceeded by the average of daily measurements taken during any 30-day period.

THM (Section 141.2). See **Trihalomethane (Section 141.2)**.

Threatened species (Section 257.3-2). See **Endangered species (Section 257.3-2)**.

Three (3)-hour period (Section 61.61 [2-10-90])* means any three consecutive 1-hour periods (each commencing on the hour), provided that the number of 3-hour periods during which the vinyl chloride concentration exceeds 10 ppm does not exceed the number of 1-hour periods during which the vinyl chloride concentration exceeds 10 ppm.

Three-process operation facility (Section 60.371) means the facility including those processes involved with plate stacking, burning or strap casting, and assembly of elements into the battery case.

Threshold Planning Quantity (Section 355.20) means, for a substance listed in Appendices A and B, the quantity listed in the column "threshold planning quantity" for that substance.

Throttle (Section 86.082-2) means the mechanical linkage which either directly or indirectly controls the fuel flow to the engine.

Throttle (Section 86.090-2 [7-26-90])* means a device used to control an engine's power output by limiting the amount of air entering the combustion chamber.

Tiering (Section 1508.28) refers to the coverage of general matters in broader environmental impact statements (such as national program or policy statements) with subsequent narrower statements or environmental analyses (such as regional or basinwide program statements or ultimately site-specific statements) incorporating by reference the general discussions and concentrating solely on the issues specific to the statement subsequently prepared. Tiering is appropriate when the sequence of statements or analyses is: (a) From a program, plan, or policy environmental impact statement to a program, plan, or policy statement or analysis of lesser scope or to a site-specific statement or analysis. (b) From an environmental impact statement on a specific action at an early stage (such as need and site selection) to a supplement (which is preferred) or a subsequent statement or analysis at a later stage (such as environmental mitigation). Tiering in such cases is appropriate when it helps the lead agency to focus on the issues which are ripe for decision and exclude from consideration issues already decided or not yet ripe.

Time period (Section 51.100) means any period of time designated by hour, month, season, calendar year, averaging time, or other suitable characteristics, for which ambient air quality is estimated.

Time-response curve (Section 797.1350) is the curve relating cumulative percentage response of a test batch of organisms, exposed to a single dose or single concentration of a chemical, to a period of exposure.

Timed Delay (Section 85.2122(a)(1)(ii)(B)) means a delayed diaphragm displacement controlled to occur within a given time period.

Tire (Section 253.4 [11/17/88]) means the following types of tires: passenger car tires, light- and heavy-duty truck tires, high speed industrial tires, bus tires, and special service tires (including Military, agricultural, off-the-road, and slow speed industrial).

Tire (Section 60.541 [9/15/87]) means any agricultural, airplane, industrial, mobile home, light-duty truck and/or passenger vehicle tire that has a bead diameter less than or equal to 0.5 meter (m) (19.7 inches) and a cross section dimension less than or equal to 0.325 m (12.8 in.), and that is mass produced in an assembly-line fashion.

Title (Section 35.6015 [6-5-90])* The valid claim to property which denotes ownership and the rights of ownership, including the rights of possession, control, and disposal of property.

Title III (Section 350.1 [7/29/88]) means Title III of the Superfund Amendments and Reauthorization Act of 1986, also titled the Emergency Planning and Community Right-to-Know Act of 1986.

Title III (Section 372.3 [2/16/88]) means Title III of the Superfund Amendments and Reauthorization Act of 1986, also titled the Emergency Planning and Community Right-To-Know Act of 1986.

Title V permit (Section 60.2 [7/21/92])* means any permit issued, renewed, or revised pursuant to Federal or State regulations established to implement title V of the Act (42 U.S.C. 7661). A title V permit issued by a State permitting authority is called a part 70 permit in this part.

Title V permit (Section 61.02 [3/16/94])* means any permit issued, renewed, or revised pursuant to Federal or State regulations established to implement title V of the Act (42 U.S.C. 7661). A title V permit issued by a State permitting authority is called a part 70 permit in this part.

TMDL (Section 130.2). See **Total maximum daily load (Section 130.2)**.

TOC (Section 401.11) means total organic carbon.

Toilet tissue (Section 250.4 [10/6/87, 6/22/88]) means a sanitary tissue paper. The principal characteristics are softness, absorbency, cleanliness, and adequate strength (considering easy disposability). It is marketed in rolls of varying sizes or in interleaved packages.

Tolerance (Section 177.3 [12-5-90]) means: (1) The amount of a pesticide residue that legally may be present in or on a raw agricultural commodity under the terms of a tolerance under FFDCA section 408 or a processed food under the terms of a food additive regulation under FFDCA section 409. Tolerances are usually expressed in terms of parts of the pesticide residue per million parts of the food (ppm), by weight.

Tolerance with regional registration (Section 180.1) means any tolerance which is established for pesticide residues resulting from the use of the pesticide pursuant to a regional registration. Such a tolerance is supported by residue data from specific growing regions for a raw agricultural commodity. Individual tolerances with regional registration are designated in separate subsections in 40 CFR 189.101 through 180.999, as appropriate. Additional residue data which are representative of the proposed use area are required to expand the geographical area of usage of a pesticide on a raw agricultural commodity having an established tolerance with regional registration. Persons seeking geographically broader registration of a crop having a tolerance with regional registration should contact the appropriate EPA product manager concerning additional residue data required to expand the use area.

Tomato-starch-cheese canned specialties (Section 407.81) shall mean canned specialties resulting from a combination of fresh and pre-processed tomatoes, starches, cheeses, spices, and other flavorings necessary to produce a variety of products similar to but not exclusively raviolis, spaghetti, tamales, and enchiladas.

Tomatoes (Section 407.61) shall mean the processing of tomatoes into canned, peeled, whole, stewed, and related piece sizes; and processing of tomatoes into the following products and product styles: Canned, peeled and unpeeled paste, concentrate, puree, sauce, juice, catsup and other similar formulated items requiring various other pre-processed food ingredients.

Tons per day (Section 245.101) means annual tonnage divided by 260 days.

Too numerous to count (Section 141.2 [7/17/92])* means that the total number of bacterial colonies exceeds 200 on a 47-mm diameter membrane filter used for coliform detection.

Top Secret (Section 11.4) refers to national security information or material which requires the highest degree of protection. The test for assigning Top Secret classification shall be whether its unauthorized disclosure could reasonably be expected to cause exceptionally grave damage to the national security. Examples of exceptionally grave damage include armed hostilities against the United States or its allies; disruption of foreign relations vitally affecting the national security; the compromise of vital national defense plans or complex cryptologic and communications intelligence systems; the revelation of sensitive intelligence operations; and the disclosure of scientific or technological developments vital to national security. This classification shall be used with the utmost restraint.

Top-blown furnace (Section 60.141a) means any BOPF in which oxygen is introduced to the bath of molten iron by means of an oxygen lance inserted from the top of the vessel.

Topcoat operation (Section 60.391) means the topcoat spray booth, flash-off area, and bake oven(s) which are used to apply and dry or cure the final coating(s) on components of automobile and light-duty truck bodies.

Total annual sales (Section 704.3 [12/22/88])* means the total annual revenue (in dollars) generated by the sale of all products of a company. Total annual sales must include the total annual sales revenue of all sites owned or controlled by that company and the total annual sales revenue of that company's subsidiaries and foreign or domestic parent company, if any.

Total Chromium (Section 410.01) shall mean hexavalent and trivalent chromium as measured by the procedures listed in 40 CFR Part 136.

Total dissolved solids (Sections 122.2; 144.3) means the total dissolved (filterable) solids as determined by use of the method specified in 40 CFR Part 136.

Total dissolved solids (TDS) (Section 146.3) means the total dissolved (filterable) solids as determined by use of the method specified in 40 CFR Part 136.

Total enclosure (Section 60.441)

means a structure or building around the coating applicator and flashoff area or the entire coating line for the purpose of confining and totally capturing fugitive VOC emissions.

Total enclosure (Section 60.711 [10/3/88]) means a structure that is constructed around a source of emissions so that all VOC emissions are collected and exhausted through a stack or duct. With a total enclosure, there will be no fugitive emissions, only stack emissions. The only openings in a total enclosure are forced makeup air and exhaust ducts and any natural draft openings such as those that allow raw materials to enter and exit the enclosure for processing. All access doors or windows are closed during routine operation of the enclosed source. Brief, occasional openings of such doors or windows to accommodate process equipment adjustments are acceptable, but, if such openings are routine or if an access door remains open during the entire operation, the access door must be considered a natural draft opening. The average inward face velocity across the natural draft openings of the enclosure must be calculated including the area of such access doors. The drying oven itself may be part of the total enclosure. A permanent enclosure that meets the requirements found in Section 60.713(b)(5)(i) is assumed to be a total enclosure. The owner or operator of a permanent enclosure that does not meet the requirements may apply to the Administrator for approval of the enclosure as a total enclosure on a case-by-case basis. Such approval shall be granted upon a demonstration to the satisfaction of the Administrator that all VOC emissions are contained and vented to the control device.

Total enclosure (Section 60.741 [9-11-89]) means a structure that is constructed around a source of emissions and operated so that all VOC emissions are collected and exhausted through a stack or duct. With a total enclosure, there will be no fugitive emissions, only stack emissions. The drying oven itself may be part of the total enclosure.

Total fluorides (Section 60.191) means elemental fluorine and all fluoride compounds as measured by reference methods specified in Section 60.195 or by equivalent or alternative methods (see Section 60.8(b)).

Total fluorides (Section 60.211) means elemental fluorine and all fluoride compounds as measured by reference methods specified in Section 60.214, or equivalent or alternative methods.

Total fluorides (Section 60.221) means elemental fluorine and all fluoride compounds as measured by reference methods specified in Section 60.224, or equivalent or alternative methods.

Total fluorides (Section 60.231) means elemental fluorine and all fluoride compounds as measured by reference methods specified in Section 60.234, or equivalent or alternative methods.

Total fluorides (Section 60.241) means elemental fluorine and all fluoride compounds as measured by reference methods specified in Section 60.244, or equivalent or alternative methods.

Total maximum daily load (TMDL) (Section 130.2). The sum of the individual WLAs for point sources and LAs for nonpoint sources and natural background. If a receiving water has only one point source discharger, the TMDL is the sum of that point source WLA plus the LAs for any nonpoint sources of pollution and natural background sources, tributaries, or adjacent segments. TMDLs can be expressed in terms of either mass per time, toxicity, or other appropriate measure. If Best Management Practices (BMPs) or other nonpoint source pollution controls make more stringent load allocations practicable, then wasteload allocations can be made less stringent. Thus, the TMDL process provides for nonpoint source control tradeoffs.

Total metal (Section 413.02) is defined as the sum of the concentration or mass of Copper (Cu), Nickel (Ni), Chromium (Cr) (total) and Zinc (Zn).

Total organic active ingredients (Section 455.21) means the sum of all organic active ingredients covered by Section 455.20(a) which are manufactured at a facility subject to this subpart.

Total organic compounds (Section 60.501) means those compounds measured according to the procedures in Section 60.503.

Total organic compounds (TOC) (Section 60.661 [6-29-90]) means those compounds measured according to the procedures in Sec. 60.664(b)(4). For the purposes of measuring molar composition as required in Sec. 60.664(d)(2)(i); hourly emissions rate as required in Sec. 60.664(d)(5) and Sec. 60.664(e); and TOC concentration as required in Sec. 60.665(b)(4) and Sec. 60.665(g)(4), those compounds which the Administrator has determined do not contribute appreciably to the formation of ozone are to be excluded. The compounds to be excluded are identified in Environmental Protection Agency's statements on ozone abatement policy for State Implementation Plans (SIP) revisions (42 FR 35314; 44 FR 32042; 45 FR 32424; 45 FR 48942).

Total Phenols (Section 464.02) shall mean total phenolic compounds as measured by the procedure listed in 40 CFR Part 136 (distillation followed by colorimetric-4AAP).

Total rated capacity (Section 52.01) means the sum of the rated capacities of all fuel-burning equipment connected to a common stack. The rated capacity shall be the maximum guaranteed by the equipment manufacturer or the maximum normally achieved during use, whichever is greater.

Total reduced sulfur (TRS) (Section 60.281) means the sum of the sulfur compounds hydrogen sulfide, methyl mercaptan, dimethyl sulfide, and dimethyl disulfide, that are released during the kraft pulping operation and measured by Reference Method 16.

Total residual chlorine (or total residual oxidants for intake water with bromides) (Section 423.11) means the value obtained using the amperometric method for total residual chlorine described in 40 CFR Part 136.

Total residual chlorine (TRC) (Section 420.02) means the value obtained by the iodometric titration with an amperometric endpoint method specified in 40 CFR 136.3.

Total residual oxidants for intake water with bromides (Section 423.11). See **Total residual chlorine (Section 423.11)**.

Total smelter charge (Section 60.161) means the weight (dry basis) of all copper sulfide ore concentrates processed at a primary copper smelter, plus the weight of all other solid materials introduced into the roasters and smelting furnaces at a primary copper smelter, except calcine, over a one-month period.

Total SO_2 equivalents (Section 60.641) means the sum of volumetric or mass concentrations of the sulfur compounds obtained by adding the quantity existing as SO_2 to the quantity of SO_2 that would be obtained if all reduced sulfur compounds were converted to SO_2 (ppmv or kg/DSCM).

Total suspended particulate (Section 51.100) means particulate matter as measured by the method described in Appendix B of Part 50 of this chapter.

Total suspended particulates (TSP) (Section 58.1) means particulate matter as measured by the method described in Appendix B of Part 50 of this chapter.

Total suspended residue (or total suspended solids) (TSS) (Section 420.02) means the value obtained by the method specified in 40 CFR 136.3.

Total suspended solids (or total suspended residue) (TSS) (Section 420.02) means the value obtained by the method specified in 40 CFR 136.3.

Total test distance (Section 86.402-78) is defined for each class of motorcycles in Section 86.427-78.

Total Toxic Organics (TTO) (Section 469.22) means the sum of the concentrations of the toxic organic compounds which are found in the discharge at a concentration greater than ten (10) micrograms per liter. (See Section 469.22(a).)

Total Toxic Organics (TTO) (Section 468.02) shall mean the sum of the masses or concentrations of each of the toxic organic compounds which is found at a concentration greater than 0.010 mg/l. (See Section 468.02(r).)

Total Toxic Organics (TTO) (Section 464.02) shall mean the sum of the mass of each of the toxic organic compounds which are found at a concentration greater than 0.010 mg/l. The specialized definitions for each subpart contain a discrete list of toxic organic compounds comprising TTO for each process segment in which TTO is regulated.

Total Toxic Organics (TTO) (Section 469.12) means the sum of the concentrations of the toxic organic compounds which are found in the discharge at a concentration greater than ten (10) micrograms per liter. (See Section 469.12(a).)

Total Toxic Organics (TTO) (Section 469.31) means the sum of the concentrations of the toxic organic compounds which are found in the discharge at a concentration greater than ten (10) micrograms per liter. (See Section 469.31(b).)

Total Toxic Organics (TTO) (Section 467.02) shall mean the sum of the masses or concentrations of each of the toxic organic compounds (listed in Section 467.02 under this definition) which are found in the discharge at a concentration greater than 0.010 mg/l.

Total Toxic Organics (TTO) (Section 464.21) is a regulated parameter under PSES (Section 464.25) and PSNS (Section 464.26) for the copper subcategory and is comprised of a discrete list of toxic organic pollutants for each process segment where it is regulated. (See Section 464.21(a) (1) through (5).)

Total Toxic Organics (TTO) (Section 464.31) is a regulated parameter under PSES (Section 464.35) and PSNS (Section 464.36) for the ferrous subcategory and is comprised of a discrete list of toxic organic pollutants for each process segment where it is regulated. (See Section 464.31(a) (1) through (7).)

Total Toxic Organics (TTO) (Section 464.41) is a regulated parameter under PSES (Section 464.45) and PSNS (Section 464.46) for the zinc subcategory and is comprised of a discrete list of toxic organic pollutants for each process segment where it is regulated. (See Section 464.41(a) (1) through (4).)

Total Toxic Organics (TTO) (Section 465.02) shall mean the sum of the mass of each of the toxic organic compounds which are found at a concentration greater than 0.010 mg/l. (See Section 462.02(j).)

Total Trihalomethanes (TTHM) (Section 141.2) means the sum of the concentration in milligrams per liter of the trihalomethane compounds (trichloromethane [chloroform], dibromochloromethane, bromodichloromethane and tribromomethane [bromoform]), rounded to two significant figures.

Totally enclosed manner (Section 761.3) means any manner that will ensure no exposure of human beings or the environment to any concentration of PCBs.

Totally enclosed treatment facility (Section 260.10) means a facility for the treatment of hazardous waste which is directly connected to an industrial production process and which is constructed and operated in a manner which prevents the release of any hazardous waste or any constituent thereof into the environment during treatment. An example is a pipe in which waste acid is neutralized.

Touch-up coat (Section 60.721 [1/29/88, 6/15/89]) means the coat applied to correct any imperfections in the finish after color or texture coats have been applied. This definition does not include conductive sensitizers or EMI/RFI shielding coatings.

Toxaphene (Section 129.4) means a material consisting of technical grade chlorinated camphene having the approximate formula of $C_{10}H_{10}Cl_8$ and normally containing 67-69 percent chlorine by weight.

Toxaphene formulator (Section 129.103) means a person who produces, prepares or processes a formulated product comprising a mixture of toxaphene and inert materials or other diluents into a product intended for application in any use registered under the Federal Insecticide, Fungicide and Rodenticide Act, as amended (7 U.S.C. 135, et seq.).

Toxaphene manufacturer (Section 129.103) means a manufacturer, excluding any source which is exclusively a toxaphene formulator, who produces, prepares or processes toxaphene or who uses toxaphene as a material in the production, preparation or processing of another synthetic organic substance.

Toxic chemical (Section 372.3 [2/16/88]) means a chemical or chemical category listed in Section 372.65.

Toxic pollutant (Section 122.2 [5/2/89])* means any pollutant listed as toxic under section 307(a)(1) or, in the case of sludge use or disposal practices, any pollutant identified in regulations implementing section 405(d) of the CWA.

Toxic pollutant (Section 233.3) means any pollutant listed as toxic under section 307(a)(1) of CWA.

Toxic pollutant (Section 501.2 [5/2/89]) means any pollutant listed as toxic under section 307(a)(1) or any pollutant identified in regulations implementing section 405(d) of the CWA.

Toxic pollutants (Section 125.58) means those substances listed in 40 CFR 401.15.

Toxic pollutants (Section 131.3) are those pollutants listed by the Administrator under section 307(a) of the Act.

Toxicity (Section 162.3) means the property of a substance or mixture of substances to cause any adverse effects.

Toxicity (Section 171.2) means the property of a pesticide to causc any adverse physiological effects.

Toxicity (Section 435.11 [3/4/93])* as applied to BAT effluent limitations and NSPS for drilling fluids and drill cuttings shall refer to the bioassay test procedure presented in Appendix 2 of 40 CFR 435, subpart A.

Toxicity curve (Section 797.1350) is the curve produced from toxicity tests when LC_{50} values are plotted against duration of exposure. (This term is also used in aquatic toxicology, but in a less precise sense, to describe the curve produced when the median period of survival is plotted against test concentrations.)

TPQ (Section 370.2 [10/15/87]) means the threshold planning quantity for an extremely hazardous substance as defined in 40 CFR Part 355.

TPQ (Seection 370.2 [7-26-90])* means the threshold planning quantity for an extremely hazardous substance as defined in 40 CFR Part 355.

Traceable (Sections 50.1; 58.1) means that a local standard has been compared and certified, either directly or via not more than one intermediate standard, to a primary standard such as a National Bureau of Standards Standard Reference Material (NBS SRM), or a USEPA/NBS-approved Certified Reference Material (CRM).

Tractor (Section 205.151) means for the purposes of this subpart, any two or three wheeled vehicle used exclusively for agricultural purposes, or for snow plowing, including self-propelled machines used exclusively in growing, harvesting or handling farm produce.

Trade name product (Section 372.3 [7-26-90])* means a chemical or mixture of chemicals that is distributed to other persons and that incorporates a toxic chemical component that is not identified by the applicable chemical name or Chemical Abstracts Service Registry number listed in Sec. 372.65.

Trade secrecy claim (Section 350.1 [7/29/88]) is a submittal under sections 303(d)(2) or (d)(3), 311, 312, or 313 of Title III in which a chemical identity is claimed as trade secret, and is accompanied by a substantiation in support of the claim of trade secrecy for chemical identity.

Trade secret (Section 350.1 [7/29/88]) means any confidential formula, pattern, process, device, information or compilation of information that is used in a submitter's business, and that gives the submitter an opportunity to obtain an advantage over competitors who do not know or use it. EPA intends to be guided by the Restatement of Torts, section 757, comment b.

Trading (Section 86.090-2 [7-26-90]) means the exchange of heavy-duty engine NOx or particulate emission credits between manufacturers.

Trainee (Section 45.115) means a student selected by the recipient organization who receives support to meet the objectives in Section 45.110.

Transfer and loading system (Section 60.251) means any facility used to transfer and load coal for shipment.

Transfer effciency (Section 60.721 [1/29/88]) means the ratio of the amount of coating solids deposited onto the surface of a plastic business machine part to the total amount of coating solids used.

Transfer efficiency (Section 60.311) means the ratio of the amount of coating solids deposited onto the surface of a part or product to the total amount of coating solids used.

Transfer efficiency (Section 60.391) means the ratio of the amount of coating solids transferred onto the surface of a part or product to the total amount of coating solids used.

Transfer efficiency (Section 60.451) means the ratio of the amount of coating solids deposited onto the surface of a large appliance part or product to the total amount of coating solids used.

Transfer facility (Section 259.10 [3/24/89]) means any transportation-related facility including loading docks, parking areas, storage areas and other similar areas where shipments of regulated medical waste are held (come to rest or are managed) during the course of transportation. For example, a location at which regulated medical waste is transferred directly between two vehicles is considered a transfer facility. A transfer facility is a transporter.

Transfer facility (Sections 260.10; 270.2) means any transportation related facility including loading docks, parking areas, storage areas and other similar areas where shipments of hazardous waste are held during the normal course of transportation.

Transfer facility Section 761.3 [12-21-89]) means any transportation-related facility including loading docks, parking areas, and other similar areas where shipments of PCB waste are held during the normal course of transportation. Transport vehicles are not transfer facilities under this definition, unless they are used for the storage of PCB waste, rather than for actual transport activities. Storage areas for PCB waste at transfer facilities are subject to the storage facility standards of Sec. 761.65, but such storage areas are exempt from the approval requirements of Sec. 761.65(d) and the recordkeeping requirements of Sec. 761.180, unless the same PCB waste is stored there for a period of more than 10 consecutive days between destinations.

Transfer point (Section 60.671) means a point in a conveying operation where the nonmetallic mineral is transferred to or from a belt conveyor except where the nonmetallic mineral is being transferred to a stockpile.

Transfer station (Section 243.101) means a site at which solid wastes are concentrated for transport to a processing facility or land disposal site. A transfer station may be fixed or mobile.

Transferee (Section 144.3 [12/3/93])* means the owner or operator receiving ownership and/or operational control of the well.

Transferor (Section 144.3 [12/3/93])* means the owner or operator transferring ownership and/or operational control of the well.

Transform (Section 82.3 [12/10/93])* means to use and entirely consume (except for trace quantities) a controlled substance in the manufacture of other chemicals for commercial purposes.

Transhipment (Section 82.3 [12/10/93])* means the continuous shipment of a controlled substance from a foreign state of origin through the United States or its territories to a second foreign state of final destination.

Transient non-community water system or TWS (Section 141.2 [7/17/92])* means a non-community water system that does not regularly serve at least 25 of the same persons over six months per year.

Transit country (Section 262.51 [7/19/88])* means any foreign country, other than a receiving country, through which a hazardous waste is transported.

Translocation (Section 797.2850) means the transference or transport of chemical from the site of uptake to other plant components.

Transmission class (Section 86.082-2) means the basic type of transmission, e.g., manual, automatic, semiautomatic.

Transmission Class (Section 600.002-85) means a group of transmissions having the following common features:

Basic transmission type (manual, automatic, or semi-automatic); number of forward gears used in fuel economy testing (e.g., manual four-speed, three-speed automatic, two-speed semi-automatic); drive system (e.g., front wheel drive, rear wheel drive, four wheel drive), type of overdrive, if applicable (e.g., final gear ratio less than 1.00, separate overdrive unit); torque converter type, if applicable (e.g., non-lockup, lockup, variable ratio); and other transmission characteristics that may be determined to be significant by the Administrator.

Transmission configuration (Section 600.002-85) means the Administrator may further subdivide within a transmission class if the Administrator determines that sufficient fuel economy differences exist. Features such as gear ratios, torque converter multiplication ratio, stall speed, shift calibration, or shift speed may be used to further distinguish characteristics within a transmission class.

Transmission configuration (Section 86.082-2) means a unique combination, within a transmission class, of the number of the forward gears and, if applicable, over-drive. The Administrator may further subdivide a transmission configuration (based on such criteria as gear ratios, torque convertor multiplication ratio, stall speed and shift calibration, etc.), if he determines that significant fuel economy or exhaust emission differences exist within that transmission configuration.

Transmissive fault or fracture (Section 148.2 [7/26/88]) is a fault or fracture that has sufficient permeability and vertical extent to allow fluids to move between formations.

Transmissive fault or fracture (Section 146.61 [7/26/88]) is a fault or fracture that has sufficient permeability and vertical extent to allow fluids to move between formations.

Transmissive fracture (Section 146.61 [7/26/88]). See **Transmissive fault (Section 146.61)**.

Transmissive fracture (Section 148.2 [7/26/88]). See **Transmissive fault (Section 148.2)**.

Transmissivity (Section 191.12) means the hydraulic conductivity integrated over the saturated thickness of an underground formation. The transmissivity of a series of formations is the sum of the individual transmissivities of each formation comprising the series.

Transport vehicle (Section 260.10) means a motor vehicle or rail car used for the transportation of cargo by any mode. Each cargo-carrying body (trailer, railroad freight car, etc.) is a separate transport vehicle.

Transport vehicle (Section 761.3) means a motor vehicle or rail car used for the transportation of cargo by any mode. Each cargo-carrying body (e.g., trailer, railroad freight car) is a separate transport vehicle.

Transportation (Section 259.10 [3/24/89]) means the shipment or conveyance of regulated medical waste by air, rail, highway, or water.

Transportation (Section 260.10) means the movement of hazardous waste by air, rail, highway, or water.

Transportation control measure (Section 51.100) means any measure that is directed toward reducing emissions of air pollutants from transportation sources. Such measures include, but are not limited to, those listed in section 108(f) of the Clean Air Act.

Transportation-related (Section 112.2 [8/25/93])* and "non-transportation-related" as applied to an onshore or offshore facility, are defined in the Memorandum of Understanding between the Secretary of Transportation and the Administrator of the Environmental Protection Agency, dated November 24, 1971, 36 FR 24080.

Transporter (Section 259.10 [3/24/89]) means a person engaged in the off-site transportation of regulated medical waste by air, rail, highway, or water.

Transporter (Sections 260.10; 270.2) means a person engaged in the off-site transportation of hazardous waste by air, rail, highway, or water.

Transporter of PCB waste (Section 761.3 [12-21-89]) means, for the purposes of subpart K of this part, any person engaged in the transportation of regulated PCB waste by air, rail, highway, or water for purposes other than consolidation by a generator.

Transuranic (Section 61 [12-15-89]) An element with an atomic number greater than the number of uranium.

Transuranic radioactive waste (Section 191.02) as used in this part, means waste containing more than 100 nanocuries of alpha-emitting transuranic isotopes, with half-lives greater than twenty years, per gram of waste, except for: (1) High-level radioactive wastes; (2) wastes that the Department has determined, with the concurrence of the Administrator, do not need the degree of isolation required by this part; or (3) wastes that the Commission has approved for disposal on a case-by-case basis in accordance with 10 CFR Part 61.

TRC (Section 420.02). See **Total residual chlorine (Section 420.02)**.

TRE index value (Section 60.661 [6-29-90]) means a measure of the supplemental total resource requirement per unit reduction of TOC associated with an individual distillation vent stream, based on vent stream flow rate, emission rate of TOC net heating value, and corrosion properties (whether or not the vent stream is halogenated), as quantified by the equation given under Sec. 60.664(e).

Tread end cementing operation (Section 60.541 [9/15/87]) means the system used to apply cement to one or both ends of the tread or combined tread/sidewall component. A tread end cementing operation consists of a cement application station and all other equipment, such as the cement supply system and feed and takeaway conveyors, necessary to apply cement to tread ends and to allow evaporation of solvent from the cemented tread ends.

Treatability Study (Section 260.10 [7/19/88]) means a study in which a hazardous waste is subjected to a treatment process to determine: (1) Whether the waste is amenable to the treatment process, (2) what pretreatment (if any) is required, (3) the optimal process conditions needed to achieve the desired treatment, (4) the efficiency of a treatment process for a specific waste or wastes, or (5) the characteristics and volumes of residuals from a particular treatment process. Also included in this definition for the purpose of the Section 261.4(e) and (f) exemptions are liner compatibility, corrosion, and other material compatibility studies and toxicological and health effects studies. A treatability study is not a means to commercially treat or dispose of hazardous waste.

Treated area (Section 170.3 [8/21/92])* means any area to which a pesticide is being directed or has been directed.

Treatment (Section 259.10 [3/24/89]) when used in the context of medical waste management means any method, technique, or process designed to change the biological character or composition of any regulated medical waste so as to reduce or eliminate its potential for causing disease. When used in the context of Section 259.30(a) of this part, treatment means either the provision of medical services or the preparation of human or animal remains for interment or cremation.

Treatment (Sections 260.10; 270.2) means any method, technique, or process, including neutralization, designed to change the physical, chemical, or biological character or composition of any hazardous waste so as to neutralize such waste, or so as to recover energy or material resources from the waste, or so as to render such waste non-hazardous, or less hazardous; safer to transport, store, or dispose of; or amenable for recovery, amenable for storage, or reduced in volume.

Treatment facility and treatment system (Section 434.11) mean all structures which contain, convey, and as necessary, chemically or physically treat coal mine drainage, coal preparation plant process wastewater, or drainage from coal preparation plant associated areas, which remove pollutants regulated by this Part from such waters. This includes all pipes, channels, ponds, basins, tanks and all other equipment serving such structures.

Treatment process (Section 61.341 [3-7-90]) means a steam stripping unit, thin-film evaporation unit, waste incinerator, or any other process used to comply with Section 61.348 of this subpart.

Treatment system (Section 434.11). See **Treatment facility (Section 434.11)**.

Treatment technique requirement (Section 142.2) means a requirement of the national primary drinking water regulations which specifies for a contaminant a specific treatment technique(s) known to the Administrator which leads to a reduction in the level of such contaminant sufficient to comply with the requirements of Part 141 of this chapter.

Treatment works (Section 35.2005 [6-29-90])* Any devices and systems for the storage, treatment, recycling, and reclamation of municipal sewage, domestic sewage, or liquid industrial wastes used to implement section 201 of the Act, or necessary to recycle or reuse water at the most economical cost over the design life of the works. These include intercepting sewers, outfall sewers, sewage collection systems, individual systems, pumping, power, and other equipment and their appurtenances; extensions, improvement, remodeling, additions, and alterations thereof; elements essential to provide a reliable recycled supply such as standby treatment units and clear well facilities; and any works, including acquisition of the land that will be an integral part of the treatment process or is used for ultimate disposal of residues resulting from such treatment (including land for composting sludge, temporary storage of such compost and land used for the storage of treated wastewater in land treatment systems before land application); or any other method or system for preventing, abating, reducing, storing, treating, separating, or disposing of municipal waste or industrial waste, including waste in combined storm water and sanitary sewer systems.

Treatment works (Section 35.905) are any devices and systems for the storage, treatment, recycling, and reclamation of municipal sewage, domestic sewage, or liquid industrial wastes used to implement section 201 of the Act, or necessary to recycle or reuse water at the most economical cost over the useful life of the works. These include intercepting sewers, outfall sewers, sewage collection systems, individual systems, pumping, power, and other equipment and their appurtenances; extensions, improvement, remodeling, additions, and alterations thereof; elements essential to provide a reliable recycled supply such as standby treatment units and clear well facilities; and any works, including site acquisition of the land that will be an integral part of the treatment process or is used for ultimate disposal of residues resulting from such treatment (including land for composting sludge, temporary storage of such compost, and land used for the storage of treated wastewater in land treatment systems before land application); or any other method or system for preventing, abating, reducing, storing, treating, separating, or disposing of municipal waste or industrial waste, including waste in combined storm water and sanitary sewer systems.

Treatment works phase or segment (Section 35.2005 [6-29-90])* A treatment works phase or segment may be any substantial portion of a facility and its interceptors described in a facilities plan under Sec. 35.2030, which can be identified as a subagreement or discrete subitem. Multiple subagreements under a project shall not be considered to be segments or phases. Completion of building of a treatment works phase or segment may, but need not in and of itself, result in an operable treatment works.

Treatment works segment (Section 35.905) may be any portion of an operable treatment works described in an approved facilities plan, under Section 35.917, which can be identified as a contract or discrete subitem or subcontract for step 1, 2, or 3 work. Completion of construction of a treatment works segment may, but need not, result in an operable treatment works.

Treatment works treating domestic sewage (Section 122.2 [5/2/89]) means a POTW or any other sewage sludge or waste water treatment devices or systems, regardless of ownership (including federal facilities), used in the storage, treatment, recycling, and reclamation of municipal or domestic sewage, including land dedicated for the disposal of sewage sludge. This definition does not include septic tanks or similar devices. For purposes of this definition, domestic sewage includes waste and waste water from humans or household operations that are discharged to or otherwise enter a treatment works. In States where there is no approved State sludge management program under section 405(f) of the CWA, the Regional Administrator may designate any person subject to the standards for sewage sludge use and disposal in 40 CFR Part 503 as a treatment works treating domestic sewage, where he or she finds that there is a potential for adverse effects on public health and the environment from poor sludge quality or poor sludge handling, use or disposal practices, or where he or she finds that such designation is necessary to ensure that such person is in compliance with 40 CFR Part 503.

Treatment works treating domestic sewage (Section 501.2 [5/2/89]) means a POTW or any other sewage sludge or wastewater treatment devices or systems, regardless of ownership (including Federal facilities), used in the storage, treatment, recycling, and reclamation of municipal or domestic sewage, including land dedicated for the disposal of sewage sludge. This definition does not include septic tanks or similar devices. For purposes of this definition, domestic sewage includes waste and waste water from humans or household operations that are discharged to or otherwise enter a treatment works.

Treatment zone (Section 260.10) means a soil area of the unsaturated zone of a land treatment unit within which hazardous constituents are degraded, transformed, or immobilized.

Trenching or burial operation (Section 257.3-6) means the placement of sewage sludge or septic tank pumpings in a trench or other natural or man-made depression and the covering with soil or other suitable material at the end of each operating day such that the wastes do not migrate to the surface.

Tribal education agency (Section 47.105 [3-9-92]) means a school or community college which is controlled by an Indian tribe, band, or nation, including any Alaska Native village, which is recognized as eligible for special programs and services provided by the United States to Indians because of their status as Indians and which is not administered by the Bureau of Indian Affairs.

Trihalomethane (THM) (Section 141.2) means one of the family of organic compounds, named as derivatives of methane, wherein three of the four hydrogen atoms in methane are each substituted by a halogen atom in the molecular structure.

Triple rinse (Section 165.1) means the flushing of containers three times, each time using a volume of the normal diluent equal to approximately ten percent of the container's capacity, and adding the rinse liquid to the spray mixture or disposing of it by a method prescribed for disposing of the pesticide.

Triple superphosphate plant (Section 60.231) means any facility manufacturing triple superphosphate by reacting phosphate rock with phosphoric acid. A run-of-pile triple superphosphate plant includes curing and storing.

Tris (Section 704.205) means tris(2,3-dibromopropyl) phosphate (also commonly named DBPP, TBPP, and Tris-BP).

TRS (Section 60.281). See **Total reduced sulfur (Section 60.281)**.

Truck dumping (Section 60.671) means the unloading of nonmetallic minerals from movable vehicles designed to transport nonmetallic minerals from one location to another. Movable vehicles include but are not limited to: trucks, front end loaders, skip hoists, and railcars.

Truck loading station (Section 60.381) means that portion of a metallic mineral processing plant where metallic minerals or metallic mineral concentrates are loaded by a conveying system into trucks.

Truck unloading station (Section 60.381) means that portion of a metallic mineral processing plant where metallic ore is unloaded from a truck into a hopper, screen, or crusher.

Trucked batteries (Section 461.2) shall mean batteries moved into or out of the plant by truck when the truck is actually washed in the plant to remove residues left in the truck from the batteries.

True vapor pressure (Section 60.111; 60.111a) means the equilibrium partial pressure exerted by a petroleum liquid as determined in accordance with methods described in American Petroleum Institute Bulletin 2517, Evaporation Loss from External Floating-Roof Tanks, Second Edition, February 1980 (incorporated by reference - see Section 60.17).

Trust Fund (Section 300.6). See **Fund (Section 300.6)**.

Trustee (Section 300.6) means any Federal natural resources management agency designated in Subpart G of this Plan, and any State agency which may pursue claims for damages under section 107(f) of CERCLA.

Trustee (Section 305.12) means any Federal natural resources management agency designated in Subpart G of the NCP, and any State agency that may prosecute claims for damages under section 111(b) of CERCLA.

Trustee (Section 306.12) means any Federal natural resources management agency designated in Subpart G of the NCP, and any State agency that may pursue claims for damages under section 111(b) of CERCLA.

TSCA (Section 704.3 [12/22/88])* means the Toxic Substances Control Act, 15 U.S.C. 2601 et seq.

TSCA (Section 707.63) means the Toxic Substances Control Act.

TSCA (Sections 712.3; 716.3; 766.3; 792.3) means the Toxic Substances Control Act, 15 U.S.C. 2601 et seq.

TSP (Section 58.1). See **Total suspended particulates (Section 58.1)**.

TSS (or total suspended solids, or total suspended residue) (Section 420.02) means the value obtained by the method specified in 40 CFR 136.3.

TSS (Section 401.11) means total

suspended non-filterable solids.

TTHM (Section 141.2). See **Total Trihalomethanes (Section 141.2)**.

TTO (Section 413.02) shall mean total toxic organics, which is the summation of all quantifiable values greater than 0.01 milligrams per liter for the toxic organics as listed in Section 413.02(i).

TTO (Section 433.11) shall mean total toxic organics, which is the summation of all quantifiable values greater than .01 milligrams per liter for the toxic organics listed in Section 433.11(e).

TTO (Section 464.02). See **Total Toxic Organics (Section 464.02)**.

TTO (Section 464.21). See **Total Toxic Organics (Section 464.21)**.

TTO (Section 464.31). See **Total Toxic Organics (Section 464.31)**.

TTO (Section 464.41). See **Total Toxic Organics (Section 464.41)**.

TTO (Section 465.02). See **Total Toxic Organics (Section 465.02)**.

TTO (Section 467.02). See **Total Toxic Organics (Section 467.02)**.

TTO (Section 468.02). See **Total Toxic Organics (Section 468.02)**.

TTO (Section 469.12). See **Total Toxic Organics (Section 469.12)**.

TTO (Section 469.22). See **Total Toxic Organics (Section 469.22)**.

TTO (Section 469.31). See **Total Toxic Organics (Section 469.31)**.

Tube reducing (Section 471.02) is an operation which reduces the diameter and wall thickness of tubing with a mandrel and a pair of rolls with tapered grooves.

Tumbling or barrel finishing (Section 471.02) is an operation in which castings, forgings, or parts pressed from metal powder are rotated in a barrel with ceramic or metal slugs or abrasives to remove scale, fins, or burrs. It may be done dry or with an aqueous solution.

Tumbling or burnishing (Section 468.02) shall mean the process of polishing, deburring, removing sharp corners, and generally smoothing parts for both cosmetic and functional purposes, as well as the process of washing the finished parts and cleaning the abrasion media.

Turbines employed in oil/gas production or oil/gas transportation (Section 60.331) means any stationary gas turbine used to provide power to extract crude oil/natural gas from the earth or to move crude oil/natural gas, or products refined from these substances through pipelines.

Twenty-four hour daily average or 24-hour daily average (Section 60.51a

[2-11-91]) means the arithmetic or geometric mean (as specified in Sec. 60.58a (e), (g), or (h) as applicable) of all hourly emission rates when the affected facility is operating and firing MSW measured over a 24-hour period between 12 midnight and the following midnight.

Twenty-five-kilotonne Party (Section 82.3 [8/12/88]) means any nation listed in Appendix D to this Part.

Twenty-five-year (25-year) Twenty-four-hour (24-hour) rainfall event (Section 422.41, 422.51) shall mean the maximum precipitation event with a probable recurrence interval of once in 25 years as defined by the National Weather Service in technical paper no. 40, "Rainfall Frequency Atlas of the United States," May, 1961, and subsequent amendments or equivalent regional or State rainfall probability information developed therefrom.

Twenty-five-year (25-year) Twenty-four-hour (24-hour) rainfall event (Section 418.11) shall mean the maximum 24-hour precipitation event with a probable recurrence interval of once in 25 years as defined by the National Weather Service in technical paper No. 40, "Rainfall Frequency Atlas of the United States," May 1961, and subsequent amendments in effect, as of the effective date of this regulation.

Twenty-four-hour (24-hour) period (Section 60.41a) means the period of time between 12:01 a.m. and 12:00 midnight.

Two or more animal feeding operations under common ownership (Section 122.23) are considered, for the purposes of these regulations, to be a single animal feeding operation if they adjoin each other or if they use a common area or system for the disposal of wastes.

Two-piece can (Section 60.491) means any beverage can that consists of a body manufactured from a single piece of steel or aluminum and a top. Coatings for a two-piece can are usually applied after fabrication of the can body.

Two-year (2-year), Twenty-four-hour (24-hour) precipitation event (Section 434.11) means the maximum 24-hour precipitation event with a probable recurrence interval of once in two years as defined by the National Weather Service and Technical Paper No. 40, "Rainfall Frequency Atlas of the U.S.," May 1961, or equivalent regional or rainfall probability information developed therefrom.

Type I Sound Level Meter (Section 205.2) means a sound level meter which meets the type I requirements of ANSI S1.4-1972 specification for sound level meters. This publication is available from the American National Standards Institute, Inc., 1430 Broadway, New York, New York 10018.

Type I Sound Level Meter (Section 204.2) means a sound level meter which meets the Type I requirements of American National Standard Specification S1.4-1971 for sound level meters. This publication is available from the American National Standards Institute, Inc., 1430 Broadway, New York, New York 10018.

Type of pesticidal product (Section 167.3 [9/8/88]) refers to each individual product as identified by: the product name; EPA Registration Number (or EPA File Symbol, if any, for planned products, or Experimental Permit Number, if the pesticide is produced under an Experimental Use Permit); active ingredients; production type (technical, formulation, repackaging, etc.); and, market for which the product was produced (domestic, foreign, etc.). In cases where a pesticide is not registered, registration is not applied for, or the pesticide is not produced under an Experimental Use Permit, the term shall also include the chemical formulation.

Type of pesticide (Section 167.1) refers to each individual product as identified by the product name; EPA Registration Number (EPA File Symbol, if any, for planned products; Experimental Permit Number if the pesticide is produced under an Experimental Use Permit); production type (technical, formulation, repackaging, etc.); product classification (fungicide, insecticide, herbicide, etc.); market produced for (domestic, foreign, etc.); and use classification. In cases where a pesticide is not registered, registration is not applied for, or is not produced under an Experimental Use Permit, the term shall also include the chemical formulation.

Type of resin (Section 61.61) means the broad classification of resin referring to the basic manufacturing process for producing that resin, including, but not limited to, the suspension, dispersion, latex, bulk, and solution processes.

U

u (Section 797.1560). See **Uptake (Section 797.1560)**.

U (Section 440.132). See **Uranium (Section 440.132)**.

UIC (Section 144.3) means the Underground Injection Control program under Part C of the Safe Drinking Water Act, including an approved State program.

UIC (Sections 124.2; 146.3; 270.2) means the Underground Injection Control program under Part C of the Safe Drinking Water Act, including an approved program.

Ultimate Biodegradability (Section 796.3100) is the breakdown of an organic compound to CO_2, water, the oxides or mineral salts of other elements and/or to products associated with normal metabolic processes of microorganisms.

Ultimate Consumer (Section 600.002-85) means the first person who purchases an automobile for purposes other than resale or leases an automobile.

Ultimate purchaser (Section 205.2) means the first person who in good faith purchases a product for purposes other than resale.

Ultimate purchaser (Section 53.1) means the first person who purchases a reference method or an equivalent method for purposes other than resale.

Ultimate purchaser (Section 85.1902) shall be given the meaning ascribed to it by section 214 of the Act.

Ultrasonic testing (Section 471.02) is a nondestructive test which applies sound, at a frequency above about 20 HJz, to metal, which has been immersed in liquid (usually water) to locate inhomogeneities or structural discontinuities.

Umbo (Sections 797.1800; 797.1830) means the narrow end (apex) of the oyster shell.

Unacceptable adverse effect (Section 231.2) means impact on an aquatic or wetland ecosystem which is likely to result in significant degradation of municipal water supplies (including surface or ground water) or significant loss of or damage to fisheries, shellfishing, or wildlife habitat or recreation areas. In evaluating the unacceptability of such impacts, consideration should be given to the relevant portions of the section 404(b)(1) guidelines (40 CFR Part 230).

Unauthorized dispersion technique (Section 57.103 [2-13-92])* refers to any dispersion technique which, under section 123 of the Act and the regulations promulgated pursuant to that section, may not be used to reduce the degree of emission limitation otherwise required in the applicable SIP.

Unbleached papers (Section 250.4 [10/6/87, 6/22/88]) means papers made of pulp that have not been treated with bleaching agents.

Uncertified person (Section 171.2) means any person who is not holding a currently valid certification document indicating that he is certified under section 4 of FIFRA in the category of the restricted use pesticide made available for use.

Uncontrolled total arsenic emissions (Section 61.161) means the total inorganic arsenic in the glass melting furnace exhaust gas preceding any add-on emission control device.

Under common control with (Section 66.3). See **Control (Section 66.3)**.

Under the direct supervision of (Section 171.2) means the act or process whereby the application of a pesticide is made by a competent person acting under the instructions and control of a certified applicator who is responsible for the actions of that person and who is available if and when needed, even though such certified applicator is not physically present at the time and place the pesticide is applied.

Underfire air (Section 240.101) means any forced or induced air, under control as to quantity and direction, that is supplied from beneath and which passes through the solid wastes fuel bed.

Underground area (Section 280.12 [9/23/88]) means an underground room, such as a basement, cellar, shaft or vault, providing enough space for physical inspection of the exterior of the tank situated on or above the surface of the floor.

Underground drinking water source (Section 257.3-4) means: (i) An aquifer supplying drinking water for human consumption, or (ii) An aquifer in which the ground water contains less than 10,000 mg/l total dissolved solids.

Underground injection (Section 260.10) means the subsurface emplacement of fluids through a bored, drilled or driven well; or through a dug well, where the depth of the dug well is greater than the largest surface dimension. (See also injection well.)

Underground injection (Sections 144.3; 146.3; 270.2) means a well injection.

Underground release (Section 280.12 [9/23/88]) means any belowground release.

Underground source of drinking water (Section 146.3) means an aquifer or its portion: (1)(i) Which supplies any public water system; or (ii) Which contains a sufficient quantity of ground water to supply a public water system; and (A) Currently supplies drinking water for human consumption; or (B) Contains fewer than 10,000 mg/l total dissolved solids; and (2) Which is not an exempted aquifer.

Underground source of drinking water (USDW) (Sections 144.3; 147.2902; 270.2) means an aquifer or its portion: (a)(1) Which supplies any public water system; or (2) Which contains a sufficient quantity of ground water to supply a public water system; and (i) Currently supplies drinking water for human consumption; or (ii) Contains fewer than 10,000 mg/l total dissolved solids; and (b) Which is not an exempted aquifer.

Underground storage tank (Section 280.1) means any one or combination of tanks (including underground pipes connected thereto) which is used to contain an accumulation of regulated substances, and the volume of which (including the volume of the underground pipes connected thereto) is 10 per centum or more beneath the surface of the ground. Such term does not include any (a) Farm or residential tank of 1,100 gallons or less capacity used for storing motor fuel for noncommercial purposes, (b) Tank used for storing heating oil for consumptive use on the premises where stored, (c) Septic tank, (d) Pipeline facility (including gathering lines): (e) Regulated under the Natural Gas Pipeline Safety Act of 1968 (49 U.S.C. App. 1671, et. seq.), or (f) Regulated under the Hazardous Liquid Pipeline Safety Act of 1979 (49 U.S.C. App. 2001, et. seq.), or (g) Which is an intrastate pipeline facility regulated under State laws comparable to the provisions of law referred to in paragraphs (e) and (f) of this definition, (h) Surface impoundment, pit, pond, or lagoon, (i) Storm water or waste water collection system, (j) Flow-through process tank, (k) Liquid trap or associated gathering lines directly related to oil or gas production and gathering operations, or (l) Storage tank situated in an underground area (such as a basement, cellar, mineworking, drift, shaft, or tunnel) if the storage tank is situated upon or above the surface of the undesignated floor. (m) Any pipes connected to any tank which is described in paragraphs (a) through (l) of this definition.

Underground storage tank or UST (Section 280.12 [9/23/88]) means any one or combination of tanks (including underground pipes connected thereto) that is used to contain an accumulation of regulated substances, and the volume of which (including the volume of underground pipes connected thereto) is 10 percent or more beneath the surface of the ground. This term does not include any: (a) Farm or residential tank of 1,100 gallons or less capacity used for storing motor fuel for noncommercial purposes; (b) Tank used for storing heating oil for consumptive use on the premises where stored; (c) Septic tank; (d) Pipeline facility (including gathering lines) regulated under: (1) The Natural Gas Pipeline Safety Act of 1968 (49 U.S.C. App. 1671, et seq.), or (2) The Hazardous Liquid Pipeline Safety Act of 1979 (49 U.S.C. App. 2001, et seq.), or (3) Which is an intrastate pipeline facility regulated under state laws comparable to the provisions of the law referred to in paragraph (d)(1) or (d)(2) of this definition; (e) Surface impoundment, pit, pond, or lagoon; (f) Storm-water or wastewater collection system; (g) Flow-through process tank; (h) Liquid trap or associated gathering lines directly related to oil or gas production and gathering operations; or (i) Storage tank situated in an underground area (such as a basement, cellar, mineworking, drift, shaft, or tunnel) if the storage tank is situated upon or above the surface of the floor. The term underground storage tank or UST does not include any pipes connected to any tank which is described in paragraphs (a) through (i) of this definition.

Underground tank (Section 260.10) means a device meeting the definition of tank in Section 260.10 whose entire surface area is totally below the surface of and covered by the ground.

Underground uranium mine (Section 61.21 [12-15-89])* means a man-made underground excavation made for the purpose of removing material containing uranium for the principal purpose of recovering uranium.

Underground uranium mine (Section 61.20) means a man-made underground excavation made for the purpose of removing material containing uranium for the principal purpose of recovering uranium.

Underlying hazardous constituent (Section 268.2 [5/24/93])* means any regulated constituent present at levels above the F039 constituent-specific treatment standard at the point of generation of the hazardous waste.

Undertread cementing operation (Section 60.541 [9/15/87]) means the system used to apply cement to a continuous strip of tread or combined tread/sidewall component. An undertread cementing operation consists of a cement application station and all other equipment, such as the cement supply system and feed and takeaway conveyors, necessary to apply cement to tread or combined tread/sidewall strips and to allow evaporation of solvent from the cemented tread or combined tread/sidewall.

Undisturbed performance (Section 191.12 [12/20/93])* means the predicted behavior of a disposal system, including consideration of the uncertainties in predicted behavior, if the disposal system is not disrupted by human intrusion or the occurrence of unlikely natural events.

Unexpended consumption allowances (Section 82.3 [12/10/93])* means consumption allowances that have not been used. At any time in any control period a person's unexpended consumption allowances are the total of the level of consumption allowances the person has authorization under this subpart to hold at that time for that control period, minus the level of controlled substances that the person has produced or imported (not including transhipments and used or recycled controlled substances) in that control period until that time.

Unexpended production allowances (Section 82.3 [12/10/93])* means production allowances that have not been used. At any time in any control period a person's unexpended production allowances are the total of the level of production allowances he has authorization under this subpart to hold at that time for that control period, minus the level of controlled substances that the person has produced in that control period until that time.

Unfit-for use tank system (Section 260.10) means a tank system that has been determined through an integrity assessment or other inspection to be no longer capable of storing or treating hazardous waste without posing a threat of release of hazardous waste to the environment.

Uniform Act (Section 4.2) means the Uniform Relocation Assistance and Real Property Acquisition Policies Act of 1970 (84 Stat. 1894; 42 U.S.C. 4601 et seq.; Pub. L. 91-646), and amendments thereto.

Uniform air quality required for the daily reporting of air quality (Part 58, Appendix G) is a modified form of the Pollutant Standards Index (PSI).

Uniforming (Section 60.721 [1/29/88]). See **Fog coating (Section 60.721)**.

Unit (Section 73.3 [12-17-91]) means a fossil fuel-fired combustion device.

Unit acquisition cost (Section 35.6015 [6-5-90])* The net invoice unit price of the property including the cost of modifications, attachments, accessories, or auxiliary apparatus necessary to make the property usable for the purpose for which it was acquired. Other charges, such as the cost of installation, transportation, taxes, duty, or protective in-transit insurance, shall be included or excluded from the unit acquisition cost in accordance with the recipient's regular accounting practices.

Unit packaging (Section 157.21) means a package that is labeled with directions to use the entire contents of the package in a single application.

United States (Section 109.2) means the States, the District of Columbia, the Commonwealth of Puerto Rico, the Canal Zone, Guam, American Samoa, the Virgin Islands, and the Trust Territory of the Pacific Islands.

United States (Section 110.1) means the States, the District of Columbia, the Commonwealth of Puerto Rico, Guam, American Samoa, the Virgin Islands, and the Trust Territory of the Pacific Islands.

United States (Section 112.2 [8/25/93])* means the States, the District of Columbia, the Commonwealth of Puerto Rico, the Canal Zone, Guam, American Samoa, the Virgin Islands, and the Trust Territory of the Pacific Islands.

United States (Section 260.10) means the 50 States, the District of Columbia, the Commonwealth of Puerto Rico, the U.S. Virgin Islands, Guam, American Samoa, and the Commonwealth of the Northern Mariana Islands.

United States (Section 300.6) when used in relation to section 311(a)(5) of the CWA, refers to the States, the District of Columbia, the Commonwealth of Puerto Rico, Guam, American Samoa, the Virgin Islands, and the Trust Territory of the Pacific Islands. United States, when used in relation to section 101(27) of CERCLA, and State, include the several States of the United States, the District of Columbia, the Commonwealth of Puerto Rico, Guam, American Samoa, the United States Virgin Islands, the Commonwealth of the Northern Marianas, and any other territory or possession over which the U.S. has jurisdiction.

United States (Section 302.3) include the several States of the United States, the District of Columbia, the Commonwealth of Puerto Rico, Guam, American Samoa, the United States Virgin Islands, the Commonwealth of the Northern Marianas, and any other territory or possession over which the United States has jurisdiction.

United States (Section 7.25) includes the states of the United States, the District of Columbia, the Commonwealth of Puerto Rico, the Virgin Islands, American Samoa, Guam, Wake Island, the Canal Zone, and all other territories and possessions of the United States; the term State includes any one of the foregoing.

United States (Section 710.2) when used in the geographic sense, means all of the States, territories, and possessions of the United States.

United States (Section 720.3) when used in the geographic sense, means all of the States.

United States (Section 762.3) has the same meaning as in 15 U.S.C. 2602.

United States (Section 763.163 [7-12-89]) has the same meaning as in section 3 of the Toxic Substances Control Act.

United States (Section 8.2) as used

herein shall include the several States, the District of Columbia, the Commonwealth of Puerto Rico, the Panama Canal Zone, and the possessions of the United States.

United States (Section 85.1502 [9/25/87]) includes the Customs territory of the United States as defined in 19 U.S.C. 1202, and the Virgin Islands, Guam, American Samoa and the Commonwealth of the Northern Mariana Islands.

Units (Section 797.1350). All concentrations are given in weight per volume (e.g., in mg/liter).

Unleaded gasoline (Section 80.2) means gasoline which is produced without the use of any lead additive and which contains not more than 0.05 gram of lead per gallon and not more than 0.005 gram of phosphorus per gallon.

Unloading leg (Section 60.301) means a device which includes a bucket-type elevator which is used to remove grain from a barge or ship.

Unproven emission control systems (Section 86.092-2 [2-28-90]) (Note: the definitions listed in this section on this date of publication apply beginning with the 1992 model year.) are emission control components or systems (and fuel metering systems) that do not qualify as proven emission control systems.

Unreasonable adverse effects on the environment (Section 166.3) means any unreasonable risk to man or the environment, taking into account the economic, social, and environmental costs and benefits of the use of any pesticide.

Unreasonable degradation of the marine environment (Section 125.121) means: (1) Significant adverse changes in ecosystem diversity, productivity and stability of the biological community within the area of discharge and surrounding biological communities, (2) Threat to human health through direct exposure to pollutants or through consumption of exposed aquatic organisms, or (3) Loss of esthetic, recreational, scientific or economic values which is unreasonable in relation to the benefit derived from the discharge.

Unreclaimable residues(Section 165.1) means residual materials of little or no value remaining after incineration.

Unsanitized (Section 350.1 [7/29/88]) means a version of a document from which information claimed as trade secret or confidential has not been withheld or omitted.

Unsaturated zone or zone of aeration (Section 260.10) means the zone between the land surface and the water table.

Unscheduled maintenance (Section

86.402-78) means any inspection, adjustment, repair, removal, disassembly, cleaning, or replacement of vehicle components or systems which is performed to correct or diagnose a part failure or vehicle malfunction which was not anticipated.

Unscheduled maintenance (Section 86.082-2) means any adjustment, repair, removal, disassembly, cleaning, or replacement of vehicle components or systems which is performed to correct a part failure or vehicle (if the engine were installed in a vehicle) malfunction.

Unscheduled maintenance (Section 86.084-2) means any adjustment, repair, removal disassembly, cleaning, or replacement of vehicle components or systems which is performed to correct a part failure or vehicle (if the engine were installed in a vehicle) malfunction which was not anticipated. This definition applies beginning with the 1984 model year.

Unsolicited proposal (Section 30.200) means an informal written offer to perform EPA funded work for which EPA did not publish a solicitation.

Upgrade (Section 280.12 [9/23/88]) means the addition or retrofit of some systems such as cathodic protection, lining, or spill and overfill controls to improve the ability of an underground storage tank system to prevent the release of product.

Upper limit (Section 86.1102-87 [11-5-90])* means the emission level for a specific pollutant above which a certificate of conformity may not be issued or may be suspended or revoked.

Uppermost aquifer (Section 258.2 [10-9-91]) means the geologic formation nearest the natural ground surface that is an aquifer, as well as, lower aquifers that are hydraulically interconnected with this aquifer within the facility's property boundary.

Uppermost aquifer (Section 260.10) means the geologic formation nearest the natural ground surface that is an aquifer, as well as lower aquifers that are hydraulically interconnected with this aquifer within the facility's property boundary.

Upset (Section 403.16) means an exceptional incident in which there is unintentional and temporary noncompliance with categorical Pretreatment Standards because of factors beyond the reasonable control of the Industrial User. An Upset does not include noncompliance to the extent caused by operational error, improperly designed treatment facilities, inadequate treatment facilities, lack of preventive maintenance, or careless or improper operation.

Uptake (Section 797.1520) is the sorption of a test substance into and onto aquatic organisms during exposure.

Uptake (Section 797.1830) is the sorption of a test chemical into and onto aquatic organisms during exposure.

Uptake (u) (Section 797.1560) is the process of sorbing testing material into and/or onto the test organisms.

Uptake phase (Section 797.1560) is the time during the test when test organisms are being exposed to the test material.

Uptake phase (Sections 797.1520; 797.1830) is the initial portion of a bioconcentration test during which the organisms are exposed to the test solution.

Uptake rate constant (k_1) (Section 797.1560) is the mathematically determined value that is used to define the uptake of test material by exposed test organisms, usually reported in units of liters/gram/hour.

Uranium (U) (Section 440.132) is measured by the procedure discussed in 40 CFR 141.25(b)(2), or an equivalent method.

Uranium byproduct material (Section 192.31) means the tailings or wastes produced by the extraction or concentration of uranium from any ore processed primarily for its source material content. Ore bodies depleted by uranium solution extraction operations and which remain underground do not constitute byproduct material for the purpose of this subpart.

Uranium byproduct material or tailings (Section 61.251) means the wastes produced by the extraction or concentration of uranium from any ore processed primarily for its source material content. Ore bodies depleted by uranium solution extractions and which remain underground do not constitute byproduct material for the purposes of this subpart.

Uranium byproduct material or tailings (Section 61.221 [12-15-89])* means the waste produced by the extraction or concentration of uranium from any ore processed primarily for its source material content. Ore bodies depleted by uranium solution extraction and which remain underground do not constitute byproduct material for the purposes of this subpart.

Uranium fuel cycle (Section 190.02) means the operations of milling of uranium ore, chemical conversion of uranium, isotopic enrichment of uranium, fabrication of uranium fuel, generation of electricity by a light-water-cooled nuclear power plant using uranium fuel, and reprocessing of spent uranium fuel, to the extent that these directly support the production of electrical power for public use utilizing nuclear energy, but excludes mining operations, operations at waste disposal sites, transportation of any radioactive material in support of these operations, and the reuse of recovered non-uranium special nuclear and by-product materials from the cycle.

Uranium Fuel Cycle (Section 61 [12-15-89]) The operations of milling of uranium ore, chemical conversion or uranium, isotopic enrichment of uranium, fabrication of uranium fuel, generation of electricity by a light-water-cooled nuclear power plant using uranium fuel, and reprocessing of spent uranium fuel, to the extent that these directly support the production of electrical power for public use utilizing nuclear energy. This definition does not include mining operations, operations at waste disposal sites, transportation of any radioactive material in support of these operations, or the reuse of recovered non-uranium special nuclear and by-product materials from the cycle.

Urban area population (Section 58.1) means the population defined in the most recent decennial U.S. Census of Population Report.

Urban bus (Section 86.091-2 [7-26-90])* means a heavy heavy-duty diesel-powered passenger-carrying vehicle with a load capacity of fifteen or more passengers and intended primarily for intra-city operation, i.e., within the confines of a city or greater metropolitan area. Urban bus operation is characterized by short rides and frequent stops. To facilitate this type of operation, more than one set of quick-operating entrance and exit doors would normally be installed. Since fares are usually paid in cash or tokens rather than purchased in advance in the form of tickets, urban buses would normally have equipment installed for collection of fares. Urban buses are also typically characterized by the absence of equipment and facilities for long distance travel, e.g., rest rooms, large luggage compartments, and facilities for stowing carry-on luggage. The useful life for urban buses is the same as the useful life for other heavy heavy-duty diesel engines.

USCG (Section 300.5) is the U.S. Coast Guard.

USDA (Section 300.5) is the U.S. Department of Agriculture.

USDW (Section 146.3). See **Underground source of drinking water (Section 146.3)**.

USDW (Sections 144.3; 147.2902; 270.2). See **Underground source of drinking water (Sections 144.3; 147.2902; 270.2)**.

Use (Section 162.3) means any act of handling or release of a pesticide, or exposure of man or the environment to a pesticide through acts, including but not limited to: (1) Application of a pesticide, including mixing and loading and any required supervisory action in or near the area of application; (2) Storage actions for pesticides and pesticide containers; and (3) Disposal actions for pesticides and pesticide containers. (Use as defined here incorporates application. However, the certification requirement for certain restricted use pesticides only applies with respect to applications of such pesticides. Many aspects of use do not include application (e.g. storage, transportation), and hence are outside the requirement for certification.)

Use (Section 372.3 [2/16/88]). See **Otherwise use (Section 372.3)**.

Use attainability analysis (Section 131.3) is a structured scientific assessment of the factors affecting the attainment of the use which may include physical, chemical, biological, and economic factors as described in Section 131.10(g).

Use in agricultural or wildlife propagation (Section 435.51) means that the produced water is of good enough quality to be used for wildlife or livestock watering or other agricultural uses and that the produced water is actually put to such use during periods of discharge.

Use of asbestos (Section 763.103) means the presence of asbestos-containing material in school buildings.

Use pattern (Section 162.3) means the manner in which a pesticide is applied and includes the following parameters of pesticide application: (1) Target pest; (2) Crop or animals treated; (3) Application site; and (4) Application technique, rate and frequency.

Use stream (Section 721.3 [7-27-89]) means all reasonably anticipated transfer, flow, or disposal of a chemical substance, regardless of physical state or concentration, through all intended operations of industrial, commercial, or consumer use.

Use-dilution (Section 162.3) means a dilution specified on the label or labeling which produces the concentration of the pesticide for a particular purpose or effect.

Used in or for the manufacturing or processing of an instant photographic or peel-apart film article (Section 723.175) when used to describe activities involving a new chemical substance, means the new chemical substance (i) is included in the article, or (ii) is an intermediate to a chemical substance included in the article or is one of a series of intermediates used to manufacture a chemical substance included in the article.

Used oil (Section 260.10 [8/18/92])* means any oil that has been refined from crude oil, or any synthetic oil, that has been used and as a result of such use in contaminated by physical or chemical impurities.

Used oil (Section 266.40) means any oil that has been refined from crude oil, used, and, as a result of such use, is contaminated by physical or chemical impurities.

Used oil fuel (Section 266.40) includes any fuel produced from used oil by processing, blending, or other treatment.

Used or recycled controlled substances (Section 82.3 [12/10/93])* means controlled substances that have been recovered from their intended use systems.

Used or reused (Section 261.1). For purposes of Sections 261.2 and 261.6, a material is used or reused if it is either: (i) Employed as an ingredient (including use as an intermediate) in an industrial process to make a product (for example, distillation bottoms from one process used as feedstock in another process). However, a material will not satisfy this condition if distinct components of the material are recovered as separate end products (as when metals are recovered from metal-containing secondary materials); or (ii) Employed in a particular function or application as an effective substitute for a commercial product (for example, spent pickle liquor used as phosphorous precipitant and sludge conditioner in wastewater treatment).

Useful life (Section 35.2005 [6-29-90])* The period during which a treatment works operates. (Not "design life" which is the period during which a treatment works is planned and designed to be operated.)

Useful life (Section 35.905) means estimated period during which a treatment works will be operated.

Useful life (Section 85.1502 [9/25/87]) means a period of time/mileage as specified in Part 86 for a nonconforming vehicle which begins at the time of resale (for a motor vehicle or motor vehicle engine owned by the ICI at the time of importation) or release to the owner (for a motor vehicle or motor vehicle engine not owned by the ICI at the time of importation) of the motor vehicle or motor vehicle engine by the ICI after modification and/or test pursuant to Section 85.1505 or Section 85.1509.

Useful life (Section 85.1902) shall be given the meaning ascribed to it by section 202(d) of the Act and regulations promulgated thereunder.

Useful life (Section 85.2102) means that period established pursuant to section 202(d) of the Act and regulations promulgated thereunder.

Useful life (Section 86.082-2) means: (1) For light-duty vehicles and light-duty trucks a period of use of 5 years or 50,000 miles, whichever first occurs. (2) For gasoline-fueled heavy-duty engines a period of use of 5 years or 50,000 miles of vehicle operation or 1,500 hours of engine operation (or an equivalent period of 1,500 hours of dynamometer operation), whichever first occurs. (3) For diesel heavy-duty engines a period of use of 5 years or

100,000 miles of vehicle operation or 3,000 hours of engine operation (or an equivalent period of 1,000 hours of dynamometer operation), whichever first occurs.

Useful life (Section 86.084-2) (applies beginning with the 1984 model year) means: (a) For light-duty vehicles a period of use of 5 years or 50,000 miles, whichever first occurs. (b)(1) For a light-duty truck engine family or heavy-duty engine family, the average period of use up to engine retirement or rebuild, whichever occurs first, as determined by the manufacturer under Section 86.084-21(b)(4)(ii)(B). (2) For a specific light-duty truck or heavy-duty engine, the period of use represented by the first occurring of the following: (i) The engine reaches the point of needing to be rebuilt, according to the criteria established by the manufacturer under Section 86.084-21(b)(4)(ii)(C), or (ii) The engine reaches its engine family's useful life. (3) If the useful life of a specific light-duty truck or heavy-duty engine is found to be less than 5 years or 50,000 miles (or the equivalent), the useful life shall be a period of use of 5 years or 50,000 miles (or the equivalent), whichever occurs first, as required by section 202(d)(2) of the Act. (4) For purpose of identification this option shall be known as the average useful-life period. (c)(1) As an option for a light-duty truck engine family, a period of use of 12 years or 130,000 miles, whichever occurs first. (2) As an option for a gasoline heavy-duty engine family, a period of use of 10 years or 120,000 miles, whichever occurs first. (3) As an option for a diesel heavy-duty engine family, a period of use of 10 years or 120,000 miles, whichever occurs first, for engines certified for use in vehicles of less than 19,500 pounds GVWR; a period of use of 10 years or 200,000 miles, whichever occurs first, for engines certified for use in vehicles of 19,501-26,000 pounds GVWR; or, a period of use of 10 years or 275,000 miles, whichever occurs first, for engines certified for use in vehicles whose GVWR exceeds 26,000 pounds. (4) As an option for both light-duty truck and heavy-duty engine families, an alternate full-life value assigned by the Administrator under Section 86.084-21(b)(4)(ii)(B)(4). (5) For purpose of identification these options shall be known as the assigned useful-life period options. (6) For those light-duty truck and heavy-duty engine families using the assigned useful-life period options, the warranty period for emissions defect warranty and emissions performance warranty shall be 5 years/50,000 miles for light-duty trucks, 5 years/50,000 miles for gasoline heavy-duty engines and for diesel heavy-duty engines certified for use in vehicle of less than 19,501 lbs. GVWR, and 5 years/100,000 miles for all other diesel heavy-duty engines. However, in no case may this period be less than the basic mechanical warranty period. (7) The assigned useful-life period options, as detailed in paragraphs (c)(1) through (c)(6) of this section, are applicable for the 1984 model year only. (d)(1) As an option for the 1984 model year and for the 1984 model year only, the useful life of light-duty trucks and heavy-duty engine families may be defined as prescribed in Section 86.077-2. (2) For purpose of identification this option shall be known as the half-life useful-life option.

Useful life (Section 86.085-2 [12/16/87])* (applies beginning with the 1985 model year) means: (a) For light-duty vehicles a period of use of 5 years or 50,000 miles, whichever first occurs. (b) For a light-duty truck engine family, a period of use of 11 years or 120,000 miles, whichever occurs first. (c) For a gasoline-fueled heavy-duty engine family (and in the case of evaporative emission regulations, for gasoline-fueled heavy-duty vehicles), a period of use of 8 years or 110,000 miles, whichever first occurs. (d) For a diesel heavy-duty engine family: (1) For light heavy-duty diesel engines, a period of use of 8 years or 110,000 miles, whichever first occurs. (2) For medium heavy-duty diesel engines, a period of use of 8 years or 185,000 miles, whichever first occurs. (3) For heavy heavy-duty diesel engines, a period of use of 8 years or 290,000 miles, whichever first occurs. (e) As an option for both light-duty truck and heavy-duty engine families, an alternative useful life period assigned by the Administrator under the provisions of paragraph (f) of Section 86.085-21. (f) The useful-life period for purposes of the emissions defect warranty and emissions performance warranty shall be a period of 5 years/50,000 miles whichever first occurs, for light-duty trucks, gasoline heavy-duty engines, and light heavy-duty diesel engines. For all other heavy-duty diesel engines the aforementioned period is 5 years/100,000 miles, whichever first occurs. However, in no case may this period be less than the manufacturer's basic mechanical warranty period for the engine family.

Useful life (Section 86.090-2 [4/11/89]) (applies beginning with the 1990 model year) means: (a) For light-duty vehicles a period of use of 5 years or 50,000 miles, whichever first occurs. (b) For a light-duty truck engine family, a period of use of 11 years or 120,000 miles, whichever occurs first. (c) For an Otto-cycle heavy-duty engine family, a period of use of 8 years or 110,000 miles, whichever first occurs. (d) For a diesel heavy-duty engine family: (1) For light heavy-duty diesel engines, period of use of 8 years or 110,000 miles, whichever first occurs. (2) For medium heavy-duty diesel engines, a period of use of 8 years or 185,000 miles, whichever first occurs. (3) For heavy heavy-duty diesel engines, a period of use of 8 years or 290,000 miles, whichever first occurs. (e) As an option for both light-duty truck and heavy-duty engine families, an alternative useful-life period assigned by the Administrator under the provisions of paragraph (f) of Section

86.090-21. (f) The useful-life period for purposes of the emissions defect warranty and emissions performance warranty shall be a period of 5 years/50,000 miles whichever first occurs, for light-duty trucks, Otto-cycle heavy-duty engines, and light heavy-duty diesel engines. For all other heavy-duty diesel engines the aforementioned period is 5 years/100,000 miles, whichever first occurs. However, in no case may this period be less than the manufacturer's basic mechanical warranty period for the engine family.

Useful life (Section 86.094-2 [6-5-91]) means: (a) For light-duty vehicles, and for model year 1994 and later light light-duty trucks not subject to the Tier 0 standards of paragraph (a) of Sec. 86.094-9, intermediate useful life and/or full useful life. Intermediate useful life is a period of use of 5 years or 50,000 miles, whichever occurs first. Full useful life is a period of use of 10 years or 100,000 miles, whichever occurs first, except as otherwise noted in Sec. 86.094-9. (b) For light light-duty trucks subject to the Tier 0 standards of paragraph (a) of Sec. 86.094-9, and for heavy light-duty truck engine families, intermediate and/or full useful life. Intermediate useful life is a period of use of 5 years or 50,000 miles, whichever occurs first. Full useful life is a period of use of 11 years or 120,000 miles, whichever occurs first. (c) For an Otto-cycle heavy-duty engine family, a period of use of 8 years or 110,000 miles, whichever first occurs. (d) For a diesel heavy-duty engine family: (1) For light heavy-duty diesel engines, period of use of 8 years or 110,000 miles, whichever first occurs. (2) For medium heavy-duty diesel engines, a period of use of 8 years or 185,000 miles, whichever first occurs. (3) For heavy heavy-duty diesel engines, a period of use of 8 years or 290,000 miles, whichever first occurs. (e) As an option for both light-duty trucks under certain conditions and heavy-duty engine families, an alternative useful life period assigned by the Administrator under the provisions of paragraph (f) of Sec. 86.094-21. (f) The useful-life period for purposes of the emissions defect warranty and emissions performance warranty shall be a period of 5 years/50,000 miles, whichever first occurs, for light-duty trucks, Otto-cycle heavy-duty engines and light heavy-duty diesel engines. For all other heavy-duty diesel engines the aforementioned period is 5 years/100,000 miles, whichever first occurs. However, in no case may this period be less than the manufacturer's basic mechanical warranty period for the engine family.

Useful life (Section 86.098-2 [4/6/94])* means: (1) For light-duty vehicles, and for light light-duty trucks not subject to the Tier 0 standards of Sec. 86.094-9(a), intermediate useful life and/or full useful life.

Useful life (Section 86.402-78) is defined for each class (see Section

86.419) of motorcycle: Class I - 5.0 years or 12,000 km (7,456 miles), whichever first occurs; Class II - 5.0 years or 18,000 km (11,185 miles), whichever first occurs; Class III - 5.0 years or 30,000 km (18,641 miles), whichever first occurs.

User (Section 403.3). See **Industrial User (Section 403.3)**.

User charge (Section 35.2005 [6-29-90])* A charge levied on users of a treatment works, or that portion of the ad valorem taxes paid by a user, for the user's proportionate share of the cost of operation and maintenance (including replacement) of such works under sections 204(b)(1)(A) and 201(h)(2) of the Act and this subpart.

User charge (Sections 35.905) means a charge levied on users of a treatment works, or that portion of the ad valorem taxes paid by a user, for the user's proportionate share of the cost of operation and maintenance (including replacement) of such works under sections 204(b)(1)(A) and 201(h)(2) of the Act and this subpart.

UST (Section 280.12 [9/23/88]). See **Underground storage tank (Section 280.12)**.

UST system or Tank system (Section 280.12 [9/23/88]) means an underground storage tank, connected underground piping, underground ancillary equipment, and containment system, if any.

Utility competitive bid solicitation (Section 72.2 [7/30/93])* is a public request from a regulated utility for offers to the utility for meeting future generating needs. A qualifying facility, independent power production facility, or new IPP may be regarded as having been "selected" in such solicitation if the utility has named the facility as a project with which the utility intends to negotiate a power sales agreement.

Utility competitive bid solicitation (Section 73.3 [12-17-91]) is a public request from a regulated electric utility for offers to the utility for meeting future capacity needs. A new independent power production facility (IPPF) may be regarded as having been "selected" in such solicitation pursuant to section 405(g)(6)(A)(iv) if the utility has named the IPPF as a project with which it intends to negotiate a power sales agreement.

Utilize (Section 60.711 [10/3/88]) refers to the use of solvent that is delivered to coating mix preparation equipment for the purpose of formulating coatings to be applied on an affected coating operation and any other solvent (e.g., dilution solvent) that is added at any point in the manufacturing process.

V

Vacuum Break (Choke Pull-off) (Section 85.2122(a)(1)(ii)(F)) means a vacuum-operated device to open the carburetor choke plate a predetermined amount on cold start.

Vacuum Leakage (Section 85.2122(a)(1)(ii)(E)) means leakage into the vacuum cavity of a vacuum break.

Vacuum Purge System (Section 85.2122(a)(1)(ii)(H)) means a vacuum system with a controlled air flow to purge the vacuum system of undesirable manifold vapors.

Valid day (Section 60.101 [8-17-89]) means a 24-hour period in which at least 18 valid hours of data are obtained. A "valid hour" is one in which at least 2 valid data points are obtained.

Valid study (Section 152.83) means a study that has been conducted in accordance with the Good Laboratory Practice standards of 40 CFR Part 160 or generally accepted scientific methodology and that EPA has not determined to be invalid.

Validated test (Section 154.3) means a test determined by the Agency to have been conducted and evaluated in a manner consistent with accepted scientific procedures.

Valuable commercial and recreational species (Section 228.2) shall mean those species for which catch statistics are compiled on a routine basis by the Federal or State agency responsible for compiling such statistics for the general geographical area impacted, or which are under current study by such Federal or State agencies for potential development for commercial or recreational use.

Value engineering (Section 35.2005 [6-29-90])* A specialized cost control technique which uses a systematic and creative approach to identify and to focus on unnecessarily high cost in a project in order to arrive at a cost saving without sacrificing the reliability or efficiency of the project.

Value engineering (Section 35.6015 [6-5-90])* A systematic and creative analysis of each contract term or task to ensure that its essential function is provided at the overall lowest cost.

Value engineering (VE) (Section 35.905) means a specialized cost control technique which uses a systematic and creative approach to identify and to focus on unnecessarily high cost in a project in order to arrive at a cost saving without sacrificing the reliability or efficiency of the project.

Value for pesticide purposes (Section 172.1) means that characteristic of a substance or mixture of substances which produces an efficacious action on a pest.

Valve height (Sections 797.1800; 797.1830) means the greatest linear dimension of the oyster as measured from the umbo to the ventral edge of the valves (the farthest distance from the umbo).

Van (Section 600.002-85) means any light truck having an integral enclosure fully enclosing the driver compartment and load-carrying device, and having no body sections protruding more than 30 inches ahead of the leading edge of the windshield.

Van (Section 86.082-2) means a light-duty truck having an integral enclosure, fully enclosing the driver compartment and load carrying device, and having no body sections protruding more than 30 inches ahead of the leading edge of the windshield.

Vapor capture system (Section 60.741 [9-11-89]) means any device or combination of devices designed to contain, collect, and route solvent vapors released from the coating mix preparation equipment or coating operation.

Vapor capture system (Section 60.581) means any device or combination of devices designed to contain, collect, and route organic solvent vapors emitted from the flexible vinyl or urethane rotogravure printing line.

Vapor collection system (Section 61.301 [3-7-90]) means any equipment located at the affected facility used for containing benzene vapors displaced during the loading of tank trucks, railcars, or marine vessels. This does not include the vapor collection system that is part of any tank truck, railcar, or marine vessel vapor collection manifold system.

Vapor collection system (Section 60.501) means any equipment used for containing total organic compounds vapors displaced during the loading of gasoline tank trucks.

Vapor incinerator (Section 264.1031 [6-21-90]) means any enclosed combustion device that is used for destroying organic compounds and does not extract energy in the form of steam or process heat.

Vapor incinerator (Section 61.131 [9-19-91]) means any enclosed combustion device that is used for destroying organic compounds and does not necessarily extract energy in the form of steam or process heat.

Vapor-mounted seal (Section 61.341 [3-7-90]) means a foam-filled primary seal mounted continuously around the perimeter of a waste management unit so there is an annular vapor space underneath the seal. The annular vapor space is bounded by the bottom of the primary seal, the unit wall, the liquid surface, and the floating roof.

Vapor processing system (Section 60.501) means all equipment used for recovering or oxidizing total organic compounds vapors displaced from the affected facility.

Vapor recovery system (Section 60.111) means a vapor gathering system capable of collecting all hydrocarbon vapors and gases discharged from the storage vessel and a vapor disposal system capable of processing such hydrocarbon vapors and gases so as to prevent their emission to the atmosphere.

Vapor-tight marine vessel (Section 61.301 [3-7-90]) means a marine vessel with a benzene product tank that has been demonstrated within the preceding 12 months to have no leaks. This demonstration shall be made using method 21 of part 60,appendix A, during the last 20 percent of loading and during a period when the vessel is being loaded at its maximum loading rate. A reading of greater than 10,000 ppm as methane shall constitute a leak. As an alternative, a marine vessel owner or operator may use the vapor-tightness test described in Sec. 61.304(f) to demonstrate vapor tightness. A marine vessel operated at negative pressure is assumed to be vapor-tight for the purpose of this standard.

Vapor-tight tank truck or vapor-tight railcar (Section 61.301 [3-7-90]) means a tank truck or railcar for which it has been demonstrated within the preceding 12 months that its product tank will sustain a pressure change of not more than 750 pascals within 5 minutes after it is pressurized to a minimum of 4,500 pascals. This capability is to be demonstrated using the pressure test procedure specified in method 27 of part 60, appendix A, and a pressure measurement device which has a precision of +/-2.5 mm water and which is capable of measuring above the pressure at which the tank truck or railcar is to be tested for vapor tightness.

Vapor-mounted seal (Section 60.111a) means a foam-filled primary seal mounted continuously around the circumference of the tank so there is an annular vapor space underneath the seal. The annular vapor space is bounded by the bottom of the primary seal, the tank wall, the liquid surface, and the floating roof.

Vapor-tight gasoline tank truck (Section 60.501) means a gasoline tank truck which has demonstrated within the 12 preceding months that its product delivery tank will sustain a pressure change of not more than 750 pascals (75 mm of water) within 5 minutes after it is pressurized to 4,500 pascals (450 mm of water). This capability is to be demonstrated using the pressure test procedure specified in Reference Method 27.

Vapor-tight tank truck, or vapor-

tight railcar (Section 61.301 [3-7-90]) means a tank truck or railcar for which it has been demonstrated within the preceding 12 months that its product tank will sustain a pressure change of not more than 750 pascals within 5 minutes after it is pressurized to a minimum of 4,500 pascals. This capability is to be demonstrated using the pressure test procedure specified in method 27 of part 60, appendix A, and a pressure measurement device which has a precision of plus or minus 2.5 mm water and which is capable of measuring above the pressure at which the tank truck or railcar is to be tested for vapor tightness.

Variance (NPDES) (Section 124.2) means any mechanism or provision under section 301 or 316 of CWA or under 40 CFR Part 125, or in the applicable effluent limitations guidelines which allows modification to or waiver of the generally applicable effluent limitation requirements or time deadlines of CWA. This includes provisions which allow the establishment of alternative limitations based on fundamentally different factors or on sections 301(c), 301(g), 301(h), 301(i), or 316(a) of CWA.

Variance (Section 122.2) means any mechanism or provision under section 301 or 316 of CWA or under 40 CFR Part 125, or in the applicable effluent limitations guidelines which allows modification to or waiver of the generally applicable effluent limitation requirements or time deadlines of CWA. This includes provisions which allow the establishment of alternative limitations based on fundamentally different factors or on sections 301(c), 301(g), 301(h), 301(i), or 316(a) of CWA.

Variance (Section 51.100) means the temporary deferral of a final compliance date for an individual source subject to an approved regulation, or a temporary change to an approved regulation as it applies to an individual source.

VE (Section 35.905). See **Value engineering (Section 35.905)**.

Vector (Section 243.101) means a carrier that is capable of transmitting a pathogen from one organism to another.

Vector (Sections 240.101; 241.101) means a carrier, usually an arthropod, that is capable of transmitting a pathogen from one organism to another.

Vegetable tan (Section 425.02) means the process of converting hides into leather using chemicals either derived from vegetable matter or synthesized to produce effects similar to those chemicals.

Vehicle (Section 160.3 [8-17-89]) means any agent which facilitates the mixture, dispersion, or solubilization of a test substance with a carrier.

Vehicle (Section 205.151) means any motorcycle regulated pursuant to this subpart.

Vehicle (Section 205.51) means any motor vehicle, machine or tractor, which is propelled by mechanical power and capable of transportation of property on a street or highway and which has a gross vehicle weight rating in excess of 10,000 pounds and a partially or fully enclosed operator's compartment.

Vehicle (Section 792.226) means any agent which facilitates the mixture, dispersion, or solubilization of a test substance with a carrier.

Vehicle (Section 792.3 [8-17-89]) means any agent which facilitates the mixture, dispersion, or solubilization of a test substance with a carrier.

Vehicle (Section 85.2102) means a light duty vehicle or a light duty truck.

Vehicle (Section 86.602) means any new production light-duty vehicle as defined in Subpart A of this part.

Vehicle configuration (Section 86.082-2) means a unique combination of basic engine, engine code, inertia weight class, transmission configuration, and axle ratio.

Vehicle configuration (Section 600.002-85) means a unique combination of basic engine, engine code, inertia weight class, transmission configuration, and axle ratio within a base level.

Vehicle curb weight (Section 86.082-2) means the actual or the manufacturer's estimated weight of the vehicle in operational status with all standard equipment, and weight of fuel at nominal tank capacity, and the weight of optional equipment computed in accordance with Section 86.082-24; incomplete light-duty trucks shall have the curb weight specified by the manufacturer.

Vehicle or Engine Configuration (Section 85.2113) means the specific subclassification unit of an engine family as determined by engine displacement, fuel system, engine code, transmission and inertia weight class as applicable.

Vent (Section 60.671) means an opening through which there is mechanically induced air flow for the purpose of exhausting from a building air carrying particulate matter emissions from one or more affected facilities.

Vent stream (Section 60.661 [6-29-90]) means any gas stream discharged directly from a distillation facility to the atmosphere or indirectly to the atmosphere after diversion through other process equipment. The vent stream excludes relief valve discharges and equipment leaks including, but not limited to, pumps, compressors, and valves.

Vented (Section 264.1031 [6-21-90]) means discharged through an opening, typically an open-ended pipe or stack, allowing the passage of a stream of liquids, gases, or fumes into the atmosphere. The passage of liquids, gases, or fumes is caused by mechanical means such as compressors or vacuum-producing systems or by process-related means such as evaporation produced by heating and not caused by tank loading and unloading (working losses) or by natural means such as diurnal temperature changes.

Very large MWC plant (Section 60.31a [2-11-91]) means an MWC plant with an MWC plant capacity greater than 1,000 megagrams per day (1,100 tons per day) of MSW.

Very low sulfur oil (Section 60.41 [12-18-89]) means an oil that contains no more than 0.5 weight percent sulfur or that, when combusted without sulfur dioxide emission control, has a sulfur dioxide emission rate equal to or less than 215 ng/J (0.5 lb/million Btu) heat input.

Very low sulfur oil (Section 60.41b [12/16/87]) means a distillate oil or residual oil that when combusted without post combustion SO_2 control has an SO_2 emission rate equal to or less than 130 ng/J (0.03 lb SO_2/million Btu).

Vessel (Section 112.2 [8/25/93])* means every description of watercraft or other artificial contrivance used, or capable of being used as a means of transportation on water, other than a public vessel.

Vessel (Section 140.1) includes every description of watercraft or other artificial contrivance used, or capable of being used, as a means of transportation on waters of the United States.

Vessel (Section 21.2) means every description of watercraft or other artificial contrivance used, or capable of being used, as a means of transportation on the navigable waters of the United States other than a vessel owned or operated by the United States or a State or a political subdivision thereof, or a foreign nation; and is used for commercial purposes by a small business concern.

Vessel (Section 260.10) includes every description of watercraft, used or capable of being used as a means of transportation on the water.

Vessel (Section 302.3) means every description of watercraft or other artificial contrivance used, or capable of being used, as a means of transportation on water.

Vessel (Sections 110.1; 112.2; 116.3) means every description of watercraft or other artificial contrivance used, or capable of being used, as a means of transportation on water other than a public vessel.

VHAP (Section 61.241). See **Volatile hazardous air pollutant (Section 61.241)**.

Viable embryos (fertility) (Section 797.2130) are eggs in which fertilization has occurred and embryonic development has begun. This is determined by candling the eggs 11 days after incubation has begun. It is difficult to distinguish between the absence of fertilization and early embryonic death. The distinction can be made by breaking out eggs that appear infertile and examining further. This distinction is especially important when a test compound induces early embryo mortality. Values are expressed as a percentage of eggs set.

Viable embryos (fertility) (Section 797.2150) are eggs in which fertilization has occurred and embryonic development has begun. This is determined by candling the eggs 14 days after incubation has begun. It is difficult to distinguish between the absence of fertilization and early embryonic death. The distinction can be made by breaking out eggs that appear infertile and examining further. This distinction is especially important when a test compound induces early embryo mortality. Values are expressed as a percentage of eggs set.

Vibration (Section 763.83 [10/30/87]) means the periodic motion of friable ACBM which may result in the release of asbestos fibers.

Vinyl chloride plant (Section 61.61) includes any plant which produces vinyl chloride by any process.

Vinyl chloride purification (Section 61.61) includes any part of the process of vinyl chloride production which follows vinyl chloride formation.

Vinyl-asbestos floor tile (Section 763.163 [7-12-89]) means an asbestos-containing product composed of vinyl resins and used as floor tile.

Violating facility (Section 30.200) means any facility that is owned, leased, or supervised by an applicant, recipient, contractor, or subcontractor that EPA lists under 40 CFR Part 15 as not in compliance with Federal, State, or local requirements under the Clean Air Act or Clean Water Act. A facility includes any building, plant, installation, structure, mine, vessel, or other floating craft.

Virgin material (Sections 246.101; 247.101) means a raw material used in manufacturing that has been mined or harvested and has not as yet become a product.

Virus (Section 141.2 [6/29/89]) means a virus of fecal origin which is infectious to humans by waterborne transmission.

Viscose process (Section 60.601) means the fiber forming process where cellulose and concentrated caustic soda are reacted to form soda or alkali cellulose. This reacts with carbon disulfide to form sodium cellulose xanthate, which is then dissolved in a solution of caustic soda. After ripening, the solution is spun into an acid coagulating bath. This precipitates the cellulose in the form of a regenerated cellulose filament.

Visibility impairment (Section 51.301) means any humanly perceptible change in visibility (visual range, contrast, coloration) from that which would have existed under natural conditions.

Visibility in any mandatory Class I Federal area (Section 51.301) includes any integral vista associated with that area.

Visibility protection area (Sections 52.26; 52.28) means any area listed in 40 CFR 81.401-81.436 (1984).

Visible emissions (Section 61.141 [11-20-90])* means any emissions, which are visually detectable without the aid of instruments, coming from RACM or asbestos-containing waste material, or from any asbestos milling, manufacturing, or fabricating operation. This does not include condensed, uncombined water vapor.

VOC (Section 58.1 [2/12/93])* means volatile organic compounds.

VOC (Section 60.481). See **Volatile organic compounds (Section 60.481)**.

VOC (Section 60.711 [10/3/88]). See **Volatile organic compounds (Section 60.711)**.

VOC (Sections 60.431; 60.441) means volatile organic compound.

VOC content (Section 60.391) means all volatile organic compounds that are in a coating expressed as kilograms of VOC per liter of coating solids.

VOC content (Section 60.461) means the quantity, in kilograms per liter of coating solids, of volatile organic compounds (VOCs) in a coating.

VOC content (Section 60.491) means all volatile organic compounds (VOC) that are in a coating. VOC content is expressed in terms of kilograms of VOC per litre of coating solids.

VOC content (Sections 60.311; 60.451) means the proportion of a coating that is volatile organic compounds (VOCs), expressed as kilograms of VOCs per liter of coating solids.

VOC content of the coating applied (Section 60.711 [10/3/88]) means the product of Method 24 VOC analyses or formulation data (if the data are demonstrated to be equivalent to Method 24 results) and the total volume of coating fed to the coating applicator. This quantity is intended to include all VOC that actually are emitted from the coating operation in the gaseous phase. Thus, for purposes of the liquid-liquid VOC material balance in Section 60.713(b)(1), any VOC (including dilution solvent) added to the coatings must be accounted for, and any VOC contained in waste coatings or retained in the final product may be measured and subtracted from the total. (These adjustments are not necessary for the gaseous emission test compliance provisions of Section 60.713(b).)

VOC emission control device (Section 60.541 [9/15/87]) means equipment that destroys or recovers VOC.

VOC emission reduction system (Section 60.541 [9/15/87]) means a system composed of an enclosure, hood, or other device for containment and capture of VOC emissions and a VOC emission control device.

VOC emissions (Section 60.311) means the mass of volatile organic compounds (VOCs), expressed as kilograms of VOCs per liter of applied coating solids, emitted from a metal furniture surface coating operation.

VOC emissions (Section 60.451) means the mass of volatile organic compounds (VOCs), expressed as kilograms of VOCs per liter of applied coating solids, emitted from a surface coating operation.

VOC emissions (Section 60.721 [1/29/88]) means the mass of VOCs emitted from the surface coating of plastic parts for business machines expressed as kilograms of VOCs per liter of coating solids applied (i.e., deposited on the surface).

VOC in the applied coating (Section 60.741 [9-11-89]) means the product of Method 24 VOC analyses or formulation data (if those data are demonstrated to be equivalent to Method 24 results) and the total volume of coating fed to the coating applicator.

VOC solvent (Section 60.431) means an organic liquid or liquid mixture consisting of VOC components.

VOC used (Section 60.741 [9-11-89]) means the amount of VOC delivered to the coating mix preparation equipment of the affected facility (including any contained in premixed coatings or other coating ingredients prepared off the plant site) for the formulation of polymeric coatings to be applied to supporting substrates at the coating operation, plus any solvent added after initial formulation is complete (e.g., dilution solvent added at the coating operation). If premixed coatings that require no mixing at the plant site are used, "VOC used" means the amount of VOC delivered to the coating applicator(s) of the affected facility.

VOL (Section 60.111b). See **Volatile organic liquid (Section 60.111b)**.

Volatile hazardous air pollutant or

VHAP (Section 61.241) means a substance regulated under this part for which a standard for equipment leaks of the substance has been proposed and promulgated. Benzene is a VHAP. Vinyl chloride is a VHAP.

Volatile organic compounds (VOC) (Section 51.100 [2-3-92]) means any compound of carbon, excluding carbon monoxide, carbon dioxide, carbonic acid, metallic carbides or carbonates, and ammonium carbonate, which participates in atmospheric photochemical reactions. (1) This includes any such organic compound other than the following, which have been determined to have negligible photochemical reactivity: Methane; ethane; methylene chloride (dichloromethane); 1,1,1-trichloroethane (methyl chloroform); 1,1,1-trichloro-2,2,2-trifluoroethane (CFC-113); trichlorofluoromethane (CFC-11); dichlorodifluoromethane (CFC-12); chlorodifluoromethane (CFC-22); trifluoromethane (FC-23); 1,2-dichloro 1,1,2,2-tetrafluoroethane (CFC-114); chloropentafluoroethane (CFC-115); 1,1,1-trifluoro 2,2-dichloroethane (HCFC-123); 1,1,1,2-tetrafluoroethane (HFC-134a); 1,1-dichloro 1-fluoroethane (HCFC-141b): 1-chloro 1,1-difluoroethane (HCFC-142b); 2-chloro-1,1,1,2-tetrafluoroethane (HCFC-124); pentafluoroethane (HFC-125); 1,1,2,2-tetrafluoroethane (HFC-134); 1,1,1-trifluoroethane (HFC-143a);1,1-difluoroethane (HFC-152a); and perfluorocarbon compounds which fall into these classes: (i) Cyclic, branched, or linear, completely fluorinated alkanes; (ii) Cyclic, branched, or linear, completely fluorinated ethers with no unsaturations; (iii) Cyclic, branched, or linear, completely fluorinated tertiary amines with no unsaturations; and (iv) Sulfur containing perfluorocarbons with no unsaturations and with sulfur bonds only to carbon and fluorine. (2) For purposes of determining compliance with emissions limits, VOC will be measured by the test methods in the approved State implementation plan (SIP) or 40 CFR part 60, appendix A, as applicable. Where such a method also measures compounds with negligible photochemical reactivity, these negligibility-reactive compounds may be excluded as VOC if the amount of such compounds is accurately quantified, and such exclusion is approved by the enforcement authority. (3) As a precondition to excluding these compounds as VOC or at any time thereafter, the enforcement authority may require an owner or operator to provide monitoring or testing methods and results demonstrating, to the satisfaction of the enforcement authority, the amount of negligibly-reactive compounds in the source's emissions. (4) For purposes of Federal enforcement for a specific source, the EPA shall use the test methods specified in the applicable EPA-approved SIP, in a permit issued pursuant to a program approved or promulgated under title V of the Act, or under 40 CFR part 51, subpart I or appendix S, or under 40 CFR parts 52 or 60. The EPA shall not be bound by any State determination as to appropriate methods for testing or monitoring negligibly-reactive compounds if such determination is not reflected in any of the above provisions.

Volatile organic compounds (VOC) (Section 60.711 [10/3/88]) means any organic compounds that participate in atmospheric photochemical reactions or that are measured by Method 18, 24, 25, or 25A or an equivalent or alternative method as defined in 40 CFR 60.2.

Volatile organic compounds (VOC) (Section 60.481) means, for the purposes of this subpart, any reactive organic compounds as defined in Section 60.2 Definitions.

Volatile organic compounds or VOC (Section 60.741 [9-11-89]) means any organic compounds that participate in atmospheric photochemical reactions; or that are measured by a reference method, an equivalent method, an alternative method, or that are determined by procedures specified under any subpart.

Volatile organic liquid (VOL) (Section 60.111b) means any organic liquid which can emit volatile organic compounds into the atmosphere except those VOLs that emit only those compounds which the Administrator has determined do not contribute appreciably to the formation of ozone. These compounds are identified in EPA statements on ozone abatement policy for SIP revisions (42 FR 35314, 44 FR 32042, 45 FR 32424, and 45 FR 48941).

Volatile Organic Compound (Section 60.2) means any organic compound which participates in atmospheric photo-chemical reactions; or which is measured by a reference method, an equivalent method, an alternative method, or which is determined by procedures specified under any subpart.

Volatility (Section 162.3) means the property of a substance or substances to convert into vapor or gas without chemical change.

Volume of process water used per year (Section 463.31) is the volume of process water that flows through a finishing water process and comes in contact with the plastics product over a period of one year.

Volume of process water used per year (Section 463.21) is the volume of process water that flows through a cleaning process and comes in contact with the plastic product over a period of one year.

Volume of process water used per year (Section 463.11) is the volume of process water that flows through a contact cooling and heating water process and comes in contact with the plastic product over a period of one year.

Voluntarily submitted information (Section 2.201) means business information in EPA's possession (1) The submission of which EPA had no statutory or contractual authority to require; and (2) The submission of which was not prescribed by statute or regulation as a condition of obtaining some benefit (or avoiding some disadvantage) under a regulatory program of general applicability, including such regulatory programs as permit, licensing, registration, or certification programs, but excluding programs concerned solely or primarily with the award or administration by EPA of contracts or grants.

Voluntary Emissions Recall (Section 85.1902) shall mean a repair, adjustment, or modification program voluntarily initiated and conducted by a manufacturer to remedy any emission-related defect for which direct notification of vehicle or engine owners has been provided.

Volunteer (Section 300.6) means any individual accepted to perform services by a Federal agency which has authority to accept volunteer services (examples: See 16 U.S.C. 742f(c)). A volunteer is subject to the provisions of the authorizing statute, and Section 300.25.

W

Waiver (Section 13.2 [9/23/88]) means the cancellation, remission, forgiveness or non-recovery of a debt or debt-related charge as permitted or required by law.

Waiver (Section 35.4010 [10/1/92])* means excusing recipients from following certain anticipated regulatory or administrative requirements if; the authority to issue a waiver is provided in the regulation itself; and the Agency believes sufficient justification exists to approve such action. The Award Official has the authority to issue a waiver. Deviation means an exemption from certain provisions of existing regulations, which may be necessary in some unforeseen instances. The Director, Grants Administration Division, is authorized under 40 CFR 30.1001(b) to approve deviations from the requirements of regulations (except for those that implement statutory or executive order requirements) when such situations warrant special consideration.

Wall insulation (Section 248.4 [2/17/89]) means a material, primarily designed to resist heat flow, which is installed within or on the walls between conditioned areas of a building and unconditioned areas of a building or the outside, as well as common wall assemblies between separately conditioned units in multiple unit structures.

Warning Device (Section 201.1) means a sound emitting device used to alert and warn people of the presence of railroad equipment.

Warranty (Section 205.151) means the warranty required by section 6(d)(1) of the Act.

Warranty (Sections 204.2; 205.2) means the warranty required by section 6(c)(1) of the Act.

Warranty Booklet (Section 85.2102) means a booklet, separate from the owner's manual, containing all warranties provided with the vehicle.

Wash-oil decanter (Section 61.131 [9-19-91])* means any vessel that functions to separate, by gravity, the condensed water from the wash oil received from a wash-oil final cooler or from a light-oil scrubber.

Wash-oil circulation tank (Section 61.131 [9-14-89]) means any vessel that functions to hold the wash oil used in light-oil recovery operations or the wash oil used in the wash-oil final cooler.

Washer (Section 60.621) means a machine which agitates fabric articles in a petroleum solvent bath and spins the articles to remove the solvent, together with the piping and ductwork used in the installation of this device.

Washout (Section 257.3-1) means the carrying away of solid waste by waters of the base flood.

Washout (Section 264.18), as used in paragraph (b)(1) of this section, means the movement of hazardous waste from the active portion of the facility as a result of flooding.

Waste (Section 191.12 [12/20/93])* as used in this subpart, means any spent nuclear fuel or radioactive waste isolated in a disposal system.

Waste (Section 60.111b) means any liquid resulting from industrial, commercial, mining or agricultural operations, or from community activities that is discarded or is being accumulated, stored, or physically, chemically, or biologically treated prior to being discarded or recycled.

Waste (Section 61.341 [3-7-90]) means any material resulting from industrial, commercial, mining or agricultural operations, or from community activities that is discarded or is being accumulated, stored, or physically, chemically, thermally, or biologically treated prior to being discarded, recycled, or discharged.

Waste (Section 704.83) means any solid, liquid, semisolid, or contained gaseous material that results from the production of a chemical substance identified in paragraph (b) of this section and which is to be disposed.

Waste form (Section 191.12 [12/20/93])* means the materials comprising the radioactive components of waste and any encapsulating or stabilizing matrix.

Waste generator (Section 61.141 [11-20-90]) means any owner or operator of a source covered by this subpart whose act or process produces asbestos-containing waste material.

Waste management unit (Section 61.341 [3-7-90]) means a piece of equipment, structure, or transport mechanism used in handling, storage, treatment, or disposal of waste. Examples of a waste management unit include a tank, surface impoundment, container, oil-water separator, individual drain system, steam stripping unit, thin-film evaporation unit, waste incinerator, and landfill.

Waste management unit boundary (Section 258.2 [10-9-91]) means a vertical surface located at the hydraulically downgradient limit of the unit. This vertical surface extends down into the uppermost aquifer.

Waste Oil (Section 761.3) means used products primarily derived from petroleum, which include, but are not limited to, fuel oils, motor oils, gear oils, cutting oils, transmission fluids, hydraulic fluids, and dielectric fluids.

Waste paper (Section 250.4 [6/22/88]) means any of the following recovered materials: (1) Postconsumer materials such as: (i) Paper, paperboard, and fibrous wastes from retail stores, office buildings, homes, and so forth, after they have passed through their end usage as a consumer item, including: Used corrugated boxes, old newspapers, old magazines, mixed waste paper, tabulating cards, and used cordage, and (ii) All paper, paperboard, and fibrous wastes that enter and are collected from municipal solid waste; and (2) Manufacturing, forest residues, and other wastes such as: (i) Dry paper and paperboard waste generated after completion of the papermaking process (that is, those manufacturing operations up to and including the cutting and trimming of the paper machine reel into smaller rolls or rough sheets) including: Envelope cuttings, bindery trimmings, and other paper and paperboard waste, resulting from printing, cutting, forming, and other converting operations; bag, box, and carton manufacturing wastes; and butt rolls, mill wrappers, and rejected unused stock; and; (ii) Finished paper and paperboard from obsolete inventories of paper and paperboard manufacturers, merchants, wholesalers, dealers, printers, converters, or others.

Waste pile or pile (Section 257.2 [3/19/93])* means any noncontainerized accumulation of solid, nonflowing waste that is used for treatment or storage.

Waste shipment record (Section 61.141 [11-20-90]) means the shipping document, required to be originated and signed by the waste generator, used to track and substantiate the disposition of asbestos-containing waste material.

Waste stream (Section 61.341 [3-7-90]) means the waste generated by a particular process unit, product tank, or waste management unit. The characteristics of the waste stream (e.g., flow rate, benzene concentration, water content) are determined at the point of waste generation. Examples of a waste stream include process wastewater, product tank drawdown, sludge and slop oil removed from waste management units, and landfill leachate.

Waste treatment systems (Section 122.1 [12/22/93])* including treatment ponds or lagoons designed to meet the requirements of CWA (other than cooling ponds as defined in 40 CFR 423.11(m) which also meet the criteria of this definition) are not waters of the United States. This exclusion applies only to manmade bodies of water which neither were originally created in waters of the United States (such as disposal area in wetlands) nor resulted from the impoundment of waters of the United States. [See Note 1 of this section.] Waters of the United States do not include prior converted cropland. Notwithstanding the determination of an area's status as prior converted cropland by any other federal agency, for the purposes of the Clean Water Act, the final authority regarding Clean Water Act jurisdiction remains with EPA.

Wasteload allocation (WLA) (Section 130.2). The portion of a receiving water's loading capacity that is allocated to one of its existing or future point sources of pollution. WLAs constitute a type of water quality-based effluent limitation.

Wastewater collection system (Section 280.12 [9/23/88]). See **Storm-water collection system (Section 280.12)**.

Wastewater system (Section 60.691 [11/23/88]) means any component, piece of equipment, or installation that receives, treats, or processes oily wastewater from petroleum refinery process units.

Wastewater treatment (Section 61.341 [3-7-90]) means any component, piece of equipment, or installation that receives, manages, or treats process wastewater, product tank drawdown, or landfill leachate prior to direct or indirect discharge in accordance with the National Pollutant Discharge Elimination System permit regulations under 40 CFR part 122. These systems typically include individual drain systems, oil-water separators, air floatation units, equalization tanks, and biological treatment units.

Wastewater treatment process (Section 61.61) includes any process which modifies characteristics such as BOD, COD, TSS, and pH, usually for the purpose of meeting effluent guidelines and standards; it does not include any process the purpose of which is to remove vinyl chloride from water to meet requirements of this subpart.

Wastewater treatment system (Section 61.341 [3-7-90]) means any component, piece of equipment, or installation that receives, manages, or treats process wastewater, product tank drawdown, or landfill leachate prior to direct or indirect discharge in accordance with the National Pollutant Discharge Elimination System permit regulations under 40 CFR part 122. These systems typically include individual drain systems, oil-water separators, air flotation units, equalization tanks, and biological treatment units.

Wastewater treatment tank (Section 280.12 [9/23/88]) means a tank that is designed to receive and treat an influent wastewater through physical, chemical, or biological methods.

Wastewater treatment unit (Section 260.10 [9/2/88])* means a device which: (1) Is part of a wastewater treatment facility that is subject to regulation under either section 402 or 307(b) of the Clean Water Act; and (2) Receives and treats or stores an influent wastewater that is a hazardous waste as defined in Section 261.3 of this chapter, or that generates and accumulates a wastewater treatment sludge that is a hazardous waste as defined in Section 261.3 of this chapter, or treats or stores a wastewater treatment sludge which is a hazardous waste as defined in Section 261.3 of this Chapter; and (3) Meets the definition of tank or tank system in Section 260.10 of this chapter.

Wastewater treatment unit (Section 270.2) means a device which: (a) Is part of a wastewater treatment facility which is subject to regulation under either section 402 or 307(b) of the Clean Water Act; and (b) Receives and treats or stores an influent wastewater which is a hazardous waste as defined in Section 261.3 of this chapter, or generates and accumulates a wastewater treatment sludge which is a hazardous waste as defined in Section 261.3 of this chapter, or treats or stores a wastewater treatment sludge which is a hazardous waste as defined in Section 261.3 of this chapter; and (c) Meets the definition of tank in Section 260.10 of this chapter.

Wastewaters (Section 268.2 [6-1-90]) are wastes that contain less than 1% by weight total organic carbon (TOC) and less than 1% by weight total suspended solids (TSS), with the following exceptions: (1) F001, F002, F003, F004, F005 solvent-water mixtures that contain less than 1% by weight TOC or less than 1% by weight total F001, F002, F003, F004, F005 solvent constituents listed in Sec. 268.41, Table CCWE. (2) K011, K013, K014 wastewaters (as generated) that contain less than 5% by weight TOC and less than 1% by weight TSS. (3) K103 and K104 wastewaters contain less than 4% by weight TOC and less than 1% by weight TSS.

Water (bulk shipment) (Section 260.10) means the bulk transportation of hazardous waste which is loaded or carried on board a vessel without containers or labels.

Water area (Section 435.11 [3/4/93])* as used in the term "site" in 40 CFR 122.29 and 122.2 shall mean the water area and ocean floor beneath any exploratory, development, or production facility where such facility is conducting its exploratory, development or production activities.

Water dumping (Section 165.1) means the disposal of pesticides in or on lakes, ponds, rivers, sewers, or other water systems as defined in Pub. L. 92-500.

Water jet weaving (Section 410.31) shall mean the internal subdivision of the low water use processing subcategory for facilities primarily engaged in manufacturing woven greige goods through the water jet weaving process.

Water Management Division Director (Section 403.3) means one of the Directors of the Water Management Divisions within the Regional offices of the Environmental Protection Agency or this person's delegated representative.

Water pollution control agency (Section 15.4) means any agency which is defined in section 502(1) or section 502(2), 33 U.S.C. 1362(1) or (2), of the CWA.

Water quality limited segment (Section 131.3) means any segment where it is known that water quality does not meet applicable water quality standards, and/or is not expected to meet applicable water quality standards, even after the application of the technology-bases effluent limitations required by sections 301(b) and 306 of the Act.

Water quality limited segment (Section 130.2). Any segment where it is known that water quality does not meet applicable water quality standards, and/or is not expected to meet applicable water quality standards, even after the application of the technology-based effluent limitations required by sections 301(b) and 306 of the Act.

Water quality management (WQM) plan (Section 130.2). A State or areawide waste treatment management plan developed and updated in accordance with the provisions of sections 205(j), 208, and 303 of the Act and this regulation.

Water quality standards (Section 125.58) means applicable water quality standards which have been approved, left in effect, or promulgated under section 303 of the Clean Water Act.

Water quality standards (Section 131.3) are provisions of State or Federal law which consist of a designated use or uses for the waters of the United States and water quality criteria for such waters based upon such uses. Water quality standards are to protect the public health or welfare, enhance the quality of the water and serve the purposes of the Act.

Water quality standards (Section 121.1) means standards established pursuant to section 10(c) of the Act, and State-adopted water quality standards for navigable waters which are not interstate waters.

Water quality standards (WQS) (Section 130.2). Provisions of State or Federal law which consist of a designated use or uses for the waters of the United States and water quality criteria for such waters based upon such uses. Water quality standards are to protect the public health or welfare, enhance the quality of water and serve the purposes of the Act.

Water reducible (Section 60.391). See **Waterborne (Section 60.391)**.

Water seal controls (Section 60.691 [11/23/88]) means a seal pot, p-leg trap, or other type of trap filled with water that has a design capability to create a water barrier between the sewer and the atmosphere.

Water seal controls (Section 61.341 [1/7/93])* means a seal pot, p-leg trap, or other type of trap filled with water (e.g., flooded sewers that maintain water levels adequate to prevent air flow through the system) that creates a water barrier between the sewer line and the atmosphere. The water level of the seal must be maintained in the vertical leg of a drain in order to be considered a water seal.

Water table (Section 241.101) means the upper water level of a body of groundwater.

Water treatment chemicals (Section 749.68 [1-3-90]) means any combination of chemical substances used to treat water in cooling systems and can include corrosion inhibitors, antiscalants, dispersants, and any other chemical substances except biocides.

Water-based green tire spray (Section 60.541 [9/15/87]) means any mold release agent and lubricant applied to the inside or outside of green tires that contains 12 percent or less, by weight, of VOC as sprayed.

Waterborne coating (Section 60.741 [9-11-89]) means a coating which contains more than 5 weight percent water in its volatile fraction.

Waterborne disease outbreak (Section 141.2 [6/29/89]) means the significant occurrence of acute infectious illness, epidemiologically associated with the ingestion of water from a public water system which is deficient in treatment, as determined by the appropriate local or State agency.

Waterborne ink systems (Section 60.431) means ink and related coating mixtures whose volatile portion consists of a mixture of VOC solvent and more than five weight percent water, as applied to the gravure cylinder.

Waterborne or water reducible (Section 60.391) means a coating which contains more than five weight percent water in its volatile fraction.

Waters of the United States (Section 232.2 [2/11/93])* means: (1) All waters which are currently used, were used in the past, or may be susceptible to us in interstate or foreign commerce, including all waters which are subject to the ebb and flow of the tide. (2) All interstate waters including interstate wetlands. (3) All other waters, such as intrastate lakes, rivers, streams (including intermittent streams), mudflats, sandflats, wetlands, sloughs, prairie potholes, wet meadows, playa lakes, or natural ponds, the use, degradation, or destruction of which would or could affect interstate or foreign ommerce including any such waters: (i) Which are or could be used by interstate or foreign travelers for recreational or other purposes; or (ii) From which fish or shellfish are or could be taken and sold in interstate or foreign commerce; or (iii) Which are used or could be used for industrial purposes by industries in interstate commerce. (4) All impoundments of waters otherwise defined as waters of the United States under this definition; (5) Tributaries of waters identified in paragraphs (g)(1)-(4) of this section; (6) The territorial sea; and (7) Wetlands adjacent to waters (other than waters that are themselves wetlands) identified in paragraphs (q)(1)-(6) of this section. Waste treatment systems, including treatment ponds or lagoons designed to meet the equirements of the Act (other than cooling ponds as defined in 40 CFR 123.11(m) which also meet the criteria of this definition) are not waters of the United States. Waters of the United States do not include prior converted cropland. Notwithstanding the determination of an area's status as prior converted cropland by any other federal agency, for the purposes of the Clean Water Act, the final authority regarding Clean Water Act jurisdiction remains with EPA.

Waters of the United States (Section 721.3 [7/27/88]) has the meaning set forth in 40 CFR 122.2.

Waters of the United States (Section 257.3-3) is defined in the Clean Water Act, as amended, 33 U.S.C. 1251 et. seq., and implementing regulations, specifically 33 CFR Part 323 (42 FR 37122, July 19, 1977).

Waters of the United States (Section 230.3 [8/25/93])* do not include prior converted cropland. Notwithstanding the determination of an area's status as prior converted cropland by any other federal agency, for the purposes of the Clean Water Act, the final authority regarding Clean Water Act jurisdiction remains with EPA.

Waters of the United States or waters of the U.S. (Section 122.2) means: (a) All waters which are currently used, were used in the past, or may be susceptible to use in interstate or foreign commerce, including all waters which are subject to the ebb and flow of the tide; (b) All interstate waters, including interstate wetlands; (c) All other waters such as intrastate lakes, rivers, streams (including intermittent streams), mudflats, sandflats, wetlands, sloughs, prairie potholes, wet meadows, playa lakes, or natural ponds the use degradation, or destruction of which would affect or could affect interstate or foreign commerce including any such waters: (1) Which are or could be used by interstate or foreign travelers for recreational or other purposes; (2) From which fish or shellfish are or could be taken and sold in interstate or foreign commerce; or (3) Which are used or could be used for industrial purposes by industries in interstate commerce; (d) All impoundments of waters otherwise defined as waters of the United States under this definition; (e) Tributaries of waters identified in paragraphs (a) through (d) of this definition; (f) The territorial sea; and (g) Wetlands adjacent to waters (other than waters that are themselves wetlands) identified in paragraphs (a) through (f) of this definition. Waste treatment systems, including treatment ponds or lagoons designed to meet the requirements of CWA (other than cooling ponds as defined in 40 CFR 423.11(m) which also meet the criteria of this definition) are not waters of the United States. This exclusion applies only to manmade bodies of water which neither were originally created in waters of the United States (such as disposal area in wetlands) nor resulted from the impoundment of waters of the United States. (See Note 1 of this section.)

Waters of the United States or waters of the U.S. (Section 233.3) means: (a) All waters which are currently used, were used in the past, or may be susceptible to use in interstate or foreign commerce, including all waters which are subject to the ebb and flow of the tide; (b) All interstate waters, including interstate wetlands; (c) All other waters such as intrastate lakes, rivers, streams (including intermittent streams), mudflats, sandflats, wetlands, sloughs, prairie potholes, wet meadows, playa lakes, or natural ponds the use, degradation, or destruction of which would affect or could affect interstate or foreign commerce including any such waters: (1) Which are or could be used by interstate or foreign travelers for recreational or other purposes; (2) From which fish or shellfish are or could be taken and sold in interstate or foreign commerce; or (3) Which are used or could be used for industrial purposes by industries in interstate commerce.

WBE (Section 35.6015 [6/5/90])* See **Women's Business Enterprise (Section 35.6015)**.

Weak nitric acid (Section 60.71) means acid which is 30 to 70 percent in strength.

Web coating (Section 60.741 [9-11-89] means the coating of products, such as fabric, paper, plastic film, metallic foil, metal coil, cord, and yarn, that are flexible enough to be unrolled from a large roll; and coated as a continuous substrate by methods including, but not limited to, knife coating, roll coating, dip coating, impregnation, rotogravure, and extrusion.

Well (Section 260.10) means any shaft or pit dug or bored into the earth, generally of a cylindrical form, and often walled with bricks or tubing to prevent the earth from caving in.

Well (Section 435.61) shall mean crude oil producing wells and shall not include gas wells or wells injecting water for disposal or for enhanced recovery of oil or gas.

Well (Sections 144.3; 146.3; 147.2902) means a bored, drilled or driven shaft, or a dug hole, whose depth is greater than the largest surface dimension.

Well completion fluids (Section 435.11 [3/4/93])* shall refer to salt solutions, weighted brines, polymers, and various additives used to prevent damage to the well bore during operations which prepare the drilled well for hydrocarbon production.

Well injection (Section 165.1) means disposal of liquid wastes through a hole or shaft to a subsurface stratum.

Well injection (Section 260.10). See **Underground injection (Section 260.10)**.

Well injection (Sections 144.3; 146.3; 147.2902) means the subsurface emplacement of fluids through a bored, drilled, or driven well; or through a dug well, where the depth of the dug well is greater than the largest surface dimension.

Well monitoring (Section 146.3) means the measurement, by on-site instruments or laboratory methods, of the quality of water in a well.

Well plug (Section 146.3) means a watertight and gastight seal installed in a borehole or well to prevent movement of fluids.

Well stimulation (Section 146.3) means several processes used to clean the well bore, enlarge channels, and increase pore space in the interval to be injected thus making it possible for wastewater to move more readily into the formation, and includes (1) surging, (2) jetting, (3) blasting, (4) acidizing, (5) hydraulic fracturing.

Well treatment fluids (Section 435.11 [3/4/93])* shall refer to any fluid used to restore or improve productivity by chemically or physically altering hydrocarbon-bearing strata after a well has been drilled.

Well workover (Section 147.2902) means any reentry of an injection well; including, but not limited to, the pulling of tubular goods, cementing or casing repairs; and excluding any routine maintenance (e.g. reseating the packer at the same depth, or repairs to surface equipment).

Wet (Section 420.41) means those steelmaking air cleaning systems that primarily use water for furnace gas cleaning.

Wet air pollution control scrubbers (Section 471.02) are air pollution control devices used to remove particulates and fumes from air by entraining the pollutants in a water spray.

Wet barking operations (Section 430.01) shall be defined to include hydraulic barking operations and wet drum barking operations which are those drum barking operations that use substantial quantities of water in either water sprays in the barking drums or in a partial submersion of the drums in a tub of water.

Wet desulfurization system (Section 420.11) means those systems which remove sulfur compounds from coke oven gases and produce a contaminated process wastewater.

Wet flue gas desulfurization technology (Section 60.41b [12/16/87]) means a sulfur dioxide control system that is located downstream of the steam generating unit and removes sulfur oxides from the combustion gases of the steam generating unit by contacting the combustion gas with an alkaline slurry or solution and forming a liquid material. This definition applies to devices where the aqueous liquid material product of this contact is subsequently converted to other forms. Alkaline reagents used in wet flue gas desulfurization technology include, but are not limited to, lime, limestone, and sodium.

Wet flue gas desulfurization technology (Section 60.41c [9-12-90]) means an SO 2 control system that is located between the steam generating unit and the exhaust vent or stack, and that removes sulfur oxides from the combustion gases of the steam generating unit by contacting the combustion gases with an alkaline slurry or solution and forming a liquid material. This definition includes devices where the liquid material is subsequently converted to another form. Alkaline reagents used in wet flue gas desulfurization systems include, but are not limited to, lime, limestone, and sodium compounds.

Wet lot (Section 412.21) shall mean a confinement facility for raising ducks which is open to the environment with a small portion of shelter area, and with open water runs and swimming areas to which ducks have free access.

Wet mixture (Section 723.175) means a water or organic solvent-based suspension, solution, dispersion, or emulsion used in the manufacture of an instant photographic or peel-apart film article.

Wet scrubber system (Section 60.41b [12/16/87])* means any emission control device that mixes an aqueous stream or slurry with the exhaust gases from a steam generating unit to control emissions of particulate matter or sulfur dioxide.

Wet scrubber system (Section 60.41c [9-12-90]) means any emission control device that mixes an aqueous stream or slurry with the exhaust gases from a steam generating unit to control emissions of particulate matter (PM) or SO 2.

Wet scrubbers (Section 467.02) are air pollution control devices used to remove particulates and fumes from air by entraining the pollutants in a water spray.

Wetlands (Part 6, Appendix A, Section 4) means those areas that are inundated by surface or ground water with a frequency sufficient to support and under normal circumstances does or would support a prevalence of vegetative or aquatic life that requires saturated or seasonally saturated soil conditions for growth and reproduction. Wetlands generally include swamps, marshes, bogs, and similar areas such as sloughs, potholes, wet meadows, river overflows, mud flats, and natural ponds.

Wetlands (Section 110.1) means those areas that are inundated or saturated by surface or ground water at a frequency or duration sufficient to support, and that under normal circumstances do support, a prevalence of vegetation typically adapted for life in saturated soil conditions. Wetlands generally include playa lakes, swamps, marshes, bogs and similar areas such as sloughs, prairie potholes, wet meadows, prairie river overflows, mudflats, and natural ponds.

Wetlands (Section 230.3 [8/25/93])* means those areas that are inundated or saturated by surface or ground water at a frequency and duration sufficient to support, and that under normal circumstances do support, a prevalence of vegetation typically adapted for life in saturated soil conditions. Wetlands generally include swamps, marshes, bogs and similar areas.

Wetlands (Section 232.2 [2/11/93])* means those areas that are inundated or saturated by surface or ground water at a frequency and duration sufficient to support, and that under normal circumstances do support, a prevalence of vegetation typically adapted for life in saturated soil conditions. Wetlands generally include swamps, marshes, bogs, and similar areas.

Wetlands (Section 257.3-3) is defined in the Clean Water Act, as amended, 33 U.S.C. 1251 et. seq., and implementing regulations, specifically 33 CFR Part 323 (42 FR 37122, July 19, 1977).

Wetlands (Section 435.41) shall mean those surface areas which are inundated or saturated by surface or ground water at a frequency and duration sufficient to support, and that under normal circumstances do support, a prevalence of vegetation typically adapted for life in saturated soil conditions. Wetlands generally include swamps, marshes, bogs, and similar areas.

Wetlands (Sections 122.2; 233.3) means those areas that are inundated or saturated by surface or ground water at a frequency and duration sufficient to support, and that under normal circumstances do support, a prevalence of vegetation typically adapted for life in saturated soil conditions. Wetlands generally include swamps, marshes, bogs, and similar areas.

Wheat (Section 406.41) shall mean wheat delivered to a plant before processing.

Whole body (Section 61.101) means all organs or tissues exclusive of the integumentary system (skin) and the cornea.

Whole body (Section 61.91) means all human organs or tissue exclusive of the integumentary system (skin) and the cornea.

Whole effluent toxicity (Section 122.2 [6/2/89]) means the aggregate toxic effect of an effluent measured directly by a toxicity test.

Wholesale purchaser-consumer (Section 80.2 [2/16/94])* means any organization that is an ultimate consumer of gasoline or diesel fuel and which purchases or obtains gasoline or diesel fuel from a supplier for use in motor vehicles and receives delivery of that product into a storage tank of at least 550-gallon capacity substantially under the control of that organization.

Wire products and fasteners (Section 420.121) means steel wire, products manufactured from steel wire, and steel fasteners manufactured from steel wire or other steel shapes.

With modified-processes (Section 60.291) means using any technique designed to minimize emissions without the use of add-on pollution controls.

Withdraw specification (Section 231.2) means to remove from designation any area already specified as a disposal site by the U.S. Army Corps of Engineers or by a state which has assumed the section 404 program, or any portion of such area.

Within the impoundment (Section 421.41(c)), for all impoundments constructed prior to the effective date of the interim final regulation (40 FR 8513), when used to calculate the volume of process wastewater which may be discharged, means the water surface area within the impoundment at maximum capacity plus the surface area of the inside and outside slopes of the impoundment dam as well as the surface area between the outside edge of the impoundment dam and any seepage ditch adjacent to the dam upon which rain falls and is returned to the impoundment. For the purpose of such calculations, the surface area allowances set forth above shall not exceed more than 30 percent of the water surface area within the impoundment dam at maximum capacity.

Within the impoundment (Section 421.61(b)), for all impoundments constructed prior to the effective date of this regulation, when used for purposes of calculating the volume of process wastewater which may be discharged shall mean the water surface area within the impoundment at maximum capacity plus the surface area of the inside and outside slopes of the impoundment dam as well as the surface area between the outside edge of the impoundment dam and any seepage ditch immediately adjacent to the dam upon which rain falls and is returned to the impoundment. For the purpose of such calculations, the surface area allowances set forth above shall not be more than 30 percent of the water surface area within the impoundment dam at maximum capacity.

Within the impoundment (Section 421.41(d)), for all impoundments constructed on or after the effective date of the interim final regulation (the interim regulation was effective February 27, 1975; 40 FR 8513, February 27, 1975), for purposes of calculating the volume of process wastewater which may be discharged, means the water surface area within the impoundment at maximum capacity.

Within the impoundment (Section 421.61(c)) for all impoundments constructed on or after the effective date of this regulation, for purposes of calculating the volume of process wastewater which may be discharged shall mean the water surface area within the impoundment at maximum capacity.

Within the impoundment (Section 421.11), for all impoundments, for purposes of calculating the volume of process wastewater which may be discharged, shall mean the surface area within the impoundment at the maximum capacity plus the area of the inside and outside slopes of the impoundment dam and the surface area between the outside edge of the impoundment dam and seepage ditches upon which rain falls and is returned to the impoundment. For the purpose of such calculations, the surface area allowance for external appurtenances to the impoundment shall not be more than 30 percent of the water surface area within the impoundment dam at maximum capacity.

WL (Section 192.11). See **Working Level (Section 192.11)**.

WLA (Section 130.2). See **Wasteload allocation (Section 130.2)**.

W_{oi} (Section 60.741 [9-11-89]) means the weight fraction of VOC in each coating (i) applied at an affected coating operation during a nominal 1-month period as determined by Method 24.

Women's business enterprise (Section 33.005) is a business which is certified as such by a State or Federal agency, or which meets the following definition: A women's business enterprise is an independent business concern which is at least 51 percent owned by a woman or women who also control and operate it. Determination of whether a business is at least 51 percent owned by a woman or women shall be made without regard to community property laws. For example, an otherwise qualified WBE which is 51 percent owned by a married woman in a community property state will not be disqualified because her husband has a 50 percent interest in her share. Similarly, a business which is 51 percent owned by a married man and 49 percent owned by an unmarried woman will not become a qualified WBE by virtue of his wife's 50 percent interest in his share of the business.

Women's Business Enterprise (WBE) (Section 35.6015 [6-5-90])* A business which is certified as a Women's Business Enterprise by a State or Federal agency, or which meets the following definition. A Women's Business Enterprise is an independent business concern which is at least 51 percent owned by a woman or women who also control and operate it. Determination of whether a business is at least 51 percent owned by a woman or women shall be made without regard to community property laws. (b) Those terms not defined in this section shall have the meanings set forth in section 101 of CERCLA, 40 CFR part 31 and 40 CFR part 300 (the National Contingency Plan).

Wood (Section 60.41b [12/16/87])* means wood, wood residue, bark, or any derivative fuel or residue thereof, in any form, including, but not limited to, sawdust, sanderdust, wood chips, scraps, slabs, millings, shavings, and processed pellets made from wood or other forest residues.

Wood (Section 60.41c [9-12-90]) means wood, wood residue, bark, or any derivative fuel or residue thereof, in any form, including but not limited to sawdust, sanderdust, wood chips, scraps, slabs, millings, shavings, and processed pellets made from wood or other forest residues.

Wood fiber furnish subdivision mills (Section 430.181) are those mills where cotton fibers are not used in the production of fine papers.

Wood heater (Section 60.531 [2/26/88]) means an enclosed, woodburning appliance capable of and intended for space heating and domestic water heating that meets all of the following criteria: (a) An air-to-fuel ratio in the combustion chamber averaging less than 35-to-1 as determined by the test procedure prescribed in Section 60.534 performed at an accredited laboratory, (b) A usable firebox volume of less than 20 cubic feet, (c) A minimum burn rate less than 5 kg/hr as determined by the test procedure prescribed in Section 60.534 performed at an accredited laboratory, and (d) A maximum weight of 800 kg. In determining the weight of an appliance for these purposes, fixtures and devices that are normally sold separately, such as flue pipe, chimney, and masonry components that are not an integral part of the appliance or heat distribution ducting, shall not be included.

Wood residue (Section 60.41) means bark, sawdust, slabs, chips, shavings, mill trim, and other wood products derived from wood processing and forest management operations.

Wool (Section 410.11) shall mean the dry raw wool as it is received by the wool scouring mill.

Wool fiberglass (Section 60.291) means fibrous glass of random texture, including fiberglass insulation, and other products listed in SIC 3296.

Wool fiberglass insulation (Section 60.681) means a thermal insulation material composed of glass fibers and made from glass produced or melted at the same facility where the manufacturing line is located.

Work (Section 61.20) means mining activity done in the usual and ordinary course of developing and operating a mine.

Work area (Section 721.3 [7-27-89]) means a room or defined space in a workplacc where a chemical substance is manufactured, processed, or used and where employees are present.

Work program (Section 35.105) means the document which identifies how and when the applicant will use program funds to produce specific outputs.

Work Program (Section 35.9010 [10-3-89]) The Scope of Work of an assistance application, which identifies how and when the applicant will use funds to produce specific outputs.

Worker (Section 170.3 [8/21/92])* means any person, including a self-employed person, who is employed for any type of compensation and who is performing activities relating to the production of agricultural plants on an agricultural establishment to which subpart B of this part applies. While persons employed by a commercial pesticide handling establishment are performing tasks as crop advisors, they are not workers covered by the requirements of subpart B of this part.

Working day (Section 129.2) means the hours during a calendar day in which a facility discharges effluents subject to this part.

Working day (Section 61.141 [11-20-90]) means Monday through Friday and includes holidays that fall on any of the days Monday through Friday.

Working day (Section 85.1502 [9/25/87]) means any day on which Federal government officcs are open for normal business. Saturdays, Sundays, and official Federal holidays are not working days.

Working day (Sections 2.201; 350.1 [7/29/88]) is any day on which Federal government offices are open for normal business. Saturdays, Sundays, and official Federal holidays are not working days; all other days are.

Working days (Section 16.2) means calendar days excluding Saturdays, Sundays, and legal public holidays.

Working face (Section 241.101)

means that portion of the land disposal site where solid wastes are discharged and are spread and compacted prior to the placement of cover material.

Working Level (WL) (Section 192.11) means any combination of short-lived radon decay products in one liter of air that will result in the ultimate emission of alpha particles with a total energy of 130 billion electron volts.

Workover fluids (Section 435.11 [3/4/93])* shall refer to salt solutions, weighted brines, polymers, or other specialty additives used in a producing well to allow for maintenance, repair or abandonment procedures.

Workplace (Section 721.3 [7-27-89]) means an establishment at one geographic location containing one or more work areas.

Worst case discharge for an onshore non-transportation-related facility (Section 112.2 [8/25/93])* means the largest foreseeable discharge in adverse weather conditions as determined using the worksheets in Appendix D to this part.

WQM plan (Section 130.2). See **Water quality management plan (Section 130.2)**.

WQS (Section 130.2). See **Water quality standards (Section 130.2)**.

Writing paper (Section 250.4 [10/6/87, 6/22/88]) means a paper suitable for pen and ink, pencil, typewriter or printing.

Written Instructions for Proper Maintenance and Use (Section 85.2102) means those maintenance and operation instructions specified in the owner's manual as being necessary to assure compliance of a vehicle with applicable emission standards for the useful life of the vehicle that are: (i) In accordance with the instructions specified for performance on the manufacturer's prototype vehicle used in certification (including those specified for vehicles used under special circumstances), and (ii) In compliance with the requirements of Section 86.XXX-38 (as appropriate for the applicable model year vehicle/engine classification), and (iii) In compliance with any other regulations promulgated by EPA governing maintenance and use instructions.

XYZ

x (Section 409.61) shall mean that fraction of the net cane harvested by the advanced harvesting systems.

X (Section 60.641) is the sulfur feed rate, i.e., the H_2S in the acid gas (expressed as sulfur) from the sweetening unit, expressed in long tons per day (LT/D) of sulfur rounded to one decimal place.

Xerographic/copy paper (Section 250.4 [10/6/87, 6/22/88]) means any grade of paper suitable for copying by the xerographic process (a dry method of reproduction).

Y (Section 60.641) is the sulfur content of the acid gas from the sweetening unit, expressed as mole percent H_2S (dry basis) rounded to one decimal place.

Year (Section 704.142) means corporate fiscal year.

Z (Section 60.641) is the minimum required sulfur dioxide (SO_2) emission reduction efficiency, expressed as percent carried to one decimal place. Z_i refers to the reduction efficiency required at the initial performance test. Z_c refers to the reduction efficiency required on a continuous basis after compliance with Z_i has been demonstrated.

Zero (0) hours (Section 86.082-2) means that point after normal assembly line operations and adjustments are completed and before ten (10) additional operating hours have been accumulated, including emission testing, if performed.

Zero (0) miles (Section 86.082-2) means that point after initial engine starting (not to exceed 100 miles of vehicle operation, or three hours of engine operation) at which normal assembly line operations and adjustments are completed, and including emission testing, if performed.

Zero device-miles (Section 610.11) means the period of time between retrofit installation and the accumulation of 100 miles of automobile operation after installation.

Zero kilometers (Section 86.402-78) means that point after normal assembly line operations and adjustments, after normal dealer setup and preride inspection operations have been completed, and before 100 kilometers of vehicle operation or three hours of engine operation have been accumulated, including emission testing if performed.

ZID (Section 125.58). See **Zone of initial dilution (Section 125.58)**.

Zinc (Section 420.02) means total zinc and is determined by the method specified in 40 CFR 136.3.

Zinc Casting (Section 464.02). The remelting of zinc or zinc alloy to form a cast intermediate or final product by pouring or forcing the molten metal into a mold, except for ingots, pigs, or other cast shapes related to nonferrous (primary) metals manufacturing (40 CFR Part 421) and nonferrous metals forming (40 CFR Part 471). Processing operations following the cooling of castings not covered under nonferrous metals forming are covered under the electroplating and metal finishing point source categories (40 CFR Parts 413 and 433).

Zone of aeration (Section 260.10). See **Unsaturated zone (Section 260.10)**.

Zone of engineering control (Section 260.10) means an area under the control of the owner/operator that, upon detection of a hazardous waste release, can be readily cleaned up prior to the release of hazardous waste or hazardous constituents to ground water or surface water.

Zone of initial dilution (ZID) (Section125.58) means the region of initial mixing surrounding or adjacent to the end of the outfall pipe or diffuser ports, provided that the ZID may not be larger than allowed by mixing zone restrictions in applicable water quality standards.

Zone of saturation (Section 260.10). See **Saturated zone (Section 260.10)**.

APPENDIX

GUIDE TO USING THE APPENDIX

I. Construction and interrelationship of the *Code of Federal Regulations* and the *Federal Register* (excerpted from the "Explanation" section of the *Code of Federal Regulations, Title 40, Protection of Environment*, Revised as of July 1, 1987):

"The *Code of Federal Regulations* is a codification of the general and permanent rules published in the *Federal Register* by the Executive departments and agencies of the Federal Government. The Code is divided into 50 titles which represent broad areas subject to Federal Regulation. Each title is divided into chapters which usually bear the name of the issuing agency. Each chapter is further subdivided into parts covering specific regulatory areas." [See "How to Use This Appendix" (II) below for a description of 40 CFR subdivisions.]

"Each volume of the Code is revised at least once each calendar year... The appropriate revision date is printed on the cover of each volume." [Title 40 is revised as of July 1 of each year.]

"The *Code of Federal Regulations* is kept up to date by the individual issues of the *Federal Register*. These two publications must be used together to determine the latest version of any given rule." [The *Federal Register* is issued daily.]

II. How to Use This Appendix

A. Description of Title 40 subdivisions:

The *Code of Federal Regulations, Title 40, Protection of Environment*, Revised as of July 1, 1987, comprises eleven volumes. The two chapters currently under this title are: Chapter I - Environmental Protection Agency (Parts 1-799); and, Chapter V - Council on Environmental Quality (Parts 1500-1599).

Chapter I is divided into "Subchapters" which identify the subjects dealt with in this chapter (e.g., programs administered by EPA, policies, guidelines and standards, acts, etc.). These subject matters are subdivided into "Parts," which cover specific regulatory areas.

"Parts" are further divided into "Subparts," which narrow the regulatory areas into definitive subcategories. The elements of a "Subpart" are dealt with in "Sections."

B. How to find the topic to which a definition applies:

Referring to the Section number appearing with a definition in this dictionary, turn to the appendix and locate the definition's Part number (this is the portion of the section number preceding the decimal point). The Part title listed in the appendix describes the regulatory area under which the definition falls. Then, scan the Subparts under this Part until you locate the Section number that appears with the definition. The Subpart under which the Section number is listed identifies the subcategory for this Section.

APPENDIX

INDEX TO 40 CFR

Title 40--Protection of the Environment

CHAPTER I--ENVIRONMENTAL PROTECTION AGENCY

SUBCHAPTER A--GENERAL

PART 1--STATEMENT OF ORGANIZATION AND GENERAL INFORMATION

Subpart A--Introduction

§ 1.1 Creation and authority.

§ 1.3 Purpose and functions.

§ 1.5 Organization and general information.

§ 1.7 Location of principal offices.

Subpart B--Headquarters

§ 1.21 General.

§ 1.23 Office of the Administrator.

§ 1.25 Staff Offices.

§ 1.27 Offices of the Associate Administrators.

§ 1.29 Office of Inspector General.

§ 1.31 Office of General Counsel.

§ 1.33 Office of Administration and Resources Management.

§ 1.35 Office of Enforcement and Compliance Monitoring.

§ 1.37 Office of External Affairs.

§ 1.39 Office of Policy, Planning and Evaluation.

§ 1.41 Office of Air and Radiation.

§ 1.43 Office of Prevention, Pesticides, and Toxic Substances.

§ 1.45 Office of Research and Development.

§ 1.47 Office of Solid Waste and Emergency Response.

§ 1.49 Office of Water.

Subpart C--Field Installations

§ 1.61 Regional Offices.

PART 2--PUBLIC INFORMATION

Subpart A--Requests for Information

§ 2.100 Definitions.

§ 2.101 Policy on disclosure of EPA records.

§ 2.102 [Reserved]

§ 2.103 Partial disclosure of records.

§ 2.104 Requests to which this subpart applies.

§ 2.105 Existing records.

§ 2.106 Where requests for agency records shall be filed.

§ 2.107 Misdirected written requests; oral requests.

§ 2.108 Form of request.

§ 2.109 Requests which do not reasonably describe records sought.

§ 2.110 Responsibilities of Freedom of Information Officers.

§ 2.111 Action by office responsible for responding to request.

§ 2.112 Time allowed for issuance of initial determination.

§ 2.113 Initial denials of requests.

§ 2.114 Appeals from initial denials; manner of making.

§ 2.115 Appeal determinations; by whom made.

§ 2.116 Contents of determination denying appeal.

§ 2.117 Time allowed for issuance of appeal determination.

§ 2.118 Exemption categories.

§ 2.119 Discretionary release of exempt documents.
§ 2.120 Fees; payment; waiver.
§ 2.121 Exclusions.

Subpart B--Confidentiality of Business Information

§ 2.201 Definitions.
§ 2.202 Applicability of subpart; priority where provisions conflict; records containing more than one kind of information.
§ 2.203 Notice to be included in EPA requests, demands, and forms; method of asserting business confidentiality claim; effect of failure to assert claim at time of submission.
§ 2.204 Initial action by EPA office.
§ 2.205 Final confidentiality determination by EPA legal office.
§ 2.206 Advance confidentiality determinations.
§ 2.207 Class determinations.
§ 2.208 Substantive criteria for use in confidentiality determinations.
§ 2.209 Disclosure in special circumstances.
§ 2.210 Nondisclosure for reasons other than business confidentiality or where disclosure is prohibited by other statute.
§ 2.211 Safeguarding of business information; penalty for wrongful disclosure.
§ 2.212 Establishment of control offices for categories of business information.
§ 2.213 Designation by business of addressee for notices and inquiries.
§ 2.214 Defense of Freedom of Information Act suits; participation by affected business.
§ 2.215 Confidentiality agreements.
Secs. 2.216--2.300 [Reserved]
§ 2.301 Special rules governing certain information obtained under the Clean Air Act.
§ 2.302 Special rules governing certain information obtained under the Clean Water Act.
§ 2.303 Special rules governing certain information obtained under the Noise Control Act of 1972.
§ 2.304 Special rules governing certain information obtained under the Safe Drinking Water Act.
§ 2.305 Special rules governing certain information obtained under the Solid Waste Disposal Act, as amended.
§ 2.306 Special rules governing certain information obtained under the Toxic Substances Control Act.
§ 2.307 Special rules governing certain information obtained under the Federal Insecticide, Fungicide and Rodenticide Act.
§ 2.308 Special rules governing certain information obtained under the Federal Food, Drug and Cosmetic Act.
§ 2.309 Special rules governing certain information obtained under the Marine Protection, Research and Sanctuaries Act of 1972.

administrative assessment of civil penalties under the Toxic Substances Control Act.

§ 22.34 Supplemental rules of practice governing the administrative assessment of civil penalties under Title II of the Clean Air Act.

§ 22.35 Supplemental rules of practice governing the administrative assessment of civil penalties under the Federal Insecticide, Fungicide, and Rodenticide Act.

§ 22.36 Supplemental rules of practice governing the administrative assessment of civil penalties and the revocation or suspension of permits under the Marine Protection, Research, and Sanctuaries Act.

§ 22.37 Supplemental rules of practice governing the administrative assessment of civil penalties under the Solid Waste Disposal Act.

§ 22.38 Supplemental rules of practice governing the administrative assessment of Class II penalties under the Clean Water Act.

§ 22.39 Supplemental rules of practice governing the administrative assessment of administrative penalties under section 109 of the Comprehensive Environmental Response, Compensation, and Liability Act of 1980, as amended.

§ 22.40 Supplemental rules of practice governing the administrative assessment of administrative penalties under section 325 of the Emergency Planning and Community Right-To-Know Act of 1986 (EPCRA).

§ 22.41 Supplemental rules of practice governing the administrative assessment of civil penalties under Title II of the Toxic Substances Control Act, enacted as section 2 of the Asbestos Hazard Emergency Response Act (AHERA).

§ 22.42 Supplemental rules of practice governing the administrative assessment of civil penalties for violations of compliance orders issued under Part B of the Safe Drinking Water Act.

§ 22.43 Supplemental rules of practice governing the administrative assessment of civil penalties under Section 113(d)(1) of the Clean Air Act.

Appendix--Addresses of EPA Regional Offices

PART 23--JUDICIAL REVIEW UNDER EPA--ADMINISTERED STATUTES

§ 23.1 Definitions.

§ 23.2 Timing of Administrator's action under Clean Water Act.

§ 23.3 Timing of Administrator's action under Clean Air Act.

§ 23.4 Timing of Administrator's action under Resource Conservation and Recovery Act.

§ 23.5 Timing of Administrator's action under Toxic Substances Control Act.

§ 23.6 Timing of Administrator's action under Federal Insecticide, Fungicide and Rodenticide Act.

§ 23.7 Timing of Administrator's action under Safe Drinking Water Act.

§ 27.8 Service of complaint.
§ 27.9 Answer.
§ 27.10 Default upon failure to file an answer.
§ 27.11 Referral of complaint and answer to the presiding officer.
§ 27.12 Notice of hearing.
§ 27.13 Parties to the hearing.
§ 27.14 Separation of functions.
§ 27.15 Ex parte contacts.
§ 27.16 Disqualification of the reviewing official or presiding officer.
§ 27.17 Rights of parties.
§ 27.18 Authority of the presiding officer.
§ 27.19 Prehearing conferences.
§ 27.20 Disclosure of documents.
§ 27.21 Discovery.
§ 27.22 Exchange of witness lists, statements and exhibits.
§ 27.23 Subpoenas for attendance at hearing.
§ 27.24 Protective order.
§ 27.25 Fees.
§ 27.26 Form, filing and service of papers.
§ 27.27 Computation of time.
§ 27.28 Motions.
§ 27.29 Sanctions.
§ 27.30 The hearing and burden of proof.
§ 27.31 Determining the amount of penalties and assessments.
§ 27.32 Location of hearing.
§ 27.33 Witnesses.
§ 27.34 Evidence.
§ 27.35 The record.
§ 27.36 Post-hearing briefs.
§ 27.37 Initial decision.
§ 27.38 Reconsideration of initial decision.
§ 27.39 Appeal to authority head.
§ 27.40 Stay ordered by the Department of Justice.
§ 27.41 Stay pending appeal.
§ 27.42 Judicial Review.
§ 27.43 Collection of civil penalties and assessments.
§ 27.44 Right to administrative offset.
§ 27.45 Deposit in Treasury of United States.
§ 27.46 Compromise or settlement.
§ 27.47 Limitations.
§ 27.48 Delegated functions.

PART 29--INTERGOVERNMENTAL REVIEW OF ENVIRONMENTAL PROTECTION AGENCY PROGRAMS AND ACTIVITIES

§ 29.1 What is the purpose of these regulations?
§ 29.2 What definitions apply to these regulations?
§ 29.3 What programs and activities of the Environmental Protection Agency are subject to these regulations?
§ 29.4 What are the Administrator's general responsibilities under the Order?

§ 29.5 What is the Administrator's obligation with respect to Federal interagency coordination?
§ 29.6 What procedures apply to the selection of programs and activities under these regulations?
§ 29.7 How does the Administrator communicate with State and local officials concerning the EPA programs and activities?
§ 29.8 How does the Administrator provide States an opportunity to comment on proposed Federal financial assistance and direct Federal development?
§ 29.9 How does the Administrator receive and respond to comments?
§ 29.10 How does the Administrator make efforts to accommodate intergovernmental concerns?
§ 29.11 What are the Administrator's obligations in interstate situations?
§ 29.12 How may a State simplify, consolidate, or substitute federally required State plans?
§ 29.13 May the Administrator waive any provision of these regulations?
SUBCHAPTER B--GRANTS AND OTHER FEDERAL ASSISTANCE
PART 30--GENERAL REGULATION FOR ASSISTANCE PROGRAMS FOR OTHER THAN STATE AND LOCAL GOVERNMENTS
Subpart A--What is the Purpose and Scope of this Regulation?
§ 30.100 What is the purpose of this regulation?
§ 30.101 What is the scope of this regulation?
§ 30.102 What laws authorize EPA to issue this regulation?
Subpart B--What Definitions Apply to this Regulation?
§ 30.200 What definitions apply to this regulation?
Subpart C--How do I Apply for and Receive Assistance?
§ 30.300 What activities does EPA fund?
§ 30.301 To whom does EPA award assistance?
§ 30.302 How do I apply for assistance?
§ 30.303 What steps must I take when filing a standard application?
§ 30.304 Is the information I submit to EPA confidential?
§ 30.305 How do I find out if EPA approved or disapproved my application?
§ 30.306 How long will I have to complete my project?
§ 30.307 How much must I contribute to the funding of my project?
§ 30.308 When may I begin incurring costs?
§ 30.309 What is the effect of accepting an assistance agreement?
Subpart D--How does EPA Pay Me?
§ 30.400 How does EPA make payments?
§ 30.405 Can I assign my payment to anyone else?
§ 30.410 How does EPA determine allowable costs?
§ 30.412 How are costs categorized?
Subpart E--How do I Manage My Award?
§ 30.500 What records must I maintain?
§ 30.501 How long must I keep these records?
§ 30.502 To whom must my contractor and I show these

records?
§ 30.503 What type of quality assurance practices am I required to have?
§ 30.505 What reports must I submit?
§ 30.510 What type of financial management system must I maintain?
§ 30.515 What restrictions on signs, surveys, and questionnaires must I observe?
§ 30.518 What are the procedures for publishing scientific, informational, and educational documents?
§ 30.520 When may I use my own employees ("force account")?
§ 30.525 How should I treat program income?
§ 30.526 How do I treat interest earned on EPA funds?
§ 30.530 May I purchase personal property using EPA assistance funds?
§ 30.531 What property management standards must I follow for nonexpendable personal property purchased with an EPA award?
§ 30.532 How do I dispose of personal property?
§ 30.535 May I purchase real property with EPA awarded funds?
§ 30.536 How do I manage federally-owned property?
§ 30.537 Are contractors required to comply with EPA property policies?
§ 30.538 May I use General Services Administration (GSA) supplies and services?
§ 30.540 Who will audit my project?
§ 30.541 What are my audit responsibilities?
§ 30.600 What Federal laws and policies affect my award?
§ 30.601 Are there restrictions on the use of assistance funds for advocacy purposes?
§ 30.603 What additional Federal laws apply to EPA assisted construction projects?
§ 30.610 What are my responsibilities for preventing and detecting fraud and other corrupt practices?
§ 30.611 Can I hire a person or agency to solicit EPA assistance for me?
§ 30.612 May an EPA employcc act as my representative?
§ 30.613 What is EPA's policy on conflict of interest?
§ 30.615 May I employ a former EPA employee and still receive assistance?
Subpart G--Can an Assistance Agreement be Changed?
§ 30.700 What changes to my assistance agreement require a formal amendment?
§ 30.705 What changes can I make to my assistance agreement without a formal amendment?
§ 30.710 Can I terminate a part or all of my assistance agreement?
Subpart H--How do I Close out my Project?
§ 30.800 What records and reports must I keep after I complete my project?
§ 30.802 Under what conditions will I owe money to EPA?

Subpart I--What Measures may EPA Take for Non-compliance?
§ 30.900 What measures may EPA take for non-compliance?
§ 30.901 What are the consequences of a stop-work order?
§ 30.902 What are the consequences of withholding payments?
§ 30.903 What are the consequences of termination for cause?
§ 30.904 What are the consequences of annulment?
§ 30.905 May I request a review of a termination or annulment?
§ 30.906 What are the consequences of suspension or debarment?
Subpart J--Can I get An Exception ("Deviation") From These Regulations?
§ 30.1001 Will EPA approve any exceptions to these regulations?
§ 30.1002 Who may request a deviation?
§ 30.1003 What information must I include in a deviation request?
§ 30.1004 Who approves or disapproves a deviation request?
§ 30.1005 May I request a review of a deviation decision?
Subpart K--What Policies Apply to Patents, Data, and Copyrights?
§ 30.1100 What assistance agreements are subject to EPA patent rules?
§ 30.1101 What Federal patent laws or policies govern my assistance agreement?
§ 30.1102 What are my invention rights and my reporting requirements if my award is other than an award under section 6914 of RCRA?
§ 30.1103 What are my invention rights and obligations if I am a profitmaking firm with an award under section 6914 of RCRA?
§ 30.1104 Can I get a waiver from section 6981(c) of RCRA?
§ 30.1106 Do the patent rules apply to subagreements?
§ 30.1108 Does EPA require any type of licensing of background patents that I own?
§ 30.1112 Are there any other patent clauses or conditions that apply to my award?
§ 30.1130 What rights in data and copyrights does EPA acquire?
Subpart L--How are Disputes Between EPA Officials and me Resolved?
§ 30.1200 What happens if an EPA official and I disagree about an assistance agreement requirement?
§ 30.1205 If I file a request for review, with whom must I file?
§ 30.1210 What must I include in my request for review or reconsideration?
§ 30.1215 What are my rights after I file a request for review or reconsideration?
§ 30.1220 If the Assistant Administrator confirms the final decision of the Headquarters disputes decision official, may I seek further administrative review?

Underground Water Source Protection (Section 1443(b))
§ 35.450 Purpose.
§ 35.455 Maximum Federal share.
§ 35.460 Limitations.
§ 35.465 Indian Tribes.
Hazardous Waste Management (Section 3011)
§ 35.500 Purpose.
§ 35.505 Maximum Federal share.
§ 35.510 Limitations.
Pesticide Enforcement (Section 23(a)(1))
§ 35.550 Purpose.
§ 35.555 Maximum Federal share.
§ 35.600 Purpose.
§ 35.605 Maximum Federal share.
Nonprofit Source Management (Sections 205(j)(5) and 319(h))
§ 35.750 Purpose.
§ 35.755 Awards to Indian Tribes.
§ 35.760 Maximum Federal share.
Subpart B--[Reserved]
Subpart C--Grants for Construction of Wastewater Treatment Works
§ 35.800 Purpose.
§ 35.801 Authority.
§ 35.805 Definitions.
§ 35.805-1 Construction.
§ 35.805-2 Intermunicipal agency.
§ 35.805-3 Interstate agency.
§ 35.805-4 Municipality.
§ 35.805-5 State.
§ 35.805-6 State water pollution control agency.
§ 35.805-7 Treatment works.
§ 35.810 Applicant eligibility.
§ 35.815 Allocation of funds.
§ 35.815-1 Allotments to States.
§ 35.815-2 Reallotment.
§ 35.820 Grant limitations.
§ 35.820-1 Exceptions.
§ 35.825 Application for grant.
§ 35.825-1 Preapplication procedures.
§ 35.825-2 Formal application.
§ 35.830 Determining the desirability of projects.
§ 35.835 Criteria for award.
§ 35.835-1 State plan and priority.
§ 35.835-2 Basin control.
§ 35.835-3 Regional and metropolitan plan.
§ 35.835-4 Adequacy of treatment.
§ 35.835-5 Industrial waste treatment.
§ 35.835-6 Design.
§ 35.835-7 Operation and maintenance.
§ 35.835-8 Operation during construction.
§ 35.835-9 Postconstruction inspection.
§ 35.840 Supplemental grant conditions.
§ 35.845 Payments.
Subpart D--Reimbursement Grants

§ 35.920-3 Contents of application.
§ 35.925 Limitations on award.
§ 35.925-1 Facilities planning.
§ 35.925-2 Water quality management plans and agencies.
§ 35.925-3 Priority determination.
§ 35.925-4 State allocation.
§ 35.925-5 Funding and other capabilities.
§ 35.925-6 Permits.
§ 35.925-7 Design.
§ 35.925-8 Environmental review.
§ 35.925-9 Civil rights.
§ 35.925-10 Operation and maintenance program.
§ 35.925-11 User charges and industrial cost recovery.
§ 35.925-12 Property.
§ 35.925-13 Sewage collection system.
§ 35.925-14 Compliance with environmental laws.
§ 35.925-15 Treatment of industrial wastes.
§ 35.925-16 Federal activities.
§ 35.925-17 Retained amounts for reconstruction and expansion.
§ 35.925-18 Limitation upon project costs incurred prior to award.
§ 35.925-19 [Reserved]
§ 35.925-20 Procurement.
§ 35.925-21 Storm sewers.
§ 35.926 Value engineering (VE).
§ 35.927 Sewer system evaluation and rehabilitation.
§ 35.927-1 Infiltration/inflow analysis.
§ 35.927-2 Sewer system evaluation survey.
§ 35.927-3 Rehabilitation.
§ 35.927-4 Sewer use ordinance.
§ 35.927-5 Project procedures.
§ 35.928 Requirements for an industrial cost recovery system.
§ 35.928-1 Approval of the industrial cost recovery system.
§ 35.928-2 Use of industrial cost recovery payments.
§ 35.928-3 Implementation of the industrial cost recovery system.
§ 35.928-4 Moratorium on industrial cost recovery payments.
§ 35.929 Requirements for user charge system.
§ 35.929-1 Approval of the user charge system.
§ 35.929-2 General requirements for all user charge systems.
§ 35.929-3 Implementation of the user charge system.
§ 35.930 Award of grant assistance.
§ 35.930-1 Types of projects.
§ 35.930-2 Grant amount.
§ 35.930-3 Grant term.
§ 35.930-4 Project scope.
§ 35.930-5 Federal share.
§ 35.930-6 Limitation on Federal share.

§ 35.935 Grant conditions.
§ 35.935-1 Grantee responsibilities.
§ 35.935-2 Procurement.
§ 35.935-3 Property.
§ 35.935-4 Step 2+3 projects.
§ 35.935-5 Davis-Bacon and related statutes.
§ 35.935-6 Equal employment opportunity.
§ 35.935-7 Access.
§ 35.935-8 Supervision.
§ 35.935-9 Project initiation and completion.
§ 35.935-10 Copies of contract documents.
§ 35.935-11 Project changes.
§ 35.935-12 Operation and maintenance.
§ 35.935-13 Submission and approval of user charge systems.
§ 35.935-14 Final inspection.
§ 35.935-15 Submission and approval of industrial cost recovery system.
§ 35.935-16 Sewer use ordinance and evaluation/ rehabilitation program.
§ 35.935-17 Training facility.
§ 35.935-18 Value engineering.
§ 35.935-19 Municipal pretreatment program.
§ 35.935-20 Innovative processes and techniques.
§ 35.936 Procurement.
§ 35.936-1 Definitions.
§ 35.936-2 Grantee procurement systems; State or local law.
§ 35.936-3 Competition.
§ 35.936-4 Profits.
§ 35.936-5 Grantee responsibility.
§ 35.936-6 EPA responsibility.
§ 35.936-7 Small and minority business.
§ 35.936-8 Privity of contract.
§ 35.936-9 Disputes.
§ 35.936-10 Federal procurement regulations.
§ 35.936-11 General requirements for subagreements.
§ 35.936-12 Documentation.
§ 35.936-13 Specifications.
§ 35.936-14 Force account work.
§ 35.936-15 Limitations on subagreement award.
§ 35.936-16 Code or standards of conduct.
§ 35.936-17 Fraud and other unlawful or corrupt practices.
§ 35.936-18 Negotiation of subagreements.
§ 35.936-19 Small purchases.
§ 35.936-20 Allowable costs.
§ 35.936-21 Delegation to State agencies; certification of procurement systems.
§ 35.936-22 Bonding and insurance.
§ 35.937 Subagreements for architectural or engineering services.
§ 35.937-1 Type of contract (subagreement).
§ 35.937-2 Public notice.

§ 35.2101 Advanced treatment.
§ 35.2102 Water quality management planning.
§ 35.2103 Priority determination.
§ 35.2104 Funding and other considerations.
§ 35.2105 Debarment and suspension.
§ 35.2106 Plan of operation.
§ 35.2107 Intermunicipal service agreements.
§ 35.2108 Phased or segmented treatment works.
§ 35.2109 Step 2+3.
§ 35.2110 Access to individual systems.
§ 35.2111 Revised water quality standards.
§ 35.2112 Marine discharge waiver applicants.
§ 35.2113 Environmental review.
§ 35.2114 Value engineering.
§ 35.2116 Collection system.
§ 35.2118 Preaward costs.
§ 35.2120 Infiltration/Inflow.
§ 35.2122 Approval of user charge system and proposed sewer use ordinance.
§ 35.2123 Reserve capacity.
§ 35.2125 Treatment of wastewater from industrial users.
§ 35.2127 Federal facilities.
§ 35.2130 Sewer use ordinance.
§ 35.2140 User charge system.
§ 35.2152 Federal share.
§ 35.2200 Grant conditions.
§ 35.2202 Step 2+3 projects.
§ 35.2203 Step 7 projects.
§ 35.2204 Project changes.
§ 35.2205 Maximum allowable project cost.
§ 35.2206 Operation and maintenance.
§ 35.2208 Adoption of sewer use ordinance and user charge system.
§ 35.2210 Land acquisition.
§ 35.2211 Field testing for Innovative and Alternative Technology Report.
§ 35.2212 Project initiation.
§ 35.2214 Grantee responsibilities.
§ 35.2216 Notice of building completion and final inspection.
§ 35.2218 Project performance.
§ 35.2250 Determination of allowable costs.
§ 35.2260 Advance purchase of eligible land.
§ 35.2262 Funding of field testing.
§ 35.2300 Grant payments.
§ 35.2350 Subagreement enforcement.
Appendix A--Determination of Allowable Costs
Appendix B--Allowance for Facilities Planning and Design
Subpart J--Construction Grants Program Delegation to States
§ 35.3000 Purpose.
§ 35.3005 Policy.
§ 35.3010 Delegation agreement.
§ 35.3015 Extent of State responsibilities.

§ 35.4130 Conflict of interest and disclosure requirements.

Subpart O--Cooperative Agreements and Superfund State Contracts for Superfund Response Actions

General

§ 35.6000 Authority.

§ 35.6005 Purpose and scope.

§ 35.6010 Eligibility.

§ 35.6015 Definitions.

§ 35.6020 Other statutory provisions.

§ 35.6025 Deviation from this subpart.

Pre-Remedial Response Cooperative Agreements

§ 35.6050 Eligibility for pre-remedial Cooperative Agreements.

§ 35.6055 State-lead pre-remedial Cooperative Agreements.

§ 35.6060 Political subdivision-lead pre-remedial Cooperative Agreements.

§ 35.6070 Indian Tribe-lead pre-remedial Cooperative Agreements.

Remedial Response Cooperative Agreements

§ 35.6100 Eligibility for remedial Cooperative Agreements.

§ 35.6105 State-lead remedial Cooperative Agreements.

§ 35.6110 Indian Tribe-lead remedial Cooperative Agreements.

§ 35.6115 Political subdivision-lead remedial Cooperative Agreements.

§ 35.6120 Notification of the out-of-State or out-of-Indian Tribal jurisdiction transfer of CERCLA wastes.

Enforcement Cooperative Agreements

§ 35.6145 Eligibility for enforcement Cooperative Agreements.

§ 35.6150 Activities eligible for funding under enforcement Cooperatve Agreements.

§ 35.6155 State, political subdivision or Indian Tribe-lead enforcement Cooperative Agreements.

Removal Response Cooperative Agreements

§ 35.6200 Eligibility for removal Cooperative Agreements.

§ 35.6205 Removal Cooperative Agreements.

Core Program Cooperative Agreements

§ 35.6215 Eligibility for Core Program Cooperative Agreements.

§ 35.6220 General.

§ 35.6225 Activities eligible for funding under Core Program Cooperative Agreements.

§ 35.6230 Application requirements.

§ 35.6235 Cost sharing.

Support Agency Cooperative Agreements

§ 35.6240 Eligibility for support agency Cooperative Agreements.

§ 35.6245 Allowable activities.

§ 35.6250 Support agency Cooperative Agreement requirements.

§ 35.6255 Cost sharing.

Financial Administration Requirements Under a Cooperative Agreement

§ 35.6270 Standards for financial management systems.
§ 35.6275 Period of availability of funds.
§ 35.6280 Payments.
§ 35.6285 Recipient payment of response costs.
§ 35.6290 Program income.

Personal Property Requirements Under a Cooperative Agreement

§ 35.6300 General personal property acquisition and use requirements.
§ 35.6305 Obtaining supplies.
§ 35.6310 Obtaining equipment.
§ 35.6315 Alternative methods for obtaining property.
§ 35.6320 Usage rate.
§ 35.6325 Title and EPA interest in CERCLA-funded property.
§ 35.6330 Title to federally owned property.
§ 35.6335 Property management standards.
§ 35.6340 Disposal of CERCLA-funded property.
§ 35.6345 Equipment disposal options.
§ 35.6350 Disposal of federally owned property.

Real Property Requirements under a Cooperative Agreement

§ 35.6400 Acquisition and transfer of interest.
§ 35.6405 Use.

Copyright Requirements under a Cooperative Agreement

§ 35.6450 General requirements.

Use of Recipient Employees ("Force Account") under a Cooperative Agreement

§ 35.6500 General requirements.

Procurement Requirements under a Cooperative Agreement

§ 35.6550 Procurement system standards.
§ 35.6555 Competition.
§ 35.6560 Master list of debarred, suspended, and voluntarily excluded persons.
§ 35.6565 Procurement methods.
§ 35.6570 Use of the same engineer during subsequent phases of response.
§ 35.6575 Restrictions on types of contracts.
§ 35.6580 Contracting with minority and women's business enterprises (MBE/WBE), small businesses, and labor surplus area firms.
§ 35.6585 Cost and price analysis.
§ 35.6590 Bonding and insurance.
§ 35.6595 Contract provisions.
§ 35.6600 Contractor claims.
§ 35.6605 Privity of contract.
§ 35.6610 Contracts awarded by a contractor.

Reports Required Under a Cooperative Agreement

§ 35.6650 Quarterly progress reports.
§ 35.6655 Notification of significant developments.
§ 35.6660 Property inventory reports.
§ 35.6665 Procurement reports.
§ 35.6670 Financial reports.

§ 39.135 Loan terms.
§ 39.140 Loan proceeds.
§ 39.145 Loan payments by borrower.
§ 39.150 Defaults and remedies.
PART 40--RESEARCH AND DEMONSTRATION GRANTS
§ 40.100 Purpose of regulation.
§ 40.105 Applicability and scope.
§ 40.110 Authority.
§ 40.115 Definitions.
§ 40.115-1 Construction.
§ 40.115-2 Intermunicipal agency.
§ 40.115-3 Interstate agency.
§ 40.115-4 Municipality.
§ 40.115-5 Person.
§ 40.115-6 State.
§ 40.120 Publication of EPA research objectives.
§ 40.125 Grant limitations.
§ 40.125-1 Limitations on duration.
§ 40.125-2 Limitations on assistance.
§ 40.130 Eligibility.
§ 40.135 Application.
§ 40.135-1 Preapplication coordination.
§ 40.135-2 Application requirements.
§ 40.140 Criteria for award.
§ 40.140-1 All applications.
§ 40.140-2 [Reserved]
§ 40.140-3 Federal Water Pollution Control Act.
§ 40.145 Supplemental grant conditions.
§ 40.145-1 Resource Conservation and Recovery Act.
§ 40.145-2 Federal Water Pollution Control Act.
§ 40.145-3 Projects involving construction.
§ 40.150 Evaluation of applications.
§ 40.155 Availability of information.
§ 40.160 Reports.
§ 40.160-1 Progress reports.
§ 40.160-2 Financial status report.
§ 40.160-3 Reporting of inventions.
§ 40.160-4 Equipment report.
§ 40.160-5 Final report.
§ 40.165 Continuation grants.
PART 45--TRAINING ASSISTANCE
§ 45.100 Purpose and scope.
§ 45.105 Authority.
§ 45.110 Objectives.
§ 45.115 Definitions.
§ 45.120 Applicant eligibility.
§ 45.125 Application requirements.
§ 45.130 Evaluation of applications.
§ 45.135 Supplemental conditions.
§ 45.140 Budget and project period.
§ 45.145 Allocability and allowability of costs.
§ 45.150 Reports.
§ 45.155 Continuation assistance.

Appendix A--Environmental Protection Agency Training Programs
PART 46--FELLOWSHIPS
§ 46.100 Purpose.
§ 46.105 Authority.
§ 46.110 Objectives.
§ 46.115 Types of fellowships.
§ 46.120 Definitions.
§ 46.125 Benefits.
§ 46.130 Eligibility.
§ 46.135 Submission of applications.
§ 46.140 Evaluation of applications.
§ 46.145 Fellowship agreement.
§ 46.150 Fellowship agreement amendment.
§ 46.155 Supplemental conditions.
§ 46.160 Acceptance of fellowship award.
§ 46.165 Duration of fellowship.
§ 46.170 Initiation of studies.
§ 46.175 Completion of studies.
§ 46.180 Payment.
Appendix A--Environmental Protection Agency Fellowship Programs
PART 47--NATIONAL ENVIRONMENTAL EDUCATION ACT GRANTS
§ 47.100 Purpose and scope.
§ 47.105 Definitions.
§ 47.110 Eligible applicants.
§ 47.115 Award amount and matching requirements.
§ 47.120 Solicitation notice and proposal procedures.
§ 47.125 Eligible and priority projects and activities.
§ 47.130 Performance of grant.
§ 47.135 Disputes.
SUBCHAPTER C--AIR PROGRAMS
PART 50--NATIONAL PRIMARY AND SECONDARY AMBIENT AIR QUALITY STANDARDS
§ 50.1 Definitions.
§ 50.2 Scope.
§ 50.3 Reference conditions.
§ 50.4 National primary ambient air quality standards for sulfur oxides (sulfur dioxide).
§ 50.5 National secondary ambient air quality standards for sulfur oxides (sulfur dioxide).
§ 50.6 National primary and secondary ambient air quality standards for particulate matter.
§ 50.7 [Reserved]
§ 50.8 National primary ambient air quality standards for carbon monoxide.
§ 50.9 National primary and secondary ambient air quality standards for ozone.
§ 50.10 [Reserved]
Subpart C--Standards for Used Oil Generators
§ 50.11 National primary and secondary ambient air quality standards for nitrogen dioxide.
§ 50.12 National primary and secondary ambient air quality standards for lead.
Appendix A--Reference Method for the Determination of Sulfur

§ 52.18 Abbreviations.
§ 52.19 Revision of plans by Administrator.
§ 52.20 Attainment dates for national standards.
§ 52.21 Prevention of significant deterioration of air quality.
§ 52.22 Maintenance of national standards.
§ 52.23 Violation and enforcement.
§ 52.24 Statutory restriction on new sources.
§ 52.25 Date for submission of Set II CTG regulations.
§ 52.26 Visibility monitoring strategy.
§ 52.27 Protection of visibility from sources in attainment areas.
§ 52.28 Protection of visibility from sources in nonattainment areas.
§ 52.29 Visibility long-term strategies.
§ 52.30 Criteria for limiting application of sanctions under section 110(m) of the Clean Air Act on a statewide basis.

Subpart B--Alabama

§ 52.50 Identification of plan.
§ 52.51 Classification of regions.
§ 52.52 Extensions.
§ 52.53 Approval status.
§ 52.54 Attainment dates for national standards.
§ 52.56 Review of new sources and modifications.
§ 52.57 Control strategy: Sulfur oxides.
§ 52.58 Control strategy: Lead.
§ 52.59 Maintenance of national standards.
§ 52.60 Significant deterioration of air quality.
§ 52.61 Visibility protection.
§ 52.62 Control strategy: sulfur oxides and particulate matter.
§ 52.63 PM10 State Implementation Plan development in group II areas.
§ 52.65 Control Strategy: Nitrogen Oxides.

Subpart C--Alaska

§ 52.70 Identification of plan.
§ 52.71 Classification of regions.
§ 52.72 Approval status.
§ 52.73 General requirements.
§ 52.74 Legal authority.
§ 52.75 Contents of the approved state-submitted implementation plan.
Secs. 52.76--52.77 [Reserved]
§ 52.78 Review of new sources and modifications
§ 52.79 [Reserved]
§ 52.80 Intergovernmental cooperation.
§ 52.81 Attainment dates for national standards.
§ 52.82 Extensions.
§ 52.83 [Reserved]
§ 52.84 Compliance schedules.
Secs. 52.85--52.94 [Reserved]
§ 52.95 Maintenance of national standards.

§ 52.228 Regulations: Particulate matter, Southeast Desert Intrastate Region.
§ 52.229 Control strategy and regulations: Photochemical oxidants (hydrocarbons), Metropolitan Los Angeles Intrastate Region.
§ 52.230 Control strategy and regulations: Nitrogen dioxide.
§ 52.231 Regulations: Sulfur oxides.
§ 52.232 Part D conditional approval.
§ 52.233 Review of new sources and modifications.
§ 52.234 Source surveillance.
§ 52.235 [Reserved]
§ 52.236 Rules and regulations.
§ 52.237 Part D disapproval.
§ 52.238 Attainment dates for national standards.
§ 52.239 Alternate compliance plans.
§ 52.240 Compliance schedules.
Secs. 52.241--52.245 [Reserved]
§ 52.246 Control of dry cleaning solvent vapor losses.
Secs. 52.247--52.251 [Reserved]
§ 52.252 Control of degreasing operations.
§ 52.253 Metal surface coating thinner and reducer.
§ 52.254 Organic solvent usage.
§ 52.255 Gasoline transfer vapor control.
§ 52.256 Control of evaporative losses from the filling of vehicular tanks.
Secs. 52.257--52.262 [Reserved]
§ 52.263 Priority treatment for buses and carpools--Los Angeles Region.
Secs. 52.264--52.266 [Reserved]
§ 52.267 Maintenance of national standards.
§ 52.268 [Reserved]
§ 52.269 Control strategy and regulations: Photochemical oxidants (hydrocarbons) and carbon monoxide.
§ 52.270 Significant deterioration of air quality.
§ 52.271 Malfunction regulations.
§ 52.272 Research operations exemptions.
§ 52.273 Open burning.
§ 52.274 California air pollution emergency plan.
§ 52.275 Particulate matter control.
§ 52.276 Sulfur content of fuels.
§ 52.277 Oxides of nitrogen, combustion gas concentration limitations.
§ 52.278 Oxides of nitrogen control.
§ 52.279 Food processing facilities.
§ 52.280 Fuel burning equipment.
§ 52.281 Visibility protection.
Subpart G--Colorado
§ 52.320 Identification of plan.
§ 52.321 Classification of regions.
§ 52.322 Extensions.
§ 52.323 Approval status.
§ 52.324 Legal authority.

§ 52.499 Significant deterioration of air quality.
Subpart K--Florida
§ 52.520 Identification of plan.
§ 52.521 Classification of regions.
§ 52.522 Approval status.
§ 52.523 Attainment dates for national standards.
§ 52.524 Compliance schedules.
§ 52.525 General requirements.
§ 52.526 Legal authority.
§ 52.527 Control strategy: General.
§ 52.528 Control strategy: Sulfur oxides and particulate matter.
§ 52.529 Maintenance of national standards.
§ 52.530 Significant deterioration of air quality.
§ 52.531 [Removed. 58 FR 37660, July 13, 1993]
§ 52.532 Extensions.
§ 52.533 Source surveillance.
§ 52.534 Visibility protection.
§ 52.535 Rules and regulations.
Subpart L--Georgia
§ 52.570 Identification of plan.
§ 52.571 Classification of regions.
§ 52.572 Approval status.
§ 52.573 Control strategy: General.
§ 52.574 Review of new sources and modifications.
§ 52.575 Attainment dates for national standards.
§ 52.576 Compliance schedules.
§ 52.577 Extensions.
§ 52.578 Control Strategy: Sulfur oxides and particulate matter.
§ 52.579 Economic feasibility considerations.
§ 52.580 Maintenance of national standards.
§ 52.581 Significant deterioration of air quality.
§ 52.582 [Reserved]
§ 52.583 Additional rules and regulations.
Subpart M--Hawaii
§ 52.620 Identification of plan.
§ 52.621 Classification of regions.
§ 52.622 Extensions.
§ 52.623 Approval status.
§ 52.624 General requirements.
§ 52.625 Legal authority.
§ 52.626 Compliance schedules.
§ 52.627 [Reserved]
§ 52.628 Attainment dates for national standards.
§ 52.629 Review of new sources and modifications.
§ 52.630 [Reserved]
§ 52.631 Maintenance of national standards.
§ 52.632 Significant deterioration of air quality.
§ 52.633 Visibility protection.
§ 52.634 Particulate matter (PM-10) Group III SIP.
Subpart N--Idaho
§ 52.670 Identification of plan.

§ 52.873 Approval status.
§ 52.874 Legal authority.
§ 52.875 [Reserved]
§ 52.876 Compliance schedules.
§ 52.877 [Reserved]
§ 52.878 Review of new sources and modifications.
§ 52.879 Attainment dates for national standards.
§ 52.880 Requests for 18-month extensions.
§ 52.881 PM10 State implementation plan development in group II areas.
[§ 52.882. Reserved.]
§ 52.883 Maintenance of national standards.
§ 52.884 Significant deterioration of air quality.

Subpart S--Kentucky

§ 52.920 Identification of plan.
§ 52.921 Classification of regions.
§ 52.922 Extensions.
§ 52.923 Approval status.
§ 52.924 Legal authority.
§ 52.925 General requirements.
§ 52.926 Attainment dates for national standards.
§ 52.927 Compliance schedules.
§ 52.928 Control strategy: Sulfur oxides.
§ 52.929 Maintenance of national standards.
§ 52.930 Control strategy: Ozone.
§ 52.931 Significant deterioration of air quality.
§ 52.932 Rules and regulations.
§ 52.933 Control Strategy: Sulfur oxides and particulate matter.
§ 52.934 VOC Rule Deficiency Correction.
§ 52.935 PM10 State implementation plan development in group II areas.
§ 52.936 Visibility protection.

Subpart T--Louisiana

§ 52.970 Identification of plan.
§ 52.971 Classification of regions.
§ 52.972 Approval status.
Secs. 52.973--52.975 [Reserved]
§ 52.976 Review of new sources and modification.
§ 52.977 [Reserved]
§ 52.978 Resources.
§ 52.979 Attainment dates for national standards.
§ 52.980 Compliance schedules.
Secs. 52.981-52.985 [Reserved]
§ 52.986 Significant deterioration of air quality.
§ 52.987 Control of hydrocarbon emissions.
§ 52.988 Rules and regulations.
§ 52.990 Stack height regulations.
§ 52.991 Small business assistance program.

Subpart U--Maine

§ 52.1020 Identification of plan.
§ 52.1021 Classification of regions.
§ 52.1022 Approval status.

§ 52.1024 Attainment dates for national standards.
§ 52.1025 Control strategy: Particulate matter.
§ 52.1026 Review of new sources and modifications.
§ 52.1027 Rules and regulations.
§ 52.1028 Maintenance of national standards.
§ 52.1029 Significant deterioration of air quality.
§ 52.1030 Control strategy: Sulfur oxides.
§ 52.1031 EPA-approved Maine regulations.
§ 52.1033 Visibility protection.
§ 52.1034 Stack height review.
Subpart V--Maryland
§ 52.1070 Identification of plan.
§ 52.1071 Classification of regions.
§ 52.1072 Extensions.
§ 52.1073 Approval status.
§ 52.1074 Legal authority.
Secs. 52.1075--52.1076 [Reserved]
§ 52.1077 Source surveillance.
§ 52.1078 Attainment dates for national standards.
§ 52.1079 [Reserved]
§ 52.1081 [Reserved]
§ 52.1082 Rules and regulations.
Secs. 52.1083--52.1085 [Reserved]
§ 52.1086 Gasoline transfer vapor control.
§ 52.1087 Control of evaporative losses from the filling of vehicular tanks.
§ 52.1088 Control of dry cleaning solvent evaporation.
Secs. 52.1089--52.1100 [Reserved]
§ 52.1110 Small business stationary source technical and environmental compliance assistance program.
Secs. 52.1111--52.1112 [Reserved]
§ 52.1101 Gasoline transfer vapor control.
§ 52.1102 Control of evaporative losses from the filling of vehicular tanks.
Secs. 52.1103--52.1106 [Reserved]
§ 52.1107 Control of dry cleaning solvent evaporation.
Secs. 52.1108--52.1109 [Reserved]
Secs. 52.1108--52.1112 [Reserved]
§ 52.1113 General requirements.
§ 52.1114 [Reserved]
§ 52.1115 Maintenance of national standards.
§ 52.1116 Significant deterioration of air quality.
§ 52.1117 Control strategy: Sulfur oxides.
§ 52.1118 Approval of bubbles in nonattainment areas lacking approved demonstrations: State assurances.
Subpart W--Massachusetts
§ 52.1120 Identification of plan.
§ 52.1121 Classification of regions.
§ 52.1122 Extensions.
§ 52.1123 Approval status.
§ 52.1124 Review of new sources and modifications.
§ 52.1126 Control strategy: Sulfur oxides.
§ 52.1127 Attainment dates for national standards.

§ 52.1389 Small business stationary source technical and environmental compliance assistance program.
Subpart CC--Nebraska
§ 52.1420 Identification of plan.
§ 52.1421 Classification of regions.
§ 52.1422 Approval status.
§ 52.1423 PM10 State implementation plan development in group II areas.
§ 52.1424 [Reserved]
§ 52.1425 Compliance schedules.
§ 52.1426 Extensions.
Secs. 52.1427--52.1430 [Reserved]
§ 52.1431 Attainment dates for national standards.
Secs. 52.1432--52.1434 [Reserved]
§ 52.1435 Maintenance of national standards.
§ 52.1436 Significant deterioration of air quality.
Subpart DD--Nevada
§ 52.1470 Identification of plan.
§ 52.1471 Classification of regions.
§ 52.1472 Approval status.
§ 52.1473 General requirements.
§ 52.1474 Part D conditional approval.
§ 52.1475 Control strategy and regulations: Sulfur oxides.
§ 52.1476 Control strategy: Particulate matter.
§ 52.1477 Nevada air pollution emergency plan.
§ 52.1478 [Reserved]
§ 52.1479 Source surveillance.
§ 52.1480 Attainment dates for national standards.
§ 52.1481 Extensions.
§ 52.1482 Compliance schedules.
§ 52.1483 Malfunction regulations.
§ 52.1484 [Reserved]
§ 52.1485 Significant deterioration of air quality.
§ 52.1486 Control strategy: Hydrocarbons and ozone.
§ 52.1487 Public hearings.
§ 52.1488 Visibility protection.
§ 52.1489 Particulate matter (PM-10) Group II SIP commitments.
Subpart EE--New Hampshire
§ 52.1520 Identification of plan.
§ 52.1521 Classification of regions.
§ 52.1522 Approval status.
§ 52.1523 Attainment dates for national standards.
§ 52.1524 Compliance schedules.
§ 52.1525 EPA-approved New Hampshire state regulations.
§ 52.1526 [Reserved. 57 FR 36607, Aug. 14, 1992]
§ 52.1527 Rules and regulations.
§ 52.1528 Maintenance of national standards.
§ 52.1529 Significant deterioration of air quality.
§ 52.1530 [Reserved. 57 FR 36607, Aug. 14, 1992]
§ 52.1531 Visibility protection.
§ 52.1532 Stack height review.
Subpart FF--New Jersey

(hydrocarbons).
§ 52.1878 Inspection and maintenance program.
§ 52.1879 Review of new sources and modifications.
§ 52.1880 Control strategy: Particulate matter.
§ 52.1881 Control strategy: Sulfur oxides (sulfur dioxide).
§ 52.1882 Compliance schedules.
§ 52.1883 Maintenance of national standards.
§ 52.1884 Significant deterioration of air quality.
§ 52.1885 Control strategy: Ozone.
§ 52.1886 [Reserved]
§ 52.1887 Control strategy: Carbon monoxide.
Subpart LL--Oklahoma
§ 52.1920 Identification of plan.
§ 52.1921 Classification of regions.
§ 52.1922 Approval status.
Secs. 52.1923--52.1924 [Reserved]
§ 52.1925 Attainment dates for national standards.
§ 52.1926 General requirements.
§ 52.1927 Maintenance of national standards.
§ 52.1928 [Reserved]
§ 52.1929 Significant deterioration of air quality.
§ 52.1930 [Reserved]
§ 52.1931 Petroleum storage tank controls.
§ 52.1932 Control strategy and regulations: ozone.
§ 52.1933 Visibility protection.
§ 52.1934 Prevention of air pollution emergency episodes.
§ 52.1935 Small business assistance program.
Subpart MM--Oregon
§ 52.1970 Identification of plan.
§ 52.1971 Classification of regions.
§ 52.1972 Approval status.
§ 52.1973 Attainment dates for national standards.
§ 52.1974 Transportation and land use controls.
§ 52.1975 Compliance schedules.
§ 52.1976 Control strategy: Particulate matter--Portland interstate air quality control region.
§ 52.1977 Content of approved State submitted implementation plan.
Secs. 52.1978--52.1980 [Reserved]
§ 52.1981 Extension.
§ 52.1982 Control strategy: Ozone.
Secs. 52.1983--52.1984 [Reserved]
§ 52.1985 Rules and regulations.
§ 52.1986 Maintenance of national standards.
§ 52.1987 Significant deterioration of air quality.
§ 52.1988 Air contaminant discharge permits.
Subpart NN--Pennsylvania
§ 52.2020 Identification of plan.
§ 52.2021 Classification of regions.
§ 52.2022 Extensions.
§ 52.2023 Approval status.
§ 52.2024 General requirements.

§ 52.2285 Control of evaporative losses from the filling of gasoline storage vessels in the Houston and San Antonio areas.
§ 52.2286 Control of evaporative losses from the filling of gasoline storage vessels in the Dallas-Fort Worth area.
Secs. 52.2287--52.2293 [Reserved]
§ 52.2294 Incentive program to reduce vehicle emissions through increased bus and carpool use.
§ 52.2295 [Reserved]
§ 52.2296 Carpool matching and promotion system.
§ 52.2297 Employer mass transit and carpool incentive program.
§ 52.2298 Monitoring transportation mode trends.
Secs. 52.2299--52.2300 [Reserved]
§ 52.2301 Federal compliance date for automobile and light-duty truck coating. Texas Air Control Board Regulation V (31 TAC chapter 115), control of air pollution from volatile organic compound, rule 115.191(1)(8)(A).
§ 52.2302 Maintenance of national standards.
§ 52.2303 Significant deterioration of air quality.
§ 52.2304 Visibility protection.
§ 52.2305 Lead control plan: Federal compliance date for requirements of Texas Air Control Board (TACB) Rule 113.53.
§ 52.2306 Particulate Matter (PM10) Group II SIP commitments.

Subpart TT--Utah
§ 52.2320 Identification of plan.
§ 52.2321 Classification of regions.
§ 52.2322 Extensions.
§ 52.2323 Approval status.
Secs. 52.2324--52.2330 [Reserved]
§ 52.2331 Attainment dates for national standards.
§ 52.2332 [Reserved]
§ 52.2333 Legal authority.
Secs. 52.2334--52.2344 [Reserved]
§ 52.2345 Maintenance of national standards.
§ 52.2346 Significant deterioration of air quality.
§ 52.2347 Stack height regulations.
§ 52.2348 Small business assistance program.

Subpart UU--Vermont
§ 52.2370 Identification of plan.
§ 52.2371 Classification of regions.
§ 52.2372 Approval status.
§ 52.2373 Legal authority.
§ 52.2374 General requirements.
§ 52.2375 Attainment dates for national standards.
§ 52.2377 Review of new sources and modifications.
§ 52.2378 Certification of no facilities.
§ 52.2379 Maintenance of national standards.
§ 52.2380 Significant deterioration of air quality.
§ 52.2381 EPA-approved Vermont state regulations.

§ 52.2490 Air bleed to intake manifold retrofit.
§ 52.2491 Exhaust gas recirculation-air bleed.
§ 52.2492 Computer carpool matching system.
§ 52.2493 Transit improvement measures.
§ 52.2494 Bike lanes and bike racks.
§ 52.2495 [Reserved]
§ 52.2496 Maintenance of national standards.
§ 52.2497 Significant deterioration of air quality.
§ 52.2498 Visibility protection.
Subpart XX--West Virginia
§ 52.2520 Identification of plan.
§ 52.2521 Classification of regions.
§ 52.2522 Approval status.
§ 52.2523 Attainment dates for national standards.
§ 52.2524 Compliance schedules.
§ 52.2525 Control strategy: Sulfur dioxide.
§ 52.2526 Maintenance of national standards.
§ 52.2527 [Reserved]
§ 52.2528 Significant deterioration of air quality.
Secs. 52.2529--52.2530 [Reserved]
§ 52.2531 Control strategy: (Hydrocarbons).
§ 52.2532 Control strategy: Particulate matter.
§ 52.2533 Visibility protection.
§ 52.2534 Stack height review.
§ 52.2560 Small business technical and environmental compliance assistance program.
Subpart YY--Wisconsin
§ 52.2569 Identification of plan--conditional approval.
§ 52.2570 Identification of plan.
§ 52.2571 Classification of regions.
§ 52.2572 Approval status.
§ 52.2573 General requirements.
§ 52.2574 Legal authority.
§ 52.2575 Control strategy: Sulfur dioxide.
§ 52.2576 [Reserved]
§ 52.2577 Attainment dates for national standards.
§ 52.2578 Compliance schedules.
§ 52.2579 Review of new sources and modifications.
§ 52.2580 Maintenance of national standards.
§ 52.2581 Significant deterioration of air quality.
§ 52.2582 Extensions.
§ 52.2583 [Reserved]
§ 52.2584 Control strategy; Particulate matter.
§ 52.2585 Control strategy: Ozone.
Subpart ZZ--Wyoming
§ 52.2620 Identification of plan.
§ 52.2621 Classification of regions.
§ 52.2622 Approval status.
§ 52.2623 Review of new sources and modifications.
§ 52.2624 [Reserved]
§ 52.2625 Compliance schedules.
§ 52.2626 [Reserved]
§ 52.2627 Attainment dates for national standards.

§ 60.45 Emission and fuel monitoring.
§ 60.46 Test methods and procedures.
§ 60.47 Innovative technology waivers; waiver of sulfur dioxide standards of performance for new stationary sources for Homer City Unit No. 3 under section 111(j) of the Clean Air Act for Multi-Steam Coal Cleaning System.
§ 60.48 [Removed. 58 FR 34375, June 25, 1993]

Subpart Da--Standards of Performance for Electric Utility Steam Generating Units for Which Construction Is Commenced After September 18, 1978

§ 60.40a Applicability and designation of affected facility.
§ 60.41a Definitions.
§ 60.42a Standard for particulate matter.
§ 60.43a Standard for sulfur dioxide.
§ 60.44a Standard for nitrogen oxides.
§ 60.45a Commercial demonstration permit.
§ 60.46a Compliance provisions.
§ 60.47a Emission monitoring.
§ 60.48a Compliance determination procedures and methods.
§ 60.49a Reporting requirements.

Subpart Db--Standards of Performance for Industrial-Commercial-Institutional Steam Generating Units

§ 60.40b Applicability and delegation of authority.
§ 60.41b Definitions.
§ 60.42b Standard for sulfur dioxide.
§ 60.43b Standard for particulate matter.
§ 60.44b Standard for nitrogen oxides.
§ 60.45b Compliance and performance test methods and procedures for sulfur dioxide.
§ 60.46b Compliance and performance test methods and procedures for particulate matter and nitrogen oxides.
§ 60.47b Emission monitoring for sulfur dioxide.
§ 60.48b Emission monitoring for particulate matter and nitrogen oxides.
§ 60.49b Reporting and recordkeeping requirements.

Subpart Dc--Standards of Performance for Small Industrial-Commercial- Institutional Steam Generating Units

§ 60.40c Applicability and delegation of authority.
§ 60.41c Definitions.
§ 60.42c Standard for sulfur dioxide.
§ 60.43c Standard for particulate matter.
§ 60.44c Compliance and performance test methods and procedures for sulfur dioxide.
§ 60.45c Compliance and performance test methods and procedures for particulate matter.
§ 60.46c Emission monitoring for sulfur dioxide
§ 60.47c Emission monitoring for particulate matter.
§ 60.48c Reporting and recordkeeping requirements.

Subpart E--Standards of Performance for Incinerators

§ 60.50 Applicability and designation of affected facility.
§ 60.51 Definitions.

§ 60.52 Standard for particulate matter.
§ 60.53 Monitoring of operations.
§ 60.54 Test methods and procedures.
§ 60.55 [Removed. 58 FR 34375, June 25, 1993]

Subpart Ea--Standards of Performance for Municipal Waste Combusters

§ 60.50a Applicability and delegation of authority.
§ 60.51a Definitions.
§ 60.52a Standard for municipal waste combustor metals.
§ 60.53a. Standard for municipal waste combustor organics.
§ 60.54a Standard for municipal waste combustor acid gases.
§ 60.55a Standard for nitrogen oxides.
§ 60.56a Standards for municipal waste combustor operating practices.
§ 60.57a [Reserved]
§ 60.58a Compliance and performance testing.
§ 60.59a Reporting and recordkeeping requirements.

Subpart F--Standards of Performance for Portland Cement Plants

§ 60.60 Applicability and designation of affected facility.
§ 60.61 Definitions.
§ 60.62 Standard for particulate matter.
§ 60.63 Monitoring of operations.
§ 60.64 Test methods and procedures.
§ 60.65 Recordkeeping and reporting requirements.
§ 60.66 Delegation of authority.

Subpart G--Standards of Performance for Nitric Acid Plants

§ 60.70 Applicability and designation of affected facility.
§ 60.71 Definitions.
§ 60.72 Standard for nitrogen oxides.
§ 60.73 Emission monitoring.
§ 60.74 Test methods and procedures.
§ 60.75 [Removed. 58 FR 34375, June 25, 1993]

Subpart H--Standards of Performance for Sulfuric Acid Plants

§ 60.80 Applicability and designation of affected facility.
§ 60.81 Definitions.
§ 60.82 Standard for sulfur dioxide.
§ 60.83 Standard for acid mist.
§ 60.84 Emission monitoring.
§ 60.85 Test methods and procedures.

Subpart I--Standards of Performance for Hot Mix Asphalt Facilities

§ 60.90 Applicability and designation of affected facility.
§ 60.91 Definitions
§ 60.92 Standard for particulate matter.
§ 60.93 Test methods and procedures.
§ 60.94 [Removed. 58 FR 34375, June 25, 1993]

Subpart J--Standards of Performance for Petroleum Refineries

§ 60.100 Applicability, designation of affected facility,

§ 60.130 Applicability and designation of affected facility.
§ 60.131 Definitions.
§ 60.132 Standard for particulate matter.
§ 60.133 Test methods and procedures.
§ 60.134 [Removed. 58 FR 34375, June 25, 1993]
Subpart N--Standards of Performance for Primary Emissions from Basic Oxygen Process Furnaces for Which Construction is Commenced After June 11, 1973
§ 60.140 Applicability and designation of affected facility.
§ 60.141 Definitions.
§ 60.142 Standard for particulate matter.
§ 60.143 Monitoring of operations.
§ 60.144 Test methods and procedures.
Subpart Na--Standards of Performance for Secondary Emissions From Basic oxygen Process Steelmaking Facilities for Which Construction Is Commenced After January 20, 1983
§ 60.140a Applicability and designation of affected facilities.
§ 60.141a Definitions.
§ 60.142a Standards for particulate matter.
§ 60.143a Monitoring of operations.
§ 60.144a Test methods and procedures.
§ 60.145a Compliance provisions.
Subpart O--Standards of Performance for Sewage Treatment Plants
§ 60.150 Applicability and designation of affected facility.
§ 60.151 Definitions.
§ 60.152 Standard for particulate matter.
§ 60.153 Monitoring of operations.
§ 60.154 Test methods and procedures.
§ 60.155 Reporting.
§ 60.156 Delegation of authority.
Subpart P--Standards of Performance for Primary Copper Smelters
§ 60.160 Applicability and designation of affected facility.
§ 60.161 Definitions.
§ 60.162 Standard for particulate matter.
§ 60.163 Standard for sulfur dioxide.
§ 60.164 Standard for visible emissions.
§ 60.165 Monitoring of operations.
§ 60.166 Test methods and procedures.
Subpart Q--Standards of Performance for Primary Zinc Smelters
§ 60.170 Applicability and designation of affected facility.
§ 60.171 Definitions.
§ 60.172 Standard for particulate matter.
§ 60.173 Standard for sulfur dioxide.
§ 60.174 Standard for visible emissions.
§ 60.175 Monitoring of operations.
§ 60.176 Test methods and procedures.
Subpart R--Standards of Performance for Primary Lead Smelters

§ 60.282 Standard for particulate matter.
§ 60.283 Standard for total reduced sulfur (TRS).
§ 60.284 Monitoring of emissions and operations.
§ 60.285 Test methods and procedures.
§ 60.286 Innovative technology waiver.

Subpart CC--Standards of Performance for Glass Manufacturing Plants

§ 60.290 Applicability and designation of affected facility.
§ 60.291 Definitions.
§ 60.292 Standards for particulate matter.
§ 60.293 Standards for particulate matter from glass melting furnace with modified-processes.
Secs. 60.294-60.295 [Reserved]
§ 60.296 Test methods and procedures.
§ 60.297 [Removed. 58 FR 34375, June 25, 1993]

Subpart DD--Standards of Performance for Grain Elevators

§ 60.300 Applicability and designation of affected facility.
§ 60.301 Definitions.
§ 60.302 Standard for particulate matter.
§ 60.303 Test methods and procedures.
§ 60.304 Modifications.
§ 60.305 [Removed. 58 FR 34375, June 25, 1993]

Subpart EE--Standards of Performance for Surface Coating of Metal Furniture

§ 60.310 Applicability and designation of affected facility.
§ 60.311 Definitions and symbols.
§ 60.312 Standard for volatile organic compounds (VOC).
§ 60.313 Performance tests and compliance provisions.
§ 60.314 Monitoring of emissions and operations.
§ 60.315 Reporting and recordkeeping requirements.
§ 60.316 Test methods and procedures.

Subpart FF--[Reserved]

Subpart GG--Standards of Performance for Stationary Gas Turbines

§ 60.330 Applicability and designation of affected facility.
§ 60.331 Definitions.
§ 60.332 Standard for nitrogen oxides.
§ 60.333 Standard for sulfur dioxide.
§ 60.334 Monitoring of operations.
§ 60.335 Test methods and procedures.
§ 60.336 [Removed. 58 FR 34375, June 25, 1993]

Subpart HH--Standards of Performance for Lime Manufacturing Plants

§ 60.340 Applicability and designation of affected facility.
§ 60.341 Definitions.
§ 60.342 Standard for particulate matter.
§ 60.343 Monitoring of emissions and operations.
§ 60.344 Test methods and procedures.

Subpart KK--Standards of Performance for Lead-Acid Battery

§ 60.482-6 Standards: Open-ended valves or lines.
§ 60.482-7 Standards: Valves in gas/vapor service in light liquid service.
§ 60.482-8 Standards: Pumps and valves in heavy liquid service, pressure relief devices in light liquid or heavy liquid service, and flanges and other connectors.
§ 60.482-9 Standards: Delay of repair.
§ 60.482-10 Standards: Closed vent systems and control devices.
§ 60.483-1 Alternative standards for valves--allowable percentage of valves leaking.
§ 60.483-2 Alternative standards for valves--skip period leak detection and repair.
§ 60.484 Equivalence of means of emission limitation.
§ 60.485 Test methods and procedures.
§ 60.486 Recordkeeping requirements.
§ 60.487 Reporting requirements.
§ 60.488 Reconstruction.
§ 60.489 List of chemicals produced by affected facilities.

Subpart WW--Standards of Performance for the Beverage Can Surface Coating Industry

§ 60.490 Applicability and designation of affected facility.
§ 60.491 Definitions.
§ 60.492 Standards for volatile organic compounds.
§ 60.493 Performance test and compliance provisions.
§ 60.494 Monitoring of emissions and operations
§ 60.495 Reporting and recordkeeping requirements.
§ 60.496 Test methods and procedures.

Subpart XX--Standards of Performance for Bulk Gasoline Terminals

§ 60.500 Applicability and designation of affected facility.
§ 60.501 Definitions.
§ 60.502 Standard for Volatile Organic Compound (VOC) emissions from bulk gasoline terminals.
§ 60.503 Test methods and procedures.
§ 60.504 [Reserved]
§ 60.505 Reporting and recordkeeping.
§ 60.506 Reconstruction.

Subpart AAA--Standards of Performance for New Residential Wood Heaters

§ 60.530 Applicability and designation of affected facility.
§ 60.531 Definitions.
§ 60.532 Standards for particulate matter.
§ 60.533 Compliance and certification.
§ 60.534 Test methods and procedures.
§ 60.535 Laboratory accreditation.
§ 60.536 Permanent label, temporary label, and owner's manual.
§ 60.537 Reporting and recordkeeping.
§ 60.538 Prohibitions.

§ 60.539 Hearing and appeal procedures.
§ 60.539a Delegation of Authority.
§ 60.539b General provisions exclusions.
Subpart BBB--Standards of Performance for the Rubber Tire Manufacturing Industry
§ 60.540 Applicability and designation of affected facilities.
§ 60.541 Definitions.
§ 60.542 Standards for volatile organic compounds.
§ 60.542a Alternate standard for volatile organic compounds.
§ 60.543 Performance test and compliance provisions.
§ 60.544 Monitoring of operations.
§ 60.545 Recordkeeping requirements.
§ 60.546 Reporting requirements.
§ 60.547 Test methods and procedures.
§ 60.548 Delegation of authority.
Subpart CCC [Reserved]
Subpart DDD--Standards of Performance for Volatile Organic Compound (VOC) Emissions from the Polymer Manufacturing Industry
§ 60.560 Applicability and designation of affected facilities.
§ 60.561 Definitions.
§ 60.562-1 Standards: Process emissions.
§ 60.562-2 Standards: Equipment leaks of VOC.
§ 60.563 Monitoring requirements.
§ 60.564 Test methods and procedures.
§ 60.565 Reporting and recordkeeping requirements.
§ 60.566 Delegation of authority.
Subpart EEE [Reserved]
Subpart FFF--Standards of Performance for Flexible Vinyl and Urethane Coating and Printing
§ 60.580 Applicability and designation of affected facility.
§ 60.581 Definitions and symbols.
§ 60.582 Standard for volatile organic compounds.
§ 60.583 Test methods and procedures.
§ 60.584 Monitoring of operations and recordkeeping requirements.
§ 60.585 Reporting requirements.
Subpart GGG--Standards of Performance for Equipment Leaks of VOC in Petroleum Refineries
§ 60.590 Applicability and designation of affected facility.
§ 60.591 Definitions.
§ 60.592 Standards.
§ 60.593 Exceptions.
§ 60.594 [Removed. 58 FR 34375, June 25, 1993]
Subpart HHH--Standards of Performance for Synthetic Fiber Production Facilities
§ 60.600 Applicability and designation of affected facility.

Compound (VOC) Emissions From Synthetic Organic Chemical Manufacturing Industry (SOCMI) Reactor Processes

§ 60.700 Applicability and designation of affected facility.
§ 60.701 Definitions.
§ 60.702 Standards.
§ 60.703 Monitoring of emissions and operations.
§ 60.704 Test methods and procedures.
§ 60.705 Reporting and recordkeeping requirements.
§ 60.706 Reconstruction.
§ 60.707 Chemicals affected by Subpart RRR.
§ 60.708 Delegation of Authority.

Subpart SSS--Standards of Performance for Magnetic Tape Coating Facilities

§ 60.710 Applicability and designation of affected facility.
§ 60.711 Definitions, symbols, and cross reference tables.
§ 60.712 Standards for volatile organic compounds.
§ 60.713 Compliance provisions.
§ 60.714 Installation of monitoring devices and recordkeeping.
§ 60.715 Test methods and procedures.
§ 60.716 Permission to use alternative means of emission limitation.
§ 60.717 Reporting and monitoring requirements.
§ 60.718 Delegation of authority.

Subpart TTT--Standards of Performance for Industrial Surface Coating: Surface Coating of Plastic Parts for Business Machines

§ 60.720 Applicability and designation of affected facility.
§ 60.721 Definitions.
§ 60.722 Standards for volatile organic compounds.
§ 60.723 Performance tests and compliance provisions.
§ 60.724 Reporting and recordkeeping requirements.
§ 60.725 Test methods and procedures.
§ 60.726 Delegation of authority.

Subpart UUU--Standards of Performance for Calciners and Dryers in Mineral Industries

§ 60.730 Applicability and designation of affected facility.
§ 60.731 Definitions.
§ 60.732 Standards for particulate matter.
§ 60.733 Reconstruction.
§ 60.734 Monitoring of emissions and operations.
§ 60.735 Recordkeeping and reporting requirements.
§ 60.736 Test methods and procedures.
§ 60.737 Delegation of authority.

Subpart VVV--Standards of Performance for Polymeric Coating of Supporting Substrates Facilities

§ 60.740 Applicability and designation of affected facility.
§ 60.741 Definitions, symbols, and cross-reference tables.

Method 6B--Determination of Sulfur Dioxide and Carbon Dioxide Daily Average Emissions From Fossil Fuel Combustion Sources

Method 6C--Determination of Sulfur Dioxide Emissions From Stationary Sources (Instrumental Analyzer Procedure)

Method 7--Determination of Nitrogen Oxide Emissions From Stationary Sources

Method 7A--Determination of Nitrogen Oxide Emissions From Stationary Sources--Ion Chromatographic Method

Method 7B--Determination of Nitrogen Oxide Emissions From Stationary Sources (Ultraviolet Spectrophotometry)

Method 7C--Determination of Nitrogen Oxide Emissions From Stationary Sources--Alkaline-Permanganate/Colorimetric Method

Method 7D--Determination of Nitrogen Oxide Emissions From Stationary Sources--Alkaline-Permanganate/Ion Chromatographic Method

Method 7E--Determination of Nitrogen Oxides Emissions From Stationary Sources (Instrumental Analyzer Procedure)

Method 8--Determination of Sulfuric Acid Mist and Sulfur Dioxide Emissions From Stationary Sources

Method 9--Visual Determination of the Opacity of Emissions From Stationary Sources

Alternate Method 1--Determination of the Opacity of Emissions From Stationary Sources Remotely by Lidar

Method 10--Determination of Carbon Monoxide Emissions From Stationary Sources

Method 10A--Determination of Carbon Monoxide Emissions in Certifying Continuous Emission Monitoring Systems at Petroleum Refineries

Method 10B--Determination of Carbon Monoxide Emissions From Stationary Sources

Method 11--Determination of Hydrogen Sulfide Content of Fuel Gas Streams in Petroleum Refineries

Method 12--Determination of Inorganic Lead Emissions From Stationary Sources

Method 13A--Determination of Total Fluoride Emissions From Stationary Sources; SPADNS Zirconium Lake Method

Method 13B--Determination of Total Fluoride Emissions From Stationary Sources--Specific Ion Electrode Method

Method 14--Determination of Fluoride Emissions from Potroom Roof Monitors for Primary Aluminum Plants

Method 15--Determination of Hydrogen Sulfide, Carbonyl Sulfide, and Carbon Disulfide Emissions from Stationary Sources

Method 15A--Determination of Total Reduced Sulfur Emissions From Sulfur Recovery Plants in Petroleum Refineries

Method 16--Semicontinuous Determination of Sulfur Emissions From Stationary Sources

Method 16A--Determination of Total Reduced Sulfur Emissions From Stationary Sources (Impinger Technique)

Method 16B--Determination of Total Reduced Sulfur Emissions From Stationary Sources

§ 61.102 Standard.
§ 61.103 Determining compliance.
§ 61.104 Reporting requirements.
§ 61.105 Recordkeeping requirements.
§ 61.106 Applications to construct or modify.
§ 61.107 Emission determination.
§ 61.108 Exemption from the reporting and testing requirements of 40 CFR 61.10.
§ 61.109 Stay of effective date.

Subpart J--National Emission Standard for Equipment Leaks (Fugitive Emission Sources) of Benzene

§ 61.110 Applicability and designation of sources.
§ 61.111 Definitions.
§ 61.112 Standards.

Subpart K--National Emission Standards for Radionuclide Emissions From Elemental Phosphorus Plants

§ 61.120 Applicability.
§ 61.121 Definitions.
§ 61.122 Emission standard.
§ 61.123 Emission testing.
§ 61.124 Recordkeeping requirements.
§ 61.125 Test methods and procedures.
§ 61.126 Monitoring of operations.
§ 61.127 Exemption from the reporting and testing requirements of 40 CFR 61.10.

Subpart L--National Emission Standard for Benzene Emissions from Coke By-Product Recovery Plants

§ 61.130 Applicability, designation of sources, and delegation of authority.
§ 61.131 Definitions.
§ 61.132 Standard: Process vessels, storage tanks, and tar-intercepting sumps.
§ 61.133 Standard: Light-oil sumps.
§ 61.134 Standard: Naphthalene processing, final coolers, and final-cooler cooling towers.
§ 61.135 Standard: Equipment leaks.
§ 61.136 Compliance provisions and alternative means of emission limitation.
§ 61.137 Test methods and procedures.
§ 61.138 Recordkeeping and reporting requirements.
§ 61.139 Provisions for alternative means for process vessels, storage tanks, and tar-intercepting sumps.

Subpart M--National Emission Standard for Asbestos

§ 61.140 Applicability.
§ 61.141 Definitions.
§ 61.142 Standard for asbestos mills.
§ 61.143 Standard for roadways.
§ 61.144 Standard for manufacturing.
§ 61.145 Standard for demolition and renovation.
§ 61.146 Standard for spraying.
§ 61.147 Standard for fabricating.
§ 61.148 Standard for insulating materials.
§ 61.149 Standard for waste disposal for asbestos mills.

§ 61.150 Standard for waste disposal for manufacturing, fabricating, demolition, renovation, and spraying operations.

§ 61.151 Standard for inactive waste disposal sites for asbestos mills and manufacturing and fabricating operations.

§ 61.152 Air-cleaning.

§ 61.153 Reporting.

§ 61.154 Standard for active waste disposal sites.

§ 61.155 Standard for operations that convert asbesto-containing waste material into nonasbestos (asbestos-free) material.

§ 61.156 Cross-reference to other asbestos regulations.

§ 61.157 Delegation of authority.

Appendix A to Subpart M--Interpretive Rule Governing Roof Removal Operations

Subpart N--National Emission Standard for Inorganic Arsenic Emissions from Glass Manufacturing Plants

§ 61.160 Applicability and designation of source.

§ 61.161 Definitions.

§ 61.162 Emission limits.

§ 61.163 Emission monitoring.

§ 61.164 Test methods and procedures.

§ 61.165 Reporting and recordkeeping requirements.

Subpart O--National Emission Standard for Inorganic Arsenic Emissions from Primary Copper Smelters

§ 61.170 Applicability and designation of source.

§ 61.171 Definitions.

§ 61.172 Standard for new and existing sources.

§ 61.173 Compliance provisions.

§ 61.174 Test methods and procedures.

§ 61.175 Monitoring requirements.

§ 61.176 Recordkeeping requirements.

§ 61.177 Reporting requirements.

Subpart P--National Emission Standard for Inorganic Arsenic Emissions From Arsenic Trioxide and Metallic Arsenic Production Facilities

§ 61.180 Applicability and designation of sources.

§ 61.181 Definitions.

§ 61.182 Standard for new and existing sources.

§ 61.183 Emission monitoring.

§ 61.184 Ambient air monitoring for inorganic arsenic.

§ 61.185 Recordkeeping requirements.

§ 61.186 Reporting requirements.

Subpart Q--National Emission Standards for Radon Emissions From Department of Energy Facilities

§ 61.190 Designation of facilities.

§ 61.191 Definitions.

§ 61.192 Standard.

§ 61.193 Exemption from the reporting and testing requirements of 40 CFR 61.10.

Subpart R--National Emission Standards for Radon Emissions From Phosphogypsum Stacks

§ 61.200 Designation of facilities.
§ 61.201 Definitions.
§ 61.202 Standard.
§ 61.203 Radon monitoring and compliance procedures.
§ 61.204 Distribution and use of phosphogypsum for agricultural purposes.
§ 61.205 Distribution and use of phosphogypsum for research and development.
§ 61.206 Distribution and use of phosphogypsum for other purposes.
§ 61.207 Radium-226 sampling and measurement procedures.
§ 61.208 Certification requirements.
§ 61.209 Required records.
§ 61.210 Exemption from the reporting and testing requirements of 40 CFR 61.10.

Subpart S--[Reserved]

Subpart T--National Emission Standards for Radon Emissions From the Disposal of Uranium Mill Tailings

§ 61.220 Designation of facilities.
§ 61.221 Definitions.
§ 61.222 Standard.
§ 61.223 Compliance procedures.
§ 61.224 Recordkeeping requirements.
§ 61.225 Exemption from the reporting and testing requirements of 40 CFR 61.10.

Subpart U--[Reserved]

Subpart V--National Emission Standard for Equipment Leaks (Fugitive Emission Sources)

§ 61.240 Applicability and designation of sources.
§ 61.241 Definitions.
§ 61.242-1 Standards: General.
§ 61.242-2 Standards: Pumps.
§ 61.242-3 Standards: Compressors.
§ 61.242-4 Standards: Pressure relief devices in gas/vapor service.
§ 61.242-5 Standards: Sampling connecting systems.
§ 61.242-6 Standards: Open-ended valves or lines.
§ 61.242-7 Standards: Valves.
§ 61.242-8 Standards: Pressure relief devices in liquid service and flanges and other connectors.
§ 61.242-9 Standards: Product accumulator vessels.
§ 61.242-10 Standards: Delay of repair.
§ 61.242-11 Standards: Closed-vent systems and control devices.
§ 61.243-1 Alternative standards for valves in VHAP service--allowable percentage of valves leaking.
§ 61.243-2 Alternative standards for valves in VHAP service--skip period leak detection and repair.
§ 61.244 Alternative means of emission limitation.
§ 61.245 Test methods and procedures.
§ 61.246 Recordkeeping requirements.
§ 61.247 Reporting requirements.

Subpart W--National Emission Standards for Radon Emissions From

Operating Mill Tailings
§ 61.250 Designation of facilities.
§ 61.251 Definitions.
§ 61.252 Standard.
§ 61.253 Determining compliance.
§ 61.254 Annual reporting requirements.
§ 61.255 Recordkeeping requirements.
§ 61.256 Exemption from the reporting and testing requirements of 40 CFR 61.10.
Subpart Y--National Emission Standard for Benzene Emissions from Benzene Storage Vessels
§ 61.270 Applicability and designation of sources.
§ 61.271 Emission standard.
§ 61.272 Compliance provisions.
§ 61.273 Alternative means of emission limitation.
§ 61.274 Initial report.
§ 61.275 Periodic report.
§ 61.276 Recordkeeping.
§ 61.277 Delegation of authority.
Subpart BB--National Emission Standard for Benzene Emissions from Benzene Transfer Operations
§ 61.300 Applicability.
§ 61.301 Definitions.
§ 61.302 Standards.
§ 61.303 Monitoring requirements.
§ 61.304 Test methods and procedures.
§ 61.305 Reporting and recordkeeping.
§ 61.306 Delegation of authority.
Subpart FF--National Emission Standard for Benzene Waste Operations
§ 61.340 Applicability
§ 61.341 Definitions.
§ 61.342 Standards: General.
§ 61.343 Standards: Tanks.
§ 61.344 Standards: Surface impoundments.
§ 61.345 Standards: Containers.
§ 61.346 Standards: Individual drain systems.
§ 61.347 Standards: Oil-water separators.
§ 61.348 Standards: Treatment processes.
§ 61.349 Standards: Closed-vent systems and control devices.
§ 61.350 Standards: Delay of repair.
§ 61.351 Alternative standards for tanks.
§ 61.352 Alternative standards for oil-water separators.
§ 61.353 Alternative means of emission limitation.
§ 61.354 Monitoring of operations.
§ 61.355 Test methods, procedures, and compliance provisions.
§ 61.356 Recordkeeping requirements.
§ 61.357 Reporting requirements.
§ 61.358 Delegation of authority.
§ 61.359 [Reserved. 58 FR 3105, Jan. 7, 1993]
Appendix A--National Emission Standards for Hazardous Air

Subpart Z--Mississippi
§ 62.6100 Identification of plan.
Sulfuric Acid Mist From Existing Sulfuric Acid Plants
§ 62.6110 Identification of sources.
Fluoride Emissions From Phosphate Fertilizer Plants
§ 62.6120 Identification of sources.
Fluoride Emissions From Primary Aluminum Reduction Plants
§ 62.6121 Identification of sources--Negative declaration.
Total Reduced Sulfur Emissions From Kraft Pulp Mills
§ 62.6122 Identification of sources.
Municipal Waste Combustors
§ 62.6123 Identification of sources--Negative declaration.
Subpart AA--Missouri
§ 62.6350 Identification of plan.
Fluoride Emissions From Existing Phosphate Fertilizer Plants
§ 62.6351 Identification of sources.
Flouride Emissions From Existing Primary Aluminum Reduction Plants
§ 62.6352 Identification of sources.
Sulfuric Acid Mist From Existing Sulfuric Acid Production Plants
§ 62.6353 Idenification of sources.
Total Reduced Sulfur Emissions From Existing Kraft Pulp Mills
§ 62.6354 Identification of plan--negative declaration.
Emissions From Existing Municipal Waste Combustors With the Capacity To Burn Greater Than 250 Tons Per Day of Municipal Solid Waste
§ 62.6355. Identification of Plan--Negative Declaration.
Subpart BB--[Reserved]
Subpart CC--Nebraska
Fluoride Emissions From Existing Phosphate Fertilizer Plants
§ 62.6850 Identification of Plan--Negative Declaration.
Sulfuric Acid Mist Emissions From Existing Sulfuric Acid Plants
§ 62.6875 Identification of plan--negative declaration.
Total Reduced Sulfur Emissions From Existing Kraft Pulp Mills
§ 62.6880 Identification of plan--negative declaration.
Fluoride Emissions From Existing Primary Aluminum Reduction Plants
§ 62.6910 Identification of plan--negative declaration.
Emissions From Existing Municipal Waste Combustors With the Capacity To Burn Greater Than 250 Tons Per Day of Municipal Solid Waste
§ 62.6911 Identification of Plan--Negative Declaration.
Subpart DD--[Reserved]
Subpart EE--New Hampshire
Plan for the Control of Designated Pollutants From Existing Facilities (Section 111(d) Plan)
§ 62.7325 Identification of plan.
Fluoride Emissions From Phosphate Fertilizer Plants
§ 62.7350 Identification of plan--negative declaration.
Sulfuric Acid Mist Emissions From Sulfuric Acid Production Units
§ 62.7375 Identification of plan--negative declaration.

PART 63--NATIONAL EMISSION STANDARDS FOR HAZARDOUS AIR POLLUTANTS FOR SOURCE CATEGORIES

Subpart A--General Provisions

§ 63.1 Applicability.

§ 63.2 Definitions.

§ 63.3 Units and abbreviations.

§ 63.4 Prohibited activities and circumvention.

§ 63.5 Construction and reconstruction.

§ 63.6 Compliance with standards and maintenance requirements.

§ 63.7 Performance testing requirements.

§ 63.8 Monitoring requirements.

§ 63.9 Notification requirements.

§ 63.10 Recordkeeping and reporting requirements.

§ 63.11 Control device requirements.

§ 63.12 State authority and delegations.

§ 63.13 Addresses of State air pollution control agencies and EPA Regional Offices.

§ 63.14 Incorporations by reference.

§ 63.15 Availability of information and confidentiality.

Subpart B--Requirements for Control Technology Determinations for Major Sources in Accordance With Clean Air Act Sections, Sections 112(g) and 112(j)

Secs. 63.40--63.49 [Reserved]

§ 63.50 Applicability.

§ 63.51 Definitions.

§ 63.52 Approval process for new and existing emission units.

§ 63.53 Application content for case-by-case MACT determinations.

§ 63.54 Preconstruction review procedures for new emission units.

§ 63.55 Maximum achievable control technology (MACT) determinations for emission units subject to case-by-case determination of equivalent emission limitations.

§ 63.56 Requirements for case-by-case determination of equivalent emission limitations after promulgation of a subsequent MACT standard.

Subpart C--List of Hazardous Air Pollutants, Petitions Process, Lesser Quantity Designations, Source Category List [Reserved]

Subpart D--Regulations Governing Compliance Extensions for Early Reductions of Hazardous Air Pollutants

§ 63.70 Applicability.

§ 63.71 Definitions.

§ 63.72 General provisions for compliance extensions

§ 63.73 Source.

§ 63.74 Demonstration of early reduction.

§ 63.75 Enforceable commitments.

§ 63.76 Review of base year emissions.

§ 63.77 Application procedures.

§ 63.78 Early reduction demonstration evaluation.

§ 63.79 Approval of applications.

§ 63.80 Enforcement.

section 113(d) (1), (3), (4) and (5) of the Act.
§ 65.351 [Reserved]
§ 65.352 [Reserved]
Subpart GG--[Reserved]
Subpart HH--[Reserved]
Subpart II--[Reserved]
Subpart JJ--[Reserved]
Subpart KK--[Reserved]
Subpart LL--[Reserved]
Subpart MM--[Reserved]
Subpart NN--Pennsylvania
§ 65.430 [Reserved]
§ 65.431 EPA approval of State delayed compliance orders issued to major stationary sources.
§ 65.432 [Reserved]
Subpart OO--[Reserved]
Subpart PP--[Reserved]
Subpart QQ--[Reserved]
Subpart RR--Tennessee
§ 65.470 [Reserved]
§ 65.471 EPA approval of State delayed compliance orders issued to major stationary sources.
§ 65.472 [Reserved]
Subpart SS--Texas
§ 65.480 [Reserved]
§ 65.481 EPA approval of State delayed compliance orders issued to major stationary sources.
§ 65.482 [Reserved]
Subpart TT--[Reserved]
Subpart UU--[Reserved]
Subpart VV--Virginia
§ 65.510 Federal delayed compliance orders issued under section 113(d) (1), (3), (4) and (5) of the Act.
Secs. 65.511-12 [Reserved]
Subpart WW--Washington
§ 65.520 [Reserved]
§ 65.521 EPA approval of State delayed compliance orders issued to major stationary sources.
§ 65.522 [Reserved]
Subpart XX--[Reserved]
Subpart YY--Wisconsin
§ 65.540 Federal delayed compliance orders issued under section 113(d) (1), (3), (4) and (5) of the Act.
§ 65.541 [Reserved]
§ 65.542 [Reserved]
Subpart ZZ--[Reserved]
Subpart AAA--Guam
§ 65.560 Federal delayed compliance orders issued under section 113(d) (1), (3), (4) and (5) of the Act.
§ 65.561 [Reserved]
§ 65.562 [Reserved]
Subpart BBB--[Reserved]
Subpart CCC--[Reserved]

commercial operation during the period from January 1, 1993 through December 31, 1995.

§ 73.19 Certain units with declining SO2 rates.
§ 73.20 Phase II early reduction credits.
§ 73.21 Phase II repowering allowances.
§ 73.22 [Reserved. 58 FR 15650, Mar. 23, 1993]
§ 73.23 [Reserved. 58 FR 15650, Mar. 23, 1993]
§ 73.24 [Reserved. 58 FR 15650, Mar. 23, 1993]
§ 73.26 Conservation and renewable energy reserve.
§ 73.25 Phase I extension reserve.
§ 73.27 Special allowance reserve.

Subpart C--Allowance Tracking System

§ 73.30 Allowance tracking system accounts.
§ 73.31 Establishment of accounts.
§ 73.32 Allowance account contents.
§ 73.33 Authorized account representative.
§ 73.34 Recordation in accounts.
§ 73.35 Compliance.
§ 73.36 Banking.
§ 73.37 Account error and dispute resolution.
§ 73.38 Closing of accounts.

Subpart D--Allowance Transfers

§ 73.50 Scope and submission of transfers.
§ 73.51 Prohibition.
§ 73.52 EPA recordation.
§ 73.53 Notification.

Subpart E--Auctions, Direct Sales, and Independent Power Producers Written Guarantee

§ 73.70 Auctions.
§ 73.71 Bidding.
§ 73.72 Direct sales.
§ 73.73 Delegation of auctions and sales and termination of auctions and sales.
§ 73.74 Independent power producers written guarantee.
§ 73.75 Application for an IPP written guarantee.
§ 73.76 Approval and exercise of the IPP written guarantee.
§ 73.77 Relationship of the independent power producers written guarantee to the direct sale subaccount.

Subpart F--Conservation and Renewable Energy Reserve

§ 73.80 Operation of allowance reserve program for conservation and renewable energy.
§ 73.81 Qualified conservation measures and renewable energy generation.
§ 73.82 Application for allowances from reserve program.
§ 73.83 Secretary of Energy's action on net income neutrality applications.
§ 73.84 Administrator's action on applications.
§ 73.85 Administrator review of the reserve program.
§ 73.86 State regulatory autonomy.

Appendix A to Subpart F--List of Qualified Energy Conservation Measures, Qualified Renewable Generation, and Measures Applicable for Reduced Utilization

Subpart G--Small Diesel Refineries

§ 73.90 Allowance allocations for small diesel refineries.

PART 75--CONTINUOUS EMISSION MONITORING

Subpart A--General

§ 75.1 Purpose and scope.

§ 75.2 Applicability.

§ 75.3 General Acid Rain Program provisions.

§ 75.4 Compliance dates.

§ 75.5 Prohibitions.

§ 75.6 Incorporation by reference.

§ 75.7 EPA Study.

§ 75.8 [Reserved]

Subpart B--Monitoring Provisions

§ 75.10 General operating requirements.

§ 75.11 Specific provisions for monitoring SO2 emissions (SO2 and flow monitors).

§ 75.12 Specific provisions for monitoring NOx emissions (NOx and diluent gas monitors).

§ 75.13 Specific provisions for monitoring CO2 emissions.

§ 75.14 Specific provisions for monitoring opacity.

§ 75.15 Specific provisions for monitoring SO2 emissions removal by qualifying Phase I technology.

§ 75.16 Special provisions for monitoring emissions from common by-pass, and multiple stacks for SO2 emissions and heat input determinations.

§ 75.17 Specific provisions for monitoring emissions from common, by- pass, and multiple stacks for NOx emission rate.

§ 75.18 Specific provisions for monitoring emissions from common and by- pass stacks for opacity.

Subpart C--Operation and Maintenance Requirements

§ 75.20 Certification and recertification procedures.

§ 75.21 Quality assurance and quality control requirements.

§ 75.22 Reference test methods.

§ 75.23 Alternatives to ASTM methods.

§ 75.24 Out-of-control periods.

Subpart D--Missing Data Substitution Procedures

§ 75.30 General provisions.

§ 75.31 Initial missing data procedures.

§ 75.32 Determination of monitor data availability for standard missing data procedures.

§ 75.33 Standard missing data procedures.

§ 75.34 Units with add-on emission controls.

Subpart E--Alternative Monitoring Systems

§ 75.40 General demonstration requirements.

§ 75.41 Precision criteria.

§ 75.42 Reliability criteria.

§ 75.43 Accessibility criteria.

§ 75.44 Timeliness criteria.

§ 75.45 Daily quality assurance criteria.

§ 75.46 Missing data substitution criteria.

§ 75.47 Criteria for a class of affected units.

Phase I Boilers

PART 77--EXCESS EMISSIONS

§ 77.1 Purpose and scope.

§ 77.2 General.

§ 77.3 Offset plans for excess emissions of sulfur dioxide.

§ 77.4 Administrator's action on proposed offset plans.

§ 77.5 Deduction of allowances to offset excess emissions of sulfur dioxide.

§ 77.6 Penalties for excess emissions of sulfur dioxide and nitrogen oxides.

PART 78--APPEAL PROCEDURES FOR ACID RAIN PROGRAM

§ 78.1 Purpose and scope.

§ 78.2 General.

§ 78.3 Petition for administrative review and request for evidentiary hearing.

§ 78.4 Filings.

§ 78.5 Limitation on filing or presenting new evidence and raising new issues.

§ 78.6 Action on petition for administrative review.

§ 78.7 Stays of contested Acid Rain requirements pending appeal.

§ 78.8 Consolidation and severance of appeals proceedings.

§ 78.9 Notice of the filing of petition for administrative review.

§ 78.10 Ex parte communications during pendency of a hearing.

§ 78.11 Intervenors.

§ 78.12 Standard of review.

§ 78.13 Scheduling orders and pre-hearing conferences.

§ 78.14 Evidentiary hearing procedure.

§ 78.15 Motions in evidentiary hearings.

§ 78.16 Record of appeal proceeding.

§ 78.17 Proposed findings and conclusions and supporting brief.

§ 78.18 Proposed decision.

§ 78.19 Interlocutory appeal.

§ 78.20 Appeal of decision of Administrator or proposed decision to the Environmental Appeals Board.

PART 79--REGISTRATION OF FUELS AND FUEL ADDITIVES

Subpart A--General Provisions

§ 79.1 Applicability.

§ 79.2 Definitions.

§ 79.3 Availability of information.

§ 79.4 Requirement of registration.

§ 79.5 Periodic reporting requirements.

§ 79.6 Requirement for testing.

§ 79.7 Samples for test purposes.

§ 79.8 Penalties.

Subpart B--Fuel Registration Procedures

§ 79.10 Application for registration by fuel manufacturer.

§ 79.11 Information and assurances to be provided by the fuel manufacturer.

§ 81.34 Metropolitan Dayton Intrastate Air Quality Control Region.
§ 81.35 Louisville Interstate Air Quality Control Region.
§ 81.36 Maricopa Intrastate Air Quality Control Region.
§ 81.37 Metropolitan Detroit-Port Huron Intrastate Air Quality Control Region.
§ 81.38 Metropolitan Houston-Galveston Intrastate Air Quality Control Region.
§ 81.39 Metropolitan Dallas-Fort Worth Intrastate Air Quality Control Region.
§ 81.40 Metropolitan San Antonio Intrastate Air Quality Control Region.
§ 81.41 Metropolitan Birmingham Intrastate Air Quality Control Region.
§ 81.42 Chattanooga Interstate Air Quality Control Region.
§ 81.43 Metropolitan Toledo Interstate Air Quality Control Region.
§ 81.44 Metropolitan Memphis Interstate Air Quality Control Region.
§ 81.45 Metropolitan Atlanta Intrastate Air Quality Control Region.
§ 81.46 U.S. Virgin Islands Air Quality Control Region.
§ 81.47 Central Oklahoma Intrastate Air Quality Control Region.
§ 81.48 Champlain Valley Interstate Air Quality Control Region.
§ 81.49 Southeast Florida Intrastate Air Quality Control Region.
§ 81.50 Metropolitan Omaha-Council Bluffs Interstate Air Quality Control Region.
§ 81.51 Portland Interstate Air Quality Control Region.
§ 81.52 Wasatch Front Intrastate Air Quality Control Region.
§ 81.53 Southern Louisiana-Southeast Texas Interstate Air Quality Control Region.
§ 81.54 Cook Inlet Intrastate Air Quality Control Region.
§ 81.55 Northeast Pennsylvania-Upper Delaware Valley Interstate Air Quality Control Region.
§ 81.57 Eastern Tennessee-Southwestern Virginia Interstate Air Quality Control Region.
§ 81.58 Columbus (Georgia)-Phenix City (Alabama) Interstate Air Quality Control Region.
§ 81.59 Cumberland-Keyser Interstate Air Quality Control Region.
§ 81.60 Duluth (Minnesota)-Superior (Wisconsin) Interstate Air Quality Control Region.
§ 81.61 Evansville (Indiana)-Owensboro-Henderson (Kentucky) Interstate Air Quality Control Region.
§ 81.62 Northeast Mississippi Intrastate Air Quality Control Region.
§ 81.63 Metropolitan Fort Smith Interstate Air Quality Control Region.
§ 81.64 Huntington (West Virginia)-Ashland (Kentucky)-

Portsmouth-Ironton (Ohio) Interstate Air Quality Control Region.
§ 81.65 Joplin (Missouri)-Northeast Oklahoma Interstate Air Quality Control Region.
§ 81.66 Southeast Minnesota-La Crosse (Wisconsin) Interstate Air Quality Control Region.
§ 81.67 Lake Michigan Intrastate Air Quality Control Region.
§ 81.68 Mobile (Alabama)-Pensacola-Panama City (Florida)-Southern Mississippi Interstate Air Quality Control Region.
§ 81.69 Paducah (Kentucky)-Cairo (Illinois) Interstate Air Quality Control Region.
§ 81.70 Parkersburg (West Virginia)-Marietta (Ohio) Interstate Air Quality Control Region.
§ 81.71 Rockford (Illinois)-Janesville-Beloit (Wisconsin) Interstate Air Quality Control Region.
§ 81.72 Tennessee River Valley (Alabama)-Cumberland Mountains (Tennessee) Interstate Air Quality Control Region.
§ 81.73 South Bend-Elkhart (Indiana)-Benton Harbor (Michigan) Interstate Air Quality Control Region.
§ 81.74 Northwest Pennsylvania-Youngstown Interstate Air Quality Control Region.
§ 81.75 Metropolitan Charlotte Interstate Air Quality Control Region.
§ 81.76 State of Hawaii Air Quality Control Region.
§ 81.77 Puerto Rico Air Quality Control Region.
§ 81.78 Metropolitan Portland Intrastate Air Quality Control Region.
§ 81.79 Northeastern Oklahoma Intrastate Air Quality Control Region.
§ 81.80 Las Vegas Intrastate Air Quality Control Region.
§ 81.81 Merrimack Valley-Southern New Hampshire Interstate Air Quality Control Region.
§ 81.82 El Paso-Las Cruces-Alamogordo Interstate Air Quality Control Region.
§ 81.83 Albuquerque-Mid Rio Grande Intrastate Air Quality Control Region.
§ 81.84 Metropolitan Fargo-Moorhead Interstate Air Quality Control Region.
§ 81.85 Metropolitan Sioux Falls Interstate Air Quality Control Region.
§ 81.86 Metropolitan Sioux City Interstate Air Quality Control Region.
§ 81.87 Metropolitan Boise Intrastate Air Quality Control Region.
§ 81.88 Billings Intrastate Air Quality Control Region.
§ 81.89 Metropolitan Cheyenne Intrastate Air Quality Control Region.
§ 81.90 Androscoggin Valley Interstate Air Quality Control Region.
§ 81.91 Jacksonville (Florida)-Brunswick (Georgia)

Interstate Air Quality Control Region.
§ 81.92 Monroe (Louisiana)--El Dorado (Arkansas) Interstate Air Quality Control Region.
§ 81.93 Hampton Roads Intrastate Air Quality Control Region.
§ 81.94 Shreveport-Texarkana-Tyler Interstate Air Quality Control Region.
§ 81.95 Central Florida Intrastate Air Quality Control Region.
§ 81.96 West Central Florida Intrastate Air Quality Control Region.
§ 81.97 Southwest Florida Intrastate Air Quality Control Region.
§ 81.98 Burlington-Keokuk Interstate Air Quality Control Region.
§ 81.99 New Mexico Southern Border Intrastate Air Quality Control Region.
§ 81.100 Eastern Washington-Northern Idaho Interstate Air Quality Control Region.
§ 81.101 Metropolitan Dubuque Interstate Air Quality Control Region.
§ 81.102 Metropolitan Quad Cities Interstate Air Quality Control Region.
§ 81.104 Central Pennsylvania Intrastate Air Quality Control Region.
§ 81.105 South Central Pennsylvania Intrastate Air Quality Control Region.
§ 81.106 Greenville-Spartanburg Intrastate Air Quality Control Region.
§ 81.107 Greenwood Intrastate Air Quality Control Region.
§ 81.108 Columbia Intrastate Air Quality Control Region.
§ 81.109 Florence Intrastate Air Quality Control Region.
§ 81.110 Camden-Sumter Intrastate Air Quality Control Region.
§ 81.111 Georgetown Intrastate Air Quality Control Region.
§ 81.112 Charleston Intrastate Air Quality Control Region.
§ 81.113 Savannah (Georgia)-Beaufort (South Carolina) Interstate Air Quality Control Region.
§ 81.114 Augusta (Georgia)-Aiken (South Carolina) Interstate Air Quality Control Region.
§ 81.115 Northwest Nevada Intrastate Air Quality Control Region.
§ 81.116 Northern Missouri Intrastate Air Quality Control Region.
§ 81.117 Southeast Missouri Intrastate Air Quality Control Region.
§ 81.118 Southwest Missouri Intrastate Air Quality Control Region.
§ 81.119 Western Tennessee Intrastate Air Quality Control Region.
§ 81.120 Middle Tennessee Intrastate Air Quality Control Region.
§ 81.121 Four Corners Interstate Air Quality Control

Region.
§ 81.122 Mississippi Delta Intrastate Air Quality Control Region.
§ 81.123 Southeastern Oklahoma Intrastate Air Quality Control Region.
§ 81.124 North Central Oklahoma Intrastate Air Quality Control Region.
§ 81.125 Southwestern Oklahoma Intrastate Air Quality Control Region.
§ 81.126 Northwestern Oklahoma Intrastate Air Quality Control Region.
§ 81.127 Central New York Intrastate Air Quality Control Region.
§ 81.128 Genesee-Finger Lakes Intrastate Air Quality Control Region.
§ 81.129 Hudson Valley Intrastate Air Quality Control Region.
§ 81.130 Southern Tier East Intrastate Air Quality Control Region.
§ 81.131 Southern Tier West Intrastate Air Quality Control Region.
§ 81.132 Abilene-Wichita Falls Intrastate Air Quality Control Region.
§ 81.133 Amarillo-Lubbock Intrastate Air Quality Control Region.
§ 81.134 Austin-Waco Intrastate Air Quality Control Region.
§ 81.135 Brownsville-Laredo Intrastate Air Quality Control Region.
§ 81.136 Corpus Christi-Victoria Intrastate Air Quality Control Region.
§ 81.137 Midland-Odessa-San Angelo Intrastate Air Quality Control Region.
§ 81.138 Central Arkansas Intrastate Air Quality Control Region.
§ 81.139 Northeast Arkansas Intrastate Air Quality Control Region.
§ 81.140 Northwest Arkansas Intrastate Air Quality Control Region.
§ 81.141 Berkshire Intrastate Air Quality Control Region.
§ 81.142 Central Massachusetts Intrastate Air Quality Control Region.
§ 81.143 Central Virginia Intrastate Air Quality Control Region.
§ 81.144 Northeastern Virginia Intrastate Air Quality Control Region.
§ 81.145 State Capital Intrastate Air Quality Control Region.
§ 81.146 Valley of Virginia Intrastate Air Quality Control Region.
§ 81.147 Eastern Mountain Intrastate Air Quality Control Region.
§ 81.148 Eastern Piedmont Intrastate Air Quality Control

Region.
§ 81.183 Eastern Connecticut Intrastate Air Quality Control Region.
§ 81.184 Northwestern Connecticut Intrastate Air Quality Control Region.
§ 81.185 Northern Washington Intrastate Air Quality Control Region.
§ 81.187 Olympic-Northwest Washington Intrastate Air Quality Control Region.
§ 81.189 South Central Washington Intrastate Air Quality Control Region.
§ 81.190 Eastern Idaho Intrastate Air Quality Control Region.
§ 81.191 Appalachian Intrastate Air Quality Control Region.
§ 81.192 Bluegrass Intrastate Air Quality Control Region.
§ 81.193 North Central Kentucky Intrastate Air Quality Control Region.
§ 81.194 South Central Kentucky Intrastate Air Quality Control Region.
§ 81.195 Central Michigan Intrastate Air Quality Control Region.
§ 81.196 South Central Michigan Intrastate Air Quality Control Region.
§ 81.197 Upper Michigan Intrastate Air Quality Control Region.
§ 81.199 East Alabama Intrastate Air Quality Control Region.
§ 81.200 Metropolitan Columbus Intrastate Air Quality Control Region.
§ 81.201 Mansfield-Marion Intrastate Air Quality Control Region.
§ 81.202 Northwest Ohio Intrastate Air Quality Control Region.
§ 81.203 Sandusky Intrastate Air Quality Control Region.
§ 81.204 Wilmington-Chillicothe-Logan Intrastate Air Quality Control Region.
§ 81.205 Zanesville-Cambridge Intrastate Air Quality Control Region.
§ 81.213 Casper Intrastate Air Quality Control Region.
§ 81.214 Black Hills-Rapid City Intrastate Air Quality Control Region.
§ 81.215 East Central Indiana Intrastate Air Quality Control Region.
§ 81.216 Northeast Indiana Intrastate Air Quality Control Region.
§ 81.217 Southern Indiana Intrastate Air Quality Control Region.
§ 81.218 Wabash Valley Intrastate Air Quality Control Region.
§ 81.219 Central Oregon Intrastate Air Quality Control Region.
§ 81.220 Eastern Oregon Intrastate Air Quality Control

Region.
§ 81.221 Southwest Oregon Intrastate Air Quality Control Region.
§ 81.226 Lincoln-Beatrice-Fairbury Intrastate Air Quality Control Region.
§ 81.230 Allegheny Intrastate Air Quality Control Region.
§ 81.231 Central West Virginia Intrastate Air Quality Control Region.
§ 81.232 Eastern Panhandle Intrastate Air Quality Control Region.
§ 81.233 Kanawha Valley Intrastate Air Quality Control Region.
§ 81.234 North Central West Virginia Intrastate Air Quality Control Region.
§ 81.235 Southern West Virginia Intrastate Air Quality Control Region.
§ 81.236 Central Georgia Intrastate Air Quality Control Region.
§ 81.237 Northeast Georgia Intrastate Air Quality Control Region.
§ 81.238 Southwest Georgia Intrastate Air Quality Control Region.
§ 81.239 Upper Rio Grande Valley Intrastate Air Quality Control Region.
§ 81.240 Northeastern Plains Intrastate Air Quality Control Region.
§ 81.241 Southwestern Mountains-Augustine Plains Intrastate Air Quality Control Region.
§ 81.242 Pecos-Permian Basin Intrastate Air Quality Control Region.
§ 81.243 Central Minnesota Intrastate Air Quality Control Region.
§ 81.244 Northwest Minnesota Intrastate Air Quality Control Region.
§ 81.245 Southwest Minnesota Intrastate Air Quality Control Region.
§ 81.246 Northern Alaska Intrastate Air Quality Control Region.
§ 81.247 South Central Alaska Intrastate Air Quality Control Region.
§ 81.248 Southeastern Alaska Intrastate Air Quality Control Region.
§ 81.249 Northwest Oregon Intrastate Air Quality Control Region.
§ 81.250 North Central Kansas Intrastate Air Quality Control Region.
§ 81.251 Northeast Kansas Intrastate Air Quality Control Region.
§ 81.252 Northwest Kansas Intrastate Air Quality Control Region.
§ 81.253 South Central Kansas Intrastate Air Quality Control Region.
§ 81.254 Southeast Kansas Intrastate Air Quality Control

Region.
§ 81.255 Southwest Kansas Intrastate Air Quality Control Region.
§ 81.256 Northeast Iowa Intrastate Air Quality Control Region.
§ 81.257 North Central Iowa Intrastate Air Quality Control Region.
§ 81.258 Northwest Iowa Intrastate Air Quality Control Region.
§ 81.259 Southwest Iowa Intrastate Air Quality Control Region.
§ 81.260 South Central Iowa Intrastate Air Quality Control Region.
§ 81.261 Southeast Iowa Intrastate Air Quality Control Region.
§ 81.262 North Central Illinois Intrastate Air Quality Control Region.
§ 81.263 East Central Illinois Intrastate Air Quality Control Region.
§ 81.264 West Central Illinois Intrastate Air Quality Control Region.
§ 81.265 Southeast Illinois Intrastate Air Quality Control Region.
§ 81.266 Alabama and Tombigbee Rivers Intrastate Air Quality Control Region.
§ 81.267 Southeast Alabama Intrastate Air Quality Control Region.
§ 81.268 Mohave-Yuma Intrastate Air Quality Control Region.
§ 81.269 Pima Intrastate Air Quality Control Region.
§ 81.270 Northern Arizona Intrastate Air Quality Control Region.
§ 81.271 Central Arizona Intrastate Air Quality Control Region.
§ 81.272 Southeast Arizona Intrastate Air Quality Control Region.
§ 81.273 Lake County Intrastate Air Quality Control Region.
§ 81.274 Mountain Counties Intrastate Air Quality Control Region.
§ 81.275 Lake Tahoe Intrastate Air Quality Control Region.
Subpart C--Section 107 Attainment Status Designations
§ 81.300 Scope.
§ 81.301 Alabama.
§ 81.302 Alaska.
§ 81.303 Arizona.
§ 81.304 Arkansas.
§ 81.305 California.
§ 81.306 Colorado.
§ 81.307 Connecticut.
§ 81.308 Delaware.
§ 81.309 District of Columbia.
§ 81.310 Florida.

§ 81.311 Georgia.
§ 81.312 Hawaii.
§ 81.313 Idaho.
§ 81.314 Illinois.
§ 81.315 Indiana.
§ 81.316 Iowa.
§ 81.317 Kansas.
§ 81.318 Kentucky.
§ 81.319 Louisiana.
§ 81.320 Maine.
§ 81.321 Maryland.
§ 81.322 Massachusetts.
§ 81.323 Michigan.
§ 81.324 Minnesota.
§ 81.325 Mississippi.
§ 81.326 Missouri.
§ 81.327 Montana.
§ 81.328 Nebraska.
§ 81.329 Nevada.
§ 81.330 New Hampshire.
§ 81.331 New Jersey.
§ 81.332 New Mexico.
§ 81.333 New York.
§ 81.334 North Carolina.
§ 81.335 North Dakota.
§ 81.336 Ohio.
§ 81.337 Oklahoma.
§ 81.338 Oregon.
§ 81.339 Pennsylvania.
§ 81.340 Rhode Island.
§ 81.341 South Carolina.
§ 81.342 South Dakota.
§ 81.343 Tennessee.
§ 81.344 Texas.
§ 81.345 Utah.
§ 81.346 Vermont.
§ 81.347 Virginia.
§ 81.348 Washington.
§ 81.349 West Virginia.
§ 81.350 Wisconsin.
§ 81.351 Wyoming.
§ 81.352 American Samoa.
§ 81.353 Guam.
§ 81.354 Northern Mariana Islands.
§ 81.355 Puerto Rico.
§ 81.356 Virgin Islands.
Subpart D--Identification of Mandatory Class I Federal Areas Where Visibility Is an Important Valuc
§ 81.400 Scope.
§ 81.401 Alabama.
§ 81.402 Alaska.
§ 81.403 Arizona.
§ 81.404 Arkansas.

§ 82.12 Transfers.
§ 82.13 Record-keeping and reporting requirements.
§ 82.14 [Removed. 57 FR 33787, July 30, 1992]
§ 82.20 [Removed. 57 FR 33787, July 30, 1992]
Appendix A to Subpart A--Class 1 Controlled Substances
Appendix B to Subpart A--Class II Controlled Substances
Appendix C to Subpart A--Annex 1-Parties to the Montreal Protocol:
Appendix D to Subpart A--Harmonized Tariff Schedule
Appendix E to Subpart A--Article 5 Parties
Appendix F to Subpart A--Listing of Ozone Depleting Chemicals

Subpart B--Servicing of Motor Vehicle Air Conditioners
§ 82.30 Purpose and scope.
§ 82.32 Definitions.
§ 82.34 Prohibitions.
§ 82.36 Approved refrigerant recycling equipment.
§ 82.38 Approved independent standards testing organizations.
§ 82.40 Technician training and certification.
§ 82.42 Certification, recordkeeping and public notification requirements.
Appendix A to Subpart B--Standard for Recycle/Recover Equipment
Appendix B to Subpart B--Standard for Recover Equipment [Reserved]

Subpart C--Ban on Nonessential Products Containing Class I Substances and Ban on Nonessential Products Containing or Manufactured With Class II Substances
§ 82.60 Purpose.
§ 82.62 Definitions.
§ 82.64 Prohibitions.
§ 82.65 Temporary exemptions.
§ 82.66 Nonessential Class I Products and Exceptions.
§ 82.68 Verification and public notice requirements.
§ 82.70 Nonessential Class II products and exceptions.

Subpart D--Federal Procurement
§ 82.80 Purpose and scope.
§ 82.82 Definitions.
§ 82.84 Requirements.
§ 82.86 Reporting requirements.

Subpart E--The Labeling of Products Using Ozone-Depleting Substances
§ 82.100 Purpose.
§ 82.102 Applicability.
§ 82.104 Definitions.
§ 82.106 Warning statement requirements.
§ 82.108 Placement of warning statement.
§ 82.110 Form of label bearing warning statement.
§ 82.112 Removal of label bearing warning statement.
§ 82.114 Compliance by manufacturers and importers with requirements for labeling of containers of controlled substances, or products containing controlled substances.
§ 82.116 Compliance by manufacturers or importers

Subpart O--Urban Bus Rebuild Requirements

§ 85.1401 General applicability.
§ 85.1402 Definitions.
§ 85.1403 Particulate standard for pre-1994 model year urban buses effective at time of engine rebuild or engine replacement.
§ 85.1404 Maintenance of records for urban bus operators; submittal of information; right of entry.
§ 85.1405 Applicability.
§ 85.1406 Certification.
§ 85.1407 Notification of intent to certify.
§ 85.1408 Objections to certification.
§ 85.1409 Warranty.
§ 85.1410 Changes after certification.
§ 85.1411 Labeling requirements.
§ 85.1412 Maintenance and submittal of records for equipment certifiers.
§ 85.1413 Decertification.
§ 85.1414 Alternative test procedures.
§ 85.1415 Treatment of confidential information.

Subpart P--Importation of Motor Vehicles and Motor Vehicle Engines

§ 85.1501 Applicability.
§ 85.1502 Definitions.
§ 85.1503 General requirements for importation of nonconforming vehicles.
§ 85.1504 Conditional admission.
§ 85.1505 Final admission of certified vehicles.
§ 85.1506 Inspection and testing of imported motor vehicles and engines.
§ 85.1507 Maintenance of certificate holder's records.
§ 85.1508 "In Use" inspections and recall requirements.
§ 85.1509 Final admission of modification and test vehicles.
§ 85.1510 Maintenance instructions, warranties, emission labeling and fuel economy requirements.
§ 85.1511 Exemptions and exclusions.
§ 85.1512 Admission of catalyst and O2 sensor-equipped vehicles.
§ 85.1513 Prohibited acts; penalties.
§ 85.1514 Treatment of confidential information.
§ 85.1515 Effective dates.

Subpart Q--[Reserved]

Subpart R--Exclusion and Exemption of Motor Vehicles and Motor Vehicle Engines

§ 85.1701 General applicability.
§ 85.1702 Definitions.
§ 85.1703 Application of section 216(2).
§ 85.1704 Who may request an exemption.
§ 85.1705 Testing exemption.
§ 85.1706 Pre-certification exemption.
§ 85.1707 Display exemption.
§ 85.1708 National security exemption.

§ 85.1709 Export exemptions.
§ 85.1710 Granting of exemptions.
§ 85.1711 Submission of exemption requests.
§ 85.1712 Treatment of confidential information.
Subpart S--Recall Regulations
§ 85.1801 Definitions.
§ 85.1802 Notice to manufacturer of nonconformity; submission of Remedial Plan.
§ 85.1803 Remedial Plan.
§ 85.1804 Approval of Plan: Implementation.
§ 85.1805 Notification to vehicle or engine owners.
§ 85.1806 Records and reports.
§ 85.1807 Public hearings.
§ 85.1808 Treatment of confidential information.
Appendix A to Subpart S--Interpretive Ruling for § 85.1803--Remedial Plans
Subpart T--Emission Defect Reporting Requirements
§ 85.1901 Applicability.
§ 85.1902 Definitions.
§ 85.1903 Emissions defect information report.
§ 85.1904 Voluntary emissions recall report; quarterly reports.
§ 85.1905 Alternative report formats.
§ 85.1906 Report filing: Record retention.
§ 85.1907 Responsibility under other legal provisions preserved.
§ 85.1908 Disclaimer of production warranty applicability.
§ 85.1909 Treatment of confidential information.
Subpart U--[Reserved]
Subpart V--Emissions Control System Performance Warranty Regulations and Voluntary Aftermarket Part Certification Program
§ 85.2101 General applicability.
§ 85.2102 Definitions.
§ 85.2103 Emission performance warranty.
§ 85.2104 Owners' compliance with instructions for proper maintenance and use.
§ 85.2105 Aftermarket parts.
§ 85.2106 Warranty claim procedures.
§ 85.2107 Warranty remedy.
§ 85.2108 Dealer certification.
§ 85.2109 Inclusion of warranty provisions in owners' manuals and warranty booklets.
§ 85.2110 Submission of owners' manuals and warranty statements to EPA.
§ 85.2111 Warranty enforcement.
§ 85.2112 Applicability.
§ 85.2113 Definitions.
§ 85.2114 Basis of certification.
§ 85.2115 Notification of intent to certify.
§ 85.2116 Objections to certification.
§ 85.2117 Warranty and dispute resolution.
§ 85.2118 Changes after certification.

§ 86.084-14 Small-volume manufacturers certification procedures.
§ 86.084-15 Emission standards for 1984 model year heavy-passenger cars.
§ 86.084-26 Mileage and service accumulation; emission measurements.
§ 86.084-40 Automatic expiration of reporting and recordkeeping requirements.
§ 86.085-1 General applicability.
§ 86.085-2 Definitions.
§ 86.085-8 Emission standards for 1985 and later model year light-duty vehicles.
§ 86.085-9 Emission standards for 1985 and later model year light-duty trucks.
§ 86.085-10 Emission standards for 1985 and later model year gasoline- fueled heavy-duty engines and vehicles.
§ 86.085-11 Emission standards for 1985 and later model year diesel heavy-duty engines.
§ 86.085-13 Alternative Durability Program.
§ 86.085-20 Incomplete vehicles, classification.
§ 86.085-21 Application for certification.
§ 86.085-22 Approval of application for certification; test fleet selections; determinations of parameters subject to adjustment for certification and selective enforcement audit, adequacy of limits, and physically adjustable ranges.
§ 86.085-23 Required data.
§ 86.085-24 Test vehicles and engines.
§ 86.085-25 Maintenance.
§ 86.085-27 Special test procedures.
§ 86.085-28 Compliance with emission standards.
§ 86.085-29 Testing by the Administrator.
§ 86.085-30 Certification.
§ 86.085-35 Labeling.
§ 86.085-37 Production vehicles and engines.
§ 86.085-38 Maintenance instructions.
§ 86.087-2 Definitions.
§ 86.087-8 Emission standards for 1987 light-duty vehicles.
§ 86.087-9 Emission standards for 1987 and later model year light-duty trucks.
§ 86.087-10 Emission standards for 1987 and later model year gasoline- fueled heavy-duty engines and vehicles.
§ 86.087-21 Application for certification.
§ 86.087-23 Required data.
§ 86.087-25 Maintenance.
§ 86.087-28 Compliance with emission standards.
§ 86.087-29 Testing by the Administrator.
§ 86.087-30 Certification.
§ 86.087-35 Labeling.
§ 86.087-38 Maintenance instructions.
§ 86.088-2 Definitions.
§ 86.088-9 Emission standards for 1988 and later model

year light-duty trucks.
§ 86.088-10 Emission standards for 1988 and 1989 model year gasoline- fueled heavy-duty engines and vehicles.
§ 86.088-11 Emission standards for 1988 and later model year diesel heavy-duty engines.
§ 86.088-21 Application for certification.
§ 86.088-23 Required data.
§ 86.088-25 Maintenance.
§ 86.088-28 Compliance with emission standards.
§ 86.088-29 Testing by the Administrator.
§ 86.088-30 Certification.
§ 86.088-35 Labeling.
§ 86.090-1 General applicability.
§ 86.090-2 Definitions.
§ 86.090-3 Abbreviations.
§ 86.090-5 General standards; increase in emissions; unsafe conditions.
§ 86.090-7 Maintenance of records; submittal of information; right of entry.
§ 86.090-8 Emission standards for 1990 and later model year light-duty vehicles.
§ 86.090-9 Emission standards for 1990 and later model year light-duty trucks.
§ 86.090-10 Emission standards for 1990 and later model year Otto-cycle heavy-duty engines and vehicles.
§ 86.090-11 Emission standards for 1990 and later model year diesel heavy-duty engines.
§ 86.090-14 Small-volume manufacturers certification procedures.
§ 86.090-15 NOX and particulate banking for heavy-duty engines.
§ 86.090-21 Application for certification.
§ 86.090-22 Approval of application for certification; test fleet selections; determinations of parameters subject to adjustment for certification and Selective Enforcement Audit, adequacy of limits, and physically adjustable ranges.
§ 86.090-23 Required data.
§ 86.090-24 Test vehicles and engines.
§ 86.090-25 Maintenance.
§ 86.090-26 Mileage and service accumulation; emission requirements.
§ 86.090-27 Special test procedures.
§ 86.090-28 Compliance with emission standards.
§ 86.090-29 Testing by the Administrator.
§ 86.090-30 Certification.
§ 86.090-35 Labeling.
§ 86.091-2 Definitions.
§ 86.091-7 Maintenance of records; submittal of information; right of entry.
§ 86.091-9 Emission standards for 1991 and later model year light-duty trucks.
§ 86.091-10 Emission standards for 1991 and later model

year Otto-cycle heavy-duty engines and vehicles.
§ 86.091-11 Emission standards for 1991 and later model year diesel heavy-duty engines.
§ 86.091-15 NOX and particulate averaging, trading, and banking for heavy-duty engines.
§ 86.091-21 Application for certification.
§ 86.091-23 Required data.
§ 86.091-28 Compliance with emission standards.
§ 86.091-29 Testing by the Administrator.
§ 86.091-30 Certification.
§ 86.091-35 Labeling.
§ 86.092-1 General applicability.
§ 86.092-2 Definitions.
§ 86.092-14 Small-volume manufacturers certification procedures.
§ 86.092-15 NOX and particulate averaging, trading, and banking for heavy-duty engines.
§ 86.092-23 Required data.
§ 86.092-24 Test vehicles and engines.
§ 86.092-26 Mileage and service accumulation; emission measurements.
§ 86.092-35 Labeling.
§ 86.093-2 Definitions.
§ 86.093-11 Emission standards for 1993 and later model year diesel heavy-duty engines.
§ 86.093-35 Labeling.
§ 86.094-1 General applicability.
§ 86.094-2 Definitions.
§ 86.094-3 Abbreviations.
§ 86.094-7 Maintenance of records; submittal of information; right of entry.
§ 86.094-8 Emission standards for 1994 and later model year light-duty vehicles.
§ 86.094-9 Emission standards for 1994 and later model year light-duty trucks.
§ 86.094-11 Emission standards for 1994 and later model year diesel heavy-duty engines.
§ 86.094-13 Light-duty exhaust durability programs.
§ 86.094-14 Small-volume manufacturers certification procedures.
§ 86.094-15 NOX and particulate averaging, trading, and banking for heavy-duty engines.
§ 86.094-16 Prohibition of defeat devices.
§ 86.094-17 Emission control diagnostic system for 1994 and later light- duty vehicles and light-duty trucks.
§ 86.094-18 Tampering prevention.
§ 86.094-21 Application for certification.
§ 86.094-22 Approval of application for certification; test fleet selections; determinations of parameters subject to adjustment for certification and Selective Enforcement Audit, adequacy of limits, and physically adjustable ranges.
§ 86.094-23 Required data.

§ 86.094-24 Test vehicles and engines.
§ 86.094-25 Maintenance.
§ 86.094-26 Mileage and service accumulation; emission requirements.
§ 86.094-28 Compliance with emission standards.
§ 86.094-30 Certification.
§ 86.094-35 Labeling.
§ 86.095-14 Small-volume manufacturers certification procedures.
§ 86.095-23 Required data.
§ 86.095-24 Test vehicles and engines.
§ 86.095-26 Mileage and service accumulation; emission measurements.
§ 86.095-30 Certification.
§ 86.095-35 Labeling.
§ 86.096-2 Definitions.
§ 86.096-3 Abbreviations.
§ 86.096-7 Maintenance of records; submittal of information; right of entry.
§ 86.096-8 Emission standards for 1996 and later model year light-duty vehicles.
§ 86.096-9 Emission standards for 1996 and later model year light-duty trucks.
§ 86.096-10 Emission standards for 1996 and later model year Otto-cycle heavy-duty engines and vehicles.
§ 86.096-11 Emission Standards for 1996 and Later Model Year Diesel Heavy-duty Engines.
§ 86.096-14 Small-volume manufacturer certification procedures.
§ 86.096-21 Application for certification.
§ 86.096-23 Required data.
§ 86.096-24 Test vehicles and engines.
§ 86.096-26 Mileage and service accumulation; emission measurements.
§ 86.096-30 Certification.
§ 86.096-35 Labeling.
§ 86.097-9 Emission standards for 1997 and later model year light-duty trucks.
§ 86.098-2 Definitions.
§ 86.098-3 Abbreviations.
§ 86.098-7 Maintenance of records; submittal of information; right of entry.
§ 86.098-8 Emission standards for 1998 and later model year light-duty vehicles.
§ 86.098-10 Emission Standards for 1998 and Later Model Year Otto-cycle Heavy-duty Engines and Vehicles.
§ 86.098-11 Emission Standards for 1998 and Later Model Year Diesel Heavy-duty Engines.
§ 86.098-14 Small-volume manufacturers certification procedures.
§ 86.098-17 Emission control diagnostic system for 1998 and later light- duty vehicles and light-duty trucks.
§ 86.098-21 Application for certification.

§ 86.098-22 Approval of application for certification; test fleet selections; determinations of parameters subject to adjustment for certification and Selective Enforcement Audit, adequacy of limits, and physically adjustable ranges.
§ 86.098-23 Required data.
§ 86.098-24 Test vehicles and engines.
§ 86.098-25 Maintenance.
§ 86.098-26 Mileage and service accumulation; emission measurements.
§ 86.098-28 Compliance with emission standards.
§ 86.098-30 Certification.
§ 86.098-35 Labeling.
§ 86.099-8 Emission standards for 1999 and later model year light-duty vehicles.
§ 86.099-9 Emission standards for 1999 and later model year light-duty trucks.
§ 86.099-10 Emission standards for 1999 and later model year Otto-cycle heavy-duty engines and vehicles.
§ 86.099-11 Emission standards for 1999 and later model year diesel heavy-duty engines and vehicles.

Subpart B--Emission Regulations for 1977 and Later Model Year New Light-Duty Vehicles and New Light-Duty Trucks; Test Procedures

§ 86.101 General applicability.
§ 86.102 Definitions.
§ 86.103 Abbreviations.
§ 86.104 Section numbering; construction.
§ 86.105 Introduction; structure of subpart.
§ 86.106-82 Equipment required; overview.
§ 86.106-90 Equipment required; overview.
§ 86.106-94 Equipment required; overview.
§ 86.106-96 Equipment required; overview.
§ 86.107-78 [Removed. 58 FR 16019, Mar. 24, 1993]
§ 86.107-90 Sampling and analytical system; evaporative emissions.
§ 86.107-96 Sampling and analytical systems; evaporative emissions.
Section 86.107-98 includes text that specifies requirements that differ from
§ 86.108-79 Dynamometer.
§ 86.109-82 Exhaust gas sampling system; gasoline-fueled vehicles.
§ 86.109-90 Exhaust gas sampling system; Otto-cycle vehicles.
§ 86.109-94 Exhaust gas sampling system; Otto-cycle vehicles not requiring particulate emission measurements.
§ 86.110-82 Exhaust gas sampling system; diesel vehicles.
§ 86.110-90 Exhaust gas sampling system; diesel vehicles.
§ 86.110-94 Exhaust gas sampling system; diesel-cycle vehicles, and Otto-cycle vehicles requiring particulate emissions measurements.
§ 86.111-82 Exhaust gas analytical system.
§ 86.111-90 Exhaust gas analytical system.

§ 86.111-94 Exhaust gas analytical system.
§ 86.112-82 Weighing chamber (or room) and microgram balance specifications.
§ 86.112-91 Weighing chamber (or room) and microgram balance specifications.
§ 86.113-82 [Removed. 58 FR 16019, Mar. 24, 1993]
§ 86.113-87 [Removed. 58 FR 16019, Mar. 24, 1993]
§ 86.113-90 [Removed. 58 FR 16019, Mar. 24, 1993]
§ 86.113-91 Fuel specifications.
§ 86.113-94 Fuel specifications.
§ 86.114-79 Analytical gases.
§ 86.114-94 Analytical gases.
§ 86.115-78 EPA urban dynamometer driving schedule.
§ 86.116-82 Calibrations, frequency and overview.
§ 86.116-94 Calibrations, frequency and overview.
§ 86.117-78 [Removed. 58 FR 16019, Mar. 24, 1993]
§ 86.117-90 Evaporative emission enclosure calibrations.
§ 86.117-96 Evaporative emission enclosure calibrations.
§ 86.118-78 Dynamometer calibration.
§ 86.119-78 CVS calibration.
§ 86.119-90 CVS calibration.
§ 86.120-82 Gas meter or flow instrumentation calibration, particulate measurement.
§ 86.121-82 Hydrocarbon analyzer calibration.
§ 86.121-90 Hydrocarbon analyzer calibration.
§ 86.122-78 Carbon monoxide analyzer calibration.
§ 86.123-78 Oxides of nitrogen analyzer calibration.
§ 86.124-78 Carbon dioxide analyzer calibration.
§ 86.125-94 Methane analyzer calibration.
§ 86.126-78 Calibration of other equipment.
§ 86.126-90 Calibration of other equipment.
§ 86.127-82 [Removed. 58 FR 16019, Mar. 24, 1993]
§ 86.127-90 Test procedures; overview.
§ 86.127-94 Test procedures; overview.
§ 86.127-96 Test procedures; overview.
§ 86.128-79 Transmissions.
§ 86.129-80 Road load power test weight and inertia weight class determination.
§ 86.129-94 Road load power test weight and inertia weight class determination.
§ 86.130-78 Test sequence; general requirements.
§ 86.130-96 Test sequence; general requirements.
§ 86.131-78 [Removed. 58 FR 16019, Mar. 24, 1993]
§ 86.131-90 Vehicle preparation.
§ 86.131-96 Vehicle preparation.
§ 86.132-82 [Removed. 58 FR 16019, Mar. 24, 1993]
§ 86.132-90 Vehicle preconditioning.
§ 86.132-96 Vehicle preconditioning.
§ 86.133-78 [Removed. 58 FR 16019, Mar. 24, 1993]
§ 86.133-90 Diurnal breathing loss test.
§ 86.133-96 Diurnal emission test.
§ 86.134-96 Running loss test.
§ 86.135-82 [Removed. 58 FR 16019, Mar. 24, 1993]

§ 86.312-79 Dynamometer and engine equipment specifications.
§ 86.313-79 Air flow measurement specifications; diesel engines.
§ 86.314-79 Fuel flow measurement specifications.
§ 86.315-79 General analyzer specifications.
§ 86.316-79 Carbon monoxide and carbon dioxide analyzer specifications.
§ 86.317-79 Hydrocarbon analyzer specifications.
§ 86.318-79 Oxides of nitrogen analyzer specifications.
§ 86.319-79 Analyzer checks and calibrations; frequency and overview.
§ 86.320-79 Analyzer bench check.
§ 86.321-79 NDIR water rejection ratio check.
§ 86.322-79 NDIR CO2 rejection ratio check.
§ 86.327-79 Quench checks; NOx analyzer.
§ 86.328-79 Leak checks.
§ 86.329-79 System response time; check procedure.
§ 86.330-79 NDIR analyzer calibration.
§ 86.331-79 Hydrocarbon analyzer calibration.
§ 86.332-79 Oxides of nitrogen analyzer calibration.
§ 86.333-79 Dynamometer calibration.
§ 86.334-79 Test procedure overview.
§ 86.335-79 Gasoline-fueled engine test cycle.
§ 86.336-79 Diesel engine test cycle.
§ 86.337-79 Information.
§ 86.338-79 Exhaust measurement accuracy.
§ 86.339-79 Pre-test procedures.
§ 86.340-79 Gasoline-fueled engine dynamometer test run.
§ 86.341-79 Diesel engine dynamometer test run.
§ 86.342-79 Post-test procedures.
§ 86.343-79 Chart reading.
§ 86.344-79 Humidity calculations.
§ 86.345-79 Emission calculations.
§ 86.346-79 Alternative NOx measurement technique.
§ 86.347-79 Alternative calculations for diesel engines.
§ 86.348-79 Alternative to fuel H/C analysis.

Subpart E--Emission Regulations for 1978 and Later New Motorcycles, General Provisions

§ 86.401-78 General applicability.
§ 86.401-90 General applicability.
§ 86.402-78 Definitions.
§ 86.403-78 Abbreviations.
§ 86.404-78 Section numbering.
§ 86.405-78 Measurement system.
§ 86.406-78 Introduction, structure of subpart, further information.
§ 86.407-78 Certificate of conformity required.
§ 86.408-78 General standards; increase in emissions; unsafe conditions.
§ 86.409-78 Defeat devices, prohibition.
§ 86.410-80 Emission standards for 1980 and later model year motorcycles.

§ 86.410-90 Emission standards for 1990 and later model year motorcycles.
§ 86.411-78 Maintenance instructions, vehicle purchaser.
§ 86.412-78 Maintenance instructions, submission to Administrator.
§ 86.413-78 Labeling.
§ 86.414-78 Submission of vehicle identification number.
§ 86.415-78 Production vehicles.
§ 86.416-80 Application for certification.
§ 86.417-78 Approval of application for certification.
§ 86.418-78 Test fleet selection.
§ 86.419-78 Engine displacement, motorcycle classes.
§ 86.420-78 Engine families.
§ 86.421-78 Test fleet.
§ 86.422-78 Administrator's fleet.
§ 86.423-78 Test vehicles.
§ 86.425-78 Test procedures.
§ 86.426-78 Service accumulation.
§ 86.427-78 Emission tests.
§ 86.428-80 Maintenance, scheduled; test vehicles.
§ 86.429-78 Maintenance, unscheduled; test vehicles.
§ 86.430-78 Vehicle failure.
§ 86.431-78 Data submission.
§ 86.432-78 Deterioration factor.
§ 86.434-78 Testing by the Administrator.
§ 86.435-78 Extrapolated emission values.
§ 86.436-78 Additional service accumulation.
§ 86.437-78 Certification.
§ 86.438-78 Amendments to the application.
§ 86.439-78 Alternative procedure for notification of additions and changes.
§ 86.440-78 Maintenance of records.
§ 86.441-78 Right of entry.
§ 86.442-78 Denial, revocation, or suspension of certification.
§ 86.443-78 Request for hearing.
§ 86.444-78 Hearings on certification.

Subpart F--Emission Regulations for 1978 and Later New Motorcycles; Test Procedures

§ 86.501-78 Applicability.
§ 86.502-78 Definitions.
§ 86.503-78 Abbreviations.
§ 86.504-78 Section numbering.
§ 86.505-78 Introduction; structure of subpart.
§ 86.508-78 Dynamometer.
§ 86.509-78 Exhaust gas sampling system.
§ 86.509-90 Exhaust gas sampling system.
§ 86.511-78 Exhaust gas analytical system.
§ 86.511-90 Exhaust gas analytical system.
§ 86.513-82 Fuel and engine lubricant specifications.
§ 86.513-87 Fuel and engine lubricant specifications.
§ 86.513-90 Fuel and engine lubricant specifications.
§ 86.514-78 Analytical gases.

§ 86.610-84 Compliance with acceptable quality level and passing and failing criteria for Selective Enforcement Audits.

§ 86.610-96 Compliance with acceptable quality level and passing and failing criteria for Selective Enforcement Audits.

§ 86.610-98 Compliance with acceptable quality level and passing and failing criteria for Selective Enforcement Audits.

§ 86.612-84 Suspension and revocation of certificates of conformity.

§ 86.614-84 Hearings on suspension, revocation, and voiding of certificates of conformity.

§ 86.615-84 Treatment of confidential information.

Subpart H--General Provisions for In-use Emission Regulations for 1994 and Later Model Year Light-Duty Vehicles and Light-Duty Trucks

§ 86.701-94 General applicability.

§ 86.702-94 Definitions.

§ 86.703-94 Abbreviations.

§ 86.704-94 Section numbering; construction.

§ 86.705-94 [Reserved]

§ 86.706-94 [Reserved]

§ 86.707-94 [Reserved]

§ 86.708-94 In-use emission standards for 1994 and later model year light duty vehicles.

§ 86.708-98 In-use emission standards for 1998 and later model year light duty vehicles.

§ 86.709-94 In-use emission standards for 1994 and later model year light-duty trucks.

§ 86.709-99 In-use emission standards for 1999 and later model year light-duty trucks.

Subpart I--Emission Regulations for New Diesel Heavy-Duty Engines; Smoke Exhaust Test Procedure

§ 86.884-1 General applicability.

§ 86.884-2 Definitions.

§ 86.884-3 Abbreviations.

§ 86.884-4 Section numbering.

§ 86.884-5 Test procedures.

§ 86.884-6 Fuel specifications.

§ 86.884-7 Dynamometer operation cycle for smoke emission tests.

§ 86.884-8 Dynamometer and engine equipment.

§ 86.884-9 Smoke measurement system.

§ 86.884-10 Information.

§ 86.884-11 Instrument checks.

§ 86.884-12 Test run.

§ 86.884-13 Data analysis.

§ 86.884-14 Calculations.

Subpart J--Fees for the Motor Vehicle and Engine Compliance Program

§ 86.901-93 Abbreviations

§ 86.902-93 Definitions.

§ 87.5 Special test procedures.
§ 87.6 Aircraft safety.
§ 87.7 Exemptions.
Subpart B--Engine Fuel Venting Emissions (New and In-Use Aircraft Gas Turbine Engines)
§ 87.10 Applicability.
§ 87.11 Standard for fuel venting emissions.
Subpart C--Exhaust Emissions (New Aircraft Gas Turbine Engines)
§ 87.20 Applicability.
§ 87.21 Standards for exhaust emissions.
Subpart D--Exhaust Emissions (In-use Aircraft Gas Turbine Engines)
§ 87.30 Applicability.
§ 87.31 Standards for exhaust emissions.
Subparts E-F--[Reserved]
Subpart G--Test Procedures for Engine Exhaust Gaseous Emissions (Aircraft and Aircraft Gas Turbine Engines)
§ 87.60 Introduction.
§ 87.61 Turbine fuel specifications.
§ 87.62 Test procedure (propulsion engines).
§ 87.63 [Reserved]
§ 87.64 Sampling and analytical procedures for measuring gaseous exhaust emissions.
Secs. 87.65--87.70 [Reserved]
§ 87.71 Compliance with gaseous emission standards.
Subpart H--Test Procedures for Engine Smoke Emissions (Aircraft Gas Turbine Engines)
§ 87.80 Introduction.
§ 87.81 Fuel specifications.
§ 87.82 Sampling and analytical procedures for measuring smoke exhaust emissions.
Secs. 87.83--87.88 [Reserved]
§ 87.89 Compliance with smoke emission standards.
PART 88--CLEAN-FUEL VEHICLES
Subpart A--Emission Standards for Clean-Fuel Vehicles.
§ 88.101-94 Definitions.
§ 88.102-94
Subpart B--California Pilot Test Program
§ 88.201-94 Scope.
§ 88.202-94 Definitions.
§ 88.203-94 Abbreviations.
§ 88.205-94 California Pilot Test Program Credits Program.
Tables to Subpart B of Part 88
Subpart C--Clean-Fuel Fleet Program
§ 88.301-93 General applicability.
§ 88.302-93 Definitions.
§ 88.302-94 Definitions.
§ 88.303-93 Abbreviations.
§ 88.304-94 Clean-fuel fleet vehicle credit program.
§ 88.305 [Reserved]
§ 88.306 [Reserved]
§ 88.307-94 Exemption from temporal transportation control measures for CFFVs.

§ 89.126-96 Denial, revocation of certificate of conformity.

§ 89.127-96 Request for hearing.

§ 89.128-96 Hearing procedures.

§ 89.129-96 Right of entry.

Subpart C--Averaging, Banking, and Trading Provisions

§ 89.201-96 Applicability.

§ 89.202-96 Definitions.

§ 89.203-96 General provisions.

§ 89.204-96 Averaging.

§ 89.205-96 Banking.

§ 89.206-96 Trading.

§ 89.207-96 Credit calculation.

§ 89.208-96 Labeling.

§ 89.209-96 Certification.

§ 89.210-96 Maintenance of records.

§ 89.211-96 End-of-year and final reports.

§ 89.212-96 Notice of opportunity for hearing.

Subpart D--Emission Test Equipment Provisions

§ 89.301-96 Scope; applicability.

§ 89.302-96 Definitions.

§ 89.303-96 Symbols/abbreviations.

§ 89.304-96 Equipment required for gaseous emissions; overview.

§ 89.305-96 Equipment measurement accuracy/calibration frequency.

§ 89.306-96 Dynamometer specifications and calibration weights.

§ 89.307-96 Dynamometer calibration.

§ 89.308-96 Sampling system requirements for gaseous emissions.

§ 89.309-96 Analyzers required for gaseous emissions.

§ 89.310-96 Analyzer accuracy and specifications.

§ 89.311-96 Analyzer calibration frequency.

§ 89.312-96 Analytical gases.

§ 89.313-96 Initial calibration of analyzers.

§ 89.314-96 Pre- and post-test calibration of analyzers.

§ 89.315-96 Analyzer bench checks.

§ 89.316-96 Analyzer leakage and response time.

§ 89.317-96 NOX converter check.

§ 89.318-96 Analyzer interference checks.

§ 89.319-96 Hydrocarbon analyzer calibration.

§ 89.320-96 Carbon monoxide analyzer calibration.

§ 89.321-96 Oxides of nitrogen analyzer calibration.

§ 89.322-96 Carbon dioxide analyzer calibration.

§ 89.323-96 NDIR analyzer calibration.

§ 89.324-96 Calibration of other equipment.

§ 89.325-96 Engine intake air temperature measurement.

§ 89.326-96 Engine intake air humidity measurement.

§ 89.327-96 Charge cooling.

§ 89.328-96 Inlet and exhaust restrictions.

§ 89.329-96 Engine cooling system.

§ 89.330-96 Lubricating oil and test fuels.

PARTS 90--92 [RESERVED]

PARTS 89-92--[Reserved]

PART 93--DETERMINING CONFORMITY OF FEDERAL ACTIONS TO STATE OR FEDERAL IMPLEMENTATION PLANS

Subpart A--Conformity to State or Federal Implementation Plans of Transportation Plans, Programs, and Projects Developed, Funded or Approved Under Title 23 U.S.C. or the Federal Transit Act

§ 93.100 Purpose.

§ 93.101 Definitions.

§ 93.102 Applicability.

§ 93.103 Priority.

§ 93.104 Frequency of conformity determinations.

§ 93.105 Consultation.

§ 93.106 Content of transportation plans.

§ 93.107 Relationship of transportation plan and TIP conformity with the NEPA process.

§ 93.108 Fiscal constraints for transportation plans and TIPs.

§ 93.109 Criteria and procedures for determining conformity of transportation plans, programs, and projects: General.

§ 93.110 Criteria and procedures: Latest planning assumptions.

§ 93.111 Criteria and procedures: Latest emissions model.

§ 93.112 Criteria and procedures: Consultation.

§ 93.113 Criteria and procedures: Timely implementation of TCMs.

§ 93.114 Criteria and procedures: Currently conforming transportation plan and TIP.

§ 93.115 Criteria and procedures: Projects from a plan and TIP.

§ 93.116 Criteria and procedures: Localized CO and PM10 violations (hot spots).

§ 93.117 Criteria and procedures: Compliance with PM10 control measures.

§ 93.118 Criteria and procedures: Motor vehicle emissions budget (transportation plan).

§ 93.119 Criteria and procedures: Motor vehicle emissions budget (TIP).

§ 93.120 Criteria and procedures: Motor vehicle emissions budget (project not from a plan and TIP).

§ 93.121 Criteria and procedures: Localized CO violations (hot spots) in the interim period.

§ 93.122 Criteria and procedures: Interim period reductions in ozone and CO areas (transportation plan).

§ 93.123 Criteria and procedures: Interim period reductions in ozone and CO areas (TIP).

§ 93.124 Criteria and procedures: Interim period reductions for ozone and CO areas (project not from a plan and TIP).

§ 93.125 Criteria and procedures: Interim period reductions for PM10 and NO2 areas (transportation plan).

§ 93.126 Criteria and procedures: Interim period

§ 104.12 Certification of record.
§ 104.13 Interlocutory and post-hearing review of rulings of the Presiding Officer; motions.
§ 104.14 Tentative and final decision by the Administrator.
§ 104.15 Promulgation of standards.
§ 104.16 Filing and time.
PART 108--EMPLOYEE PROTECTION HEARINGS
§ 108.1 Applicability.
§ 108.2 Definitions.
§ 108.3 Request for investigation.
§ 108.4 Investigation by Regional Administrator.
§ 108.5 Procedure.
§ 108.6 Recommendations.
§ 108.7 Hearing before Administrator.
PART 109--CRITERIA FOR STATE, LOCAL AND REGIONAL OIL REMOVAL CONTINGENCY PLANS
§ 109.1 Applicability.
§ 109.2 Definitions.
§ 109.3 Purpose and scope.
§ 109.4 Relationship to Federal response actions.
§ 109.5 Development and implementation criteria for State, local and regional oil removal contingency plans.
§ 109.6 Coordination.
PART 110--DISCHARGE OF OIL
§ 110.1 Definitions.
§ 110.2 Applicability.
§ 110.3 Discharge into navigable waters of such quantities as may be harmful.
§ 110.4 Discharge into contiguous zone of such quantities as may be harmful.
§ 110.5 Discharge beyond contiguous zone of such quantities as may be harmful.
§ 110.6 Discharge prohibited.
§ 110.7 Exception for vessel engines.
§ 110.8 Dispersants.
§ 110.9 Demonstration projects.
§ 110.10 Notice.
§ 110.11 Discharge at deepwater ports.
PART 112--OIL POLLUTION PREVENTION
§ 112.1 General applicability.
§ 112.2 Definitions.
§ 112.3 Requirements for preparation and implementation of Spill Prevention Control and Countermeasure Plans.
§ 112.4 Amendment of SPCC Plans by Regional Administrator.
§ 112.5 Amendment of Spill Prevention Control and Countermeasure Plans by owners or operators.
§ 112.6 Civil penalties for violation of oil pollution prevention regulations.
§ 112.7 Guidelines for the preparation and implementation of a Spill Prevention Control and Countermeasure Plan.
§ 112.20 Facility response plans.
§ 112.21 Facility response training and drills/exercises.

§ 117.13 Applicability to discharges from publicly owned treatment works and their users.

§ 117.14 Demonstration projects.

Subpart C--Notice of Discharge of a Reportable Quantity

§ 117.21 Notice.

§ 117.22 Penalties.

§ 117.23 Liabilities for removal.

PART 121--STATE CERTIFICATION OF ACTIVITIES REQUIRING A FEDERAL LICENSE OR PERMIT

Subpart A--General

§ 121.1 Definitions.

§ 121.2 Contents of certification.

§ 121.3 Contents of application.

Subpart B--Determination of Effect on Other States

§ 121.11 Copies of documents.

§ 121.12 Supplemental information.

§ 121.13 Review by Regional Administrator and notification.

§ 121.14 Forwarding to affected State.

§ 121.15 Hearings on objection of affected State.

§ 121.16 Waiver.

Subpart C--Certification by the Administrator

§ 121.21 When Administrator certifies.

§ 121.22 Applications.

§ 121.23 Notice and hearing.

§ 121.24 Certification.

§ 121.25 Adoption of new water quality standards.

§ 121.26 Inspection of facility or activity before operation.

§ 121.27 Notification to licensing or permitting agency.

§ 121.28 Termination of suspension.

Subpart D--Consultations

§ 121.30 Review and advice.

PART 122--EPA ADMINISTERED PERMIT PROGRAMS: THE NATIONAL POLLUTANT DISCHARGE ELIMINATION SYSTEM

Subpart A--Definitions and General Program Requirements

§ 122.1 Purpose and scope.

§ 122.2 Definitions.

§ 122.3 Exclusions.

§ 122.4 Prohibitions (applicable to State NPDES programs, see § 123.25).

§ 122.5 Effect of a permit.

§ 122.6 Continuation of expiring permits.

§ 122.7 Confidentiality of information.

Subpart B--Permit Application and Special NPDES Program Requirements

§ 122.21 Application for a permit (applicable to State programs, see § 123.25).

§ 122.22 Signatories to permit applications and reports (applicable to State programs, see § 123.25).

§ 122.23 Concentrated animal feeding operations (applicable to State NPDES programs, see § 123.25).

§ 122.24 Concentrated aquatic animal production facilities

(applicable to State NPDES programs, see § 123.25).

§ 122.25 Aquaculture projects (applicable to State NPDES programs, see § 123.25).

§ 122.26 Storm water discharges (applicable to State NPDES programs, see § 123.25).

§ 122.27 Silvicultural activities (applicable to State NPDES programs, see § 123.25).

§ 122.28 General permits (applicable to State NPDES programs, see § 123.25).

§ 122.29 New sources and new dischargers.

Subpart C--Permit Conditions

§ 122.41 Conditions applicable to all permits (applicable to State programs, see § 123.25).

§ 122.42 Additional conditions applicable to specified categories of NPDES permits (applicable to State NPDES programs, see § 123.25).

§ 122.43 Establishing permit conditions (applicable to State programs, see § 123.25).

§ 122.44 Establishing limitations, standards, and other permit conditions (applicable to State NPDES programs, see § 123.25).

§ 122.45 Calculating NPDES permit conditions (applicable to State NPDES programs, see § 123.25).

§ 122.46 Duration of permits (applicable to State programs, see § 123.25).

§ 122.47 Schedules of compliance.

§ 122.48 Requirements for recording and reporting of monitoring results (applicable to State programs, see § 123.25).

§ 122.49 Considerations under Federal law.

§ 122.50 Disposal of pollutants into wells, into publicly owned treatment works or by land application (applicable to State NPDES programs, see § 123.25).

Subpart D--Transfer, Modification, Revocation and Reissuance, and Termination of Permits

§ 122.61 Transfer of permits (applicable to State programs, see § 123.25).

§ 122.62 Modification or revocation and reissuance of permits (applicable to State programs, see § 123.25).

§ 122.63 Minor modifications of permits.

§ 122.64 Termination of permits (applicable to State programs, see § 123.25).

Appendix A--NPDES Primary Industry Categories

Appendix B--Criteria for Determining a Concentrated Animal Feeding Operation (§ 122.23)

Appendix C--Criteria for Determining a Concentrated Aquatic Animal Production Facility (§ 122.24)

Appendix D--NPDES Permit Application Testing Requirements (§ 122.21)

Appendix E to Part 122--Rainfall Zones of the United States

Appendix F to Part 122--Incorporated Places With Populations Greater Than 250,000 According to Latest Decennial Census by Bureau of Census.

§ 124.4 Consolidation of permit processing.
§ 124.5 Modification, revocation and reissuance, or termination of permits.
§ 124.6 Draft permits.
§ 124.7 Statement of basis.
§ 124.8 Fact sheet.
§ 124.9 Administrative record for draft permits when EPA is the permitting authority.
§ 124.10 Public notice of permit actions and public comment period.
§ 124.11 Public comments and requests for public hearings.
§ 124.12 Public hearings.
§ 124.13 Obligation to raise issues and provide information during the public comment period.
§ 124.14 Reopening of the public comment period.
§ 124.15 Issuance and effective date of permit.
§ 124.16 Stays of contested permit conditions.
§ 124.17 Response to comments.
§ 124.18 Administrative record for final permit when EPA is the permitting authority.
§ 124.19 Appeal of RCRA, UIC, and PSD permits.
§ 124.20 Computation of time.
§ 124.21 Effective date of Part 124.

Subpart B--Specific Procedures Applicable to RCRA Permits [Reserved]

Subpart C--Specific Procedures Applicable to PSD Permits

§ 124.41 Definitions applicable to PSD permits.
§ 124.42 Additional procedures for PSD permits affecting Class I areas.

Subpart D--Specific Procedures Applicable to NPDES Permits

§ 124.51 Purpose and scope.
§ 124.52 Permits required on a case-by-case basis.
§ 124.53 State certification.
§ 124.54 Special provisions for State certification and concurrence on applications for section 301(h) variances.
§ 124.55 Effect of State certification.
§ 124.56 Fact sheets.
§ 124.57 Public notice.
§ 124.58 Special procedures for EPA-issued general permits for point sources other than separate storm sewers.
§ 124.59 Conditions requested by the Corps of Engineers and other government agencies.
§ 124.60 Issuance and effective date and stays of NPDES permits.
§ 124.61 Final environmental impact statement.
§ 124.62 Decision on variances.
§ 124.63 Procedures for variances when EPA is the permitting authority.
§ 124.64 Appeals of variances.
§ 124.65 [Reserved]
§ 124.66 Special procedures for decisions on thermal variances under section 316(a).

Subpart E--Evidentiary Hearings for EPA-Issued NPDES Permits and

§ 129.6 Adjustment of effluent standard for presence of toxic pollutant in the intake water.

§ 129.7 Requirement and procedure for establishing a more stringent effluent limitation.

§ 129.8 Compliance date.

Secs. 129.9--129.99 [Reserved]

§ 129.100 Aldrin/dieldrin.

§ 129.101 DDT, DDD and DDE.

§ 129.102 Endrin.

§ 129.103 Toxaphene.

§ 129.104 Benzidine.

§ 129.105 Polychlorinated biphenyls (PCBs).

PART 130--WATER QUALITY PLANNING AND MANAGEMENT

§ 130.0 Program summary and purpose.

§ 130.1 Applicability.

§ 130.2 Definitions.

§ 130.3 Water quality standards.

§ 130.4 Water quality monitoring.

§ 130.5 Continuing planning process.

§ 130.6 Water quality management plans.

§ 130.7 Total maximum daily loads (TMDL) and individual water quality- based effluent limitations.

§ 130.8 Water quality report.

§ 130.9 Designation and de-designation.

§ 130.10 State submittals to EPA.

§ 130.11 Program management.

§ 130.12 Coordination with other programs.

§ 130.15 Processing application for Indian tribes.

PART 131--WATER QUALITY STANDARDS

Subpart A--General Provisions

§ 131.1 Scope.

§ 131.2 Purpose.

§ 131.3 Definitions.

§ 131.4 State authority.

§ 131.5 EPA authority.

§ 131.6 Minimum requirements for water quality standards submission.

§ 131.7 Dispute resolution mechanism.

§ 131.8 Requirements for Indian Tribes to be treated as States for purposes of water quality standards.

Subpart B--Establishment of Water Quality Standards

§ 131.10 Designation of uses.

§ 131.11 Criteria.

§ 131.12 Antidegradation policy.

§ 131.13 General policies.

Subpart C--Procedures for Review and Revision of Water Quality Standards

§ 131.20 State review and revision of water quality standards.

§ 131.21 EPA review and approval of water quality standards.

§ 131.22 EPA promulgation of water quality standards.

Subpart D--Federally Promulgated Water Quality Standards

§ 141.87 Monitoring requirements for water quality parameters.

§ 141.88 Monitoring requirements for lead and copper in source water.

§ 141.89 Analytical methods.

§ 141.90 Reporting requirements.

§ 141.91 Recordkeeping requirements.

Subpart J--Use of Non-Centralized Treatment Devices

§ 141.100 Criteria and procedures for public water systems using point- of-entry devices.

§ 141.101 Use of other non-centralized treatment devices.

Subpart K--Treatment Techniques

§ 141.110 General requirements.

§ 141.111 Treatment techniques for acrylamide and epichlorohydrin.

PART 142--NATIONAL PRIMARY DRINKING WATER REGULATIONS IMPLEMENTATION

Subpart A--General Provisions

§ 142.1 Applicability.

§ 142.2 Definitions.

§ 142.3 Scope.

§ 142.4 State and local authority.

Subpart B--Primary Enforcement Responsibility

§ 142.10 Requirements for a determination of primary enforcement responsibility.

§ 142.11 Initial determination of primary enforcement responsibility.

§ 142.12 Revision of State programs.

§ 142.13 Public hearing.

§ 142.14 Records kept by States.

§ 142.15 Reports by States.

§ 142.16 Special primacy requirements.

§ 142.17 Review of State programs and procedures for withdrawal of approved primacy programs.

§ 142.18 EPA review of State monitoring determinations.

§ 142.19 EPA review of State implementation of national primary drinking water regulations for lead and copper.

Subpart C--Review of State-Issued Variances and Exemptions

§ 142.20 State-issued variances and exemptions.

§ 142.21 State consideration of a variance or exemption request.

§ 142.22 Review of State variances, exemptions and schedules.

§ 142.23 Notice to State.

§ 142.24 Administrator's rescission.

Subpart D--Federal Enforcement

§ 142.30 Failure by State to assure enforcement.

§ 142.31 [Reserved]

§ 142.32 Petition for public hearing.

§ 142.33 Public hearing.

§ 142.34 Entry and inspection of public water systems.

Subpart E--Variances Issued by the Administrator

§ 142.40 Requirements for a variance.

§ 142.41 Variance request.

§ 144.41 Minor modifications of permits.

Subpart E--Permit Conditions

§ 144.51 Conditions applicable to all permits.

§ 144.52 Establishing permit conditions.

§ 144.53 Schedule of compliance.

§ 144.54 Requirements for recording and reporting of monitoring results.

§ 144.55 Corrective action.

Subpart F--Financial Responsibility: Class I Hazardous Waste Injection Wells

§ 144.60 Applicability.

§ 144.61 Definitions of terms as used in this subpart.

§ 144.62 Cost estimate for plugging and abandonment.

§ 144.63 Financial assurance for plugging and abandonment.

§ 144.64 Incapacity of owners or operators, guarantors, or financial institutions.

§ 144.65 Use of State-required mechanisms.

§ 144.66 State assumption of responsibility.

§ 144.70 Wording of the instruments.

PART 145--STATE UIC PROGRAM REQUIREMENTS

Subpart A--General Program Requirements

§ 145.1 Purpose and scope.

§ 145.2 Definitions.

Subpart B--Requirements for State Programs

§ 145.11 Requirements for permitting.

§ 145.12 Requirements for compliance evaluation programs.

§ 145.13 Requirements for enforcement authority.

§ 145.14 Sharing of information.

Subpart C--State Program Submissions

§ 145.21 General requirements for program approvals.

§ 145.22 Elements of a program submission.

§ 145.23 Program description.

§ 145.24 Attorney General's statement.

§ 145.25 Memorandum of Agreement with the Regional Administrator.

Subpart D--Program Approval, Revision and Withdrawal

§ 145.31 Approval process.

§ 145.32 Procedures for revision of State programs.

§ 145.33 Criteria for withdrawal of State programs.

§ 145.34 Procedures for withdrawal of State programs.

Subpart E--Treatment of Indian Tribes as States

§ 145.52 Requirements for treatment as a State.

§ 145.56 Request by an Indian Tribe for a determination of treatment as a State.

§ 145.58 Procedure for processing an Indian Tribe's application for treatment as a State.

PART 146--UNDERGROUND INJECTION CONTROL PROGRAM: CRITERIA AND STANDARDS

Subpart A--General Provisions

§ 146.1 Applicability and scope.

§ 146.2 Law authorizing these regulations.

§ 146.3 Definitions.

§ 146.4 Criteria for exempted aquifers.

§ 147.2 Severability of provisions.
Subpart B--Alabama
§ 147.50 State-administered program--Class II wells.
§ 147.51 State-administered program--Class I, III, IV, and V wells.
§ 147.60 EPA-administered program--Indian lands.
Subpart C--Alaska
§ 147.100 State-administered program--Class II wells.
§ 147.101 EPA-administered program.
§ 147.102 Aquifer exemptions.
§ 147.103 Existing Class I, II (except enhanced recovery and hydrocarbon storage) and III wells authorized by rule.
§ 147.104 Existing Class II enhanced recovery and hydrocarbon storage wells authorized by rule.
Subpart D--Arizona
§ 147.150 State-administered program. [Reserved]
§ 147.151 EPA-administered program.
§ 147.152 Aquifer exemptions. [Reserved]
Subpart E--Arkansas
§ 147.200 State-administered program--Class I, III, IV, and V wells.
§ 147.201 State-administered program--Class II wells. [Reserved]
§ 147.205 EPA-administered program--Indian lands.
Subpart F--California
§ 147.250 State-administered program--Class II wells.
§ 147.251 EPA-administered program--Class I, III, IV and V wells and Indian lands.
§ 147.252 Aquifer exemptions. [Reserved]
§ 147.253 Existing Class I, II (except enhanced recovery and hydrocarbon storage) and III wells authorized by rule.
Subpart G--Colorado
§ 147.300 State-administered program--Class II wells.
§ 147.301 EPA-administered program--Class I, III, IV, V wells and Indian lands.
§ 147.302 Aquifer exemptions.
§ 147.303 Existing Class I, II (except enhanced recovery and hydrocarbon storage) and III wells authorized by rule.
§ 147.304 Existing Class II enhanced recovery and hydrocarbon storage wells authorized by rule.
§ 147.305 Requirements for all wells.
Subpart H--Connecticut
§ 147.350 State-administered program.
§ 147.351 [Reserved. 56 FR 9413, Mar. 6, 1991]
§ 147.352 [Reserved. 56 FR 9413, Mar. 6, 1991]
§ 147.353 EPA-administered program--Indian lands.
Secs. 147.354--147.359 [Reserved]
Subpart I--Delaware
§ 147.400 State-administered program.
§ 147.401 [Reserved. 56 FR 9413, Mar. 6, 1991]
§ 147.402 [Reserved. 56 FR 9413, Mar. 6, 1991]
§ 147.403 EPA-administered program--Indian lands.
Secs. 147.404--147.449 [Reserved]

§ 152.85 Formulators' exemption.
§ 152.86 The cite-all method.
§ 152.90 The selective method.
§ 152.91 Waiver of a data requirement.
§ 152.92 Submission of a new valid study.
§ 152.93 Citation of a previously submitted valid study.
§ 152.94 Citation of a public literature study or study generated at government expense.
§ 152.95 Citation of all studies in the Agency's files pertinent to a specific data requirement.
§ 152.96 Documentation of a data gap.
§ 152.97 Rights and obligations of data submitters.
§ 152.98 Procedures for transfer of exclusive use or compensation rights to another person.
§ 152.99 Petitions to cancel registration.

Subpart F--Agency Review of Applications

§ 152.100 Scope.
§ 152.102 Publication.
§ 152.104 Completeness of applications.
§ 152.105 Incomplete applications.
§ 152.107 Review of data.
§ 152.108 Review of labeling.
§ 152.110 Time for Agency review.
§ 152.111 Choice of standards for review of applications.
§ 152.112 Approval of registration under FIFRA sec. 3(c)(5).
§ 152.113 Approval of registration under FIFRA sec. 3(c)(7)--Products that do not contain a new active ingredient.
§ 152.114 Approval of registration under FIFRA sec. 3(c)(7)--Products that contain a new active ingredient.
§ 152.115 Conditions of registration.
§ 152.116 Notice of intent to register to original submitters of exclusive use data.
§ 152.117 Notification to applicant.
§ 152.118 Denial of application.
§ 152.119 Availability of material in support of registration.

Subpart G--Obligations and Rights of Registrants

§ 152.122 Currency of address of record and authorized agent.
§ 152.125 Submission of information pertaining to adverse effects.
§ 152.130 Distribution under approved labeling.
§ 152.132 Supplemental distribution.
§ 152.135 Transfer of registration.
§ 152.138 Voluntary cancellation.

Subpart H--Agency Actions Affecting Registrations

§ 152.140 Classification of pesticide products.
§ 152.142 Submission of information to maintain registration in effect.
§ 152.144 Reregistration.
§ 152.146 Special review of pesticides.

§ 160.41 General.
§ 160.43 Test system care facilities.
§ 160.45 Test system supply facilities.
§ 160.47 Facilities for handling test, control, and reference substances.
§ 160.49 Laboratory operation areas.
§ 160.51 Specimen and data storage facilities.
Subpart D--Equipment
§ 160.61 Equipment design
§ 160.63 Maintenance and calibration of equipment.
Subpart E--Testing Facilities Operation
§ 160.81 Standard operating procedures.
§ 160.83 Reagents and solutions.
§ 160.90 Animal and other test system care.
Subpart F--Test, Control, and Reference Substances
§ 160.105 Test, control, and reference substance characterization.
§ 160.107 Test, control, and reference substance handling.
§ 160.113 Mixtures of substances with carriers.
Subpart G--Protocol for and Conduct of a Study
§ 160.120 Protocol.
§ 160.130 Conduct of a study.
§ 160.135 Physical and chemical characterization studies.
Subparts H and I [Reserved]
Subpart J--Records and Reports
§ 160.185 Reporting of study results.
§ 160.190 Storage and retrieval of records and data.
§ 160.195 Retention of records.
PART 162--STATE REGISTRATION OF PESTICIDE PRODUCTS
Subparts A--C [Reserved]
Subpart D--Regulations Pertaining to State Registration of Pesticides To Meet Special Local Needs
§ 162.150 General.
§ 162.151 Definitions.
§ 162.152 State registration authority.
§ 162.153 State registration procedures.
§ 162.154 Disapproval of State registrations.
§ 162.155 Suspension of State registration authority.
§ 162.156 General requirements.
Subpart E--Reserved
PART 163--CERTIFICATION OF USEFULNESS OF PESTICIDE CHEMICALS
§ 163.1 Words in the singular form.
§ 163.2 Definitions.
§ 163.3 Administration.
§ 163.4 Filing of requests for certification.
§ 163.5 Material in support of the request for certification.
§ 163.6 Certification limited to economic poison uses.
§ 163.7 Factors considered in determining usefulness.
§ 163.8 Basis for determination of usefulness.
§ 163.9 Proposed certification; notice; request for hearing.
§ 163.10 Withdrawal of request for certification pending

clarification or completion.
§ 163.11 Registration under the Federal Insecticide, Fungicide, and Rodenticide Act.
§ 163.12 Opinion as to residue.
PART 164--RULES OF PRACTICE GOVERNING HEARINGS, UNDER THE FEDERAL INSECTICIDE, FUNGICIDE, AND RODENTICIDE ACT, ARISING FROM REFUSALS TO REGISTER, CANCELLATIONS OF REGISTRATIONS, CHANGES OF CLASSIFICATIONS, SUSPENSIONS OF REGISTRATIONS AND OTHER HEARINGS CALLED PURSUANT TO SECTION 6 OF THE ACT
Subpart A--General
§ 164.1 Number of words.
§ 164.2 Definitions.
§ 164.3 Scope and applicability of this part.
§ 164.4 Arrangements for examining Agency records, transcripts, orders, and decisions.
§ 164.5 Filing and service.
§ 164.6 Time.
§ 164.7 Ex parte discussion of proceeding.
§ 164.8 Publication.
Subpart B--General Rules of Practice Concerning Proceedings (Other Than Expedited Hearings)
Commencement of Proceeding
§ 164.20 Commencement of proceeding.
§ 164.21 Contents of a denial of registration, notice of intent to cancel a registration, or notice of intent to change a classification.
§ 164.22 Contents of document setting forth objections.
§ 164.23 Contents of the statement of issues to accompany notice of intent to hold a hearing.
§ 164.24 Response to the Administrator's notice of intention to hold a hearing.
§ 164.25 Filing copies of notification of intent to cancel registration or change classification or refusal to register, and statement of issues.
Appearances, Intervention, and Consolidation
§ 164.30 Appearances.
§ 164.31 Intervention.
§ 164.32 Consolidation.
Administrative Law Judge
§ 164.40 Qualifications and duties of Administrative Law Judge.
Prehearing Procedures and Discovery
§ 164.50 Prehearing conference and primary discovery.
§ 164.51 Other discovery.
Motions
§ 164.60 Motions.
Subpoenas and Witness Fees
§ 164.70 Subpoenas.
§ 164.71 Fees of witnesses.
The Hearings
§ 164.80 Order of proceeding and burden of proof.
§ 164.81 Evidence.
§ 164.82 Transcripts.

Initial or Accelerated Decision

§ 164.90 Initial decision.

§ 164.91 Accelerated decision.

Appeals

§ 164.100 Appeals from or review of interlocutory orders or rulings.

§ 164.101 Appeals from or review of initial decisions.

§ 164.102 Appeals from accelerated decisions.

§ 164.103 Final decision or order on appeal or review.

§ 164.110 Motion for reopening hearings; for rehearing; for reargument of any proceeding; or for reconsideration of order.

§ 164.111 Procedure for disposition of motions.

Subpart C--General Rules of Practice for Expedited Hearings

§ 164.120 Notification.

§ 164.121 Expedited hearing.

§ 164.122 Final order and order of suspension.

§ 164.123 Emergency order.

Subpart D--Rules of Practice for Applications Under Sections 3 and 18 To Modify Previous Cancellation or Suspension Orders

§ 164.130 General.

§ 164.131 Review by Administrator.

§ 164.132 Procedures governing hearing.

§ 164.133 Emergency waiver of hearing.

PART 165--REGULATIONS FOR THE ACCEPTANCE OF CERTAIN PESTICIDES AND RECOMMENDED PROCEDURES FOR THE DISPOSAL AND STORAGE OF PESTICIDES AND PESTICIDES CONTAINERS

Subpart A--General

§ 165.1 Definitions.

§ 165.2 Authorization and scope.

Subpart B--Acceptance Regulations

§ 165.3 Acceptable pesticides.

§ 165.4 Request for acceptance.

§ 165.5 Delivery.

§ 165.6 Disposal.

Subpart C--Pesticides and Containers

§ 165.7 Procedures not recommended.

§ 165.8 Recommended procedures for the disposal of pesticides.

§ 165.9 Recommended procedures for the disposal of pesticide containers and residues.

§ 165.10 Recommended procedures and criteria for storage of pesticides and pesticide containers.

Subpart D--Pesticide-Related Wastes

§ 165.11 Procedures for disposal and storage of pesticide-related wastes.

PART 166--EXEMPTION OF FEDERAL AND STATE AGENCIES FOR USE OF PESTICIDES UNDER EMERGENCY CONDITIONS

Subpart A--General Provisions

§ 166.1 Purpose and organization.

§ 166.2 Types of exemptions.

§ 166.3 Definitions.

§ 166.7 User notification; advertising.

§ 171.9 Submission and approval of Government Agency Plan.

§ 171.10 Certification of applicators on Indian Reservations.

§ 171.11 Federal certification of pesticide applicators in States or on Indian Reservations where there is no approved State or Tribal certification plan in effect.

PART 172--EXPERIMENTAL USE PERMITS

Subpart A--Federal Issuance of Experimental Use Permits

§ 172.1 Definitions.

§ 172.2 General.

§ 172.3 Scope of requirement.

§ 172.4 Applications.

§ 172.5 The permit.

§ 172.6 Labeling.

§ 172.7 Importation of technical material.

§ 172.8 Program surveillance and reporting of data.

§ 172.9 Renewals.

§ 172.10 Refusals to issue and revocation.

§ 172.11 Publication.

Subpart B--State Issuance of Experimental Use Permits

§ 172.20 Scope.

§ 172.21 Definitions.

§ 172.22 General.

§ 172.23 State plans.

§ 172.24 State issuance of permits.

§ 172.25 Administration of State programs.

§ 172.26 EPA review of permits.

PART 173--PROCEDURES GOVERNING THE RESCISSION OF STATE PRIMARY ENFORCEMENT RESPONSIBILITY FOR PESTICIDE USE VIOLATIONS

§ 173.1 Applicability.

§ 173.2 Definitions.

§ 173.3 Initiation of rescission proceedings.

§ 173.4 Informal conference and settlement.

§ 173.5 Request for hearing.

§ 173.6 Publication of the notice; scheduling the hearing.

§ 173.7 Hearing and recommended decision.

§ 173.8 Final order.

§ 173.9 Judicial review.

PART 177--ISSUANCE OF FOOD ADDITIVE REGULATIONS

Subpart A--General Provisions

§ 177.1 Scope and applicability.

§ 177.3 Definitions.

Subparts B-D--[Reserved]

Subpart E -- Procedures for Filing Petitions

§ 177.81 Petition for establishment, modification, or revocation of a food additive regulation.

§ 177.84 Deficient or incomplete petitions.

§ 177.86 Acceptance for review.

§ 177.88 Publication of notice.

§ 177.92 Amendments or supplements to petitions.

§ 177.98 Withdrawal of petitions.

§ 177.99 Demand for action.

Subpart F--Submission of Scientific and Technical Information

§ 180.138 [Removed. 58 FR 46088, Sept. 1, 1993]
§ 180.139 1,1-Dichloro-2,2-bis (r-ethylphenyl) ethane; tolerances for residues.
§ 180.141 Biphenyl; tolerances for residues.
§ 180.142 2,4-D; tolerances for residues.
§ 180.143 Dipropyl isocinchomeronate; tolerances for residues.
§ 180.144 Cyhexatin; tolerances for residues.
§ 180.145 Fluorine compounds; tolerances for residues.
§ 180.148 b-Naphthoxyacetic acid; tolerances for residues.
§ 180.149 Mineral oil; tolerances for residues.
§ 180.150 Dalapon; tolerances for residues.
§ 180.151 Ethylene oxide; tolerances for residues.
§ 180.152 Sodium dimethyldithiocarba-mate; tolerance for residues.
§ 180.153 Diazinon; tolerances for residues.
§ 180.154 O,O-Dimethyl S-[(4-oxo-1,2,3-benzotriazin-3(4H)-yl)methyl] phosphorodithioate; tolerances for residues.
§ 180.154a O,O-Dimethyl S-[(4-oxo-1,2, 3-benzotriazin-3 (4H)-yl)methyl] phosphorodithioate residues and/or its metabolites in milk.
§ 180.155 1-Naphthaleneacetic acid; tolerances for residues.
§ 180.156 Carbophenothion; tolerances for residues.
§ 180.157 Methyl 3-[(dimethoxyphosphinyl) oxy]butenoate, alpha and beta isomers; tolerances for residues.
§ 180.158 2,4-Dichloro-6-o-chloroanilino-s-triazine; tolerances for residues.
§ 180.159 Sodium dehydroacetate; tolerances for residues.
§ 180.160 [Removed. 56 FR 14472, Apr. 10, 1991]
§ 180.161 Manganous dimethyldithio-carbamate; tolerance for residues.
§ 180.162 Tetraiodoethylene; tolerance for residues.
§ 180.163 1,1-Bis(p-chlorophenyl)-2,2,2-trichloroethanol; tolerances for residues.
§ 180.165 [Removed. 58 FR 32300, June 9, 1993]
§ 180.167 Nicotine-containing compounds; tolerances for residues.2
§ 180.167a Nicotine; tolerances for residues.
§ 180.169 Carbaryl; tolerances for residues.
§ 180.170 Temephos; tolerances for residues.
§ 180.171 Dioxathion; tolerances for residues.
§ 180.172 Dodine; tolerances for residues.
§ 180.173 Ethion; tolerances for residues.
§ 180.174 Tetradifon; tolerances for residues.
§ 180.175 Maleic hydrazide; tolerances for residues.
§ 180.176 Coordination product of zinc ion and maneb; tolerances for residues.
§ 180.177 [Removed. 59 FR 13659, Mar. 23, 1994]
§ 180.178 Ethoxyquin; tolerances for residues.
§ 180.179 Tartar emetic; tolerances for residues.2
§ 180.180 Orthoarsenic acid.
§ 180.181 CIPC; tolerances for residues.

§ 180.182 Endosulfan; tolerances for residues.
§ 180.183 O,O-Diethyl S-[2-(ethylthio)ethyl] phosphorodithioate; tolerances for residues.
§ 180.184 Linuron; tolerances for residues.
§ 180.185 Dimethyl tetrachlorotereph thalate; tolerances for residues.
§ 180.186 Tributylphosphorotrithioite; tolerance for residues.
§ 180.188 Ammonium sulfamate; tolerances for residues.
§ 180.189 Coumaphos; tolerances for residues.
§ 180.190 Diphenylamine; tolerances for residues.
§ 180.191 Folpet; tolerances for residues.
§ 180.192 [Removed. 56 FR 13595, Apr. 3, 1991]
§ 180.194 [Removed. 56 FR 13594, Apr. 3, 1991]
§ 180.198 Dimethyl (2,2,2-trichloro-1-hydroxyethyl) phosphonate; tolerances for residues.
§ 180.199 Inorganic bromides resulting from soil treatment with combinations of chloropicrin, methyl bromide, and propargyl bromide; tolerances for residues.
§ 180.200 2,6-Dichloro-4-nitroaniline; tolerances for residues.
§ 180.201 Chlorosulfamic acid; tolerances for residues.
§ 180.202 p-Chlorophenoxyacetic acid; tolerances for residues.
§ 180.203 1,2,4,5-Tetrachloro-3-nitro-benzene; tolerances for residues.
§ 180.204 Dimethoate including its oxygen analog; tolerances for residues.
§ 180.205 Paraquat; tolerances for residues.
§ 180.206 Phorate; tolerances for residues.
§ 180.207 Trifluralin; tolerances for residues.
§ 180.208 N-Butyl-N-ethyl-<alpha>,<alpha>,<alpha>-trifluoro-2,6-dinitro- p-toluidine; tolerances for residues
§ 180.209 Terbacil; tolerances for residues.
§ 180.210 Bromacil; tolerances for residues.
§ 180.211 2-Chloro-N-isopropylacetanilide; tolerances for residues.
§ 180.212 S-Ethyl cyclohexylethylthiocarbamate; tolerances for residues.
§ 180.213 Simazine; tolerances for residues.
§ 180.213a Simazine; tolerances for residues.
§ 180.214 Fenthion; tolerances for residues.
§ 180.215 Naled; tolerances for residues.
§ 180.216 Chloroxuron; tolerances for residues.
§ 180.217 Ammoniates for [ethylenebis-(dithiocarbamato)] zinc and ethylenebis [dithiocarbamic acid] bimolecular and trimolecular cyclic anhydrosulfides and disulfides; tolerances for residues.
§ 180.219 2,3,5-Triiodobenzoic acid; tolerances for residues.
§ 180.220 Atrazine; tolerances for residues.
§ 180.221 O-Ethyl S-phenyl ethylphosphonodithioate;

tolerances for residues.
§ 180.222 Prometryn; tolerances for residues.
§ 180.224 Gibberellins; tolerances for residues.
§ 180.225 Aluminum phosphide; tolerances for residues.
§ 180.226 Diquat; tolerances for residues.
§ 180.227 Dicamba; tolerances for residues
§ 180.228 S-Ethyl hexahydro-1H-azepine-1-carbothioate; tolerances for residues.
§ 180.229 Fluometuron; tolerances for residues.
§ 180.230 Diphenamid; tolerances for residues.
§ 180.231 Dichlobenil; tolerances for residues.
§ 180.232 S-Ethyl diisobutylthiocarbamate; tolerances for residues.
§ 180.233 O,O-Dimethyl O-p-(dimethylsulfamoyl) phenyl phosphorothioate including its oxygen analog; tolerances for residues.
§ 180.234 [Removed. 58 FR 60559, Nov. 17, 1993]
§ 180.235 2,2-Dichlorovinyl dimethyl phosphate; tolerances for residues.
§ 180.236 Triphenyltin hydroxide; tolerances for residues.
§ 180.237 [Removed. 56 FR 14472, Apr. 10, 1991]
§ 180.238 S-Propyl butylethylthiocarbamate; tolerances for residues.
§ 180.239 Phosphamidon; tolerances for residues.
§ 180.240 S-Propyl dipropylthiocarbamate; tolerances for residues.
§ 180.241 S-(O,O-Diisopropyl phosphorodithioate) of N-(2-mercaptoethyl) benzenesulfonamide; tolerances for residues.
§ 180.242 Thiabendazole; tolerances for residues.
§ 180.243 Propazine; tolerances for residues.
§ 180.244 Basic zinc sulfate; tolerances for residues.
§ 180.245 Streptomycin; tolerances for residues.
§ 180.246 Daminozide; tolerances for residues.
§ 180.247 2-Chloroallyldiethyldithiocar-bamate; tolerances for residues.
§ 180.249 Alachlor; tolerances for residues.
§ 180.250 Metobromuron; tolerances for residues.
§ 180.252 2-Chloro-1,(2,4,5-trichlorophenyl)- vinyl dimethyl phosphate; tolerances for residues.
§ 180.253 Methomyl; tolerances for residues.
§ 180.254 Carbofuran; tolerances for residues.
§ 180.255 [Removed. 58 FR 32303, June 9, 1993]
§ 180.257 Chloroneb; tolerances for residues.
§ 180.258 Ametryn; tolerances for residues.
§ 180.259 Propargite; tolerances for residues.
§ 180.260 Norea; tolerances for residues.
§ 180.261 N-(Mercaptomethyl)phthalimide S-(O,O-dimethyl phosphorodithioate) and its oxygen analog; tolerances for residues.
§ 180.262 Ethoprop; tolerances for residues.
§ 180.263 Phosalone; tolerances for residues.
§ 180.265 Terbutryn; tolerances for residues.

§ 180.310 Sodium trichloroacetate; tolerances for residues.
§ 180.311 Cacodylic acid; tolerances for residues.
§ 180.312 4-Aminopyridine; tolerances for residues.
§ 180.313 [Removed. 56 FR 26915, June 12, 1991]
§ 180.314 S-2,3,3-trichloroallyl diisopropylthiocarbamate; tolerances for residues.
§ 180.315 Methamidophos; tolerances for residues.
§ 180.316 Pyrazon; tolerances for residues.
§ 180.317 3,5-Dichloro-N-(1,1-dimethyl-2-propynyl)benzamide; tolerances for residues.
§ 180.318 4 - (2 - Methyl - 4 - chlorophenoxy) butyric acid; tolerances for residues.
§ 180.319 Interim tolerances.
§ 180.320 3,5-Dimethyl-4-(methylthio)phenyl methylcarbamate; tolerances for residues.
§ 180.321 sec-Butylamine; tolerances for residues.
§ 180.322 2-Chloro-1-(2,4-dichlorophenyl) vinyl diethyl phosphate; tolerances for residues.
§ 180.324 Bromoxynil; tolerances for residues.
§ 180.325 2-(m-Chlorophenoxy) propionic acid; tolerances for residues.
§ 180.326 Dialifor; tolerances for residues.
§ 180.327 Dinitramine; tolerances for residues.
§ 180.328 N,N-Diethyl-2-(1-naphthalenyloxy)-propionamide; tolerances for residues.
§ 180.329 Dipropetryn; tolerance for residues.
§ 180.330 S-[2-(Ethylsulfinyl)ethyl] O, O-dimethyl phosphorothioate; tolerances for residues.
§ 180.331 4-(2,4-Dichlorophenoxy) butyric acid; tolerances for residues.
§ 180.332 4-Amino-6-(1,1-dimethylethyl)-3-(methylthio)-1,2,4-triazin- 5(4H)-one; tolerances for residues.
§ 180.333 [Removed. 57 FR 30132, July 8, 1992]
§ 180.335 [Removed. 58 FR 39154, July 22, 1993]
§ 180.336 Cycloheximide; tolerances for residues.
§ 180.337 Oxytetracycline; tolerances for residues.
§ 180.338 6-methyl-1,3-dithiolo [4,5-b] quinoxalin-2-one, tolerances for residues.
§ 180.339 2-methyl-4-chlorophenoxyacetic acid; tolerances for residues.
§ 180.340 [Removed. 58 FR 33212, June 16, 1993]
§ 180.341 2,4-Dinitro-6-octylphenyl crotonate and 2,6-dinitro-4- octylphenyl crotonate; tolerances for residues.
§ 180.342 Chlorpyrifos; tolerances for residues.
§ 180.344 4,6-Dinitro-o-cresol and its sodium salt; tolerance for residues.
§ 180.345 Ethofumesate; tolerances for residues.
§ 180.346 Oxadiazon; tolerances for residues.
§ 180.347 Tetraethyl pyrophosphate; tolerances for residues.
§ 180.348 [Removed. 58 FR 32299, June 9, 1993]
§ 180.349 Ethyl 3-methyl-4-(methylthio)phenyl (1-

methylethyl) phosphoramidate; tolerances for residues.
§ 180.350 Nitrapyrin; tolerances for residues.
§ 180.351 Bifenox; tolerances for residues.
§ 180.352 Terbufos; tolerances for residues.
§ 180.353 Desmedipham; tolerances for residues.
§ 180.355 Bentazon; tolerances for residues.
§ 180.356 Norflurazon; tolerances for residues.
§ 180.357 Methazole, tolerances for residues.
§ 180.358 Butralin; tolerances for residues.
§ 180.359 Methoprene; tolerances for residues.
§ 180.360 Asulam; tolerances for residues.
§ 180.361 Pendimethalin; tolerances for residues.
§ 180.362 Hexakis (2-methyl-2-phenylpropyl)distannoxane; tolerances for residues.
§ 180.363 Fluchloralin: tolerances for residues.
§ 180.364 Glyphosate; tolerances for residues.
§ 180.366 Octhilinone; tolerances for residues.
§ 180.367 n-Octyl bicycloheptenedicarboximide; tolerances for residues.
§ 180.368 Metolachlor; tolerances for residues.
§ 180.369 Difenzoquat; tolerances for residues.
§ 180.370 5-Ethoxy-3-(trichloromethyl)-1,2,4-thiadiazole; tolerances for residues.
§ 180.371 Thiophanate-methyl; tolerances for residues.
§ 180.372 2,6-dimethyl-4-tridecylmorpho-line; tolerances for residues.
§ 180.373 [Reserved]
§ 180.374 O - Ethyl O - [4-(methylthio) phenyl] S-propyl phosphorodithioate; tolerances for residues.
§ 180.375 Magnesium phosphide; tolerances for residues.
§ 180.376 6-Benzyladenine; tolerances for residues.
§ 180.377 Diflubenzuron; tolerances for residues.
§ 180.378 Permethrin; tolerances for residues.
§ 180.379 Cyano(3-phenoxyphenyl)methyl-4-chloro-<alpha>-(1-methylethyl) benzeneacetate; tolerances for residues.
§ 180.380 3-(3,5-Dichlorophenyl)-5-ethenyl-5-methyl-2,4 oxazolidinedione; tolerances for residues.
§ 180.381 Oxyfluorfen; tolerances for residues.
§ 180.382 Triforine; tolerances for residues.
§ 180.383 Sodium salt of acifluorfen; tolerances for residues.
§ 180.384 N,N-Dimethylpiperidinium chloride; tolerances for residues.
§ 180.385 Diclofop-methyl; tolerances for residues.
§ 180.386 Mefluidide; tolerances for residues.
§ 180.387 1-Methyl 2-[[ethoxy-[(1-methylethyl) amino] phosphinothioyl)oxy)benzoate.
Secs. 180.388--180.389 [Reserved]
§ 180.390 Tebuthiuron; tolerances for residues.
§ 180.395 Tetrahydro-5,5-dimethyl-2(1H)-pyrimidinone(3-(4-trifluoromethyl)phenyl)-1-(2-(4-(trifluoromethyl)phenyl)ethenyl)-2-propenylidene)hydrazone; tolerances for residues.

§ 180.396 Hexazinone; tolerances for residues.
§ 180.397 [Removed. 58 FR 65556, Dec. 15, 1993]
§ 180.398 Chlorthiophos; tolerances for residues.
§ 180.399 Iprodione; tolerances for residues.
§ 180.400 Flucythrinate; tolerances for residues.
§ 180.401 Thiobencarb; tolerances for residues.
§ 180.402 Diethatyl-ethyl; tolerances for residues.
§ 180.403 Thidiazuron; tolerances for residues.
§ 180.404 Profenofos; tolerances for residues.
§ 180.405 Chlorsulfuron; tolerances for residues.
§ 180.406 Dimethipin; tolerances for residues.
§ 180.407 Thiodicarb; tolerances for residue.
§ 180.408 Metalaxyl; tolerances for residues.
§ 180.409 Pirimiphos-methyl; tolerances for residues.
§ 180.410 1-(4-Chlorophenoxy)-3,3-dimethyl-1(1H-1,2,4-triazol-1-yl)-2- butanone; tolerances for residues.
§ 180.411 Fluazifop-butyl; tolerances for residues.
§ 180.412 2-[1-(Ethoxyimino)butyl]-5-[2-(ethylthio)propyl]-3-hydroxy-2- cyclohexene-1-one; tolerances for residues.
§ 180.413 Imazalil; tolerances for residues.
§ 180.414 Cyromazine; tolerances for residues.
§ 180.415 Aluminum tris (O-ethylphosphonate); tolerances for residues.
§ 180.416 Ethalfluralin; tolerances for residues.
§ 180.417 Triclopyr; tolerances for residues.
§ 180.418 Cypermethrin; tolerances for residues.
§ 180.419 Chlorpyrifos-methyl.
§ 180.420 Fluridone; tolerances for residues.
§ 180.421 Fenarimol; tolerances for residues.
§ 180.422 Tralomethrin; tolerances for residues.
§ 180.423 Fenridazon, potassium salt; tolerances for residues.
§ 180.424 2-(3,5-Dichlorophenyl)-2-(2,2,2-trichloroethyl)-oxirane; tolerances for residues.
§ 180.425 2-(2-Chlorophenyl)methyl-4,4-dimethyl-3-isoxazolidinone; tolerances for residues.
§ 180.426 2-[4,5-Dihydro-4-methyl-4-(1-methylethyl)-5-oxo-1H-imidazol-2- yl]-3-quinoline carboxylic acid; tolerance for residues.
§ 180.427 (Alpha RS,2R)-fluvalinate [(RS)-alpha-cyano-3-phenoxybenzyl(R)- 2-[2-chloro-4-(trifluoromethyl)anilino]-3-methylbutanoate; tolerances for residues.
§ 180.428 Metsulfuron methyl; tolerances for residues.
§ 180.429 Chlorimuron ethyl; tolerance for residues.
§ 180.430 Fenoxaprop-ethyl; tolerances for residues.
§ 180.431 Clopyralid; tolerances for residues.
§ 180.432 Lactofen; tolerance for residues.
§ 180.433 Sodium salt of fomesafen; tolerance for residues.
§ 180.434 1-[[2-(2,4-dichlorophenyl)-4-propyl-1,3-dioxolan-2-yl]methyl]- 1H-1,2,4-triazole; tolerances for residues.

§ 180.435 Deltamethrin; tolerance for residues.
§ 180.436 Cyfluthrin; tolerances for residues.
§ 180.437 Methyl 2-(4-isopropyl-4-methyl-5-oxo-2-imidazolin-2-yl)-p- toluate and methyl 6-(4-isopropyl-4-methyl-5-oxo-2-imidazolin-2-yl)-m- toluate; tolerances for residues.
§ 180.438 [1 alpha-(S*),3 alpha(Z)]-(+/-)-cyano(3-phenoxyphenyl)methyl 3- (2-chloro-3,3,3-trifluoro-1-propenyl)-2,2-dimethylcyclopropanecarboxylate; tolerances for residues.
§ 180.439 Thifensulfuron methyl (methy-3-[[[[(4-methoxy-6-methyl-1,3,5- triazin-2-yl) amino]carbonyl]amino]sulfonyl]-2-thiophene carboxylate); tolerances for residues.
§ 180.440 Tefluthrin; tolerances for residues.
§ 180.441 Quizalofop ethyl; tolerances for residues.
§ 180.442 Bifenthrin; tolerances for residues.
§ 180.443 Myclobutanil; tolerances for residues.
§ 180.444 Sulfur dioxide; tolerances for residues.
§ 180.445 Bensulfuron methyl ester; tolerances for residues.
§ 180.446 Clofentezine; tolerances for residues.
§ 180.447 Imazethapyr, ammonium salt; tolerance for residues.
§ 180.448 Hexythiazox; tolerance for residues.
§ 180.449 Avermectin B1 and its delta-8,9-isomer; tolerances for residues.
§ 180.450 Beta-(4-Chlorophenoxy)-alpha-(1,1-dimethylethyl)-1H-1,2,4- triazole-1-ethanol; tolerances for residues.
§ 180.451 Tribenuron methyl (methy-2-[[[[N-(4-methoxy-6-methyl-1,3,5- triazin-2-yl) methylamino] carbonyl]amino]sulfonyl]benzoate); tolerances for residues.
§ 180.452 Primisulfuron-methyl; tolerances for residues.
§ 180.453 [Removed. 55 FR 33695, Aug. 17, 1990]
§ 180.454 Nicosulfuron, [3-pyridinecarboxamide, 2-((((4,6-dimethoxypyrimidin-2-yl)aminocarbonyl)aminosulfonyl))-N,N-dimethyl]; tolerances for residues.
§ 180.455 Procymidone; tolerances for residues.
§ 180.456 Oxadixyl; tolerances for residues.
§ 180.457 Beta-([1,1'-biphenyl]-4-yloxy)-alpha-(1,1-dimethylethyl)-1H- 1,2,4-triazole-1-ethanol; tolerances for residues.
§ 180.458 Clethodim ((E)-(+/-)-2-[1-[[(3-chloro-2-propenyl)oxy]imino]propyl]-5-[2-(ethylthio)propyl]-3-hydroxy-2-cyclohexen-1- one); tolerances for residues.
§ 180.459 Triasulfuron; tolerances for residues.
§ 180.460 4-(Dichloroacetyl)-3,4-dihydro-3-methyl-2H-1,4-benzoxazine; tolerances for residues.
§ 180.461 Cadusafos; tolerances for residues.
§ 180.462 Pyridate; tolerances for residues.
§ 180.463 3,7-Dichloro-8-quinoline carboxylic acid;

requirement of a tolerance.
§ 180.1023 Propionic acid; exemptions from the requirement of a tolerance.
§ 180.1024 Paraformaldehyde; exemption from the requirement of a tolerance.
§ 180.1025 Xylene; exemption from the requirement of a tolerance.
§ 180.1026 N,N-Diallyl dichloroacetamide; exemption from the requirement of a tolerance.
§ 180.1027 Nuclear polyhedrosis virus of Heliothis zea; exemption from the requirement of a tolerance.
§ 180.1028 Cross-linked nylon-type encapsulating polymer, exemption from the requirement of a tolerance.
§ 180.1029 [Removed. 58 FR 47215, Sept. 8, 1993]
§ 180.1030 Isobutyric acid; exemption from the requirement of a tolerance.
§ 180.1031 Acetaldehyde; exemption from the requirement of a tolerance.
§ 180.1032 Formaldehyde, exemption from the requirement of a tolerance.
§ 180.1033 Methoprene; exemption from the requirement of a tolerance.
§ 180.1034 Butanoic anhydride; exemption from the requirement of a tolerance.
§ 180.1035 Pine oil; exemption from the requirement of a tolerance.
§ 180.1036 Hydrogenated castor oil; exemption from the requirement of a tolerance.
§ 180.1037 Polybutenes; exemption from the requirement of a tolerance.
§ 180.1038 Polyoxymethylene copolymer; exemption from the requirement of a tolerance.
§ 180.1039 Cross-linked polyurea-type encapsulating polymer; exemption from the requirement of a tolerance.
§ 180.1040 Ethylene glycol; exemption from the requirement of a tolerance.
§ 180.1041 Nosema locustae; exemption from the requirement of a tolerance.
§ 180.1042 Aqueous extract of seaweed meal; exemption from the requirement of a tolerance.
§ 180.1043 Gossyplure; exemption from the requirement of a tolerance.
§ 180.1045 Chlorotoluene; exemption from the requirement of a tolerance.
§ 180.1046 Dimethylformamide; exemption from the requirement of a tolerance.
§ 180.1049 Carbon dioxide; exemption from the requirement of a tolerance.
§ 180.1050 Nitrogen; exemption from the requirement of a tolerance.
§ 180.1051 Combustion product gas; exemption from the requirements of a tolerance.
§ 180.1052 2,2,5-trimethyl-3-dichloroacetyl-1,3-

oxazolidine; exemption from the requirement of a tolerance.
§ 180.1053 Polyamide polymer derived from sebacic acid; exemption from requirement of tolerance.
§ 180.1054 Calcium hypochlorite; exemption from the requirement of a tolerance.
§ 180.1055 (E,Z)-3,13-octadecadien-1-ol acetate and (Z,Z)-3,13- octadecadien-1-ol acetate; exemption from the requirement of a tolerance.
§ 180.1056 Boiled linseed oil; exemption from requirement of tolerance.
§ 180.1057 Phytophthora palmivora; exemption from requirement of tolerance.
§ 180.1058 Sodium diacetate; exemption from the requirement of a tolerance.
§ 180.1059 Methyl alpha-eleosterate; exemption from the requirement of a tolerance.
§ 180.1060 Polyvinyl chloride; exemption from requirement of a tolerance.
§ 180.1061 Hirsutella thompsonii; exemption from the requirement of a tolerance.
§ 180.1062 Butyl benzyl phthalate; exemption from the requirement tolerance.
§ 180.1063 Kontrol H. V.; exemption from the requirement of a tolerance.
§ 180.1064 Tomato pinworm insect pheromone; exemption from the requirement of a tolerance.
§ 180.1065 2-Amino-4,5-dihydro-6-methyl-4-propyl-s-triazolo(1,5- alpha)pyrimidin-5-one; exemption from the requirement of a tolerance.
§ 180.1066 O-O-Diethyl-O-phenylphosphorothioate; exemption from the requirement of a tolerance.
§ 180.1067 Methyl eugenol and malathion combination; exemption from the requirement of a tolerance.
§ 180.1068 Potassium oleate and related C12-C18 fatty acid potassium salts; exemption from the requirement of a tolerance.
§ 180.1069 (Z)-11-Hexadecenal; exemption from the requirement of a tolerance.
§ 180.1070 Sodium chlorite; exemption from the requirement of a tolerance.
§ 180.1071 Egg solids (whole); exemption from the requirement of a tolerance.
§ 180.1072 Poly-D-glucosamine (chitosan); exemption from the requirement of a tolerance.
§ 180.1073 Isomate-M; exemption from the requirement of a tolerance.
§ 180.1074 F.D. & C Blue No. 1; exemption from the requirement of a tolerance.
§ 180.1075 Colletotrichum gloeosporioides f. sp. aeschynomene; exemption from the requirement of a tolerance.
§ 180.1076 Viable spores of the microorganism Bacillus

popilliae; exemption from the requirement of a tolerance.
§ 180.1077 2,2-Dichloro-N-(1,3-dioxolan-2-ylmethyl)-N-2-propenylacetamide; exemption from the requirement of a tolerance.
§ 180.1078 Poly(oxy-1,2-ethanediyl), alpha-isooctadyl-omega-hydroxy; exemption from the requirement of a tolerance.
§ 180.1079 1-(8-Methoxy-4,8-dimethylnonyl)-4-(1-methylethyl) benzene; exemptions from the requirement of a tolerance.
§ 180.1080 Plant volatiles and pheromone; exemptions from the requirement of a tolerance.
§ 180.1081 1-Triacontanol; exemption from the requirement of a tolerance.
§ 180.1082 Cross-linked polyurea-type encapsulating polymer (Alachlor); exemption from the requirement of a tolerance.
§ 180.1083 Dimethyl sulfoxide; exemption from the requirement of a tolerance.
§ 180.1084 Monocarbamide dihydrogen sulfate; exemption from the requirement of a tolerance.
§ 180.1085 Potassium ricinoleate and related C12-C18 fatty acid potassium salts; exemption from the requirement of a tolerance.
§ 180.1086 3,7,11-Trimethyl-1,6,10-dodecatriene-1-ol and 3,7,11- trimethyl-2,6,10-dodecatriene-3-ol; exemption from the requirement of a tolerance.
§ 180.1087 Sesame stalks; exemption from the requirement of a tolerance.
§ 180.1088 Pseudomonas fluorescens EG-1053; exemption from the requirement of tolerance.
§ 180.1089 Poly-N-acetyl-D-glucosamine; exemption from the requirement of tolerance.
§ 180.1090 Lactic acid; exemption from the requirement of a tolerance.
§ 180.1091 Aluminum isopropoxide and aluminum secondary butoxide; exemption from the requirement of a tolerance.Aluminum isopropoxide (CAS Reg. No. 555--31--7) and aluminum secondary butoxide (CAS Reg. No. 2269-- 22--9) are exempted from the requirement of a tolerance when used in accordance with good agricultural practices as stabilizers in formulations of the
§ 180.1092 Menthol; exemption from the requirement of a tolerance.
§ 180.1095 Chlorine gas; exemption from the requirement of a tolerance
§ 180.1097 GBM-ROPE; exemption from the requirement of a tolerance.
§ 180.1098 Gibberellins (GA3); exemption from the requirement of a tolerance.
§ 180.1099 Indole butyric acid (IBA); exemption from the requirement of tolerance
§ 180.1100 Gliocladium virens GL-21; exemption from the

requirement of a tolerance.
§ 180.1101 Parasitic (parasitoid) and predatory insects; exemption from the requirement of a tolerance.
§ 180.1102 Trichoderma harzianum, Rifai Strain KRL-AG2; exemption from the requirement of a tolerance.
§ 180.1103 Isomate-C; exemption from the requirement of a tolerance.
§ 180.1104 Poly(vinylpyrrolidone/1-eicosene); exemption from the requirement of a tolerance.
§ 180.1105 Poly(vinylpyrrolidone/1-hexadecene); exemption from the requirement of a tolerance.
§ 180.1106 Vinylpyrrolidone-vinyl acetate copolymer; exemption from the requirement of a tolerance.
§ 180.1107 Delta endotoxin of Bacillus thuringiensis variety kurstaki encapsulated into killed Pseudomonas fluorescens; exemption from the requirement of a tolerance.
§ 180.1108 Delta endotoxin of Bacillus thuringiensis variety San Diego encapsulated into killed Pseudomonas fluorescens; exemption from the requirement of a tolerance.
§ 180.1110 3-Carbamyl-2,4,5-trichlorobenzoic acid; exemption from the requirement of a tolerance.
§ 180.1111 Bacillus subtilis GB03; exemption from the requirement of a tolerance.
§ 180.1112 Alkyl acrylate/methacrylate copolymers; exemptions from the requirement of a tolerance.
§ 180.1113 Lagenidium giganteum; exemption from the requirement of a tolerance.
§ 180.1114 Pseudomonas fluorescens A506, Pseudomonas fluorescens 1629RS, and Pseudomonas syringae 742RS; exemptions from the requirement of a tolerance.
§ 180.1115 Pseudomonas cepacia type Wisconisn; exemption from the requirement of a tolerance.
§ 180.1116 Metarhizium anisopliae strain ESF1; exemption from the requirement of a tolerance.
§ 180.1118 Spodoptera exigua nuclear polyhedrosis virus; exemption from the requirement of a tolerance.
§ 180.1119 Azadirachtin; exemption from the requirement of a tolerance.
§ 180.1120 Streptomyces sp. strain K61; exemption from the requirement of a tolerance.
§ 180.1121 Boric acid and its salts, borax (sodium borate decahydrate), disodium octaborate tetrahydrate, boric oxide (boric anhydride), sodium borate and sodium metaborate; exemptions from the requirement of a tolerance.
§ 180.1122 Inert ingredicnts of semiochemical dispensers; exemptions from the requirement of a tolerance.
§ 180.1123 Puccinia canaliculata (ATCC 40199); exemption from the requirement of a tolerance.
§ 180.1124 Arthropod pheromones; exemption from the requirement of a tolerance.

§ 180.1125 Polyhedral occlusion bodies of Autographa californica nuclear polyhedrosis virus; exemption from the requirement of a tolerance.

§ 180.1126 Codlure, (E,E)-8,10-Dodecadien-1-ol; exemption from the requirement of a tolerance.

§ 180.1127 Biochemical pesticide plant floral volatile attractant compounds: cinnamaldehyde, cinnamyl alcohol, 4-methoxy cinnamaldehyde, 3- phenyl propanol, 4-methoxy phenethyl alcohol, indole, and 1,2,4- trimethoxybenzene; exemptions from the requirement of a tolerance.

§ 180.1128 Bacillus subtilis MBI 600; exemption from the requirement of a tolerance.

§ 180.1130 N-(n-octyl)-2-pyrrolidone and N-(n-dodecyl)-2-pyrrolidone; exemptions from the requirement of a tolerance.

§ 180.1131 Ampelomyces quisqualis isolate M10; exemption from the requirement of a tolerance.

PART 185--TOLERANCES FOR PESTICIDES IN FOOD

Subpart A--[Reserved]

Subpart B--Food Additives Permitted in Food for Human Consumption

§ 185.100 Acephate.

§ 185.150 Aldicarb.

§ 185.200 Aluminum phosphide.

§ 185.250 4-Amino-6-(1,1-dimethylethyl)-3-methylthio)-1,2,4-triazin- 5(4H)-one.

§ 185.300 Avermectin B1 and its delta-8,9-isomer; tolerances for residues.

§ 185.350 [Removed. 59 FR 33694, June 30, 1994]

§ 185.410 [Removed. 59 FR 10997, Mar. 9, 1994]

§ 185.425 Bromide ion and residual bromine.

§ 185.500 Captan.

§ 185.600 Carbofuran; tolerances for residues.

§ 185.650 Carbon dioxide.

§ 185.700 Carbophenothion.

§ 185.800 1-(4-Chlorophenoxy)-3,3-dimethyl-1-(1H-1,2,4-triazol-1-yl)-2- butanone.

§ 185.1000 Chlorpyrifos.

§ 185.1050 Chlorpyrifos-methyl.

§ 185.1100 Clopyralid.

§ 185.1150 Combustion product gas.

§ 185.1200 Copper.

§ 185.1250 Cyfluthrin.

§ 185.1300 Cyano(3-phenoxyphenyl)methyl-4-chloro-alpha-(1-methylethyl)benzeneacetate and its S,S isomer.

§ 185.1310 [1 alpha (S*),3 alpha(Z)]-(+/-)-cyano(3-phenoxyphenyl)methyl 3- (2-chloro-3,3,3-trifluoro-1-propenyl)-2,2-dimethylcyclopropanecarboxylate.

§ 185.1350 Cyhexatin.

§ 185.1450 2,4-D.

§ 185.1500 Dalapon.

§ 185.1550 [Removed. 55 FR 10222, Mar. 19, 1990].

§ 185.1580 Deltamethrin.

§ 185.1650 Dialifor.

§ 185.1700 Diatomaceous earth.
§ 185.1750 Diazinon.
§ 185.1800 Dicamba.
§ 185.1850 3-(3,5-Dichlorophenyl)-5-ethenyl-5-methyl-2,4-oxazolidinedione.
§ 185.1900 2,2-Dichlorovinyl dimethyl phosphate.
§ 185.2150 2,2-dimethyl-1,3-benzodioxol-4-ol methylcarbamate.
§ 185.2200 O,O-Dimethyl O-(4-nitro-m-tolyl) phosphorothioate.
§ 185.2225 O,O-Dimethyl S-[4-oxo-1,2,3-benzotriazin-3 (4H)-ylmethyl] phosphorodithioate.
§ 185.2250 Dimethyl phosphate of 3-hydroxy-N-methyl-cis-crotonamide.
§ 185.2275 N,N-dimethylpiperidinium chloride
§ 185.2500 Diquat.
§ 185.2600 Endosulfan.
§ 185.2650 Endothall.
§ 185.2700 Ethephon.
§ 185.2750 Ethion.
§ 185.2800 2-[1-(Ethoxyimino)butyl]-5-[2-(ethylthio)propyl]-3-hydroxy-2- cyclohexene-1-one.
§ 185.2850 Ethylene oxide.
§ 185.2900 Ethyl formate.
§ 185.2950 Ethyl 3-methyl-4-(methylthio)phenyl (1-methylethyl)- phosphoramidate.
§ 185.3000 O-Ethyl O-[4-(methylthio) phenyl] S-propyl phosphorodithioate.
§ 185.3200 Fenarimol.
§ 185.3225 Fenpropathrin.
§ 185.3250 Fluazifop-butyl.
§ 185.3300 Flucythrinate.
§ 185.3450 Formetanate hydrochloride.
§ 185.3475 Fumigants for grain-mill machinery.
§ 185.3480 Fumigants for processed grains used in production of fermented malt beverages.
§ 185.3500 Glyphosate.
§ 185.3550 Hexakis.
§ 185.3600 Hydrogen cyanide.
§ 185.3625 Hydroprene; tolerances for residues.
§ 185.3650 Imazalil.
§ 185.3700 Inorganic bromide.
§ 185.3750 Iprodione.
§ 185.3800 Magnesium phosphide.
§ 185.3850 Malathion.
§ 185.3900 Maleic hydrazide.
§ 185.3950 [Removed. 59 FR 33694, June 30, 1994]
§ 185.4000 Metalaxyl.
§ 185.4025 Metaldehyde.
§ 185.4035 Metarhizium anisopliae strain ESF1.
§ 185.4100 Methomyl.
§ 185.4150 Methoprene.
§ 185.4200 1-Methoxycarbonyl-1-propen-2-yl

§ 186.850 2-(m-Chlorophenoxy) propionic acid.
§ 186.950 2-Chloro-1-(2,4,5-trichloro-phenyl)vinyl dimethyl phosphate.
§ 186.1000 Chlorpyrifos.
§ 186.1050 Chlorpyrifos-methyl.
§ 186.1075 Clethodim ((E)-(+/-)-2-[1-[[(3-chloro-2-propenyl)oxy]imino]propyl]-5-[2-(ethylthio)propyl]-3-hydroxy-2-cyclohexen-1- one); tolerances for residues.
§ 186.1100 Clopyralid.
§ 186.1250 Cyfluthrin.
§ 186.1300 Cyano(3-phenoxyphenyl)methyl 4-chloro-alpha-(1-methylethyl)benzeneacetate.
§ 186.1350 Cyhexatin.
§ 186.1400 Cyromazine.
§ 186.1450 2,4-D.
§ 186.1500 Dalapon.
§ 186.1550 [Removed. 55 FR 10222, Mar. 19, 1990].
§ 186.1600 Demeton.
§ 186.1650 Dialifor.
§ 186.1700 Diatomaceous earth.
§ 186.1750 Diazinon.
§ 186.1800 Dicamba.
§ 186.1850 3-(3,5-Dichlorophenyl)-5-ethenyl-5-methyl-2,4-oxazolidinedione.
§ 186.1860 3,7-Dichloro-8-quinoline carboxylic acid.
§ 186.1875 Propanil.
§ 186.1950 O,O-Diethyl S-2-(ethyl-thio)ethyl phosphorodithioate.
§ 186.2000 Diflubenzuron.
§ 186.2050 Dimethipin.
§ 186.2100 Dimethoate including its oxygen analog.
§ 186.2150 2,2-Dimethyl-1,3-benzodioxol-4-ol methylcarbamate.
§ 186.2225 O,O-Dimethyl S-[4-oxo-1,2,3-benzotriazin-3(4H)-ylmethyl] phosphorodithioate.
§ 186.2275 N,N-Dimethylpiperidinium chloride.
§ 186.2325 O,O-Dimethyl 2,2,2-trichloro-1-hydroxyethyl phosphonate.
§ 186.2400 2,4-Dinitro-6-octylphenyl crotonate and 2,6-dinitro-4- octylphenyl crotonate.
§ 186.2450 Dioxathion.
§ 186.2500 Diquat.
§ 186.2550 Diuron.
§ 186.2700 Ethephon.
§ 186.2750 Ethion.
§ 186.2775 Ethofumesate.
§ 186.2800 2-[1-(Ethoxyimino)butyl]-5-[2-(ethylthio)propyl]-3-hydroxy-2- cyclohexene-1-one.
§ 186.2950 Ethyl 3-methyl-4-(methylthio)phenyl (1-methylethyl)- phosphoramidate.
§ 186.3000 O-Ethyl O-[4-(methylthio)phenyl] S-propyl phosphorodithioate.
§ 186.3050 S-[2-(ethylsulfinyl)ethyl] O, O-dimethyl

phosphorothioate.
§ 186.3200 Fenarimol.
§ 186.3225 Fenpropathrin
§ 186.3250 Fluazifop-butyl.
§ 186.3300 Flucythrinate.
§ 186.3350 Fluometuron.
§ 186.3375 Fluorine compounds.
§ 186.3400 (Alpha RS,2R)-fluvalinate [(RS)-alpha-cyano-3phenoxybenzyl (R)-2-[2-chloro-4-(trifluoromethyl) anilino]-3-methylbutanoate].
§ 186.3415 Fluzilazol.A feed additive regulation is established to permit residues of the fungicide fluzilazol, bis(4-fluorophenyl)methyl (1H- 1,2,4-triazol-1-ylmethyl) silane, in or on the following processed feed when present thehrein as a result of application to apples in connection with an experimental use program that expires October 9, 1989:
§ 186.3450 Formetanate hydrochloride.
§ 186.3500 Glyphosate.
§ 186.3550 Hexakis (2-methyl-2-phenylpropyl)distannoxane.
§ 186.3650 Imazalil.
§ 186.3700 Inorganic bromides.
§ 186.3750 Iprodione.
§ 186.3800 Magnesium phosphide.
§ 186.3850 Malathion.
§ 186.4000 Metalaxyl.
§ 186.4035 Metarhizium anisopliae strain ESF1.
§ 186.4050 Methanearsonic acid.
§ 186.4150 Methoprene.
§ 186.4350 Myclobutanil.
§ 186.4450 Norflurazon.
§ 186.4575 Oxamyl.
§ 186.4700 Paraquat.
§ 186.4725 Pentyl 2-chloro-4-fluoro-5-(3,4,5,6-tetrahydrophthalimido) phenoxyacetate.
§ 186.4750 Phorate.
§ 186.4800 Phosalone.
§ 186.4850 Picloram.
§ 186.4900 Piperonyl butoxide.
§ 186.4950 Pirimiphos-methyl.
§ 186.4975 Profenofos.
§ 186.5000 Propargite.
§ 186.5100 Propetamphos.
§ 186.5200 Pyrethrins.
§ 186.5225 Quinclorac.
§ 186.5250 Quizalofop ethyl.
§ 186.5350 Simazine.
§ 186.5400 Synthetic isoparaffinic petroleum hydrocarbons.
§ 186.5550 Thiabendazole.
§ 186.5600 Thidiazuron.
§ 186.5650 Thiodicarb.
§ 186.5700 Thiophanate-methyl.
§ 186.5800 S,S,S-Tributyl phosphorotrithioate.

distance of the noise from locomotive and rail car operations and locomotive load cell test stands.

§ 201.24 Procedures for measurement at a 30 meter (100 feet) distance of the noise from locomotive and rail car operations and locomotive load cell test stands.

§ 201.25 Measurement location and weather conditions for measurement on receiving property of the noise of retarders, car coupling, locomotive load cell test stands, and stationary locomotives.

§ 201.26 Procedures for the measurement on receiving property of retarder and car coupling noise.

§ 201.27 Procedures for: (1) Determining applicability of the locomotive load cell test stand standard and switcher locomotive standard by noise measurement on a receiving property; (2) measurement of locomotive load cell test stands more than 120 meters (400 feet) on a receiving property.

§ 201.28 Testing by railroad to determine probable compliance with the standard.

PART 202--MOTOR CARRIERS ENGAGED IN INTERSTATE COMMERCE

Subpart A--General Provisions

§ 202.10 Definitions.

§ 202.11 Effective date.

§ 202.12 Applicability.

Subpart B--Interstate Motor Carrier Operations Standards

§ 202.20 Standards for highway operations.

§ 202.21 Standard for operation under stationary test.

§ 202.22 Visual exhaust system inspection.

§ 202.23 Visual tire inspection.

PART 203--LOW-NOISE-EMISSION PRODUCTS

§ 203.1 Definitions.

§ 203.2 Application for certification.

§ 203.3 Test procedures.

§ 203.4 Low-noise-emission product determination.

§ 203.5 Suitable substitute decision.

§ 203.6 Contracts for low-noise-emission products.

§ 203.7 Post-certification testing.

§ 203.8 Recertification.

PART 204--NOISE EMISSION STANDARDS FOR CONSTRUCTION EQUIPMENT

Subpart A--General Provisions

§ 204.1 General applicability.

§ 204.2 Definitions.

§ 204.3 Number and gender.

§ 204.4 Inspection and monitoring.

§ 204.5 Exemptions.

§ 204.5-1 Testing exemption.

§ 204.5-2 National security exemptions.

§ 204.5-3 Export exemptions.

Subpart B--Portable Air Compressors

§ 204.50 Applicability.

§ 204.51 Definitions.

§ 204.52 Portable air compressor noise emission standard.

§ 204.54 Test procedures.

§ 223.1 Contents of special, interim, emergency, general and research permits; posting requirements.

Subpart B--Procedures for Revision, Revocation or Limitation of Ocean Dumping Permits Under Section 104(d) of the Act

§ 223.2 Scope of these rules.

§ 223.3 Preliminary determination; notice.

§ 223.4 Request for, scheduling and conduct of public hearing; determination.

§ 223.5 Request for, scheduling and conduct of adjudicatory hearing; determination.

PART 224--RECORDS AND REPORTS REQUIRED OF OCEAN DUMPING PERMITTEES UNDER SECTION 102 OF THE ACT

§ 224.1 Records of permittees.

§ 224.2 Reports.

PART 225--CORPS OF ENGINEERS DREDGED MATERIAL PERMITS

§ 225.1 General.

§ 225.2 Review of Dredged Material Permits.

§ 225.3 Procedure for invoking economic impact.

§ 225.4 Waiver by Administrator.

PART 227--CRITERIA FOR THE EVALUATION OF PERMIT APPLICATIONS FOR OCEAN DUMPING OF MATERIALS

Subpart A--General

§ 227.1 Applicability.

§ 227.2 Materials which satisfy the environmental impact criteria of Subpart B.

§ 227.3 Materials which do not satisfy the environmental impact criteria set forth in Subpart B.

Subpart B--Environmental Impact

§ 227.4 Criteria for evaluating environmental impact.

§ 227.5 Prohibited materials.

§ 227.6 Constituents prohibited as other than trace contaminants.

§ 227.7 Limits established for specific wastes or waste constituents.

§ 227.8 Limitations on the disposal rates of toxic wastes.

§ 227.9 Limitations on quantities of waste materials.

§ 227.10 Hazards to fishing, navigation, shorelines or beaches.

§ 227.11 Containerized wastes.

§ 227.12 Insoluble wastes.

§ 227.13 Dredged materials.

Subpart C--Need for Ocean Dumping

§ 227.14 Criteria for evaluating the need for ocean dumping and alternatives to ocean dumping.

§ 227.15 Factors considered.

§ 227.16 Basis for determination of need for ocean dumping.

Subpart D--Impact of the Proposed Dumping on Esthetic, Recreational and Economic Values

§ 227.17 Basis for determination.

§ 227.18 Factors considered.

§ 227.19 Assessment of impact.

Subpart E--Impact of the Proposed Dumping on Other Uses of the

restrictions on discharge.

Subpart C--Potential Impacts on Physical and Chemical Characteristics of the Aquatic Ecosystem

§ 230.20 Substrate.
§ 230.21 Suspended particulates/turbidity.
§ 230.22 Water.
§ 230.23 Current patterns and water circulation.
§ 230.24 Normal water fluctuations.
§ 230.25 Salinity gradients.

Subpart D--Potential Impacts on Biological Characteristics of the Aquatic Ecosystem

§ 230.30 Threatened and endangered species.
§ 230.31 Fish, crustaceans, mollusks, and other aquatic organisms in the food web.
§ 230.32 Other wildlife.

Subpart E--Potential Impacts on Special Aquatic Sites

§ 230.40 Sanctuaries and refuges.
§ 230.41 Wetlands.
§ 230.42 Mud flats.
§ 230.43 Vegetated shallows.
§ 230.44 Coral reefs.
§ 230.45 Riffle and pool complexes.

Subpart F--Potential Effects on Human Use Characteristics

§ 230.50 Municipal and private water supplies.
§ 230.51 Recreational and commercial fisheries.
§ 230.52 Water-related recreation.
§ 230.53 Aesthetics.
§ 230.54 Parks, national and historical monuments, national seashores, wilderness areas, research sites, and similar preserves.

Subpart G--Evaluation and Testing

§ 230.60 General evaluation of dredged or fill material.
§ 230.61 Chemical, biological, and physical evaluation and testing.

Subpart H--Actions To Minimize Adverse Effects

§ 230.70 Actions concerning the location of the discharge.
§ 230.71 Actions concerning the material to be discharged.
§ 230.72 Actions controlling the material after discharge.
§ 230.73 Actions affecting the method of dispersion.
§ 230.74 Actions related to technology.
§ 230.75 Actions affecting plant and animal populations.
§ 230.76 Actions affecting human use.
§ 230.77 Other actions.

Subpart I--Planning To Shorten Permit Processing Time

§ 230.80 Advanced identification of disposal areas.

PART 231--SECTION 404(c) PROCEDURES

§ 231.1 Purpose and scope.
§ 231.2 Definitions.
§ 231.3 Procedures for proposed determinations.
§ 231.4 Public comments and hearings.
§ 231.5 Recommended determination.
§ 231.6 Administrator's final determinations.
§ 231.7 Emergency procedure.

§ 249.31 Recommendations for documentation.
§ 249.32 Quality control.
§ 249.33 Date recommendations.
PART 250--GUIDELINE FOR FEDERAL PROCUREMENT OF PAPER AND PAPER PRODUCTS CONTAINING RECOVERED MATERIALS
Subpart A--General
§ 250.1 Purpose.
§ 250.2 Designation.
§ 250.3 Applicability.
§ 250.4 Definitions.
Subpart B--Revisions and Additions to Paper and Paper Product Specifications
§ 250.10 Introduction.
§ 250.11 Elimination of recovered materials exclusion.
§ 250.12 Requirement of recovered materials content.
§ 250.13 Exclusion of products containing recovered materials that do not meet reasonable performance standards.
§ 250.14 New specifications.
Subpart C--Affirmative Procurement Program
§ 250.20 General.
§ 250.21 Recovered materials preference program.
§ 250.22 Promotion program.
§ 250.23 Estimates, certification, and verification.
§ 250.24 Annual review and monitoring.
§ 250.25 Implementation.
PART 252--GUIDELINE FOR FEDERAL PROCUREMENT OF LUBRICATING OILS CONTAINING RE-REFINED OIL
Subpart A--General
§ 252.1 Purpose.
§ 252.2 Designation.
§ 252.3 Applicability.
§ 252.4 Definitions.
Subpart B--Specifications
§ 252.10 Revisions.
§ 252.11 Recommendations.
Subpart C--Affirmative Procurement Program
§ 252.20 General.
§ 252.21 Preference program.
§ 252.22 Promotion program.
§ 252.23 Estimates, certification, and verification.
§ 252.24 Annual review and monitoring.
§ 252.25 Implementation.
PART 253--GUIDELINE FOR FEDERAL PROCUREMENT OF RETREAD TIRES
Subpart A--General
§ 253.1 Purpose.
§ 253.2 Designation.
§ 253.3 Applicability.
§ 253.4 Definitions.
Subpart B--Specifications
§ 253.10 Revisions.
§ 253.11 Exclusions.
Subpart C--Affirmative Procurement Program

PART 259--STANDARDS FOR THE TRACKING AND MANAGEMENT OF MEDICAL WASTE

Subpart A--General

§ 259.1 Purpose, scope, and applicability.

§ 259.2 Effective dates and duration of the demonstration program.

Subpart B--Definitions

§ 259.10 Definitions.

Subpart C--Covered States

Subpart C--Covered States

§ 259.20 States included in the demonstration program.

Subpart D--Regulated Medical Waste

§ 259.30 Definition of regulated medical waste.

§ 259.31 Mixtures.

Subpart E--Pre-Transport Requirements

§ 259.39 Applicability.

§ 259.40 Segregation requirements.

§ 259.41 Packaging requirements.

§ 259.42 Storage of regulated medical waste prior to transport, treatment, destruction, or disposal.

§ 259.43 Decontamination standards for reusable containers.

§ 259.44 Labeling requirements.

§ 259.45 Marking (identification) requirements.

Subpart F--Generator Standards

§ 259.50 Applicability and general requirements.

§ 259.51 Exemptions.

§ 259.52 Use of the tracking form.

§ 259.53 Generators exporting regulated medical waste.

§ 259.54 Recordkeeping.

§ 259.55 Exception Reporting.

§ 259.56 Additional Reporting.

Subpart G--On-Site Incinerators

§ 259.60 Applicability.

§ 259.61 Recordkeeping.

§ 259.62 Reporting.

Subpart H--Transporter Requirements

§ 259.70 Applicability.

§ 259.71 Transporter acceptance of regulated medical waste.

§ 259.72 Transporter notification.

§ 259.73 Vehicle requirements.

§ 259.74 Tracking form requirements.

§ 259.75 Compliance with the tracking form.

§ 259.76 Consolidating or remanifesting waste to a new tracking form.

§ 259.77 Recordkeeping.

§ 259.78 Reporting.

§ 259.79 Additional reporting.

Subpart I--Treatment, Destruction, and Disposal Facilities

§ 259.80 Applicability.

§ 259.81 Use of the tracking form.

§ 259.82 Tracking form discrepancies.

§ 259.83 Recordkeeping.

§ 262.50 Applicability.
§ 262.51 Definitions.
§ 262.52 General requirements.
§ 262.53 Notification of intent to export.
§ 262.54 Special manifest requirements.
§ 262.55 Exception reports.
§ 262.56 Annual reports.
§ 262.57 Recordkeeping.
§ 262.58 International agreements. [(Reserved)]
Subpart F--Imports of Hazardous Waste
§ 262.60 Imports of hazardous waste.
Subpart G--Farmers
§ 262.70 Farmers.
Appendix--Uniform Hazardous Waste Manifest and Instructions (EPA Forms 8700- 22 and 8700-22A and Their Instructions)
PART 263--STANDARDS APPLICABLE TO TRANSPORTERS OF HAZARDOUS WASTE
Subpart A--General
§ 263.10 Scope.
§ 263.11 EPA identification number.
§ 263.12 Transfer facility requirements.
Subpart B--Compliance With the Manifest System and Recordkeeping
§ 263.20 The manifest system.
§ 263.21 Compliance with the manifest.
§ 263.22 Recordkeeping.
Subpart C--Hazardous Waste Discharges
§ 263.30 Immediate action.
§ 263.31 Discharge clean up.
PART 264--STANDARDS FOR OWNERS AND OPERATORS OF HAZARDOUS WASTE TREATMENT, STORAGE, AND DISPOSAL FACILITIES
Subpart A--General
§ 264.1 Purpose, scope and applicability.
§ 264.2 [Reserved]
§ 264.3 Relationship to interim status standards.
§ 264.4 Imminent hazard action.
Subpart B--General Facility Standards
§ 264.10 Applicability.
§ 264.11 Identification number.
§ 264.12 Required notices.
§ 264.13 General waste analysis.
§ 264.14 Security.
§ 264.15 General inspection requirements.
§ 264.16 Personnel training.
§ 264.17 General requirements for ignitable, reactive, or incompatible wastes.
§ 264.18 Location standards.
§ 264.19 Construction quality assurance program.
Subpart C--Preparedness and Prevention
§ 264.30 Applicability.
§ 264.31 Design and operation of facility.
§ 264.32 Required equipment.
§ 264.33 Testing and maintenance of equipment.
§ 264.34 Access to communications or alarm system.
§ 264.35 Required aisle space.

§ 264.36 [Reserved]
§ 264.37 Arrangements with local authorities.
Subpart D--Contingency Plan and Emergency Procedures
§ 264.50 Applicability.
§ 264.51 Purpose and implementation of contingency plan.
§ 264.52 Content of contingency plan.
§ 264.53 Copies of contingency plan.
§ 264.54 Amendment of contingency plan.
§ 264.55 Emergency coordinator.
§ 264.56 Emergency procedures.
Subpart E--Manifest System, Recordkeeping, and Reporting
§ 264.70 Applicability.
§ 264.71 Use of manifest system.
§ 264.72 Manifest discrepancies.
§ 264.73 Operating record.
§ 264.74 Availability, retention, and disposition of records.
§ 264.75 Biennial report.
§ 264.76 Unmanifested waste report.
§ 264.77 Additional reports.
Subpart F--Releases From Solid Waste Management Units
§ 264.90 Applicability.
§ 264.91 Required programs.
§ 264.92 Ground-water protection standard.
§ 264.93 Hazardous constituents.
§ 264.94 Concentration limits.
§ 264.95 Point of compliance.
§ 264.96 Compliance period.
§ 264.97 General ground-water monitoring requirements.
§ 264.98 Detection monitoring program.
§ 264.99 Compliance monitoring program.
§ 264.100 Corrective action program.
§ 264.101 Corrective action for solid waste management units.
Subpart G--Closure and Post-Closure
§ 264.110 Applicability.
§ 264.111 Closure performance standard.
§ 264.112 Closure plan; amendment of plan.
§ 264.113 Closure; time allowed for closure.
§ 264.114 Disposal or decontamination of equipment, structures and soils.
§ 264.115 Certification of closure.
§ 264.116 Survey plat.
§ 264.117 Post-closure care and use of property.
§ 264.118 Post-closure plan; amendment of plan.
§ 264.119 Post-closure notices.
§ 264.120 Certification of completion of post-closure care.
Subpart H--Financial Requirements
§ 264.140 Applicability.
§ 264.141 Definitions of terms as used in this subpart.
§ 264.142 Cost estimate for closure.
§ 264.143 Financial assurance for closure.

§ 264.144 Cost estimate for post-closure care.
§ 264.145 Financial assurance for post-closure care.
§ 264.146 Use of a mechanism for financial assurance of both closure and post-closure care.
§ 264.147 Liability requirements.
§ 264.148 Incapacity of owners or operators, guarantors, or financial institutions.
§ 264.149 Use of State-required mechanisms.
§ 264.150 State assumption of responsibility.
§ 264.151 Wording of the instruments.

Subpart I--Use and Management of Containers

§ 264.170 Applicability.
§ 264.171 Condition of containers.
§ 264.172 Compatibility of waste with containers.
§ 264.173 Management of containers.
§ 264.174 Inspections.
§ 264.175 Containment.
§ 264.176 Special requirements for ignitable or reactive waste.
§ 264.177 Special requirements for incompatible wastes.
§ 264.178 Closure.

Subpart J--Tank Systems

§ 264.190 Applicability.
§ 264.191 Assessment of existing tank system's integrity.
§ 264.192 Design and installation of new tank systems or components.
§ 264.193 Containment and detection of releases.
§ 264.194 General operating requirements.
§ 264.195 Inspections.
§ 264.196 Response to leaks or spills and disposition of leaking or unfit-for-use tank systems.
§ 264.197 Closure and post-closure care.
§ 264.198 Special requirements for ignitable or reactive wastes.
§ 264.199 Special requirements for incompatible wastes.

Subpart K--Surface Impoundments

§ 264.220 Applicability.
§ 264.221 Design and operating requirements.
§ 264.222 Action leakage rate.
§ 264.223 Response actions.
§ 264.224 [Reserved]
§ 264.225 [Reserved]
§ 264.226 Monitoring and inspection.
§ 264.227 Emergency repairs; contingency plans.
§ 264.228 Closure and post-closure care.
§ 264.229 Special requirements for ignitable or reactive waste.
§ 264.230 Special requirements for incompatible wastes.
§ 264.231 Special requirements for hazardous wastes FO20, FO21, FO22, FO23, FO26, and FO27.

Subpart L--Waste Piles

§ 264.250 Applicability.
§ 264.251 Design and operating requirements.

§ 264.252 Action leakage rate.
§ 264.253 Response actions.
§ 264.254 Monitoring and inspection.
§ 264.255 [Reserved]
§ 264.256 Special requirements for ignitable or reactive waste.
§ 264.257 Special requirements for incompatible wastes.
§ 264.258 Closure and post-closure care.
§ 264.259 Special requirements for hazardous wastes FO20, FO21, FO22, FO23, FO26, and FO27.

Subpart M--Land Treatment

§ 264.270 Applicability.
§ 264.271 Treatment program.
§ 264.272 Treatment demonstration.
§ 264.273 Design and operating requirements.
Secs. 264.274--264.275 [Reserved]
§ 264.276 Food-chain crops.
§ 264.277 [Reserved]
§ 264.278 Unsaturated zone monitoring.
§ 264.279 Recordkeeping.
§ 264.280 Closure and post-closure care.
§ 264.281 Special requirements for ignitable or reactive waste.
§ 264.282 Special requirements for incompatible wastes.
§ 264.283 Special requirements for hazardous wastes FO20, FO21, FO22, FO23, FO26, and FO27.

Subpart N--Landfills

§ 264.300 Applicability.
§ 264.301 Design and operating requirements.
§ 264.302 Action leakage rate.
§ 264.303 Monitoring and inspection.
§ 264.304 Response actions.
§ 264.305--264.308 [Reserved]
§ 264.309 Surveying and recordkeeping.
§ 264.310 Closure and post-closure care.
§ 264.311 [Reserved]
§ 264.312 Special requirements for ignitable or reactive waste.
§ 264.313 Special requirements for incompatible wastes.
§ 264.314 Special requirements for bulk and containerized liquids.
§ 264.315 Special requirements for containers.
§ 264.316 Disposal of small containers of hazardous waste in overpacked drums (lab packs).
§ 264.317 Special requirements for hazardous wastes FO20, FO21, FO22, FO23, FO26, and FO27.

Subpart O--Incinerators

§ 264.340 Applicability.
§ 264.341 Waste analysis.
§ 264.342 Principal organic hazardous constituents (POHCs).
§ 264.343 Performance standards.
§ 264.344 Hazardous waste incinerator permits.

§ 264.345 Operating requirements.
§ 264.346 [Reserved]
§ 264.347 Monitoring and inspections.
Secs. 264.348--264.350 [Reserved]
§ 264.351 Closure.

Subparts P-R--[Reserved]

Subpart S--Corrective Action for Solid Waste Management Units

§ 264.552 Corrective Action Management Units (CAMU).
§ 264.553 Temporary Units (TU).

Subparts T-V--[Reserved]

Subpart W--Drip Pads

§ 264.570 Applicability.
§ 264.571 Assessment of existing drip pad integrity.
§ 264.572 Design and installation of new drip pads.
§ 264.573 Design and operating requirements.
§ 264.574 Inspections.
§ 264.575 Closure.

Subpart X--Miscellaneous Units

§ 264.600 Applicability.
§ 264.601 Environmental performance standards.
§ 264.602 Monitoring, analysis, inspection, response, reporting, and corrective action.
§ 264.603 Post-closure care.

Subpart AA--Air Emission Standards for Process Vents

§ 264.1030 Applicability.
§ 264.1031 Definitions.
§ 264.1032 Standards: Process vents.
§ 264.1033 Standards: Closed-vent systems and control devices.
§ 264.1034 Test methods and procedures.
§ 264.1035 Recordkeeping requirements.
§ 264.1036 Reporting requirements.
Secs. 264.1037-264.1049 [Reserved].

Subpart BB--Air Emission Standards for Equipment Leaks

§ 264.1050 Applicability.
§ 264.1051 Definitions.
§ 264.1052 Standards: Pumps in light liquid service.
§ 264.1053 Standards: Compressors.
§ 264.1054 Standards: Pressure relief devices in gas/vapor service.
§ 264.1055 Standards: Sampling connecting systems.
§ 264.1056 Standards: Open-ended valves or lines.
§ 264.1057 Standards: Valves in gas/vapor service or in light liquid service.
§ 264.1058 Standards: Pumps and valves in heavy liquid service, pressure relief devices in light liquid or heavy liquid service, and flanges and other connectors.
§ 264.1059 Standards: Delay of repair.
§ 264.1060 Standards: Closed-vent systems and control devices.
§ 264.1061 Alternative standards for valves in gas/vapor service or in light liquid service: percentage of valves allowed to leak.

§ 265.54 Amendment of contingency plan.
§ 265.55 Emergency coordinator.
§ 265.56 Emergency procedures.
Subpart E--Manifest System, Recordkeeping, and Reporting
§ 265.70 Applicability.
§ 265.71 Use of manifest system.
§ 265.72 Manifest discrepancies.
§ 265.73 Operating record.
§ 265.74 Availability, retention, and disposition of records.
§ 265.75 Biennial report.
§ 265.76 Unmanifested waste report.
§ 265.77 Additional reports.
Subpart F--Ground-Water Monitoring
§ 265.90 Applicability.
§ 265.91 Ground-water monitoring system.
§ 265.92 Sampling and analysis.
§ 265.93 Preparation, evaluation, and response.
§ 265.94 Recordkeeping and reporting.
Subpart G--Closure and Post-Closure
§ 265.110 Applicability.
§ 265.111 Closure performance standard.
§ 265.112 Closure plan; amendment of plan.
§ 265.113 Closure; time allowed for closure.
§ 265.114 Disposal or decontamination of equipment, structures and soils.
§ 265.115 Certification of closure.
§ 265.116 Survey plat.
§ 265.117 Post-closure care and use of property.
§ 265.118 Post-closure plan; amendment of plan.
§ 265.119 Post-closure notices.
§ 265.120 Certification of completion of post-closure care.
Subpart H--Financial Requirements
§ 265.140 Applicability.
§ 265.141 Definitions of terms as used in this subpart.
§ 265.142 Cost estimate for closure.
§ 265.143 Financial assurance for closure.
§ 265.144 Cost estimate for post-closure care.
§ 265.145 Financial assurance for post-closure care.
§ 265.146 Use of a mechanism for financial assurance of both closure and post-closure care.
§ 265.147 Liability requirements.
§ 265.148 Incapacity of owners or operators, guarantors, or financial institutions.
§ 265.149 Use of State-required mechanisms.
§ 265.150 State assumption of responsibility.
Subpart I--Use and Management of Containers
§ 265.170 Applicability.
§ 265.171 Condition of containers.
§ 265.172 Compatibility of waste with container.
§ 265.173 Management of containers.
§ 265.174 Inspections.

§ 265.175 [Reserved]

§ 265.176 Special requirements for ignitable or reactive waste.

§ 265.177 Special requirements for incompatible wastes.

Subpart J--Tank Systems

§ 265.190 Applicability.

§ 265.191 Assessment of existing tank system's integrity.

§ 265.192 Design and installation of new tank systems or components.

§ 265.193 Containment and detection of releases.

§ 265.194 General operating requirements.

§ 265.195 Inspections.

§ 265.196 Response to leaks or spills and disposition of leaking or unfit-for-use tank systems.

§ 265.197 Closure and post-closure care.

§ 265.198 Special requirements for ignitable or reactive wastes.

§ 265.199 Special requirements for incompatible wastes.

§ 265.200 Waste analysis and trial tests.

§ 265.201 Special requirements for generators of between 100 and 1,000 kg/mo that accumulate hazardous waste in tanks.

Subpart K--Surface Impoundments

§ 265.220 Applicability.

§ 265.221 Design and operating requirements.

§ 265.222 Action leakage rate.

§ 265.223 Containment system.

§ 265.224 Response actions.

§ 265.225 Waste analysis and trial tests.

§ 265.226 Monitoring and inspection.

§ 265.227 [Reserved]

§ 265.228 Closure and post-closure care.

§ 265.229 Special requirements for ignitable or reactive waste.

§ 265.230 Special requirements for incompatible wastes.

Subpart L--Waste Piles

§ 265.250 Applicability.

§ 265.251 Protection from wind.

§ 265.252 Waste analysis.

§ 265.253 Containment.

§ 265.254 Design and operating requirements.

§ 265.255 Action leakage rates

§ 265.256 Special requirements for ignitable or reactive waste.

§ 265.257 Special requirements for incompatible wastes.

§ 265.258 Closure and post-closure care.

§ 265.259 Response actions.

§ 265.260 Monitoring and inspection.

Subpart M--Land Treatment

§ 265.270 Applicability.

§ 265.271 [Reserved]

§ 265.272 General operating requirements.

§ 265.273 Waste analysis.

Secs. 265.274--265.275 [Reserved]
§ 265.276 Food chain crops.
§ 265.277 [Reserved]
§ 265.278 Unsaturated zone (zone of aeration) monitoring.
§ 265.279 Recordkeeping.
§ 265.280 Closure and post-closure.
§ 265.281 Special requirements for ignitable or reactive waste.
§ 265.282 Special requirements for incompatible wastes.

Subpart N--Landfills

§ 265.300 Applicability.
§ 265.301 Design and operating requirements.
§ 265.302 Action leakage rate.
§ 265.303 Response actions.
§ 265.304 Monitoring and inspection.
§ 265.305--265.308 [Reserved]
§ 265.309 Surveying and recordkeeping.
§ 265.310 Closure and post-closure care.
§ 265.311 [Reserved]
§ 265.312 Special requirements for ignitable or reactive waste.
§ 265.313 Special requirements for incompatible wastes.
§ 265.314 Special requirements for bulk and containerized liquids.
§ 265.315 Special requirements for containers.
§ 265.316 Disposal of small containers of hazardous waste in overpacked drums (lab packs).

Subpart O--Incinerators

§ 265.340 Applicability.
§ 265.341 Waste analysis.
Secs. 265.342--265.344 [Reserved]
§ 265.345 General operating requirements.
§ 265.346 [Reserved]
§ 265.347 Monitoring and inspections.
Secs. 265.348--265.350 [Reserved]
§ 265.351 Closure.
§ 265.352 Interim status incinerators burning particular hazardous wastes.
Secs. 265.353--265.369 [Reserved]

Subpart P--Thermal Treatment

§ 265.370 Other thermal treatment.
Secs. 265.371--265.372 [Reserved]
§ 265.373 General operating requirements.
§ 265.374 [Reserved]
§ 265.375 Waste analysis.
§ 265.376 [Reserved]
§ 265.377 Monitoring and inspections.
Secs. 265.378--265.380 [Reserved]
§ 265.381 Closure.
§ 265.382 Open burning; waste explosives.
§ 265.383 Interim status thermal treatment devices burning particular hazardous waste.

Subpart Q--Chemical, Physical, and Biological Treatment

Subpart DD--Containment Buildings
§ 265.1100 Applicability.
§ 265.1101 Design and operating standards.
§ 265.1102 Closure and post-closure care.
Secs. 265.1103--265.1110 [Reserved]
Appendix I--Recordkeeping Instructions
Appendix II--[Reserved]
Appendix III--EPA Interim Primary Drinking Water Standards
Appendix IV--Tests for Significance
Appendix V--Examples of Potentially Incompatible Waste

PART 266--STANDARDS FOR THE MANAGEMENT OF SPECIFIC HAZARDOUS WASTES AND SPECIFIC TYPES OF HAZARDOUS WASTE MANAGEMENT FACILITIES

Subparts A--B [Reserved]
Subpart C--Recyclable Materials Used in a Manner Constituting Disposal
§ 266.20 Applicability.
§ 266.21 Standards applicable to generators and transporters of materials used in a manner that constitute disposal.
§ 266.22 Standards applicable to storers of materials that are to be used in a manner that constitutes disposal who are not the ultimate users.
§ 266.23 Standards applicable to users of materials that are used in a manner that constitutes disposal.
[Subpart D--Reserved. 56 FR 7208, Feb. 21, 1991]
Subpart E--[Reserved. 57 FR 41612, Sept. 10, 1992]
Subpart F--Recyclable Materials Utilized for Precious Metal Recovery
§ 266.70 Applicability and requirements.
Subpart G--Spent Lead-Acid Batteries Being Reclaimed
§ 266.80 Applicability and requirements.
Subpart H--Hazardous Waste Burned in Boilers and Industrial Furnaces
§ 266.100 Applicability.
§ 266.101 Management prior to burning.
§ 266.102 Permit standards for burners.
§ 266.103 Interim status standards for burners.
§ 266.104 Standards to control organic emissions.
§ 266.105 Standards to control particulate matter.
§ 266.106 Standards to control metals emissions.
§ 266.107 Standards to control hydrogen chloride (HCl) and chlorine gas (Cl2) emissions.
§ 266.108 Small quantity on-site burner exemption.
§ 266.109 Low risk waste exemption.
§ 266.110 Waiver of DRE trial burn for boilers.
§ 266.111 Standards for direct transfer.
§ 266.112 Regulation of residues.
Appendix I.--Tier I and Tier II Feed Rate and Emissions Screening Limits for Metals
Appendix II.--Tier I Feed Rate Screening Limits for Total Chlorine
Appendix III.--Tier II Emission Rate Screening Limits for Free Chlorine and Hydrogen Chloride

§ 271.7 Attorney General's statement.

§ 271.8 Memorandum of Agreement with the Regional Administrator.

§ 271.9 Requirements for identification and listing of hazardous wastes.

§ 271.10 Requirements for generators of hazardous wastes.

§ 271.11 Requirements for transporters of hazardous wastes.

§ 271.12 Requirements for hazardous waste management facilities.

§ 271.13 Requirements with respect to permits and permit applications.

§ 271.14 Requirements for permitting.

§ 271.15 Requirements for compliance evaluation programs.

§ 271.16 Requirements for enforcement authority.

§ 271.17 Sharing of information.

§ 271.18 Coordination with other programs.

§ 271.19 EPA review of State permits.

§ 271.20 Approval process.

§ 271.21 Procedures for revision of State programs.

§ 271.22 Criteria for withdrawing approval of State programs.

§ 271.23 Procedures for withdrawing approval of State programs.

§ 271.24 Interim authorization under section 3006(g) of RCRA.

§ 271.25 HSWA requirements.

§ 271.26 Requirements for used oil management.

Subpart B--Requirements for Interim Authorization

§ 271.121 Purpose and scope.

§ 271.122 Schedule.

§ 271.123 Elements of a program submission.

§ 271.124 Program description.

§ 271.125 Attorney General's statement.

§ 271.126 Memorandum of Agreement with the Regional Administrator.

§ 271.127 Authorization plan.

§ 271.128 Program requirements for interim authorization for Phase I.

§ 271.129 Additional program requirements for interim authorization for Phase II.

§ 271.130 Interstate movement of hazardous waste.

§ 271.131 Progress reports.

§ 271.132 Sharing of information.

§ 271.133 Coordination with other programs.

§ 271.134 EPA review of State permits.

§ 271.135 Approval process.

§ 271.136 Withdrawal of State programs.

§ 271.137 Reversion of State programs.

§ 271.138 Interim authorization under section 3006(g) of RCRA.

PART 272--APPROVED STATE HAZARDOUS WASTE MANAGEMENT PROGRAMS

Subpart A--General Provisions

§ 279.12 Prohibitions.
§ 279.20 Applicability.
§ 279.21 Hazardous waste mixing.
§ 279.22 Used oil storage.
§ 279.23 On-site burning in space heaters.
§ 279.24 Off-site shipments.

Subpart D--Standards for Used Oil Collection Centers and Aggregation Points

§ 279.30 Do-it-yourselfer used oil collection centers.
§ 279.31 Used oil collection centers.
§ 279.32 Used oil aggregation points owned by the generator.

Subpart E--Standards for Used Oil Transporter and Transfer Facilities

§ 279.40 Applicability.
§ 279.41 Restrictions on transporters who are not also processors or re- refiners.
§ 279.42 Notification.
§ 279.43 Used oil transportation.
§ 279.44 Rebuttable presumption for used oil.
§ 279.45 Used oil storage at transfer facilities.
§ 279.46 Tracking.
§ 279.47 Management of residues.

Subpart F--Standards for Used Oil Processors and Re-Refiners

§ 279.50 Applicability.
§ 279.51 Notification.
§ 279.52 General facility standards.
§ 279.53 Rebuttable presumption for used oil.
§ 279.54 Used oil management.
§ 279.55 Analysis plan.
§ 279.56 Tracking.
§ 279.57 Operating record and reporting.
§ 279.58 Off-site shipments of used oil.
§ 279.59 Management of residues.

Subpart G--Standards for Used Oil Burners Who Burn Off-Specification Used Oil for Energy Recovery

§ 279.60 Applicability.
S%9m:;Kr2
Wk.Z,]ZKo4K<}9ning.
§ 279.62 Notification
§ 279.63 Rebuttable presumption for used oil.
§ 279.64 Used oil storage.
§ 279.65 Tracking.
§ 279.66 Notices.
§ 279.67 Management of residues.

Subpart H--Standards for Used Oil Fuel Marketers

§ 279.70 Applicability.
§ 279.71 Prohibitions.
§ 279.72 On-specification used oil fuel.
§ 279.73 Notification.
§ 279.74 Tracking.
§ 279.75 Notices.

Subpart I--Standards for Use as a Dust Suppressant and Disposal

§ 300.105 General organization concepts.
§ 300.110 National Response Team.
§ 300.115 Regional Response Teams.
§ 300.120 On-scene coordinators and remedial project managers: general responsibilities.
§ 300.125 Notification and communications.
§ 300.130 Determinations to initiate response and special conditions.
§ 300.135 Response operations.
§ 300.140 Multi-regional responses.
§ 300.145 Special teams and other assistance available to OSCs/RPMs.
§ 300.150 Worker health and safety.
§ 300.155 Public information and community relations.
§ 300.160 Documentation and cost recovery.
§ 300.165 OSC reports.
§ 300.170 Federal agency participation.
§ 300.175 Federal agencies: additional responsibilities and assistance.
§ 300.180 State and local participation in response.
§ 300.185 Nongovernmental participation.

Subpart C--Planning and Preparedness

§ 300.200 General.
§ 300.205 Planning and coordination structure.
§ 300.210 Federal contingency plans.
§ 300.215 Title III local emergency response plans.
§ 300.220 Related Title III issues.

Subpart D--Operational Responses Phases for Oil Removal

§ 300.300 Phase I--Discovery or notification.
§ 300.305 Phase II--Preliminary assessment and initiation of action.
§ 300.310 Phase III--Containment, countermeasures, cleanup, and disposal.
§ 300.315 Phase IV--Documentation and cost recovery.
§ 300.320 General pattern of response.
§ 300.330 Wildlife conservation.
§ 300.335 Funding.

Subpart E--Hazardous Substance Response

§ 300.400 General.
§ 300.405 Discovery or notification.
§ 300.410 Removal site evaluation.
§ 300.415 Removal action.
§ 300.420 Remedial site evaluation.
§ 300.425 Establishing remedial priorities.
§ 300.430 Remedial investigation/feasibility study and selection of remedy.
§ 300.435 Remedial design/remedial action, operation and maintenance.
§ 300.440 Procedures for planning and implementing off-site response actions.

Subpart F--State Involvement in Hazardous Substance Response

§ 300.500 General.
§ 300.505 EPA/State Superfund Memorandum of Agreement

available.
§ 405.63 [Reserved]
§ 405.64 Pretreatment standards for existing sources.
§ 405.65 Standards of performance for new sources.
§ 405.66 Pretreatment standards for new sources.
§ 405.67 Effluent limitations guidelines representing the degree of effluent reduction attainable by the application of the best conventional pollutant control technology (BCT).

Subpart G--Fluid Mix for Ice Cream and Other Frozen Desserts Subcategory

§ 405.70 Applicability; description of the fluid mix for ice cream and other frozen desserts subcategory.
§ 405.71 Specialized definitions.
§ 405.72 Effluent limitations guidelines representing the degree of effluent reduction attainable by the application of the best practicable control technology currently available.
§ 405.73 [Reserved]
§ 405.74 Pretreatment standards for existing sources.
§ 405.75 Standards of performance for new sources.
§ 405.76 Pretreatment standards for new sources.
§ 405.77 Effluent limitations guidelines representing the degree of effluent reduction attainable by the application of the best conventional pollutant control technology (BCT).

Subpart H--Ice Cream, Frozen Desserts, Novelties and Other Dairy Desserts Subcategory

§ 405.80 Applicability; description of the ice cream, frozen desserts, novelties and other dairy desserts subcategory.
§ 405.81 Specialized definitions.
§ 405.82 Effluent limitations guidelines representing the degree of effluent reduction attainable by the application of the best practicable control technology currently available.
§ 405.83 [Reserved]
§ 405.84 Pretreatment standards for existing sources.
§ 405.85 Standards of performance for new sources.
§ 405.86 Pretreatment standards for new sources.
§ 405.87 Effluent limitations guidelines representing the degree of effluent reduction attainable by the application of the best conventional pollutant control technology (BCT).

Subpart I--Condensed Milk Subcategory

§ 405.90 Applicability; description of the condensed milk subcategory.
§ 405.91 Specialized definitions.
§ 405.92 Effluent limitations guidelines representing the degree of effluent reduction attainable by the application of the best practicable control technology currently available.
§ 405.93 [Reserved]

§ 405.94 Pretreatment standards for existing sources.
§ 405.95 Standards of performance for new sources.
§ 405.96 Pretreatment standards for new sources.
§ 405.97 Effluent limitations guidelines representing the degree of effluent reduction attainable by the application of the best conventional pollutant control technology (BCT).

Subpart J--Dry Milk Subcategory

§ 405.100 Applicability; description of the dry milk subcategory.
§ 405.101 Specialized definitions.
§ 405.102 Effluent limitations guidelines representing the degree of effluent reduction attainable by the application of the best practicable control technology currently available.
§ 405.103 [Reserved]
§ 405.104 Pretreatment standards for existing sources.
§ 405.105 Standards of performance for new sources.
§ 405.106 Pretreatment standards for new sources.
§ 405.107 Effluent limitations guidelines representing the degree of effluent reduction attainable by the application of the best conventional pollutant control technology (BCT).

Subpart K--Condensed Whey Subcategory

§ 405.110 Applicability; description of the condensed whey subcategory.
§ 405.111 Specialized definitions.
§ 405.112 Effluent limitations guidelines representing the degree of effluent reduction attainable by the application of the best practicable control technology currently available.
§ 405.113 [Reserved]
§ 405.114 Pretreatment standards for existing sources.
§ 405.115 Standards of performance for new sources.
§ 405.116 Pretreatment standards for new sources.
§ 405.117 Effluent limitations guidelines representing the degree of effluent reduction attainable by the application of the best conventional pollutant control technology (BCT).

Subpart L--Dry Whey Subcategory

§ 405.120 Applicability; description of the dry whey subcategory.
§ 405.121 Specialized definitions.
§ 405.122 Effluent limitations guidelines representing the degree of effluent reduction attainable by the application of the best practicable control technology currently available.
§ 405.123 [Reserved]
§ 405.124 Pretreatment standards for existing sources.
§ 405.125 Standards of performance for new sources.
§ 405.126 Pretreatment standards for new sources.
§ 405.127 Effluent limitations guidelines representing the degree of effluent reduction attainable by the application

§ 406.40 Applicability; description of the bulgur wheat flour milling subcategory.

§ 406.41 Specialized definitions.

§ 406.42 Effluent limitations guidelines representing the degree of effluent reduction attainable by the application of the best practicable control technology currently available.

§ 406.43 [Reserved]

§ 406.44 Pretreatment standards for existing sources.

§ 406.45 Standards of performance for new sources.

§ 406.46 Pretreatment standards for new sources.

§ 406.47 Effluent limitations guidelines representing the degree of effluent reduction attainable by the application of the best conventional pollutant control technology (BCT).

Subpart E--Normal Rice Milling Subcategory

§ 406.50 Applicability; description of the normal rice milling subcategory.

§ 406.51 Specialized definitions.

§ 406.52 Effluent limitations guidelines representing the degree of effluent reduction attainable by the application of the best practicable control technology currently available.

§ 406.53 Effluent limitations guidelines representing the degree of effluent reduction attainable by the application of the best available technology economically achievable.

§ 406.54 Pretreatment standards for existing sources.

§ 406.55 Standards of performance for new sources.

§ 406.56 Pretreatment standards for new sources.

§ 406.57 Effluent limitations guidelines representing the degree of effluent reduction attainable by the application of the best conventional pollutant control technology (BCT).

Subpart F--Parboiled Rice Processing Subcategory

§ 406.60 Applicability; description of the parboiled rice processing subcategory.

§ 406.61 Specialized definitions.

§ 406.62 Effluent limitations guidelines representing the degree of effluent reduction attainable by the application of the best practicable control technology currently available.

§ 406.63 [Reserved]

§ 406.64 Pretreatment standards for existing sources.

§ 406.65 Standards of performance for new sources.

§ 406.66 Pretreatment standards for new sources.

§ 406.67 Effluent limitations guidelines representing the degree of effluent reduction attainable by the application of the best conventional pollutant control technology (BCT).

Subpart G--Animal Feed Subcategory

§ 406.70 Applicability; description of the animal feed subcategory.

§ 406.71 Specialized definitions.

§ 406.72 Effluent limitations guidelines representing the degree of effluent reduction attainable by the application of the best practicable control technology currently available.
§ 406.73 Effluent limitations guidelines representing the degree of effluent reduction attainable by the application of the best available technology economically achievable.
§ 406.74 [Reserved]
§ 406.75 Standards of performance for new sources.
§ 406.76 Pretreatment standards for new sources.
§ 406.77 Effluent limitations guidelines representing the degree of effluent reduction attainable by the application of the best conventional pollutant control technology (BCT).

Subpart H--Hot Cereal Subcategory
§ 406.80 Applicability; description of the hot cereal subcategory.
§ 406.81 Specialized definitions.
§ 406.82 Effluent limitations guidelines representing the degree of effluent reduction attainable by the application of the best practicable control technology currently available.
§ 406.83 Effluent limitations guidelines representing the degree of effluent reduction attainable by the application of the best available technology economically achievable.
§ 406.84 [Reserved]
§ 406.85 Standards of performance for new sources.
§ 406.86 Pretreatment standards for new sources.
§ 406.87 Effluent limitations guidelines representing the degree of effluent reduction attainable by the application of the best conventional pollutant control technology (BCT).

Subpart I--Ready-To-Eat Cereal Subcategory
§ 406.90 Applicability; description of the ready-to-eat cereal subcategory.
§ 406.91 Specialized definitions.
§ 406.92 Effluent limitations guidelines representing the degree of effluent reduction attainable by the application of the best practicable control technology currently available.
Secs. 406.93--406.94 [Reserved]
§ 406.95 Standards of performance for new sources.
§ 406.96 Pretreatment standards for new sources.
§ 406.97 Effluent limitations guidelines representing the degree of effluent reduction attainable by the application of the best conventional pollutant control technology (BCT).

Subpart J--Wheat Starch and Gluten Subcategory
§ 406.100 Applicability; description of the wheat starch and gluten subcategory.
§ 406.101 Specialized definitions.
§ 406.102 Effluent limitations guidelines representing the degree of effluent reduction attainable by the application

of the best practicable control technology currently available.

Secs. 406.103--406.104 [Reserved]

§ 406.105 Standards of performance for new sources.

§ 406.106 Pretreatment standards for new sources.

§ 406.107 Effluent limitations guidelines representing the degree of effluent reduction attainable by the application of the best conventional pollutant control technology (BCT).

PART 407--CANNED AND PRESERVED FRUITS AND VEGETABLES PROCESSING POINT SOURCE CATEGORY

Subpart A--Apple Juice Subcategory

§ 407.10 Applicability; description of the apple juice subcategory.

§ 407.11 Specialized definitions.

§ 407.12 Effluent limitations guidelines representing the degree of effluent reduction attainable by the application of the best practicable control technology currently available.

§ 407.13 [Reserved]

§ 407.14 Pretreatment standards for existing sources.

§ 407.15 Standards of performance for new sources.

§ 407.16 Pretreatment standards for new sources.

§ 407.17 Effluent limitations guidelines representing the degree of effluent reduction attainable by the application of the best conventional pollutant control technology (BCT).

Subpart B--Apple Products Subcategory

§ 407.20 Applicability; description of the apple products subcategory.

§ 407.21 Specialized definitions.

§ 407.22 Effluent limitations guidelines representing the degree of effluent reduction attainable by the application of the best practicable control technology currently available.

§ 407.23 [Reserved]

§ 407.24 Pretreatment standards for existing sources.

§ 407.25 Standards of performance for new sources.

§ 407.26 Pretreatment standards for new sources.

§ 407.27 Effluent limitations guidelines representing the degree of effluent reduction attainable by the application of the best conventional pollutant control technology (BCT).

Subpart C--Citrus Products Subcategory

§ 407.30 Applicability; description of the citrus products subcategory.

§ 407.31 Specialized definitions.

§ 407.32 Effluent limitations guidelines representing the degree of effluent reduction attainable by the application of the best practicable control technology currently available.

§ 407.33 [Reserved]

§ 407.34 Pretreatment standards for existing sources.

§ 407.35 Standards of performance for new sources.
§ 407.36 Pretreatment standards for new sources.
§ 407.37 Effluent limitations guidelines representing the degree of effluent reduction attainable by the application of the best conventional pollutant control technology (BCT).

Subpart D--Frozen Potato Products Subcategory
§ 407.40 Applicability; description of the frozen potato products subcategory.
§ 407.41 Specialized definitions.
§ 407.42 Effluent limitations guidelines representing the degree of effluent reduction attainable by the application of the best practicable control technology currently available.
§ 407.43 [Reserved]
§ 407.44 Pretreatment standards for existing sources.
§ 407.45 Standards of performance for new sources.
§ 407.46 Pretreatment standards for new sources.
§ 407.47 Effluent limitations guidelines representing the degree of effluent reduction attainable by the application of the best conventional pollutant control technology (BCT).

Subpart E--Dehydrated Potato Products Subcategory
§ 407.50 Applicability; description of the dehydrated potato products subcategory.
§ 407.51 Specialized definitions.
§ 407.52 Effluent limitations guidelines representing the degree of effluent reduction attainable by the application of the best practicable control technology currently available.
§ 407.53 [Reserved]
§ 407.54 Pretreatment standards for existing sources.
§ 407.55 Standards of performance for new sources.
§ 407.56 Pretreatment standards for new sources.
§ 407.57 Effluent limitations guidelines representing the degree of effluent reduction attainable by the application of the best conventional pollutant control technology (BCT).

Subpart F--Canned and Preserved Fruits Subcategory
§ 407.60 Applicability; description of the canned and preserved fruits subcategory.
§ 407.61 Specialized definitions.
§ 407.62 Effluent limitations guidelines representing the degree of effluent reduction attainable by the application of the best practicable control technology currently available.
§ 407.63 [Reserved]
§ 407.64 Pretreatment standards for existing sources.
§ 407.65 [Reserved]
§ 407.66 Pretreatment standards for new sources.
§ 407.67 Effluent limitations guidelines representing the degree of effluent reduction attainable by the application of the best conventional pollutant control technology

(BCT).

Subpart G--Canned and Preserved Vegetables Subcategory

§ 407.70 Applicability; description of the canned and preserved vegetables subcategory.

§ 407.71 Specialized definitions.

§ 407.72 Effluent limitations guidelines representing the degree of effluent reduction attainable by the application of the best practicable control technology currently available.

§ 407.73 [Reserved]

§ 407.74 Pretreatment standards for existing sources.

§ 407.75 [Reserved]

§ 407.76 Pretreatment standards for new sources.

§ 407.77 Effluent limitations guidelines representing the degree of effluent reduction attainable by the application of the best conventional pollutant control technology (BCT).

Subpart H--Canned and Miscellaneous Specialties Subcategory

§ 407.80 Applicability; description of the canned and miscellaneous specialties subcategory.

§ 407.81 Specialized definitions.

§ 407.82 Effluent limitations guidelines representing the degree of effluent reduction attainable by the application of the best practicable control technology currently available.

§ 407.83 [Reserved]

§ 407.84 Pretreatment standards for existing sources.

§ 407.85 [Reserved]

§ 407.86 Pretreatment standards for new sources.

§ 407.87 Effluent limitations guidelines representing the degree of effluent reduction attainable by the application of the best conventional pollutant control technology (BCT).

PART 408--CANNED AND PRESERVED SEAFOOD PROCESSING POINT SOURCE CATEGORY

Subpart A--Farm-Raised Catfish Processing Subcategory

§ 408.10 Applicability; description of the farm-raised catfish processing subcategory.

§ 408.11 Specialized definitions.

§ 408.12 Effluent limitations guidelines representing the degree of effluent reduction attainable by the application of the best practicable control technology currently available.

§ 408.13 [Reserved]

§ 408.14 Pretreatment standards for existing sources.

§ 408.15 Standards of performance for new sources.

§ 408.16 Pretreatment standards for new sources.

§ 408.17 Effluent limitations guidelines representing the degree of effluent reduction attainable by the application of the best conventional pollutant control technology (BCT).

Subpart B--Conventional Blue Crab Processing Subcategory

§ 408.20 Applicability; description of the conventional

available.

§ 408.83 [Reserved]

§ 408.84 Pretreatment standards for existing sources.

§ 408.85 Standards of performance for new sources.

§ 408.86 Pretreatment standards for new sources.

§ 408.87 Effluent limitations guidelines representing the degree of effluent reduction attainable by the application of the best conventional pollutant control technology (BCT).

Subpart I--Non-Remote Alaskan Shrimp Processing Subcategory

§ 408.90 Applicability; description of the non-remote Alaskan shrimp processing subcategory.

§ 408.91 Specialized definitions.

§ 408.92 Effluent limitations guidelines representing the degree of effluent reduction attainable by the application of the best practicable control technology currently available.

§ 408.93 [Reserved]

§ 408.94 Pretreatment standards for existing sources.

§ 408.95 Standards of performance for new sources.

§ 408.96 Pretreatment standards for new sources.

§ 408.97 Effluent limitations guidelines representing the degree of effluent reduction attainable by the application of the best conventional pollutant control technology (BCT).

Subpart J--Remote Alaskan Shrimp Processing Subcategory

§ 408.100 Applicability; description of the remote Alaskan shrimp processing subcategory.

§ 408.101 Specialized definitions.

§ 408.102 Effluent limitations guidelines representing the degree of effluent reduction attainable by the application of the best practicable control technology currently available.

§ 408.103 [Reserved]

§ 408.104 Pretreatment standards for existing sources.

§ 408.105 Standards of performance for new sources.

§ 408.106 Pretreatment standards for new sources.

§ 408.107 Effluent limitations guidelines representing the degree of effluent reduction attainable by the application of the best conventional pollutant control technology (BCT).

Subpart K--Northern Shrimp Processing in the Contiguous States Subcategory

§ 408.110 Applicability; description of the Northern shrimp processing in the contiguous States subcategory.

§ 408.111 Specialized definitions.

§ 408.112 Effluent limitations guidelines representing the degree of effluent reduction attainable by the application of the best practicable control technology currently available.

§ 408.113 [Reserved]

§ 408.114 Pretreatment standards for existing sources.

§ 408.115 Standards of performance for new sources.

degree of effluent reduction attainable by the application of the best conventional pollutant control technology (BCT).

Subpart O--Fish Meal Processing Subcategory

§ 408.150 Applicability; description of the fish meal processing subcategory.

§ 408.151 Specialized definitions.

§ 408.152 Effluent limitations guidelines representing the degree of effluent reduction attainable by the application of the best practicable control technology currently available.

§ 408.153 [Reserved]

§ 408.154 Pretreatment standards for existing sources.

§ 408.155 Standards of performance for new sources.

§ 408.156 Pretreatment standards for new sources.

§ 408.157 Effluent limitations guidelines representing the degree of effluent reduction attainable by the application of the best conventional pollutant control technology (BCT).

Subpart P--Alaskan Hand-Butchered Salmon Processing Subcategory

§ 408.160 Applicability; description of the Alaskan hand-butchered salmon processing subcategory.

§ 408.161 Specialized definitions.

§ 408.162 Effluent limitations guidelines representing the degree of effluent reduction attainable by the application of the best practicable control technology currently available.

§ 408.163 [Reserved]

§ 408.164 Pretreatment standards for existing sources.

§ 408.165 Standards of performance for new sources.

§ 408.166 Pretreatment standards for new sources.

§ 408.167 Effluent limitations guidelines representing the degree of effluent reduction attainable by the application of the best conventional pollutant control technology.

Subpart Q--Alaskan Mechanized Salmon Processing Subcategory

§ 408.170 Applicability; description of the Alaskan mechanized salmon processing subcategory.

§ 408.171 Specialized definitions.

§ 408.172 Effluent limitations guidelines representing the degree of effluent reduction attainable by the application of the best practicable control technology currently available.

§ 408.173 [Reserved]

§ 408.174 Pretreatment standards for existing sources.

§ 408.175 Standards of performance for new sources.

§ 408.176 Pretreatment standards for new sources.

§ 408.177 Effluent limitations guidelines representing the degree of effluent reduction attainable by the application of the best conventional pollutant control technology (BCT).

Subpart R--West Coast Hand-Butchered Salmon Processing Subcategory

§ 408.180 Applicability; description of the West Coast

hand-butchered salmon processing subcategory.
§ 408.181 Specialized definitions.
§ 408.182 Effluent limitations guidelines representing the degree of effluent reduction attainable by the application of the best practicable control technology currently available.
§ 408.183 [Reserved]
§ 408.184 Pretreatment standards for existing sources.
§ 408.185 Standards of performance for new sources.
§ 408.186 Pretreatment standards for new sources.
§ 408.187 Effluent limitations guidelines representing the degree of effluent reduction attainable by the application of the best conventional pollutant control technology (BCT).

Subpart S--West Coast Mechanized Salmon Processing Subcategory
§ 408.190 Applicability; description of the West Coast mechanized salmon processing subcategory.
§ 408.191 Specialized definitions.
§ 408.192 Effluent limitations guidelines representing the degree of effluent reduction attainable by the application of the best practicable control technology currently available.
§ 408.193 [Reserved]
§ 408.194 Pretreatment standards for existing sources.
§ 408.195 Standards of performance for new sources.
§ 408.196 Pretreatment standards for new sources.
§ 408.197 Effluent limitations guidelines representing the degree of effluent reduction attainable by the application of the best conventional pollutant control technology (BCT).

Subpart T--Alaskan Bottom Fish Processing Subcategory
§ 408.200 Applicability; description of the Alaskan bottom fish processing subcategory.
§ 408.201 Specialized definitions.
§ 408.202 Effluent limitations guidelines representing the degree of effluent reduction attainable by the application of the best practicable control technology currently available.
§ 408.203 [Reserved]
§ 408.204 Pretreatment standards for existing sources.
§ 408.205 Standards of performance for new sources.
§ 408.206 Pretreatment standards for new sources.
§ 408.207 Effluent limitations guidelines representing the degree of effluent reduction attainable by the application of the best conventional pollutant control technology.

Subpart U--Non-Alaskan Conventional Bottom Fish Processing Subcategory
§ 408.210 Applicability; description of the non-Alaskan conventional bottom fish processing subcategory.
§ 408.211 Specialized definitions.
§ 408.212 Effluent limitations guidelines representing the degree of effluent reduction attainable by the application of the best practicable control technology currently

available.

§ 408.213 [Reserved]

§ 408.214 Pretreatment standards for existing sources.

§ 408.215 Standards of performance for new sources.

§ 408.216 Pretreatment standards for new sources.

§ 408.217 Effluent limitations guidelines representing the degree of effluent reduction attainable by the application of the best conventional pollutant control technology (BCT).

Subpart V--Non-Alaskan Mechanized Bottom Fish Processing Subcategory

§ 408.220 Applicability; description of the non-Alaskan mechanized bottom fish processing subcategory.

§ 408.221 Specialized definitions.

§ 408.222 Effluent limitations guidelines representing the degree of effluent reduction attainable by the application of the best practicable control technology currently available.

§ 408.223 [Reserved]

§ 408.224 Pretreatment standards for existing sources.

§ 408.225 Standards of performance for new sources.

§ 408.226 Pretreatment standards for new sources.

§ 408.227 Effluent limitations guidelines representing the degree of effluent reduction attainable by the application of the best conventional pollutant control technology (BCT).

Subpart W--Hand-Shucked Clam Processing Subcategory

§ 408.230 Applicability; description of the hand-shucked clam processing subcategory.

§ 408.231 Specialized definitions.

§ 408.232 Effluent limitations guidelines representing the degree of effluent reduction attainable by the application of the best practicable control technology currently available.

§ 408.233 [Reserved]

§ 408.234 Pretreatment standards for existing sources.

§ 408.235 Standards of performance for new sources.

§ 408.236 Pretreatment standards for new sources.

§ 408.237 Effluent limitations guidelines representing the degree of effluent reduction attainable by the application of the best conventional pollutant control technology (BCT).

Subpart X--Mechanized Clam Processing Subcategory

§ 408.240 Applicability; description of the mechanized clam processing subcategory.

§ 408.241 Specialized definitions.

§ 408.242 Effluent limitations guidelines representing the degree of effluent reduction attainable by the application of the best practicable control technology currently available.

§ 408.243 [Reserved]

§ 408.244 Pretreatment standards for existing sources.

§ 408.245 Standards of performance for new sources.

§ 408.246 Pretreatment standards for new sources.
§ 408.247 Effluent limitations guidelines representing the degree of effluent reduction attainable by the application of the best conventional pollutant control technology (BCT).

Subpart Y--Pacific Coast Hand-Shucked Oyster Processing Subcategory

§ 408.250 Applicability; description of the Pacific Coast hand-shucked oyster processing subcategory.
§ 408.251 Specialized definitions.
§ 408.252 Effluent limitations guidelines representing the degree of effluent reduction attainable by the application of the best practicable control technology currently available.
§ 408.253 [Reserved]
§ 408.254 Pretreatment standards for existing sources.
§ 408.255 Standards of performance for new sources.
§ 408.256 Pretreatment standards for new sources.
§ 408.257 Effluent limitations guidelines representing the degree of effluent reduction attainable by the application of the best conventional pollutant control technology.

Subpart Z--Atlantic and Gulf Coast Hand-Shucked Oyster Processing Subcategory

§ 408.260 Applicability; description of the Atlantic and Gulf Coast hand-shucked oyster processing subcategory.
§ 408.261 Specialized definitions.
§ 408.262 Effluent limitations guidelines representing the degree of effluent reduction attainable by the application of the best practicable control technology currently available.
§ 408.263 [Reserved]
§ 408.264 Pretreatment standards for existing sources.
§ 408.265 Standards of performance for new sources.
§ 408.266 Pretreatment standards for new sources.
§ 408.267 Effluent limitations guidelines representing the degree of effluent reduction attainable by the application of the best conventional pollutant control technology.

Subpart AA--Steamed and Canned Oyster Processing Subcategory

§ 408.270 Applicability; description of the steamed and canned oyster processing subcategory.
§ 408.271 Specialized definitions.
§ 408.272 Effluent limitations guidelines representing the degree of effluent reduction attainable by the application of the best practicable control technology currently available.
§ 408.273 [Reserved]
§ 408.274 Pretreatment standards for existing sources.
§ 408.275 Standards of performance for new sources.
§ 408.276 Pretreatment standards for new sources.
§ 408.277 Effluent limitations guidelines representing the degree of effluent reduction attainable by the application of the best conventional pollutant control technology (BCT).

Subpart AB--Sardine Processing Subcategory

§ 408.280 Applicability; description of the sardine processing subcategory.

§ 408.281 Specialized definitions.

§ 408.282 Effluent limitations guidelines representing the degree of effluent reduction attainable by the application of the best practicable control technology currently available.

§ 408.283 [Reserved]

§ 408.284 Pretreatment standards for existing sources.

§ 408.285 Standards of performance for new sources.

§ 408.286 Pretreatment standards for new sources.

§ 408.287 Effluent limitations guidelines representing the degree of effluent reduction attainable by the application of the best conventional pollutant control technology (BCT).

Subpart AC--Alaskan Scallop Processing Subcategory

§ 408.290 Applicability; description of the Alaskan scallop processing subcategory.

§ 408.291 Specialized definitions.

§ 408.292 Effluent limitations guidelines representing the degree of effluent reduction attainable by the application of the best practicable control technology currently available.

§ 408.293 [Reserved]

§ 408.294 Pretreatment standards for existing sources.

§ 408.295 Standards of performance for new sources.

§ 408.296 Pretreatment standards for new sources.

§ 408.297 Effluent limitations guidelines representing the degree of effluent reduction attainable by the application of the best conventional pollutant control technology.

Subpart AD--Non-Alaskan Scallop Processing Subcategory

§ 408.300 Applicability; description of the non-Alaskan scallop processing subcategory.

§ 408.301 Specialized definitions.

§ 408.302 Effluent limitations guidelines representing the degree of effluent reduction attainable by the application of the best practicable control technology currently available.

§ 408.303 [Reserved]

§ 408.304 Pretreatment standards for existing sources.

§ 408.305 Standards of performance for new sources.

§ 408.306 Pretreatment standards for new sources.

§ 408.307 Effluent limitations guidelines representing the degree of effluent reduction attainable by the application of the best conventional pollutant control technology.

Subpart AE--Alaskan Herring Fillet Processing Subcategory

§ 408.310 Applicability; description of the Alaskan herring fillet processing subcategory.

§ 408.311 Specialized definitions.

§ 408.312 Effluent limitations guidelines representing the degree of effluent reduction attainable by the application of the best practicable control technology currently

Subpart A--Wool Scouring Subcategory

§ 410.10 Applicability; description of the wool scouring subcategory.

§ 410.11 Specialized definitions.

§ 410.12 Effluent limitations representing the degree of effluent reduction attainable by the application of the best practicable control technology currently available (BPT).

§ 410.13 Effluent limitations representing the degree of effluent reduction attainable by the application of the best available technology economically achievable (BAT).

§ 410.14 Pretreatment standards for existing sources (PSES).

§ 410.15 New source performance standards (NSPS).

§ 410.16 Pretreatment standards for new sources (PSNS).

§ 410.17 Effluent limitations representing the degree of effluent reduction attainable by the application of the best conventional pollutant control technology (BCT). [Reserved]

Subpart B--Wool Finishing Subcategory

§ 410.20 Applicability; description of the wool finsihing subcategory.

§ 410.21 Specialized definitions.

§ 410.22 Effluent limitations representing the degree of effluent reduction attainable by the application of the best practicable control technology currently available (BPT).

§ 410.23 Effluent limitation representing the degree of effluent reduction attainable by the application of the best available technology economically achievable (BAT).

§ 410.24 Pretreatment standards for existing sources (PSES).

§ 410.25 New source performance standards (NSPS).

§ 410.26 Pretreatment standards for new sources (PSNS).

§ 410.27 Effluent limitations representing the degree of effluent reduction attainable by the application of the best conventional pollutant control technology (BCT). [Reserved]

Subpart C--Low Water Use Processing Subcategory

§ 410.30 Applicability; description of the low water use processing subcategory.

§ 410.31 Specialized definitions.

§ 410.32 Effluent limitations representing the degree of effluent reduction attainable by the application of the best practicable control technology currently available (BPT).

§ 410.33 Effluent limitations representing the degree of effluent reduction attainable by the application of the best available technology economically achievable (BAT).

§ 410.34 Pretreatment standards for existing sources (PSES).

§ 410.35 New source performance standards (NSPS).

§ 410.36 Pretreatment standards for new sources (PSNS).

§ 410.37 Effluent limitations representing the degree of effluent reduction attainable by the application of the best conventional pollutant control technology (BCT). [Reserved]

Subpart D--Woven Fabric Finishing Subcategory

§ 410.40 Applicability; description of the woven fabric finishing subcategory.

§ 410.41 Specialized definitions.

§ 410.42 Effluent limitations representing the degree of effluent reduction attainable by the application of the best practicable control technology currently available (BPT).

§ 410.43 Effluent limitations representing the degree of effluent reduction attainable by the application of the best available technology economically achievable (BAT).

§ 410.44 Pretreatment standards for existing sources (PSES).

§ 410.45 New source performance standards (NSPS).

§ 410.46 Pretreatment standards for new sources (PSNS).

§ 410.47 Effluent limitations representing the degree of effluent reduction attainable by the application of the best conventional pollutant control technology (BCT). [Reserved]

Subpart E--Knit Fabric Finishing Subcategory

§ 410.50 Applicability; description of the knit fabric finishing subcategory.

§ 410.51 Specialized definitions.

§ 410.52 Effluent limitations representing the degree of effluent reduction attainable by the application of the best practicable control technology currently available (BPT).

§ 410.53 Effluent limitations representing the degree of effluent reduction attainable by the application of the best available technology economically achievable (BAT).

§ 410.54 Pretreatment standards for existing sources (PSES).

§ 410.55 New source performance standards (NSPS).

§ 410.56 Pretreatment standards for new sources (PSNS).

§ 410.57 Effluent limitations representing the degree of effluent reduction attainable by the application of the best conventional pollutant control technology (BCT). [Reserved]

Subpart F--Carpet Finishing Subcategory

§ 410.60 Applicability; description of the carpet finishing subcategory.

§ 410.61 Specialized definitions.

§ 410.62 Effluent limitations representing the degree of effluent reduction attainable by the application of the best practicable control technology currently available (BPT).

§ 410.63 Effluent limitations representing the degree of effluent reduction attainable by the application of the best available technology economically achievable (BAT).

best available technology economically achievable (BAT).
§ 414.24 New source performance standards (NSPS).
§ 414.25 Pretreatment standards for existing sources (PSES).
§ 414.26 Pretreatment standards for new sources (PSNS).

Subpart C--Other Fibers

§ 414.30 Applicability; description of the other fibers subcategory.
§ 414.31 Effluent limitations representing the degree of effluent reduction attainable by the application of the best practicable control technology currently available (BPT).
§ 414.32 Effluent limitations representing the degree of effluent reduction attainable by the application of the best conventional pollutant control technology (BCT). [Reserved]
§ 414.33 Effluent limitations representing the degree of effluent reduction attainable by the application of the best available technology economically achievable (BAT).
§ 414.34 New source performance standards (NSPS).
§ 414.35 Pretreatment standards for existing sources (PSES).
§ 414.36 Pretreatment standards for new sources (PSNS).

Subpart D--Thermoplastic Resins

§ 414.40 Applicability; description of the thermoplastic resins subcategory.
§ 414.41 Effluent limitations representing the degree of effluent reduction attainable by the application of the best practicable control technology currently available (BPT).
§ 414.42 Effluent limitations representing the degree of effluent reduction attainable by the application of the best conventional pollutant control technology (BCT). [Reserved]
§ 414.43 Effluent limitations representing the degree of effluent reduction attainable by the application of the best available technology economically achievable (BAT).
§ 414.44 New source performance standards (NSPS).
§ 414.45 Pretreatment standards for existing sources (PSES).
§ 414.46 Pretreatment standards for new sources (PSNS).

Subpart E--Thermosetting Resins

§ 414.50 Applicability; description of the thermosetting resins subcategory.
§ 414.51 Effluent limitations representing the degree of effluent reduction attainable by the application of the best practicable control technology currently available (BPT).
§ 414.52 Effluent limitations representing the degree of effluent reduction attainable by the application of the best conventional pollutant control technology (BCT). [Reserved]
§ 414.53 Effluent limitations representing the degree of

effluent reduction attainable by the application of the best available technology economically achievable (BAT).

§ 414.54 New source performance standards (NSPS).

§ 414.55 Pretreatment standards for existing sources (PSES).

§ 414.56 Pretreatment standards for new sources (PSNS).

Subpart F--Commodity Organic Chemicals

§ 414.60 Applicability; description of the commodity organic chemicals subcategory.

§ 414.61 Effluent limitations representing the degree of effluent reduction attainable by the application of the best practicable control technology currently available (BPT).

§ 414.62 Effluent limitations representing the degree of effluent reduction attainable by the application of the best conventional pollutant control technology (BCT). [Reserved]

§ 414.63 Effluent limitations representing the degree of effluent reduction attainable by the application of the best available technology economically achievable (BAT).

§ 414.64 New source performance standards (NSPS)

§ 414.65 Pretreatment standards for existing sources (PSES).

§ 414.66 Pretreatment standards for new sources (PSNS).

Subpart G--Bulk Organic Chemicals

§ 414.70 Applicability; description of the bulk organic chemicals subcategory.

§ 414.71 Effluent limitations representing the degree of effluent reduction attainable by the application of the best practicable control technology currently available (BPT).

§ 414.72 Effluent limitations representing the degree of effluent reduction attainable by the application of the best conventional pollutant control technology (BCT). [Reserved]

§ 414.73 Effluent limitations representing the degree of effluent reduction attainable by the application of the best available technology economically achievable (BAT).

§ 414.74 New source performance standards (NSPS)

§ 414.75 Pretreatment standards for existing sources (PSES).

§ 414.76 Pretreatment standards for new sources (PSNS).

Subpart H--Specialty Organic Chemicals

§ 414.80 Applicability; description of the specialty organic chemicals subcategory.

§ 414.81 Effluent limitations representing the degree of effluent reduction attainable by the application of the best practicable control technology currently available (BPT).

§ 414.82 Effluent limitations representing the degree of effluent reduction attainable by the application of the best conventional pollutant control technology (BCT). [Reserved]

(BAT).

§ 415.24 Pretreatment standards for existing sources (PSES).

§ 415.25 New source performance standards (NSPS).

§ 415.26 Pretreatment standards for new sources (PSNS).

Subpart C--Calcium Carbide Production Subcategory

§ 415.30 Applicability; description of the calcium carbide production subcategory.

§ 415.31 Specialized definitions. [Reserved]

§ 415.32 Effluent limitations guidelines representing the degree of effluent reduction attainable by the application of the best practicable control technology currently available (BPT).

§ 415.33 Effluent limitations guidelines representing the degree of effluent reduction attainable by the application of the best available technolgy economically achievable (BAT).

§ 415.34 [Reserved]

§ 415.35 New source performance standards (NSPS).

§ 415.36 Pretreatment standards for new sources (PSNS).

Subpart D--Calcium Chloride Production Subcategory

§ 415.40 Applicability; description of the calcium chloride production subcategory.

§ 415.41 Specialized definitions.

§ 415.42 Effluent limitations guidelines representing the degree of effluent reduction attainable by the application of the best practicable control technology currently available (BPT).

§ 415.43 Effluent limitations guidelines representing the degree of effluent reduction attainable by the application of the best available technology economically achievable (BAT).

§ 415.44 [Reserved]

§ 415.45 New source performance standards (NSPS).

§ 415.46 Pretreatment standards for new sources (PSNS).

Subpart E--Calcium Oxide Production Subcategory

§ 415.50 Applicability; description of the calcium oxide production subcategory.

§ 415.51 Specialized definitions. [Reserved]

§ 415.52 Effluent limitations guidelines representing the degree of effluent reduction attainable by the application of the best practicable control technology currently available (BPT).

§ 415.53 Effluent limitations guidelines representing the degree of effluent reduction attainable by the application of the best available technology economically achievable (BAT).

§ 415.54 [Reserved]

§ 415.55 New source performance standards (NSPS).

§ 415.56 Pretreatment standards for new sources (PSNS).

Subpart F--Chlor-alkali Subcategory (Chlorine and Sodium or Potassium Hydroxide Production)

§ 415.60 Applicability; description of the chlorine and

sodium or potassium hydroxide production subcategory.
§ 415.61 Specialized definitions.
§ 415.62 Effluent limitations guidelines representing the degree of effluent reduction attainable by the application of the best practicable control technology currently available (BPT).
§ 415.63 Effluent limitations guidelines representing the degree of effluent reduction attainable by the application of the best available technology economically achievable (BAT).
§ 415.64 Pretreatment standards for existing sources (PSES).
§ 415.65 New source performance standards (NSPS).
§ 415.66 Pretreatment standards for new sources (PSNS).
§ 415.67 Effluent limitations guidelines representing the degree of effluent reduction attainable by the application of the best conventional pollutant control technology (BCT).

Subpart G--Hydrochloric Acid Production Subcategory [Reserved]

Subpart H--Hydrofluoric Acid Production Subcategory

§ 415.80 Applicability; description of the hydrofluoric acid production subcategory.
§ 415.81 Specialized definitions. [Reserved]
§ 415.82 Effluent limitations guidelines representing the degree of effluent reduction attainable by the application of the best practicable control technology currently available (BPT).
§ 415.83 Effluent limitations guidelines representing the degree of effluent reduction attainable by the application of the best available technology economically achievable (BAT).
§ 415.84 [Reserved]
§ 415.85 New source performance standards (NSPS).
§ 415.86 Pretreatment standards for new sources (PSNS).
§ 415.87 [Reserved]

Subpart I--Hydrogen Peroxide Production Subcategory

§ 415.90 Applicability; description of the hydrogen peroxide production subcategory.
§ 415.91 Specialized definitions.
§ 415.92 Effluent limitations guidelines representing the degree of effluent reduction attainable by the application of the best practicable control technology currently available (BPT).

Subpart J--Nitric Acid Production Subcategory [Reserved]

Subpart K--Potassium Metal Production Subcategory

§ 415.110 Applicability; description of the potassium metal production subcategory.
§ 415.111 Specialized definitions. [Reserved]
§ 415.112 Effluent limitations guidelines representing the degree of effluent reduction attainable by the application of the best practicable control technology currently available (BPT).
§ 415.113 Effluent limitations guidelines representing the

Subpart P--Sodium Chloride Production Subcategory

§ 415.160 Applicability; description of the sodium chloride production subcategory.

§ 415.161 Specialized definitions.

§ 415.162 Effluent limitations guidelines representing the degree of effluent reduction attainable by the application of the best practicable control technology currently available (BPT).

§ 415.163 Effluent limitations guidelines representing the degree of effluent reduction attainable by the application of the best available technology economically achievable (BAT).

§ 415.164 [Reserved]

§ 415.165 New source performance standards (NSPS).

§ 415.166 Pretreatment standards for new sources (PSNS).

Subpart Q--Sodium Dichromate and Sodium Sulfate Production Subcategory

§ 415.170 Applicability; description of the sodium dichromate and sodium sulfate production subcategory.

§ 415.171 Specialized definitions.

§ 415.172 Effluent limitations guidelines representing the degree of effluent reduction attainable by the application of the best practicable control technology currently available (BPT).

§ 415.173 Effluent limitations guidelines representing the degree of effluent reduction attainable by the application of the best available technology economically achievable (BAT).

§ 415.174 [Reserved]

§ 415.175 New source performance standards (NSPS).

§ 415.176 Pretreatment standards for new sources (PSNS).

§ 415.177 Effluent limitations guidelines representing the degree of effluent reduction attainable by the application of the best conventional pollutant control technology (BCT).

Subpart R--Sodium Metal Production Subcategory [Reserved]

Subpart S--Sodium Silicate Production Subcategory [Reserved]

Subpart T--Sodium Sulfite Production Subcategory

§ 415.200 Applicability; description of the sodium sulfite production subcategory.

§ 415.201 Specialized definitions.

§ 415.202 Effluent limitations guidelines representing the degree of effluent reduction attainable by the application of the best practicable control technology currently available (BPT).

§ 415.203 Effluent limitations guidelines representing the degree of effluent reduction attainable by the application of the best available technology economically achievable (BAT).

§ 415.204 [Reserved]

§ 415.205 New source performance standards (NSPS).

§ 415.206 Pretreatment standards for new sources (PSNS).

§ 415.207 Effluent limitations guidelines representing the

degree of effluent reduction attainable by the application of the best conventional pollutant control technology (BCT).

Subpart U--Sulfuric Acid Production Subcategory [Reserved]

Subpart V--Titanium Dioxide Production Subcategory

§ 415.220 Applicability; description of the titanium dioxide production subcategory.

§ 415.221 Specialized definitions.

§ 415.222 Effluent limitations guidelines representing the degree of effluent reduction attainable by the application of the best practicable control technology currently available (BPT).

§ 415.223 Effluent limitations guidelines representing the degree of effluent reduction attainable by the application of the best available technology economically achievable (BAT).

§ 415.224 [Reserved]

§ 415.225 New source performance standards (NSPS).

§ 415.226 Pretreatment standards for new sources (PSNS).

§ 415.227 Effluent limitations guidelines representing the degree of effluent reduction attainable by the application of the best conventional pollutant control technology (BCT).

Subpart W--Aluminum Fluoride Production Subcategory

§ 415.230 Applicability; description of the aluminum fluoride production subcategory.

§ 415.231 Specialized definitions.

§ 415.232 Effluent limitations guidelines representing the degree of effluent reduction attainable by the application of the best practicable control technology currently available (BPT).

§ 415.233 Effluent limitations guidelines representing the degree of effluent reduction attainable by the application of the best available technology economically achievable (BAT).

§ 415.234 [Reserved]

§ 415.235 New source performance standards (NSPS).

§ 415.236 [Reserved]

§ 415.237 Effluent limitations guidelines representing the degree of effluent reduction attainable by the application of the best conventional pollutant control technology (BCT).

Subpart X--Ammonium Chloride Production Subcategory

§ 415.240 Applicability; description of the ammonium chloride production subcategory.

§ 415.241 Specialized definitions.

§ 415.242 Effluent limitations guidelines representing the degree of effluent reduction attainable by the application of the best practicable control technology currently available (BPT).

Subpart Y--Ammonium Hydroxide Production Subcategory [Reserved]

Subpart Z--Barium Carbonate Production Subcategory [Reserved]

Subpart AA--Borax Production Subcategory

§ 415.270 Applicability; description of the borax production subcategory.

§ 415.271 Specialized definitions. [Reserved]

§ 415.272 Effluent limitations guidelines representing the degree of effluent reduction attainable by the application of the best practicable control technology currently available (BPT).

Secs. 415.273--415.275 [Reserved]

§ 415.276 Pretreatment standards for new sources (PSNS).

Subpart AB--Boric Acid Production Subcategory

§ 415.280 Applicability; description of the boric acid production subcategory.

§ 415.281 Specialized definitions.

§ 415.282 Effluent limitations guidelines representing the degree of effluent reduction attainable by the application of the best practicable control technology currently available (BPT).

Subpart AC--Bromine Production Subcategory

§ 415.290 Applicability; description of the bromine production subcategory.

§ 415.291 Specialized definitions. [Reserved]

§ 415.292 Effluent limitations guidelines representing the degree of effluent reduction attainable by the application of the best practicable control technology currently available (BPT).

Secs. 415.293--415.295 [Reserved]

§ 415.296 Pretreatment standards for new sources (PSNS).

Subpart AD--Calcium Carbonate Production Subcategory

§ 415.300 Applicability; description of the calcium carbonate production subcategory.

§ 415.301 Specialized definitions.

§ 415.302 Effluent limitations guidelines representing the degree of effluent reduction attainable by the application of the best practicable control technology currently available (BPT).

Subpart AE--Calcium Hydroxide Production Subcategory

§ 415.310 Applicability; description of the calcium hydroxide production subcategory.

§ 415.311 Specialized definitions.

§ 415.312 Effluent limitations guidelines representing the degree of effluent reduction attainable by the application of the best practicable control technology currently available (BPT).

Secs. 415.313--415.315 [Reserved]

§ 415.316 Pretreatment standards for new sources (PSNS).

Subpart AF--Carbon Dioxide Production Subcategory [Reserved]

Subpart AG--Carbon Monoxide and By-Product Hydrogen Production Subcategory

§ 415.330 Applicability; description of the carbon monoxide and by- product hydrogen production subcategory.

§ 415.331 Specialized definitions.

§ 415.332 Effluent limitations guidelines representing the degree of effluent reduction attainable by the application

Subpart AL--Ferric Chloride Production Subcategory

§ 415.380 Applicability; description of the ferric chloride production subcategory.

§ 415.381 Specialized definitions.

§ 415.382 Effluent limitations guidelines representing the degree of effluent reduction attainable by the application of the best practicable control technology currently available (BPT).

§ 415.383 [Reserved]

§ 415.384 Pretreatment standards for existing sources (PSES).

§ 415.385 [Reserved]

§ 415.386 Pretreatment standards for new sources (PSNS).

Subpart AM--Ferrous Sulfate Production Subcategory [Reserved]

Subpart AN--Fluorine Production Subcategory

§ 415.400 Applicability; description of the fluorine production subcategory.

§ 415.401 Specialized definitions.

§ 415.402 Effluent limitations guidelines representing the degree of effluent reduction attainable by the application of the best practicable control technology currently available (BPT).

Secs. 415.403--415.405 [Reserved]

§ 415.406 Pretreatment standards for new sources (PSNS).

Subpart AO--Hydrogen Production Subcategory

§ 415.410 Applicability; description of the hydrogen production subcategory.

§ 415.411 Specialized definitions.

§ 415.412 Effluent limitations guidelines representing the degree of effluent reduction attainable by the application of the best practicable control technology currently available (BPT).

Subpart AP--Hydrogen Cyanide Production Subcategory

§ 415.420 Applicability; description of the hydrogen cyanide production subcategory.

§ 415.421 Specialized definitions.

§ 415.422 Effluent limitations guidelines representing the degree of effluent reduction attainable by the application of the best practicable control technology currently available (BPT).

§ 415.423 Effluent limitations guidelines representing the degree of effluent reduction attainable by the application of the best available technology economically achievable (BAT).

§ 415.424 [Reserved]

§ 415.425 New source performance standards (NSPS).

§ 415.426 Pretreatment standards for new sources (PSNS).

§ 415.427 Effluent limitations guidelines representing the degree of effluent reduction attainable by the application of the best conventional pollutant control technology (BCT).

Subpart AQ--Iodine Production Subcategory

§ 415.430 Applicability; description of the iodine

production subcategory.
§ 415.431 Specialized definitions.
§ 415.432 Effluent limitations guidelines representing the degree of effluent reduction attainable by the application of the best practicable control technology currently available (BPT).
Secs. 415.433--415.435 [Reserved]
§ 415.436 Pretreatment standards for new sources (PSNS).
Subpart AR--Lead Monoxide Production Subcategory
§ 415.440 Applicability; description of the lead monoxide production subcategory.
§ 415.441 Specialized definitions.
§ 415.442 Effluent limitations quidelines respresenting the degree of effluent reduction attainable by the application of the best practicable control technology currently available (BPT).
§ 415.443 [Reserved]
§ 415.444 Pretreatment standards for existing sources (PSES).
Secs. 415.445 [Reserved]
§ 415.446 Pretreatment standards for new sources (PSNS).
Subpart AS--Lithium Carbonate Production Subcategory
§ 415.450 Applicability; description of the lithium carbonate production subcategory.
§ 415.451 Specialized definitions.
§ 415.452 Effluent limitations guidelines representing the degree of effluent reduction attainable by the application of the best practicable control technology currently available (BPT).
Subpart AT--Manganese Sulfate Production Subcategory [Reserved]
Subpart AU--Nickel Salts Production Subcategory
§ 415.470 Applicability; description of the nickel salts production subcategory.
§ 415.471 Specialized definitions.
§ 415.472 Effluent limitations guidelines representing the degree of effluent reduction attainable by the application of the best practicable control technology currently available (BPT).
§ 415.473 Effluent limitations guidelines representing the degree of effluent reduction attainable by the application of the best available technology economically achievable (BAT).
§ 415.474 Pretreatment standards for existing sources (PSES).
§ 415.475 New source performance standards (NSPS).
§ 415.476 Pretreatment standards for new sources (PSNS).
§ 415.477 Effluent limitations guidelines representing the degree of effluent reduction attainable by the application of the best conventional pollutant control technology (BCT).
Subpart AV--Strong Nitric Acid Production Subcategory--[Reserved]
Subpart AW--Oxygen and Nitrogen Production Subcategory
§ 415.490 Applicability; description of the oxygen and

nitrogen production subcategory.
§ 415.491 Specialized definitions. [Reserved]
§ 415.492 Effluent limitations guidelines representing the degree of effluent reduction attainable by the application of the best practicable control technology currently available (BPT).

Subpart AX--Potassium Chloride Production Subcategory
§ 415.500 Applicability; description of the potassium chloride production subcategory.
§ 415.501 Specialized definitions. [Reserved]
§ 415.502 Effluent limitations guidelines representing the degree of effluent reduction attainable by the application of the best practicable control technology currently available (BPT).
Secs. 415.503--415.505 [Reserved]
§ 415.506 Pretreatment standards for new sources (PSNS).

Subpart AY--Potassium Iodide Production Subcategory
§ 415.510 Applicability; description of the potassium iodide production subcategory.
§ 415.511 Specialized definitions.
§ 415.512 Effluent limitations guidelines representing the degree of effluent reduction attainable by the application of the best practicable control technology currently available (BPT).

Subpart AZ--Potassium Permanganate Production Subcategory [Reserved]

Subpart BA--Silver Nitrate Production Subcategory
§ 415.530 Applicability; description of the silver nitrate production subcategory.
§ 415.531 Specialized definitions.
§ 415.532 Effluent limitations guidelines representing the degree of effluent reduction attainable by the application of the best practicable control technology currently available (BPT).
§ 415.533 [Reserved]
§ 415.534 Pretreatment standards for existing sources (PSES).

Subpart BB--Sodium Bisulfite Production Subcategory
§ 415.540 Applicability; description of the sodium bisulfite production subcategory.
§ 415.541 Specialized definitions.
§ 415.542 Effluent limitations guidelines representing the degree of effluent reduction attainable by the application of the best practicable control technology currently available (BPT).
§ 415.543 Effluent limitations guidelines representing the degree of effluent reduction attainable by the application of the best available technology economically achievable (BAT).
§ 415.544 [Reserved]
§ 415.545 New source performance standards (NSPS).
§ 415.546 Pretreatment standards for new sources (PSNS).
§ 415.547 Effluent limitations guidelines representing the

degree of effluent reduction attainable by the application of the best conventional pollutant control technology (BCT).

Subpart BC--Sodium Fluoride Production Subcategory

§ 415.550 Applicability; description of the sodium fluoride production subcategory.

§ 415.551 Specialized definitions.

§ 415.552 Effluent limitations guidelines representing the degree of effluent reduction attainable by the application of the best practicable control techology currently available (BPT).

§ 415.553 [Reserved]

§ 415.554 Pretreatment standards for existing sources (PSES).

§ 415.555 [Reserved]

§ 415.556 Pretreatment standards for new sources (PSNS).

Subpart BD--Sodium Hydrosulfide Production Subcategory [Reserved]

Subpart BE--Sodium Hydrosulfite Production Subcategory [Reserved]

Subpart BF--Sodium Silicofluoride Production Subcategory [Reserved]

Subpart BG--Sodium Thiosulfate Production Subcategory [Reserved]

Subpart BH--Stannic Oxide Production Subcategory

§ 415.600 Applicability; description of the stannic oxide production subcategory.

§ 415.601 Specialized definitions.

§ 415.602 Effluent limitations guidelines representing the degree of effluent reduction attainable by the application of the best practicable control technology currently available (BPT).

Secs. 415.603--415.605 [Reserved]

§ 415.606 Pretreatment standards for new sources (PSNS).

Subpart BI--Sulfur Dioxide Production Subcategory [Reserved]

Subpart BJ--Zinc Oxide Production Subcategory [Reserved]

Subpart BK--Zinc Sulfate Production Subcategory

§ 415.630 Applicability; description of the zinc sulfate production subcategory.

§ 415.631 Specialized definitions.

§ 415.632 Effluent limitations guidelines representing the degree of effluent reduction attainable by the application of the best practicable control technology currently available (BPT).

Secs. 415.633--415.635 [Reserved]

§ 415.636 Pretreatment standards for new sources (PSNS).

Subpart BL--Cadmium Pigments and Salts Production Subcategory

§ 415.640 Applicability; description of the cadmium pigments and salts production subcategory.

§ 415.641 Specialized definitions.

§ 415.642 Effluent limitations guidelines representing the degree of effluent reduction attainable by the application of the best practicable control technology currently available (BPT).

§ 415.643 Effluent limitations guidelines representing the degree of effluent reduction attainable by the application

of the best available technology economically achievable (BAT).

§ 415.644 Pretreatment standards for existing sources (PSES).

§ 415.645 New source performance standards (NSPS).

§ 415.646 Pretreatment standards for new sources (PSNS).

§ 415.647 Effluent limitations guidelines representing the degree of effluent reduction attainable by the application of the best conventional pollutant control technology (BCT).

Subpart BM--Cobalt Salts Production Subcategory

§ 415.650 Applicability; description of the cobalt salts production subcategory.

§ 415.651 Specialized definitions.

§ 415.652 Effluent limitations guidelines representing the degree of effluent reduction attainable by the application of the best practicable control technology currently available (BPT).

§ 415.653 Effluent limitations guidelines representing the degree of effluent reduction attainable by the application of the best available technology economically achievable (BAT).

§ 415.654 Pretreatment standards for existing sources (PSES).

§ 415.655 New source performance standards (NSPS).

§ 415.656 Pretreatment standards for new sources (PSNS).

§ 415.657 Effluent limitations guidelines representing the degree of effluent reduction attainable by the application of the best conventional pollutant control technology (BCT).

Subpart BN--Sodium Chlorate Production Subcategory

§ 415.660 Applicability; description of the sodium chlorate production subcategory.

§ 415.661 Specialized definitions.

§ 415.662 Effluent limitations guidelines representing the degree of effluent reduction attainable by the application of the best practicable control technology currently available (BPT).

§ 415.663 Effluent limitations guidelines representing the degree of effluent reduction attainable by the application of the best available technology economically achievable (BAT).

§ 415.664 Pretreatment standards for existing sources (PSES). [Reserved]

§ 415.665 New source performance standards (NSPS).

§ 415.666 Pretreatment standards for new sources (PSNS).

§ 415.667 Effluent limitations guidelines representing the degree of effluent reduction attainable by the application of the best conventional pollutant control technology (BCT).

Subpart BO--Zinc Chloride Production Subcategory

§ 415.670 Applicability; description of the zinc chloride production subcategory.

degree of effluent reduction attainable by the application of the best practicable control technology currently available.

§ 417.33 Effluent limitations guidelines representing the degree of effluent reduction attainable by the application of the best available technology economically achievable.

§ 417.34 Pretreatment standards for existing sources.

§ 417.35 Standards of performance for new sources.

§ 417.36 Pretreatment standards for new sources.

Subpart D--Glycerine Concentration Subcategory

§ 417.40 Applicability; description of the glycerine concentration subcategory.

§ 417.41 Specialized definitions.

§ 417.42 Effluent limitations guidelines representing the degree of effluent reduction attainable by the application of the best practicable control technology currently available.

§ 417.43 Effluent limitations guidelines representing the degree of effluent reduction attainable by the application of the best available technology economically achievable.

§ 417.44 Pretreatment standards for existing sources.

§ 417.45 Standards of performance for new sources.

§ 417.46 Pretreatment standards for new sources.

Subpart E--Glycerine Distillation Subcategory

§ 417.50 Applicability; description of the glycerine distillation subcategory.

§ 417.51 Specialized definitions.

§ 417.52 Effluent limitations guidelines representing the degree of effluent reduction attainable by the application of the best practicable control technology currently available.

§ 417.53 Effluent limitations guidelines representing the degree of effluent reduction attainable by the application of the best available technology economically achievable.

§ 417.54 Pretreatment standards for existing sources.

§ 417.55 Standards of performance for new sources.

§ 417.56 Pretreatment standards for new sources.

Subpart F--Manufacture of Soap Flakes and Powders Subcategory

§ 417.60 Applicability; description of the manufacture of soap flakes and powders subcategory.

§ 417.61 Specialized definitions.

§ 417.62 Effluent limitations guidelines representing the degree of effluent reduction attainable by the application of the best practicable control technology currently available.

§ 417.63 Effluent limitations guidelines representing the degree of effluent reduction attainable by the application of the best available technology economically achievable.

§ 417.64 Pretreatment standards for existing sources.

§ 417.65 Standards of performance for new sources.

§ 417.66 Pretreatment standards for new sources.

Subpart G--Manufacture of Bar Soaps Subcategory

§ 417.70 Applicability; description of the manufacture of

bar soaps subcategory.

§ 417.71 Specialized definitions.

§ 417.72 Effluent limitations guidelines representing the degree of effluent reduction attainable by the application of the best practicable control technology currently available.

§ 417.73 Effluent limitations guidelines representing the degree of effluent reduction attainable by the application of the best available technology economically achievable.

§ 417.74 Pretreatment standards for existing sources.

§ 417.75 Standards of performance for new sources.

§ 417.76 Pretreatment standards for new sources.

Subpart H--Manufacture of Liquid Soaps Subcategory

§ 417.80 Applicability; description of the manufacture of liquid soaps subcategory.

§ 417.81 Specialized definitions.

§ 417.82 Effluent limitations guidelines representing the degree of effluent reduction attainable by the application of the best practicable control technology currently available.

§ 417.83 Effluent limitations guidelines representing the degree of effluent reduction attainable by the application of the best available technology economically achievable.

§ 417.84 Pretreatment standards for existing sources.

§ 417.85 Standards of performance for new sources.

§ 417.86 Pretreatment standards for new sources.

Subpart I--Oleum Sulfonation and Sulfation Subcategory

§ 417.90 Applicability; description of the oleum sulfonation and sulfation subcategory.

§ 417.91 Specialized definitions.

§ 417.92 Effluent limitations guidelines representing the degree of effluent reduction attainable by the application of the best practicable control technology currently available.

§ 417.93 Effluent limitations guidelines representing the degree of effluent reduction attainable by the application of the best available technology economically achievable.

§ 417.94 Pretreatment standards for existing sources.

§ 417.95 Standards of performance for new sources.

§ 417.96 Pretreatment standards for new sources.

Subpart J--Air--SO3 Sulfation and Sulfonation Subcategory

§ 417.100 Applicability; description of the air--SO3 sulfation and sulfonation subcategory.

§ 417.101 Specialized definitions.

§ 417.102 Effluent limitations guidelines representing the degree of effluent reduction attainable by the application of the best practicable control technology currently available.

§ 417.103 Effluent limitations guidelines representing the degree of effluent reduction attainable by the application of the best available technology economically achievable.

§ 417.104 Pretreatment standards for existing sources.

§ 417.105 Standards of performance for new sources.

degree of effluent reduction attainable by the application of the best available technology economically achievable.

§ 417.144 Pretreatment standards for existing sources.

§ 417.145 Standards of performance for new sources.

§ 417.146 Pretreatment standards for new sources.

Subpart O--Manufacture of Spray Dried Detergents Subcategory

§ 417.150 Applicability; description of the manufacture of spray dried detergents subcategory.

§ 417.151 Specialized definitions.

§ 417.152 Effluent limitations guidelines representing the degree of effluent reduction attainable by the application of the best practicable control technology currently available.

§ 417.153 Effluent limitations guidelines representing the degree of effluent reduction attainable by the application of the best available technology economically achievable.

§ 417.154 [Reserved]

§ 417.155 Standards of performance for new sources.

§ 417.156 Pretreatment standards for new sources.

Subpart P--Manufacture of Liquid Detergents Subcategory

§ 417.160 Applicability; description of the manufacture of liquid detergents subcategory.

§ 417.161 Specialized definitions.

§ 417.162 Effluent limitations guidelines representing the degree of effluent reduction attainable by the application of the best practicable control technology currently available.

§ 417.163 Effluent limitations guidelines representing the degree of effluent reduction attainable by the application of the best available technology economically achievable.

§ 417.164 [Reserved]

§ 417.165 Standards of performance for new sources.

§ 417.166 Pretreatment standards for new sources.

Subpart Q--Manufacture of Detergents by Dry Blending Subcategory

§ 417.170 Applicability; description of the manufacture of detergents by dry blending subcategory.

§ 417.171 Specialized definitions.

§ 417.172 Effluent limitations guidelines representing the degree of effluent reduction attainable by the application of the best practicable control technology currently available.

§ 417.173 Effluent limitations guidelines representing the degree of effluent reduction attainable by the application of the best available technology economically achievable.

§ 417.174 [Reserved]

§ 417.175 Standards of performance for new sources.

§ 417.176 Pretreatment standards for new sources.

Subpart R--Manufacture of Drum Dried Detergents Subcategory

§ 417.180 Applicability; description of the manufacture of drum dried detergents subcategory.

§ 417.181 Specialized definitions.

§ 417.182 Effluent limitations guidelines representing the degree of effluent reduction attainable by the application

§ 418.25 Standards of performance for new sources.
§ 418.26 Pretreatment standards for new sources.
§ 418.27 Effluent limitations guidelines representing the degree of effluent reduction attainable by the application of the best conventional pollutant control technology.

Subpart C--Urea Subcategory

§ 418.30 Applicability; description of the urea subcategory.
§ 418.31 Specialized definitions.
§ 418.32 Effluent limitations and guidelines representing the degree of effluent reduction attainable by the application of the best practicable control technology currently available.
§ 418.33 Effluent limitations and guidelines representing the degree of effluent reduction attainable by the application of the best available technology economically achievable.
§ 418.34 [Reserved]
§ 418.35 Standards of performance for new sources.
§ 418.36 Pretreatment standards for new sources.

Subpart D--Ammonium Nitrate Subcategory

§ 418.40 Applicability; description of the ammonium nitrate subcategory.
§ 418.41 Specialized definitions.
§ 418.42 Effluent limitations and guidelines representing the degree of effluent reduction attainable by the application of the best practicable control technology currently available.
§ 418.43 Effluent limitations and guidelines representing the degree of effluent reduction attainable by the application of the best available technology economically achievable.
§ 418.44 [Reserved]
§ 418.45 Standards of performance for new sources.
§ 418.46 Pretreatment standards for new sources.

Subpart E--Nitric Acid Subcategory

§ 418.50 Applicability; description of the nitric acid subcategory.
§ 418.51 Specialized definitions.
§ 418.52 Effluent limitations guidelines representing the degree of effluent reduction attainable by the application of the best practicable control technology currently available.
§ 418.53 Effluent limitations guidelines representing the degree of effluent reduction attainable by the application of the best available technology economically achievable.
§ 418.54 [Reserved]
§ 418.55 Standards of performance for new sources.
§ 418.56 Pretreatment standards for new sources.

Subpart F--Ammonium Sulfate Production Subcategory

§ 418.60 Applicability; description of the ammonium sulfate production subcategory.
§ 418.61 Specialized definitions.

§ 418.62 Effluent limitations guidelines representing the degree of effluent reduction attainable by the application of the best practicable control technology currently available.

§ 418.63 Effluent limitations guidelines representing the degree of effluent reduction attainable by the application of the best available technology economically achievable.

§ 418.64 [Reserved]

§ 418.65 Standards of performance for new sources.

§ 418.66 Pretreatment standard for new sources.

§ 418.67 Effluent limitations guidelines representing the degree of effluent reduction attainable by the application of the best conventional pollutant control technology.

Subpart G--Mixed and Blend Fertilizer Production Subcategory

§ 418.70 Applicability; description of the mixed and blend fertilizer production subcategory.

§ 418.71 Specialized definitions.

§ 418.72 Effluent limitations guidelines representing the degree of effluent reduction attainable by the application of the best practicable control technology currently available.

§ 418.73 Effluent limitations guidelines representing the degree of effluent reduction attainable by the application of the best available technology economically achievable.

§ 418.74 [Reserved]

§ 418.75 Standards of performance for new sources.

§ 418.76 Pretreatment standard for new sources.

§ 418.77 Effluent limitations guidelines representing the degree of effluent reduction attainable by the application of the best conventional pollutant control technology.

PART 419--PETROLEUM REFINING POINT SOURCE CATEGORY

Subpart A--Topping Subcategory

§ 419.10 Applicability; description of the topping subcategory.

§ 419.11 Specialized definitions.

§ 419.12 Effluent limitations guidelines representing the degree of effluent reduction attainable by the application of the best practicable control technology currently available (BPT).

§ 419.13 Effluent limitations guidelines representing the degree of effluent reduction attainable by the application of the best available technology economically achievable (BAT).

§ 419.14 Effluent limitations guidelines representing the degree of effluent reduction attainable by the application of the best conventional pollutant control technology (BCT).

§ 419.15 Pretreatment standards for existing sources (PSES).

§ 419.16 Standards of performance for new sources (NSPS).

§ 419.17 Pretreatment standards for new sources (PSNS).

Subpart B--Cracking Subcategory

§ 419.20 Applicability; description of the cracking

subcategory.
§ 419.21 Specialized definitions.
§ 419.22 Effluent limitations guidelines representing the degree of effluent reduction attainable by the application of the best practicable control technology currently available (BPT).
§ 419.23 Effluent limitations guidelines representing the degree of effluent reduction attainable by the application of the best available technology economically achievable (BAT).
§ 419.24 Effluent limitations guidelines representing the degree of effluent reduction attainable by the application of the best conventional pollutant control technology (BCT).
§ 419.25 Pretreatment standards for existing sources (PSES).
§ 419.26 Standards of performance for new sources (NSPS).
§ 419.27 Pretreatment standards for new sources (PSNS).

Subpart C--Petrochemical Subcategory
§ 419.30 Applicability; description of the petrochemical subcategory.
§ 419.31 Specialized definitions.
§ 419.32 Effluent limitations guidelines representing the degree of effluent reduction attainable by the application of the best practicable control technology currently available.
§ 419.33 Effluent limitations guidelines representing the degree of effluent reduction attainable by the application of the best available technology economically achievable (BAT).
§ 419.34 Effluent limitations guidelines representing the degree of effluent reduction attainable by the application of the best conventional pollutant control technology (BCT).
§ 419.35 Pretreatment standards for existing sources (PSES).
§ 419.36 Standards of performance for new sources (NSPS).
§ 419.37 Pretreatment standards for new sources (PSNS).

Subpart D--Lube Subcategory
§ 419.40 Applicability; description of the lube subcategory.
§ 419.41 Specialized definitions.
§ 419.42 Effluent limitations guidelines representing the degree of effluent reduction attainable by the application of the best practicable control technology currently available (BPT).
§ 419.43 Effluent limitations guidelines representing the degree of effluent reduction attainable by the application of the best available technology economically achievable (BAT).
§ 419.44 Effluent limitations guidelines representing the degree of effluent reduction attainable by the application of the best conventional pollutant control technology

effluent reduction attainable by the application of the best conventional technology (BCT).

Subpart B--Sintering Subcategory

§ 420.20 Applicability; description of the sintering subcategory.

§ 420.21 [Reserved]

§ 420.22 Effluent limitations representing the degree of effluent reduction attainable by the application of the best practicable control technology currently available (BPT).

§ 420.23 Effluent limitations representing the degree of effluent reduction attainable by the application of the best available technology economically achievable (BAT).

§ 402.24 New source performance standards (NSPS).

§ 420.25 Pretreatment standards for existing sources (PSES).

§ 420.26 Pretreatment standards for new sources (PSNS).

§ 420.27 [Reserved]

Subpart C--Ironmaking Subcategory

§ 420.30 Applicability; description of the ironmaking subcategory.

§ 420.31 Specialized definitions.

§ 420.32 Effluent limitations representing the degree of effluent reduction attainable by the application of the best practicable control technology currently available (BPT).

§ 420.33 Effluent limitations representing the degree of effluent reduction attainable by the application of the best available technology economically achievable (BAT).

§ 420.34 New source performance standards (NSPS).

§ 420.35 Pretreatment standards for existing sources (PSES).

§ 420.36 Pretreatment standards for new sources (PSNS).

§ 420.37 [Reserved]

Subpart D--Steelmaking Subcategory

§ 420.40 Applicability; description of the steelmaking subcategory.

§ 420.41 Specialized definitions.

§ 420.42 Effluent limitations representing the degree of effluent reduction attainable by the application of the best practicable control technology currently available (BPT).

§ 420.43 Effluent limitations representing the degree of effluent reduction attainable by the application of the best available technology economically achievable (BAT).

§ 420.44 New source performance standards (NSPS).

§ 420.45 Pretreatment standards for existing sources (PSES).

§ 420.46 Pretreatment standards for new sources (PSNS).

§ 420.47 Effluent limitations representing the degree of effluent reduction attainable by the application of the best conventional control technology (BCT).

Subpart E--Vacuum Degassing Subcategory

§ 420.50 Applicability; description of the vacuum degassing subcategory.
§ 420.51 [Reserved]
§ 420.52 Effluent limitations representing the degree of effluent reduction attainable by the application of the best practicable control technology currently available (BPT).
§ 420.53 Effluent limitations representing the degree of effluent reduction attainable by the application of the best available technology economically achievable (BAT).
§ 420.54 New source performance standards (NSPS).
§ 420.55 Pretreatment standards for existing sources (PSES).
§ 420.56 Pretreatment standards for new sources (PSNS).
§ 420.57 [Reserved]

Subpart F--Continuous Casting Subcategory

§ 420.60 Applicability; description of the continuous casting subcategory.
§ 420.61 [Reserved]
§ 420.62 Effluent limitations representing the degree of effluent reduction attainable by the application of the best practicable control technology currently available (BPT).
§ 420.63 Effluent limitations representing the degree of effluent reduction attainable by the application of the best available technology economically achievable (BAT).
§ 420.64 New source performance standards (NSPS).
§ 420.65 Pretreatment standards for existing sources (PSES).
§ 420.66 Pretreatment standards for new sources (PSNS).
§ 420.67 [Reserved]

Subpart G--Hot Forming Subcategory

§ 420.70 Applicability; description of the hot forming subcategory.
§ 420.71 Specialized definitions.
§ 420.72 Effluent limitations representing the degree of effluent reduction attainable by the application of the best practicable control technology currently available (BPT).
§ 420.73 Effluent limitations representing the degree of effluent reduction attainable by the application of the best available technology economically achievable (BAT).
§ 420.74 New source performance standards (NSPS).
§ 420.75 Pretreatment standards for existing sources (PSES).
§ 420.76 Pretreatment standards for new sources (PSNS).
§ 420.77 Effluent limitations representing the degree of effluent reduction attainable by the application of the best conventional technology (BCT).

Subpart H--Salt Bath Descaling Subcategory

§ 420.80 Applicability; description of the salt bath descaling subcategory.
§ 420.81 Specialized definitions.

§ 420.82 Effluent limitations representing the degree of effluent reduction attainable by the application of the best practicable control technology currently available (BPT).

§ 420.83 Effluent limitations representing the degree of effluent reduction attainable by the application of the best available technology economically achievable (BAT).

§ 420.84 New source performance standards (NSPS).

§ 420.85 Pretreatment standards for existing sources (PSES).

§ 420.86 Pretreatment standards for new sources (PSNS).

§ 420.87 Effluent limitations representing the degree of effluent reduction attainable by the application of the best conventional technology (BCT).

Subpart I--Acid Pickling Subcategory

§ 420.90 Applicability; description of the acid pickling subcategory.

§ 420.91 Specialized definitions.

§ 420.92 Effluent limitations representing the degree of effluent reduction attainable by the application of the best practicable control technology currently available (BPT).

§ 420.93 Effluent limitations representing the degree of effluent reduction attainable by the application of the best available technology economically achievable (BAT).

§ 420.94 New source performance standards (NSPS).

§ 420.95 Pretreatment standards for existing sources (PSES).

§ 420.96 Pretreatment standards for new sources (PSNS).

§ 420.97 Effluent limitations representing the degree of effluent reduction attainable by the application of the best conventional technology (BCT).

Subpart J--Cold Forming Subcategory

§ 420.100 Applicability; description of the cold forming subcategory.

§ 420.101 Specialized definitions.

§ 420.102 Effluent limitations representing the degree of effluent reduction attainable by the application of the best practicable control technology currently available (BPT).

§ 420.103 Effluent limitations representing the degree of effluent reduction attainable by the application of the best available technology economically achievable (BAT).

§ 420.104 New source performance standards (NSPS).

§ 420.105 Pretreatment standards for existing sources (PSES).

§ 420.106 Pretreatment standards for new sources (PSNS).

§ 420.107 Effluent limitations representing the degree of effluent reduction attainable by the application of the best conventional technology (BCT).

Subpart K--Alkaline Cleaning Subcategory

§ 420.110 Applicability; description of the alkaline cleaning subcategory.

§ 420.111 Specialized definitions.
§ 420.112 Effluent limitations representing the degree of effluent reduction attainable by the application of the best practicable control technology currently available (BPT).
§ 420.113 Effluent limitations representing the degree of effluent reduction attainable by the application of the best available technology economically achievable (BAT).
§ 420.114 New source performance standards (NSPS).
§ 420.115 Pretreatment standards for existing sources (PSES).
§ 420.116 Pretreatment standards for new sources (PSNS).
§ 420.117 Effluent limitations representing the degree of effluent reduction attainable by the application of the best conventional technology (BCT).

Subpart L--Hot Coating Subcategory

§ 420.120 Applicability; description of the hot coating subcategory.
§ 420.121 Specialized definitions.
§ 420.122 Effluent limitations representing the degree of effluent reduction attainable by the application of the best practicable control technology currently available (BPT).
§ 420.123 Effluent limitations representing the degree of effluent reduction attainable by the application of the best available technology economically achievable (BAT).
§ 420.124 New source performance standards (NSPS).
§ 420.125 Pretreatment standards for existing sources (PSES).
§ 420.126 Pretreatment standards for new sources (PSNS).
§ 420.127 Effluent limitations representing the degree of effluent reduction attainable by the application of the best conventional technology (BCT).

PART 421--NONFERROUS METALS MANUFACTURING POINT SOURCE CATEGORY

General Provisions

§ 421.1 Applicability.
§ 421.2 [Reserved]
§ 421.3 Monitoring and reporting requirements.
§ 421.4 Compliance date for pretreatment standards for existing sources (PSES).
§ 421.5 Removal allowances for pretreatment standards.

Subpart A--Bauxite Refining Subcategory

§ 421.10 Applicability; description of the bauxite refining subcategory.
§ 421.11 Specialized definitions.
§ 421.12 Effluent limitations guidelines representing the degree of effluent reduction attainable by the application of the best practicable control technology currently available.
§ 421.13 Effluent limitations guidelines representing the degree of effluent reduction attainable by the application of the best available technology economically achievable.
§ 421.14 [Reserved]

of the best practicable control technology currently available.

§ 421.53 Effluent limitations guidelines representing the degree of effluent reduction attainable by the application of the best available technology economically achievable.

§ 421.54 Standards of performance for new sources.

§ 421.55 [Reserved]

§ 421.56 Pretreatment standards for new sources.

§ 421.57 [Reserved]

Subpart F--Secondary Copper Subcategory

§ 421.60 Applicability: Description of the secondary copper subcategory.

§ 421.61 Specialized definitions.

§ 421.62 Effluent limitations guidelines representing the degree of effluent reduction attainable by the application of the best practicable control technology currently available.

§ 421.63 Effluent limitations guidelines representing the degree of effluent reduction attainable by the application of the best available technology economically achievable.

§ 421.64 Standards of performance for new sources.

§ 421.65 Pretreatment standards for existing sources.

§ 421.66 Pretreatment standards for new sources.

§ 421.67 [Reserved]

Subpart G--Primary Lead Subcategory

§ 421.70 Applicability: Description of the primary lead subcategory.

§ 421.71 Specialized definitions.

§ 421.72 Effluent limitations guidelines representing the degree of effluent reduction attainable by the application of the best practicable control technology currently available.

§ 421.73 Effluent limitations guidelines representing the degree of effluent reduction attainable by the application of the best available technology economically achievable.

§ 421.74 Standards of performance for new sources.

§ 421.75 Pretreatment standards for existing sources.

§ 421.76 Pretreatment standards for new sources.

§ 421.77 [Reserved]

Subpart H--Primary Zinc Subcategory

§ 421.80 Applicability: Description of the primary zinc subcategory.

§ 421.81 Specialized definitions.

§ 421.82 Effluent limitations guidelines representing the degree of effluent reduction attainable by the application of the best practicable control technology currently available.

§ 421.83 Effluent limitations guidelines representing the degree of effluent reduction attainable by the application of the best available technology economically achievable.

§ 421.84 Standards of performance for new sources.

§ 421.85 Pretreatment standards for existing sources.

§ 421.86 Pretreatment standards for new sources.

§ 421.87 [Reserved]

Subpart I--Metallurgical Acid Plants Subcategory

§ 421.90 Applicability: Description of the metallurgical acid plants subcategory.

§ 421.91 Specialized definitions.

§ 421.92 Effluent limitations guidelines representing the degree of effluent reduction attainable by the application of the best practicable control technology currently available.

§ 421.93 Effluent limitations guidelines representing the degree of effluent reduction attainable by the application of the best available technology economically achievable.

§ 421.94 Standards of performance for new sources.

§ 421.95 Pretreatment standards for existing sources.

§ 421.96 Pretreatment standards for new sources.

§ 421.97 [Reserved]

Subpart J--Primary Tungsten Subcategory

§ 421.100 Applicability: Description of the primary tungsten subcategory.

§ 421.101 Specialized definitions.

§ 421.102 Effluent limitations guidelines representing the degree of effluent reduction attainable by the application of the best practicable control technology currently available.

§ 421.103 Effluent limitations guidelines representing the degree of effluent reduction attainable by the application of the best available technology economically achievable.

§ 421.104 Standards of performance for new sources.

§ 421.105 Pretreatment standards for existing sources.

§ 421.106 Pretreatment standards for new sources.

§ 421.107 [Reserved]

Subpart K--Primary Columbium-Tantalum Subcategory

§ 421.110 Applicability: Description of the primary columbium-tantalum subcategory.

§ 421.111 Specialized definitions.

§ 421.112 Effluent limitations guidelines representing the degree of effluent reduction attainable by the application of the best practicable control technology currently available.

§ 421.113 Effluent limitations guidelines representing the degree of effluent reduction attainable by the application of the best available technology economically achievable.

§ 421.114 Standards of performance for new sources.

§ 421.115 Pretreatment standards for existing sources.

§ 421.116 Pretreatment standards for new sources.

§ 421.117 [Reserved]

Subpart L--Secondary Silver Subcategory

§ 421.120 Applicability: Description of the secondary silver subcategory.

§ 421.121 Specialized definitions.

§ 421.122 Effluent limitations guidelines representing the degree of effluent reduction attainable by the application of the best practicable control technology currently

available.

§ 421.123 Effluent limitations guidelines representing the degree of effluent reduction attainable by the application of the best available technology economically achievable.

§ 421.124 Standards of performance for new sources.

§ 421.125 Pretreatment standards for existing sources.

§ 421.126 Pretreatment standards for new sources.

§ 421.127 [Reserved]

Subpart M--Secondary Lead Subcategory

§ 421.130 Applicability: Description of the secondary lead subcategory.

§ 421.131 Specialized definitions.

§ 421.132 Effluent limitations guidelines representing the degree of effluent reduction attainable by the application of the best practicable control technology currently available.

§ 421.133 Effluent limitations guidelines representing the degree of effluent reduction attainable by the application of the best available technology economically achievable.

§ 421.134 Standards of performance for new sources.

§ 421.135 Pretreatment standards for existing sources.

§ 421.136 Pretreatment standards for new sources.

§ 421.137 [Reserved]

Subpart N--Primary Antimony Subcategory

§ 421.140 Applicability: Description of the primary antimony subcategory.

§ 421.141 Specialized definitions.

§ 421.142 Effluent limitations guidelines representing the degree of effluent reduction attainable by the application of the best practicable control technology currently available.

§ 421.143 Effluent limitations guidelines representing the degree of effluent reduction attainable by the application of the best available technology economically achievable.

§ 421.144 Standards of performance for new sources.

§ 421.145 [Reserved]

§ 421.146 Pretreatment standards for new sources.

§ 421.147 [Reserved]

Subpart O--Primary Beryllium Subcategory

§ 421.150 Applicability: Description of the primary beryllium subcategory.

§ 421.151 Specialized definitions.

§ 421.152 Effluent limitations guidelines representing the degree of effluent reduction attainable by the application of the best practicable control technology currently available.

§ 421.153 Effluent limitations guidelines representing the degree of cffluent reduction attainable by the application of the best available technology economically achievable.

§ 421.154 Standards of performance for new sources.

§ 421.155 [Reserved]

§ 421.156 Pretreatment standards for new sources.

§ 421.157 [Reserved]

Subpart P--Primary and Secondary Germanium and Gallium Subcategory

§ 421.180 Applicability: Description of the primary and secondary germanium and gallium subcategory.

§ 421.181 Specialized definitions.

§ 421.182 Effluent limitations guidelines representing the degree of effluent reduction attainable by the application of the best practicable control technology currently available.

§ 421.183 Effluent limitations guidelines representing the degree of effluent reduction attainable by the application of the best available technology economically achievable.

§ 421.184 Standards of performance for new sources.

§ 421.185 Pretreatment standards for existing sources.

§ 421.186 Pretreatment standards for new sources.

§ 421.187 [Reserved]

Subpart Q--Secondary Indium Subcategory

§ 421.190 Applicability: Description of the secondary indium subcategory.

§ 421.191 Specialized definitions.

Secs. 421.192--421.193 [Reserved]

§ 421.194 Standards of performance for new sources.

§ 421.195 Pretreatment standards for existing sources.

§ 421.196 Pretreatment standards for new sources.

§ 421.197 [Reserved]

Subpart R--Secondary Mercury Subcategory

§ 421.200 Applicability: Description of the secondary mercury subcategory.

§ 421.201 Specialized definitions.

Secs. 421.202--421.203 [Reserved]

§ 421.204 Standards of performance for new sources.

§ 421.205 [Reserved]

§ 421.206 Pretreatment standards for new sources.

§ 421.207 [Reserved]

Subpart S--Primary Molybdenum and Rhenium Subcategory

§ 421.210 Applicability: Description of the primary molybdenum and rhenium subcategory.

§ 421.211 Specialized definitions.

§ 421.212 Effluent limitations guidelines representing the degree of effluent reduction attainable by the application of the best practicable control technology currently available.

§ 421.213 Effluent limitations guidelines representing the degree of effluent reduction attainable by the application of the best available technology economically achievable.

§ 421.214 Standards of performance for new sources.

§ 421.215 [Reserved]

§ 421.216 Pretreatment standards for new sources.

§ 421.217 [Reserved]

Subpart T--Secondary Molybdenum and Vanadium Subcategory

§ 421.220 Applicability: Description of the secondary molybdenum and vanadium subcategory.

§ 421.221 Specialized definitions.

§ 421.222 Effluent limitations guidelines representing the degree of effluent reduction attainable by the application of the best practicable control technology currently available.

§ 421.223 Effluent limitations guidelines representing the degree of effluent reduction attainable by the application of the best available technology economically achievable.

§ 421.224 Standards of performance for new sources.

§ 421.225 [Reserved]

§ 421.226 Pretreatment standards for new sources.

§ 421.227 [Reserved]

Subpart U--Primary Nickel and Cobalt Subcategory

§ 421.230 Applicability: Description of the primary nickel and cobalt subcategory.

§ 421.231 Specialized definitions.

§ 421.232 Effluent limitations guidelines representing the degree of effluent reduction attainable by the application of the best practicable control technology currently available.

§ 421.233 Effluent limitations guidelines representing the degree of effluent reduction attainable by the application of the best available technology economically achievable.

§ 421.234 Standards of performance for new sources.

§ 421.235 [Reserved]

§ 421.236 Pretreatment standards for new sources.

§ 421.237 [Reserved]

Subpart V--Secondary Nickel Subcategory

§ 421.240 Applicability: Description of the secondary nickel subcategory.

§ 421.241 Specialized definitions.

Secs. 421.242--421.243 [Reserved]

§ 421.244 Standards of performance for new sources.

§ 421.245 Pretreatment standards for existing sources.

§ 421.246 Pretreatment standards for new sources.

§ 421.247 [Reserved]

Subpart W--Primary Precious Metals and Mercury Subcategory

§ 421.250 Applicability: Description of the primary precious metals and mercury subcategory.

§ 421.251 Specialized definitions.

§ 421.252 Effluent limitations guidelines representing the degree of effluent reduction attainable by the application of the best practicable control technology currently available.

§ 421.253 Effluent limitations guidelines representing the degree of effluent reduction attainable by the application of the best available technology economically achievable.

§ 421.254 Standards of performance for new sources.

§ 421.255 [Reserved]

§ 421.256 Pretreatment standards for new sources.

§ 421.257 [Reserved]

Subpart X--Secondary Precious Metals Subcategory

§ 421.260 Applicability: Description of the secondary precious metals subcategory.

§ 421.261 Specialized definitions.
§ 421.262 Effluent limitations guidelines representing the degree of effluent reduction attainable by the application of the best practicable control technology currently available.
§ 421.263 Effluent limitations guidelines representing the degree of effluent reduction attainable by the application of the best available technology economically achievable.
§ 421.264 Standards of performance for new sources.
§ 421.265 Pretreatment standards for existing sources.
§ 421.266 Pretreatment standards for new sources.
§ 421.267 [Reserved]

Subpart Y--Primary Rare Earth Metals Subcategory

§ 421.270 Applicability: Description of the primary rare earth metals subcategory.
§ 421.271 Specialized definitions.
Secs. 421.272--421.273 [Reserved]
§ 421.274 Standards of performance for new sources.
§ 421.275 Pretreatment standards for existing sources.
§ 421.276 Pretreatment standards for new sources.
§ 421.277 [Reserved]

Subpart Z--Secondary Tantalum Subcategory

§ 421.280 Applicability: Description of the secondary tantalum subcategory.
§ 421.281 Specialized definitions.
§ 421.282 Effluent limitations guidelines representing the degree of effluent reduction attainable by the application of the best practicable control technology currently available.
§ 421.283 Effluent limitations guidelines representing the degree of effluent reduction attainable by the application of the best available technology economically achievable.
§ 421.284 Standards of performance for new sources.
§ 421.285 [Reserved]
§ 421.286 Pretreatment standards for new sources.
§ 421.287 [Reserved]

Subpart AA--Secondary Tin Subcategory

§ 421.290 Applicability: Description of the secondary tin subcategory.
§ 421.291 Specialized definitions.
§ 421.292 Effluent limitations guidelines representing the degree of effluent reduction attainable by the application of the best practicable control technology currently available.
§ 421.293 Effluent limitations guidelines representing the degree of effluent reduction attainable by the application of the best available technology economically achievable.
§ 421.294 Standards of performance for new sources.
§ 421.295 Pretreatment standards for existing sources.
§ 421.296 Pretreatment standards for new sources.
§ 421.297 [Reserved]

Subpart AB--Primary and Secondary Titanium Subcategory

§ 421.300 Applicability: Description of the primary and

secondary titanium subcategory.
§ 421.301 Specialized definitions.
§ 421.302 Effluent limitations guidelines representing the degree of effluent reduction attainable by the application of the best practicable control technology currently available.
§ 421.303 Effluent limitations guidelines representing the degree of effluent reduction attainable by the application of the best available technology economically achievable.
§ 421.304 Standards of performance for new sources.
§ 421.305 Pretreatment standards for existing sources.
§ 421.306 Pretreatment standards for new sources.
§ 421.307 [Reserved]

Subpart AC--Secondary Tungsten and Cobalt Subcategory
§ 421.310 Applicability: Description of the secondary tungsten and cobalt subcategory.
§ 421.311 Specialized definitions.
§ 421.312 Effluent limitations guidelines representing the degree of effluent reduction attainable by the application of the best practicable control technology currently available.
§ 421.313 Effluent limitations guidelines representing the degree of effluent reduction attainable by the application of the best available technology economically achievable.
§ 421.314 Standards of performance for new sources.
§ 421.315 Pretreatment standards for existing sources.
§ 421.316 Pretreatment standards for new sources.
§ 421.317 [Reserved]

Subpart AD--Secondary Uranium Subcategory
§ 421.320 Applicability: Description of the secondary uranium subcategory.
§ 421.321 Specialized definitions.
§ 421.322 Effluent limitations guidelines representing the degree of effluent reduction attainable by the application of the best practicable control technology currently available.
§ 421.323 Effluent limitations guidelines representing the degree of effluent reduction attainable by the application of the best available technology economically achievable.
§ 421.324 Standards of performance for new sources.
§ 421.325 [Reserved]
§ 421.326 Pretreatment standards for new sources.
§ 421.327 [Reserved]

Subpart AE--Primary Zirconium and Hafnium Subcategory
§ 421.330 Applicability: Description of the primary zirconium and hafnium subcategory.
§ 421.331 Specialized definitions.
§ 421.332 Effluent limitations guidelines representing the degree of effluent reduction attainable by the application of the best practicable control technology currently available.
§ 421.333 Effluent limitations guidelines representing the degree of effluent reduction attainable by the application

of the best available technology economically achievable.
§ 421.334 Standards of performance for new sources.
§ 421.335 [Reserved]
§ 421.336 Pretreatment standards for new sources.
§ 421.337 [Reserved]

PART 422--PHOSPHATE MANUFACTURING POINT SOURCE CATEGORY

Subpart A--Phosphorus Production Subcategory
§ 422.10 Applicability; description of the phosphorus production subcategory.

Subpart B--Phosphorus Consuming Subcategory
§ 422.20 Applicability; description of the phosphorus consuming subcategory.

Subpart C--Phosphate Subcategory
§ 422.30 Applicability; description of the phosphate subcategory.

Subpart D--Defluorinated Phosphate Rock Subcategory
§ 422.40 Applicability; description of the defluorinated phosphate rock subcategory.
§ 422.41 Specialized definitions.
§ 422.42 Effluent limitations and guidelines representing the degree of effluent reduction attainable by the application of the best practicable control technology currently available.
§ 422.43 Effluent limitations and guidelines representing the degree of effluent reduction attainable by the application of the best available technology economically achievable.
§ 422.44 [Reserved]
§ 422.45 Standards of performance for new sources.
§ 422.46 [Reserved]
§ 422.47 Effluent limitations guidelines representing the degree of effluent reduction attainable by the application of the best conventional pollutant control technology.

Subpart E--Defluorinated Phosphoric Acid Subcategory
§ 422.50 Applicability; description of the defluorinated phosphoric acid subcategory.
§ 422.51 Specialized definitions.
§ 422.52 Effluent limitations and guidelines representing the degree of effluent reduction attainable by the application of the best practicable control technology currently available.
§ 422.53 Effluent limitations and guidelines representing the degree of effluent reduction attainable by the application of the best available technology economically achievable.
§ 422.54 [Reserved]
§ 422.55 Standards of performance for new sources.
§ 422.56 [Reserved]
§ 422.57 Effluent limitations guidelines representing the degree of effluent reduction attainable by the application of the best conventional pollutant control technology.

Subpart F--Sodium Phosphates Subcategory
§ 422.60 Applicability; description of the sodium

phosphates subcategory.
§ 422.61 Specialized definitions.
§ 422.62 Effluent limitations and guidelines representing the degree of effluent reduction attainable by the application of the best practicable control technology currently available.
§ 422.63 Effluent limitations guidelines representing the degree of effluent reduction attainable by the application of the best available technology economically achievable.
§ 422.64 [Reserved]
§ 422.65 Standards of performance for new sources.
§ 422.66 [Reserved]
§ 422.67 Effluent limitations guidelines representing the degree of effluent reduction attainable by the application of the best conventional pollutant control technology.

PART 423--STEAM ELECTRIC POWER GENERATING POINT SOURCE CATEGORY

§ 423.10 Applicability.
§ 423.11 Specialized definitions.
§ 423.12 Effluent limitations guidelines representing the degree of effluent reduction attainable by the application of the best practicable control technology currently available (BPT).
§ 423.13 Effluent limitations guidelines representing the degree of effluent reduction attainable by the application of the best available technology economically achievable (BAT).
§ 423.14 Effluent limitations guidelines representing the degree of effluent reduction attainable by the application of the best conventional pollutant control technology (BCT). [Reserved]
§ 423.15 New source performance standards (NSPS).
§ 423.16 Pretreatment standards for existing sources (PSES).
§ 423.17 Pretreatment standards for new sources (PSNS).
Appendix A--126 Priority Pollutants

PART 424--FERROALLOY MANUFACTURING POINT SOURCE CATEGORY

Subpart A--Open Electric Furnaces With Wet Air Pollution Control Devices Subcategory

§ 424.10 Applicability; description of the open electric furnaces with wet air pollution control devices subcategory.
§ 424.11 Specialized definitions.
§ 424.12 Effluent limitations guidelines representing the degree of effluent reduction attainable by the application of the best practicable control technology currently available.
§ 424.13 Effluent limitations guidelines representing the degree of effluent reduction attainable by the application of the best available technology economically achievable.
§ 424.14 [Reserved]
§ 424.15 Standards of performance for new sources.
§ 424.16 Pretreatment standards for new sources.
§ 424.17 Effluent limitations guidelines representing the

degree of effluent reduction attainable by the application of the best conventional pollutant control technology.

Subpart E--Other Calcium Carbide Furnaces Subcategory

§ 424.50 Applicability; description of the other calcium carbide furnaces subcategory.

§ 424.51 Specialized definitions.

§ 424.52 Effluent limitations guidelines representing the degree of effluent reduction attainable by the application of the best practicable control technology currently available.

§ 424.53 Effluent limitations guidelines representing the degree of effluent reduction attainable by the application of the best available technology economically achievable.

Secs. 424.54--424.56 [Reserved]

§ 424.57 Effluent limitations guidelines representing the degree of effluent reduction attainable by the application of the best conventional pollutant control technology.

Subpart F--Electrolytic Manganese Products Subcategory

§ 424.60 Applicability; description of the electrolytic manganese products subcategory.

§ 424.61 Specialized definitions.

§ 424.62 Effluent limitations guidelines representing the degree of effluent reduction attainable by the application of the best practicable control technology currently available.

§ 424.63 Effluent limitations guidelines representing the degree of effluent reduction attainable by the application of the best available technology economically achievable.

Secs. 424.64--424.66 [Reserved]

§ 424.67 Effluent limitations guidelines representing the degree of effluent reduction attainable by the application of the best conventional pollutant control technology.

Subpart G--Electrolytic Chromium Subcategory

§ 424.70 Applicability; description of the electrolytic chromium subcategory.

§ 424.71 Specialized definitions.

§ 424.72 Effluent limitations guidelines representing the degree of effluent reduction attainable by the application of the best practicable control technology currently available.

§ 424.73 Effluent limitations guidelines representing the degree of effluent reduction attainable by the application of the best available technology economically achievable.

Secs. 424.74--424.76 [Reserved]

§ 424.77 Effluent limitations guidelines representing the degree of effluent reduction attainable by the application of the best conventional pollutant control technology.

PART 425--LEATHER TANNING AND FINISHING POINT SOURCE CATEGORY

General Provisions

§ 425.01 Applicability.

§ 425.02 General definitions.

§ 425.03 Sulfide analytical methods and applicability.

§ 425.04 Applicability of sulfide pretreatment standards.

§ 425.36 Pretreatment standards for new sources (PSNS).

Subpart D--Retan-Wet Finish-Sides Subcategory

§ 425.40 Applicability; description of the retan-wet finish-sides subcategory.

§ 425.41 Effluent limitations representing the degree of effluent reduction attainable by the application of the best practicable control technology currently available (BPT).

§ 425.42 Effluent limitations representing the degree of effluent reduction attainable by the application of the best conventional pollutant control technology (BCT).

§ 425.43 Effluent limitations representing the degree of effluent reduction attainable by the application of the best available technology economically achievable (BAT).

§ 425.44 New source performance standards (NSPS).

§ 425.45 Pretreatment standards for existing sources (PSES).

§ 425.46 Pretreatment standards for new sources (PSNS).

Subpart E--No Beamhouse Subcategory

§ 425.50 Applicability; description of the no beamhouse subcategory.

§ 425.51 Effluent limitations representing the degree of effluent reduction attainable by the application of the best practicable control technology currently available (BPT).

§ 425.52 Effluent limitations representing the degree of effluent reduction attainable by the application of the best conventional pollutant control technology (BCT).

§ 425.53 Effluent limitations representing the degree of effluent reduction attainable by the application of the best available technology economically achievable (BAT).

§ 425.54 New source performance standards (NSPS).

§ 425.55 Pretreatment standards for existing sources (PSES).

§ 425.56 Pretreatment standards for new sources (PSNS).

Subpart F--Through-the-Blue Subcategory

§ 425.60 Applicability; description of the through-the-blue subcategory.

§ 425.61 Effluent limitations representing the degree of effluent reduction attainable by the application of the best practicable control technology currently available (BPT).

§ 425.62 Effluent limitations representing the degree of effluent reduction attainable by the application of the best conventional pollutant control technology (BCT).

§ 425.63 Effluent limitations representing the degree of effluent reduction attainable by the application of the best available technology economically achievable (BAT).

§ 425.64 New source performance standards (NSPS).

§ 425.65 Pretreatment standards for existing sources (PSES).

§ 425.66 Pretreatment standards for new sources (PSNS).

Subpart G--Shearling Subcategory

of the best conventional pollutant control technology.

Subpart D--Plate Glass Manufacturing Subcategory

§ 426.40 Applicability; description of the plate glass manufacturing subcategory.

§ 426.41 Specialized definitions.

§ 426.42 Effluent limitations guidelines representing the degree of effluent reduction attainable by the application of the best practicable control technology currently available.

§ 426.43 [Reserved]

§ 426.44 Pretreatment standards for existing sources.

§ 426.45 Standards of performance for new sources.

§ 426.46 Pretreatment standards for new sources.

§ 426.47 Effluent limitations guidelines representing the degree of effluent reduction attainable by the application of the best conventional pollutant control technology (BCT).

Subpart E--Float Glass Manufacturing Subcategory

§ 426.50 Applicability; description of the float glass manufacturing subcategory.

§ 426.51 Specialized definitions.

§ 426.52 Effluent limitations guidelines representing the degree of effluent reduction attainable by the application of the best practicable control technology currently available.

§ 426.53 Effluent limitations guidelines representing the degree of effluent reduction attainable by the application of the best available technology economically achievable.

§ 426.54 [Reserved]

§ 426.55 Standards of performance for new sources.

§ 426.56 Pretreatment standards for new sources.

§ 426.57 Effluent limitations guidelines representing the degree of effluent reduction attainable by the application of the best conventional pollutant control technology.

Subpart F--Automotive Glass Tempering Subcategory

§ 426.60 Applicability; description of the automotive glass tempering subcategory.

§ 426.61 Specialized definitions.

§ 426.62 Effluent limitations guidelines representing the degree of effluent reduction attainable by the application of the best practicable control technology currently available.

§ 426.63 [Reserved]

§ 426.64 Pretreatment standards for existing sources.

§ 426.65 Standards of performance for new sources.

§ 426.66 Pretreatment standards for new sources.

§ 426.67 Effluent limitations guidelines representing the degree of effluent reduction attainable by the application of the best conventional pollutant control technology.

Subpart G--Automotive Glass Laminating Subcategory

§ 426.70 Applicability; description of the automotive glass laminating subcategory.

§ 426.71 Specialized definitions.

§ 426.72 Effluent limitations guidelines representing the degree of effluent reduction attainable by the application of the best practicable control technology currently available.
§ 426.73 Effluent limitations guidelines representing the degree of effluent reduction attainable by the application of the best available technology economically achievable.
§ 426.74 [Reserved]
§ 426.75 Standards of performance for new sources.
§ 426.76 Pretreatment standards for new sources.
§ 426.77 Effluent limitations guidelines representing the degree of effluent reduction attainable by the application of the best conventional pollutant control technology.
Subpart H--Glass Container Manufacturing Subcategory
§ 426.80 Applicability; description of the glass container manufacturing subcategory.
§ 426.81 Specialized definitions.
§ 426.82 Effluent limitations guidelines representing the degree of effluent reduction attainable by the application of the best practicable control technology currently available.
Secs. 426.83--426.84 [Reserved]
§ 426.85 Standards of performance for new sources.
§ 426.86 Pretreatment standards for new sources.
§ 426.87 Effluent limitations guidelines representing the degree of effluent reduction attainable by the application of the best conventional pollutant control technology.
Subpart I--Machine Pressed and Blown Glass Manufacturing Subcategory [Reserved]
Subpart J--Glass Tubing (Danner) Manufacturing Subcategory
§ 426.100 Applicability; description of the glass tubing (Danner) manufacturing subcategory.
§ 426.101 Specialized definitions.
§ 426.102 Effluent limitations guidelines representing the degree of effluent reduction attainable by the application of the best practicable control technology currently available.
Secs. 426.103--426.104 [Reserved]
§ 426.105 Standards of performance for new sources.
§ 426.106 Pretreatment standards for new sources.
§ 426.107 Effluent limitations guidelines representing the degree of effluent reduction attainable by the application of the best conventional pollutant control technology.
Subpart K--Television Picture Tube Envelope Manufacturing Subcategory
§ 426.110 Applicability; description of the television picture tube envelope manufacturing subcategory.
§ 426.111 Specialized dcfinitions.
§ 426.112 Effluent limitations guidelines representing the degree of effluent reduction attainable by the application of the best practicable control technology currently available.
§ 426.113 Effluent limitations guidelines representing the

degree of effluent reduction attainable by the application of the best available technology economically achievable.
§ 426.114 [Reserved]
§ 426.115 Standards of performance for new sources.
§ 426.116 Pretreatment standards for new sources.
§ 426.117 Effluent limitations guidelines representing the degree of effluent reduction attainable by the application of the best conventional pollutant control technology.

Subpart L--Incandescent Lamp Envelope Manufacturing Subcategory

§ 426.120 Applicability; description of the incandescent lamp envelope manufacturing subcategory.
§ 426.121 Specialized definitions.
§ 426.122 Effluent limitations guidelines representing the degree of effluent reduction attainable by the application of the best practicable control technology currently available.
§ 426.123 Effluent limitations guidelines representing the degree of effluent reduction attainable by the application of the best available technology economically achievable.
§ 426.124 [Reserved]
§ 426.125 Standards of performance for new sources.
§ 426.126 Pretreatment standards for new sources.
§ 426.127 Effluent limitations guidelines representing the degree of effluent reduction attainable by the application of the best conventional pollutant control technology.

Subpart M--Hand Pressed and Blown Glass Manufacturing Subcategory

§ 426.130 Applicability; description of the hand pressed and blown glass manufacturing subcategory.
§ 426.131 Specialized definitions.
§ 426.132 Effluent limitations guidelines representing the degree of effluent reduction attainable by the application of the best practicable control technology currently available.
§ 426.133 Effluent limitations guidelines representing the degree of effluent reduction attainable by the application of the best available technology economically achievable.
§ 426.134 [Reserved]
§ 426.135 Standards of performance for new sources.
§ 426.136 Pretreatment standards for new sources.
§ 426.137 [Reserved]

PART 427--ASBESTOS MANUFACTURING POINT SOURCE CATEGORY

Subpart A--Asbestos-Cement Pipe Subcategory

§ 427.10 Applicability; description of the asbestos-cement pipe subcategory.
§ 427.11 Specialized definitions.
§ 427.12 Effluent limitations guidelines representing the degree of effluent reduction attainable by the application of the best practicable control technology currently available.
§ 427.13 Effluent limitations guidelines representing the degree of effluent reduction attainable by the application of the best available technology economically achievable.
§ 427.14 Pretreatment standards for existing sources.

§ 427.15 Standards of performance for new sources.
§ 427.16 Pretreatment standards for new sources.

Subpart B--Asbestos-Cement Sheet Subcategory

§ 427.20 Applicability; description of the asbestos-cement sheet subcategory.
§ 427.21 Specialized definitions.
§ 427.22 Effluent limitations guidelines representing the degree of effluent reduction attainable by the application of the best practicable control technology currently available.
§ 427.23 Effluent limitations guidelines representing the degree of effluent reduction attainable by the application of the best available technology economically achievable.
§ 427.24 Pretreatment standards for existing sources.
§ 427.25 Standards of performance for new sources.
§ 427.26 Pretreatment standards for new sources.

Subpart C--Asbestos Paper (Starch Binder) Subcategory

§ 427.30 Applicability; description of the asbestos paper (starch binder) subcategory.
§ 427.31 Specialized definitions.
§ 427.32 Effluent limitations guidelines representing the degree of effluent reduction attainable by the application of the best practicable control technology currently available.
§ 427.33 Effluent limitations guidelines representing the degree of effluent reduction attainable by the application of the best available technology economically achievable.
§ 427.34 Pretreatment standards for existing sources.
§ 427.35 Standards of performance for new sources.
§ 427.36 Pretreatment standards for new sources.

Subpart D--Asbestos Paper (Elastomeric Binder) Subcategory

§ 427.40 Applicability; description of the asbestos paper (elastomeric binder) subcategory.
§ 427.41 Specialized definitions.
§ 427.42 Effluent limitations guidelines representing the degree of effluent reduction attainable by the application of the best practicable control technology currently available.
§ 427.43 Effluent limitations guidelines representing the degree of effluent reduction attainable by the application of the best available technology economically achievable.
§ 427.44 Pretreatment standards for existing sources.
§ 427.45 Standards of performance for new sources.
§ 427.46 Pretreatment standards for new sources.

Subpart E--Asbestos Millboard Subcategory

§ 427.50 Applicability; description of the asbestos millboard subcategory.
§ 427.51 Specialized definitions.
§ 427.52 Effluent limitations guidelines representing the degree of effluent reduction attainable by the application of the best practicable control technology currently available.
§ 427.53 Effluent limitations guidelines representing the

degree of effluent reduction attainable by the application of the best available technology economically achievable.

§ 427.54 Pretreatment standards for existing sources.

§ 427.55 Standards of performance for new sources.

§ 427.56 Pretreatment standards for new sources.

Subpart F--Asbestos Roofing Subcategory

§ 427.60 Applicability; description of the asbestos roofing subcategory.

§ 427.61 Specialized definitions.

§ 427.62 Effluent limitations guidelines representing the degree of effluent reduction attainable by the application of the best practicable control technology currently available.

§ 427.63 Effluent limitations guidelines representing the degree of effluent reduction attainable by the application of the best available technology economically achievable.

§ 427.64 Pretreatment standards for existing sources.

§ 427.65 Standards of performance for new sources.

§ 427.66 Pretreatment standards for new sources.

Subpart G--Asbestos Floor Tile Subcategory

§ 427.70 Applicability; description of the asbestos floor tile subcategory.

§ 427.71 Specialized definitions.

§ 427.72 Effluent limitations guidelines representing the degree of effluent reduction attainable by the application of the best practicable control technology currently available.

§ 427.73 Effluent limitations guidelines representing the degree of effluent reduction attainable by the application of the best available technology economically achievable.

§ 427.74 Pretreatment standards for existing sources.

§ 427.75 Standards of performance for new sources.

§ 427.76 Pretreatment standards for new sources.

Subpart H--Coating or Finishing of Asbestos Textiles Subcategory

§ 427.80 Applicability; description of the coating or finishing of asbestos textiles subcategory.

§ 427.81 Specialized definitions.

§ 427.82 Effluent limitations guidelines representing the degree of effluent reduction attainable by the application of the best practicable control technology currently available.

§ 427.83 Effluent limitations guidelines representing the degree of effluent reduction attainable by the application of the best available technology economically achievable.

§ 427.84 [Reserved]

§ 427.85 Standards of performance for new sources.

§ 427.86 Pretreatment standards for new sources.

Subpart I--Solvent Recovery Subcategory

§ 427.90 Applicability; description of the solvent recovery subcategory.

§ 427.91 Specialized definitions.

§ 427.92 Effluent limitations guidelines representing the degree of effluent reduction attainable by the application

§ 428.16 Pretreatment standards for new sources.

Subpart B--Emulsion Crumb Rubber Subcategory

§ 428.20 Applicability; description of the emulsion crumb rubber subcategory.

§ 428.21 Specialized definitions.

§ 428.22 Effluent limitations guidelines representing the degree of effluent reduction attainable by the application of the best practicable control technology currently available.

§ 428.23 Effluent limitations guidelines representing the degree of effluent reduction attainable by the application of the best available technology economically achievable.

§ 428.24 [Reserved]

§ 428.25 Standards of performance for new sources.

Subpart C--Solution Crumb Rubber Subcategory

§ 428.30 Applicability; description of the solution crumb rubber subcategory.

§ 428.31 Specialized definitions.

§ 428.32 Effluent limitations guidelines representing the degree of effluent reduction attainable by the application of the best practicable control technology currently available.

§ 428.33 Effluent limitations guidelines representing the degree of effluent reduction attainable by the application of the best available technology economically achievable.

§ 428.34 [Reserved]

§ 428.35 Standards of performance for new sources.

Subpart D--Latex Rubber Subcategory

§ 428.40 Applicability; description of the latex rubber subcategory.

§ 428.41 Specialized definitions.

§ 428.42 Effluent limitations guidelines representing the degree of effluent reduction attainable by the application of the best practicable control technology currently available.

§ 428.43 Effluent limitations guidelines representing the degree of effluent reduction attainable by the application of the best available technology economically achievable.

§ 428.44 [Reserved]

§ 428.45 Standards of performance for new sources.

§ 428.46 Pretreatment standards for new sources.

Subpart E--Small-Sized General Molded, Extruded, and Fabricated Rubber Plants Subcategory

§ 428.50 Applicability; description of the small-sized general molded, extruded, and fabricated rubber plants subcategory.

§ 428.51 Specialized definitions.

§ 428.52 Effluent limitations guidelines representing the degree of effluent reduction attainable by the application of the best practicable control technology currently available.

§ 428.53 Effluent limitations guidelines representing the degree of effluent reduction attainable by the application

of the best available technology economically achievable.
§ 428.54 [Reserved]
§ 428.55 Standards of performance for new sources.
§ 428.56 Pretreatment standards for new sources.
Subpart F--Medium-Sized General Molded, Extruded, and Fabricated Rubber Plants Subcategory
§ 428.60 Applicability; description of the medium-sized general molded, extruded, and fabricated rubber plants subcategory.
§ 428.61 Specialized definitions.
§ 428.62 Effluent limitations guidelines representing the degree of effluent reduction attainable by the application of the best practicable control technology currently available.
§ 428.63 Effluent limitations guidelines representing the degree of effluent reduction attainable by the application of the best available technology economically achievable.
§ 428.64 [Reserved]
§ 428.65 Standards of performance for new sources.
§ 428.66 Pretreatment standards for new sources.
Subpart G--Large-Sized General Molded, Extruded, and Fabricated Rubber Plants Subcategory
§ 428.70 Applicability; description of the large-sized general molded, extruded, and fabricated rubber plants subcategory.
§ 428.71 Specialized definitions.
§ 428.72 Effluent limitations guidelines representing the degree of effluent reduction attainable by the application of the best practicable control technology currently available.
§ 428.73 Effluent limitations guidelines representing the degree of effluent reduction attainable by the application of the best available technology economically achievable.
§ 428.74 [Reserved]
§ 428.75 Standards of performance for new sources.
§ 428.76 Pretreatment standards for new sources.
Subpart H--Wet Digestion Reclaimed Rubber Subcategory
§ 428.80 Applicability; description of the wet digestion reclaimed rubber subcategory.
§ 428.81 Specialized definitions.
§ 428.82 Effluent limitations guidelines representing the degree of effluent reduction attainable by the application of the best practicable control technology currently available.
§ 428.83 Effluent limitations guidelines representing the degree of effluent reduction attainable by the application of the best available technology economically achievable.
§ 428.84 [Reserved]
§ 428.85 Standards of performance for new sources.
§ 428.86 Pretreatment standards for new sources.
Subpart I--Pan, Dry Digestion, and Mechanical Reclaimed Rubber Subcategory
§ 428.90 Applicability; description of the pan, dry

§ 429.22 Effluent limitations representing the degree of effluent reduction attainable by the application of the best conventional pollutant control technology (BCT). [Reserved]

§ 429.23 Effluent limitations representing the degree of effluent reduction attainable by the application of the best available technology economically achievable (BAT). [Reserved]

§ 429.24 New source peformance standards (NSPS).

§ 429.25 Pretreatment standards for existing sources (PSES).

§ 429.26 Pretreatment standards for new sources (PSNS).

Subpart B--Veneer Subcategory

§ 429.30 Applicability; description of the veneer subcategory.

§ 429.31 Effluent limitations representing the degree of effluent reduction attainable by the application of the best practicable control technology currently available (BPT).

§ 429.32 Effluent limitations representing the degree of effluent reduction attainable by the application of the best conventional pollutant control technology (BCT). [Reserved]

§ 429.33 Effluent limitations representing the degree of effluent reduction attainable by the application of the best available technology economically achievable (BAT).

§ 429.34 New source performance standards (NSPS).

§ 429.35 Pretreatment standards for existing sources (PSES).

§ 429.36 Pretreatment standards for new sources (PSNS).

Subpart C--Plywood Subcategory

§ 429.40 Applicability; description of the plywood subcategory.

§ 429.41 Effluent limitations representing the degree of effluent reduction attainable by the application of the best practicable control technology currently available (BPT).

§ 429.42 Effluent limitations representing the degree of effluent reduction attainable by the application of the best conventional pollutant control technology (BCT). [Reserved]

§ 429.43 Effluent limitations representing the degree of effluent reduction attainable by the application of the best available technology economically achievable (BAT).

§ 429.44 New source performance standards (NSPS).

§ 429.45 Pretreatment standards for existing sources (PSES).

§ 429.46 Pretreatment standards for new sources (PSNS).

Subpart D--Dry Process Hardboard Subcategory

§ 429.50 Applicability; description of the dry process hardboard subcategory.

§ 429.51 Effluent limitations representing the degree of effluent reduction attainable by the application of the

best practicable control technology currently available (BPT).

§ 429.52 Effluent limitations representing the degree of effluent reduction attainable by the application of the best conventional pollutant control technology (BCT). [Reserved]

§ 429.53 Effluent limitations representing the degree of effluent reduction attainable by the application of the best available technology economically achievable (BAT).

§ 429.54 New source performance standards (NSPS).

§ 429.55 Pretreatment standards for existing sources (PSES).

§ 429.56 Pretreatment standards for new sources (PSNS).

Subpart E--Wet Process Hardboard Subcategory

§ 429.60 Applicability; description of the wet process hardboard subcategory.

§ 429.61 Effluent limitations representing the degree of effluent reduction attainable by the application of the best practicable control technology currently available (BPT).

§ 429.62 [Reserved]

§ 429.63 Effluent limitations representing the degree of effluent reduction attainable by the application of the best available technology economically achievable (BAT). [Reserved]

§ 429.64 New source performance standards (NSPS).

§ 429.65 Pretreatment standards for existing sources (PSES).

§ 429.66 Pretreatment standards for new sources (PSNS).

Subpart F--Wood Preserving--Water Borne or Nonpressure Subcategory

§ 429.70 Applicability; description of the wood preserving-water borne or nonpressure subcategory.

§ 429.71 Effluent limitations representing the degree of effluent reduction attainable by the application of the best practicable control technology currently available (BPT).

§ 429.72 Effluent limitations representing the degree of effluent reduction attainable by the application of the best conventional pollutant control technology (BCT). [Reserved]

§ 429.73 Effluent limitations representing the degree of effluent reduction attainable by the application of the best available technology economically achievable (BAT).

§ 429.74 New source performance standards (NSPS).

§ 429.75 Pretreatment standards for existing sources (PSES).

§ 429.76 Pretreatment standards for new sources (PSNS).

Subpart G--Wood Preserving Steam Subcategory

§ 429.80 Applicability; description of the wood preserving--steam subcategory.

§ 429.81 Effluent limitations representing the degree of effluent reduction attainable by the application of the

best practicable control technology currently available (BPT).

§ 429.82 Effluent limitations representing the degree of effluent reduction attainable by the application of the best conventional pollutant control technology (BCT). [Reserved]

§ 429.83 Effluent limitations representing the degree of effluent reduction attainable by the application of the best available technology economically achievable (BAT). [Reserved]

§ 429.84 New source performance standards (NSPS).

§ 429.85 Pretreatment standards for existing sources (PSES).

§ 429.86 Pretreatment standards for new sources (PSNS).

Subpart H--Wood Preserving--Boulton Subcategory

§ 429.90 Applicability; description of the wood preserving--Boulton subcategory.

§ 429.91 Effluent limitations representing the degree of effluent reduction attainable by the application of the best practicable control technology currently available (BPT).

§ 429.92 Effluent limitations representing the degree of effluent reduction attainable by the application of the best conventional pollutant control technology (BCT). [Reserved]

§ 429.93 Effluent limitations representing the degree of effluent reduction attainable by the application of the best available technology economically achievable (BAT).

§ 429.94 New source performance standards (NSPS).

§ 429.95 Pretreatment standards for existing sources (PSES).

§ 429.96 Pretreatment standards for new sources (PSNS).

Subpart I--Wet Storage Subcategory

§ 429.100 Applicability; description of the wet storage subcategory.

§ 429.101 Effluent limitations representing the degree of effluent reduction attainable by the application of the best practicable control technology currently available (BPT).

§ 429.102 Effluent limitations representing the degree of effluent reduction attainable by the application of the best conventional pollutant control technology (BCT). [Reserved]

§ 429.103 Effluent limitations representing the degree of effluent reduction attainable by the application of the best available technology economically achievable (BAT).

§ 429.104 New source performance standards (NSPS).

§ 429.105 Pretreatment standards for existing sources (PSES).

§ 429.106 Pretreatment standards for new sources (PSNS).

Subpart J--Log Washing Subcategory

§ 429.110 Applicability; description of the log washing subcategory.

§ 429.111 Effluent limitations representing the degree of effluent reduction attainable by the application of the best practicable control technology currently available (BPT).

§ 429.112 Effluent limitations representing the degree of effluent reduction attainable by the application of the best conventional pollutant control technology (BCT). [Reserved]

§ 429.113 Effluent limitations representing the degree of effluent reduction attainable by the application of the best available technology economically achievable (BAT).

§ 429.114 New source performance standards (NSPS).

§ 429.115 Pretreatment standards for existing sources (PSES).

§ 429.116 Pretreatment standards for new sources (PSNS).

Subpart K--Sawmills and Planing Mills Subcategory

§ 429.120 Applicability; description of the sawmills and planing mills subcategory.

§ 429.121 Effluent limitations representing the degree of effluent reduction attainable by the application of the best practicable control technology currently available (BPT).

§ 429.122 Effluent limitations representing the degree of effluent reduction attainable by the application of the best conventional pollutant control technology (BCT). [Reserved]

§ 429.123 Effluent limitations representing the degree of effluent reduction attainable by the application of the best available technology economically achievable (BAT).

§ 429.124 New source performance standards (NSPS).

§ 429.125 Pretreatment standards for existing sources (PSES).

§ 429.126 Pretreatment standards for new sources (PSNS).

Subpart L--Finishing Subcategory

§ 429.130 Applicability; description of the finishing subcategory.

§ 429.131 Effluent limitations representing the degree of effluent reduction attainable by the application of the best practicable control technology currently available (BPT).

§ 429.132 Effluent limitations representing the degree of effluent reduction attainable by the application of the best conventional pollutant control technology (BCT). [Reserved]

§ 429.133 Effluent limitations representing the degree of effluent reduction attainable by the application of the best available technology economically achievable (BAT).

§ 429.134 New source performance standards (NSPS).

§ 429.135 Pretreatment standards for existing sources (PSES).

§ 429.136 Pretreatment standards for new sources (PSNS).

Subpart M--Particleboard Manufacturing Subcategory

§ 429.140 Applicability; description of the particleboard

manufacturing subcategory.

§ 429.141 Effluent limitations representing the degree of effluent reduction attainable by the application of the best practicable control technology currently available (BPT).

§ 429.142 Effluent limitations representing the degree of effluent reduction attainable by the application of the best conventional pollutant control technology (BCT). [Reserved]

§ 429.143 Effluent limitations representing the degree of effluent reduction attainable by the application of the best available technology economically achievable (BAT).

§ 429.144 New source performance standards (NSPS).

§ 429.145 Pretreatment standards for existing sources (PSES).

§ 429.146 Pretreatment standards for new sources (PSNS).

Subpart N--Insulation Board Subcategory

§ 429.150 Applicability; description of the insulation board subcategory.

§ 429.151 Effluent limitations representing the degree of effluent reduction attaintable by the application of the best practicable control technology currently available (BPT).

§ 429.152 [Reserved]

§ 429.153 Effluent limitations representing the degree of effluent reduction attainable by the application of the best available technology economically achievable (BAT). [Reserved]

§ 429.154 New source performance standards (NSPS).

§ 429.155 Pretreatment standards for existing sources (PSES)

§ 429.156 Pretreatment standards for new sources (PSNS).

Subpart O--Wood Furniture and Fixture Production Without Water Wash Spray Booth(s) or Without Laundry Facilities Subcategory

§ 429.160 Applicability; description of the wood furniture and fixture production without water wash spray booth(s) or without laundry facilities subcategory.

§ 429.161 Effluent limitations representing the degree of effluent reduction attainable by the application of the best practicable control technology currently available (BPT).

§ 429.162 Effluent limitations representing the degree of effluent reduction attainable by the application of the best conventional pollutant control technology (BCT). [Reserved]

§ 429.163 Effluent limitations representing the degree of effluent reduction attainable by the application of the best available technology economically achievable (BAT).

§ 429.164 New source performance standards (NSPS).

§ 429.165 Pretreatment standards for existing sources (PSES).

§ 429.166 Pretreatment standards for new sources (PSNS).

Subpart P--Wood Furniture and Fixture Production With Water Wash

Spray Booth(s) or With Laundry Facilities Subcategory

§ 429.170 Applicability; description of the wood furniture and fixture production with water wash spray booth(s) or with laundry facilities subcategory.

§ 429.171 Effluent limitations representing the degree of effluent reduction attainable by the application of the best practicable control technology currently available (BPT).

§ 429.172 Effluent limitations representing the degree of effluent reduction attainable by the application of the best conventional pollutant control technology (BCT). [Reserved]

§ 429.173 Effluent limitations representing the degree of effluent reduction attainable by the application of the best available technology economically achievable (BAT).

§ 429.174 New source performance standards (NSPS).

§ 429.175 Pretreatment standards for existing sources (PSES).

§ 429.176 Pretreatment standards for new sources (PSNS).

PART 430--PULP, PAPER, AND PAPERBOARD POINT SOURCE CATEGORY

General Provisions

§ 430.00 Applicability.

§ 430.01 General definitions.

§ 430.02 Monitoring requirements. [Reserved]

Subpart A--Unbleached Kraft Subcategory

§ 430.10 Applicability; description of the unbleached kraft subcategory.

§ 430.11 Specialized definitions.

§ 430.12 Effluent limitations representing the degree of effluent reduction attainable by the application of the best practicable control technology currently available (BPT).

§ 430.13 Effluent limitations guidelines representing the degree of effluent reduction attainable by the application of the best conventional pollutant control technology (BCT).

§ 430.14 Effluent limitations representing the degree of effluent reduction attainable by the application of the best available technology economically achievable (BAT).

§ 430.15 New source performance standards (NSPS).

§ 430.16 Pretreatment standards for existing sources (PSES).

§ 430.17 Pretreatment standards for new sources (PSNS).

Subpart B--Semi-Chemical Subcategory

§ 430.20 Applicability; description of the semi-chemical subcategory.

§ 430.21 Specialized definitions.

§ 430.22 Effluent limitations representing the degree of effluent reduction attainable by the application of the best practicable control technology currently available (BPT).

§ 430.23 Effluent limitations guidelines representing the degree of effluent reduction attainable by the application

of the best conventional pollutant control technology (BCT).

§ 430.24 Effluent limitations representing the degree of effluent reduction attainable by the application of the best available technology economically achievable (BAT).

§ 430.25 New source performance standards (NSPS).

§ 430.26 Pretreatment standards for existing sources (PSES).

§ 430.27 Pretreatment standards for new sources (PSNS).

Subpart C--[Reserved]

Subpart D--Unbleached Kraft--Neutral Sulfite Semi-Chemical (Cross Recovery) Subcategory

§ 430.40 Applicability; description of the unbleached kraft-neutral sulfite semi-chemical (cross recovery) subcategory.

§ 430.41 Specialized definitions.

§ 430.42 Effluent limitations representing the degree of effluent reduction attainable by the application of the best practicable control technology currently available (BPT).

§ 430.43 Effluent limitations guidelines representing the degree of effluent reduction attainable by the application of the best conventional pollutant control technology (BCT).

§ 430.44 Effluent limitations representing the degree of effluent reduction attainable by the application of the best available technology economically achievable (BAT).

§ 430.45 New source performance standards (NSPS).

§ 430.46 Pretreatment standards for existing sources (PSES).

§ 430.47 Pretreatment standards for new sources (PSNS).

Subpart E--Paperboard From Wastepaper Subcategory

§ 430.50 Applicability; description of the paperboard from wastepaper subcategory.

§ 430.51 Specialized definitions.

§ 430.52 Effluent limitations representing the degree of effluent reduction attainable by the application of the best practicable control technology currently available (BPT).

§ 430.53 Effluent limitations guidelines representing the degree of effluent reduction attainable by the application of the best conventional pollutant control technology (BCT).

§ 430.54 Effluent limitations representing the degree of effluent reduction attainable by the application of the best available technology economically achievable (BAT).

§ 430.55 New source performance standards (NSPS).

§ 430.56 Pretreatment standards for existing sources (PSES).

§ 430.57 Pretreatment standards for new sources (PSNS).

Subpart F--Dissolving Kraft Subcategory

§ 430.60 Applicability; description of the dissolving kraft subcategory.

§ 430.61 Specialized definitions.

§ 430.62 Effluent limitations representing the degree of effluent reduction attainable by the application of the best practicable control technology currently available (BPT).

§ 430.63 Effluent limitations guidelines representing the degree of effluent reduction attainable by the application of the best conventional pollutant control technology (BCT).

§ 430.64 Effluent limitations representing the degree of effluent reduction attainable by the application of the best available technology economically achievable (BAT).

§ 430.65 New source performance standards (NSPS).

§ 430.66 Pretreatment standards for existing sources (PSES).

§ 430.67 Pretreatment standards for new sources (PSNS).

Subpart G--Market Bleached Kraft Subcategory

§ 430.70 Applicability; description of the market bleached kraft subcategory.

§ 430.71 Specialized definitions.

§ 430.72 Effluent limitations representing the degree of effluent reduction attainable by the application of the best practicable control technology currently available (BPT).

§ 430.73 Effluent limitations guidelines representing the degree of effluent reduction attainable by the application of the best conventional pollutant control technology (BCT).

§ 430.74 Effluent limitations representing the degree of effluent reduction attainable by the application of the best available technology economically achievable (BAT).

§ 430.75 New source performance standards (NSPS).

§ 430.76 Pretreatment standards for existing sources (PSES).

§ 430.77 Pretreatment standards for new sources (PSNS).

Subpart H--BCT Bleached Kraft Subcategory

§ 430.80 Applicability; description of the BCT bleached kraft subcategory.

§ 430.81 Specialized definitions.

§ 430.82 Effluent limitations representing the degree of effluent reduction attainable by the application of the best practicable control technology currently available (BPT).

§ 430.83 Effluent limitations guidelines representing the degree of effluent reduction attainable by the application of the best conventional pollutant control technology (BCT).

§ 430.84 Effluent limitations representing the degree of effluent reduction attainable by the application of the best available technology economically achievable (BAT).

§ 430.85 New source performance standards (NSPS).

§ 430.86 Pretreatment standards for existing sources (PSES).

§ 430.87 Pretreatment standards for new sources (PSNS).

Subpart I--Fine Bleached Kraft Subcategory

§ 430.90 Applicability; description of the fine bleached kraft subcategory.

§ 430.91 Specialized definitions.

§ 430.92 Effluent limitations representing the degree of effluent reduction attainable by the application of the best practicable control technology currently available (BPT).

§ 430.93 Effluent limitations guidelines representing the degree of effluent reduction attainable by the application of the best conventional pollutant control technology (BCT).

§ 430.94 Effluent limitations representing the degree of effluent reduction attainable by the application of the best available technology economically achievable (BAT).

§ 430.95 New source performance standards (NSPS).

§ 430.96 Pretreatment standards for existing sources (PSES).

§ 430.97 Pretreatment standards for new sources (PSNS).

Subpart J--Papergrade Sulfite (Blow Pit Wash) Subcategory

§ 430.100 Applicability; description of the papergrade sulfite (blow pit wash) subcategory.

§ 430.101 Specialized definitions.

§ 430.102 Effluent limitations representing the degree of effluent reduction attainable by the application of the best practicable control technology currently available (BPT).

§ 430.103 Effluent limitations guidelines representing the degree of effluent reduction attainable by the application of the best conventional pollutant control technology (BCT).

§ 430.104 Effluent limitations representing the degree of effluent reduction attainable by the application of the best available technology economically achievable (BAT).

§ 430.105 New source performance standards (NSPS).

§ 430.106 Pretreatment standards for existing sources (PSES).

§ 430.107 Pretreatment standards for new sources (PSNS).

Subpart K--Dissolving Sulfite Pulp Subcategory

§ 430.110 Applicability; description of the dissolving sulfite pulp subcategory.

§ 430.111 Specialized definitions.

§ 430.112 Effluent limitations representing the degree of effluent reduction attainable by the application of the best practicable control technology currently available (BPT).

§ 430.113 Effluent limitations guidelines representing the degree of effluent reduction attainable by the application of the best conventional pollutant control technology (BCT).

§ 430.114 Effluent limitations representing the degree of effluent reduction attainable by the application of the

best available technology economically achievable (BAT).
§ 430.115 New source performance standards (NSPS).
§ 430.116 Pretreatment standards for existing sources (PSES).
§ 430.117 Pretreatment standards for new sources (PSNS).

Subpart L--Groundwood-Chemi-Mechanical Subcategory

§ 430.120 Applicability; description of the groundwood-chemi-mechanical subcategory.
§ 430.121 Specialized definitions.
§ 430.122 Effluent limitations representing the degree of effluent reduction attainable by the application of the best practicable control technology currently available (BPT).
§ 430.123 Effluent limitations representing the degree of effluent reduction attainable by the application of the best conventional pollutant control technology (BCT). [Reserved]
§ 430.124 Effluent limitations representing the degree of effluent reduction attainable by the application of the best available technology economically achievable (BAT). [Reserved]
§ 430.125 New source performance standards (NSPS). [Reserved]
§ 430.126 Pretreatment standards for existing sources (PSES). [Reserved]
§ 430.127 Pretreatment standards for new sources (PSNS). [Reserved]

Subpart M--Groundwood--Thermo--Mechanical Subcategory

§ 430.130 Applicability; description of the groundwood-thermo-mechanical subcategory.
§ 430.131 Specialized definitions.
§ 430.132 Effluent limitations representing the degree of effluent reduction attainable by the application of the best practicable control technology currently available (BPT).
§ 430.133 Effluent limitations representing the degree of effluent reduction attainable by the application of the best conventional pollutant control technology (BCT). [Reserved]
§ 430.134 Effluent limitations representing the degree of effluent reduction attainable by the application of the best available technology economically achievable (BAT).
§ 430.135 New source performance standards (NSPS).
§ 430.136 Pretreatment standards for existing sources (PSES).
§ 430.137 Pretreatment standards for new sources (PSNS).

Subpart N--Groundwood-CMN Papers Subcategory

§ 430.140 Applicability; description of the groundwood-CMN papers subcategory.
§ 430.141 Specialized definitions.
§ 430.142 Effluent limitations representing the degree of effluent reduction attainable by the application of the best practicable control technology currently available

§ 430.197 Pretreatment standards for new sources (PSNS).

Subpart T--Tissue from Wastepaper Subcategory

§ 430.200 Applicability; description of the tissue from wastepaper subcategory.

§ 430.201 Specialized definitions.

§ 430.202 Effluent limitations representing the degree of effluent reduction attainable by the application of the best practicable control technology currently available (BPT).

§ 430.203 Effluent limitations guidelines representing the degree of effluent reduction attainable by the application of the best conventional pollutant control technology (BCT).

§ 430.204 Effluent limitations representing the degree of effluent reduction attainable by the application of the best available technology economically achievable (BAT).

§ 430.205 New source performance standards (NSPS).

§ 430.206 Pretreatment standards for existing sources (PSES).

§ 430.207 Pretreatment standards for new sources (PSNS).

Subpart U--Papergrade Sulfite (Drum Wash) Subcategory

§ 430.210 Applicability; description of the papergrade sulfite (drum wash) subcategory.

§ 430.211 Specialized definitions.

§ 430.212 Effluent limitations representing the degree of effluent reduction attainable by the application of the best practicable control technology currently available (BPT).

§ 430.213 Effluent limitations guidelines representing the degree of effluent reduction attainable by the application of the best conventional pollutant control technology (BCT).

§ 430.214 Effluent limitations representing the degree of effluent reduction attainable by the application of the best available technology economically achievable (BAT).

§ 430.215 New source performance standards (NSPS).

§ 430.216 Pretreatment standards for existing sources (PSES).

§ 430.217 Pretreatment standards for new sources (PSNS).

Subpart V--Unbleached Kraft and Semi-Chemical Subcategory

§ 430.220 Applicability; description of the unbleached kraft and semi- chemical subcategory.

§ 430.221 Specialized definitions.

§ 430.222 Effluent limitations representing the degree of effluent reduction attainable by the application of the best practicable control technology currently available (BPT). [Reserved]

§ 430.223 Effluent limitations guidelines representing the degree of effluent reduction attainable by the application of the best conventional pollutant control technology (BCT).

§ 430.224 Effluent limitations representing the degree of effluent reduction attainable by the application of the

best available technology economically achievable (BAT).

§ 430.225 New source performance standards (NSPS).

§ 430.226 Pretreatment standards for existing sources (PSES).

§ 430.227 Pretreatment standards for new sources (PSNS).

Subpart W--Wastepaper-Molded Products Subcategory

§ 430.230 Applicability; description of the wastepaper-molded products subcategory.

§ 430.231 Specialized definitions.

§ 430.232 Effluent limitations representing the degree of effluent reduction attainable by the application of the best practicable control technology currently available (BPT).

§ 430.233 Effluent limitations guidelines representing the degree of effluent reduction attainable by the application of the best conventional pollutant control technology (BCT).

§ 430.234 Effluent limitations representing the degree of effluent reduction attainable by the application of the best available technology economically achievable (BAT).

§ 430.235 New source performance standards (NSPS).

§ 430.236 Pretreatment standards for existing sources (PSES).

§ 430.237 Pretreatment standards for new sources (PSNS).

Subpart X--Nonintegrated-Lightweight Papers Subcategory

§ 430.240 Applicability; description of the nonintegrated-lightweight papers subcategory.

§ 430.241 Specialized definitions.

§ 430.242 Effluent limitations representing the degree of effluent reduction attainable by the application of the best practicable control technology currently available (BPT).

§ 430.243 Effluent limitations guidelines representing the degree of effluent reduction attainable by the application of the best conventional pollutant control technology (BCT).

§ 430.244 Effluent limitations representing the degree of effluent reduction attainable by the application of the best available technology economically achievable (BAT).

§ 430.245 New source performance standards (NSPS).

§ 430.246 Pretreatment standards for existing sources (PSES).

§ 430.247 Pretreatment standards for new sources (PSNS).

Subpart Y--Nonintegrated-Filter and Nonwoven Papers Subcategory

§ 430.250 Applicability; description of the nonintegrated-filter and nonwoven papers subcategory.

§ 430.251 Specialized definitions.

§ 430.252 Effluent limitations representing the degree of effluent reduction attainable by the application of the best practicable control technology currently available (BPT).

§ 430.253 Effluent limitations guidelines representing the degree of effluent reduction attainable by the application

§ 432.12 Effluent limitations guidelines representing the degree of effluent reduction attainable by the application of the best practicable control technology currently available.

§ 432.13 [Reserved]

§ 432.14 Pretreatment standards for existing sources.

§ 432.15 Standards of performance for new sources.

§ 432.16 Pretreatment standards for new sources.

§ 432.17 Effluent limitations guidelines representing the degree of effluent reduction attainable by the application of the best conventional pollutant control technology.

Subpart B--Complex Slaughterhouse Subcategory

§ 432.20 Applicability; description of the complex slaughterhouse subcategory.

§ 432.21 Specialized definitions.

§ 432.22 Effluent limitations guidelines representing the degree of effluent reduction attainable by the application of the best practicable control technology currently available.

§ 432.23 [Reserved]

§ 432.24 Pretreatment standards for existing sources.

§ 432.25 Standards of performance for new sources.

§ 432.26 Pretreatment standards for new sources.

§ 432.27 Effluent limitations guidelines representing the degree of effluent reduction attainable by the application of the best conventional pollutant control technology.

Subpart C--Low-Processing Packinghouse Subcategory

§ 432.30 Applicability; description of the low-processing packinghouse subcategory.

§ 432.31 Specialized definitions.

§ 432.32 Effluent limitations guidelines representing the degree of effluent reduction attainable by the application of the best practicable control technology currently available.

§ 432.33 [Reserved]

§ 432.34 Pretreatment standards for existing sources.

§ 432.35 Standards of performance for new sources.

§ 432.36 Pretreatment standards for new sources.

§ 432.37 Effluent limitations guidelines representing the degree of effluent reduction attainable by the application of the best conventional pollutant control technology.

Subpart D--High-Processing Packinghouse Subcategory

§ 432.40 Applicability; description of the high-processing packinghouse subcategory.

§ 432.41 Specialized definitions.

§ 432.42 Effluent limitations guidelines representing the degree of effluent reduction attainable by the application of the best practicable control technology currently available.

§ 432.43 [Reserved]

§ 432.44 Pretreatment standards for existing sources.

§ 432.45 Standards of performance for new sources.

§ 432.46 Pretreatment standards for new sources.

§ 432.47 Effluent limitations guidelines representing the degree of effluent reduction attainable by the application of the best conventional pollutant control technology.

Subpart E--Small Processor Subcategory

§ 432.50 Applicability; description of the small processor subcategory.

§ 432.51 Specialized definitions.

§ 432.52 Effluent limitations guidelines representing the degree of effluent reduction attainable by the application of the best practicable control technology currently available.

Secs. 432.53--432.54 [Reserved]

§ 432.55 Standards of performance for new sources.

§ 432.56 Pretreatment standards for new sources.

§ 432.57 Effluent limitations guidelines representing the degree of effluent reduction attainable by the application of the best conventional pollutant control technology.

Subpart F--Meat Cutter Subcategory

§ 432.60 Applicability; description of the meat cutter subcategory.

§ 432.61 Specialized definitions.

§ 432.62 Effluent limitations guidelines representing the degree of effluent reduction attainable by the application of the best practicable control technology currently available.

§ 432.63 Effluent limitations guidelines representing the degree of effluent reduction attainable by the application of the best available technology economically achievable.

§ 432.64 [Reserved]

§ 432.65 Standards of performance for new sources.

§ 432.66 Pretreatment standards for new sources.

§ 432.67 Effluent limitations guidelines representing the degree of effluent reduction attainable by the application of the best conventional pollutant control technology.

Subpart G--Sausage and Luncheon Meats Processor Subcategory

§ 432.70 Applicability; description of the sausage and luncheon meat processor subcategory.

§ 432.71 Specialized definitions.

§ 432.72 Effluent limitations guidelines representing the degree of effluent reduction attainable by the application of the best practicable control technology currently available.

§ 432.73 Effluent limitations guidelines representing the degree of effluent reduction attainable by the application of the best available technology economically achievable.

§ 432.74 [Reserved]

§ 432.75 Standards of performance for new sources.

§ 432.76 Pretreatment standards for new sources.

§ 432.77 Effluent limitations guidelines representing the degree of effluent reduction attainable by the application of the best conventional pollutant control technology.

Subpart H--Ham Processor Subcategory

§ 432.80 Applicability; description of the ham processor

§ 434.40 Applicability; description of the alkaline mine drainage subcategory.

§ 434.41 [Reserved]

§ 434.42 Effluent limitations guidelines representing the degree of effluent reduction attainable by the application of the best practicable control technology currently available (BPT).

§ 434.43 Effluent limitations guidelines representing the degree of effluent reduction attainable by application of the best available technology economically achievable (BAT).

§ 434.44 Effluent limitations guidelines representing the degree of effluent reduction attainable by the application of the best conventional pollutant control technology (BCT). [Reserved]

§ 434.45 New source performance standards (NSPS).

Subpart E--Post-Mining Areas

§ 434.50 Applicability.

§ 434.51 [Reserved]

§ 434.52 Effluent limitations quidelines representing the degree of effluent reduction attainable by the application of the best practicable control technology currently available (BPT).

§ 434.53 Effluent limitations guidelines representing the degree of effluent reduction attainable by application of the best available technology economically achievable (BAT).

§ 434.54 Effluent limitations guidelines representing the degree of effluent reduction attainable by the application of the best conventional pollutant control technology (BCT). [Reserved]

§ 434.55 New source performance standards (NSPS).

Subpart F--Miscellaneous Provisions

§ 434.60 Applicability.

§ 434.61 Commingling of waste streams.

§ 434.62 Alternate effluent limitation for pH.

§ 434.63 Effluent limitations for precipitation events.

§ 434.64 Procedure and method detection limit for measurement of settleable solids.

§ 434.65 Modification of NPDES permits for new sources.

PART 435--OIL AND GAS EXTRACTION POINT SOURCE CATEGORY

Subpart A--Offshore Subcategory

§ 435.10 Applicability; description of the offshore subcategory.

§ 435.11 Specialized definitions.

§ 435.12 Effluent limitations guidelines representing the degree of effluent reduction attainable by the application of the best practicable control technology currently available (BPT).

§ 435.13 Effluent limitations guidelines representing the degree of effluent reduction attainable by the application of the best available technology economically achievable (BAT).

of the best practicable control technology currently available.

Subpart D--Industrial Sand Subcategory

§ 436.40 Applicability; description of the industrial sand subcategory.

§ 436.41 Specialized definitions.

§ 436.42 Effluent limitations guidelines representing the degree of effluent reduction attainable by the application of the best practicable control technology currently available.

Subpart E--Gypsum Subcategory

§ 436.50 Applicability; description of the gypsum subcategory.

§ 436.51 Specialized definitions.

§ 436.52 Effluent limitations guidelines representing the degree of effluent reduction attainable by the application of the best practicable control technology currently available.

Subpart F--Asphaltic Mineral Subcategory

§ 436.60 Applicability; description of the asphaltic mineral subcategory.

§ 436.61 Specialized definitions.

§ 436.62 Effluent limitations guidelines representing the degree of effluent reduction attainable by the application of the best practicable control technology currently available.

Subpart G--Asbestos and Wollastonite Subcategory

§ 436.70 Applicability; description of the asbestos and wollastonite subcategory.

§ 436.71 Specialized definitions.

§ 436.72 Effluent limitations guidelines representing the degree of effluent reduction attainable by the application of the best practicable control technology currently available.

Subpart H--Lightweight Aggregates Subcategory [Reserved]

Subpart I--Mica and Sericite Subcategory [Reserved]

Subpart J--Barite Subcategory

§ 436.100 Applicability; description of the barite subcategory.

§ 436.101 Specialized definitions.

§ 436.102 Effluent limitations guidelines representing the degree of effluent reduction attainable by the application of the best practicable control technology currently available.

Subpart K--Fluorspar Subcategory

§ 436.110 Applicability; description of the fluorspar subcategory.

§ 436.111 Specialized definitions.

§ 436.112 Effluent limitations guidelines representing the degree of effluent reduction attainable by the application of the best practicable control technology currently available.

Subpart L--Salines From Brine Lakes Subcategory

§ 436.120 Applicability; description of the salines from brine lakes subcategory.
§ 436.121 Specialized definitions.
§ 436.122 Effluent limitations guidelines representing the degree of effluent reduction attainable by the application of the best practicable control technology currently available.

Subpart M--Borax Subcategory

§ 436.130 Applicability; description of the borax subcategory.
§ 436.131 Specialized definitions.
§ 436.132 Effluent limitations guidelines representing the degree of effluent reduction attainable by the application of the best practicable control technology currently available.

Subpart N--Potash Subcategory

§ 436.140 Applicability; description of the potash subcategory.
§ 436.141 Specialized definitions.
§ 436.142 Effluent limitations guidelines representing the degree of effluent reduction attainable by the application of the best practicable control technology currently available.

Subpart O--Sodium Sulfate Subcategory

§ 436.150 Applicability; description of the sodium sulfate subcategory.
§ 436.151 Specialized definitions.
§ 436.152 Effluent limitations guidelines representing the degree of effluent reduction attainable by the application of the best practicable control technology currently available.

Subpart P--Trona Subcategory [Reserved]

Subpart Q--Rock Salt Subcategory [Reserved]

Subpart R--Phosphate Rock Subcategory

§ 436.180 Applicability; description of the phosphate rock subcategory.
§ 436.181 Specialized definitions.
§ 436.182 Effluent limitations guidelines representing the degree of effluent reduction attainable by the application of the best practicable control technology currently available.
Secs. 436.183-436.184 [Reserved]
§ 436.185 Standards of performance for new sources.

Subpart S--Frasch Sulfur Subcategory

§ 436.190 Applicability; description of the Frasch sulfur subcategory.
§ 436.191 Specialized definitions.
§ 436.192 Effluent limitations guidelines representing the degree of effluent reduction attainable by the application of the best practicable control technology currently available.

Subpart T--Mineral Pigments Subcategory [Reserved]

Subpart U--Lithium Subcategory [Reserved]

Subpart V--Bentonite Subcategory

§ 436.220 Applicability; description of the bentonite subcategory.

§ 436.221 Specialized definitions.

§ 436.222 Effluent limitations guidelines representing the degree of effluent reduction attainable by the application of the best practicable control technology currently available.

Subpart W--Magnesite Subcategory

§ 436.230 Applicability; description of the magnesite subcategory.

§ 436.231 Specialized definitions.

§ 436.232 Effluent limitations guidelines representing the degree of effluent reduction attainable by the application of the best practicable control technology currently available.

Subpart X--Diatomite Subcategory

§ 436.240 Applicability; description of the diatomite subcategory.

§ 436.241 Specialized definitions.

§ 436.242 Effluent limitations guidelines representing the degree of effluent reduction attainable by the application of the best practicable control technology currently available.

Subpart Y--Jade Subcategory

§ 436.250 Applicability; description of the jade subcategory.

§ 436.251 Specialized definitions.

§ 436.252 Effluent limitations guidelines representing the degree of effluent reduction attainable by the application of the best practicable control technology currently available.

Subpart Z--Novaculite Subcategory

§ 436.260 Applicability; description of the novaculite subcategory.

§ 436.261 Specialized definitions.

§ 436.262 Effluent limitations guidelines representing the degree of effluent reduction attainable by the application of the best practicable control technology currently available.

Subpart AA--Fire Clay Subcategory [Reserved]

Subpart AB--Attapulgite and Montmorillonite Subcategory [Reserved]

Subpart AC--Kyanite Subcategory [Reserved]

Subpart AD--Shale and Common Clay Subcategory [Reserved]

Subpart AE--Aplite Subcategory [Reserved]

Subpart AF--Tripoli Subcategory

§ 436.310 Applicability; description of the tripoli subcategory.

§ 436.321 Specialized definitions.

§ 436.322 Effluent limitations guidelines representing the degree of effluent reduction attainable by the application of the best practicable control technology currently

available.

Subpart AG--Kaolin Subcategory [Reserved]

Subpart AH--Ball Clay Subcategory [Reserved]

Subpart AI--Feldspar Subcategory [Reserved]

Subpart AJ--Talc, Steatite, Soapstone and Pyrophyllite Subcategory [Reserved]

Subpart AK--Garnet Subcategory [Reserved]

Subpart AL--Graphite Subcategory

§ 436.380 Applicability; description of the graphite subcategory.

§ 436.381 Specialized definitions.

§ 436.382 Effluent limitations guidelines representing the degree of effluent reduction attainable by the application of the best practicable control technology currently available.

PART 439--PHARMACEUTICAL MANUFACTURING POINT SOURCE CATEGORY

General Provisions

§ 439.0 Applicability.

§ 439.1 General definitions.

§ 439.2 Monitoring requirements.

Subpart A--Fermentation Products Subcategory

§ 439.10 Applicability; description of the fermentation products subcategory.

§ 439.11 Specialized definitions.

§ 439.12 Effluent limitations representing the degree of effluent reduction attainable by the application of the best practicable control technology currently available (BPT).

§ 439.13 Effluent limitations representing the degree of effluent reduction attainable by the application of the best conventional pollutant control technology (BCT).

§ 439.14 Effluent limitations representing the degree of effluent reduction attainable by the application of the best available technology economically achievable (BAT).

§ 439.15 New source performance standards (NSPS).

§ 439.16 Pretreatment standards for existing sources (PSES).

§ 439.17 Pretreatment standards for new sources (PSNS).

Subpart B--Extraction Products Subcategory

§ 439.20 Applicability; description of the extraction products subcategory.

§ 439.21 Specialized definitions.

§ 439.22 Effluent limitations representing the degree of effluent reduction attainable by the application of the best practicable control technology currently available (BPT).

§ 439.23 Effluent limitations representing the degree of effluent reduction attainable by the application of the best conventional pollutant control technology (BCT).

§ 439.24 Effluent limitations representing the degree of effluent reduction attainable by the application of the best available technology economically achievable (BAT).

§ 439.25 New source performance standards (NSPS).

§ 439.26 Pretreatment standards for existing sources (PSES).

§ 439.27 Pretreatment standards for new sources (PSNS).

Subpart C--Chemical Synthesis Products Subcategory

§ 439.30 Applicability; description of the chemical synthesis products subcategory.

§ 439.31 Specialized definitions.

§ 439.32 Effluent limitations representing the degree of effluent reduction attainable by the application of the best practicable control technology currently available (BPT).

§ 439.33 Effluent limitations representing the degree of effluent reduction attainable by the application of the best conventional pollutant control technology (BCT).

§ 439.34 Effluent limitations representing the degree of effluent reduction attainable by the application of the best available technology economically achievable (BAT).

§ 439.35 New source performance standards (NSPS).

§ 439.36 Pretreatment standards for existing sources (PSES).

§ 439.37 Pretreatment standards for new sources (PSNS).

Subpart D--Mixing/Compounding and Formulation Subcategory

§ 439.40 Applicability; description of the mixing/ compounding and formulation subcategory.

§ 439.41 Specialized definitions.

§ 439.42 Effluent limitations representing the degree of effluent reduction attainable by the application of the best practicable control technology currently available (BPT).

§ 439.43 Effluent limitations representing the degree of effluent reduction attainable by the application of the best conventional pollutant control technology (BCT).

§ 439.44 Effluent limitations representing the degree of effluent reduction attainable by the application of the best available technology economically achievable (BAT).

§ 439.45 New source performance standards (NSPS).

§ 439.46 Pretreatment standards for existing sources (PSES).

§ 439.47 Pretreatment standards for new sources (PSNS).

Subpart E--Research Subcategory

§ 439.50 Applicability; description of the research subcategory.

§ 439.51 Specialized definitions.

§ 439.52 Effluent limitations representing the degree of effluent reduction attainable by the application of the best practicable control technology currently available (BPT).

§ 439.53 Effluent limitations representing the degree of effluent reduction attainable by the application of the best conventional pollutant control technology (BCT). [Reserved]

§ 439.54 Effluent limitations representing the degree of effluent reduction attainable by the application of the

best available technology economically achievable (BAT). [Reserved]

§ 439.55 New source performance standards (NSPS). [Reserved]

§ 439.56 Pretreatment standards for existing sources (PSES). [Reserved]

§ 439.57 Pretreatment standards for new sources (PSNS). [Reserved]

PART 440--ORE MINING AND DRESSING POINT SOURCE CATEGORY

Subpart A--Iron Ore Subcategory

§ 440.10 Applicability; description of the iron ore subcategory.

§ 440.11 [Reserved]

§ 440.12 Effluent limitations representing the degree of effluent reduction attainable by the application of the best practicable control technology currently available (BPT).

§ 440.13 Effluent limitations representing the degree of effluent reduction attainable by the application of the best available technology economically achievable (BAT).

§ 440.14 New source performance standards (NSPS).

§ 440.15 Effluent limitations representing the degree of effluent reduction attainable by the application of the best conventional pollutant control technology (BCT). [Reserved]

Subpart B--Aluminum Ore Subcategory

§ 440.20 Applicability; description of the aluminum ore subcategory.

§ 440.21 [Reserved]

§ 440.22 Effluent limitations representing the degree of effluent reduction attainable by the application of the best practicable control technology currently available (BPT).

§ 440.23 Effluent limitations representing the degree of effluent reduction attainable by the application of the best available technology economically achievable (BAT).

§ 440.24 New Source performance standards (NSPS).

§ 440.25 Effluent limitations representing the degree of effluent reduction attainable by the application of the best conventional pollutant control technology (BCT). [Reserved]

Subpart C--Uranium, Radium and Vanadium Ores Subcategory

§ 440.30 Applicability; description of the uranium, radium and vanadium ores subcategory.

§ 440.31 [Reserved]

§ 440.32 Effluent limitations representing the degree of effluent reduction attainable by the application of the best practicable control technology currently available (BPT).

§ 440.33 Effluent limitations representing the degree of effluent reduction attainable by the application of the best available technology economically achievable (BAT).

§ 440.34 New source performance standards (NSPS).

§ 440.35 Effluent limitations representing the degree of effluent reduction attainable by the application of the best conventional pollutant control technology (BCT). [Reserved]

Subpart D--Mercury Ore Subcategory

§ 440.40 Applicability; description of the mercury ore subcategory.

§ 440.41 [Reserved]

§ 440.42 Effluent limitations representing the degree of effluent reduction attainable by the application of the best practicable control technology currently available (BPT).

§ 440.43 Effluent limitations representing the degree of effluent reduction attainable by the application of the best available technology economically achievable (BAT).

§ 440.44 New source performance standards (NSPS).

§ 440.45 Effluent limitations representing the degree of effluent reduction attainable by the application of the best conventional pollutant control technology (BCT). [Reserved]

Subpart E--Titanium Ore Subcategory

§ 440.50 Applicability; description of the titanium ore subcategory.

§ 440.51 [Reserved]

§ 440.52 Effluent limitations guidelines representing the degree of effluent reduction attainable by the application of the best practicable control technology currently available (BPT).

§ 440.53 Effluent limitations representing the degree of effluent reduction attainable by the application of the best available technology economically achievable (BAT).

§ 440.54 New source performance standards (NSPS).

§ 440.55 Effluent limitations representing the degree of effluent reduction attainable by the application of the best conventional pollutant control technology (BCT). [Reserved]

Subpart F--Tungsten Ore Subcategory

§ 440.60 Applicability; description of the tungsten ore subcategory.

§ 440.61 [Reserved]

§ 440.62 Effluent limitations representing the degree of effluent reduction attainable by the application of the best practicable control technology currently available (BPT).

§ 440.63 Effluent limitations representing the degree of effluent reduction attainable by the application of the best available technology economically achievable (BAT).

§ 440.64 New source performance standards (NSPS).

§ 440.65 Effluent limitations representing the degree of effluent reduction attainable by the application of the best conventional pollutant control technology (BCT). [Reserved]

Subpart G--Nickel Ore Subcategory

§ 440.70 Applicability; description of the nickel ore subcategory.
§ 440.71 [Reserved]
§ 440.72 Effluent limitations representing the degree of effluent reduction attainable by the application of the best practicable control technology currently available (BPT).
§ 440.73 Effluent limitations representing the degree of effluent reduction attainable by the application of the best available technology economically achievable (BAT). [Reserved]
§ 440.74 New source performance standards (NSPS). [Reserved]
§ 440.75 Effluent limitations representing the degree of effluent reduction attainable by the application of the best conventional pollutant control technology (BCT). [Reserved]

Subpart H--Vanadium Ore Subcategory (Mined Alone and Not as a Byproduct)

§ 440.80 Applicability; description of the vanadium ore subcategory.
§ 440.81 [Reserved]
§ 440.82 Effluent limitations representing the degree of effluent reduction attainable by the application of the best practicable control technology currently available (BPT).
§ 440.83 Effluent limitations representing the degree of effluent reduction attainable by the application of the best available technology economically achievable (BAT). [Reserved]
§ 440.84 New source performance standards (NSPS). [Reserved]
§ 440.85 Effluent limitations representing the degree of effluent reduction attainable by the application of the best conventional pollutant control technology (BCT). [Reserved]

Subpart I--Antimony Ore Subcategory

§ 440.90 Applicability; description of the antimony ore subcategory.
§ 440.91 [Reserved]
§ 440.92 Effluent limitations representing the degree of effluent reduction attainable by the application of the best practicable control technology currently available (BPT). [Reserved]
§ 440.93 Effluent limitations representing the degree of effluent reduction attainable by the application of the best available technology economically achievable (BAT). [Reserved]
§ 440.94 New source performance standards (NSPS). [Reserved]
§ 440.95 Effluent limitations representing the degree of effluent reduction attainable by the application of the best conventional pollutant control technology (BCT).

[Reserved]

Subpart J--Copper, Lead, Zinc, Gold, Silver, and Molybdenum Ores Subcategory

§ 440.100 Applicability; description of the copper, lead, zinc, gold, silver, and molybdenum ores subcategory.

§ 440.101 [Reserved]

§ 440.102 Effluent limitations representing the degree of effluent reduction attainable by the application of the best practicable control technology (BPT).

§ 440.103 Effluent limitations representing the degree of effluent reduction attainable by the application of the best available technology economically achievable (BAT).

§ 440.104 New source performance standards (NSPS).

§ 440.105 Effluent limitations representing the degree of effluent reduction attainable by the application of the best conventional pollutant control technology (BCT). [Reserved]

Subpart K--Platinum Ores Subcategory

§ 440.110 Applicability; description of the platinum ore subcategory.

§ 440.111 [Reserved]

§ 440.112 Effluent limitations representing the degree of effluent reduction attainable by the application of the best practicable control technology currently available (BPT). [Reserved]

§ 440.113 Effluent limitations representing the degree of effluent reduction attainable by the application of the best available technology economically achievable (BAT).

§ 440.114 New source performance standards (NSPS). [Reserved]

§ 440.115 Effluent limitations representing the degree of effluent reduction attainable by the application of the best conventional pollutant control technology (BTC). [Reserved]

Subpart L--General Provisions and Definitions

§ 440.130 Applicability.

§ 440.131 General provisions.

§ 440.132 General definitions.

Subpart M--Gold Placer Mine Subcategory

§ 440.140 Applicability; description of the gold placer mine subcategory.

§ 440.141 Specialized definitions and provisions.

§ 440.142 Effluent limitations representing the degree of effluent reduction attainable by the application of the best practicable control technology currently available (BPT).

§ 440.143 Effluent limitations representing the degree of effluent reduction attainable by the application of the best available technology economically achievable (BAT).

§ 440.144 New Source Performance Standards (NSPS).

Secs. 440.145--440.147 [Reserved]

§ 440.148 Best Management Practices (BMP).

PART 443--EFFLUENT LIMITATIONS GUIDELINES FOR EXISTING SOURCES AND

STANDARDS OF PERFORMANCE AND PRETREATMENT STANDARDS FOR NEW SOURCES FOR THE PAVING AND ROOFING MATERIALS (TARS AND ASPHALT) POINT SOURCE CATEGORY

§ 443.43 Effluent limitations guidelines representing the degree of effluent reduction attainable by the application of the best available technology economically achievable.
§ 443.44 [Reserved]
§ 443.45 Standards of performance for new sources.
§ 443.46 Pretreatment standard for new sources.

PART 446--PAINT FORMULATING POINT SOURCE CATEGORY

Subpart A--Oil-Base Solvent Wash Paint Subcategory

§ 446.10 Applicability; description of the oil-base solvent wash paint subcategory.
§ 446.11 Specialized definitions.
§ 446.12 Effluent limitations guidelines representing the degree of effluent reduction attainable by the application of the best practicable control technology currently available.
§ 446.13 Effluent limitations guidelines representing the degree of effluent reduction attainable by the application of the best available technology economically achievable.
§ 446.14 [Reserved]
§ 446.15 Standards of performance for new sources.
§ 446.16 Pretreatment standard for new sources.

PART 447--INK FORMULATING POINT SOURCE CATEGORY

Subpart A--Oil-Base Solvent Wash Ink Subcategory

§ 447.10 Applicability; description of the oil-base solvent wash ink subcategory.
§ 447.11 Specialized definitions.
§ 447.12 Effluent limitations guidelines representing the degree of effluent reduction attainable by the application of the best practicable control technology currently available.
§ 447.13 Effluent limitations guidelines representing the degree of effluent reduction attainable by the application of the best available technology economically achievable.
§ 447.14 [Reserved]
§ 447.15 Standards of performance for new sources.
§ 447.16 Pretreatment standard for new sources.

PART 454--GUM AND WOOD CHEMICALS MANUFACTURING POINT SOURCE CATEGORY

Subpart A--Char and Charcoal Briquets Subcategory

§ 454.10 Applicability; description of the manufacture of char and charcoal briquets subcategory.
§ 454.11 Specialized definitions.
§ 454.12 Effluent limitations and guidelines representing the degree of effluent reduction attainable by the application of the best practicable control technology currently available.

Subpart B--Gum Rosin and Turpentine Subcategory

§ 454.20 Applicability; description of the manufacture of gum rosin and turpentine subcategory.
§ 454.21 Specialized definitions.
§ 454.22 Effluent limitations and guidelines representing the degree of effluent reduction attainable by the application of the best practicable control technology currently available.

Subpart C--Wood Rosin, Turpentine and Pine Oil Subcategory

§ 454.30 Applicability; description of the manufacture of wood rosin, turpentine and pine oil subcategory.

§ 454.31 Specialized definitions.

§ 454.32 Effluent limitations and guidelines representing the degree of effluent reduction attainable by the application of the best practicable control technology currently available.

Subpart D--Tall Oil Rosin, Pitch and Fatty Acids Subcategory

§ 454.40 Applicability; description of manufacture of tall oil rosin, pitch and fatty acids subcategory.

§ 454.41 Specialized definitions.

§ 454.42 Effluent limitations and guidelines representing the degree of effluent reduction attainable by the application of the best practicable control technology currently available.

Subpart E--Essential Oils Subcategory

§ 454.50 Applicability; description of the essential oils subcategory.

§ 454.51 Specialized definitions.

§ 454.52 Effluent limitations and guidelines representing the degree of effluent reduction attainable by the application of the best practicable control technology currently available.

Subpart F--Rosin-Based Derivatives Subcategory

§ 454.60 Applicability; description of manufacture of rosin-based derivatives subcategory.

§ 454.61 Specialized definitions.

§ 454.62 Effluent limitations and guidelines representing the degree of effluent reduction attainable by the application of the best practicable control technology currently available.

PART 455--PESTICIDE CHEMICALS

§ 455.10 General definitions.

§ 455.11 Compliance date for pretreatment standards for existing sources (PSES).

Subpart A--Organic Pesticide Chemicals Manufacturing Subcategory

§ 455.20 Applicability; description of the organic pesticide chemicals manufacturing subcategory.

§ 455.21 Specialized definitions.

§ 455.22 Effluent limitations guidelines representing the degree of effluent reduction attainable by the application of the best practicable control technology currently available.

§ 455.23 Effluent limitations guidelines representing the degree of effluent reduction attainable by the application of the best conventional pollutant control technology (BCT).

§ 455.24 Effluent limitations guidelines representing the degree of effluent reduction attainable by the application of the best available control technology economically achievable (BAT).

§ 455.25 New source performance standards (NSPS).

best available technology economically achievable (BAT).
§ 461.13 New source performance standards (NSPS).
§ 461.14 Pretreatment standards for existing sources (PSES).
§ 461.15 Pretreatment standards for new sources (PSNS).

Subpart B--Calcium Subcategory
§ 461.20 Applicability; description of the calcium subcategory.
Secs. 461.21-461.22 [Reserved]
§ 461.23 New source performance standards (NSPS).
§ 461.24 [Reserved]
§ 461.25 Pretreatment standards for new sources (PSNS).

Subpart C--Lead Subcategory
§ 461.30 Applicability; description of the lead subcategory.
§ 461.31 Effluent limitations representing the degree of effluent reduction attainable by the application of the best practicable control technology currently available (BPT).
§ 461.32 Effluent limitations representing the degree of effluent reduction attainable by the application of the best available technology economically achievable (BAT).
§ 461.33 New source performance standards (NSPS).
§ 461.34 Pretreatment standards for existing sources (PSES).
§ 461.35 Pretreatment standards for new sources (PSNS).

Subpart D--Leclanche Subcategory
§ 461.40 Applicability; description of the Leclanche subcategory.
Secs. 461.41-461.42 [Reserved]
§ 461.43 New source performance standards (NSPS).
§ 461.44 Pretreatment standards for existing sources (PSES).
§ 461.45 Pretreatment standards for new sources (PSNS).

Subpart E--Lithium Subcategory
§ 461.50 Applicability; description of the lithium subcategory.
Secs. 461.51-461.52 [Reserved]
§ 461.53 New source performance standards (NSPS).
§ 461.54 [Reserved]
§ 461.55 Pretreatment standards for new sources (PSNS).

Subpart F--Magnesium Subcategory
§ 461.60 Applicability; description of the magnesium subcategory.
Secs. 461.61-461.62 [Reserved]
§ 461.63 New source performance standards (NSPS).
§ 461.64 Pretreatment standards for existing sources (PSES).
§ 461.65 Pretreatment standards for new sources (PSNS).

Subpart G--Zinc Subcategory
§ 461.70 Applicability; description of the zinc subcategory.
§ 461.71 Effluent limitations representing the degree of

§ 463.31 Specialized definitions.
§ 463.32 Effluent limitations guidelines representing the degree of effluent reduction attainable by the application of the best practicable control technology currently available.
§ 463.33 Effluent limitations guidelines representing the degree of effluent reduction attainable by the application of the best available technology economically achievable.
§ 463.34 New source performance standards.
§ 463.35 Pretreatment standards for existing sources.
§ 463.36 Pretreatment standards for new sources.
§ 463.37 Effluent limitations guidelines representing the degree of effluent reduction attainable by the application of the best conventional pollutant control technology. [Reserved]

PART 464--METAL MOLDING AND CASTING POINT SOURCE CATEGORY

General Provisions

§ 464.01 Applicability.
§ 464.02 General definitions.
§ 464.03 Monitoring and reporting requirements.
§ 464.04 Compliance date for PSES.

Subpart A--Aluminum Casting Subcategory

§ 464.10 Applicability; description of the aluminum casting subcategory.
§ 464.11 Specialized definitions.
§ 464.12 Effluent limitations guidelines representing the degree of effluent reduction attainable by the application of the best practicable control technology currently available.
§ 464.13 Effluent limitations guidelines representing the degree of effluent reduction attainable by the application of the best available technology economically achievable.
§ 464.14 New source performance standards.
§ 464.15 Pretreatment standards for existing sources.
§ 464.16 Pretreatment standards for new sources.
§ 464.17 Effluent limitations guidelines representing the degree of effluent reduction attainable by the application of the best conventional pollutant control technology. [Reserved]

Subpart B--Copper Casting Subcategory

§ 464.20 Applicability; description of the copper casting subcategory.
§ 464.21 Specialized definitions.
§ 464.22 Effluent limitations guidelines representing the degree of effluent reduction attainable by the application of the best practicable control technology currently available.
§ 464.23 Effluent limitations guidelines representing the degree of effluent reduction attainable by the application of the best available technology economically achievable.
§ 464.24 New source performance standards.
§ 464.25 Pretreatment standards for existing sources.
§ 464.26 Pretreatment standards for new sources.

§ 464.27 Effluent limitations guidelines representing the degree of effluent reduction attainable by the application of the best conventional pollutant control technology. [Reserved]

Subpart C--Ferrous Casting Subcategory

§ 464.30 Applicability; description of the ferrous casting subcategory.

§ 464.31 Specialized definitions.

§ 464.32 Effluent limitations guidelines representing the degree of effluent reduction attainable by the application of the best practicable control technology currently available.

§ 464.33 Effluent limitations guidelines representing the degree of effluent reduction attainable by the application of the best available technology economically achievable.

§ 464.34 New source performance standards.

§ 464.35 Pretreatment standards for existing sources.

§ 464.36 Pretreatment standards for new sources.

§ 464.37 Effluent limitations guidelines representing the degree of effluent reduction attainable by the application of the best conventional pollutant control technology. [Reserved]

Subpart D--Zinc Casting Subcategory

§ 464.40 Applicability; description of the zinc casting subcategory.

§ 464.41 Specialized definitions.

§ 464.42 Effluent limitations guidelines representing the degree of effluent reduction attainable by the application of the best practicable control technology currently available.

§ 464.43 Effluent limitations guidelines representing the degree of effluent reduction attainable by the application of the best available technology economically achievable.

§ 464.44 New source performance standards.

§ 464.45 Pretreatment standards for existing sources.

§ 464.46 Pretreatment standards for new sources.

§ 464.47 Effluent limitations guidelines representing the degree of effluent reduction attainable by the application of the best conventional pollutant control technology. [Reserved]

PART 465--COIL COATING POINT SOURCE CATEGORY

General Provisions

§ 465.01 Applicability.

§ 465.02 General definitions.

§ 465.03 Monitoring and reporting requirements.

§ 465.04 Compliance date for PSES.

Subpart A--Steel Basis Material Subcategory

§ 465.10 Applicability; description of the steel basis material subcategory.

§ 465.11 Effluent limitations representing the degree of effluent reduction attainable by the application of the best practicable control technology currently available.

§ 465.12 Effluent limitations representing the degree of

effluent reduction attainable by the application of the best available technology economically achievable.

§ 465.13 New source performance standards.

§ 465.14 Pretreatment standards for existing sources.

§ 465.15 Pretreatment standards for new sources.

Subpart B--Galvanized Basis Material Subcategory

§ 465.20 Applicability; description of the galvanized basis material subcategory.

§ 465.21 Effluent limitations representing the degree of effluent reduction attainable by the application of the best practicable control technology currently available.

§ 465.22 Effluent limitations representing the degree of effluent reduction attainable by the application of the best available technology economically achievable.

§ 465.23 New source performance standards.

§ 465.24 Pretreatment standards for existing sources.

§ 465.25 Pretreatment standards for new sources.

Subpart C--Aluminum Basis Material Subcategory

§ 465.30 Applicability; description of the aluminum basis material subcategory.

§ 465.31 Effluent limitations representing the degree of effluent reduction attainable by the application of the best practicable control technology currently available.

§ 465.32 Effluent limitations representing the degree of effluent reduction attainable by the application of the best available technology economically achievable.

§ 465.33 New source performance standards.

§ 465.34 Pretreatment standards for existing sources.

§ 465.35 Pretreatment standards for new sources.

Subpart D--Canmaking Subcategory

§ 465.40 Applicability; description of the canmaking subcategory.

§ 465.41 Effluent limitations representing the degree of effluent reduction attainable by the application of the best practicable control technology currently available.

§ 465.42 Effluent limitations representing the degree of effluent reduction attainable by the application of the best available technology economically achievable.

§ 465.43 New source performance standards.

§ 465.44 Pretreatment standards for existing sources.

§ 465.45 Pretreatment standards for new sources.

§ 465.46 Effluent limitations representing the degree of effluent reduction attainable by the application of the best conventional pollutant control technology. [Reserved]

PART 466--PORCELAIN ENAMELING POINT SOURCE CATEGORY

General Provisions

§ 466.01 Applicability.

§ 466.02 General definitions.

§ 466.03 Monitoring and reporting requirements.

§ 466.04 Compliance date for PSES.

Subpart A--Steel Basis Material Subcategory

§ 466.10 Applicability; description of the steel basis material.

best practicable control technology currently available.
§ 467.13 Effluent limitations representing the degree of effluent reduction attainable by the application of the best available technology economically achievable.
§ 467.14 New source performance standards.
§ 467.15 Pretreatment standards for existing sources.
§ 467.16 Pretreatment standards for new sources.
§ 467.17 Effluent limitations representing the degree of effluent reduction attainable by the application of the best conventional pollutant control technology. [Reserved]

Subpart B--Rolling With Emulsions Subcategory
§ 467.20 Applicability; description of the rolling with emulsions subcategory.
§ 467.21 Specialized definitions.
§ 467.22 Effluent limitations representing the degree of effluent reduction attainable by the application of the best practicable control technology currently available.
§ 467.23 Effluent limitations representing the degree of effluent reduction attainable by the application of the best available technology economically achievable.
§ 467.24 New source performance standards.
§ 467.25 Pretreatment standards for existing sources.
§ 467.26 Pretreatment standards for new sources.
§ 467.27 Effluent limitations representing the degree of effluent reduction attainable by the application of the best conventional pollutant control technology. [Reserved]

Subpart C--Extrusion Subcategory
§ 467.30 Applicability; description of the extrusion subcategory.
§ 467.31 Specialized definitions.
§ 467.32 Effluent limitations representing the degree of effluent reduction attainable by the application of the best practicable control technology currently available.
§ 467.33 Effluent limitations representing the degree of effluent reduction attainable by the application of the best available technology economically achievable.
§ 467.34 New source performance standards.
§ 467.35 Pretreatment standards for existing sources.
§ 467.36 Pretreatment standards for new sources.
§ 467.37 Effluent limitations representing the degree of effluent reduction attainable by the application of the best conventional pollutant control technology. [Reserved]

Subpart D--Forging Subcategory
§ 467.40 Applicability; description of the forging subcategory.
§ 467.41 Specialized definitions.
§ 467.42 Effluent limitations representing the degree of effluent reduction attainable by the application of the best practicable control technology currently available. [Reserved]
§ 467.43 Effluent limitations representing the degree of effluent reduction attainable by the application of the best available technology economically achievable.

[Reserved]

Subpart C--Nickel-Cobalt Forming Subcategory

§ 471.30 Applicability; description of the nickel-cobalt forming subcategory.

§ 471.31 Effluent limitations representing the degree of effluent reduction attainable by the application of the best practicable control technology currently available (BPT).

§ 471.32 Effluent limitations representing the degree of effluent reduction attainable by the application of the best available technology economically achievable (BAT).

§ 471.33 New source performance standards (NSPS).

§ 471.34 Pretreatment standards for existing sources (PSES).

§ 471.35 Pretreatment standards for new sources (PSNS).

§ 471.36 Effluent limitations representing the degree of effluent reduction attainable by the application of the best conventional pollutant control technology (BCT). [Reserved]

Subpart D--Precious Metals Forming Subcategory

§ 471.40 Applicability; description of the precious metals forming subcategory.

§ 471.41 Effluent limitations representing the degree of effluent reduction attainable by the application of the best practicable control technology currently available (BPT).

§ 471.42 Effluent limitations representing the degree of effluent reduction attainable by the application of the best available technology economically achievable (BAT).

§ 471.43 New source performance standards (NSPS).

§ 471.44 Pretreatment standards for existing sources (PSES).

§ 471.45 Pretreatment standards for new sources (PSNS).

§ 471.46 Effluent limitations representing the degree of effluent reduction attainable by the application of the best conventional pollutant control technology (BCT). [Reserved]

Subpart E--Refractory Metals Forming Subcategory

§ 471.50 Applicability; description of the refractory metals forming subcategory.

§ 471.51 Effluent limitations representing the degree of effluent reduction attainable by the application of the best practicable control technology currently available (BPT).

§ 471.52 Effluent limitations representating the degree of effluent reduction attainable by the application of the best available technology economically achievable (BAT).

§ 471.53 New source performance standards (NSPS).

§ 471.54 Pretreatment standards for existing sources (PSES).

§ 471.55 Pretreatment standards for new sources (PSNS).

§ 471.56 Effluent limitations representing the degree of effluent reduction attainable by the application of the

best conventional pollutant control technology (BCT). [Reserved]

Subpart F--Titanium Forming Subcategory

§ 471.60 Applicability; description of the titanium forming subcategory.

§ 471.61 Effluent limitations representing the degree of effluent reduction attainable by the application of the best practicable control technology currently available (BPT).

§ 471.62 Effluent limitations representing the degree of effluent reduction attainable by the application of the best available technology economically achievable (BAT).

§ 471.63 New source performance standards (NSPS).

§ 471.64 Pretreatment standards for existing sources (PSES).

§ 471.65 Pretreatment standards for new sources (PSNS).

§ 471.66 Effluent limitations representing the degree of effluent reduction attainable by the application of the best conventional pollutant control technology (BCT). [Reserved]

Subpart G--Uranium Forming Subcategory

§ 471.70 Applicability; description of the uranium forming subcategory.

§ 471.71 Effluent limitations representing the degree of effluent reduction attainable by the application of the best practicable control technology currently available (BPT).

§ 471.72 Effluent limitations representing the degree of effluent reduction attainable by the application of the best available technology economically achievable (BAT).

§ 471.73 New source performance standards (NSPS).

§ 471.74 Pretreatment standards for existing sources (PSES). [Reserved]

§ 471.75 Pretreatment standards for new sources (PSNS).

§ 471.76 Effluent limitations representing the degree of effluent reduction attainable by the application of the best conventional pollutant control technology (BCT). [Reserved]

Subpart H--Zinc Forming Subcategory

§ 471.80 Applicability; description of the zinc forming subcategory.

§ 471.81 Effluent limitations representing the degree of effluent reduction attainable by the application of the best practicable control technology currently available (BPT).

§ 471.82 Effluent limitations representing the degree of effluent reduction attainable by the application of the best available technology economically achievable (BAT).

§ 471.83 New source performance standards (NSPS).

§ 471.84 Pretreatment standards for existing sources (PSES). [Reserved]

§ 471.85 Pretreatment standards for new sources (PSNS).

§ 471.86 Effluent limitations representing the degree of

effluent reduction attainable by the application of the best conventional pollutant control technology (BCT). [Reserved]

Subpart I--Zirconium-Hafnium Forming Subcategory

§ 471.90 Applicability; description of the zirconium-hafnium forming subcategory.

§ 471.91 Effluent limitations representing the degree of effluent reduction attainable by the application of the best practicable control technology currently available (BPT).

§ 471.92 Effluent limitations representing the degree of effluent reduction attainable by the application of the best available technology economically achievable (BAT).

§ 471.93 New source performance standards (NSPS).

§ 471.94 Pretreatment standards for existing sources (PSES).

§ 471.95 Pretreatment standards for new sources (PSNS).

§ 471.96 Effluent limitations representing the degree of effluent reduction attainable by the application of the best conventional pollutant control technology (BCT). [Reserved]

Subpart J--Metals Powders Subcategory

§ 471.100 Applicability; description of the powder metals subcategory.

§ 471.101 Effluent limitations representing the degree of effluent reduction attainable by the application of the best practicable control technology currently available (BPT).

§ 471.102 Effluent limitations representing the degree of effluent reduction attainable by the application of the best available technology economically achievable (BAT).

§ 471.103 New source performance standards (NSPS).

§ 471.104 Pretreatment standards for existing sources (PSES).

§ 471.105 Pretreatment standards for new sources (PSNS).

§ 471.106 Effluent limitations representing the degree of effluent reduction attainable by the application of the best conventional pollutant control technology (BCT). [Reserved]

SUBCHAPTER O--SEWAGE SLUDGE

PART 501--STATE SLUDGE MANAGEMENT PROGRAM REGULATIONS

Subpart A--Purpose, Scope and General Program Requirements

§ 501.1 Purpose and scope.

§ 501.2 Definitions.

§ 501.3 Coordination with other programs.

Subpart B--Development and Submission of State Programs

§ 501.11 Elements of a sludge management program submission.

§ 501.12 Program description.

§ 501.13 Attorney General's statement.

§ 501.14 Memorandum of Agreement with the Regional Administrator.

§ 501.15 Requirements for Permitting.

§ 600.112-78 Exhaust sample analysis.
§ 600.113-78 Fuel economy calculations.
§ 600.113-88 Fuel economy calculations.
Subpart C--Fuel Economy Regulations for 1977 and Later Model Year Automobiles--Procedures for Calculating Fuel Economy Values
§ 600.201-86 General applicability.
§ 600.202-77 Definitions.
§ 600.203-77 Abbreviations.
§ 600.204-77 Section numbering, construction.
§ 600.205-77 Recordkeeping.
§ 600.206-86 Calculation and use of fuel economy values for gasoline- fueled, diesel, and electric vehicle configurations.
§ 600.207-86 Calculation of fuel economy values for a model type.
§ 600.208-77 Sample calculation.
§ 600.209-85 Calculation of fuel economy values for labeling.
Subpart D--Fuel Economy Regulations for 1977 and Later Model Year Automobiles--Labeling
§ 600.301-86 General applicability.
§ 600.302-77 Definitions.
§ 600.303-77 Abbreviations.
§ 600.304-77 Section numbering, construction.
§ 600.305-77 Recordkeeping.
§ 600.306-86 Labeling requirements.
§ 600.307-86 Fuel economy label format requirements.
§ 600.310-86 Labeling of high altitude vehicles.
§ 600.311-86 Range of fuel economy for comparable automobiles.
§ 600.312-86 Labeling, reporting, and recordkeeping; Administrator reviews.
§ 600.313-86 Timetable for data and information submittal and review.
§ 600.314-86 Updating label values, annual fuel cost, Gas Guzzler Tax, and range of fuel economies for comparable automobiles.
§ 600.315-82 Classes of comparable automobiles.
§ 600.316-78 Multistage manufacture.
Subpart E--Fuel Economy Regulations for 1977 and Later Model Year Automobiles--Dealer Availability of Fuel Economy Information
§ 600.401-77 General applicability.
§ 600.402-77 Definitions.
§ 600.403-77 Abbreviations.
§ 600.404-77 Section numbering, construction.
§ 600.405-77 Dealer requirements.
§ 600.406-77 [Reserved]
§ 600.407-77 Booklets displayed by dealers.
Subpart F--Fuel Economy Regulations for Model Year 1978 Passenger Automobiles and for 1979 and Later Model Year Automobiles (Light Trucks and Passenger Automobiles)--Procedures for Determining Manufacturer's Average Fuel Economy
§ 600.501-85 General applicability.

§ 720.22 Persons who must report.
§ 720.25 Determining whether a chemical substance is on the Inventory.
§ 720.30 Chemicals not subject to notification requirements.
§ 720.36 Exemption for research and development.
§ 720.38 Exemptions for test marketing.

Subpart C--Notice Form
§ 720.40 General.
§ 720.45 Information that must be included in the notice form.
§ 720.50 Submission of test data and other data concerning the health and environmental effects of a substance.
§ 720.57 Imports.

Subpart D--Disposition of Notices
§ 720.60 General.
§ 720.62 Notice that notification is not required.
§ 720.65 Acknowledgment of receipt of a notice; errors in the notice; incomplete submissions; false and misleading statements.
§ 720.70 Notice in the Federal Register.
§ 720.75 Notice review period.
§ 720.78 Recordkeeping.

Subpart E--Confidentiality and Public Access to Information
§ 720.80 General provisions.
§ 720.85 Chemical identity.
§ 720.87 Categories or proposed categories of uses of a new chemical substance.
§ 720.90 Data from health and safety studies.
§ 720.95 Public file.

Subpart F--Commencement of Manufacture or Import
§ 720.102 Notice of commencement of manufacture or import.

Subpart G--Compliance and Inspections
§ 720.120 Compliance.
§ 720.122 Inspections.

Appendix A--Premanufacture Notice for New Chemical Substances

PART 721--SIGNIFICANT NEW USES OF CHEMICAL SUBSTANCES

Subpart A--General Provisions
§ 721.1 Scope and applicability.
§ 721.3 Definitions.
§ 721.5 Persons who must report.
§ 721.11 Applicability determination when the specific chemical identity is confidential.
§ 721.20 Exports and imports.
§ 721.25 Notice requirements and procedures.
§ 721.30 EPA approval of alternative control measures.
§ 721.35 Compliance and enforcement.
§ 721.40 Recordkeeping.
§ 721.45 Exemptions.
§ 721.47 Conditions for research and development exemption.

Subpart B--Certain Significant New Uses
§ 721.50 Applicability.

§ 721.63 Protection in the workplace.
§ 721.72 Hazard communication program.
§ 721.80 Industrial, commercial, and consumer activities.
§ 721.85 Disposal.
§ 721.90 Release to water.
§ 721.91 Computation of estimated surface water concentrations: instructions.

Subpart C--Recordkeeping Requirements

§ 721.100 Applicability.
§ 721.125 Recordkeeping requirements.

Subpart D--Expedited Process for Issuing Significant New Use Rules for Selected Chemical Substances and Limitation or Revocation of Selected Significant New Use Rules

§ 721.160 Notification requirements for new chemical substances subject to section 5(e) orders.
§ 721.170 Notification requirements for selected new chemical substances that have completed premanufacture review.
§ 721.185 Limitation or revocation of certain notification requirements.

Subpart E--Significant New Uses for Specific Chemical Substances

§ 721.224 [Redesignated. 58 FR 29946, May 24, 1993]
§ 721.225 2-Chloro-N-methyl-N-substituted acetamide (generic name).
§ 721.235 [Redesignated. 58 FR 29946, May 24, 1993]
§ 721.263 [Redesignated. 58 FR 29946, May 24, 1993]
§ 721.264 [Redesignated. 58 FR 29946, May 24, 1993]
§ 721.266 [Redesignated. 58 FR 29946, May 24, 1993]
§ 721.266 [Redesignated. 55 FR 52276, Dec. 21, 1990]
§ 721.270 [Redesignated. 58 FR 29946, May 24, 1993]
§ 721.273 [Redesignated. 58 FR 29946, May 24, 1993]
§ 721.275 Halogenated-N-(2-propenyl)-N-(substituted phenyl) acetamide.
§ 721.278 [Redesignated. 58 FR 29946, May 24, 1993]
§ 721.285 Acetamide, N-[4-(pentyloxy)phenyl]-, acetamide, N-[2-nitro-4- (pentyloxy)phenyl]-, and acetamide, N-[2-amino-4-(pentyloxy)phenyl]-.
§ 721.287 [Redesignated. 58 FR 29946, May 24, 1993]
§ 721.288 [Removed. 56 FR 43877, Sept. 5, 1991]
§ 721.289 [Redesignated. 58 FR 29946, May 24, 1993]
§ 721.290 [Redesignated. 58 FR 29946, May 24, 1993]
§ 721.291 [Redesignated. 58 FR 29946, May 24, 1993]
§ 721.293 [Redesignated. 58 FR 29946, May 24, 1993]
§ 721.295 [Redesignated. 58 FR 29946, May 24, 1993]
§ 721.296 [Redesignated. 58 FR 29946, May 24, 1993]
§ 721.305 [Redesignated. 58 FR 29946, May 24, 1993]
§ 721.315 [Redesignated. 58 FR 29946, May 24, 1993]
§ 721.320 Acrylamide-substituted epoxy.
§ 721.323 Substituted acrylamide.
§ 721.325 Certain acrylates.
§ 721.350 [Redesignated. 58 FR 29946, May 24, 1993]
§ 721.370 Substituted diacrylate.
§ 721.377 [Redesignated. 58 FR 29946, May 24, 1993]

§ 721.390 Monoacrylate.
§ 721.400 Polyalkylpolysilazane, bis(substituted acrylate).
§ 721.415 Aliphatic diurethane acrylate ester.
§ 721.425 [Redesignated. 58 FR 29946, May 24, 1993]
§ 721.430 Oxo-substituted aminoalkanoic acid derivative.
§ 721.435 [Redesignated. 58 FR 29946, May 24, 1993]
§ 721.440 [Redesignated. 58 FR 29946, May 24, 1993]
4. By adding new § 721.445 to subpart E to read as follows:
§ 721.450 [Redesignated. 58 FR 29946, May 24, 1993]
§ 721.454 [Redesignated. 58 FR 29946, May 24, 1993]
§ 721.460 Amino acrylate monomer.
§ 721.462 [Reserved. 58 FR 29946, May 24, 1993]
§ 721.464 [Redesignated. 58 FR 29946, May 24, 1993]
§ 721.466 [Redesignated. 58 FR 29946, May 24, 1993]
§ 721.467 [Redesignated. 58 FR 29946, May 24, 1993]
§ 721.470 Aliphatic difunctional acrylic acid ester.
§ 721.490 Modified acrylic ester (generic name).
§ 721.500 [Redesignated. 58 FR 29946, May 24, 1993]
§ 721.505 Halogenated acrylonitrile.
§ 721.520 Alanine, N-(2-carboxyethyl)-N-alkyl-, salt.
§ 721.523 [Redesignated. 58 FR 29946, May 24, 1993]
§ 721.530 Substituted aliphatic acid halide (generic name).
§ 721.536 Halogenated phenyl alkane.
§ 721.540 Alkylphenoxypolyalkoxyamine (generic name).
§ 721.550 Alkyl alkenoate, azobis-.
§ 721.555 [Redesignated. 58 FR 29946, May 24, 1993]
§ 721.557 [Redesignated. 58 FR 29946, May 24, 1993]
§ 721.564 [Redesignated. 58 FR 29946, May 24, 1993]
§ 721.566 [Redesignated. 58 FR 29946, May 24, 1993]
§ 721.567 [Redesignated. 58 FR 29946, May 24, 1993]
§ 721.570 [Redesignated. 58 FR 29946, May 24, 1993]
§ 721.575 Substituted alkyl halide.
§ 721.580 [Redesignated. 58 FR 29946, May 24, 1993]
§ 721.586 [Redesignated. 58 FR 29946, May 24, 1993]
§ 721.600 3-Alkyl-2-(2-anilino)vinyl thiazolinium salt (generic name).
§ 721.605 [Redesignated. 58 FR 29946, May 24, 1993]
§ 721.607 [Redesignated. 58 FR 29946, May 24, 1993]
§ 721.609 [Redesignated. 58 FR 29946, May 24, 1993]
§ 721.611 [Redesignated. 58 FR 29946, May 24, 1993]
§ 721.612 [Redesignated. 58 FR 29946, May 24, 1993]
§ 721.617 [Redesignated. 58 FR 29946, May 24, 1993]
§ 721.625 Alkylated diarylamine, sulfurized (generic name).
§ 721.648 [Redesignated. 58 FR 29946, May 24, 1993]
§ 721.650 11-Aminoundecanoic acid.
§ 721.660 [Redesignated. 58 FR 29946, May 24, 1993]
§ 721.700 Methylenebistrisubstituted aniline (generic name).
§ 721.715 Trisubstituted anthracene.

§ 721.740 [Redesignated. 58 FR 29946, May 24, 1993]
§ 721.750 Aromatic amine compound.
§ 721.756 [Redesignated. 58 FR 29946, May 24, 1993]
§ 721.757 Polyoxyalkylene substituted aromatic azo colorant.
§ 721.759 [Redesignated. 58 FR 29946, May 24, 1993]
§ 721.760 [Redesignated. 58 FR 29946, May 24, 1993]
§ 721.766 [Removed. 58 FR 26691, May 5, 1993]
§ 721.767 [Redesignated. 58 FR 29946, May 24, 1993]
§ 721.770 [Removed. 58 FR 27206, May 7, 1993]
§ 721.775 Brominated aromatic compound (generic name).
§ 721.782 [Removed. 56 FR 12852, Mar. 28, 1991]
§ 721.783 [Redesignated. 58 FR 29946, May 24, 1993]
§ 721.792 [Redesignated. 58 FR 29946, May 24, 1993]
§ 721.800 [Redesignated. 58 FR 29946, May 24, 1993]
§ 721.805 Benzenamine, 4,4'-[1,3-phenylenebis(1-methylethyl idene)]bis[2,6-dimethyl-.
§ 721.818 [Redesignated. 58 FR 29946, May 24, 1993]
§ 721.821 [Redesignated. 58 FR 29946, May 24, 1993]
§ 721.840 Alkyl substituted diaromatic hydrocarbons.
§ 721.850 [Redesignated. 58 FR 29946, May 24, 1993]
§ 721.853 [Redesignated. 58 FR 29946, May 24, 1993]
§ 721.875 Aromatic nitro compound.
§ 721.880 [Redesignated. 58 FR 29946, May 24, 1993]
§ 721.925 Substituted aromatic (generic).
§ 721.950 Sodium salt of an alkylated, sulfonated aromatic (generic name).
§ 721.953 [Redesignated. 58 FR 29946, May 24, 1993]
§ 721.956 [Redesignated. 58 FR 29946, May 24, 1993]
§ 721.960 [Redesignated. 58 FR 29946, May 24, 1993]
§ 721.975 [Redesignated. 58 FR 29946, May 24, 1993]
§ 721.976 [Redesignated. 58 FR 29946, May 24, 1993]
§ 721.977 [Redesignated. 58 FR 29946, May 24, 1993]
§ 721.978 [Redesignated. 58 FR 29946, May 24, 1993]
§ 721.979 [Redesignated. 58 FR 29946, May 24, 1993]
§ 721.980 [Redesignated. 58 FR 29946, May 24, 1993]
§ 721.982 Calcium, bis(2,4-pentanedionato-O,O').
§ 721.983 [Redesignated. 58 FR 29946, May 24, 1993]
§ 721.990 [Redesignated. 58 FR 29946, May 24, 1993]
§ 721.1000 Benzenamine, 3-chloro-2,6-dinitro-N,N-dipropyl-4- (trifluoromethyl)-.
§ 721.1005 [Redesignated. 58 FR 29946, May 24, 1993]
§ 721.1006 [Removed. 56 FR 47677, Sept. 20, 1991]
§ 721.1007 [Redesignated. 58 FR 29946, May 24, 1993]
§ 721.1025 Benzenamine, 4-chloro-2-methyl-; benzenamine, 4-chloro-2- methyl-, hydrochloride; and benzenamine, 2-chloro-6-methyl-.
§ 721.1027 [Redesignated. 58 FR 29946, May 24, 1993]
§ 721.1028 [Redesignated. 58 FR 29946, May 24, 1993]
§ 721.1029 [Redesignated. 58 FR 29946, May 24, 1993]
§ 721.1030 [Redesignated. 58 FR 29946, May 24, 1993]
§ 721.1032 [Redesignated. 58 FR 29946, May 24, 1993]
§ 721.1033 [Redesignated. 58 FR 29946, May 24, 1993]

§ 721.1036 [Redesignated. 58 FR 29946, May 24, 1993]
§ 721.1040 [Redesignated. 58 FR 29946, May 24, 1993]
§ 721.1045 [Redesignated. 58 FR 29946, May 24, 1993]
§ 721.1050 Benzenamine, 2,5-dibutoxy-4-(4-morpholinyl)-, sulfate
§ 721.1054 [Redesignated. 58 FR 29946, May 24, 1993]
§ 721.1060 [Removed. 57 FR 9983, Mar. 23, 1992]
§ 721.1064 [Redesignated. 58 FR 29946, May 24, 1993]
§ 721.1068 Benzenamine, 4-isocyanato-N,N-bis(4-isocyanatophenyl)-2,5- dimethoxy-.
§ 721.1075 Benzenamine, 4-(1-methylbutoxy)-, hydrochloride.
§ 721.1078 [Redesignated. 58 FR 29946, May 24, 1993]
§ 721.1082 [Redesignated. 58 FR 29946, May 24, 1993]
§ 721.1100 [Redesignated. 58 FR 29946, May 24, 1993]
§ 721.1105 [Redesignated. 58 FR 29946, May 24, 1993]
§ 721.1120 Benzenamine, 4,4'-[1,4-phenylenebis(1-methylethylidene)]bis[2,6-dimethyl-.
§ 721.1125 [Redesignated. 58 FR 29946, May 24, 1993]
§ 721.1130 [Redesignated. 58 FR 29946, May 24, 1993]
§ 721.1137 [Redesignated. 58 FR 29946, May 24, 1993]
§ 721.1140 [Redesignated. 58 FR 29946, May 24, 1993]
§ 721.1143 [Redesignated. 58 FR 29946, May 24, 1993]
§ 721.1150 Substituted polyglycidyl benzeneamine.
§ 721.1175 Benzene, substituted, alkyl acrylate derivative (generic name).
§ 721.1200 [Redesignated. 58 FR 29946, May 24, 1993]
§ 721.1204 [Redesignated. 58 FR 29946, May 24, 1993]
§ 721.1208 [Redesignated. 58 FR 29946, May 24, 1993]
§ 721.1210 Benzene, (2-chloroethoxy)-.
§ 721.1225 Benzene, 1,2-dimethyl-, polypropene derivatives, sulfonated, potassium salts.
§ 721.1232 [Redesignated. 58 FR 29946, May 24, 1993]
§ 721.1233 [Redesignated. 58 FR 29946, May 24, 1993]
§ 721.1234 [Redesignated. 58 FR 29946, May 24, 1993]
§ 721.1235 [Redesignated. 58 FR 29946, May 24, 1993]
§ 721.1237 [Redesignated. 58 FR 29946, May 24, 1993]
§ 721.1243 [Redesignated. 58 FR 29946, May 24, 1993]
§ 721.1245 [Removed. 58 FR 27207, May 7, 1993]
§ 721.1247 [Redesignated. 58 FR 29946, May 24, 1993]
§ 721.1250 [Redesignated. 58 FR 29946, May 24, 1993]
§ 721.1261 [Redesignated. 58 FR 29946, May 24, 1993]
§ 721.1265 [Redesignated. 58 FR 29946, May 24, 1993]
§ 721.1272 [Redesignated. 58 FR 29946, May 24, 1993]
§ 721.1282 [Redesignated. 58 FR 29946, May 24, 1993]
§ 721.1285 [Redesignated. 58 FR 29946, May 24, 1993]
§ 721.1287 [Redesignated. 58 FR 29946, May 24, 1993]
§ 721.1290 [Redesignated. 58 FR 29946, May 24, 1993]
§ 721.1296 [Redesignated. 58 FR 29946, May 24, 1993]
§ 721.1298 [Redesignated. 58 FR 29946, May 24, 1993]
§ 721.1300 [(Dinitrophenyl)azo]-[2,4-diamino-5-methoxybenzene] derivatives.
§ 721.1325 Benzene, 1-(1-methylbutoxy)-4-nitro-.

§ 721.1350 Benzene, (1-methylethyl)(2-phenylethyl)-.
§ 721.1372 Substituted nitrobenzene.
§ 721.1375 Disubstituted nitrobenzene (generic name).
§ 721.1390 [Redesignated. 58 FR 29946, May 24, 1993]
§ 721.1395 [Redesignated. 58 FR 29946, May 24, 1993]
§ 721.1425 Pentabromoethylbenzene.
§ 721.1430 Pentachlorobenzene.
§ 721.1435 1,2,4,5-Tetrachlorobenzene.
§ 721.1440 1,3,5-Trinitrobenzene.
§ 721.1450 1,3-Benzenediamine, 4-(1,1-dimethylethyl)-ar-methyl.
§ 721.1454 [Redesignated. 58 FR 29946, May 24, 1993]
§ 721.1456 [Redesignated. 58 FR 29947, May 24, 1993]
§ 721.1460 [Redesignated. 58 FR 29947, May 24, 1993]
§ 721.1465 [Redesignated. 58 FR 29947, May 24, 1993]
§ 721.1470 [Redesignated. 58 FR 29947, May 24, 1993]
§ 721.1475 [Redesignated. 55 FR 33305, Aug. 15, 1990; 56 FR 29903, July 1, 1991]
§ 721.1475 [Redesignated. 58 FR 29947, May 24, 1993]
§ 721.1477 [Redesignated. 58 FR 29947, May 24, 1993]
§ 721.1478 [Redesignated. 58 FR 29947, May 24, 1993]
§ 721.1483 [Redesignated. 58 FR 29947, May 24, 1993]
§ 721.1488 [Redesignated. 58 FR 29947, May 24, 1993]
§ 721.1489 [Redesignated. 58 FR 29947, May 24, 1993]
§ 721.1490 [Redesignated. 58 FR 29947, May 24, 1993]
§ 721.1491 [Redesignated. 58 FR 29947, May 24, 1993]
§ 721.1495 [Redesignated. 58 FR 29947, May 24, 1993]
§ 721.1497 [Redesignated. 58 FR 29947, May 24, 1993]
§ 721.1550 1,2-Benzenediamine, 4-ethoxy, sulfate.
§ 721.1502 [Redesignated. 58 FR 29947, May 24, 1993]
§ 721.1504 [Redesignated. 58 FR 29947, May 24, 1993]
§ 721.1515 [Redesignated. 58 FR 29447, May 24, 1993]
§ 721.1525 Mixture of: 1,3-benzenediamine, 2-methyl-4,6-bis(methylthio)- (CAS NO. 104983-85-9) and 1,3-benzenediamine, 4-methyl-2,6-bis(methylthio)- (CAS NO. 102093-68-5).
§ 721.1536 [Removed. 58 FR 27208, May 7, 1993]
§ 721.1537 [Redesignated. 58 FR 29947, May 24, 1993]
§ 721.1538 [Redesignated. 58 FR 29947, May 24, 1993]
§ 721.1540 [Redesignated. 58 FR 29947, May 24, 1993]
§ 721.1541 [Redesignated. 58 FR 29947, May 24, 1993]
§ 721.1542 [Redesignated. 58 FR 29947, May 24, 1993]
§ 721.1544 [Redesignated. 58 FR 29947, May 24, 1993]
§ 721.1550 Benzenediazonium, 4-(dimethylamino)-, salt with 2-hydroxy-5- sulfobenzoic acid (1:1).
§ 721.1555 Substituted phenyl azo substituted benzenediazonium salt.
§ 721.1560 [Redesignated. 58 FR 29947, May 24, 1993]
§ 721.1565 [Redesignated. 58 FR 29947, May 24, 1993]
§ 721.1568 Substituted benzenediazonium.
§ 721.1575 Substituted benzenedicarboxylic acid, poly(alkyl acrylate) derivative.
§ 721.1582 [Redesignated. 58 FR 29947, May 24, 1993]

§ 721.1585 [Redesignated. 58 FR 29947, May 24, 1993]
§ 721.1590 [Redesignated. 58 FR 29947, May 24, 1993]
§ 721.1600 [Removed. 59 FR 17491, Apr. 13, 1994]
§ 721.1608 [Redesignated. 58 FR 29947, May 24, 1993]
§ 721.1610 [Redesignated. 58 FR 29947, May 24, 1993]
§ 721.1611 [Redesignated. 58 FR 29947, May 24, 1993]
§ 721.1612 Substituted 2-nitro- and 2-aminobenzesulfonamide.
§ 721.1614 [Redesignated. 58 FR 29947, May 24, 1993]
§ 721.1616 [Redesignated. 58 FR 29947, May 24, 1993]
§ 721.1617 [Redesignated. 58 FR 29947, May 24, 1993]
§ 721.1619 [Redesignated. 58 FR 29947, May 24, 1993]
§ 721.1620 [Redesignated. 58 FR 29947, May 24, 1993]
§ 721.1621 [Removed. 58 FR 26692, May 5, 1993]
§ 721.1622 [Redesignated. 58 FR 29947, May 24, 1993]
§ 721.1624 [Redesignated. 58 FR 29947, May 24, 1993]
§ 721.1625 Alkylbenzene sulfonate, amine salt.
§ 721.1630 1,2-Ethanediol bis(4-methylbenzenesulfonate); 2,2-oxybis- ethane bis(4-methylbenzenesulfonate); ethanol, 2,2'-[oxybis(2,1-ethanediyl oxy)]bis-, bis(4-methylbenzenesulfonate); ethanol, 2,2'-[oxybis (2,1-ethane diyloxy)] bis-, bis(4-methylbenzenesulfonate); ethanol, 2,2'-[[1-[(2- propenyloxy) methyl]-1,2-ethanediyl] bis(oxy)]bis-, bis(4-methylbenzene sulfonate); and etha
§ 721.1632 [Redesignated. 58 FR 29947, May 23, 1993]
§ 721.1634 [Redesignated. 58 FR 29947, May 24, 1993]
§ 721.1637 1,2-Propanediol, 3-(2-propenyloxy)-, bis(4-methylbenzene sulfonate); 2-propanol, 1-[2-[[(4-methylphenyl)sulfonyl] oxy]ethoxy]-3-(2- propenyloxy)-4-methylbenzenesulfonate; and 2-propanol, 1-[2-[2-[[(4-methylphenyl)sulfonyl]oxy] ethoxy]ethoxy]-3-(2-propenyloxy)-, 4- methylbenzenesulfonate.
§ 721.1638 [Redesignated. 58 FR 29947, May 24, 1993]
§ 721.1640 3,6,9,12,-Tetraoxatetradecane-1,14-diol, bis(4-methylbenzenesulfonate; 3,6,9,13-tetraoxahexadec-15-ene-1,11-diol, bis(4- methylbenzenesulfonate); 3,6,9,12,16-pentaoxanonadec-18-ene-1,14-diol, bis(4- methyl benzenesulfonate); and 3,6,9,12-tetraoxatetradecane-1,14-diol, 7-[(2- propenyloxy)methyl]-, bis(4-methylbenzenesulfonate).
§ 721.1641 [Redesignated. 58 FR 29947, May 24, 1993]
§ 721.1643 [Redesignated. 58 FR 29947, May 24, 1993]
§ 721.1645 Benzenesulfonic acid, 4-methyl-, reaction products with oxirane mono[(C10-16-alkyloxy)methyl] derivatives and 2,2,4(or 2,4,4)- trimethyl-1,6-hexanediamine.
§ 721.1646 [Redesignated. 58 FR 29947, May 24, 1993]
§ 721.1648 [Redesignated. 58 FR 29947, May 24, 1993]
§ 721.1650 Alkylbenzenesulfonic acid and sodium salts.
§ 721.1675 Disulfonic acid rosin amine salt of a benzidine derivative (generic name).
§ 721.1700 Halonitrobenzoic acid, substituted (generic name).

§ 721.1702 [Redesignated. 58 FR 29947, May 24, 1993]
§ 721.1704 [Redesignated. 58 FR 29947, May 24, 1993]
§ 721.1706 [Redesignated. 58 FR 29947, May 24, 1993]
§ 721.1708 [Redesignated. 58 FR 29947, May 24, 1993]
§ 721.1710 [Redesignated. 58 Fr 29947, May 24, 1993]
§ 721.1711 [Redesignated. 58 FR 29947, May 24, 1993]
§ 721.1712 [Redesignated. 58 FR 29947, May 24, 1993]
§ 721.1715 [Redesignated. 58 FR 29947, May 24, 1993]
§ 721.1725 Benzoic acid, 3,3'-methylenebis $6 amino-, di-2-propenyl ester.
§ 721.1728 Benzoic acid, 2-(3-phenylbutylidene)amino-, methyl ester.
§ 721.1732 Nitrobenzoic acid octyl ester.
§ 721.1735 Alkylbisoxyalkyl (substituted-1,1-dimethylethylphenyl) benzotriazole (generic name).
§ 721.1740 Substituted dichlorobenzothiazoles.
§ 721.1745 Ethoxybenzothiazole disulfide.
§ 721.1750 1H-Benzotriazole, 5-(pentyloxy)- and 1H-benzotriazole, 5- (pentyloxy)-, sodium and potassium salts.
§ 721.1760 [Redesignated. 58 FR 29947, May 24, 1993]
§ 721.1763 [Redesignated. 58 FR 29947, May 24, 1993]
§ 721.1765 2-Substituted benzotriazole.
§ 721.1775 6-Nitro-2(3H)-benzoxazolone.
§ 721.1778 [Redesignated. 58 FR 29947, May 24, 1993]
§ 721.1780 [Redesignated. 58 FR 29947, May 24, 1993]
§ 721.1790 Polybrominated biphenyls.
§ 721.1795 [Redesignated. 58 FR 29947, May 24, 1993]
§ 721.1796 [Redesignated. 58 FR 29947, May 24, 1993]
§ 721.1797 [Redesignated. 58 FR 29947, May 24, 1993]
§ 721.1798 [Redesignated. 58 FR 29947, May 24, 1993]
§ 721.1800 3,3',5,5'-Tetramethylbiphenyl-4,4'-diol.
§ 721.1805 [Redesignated. 58 FR 29947, May 24, 1993]
§ 721.1810 [Redesignated. 58 FR 29947, May 24, 1993]
§ 721.1814 [Redesignated. 58 FR 29947, May 24, 1993]
§ 721.1815 [Redesignated. 58 FR 29947, May 24, 1993]
§ 721.1816 [Redesignated. 58 FR 29947, May 24, 1993]
§ 721.1817 [Redesignated. 58 FR 29947, May 24, 1993]
§ 721.1818 [Redesignated. 58 FR 29947, May 24, 1993]
§ 721.1820 Bisphenol derivative.
§ 721.1822 [Redesignated. 58 FR 29947, May 24, 1993]
§ 721.1824 [Redesignated. 58 FR 29947, May 24, 1993]
§ 721.1825 Bisphenol A, epichlorohydrin, polyalkylenepolyol and polyisocyanato derivative.
§ 721.1828 [Redesignated. 58 FR 29947, May 24, 1993]
§ 721.1830 [Redesignated. 58 FR 29947, May 24, 1993]
§ 721.1832 [Redesignated. 58 FR 29947, May 24, 1993]
§ 721.1835 [Redesignated. 58 FR 29947, May 24, 1993]
§ 721.1840 [Redesignated. 58 FR 29947, May 24, 1993]
§ 721.1845 [Redesignated. 58 FR 29947, May 24, 1993]
§ 721.1850 Toluene sulfonamide bisphenol A epoxy adduct.
§ 721.1858 [Redesignated. 58 FR 29947, May 24, 1993]
§ 721.1875 Boric acid, alkyl and substituted alkyl esters

(generic name).
§ 721.1880 [Redesignated. 58 FR 29947, May 24, 1993]
§ 721.1883 [Redesignated. 58 FR 29947, May 24, 1993]
§ 721.1886 [Redesignated. 58 FR 29947, May 24, 1993]
§ 721.1887 [Redesignated. 58 FR 29947, May 24, 1993]
§ 721.1888 [Redesignated. 58 FR 29947, May 24, 1993]
§ 721.1889 [Redesignated. 58 FR 29947, May 24, 1993]
§ 721.1890 [Removed. 57 FR 9984, Mar. 23, 1992]
§ 721.1895 [Redesignated. 58 FR 29947, May 24, 1993]
§ 721.1896 [Redesignated. 58 FR 29947, May 24, 1993]
§ 721.1897 [Redesignated. 58 FR 29947, May 24, 1993]
§ 721.1898 [Redesignated. 58 FR 29947, May 24, 1993]
§ 721.1900 Substituted bromothiophene.
§ 721.1920 1,4-Bis(3-hydroxy-4-benzoylphenoxy)butane.
§ 721.1925 Substituted carboheterocyclic butane tetracarboxylate.
§ 721.1950 2-Butenedioic acid (Z), mono(2-((1-oxopropenyloxy)ethyl) ester.
§ 721.2000 [Removed. 59 FR 17489, Apr. 13, 1994]
§ 721.2025 Substituted phenylimino carbamate derivative.
§ 721.2050 Carbamic acid, (trialkyloxy silyalkyl)-substituted acrylate ester.
§ 721.2070 [Redesignated. 58 FR 29947, May 24, 1993]
§ 721.2075 Carbamodithioic acid, methyl-, compound with methanamine (1:1).
§ 721.2084 Carbon oxyfluoride (Carbonic difluoride).
§ 721.2085 [Redesignated. 58 FR 29947, May 24, 1993]
§ 721.2086 Coco acid triamine condensate, polycarboxylic acid salts.
§ 721.2092 3-Methylcholanthrene.
§ 721.2094 [Redesignated. 58 FR 29947, May 24, 1993]
§ 721.2100 [Redesignated. 58 FR 29947, May 24, 1993]
§ 721.2120 Cyclic amide.
§ 721.2132 [Redesignated. 58 FR 29947, May 24, 1993]
§ 721.2140 Carbopolycyclicol azoalkylaminoalkylcarbomonocyclic ester, halogen acid salt.
§ 721.2150 [Redesignated. 58 FR 29947, May 24, 1993]
§ 721.2155 [Redesignated. 58 FR 29947, May 24, 1993]
§ 721.2170 Cyclic phosphazene, methacrylate derivative.
§ 721.2175 Salt of cyclodiamine and mineral acid.
§ 721.2180 [Redesignated. 58 FR 29947, May 24, 1993]
§ 721.2184 [Redesignated. 58 FR 29947, May 24, 1993]
§ 721.2188 [Redesignated. 58 FR 29947, May 24, 1993]
§ 721.2192 [Redesignated. 58 FR 29947, May 24, 1993]
§ 721.2194 [Redesignated. 58 FR 29947, May 24, 1993]
§ 721.2196 [Redesignated. 58 FR 29947, May 24, 1993]
§ 721.2198 [Redesignated. 58 FR 29947, May 24, 1993]
§ 721.2200 [Redesignated. 58 FR 29947, May 24, 1993]
§ 721.2225 Cyclohexanecarbonitrile, 1,3,3-trimethyl-5-oxo-

§ 721.2250 1,4-Cyclohexanediamine, cis- and trans-.
§ 721.2260 1,2-Cyclohexanedicarboxylic acid, 2,2-bis[[[2-

§ 721.3080 Substituted phosphate ester (generic).
§ 721.3100 Oligomeric silicic acid ester compound with a hydroxylalkylamine.
§ 721.3120 Propenoate-terminated alkyl substituted silyl ester.
§ 721.3140 Vinyl epoxy ester.
§ 721.3160 1-Chloro-2-bromoethane.
§ 721.3180 Ethane, 2-chloro-1,1,1,2-tetrafluoro-.
§ 721.3200 Ethane, 1,1-dichloro-1-fluoro-.
§ 721.3220 Pentachloroethane.
§ 721.3240 Ethane, 1,1,1,2,2-pentafluoro-.
§ 721.3248 Ethane, 1,2,2-trichlorodifluoro-.
§ 721.3254 Ethane, 1,1,1 trifluoro-.
§ 721.3260 Ethanediimidic acids.
§ 721.3320 Ethanol, 2-amino-, compound with N-hydroxy-N-nitrosobenzenamine (1:1).
§ 721.3340 Ethanol, 2,2'-(hexylamino)bis-.
§ 721.3350 N-Nitrosodiethanolamine.
§ 721.3360 Substituted ethanolamine.
§ 721.3364 Aliphatic ether.
§ 721.3367 Alkenyl ether of alkanetriol polymer.
§ 721.3374 Alkylenediolalkyl ether.
§ 721.3380 Anilino ether.
§ 721.3390 [Removed. 59 FR 29203, June 6, 1994]
§ 721.3420 Brominated arylalkyl ether.
§ 721.3430 4-Bromophenyl phenyl ether.
§ 721.3435 Butoxy-substituted ether alkane.
§ 721.3440 Haloalkyl substituted cyclic ethers.
§ 721.3460 Diglycidyl ether of disubstituted carbopolycyle (generic name).
§ 721.3480 Halogenated biphenyl glycidyl ethers.
§ 721.3500 Perhalo alkoxy ether.
§ 721.3520 Aliphatic polyglycidyl ether.
§ 721.3540 [Removed. 59 FR 17489, Apr. 13, 1994]
§ 721.3560 Derivative of tetrachloroethylene.
§ 721.3580 Substituted ethylene diamine, methyl sulfate quaternized (generic name).
§ 721.3620 Fatty acid amine condensate, polycarboxylic acid salts.
§ 721.3625 Fatty acid amine salt (generic name).
§ 721.3629 Triethanolamine salts of fatty acids.
§ 721.3640 Trimethylolpropane fatty acid diacrylate.
§ 721.3680 Ethylene oxide adduct of fatty acid ester with pentaerythritol.
§ 721.3700 Fatty acid, ester with styrenated phenol, ethylene oxide adduct.
§ 721.3720 Fatty amide.
§ 721.3740 Bisalkylated fatty alkyl amine oxide
§ 721.3764 Fluorene substituted aromatic amine.
§ 721.3800 Formaldehyde, condensated polyoxyethylene fatty acid, ester with styrenated phenol, ethylene oxide adduct.
§ 721.3840 Tetraglycidalamines (generic name).
§ 721.3860 Glycol monobenzoate.

§ 721.3870 Monomethoxy neopentyl glycol propoxylate monoacrylate.
§ 721.3880 Polyalkylene glycol substituted acetate.
§ 721.3900 Alkyl polyethylene glycol phosphate, potassium salt.
§ 721.4000 Polyoxy alkylene glycol amine.
§ 721.4020 Polyalkylene glycol alkyl ether acrylate.
§ 721.4040 Glycols, polyethylene-, 3-sulfo-2-hydroxypropyl-p-(1,1,3,3- tetramethylbutyl)phenyl ether, sodium salt.
§ 721.4060 Alkylene glycol terephthalate and substituted benzoate esters (generic name).
§ 721.4080 MNNG (N-methyl-N'-nitro-N-nitrosoguanidine).
§ 721.4100 Tris(disubstituted alkyl) heterocycle.
§ 721.4128 Dimethyl-3-substituted heteromonocycle.
§ 721.4133 Dimethyl-3-substituted heteromonocyclic amine.
§ 721.4140 Hexachloronorbornadiene.
§ 721.4155 Hexachloropropene.
§ 721.4160 Hexafluoropropylene oxide.
§ 721.4180 Hexamethylphosphoramide.
§ 721.4200 Substituted alkyl peroxyhexane carboxylate (mixed isomers) (generic name).
§ 721.4215 Hexanedioic acid, diethenyl ester.
§ 721.4220 Hexanedioic acid, polymer with 1,2-ethanediol and 1,6- diisocyanato-2,2,4(or 2,4,4)-trimethylhexane, 2-hydroxyethyl-acrylate- blocked.
§ 721.4240 Alkyl peroxy-2-ethyl hexanoate.
§ 721.4250 Hexanoic acid, 2-ethyl-, ethenyl ester.
§ 721.4255 1,4,7,10,13,16-Hexaoxacyclooctadecane, 2-[(2-propenyl oxy)methyl]-.
§ 721.4260 Hydrazine, [4-(1-methylbutoxy)phenyl]-, monohydrochloride.
§ 721.4270 Nitrophenoxylalkanoic acid substituted thiazino hydrazide (generic name).
§ 721.4280 Substituted hydrazine.
§ 721.4300 Hydrazinecarboxamide, N,N'-1,6-hexanediylbis [2,2-dimethyl-
§ 721.4320 Hydrazinecarboxamide, N,N'-(methylenedi-4,1-phenylene)bis[2,2-dimethyl-.
§ 721.4340 Substituted thiazino hydrazine salt (generic name).
§ 721.4360 Certain hydrogen containing chlorofluorocarbons.
§ 721.4380 Modified hydrocarbon resin.
§ 721.4390 Trisubstituted hydroquinone diester.
§ 721.4400 Substituted hydroxyalkyl alkenoate, [(1-oxo-2-propenyl)oxy]alkoxy] carbonylamino] substituted] aminocarbonyl]oxy-.
§ 721.4420 Substituted hydroxylamine.
§ 721.4460 Amidinothiopropionic acid hydrochloride.
§ 721.4480 2-Imino-1,3-thiazin-4-one-5,6-dihydromonohydrochloride.
§ 721.4490 Capped aliphatic isocyanate.

§ 721.4500 Isopropylamine distillation residues and ethylamine distillation residues.
§ 721.4520 Isopropylidene, bis(1,1-dimethylpropyl) derivative.
§ 721.4550 Diperoxy ketal.
§ 721.4568 Methylpolychloro aliphatic ketone.
§ 721.4585 Lecithins, phospholipase A2-hydrolyzed.
§ 721.4590 Mannich-based adduct.
§ 721.4600 Recovered metal hydroxide.
§ 721.4620 Dialkylamino alkanoate metal salt.
§ 721.4640 Substituted benzenesulfonic acid, alkali metal salt.
§ 721.4660 Alcohol, alkali metal salt.
§ 721.4680 Metal salts of complex inorganic oxyacids (generic name).
§ 721.4700 Metalated alkylphenol copolymer (generic name).
§ 721.4720 Disubstituted phenoxazine, chlorometalate salt.
§ 721.4740 Alkali metal nitrites.
§ 721.4780 Hydroxyalkyl methacrylate, alkyl ester.
§ 721.4790 2-(2-Hydroxy-3-tert-butyl-5-methylbenzyl)-4-methyl-6-tert- butylphenyl methacrylate.
§ 721.4794 Polypiperidinol-acrylate methacrylate.
§ 721.4800 Methacrylic ester.
§ 721.4820 Methane, bromodifluoro- .
§ 721.4840 Substituted triphenylmethane.
§ 721.4880 Methanol, trichloro-, carbonate (2:1).
§ 721.4925 Methyl n-butyl ketone.
§ 721.5050 2,2'-[(1-Methylethylidene)bis[4,1-phenyloxy[1-(butoxymethyl)- (2,1-ethanediyl]oxymethylene]]bisoxirane, reaction product with a diamine.
§ 721.5075 Mixed methyltin mercaptoester sulfides.
§ 721.5175 Mitomycin C.
§ 721.5200 Disubstituted phenylazo trisubstituted naphthalene.
§ 721.5225 Naphthalene,1,2,3,4-tetrahydro(1-phenylethyl) (specific name).
§ 721.5250 Trimethyl spiropolyheterocyclic naphthalene compound.
§ 721.5275 2-Napthalenecarboxamide-N-aryl-3-hydroxy-4-arylazo (generic name).
§ 721.5285 Ethoxylated substituted naphthol.
§ 721.5300 Neodecaneperoxoic acid, 1,1,3,3-tetramethylbutyl ester.
§ 721.5310 Neononanoic acid, ethenyl ester.
§ 721.5325 Nickel acrylate complex.
§ 721.5330 Nickel salt of an organo compound containing nitrogen.
§ 721.5350 Substituted nitrile (generic name).
§ 721.5375 Nitrothiophenecarboxylic acid, ethyl ester, bis[[[[(substituted)]amino]alkylphenyl]azo] (generic name).
§ 721.5385 Octanoic acid, hydrazide.
§ 721.5400 3,6,9,12,15,18,21-Heptaoxatetratriaoctanoic

acid, sodium salt.
§ 721.5425 a-Olefin sulfonate, potassium salts.
§ 721.5450 a-Olefin sulfonate, sodium salt.
§ 721.5475 1-Oxa-4-azaspiro[4.5]decane, 4-dichloroacetyl-.
§ 721.5500 7-Oxabicyclo[4.1.0]heptane, 3-ethenyl, homopolymer, ether with 2-ethyl-2-(hydroxymethyl)-1,3-propanediol (3:1), epoxidized.
§ 721.5525 Substituted spiro oxazine.
§ 721.5550 Substituted dialkyl oxazolone (generic name).
§ 721.5575 Oxirane, 2,2'-(1,6-hexanediylbis (oxymethylene)) bis-.
§ 721.5600 Substituted oxirane.
§ 721.5625 Oxiranemethanamine, N,N'-[methylenebis(2-ethyl-4,1- phenylene)]bis[N-(oxiranylmethyl)]-.
§ 721.5660 Pentaerythritol, mixed esters with carboxylic acids.
§ 721.5700 Pentanenitrile, 3-amino-.
§ 721.5705 2,5,8,10,13-Pentaoxahexadec-15-enoic acid, 9,14-dioxo-2-[(1- oxo-2-propenyl)oxy]ethyl ester.
§ 721.5710 Phenacetin.
§ 721.5740 Phenol, 4,4'-methylenebis(2,6-dimethyl-.
§ 721.5760 Phenol, 4,4'-[methylenebis(oxy-2,1-ethanediylthio)]bis-.
§ 721.5780 Phenol, 4,4'-(oxybis(2,1-ethanediylthio)bis-.
§ 721.5800 Sulfurized alkylphenol.
§ 721.5820 Aminophenol.
§ 721.5840 Ethylated aminophenol.
§ 721.5860 Methylphenol, bis(substituted)alkyl.
§ 721.5880 Sulfur bridged substituted phenols (generic name).
§ 721.5900 Trisubstituted phenol (generic name).
§ 721.5910 Acrylated epoxy phenolic resin.
§ 721.5915 Polysubstituted phenylazopolysubstitutedphenyl dye.
§ 721.5920 Phenyl(disubstitutedpolycyclic).
§ 721.5960 N,N'-Bis(2-(2-(3-alkyl)thiazoline) vinyl)-1,4-phenylenediamine methyl sulfate double salt (generic name).
§ 721.5970 Phosphated polyarylphenol ethoxylate, potassium salt.
§ 721.5980 Dialkyl phosphorodithioate phosphate compounds.
§ 721.5990 Halogenated phosphate ester.
§ 721.6000 Tris (2,3-dibromopropyl) phosphate.
§ 721.6020 Phosphine, dialkylyphenyl.
§ 721.6060 Alkylaryl substituted phosphite.
§ 721.6070 Alkyl phosphonate ammonium salts.
§ 721.6080 Phosphonium salt (generic name).
§ 721.6085 Phosphonocarboxylate salts.
§ 721.6090 Phosphoramide.
§ 721.6100 Phosphoric acid, C6-12-alkyl esters, compounds with 2- (dibutylamino) ethanol.
§ 721.6120 Phosphoric acid, 1,2-ethanediyl tetrakis(2-chloro-1- methylethyl) ester.

§ 721.6140 Dialkyldithiophosphoric acid, aliphatic amine salt.
§ 721.6160 Piperazinone, 1,1',1"-[1,3,5-triazine-2,4,6-triyltris[(cyclohexylimino)-2,1-ethanediyl]]tris-[3,3,4,5,5-pentamethyl]-.
§ 721.6180 [Removed. 59 FR 29204, June 6, 1994]
§ 721.6186 Polyamine dithiocarbamate.
§ 721.6193 Polyalkylene polyamine.
§ 721.6200 Fatty acid polyamine condensate, phosphoric acid ester salts.
§ 721.6220 Aryl sulfonate of a fatty acid mixture, polyamine condensate.
§ 721.6440 Polyamine ureaformaldehyde condensate (specific name).
§ 721.6470 Polyaminopolyacid.
§ 721.6480 [Removed. 59 FR 17490, Apr. 13, 1994]
§ 721.6500 Polymer.
§ 721.6520 Acrylamide, polymer with substituted alkylacrylamide salt (generic name).
§ 721.6540 Acrylamide, polymers with tetraalkyl ammonium salt and polyalkyl, aminoalkyl methacrylamide salt.
§ 721.6560 Acrylic acid, polymer with substituted ethene.
§ 721.6580 Polymer of adipic acid, alkanepolyol, alkyldiisocyanatocarbomonocycle, hydroxyalkyl acrylate ester.
§ 721.6600 [Removed. 59 FR 29205, June 6, 1994]
§ 721.6620 Alkanaminium, polyalkyl-[(2-methyl-1-oxo-2-propenyl)oxy] salt, polymer with acrylamide and substituted alkyl methacrylate.
§ 721.6625 [Redesignated. 58 FR 29947, May 24, 1993]
§ 721.6640 Polymer of alkanedioic acid, methylenebiscarbomonocyclic diisocyanate, and alkylene glycols, hydroxyalkyl acrylate ester.
§ 721.6660 Polymer of alkanepolyol and polyalkylpolyisocyanatocarbomonocycle, acetone oxime-blocked (generic name).
§ 721.6680 Alkanoic acid, butanediol and cyclohexanealkanol polymer (generic name).
§ 721.6700 Polymer of alkenoic acid, substituted alkylacrylate sodium salt (generic name).
§ 721.6720 Alkyldicarboxylic acids, polymers with alkanepolyol and TDI, alkanol blocked, acrylate.
§ 721.6740 Polymer of alkyl carbomonocycle diisocyanate with alkanepolyol polyacrylate.
§ 721.6760 Alkylenebis (substituted carbomonocycle), epichlorohydrin, disubstituted heteromonocycle, acrylate polymer.
§ 721.6780 Polymer of substituted alkylphenol formaldehyde and phthalic anhydride, acrylate (generic name).
§ 721.6820 [Removed. 58 FR 45842, Aug. 31, 1993]
§ 721.6840 Substituted bis(hydroxyalkane) polymer with epichlorohydrin, acrylate.
§ 721.6880 Bisphenol A, epichlorohydrin, methylenebis

(substituted carbomonocycle), polyalkylene glycol, alkanol, methacrylate polymer.

§ 721.6900 Polymer of bisphenol A diglycidal ether, substituted alkenes, and butadiene.

§ 721.6920 Butyl acrylate, polymer with substituted methyl styrene, methyl methacrylate, and substituted silane.

§ 721.6940 Caprolactone, polymer with hexamethylene diisocyanate, hydroxyalkyl acrylate ester, reaction products with substituted alkanoic acid and metal heteromonocycle.

§ 721.6960 E-Caprolactone modified 2-hydroxyethyl acrylate monomer.

§ 721.6980 Dimer acids, polymer with polyalkylene glycol, bisphenol A- diglycidyl ether, and alkylenepolyols polyglycidyl ethers (generic name).

§ 721.7000 Polymer of disodium maleate, allyl ether, and ethylene oxide.

§ 721.7020 Distillates (petroleum), C(3-6), polymers with styrene and mixed terpenes (generic name).

§ 721.7040 Formaldehyde, polymer with (chloromethyl)oxirane, 4,4'-(1- methyl ethylidene)bis[2,6-dibromophenol] and phenol, 2-methyl-2-propenoate.

§ 721.7080 Polymer of hydroxyethyl acrylate and polyisocyanate.

§ 721.7100 Polymer of isophorone diisocyanate, trimethylolpropane, polyalkylenepolyol, disubstituted alkanes and hydroxyethyl acrylate.

§ 721.7140 Methylenebis(4-isocyanato benzene), polymer with polycaprolactone triol and alkoxylated alkanepolyol, hydroxyalkyl methacrylate ester.

§ 721.7160 2-Oxepanone, polymer with 4,4'-(1-methylethylidene)bisphenol and 2,2-[(1-methylethylidene)bis(4,1-phenyleneoxymethylene)]bisoxirane, graft.

§ 721.7180 Substituted oxide-alkylene polymer, methacrylate.

§ 721.7200 Perfluoroalkyl aromatic carbamate modified alkyl methacrylate copolymer.

§ 721.7210 Epoxidized copolymer of phenol and substituted phenol.

§ 721.7220 Polymer of substituted phenol, formaldehyde, epichlorohydrin, and disubstituted benzene.

§ 721.7240 Polymer of disubstituted phthalate, dioxoheteropolycycle, and methacrylic acid.

§ 721.7260 Polymer of polyethylenepolyamine and alkanediol diglycidyl ether.

§ 721.7280 1,3-Propanediamine, N,N'-1,2-ethanediylbis-, polymer with 2,4,6-trichloro-1,3,5-triazine, reaction products with N-butyl-2,2,6,6- tetramethyl-4-piperidinamine.

§ 721.7300 2-Propenenitrile, polymer with 1,3-butadiene, 3-carboxy-1- cyano-1-methylpropyl-terminated, polymers with bisphenol A, epichlorohydrin, and 4,4'-

(1methylethylidene)bis[2,6-dibromophenol], dimethacrylate.
§ 721.7320 2-Propenenitrile, polymer with 1,3-butadiene, 3-carboxy-1- cyano-1-methylpropyl-terminated, polymers with epichlorohydrin, formaldehyde, 4,4'-(1-methylethylidene)bis[2,6-dibromophenol], and phenol, 2-methyl-2-propenoate.
§ 721.7340 Polymer of styrene, substituted alkyl methacrylates, 2- ethylhexyl acrylate, methacrylic acid and substituted bis(benzene).
§ 721.7360 Terpenes and terpenoids, limonene fraction, polymer with substituted carbopolycycles (generic name).
§ 721.7370 Acrylates of aliphatic polyol.
§ 721.7400 Di(alkanepolyol) ether, polyacrylate.
§ 721.7420 Oxyalkanepolyol polyacrylate.
§ 721.7440 Polyalkylenepolyol alkylamine. (generic name).
§ 721.7450 Aromatic amine polyols.
§ 721.7460 Polyol carboxylate ester.
§ 721.7480 Isocyanate terminated polyols.
§ 721.7500 Nitrate polyether polyol (generic name).
§ 721.7540 Polysubstituted polyol.
§ 721.7560 Alkoxylated alkane polyol, polyacrylate ester.
§ 721.7580 Substituted acrylated alkoxylated aliphatic polyol.
§ 721.7600 Alkyl(heterocyclicyl) phenylazohetero monocyclic polyone (generic name).
§ 721.7620 Alkyl(heterocyclicyl) phenylazohetero monocyclic polyone, ((alkylimidazolyl) methyl) derivative (generic name).
§ 721.7655 Alkylsulfonium salt.
§ 721.7660 Poly(oxy-1,4-butanediyl), a-(1-oxo-2-propenyl)-v-[(1-oxo-2- propenyl)oxy].
§ 721.7680 Poly(oxy-1,2-ethanediyl), .a.-hydro-.v.hydroxy-, ether with 2- ethyl-2-(hydroxymethyl)-1,3-propanediol (3:1) di-2-propenoate, methyl ether
§ 721.7700 Poly(oxy-1,2-ethanediyl), a-hydro-v-(oxiranylmethoxy)-, ether with 2-ethyl-2-(hydroxymethyl)-1,3-propanediol (3:1).
§ 721.7710 Polyepoxy polyol.
§ 721.7720 Poly(oxy-1,2-ethanediyl), a,a'-[(1-methylethylidene) di-4,1- phenylene] bis [v-(oxiranylmethoxy)-.
§ 721.7740 Poly(oxy-1,2-ethanediyl), a-(2-methyl-1-oxo-2-propenyl)-v- hydroxy-, C10-16-alkyl ethers.
§ 721.7760 Poly(oxy-1,2-ethanediyl), a-(1-oxo-2-propenyl)-v-hydroxy-, C10-16-alkyl ethers.
§ 721.7770 Alkylphenoxypoly(oxyethylene) sulfuric acid ester, substituted amine salt.
§ 721.7780 Poly[oxy(methyl-1,2-ethanediyl)], a,a'-(2,2-dimethyl-1,3- propanediyl)bis[v-(oxiranymethoxy)-.
§ 721.8075 Polyurethane.
§ 721.8082 Polyester polyurethane acrylate.
§ 721.8100 Potassium N,N-bis (hydroxyethyl) cocoamine oxide phosphate, and potassium N,N-bis (hydroxyethyl)

tallowamine oxide phosphate.
§ 721.8125 Propane, 1,1,1,2,3,3,3-heptafluoro-.
§ 721.8160 Propanoic acid, 2,2-dimethyl-, ethenyl ester.
§ 721.8225 2-Propenamide, N-[3-dimethylamino)propyl]-.
§ 721.8250 1-Propanol, 3,3'-oxybis[2,2-bis(bromomethyl)
§ 721.8265 2-Propenoic acid, C18-26 and C>20 alkyl esters.
§ 721.8275 2-Propenoic acid, 3-(dimethylamino)-2,2-dimethylpropyl ester.
§ 721.8290 2-Propenoic acid, docosyl ester.
§ 721.8300 2-Propenoic acid, 2-hydroxybutyl ester.
§ 721.8325 2-Propenoic acid, 1-(hydroxymethyl) propyl ester.
§ 721.8335 2-Propenoic acid, 2-[[(1-methylethoxy)carbonyl]amino]ethyl ester.
§ 721.8350 2-Propenoic acid, 7-oxabicyclo[4.1.0]hept-3-ylmethyl ester.
§ 721.8375 2-Propenoic acid, 2-(2-oxo-3-oxazolidinyl)ethyl ester.
§ 721.8400 2-Propenoic acid, 3,3,5-trimethylcyclohexyl ester.
§ 721.8425 2-Propenoic acid, 2-[[[[[1,3,3-trimethyl-5-[[[2-[(1-oxo-2- propenyl)oxy] ethoxy]carbonyl]amino]cyclohexyl]methyl] amino]carbonyl]oxy]ethyl ester.
§ 721.8450 2-Propenoic acid, 2-methyl-, 2-[3-(2H-benzotriazol-2-yl)-4- hydroxyphenyl]ethyl ester.
§ 721.8475 2-Propenoic acid, 2-methyl-, 1,1-dimethylethyl ester.
§ 721.8500 2-Propenoic acid, 2-methyl-, 7-oxabicyclo [4.1.0]hept-3- ylmethyl ester.
§ 721.8525 2-Propenoic acid, 2-methyl-, 3,3,5-trimethylcyclohexyl ester.
§ 721.8550 2-Propenoic acid, 2-methyl-, 7,7,9-trimethyl-4,13-dioxo-3,14- dioxo-5,12-diazahexadecane, 1,16-diyl ester.
§ 721.8575 2-Propenoic acid [octahydro-4,7-methano-1H-indene-1, 5(1,6 or 2,5)-diyl]bis(methylene) ester.
§ 721.8600 2-Propenoic acid, octahydro-4, 7-methano-1H-indenyl ester.
§ 721.8650 2-Propenoic acid, reaction product with 2-oxepanone and alkyltriol.
§ 721.8654 2-Propenoic acid 3-(trimethoxy silyl)propyl ester.
§ 721.8675 Halogenated pyridines.
§ 721.8700 Halogenated alkyl pyridine.
§ 721.8750 Halogenated substituted pyridine.
§ 721.8775 Substituted pyridines.
§ 721.8825 Substituted methylpyridine and substituted 2-phenoxypyridine.
§ 721.8850 Disubstituted halogenated pyridinol.
§ 721.8875 Substituted halogenated pyridinol.
§ 721.8900 Substituted halogenated pyridinol, alkali salt.
§ 721.8965 1H-Pyrole-2, 5-dione, 1-(2,4,6-tribromophenyl)-

§ 721.9000 N-Nitrosopyrrolidine.
§ 721.9075 Quaternary ammonium salt of fluorinated alkylaryl amide.
§ 721.9100 Substituted quinoline.
§ 721.9220 Reaction products of secondary alkyl amines with a substituted benzenesulfonic acid and sulfuric acid (generic name).
§ 721.9240 Reaction product of alkyl carboxylic acids, alkane polyols, alkyl acrylate, and isophorone diisocyanate.
§ 721.9260 Reaction product of alkylphenol, tetraalkyl titanate and tin complex.
§ 721.9280 Reaction product of ethoxylated fatty acid oils and a phenolic pentaerythritol tetraester.
§ 721.9300 Reaction products of substituted hydroxyalkanes and polyalkylpolyisocyanatocarbomonocycle.
§ 721.9320 Reaction product of hydroxyethyl acrylate and methyl oxirane.
§ 721.9360 Reaction product of a monoalkyl succinic anhydride with an v- hydroxy methacrylate.
§ 721.9400 Reaction product of phenolic pentaerythritol tetraesters with fatty acid esters and oils, and glyceride triesters.
§ 721.9420 Polymethylcarbomonocycle, reaction product with 2- hydroxyethyl acrylate.
§ 721.9460 Tall oil fatty acids, reaction products with polyamines, alkyl substituted.
§ 721.9470 Reserpine.
§ 721.9480 Resorcinol, formaldehyde substituted carbomonocycle resin.
§ 721.9500 Silane, (1,1-dimethylethoxy)dimethoxy(2-methyl propyl)-.
§ 721.9510 Silicone ester polyacrylate.
§ 721.9525 Acrylate substituted siloxanes and silicones.
§ 721.9527 Bis(1,2,2,6,6-pentamethyl-4-piperidin-4-ol) ester of cycloaliphatic spiroketal.
§ 721.9530 Bis(2,2,6,6-tetramethylpiperidinyl) ester of cycloalkyl spiroketal.
§ 721.9550 Sulfonamide.
§ 721.9570 Halophenyl sulfonamide salt.
§ 721.9580 Ethyl methanesulfonate.
§ 721.9620 Aromatic sulfonic acid compound with amine.
§ 721.9630 Polyfluorosulfonic acid salt.
§ 721.9650 Tetramethylammonium salts of alkylbenzenesulfonic acid.
§ 721.9660 Methylthiouracil.
§ 721.9665 Organotin catalysts.
§ 721.9668 Organotin lithium compound.
§ 721.9675 Titanate [Ti6O13 (2-)], dipotassium.
§ 721.9700 Monosubstituted alkoxyaminotrazines (generic name).
§ 721.9720 Disubstituted alkyl triazines (generic name).

Appendix A to Subpart F--Interim Method of the Determination of Asbestos in Bulk Insulation Samples

Subpart G--Asbestos Abatement Projects

§ 763.120 Scope.

§ 763.121 Regulatory requirements.

Appendix A To § 763.121--EPA/OSHA Reference Method--Mandatory

Appendix B to § 763.121--Detailed Procedure for Asbestos Sampling and Analysis--Non-Mandatory

Appendix C to § 763.121--Qualitative and Quantitative Fit Testing Procedures--Mandatory

Appendix D to § 763.121--Medical Questionnaires--Mandatory

Appendix E to § 763.121--Interpretation and Classification of Chest Roentgenograms--Mandatory

§ 763.122 Exclusions for States.

§ 763.124 Reporting.

§ 763.125 Enforcement.

§ 763.126 Inspections.

Subpart I--Prohibition of the Manufacture, Importation, Processing, and Distribution in Commerce of Certain Asbestos-Containing Products; Labeling Requirements

§ 763.160 Scope.

§ 763.163 Definitions.

§ 763.165 Manufacture and importation prohibitions.

§ 763.167 Processing prohibitions.

§ 763.169 Distribution in commerce prohibitions.

§ 763.171 Labeling requirements.

§ 763.173 Exemptions.

§ 763.175 Enforcement.

§ 763.176 Inspections.

§ 763.178 Recordkeeping.

§ 763.179 Confidential business information claims.

PART 766--DIBENZO-PARA-DIOXINS/DIBENZOFURANS

Subpart A--General Provisions

§ 766.1 Scope and purpose.

§ 766.2 Applicability and duration of this part.

§ 766.3 Definitions.

§ 766.5 Compliance.

§ 766.7 Submission of information.

§ 766.10 Test standards.

§ 766.12 Testing guidelines.

§ 766.14 Contents of protocols.

§ 766.16 Developing the analytical test method.

§ 766.18 Method sensitivity.

Subpart B--Specific Chemical Testing/Reporting Requirements

§ 766.20 Who must test.

§ 766.25 Chemical substances for testing.

§ 766.27 Congeners and LOQs for which quantitation is required.

§ 766.28 Expert review of protocols.

§ 766.32 Exclusions and waivers.

§ 766.35 Reporting requirements.

§ 796.1550 Partition coefficient (n-Octanol/water).
§ 796.1570 Partition coefficient (n-Octanol/water)--Estimation by liquid chromatography.
§ 796.1720 Octanol/water partition coefficient, generator column method.
§ 796.1840 Water solubility.
§ 796.1860 Water solubility (generator column method).
§ 796.1950 Vapor pressure.

Subpart C--Transport Processes

§ 796.2700 Soil thin-layer chromatography.
§ 796.2750 Sediment and soil adsorption isotherm.

Subpart D--Transformation Processes

§ 796.3100 Aerobic aquatic biodegradation.
§ 796.3140 Anaerobic biodegradability of organic chemicals.
§ 796.3180 Ready biodegradability: Modified AFNOR test.
§ 796.3200 Ready biodegradability: Closed bottle test.
§ 796.3220 Ready biodegradability: Modified MITI test (I).
§ 796.3240 Ready biodegradability: Modified OECD screening test.
§ 796.3260 Ready biodegradability: Modified Sturm test.
§ 796.3300 Simulation test--aerobic sewage treatment: Coupled units test.
§ 796.3340 Inherent biodegradability: Modified SCAS test.
§ 796.3360 Inherent biodegradability: Modified Zahn-Wellens test.
§ 796.3400 Inherent biodegradability in soil.
§ 796.3480 Complex formation ability in water.
§ 796.3500 Hydrolysis as a function of pH at 25 deg.C.
§ 796.3700 Photolysis in aqueous solution in sunlight.
§ 796.3780 Laboratory determination of the direct photolysis reaction quantum yield in aqueous solution and sunlight photolysis.
§ 796.3800 Gas phase absorption spectra and photolysis.

PART 797--ENVIRONMENTAL EFFECTS TESTING GUIDELINES

Subpart A--[Reserved]

Subpart B--Aquatic Guidelines

§ 797.1050 Algal acute toxicity test.
§ 797.1060 Freshwater algae acute toxicity test.
§ 797.1075 Freshwater and marine algae acute toxicity test.
§ 797.1160 Lemna acute toxicity test.
§ 797.1300 Daphnid acute toxicity test.
§ 797.1330 Daphnid chronic toxicity test.
§ 797.1350 Daphnid chronic toxicity test.
§ 797.1400 Fish acute toxicity test.
§ 797.1440 Fish acute toxicity test.
§ 797.1520 Fish bioconcentration test.
§ 797.1560 Fish bioconcentration test.
§ 797.1600 Fish early life stage toxicity test.
§ 797.1800 Oyster acute toxicity test.
§ 797.1830 Oyster bioconcentration test.
§ 797.1930 Mysid shrimp acute toxicity test.

§ 797.1950 Mysid shrimp chronic toxicity test.
§ 797.1970 Penaeid shrimp acute toxicity test.
Subpart C--Terrestrial Guidelines
§ 797.2050 Avian dietary toxicity test.
§ 797.2130 Bobwhite reproduction test.
§ 797.2150 Mallard reproduction test.
§ 797.2175 Avian acute oral toxicity test.
§ 797.2750 Seed germination/root elongation toxicity test.
§ 797.2800 Early seedling growth toxicity test.
§ 797.2850 Plant uptake and translocation test.
PART 798--HEALTH EFFECTS TESTING GUIDELINES
Subpart A--[Reserved]
Subpart B--General Toxicity Testing
§ 798.1100 Acute dermal toxicity.
§ 798.1150 Acute inhalation toxicity.
§ 798.1175 Acute oral toxicity.
Subpart C--Subchronic Exposure
§ 798.2250 Dermal toxicity.
§ 798.2450 Inhalation toxicity.
§ 798.2650 Oral toxicity.
§ 798.2675 Oral toxicity with satellite reproduction and fertility study.
Subpart D--Chronic Exposure
§ 798.3260 Chronic toxicity.
§ 798.3300 Oncogenicity.
§ 798.3320 Combined chronic toxicity/oncogenicity.
Subpart E--Specific Organ/Tissue Toxicity
§ 798.4100 Dermal sensitization.
§ 798.4350 Inhalation developmental toxicity study.
§ 798.4420 Preliminary developmental toxicity screen.
§ 798.4470 Primary dermal irritation.
§ 798.4500 Primary eye irritation.
§ 798.4700 Reproduction and fertility effects.
§ 798.4900 Developmental toxicity study.
Subpart F--Genetic Toxicity
§ 798.5100 Escherichia coli WP2 and WP2 uvrA reverse mutation assays.
§ 798.5140 Gene mutation in aspergillus nidulans.
§ 798.5195 Mouse biochemical specific locus test.
§ 798.5200 Mouse visible specific locus test.
§ 798.5250 Gene mutation in neurospora crassa.
§ 798.5265 The salmonella typhimurium reverse mutation assay.
§ 798.5275 Sex-linked recessive lethal test in drosophila melanogaster.
§ 798.5300 Detection of gene mutations in somatic cells in culture.
§ 798.5375 In vitro mammalian cytogenetics.
§ 798.5385 In vivo mammalian bone marrow cytogenetics tests: Chromosomal analysis.
§ 798.5395 In vivo mammalian bone marrow cytogenetics tests: Micronucleus assay.
§ 798.5450 Rodent dominant lethal assay.

§ 798.5460 Rodent heritable translocation assays.
§ 798.5500 Differential growth inhibition of repair proficient and repair deficient bacteria: "Bacterial DNA damage or repair tests."
§ 798.5550 Unscheduled DNA synthesis in mammalian cells in culture.
§ 798.5575 Mitotic gene conversion in Saccharomyces cerevisiae.
§ 798.5900 In vitro sister chromatid exchange assay.
§ 798.5915 In vivo sister chromatid exchange assay.
§ 798.5955 Heritable translocation test in drosophila melanogaster.
Subpart G--Neurotoxicity
§ 798.6050 Functional observational battery.
§ 798.6200 Motor activity.
§ 798.6400 Neuropathology.
§ 798.6450 NTE neurotox assay.
§ 798.6500 Schedule-controlled operant behavior.
§ 798.6540 Acute delayed neurotoxicity of organophosphorus substances.
§ 798.6560 Subchronic delayed neuro-toxicity of organophosphorus substances.
§ 798.6850 Peripheral nerve function.
Subpart H--Special Studies
§ 798.7100 Metabolism.
PART 799--IDENTIFICATION OF SPECIFIC CHEMICAL SUBSTANCE AND MIXTURE TESTING REQUIREMENTS
Subpart A--General Provisions
§ 799.1 Scope and purpose.
§ 799.2 Applicability.
§ 799.3 Definitions.
§ 799.5 Submission of information.
§ 799.10 Test standards.
§ 799.11 Availability of test guidelines.
§ 799.12 Test results.
§ 799.17 Effects of non-compliance.
§ 799.19 Chemical imports and exports.
Subpart B--Specific Chemical Test Rules
§ 799.500 Anthraquinone.
§ 799.925 Biphenyl.
§ 799.940 Bisphenol A.
§ 799.1051 Monochlorobenzene.
§ 799.1052 Dichlorobenzenes.
§ 799.1053 Trichlorobenzenes.
§ 799.1054 1,2,4,5-Tetrachlorobenzene.
§ 799.1250 Cresols.
§ 799.1285 Cumene.
§ 799.1550 1,2-Dichloropropane.
§ 799.1560 Diethylene glycol butyl ether and diethylene glycol butyl ether acetate.
§ 799.1575 Diethylenetriamine (DETA).
§ 799.1645 2-Ethylhexanol.
§ 799.1650 2-Ethylhexanoic acid.

ACRONYMS & ABBREVIATIONS

ACRONYMS & ABBREVIATIONS

301(H)APPS	Applications for Variance from Secondary Treatment Requirements (File)
301(H)INFO 301(h)	Application Tracking System
403C	Section 403(c) Information
5SEG	Spatially Segmented Phytoplankton Model
A & C	Abatement & Control
A&R	Air and Radiation
A/WPR	Air/Water Pollution Report
AA	Accountable Area
AA	Adverse Action
AA	Advices of Allowance
AA	Assistant Administrator
AA	Associate Administrator
AA	Atomic Absorption
AAA	American Arbitration Association
AAA	American Automobile Association
AAAS	American Association for the Advancement of Science
AAEE	American Academy of Environmental Engineers
AAEM	American Academy of Environmental Medicine
AAES	American Association of Engineering Societies
AANWR	Alaskan Arctic National Wildlife Refuge
AAOHN	American Association of Occupational Health Nurses
AAP	Affirmative Action Plan
AAP	Affirmative Action Program
AAP	Asbestos Action Program
AAPCO	Association of American Pesticides Control Officers
AAR/BOE	Association of American Railroads/Bureau of Explosives
AARC	Alliance for Acid Rain Control
AARP	American Association of Retired Persons
ABA	American Bar Association
ABAG	Association of Bay Area Governments
ABES	Alliance for Balanced Environmental Solutions
ABMA	American Boiler Manufacturers Association
ABTRES	Abatement and Residual Forecasting Model
AC	Actual Commitment
AC	Advisory Circular
AC	Alternating Current
ACA	American Conservation Association, Inc.
ACBM	Asbestos-Containing Building Material
ACE	Alliance for Clean Energy
ACEC	American Consulting Engineers Council
ACEEE	American Council for an Energy Efficient Economy

ACFM	Actual Cubic Feet Per Minute
ACGIH	American Conference of Governmental Industrial Hygienists
ACI	Association for Conservation Information
ACL	Alternative Concentration Limits
ACL	Analytical Chemistry Laboratory
ACM	Asbestos-Containing Material
ACP	Air Carcinogen Policy
ACQR	Air Quality Control Region
ACQUIRE	Aquatic Information Retrieval
ACR	Agency Confirmation Agreement
ACS	American Chemical Society
ACSH	American Council on Science and Health
ACT	Action
ACTS	Asbestos Contractor Tracking System
ACWA	American Clean Water Association
ADABAS	Adaptable Data Base
ADAPT	Adapt II Structural Activities
ADARD	Acid Deposition and Atmospheric Research Division (ORD)
ADAS	Acid Deposition Assessment Staff (ORD)
ADB	Applications Data Base
ADBA	Adabas Administrator
ADCO	Alternate Document Control Officer
ADCR	Automated Document Control Register
ADCRMG	Automated Document Control Register Management Group
ADI	Acceptable Daily Intake
ADMIN	ERL-Athens Administrative System
ADP	Applications Data Base
ADP	Automated Data Processing
ADPCE	Arkansas Department of Pollution Control and Ecology
ADPCETS	ADP Capital Equipment Inventory System
ADPE	Automated Data Processing Equipment
ADPS	Acid Deposition Planning Staff (ORD)
ADQ	Audits of Data Quality
ADR	Alternative Dispute Resolution
ADSS	Air Data Screening System
ADT	Average Daily Traffic
ADTRACS	Assistance Disputes
AEA	Atomic Energy Act
AEC	Associate Enforcement Counsel (OECM)
AED	Air Enforcement Division (OECM)
AED	Analysis and Evaluation Division (OW)
AEE	Alliance for Environmental Education
AEERL	Air and Energy Engineering Research Laboratory (ORD)

AEM	Acoustic Emission Monitoring
AERE	Association of Environmental & Resource Economists
AEROS	Aerometric and Emissions Reporting System
AES	Air and Energy Staff (ORD)
AES	American Electroplating Society
AES	Analysis and Evaluation Staff
AES	Auger Electron Spectrometry
AESA	Association of Environmental Scientists and Administrators
AFA	American Forestry Association
AFBF	American Farm Bureau Federation
AFCA	Area Fuel Consumption Allocation
AFDO	Award Fee Determination Official
AFGE	American Federation of Government Employees
AFI	American Forest Institute
AFRCE	Air Force Regional Civil Engineers
AFS	AIRS Facility Subsystem (OAR)
AFUG	AIRS Facility Users Group (OAR)
AG	Attorney General
AGA	American Gas Association, Inc.
AGC	Associate General Counsels (OGC)
AGCA	Associated General Contractors of America
AGDS	Automated Grants Documentation System
AH	Allowance Holder
AHERA	Asbestos Hazard Emergency Response Act
AHM	Allowance Monthly Holder
AI	Artificial Intelligence
AIA	American Institute of Architects
AIA	Asbestos Information Association
AIADA	American International Automobile Dealers Association
AIC	Acceptable Intake for Chronic Exposures
AIChE	American Institute of Chemical Engineers
AICPA	American Institute of Certified Public Accountants
AICR	Alternative Internal Control Review
AICUZ	Air Installation Compatible Use Zones
AID	Agency for International Development
AIF	Atomic Industrial Health Forum, Inc.
AIG	Assistant Inspector General
AIHC	American Industrial Health Council
AIME	American Institute of Metallurgical, Mining and Petroleum Engineers
AIP	Auto Ignition Point
AIRDOS-EPA	Atmospheric Dispersion of Radionuclides
AIRS	Accident and Illness Reporting System
AIRS	Aerometric Information Retrieval System

AIRS	Air Quality Subsystem
AIS	Acceptable Intake for Subchronic Exposures
AIS	Asbestos Information System
AISCM	Advanced Information System for Career Management
AISI	American Iron & Steel Institute
AL	Acceptable Level
AL	Administrative Leave
AL	Annual Leave
ALA	American League of Anglers, Inc.
ALA	American Lung Association
ALA	Delta-Aminolevulinic Acid
ALAPCO	Association of Local Air Pollution Control Officials
ALARA	As Low As Reasonably Achievable
ALC	Application Limiting Constituent
ALD-O	Delta-Aminolevulinic Acid Dehydrates
ALEC	American Legislative Exchange Council
ALJ	Administrative Law Judge
ALMS	TALMS Without the Tunable
ALPS	ERL-Athens Lab Planning System
ALR	Action Leakage Rate
ALS	American Littoral Society
ALT-SEA	Assembly Line Test and Selective Enforcement Audit Data
AMA	American Medical Association
AMBIENS	Atmospheric Mass Balance of Industrially Emitted and Natural Sulfur
AMC	American Mining Congress
AMC	Army Material Command (DOD)
AMD	Air Management Division (regional)
AMIS	Air Management Information System
AMPS	Automatic Mapping and Planning System
AMS	Administrative Management Staff (OAR)(ORD)
AMS	American Meteorological Society
AMS	Army Map Service
AMSA	Association of Metropolitan Sewage Agencies
AMSD	Administrative and Management Services Division (OEA)
ANEC	American Nuclear Energy Council
ANPR	Advance Notice of Proposed Rulemaking
ANRHRD	Air, Noise, and Radiation Health Research Division (ORD)
ANSI	American National Standards Institute
ANSS	American Nature Study Society
AO	Administrative Officer
AO	Administrative Order
AO	Administrator's Officer
AO	Area Office

AO	Awards and Obligations
AOC	Abnormal Operating Conditions
AOD	Argon-Oxygen Decarbonization
AOML	Atlantic Oceanographic and Meteorological Laboratory
AOO	Accounting Operations Office
AOO	American Oceanic Organization
AOS	Audit Operations Staff (OEA)
AP	Accounting Point
APA	Administrative Procedure Act
APA	American Planning Association
APCA	Air Pollution Control Association
APCD	Air Pollution Control District
APDS	Automatic Procurement Documentation Systems
APER	Air Pollution Emissions Report
APGR	Ann Arbor AP-42 Program
APHA	American Public Health Association
API	American Paper Institute
API	American Petroleum Institute
APPA	American Public Power Association
APRAC	Urban Diffusion Model for Carbon Monoxide from Motor Vehicle Traffic
APS	ADP Planning System
APS	Automated Personnel System
APT	Associated Pharmacologists & Toxicologists
APTI	Air Pollution Training Institute
APTMD	Air, Pesticides, and Toxics Management Division
APWA	American Public Works Association
AQ-7	Nonreactive Pollutant Modeling
AQCCT	Air Quality Criteria and Control Techniques
AQCR	Air Quality Control Region (CAA)
AQD	Air Quality Digest
AQDHS	Air Quality Data Handling System
AQDHS	Air Quality Data Handling System II
AQDM	Air Quality Display Model
AQM1	Region 1 Air Quality Models
AQM2	Region 2 Puerto Rico EQB Air Quality Model
AQMA	Air Quality Maintenance Area
AQMD	Air Quality Management Division (OAR)
AQMP	Air Quality Maintenance Plan
AQMP	Air Quality Management Plan
AQSM	Air Quality Simulation Model
AQSY	Ann Arbor Air Quality System
AQTAD	Air Quality Technical Assistance Demonstration
AQUIFR	Artificial Aquifer Data Collection System

AQUIRE	Aquatic Information Retrieval
Ar	Argon
ARA	Assistant Regional Administrator
ARA	Associate Regional Administrator
ARAR	Applicable Relevant and Appropriate Requirements (CERCLA)
ARB	Air Resources Board
ARC	Agency Ranking Committee
ARCC	American Rivers Conservation Council
ARD	Air & Radiation Division (OGC)
ARD	Aquatic Resource Division
ARG	American Resources Group
ARIP	Accidental Release Information Program
ARL	Air Resources Laboratory
ARM	Air Resources Management
ARO	Alternative Regulatory Option
ARPO	Acid Rain Policy Office
ARPS	Atmospheric Research Program Staff (ORD)
ARRP	Acid Rain Research Program
ARRPA	Air Resources Regional Pollution Assessment Model
ARZ	Auto-Restricted Zone
AS	Area Source
ASA	American Society of Agronomy
ASAE	American Society of Agriculture
ASAP	As Soon as Possible
ASB	Ambient Standards Branch
ASBESTOS	Region 7 Asbestos in Schools
ASC	Area Source Category
ASCE	American Society of Civil Engineers
ASCII	American Standard Code for Information Interchange
ASCP	American Society of Consulting Planners
ASCS	Agricultural Stabilization and Conservation Services
ASD	Administrative Services Division (regional)
ASD	Analysis and Support Division (OA&R)
ASDWA	Association of State Drinking Water Administrators
ASHAA	Asbestos in Schools Hazard Abatement Act of 1984
ASHAAIS	Asbestos in Schools Hazard Abatement Automated Information System
ASIWPCA	Association of State and Interstate Water Pollution Control Administrators
ASMDHS	Airshed Model Data Handling System
ASME	American Society of Mechanical Engineers
ASN	American Society of Naturalists
ASPA	American Society of Public Administration
ASRL	Atmospheric Sciences Research Laboratory (ORD)
ASSE	American Society of Sanitary Engineers

ASTHO	Association of State and Territorial Health Officials
ASTM	American Society for Testing and Materials
ASTS	Asbestos in Schools Tracking System
ASTSWMO	Association of State & Territorial Solid Wastes Management Officials
ASUS	Administrative Support/Utilization
AT	Advanced Treatment (water)
ATA	American Trucking Association
ATC	Area Training Center
ATCS	Audit Tracking and Control System
ATD	Air and Toxics Division
ATERIS	Air Toxics Exposure and Risk Information System
ATMI	American Textile Manufacturing Institute
ATP	Antitampering Program (FOSD, OMS)
ATR	Agency Technical Representative (GSA Program)
ATRMRD	Air Toxics and Radiation Monitoring Research Division (ORD)
ATRS	Air, Toxics, and Radiation Staff (ORD)
ATS	Action Tracking System
ATS	Administrator's Tracking System
ATS	Assignment Tracking System
ATSDR	Agency for Toxic Substances and Disease Registry
ATTF	Air Toxics Task Force
ATTS	Agency Technology Transfer Staff
AUSA	Assistant U.S. Attorney
AUSM	Advanced Utility Simulation Model
AVD	Audio Visual Division (OEA)
AWI	Animal Welfare Institute
AWISE	Association of Women in Science and Engineering
AWMD	Air and Waste Management Division (regional)
AWOL	Absent Without Official Leave
AWPI	American Wood Preservers' Institute
AWRA	American Water Resources Association
AWWA	American Water Works Association
AWWARF	American Water Works Association Research Foundation
AWWUC	American Water Works Utility Council
AX	Administrator's Office
B & F	Building and Facilities
BAA	Board of Assistance Appeals (OGC)
BAAQMD	Bay Area Air Quality Management District
BAC	Biotechnology Advisory Committee
BACER	Biological and Climatological Effects Research
BACT	Best Available Control Technology
BADT	Best Available Demonstrated Technology

BAP	Benefits Analysis Program
BaP	Benzo(a)Pyrene
BARCDECAL	BARCODE/DECAL Systems
BARCIS	Barcode Information System
BARF	Best Available Retrofit Facility
BART	Best Available Retrofit Technology
BAS	Branch Accounting System
BASIS	Battelle's Automated Search Information System
BAT	Best Available Technology
BAT	Best Available Treatment
BATEA	Best Available Technology Economically Achievable
BBI	OA-Cinci (EMSAC) Foreign Tape
BBS	Bulletin Board System (WIC)
BCC	Blind Carbon Copy
BCCM	Board for Certified Consulting Meteorologists
BCF	Bioconcentration Factor
BCPT	Best Conventional Pollutant Technology
BCT	Best Control Technology
BCT	Best Conventional Pollutant Control Technology
BD	Budget Division (OARM)
BDAT	Best Demonstrated Available Technology (RCRA)
BEA	Bureau of Economic Advisors
BEEP	Benignus EEG Evoked Potential
BEJ	Best Expert Judgment
BEP	Black Employment Program
BG	Billion Gallons
BI	Background Information
BI	Brookings Institution
BIA	Bureau of Indian Affairs
BID	Background Information Document
BID	Buoyancy Induced Dispersion
BIOLOGS	Biological Data Management System (White River)
BIOPLUME	Model to Predict the Maximum Extent of Existing Plumes
BIOS	Natural Biological Information System
BIOSTU	Bioassay Studies
BLIS	BACT/LAER Determinations
BLM	Bureau of Land Management (DOI)
BLOB	Biologically Liberated Organo-Beasties
BLP	Buoyant Line and Point Source Model
BLS	Bureau of Labor Statistics
BMP	Best Management Practices
BMR	Baseline Monitoring Report (CWA)
BNA	Bureau of National Affairs

BOA	Basic Ordering Agreement
BOAC	Bill Office Address Code
BOD	Biochemical Oxygen Demand
BOD	Biological Oxygen Demand
BOF	Basic Oxygen Furnace
BOM	Bureau of Mines
BOP	Basic Oxygen Process
BOPF	Basic Oxygen Process Furnace
BOULD	Boulder Remote Data Collection and Control System
BOY	Beginning of Year Violator (CAA)
BOYSNC	Beginning of Year Significant Noncompliers
BP	Boiling Point
BPA	Blanket Purchase Agreement
BPJ	Best Professional Judgment (CWA)
BPT	Best Practicable Control Technology
BPT	Best Practicable Technology
BPT	Best Practicable Treatment
BR	Business Roundtable
BRS	Bibliographic Retrieval Service
BS	Bilateral Staff (OIA)
BSAC	Biotechnology Science Advisory Committee
BSO	Benzene Soluble Organics
BTU	British Thermal Units
BTZ	Below the Treatment Zone
BU	Bargaining Unit
BUBBLE	Use of Alternative Emission Limits To Meet SIPS/NSPS Requirements
BUD	Benefits and Use Division (OPTS)
BUN	Blood Urea Nitrogen
BY	Budget Year
C	Celsius (degrees)
C/O	Carry Over Funds
CA	Citizen Act
CA	Competition Advocate
CA	Cooperative Agreements
CAA	Clean Air Act
CAA	Compliance Assurance Agreement
CAAA	Clean Air Act Amendments
CAASE	Computer Assisted Area Source Emissions
CAB	Civil Aeronautics Board
CAD	Characterization and Assessment Division (OSWER)
CAD	Computer Aided Design
CAER	Chemical Awareness and Emergency Response Program (CMA)

CAER	Community Awareness and Emergency Response Program
CAFE	Corporate Average Fuel Economy
CAFO	Consent Agreement/Final Order
CAG	Carcinogen Assessment Group (ORD)
CAIR	Comprehensive Assessment Information Rule
CAIRD	Cohort Analysis of Increased Risks of Deaths Model
CALINE	California Line Source Model
CAMEO	Computer Aided Management of Emergency Operations
CAMP	Continuous Air Monitoring Program
CAN	Common Account Number
CANAL	Love Canal & Data Handling System
CAO	Corrective Action Order
CAOO	Cincinnati Accounting Operations Office (OARM)
CAP	Corrective Action Plan
CAP	Cost Allocation Procedure
CAP	Criteria Air Pollutants
CAPCA	Carolinas Air Pollution Control Association
CAPCOA	California Air Pollution Control Officers Association
CAPDET	Procedure for Design and Evaluation of TWKS
CAR	Corrective Action Report
CARB	California Air Resources Board
CARPOOL	Carpool System
CARPS	Computerized Accidental Release Planning System
CARS	Carcinogen System
CAS	Center for Automotive Safety
CAS	Chemical Abstracts Service
CASAC	Clean Air Scientific Advisory Committee (CAA)
CASEREP	Field Office Inspection Data Base
CASETRK	FIFRA and TSCA Case Tracking System
CASLP	Conference on Alternative State and Local Policies
CASU	Cooperative Administrative Support Units
CATS	Corrective Action Tracking System
CAU	Carbon Absorption Unit
CAU	Command Arithmetic Unit
CB	Continuous Bubbler
CBA	Central Business Area
CBA	Chesapeake Bay Agreement
CBA	Cost Benefit Analysis
CBB	Chesapeake Bay Basin
CBD	Central Business District
CBD	Commerce Business Daily
CBI	Compliance Biomonitoring Inspection (CWA)
CBI	Confidential Business Information

CBO	Congressional Budget Office
CBOD	Carbonaceous Biochemical Oxygen Demand
CBP	Chesapeake Bay Program
CBP	County Business Patterns
CBT	Computer Based Training
CC	Carbon Copy
CC	Common Cause
CC/RTS	Chemical Collection/Request Tracking System
CCA	Competition in Contracting Act
CCAA	Canadian Clean Air Act
CCAP	Center for Clean Air Policy
CCD	Chemical Control Division (OPTS)
CCDH	Clark County Department of Health
CCEA	Conventional Combustion Environmental Assessment
CCH	Commerce Clearing House
CCHW	Citizens Clearinghouse for Hazardous Wastes
CCID	Confidential Chemicals Identification System
CCMS	Committee on the Challenges of a Modern Society (NATO)
CCP	Composite Correction Plan (CWA)
CCS	Chemical Coordination Staff (OPTS)
CCS/RTS	Chemical Collection System
CCTP	Clean Coal Technology Program
CCU	Correspondence Control Unit (OECM)
CD	Certification Division (OA&R, Ann Arbor, MI)
CD	Climatological Data
CD	Compliance Division (OPTS)
CDB	Waste Management Data Base System
CDBA	Central Data Base Administrator
CDC	Centers for Disease Control (HHS)
CDD	Chlorinated dibenzo-p-dioxin
CDETS	Consent Decree Tracking System
CDF	Chlorinated dibenzofuran
CDHS	Comprehensive Data Handling System
CDI	Chronic Daily Intake
CDM	Climatological Dispersion Model
CDM	Comprehensive Data Management
CDMQC	Climatological Dispersion Model with Calibration and Source Contribution
CDMS	Cost Development Management System
CDNS	Climatological Data National Summary
CDOTS	Contract Delivery Order Tracking System
CDP	Census Designated Places
CDS	Compliance Data System (CAA)
CE	Categorical Exclusion

CE	Cost Effectiveness
CEA	Cooperative Enforcement Agreement
CEA	Cost and Economic Assessment
CEA	Council of Economic Advisors
CEAM	Center for Exposure Assessment Modeling
CEARC	Canadian Environmental Assessment Research Council
CEAS	Coastal Environmental Assessment Studies
CEAT	Contractor Evidence Audit Team
CEB	Chemical Element Balance
CEB	Commission of European Communities
CECATS	CSB Existing Chemicals Assessment Tracking System (OPTS)
CED	CERCLA Enforcement Division (OSWER)
CED	Criminal Enforcement Division (OECM)
CEE	Center for Environmental Education, Inc.
CEEM	Center for Energy and Environmental Management
CEI	Compliance Evaluation Inspection (CWA)
CELRF	Canadian Environmental Law Research Foundation
CEM	Continuous Emission Monitoring (CAA)
CEM	Cooperative Environmental Management
CEMS	Continuous Emission Monitoring System
CEMS	Continuous Emissions Monitoring Subset
CEO	Chief Executive Officer
CEP	Council on Economic Priorities
CEPP	Chemical Emergency Preparedness Program
CEQ	Council on Environmental Quality
CEQA	California Environmental Quality Act
CERCLA	Comprehensive Environmental Response, Compensation and Liability Act of 1980
CERCLIS	Comprehensive Environmental Response, Compensation and Liability Information System
CERI	Center for Environmental Research Information
CERT	Certificate of Eligibles
CERTAPPL	Applications for Certification
CEU	Continuing Education Units
CF	Conservation Foundation
CFA	Consumer Federation of America
CFC	Chlorofluorocarbons
CFC	Combined Federal Campaign
CFM	Chlorofluoromethanes
CFM	Cubic Feet per Minute
CFMC	Cincinnati Financial Management Center (FMD)
CFR	Code of Federal Regulations
CFS	Command File System

CFS	Cubic Feet per Second
CFSG	Citizen Forum on Self Government (NML)
CGGICS	Construction Grants GICS
CGPRM	Construction Grants Resource Model
CHABA	Committee on Hearing and Bio Acoustics
CHAMP	Community Health Air Monitoring Program
CHEMD	OTS Chemical Directory
CHEMNET	A Mutual Aid Network of Chemical Shippers and Manufacturers and Responders
CHEMTREC	Chemical Transportation Emergency Center
CHESS	Community Health and Environmental Surveillance System
CHIP	Chemical Hazard Information Profile (TSCA)
CHIPS	Chemical Hazard Information Profile System (OPTS)
CHLOREP	A Mutual Aid Group Comprised of Shippers and Carriers
CHRIS/HACS	Chemical Hazards Response Information System/Hazard Communication System
CI	Compression Ignition
CI	Confidence Interval
CIAQ	Council on Indoor Air Quality (Interagency)
CIBL	Convective Internal Boundary Layer
CIBO	Council of Industrial Boiler Owners
CICA	Competition in Contracting Act
CICIS	Chemicals in Commerce Information System
CICS	Customer Information Control System
CIDB	Ann Arbor Certification Information and Fuel Economy Data Base
CIDRS	Cascade Impactor Data Reduction System
CII	Criminal Investigation Index
CIMI	Committee on Integrity and Management Improvement
CIS	Chemical Information System
CIS	Contract Information System
CISR	Chemical Inventory System
CIVP	Region 4, Civil Penalties System
CJE	Critical Job Element
CJO	Chief Judicial Officer (OA)
CLC	Capacity Limiting Constituents
CLCL	HERL-RTP CLEANS CLEVER Clinical Studies
CLEANS	Clinical Laboratory for Evaluation and Assessment of Noxious Substances
CLEVER	Clinical Laboratory for Evaluation and Validation of Epidemiologic Research
CLF	Conservation Law Foundation
CLIPS	Chemical List and Information Pointer System (OIRM)
CLIPS	Chemical List Indexing and Processing System
CLP	Contract Laboratory Program
CLPQA	Contract Laboratory Program Quality Assurance

CLPS	Contract Lab System
CLS	Community Liaison Staff
CLSP	Center for Law & Social Policy
CM	Corrective Measure
CM	Crystal Mall
CMA	Chemical Manufacturers Association
CMAS	Cross-Media Analysis Staff
CMB	Chemical Mass Balance
CMD	Contracts Management Division (OARM)
CME	Comprehensive (ground water) Monitoring Evaluation
CMEL	Comprehensive (ground water) Monitoring Evaluation Log
CMEP	Critical Mass Energy Project
CMO	Contract Management Office
CMS	Case Management System
CNC	Condensation Nucleus Counter
CNG	Coalition of Northeastern Governors
CNG	Compressed Natural Gas
CNR	Composite Noise Rating
CNS	Grant Administration Division Congressional (92-500) Notification
CO	Carbon Monoxide
CO	Change Order
CO	Commissioned Officer
CO	Contracting Officer
CO	Custodial Officer
CO_2	Carbon Dioxide
COA	Commissioned Officers Association
COA	Construction Quality Assurance
COB	Close of Business
COBOL	Common Business Oriented Language
COCO	Contractor-Owned/Contractor-Operated
COD	Chemical Oxygen Demand
CODES	Commitments & Obligations Data Entry System
COE	Corps of Engineers (DOD)
COG	Compliance Order Guidance
COH	Coefficient of Haze
COLA	Cost of Living Adjustment
COM	Continuous Opacity Monitor
COMPLEX	Complex Terrain Screening Model
COMPTER	Multiple Source Air Quality Model
CON	Selected Contractor or Awardee
CONG	Congressional Committee
CONUS	Continental United States
CORPS	Army Corps of Engineers

CORR	Chemicals on Reporting Rules
COS	Conservative Opportunity Society
COS	Cost Accounting System
COWPS	Council on Wage and Price Stability
CPA	Certified Public Accountant
CPA	Contract Property Administrator
CPAF	Cost Plus Award Fee
CPDD	Control Programs Development Division (OAR)
CPE	Carcinogenic Potency Factor
CPF	Cancer Potency Factor
CPFF	Cost Plus Fixed Fee
CPG 1-3	Federal Assistance Handbook Emergency Management and Planning
CPG 1-8	Guide for Development of State and Local Emergency Operations Plans
CPG 1-8A	FEMA Planning Guide for State and Local Emergency Operations Plans
CPI	Consumer Price Index
CPIF	Cost Plus Incentive Fee
CPL	Chemistry and Physics Laboratory
CPM	Continuous Particle Monitor
CPO	Certified Project Officer
CPP	Compliance Policy and Planning (OECM)
CPR	Center for Public Resources
CPR	Coalition for Pesticide Reform
CPS	Compliance Program and Schedule
CPS	Contract Payment System
CPSC	Consumer Products Safety Commission
CPSDAA	Compliance and Program Staff to the Deputy Assistant Administrator
CPU	Central Processing Unit
CR	Community Relations
CRA	Civil Rights Act
CRA	Classification Review Area
CRAVE	Carcinogen Risk Assessment Verification Exercise
CRC	Community Relations Coordinator
CRD	Community Relations Division
CRF	Community Research Facility
CRGS	Chemical Regulations and Guidelines System
CRIB	Criteria Reference Information Bank
CRIMDOCK	Criminal Docket System
CRISP	Comprehensive Risk Information Structure Project
CRL	Central Regional Laboratory
CROP	Consolidated Rules of Practice
CRP	Community Relations Plan
CRQL	Contract Required Quantitation Limit
CRR	Center for Renewable Resources

CRS	Community Relations Staff (OEA)
CRS	Congressional Research Service
CRSTER	Single Source Model
CRSTER 2	Multisource CRSTER
CRT	Cathode Ray Tube
CS	Compliance Staff (GAD)
CS	Contract Specialist
CSB	Chemical Species Balance
CSD	Criteria and Standards Division (OW)
CSEU	OA-Cinci ADP Timeshare Reporting System
CSG	Council of State Governments
CSHEM	Conference of State Health and Environmental Managers
CSI	Chemical Substances Inventory
CSI	Clean Sites, Inc.
CSI	Compliance Sampling Inspection (CWA)
CSIN	Chemical Substances Information Network (TSCA)
CSMA	Chemical Specialties Manufacturers Association
CSO	Combined Sewer Overflow
CSPA	Council of State Planning Agencies
CSPD	Chemicals and Statistical Policy Division (OPPE)
CSPI	Center for Science in the Public Interest
CSRA	Civil Service Reform Act
CSRL	Center for the Study of Responsive Law
CSRS	Civil Service Retirement System
CSS	Clerical Support Staff
CSSD	Computer Services and Systems Division (OARM)
CSSE	Conference of State Sanitary Engineers
CTARC	Chemical Testing and Assessment Research Commission
CTD	Control Technology Document
CTG	Control Technology Guidelines
CTGD	Control Techniques Guideline Document
CTM	Complex Terrain Data Base
CTO	Control Technology Office
CTS	Correspondence Tracking System
CULP	Estimating Water Treatment Costs
CURE	Chemical Unit Record Estimates Data Base (ECAO CM)
CUS	Chemical Update System
CVFM	ERL-CORV Financial Management
CVLB	ERL-CORV Library Circulation System
CVPM	ERL-CORV Personnel Management
CVS	Cardiovascular System
CW	Congress Watch
CWA	Clean Water Act (aka FWPCA)

CWAP	Clean Water Act Project
CWS	Compressed Work Schedule
CWTC	Chemical Waste Transportation Council
CY	Calendar Year
CY	Current Year
CZM	Coastal Zone Management
CZMA	Coastal Zone Management Act
D & F	Determination and Findings
DA	Deputy Administrator (AO)
DA	Designated Agent
DAA	Deputy Assistant Administrator
DAIG	Deputy Assistant Inspector General
DAMDF	Durham Air Monitoring Demonstration Facility
DAPD	AEERL Dual Alkali FGD Process Demonstration
D APSS	Document and Personnel Security System
DAR	Direct Assistance Request
DARTAB	Dose and Risk Assessment Tabulation
DAS	Data Analysis System [Geographical Information System (GIS)]
DASD	Direct Access Storage Drive
DB	Decibel
DB	Dry Bulb
DBA	Data Base Administrator
DBCP	Dibromochloropropane
DBM	Data Base Manager
DBMS	Data Base Management System
DC	Direct Current
DCA	Document Control Assistant
DCA	Washington International Airport
DCIS	Data Call-In Staff (OPTS)
DCL1	Region 1 Library Document Control System
DCMA	Dry Color Manufacturers Association
DCN	Document Control Number
DCO	Delayed Compliance Order (CAA)
DCO	Document Control Officer
DCP	Discrimination Complaints Program
DCPU	Data Center Policy and Usage
DCR	Document Control Register
DCR REG5	Document Control
DCS	Developing Countries Staff (OIA)
DCS	Region 10 Library Document Control
DD	Deputy Director
DDT	D(Ichloro)D(Iphebyl)T(Richloroethane)

DE	Department of Education
DE	Destruction Efficiency
DEA	Drug Enforcement Administration
DEC	Department of Environmental Conservation
DEEP	Dyer EEG Evoked Potential
DEFENSIVE	General Counsel Defense Docket System
DEM	ERL-Athens Dynamic Estuary Model
DEMA	Diesel Engine Manufacturers Association
DEP	Displaced Employee Program
DEPS	Data Entry and Payment System
DES	Diethylistilbesterol
DHHS	Department of Health and Human Services
DI	Diagnostic Inspection (CWA)
DIDS	Domestic Information Display System
DIG	Deputy Inspector General
DIPS	Department of Interior Payroll System
DIS	Defense Investigative Service
DISCO	Defense Investigative Service Cognizant Office
DIVPRTR	EMSL-LV Expenditure System
DL	Detection Limit
DMARS	Deposit Message Retrieval System (part of Treasury's TFCS)
DMIS	Duns Marketing Identification System
DMR	Discharge Monitoring Report (CWA)
DMR	NPDES Discharge Monitoring Report
DMR-QA	Discharge Monitoring Report-Quality Assurance Studies
DMRLS	Data Management and Research Liaison Staff (OW)
DNA	Deoxyribonucleic Acid
DNR	Department of Natural Resources
DO	Dissolved Oxygen
DO	Duty Officer
DOB	Date of Birth
DOC	U.S. Department of Commerce
DOC	Region 4 Library Tracking System
DOCKET	Enforcement Docket System
DOD	U.S. Department of Defense
DOE	U.S. Department of Ecology
DOE	U.S. Department of Energy
DOI	U.S. Department of the Interior
DOIG	Divisional Offices of the Inspector General
DOJ	U.S. Department of Justice
DOL	U.S. Department of Labor
DOPO	Delivery Order Project Officer
DOS	U.S. Department of State

DOS	Disk Operating System
DOT	U.S. Department of Transportation
DOW	Defenders of Wildlife
DPA	Deepwater Ports Act
DPC	Domestic Policy Council
DQO	Data Quality Objective
DRA	Deputy Regional Administrator
DRC	Deputy Regional Counsel
DRE	Destruction/Removal Efficiency (TSCA/RCRA)
DRMO	Defense Reutilization and Marketing Office
DRMS	Defense Reutilization and Marketing Service
DS	Dichotomous Sampler
DSAP	Data Self Auditing Program
DSCF	Dry Standard Cubic Feet
DSCM	Dry Standard Cubic Meter
DSS	Data Systems Staff (OAR)
DSS	Decision Support System
DSS	Domestic Sewage Study
DT	Declaration of Taking
DT	Detention Time
DU	Decision Unit
DU	Ducks Unlimited
DUC	Decision Unit Coordinator
DWS	Drinking Water Standards
DYNHYD4	Hydrodynamic Model
DYNTOX	Dynamic Toxics Model
E	Exposure Level
E-MAIL	EPA's Electronic Mail System
EA	Enforcement Agreement
EA	Environmental Action
EA	Environmental Assessment (NEPA)
EA	Environmental Auditing
EAD	Economic Analysis Division (OPPE)
EAD	Energy and Air Division (ORD)
EADS	Environmental Assessment Data System
EAF	Electric Arc Furnace
EAG	Exposure Assessment Group (ORD)
EAP	Environmental Action Plan
EAR	Environmental Auditing Round Table
EAS	Economic Analysis Staff
EB	Emissions Balancing
EBCDIC	Extended Binary Coded Decimal Interchange Code

EC	Education Center
EC	Effective Concentration
EC	Environment Canada
EC	European Community (Common Market)
ECA	Economic Community for Africa
ECAD	Existing Chemical Assessment Division (OPTS)
ECAO	Environmental Criteria and Assessment Office (ORD)
ECAP	Employee Counseling and Assistance Program
ECC	Executive and Congressional Communications (OA)
ECD	Electron Capture Detector
ECDB	Emissions Certification Data Base
ECE	Economic Commission for Europe
ECHD	Ann Arbor ECTD HD System
ECHH	Electro-Catalytic Hyper-Heaters
ECHO	Each Community Helps Others
ECL	Environmental Chemistry Laboratory
ECL	Executive Control Language
ECLA	Economic Commission for Latin America
ECP	External Compliance Programs (OCR, AO)
ECRA	Economic Cleanup Responsibility Act
ECTD	Emission Control Technology Division (OA&R, Ann Arbor, MI)
ECTS	Executive Correspondence Tracking System
ECU	Environmental Crimes Unit (DOJ)
ED	Department of Education
ED	Editorial Division (OEA)
ED	Effective Dose
ED	Enforcement Division (OW)
EDA	Economic Development Administration
EDA	Emergency Declaration Area
EDASS	EMSL-Cinci Equivalency Statistical System
EDB	Ethylene Dibromide
EDC	Ethylene Dichloride
EDD	Enforcement Decision Document
EDF	Environmental Defense Fund, Inc.
EDP	Electronic Data Processing
EDRS	Enforcement Document Retrieval System
EDS	Electronic Data Systems
EDS	Energy Data System
EDT	Edit Data Transmission
EDTA	Ethylene Diamine Triacetic Acid
ED10	Ten Percent Effective Dose
EDTS	Ann Arbor Evaluation and Development Test System
EDZ	Emission Density Zoning

EEA	Energy and Environmental Analysis
EEC	European Economic Community
EED	Exposure Evaluation Division (OPTS)
EEG	Electroencephalogram
EEI	Edison Electric Institute
EEMS	Emissions Elements Needs Survey (RCRA)
EENET	Emergency Education Network (IFEMA)
EEO	Equal Employment Opportunity
EEOC	Equal Employment Opportunity Commission
EER	Excess Emission Report
EERF	Eastern Environmental Radiation Facilities (EERF) Sample Data Base
EERF	Eastern Environmental Radiation Facility (OA&R, Montgomery, AL)
EERL	Eastern Environmental Radiation Laboratory (OA&R)
EERU	Environmental Emergency Response Unit
EESI	Environment and Energy Study Institute
EESL	Environmental Ecological and Support Laboratory
EETFC	Environmental Effects, Transport and Fate Committee (SAB)
EF	Emission Factor
EFE	Early Fuel Evaporative System
EFI	Electronic Fuel Injection Systems
EFO	Equivalent Field Office
EFTC	European Fluorocarbon Technical Committee
EGD	Effluent Guidelines Division (OW)
EGR	Exhaust Gas Recirculation Systems
EH	Redox Potential
EHC	Environmental Health Committee (SAB)
EHIS	Emission History Information System
EHRS	Environmental Health Research Staff (ORD)
EHS	Extremely Hazardous Substance
EI	Emission Inventory
EIA	Economic Impact Assessment
EIA	Environmental Impact Assessment
EIC	Environmental Industry Council
EIL	Environmental Impairment Liability
EIN	Employer Identification Number
EINDES	Employer ID No. Data Entry System on the PDP 11/70s
EIR	Endangerment Information Report
EIR	Environmental Impact Rep
EIS	Emissions Inventory System
EIS	Environmental Impact Statement (Environmental Review Tracking System)
EIS	Environmental Impact Statement (NEPA)
EIS	Environmental Inventory System
EIS/AS	Emissions Inventory System/Area Source

EIS/PS	Emissions Inventory System/Point Source
EIS/PS&AS	Emissions Inventory Subsystem/Point Source and Area Source
EIS7	Region 7 EIS 404 Program
EKMA	Empirical Kinetic Modeling Approach
EL	Exposure Level
ELI	Environmental Law Institute
ELR	Environmental Law Reporter
EM	Electron Microscope
EMA	Emergency Management Agency
EMAS	Enforcement Management and Accountability System (OECM)
EMI	Emergency Management Institute
EML	Emission Measurement Laboratory
EMR	Environmental Management Report
EMS	Enforcement Management Subsystem
EMS	Enforcement Management System
EMS	Environmental Mutagen Society
EMSD	Environmental Monitoring Systems Division (ORD)
EMSL	Environmental Monitoring Support Laboratory (ORD)
EMSL	Environmental Monitoring Systems Laboratory
EMTS	Environmental Methods Testing Site
EMTS	Exposure Monitoring Test Site
ENF5	Reg. 5 Enforcement Tracking System
ENFOMAIN	Enforcement Mail Computer
ENG-AUDIT	GICS Engineer Audit Tracking System
EO	Ethylene Oxide
EO	Executive Officer
EO	Executive Order
EOB	Executive Office Building
EOC	Emergency Operating Center
EOD	Engineering Operations Division (OA&R, Ann Arbor, MI)
EOD	Entrance on Duty
EOE	Equal Opportunity Employer
EOF	Emergency Operations Facility (RTP)
EOJ	End of Job
EOP	Emergency Operations Plan
EOT	Emergency Operations Team
EOY	End of Year
EP	Earth Protectors
EP	Emergency Preparedness
EP	Environmental Profiles
EP	Extraction Procedure
EPA	Environmental Protection Agency
EPAA	Environmental Programs Assistance Act of 1984

EPAAR	EPA Acquisition Regulations
EPAC	Emergency Preparedness Advisory Committee
EPACASR	EPA Chemical Activities Status Report
EPACIR	Circulation System
EPADOC	Document Control System
EPALIT	ERL-Gulf Breeze Text Data Management
EPANTS	NTIS/EPA Report System
EPATR	Translation System
EPAYS	EPA Payroll System
EPC	Economic Policy Council
EPC	Emergency Preparedness Coordinator
EPCA	Energy Policy and Conservation Act of 1975
EPD	Emergency Planning District
EPI	Environmental Policy Institute
EPIC	Environmental Photographic Interpretation Center
EPID	Epidemiological Studies
EPNL	Effective Perceived Noise Level
EPO	Estuarine Programs Office (NOAA)
EPRI	Electric Power Research Institute
EPTC	Extraction Procedure Toxicity Characteristic
ER	Electrical Resistivity
ERA	Economic Regulatory Agency
ERA	Equal Rights Amendment
ERAD	Economic and Regulatory Analysis Division (OPPE)
ERAMS	Environmental Radiation Ambient Monitoring System (OA&R)
ERC	Emergency Response Commission
ERC	Emission Reduction Credit
ERC	Environmental Research Center
ERCS	Emergency Response Cleanup Services
ERD	Emergency Response Division (OSWER)
ERD&DAA	Environmental Research, Development and Demonstration Authorization Act
ERDA	Energy Research and Development Administration
ERDB	Environmental Radiofrequency Data Base
ERFD	Airborne Particulate and Precipitation Data
ERIS	Enforcement Case Support Expert Resources Inventory System
ERL	Environmental Research Laboratory (ORD)
ERNS	Emergency Response Notification System
ERP	Enforcement Response Policy
ERRD	Emergency and Remedial Response Division
ERS	Economic Research Service
ERSS	Establishment Registration Support System
ERT	Environmental Response Team
ERTAQ	ERT Air Quality Model

ES	Enforcement Strategy
ES	Engineering Staff (OA)
ES	Expert System
ES&H	Environmental Safety and Health
ES001	Estuarine Water Quality Model
ESA	Ecological Society of America
ESA	Endangered Species Act
ESA	Environmentally Sensitive Area
ESC	Endangered Species Committee
ESCA	Electron Spectroscopy for Chemical Analysis
ESCAP	Economic and Social Commission for Asia and the Pacific
ESCP	Employee Counseling Services Program
ESD	Emission Standards Division (OAR)
ESD	Environmental Services Division (regional)
ESE	Environmental Science and Engineering
ESECA	Energy Supply and Environmental Coordination Act of 1974
ES&H	Environmental Safety and Health
ESO	Enforcement Specialist Office (NEIC)
ESP	Electrostatic Precipitator
ESRL	Environmental Sciences Research Laboratory
ET	Emissions Trading
ETA	Energy Tax Act
ETD	Economics and Technology Division (OPTS)
ETHOX	Summary of Ecotox Data on Ethoxylated Surfactants
ETP	Emissions Trading Policy
ETS	Environmental Tobacco Smoke
ETS	Extramural Tracking System
ETS	Region 4 Ref and Administrative Orders Tracking
EUP	Environmental Use Permit
EUTR04	Eutrophication Model
EWCC	Environmental Work Force Coordinating Committee
EX	Executive Level Appointments
EXAMS	Exposure Analysis Modeling System
EXAMSII	Exposure Analysis Modeling System II
ExEx	Expected Exceedance
EXL	Executive Control Language
EZP2	Region 2, EZPLOT-User Operated Business Graphics Package
EZPLOT	EZPLOT
F	Fahrenheit (Degrees)
f/cc	Fibers per cubic centimeters (of air)
F/M	Food to Microorganism Ratio
FAA	Federal Aviation Administration

FAC	Facility Advisory Committee
FACA	Federal Advisory Committee Act
FACM	Friable Asbestos-Containing Material
FACT	ERL-Gulf Breeze Financial Data Management
FALD	Fahrenheit Agency Liaison Division (OEA)
FAM	Friable Asbestos Material
FAME	Framework for Achieving Managerial Excellence (AX)
FAN	Fixed Account Number
FAO	Food and Agriculture Organization
FAR	Federal Acquisition Regulations
FAS	Frontera Audobon Society
FASB	Financial Accounting Standards Board
FAST	Fugitive Assessment Sampling Train
FATES	FIFRA and TSCA Enforcement System
FBANK3	EMSL-RTP National Filter Analysis Network
FBC	Fluidized Bed Combustion
FBI	Federal Bureau of Investigation
FCC	Federal Communications Commission
FCC	Fluid Catalytic Converter
FCCU	Fluid Catalytic Cracking Unit
FCO	Federal Coordinating Officer (in disaster areas)
FCO	Forms Control Officer
FCQAS	Financial Compliance and Quality Assurance Staff (FMD)
FDA	Food and Drug Administration
FDF	Fundamentally Different Factors
FDIC	Federal Deposit Insurance Corporation
FDL	Final Determination Letter
FDO	Fee Determination Official
FE	Fugitive Emissions
FEA	Federal Energy Administration
FEB	Federal Executive Board
FEC	Federal Executive Council
FEDS	Federal Energy Data System
FEGLI	Federal Employee Group Life Insurance
FEHB	Federal Employees Health Benefits
FEI	Federal Executive Institute
FEIS	Fugitive Emissions Information System
FEL	Frank Effect Level
FEMA	Federal Emergency Management Agency
FEMA-10	Federal Emergency Management Agency's Planning Guide & Checklist for Hazardous Materials Contingency Plans
FEMA-REP-1	Response Plans and Preparedness in Support of Nuclear Power Plants

FEMA-REP-5	Guidance for Developing State and Local Radiological Emergency Response Plans
FEME-RFP-2	Guidance for Developing State and Local Radiological Emergency Response Plans
FEMIS	Federal Emergency Management Information System
FEPCA	Federal Environmental Pesticides Control Act
FERC	Federal Energy Regulatory Commission (DOE)
FERS	Federal Employees Retirement System
FERSA	Federal Employees Retirement System Act
FES	Factor Evaluation System
FET	Foundation on Economic Trends
FEV	Forced Expiratory Volume
FEV1	Forced Expiratory Volume— one second
FEVI	Front End Volatility Index
FEW	Federally Employed Women
FF	Federal Facilities
FFAR	Fuel and Fuel Additive Registration
FFARS	Fuel and Fuel Additives Registration System
FFCS	Federal Facilities Compliance Staff (OEA)
FFDCA	Federal Food, Drug and Cosmetic Act
FFF	Firm Financial Facility
FFFSG	Fossil Fuel Fired Steam Generator
FFI	Full Field Investigation
FFIS	Federal Facilities Information System
FFMC	Federal Financial Managers' Council
FFP	Firm Fixed Price
FFTF	Future Framework Task Force
FGD	Fuel Gas Desulfurization
FGDIS	Fuel Gas Desulfurization Information System
FHA	Farmers Home Administration
FHA	Federal Housing Administration
FHLBB	Federal Home Loan Bank Board
FHwA	Federal Highway Administration
FIA	Federal Insurance Administration
FIATS	Freedom of Information Action Tracking System
FIC	Federal Information Center
FICA	Federal Insurance Contributions Act
FID	Flame Ionization Detector
FIFO	First In/First Out
FIFR	Region 7 FIFRA Neutral Inspection Selection System
FIFRA	Federal Insecticide, Fungicide, and Rodenticide Act
FILS	Federal Information Locator System
FIM	Friable Insulation Material

FINDS	Facility Index System (OIRM)
FIP	Federal Implementation Plan
FIP	Federal Information Plan
FIP	Final Implementation Plan
FIPS	Federal Information Procedures System
FIPS	Federal Information Processing Standards
FIRMIS	OSWER Info System
FIRMR	Federal Info Resources Management Regulation
FIRST-UP	Financial Information Register Satellite Terminal Users Package
FISHTEMP	National Compendium of Freshwater Fish & Water Temperature Data
FIT	Field Investigation Team
FLETC	Federal Law Enforcement Training Center
FLM	Federal Land Manager
FLP	Flash Point
FLPMA	Federal Land Policy and Management Act
FLRA	Federal Labor Relations Authority
FLRC	Federal Labor Relations Council
FLSA	Fair Labor Standards Act
FM	Friable Material
FMC	Federal Maritime Commission
FMCS	Federal Mediation and Conciliation Service
FMD	Financial Management Division (OARM)
FMFIA	Federal Managers' Financial Integrity Act
FML	Flexible Membrane Liner
FMO	Financial Management Officer
FMP	Facility Management Plan
FMP	Financial Management Plan
FMS	ERL-Ada Financial System Management
FMS	Financial Management System
FMSD	Facilities Management and Services Division (OARM)
FMSMG	Financial Management System Management Group
FMSR	Facilities Management System
FMSTI	FMS Transaction Input System
FMSTIU	FMS Transaction Input & Update System
FMVCP	Federal Motor Vehicle Control Program
FMWOS	Facility Management Work Order System
FO	Facilities Office
FOCUS	For On-Line Computer Users (report writing software)
FOE	Friends of the Earth
FOI	Freedom of Information
FOIA	Freedom of Information Act
FOIA	OWPE Freedom of Information Act System
FOISD	Fiber Optic Isolated Spherical Dipole Antenna

FONSI	Finding of No Significant Impact (NEPA)
FORAST	Forest Response to Anthropogenic Stress
FORTRAN	Formula Translation
FOSD	Field Operations and Support Division (OA&R)
FP	Fine Particulate Matter
FPA	Federal Pesticide Act
FPC	Federal Power Commission
FPD	Flame Photometric Detector
FPEIS	Fine Particulate Emissions Information System
FPI	Federal Prison Industries
FPM	Federal Personnel Manual
FPO	Federal Protective Officer
FPPB	Fiscal Policies and Procedures Branch (FMD)
FPR	Federal Procurement Regulation
FPRS	Federal Program Resources Statement
FPRS	Formal Planning and Reporting System
FPS	Federal Protective Service (GSA)
FR	Federal Register
FR	Final Rulemaking
FRA	Federal Register Act
FRAB	Financial Reports and Analysis Branch (FMD)
FRB	Federal Reserve Board
FRC	Federal Records Center
FRCS	Federal Register Chargeback System
FRD	Facility Requirements Division (OW)
FRDS	Federal Reporting Data System
FREDS	Flexible Regional Emissions Data System
FREE	Fund for Renewable Energy and the Environment
FRES	Forest Range Environmental Study
FRM	Federal Reference Methods
FRM	Final Rule Making
FRN	Final Rulemaking Notice
FRO	Federal Register Office
FRS	Formal Reporting System
FRTIB	Federal Retirement Thrift Investment Board
FS	Feasibility Study
FS	Forest Service
FSA	Food Security Act
FSB	Financial Systems Branch (FMD)
FSC	Facilities Service Center
FSIP	Federal Service Impasse Panel
FSOD	MERL-Cinci Field Scale Organics (IBM)
FSP	Field Sampling Plan

FSS	Facility Status Sheet
FSS	Federal Supply Schedule
FSSD	Facilities and Support Services Division (OARM)
FT	Full Time
FTA	Fairchild Tenants Association
FTC	Federal Trade Commission
FTE	Full-Time Equivalent
FTP	Federal Test Procedure
FTR	Federal Travel Regulations
FTS	Federal Telecommunications System
FTS	Field Transfer Service
FTT	Full Time Temporary
FTTA	Federal Technology Transfer Act
FUA	Fuel Use Act
FUELDB	Fuels Inspection Data Base
FURS	Federal Underground Injection Control Reporting System
FUSRAP	Formally Used Sites Remedial Action Plan (NWPA)
FVC	Forced Vital Capacity
FVMP	Federal Visibility Monitoring Program
FWCA	Fish and Wildlife Coordination Act
FWP	Federal Women's Program
FWPAC	FWP Advisory Committee (OCR)
FWPCA	Federal Water Pollution Control Act (aka CWA)
FWQA	Federal Water Quality Association
FWS	Fish and Wildlife Service (DOI)
FY	Fiscal Year
FYI	For Your Information
g/mi	Grams Per Mile
GAAP	Generally Accepted Accounting Principles
GAARP	Grants ADABAS Access and Retrieval Package
GAC	Granular Activated Carbon
GAC	Groundwater Activated Carbon
GACT	Granular Activated Carbon Treatment
GAD	Grants Administration Division (OARM)
GADMIS	Grants Administration Division Management Information System
GADOSAG	Georgia DOSAG
GAO	General Accounting Office (U.S. Congress)
GASP	General ADP Support-PDP 11/70
GAUD	Region 4 Grants Audit System
GBL	Government Bill of Lading
GC	Gas Chromatography
GC	General Counsel

GC/MS	Gas Chromatograph/Mass Spectrograph
GCGLD	Grants, Contracts, and General Law Division (OGC)
GCWR	Gross Combination Weight Rating
GEA	Glossary of EPA Acronyms
GEI	Geographic Enforcement Initiative
GEMS	Global Environmental Monitoring System
GEMS	Graphic Exposure Modeling System (OTS)
GEP	Good Engineering Practice
GF	General Files
GFF	Glass Fiber Filter
GFP	Government-Furnished Property
GI	Gastrointestinal
GI	Global Indexing System
GICS	Grant Information and Control System
GIS	Geographic Information Systems
GIS	Global Indexing System
GIS	Guidelines Implementation Staff (QW)
GLC	Gas Liquid Chromatography
GLERL	Great Lakes Environmental Research Laboratory
GLNPO	Great Lakes National Program Office
GLOW	Greater Leadership Opportunity for Women
GLP	Good Laboratory Practices
GLWQA	Great Lakes Water Quality Agreement
GMA	Grocery Manufacturers Association
GMCC	Global Monitoring for Climatic Change
GMDI	Geophysical Models for Data Interpretation
GMT	Greenwich Mean Time
GNP	Gross National Product
GOCM	Goals, Objectives, Commitments and Measures
GOCO	Government-Owned/Contractor-Operated
GOGO	Government-Owned/Government-Operated
GOLE	GICS On-Line Data Entry System
GOMS	Grants Obligations Management System
GOP	General Operating Procedures
GOPO	Government-Owned/Privately-Operated
GPAD	Gallons per Acre Per Day
GPG	Gram per Gallon
GPO	Government Printing Office
GPS	Groundwater Protection Strategy
GRCDA	Government Refuse Collection & Disposal Association
GREAT	General Record of Enforcement Actions Tracked
GRGL	Groundwater Residue Guidance Level
GRPH	Region 4 Graphics System

GS	General Schedule
GSA	General Services Administration
GSDP	Geophysical Survey Data Processing System
GTDMIS	GTD Bioassay System
GTN	Global Trends Network
GTR	Government Transportation Request
GTTS	Grants Treasury Tape System
GVM	Gross Vehicle Weight
GVP	Gasoline Vapor Pressure
GVW	Gross Vehicle Weight
GVWR	Gross Vehicle Weight Rating
GW	Ground Water
GWM	Ground Water Monitoring
GWPMS	Ground Water Policy and Management Staff (OW)
GWPS	General Word Processor Support
GWPS	Ground Water Protection Standard
GWPS	Ground Water Protection Strategy
GWTF	Ground Water Task Force (OSWER)
H_20	Water
H_20_2	Hydrogen Peroxide
H_2S	Hydrogen Sulfide
HA	Hatch Act
HAD	Health Assessment Document
HAOB	Headquarters Accounting Operations Branch (FMD)
HAOS	Houston Area Oxidant Study
HAP	Hazardous Air Pollutant
HAPEMS	Hazardous Air Pollutants Enforcement Management System
HAPPS	Hazardous Air Pollutant Prioritization System
HAR03	Region 2 Water Quality Models
HATREMS	Hazardous and Trace Emissions System
HAZARD	Hazardous Waste Data Base
HAZMAT	Hazardous Material
HAZMAT	Hazardous Materials
HAZOP	Hazard and Operability Study
HB	Health Benefits
HBEP	Hispanic and Black Employment Programs
HC	Hazardous Constituents
HC	Hydrocarbons
HCCPD	Hexachlorocyclopentadiene
HCl	Hydrogen Chloride
HCP	Hydrothermal Coal Process
HDD	Heavy-Duty Diesel

HDECERT	Heavy-Duty Engine Certification Data
HDG	Heavy-Duty Gasoline Powered Vehicle
HDPE	High Density Polyethylene
HDT	Heavy-Duty Truck
HDV	Heavy-Duty Vehicle
HEA	Health Effects Assessment
HEAL	Human Exposure Assessment Location
HEAST	Health Effects Assessment Summary Tables
HECC	House Energy and Commerce Committee
HED	Hazard Evaluation Division (OPTS)
HEED	Health and Environmental Effects Document
HEEP	Health and Environmental Effects Profile
HEGP	Gases and Particles
HEHP	Hazardous Pollutants Research
HEI	Health Effects Institute
HELP MODEL	Hydrologic Evaluation of Landfill Performance Model
HEM	Human Exposure Modeling
HEOX	Oxidants
HEP	Hispanic Employment Program
HEP	Household Evaluation Program (ORP)
HEPA	High-Efficiency Particulate Air
HEPS	Pesticides Research
HERD	Health and Environmental Review Division (OPTS)
HERL	Health Effects Research Laboratory
HERS	Hyperion Energy Recovery System
HESAP	Health and Environmental Study Audit Program
HETC	Toxic Substances Research
HEX-BCH	Hexachloronorbornadiene
HFOS	HERL-RTP Forced Oscillation System
HGAS	HERL-RTP Gas/Aerosol System
HHE	Human Health and the Environment
HHS	Department of Health and Human Services
HHV	Higher Heating Value
HHW	Household Hazardous Waste
HI	Hazard Index
HI-VOL	High-Volume Sampler
HIDE	Heavy-Duty Engine
HISLIB	Effluent Guidelines GC/MS Screening Analysis Data Base
HITS	Headquarters Invoice Tracking System
HIWAY	Line Source Model for Gaseous Pollutants
HIWS	High-Level Waste and Standards
HMIS	Hazardous Materials Information System
HMIS	HERL-RTP Management Information System

HMS	Highway Mobile Source
HMT	Hazardous Materials Table
HMTA	Hazardous Materials Transportation Act
HMTR	Hazardous Materials Transportation Regulations
HO	Headquarters Offices
HOC	Hazardous Organic Constituents (TSCA/RCRA)
HON	Hazardous Organic NESHAP
HOV	High Occupancy Vehicle
HP	Horse Power
HPLC	High Performance Liquid Chromatography
HPLX	HERL-RTP Plexiglas System
HPV	High Priority Violator
HQ	Headquarters
HQCDO	Headquarters Case Development Officer
HRC	Human Resources Council
HRDB	Human Resources Development Branch (OARM)
HRPS	High Risk Point Sources
HRS	Hazard Ranking System
HRSD	Hazardous Response Support Division (OSWER)
HRUP	High Risk Urban Problem
HSCD	Hazardous Site Control Division (OSWER)
HSDB	Hazardous Substance Data Base
HSIA	Halogenated Solvent Industry Alliance
HSIRS	Health and Safety Inspection Report System
HSL	Hazardous Substance List
HSPF	Hydrological Simulation Program Fortran
HSWA	Hazardous and Solid Waste Amendments of 1984
HT	Hydrothermally Treated
HTP	High Temperature and Pressure
HTRN	HERL-RTP Training System
HUD	Department of Housing and Urban Development
HVAC	Heating, Ventilation, and Air Conditioning (System)
HVIO	High Volume Industrial Organics
HVLD	HERL-RTP Validation System
HW	Hazardous Waste
HW-FW	Half Wave/Full Wave (Electrical Distribution)
HWD2	Region 2 RCRA Facilities Hazard Rating Model
HWDMS	Hazardous Waste Data Management System (OSWER)
HWED	Hazardous Waste Enforcement Division (OECM)
HWERL	Hazardous Waste Engineering Research Laboratory (ORD)
HWFW	Half Wave/Full Wave
HWGTF	Hazardous Waste Groundwater Task Force
HWLT	Hazardous Waste Land Treatment

HWM	Hazardous Waste Management
HWMD	Hazardous Waste Management Division
HWRTF	Hazardous Waste Restrictions Task Force (OW)
HWSA	Hazardous Waste Services Association
HWSD	Hazardous Waste Site Data Base (Indicator Parameters)
HWSS	Hazardous Waste and Superfund Staff (ORD)
HWTC	Hazardous Waste Treatment Council
I/M	Inspection and Maintenance
IA	Interagency Agreement
IAAC	Interagency Assessment Advisory Committee
IAD	Internal Audit Division
IADB	Innovative/Alternative Pollution Control Technology Facility File Data Base
IAEA	International Atomic Energy Agency
IAG	Interagency Agreement
IAG	Interagency Group
IAP	Incentive Awards Program
IAP	Indoor Air Pollution
IARC	International Agency for Research on Cancer
IARDB	Interim Air Toxics Data Base
IAS	Incineration at Sea Site Monitoring and Permits File
IATDB	Interim Air Toxics Data Base
IBA	Industrial Biotechnology Association
IBR	Incorporation by Reference
IBRD	International Bank for Reconstruction and Development
ICAIR	Interdisciplinary Planning and Information Research
ICAP	Inductively Coupled Argon Plasma
ICBEN	International Commission on the Biological Effects of Noise
ICC	Interstate Commerce Commission
ICE	Industrial Combustion Emissions Model
ICE	Internal Combustion Engine
ICE MODEL	Industrial Combustion Emissions Model
ICP	Inductively Coupled Plasma
ICR	Information Collection Request
ICRA	Industrial Chemical Research Association
ICRE	Ignitability, Corrosivity, Reactivity, Extraction (Characteristics)
ICRP	International Commission on Radiological Protection
ICS	Incident Command System
ICS	Institute for Chemical Studies
ICS	Intermittent Control Strategies (CAA)
ICS	Intermittent Control System (CAA)
ICWM	Institute for Chemical Waste Management
ICWP	Interstate Conference on Water Problems
ID	Inside Diameter

IDBS	Imports Data Base
IDL	Instrument Detection Limit
IDLH	Immediately Dangerous to Life and Health
IDP	Individual Development Plan
IEB	International Environment Bureau
IEMD	Integrated Environmental Management Division (OPPE)
IEMP	Integrated Environmental Management Project
IEMS	Integrated Emergency Management System
IEPD	Industrial and Extractive Processes Division (ORD)
IERL	Industrial Environmental Research Laboratory
IES	Institute of Environmental Sciences
IFB	Invitation for Bid
IFCAM	Industrial Fuel Choice Analysis Model
IFIS	Industry File Information System
IFMS	Integrated Financial Management System
IFPP	Industrial Fugitive Process Particulate
IFR	Interim Final Rule
IG	Inspector General
IGCI	Industrial Gas Cleaning Institute
IGD	Inspector General Division (OGC)
IHEU	In-House Equipment Utilization
IHS	Indian Health Service
IIS	Inflationary Impact Statement
IJC	International Joint Commission
ILS	Intergovernmental Liaison Staff
IMAGERY	Multispectral Scanner and Photographic Imagery
IMAN	NEIC Image Analysis System
IMD	Information Management Division (OPTS)
IMIS	IERL-RTP Management Information System
IMM	Intersection Midblock Model
IMPACT	Integrated Model of Plumes and Atmosphere in Complex Terrain
IMPREST	IMPREST System
IMPROVE	Interagency Monitoring of Protected Visual Environment
IMS	Information Management Staff (OSWER)
IMSD	Information Management and Services Division (OARM)
INCE	Institute of Noise Control Engineers
INDX	Enforcement Document Retrieval System
INPUFF	A Gaussian Puff Dispersion Model
INT	Intermittent
INVITRO	HERL-RTP In Vitro System
IO	Immediate Office
IOAA	Immediate Office of the Assistant Administrator
IOAU	Input/Output Arithmetic Unit

IOB	Iron Ore Benefication
IOTV	Interoffice Transfer Voucher
IOU	Input/Output Unit
IP	Inhalable Particulates
IPA	Intergovernmental Personnel Act
IPA	Intergovernmental Personnel Agreement
IPM	Inhalable Particulate Matter
IPM	Integrated Pest Management
IPMN	Inhalable Particulate Network
IPMPCS	Integrated Pest Management and Program Coordination Staff (OPTS)
IPP	Implementation Planning Program
IPP	Integrated Plotting Package
IR	Infrared
IRD	Information Resources Directory (OIRM)
IRG	Interagency Review Group
IRIS	Instructional Resources Information System
IRIS	Integrated Risk Information System
IRLG	Interagency Regulatory Liaison Group
IRM	Information Resources Management
IRM	Interim Remedial Measures (CERCLA)
IRMC	Interagency Risk Management Council
IRP	Installation Restoration Program
IRPTC	International Register of Potentially Toxic Chemicals
IRR	Institute of Resource Recovery
IRR	Inventory Reporting Requirement System
IRS	Intergovernmental Relations Staff (OEA)
IRS	Intermedia Ranking Staff
IRS	Internal Revenue Service
IRS	International Referral System
IRSD	Information and Regulatory Systems Division (OPPE)
IS	Indicator Score
IS	Interim Status
ISAM	Indexed Sequential File Access Method
ISC	Industrial Source Complex
ISCL	Interim Status Compliance Letter
ISCLT	Industrial Source Complex Long Term Model
ISCST	Industrial Source Complex Short Term Model
ISD	Information Systems Division (OARM)
ISD	Interim Status Document (RCRA)
ISDB	Industry Studies Data Base
ISE	Ion-Specific Electrode
ISI	EPA Information Systems Inventory
ISIS	Industry File Indexing System

ISMAP	Indirect Source Model for Air Pollution
ISS	Information Security Specialist
ISS	Information Systems Staff (ORD)
ISS	Interim Status Standards
ITC	International Trade Commission
ITD	Industrial Technology Division (OW)
ITD	Inhalation Toxicology Division
ITDP	Individual Training and Development Plan
ITP	Individual Training Plan
ITS	Interagency Testing Committee Tracking System
ITSS	Integrated Technical Support Services System
IUFED	In-Use Vehicle Fuel Economy Data
IUTA	In-Use Technology Assessment
IUTD	Ann Arbor In-Use Test Data System
IVTC	International Visitors and Travel Coordinator (OIA)
IWC	In-Stream Waste Concentration (CWA)
IWS	Ionizing Wet Scrubber
JAPCA	Journal of Air Pollution Control Association
JATS	Job Application Tracking System
JCL	Job Control Language
JEC	Joint Economic Committee
JFMIP	Joint Financial Management Improvement Program
JLC	Justification for Limited Competition
JNCP	Justification for Noncompetitive Procurement
JOFOC	Justification for Other Than Full and Open Competition
JPA	Joint Permitting Agreement
JSD	Jackson Structured Design
JSP	Jackson Structured Programming
JTU	Jackson Turbidity Unit
JUDO	Judicial Officer Case Tracking System
JV	Journal Voucher
KW	Kilowatt
KWH	Kilowatt Hour
KWIC/UNVAC	OA-RTP Key Word in Context Index
LAA	Lead Agency Attorney
LABELS	Region 3 Mail Labels System
LABPROP	Laboratory Property Management System
LACS	Los Angeles Catalytic Study
LAER	Lowest Achievable Emission Rate
LAI	Laboratory Audit Inspection

LAMP	Lake Acidification Mitigation Project (EPRI)
LAMS	Lake Analysis Management System
LAN	Local Area Network
LAST	Labor and Sample Tracking
LBAU	Lab Automation System
LBI	Limited Background Investigation
LC	Lethal Concentration
LC	Liquid Chromatography
LCD	Local Climatological Data
LCL	Lower Control Limit
LCM	Life Cycle Management
LCRS	Leachate Collection and Removal System
LCS	Ann Arbor Laboratory Computer System
LCS	OA-Cinci (EMSAC) Library Circulation System
LCS	Region 10 Library Circulation System
LD	Light Duty
LD50	Low Dose Where Fifty Percent of Animals Die
LD50	Median Lethal Dose
LDC	London Dumping Convention
LDCRS	Leachate Detection, Collection, and Removal System
LDD	Light Duty Diesel
LDEQ	Louisiana Department of Environmental Quality
LDIP	Laboratory Data Integrity Program
LDMS	Region 2, Lab Data Management System
LDPE	Low Density Polyethylene
LDPHDN	Lead Additive Report for Refineries and Importers and for Manufacturing
LDR	Land Disposal Restrictions
LDRRDDB	Land Disposal Restrictions (Rule Development)
LDRTF	Land Disposal Restrictions Task Force
LDS	Light-Duty Vehicle/Truck Certification
LDSFE	Fuel Economy
LDT	Light Duty Truck
LDV	Light Duty Vehicle
LEL	Lower Explosive Limit
LEP	Laboratory Evaluation Program
LEPC	Local Emergency Planning Committee
LEPD	Legal Enforcement Policy Division (OECM)
LERC	Local Emergency Response Committee
LEVEL8(A)	TSCA 8(a) Level A Information System
LFL	Lower Flammability Limit
LIBBKS	Headquarters Book System
LIBR	RTP Library
LIDAR	NEIC Light Detection and Ranging System

LIFO	Last In/First Out
LIMB	Limestone-Injection, Multi-Stage Burner
LIRAQ	Livermore Regional Air Quality Model
LITS	Litigation Support System
LLRW	Low-Level Radioactive Waste
LLWPA	Low-Level Waste Policy Act
LMF	Logical Mainframe
LMFBR	Liquid Metal Fast Breeder Reactor
LMR	Labor Management Relations
LNEP	Low Noise Emission Product
LNG	Liquified Natural Gas
LNRD	Land and Natural Resources Division (DOJ)
LOAFL	Lowest Observed Adverse Effect Level
LOC	Library of Congress
LOC-TFCS	Letter of Credit—(Dept.of) Treasury Financial Communications System
LOCATOR	Mail/Locator
LOE	Level of Effort
LOEL	Lowest Observed Effect Level
LOIS	Loss of Interim Status (SDWA)
LONGZ	Long Term Terrain Model
LOOK-UP	On-Line Account Status and Payment History
LOQ	Level of Quantification
LP	Legislative Proposal
LPG	Liquified Petroleum Gas
LRMS	Low Resolution Mass Spectroscopy
LRO	Labor Relations Officer
LRTAP	Long-Range Transportation of Air Pollution
LSD	Laboratory Services Division (NEIC)
LSERA	ERL-Duluth Financial Management Package
LSERB	ERL-Duluth Personnel and Payroll
LSI	Legal Support Inspection (CWA)
LSL	Lump Sum Leave
LST	Low-Solvent Technology
LTA	Lead Trial Attorney
LTC	Long-Term Concentration
LTD	Land Treatment Demonstration
LTO	Landing-Takeoff Cycle
LTOP	Lease to Purchase
LTR	Lead Technical Representative
LTU	Land Treatment Unit
LUCIFER	Listing of Organic Compounds Identified in Region 4
LUST	Leaking Underground Storage Tanks
LVAOO	Las Vegas Accounting Operations Office (OARM)

LVFMC	Las Vegas Financial Management Center (FMD)
LVRO	Las Vegas Radiation Operations
LVS	Laboratory Ventilation Data System
LWCF	Land and Water Conservation Fund
LWOP	Lease with Option to Purchase
LWOP	Leave without Pay
M&IE	Meals and Incidental Expenses
MAB	Man and Biosphere Program
MAC	Management Advisory Committee
MADCAP	Model of Advection, Diffusion and Chemistry for Air Pollution
MAER	Maximum Allowable Emission Rate
MAG	Management Advisory Group
MAP	Management Assistance Program
MAP3S	Multistate Atmospheric Power Production Pollution Study
MAPC	Manpower Model for Control Agencies
MAPPER	Maintaining, Preparing, and Producing Executive Reports
MAPS	Region 2 Environmental Map Catalog System
MAPSIM	Mesoscale Air Pollution Simulation Model
MAR	Management Assistance Review
MARC	Mining and Reclamation Council
MASBAL	RCRA Mass Balance System
MATC	Maximum Allowable Toxicant Concentration
MAXDOSE	Maximum Individual Dose Model
MBDA	Minority Business Development Agency
MBE	Minority Business Enterprises
MBE/WBE	Certification Questionnaire-Minority/Women Business Enterprise
MBEP	Region 4 Minority Business Tracking
MBER	Minority Business Enterprise Representative
MBO	Management by Objectives
MCA	Manufacturing Chemists Association
MCD	Municipal Construction Division (OW)
MCDF	Master Code Descriptor File
MCEF	Mixed Cellulose Ester Filter
MCIA	Methyl Chloride Industry Alliance
MCL	Maximum Contaminant Level (SDWA)
MCLG	Maximum Contaminant Level Goals
MCP	Municipal Compliance Plan (CWA)
MD	Mail Drop
MD	Management Division (regional)
MDA	Methylenedianilline
MDAAQS	Miscellaneous Data Analysis and Air Quality Simulation Studies
MDAD	Monitoring and Data Analysis Division (OA&R)

MDD	Management Division Director
MDEQ	Massachusetts Department of Environmental Quality
MDL	Method Detection Limit
MDSD	Monitoring and Data Support Division (OW)
MECA	Manufacturers of Emission Controls Association
MED	Minimum Effective Dose
MEFS	Midterm Energy Forecasting System
MEI	Maximum Exposed Individual (TSCA)
MEK	Methyl Ethyl Ketone
MEM	Modal Emission Model
MENS	Mission Element Needs Statement
MEP	Multiple Extraction Procedure
MERL	Municipal Environmental Research Laboratory
MES	Modal Emission Model
MESOPAC	Mesoscale Meteorological Preprocessor Program
MESOPLUME	Mesoscale "Bent Plume" Model
MESOPUFF	Mesoscale Puff Model
MESS	Model Evaluation Support System
METL	Metals Data Base
METL	Region 4 Metals System
MEXAMS	Metals Exposure Analysis Modeling System
MFBI	Major Fuel Burning Installation
MFC	Metal Finishing Category
MFD	Municipal Facilities Division (OW)
MGD	Millions of Gallons per Day
MH	Man Hours
MHD	Magnethydrodynamics
MIBK	Methyl Isobutyl Ketone
MIC	Master Item Code
MIC	Methyl Isocyanate
MICAD	Micro-Installation, Inc.
MICE	Management Information Capability for Enforcement
MICROMORT	A One in a Million Chance of Death from Environmental Hazards
MIDSD	Management Information and Data Systems Division
MINTEQ	Geochemical Model
MIPS	Millions of Instructions per Second
MIS	Mineral Industry Surveys
MISTT	Midwest Interstate Sulfur Transformation and Transport
MITS	Management Information Tracking System (New Chemicals)
ML	Meteorology Laboratory
ML	Military Leave
MLAP	Migrant Legal Action Program
MLSS	Mixed Liquor Suspended Solids

MLVSS	Mixed Liquor Volatile Suspended Solids
MMD	Mass Median Diameter
MMS	Minerals Management Service (DOI)
MMT	Million Metric Tons
MOA	Memorandum of Agreement
MOBILE	Mobile Source Emissions Model
MOCERT	Motorcycle Certification Data
MOCS	MacPhail Operant Chambers
MOD	Management and Organization Division (OARM)
MOD	Manufacturers Operations Division (OA&R)
MOD	Marine Operations Division (OW)
MOD	Miscellaneous Obligation Document
MOD	Modification
MODHIWAY	Modified Highway Program
MOI	Memorandum of Intent
MOI	Memorandum of Information (OARM)
MORT	HERL-RTP Mortality Data Base
MOS	Management Operations Staff (OFC)
MOS	Margin of Safety
MOTRON	MacPhail Motron System
MOU	Memorandum of Understanding
MP	Melting Point
MPD	Metropolitan Police Department
MPES	Management Planning and Evaluation Staff (OIRM)
MPO	Metropolitan Planning Organization
MPP	Merit Promotion Plan
MPRSA	Marine Protection, Research, and Sanctuaries Act
MPS2	Region 2 Merit Pay System
MPTDS	MPTER Model with Deposition and Settling of Pollutants
MPTER	Multiple Point Source Model with Terrain
MRA	Minimum Retirement Age
MRAM	MacPhail Radial Arm Maze
MRI	Midwest Research Institute
MRP	Multiroller Press (in sludge drying unit)
MS	Mail Stop
MS	Management Staff (OPTS)
MS	Mass Spectrometry
MS	Multilateral Staff (OIA)
MSA	Management System Audits
MSA	Metropolitan Statistical Areas
MSAM	Multikeyed Indexed Sequential File Access Method
MSD	Management Systems Division (OPPE)
MSDS	Material Safety Data Sheet

MSEE	Major Source Enforcement Effort
MSHA	Mine Safety and Health Administration (DOL)
MSIS	Model State Information System
MSL	Mean Sea Level
MSPB	Merit Systems Protection Board
MSRM	Mixture and Systemic Toxicant Risk Model
MSS	Management Support Staff (OPTS)
MSS	Management Systems Staff (OGC)
MTB	Materials Transportation Bureau
MTBE	Methyl Tertiary Butyl Ether
MTD	Maximum Tolerated Dose
MTDDIS	Mesoscale Transport Diffusion and Deposition Model for Industrial Sciences
MTG	Media Task Group
MTS	Management Tracking System (OW)
MTSL	Monitoring and Technical Support Laboratory
MTU	Mobile Treatment Unit
MUSWTCH	Mussel Watch
MVA	Multivariate Analysis
MVAPCA	Motor Vehicle Air Pollution Control Act
MVEL	Motor Vehicle Emission Laboratory
MVI/M	Motor Vehicle Inspection/Maintenance
MVICSA	Motor Vehicle Information and Cost Savings Act
MVMA	Motor Vehicle Manufacturers Association
MVRS	Marine Vapor Recovery System
MVS	Multiple Virtual System
MVTS	Motor Vehicle Tampering Survey (FISD, OMS)
MW	Megawatt
MW	Molecular Weight
MWC	Municipal Waste Combustor
MWG	Model Work Group
MWL	Municipal Waste Leachate
MYDP	Multiyear Development Plans
N/A	Not Applicable
N/A	Not Available
NA	National Archives
NA	Nonattainment
NAA	Nonattainment Area
NAAG	National Association of Attorneys General
NAAQS	National Ambient Air Quality Standards Program (CAA)
NAAS	National Air Audit System (OAR)
NABN	EMSL-RTP National Atmospheric Background Network
NAC	National Agency Check

NAC	National Asbestos Council
NACA	National Agricultural Chemicals Association
NACD	National Association of Conservation Districts
NACI	National Agency Check and Inquiry
NACO	National Association of Counties
NADB	National Aerometric Data Bank
NADP	National Atmospheric Deposition Program
NADPSC	National ADP Service Center
NAE	National Academy of Engineering
NAEP	National Association of Environmental Professionals
NAIS	Neutral Administrative Inspection Scheme
NALD	Nonattainment Areas Lacking Demonstrations
NAM	National Association of Manufacturers
NAMA	National Air Monitoring Audits
NAMF	National Association of Metal Finishers
NAMS	National Air Monitoring Station
NAMS/MIS	National Air Monitoring Stations Management Information System
NANCO	National Association of Noise Control Officials
NAPA	National Academy of Public Administration
NAPAP	National Acid Precipitation Assessment Program
NAPAP	National Acid Precipitation Program Emission Inventories
NAPBN	National Air Pollution Background Network
NAPCTAC	National Air Pollution Control Techniques Advisory Committee
NAR	National Asbestos Registry
NARA	National Air Resources Act
NARA	National Archives and Records Administration
NARS	National Asbestos-Contractor Registry System
NAS	National Academy of Sciences
NAS	National Audubon Society
NASA	National Aeronautics and Space Administration
NASLR	National Association of State Land Reclamationists
NASN	National Air Sampling Network
NASN	National Air Surveillance Network
NASR	National Association of Solvent Recyclers
NATICH	National Air Toxics Information Clearinghouse
NATO	North Atlantic Treaty Organization
NATS	National Air Toxics Strategy
NAWC	National Association of Water Companies
NAWDEX	National Water Data Exchange
NBAR	Nonbinding Allocation of Responsibility
NBS	National Bureau of Standards
NCA	National Coal Association
NCA	Noise Control Act

NCAC	National Clean Air Coalition
NCAF	National Clean Air Fund
NCAMP	National Coalition Against the Misuse of Pesticides
NCAQ	National Commission on Air Quality
NCAR	National Center for Atmospheric Research
NCASI	National Council of the Paper Industry for Air and Stream Improvements, Inc.
NC	Noncarcinogen
NCC	National Climatic Center
NCC	National Computer Center
NCCMAG	National Computer Center Management Advisory Group
NCF	Network Control Facility
NCHS	National Center for Health Statistics (NIH)
NCI	National Cancer Institute
NCIC	National Crime Information Center
NCL	National Consumers League
NCLAN	National Crop Loan Assessment Network
NCLEHA	Natural Conference of Local Environmental Health Administrators
NCLP	National Contract Laboratory Program
NCLS	NEIC Library System
NCM	National Coal Model
NCM	Notice of Commencement of Manufacture (TSCA)
NCO	Negotiated Consent Order
NCP	National Contingency Plan (CERCLA)
NCP	Noncompliance Penalties (CAA)
NCP	Nonconformance Penalty
NCPIASI	National Council of the Paper Industry for Air and Stream Improvements
NCR	Noncompliance Report (CWA)
NCR	Nonconformance Report
NCRIC	National Chemical Response and Information Center
NCRP	National Council on Radiation Protection and Measurements
NCS	National Compliance Strategy
NCSL	National Conference of State Legislatures
NCV	Nerve Conduction Velocity
NCVECS	National Center for Vehicle Emissions Control and Safety
NCWM	National Conference of Weights and Measures
NCWQ	National Commission on Water Quality
ND	Nondetect
NDD	Negotiation Decision Document
NDDN	National Dry Deposition Network
NDIR	Nondispersive Infrared Analysis
NDPD	National Data Processing Division (OARM)
NDS	National Dioxin Study
NDS	National Disposal Site

NDWAC	National Drinking Water Advisory Council
NEA	National Energy Act
NECRMP	Northeast Corridor Regional Modeling Project
NEDA	National Environmental Development Association
NEDS	National Emissions Data System (CAA)
NEEC	National Environmental Enforcement Council (NAAG)
NEEDS	Needs Survey
NEEJ	National Environmental Enforcement Journal (NAAG)
NEHA	National Environmental Health Association
NEIC	National Enforcement Investigations Center (OECM)
NEMA	National Electrical Manufacturers Association
NEP	National Energy Plan
NEP	National Estuary Program
NEPA	National Environmental Policy Act of 1969
NER	National Emissions Report
NERA	National Economic Research Associates
NERO	National Energy Resources Organization
NEROS	ASRL-RTP Northeast Regional Oxidant Study
NES PHYTO	National Phytoplankton Data Base (in Lakes)
NESCAUM	Northeast States for Coordinated Air Use Management
NESHAPS	National Emissions Standards for Hazardous Air Pollutants (CAA)
NETA	National Environmental Training Association
NETC	National Emergency Training Center
NETTING	Emission Trading Used to Avoid PSD/NSR Permit Review Requirements
NFA	National Fire Academy
NFAN	National Filter Analysis Network
NFFE	National Federation of Federal Employees
NFIB	National Federation of Independent Business
NFIP	National Flood Insurance Program
NFMA	National Forest Management Act
NFPA	National Fire Protection Association
NFPA	National Forest Products Association
NFS	National Forest Service
NFWF	National Fish and Wildlife Foundation
NGA	National Governors Association
NGA	Natural Gas Association
NGPA	Natural Gas Policy Act
NGWIC	National Ground Water Information Center
NHANES	National Health and Nutrition Examination Study
NHIS	National Health Interview Survey
NHMP	National Human Milk Monitoring Program
NHPA	National Historic Preservation Act
NHTSA	National Highway Traffic Safety Act

NHTSA	National Highway Traffic Safety Administration (DOT)
NHWP	Northeast Hazardous Waste Project
NICS	National Institute for Chemical Studies
NICT	National Incident Coordination Team
NIEHS	National Institute of Environmental Health Sciences
NIEI	National Indoor Environmental Institute
NIH	National Institutes of Health
NIM	National Impact Model
NIMBY	Not In My Back Yard
NIOSH	National Institute for Occupational Safety and Health
NIPDWR	National Interim Primary Drinking Water Regulations
NIS	Noise Information System
NISAC	National Industrial Security Advisory Committee
NIST	National Institute for Standards and Technology (Formerly NBS)
NITEP	National Incinerator Testing and Evaluation Program
NJDEP	New Jersey Department of Environmental Protection
NLA	National Lime Association
NLAP	National Lab Audit Program
NLC	National League of Cities
NLETS	National Law Enforcement Teletype System
NLM	National Library of Medicine
NLT	Not Later Than
NMC	National Meteorological Center
NMFS	National Marine Fisheries Service (DOC)
NMFWF	National Fish and Wildlife Foundation
NMHC	Nonmethane Hydrocarbons
NMI	Northeast Midwest Institute
NML	National Municipal League
NMOC	Nonmethane Organic Compounds
NMP	National Municipal Policy
NMR	Nuclear Magnetic Resonance
NNC	Notice of Noncompliance (TSCA)
NNEMS	National Network for Environmental Management Studies
NNPSPP	National Nonpoint Source Pollution Program
NO	Nitric Oxide
NO_2	Nitrogen Dioxide
NOA	New Obligation Authority
NOAA	National Oceanic and Atmospheric Administration (DOC)
NOAEL	No Observed Adverse Effect Level
NOC	Notice of Commencement
NOD	Notice of Deficiency (RCRA)
NOEL	No Observable Effects Level
NOHSCP	National Oil and Hazardous Substances Contingency Plan

NOJC	National Oil Jobbers Council
NON	Notice of Noncompliance
NON	Notice of Noncompliance (TSCA)
NOPES	Nonoccupational Pesticide Exposure Study
NORA	National Oil Recyclers Association
NOS	National Ocean Survey (NOAA)
NOV	Notice of Violation (CAA, CWA, FIFRA)
NOV/CD	Notice of Violation/Compliance Demand
NO_x	Nitrogen Oxides
NP&AA	Noise Pollution and Abatement Act of 1970
NPAA	Noise Pollution and Abatement Act
NPCA	National Parks and Conservation Association
NPDES	Laboratory Performance Evaluation (NPDES)
NPDES	National Pollutant Discharge Elimination System (CWA)
NPDES FILE	National Pollutant Discharge Elimination System (NPDES) Compliance Files
NPIRS	National Pesticide Information Retrieval System
NPL	National Priority List (CERCLA)
NPM	National Program Manager
NPN	National Particulate Network
NPR	Notice of Proposed Rulemaking
NPS	National Park Service
NPS	National Permit Strategy
NPS	National Pesticide Survey (OW)
NPTN	National Pesticides Telecommunications Network
NPUG	National Prime User Group
NRA	National Recreation Area
NRC	National Referral Center
NRC	National Research Council
NRC	National Response Center
NRC	Nonreusable Container (DOT)
NRC	U.S. Nuclear Regulatory Commission
NRDC	Natural Resources Defense Council
NRT	National Response Team
NRWA	National Rural Water Association
NSA	National Security Agency
NSC	National Security Council
NSDWR	National Secondary Drinking Water
NSEC	National System for Emergency Coordination
NSEP	National System for Emergency Preparedness
NSF	National Sanitation Foundation
NSF	National Science Foundation
NSF	National Strike Force
NSI	National Security Information

NSO	Nonferrous Smelter Order (CAA)
NSPE	National Society for Professional Engineers
NSPS	New Source Performance Standards (CAA)
NSR	New Source Review (CAA)
NSR/PSD	New Source Review and Prevention of Significant Deterioration Permitting
NSTL	National Space Technology Laboratory
NSWMA	National Solid Waste Management Association
NSWS	National Surface Water Survey
NTE	Not To Exceed
NTGS	National Technical Guidance Studies
NTIS	National Technical Information Service
NTN	National Trends Network
NTP	National Toxicology Program
NTSB	National Transportation Safety Board
NUCA	National Utility Contractors Association
NUREG	Criteria for Preparation and Evaluation of Radiological Regulations and Guides
NURF	NAPAP Utility Reference File
NVPP	National Vehicle Population Poll
NWA	National Water Alliance
NWF	National Wildlife Federation
NWPA	Nuclear Waste Policy Act
NWRC	National Weather Records Center
NWS	National Weather Service
O&G	Oil and Gas
O&M	Operation and Maintenance
O&M	Region 4 O&M Municipal Inventory
0_3	Ozone
OA	Office of Administration (OARM)
OA	Office of Audits
OA	Office of the Administrator
OA&R	Office of Air and Radiation
OADEMQA	Office of Acid Deposition, Environmental Monitoring, and Quality Assurance
OAE	Office of Analysis and Evaluation (OW)
OALJ	Office of Administrative Law Judges (AO)
OAQPS	Office of Air Quality Planning and Standards (OA&R, RTP)
OAR	Office of Air and Radiation
OARM	Office of Administration and Resources Management
OASDI	Old-Age Survivors and Disability Insurance
OASIS	OERR Info System
OC	Object Class
OC	Office of the Comptroller (OARM)

OCAPO	Office of Compliance Analysis and Program Operations
OCD	Offshore and Coastal Dispersion Model
OCE	Office of Criminal Enforcement (OECM)
OCEM	Office of Cooperative Environmental Management
OCESL	Office of Criminal Enforcement and Special Litigation
OCI	Office of Criminal Investigation (NEIC)
OCI	Organizational Conflicts of Interest
OCIL	Office of Community and Intergovernmental Liaison
OCIR	Office of Community and Intergovernmental Relations (OEA)
OCIS	On-Line Chemical Information System
OCL	Office of Congressional Liaison (OEA)
OCM	Office of Compliance Monitoring (OPTS)
OCR	Office of Civil Rights (AO)
OCR	Office of Community Relations
OCR	Optical Character Reader
OCS	Outer Continental Shelf
OCSLA	Outer Continental Shelf Lands Act
OD	Office of the Director
OD	Operations Division (NEIC)
OD	Organizational Development
OD	Outside Diameter
ODA	Office of the Deputy Administrator
ODATS	Office Director Assignment Tracking System
ODES	Ocean Data Evaluation System
ODN	Obligation Document Number
ODW	Office of Drinking Water (OW)
OE	Office of Enforcement
OEA	Office of External Affairs
OECD	Organization for Economic Cooperation and Development
OECM	Office of Enforcement and Compliance Monitoring
OEET	Office of Environmental Engineering and Technology (ORD)
OEOB	Old Executive Office Building
OEP	Office of Enforcement Policy (OECM)
OEP	Office of External Programs
OEPER	Office of Environmental Processes and Effects Research (ORD)
OER	Office of Exploratory Research (ORD)
OERR	Office of Emergency and Remedial Response (OSWER)
OES	Office of Executive Support (AO)
OF	Optional Form
OFA	Office of Federal Activities (OEA)
OFPPA	Office of Federal Procurement Policy Act
OFSPS	Office of Federal Statistical Policy and Standards
OGC	Office of General Counsel

OGE	Office of Government Ethics
OGR	Office of Government Relations (Regional)
OGWP	Office of Ground Water Protection (OW)
OHEA	Office of Health and Environmental Assessment (ORD)
OHEA	Office of Health Effects Assessment
OHMTADS	Oil and Hazardous Materials Technical Assistance Data System
OHR	Office of Health Research (ORD)
OHRM	Office of Human Resources Management (OARM)
OHSS	Occupation Health and Safety Staff (OA)
OI	Office of Investigations
OIA	Office of International Activities
OIG	Office of Inspector General
OIL	Office of Intergovernmental Liaison (OEA)
OILHM	Oil and Hazardous Material Information System
OIRA	Office of Information and Regulatory Affairs
OIRM	Office of Information Resources Management (OARM)
OLA	Office of Legislative Analysis (OEA)
OMB	Office of Management and Budget
OMEP	Office of Marine and Estuarine Protection (OW)
OMMSQA	Office of Modeling, Monitoring Systems and Quality Assurance (ORD)
OMPC	Office of Municipal Pollution Control (OW)
OMPE	Office of Management Planning and Evaluation (OPPE)
OMS	Office of Management Support (AO)
OMS	Office of Mobile Sources (OA&R)
OMSE	Office of Management Systems and Evaluation (OPPE)
OMTA	Office of Management and Technical Assessment
OO	Operation Office (ORD)
OP	Operating Plan
OPA	Office of Policy Analysis (OPPE)
OPA	Office of Public Affairs (OEA)
OPAC	Overall Performance Appraisal Certification
OPAR	Office of Policy Analysis and Review (OAR)
OPD	Office of Program Development (OAR)
OPF	Official Personnel Folder
OPFT	Other Than Permanent Full Time
OPFTE	Other Than Permanent Full Time Equivalent
OPM	Office of Program Management (OSWER)
OPME	Office of Program Management and Evaluation (OPTS)
OPMO	Office of Program Management Operations (OPTS)(OAR)
OPMS	Office of Program Management and Support (OSWER)
OPMT	Office of Program Management and Technology (OSWER)
OPP	Office of Pesticide Programs (OPTS)
OPPAS	OPP Administrative Support Systems

OPPE	Office of Policy, Planning and Evaluation
OPPI	Office of Policy, Planning and Information (OSWER)
OPPM	Office of Policy and Program Management (OSWER)
OPTS	Office of Pesticides and Toxic Substances
OPTS RTS	OPTS Regulation Tracking System
ORALTOX	Acute Oral Toxicity for Birds, Mice, Rats
ORC	Office of Regional Counsel
ORD	Office of Research and Development
ORDIS	Office of Research and Development Information Systems
ORM	Other Regulated Material
ORMS	Office Resources Management System
ORNL	Oak Ridge National Laboratory
ORO	Office of Regional Operations
ORP	Office of Radiation Programs (OA&R)
ORPM	Office of Research Program Management (ORD)
ORS	Office of Regulatory Support
ORSSA	Office of Regulatory Support and Scientific Analysis (ORA)
ORV	Off-Road Vehicle
OS/VS	Operating System/Virtual Storage
OSC	On-Scene Coordinator
OSC	Options Selection Committee
OSDBU	Office of Small and Disadvantaged Business Utilization (AO)
OSDH	Oklahoma State Department of Health
OSHA	Occupational Safety and Health Administration
OSM	Office of Surface Mining (DOI)
OSR	Office of Standards and Regulations (OPPE)
OSTP	Office of Science and Technology Policy
OSW	Office of Solid Waste (OSWER)
OSWER	Office of Solid Waste and Emergency Response
OT	Overtime
OTA	Office of Technology Assessment (U.S. Congress)
OTP	Office of Territorial Programs
OTS	Office of Toxic Substances (OPTS)
OTS MTS	OTS Milestone Tracking System
OUO	Official Use Only
OUST	Office of Underground Storage Tanks (OSWER)
OW	Office of Water
OWCP	Office of Workers Compensation Programs
OWEP	Office of Water Enforcement and Permits (OW)
OWEP	Oily Waste Extraction Program
OWP	Office of Wetlands Protection (OW)
OWPE	Office of Waste Programs Enforcement (OSWER)
OWPO	Office of Water Program Operations (OW)

OWRS	Office of Water Regulations and Standards (OW)
O_x	Total Oxidants
OY	Operating Year
OYG	Operating Year Guidance
OZIPP	Kinetics Model and Ozone Isopleth Plotting Package
OZIPPM	Modified Ozone Isopleth Plotting Package

P&A	Precision & Accuracy
PA	Policy Analyst (OMS)
PA	Preliminary Assessment
PA	Property Administrator
PA/SI	Preliminary Assessment/Site Inspection
PAA	Priority Abatement Areas
PAAS	Region 10 External Affairs Labels System
PAD	Planning and Analysis Division (OW)
PADRE	Particulate Data Reduction
PAGM	Permit Applicants Guidance Manual
PAH	Polycyclic Aromatic Hydrocarbon
PAHO	Pan American Health Organization
PAI	Performance Audit Inspection (CWA)
PAID	Plutonium Air Inhalation Dose
PAIR	Preliminary Assessment Information Rule
PAL	Point, Area, and Line Source Air Quality Model
PALDS	PAL Model with Deposition and Settling of Pollutants
PAN	Peroxyacetyl Nitrate
PAO	Property Accountable Officer
PAPR	Powered Air Purifying Respirator
PARACDS	Missing Parameter Codes
PARS	Precision and Accuracy Reporting System
PAS	Policy Analysis Staff
PASS	Procurement Automated Source System (SBA)
PAT	Permit Assistance Team (RCRA)
PBB	Polybrominated biphenyl
PBL	Planetary Boundary Layer
PBLSQ	The Lead Line Source (PBLSQ) Model
PC	Personal Computer
PC	Planned Commitment
PC	Position Classification
PC	Potential Carcinogen
PC	Pulverized Coal
PC&B	Personnel Compensation and Benefits
PCA	Principal Component Analysis
PCB	Polychlorinated biphenyl

PCCW	Public Citizens Congress Watch
PCDD	Polychlorinated dibenzodioxin
PCDF	Polychlorinated dibenzofuran
PCE	Perchloroethylene
PCEE	President's Commission on Executive Exchange
pCi/l	Piocuries per Liter
PCIE	President's Council on Integrity and Efficiency in Government
PCIO	PC Information Officer
PCIOS	Processor Common Input/Output System
PCM	Phase Contrast Microscopy
PCMD	Procurement and Contracts Management Division (OARM)
PCMI	President's Council on Management Improvement
PCO	Printing Control Officer
PCON	Potential Contractor
PCP	Pentachlorophenyl
PCS	Permanent Change of Station
PCS	Permit Compliance System (CWA)
PCS	Program Coordination Staff (OPTS)(ORD)
PCSC	PC Site Coordinator
PCV	Positive Crankcase Ventilation System
PD	Permits Division (OW)
PD	Position Description
PD	Position Document
PD	Press Division (OEA)
PDAS	Physiological Data Acquisition System
PDED	Program Development and Evaluation Division (OW)
PDFID	Preconcentration Direct Flame Ionization Detection
PDM	Probabilistic Dilution Model
PDMS	Pesticide Document Management System (OPP)
PDR	Particulate Data Reduction
PE	Performance Evaluation
PE	Program Element
PEARL	Portable Environmental Assessment and Research Laboratory
PEAS	Policy and External Affairs Staff (POSWER)
PED	Program Evaluation Division (OPPE)
PEFOS	Program Evaluation and Field Operations Staff
PEI	Petroleum Equipment Institute
PEL	Permissible Exposure Limit
PEL	Personal Exposure Limit
PEM	Partial Equilibrium Multimarket Model
PEM	Personal Exposure Model
PEPE	Prolonged Elevated Pollution Episode
PER7	Region 7 Personnel System

PERF	Police Executive Research Forum
PERMDATA	PERMDATA Management System
PERS	ERL-Ada Personnel System
PERSPROP	Personal Property Data Entry and Report System
PES	Planning and Evaluation Staff (ORD)
PESC	EMSL-Cinci Performance Evaluation System
PEST	Region 7 Nebraska Pesticide
PESTAN	Pesticides Analytical Transport Solution
PF	Potency Factor
PF	Protection Factor
PFLT	Paint Filter Liquids Test
PFSS	Pesticide Farm Worker Safety Staff
PFT	Permanent Full Time
PFTE	Permanent Full Time Equivalent
PGD	Policy and Grants Division (OPTS)
PHC	Principal Hazardous Constituent
PHE	Public Health Evaluation
PHEM	Public Health Evaluation Manual
PHF	Public Health Foundation
PHN	Public Health Network
PHONEBOOK	Region 3 Telephone Directory
PHRED	Public Health Risk Evaluation Data Base (OERR)
PHS	Public Health Service
PHSA	Public Health Service Act
PI	Preliminary Injunction
PI	Program Information
PIAT	Public Information Assist Team
PIC	Pressurized Ion Chamber
PIC	Products of Incomplete Combustion
PIC	Public Information Center
PICO	Pacific Islands Contact Office
PIGS	Pesticides in Groundwater Strategy
PIMS	Pesticide Incident Monitoring System
PIN	Procurement Information Notice
PIP	Public Involvement Program
PIPQUIC	Program Integration Project Queries Use in Interactive Command
PIRG	Public Interest Research Group
PIRT	Pretreatment Implementation Review Task Force
PIRU	Public Information Reference Unit
PIS	Public Information Specialist
PITS	Project Information Tracking System (OTS)
PKSDD	Probability Kriging to Spatially Distributed Data System
PLIRRA	Pollution Liability Insurance and Risk Retention Act

PLM	Polarized Light Microscopy
PLPT	Pretreatment Local Program Tracking System
PLS	Program Liaison Staff (ORD)
PLUME	Outfall PLUME Model
PLUME2D	Two-Dimensional Plumes in Uniform Ground Water Flow
PLUME3D	Three-Dimensional Plumes in Uniform Ground Water Flow
PLUVUE	Plume Visibility Model
PM	Particulate Matter
PM	Product Manager
PM10	Particulate Matter Nominally 10 m and Less
PM15	Particulate Matter Nominally 15 m and Less
PMD	Personnel Management Division (OARM)
PMD	Planning and Management Division (regional)
PMEL	Pacific Marine Environment Laboratory
PMI	Presidential Management Intern
PMIP	Presidential Management Intern Program
PMIS	Personnel Management Information System
PMN	Premanufacture Notification (TSCA)
PMNF	Premanufacture Notification Form (TSCA)
PMO	Program Management Office (OAR)
PMOS	Program Management and Operation Staff (OSWER)
PMR	Pollutant Mass Rate
PMRK	NEIC Permit Ranking System
PMRS	Merit Pay System
PMRS	Performance Management and Recognition System
PMRS	Performance Monitoring and Reporting System
PMS	Performance Management System
PMS	Personnel Management Specialist
PMS	Planning and Management Staff (OAQPS)
PMS	Program Management Staff (OPPE)(OW)
PMS	Region 4 Position Management System
PMSD	Program Management and Support Division (OPTS)
PMSO	Program Management and Support Office (OSWER)
PMSS	Policy and Management Support Staff (OW)
PNA	Polynuclear Aromatic Hydrocarbons
PO	Program Operations (OAR)
PO	Project Officer
PO	Purchase Order
POA	Program Office Approvals
POC	Point of Compliance
POC	Program Office Contacts
POE	Point of Exposure
POGO	Privately-Owned/Government-Operated

POHC	Principal Organic Hazardous Constituent
POI	Point of Interception
POLRE	Pollution Report
POM	Particulate Organic Matter
POM	Polycyclic Organic Matter
POMO	Program Operations and Management Office
POR	Program of Requirements
PORS	Project Officer Record System
POS	Program Operations Staff (ORD)
POSS	Program Operations Support Staff (OARM)
POTRK	Region 3 Purchase Order Tracking System
POTW	Publicly-Owned Treatment Works
POV	Privately Owned Vehicle
PP	Pay Period
PP	Program Planning
PPA	Pesticide Producers Association
PPA	Planned Program Accomplishment
PPAS	Personal Property Accountability System
PPB	Parts per Billion
PPBS	Program Planning and Budget Staff (OGC)
PPC	Personal Protective Clothing
PPE	Personal Protective Equipment
PPIS	Pesticide Product Information System
PPM	Parts per Million
PPMAP	Power Planning Modeling Application Procedure
PPMS	Personal Property Management System
PPRS	Program Planning and Review Staff
PPSP	Power Plant Siting Program
ppt	Parts per Trillion
PPT	Permanent Part Time
ppth	Parts per Thousand
PPU	Pollution Prevention Office
PR	Preliminary Review
PR	Procurement Request
PRA	Paperwork Reduction Act
PRA	Planned Regulatory Action
PRAMS	Paperwork Reduction Act Management System (PRAMS)
PREPRZM	Pesticide Root Zone Model
PRESTO-EPA	A Low-Level Radioactive Waste Environmental Transport and Risk
PRI	Periodic Reinvestigation
PRM	Prevention Reference Manuals
PRMS	Plans Review Management System
PROF	Premixed One-Dimensional Flame Code

PRP	Potentially Responsible Parties (CERCLA)
PRS	Procurement Tracking System
PRTYPOLS	Priority Pollutants
PRZM	Pesticide Root Zone Model
PS	Planning Staff (OPTS) (ORD)
PS	Point Source
PS	Preparedness Staff (OSWER)
PSAM	Point Source Ambient Monitoring
PSD	Prevention of Significant Deterioration (CAA)
PSD	Program Systems Division (OIRM)
PSDL	Region 4 PSD Log System
PSE	Program Sub-Element
PSES	Pretreatment Standards for Existing Sources
PSEUDO-HWD	HWDMS-SAS
PSI	Pollutant Standards Index
PSI	Pounds per Square Inch (Pressure)
PSI	Pressure per Square Inch
PSIG	Pressure per Square Inch Gauge
PSM	Point Source Modeling
PSNS	Pretreatment Standards for New Sources
PSP	Payroll Savings Plan
PSPD	Permits and State Programs Division (OSWER)
PSS	Personnel Security Staff (OEA)
PSS	Personnel Staffing Specialist
PSS	Physical Security Specialist
PSS	Program Support Staff
PSTN	Pesticide Safety Team Network
PT	Part Time
PTAT	Pesticides, Toxics and Air Team (ORD)
PTDIS	Single Stack Meteorological Model in EPA UNAMAP Series
PTE	Potential to Emit
PTFE	Polytetrafluoroethylene (Teflon)
PTMAX	Single Stack Meteorological Model in EPA UNAMAP Series
PTMTP	Multistack Meteorological Model in EPA UNAMAP Series
PTPLU	Point Source Gaussian Diffusion Model
PTS	Project Tracking System
PTSD	Pesticides and Toxic Substances Division (OGS)
PTSED	Pesticides and Toxic Substances Enforcement Division (ORD)
PUC	Public Utility Commission
PV	Project Verification
PVC	Polyvinyl Chloride
PWEP	Oily Waste Extraction Program
PWS	Public Water Supply

PWS	Public Water System (SDWA)
PWSS	Public Water Supply System (SDWA)
PY	Prior Year
QA	Quality Assurance
QA/QC	Quality Assistance/Quality Control
QAC	Quality Assurance Coordinator
QAFILE	Ambient Quality Assurance Data Base
QAMIS	Quality Assurance Management and Information System
QAMS	Quality Assurance Management Staff (ORD)
QAO	Quality Assurance Officer (OAR)
QAPP	Quality Assurance Project Plan
QASRCE	EMSL-RTP Quality Assurance Data System
qBtu	Quadrillion British Thermal Units
QC	Quality Control
QCA	Quiet Communities Act
QCI	Quality Control Index
QCP	Quiet Community Program
QCSDS	Quality Control Sample Distribution
QL	Quantitation Limit
QNCR	Quarterly Noncompliance Report
QSI	Quality Step Increase
QUAL-II	Stream Quality Model
QUAL2E	Enhanced Stream Water Quality Model
QUAL2EU	Enhanced Stream Water Quality Model with Uncertainty Analysis
R&D	Research and Development
RA	Reasonable Alternative
RA	Regional Administrator
RA	Regulatory Alternative
RA	Regulatory Analysis
RA	Remedial Action
RA	Resource Allocation
RA	Risk Analysis
RA	Risk Assessment
RAATS	RCRA Administrative Action Tracking System
RAC	Radiation Advisory Committee (SAB)
RAC	Regional Asbestos Coordinator
RACF	Resource Access Control Facility (NCC systems security)
RACM	Reasonably Available Control Measures
RACS	Reiter Acoustic Startle
RACT	Reasonably Available Control Technology
RAD	Radiation

RADM	Random Walk Advection and Dispersion Model
RADM	Regional Acid Deposition Model
RADRISK	Radionuclide Dose Rate/Risk
RADX	ORD-Cinci Radiation Exposure Tracking System
RAM	Urban Air Quality Model for Point and Area Sources in EPA UNAMAP Series
RAMP	Rural Abandoned Mine Program
RAMS	Regional Air Monitoring System
RAP	Radon Action Program
RAP	Remedial Accomplishment Plan
RAPS	ASRL-RTP Regional Air Pollution Study
RARG	Regulatory Analysis Review Group
RAS	Routine Analytical Services
RAT	Relative Accuracy Test
RB	Red Border
RBC	Red Blood Cells
RC	Regional Counsel
RC	Responsibility Center
RCA	Responsibility Center A (OAA/OPMO)
RCB	Responsibility Center B (OTLIP)
RCC	Radiation Coordinating Council
RCDO	Regional Case Development Officer
RCF	Responsibility Center F (OPA)
RCM	Responsibility Center Monthly
RCP	Research Centers Program (ORD)
RCP	Regional Contingency Plan
RCRA	Resource Conservation and Recovery Act
RCRC	Regional Cost Recovery Coordinator
RCRIS	Resource Conservation & Recovery Information System (RCRA)
RCS	Computer Timesharing Resource Control System
RD	Registration Division (OPTS)
RD	Remedial Design (CERCLA)
RD&D	Research Development and Demonstration
RDF	Refuse-Derived Fuel
rDNA	Recombinant DNA
RDS-HWDMS	National Hazardous Waste Data Management Systems
RDU	Regional Decision Unit
RE	Reasonable Efforts
RE	Reportable Event
READ	Regulatory and Economics Analysis Division
REAG	Reproductive Effects Assessment Group (ORD)
REAP	Regional Enforcement Activities Plan
REAS	Regional Enforcement and Superfund System
RECALLDB	Recall Data Base

RECEIV-II	Receiving Water Model
RED	RCRA Enforcement Division (OSWER)
REE	Rare Earth Elements
REEP	Reasonable Extra Efforts Program
REEP	Review of Environmental Effects of Pollutants
REF	Reference
REIT	HERL-RTP Reiter Maze System
REM	Roentgen Equivalent, Man
REM/FIT	Remedial/Field Investigation Team
REMARC	Resource Entry, Management, Accountability and Reporting for Contracts
REMS	RCRA Enforcement Management System
REP	Reasonable Efforts Program
REPR	Real Estate Planning Report
REPRISK	High Level Radioactive Waste-Repository Risk Model
REPS	Regional Emissions Projection System
RESOLVE	Center for Environmental Conflict Resolution (CF)
REXS	Report on Executive Staffing
RF	Radio Frequency
RF	Response Factor
RFA	RCRA Facility Assessment (RCRA)
RFA	Regulatory Flexibility Act
RFB	Request for Bids
RfD	Reference Dose
RfDdt	Developmental RfD
RFD	Reference Dose Values
RFI	Remedial Facility Investigation
RFP	Reasonable Further Progress
RFP	Request for Proposal
RFQ	Request for Quote
RGICS	Region 7 Version of Region 5 Grants Information and Control System
RGS	Research Grants Staff (ORD)
RI	Reconnaissance Inspection (CWA)
RI	Remedial Investigation
RI/FS	Remedial Investigation/Feasibility Study (CERCLA)
RIA	Regulatory Impact Analysis
RIA	Regulatory Impact Analysis Facility Profile
RIA	Regulatory Impact Assessment
RIC	Radon Information Council
RIC	RTP Information Center
RICC	Retirement Information and Counseling Center
RICO	Racketeer Influenced and Corrupt Organizations Act
RID	Regulatory Integration Division (OPPE)
RIF	Reduction in Force

RIM	Regulatory Interpretation Memorandum
RIMD	Regulation and Information Management Division (OPPE)
RIN	Regulatory Identifier Number
RIP	RCRA Implementation Plan
RIS	Regulatory Innovations Staff (OPPE)
RISC	Regulatory Information Service Center (OMB)
RITA	Relocation Income Tax Allowance
RITZ	Regulator and Treatment Zone Model
RJE	Remote Job Entry
RLAB	Region 4 Labels System
RLL	Rapid and Large Leakage (Rate)
RMAO	Resources Management and Administrative Office (OW)
RMCL	Recommended Maximum Contaminant Levels (SDWA)
RMDHS	Regional Model Data Handling System
RMDS	Resource Management Directives System
RME	Reasonable Maximum Exposure
RMIS	Resources Management Information System
RMIS/BARS	Resource Management Information System/Budget Analysis System
RMO	Records Management Officer
RMP	Radon/Radon Progeny Measurement Proficiency Program (ORD)
RMS	Region 10 Resource Management System
RMS	Resource Management Staff (OSWER)
RNA	Ribonucleic Acid
RO	Regional Office
ROADCHEM	ROADWAY Version (that Includes Chemical Reactions of NO, NO, and O_3)
ROADWAY	A Model to Predict Pollutant Concentrations Near a Roadway
ROC	Record of Communication
ROD	Record of Decisions
ROG	Reactive Organic Gases
ROLLBACK	A Proportional Reduction Model
ROM	Regional Oxidant Model
ROMCOE	Rocky Mountain Center on Environment
ROP	Regional Oversight Policy
ROPA	Record of Procurement Action
RP	Respirable Particulates
RP	Responsible Party
RPAR	Rebuttable Presumption Against Registration (FIFRA)
RPD	Regulatory Policy Division (OPPE)
RPIO	Responsible Planning and Implementation Officer
RPM	Reactive Plume Model
RPM	Remedial Project Manager (CERCLA)
RPM	Revolutions Per Minute
RPO	Regional Planning Officer

RPO	Regional Program Officer (RCRA/CERCLA)
RPS	Remedial Planning Staff (OSWER)
RPTS	Regional Priority Tracking System
RQs	Reportable Quantities
RQ	Reportable Quantity
RRC	Regional Response Center
RRS	Regulatory Reform Staff (OPPE)
RRT	Regional Response Team
RRT	Requisite Remedial Technology
RSCC	Regional Sample Control Center
RSKERL	Robert S. Kerr Environmental Research Laboratory (ORD)
RSMD	Resource Systems Management Division (OC)
RSPA	Research and Special Programs Administration
RSS	Regional Services Staff (ORD)
RT	Regional Total
RTCM	Reasonable Transportation Control Measures
RTD	Return to Duty
RTDM	Rough Terrain Diffusion Model
RTECS	Registry of Toxic Effects of Chemical Substances
RTEL	OA-RTP FTS Telephone System
RTM	Regional Transport Model
RTP	Research Triangle Park (North Carolina)
RTS	Regulation Tracking System
RUP	Restricted Use Pesticide (FIFRA)
RVP	Reid Vapor Pressure
RWC	Residential Wood Combustion
RX75	Region 10 Laboratory Management System
S&A	Surveillance and Analysis
S&A	Sampling and Analysis
S&E	Salaries and Expenses
S/TCAC	Scientific/Technical Careers Advisory Committee
SA	Special Assistant
SA	Sunshine Act
SAAMS	Sample Analyses and Management System, Region 4
SAB	Science Advisory Board (AO)
SAC	Secretarial Advisory Committee
SAC	Support Agency Coordinator
SAC	Suspended and Canceled Pesticides (FIFRA)
SAD	EMSL-RTP Acid Rain System
SADAA	Science Assistant to the Deputy Assistant Administrator
SAE	Society of Automotive Engineers
SAEWG	Standing Air Emission Work Group (OAR)

SAIC	Special-Agents-in-Charge (NEIC)
SAIP	Systems Acquisition and Implementation Program
SAMAC	Standing AIRS Management Advisory Committee (OAR)
SAMWG	Standing Air Monitoring Work Group
SANE	Sulfur and Nitrogen Emissions
SANSS	Structure and Nomenclature Search System
SAP	Sampling and Analysis Plan
SAP	Scientific Advisory Panel
SAP	Special Access Program
SAPCD	Sacramento Air Pollution Control District
SAR	Start Action Request
SAR	Structural Activity Relationship
SARA	Superfund Amendments and Reauthorization Act of 1986
SAROAD	Storage and Retrieval of Aerometric Data/AIRS
SAS	Special Analytical Services
SAS	Statistical Analysis System
SASD	Strategies and Air Standards Division (OAR)
SASS	Source Assessment Sampling System
SATO	Scheduled Airline Traffic Office
SBA	Small Business Act
SBA	Small Business Administration
SBI	Special Background Investigation
SBO	Senior Budget Officer
SBO	Small Business Ombudsman (OSDBU)
SC	Sierra Club
SC	Steering Committee
SCAB	South Coast Air Basin
SCAC	Support Careers Advisory Committee
SCAP	Superfund Comprehensive Accomplishments Plan (CERCLA)
SCBA	Self-Contained Breathing Apparatus
SCC	Source Classification Code
SCFM	Standard Cubic Feet Per Minute
SCI	Sensitive Compartmented Information
SCLDF	Sierra Club Legal Defense Fund
SCORPIO	Subject Content Oriented Retriever for Processing Information On-Line
SCR	Selective Catalytic Reduction
SCRAM	State Consolidated RCRA Authorization Manual (RCRA)
SCRC	Superfund Community Relations Coordinator (OSWER)
SCRP	Superfund Community Relations Program (OSWER)
SCS	Soil Conservation Service
SCS	Supplementary Control Strategy
SCS	Supplementary Control System
SCSA	Soil Conservation Society of America

SCSP	Storm and Combined Sewer Program
SCW	Supercritical Water Oxidation
SD	Standard Deviation
SDBE	Small and Disadvantaged Business Enterprise
SDBUS	Small and Disadvantaged Business Utilization Specialist (OSDBU)
SDC	Systems Decision Plan
SDCM	Dry standard cubic meter
SDI	Subchronic Daily Intake
SDWA	Safe Drinking Water Act
SEA	State Enforcement Agreement
SEA	State/EPA Agreement
SEAM	Superfund Exposure Assessment Manual
SEAM	Surface, Environment, and Mining
SEAS	Strategic Environmental Assessment System
SEB	Source Evaluation Board
SEC	Securities and Exchange Commission
SEC	Senior Enforcement Counsel (OECM)
SEE	Senior Environmental Employment
SEIA	Socioeconomic Impact Analysis
SEM	Scanning Electron Microscope
SEP	Special Emphasis Program (OCR)
SEPWC	Senate Environment and Public Works Committee
SER	Society for Epidemiologic Research
SERC	State Emergency Response Commission
SERF	Systems Enhancement Request Form
SERS	SES and Executive Resources Staff (OHRM)
SES	Secondary Emissions Standard
SES	Senior Executive Service
SES	Socioeconomic Status
SETS	Site Enforcement Tracking System
SETS	Superfund Enforcement Tracking System
SF	Slope Factor
SF	Standard Form
SF	Superfund
SF-52	SF-52 Personnel Management Tracking System
SFA	Spectral Flame Analyzers
SFFAS	Superfund Financial Assessment System
SFIREG	State FIFRA Issues Research and Evaluation Group (FIFRA)
SFO	Servicing Finance Office
SGL	Standard General Ledger
SHORTZ	Short Term Terrain Model
SHWL	Seasonal High Water Level
SI	International Systems of Units

SI	Site Inspection
SI	Spark Ignition
SIBAC	Simplified Intragovernmental Billing and Collection System
SIC	Standard Industrial Classification
SIC	Standard Industrial Code
SICEA	Steel Industry Compliance Extension Act
SIEFA	Source Inventory Emission Factor Analysis
SIMS	Secondary Ion-Mass Spectrometry
SIP	State Implementation Plan (CAA)
SIRMO	Senior Information Resources Management Officer
SIS	Science Integration Staff (OPTS)
SIS	Secretarial Information System
SIS	Stay in School
SITE	Superfund Innovative Technology Evaluation
SITS	Site Investigation Tracking System
SL	Sick Leave
SLAMS	State/Local Air Monitoring Station
SLANG	Selected Letter and Abbreviated Name Guide
SLAPS	St. Louis Air Pollution Study
SLD	Special Litigation Division (OECM)
SLPD	State and Local Planning Division
SLS	Superfund Label System
SLSM	Simple Line Source Model
SM/HD	Survey Meter and Historical Dosimetry Data Base
SMCL	Secondary Maximum Contaminant Level
SMCRA	Surface Mining Control and Reclamation Act of 1977
SME	Subject Matter Expert
SMO	Sample Management Office
SMOA	Superfund Memorandum of Agreement
SMS	Security Management System
SMSA	Standard Metropolitan Statistical Area
SNA	System Network Architecture
SNAAQS	Secondary National Ambient Air Quality Standards
SNAP	Significant Noncompliance Action Program
SNARL	Suggested No Adverse Response Level
SNARS	Spill Notification and Response System
SNC	Significant Noncompliers
SNUR	Significant New Use Rule (TSCA)
$S0_2$	Sulfur Dioxide
SOC	Synthetic Organic Chemicals
SOCMA	Synthetic Organic Chemical Manufacturers Association
SOCMI	Synthetic Organic Chemical Manufacturing Industry
SOP	Standard Operating Procedures

SORTST	Region I Source Test File
SOTDAT	Source Test Data
SOV	Single Occupancy Vehicle
SOW	Scope of Work
SOW	Statement of Work
SO_x	Sulfur Oxides
SPAD	Special Programs and Analysis Division (OEA)
SPAR	Status of Permit Application Report
SPATS	Small Purchase Automated Tracking System
SPCC	Spill Prevention, Containment and Countermeasures (CWA)
SPD	State Programs Division (OW)
SPE	Secondary Particulate Emissions
SPECS	Specifications
SPF	Structured Programming Facility
SPHEM	Superfund Public Health Evaluation Manual
SPHI	OSW Sludge Program-Health Impacts
SPI	Strategic Planning Initiative
SPLMD	Soil-Pore Liquid Monitoring Device
SPM	Special Purpose Monitoring
SPMB	Security and Property Management Branch
SPMS	Special Purpose Monitoring Stations
SPMS	Strategic Planning and Management System
SPO	Servicing Personnel Officers
SPOC	Single Points of Contact
SPS	Safety Plan System
SPS	State Permit System
SPSS	Statistical Package for the Social Sciences
SPTS	Small Purchase Tracking System
SPUR	Software Package for Unique Reports
SQAO	Reg 5 Toxics Monitoring System
SQBE	Small Quantity Burner Exemption (RCRA)
SQG	Small Quantity Generator (RCRA)
SRAP	Superfund Remedial Accomplishment Plan
SRC	Solvent Refined Coal
SRM	Standard Reference Method
SRU	System Resource Unit
SS	Settleable Solids
SS	Superfund Surcharge
SSA	Sole Source Aquifer
SSAC	Soil Site Assimilated Capacity
SSAN	Social Security Account Number
SSC	Scientific Support Coordinator
SSC	State Superfund Contracts (OSWER)

SSCD	Stationary Source Compliance Division (OA&R, RTP)
SSD	Standards Support Document
SSEIS	Standard Support and Environmental Impact Statement
SSEIS	Stationary Source Emissions and Inventory System
SSI	Size Selective Inlet
SSIS	Store Stock Inventory/Accounting System
SSMS	Spark Source Mass Spectroscopy
SSN	Social Security Number
SSO	Source Selection Official
SSO	Support Services Office
SSS	Security Support Staff
SSS	Strategic Studies Staff (OPPE)
SST	Supersonic Transport
SSURO	Stop Sale, Use and Removal Order (FIFRA)
STALAPCO	State and Local Air Pollution Control Officials
STAPPA	State and Territorial Air Pollution Program Administrators
STAR	Stability Wind Rose
STAR	State Acid Rain Projects
STARA	Studies on Toxicity Applicable to Risk Assessment
STARS	Superfund Transactions Automated Retrieval System
STATUS	Region 4 Status of GSA Orders
STC	Short-Term Concentration
STDMS	Sample Tracking and Data Management System
STEL	Short-Term Exposure Limit
STEM	Scanning Transmission-Electron Microscopy
STN	Scientific and Technical Information Network
STORAGE	CSSD-Cinci(EMSAC) Disk Storage Annual Report
STORET	Storage and Retrieval of Water Quality Information
STP	Sewage Treatment Plant
STP	Standard Temperature and Pressure
SUP	Standard Unit of Processing
SUPTRK	Superfund Site Tracking Information System
SURE	Sulfate Regional Experiment Program
SV	Sampling Visit
SVE	Society of Vector Ecologists
SVOC	Semivolatile Organic Chemical
SW	Slow Wave
SWAG	Simulated Waste Access To Ground Water
SWC	Settlement With Conditions
SWDA	Solid Waste Disposal Act
SWE	Society of Women Engineers
SWERD	Solid Waste and Emergency Response Division (OGC)
SWETS	Safe Water Enforcement Tracking

SWIE	Southern Waste Information Exchange
SWMM	Storm Water Management Model
SWMU	Solid Waste Management Unit
SYSOP	Systems Operator
T-R	Transformer-Rectifier
T&A	Time and Attendance
TA	Travel Authorization
TACB	Texas Air Control Board
TAFPD	Technical Assessment and Fraud Prevention Division (OEA)
TALMS	Tunable Atomic Line Molecular Spectroscopy
TAMPER	Antitampering and Fuel-Switching Information System
TAMS	Toxic Air Monitoring System
TAMTAC	Toxic Air Monitoring Technical Advisory Committee
TAO	TSCA Assistance Office (OPTS)
TAP	Technical Assistance Program
TAPDS	Toxic Air Pollutant Data System
TAPP	Time and Attendance, Personnel, Payroll
TAPPI	Technical Association of the Pulp and Paper Industry
TAS	Terminal Access System
TAT	Technical Assistance Team
TBI	Throttle Body Injection Systems
TBT	Tributyltin
TC	Target Concentration
TC	Technical Center
TC	Toxic Concentration
TCAS	Telephone Call Analysis System
TCDD	Dioxin (Tetrachlorodibenzo-p-dioxin)
TCDF	Tetrachlorodibenzofurans
TCE	Trichloroethylene
TCL	Target Compound List
TCLP	Toxicity Characteristic Leaching Procedure (RCRA)
TCM	Texas Climatological Model
TCM	Transportation Control Measure
TCP	Transportation Control Plan
TCP	Trichloroethylene
TCP	Trichloropropane
TCRI	Toxic Chemical Release Inventory
TCTS	EMSL-Vegas Time Card Tracking System
TD	Toxic Dose
TDH	Texas Department of Health
TDS	Total Dissolved Solids
TDTOX	Tetradichloroxylene

TDY	Temporary Duty
TEAM	Total Exposure Assessment Methodology
TEC	Technical Evaluation Committee
TEG	Tetraethylene Glycol
TEGD	Technical Enforcement Guidance Document
TEM	Texas Episodic Model
TEM	Transmission Electron Microscopy
TEP	Technical Evaluation Panel
Term	Definition/Use
TES	Technical Enforcement Support
TEXIN	Texas Intersection Air Quality Model
TFCS	Treasury Financial Communications System
TFMS	Treasury File Management System
TFT	Temporary Full Time
TFTE	Temporary Full Time Equivalent
TGO	Total Gross Output
THC	Total Hydrocarbons
THM	Trihalomethane
TI	Temporary Intermittent
TIBL	Thermal Internal Boundary Layer
TIC	Technical Information Coordinator
TIC	Tentatively Identified Compounds
TIM	Technical Information Manager
TIP	Transportation Improvement Program
TIS	Technical Information Staff
TISE	Take It Somewhere Else (Solid Waste Syndrome. See NIMBY)
TITC	TSCA Interagency Testing Committee
TLD	Thermoluminescent Dosimetry Data Base
TLD	Toxics Litigation Division (OECM)
TLV	Threshold Limit Value
TMAC	Time Accounting System
TMC	Travel Management Center
TMI	Three Mile Island
TMI RAD	Three Mile Island Environmental Radiation
TMS	FMD Travel Management System
TNSS	Region 7 TOSCA Neutral Selection
TNT	Trinitrotoluene
TO	Task Order
TO	Travel Order
TOA	Trace Organic Analysis
TOC	Total Organic Carbon
TOPSY	Toxics and Pesticides Management System

TOT	Time-of-Travel
TOX	Total Organic Halogens
TOX	Tetradichloroxylene
TOXFLO	Urban Wastewater Toxics Flow Model
TPC	Testing Priorities Committee
TPD	Technical Programs Division (ORD)
TPD	Toxics and Pesticide Division (ORD)
TPI	Technical Proposal Instructions
TPQ	Threshold Planning Quantity
TPSIS	Transportation Planning Support Information System
TPTH	Triphenyltinhydroxide
TPY	Tons per year
TQM	Total Quality Management
TR	Transportation Request
TRAV	Travel System
TRC	Technical Review Committee
TRD	Technical Resources Document
TRI	Toxic Chemical Release Inventory
TRI	Toxic Release Inventory
TRIANA	Triana Medical Claims Information System
TRIP	Toxic Release Inventory Program
TRIS	Toxic Chemical Release Inventory System
TRLN	Triangle Research Library Network
TRO	Temporary Restraining Order
TRRP	Trends Report
TRS	Total Reduced Sulfur
TRSC	Total Reduced Sulfur Compounds
TS	Toxic Substances
TSA	Technical Systems Audit
TSC	Toxic Substances Coordinator
TSC	Transportation Systems Center
TSCA	Toxic Substances Control Act
TSCATS	TSCA Test Submissions Database (OTS)
TSCC	Toxic Substances Coordinating Committee
TSD	Technical Support Division (OQAPS)
TSD	Technical Support Document
TSD	Treatment, Storage, and Disposal Facilities
TSDF	Air Emissions from Treatment Storage and Disposal Facilities for Hazardous Waste
TSDF	Treatment, Storage & Disposal Facility (RCRA)
TSDG	Toxic Substances Dialogue Group
TSM	Transportation System Management
TSO	Time Sharing Option

TSP	Teleprocessing Services Program
TSP	Thrift Savings Plan
TSP	Total Suspended Particulates
TSS	Technical Services Staff (OEA)
TSS	Terminal Security System
TSS	Total Suspended Solids
TSS	Technical Support Staff (OAR)
TSSMS	Time Sharing Services Management System
TTFA	Target Transformation Factor Analysis
TTHM	Total Trihalomethane
TTO	Total Toxic Organics
TTS	TRI Tracking System
TTY	Teletypewriter
TUCC	Triangle University Computer Center
TV	Travel Voucher
TVA	Tennessee Valley Authority
TWA	Time Weighted Authority
TWC	Texas Water Commission
TWMD	Toxics and Waste Management Division (regional)
TZ	Treatment Zone
UAC	User Advisory Committee
UAM	Urban Airshed Model
UAPSP	Utility Acid Precipitation Study Program
UAQI	Uniform Air Quality Index
UARG	Utility Air Regulatory Group
UCC	Ultra Clean Coal
UCL	Upper Control Limit
UDMH	Unsymmetrical Dimethyl Hydrazine
UEL	Upper Explosive Limit
UFL	Upper Flammability Limit
UIC	Underground Injection Control (SDWA)
UICTS	UIC Tracking System
UL	Underwriter's Laboratories
ULP	Unfair Labor Practice
UMP	Upward Mobility Program
UMTA	Urban Mass Transportation Administration
UMTRCA	Uranium Mill Tailings Radiation Control Act
UN	United Nations
UNAMAP	Users' Network for Applied Modeling of Air Pollution
UNEP	United Nations Environment Program
UNESCO	United Nations Educational, Scientific and Cultural Organization
UNIDO	United National Industrial Development Organization

UO	Utilization Officer
UPCONE	Upconing of a Salt-Water/Fresh-Water Interface Below a Pumping Well
USA	United States Attorney
USAO	United States Attorney's Office
USBM	United States Bureau of Mines
USBS	United States Bureau of Standards
USC	Unified Soil Classification
USC	United States Code
USCA	United States Code Annotated
USCG	United States Coast Guard
USCP	United States Capitol Police
USDA	United States Department of Agriculture
USDOI	United States Department of the Interior
USDW	Underground Source of Drinking Water
USEPA	United States Environmental Protection Agency
USFS	United States Forest Service
USGS	United States Geological Survey (DOI)
USNRC	United States Nuclear Regulatory Commission
USPHS	United States Public Health Service
USPP	United States Park Police
USPS	United States Postal Service
USSS	United States Secret Service
UST	Underground Storage Tank
UTM	Universal Transverse Mercator
UTP	Urban Transportation Planning
UV	Ultraviolet
UZM	Unsaturated Zone Monitoring

VA	Veterans Administration
VALLEY	Meteorological Model to Calculate Concentrations on Elevated Terrain
VAN	HERL-RTP Van System
VAT	Value Added Tax
VBOK	EMSL-Vegas Library Book System
VCEM	EMSL-Vegas Chemical Inventory
VCM	Vinyl Chloride Monomer
VDT	Video Display Terminal
VE	Visual Emissions
VEO	Visible Emission Observation
VHS	Vertical and Horizontal Spread Model
VHT	Vehicle-Hours of Travel
VIEW	Visibility Investigative Experiment in the West
VISTA	Employee I.D. No. Data Entry at Parklawn Computer Center
VISTTA	Visibility Impairment from Sulfur Transformation & Transport in the Atmosphere

VKT	Vehicle Kilometers Traveled
VLIB	EMSL-Vegas Library Inventory
VMAL	EMSL-Vegas Mail List
VMT	Vehicle Miles Traveled
VOC	Volatile Organic Compounds
VOLSTORAGE	Volatile Organic Liquid Storage Facilities
VOS	Vehicle Operating Survey
VOST	Volatile Organic Sampling Train
VP	Vapor Pressure
VSAM	Virtual Storage Access Method
VSD	Virtually Safe Dose
VSI	Visual Site Inspection
VSPC	EMSL-Vegas Species System
VSS	Volatile Suspended Solids
VSS	Voucher and Scheduling System
WA	Work Assignment
WACO	Region 7 NPDES Water Compliance System
WADTF	Western Atmospheric Deposition Task Force
WAM	Work Assignment Manager
WAP	Waste Analysis Plan (RCRA)
WASP	Water Quality Analysis Simulation Program, Version 3.1
WATER	Economic Analysis of Water Supply
WATQ	Region 2 Water Quality Analysis Graphics System
WB	Wet Bulb
WB	World Bank
WBC	White Blood Cells
WBC	Whole-Body Count and Bioassay
WBE	Women's Business Enterprise
WCC	Washington Computer Center
WCED	World Commission on Environment and Development
WD	Water Division (OGC)
WDROP	Distribution Register of Organic Pollutants in Water
WED	Water Enforcement Division (OECM)
WED	Women's Equality Day
WENDB	Water Enforcement National Data Base
WERL	Water Engineering Research Laboratory (ORD)
WESPDOSE2	High Level Radioactive Waste Risk Model Environmental Pathways
WFPM	Western Fine Particulate Monitoring
WG	Wage Grade
WG	Work Group
WGI	Within Grade Increase
WHO	World Health Organization (UN)

WHO-WMO	International Air Data Base
WHWT	Water and Hazardous Waste Team
WIC	Washington Information Center
WICEM	World Industry Conference on Environmental Management
WISE	Women in Science and Engineering
WL	Warning Letter
WL	Working Levels
WLA/TMDL	Wasteload Allocation/Total Maximum Daily Load
WLD	Water and Land Division (ORD)
WLM	Working Level Months
WMD	Waste Management Division (ORD)
WMD	Water Management Division (Regional)
WMED	Waste Management and Economics Division (OSWER)
WMO	World Meteorological Organization
WMS	Work Force Management Staff (OHRM)
WOW	We-Cycle Office Wastepaper
WP	Water Pollution Lab Performance Evaluation Studies
WPC	Word Processing Center
WPCF	Water Pollution Control Federation
WPD	Water Planning Division (OW)
WPI	Wholesale Price Index
WPO	Water Policy Office (OW)
WPPISDC	Wisconsin Power Plant Impact Study Data Center
WQC	Water Quality Criteria
WQM4	Region 4 Water Quality Modeling
WQMSGL	Water Quality Modeling System for the Great Lakes
WRC	Water Resources Congress
WRC	Water Resources Council
WRDA	Water Resources Development Act
WRI	World Resources Institute
WRS	Reg 5 ADP Workload Reporting System
WS	Water Supply Evaluation Studies Lab Performance
WS	Work Status
WSAP	Weighted Sensitivity Analysis Program
WSF	Water Soluble Fraction
WSM	Waterside Mall
WSRA	Wild and Scenic Rivers Act
WSSM	Water Supply Simulation Model
WSTB	Water Science and Technology Board
WSVS	Region 2 Water Supply Violation Assessment Graphics System
WTPS	Water, Toxics and Pesticides Staff (ORD)
WTS	HPOB Automated Workload Management System
WTSHRD	Water and Toxic Substances Health Research Division (ORD)

WWEMA	Waste and Wastewater Equipment Manufacturers Association
WWF	World Wildlife Fund
WWMMRD	Water and Waste Management Monitoring Research Division (ORD)
WWMS	Water and Waste Management Staff (ORD)
WWTP	Wastewater Treatment Plant
YTD	Year to Date
ZBB	Zero Base Budgeting
ZHE	Zero Headspace Extractor
ZOI	Zone of Incorporation
ZRL	Zero Risk 3 Level

BIBLIOGRAPHY

All definitions in this dictionary have been reproduced from the *Code of Federal Regulations, Title 40, Protection of Environment* (40 CFR), Revised as of July 1, 1987, and the *Federal Register* (updates pertaining to 40 CFR) covering the period from July 1, 1987 through June 30, 1994.

There are no restrictions on the republication of material appearing in the *Code of Federal Regulations.*

The user of this dictionary should consult 40 CFR and the *Federal Register* for bibliographical details on material in the definitions that is reproduced from other publications.

The Acronyms and Abbreviations section are from U.S. E.P.A. publication *ENVIRONMENTAL ACRONYMS, ABBREVIATIONS, AND GLOSSARY OF TERMS*, Source: United States Environmental Protection Agency Information Resources Directory, Fall 1989 OCPA19M-4001.

ABOUT THE AUTHOR

Mr. King is a consultant, author, and lecturer on the overall impacts of environmental regulation on business, communities, and federal facilities. He has a wide variety of consulting and project management experiences in all aspects of the regulatory and legislative process. With this background he provide clients with in-depth environmental liability assessments, management systems development (i.e., recordkeeping and documentation issues), environmental program implementation and training as well as evaluating prospective financial risk management activities in the form of environmental liability risk assessments for financing and insuring real property. Recently, Mr. King has established a total quality management approach to environmental compliance which is a necessary component of cost-effective problem solving for several major clients.

Mr. King is co-author with John P. Woodyard of the *PCB Management Handbook*, also published by John Wiley & Sons, Inc.